2, 35, 36, 17, 16 T...

18, 19, 20, 21, 22, 23, 24, 25, 27, 28 TEST 2

Materials and Processes in Manufacturing

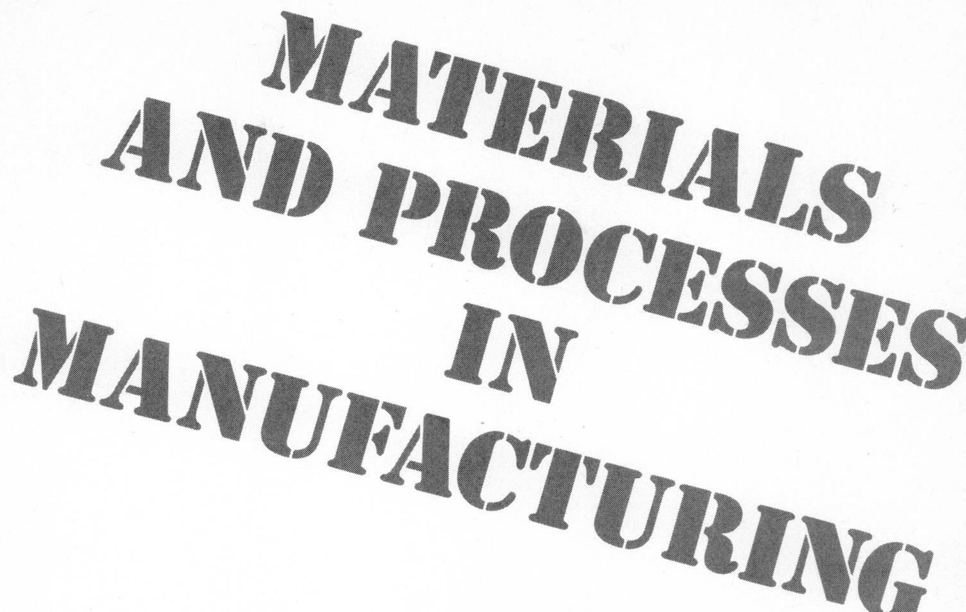

fifth edition

E. Paul DeGarmo, P.E.

Professor of Industrial Engineering
and Mechanical Engineering, Emeritus
University of California, Berkeley

MACMILLAN PUBLISHING CO., INC.
New York
COLLIER MACMILLAN PUBLISHERS
London

Printed in the United States of America

Macmillan Publishing Co., Inc.
866 Third Avenue, New York, New York 10022

Collier Macmillan Canada, Ltd.

Library of Congress Cataloging in Publication Data

DeGarmo, E. Paul, (date)
Materials and processes in manufacturing.

Bibliography: p.
Includes index.
1. Manufacturing processes. 2. Materials.
I. Title.
TS183.D4 1979 671 78-3456
ISBN 0-02-328120-0 (Hardbound)
ISBN 0-02-978760-2 (International Edition)

Printing: 2 3 4 5 6 7 8 Year: 9 0 1 2 3 4 5

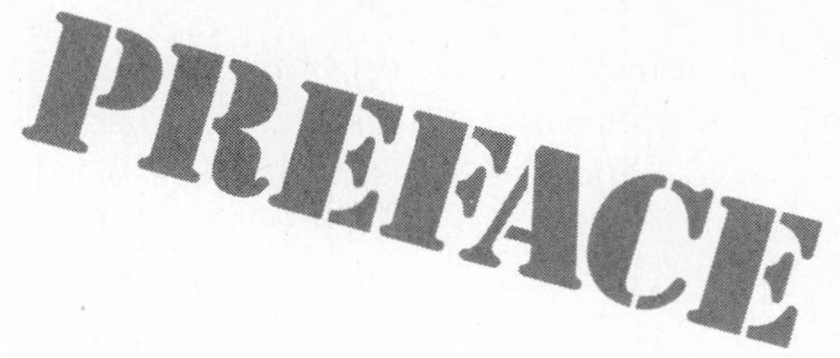

Since the publication of the fourth edition of this book in 1974, six developments have occurred that have profoundly affected manufacturing throughout the world, especially in the United States. These are (1) the availability of low-cost, solid-state microprocessors, making possible minicomputers that have resulted in the widespread adoption of computer-assisted and computer-controlled machine tools; (2) machine tools that carry out several basic machining operations, thereby permitting several operations to be performed with a single setup of the workpiece; (3) what essentially are single-purpose, automatic machine tools, such as screw machines, on which all the control functions required for producing a new part can be quickly and easily set by pushing buttons or turning dials, thereby eliminating the costly and time-consuming requirement of designing and making special control cams or templates; (4) a more widespread understanding and acceptance of the importance of productivity in determining the economic progress of a country and the standard of living of its people; (5) the great increase in the attention that manufacturing industries have had to give to product reliability in order to reduce and avoid costly product-liability claims and awards, and (6) the adoption of the international system (SI) of units by the United States government.

This edition reflects all these changes in the hope that as it continues to serve as a widely used text in engineering schools and colleges and in industry throughout the United States and numerous foreign countries, it will contribute significantly to needed progress. Both SI and the corresponding English units are given throughout the textual portions of this edition. SI units also are used in some of the tables and illustrations that relate to matters where they are likely to be encountered during the next few years.

Engineers are involved in the total manufacturing process in a variety of ways—the development of materials, the development and design of processing equipment, and the processing of materials. However, the vast majority of engineers are concerned with materials and processes as the means whereby their designs are brought into reality. At the design stage many decisions are made that determine which materials are to be used and what processes must be employed to process them. At some stage in the design–material selection–processing sequence *someone* must make decisions regarding the material to be used and the processing that will be employed. Because these decisions always affect the cost of the product—and they may vitally affect its functioning—it is most desirable that the designer either make these decisions or, at least, be in on the making of them. Otherwise, either costs or functioning, or both, may suffer. The design, material, and processing constitute a system that should be considered as an entity.

While new, and usually more specialized, materials continue to be developed, the fact that the supply of materials on our earth is limited is now generally recognized. In earlier years the availability of materials often limited the designer. In more recent years new materials have been created to meet special requirements as they arose. Often these require special, and more precise, processing in order for their properties to be utilized effectively. Now, and in the future, the economics of scarcity and the necessity for recycling will require efficient use of available materials, and these factors may increasingly be important in their selection. Similarly, more sophisticated and more versatile machines have become more commonplace. Yet, in most cases, their full capabilities can be utilized only, or at least more effectively, if the designer has an understanding of their possibilities and limitations. Therefore, although the basic objective of this edition remains the same as for prior editions, even more attention is given to the design–material selection–processing relationship. The chapter on "Material Selection" has been given even more emphasis. The basic processes continue to be stressed, but with even more emphasis on the manner in which they are accomplished in modern, multifunctional machine tools. The use of digital readout controls and numerical-, tape-, and computer-control systems is emphasized. Special attention is given to the processes that permit components to be formed into final, or semifinal, shape with little or no waste of materials.

Although considerable effort has been made to include all the significant and promising new developments in both materials and processes, the major emphasis remains on fundamentals, which provide an enduring basis for understanding both existing phenomena and those not yet in use. Thus the chapters on materials are designed to emphasize *why* they are suitable for certain applications, *why* they react as they do when subjected to certain processing, and *why* they must be treated in a specific manner to obtain desired results. *How* they are processed is an important, but secondary, objective. Similarly, with respect to machine tools, the primary emphasis is on *what* they will do, *how* they do it, their accuracy, and their relative advantages and limitations—particularly economic. Although some attention must be given to details regarding their construction and operation, this is solely for the purpose of providing a better understanding of the relationship of the tools to the foregoing objectives.

The thirty case studies that occur at the ends of various chapters are an important addition to this edition. With few exceptions, these are actual examples, or based on actual examples, from industry. It is emphasized that they do not necessarily involve only the subject matter dealt with in the chapter that they follow. Rather, all the subject matter required for their solution is contained in chapters up to that point. (In some cases, the student will have to consult standard data sources for needed information, as would be necessary in actual practice.) Such case examples are extremely useful in making students aware of the great importance of properly coordinating design, material selection, and manufacturing in order to achieve a satisfactory and failure-free product.

As in previous editions, great care has been given to the illustrations. The photographs have been selected to instruct, not to advertise a particular product. Many were made especially for this text, and numerous companies have been most cooperative in this regard. However, it should be understood that in many instances safety guards have been removed from equipment so as to show important details, and the personnel shown are not wearing certain items of safety apparel that would be worn in normal operation.

The book continues to be organized so that it can be used either in courses that cover both materials and processes, or in courses covering only manufacturing processes. For the combined type of course, the use of all chapters will give comprehensive coverage of both materials and processes. For use in courses covering only manufacturing processes, Chapters 2 through 10 can be omitted. Yet they are available as a ready reference for explanation of *why* materials behave as they do when processed.

The author wishes to acknowledge the very substantial contribution of Dr. Ronald A. Kohser, Assistant Professor of Metallurgical Engineering, University of Missouri—Rolla, who assumed primary responsibility for the revision of Chapters 2 through 8 and 14 and 15. His help is most sincerely appreciated. Additionally appreciated are comments and suggestions that have come from many who have used the previous editions. If future users find this edition equally useful, my goal will have been achieved.

E. Paul DeGarmo

Part One
MATERIALS 1

Part Two
CASTING AND FORMING PROCESSES 255

Part Three
MACHINING PROCESSES 457

Part Four
WELDING AND ALLIED PROCESSES 829

Part Five
PROCESSES AND TECHNIQUES RELATED TO MANUFACTURING 937

Case Studies

Materials

Introduction

Materials, manufacturing, and the standard of living. The standard of living in any civilization is determined, primarily, by the goods and services that are available to its people. In most cases, materials are utilized in the form of manufactured products. The efficient production of agricultural products, which are consumed as food, is greatly dependent on the use of manufactured products. Likewise, many services depend on the use of converted materials and manufactured devices; health-care services are notable examples. Thus the production and modification of materials into useful forms, through the utilization of various manufacturing processes, is of tremendous importance in modern civilization.

It follows, as a corollary, that the more efficiently and effectively materials can be produced, selected, and converted into products by manufacturing—from the viewpoint of avoiding wastage of both material and human resources and of assuring minimum cost within desired quality constraints—

the higher the standard of living can be. To a very considerable degree, the difference in the standards of living in the highly developed countries, such as the United States, and the underdeveloped countries is a reflection of the effectiveness with which available resources are utilized through manufacturing.

Although man's knowledge of materials dates back many centuries, more materials, and more knowledge about them, have come into being in the past 100 years than during all previous recorded history. The ancients dealt almost exclusively with wood, iron, brass, lead, and copper, and they knew almost nothing about why their materials behaved as they did. In certain instances they did very remarkable things with these materials. However, they developed no new materials, and most of their processing was based on accidental experience or superstition. For example, it is reported that a portion of the process used in making the famous Damascus blades was the plunging of the blade through the belly of a Nubian slave after it had been heated to the color of the rising sun as observed from a certain location in the desert. Most of the modern developments in, and understanding regarding, metals have occurred since the work of Sorby in 1861.

Although we no longer are dependent on using materials only in their natural forms or on merely modifying them in minor ways, we at last have had to face up to the fact that there is an absolute limit to the amounts of many materials that are available on spaceship earth. Thus, while research has made it possible to create completely new materials to fill specific needs, efficient use of all materials and recycling of them are both absolute and economic necessities. Concurrent has been the development of new manufacturing processes and equipment which permit old and new materials to be converted into desired forms and products with greater efficiency and, frequently, with less wastage. In countries where these techniques are planned and controlled by progressive, modern management and free, individual initiative, economies of abundance have resulted.

Materials, men, and equipment are interrelated factors in manufacturing that must be combined properly in order to achieve economical production. This important concept is indicated in Figure 1-1. What may be a proper combination for one product may not be optimal for another. The optimal combination for producing a small quantity of a given product may be very inefficient for a larger quantity of the same product. The proper combination for one product may be entirely wrong for a different product. Consequently, a systems approach, taking into account all the factors, must be used. This requires a sound and broad understanding of materials and processes and equipment on the part of those who make all the decisions involved.

The roles of engineers in manufacturing. Most engineers have as their function the designing of products that are to be brought into reality through the processing or fabrication of materials. In this capacity they are a key factor in the material selection–manufacturing procedure. A design engineer, better than any other person, should know what he or she wants a design to ac-

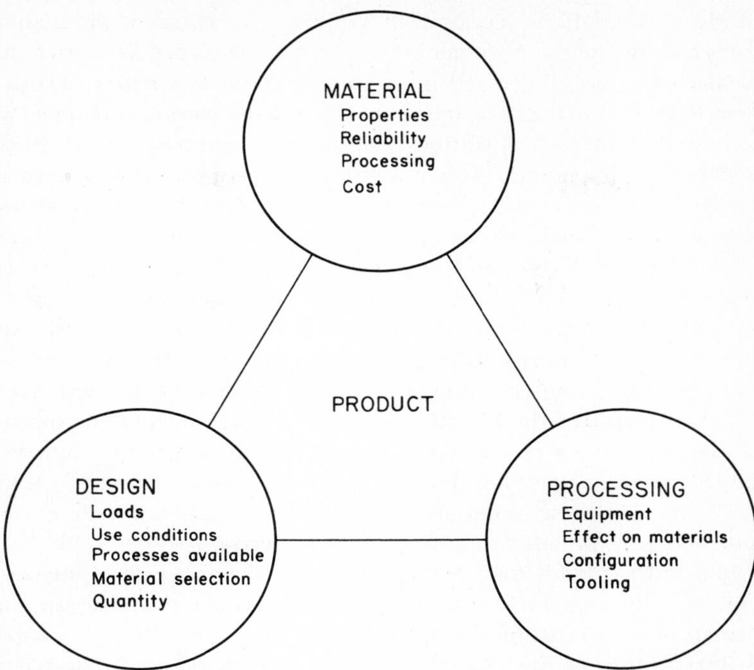

FIGURE 1-1. Interrelationships among materials, design, and processing in manufacturing a product.

complish. He knows what assumptions he has made about service loads and requirements, what service environment the product must withstand, and what appearance he wants the final product to have. In order to meet these requirements he must select and specify the material(s) to be used. In most cases, in order to utilize the material and to enable the product to have the desired form, he knows that certain manufacturing processes will have to be employed. In many instances, the selection of a specific material may dictate what processing must be used. At the same time, when certain processes are to be used, the design may have to be modified in order for the process to be utilized effectively and economically. Certain dimensional tolerances can dictate the processing, and some processes require certain tolerances. In any case, in the sequence of converting the design into reality, such decisions must be made by someone. In most instances they can be made most effectively at the design stage, by the designer if he has a reasonably adequate knowledge concerning materials and manufacturing processes. Otherwise, decisions may be made that will detract from the effectiveness of the product, or the product may be needlessly costly. It thus is apparent that the design engineer is a vital factor in the manufacturing process.

Some engineers are involved directly in manufacturing, selecting, and coordinating the specific processes and equipment to be used, or in supervising and managing their use. Some design special tooling that is used so that standard machines can be utilized in producing specific products. These engineers must have a broad knowledge of machine and process capabilities and of materials, so that desired operations can be done effectively and efficiently without overloading or damaging machines and without adversely affecting the materials being processed. These tool or manufacturing engineers also play an important role in manufacturing.

A relatively small group of engineers design the machines and equipment used in manufacturing. They obviously are design engineers and, relative to their products, they have the same concerns of the interrelationship of design, materials, and manufacturing processes. However, they have an even greater concern regarding the properties of the materials that their machines are going to process and the interreaction of the materials and the machines.

Still another group of engineers—the materials engineers—devote their major efforts toward developing new and better materials. They, too, must be concerned with how these materials can be processed and with the effects the processing will have on the properties of the materials.

Although their roles may be quite different, it is apparent that a large proportion of engineers must concern themselves with the interrelationship between materials and manufacturing processes.

As an example of the close interrelationship of design, material selection, and the selection and use of manufacturing processes, consider the electrical appliance plug shown in Figure 1-2. This plug was purchased at a retail store for 40 cents, and the manufacturer probably received about 25 cents for it. As shown in Figure 1-2, it consists of 10 parts. Thus the manufacturer had to produce, assemble, and sell the 10 parts for less than 25 cents—an average of 2½ cents per part—in order to make a profit. Only by giving a great amount of attention to design, selection of materials, selection of processes, selection of the equipment used for manufacturing (tooling), and utilization of personnel could such a result be achieved.

FIGURE 1-2. Appliance plug, assembled and disassembled.

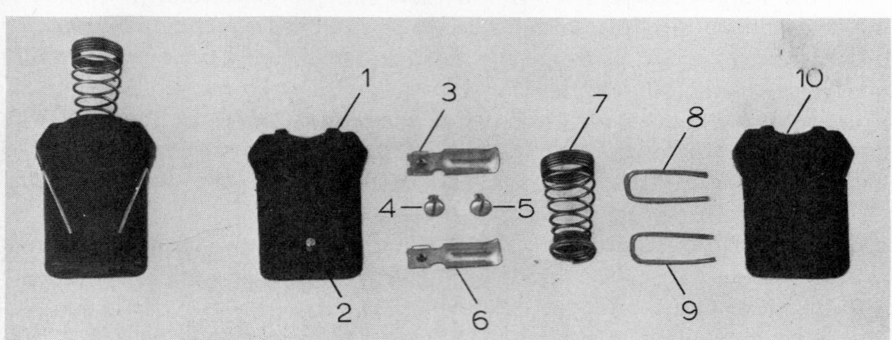

The appliance plug is a relatively simple product, yet the problems involved in its manufacture are typical of those with which manufacturing industries must deal. The elements of design, materials, and processes mentioned are all closely related. Each has its effect on the others. For example, if the two plastic shell components were to be fastened together by screws and nuts, instead of by the two U-shaped clips, entirely different machines, processes, and assembly procedures would be required. Similarly, the success of the plug is dependent on the proper material being selected for the clips. The material had to be sufficiently ductile to permit it to be bent without breaking, yet it had to be sufficiently strong and stiff to act as a spring and cause the clips to hold the shells together firmly. It is apparent that both the material and the processing had to be considered when the plug and the clips were designed in order to assure a satisfactory product that could be manufactured economically.

A list of all the problems that had to be solved to enable the appliance plug to be produced for 25 cents would be quite long. Imagine the magnitude of a similar list for an automobile or a space rocket. The fact that a modern automobile can be purchased for under $4000 is one proof that industry has learned to deal effectively with the multitude of problems that attend the design and production of complex, modern products. Solutions to these problems require engineers who have a fundamental and comprehensive knowledge concerning materials and manufacturing processes and their interrelationship, and the application of this knowledge at all stages, from the conceptual design through the supervision of the production equipment and facilities.

Basic types of manufacturing. Manufacturing frequently is classified as being *job-shop, mass production,*[1] or *process.* A job-shop plant consists of a group of general-purpose machines that, when operated by highly skilled labor, permit a wide variety of products to be produced. But such a shop, although it provides flexibility, seldom can produce large quantities economically. Work or products that require operations that are outside the scope of its machines must be subcontracted or purchased from other sources. However, for small quantities a job shop may be more economical than any other type.

Mass-production plants are composed of specialized equipment that is designed and selected to, in combination, manufacture a certain product or group of related products. The work force, as a whole, is less skilled than in a job shop, or is skilled only in a very limited range of operations. Such a plant is designed to produce its particular products very economically, *provided a sufficient quantity is produced.* If it is operated at a low volume, it will be very uneconomical.

[1] Mass production is not merely the production of large quantities of a product. Rather, it involves the manufacture of large quantities of a *standardized* product through the use of division, or *specialization,* of labor and the principle of *interchangeability.* Large quantities can be produced by other means, but usually not economically.

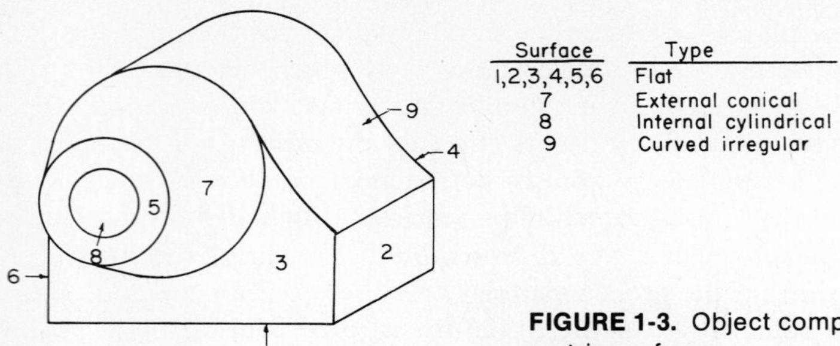

Surface	Type
1,2,3,4,5,6	Flat
7	External conical
8	Internal cylindrical
9	Curved irregular

FIGURE 1-3. Object composed of seven geometric surfaces.

Process-type plants are constructed around specific processes, usually chemical in nature, such as refining or distillation. They primarily are the concern of chemical engineers and will not be considered in this book.

Configuration analysis. In the manufacturing of *hardware,* the primary objective is to produce a component having a desired configuration, size, and finish. Every component has a shape that is bounded by various types of surfaces of certain sizes that are spaced and arranged relative to each other. Consequently, a component is manufactured by producing the surfaces that bound the shape. Surfaces may be

1. Plane or flat.
2. Cylindrical: external or internal.
3. Conical: external or internal.
4. Irregular: curved or warped.

Figure 1-3 illustrates how a shape can be analyzed and broken up into these basic, bounding surfaces. Parts are manufactured by using processes that will either (1) remove portions of a rough block of material so as to produce and leave the desired bounding surfaces, or (2) cause material to form into a stable configuration that has the required bounding surfaces. Consequently, in designing an object, one delineates and specifies the shape, size, and arrangement of the bounding surfaces. Next, the designed configuration must be analyzed to determine what materials will provide the desired properties and what processes can best be employed to obtain the required bounding surfaces from the material selected. The major objective of this book is directed toward the solution of these last two design steps.

The basic manufacturing processes. Manufacturing processes can be classified into seven types:

1. Casting or molding
2. Forming and shearing
3. Machining (material removal)
    Chip making

Chipless
Heat cutting
4. Heat treating
5. Finishing
6. Assembly
7. Inspection

These types are not completely exclusive. For example, some finishing operations involve a small amount of material removal. However, the material removal is minor, and it is not the purpose of the operation. Similar situations exist relative to some of the other types.

Casting and *molding* involve introducing liquid, granular, or powdered material into a previously prepared mold cavity. Liquid material (usually mol-

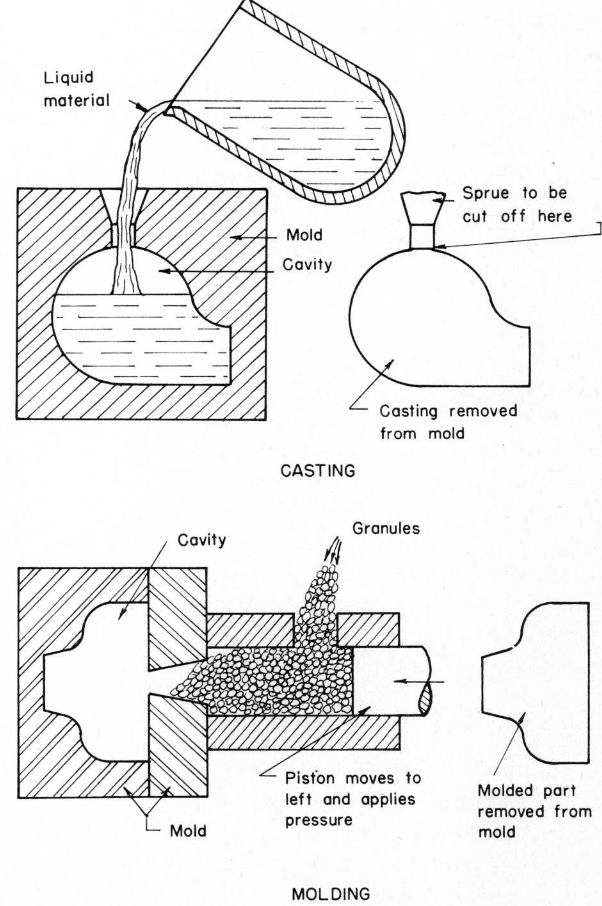

FIGURE 1-4. Casting and molding illustrated.

ten metal) takes the shape of the cavity and solidifies; it retains the desired shape of the mold cavity after it is removed, either by the mold being opened or broken away. Where granular or powdered material is involved, the application of considerable pressure is required in order to cause it to conform to the shape of the mold cavity and to acquire the desired density. Heat often is applied in addition to pressure. When the material has permanently attained the desired shape and density, the mold is opened and the part is removed.

An important advantage of casting and molding is that, in a single step, materials can be converted from a crude form into a desired shape. In most cases, there is a secondary advantage that excess, or scrap, material can easily be recycled. Figure 1-4 illustrates, schematically, the basic concepts of these processes.

Casting processes commonly are classified into two types, based on whether the mold is permanent and can be used repeatedly or nonpermanent so that a new mold must be prepared for each casting made.

Nonpermanent Molds	Permanent Molds
Sand	Nonpressure
Shell	Pressure
Plaster	Die
Investment	Centrifugal

These processes will be discussed in detail in Chapter 11.

Molding processes usually are classified according to the material being molded.

Powdered Metal (also mixed with nonmetals)	Plastics
Powder metallurgy process	Compression molding
	Transfer molding
	Injection molding
	Extrusion
	Laminating

The powder metallurgy process is discussed in Chapter 12 and the molding of plastics in Chapter 9.

Forming and *shearing* operations are extensive in number and utilize material (metal or plastics) that previously has been cast or molded. In many cases the materials pass through a series of forming or shearing operations, so the form of the material for a specific operation may be the result of all the prior operations. The basic purpose of forming and shearing is to modify the shape and size and/or physical properties of the material. Often the concurrent modifications of the properties that may occur are not desired since, as will be discussed in a later chapter, they may limit what can be accomplished in a process or may make additional processing necessary.

The basic forming and shearing processes are as follows:

Forming		Shearing
Rolling	Deep drawing	Line shearing
Forging	Stretching	Piercing and blanking
Extrusion	Swaging	Trimming
Drawing	Bending	
Wire	Coining	
Tube	Spinning	

Most of these processes are illustrated schematically in Figure 1-5. Trimming is a modification of blanking; it may or may not involve a straight line. It is used to trim off excess metal that has been left on the edges of parts that have been formed previously by drawing, stretching, or spinning.

FIGURE 1-5. Common forming and shearing processes.

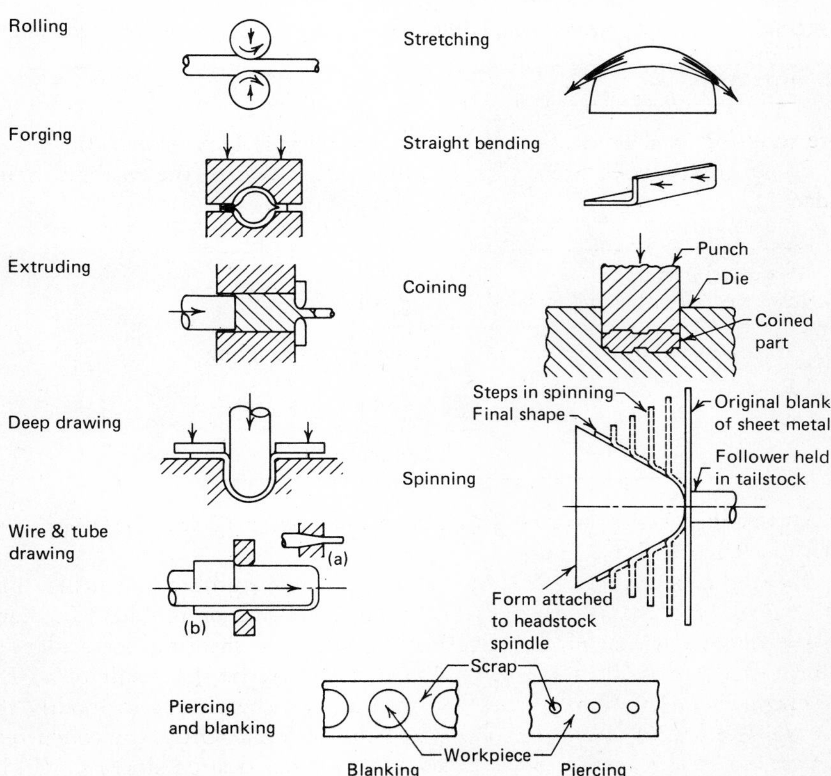

All the forming and shearing operations are done both hot and cold except coining; it always is done cold. All the fundamental surfaces can be produced by various forming and shearing processes. Some type of die or shaping equipment (tooling) is required in all cases. For most of the processes, heavy and costly equipment is required. Consequently, numerous variations of the processes have been developed in order to obtain essentially the same results with less cost for tooling and equipment.

Machining is the removal of selected portions from a piece of material in order to leave a desired shape. There are three basic types. The first, and most common, involves removal of the material in the form of *chips.* Chips are formed by the action of a cutting tool that applies compressive forces sufficient to produce shear stresses that exceed the shear strength of the material, resulting in a chip separating from the workpiece. Figure 1-6 shows a chip being formed by a single-edge cutting tool. A number of chip-type machining operations have been developed, seven of which are considered to be basic in that each involves a distinctly different type of cutting tool or tool-work-related motions. These seven, with their related variations are as follows:

FIGURE 1-6. Chip being formed by a cutting tool. (*Courtesy Cincinnati Milacron Inc.*)

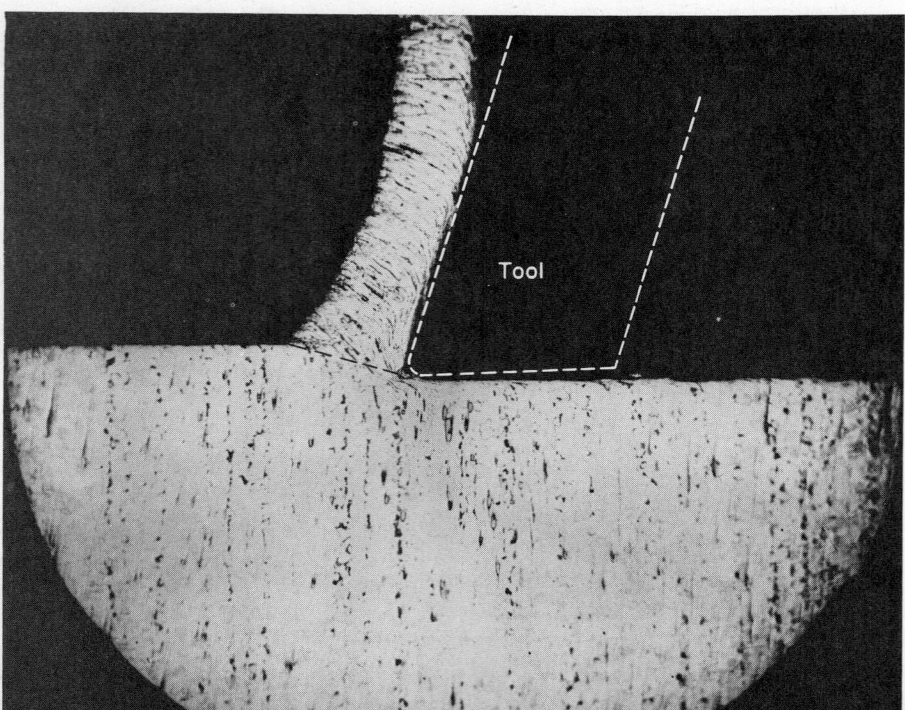

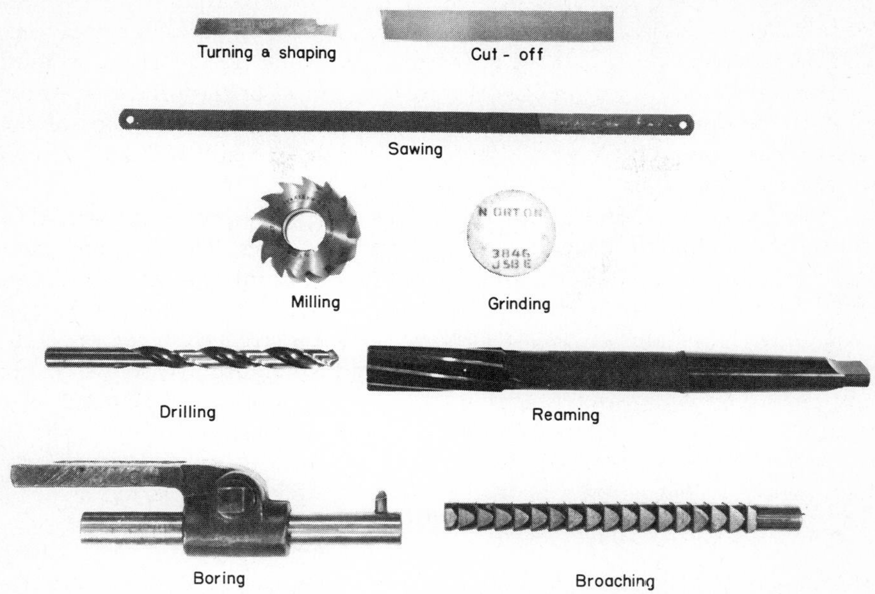

FIGURE 1-7. Typical cutting tools used in common machining operations.

1. Shaping (or planing)
2. Drilling
   a. Reaming
3. Turning
   a. Facing
   b. Boring
4. Milling
5. Sawing
   a. Filing
6. Broaching
7. Abrasive machining (or grinding)
   a. Lapping
   b. Honing
   c. Ultrasonic

The cutting tools used in these processes often appear to be quite different but are, in fact, just different arrangements of one or more cutting edges that remove material by the same basic chip-forming process.

Typical cutting tools used for various chip-type machining processes are shown in Figure 1-7. Except for those used for the abrasive (grinding) processes, they are solid and have cutting edges made from tool steel, tungsten or other metal carbides, or special ceramic material. The cutting edges of those used for the abrasive processes are small particles of aluminum oxide (Al_2O_3) or silicon carbide (SiC), and the particles either are bonded into a wheel or stone or are used to "charge" a soft metal lap or, in the case of ultrasonic machining, used in the form of a liquid slurry.

The chips from the various chip-type machining processes vary greatly in size, from so small that they can be seen only with a magnifying glass, in the case of some of the abrasive processes, to as large as 1 square inch in cross section when made with very large single-point tools. Figure 1-8 illustrates the

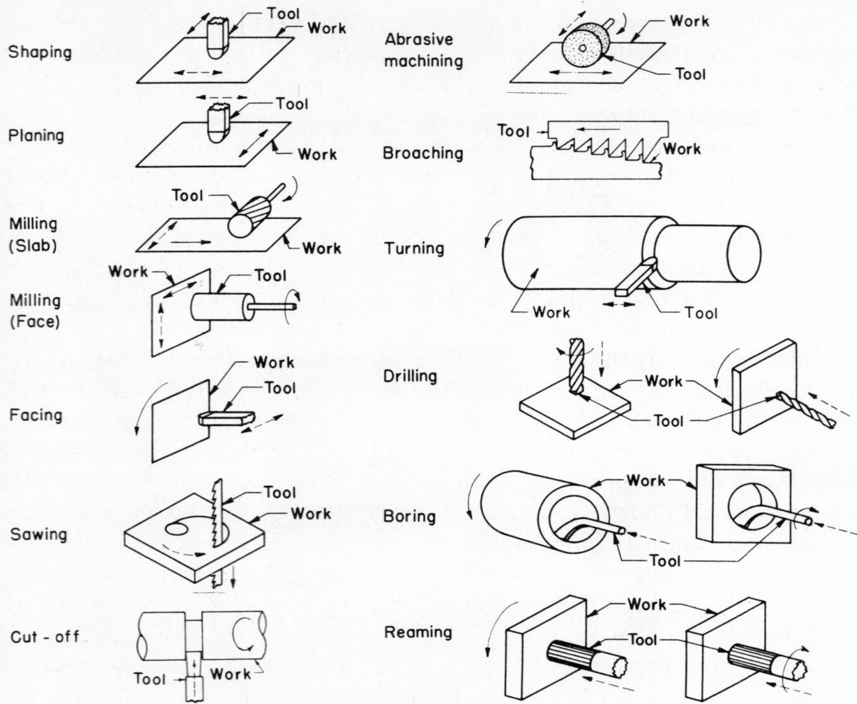

FIGURE 1-8. Schematic representation of the basic machining processes.

chip-type machining processes schematically, showing the type of cutting tool used, the relative motion between the tool and the workpiece, and the type of surface produced.

Increasing attention has been given, in recent years, to the development of the second, *chipless* group of machining processes. In these, metal is removed either by chemical or electrochemical action or by the erosion of the metal workpiece by the action of a high-voltage spark. They have added advantages that materials too hard to be cut effectively by a cutting tool can be machined and that there are no high reacting forces between the tool and the workpiece which tend to distort the latter and alter its properties. Consequently, very accurate work can be done on delicate parts, although in some cases large amounts of metal are removed at very high removal rates from large parts.

Heat cutting is accomplished either by heating metal to a sufficiently high temperature so that it can be oxidized very rapidly by a stream of oxygen, or by heating it to its melting temperature so that it can be blown away by a stream of air. A gas–oxygen flame is used in flame cutting, in conjunction with a stream of oxygen, for cutting ferrous and nonferrous metals of almost any thickness. Both ferrous and nonferrous metals can be cut either by the arc air process, in which the metal is melted by an electric arc and then blown away by a high-pressure air stream, or by means of a plasma torch, which produces extremely hot plasmas that impinge on the metal to melt it and blow it out of the kerf.

The various heat-cutting processes, one of which is illustrated in Figure 1-9, can produce flat, internal and external cylindrical and conical, and irregular surfaces. Although the accuracy is not as good as can be obtained by most of the chip-type or chipless machining processes, accuracies of ± 0.79 mm ($^1/_{32}$ inch) are readily obtainable. Most heat cutting is done to rough-prepare metals for further machining or assembly operations, but in many instances parts are heat-cut to final form.

Heat treatment is the heating and cooling of a metal for the specific purpose of altering its metallurgical and mechanical properties. Because the changing and controlling of these properties is so important in the processing and performance of metals, heat treatment is a very important manufacturing process. Each type of metal reacts differently to heat treatment. Consequently, a designer should know not only how a selected metal can be altered by heat treatment but, equally important, how a selected metal will react, favorably or unfavorably, to any heating or cooling that may be incidental to a manufacturing process. Through proper knowledge and use of heat treatment, less expensive metals can often be substituted in place of more costly materials, adverse effects from processing can be avoided, or less costly processing can be employed.

FIGURE 1-9. Three cuts being made simultaneously in a steel plate by means of an oxyacetylene flame. (*Courtesy Linde Division of Union Carbide Corporation.*)

Finishing processes are used for cleaning, deburring, or providing a protective and/or decorative surface on workpieces. These processes include

1. Cleaning
   a. Chemical
   b. Mechanical
2. Deburring
3. Painting
4. Plating
5. Buffing
6. Galvanizing
7. Anodyzing

Cleaning removes such foreign substances as dirt, grease, and scale, which result from various manufacturing operations or handling. Frequently, it must be done in preparation for subsequent finishing or manufacturing operations. Machining, casting, and shearing operations often leave sharp, and possibly dangerous, edges that are removed by *deburring.*

Buffing, sometimes called *polishing,* reduces the roughness of a surface by actual smearing of the material, reducing microscopic protrusions and filling in small hollows. Cleaning, deburring, and buffing primarily improve the appearance of workpieces. However, although customers usually prefer such improved appearance, they seldom will consciously pay an extra price for it, and these operations must be done at minimum cost. As pointed out in Chapter 37, the designer can do much to eliminate the necessity for such operations or to permit them to be done economically.

Galvanizing and *anodyzing* are done almost exclusively to provide corrosion resistance, although anodyzing, in some cases, is used to provide an improved surface for later painting. In galvanizing, a coating of zinc is built up on the surface of steel, either by dipping in a molten zinc bath or by electrolytic deposition. In anodyzing, on the other hand, the surface of the metal is converted, extending into the metal for a few thousandths of an inch. Consequently, it causes no appreciable change in dimensions. It most commonly is used on aluminum alloys, providing a surface that is corrosion-resistant in numerous media.

Painting and *plating* are, of course, the adding of protective and decorative materials to the surface of a workpiece. Although paints and lacquers can be applied by brush, in manufacturing they almost always are applied by dipping, spraying, or by an electrolytic process. Metals are plated on to surfaces by the electroplating process indicated in Figure 1-10. However, some plating is done by a process wherein metal is melted in an arc or an oxyacetylene flame and caused to impinge on the surface of the workpiece.

A modification of the electroplating process can be used to produce certain desired shapes by plating thick deposits, up to ¾ inch, on a mandrel that has an external shape coinciding with the desired inside shape of the workpiece. After the desired thickness of the plated material has been obtained, the mandrel is removed. This process is called *electroforming.*

Assembly processes are those that fasten together component parts, either temporarily or permanently. These include the following:

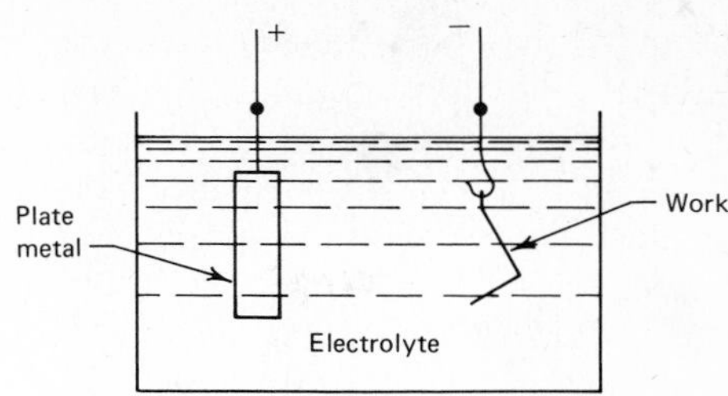

FIGURE 1-10. Basic electroplating circuit.

1. Mechanical fastening
2. Soldering and brazing
3. Welding
4. Press fitting
5. Shrink fitting
6. Adhesive bonding

Mechanical fastening includes semipermanent methods, such as bolting, and permanent methods, such as riveting and staking. A wide variety of special devices has been developed to meet specific needs.

Soldering and *brazing* are used to make semipermanent connections between metal parts by means of solders or braze metals that have a lower melting point than the metals being joined. When sufficient heat is applied to the base metal to melt the solder or braze metal, they form an alloy with the surface of the parent metal and, upon solidification, form a joint. Such a joint is commonly thought to be permanent, but it must be remembered that the parts can be unfastened, purposely or accidentally, by reheating; the parent parts are not destroyed.

Welding, of both metals and plastics, involves either melting the materials being joined at the interface or combining temperature and pressure so as to cause localized coalescence. Consequently, in most instances higher temperatures are involved than for brazing and soldering and the union is permanent. The heating is provided by gas–oxygen flames, electric arcs, or by the electrical resistance of the metals and interface, with four exceptions. In one case a beam of high-energy electrons is used. In another method heat is obtained from the combustion of fine aluminum powder. A third exception obtains heat from friction, and in a fourth exception there is essentially no heat involved—ultrasonic vibrations break up the impeding surface impurities so that coalescence is established across the interface with the application of moderate pressure.

In *press fitting,* mating parts, on which the dimension of the interior member is the same as or slightly greater than the interior dimension of the exterior member, are forced together. Obviously, there is a slight displacement of metal on the two parts. Such a joint normally will be permanent unless a suitable amount of force is used to press the parts apart. In *shrink fitting,* on the other hand, there is a substantial amount of interference between the interior and exterior parts, and they can be joined only by expanding the exterior part by heating or by contracting the interior part by cooling. Once the joint has been made, it ordinarily can be disassembled only by corresponding heating or cooling of one part.

Adhesive bonding produces joints by means of various adhesive agents. Most such joints are permanent unless too great loads are applied so as to produce failure. The use of adhesive bonding has increased greatly in recent years.

Inspection processes do not contribute directly to obtaining a desired shape or appearance but, because they determine whether desired objectives have been attained, they are very important. Consequently, they can be considered as manufacturing processes, without which the other processes likely would be of little value.

Historical development of manufacturing processes. The development of manufacturing processes, to a great extent, parallels the development of machine tools and methods of measurement to facilitate the production of desired products. Probably the earliest machine tools were crude lathes for making Etruscan wooden bowls, about 700 B.C. Such early machine tools were made by hand and were developed for a single purpose. Manufacturing, in a more real sense, followed the development of machine tools that not only could make a specific product but also could produce themselves and make the prime movers that were required to produce the power that was required to drive them.

Modern machine tools date from about 1775, when John Wilkinson, in England, constructed a horizontal boring machine for machining internal cylindrical surfaces, such as in piston-type pumps. In Wilkinson's machine, a model of which is shown in Figure 1-11, the boring bar extended through the casting to be machined and was supported at its outer end by a bearing. Modern boring machines still employ this basic design.[2] Wilkinson reported that with his machine he could bore a 57-inch-diameter cylinder to such accuracy

FIGURE 1-11. Model of Wilkinson's horizontal boring machine. (*British Crown Copyright, Science Museum, London.*)

[2] See Figure 21-7.

FIGURE 1-12. Maudsley's screw-cutting lathe. (*British Crown Copyright, Science Museum, London.*)

that nothing greater than an English shilling (about $^{1}/_{16}$ inch or 1.59 mm) could be inserted between the piston and the cylinder.

In 1794, Henry Maudsley developed an engine lathe with a practical slide tool rest. This machine tool, shown in Figure 1-12, was the forerunner of the modern engine lathe. The lead screw and change gears, which enabled threads to be cut, were added about 1800.

Joseph Whitworth, starting about 1830, speeded the wider use of Wilkinson's and Maudsley's machine tools by developing precision measuring methods. He first made three true surface plates; three were necessary so that two could not fit together exactly with spherical as well as plane surfaces. Later he developed a famous measuring machine using a large micrometer screw. Still later he worked toward establishing thread standards and made plug and ring gages. His work was of inestimable value because precise methods of measurement were a prerequisite for developing interchangeable manufacture, a requirement for later mass production.

While the early work in machine tools and precision measurement was done in England, the earliest attempts at interchangeable manufacture apparently occurred almost simultaneously in Europe and the United States. These, basically, involved the use of filing jigs, with which duplicate parts could be hand-filed to substantially identical dimensions. In 1798, Eli Whitney, using this technique, was able to obtain and fulfill a contract from the United States government to produce 10,000 army muskets, the parts of each being interchangeable. However, this truly remarkable achievement was accomplished primarily by painstaking handwork and not by specialized machines.

Relatively few basic developments in machine tools occurred between 1875 and 1920. Primarily, basic tools were enlarged and made more accurate. Following 1920, machine tools were developed for specialized work. From about 1930 through 1950, more powerful and more rigid machine tools became available to more effectively utilize the greatly improved cutting materials that were developed. These specialized machine tools and other equipment made it possible to manufacture standardized products very economically, using relatively unskilled labor. However, they lacked flexibility and were not adaptable to a variety of products or modest variations in standard products. The obvious need for greater flexibility and accuracy, without the necessity for highly skilled labor, led to the development, during the last 25 years, of highly versatile and very accurate numerically controlled machine tools. These have been adapted to tape and computer control and have made possible the economical manufacture of products in both small and large quantities, often involving complexities of design that would not have been attempted only a few years ago. Such machine tools, which will be discussed in Chapter 39, have achieved unexpected acceptance in the past 10 years, and their use is now quite common, even in small job shops.

The problem of continually reducing production costs does, and always will, remain. New materials and processes are constantly being sought. Despite the great advances of recent years, even greater progress may be expected in the future. More attention will be given to eliminating the wastage of materials, particularly through the development and use of processes that inherently do not produce scrap, to the recycling of materials, and to avoiding pollution of the environment.

Automation. Automation involves machines, or integrated groups of machines, which automatically perform required machining, forming, assembly, handling, and inspection operations and, through sensing and feedback devices, automatically make necessary corrective adjustments. There are relatively few completely automated production units, but there are numerous examples of highly automated individual machines. The potential advantages of a completely automated plant are great, but the inherent limitations are such that, in most cases, step-by-step automation of individual operations occurs. A very important limitation in the usage of automation is the economic factor: a very high initial investment is involved. As a consequence, very high use factors must be obtained. Another difficulty is that of obtaining flexibility with such equipment. The linking of computers to numerically controlled equipment provides many possibilities. Such developments will be discussed in Chapter 39.

Planning for production. Low-cost manufacture does not just happen. There is such a close and interdependent relationship between the design of a product, selection of materials, processes, and equipment, and tooling selection and design, that each of these steps must be carefully considered, planned, and co-

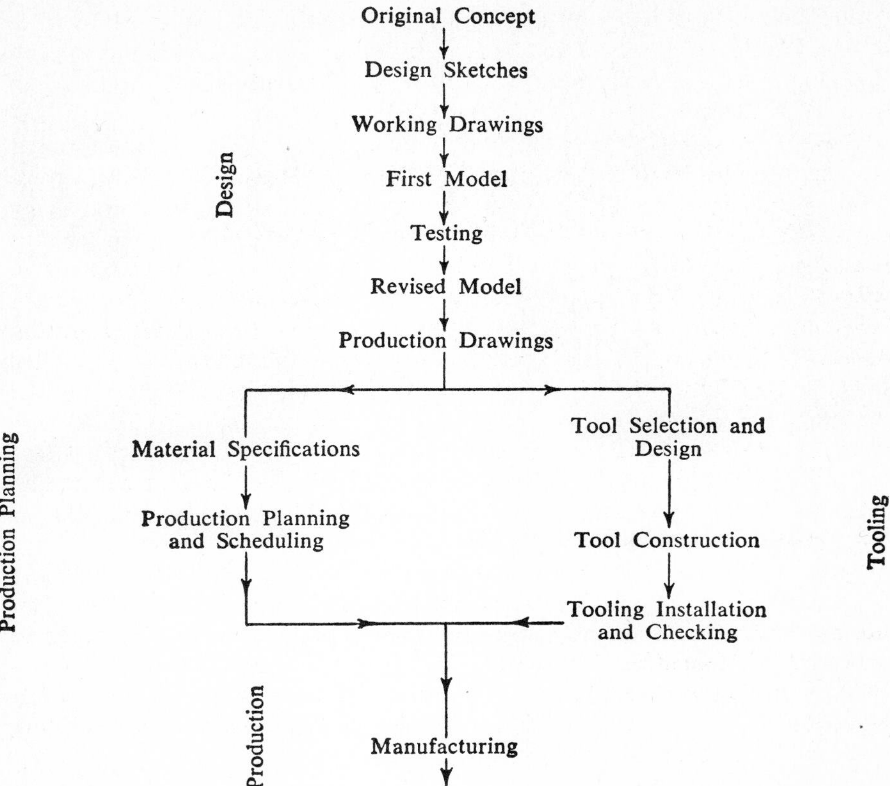

FIGURE 1-13. Steps required to convert an idea into a finished product.

ordinated before manufacturing starts. This *lead time,* particularly for complicated products, may take months, even years, and the expenditure of large amounts of money may be involved. Typically, the lead time for a completely new model of an automobile is about 2 years; for a modern aircraft it may be 4 years.

Figure 1-13 shows the steps involved in getting one product from the original idea stage to the point where it was coming off the assembly line. Note that most of the steps are closely related to the others. For example, the design of the tooling was conditioned by the design of the parts to be produced. It often is possible to simplify the tooling if certain changes are made in the design of the parts to be produced. Similarly, the selection of the materials to be used may affect the design of the tooling. On the other hand, it frequently is desirable to change the design of a part so as to enable it to be produced with tooling already on hand and thus avoid the purchase of new equipment. Close coordination of all the various phases of manufacture is essential if economy is to result. All mistakes and "bugs" should be eliminated

FIGURE 1-14. Making a full-scale clay mock-up of an automobile body. (*Courtesy Fisher Body Division, General Motors Corporation.*)

during the preliminary phases, because changes become more and more costly as work progresses.

When products are complex, involving surfaces of double curvature or requiring numerous parts, wiring, or piping to fit precisely within a limited space, full-scale models or mock-ups often are used before the design is "frozen" and tooling started. Figure 1-14 shows a typical mock-up of an automobile body. It now is possible, through the use of computers and tape-controlled machines, to machine forming dies directly from full-scale mock-ups without having to make detail drawings. Some dies also can be made directly from mathematical expressions for the desired shapes, eliminating both drawings and mock-ups.

Organization for production. One of the most important factors in economical, and successful, manufacturing is the manner in which the resources—men, materials, and capital—are organized and controlled so as to provide effective coordination, responsibility, and control. Figure 1-15 shows the organization chart of one company, indicating the relationship between the various departments and the people involved. In this company, coordination at the design stage is provided by a committee composed of all the vice presidents, the manager of planning and scheduling, the production engineer, and the chief industrial engineer. Each can provide vital input as to whether a new product should be made or an existing one altered or discontinued. The vice president (manufacturing) must know that the product can be manufactured economically and what equipment will be required. The manager of planning and scheduling must know what materials will be required and at what time.

The production engineer will know what special tooling and equipment will be required, if any, and can assure its availability when needed, and often he can suggest design modifications to reduce and simplify tooling requirements. The chief industrial engineer must be able to predict the labor costs, determine the methods to be used, and plan the layout of the equipment. With such coordination, costly errors can be avoided and changes made on paper, rather than in the factory, and a better and less costly product usually results.

Each person on such a committee obviously will not know all about the regular work of the other members. But it is important that he understand how their functions are interrelated and how decisions within their own departments will affect the operations of the others. For example, in designing the special tooling, the production engineer must keep in mind that, usually, the tools will be operated by people, and ease and speed of operation should be given just as much consideration as functional performance.

The type of cooperation and coordination that has just been described calls for engineers who are more than specialists in a given field. They also must possess a broad fundamental knowledge of design, metallurgy, processing, economics, accounting, and human relations. Without this a team of nine pitchers is apt to result. No ball player can be the best possible at all the nine positions. He must know all the tricks of playing his particular position, but he must also be able to throw, catch, hit, and run, and know what the other eight men are apt to be doing and where they will be in any situation the game may present. In the manufacturing game, low-cost mass production is the result of a team that cooperates to operate a plant as a coordinated unit. This is the key to producing more goods of better quality at less cost.

FIGURE 1-15. Organization chart for a typical modern business.

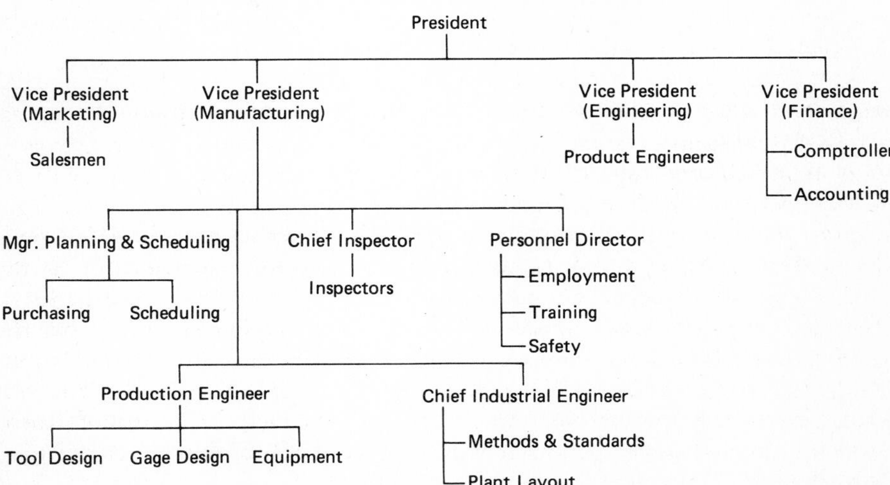

Planning manufacturing operations. As has just been indicated, successful manufacturing requires careful and extensive planning at both the product-design stage and when a decision to manufacture has been made. Extensive discussion of this latter aspect is contained in Chapter 40, "Planning Manufacturing Operations." Many readers may wish to refer to this chapter before reading Chapters 2 through 39. In such cases, it is suggested that a rereading at the conclusion of study of the book will be found beneficial.

Review questions

1. What role does manufacturing play relative to the standard of living of a country?
2. Explain the differences between job-shop, mass-production, and process manufacturing.
3. What are three ways in which an engineer may be related to manufacturing?
4. What is meant by configuration analysis?
5. Explain why the production of surfaces is of great importance in manufacturing "hard" items.
6. What are seven basic manufacturing processes?
7. Why would it be advantageous if casting could be used to produce a complex-shaped part to be made from a hard-to-machine metal?
8. What are the basic factors that distinguish mass production?
9. Why may large-scale production not be the same as mass production?
10. List eight forming processes.
11. Define "machining."
12. It is acknowledged that chip-type machining, basically, is an inefficient process, yet it probably is used more than any other to produce desired shapes. Why?
13. What are 10 basic chip-type machining operations?
14. List three purposes of finishing operations.
15. What is the basic difference between galvanizing and anodyzing?
16. List five assembly processes.
17. How is electroforming related to electroplating?
18. How does brazing differ basically from welding?
19. What is a basic difference between mechanization and automation?
20. For what types of products are full-scale mockups advantageous? Name two specific products.
21. What difficulties might result if the steps "Revised Model" and "Production Planning and Scheduling" were omitted from the procedure shown in Figure 1-13?
22. A company is considering making automobile bumpers from aluminum instead of from steel. List some of the factors it would have to consider in arriving at its decision.

23. Discuss briefly the relationship of design to production.
24. It has been said that low-cost products are more likely to be more carefully designed than high-priced items. Do you think this is true, and why?
25. Obtain from a store some mass-produced product selling for less than $1. Disassemble, if possible, and list the features that make it easy to manufacture. Can you determine any way in which it could be further simplified without detracting from its functional worth or quality?

CHAPTER 2

Properties of materials

In selecting a material, the primary concern of engineers is to match the material properties to the service requirements of the component. Knowing the conditions of load and environment under which a component must operate, engineers must select an appropriate material, using tabulated test data as their primary guide. They must know what properties they want to consider, how these are determined, and what restrictions or limitations should be placed on their application. Only by having a familiarity with test procedures, capabilities, and limitations can engineers determine whether the listed values of specific properties are, or are not, directly applicable to the problem at hand, and then use them intelligently to select a material.

Metallic and nonmetallic materials. Perhaps the most common classification that is encountered in materials selection is whether the material is *metallic* or *nonmetallic*. The common metallic materials are such metals as iron, copper, aluminum, magnesium, nickel, titanium, lead, tin, and zinc and the alloys of these metals, such as steel, brass, and bronze. They possess the *metallic properties* of luster, thermal conductivity, and electrical conductivity; are relatively ductile; and some have good magnetic properties. The common nonmetals are wood, brick, concrete, glass, rubber, and plastics. Their properties vary widely, but they generally tend to be less ductile, weaker, and less dense than the metals, and they have no electrical conductivity and poor thermal conductivity.

Although it is likely that metals always will be the more important of the two groups, the relative importance of the nonmetallic group is increasing rapidly, and since new nonmetals are being created almost continuously, this trend is certain to continue. In many cases the selection between a metal and nonmetal is determined by a consideration of required properties. Where the required properties are available in both, total cost becomes the determining factor.

Physical and mechanical properties. One material can often be distinguished from another by means of *physical properties,* such as color, density, specific heat, coefficient of thermal expansion, thermal and electrical conductivity, magnetic properties, and melting point. Some of these, for example thermal conductivity, electrical conductivity, and density, may be of prime importance in selecting material for certain specific uses. Those properties that describe how a material reacts to mechanical usage, however, are often more important to the engineer in selecting materials in connection with design. These *mechanical properties* relate to how the material will react to the various loadings during service.

The mechanical properties of materials are determined by subjecting them to standard laboratory tests, so that their reaction to changes in the controlled, influencing conditions can be determined. In using the results of such tests, however, the engineer must remember that they apply *only* to the specific test conditions. Caution should be exercised in the application of results, for the actual service conditions rarely duplicate the conditions of testing.

Stress and strain. When a load is applied to a mechanism or structure, the material is deformed (*strained*) and internal reactive forces (*stresses*) are set up to resist the applied force. For example, if a weight, W, is suspended from a bar of uniform cross section, as in Figure 2-1, the bar will elongate slightly by an amount ΔL. For a given weight, W, the magnitude of the *elongation,* ΔL, will depend upon the original length of the bar. The amount of deformation of each unit length of the bar, expressed as $e = \Delta L/L$, is called the *unit strain.* Although it is a ratio of a length to another length and is a dimensionless number, it is usually expressed in terms of millimeter per meter, inch per inch, or as a percentage.

Application of the load W also produces reactive stresses within the bar, through which the load is transmitted to the supports. *Stress* is defined as the

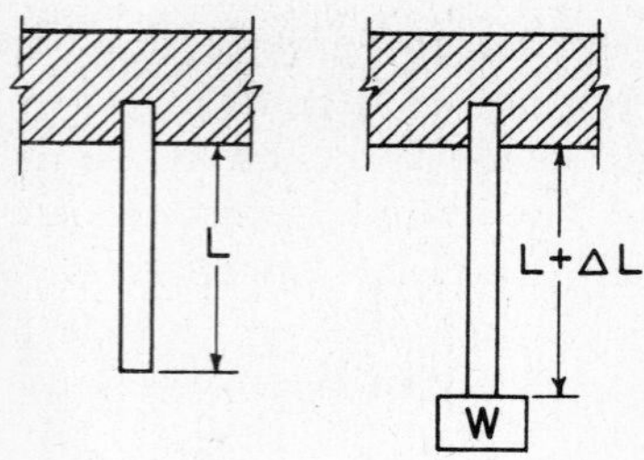

FIGURE 2-1. Tension loading and resulting elongation.

force or load being transmitted divided by the cross-sectional area transmitting the load. Thus, in Figure 2-1, the stress is $s = W/A$, where A is the cross-sectional area of the supporting bar. It ordinarily is expressed in terms of megapascals (in SI units) or pounds per square inch (in the English system).

In Figure 2-1, the weight tends to stretch or lengthen the bar, so the strain is known as a *tensile strain* and the stress as a *tensile stress.* Other types of loadings produce other types of stress and strain, as illustrated in Figure 2-2. *Compressive forces* tend to shorten the material and produce *compressive stresses and strains. Shearing stresses and strains* result from two forces acting on an element of material at an angle with respect to each other.

Static properties. When the loads applied to a material are constant and stationary, or nearly so, they are said to be *static.* In many uses the load conditions are essentially static, so it is important to characterize the behavior of materials under such conditions. Consequently, a number of standardized tests have been developed as a means to determine and report the *static properties* of materials. The documented test results can be used to select materials, provided that the service conditions are essentially the same as those of testing. Even when the service conditions differ from those of testing, the results can be used to qualitatively rate and compare various materials.

The tensile test. Considerable information about the properties of a material can be obtained from a uniaxial *tensile test.* A standard specimen is loaded in

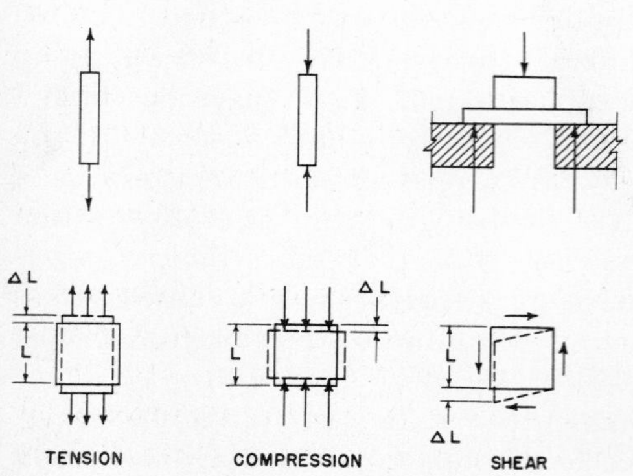

FIGURE 2-2. Examples of tension, compression, and shear loading and their strain response.

FIGURE 2-3. Hydraulic-type tension and compression testing machine. (*Courtesy Baldwin-Hamilton Co.*)

tension in a testing machine such as the one shown in Figure 2-3. Standard test conditions assure meaningful and reproducible test results. Standard specimens, the two most common of which are shown in Figure 2-4, are designed to produce uniform uniaxial tension in the central test portion and assure reduced stresses in the sections that are gripped.

A load, W, is applied and measured by the testing machine, while the elongation (ΔL) or strain over the specified gage length is determined by an external measuring device attached to the specimen. The result is a plot of coordinated load-elongation points, producing a curve of the form in Figure 2-5. Since characteristic loads will differ with different-size specimens and elongations will vary with different gage lengths, it becomes desirable to remove the size effects and establish a plot that is characteristic of the material's response to the test conditions. If the load is divided by the *original* cross-sectional area and the elongation is divided by the *original* gage length, the size effects are eliminated and the plot becomes known as an *engineering stress–strain curve,* as shown in Figure 2-5. This curve is simply a load-elongation curve with the scales of both axes modified to remove size effects.

In Figure 2-5 it will be noted that, up to a certain stress, the strain is directly proportional to the stress. The stress at which this proportionality ceases to exist is known as the *proportional limit.* Up to the proportional limit, the material obeys *Hooke's law,* which states that stress is directly proportional to strain. The proportionality constant, or ratio of stress to strain in this

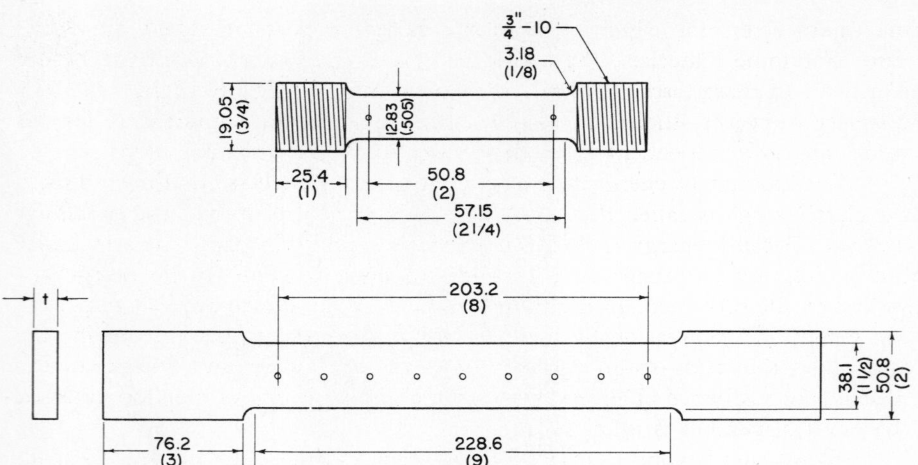

FIGURE 2-4. Two common types of standard tensile test specimens: round (*upper*) and flat (*lower*).

region, is known as *Young's modulus* or the *modulus of elasticity*. It is an inherent and constant property of a given material and is of considerable importance. As a measure of stiffness, it indicates the ability of a given material, for a given cross section, to resist deflection when loaded. It is commonly designated by the symbol E.

Up to a certain stress, if the applied load is removed, the specimen will return to its original length. Thus, from zero stress up to this point, the behavior is elastic and this region of the curve is known as the *elastic region*. The

FIGURE 2-5. Stress–strain diagram for a low-carbon steel.

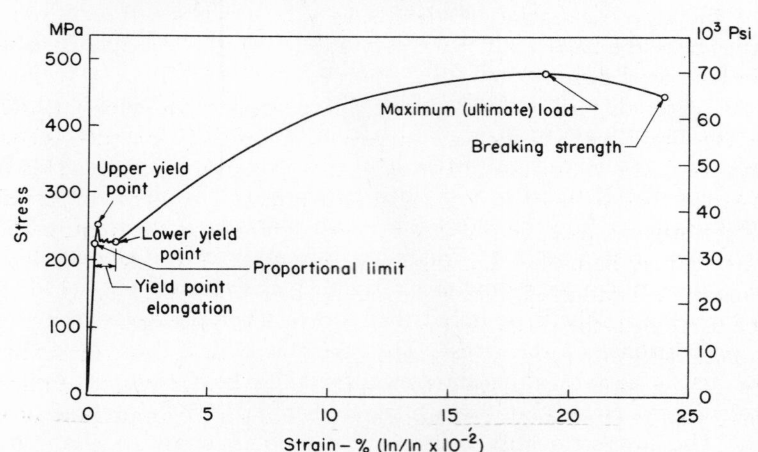

maximum stress for which truly elastic behavior exists is called the *elastic limit.* For some materials, the elastic limit and proportional limit are almost identical. In most cases, however, the elastic limit is slightly higher. Neither quantity, however, should be assigned great engineering significance, for the values are quite dependent upon the sensitivity of the test equipment.

The amount of energy that a unit volume of material can absorb within the elastic range is called the *resilience* or, in quantitative terms, the *modulus of resilience.* Because energy is the product of force times distance, the area under the load-elongation curve up to the elastic limit is equal to the energy absorbed by the specimen. In dividing load by original area to produce engineering stress, and elongation by gage length to produce engineering strain, the area under the stress–strain curve becomes the energy per unit volume or the modulus or resilience. This energy is potential energy and is therefore released whenever a member is unloaded.

Elongation beyond the elastic limit becomes unrecoverable and is known as *permanent set* or *plastic deformation.* Upon removal of all loads, the specimen will retain a permanent change in shape. Because the engineer is usually interested in either elastic or plastic response (but rarely both), the location of the elastic-to-plastic transition becomes an important material property. For most components, plastic flow, except for a slight amount to permit the redistribution of stresses, represents failure of the component through loss of dimensional and tolerance control. In manufacturing where plastic deformation is used to shape a product, design stresses must be sufficient to put the workpiece into the plastic region. Thus some means is desired to determine the transition from elastic behavior to plastic flow.

Beyond the elastic limit, increases in strain do not require proportionate increases in stress. In some materials, a point may be reached where additional strain occurs without any increase in stress, this point being known as the *yield point* or *yield-point stress.* For low-carbon steels, as in Figure 2-5, two distinct points are significant: the highest stress preceding extensive strain, known as the *upper yield point,* and the lower, relatively constant, "run-out" value, known as the *lower yield point.* The lower value is the one that would appear in tabulated data.

Most materials, however, do not have a well-defined yield point, but have a stress–strain curve of the form shown in Figure 2-6. For such materials, the elastic-to-plastic transition is *defined* by the *offset yield* strength. This is the value of stress that will produce a given and tolerable amount of permanent strain. Deformations used are usually 0.2 per cent or 0.1 per cent, although 0.02 per cent may be used when minute amounts of plastic deformation may lead to component failure. Offset yield strength is then determined by drawing a line parallel to the elastic line, displaced by the offset strain, and reporting the point where it intersects the stress–strain curve, as illustrated in Figure 2-6. The value is reproducible and is independent of equipment sensitivity, but it is meaningless unless reported with the amount of offset used.

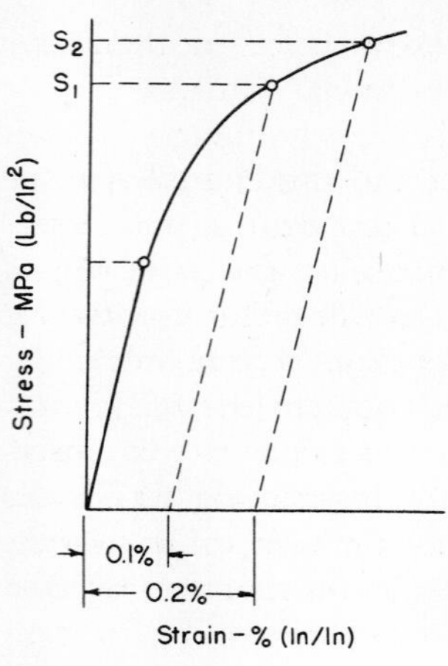

FIGURE 2-6. Stress–strain diagram for a material not having a well-defined yield point, showing the offset method for determining yield strength.

As the straining of the material continues into the plastic range, we find that the material gains increasing load-bearing ability. Since load-bearing ability is equal to strength times cross-sectional area, and the cross-sectional area is decreasing with tensile stretching of the specimen, the material must be increasing in strength. When the mechanism for this strengthening is discussed, in Chapter 3, it will be seen that the strength will always continue to increase with deformation. In the tensile test, however, a point is reached where the drop in area with increased strain dominates the increase in strength and the overall load-bearing capacity peaks and begins to diminish, as in Figure 2-5. The value of this point on the stress–strain curve is known as the *ultimate strength* or *tensile strength* of the material. The weakest point of the tensile bar at that time continues to be the weakest point by virtue of the decrease in area, and deformation becomes localized. This localized reduction of the cross-sectional area, known as *necking,* is shown in Figure 2-7. It is accompanied by a reduction in the amount of load required to produce additional straining, and the stress–strain curve drops off.

If straining is continued far enough, the tensile specimen will ultimately fracture. The stress at which this occurs is called the *breaking strength* or *fracture strength.* For relatively ductile materials, the breaking strength is less than the ultimate tensile strength and necking precedes fracture. For a brittle ma-

FIGURE 2-7. Standard 0.505-inch-diameter tensile specimen showing a necked portion developed prior to failure.

terial, fracture usually terminates the stress–strain curve before necking and possibly before the onset of plastic flow.

Ductility and brittleness. The extent to which a material exhibits *plasticity* is significant in evaluating its suitability to certain manufacturing processes. Metal deformation processes, for example, require plasticity; the more plastic a material is, the more it can be deformed without rupture. This ability of a material to be deformed plastically without fracture is known as *ductility.*

One of the major ways of evaluating ductility is to consider the *per cent elongation* of a tensile-test specimen. As shown in Figure 2-8, however, materials do not elongate uniformly along their entire length when loaded beyond necking, even though the specimen is originally of uniform cross section. Thus it has become common practice to report ductility in terms of per cent elongation of a specified gage length. The actual gage length used in the evaluation is of great significance. For the entire 8-inch gage length of Figure 2-8, the elongation becomes 31 per cent. If the center 2-inch segment is considered, elongation becomes 60 per cent. Thus quantitative comparison of material ductility through elongation requires testing of identical specimens. A more meaningful measure of ductility may be the *uniform elongation* or *per cent elongation prior to necking,* but elongation at fracture is the commonly reported value.

Another indication of ductility is the *per cent reduction in area* that occurs in the necked region of the specimen. It is computed as

$$\frac{A_o - A_f}{A_o} \times 100\%$$

where A_o is the original cross-sectional area and A_f is the smallest area in the necked region, independent of gage length.

Other terms that are often related to the ductility of materials include *malleability, workability,* and *formability.* These terms relate to the ability of a metal to undergo mechanical working processes without rupture. Although

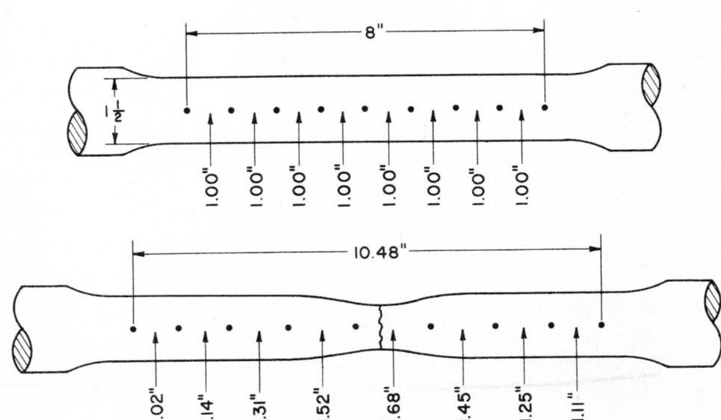

FIGURE 2-8. Elongation in various sections of a tensile-test specimen.

plasticity is the controlling property, the terms above relate to the material response to specific processes and as such do not describe a material property.

If the material fails with little or no ductility, it is said to be *brittle.* Thus brittleness can be viewed as the opposite of ductility. Brittleness should not be considered as the lack of strength, however, but simply the lack of significant plasticity.

Toughness. *Toughness* is defined as the work per unit volume required to fracture a material and is commonly expressed as a *modulus of toughness.* One means of measuring toughness is through the tensile test, for the total area under the stress–strain curve represents the energy required to produce a fracture in a unit volume of material.

Caution should be exercised in the use of toughness data, however, for the values can vary markedly with different conditions of testing. As will be seen later, variation in temperature and load-application rate can markedly change the nature of a material's stress–strain curve and, hence, the toughness. Toughness is commonly associated with impact or shock loadings. The values obtained from impact tests, however, often fail to correlate with those from static-type tests.

True stress/true strain curves. The stress–strain curve of Figure 2-5 is a plot of *engineering stress, s,* versus *engineering strain, e,* where s is computed as load (W) divided by the *original* cross-sectional area (A) and e is the elongation (ΔL) divided by the *original* gage length (L_o). As noted previously and illustrated in Figures 2-7 and 2-8, the cross section of the test bar changes as the test proceeds, first uniformly and then nonuniformly after necking begins. The actual stress within the specimen, therefore, should be based on the instantaneous cross-sectional area and not the original and will be greater than the engineering stress shown in Figure 2-5. True stress, σ, can be computed by taking simultaneous readings of load and minimum specimen diameter. The true area can be computed and true stress determined as

$$\sigma = \frac{W}{A}$$

as opposed to the engineering stress,

$$s = \frac{W}{A_o}$$

The determination of the true strain is somewhat more complex. What is desired is the strain at a point, since we have already seen strain variation throughout the specimen. One way is to utilize the engineering strain expression with an infinitesimal gage length, l_o, such that

$$e' = \frac{l - l_o}{l_o} = \frac{l}{l_o} - 1$$

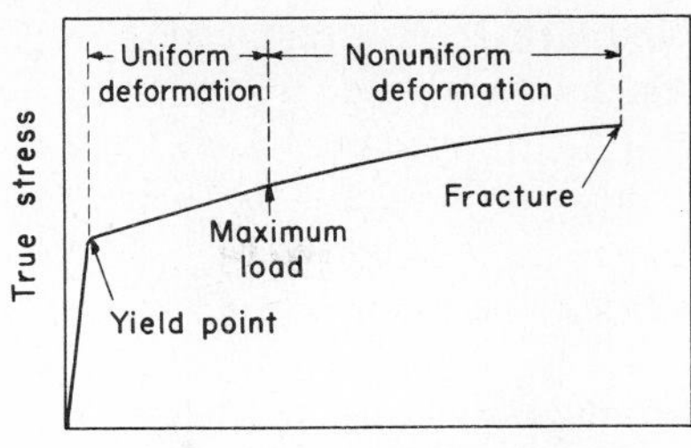

FIGURE 2-9. True stress/true strain curve.

This strain, e', is often called the *mean strain,* or, more correctly, the *zero gage length strain.* Although the infinitesimal gage length, l_0, cannot be measured directly, volume constancy enables determination through measurements of diameter and is valid up to necking.

$$\frac{l}{l_0} = \frac{A_0}{A} = \frac{D_0^2}{D^2}$$

A more precise and useful measure for true strain is the *natural strain* or *logarithmic strain,* which is the integral summation of the incremental elements, expressed as

$$\epsilon = \int_{l_0}^{l} \frac{de}{l} = \ln\frac{l}{l_0} = 2\ \ln\frac{D_0}{D}$$

Before the onset of necking, where strain is uniform, the natural strain can be related to the engineering strain by

$$\epsilon = \ln\ (e + 1)$$

Figure 2-9 shows the type of curve that results when the uniaxial tensile test data are plotted in the form of true stress vs. true strain. It should be noted that the true stress of the material, a measure of the material strength at that point, continues to rise throughout the test, even after necking. Data beyond the point of necking should be used with extreme caution, however, for once the geometry of the neck begins to form, the stress state in that region becomes a triaxial tension instead of the uniaxial tension assumed for the test. Voids or cracks, such as in Figure 2-10, tend to open in the necked region as a preface to failure. Diameter measurements no longer reflect the true load-bearing area and the data are further distorted.

Repeated static loads. In many applications, static loads are applied more than once. In order to understand how a ductile metal, such as steel, behaves when subjected to such repeated slow loading and unloading, refer to the stress–strain diagram of Figure 2-11. Unloading and reloading within the elastic range results in simply cycling up and down the linear portion of the

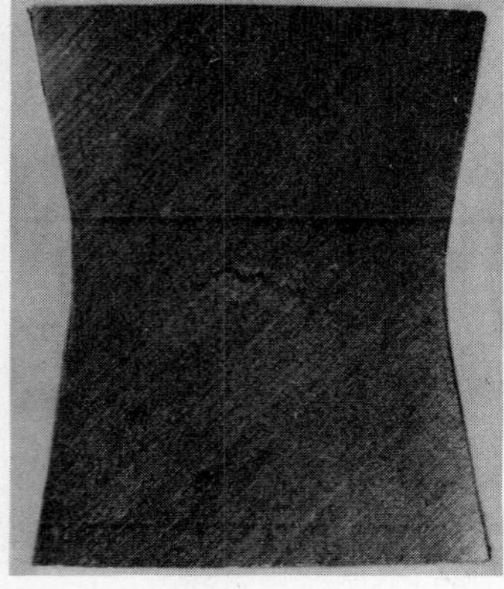

FIGURE 2-10. Section of a tensile-test specimen stopped just prior to failure, showing a crack already started in the necked region. (*Courtesy E. R. Parker.*)

diagram between O and A. However, if unloading takes place from point B in the plastic region, the unloading curve follows the path BeC, which is approximately parallel to OA. The permanent set at this point would be OC. Reloading from C would tend to follow the curve CfD, a slightly different path from that of unloading. The area within the loop formed by the two paths is called a *hysteresis loop.* The energy represented by this loop corresponds to the energy per unit volume that was transformed to heat within the material during the unloading and reloading cycle.

When most materials are plastically deformed, they *work-harden;* that is, they become harder and the yield-point stress is raised. This is a progressive phenomenon, with the result that, as the applied load is increased to produce plastic deformation, a greater load will be required to produce further deformation. In our example above, the work-hardening properties of the material have raised the yield point to D and elastic behavior is experienced up to this point upon reloading. Beyond the new yield point, D, additional plastic deformation takes place. If unloading then took place at E, with subsequent reloading as indicated, another hysteresis loop would be formed, and further

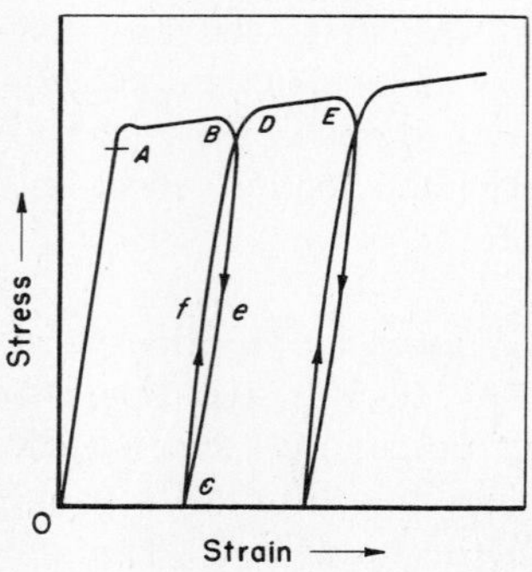

FIGURE 2-11. Stress–strain diagram obtained by unloading and reloading a specimen.

raising of the yield point would occur. From this example it may be seen that beyond the elastic region the true stress/true strain curve actually represents the locus of the yield stress for various amounts of strain.

Damping capacity. The hysteresis loop, discussed in the previous section, was caused by some of the mechanical energy that was put into the material during the loading and unloading cycle being converted into heat energy. This process produces *mechanical damping,* and materials that possess this property to a high degree are able to absorb mechanical vibrations or damp them out rapidly. This is an important property of materials for certain uses, such as crankshafts and engine bases. Gray cast iron is used in many applications because of its high damping capacity. Materials with lower damping capacities, such as brass and steel, will continue to ring when struck by a blow.

Hardness. *Hardness* is a very important, yet difficult to define, property of materials. Numerous tests have been developed around the definition of resistance to permanent indentation under static or dynamic loading. Other tests evaluate resistance to scratching, energy absorption under impact loading, wear resistance, or even resistance to cutting or drilling. Clearly, these phenomena are not the same. Thus, although hardness can be measured by a variety of well-standardized tests, there may be no correlation between the results obtained from the various tests. Caution should be exercised so that the test selected clearly evaluates the phenomenon of interest.

Brinell hardness test. One of the earlier standardized methods of measuring hardness was the *Brinell test.* A hardened steel ball 1 centimeter in diameter is pressed into a smooth surface of material by a standard load of 500, 1500, or 3000 kilograms. The load and ball are removed and the diameter of the spherical indentation is measured, usually by means of a special grid or traveling microscope. The numerical Brinell hardness number is equal to the load divided by the spherical surface area of the indentation expressed in kilograms per square millimeter:

$$\text{Brinell hardness number (BHN)} = \frac{\text{load}}{\text{surface area of indentation}}$$

In actual practice, the Brinell hardness number is determined from tables that tabulate the number versus the diameter of the indentation.

The Brinell test is subject to several limitations:

1. It cannot be used on very hard or very soft materials.
2. The test may not be valid for thin specimens. Preferably, the thickness of the material should be at least 10 times the depth of the indentation. Standards specify minimum hardnesses for which tests on thin specimens are valid.
3. The test is not valid for case-hardened surfaces.

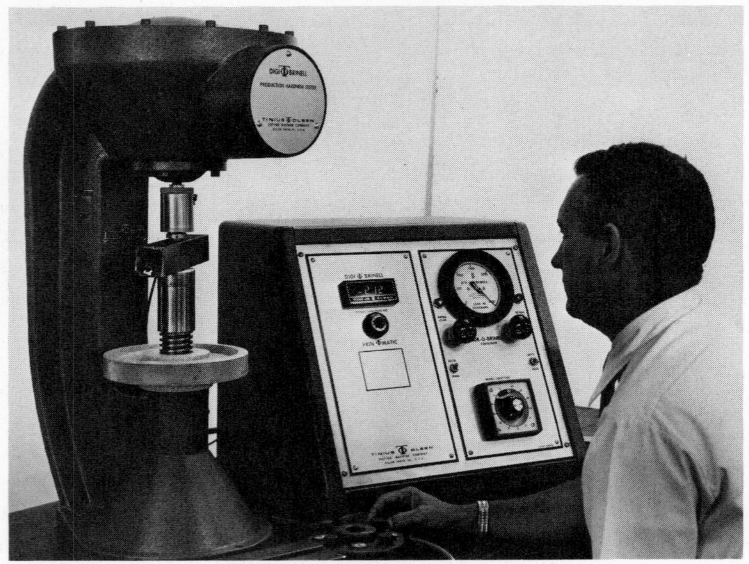

FIGURE 2-12. Brinell hardness tester. (*Courtesy Tinius Olsen Testing Machine Co., Inc.*)

4. The test should be conducted on a location far enough removed from the edge of the material so that no edge bulging results.
5. The noticeable indentation may be objectionable on finished parts.
6. The edge of the indentation may not always be clearly defined or may be rather difficult to see on material of certain colors.

Nevertheless, the test is not difficult to conduct and has the advantage that it measures the hardness over a relatively large area and, therefore, does not reflect small-scale variations. It is used to a large extent on iron and steel castings. Figure 2-12 shows a standard Brinell tester. Relatively small, portable testers are also available.

Rockwell hardness test. The widely used *Rockwell hardness test* is similar to the Brinell test in that the hardness value is a function of the indentation of a test piece by an indentor under static load. The nature of the test can be explained in connection with Figure 2-13. A small indentor, either a $^{1}/_{16}$-inch ball or a diamond cone called a *brale,* is first seated firmly in the material by the application of a "minor" load of 10 kilograms, causing a very small indentation. The indicator on the dial of the tester, shown in Figure 2-14, is set at zero and a "major" load is then applied to the indentor to produce a deeper indentation. After the indicating pointer has come to rest, the major load is removed. With the major load removed, the pointer now indicates the Rock-

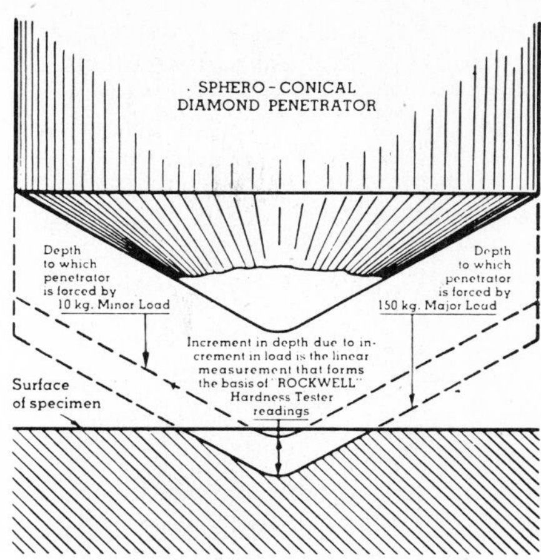

FIGURE 2-13. Operating principle of the Rockwell hardness tester. (*Courtesy Wilson Instrument Division, ACCO, Inc.*)

well hardness number on the appropriate scale of the dial. This number is a measure of the *plastic* or *permanent penetration* produced by the major load.

Different combinations of major loads and indentors are available and are used for materials of various degrees of hardness. Table 2-1 provides a partial

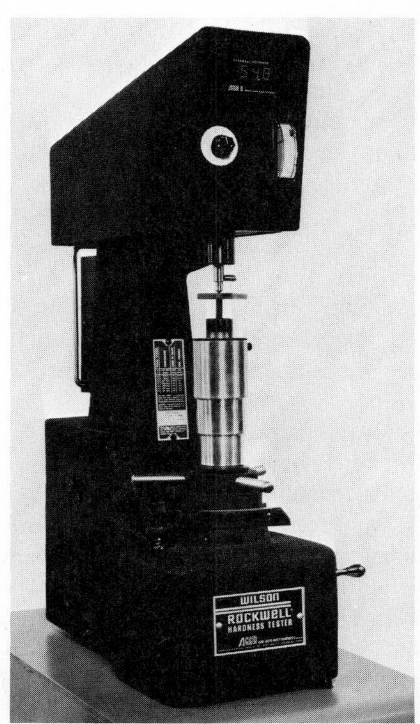

FIGURE 2-14. Rockwell hardness tester having digital readout. (*Courtesy Wilson Instrument Division, ACCO, Inc.*)

TABLE 2-1. Loads and indentors for Rockwell hardness tests

Test	Load (kg)	Indentor
A	60	Brale
B	100	$^1/_{16}$-in. ball
C	150	Brale
D	100	Brale
F	60	$^1/_{16}$-in. ball
G	150	$^1/_{16}$-in. ball

listing of available Rockwell hardness scales. Since multiple scales exist, Rockwell hardness numbers must always be accompanied by a letter indicating the particular combination of load and indentor used in the test. The notation R_C 60 (Rockwell C 60) indicates that the brale indentor was used with a major load of 150 kilograms and a reading of 60 was obtained. The B and C scales are used more extensively than the others.

The standard Rockwell tester should not be used on materials less than $^1/_6$ inch thick, on rough surfaces, or on materials that are not homogeneous, such as gray cast iron. As a result of the small size of the indentation, localized variations of roughness, composition, or structure can greatly influence the results. For thinner materials or purposes where a very shallow indentation is desired, the *Rockwell superficial-hardness test* is used. Operating on the same Rockwell principle, the test employs smaller major and minor loads and uses a more sensitive depth-measuring device. The test was designed primarily for determining the hardness of thin sheet metal and the surface hardness of materials that have received such surface treatments as nitriding or carburizing.

In comparison with the Brinell test, the Rockwell test offers the benefit of direct readings in a single step. Since it can be conducted rapidly, it is suitable for routine tests of hardness in mass production. Furthermore, it has the additional advantage that the smaller indentation is often not objectionable to the appearance of the component or is more easily removed in a later operation.

Vickers hardness test. The *Vickers hardness test* is similar to the Brinell test, with a square-based diamond pyramid being used as the indentor. As in the Brinell test, the Vickers hardness number is the ratio of the load to the surface area of the indentation in kilograms per square millimeter. An advantage of the Vickers machine, shown in Figure 2-15, is the increased accuracy in determining the diagonal of a square as opposed to the diameter of a circle.

Numerous advantages make the indentation hardness methods quite popular: (1) simple to conduct, (2) little time involved, (3) little surface preparation required, (4) can be done on location, (5) is relatively inexpensive, and

FIGURE 2-15. Vickers hardness tester. (*Courtesy Riehle Testing Machines Division, American Machine and Metals, Inc.*)

(6) often provides results that can be correlated to quality control, material strength, and so on.

Ultrasonic hardness tests. The *Sonodur hardness tester* operates on a completely different principle, making it rather useful for certain conditions. A magnetostrictive, diamond-tipped rod, vibrating at ultrasonic frequencies, is brought into contact with the metal at a load of 1½ pounds. Tip penetration, which is a function of the hardness of the metal, causes the vibration frequency of the rod to change. This change is then read on a calibrated scale in terms of either Rockwell or Vickers hardness numbers. The equipment, as shown in Figure 2-16, is readily portable, and the shallowness of the indentation—less than 0.013 millimeter—makes it particularly suitable for use where a larger indentation would be objectionable. Unfortunately, its use is restricted primarily to steel.

Microhardness tests. Various microhardness tests are available for use when it is necessary to determine hardness over a very small area of a material. The *Tukon tester,* shown in Figure 2-17, is typical. The position for the test is selected under high magnification. A small diamond penetrator is then loaded with a predetermined load of from 25 to 3600 grams. The hardness number, known as a *Knoop hardness number,* is then obtained by dividing the load (in kilograms) by the projected area of the now diamond-shaped indentation (in square millimeters). The length of the indentation is determined using a microscope, because the mark is very small.

Durometer hardness test. For testing very soft, elastic materials, such as rubbers and nonrigid plastics, a *Durometer* is often used. This instrument, shown in Figure 2-18, measures the resistance of the material to elastic penetration

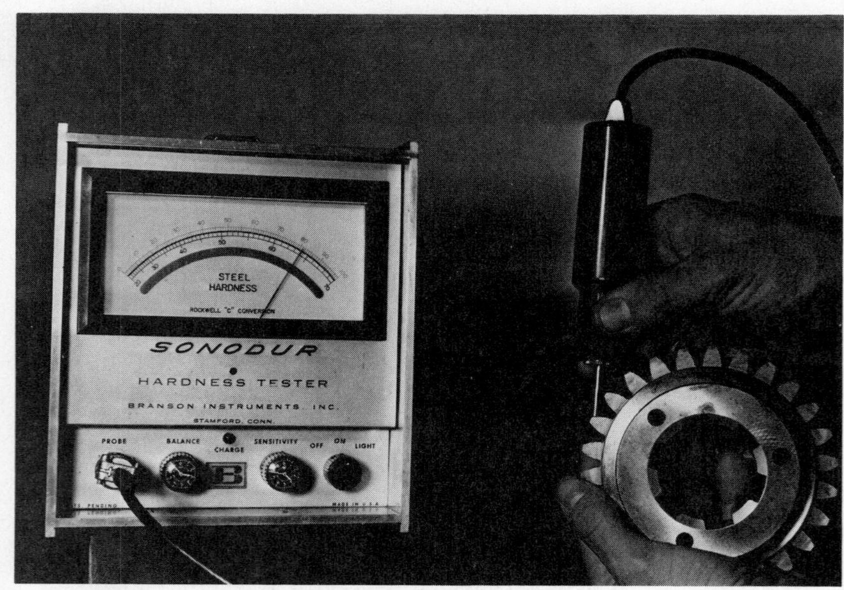

FIGURE 2-16. Sonodur ultrasonic hardness tester checking the hardness of gear teeth. (*Courtesy Branson Instruments, Inc.*)

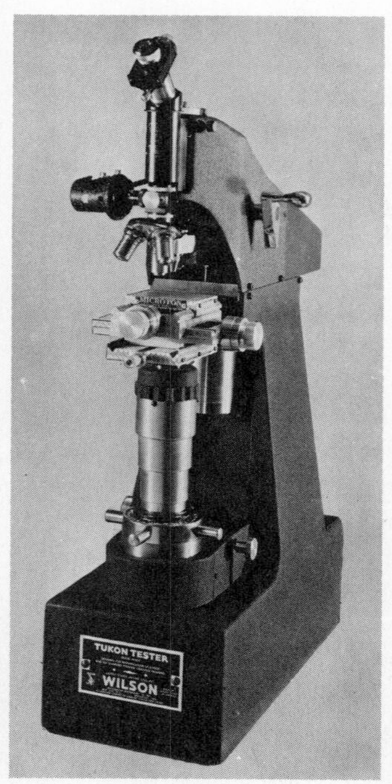

FIGURE 2-17. Tukon microhardness tester. (*Courtesy Wilson Instrument Division, ACCO, Inc.*)

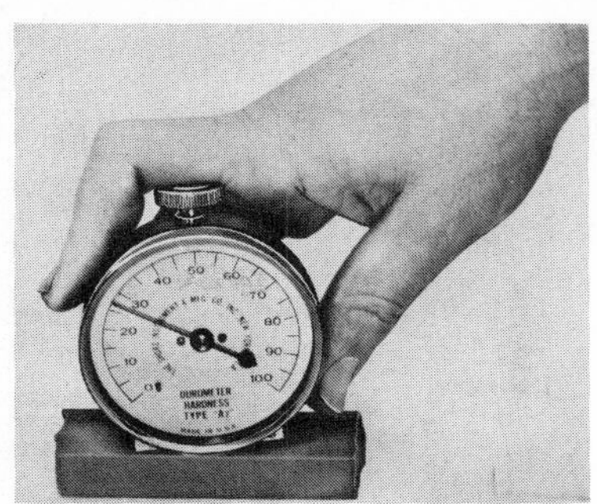

FIGURE 2-18. Durometer hardness tester. (*Courtesy The Shore Instrument & Mfg. Company, Inc.*)

by a spring-loaded conical steel indentor. No permanent deformation occurs. A similar test is used to evaluate molding sands in the casting industry.

Scleroscope hardness test—a dynamic test. In the *Scleroscope test,* hardness is measured by the rebound of a small, diamond-tipped "hammer" that is dropped from a fixed height onto the surface of a material being tested, using a test instrument such as in Figure 2-19. Obviously, this test measures the resilience of the material, and the surface on which the test is made must have a fairly high polish to obtain good results. Scleroscope hardness numbers are comparable only among similar materials. A comparison between steel and rubber, therefore, would not be valid.

FIGURE 2-19. Two types of Scleroscope hardness testers. (*Courtesy The Shore Instrument & Mfg. Company, Inc.*)

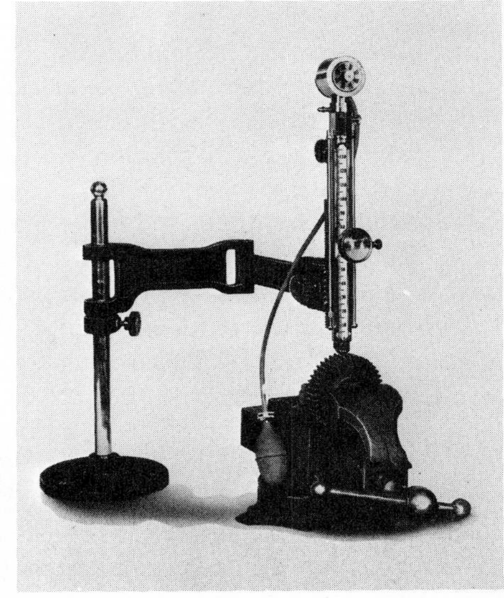

Scratch hardness tests. As mentioned previously, hardness can also be defined as the ability of a material to resist being scratched. At least two types of tests have been developed to determine hardness by this method.

The *Mohs hardness scale* arranges 10 minerals in order of ascending hardness as follows:

1. Talc
2. Gypsum
3. Calcite
4. Fluorite
5. Apatite
6. Feldspar
7. Quartz
8. Topaz
9. Sapphire or corundum
10. Diamond

According to this scale, a given material should be able to scratch any material having a lower Mohs number. In this manner, any substance can be assigned an approximate number on the Mohs scale. Ordinary glass, for example, would be about 5.5; hardened steel, about 6.5. This test is sufficiently crude that it is not suitable for manufacturing purposes, but it is quite useful in mineral identification.

Another crude, but often useful, hardness test is the *file test,* wherein one determines whether the material can be cut with a file. This test can be either a pass–fail test using a single file, or a semiquantitative evaluation using a series of files pretreated to various levels of known hardness.

Relationships among the various hardness tests. Since the various tests tend to evaluate somewhat different material phenomena, there is no simple relationship between the several types of hardness numbers that can be determined. Approximate relationships have been developed, however, by testing the same material on various devices. Table 2-2 gives a comparison of hardness values for plain carbon and low-alloy steels. It may be noted that for Rockwell C numbers above 20, the Brinell numbers are approximately 10 times the Rockwell numbers. Also, for hardnesses below 320 Brinell, the Vickers and Brinell numbers agree closely. Since the relationships will vary with material, mechanical treatment, and heat treatment, tables such as Table 2-2 should be used with caution.

Relationship of hardness to tensile strength. Table 2-2 also shows a comparison of hardness and tensile strength for steel. For plain carbon and very low alloy steels, the tensile strength (in psi) can be determined fairly well by multiplying the Brinell hardness number by 500. This provides a simple, and very useful, method of determining the approximate tensile strength of a steel by means of a hardness test. For other materials, the relationship may be too variable to be dependable. For example, for duraluminum the ratio is about 600, whereas for soft brass it is about 800.

Compression tests. When a material is subjected to compressive forces, the relationships between stress and strain are similar to those for a tension test.

Up to a certain value of stress, the material behaves elastically; beyond it plastic flow occurs. In general, however, the compression test is more difficult and more complex than the standard tensile test. Test specimens must have larger cross-sectional areas to resist bending or buckling. As deformation proceeds,

TABLE 2-2. Hardness conversion table

Brinell Number	Vickers Number	Rockwell Number		Scleroscope Number	Tensile Strength	
		C	B		MPa	1000 psi
	940	68		97	2537	368
757[a]	860	66		92	2427	352
722[a]	800	64		88	2324	337
686[a]	745	62		84	2234	324
660[a]	700	60		81	2144	311
615[a]	655	58		78	2055	298
559[a]	595	55		73	1903	276
500	545	52		69	1765	256
475	510	50		67	1703	247
452	485	48		65	1641	238
431	459	46		62	1462	212
410	435	44		58	1407	204
390	412	42		56	1351	196
370	392	40		53	1303	189
350	370	38	110	51	1213	176
341	350	36	109	48	1138	165
321	327	34	108	45	1069	155
302	305	32	107	43	1007	146
285	287	30	105	40	951	138
277	279	28	104	39	924	134
262	263	26	103	37	883	128
248	248	24	102	36	841	122
228	240	20	98	34	800	116
210	222	17	96	32	738	107
202	213	14	94	30	683	99
192	202	12	92	29	655	95
183	192	9	90	28	627	91
174	182	7	88	26	600	87
166	175	4	86	25	572	83
159	167	2	84	24	552	80
153	162		82	23	524	76
148	156		80	22	510	74
140	148		78	22	490	71
135	142		76	21	469	68
131	137		74	20	455	66
126	132		72	20	441	64
121	121		70		427	62
112	114		66			58

[a] Tungsten carbide ball; others standard ball.

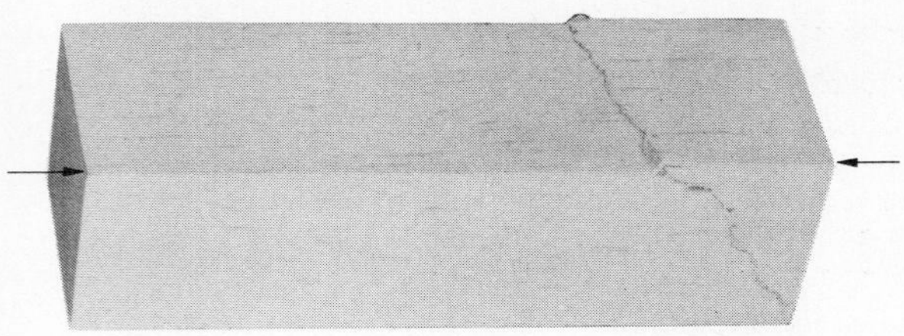

FIGURE 2-20. Failure of wood under compressive loading.

the cross section of the specimen tends to increase producing a substantial increase in required load (true stress versus true strain is substantially the same for both cases). Frictional effects between the testing machine surfaces and the end surfaces of the specimen will tend to alter the results if not properly considered. The selection of the tension or compression mode of testing, however, is largely determined by the type of service to which the material is to be subjected.

Failure under compressive loading is generally by buckling or by shear along a plane at 45° to the axis of loading. Figure 2-20 shows a compression failure of a wood specimen.

Dynamic properties. In many engineering applications, materials are subjected to dynamic loadings. Such cases may involve components that (1) experience sudden loads or loads which rapidly vary in magnitude, (2) are loaded and unloaded repeatedly, or (3) undergo frequent changes in loading modes, such as from tension to compression. For such service conditions, the engineer is concerned with properties other than those determined by the static tests.

Unfortunately, the dynamic tests are not as well standardized, or controlled as easily, as the static tests. In addition, many of the dynamic tests do not give results that can be used directly in design. In such cases, the tests merely classify materials relative to each other as to their behavior when subjected to certain loading conditions. Nevertheless, they can serve a very useful purpose, provided that one remembers their limitations.

The impact test. Several tests have been developed to evaluate the fracture resistance of a material under rapidly applied dynamic loads, or *impacts.* Of those tests that have become common, two basic types have emerged: (1) bending impact tests, which include the standard Charpy and Izod tests, and (2) tension impact tests.

The bending impact tests utilize specimens that are supported as beams. In the *Charpy test,* the specimen contains either a V, keyhole, or U notch—the keyhole and V, as shown in Figures 2-21 and 2-22, being most common. For the Charpy test, the specimen is supported as a simple beam, and the impact is applied to the center, behind the notch, to complete a three-point bending. The *Izod test* specimen is supported as a cantilever beam and is impacted on the end as in Figure 2-22. Standard testing machines, such as the one shown

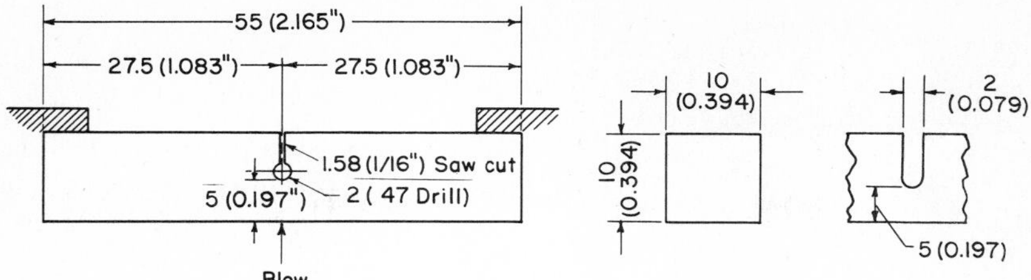

FIGURE 2-21. Standard Charpy impact specimens and mode of loading.

in Figure 2-23, apply a predetermined impact load of up to 162.7 joules (120 ft-lb) of energy by means of a swinging pendulum. After breaking or deforming the specimen, the pendulum continues to swing with an energy equal to its original minus that absorbed by the broken specimen. This loss is measured by the angle attained by the pendulum on its upward swing.

Bending impact specimens must be prepared with careful precision to assure consistent and reproducible results. Notch profile, particularly the radius at the root of V-notch specimens, is extremely critical, for the test measures the energy required to both initiate and propagate a fracture. The effect of notch profile is shown dramatically in Figure 2-24. Here, two specimens have been made from the same piece of steel with the same reduced cross-sectional area. The one with the keyhole notch fractures and absorbs only 58 joules (43 ft-lbs) of energy, while the other specimen resists fracture and absorbs 88 joules (65 ft-lbs) during impact.

Additional cautions should be placed on the use of impact test data for design purposes. These tests only indicate the impact resistance of materials that contain a standardized notch. Changes in the form of the notch or minor variations from standard geometry can produce significant changes in the results. The test also evaluates a standard specimen under only one condition of impact rate. Under modified test conditions using faster rates of loading or wider specimens with a higher degree of constraint, many materials will behave in a more brittle fashion. Bending impact tests are valuable in ranking materials as to their sensitivity to notches and the multiaxial stresses that exist

FIGURE 2-22. Izod impact specimen and mode of loading.

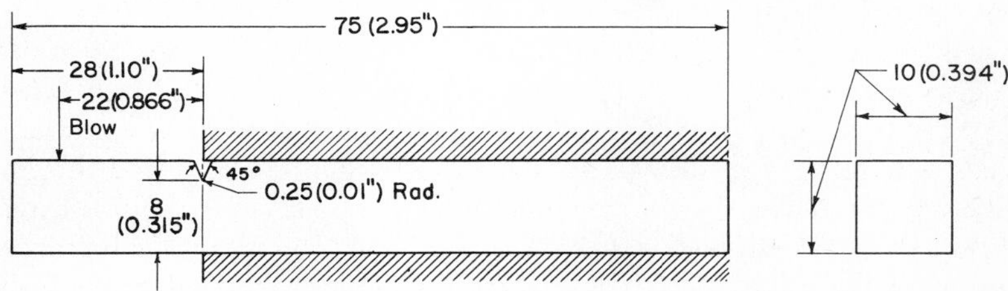

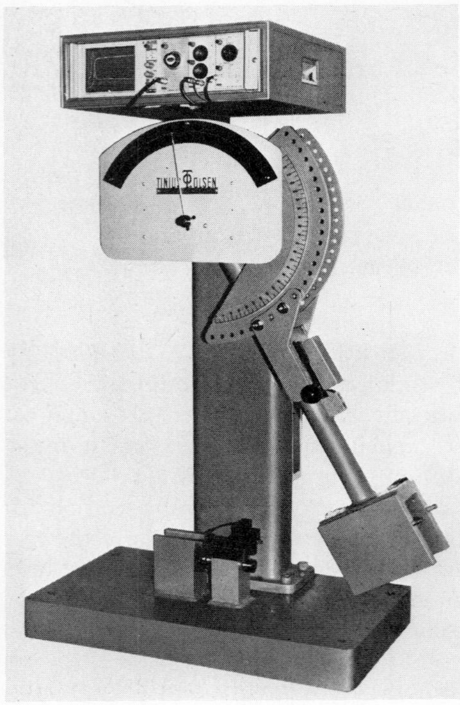

FIGURE 2-23. Impact testing machine. (*Courtesy Tinius Olsen Testing Machine Co., Inc.*)

around a notch. In addition, testing temperature can be varied to enable evaluation of the fracture resistance of a material as a function of temperature. Such information can provide invaluable information to the engineer involved in material selection.

FIGURE 2-24. Notched and unnotched impact specimens before and after testing. Both specimens had the same cross-sectional area.

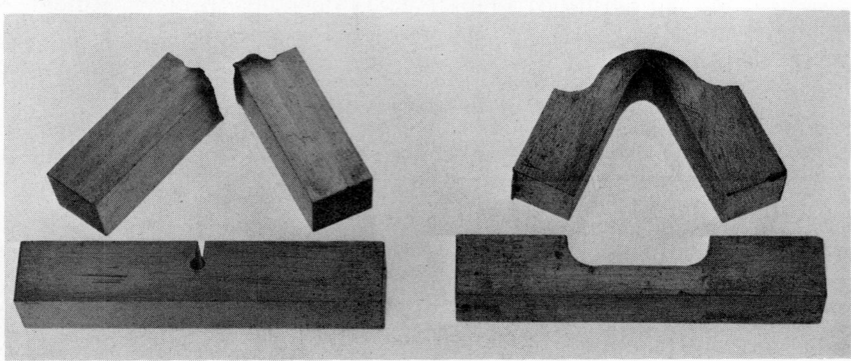

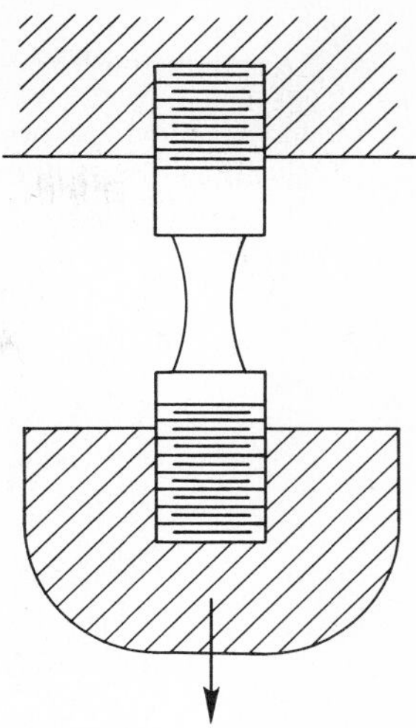

FIGURE 2-25. Tensile impact test schematic.

The *tensile impact test,* illustrated schematically in Figure 2-25, avoids many of the objections inherent in the Charpy and Izod tests, but is more difficult to perform. The behavior of ductile materials under uniaxial impact loading can be studied without the complications introduced by the use of a notched specimen. Various methods, such as drop-weight, modified pendulum, and variable-speed flywheel, have been used to supply the impact.

Metal fatigue and endurance limit. Metals may also fracture when subjected to repeated variations in applied stress, even though all applied stresses lie below the ultimate tensile strength and usually below the yield strength determined by a tensile specimen. This phenomenon, known as *metal fatigue,* may result from a repetition of a particular loading cycle or from an entirely random variation of stress. Since such failures probably account for more than 90 percent of all mechanical fractures, it is important for the engineer to know how metals will react to fatigue conditions.

Although there are an infinite number of possible repeated loadings, the periodic, sinusoidal mode is most suitable to experimental reproduction and subsequent analysis. Restricting conditions even further by considering only equal-magnitude tension–compression reversals, curves such as that of Figure 2-26 can be developed. If this material were subjected to a normal static tensile test, it would break at about 480 MPa (70,000 psi). However, if it were repeatedly subjected to a reversing stress of ±380 MPa (55,000 psi)—considerably below its breaking stress—it would fail when the load was repeated about 100,000 times. Similarly, if a stress of ±350 MPa (51,000 psi) were applied, 1,000,000 cycles could be sustained prior to failure. By reducing the applied stress to below ±340 MPa (49,000 psi), the metal will not fail regardless of the number of stress applications. Such a curve is known as a stress

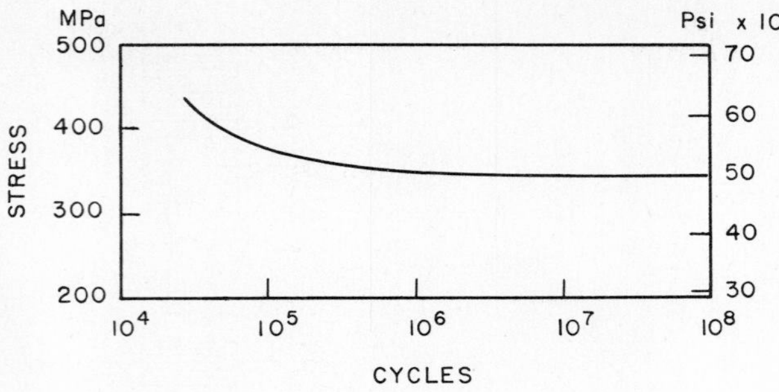

FIGURE 2-26. Typical *S–N* or endurance limit curve for steel.

versus number of cycles or *S–N curve.* Any point on the curve is the *fatigue strength* corresponding to the given number of cycles of loading. The limiting stress level, below which the metal will not fail regardless of the number of cycles of loading, is known as the *endurance limit* or *endurance strength* and is an important criterion in many design applications.

A different number of cycles of loading is required to reveal the endurance limit of different materials. For steels, about 10 million (10^7) cycles are usually sufficient. Several of the nonferrous metals require 500 million (5×10^8) cycles. Some aluminum alloys require an even greater number, such that no endurance limit is apparent under typical test conditions.

The apparent fatigue strength of materials may be affected by several factors. One of the most important of these is the presence of stress raisers such as small surface cracks, machine cracks, and so on. Data for *S–N* curves are obtained from polished specimens, and the observed lifetime is the cumulative number of cycles required to initiate a fatigue crack and propagate it to failure. If a part contains a surface crack or flaw, the cycles required for crack initiation can be markedly reduced. In addition, stresses concentrate at the tip of the crack producing an accelerated rate of crack growth. Great care should be taken to eliminate stress raisers and surface flaws on parts subjected to cyclic loadings. Proper design and manufacturing practices are often more critical than material selection and heat treatment for fatigue applications.

Another factor worthy of consideration is the temperature of testing. Figure 2-27 shows the shifts in the *S–N* curve for Inconel alloy 625 (Ni–Cr–Fe alloy) as temperature is varied. Since most test data are generated at room temperature, caution should be exercised when the application involves elevated service temperatures.

Fatigue lifetime may be further shortened when metals are subjected to corrosion simultaneously with repeated loading, a situation known as *corrosion fatigue.* With the exception of a few high-alloy steels, this condition results in very low endurance limits. Where corrosion conditions are significant, special corrosion resistant coatings, such as zinc or cadmium, may be employed.

When the magnitude of the applied stress is varied during service, a condition common to many components, the fatigue response of the metal

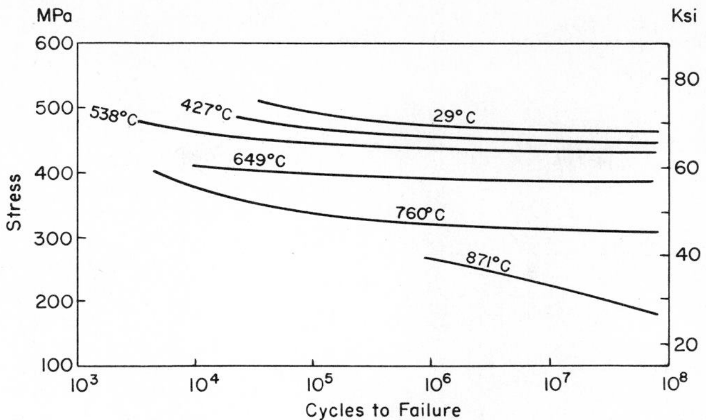

FIGURE 2-27. Fatigue strength of Inconel alloy 625 at various temperatures. (*Courtesy Huntington Alloy Products Division, The International Nickel Company, Inc.*)

becomes quite complex. Low-stress cycles are less damaging to the material. A few high-stress cycles may substantially reduce the expected lifetime. Such variations and their response are of significant importance to design engineers.

Table 2-3 shows the approximate value of the endurance limit to ultimate tensile strength ratios for several metals.

Fatigue failures. Metal components that fail as the result of repeated applications of load and the fatigue phenomenon are commonly called *fatigue failures.* These fractures are a major part of a larger classification known as *progressive fractures.* If the fracture surface of Figure 2-28 is examined closely, two points of fracture initiation can be located. These points usually correspond to discontinuities, in the form of a fine surface crack, a sharp corner, machining marks, or even "metallurgical notches" such as the abrupt change in metal

TABLE 2-3. Ratio of endurance limit to tensile strength for various materials

Material	Ratio
Steel, AISI 1035	0.46
Steel, screw stock	0.44
Steel, AISI 4140 normalized	0.54
Wrought iron	0.63
Copper, hard	0.33
Beryllium copper (heat-treated)	0.29
Aluminum	0.38
Magnesium	0.38

FIGURE 2-28. Progressive fracture of an axle within a ball-bearing ring, starting from two points (arrows).

structure which led to the initiation of the fracture in Figure 2-28. Once started, the crack propagates through the metal upon repeated application of load, crack growth being due to the stress at the tip of the crack exceeding the strength of the material. Crack propagation continues until the remaining section of metal no longer has sufficient area to sustain the applied load, at which time complete failure of the remaining section occurs. The section of metal involved in this final failure will have a relatively coarse, granular appearance; whereas the section between the origin of the crack and the coarse-appearing area will be relatively smooth. This smoothness may be attributable to the fact that the two free surfaces of the crack rubbed together under the repeated cyclic loadings.

The smooth areas of the fracture may often contain a series of crescent shaped ridges radiating outward from the origin of the crack. These markings, however, may not be observable under ordinary visual examination. They may be very fine, they may have been obliterated by the rubbing action, or there may only be a few such marks if failure occurred after only a few cycles of loading ("low-cycle fatigue"). Electron microscope studies of the fracture surface can often reveal these small parallel ridges, or *striations,* which are characteristic of progressive failure. Figure 2-29 presents a typical fatigue fracture at high magnification.

Because the final area of fracture has a crystalline appearance, it has often been said that such failures are due to the metal having "crystallized." Since solid metals are always crystalline, such a conclusion is obviously erroneous and the term should not be applied.

Another common misnomer is to apply the term "fatigue failure" to all fractures having the characteristic progressive failure appearance. The general appearance of fractures where fatigue was a major factor is often the same as for other fractures where fatigue may be only a minor contributor. Also, the same fracture phenomena may lead to different general appearances depending on the specific conditions of load magnitude, load type (torsional, bending,

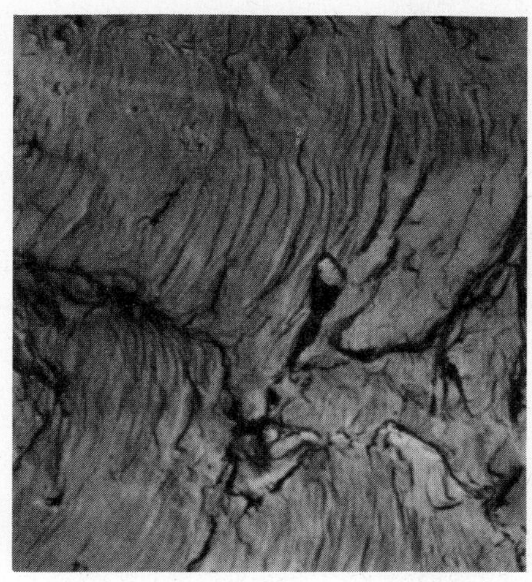

FIGURE 2-29. Electron fractograph of AISI 1018 steel fractured in fatigue, showing the fatigue striations (5000×).

tension), temperature, and so on. Correct failure analysis requires far more information than can be provided by the examination of a fracture surface.

A final fact regarding failure by fatigue relates to the misconception that failure is time-dependent. The failure of materials under repeated loads below their static strengths is not a function of time, but is dependent upon the history of loading. High cyclic frequencies can produce failure in relatively short time intervals.

Temperature effects. It cannot be overemphasized that test data used in design and engineering decisions should be obtained under conditions that best simulate the conditions of service. Engineers are frequently being confronted with the design of structures, such as aircraft, space vehicles, gas turbines, and nuclear power plants, that require operation under temperatures as low as −130°C (−200°F) or as high as 1250°C (2300°F). Consequently, it is imperative for the designer to know both the short-range and long-range effects of temperature on the mechanical and physical properties of a material being considered for such applications. From a manufacturing viewpoint, the effects of temperature variations are equally important. Since numerous manufacturing processes involve the use of heat, the processing may tend to alter the properties in a favorable or unfavorable manner. Often a material can be processed successfully, or economically, only because its properties can be changed by heating or cooling.

From a manufacturing viewpoint, the most important effects of temperature on materials are probably those relating to the tensile properties. Figure 2-30 illustrates changes in key data for the case of a medium-carbon steel. Similar effects are shown for magnesium in Figure 2-31. In general terms, an increase in temperature tends to promote a drop in strength properties and an

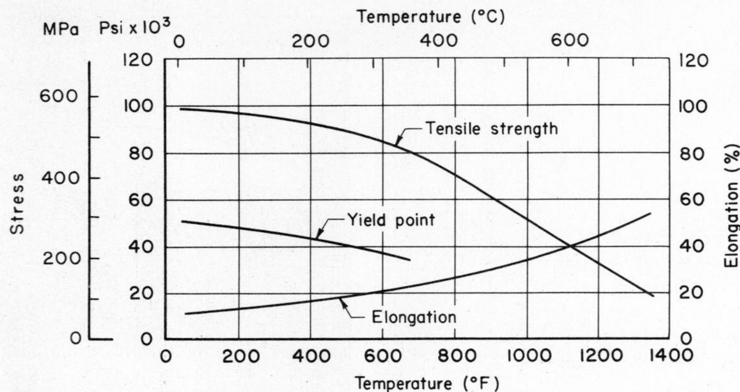

FIGURE 2-30. Some effects of temperature on the tensile properties of a medium-carbon steel.

increased elongation. In forming operations, these trends are of considerable importance because they permit forming to be done more readily at elevated temperatures. The material is weaker and more ductile.

Figure 2-32 adds another dimension by showing the effects of both temperature and strain rate on the ultimate tensile stress. From this graph it can be clearly seen that the rate of deformation can strongly influence mechanical properties. Room-temperature, standard-rate tensile test data will be of little use to the engineer concerned with the properties of a material being hot rolled at speeds of 1200 meters per minute (4600 feet per minute). The effects of strain rate on the more important yield strength value are more difficult to evaluate, but follow the same trends as tensile strength.

The effect of temperature on impact properties became the subject of in-

FIGURE 2-31. Effects of temperature on the tensile properties of magnesium.

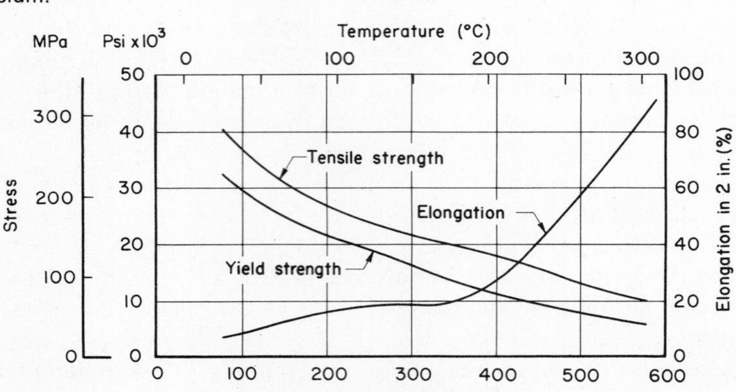

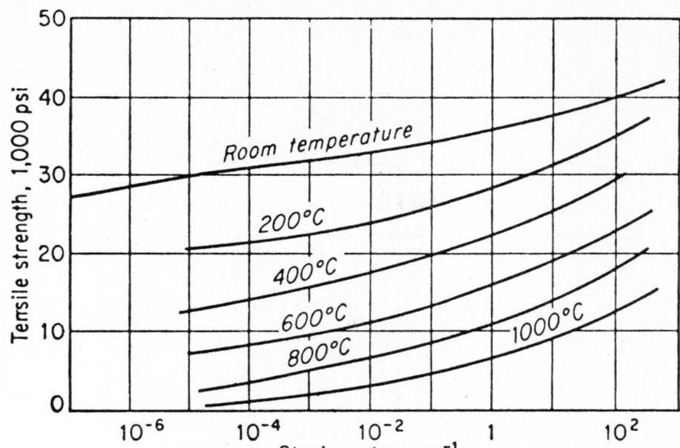

FIGURE 2-32. Effects of temperature and strain rate on the tensile strength of copper. (*From A. Nadai and M. J. Manjoine, J. Appl. Mech., Vol. 8, p. A82, 1941, courtesy ASME.*)

tense study when ships, structures, and components fractured unexpectedly in cold environments. Figure 2-33 shows the effect of decreasing temperature on the impact properties of two low-carbon steels. Although of similar compositions, the steels show distinctly different response. The steel indicated by the solid line becomes brittle at temperatures below −4°C (25°F), while the other steel retains its fracture resistance down to −26°C (−15°F). The temperature at which the response goes from high to low energy absorbtion is known as the *transition temperature* and is useful in evaluating the suitability of materials for certain applications. All steels tend to exhibit the rapid transition in impact strength when temperature is decreased, but the temperature at which it occurs varies with the material. Special steels with high nickel contents and several other alloys have been developed for cryogenic applications requiring retention of impact resistance to −195°C (−320°F).

Creep. The long-term effect of temperature is manifest in a phenomenon known as *creep.* If a tensile-type specimen is subjected to a fixed load at an

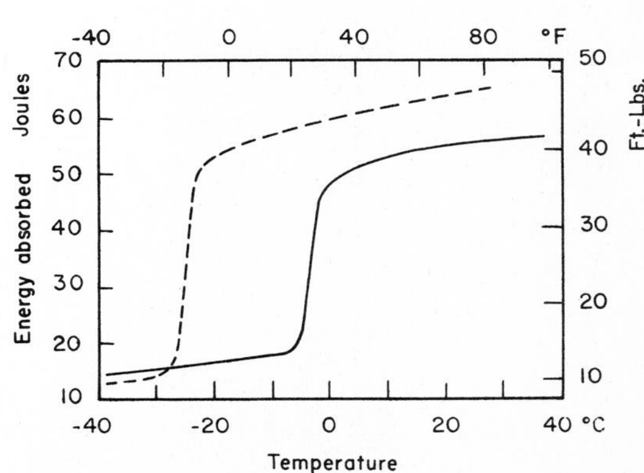

FIGURE 2-33. Effect of temperature on the impact properties of two low-carbon steels.

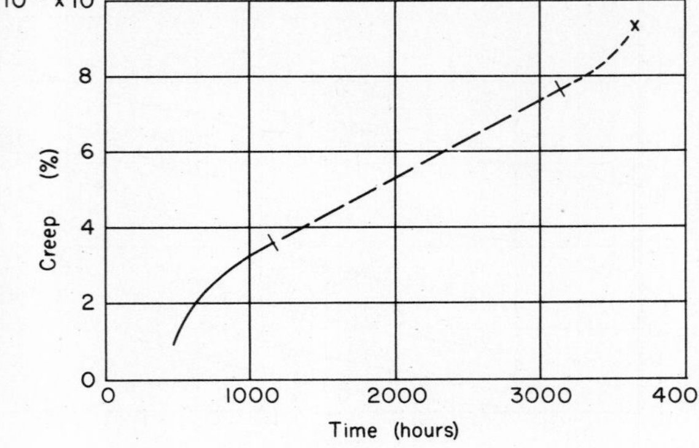

FIGURE 2-34. Creep curve for a single specimen at a fixed temperature, showing three stages of creep.

elevated temperature, it will elongate continuously until rupture occurs, even though the applied stress is below the yield strength of the material at the temperature of testing. Although the rate of elongation is small, it is sufficient to be of great importance in the design of equipment such as steam or gas turbines, power plants, and high-temperature pressure vessels that operate at high temperatures for long periods of time.

If a single specimen is tested under fixed load and fixed temperature, a curve such as that of Figure 2-34 is generated. The curve contains three distinct stages: a short-lived initial stage, a rather long second stage of rather linear elongation rate, and a short-lived third stage leading to fracture. From each test, two significant pieces of engineering data are obtained, the rate of elongation in the second stage, or *creep rate,* and the total elapsed time to rupture. Tests conducted at higher temperatures or with higher applied loads would produce higher creep rates and shorter rupture times.

A very useful engineering tool where creep is a significant factor is a *stress–rupture diagram,* such as the one in Figure 2-35. Rupture-time data from a number of tests at various temperatures and stresses are plotted on a single diagram. Creep-rate data can also be plotted to show the effects of temperature and stress as in Figure 2-36.

In general, the alloying elements of nickel, manganese, molybdenum,

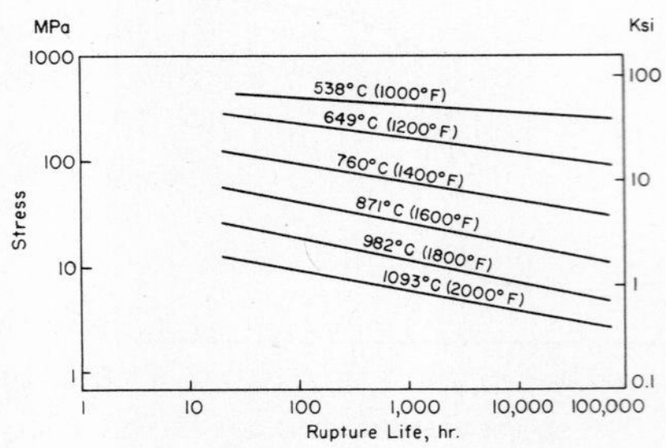

FIGURE 2-35. Stress-rupture diagram of a solution-annealed Incoloy alloy 800 (Fe–Ni–Cr alloy). (*Courtesy Huntington Alloy Products Division, The International Nickel Company, Inc.*)

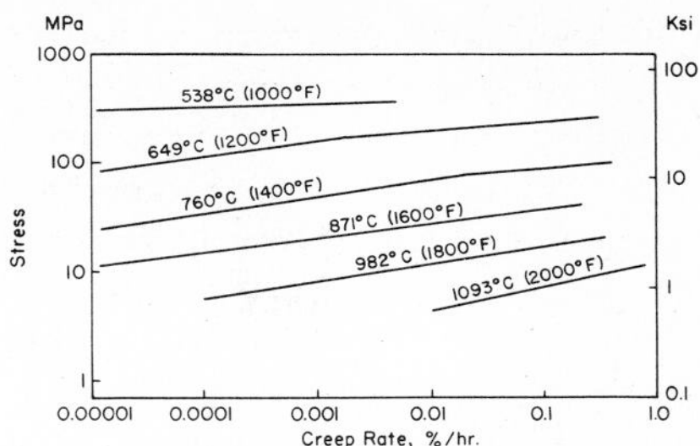

FIGURE 2-36. Creep-rate properties of solution-annealed Incoloy alloy 800. (*Courtesy Huntington Alloy Products Division, The International Nickel Company, Inc.*)

tungsten, vanadium, and chromium are helpful in lowering the creep rate of steel. At high temperatures, coarse-grained steels seem to be more creep-resistant than fine-grained steels. Below the lowest recrystallization temperature (a property discussed in Chapter 3), however, the reverse is true, with a fine-grained structure being preferred. Killed steels show superior creep-resisting properties when compared to rimmed steels. Steels having about 0.4 per cent carbon tend to be more creep-resistant than those having higher or lower carbon contents.

Machinability. Material properties relating to the response of a material to various processing techniques are some of the most difficult properties to define. The terms "malleability," "workability," and "formability" have already been introduced as measures of a material's suitability to plastic deformation processes. *Machinability* is another such term, important from the production viewpoint, but hard to define in that it depends not only on the material involved, but also on the specific process and aspect of interest. In some cases one is concerned with how easily a material cuts, with little regard for surface finish. At other times surface finish may be of prime importance. The formation of fine chips, so as to facilitate chip removal, may be another desirable feature. Also, tool life may be a dominant consideration. Thus, the general term "machinability" may involve several properties of a material, each of varying importance.

Perhaps the best approach to machinability is to consider that one is usually interested in removing the greatest amount of material in the shortest time, without requiring tool redressing or replacement, producing a satisfactory surface finish, and maintaining a low overall cost. Thus, good machinability is associated with the removal of material with moderate forces, formation of rather small chips, minimization of tool abrasion, and production of good surface finish. Chapter 17 will deal more fully with the relationships between machinability and the properties of materials.

The fracture mechanics approach. This chapter would not be complete without mention of the many tests and design concepts based on the fracture mechanics approach. Using the premise that all materials contain flaws, material characterization tends to focus on three quantities: (1) the size of the largest or most critical flaw, (2) the applied stress, and (3) the *fracture toughness*—a property that describes the conditions necessary for flaw growth or propagation leading to fracture. If nondestructive testing or quality control checks are applied, the size of the largest flaw that might go undetected can be determined. Assuming this flaw to be in the most critical location and knowing the stress in that location, the designer can specify a material with sufficient fracture toughness that the flaw would not propagate to a failure in service. Conversely, if the material and stress conditions were defined, the size of the maximum permissible flaw that would not propagate could be determined. The approach has proved invaluable in many areas where fracture could be catastrophic and has shown great refinement and increased acceptance in recent years.

Review questions

1. Why is it important to know what test conditions were utilized in determining the listed values of mechanical properties of materials?
2. Why is the designer usually more concerned with the mechanical properties of materials than with their physical properties?
3. What is stress? How does it differ from load or force?
4. What is strain? How is it normally expressed?
5. Why is a plot of stress versus strain preferable to one of load versus elongation?
6. Is strain proportional to stress for all stress values?
7. What is the significance of the modulus of elasticity of a material to a designer?
8. Can the modulus of elasticity of a given material be changed significantly?
9. Why are the proportionality limit and elastic limit not used extensively in engineering applications?
10. What is resilience? Where might it be important?
11. Why is it important to define the conditions where elastic response terminates and plastic flow begins?
12. What is a yield point? Do all materials exhibit the yield point phenomenon?
13. What is the offset yield strength? How is it determined?
14. What is the tensile strength of a material? Why might it be less significant than the yield strength?
15. At what point on the stress–strain curve does necking occur?
16. What is the difference between ductile and brittle response in a material?
17. How can the tensile test provide an indication of material ductility?
18. Does the absence of necking in a piece of matérial that has been loaded in tension indicate that no plastic deformation has occurred?

19. Why is it important that toughness be evaluated under conditions approximating those of service?
20. What is the difference between true stress and engineering stress?
21. Although true stress–true strain curves are more difficult to determine than their engineering counterpart, they more accurately reflect material properties. Why?
22. What is work hardening? How might it be important in plastic deformation processes?
23. Why is damping capacity of importance in selecting material for certain applications such as crankshafts and equipment bases?
24. Explain the lack of correlation beteween the values obtained by the various hardness tests.
25. What are some of the limitations of the Brinell hardness test?
26. Under what conditions might the Brinell hardness test be better than the Rockwell test?
27. Why must Rockwell test numbers always include a letter, such as A, B, or C?
28. What are some of the attractive features of indentation hardness tests?
29. What is the basic premise of the scratch hardness tests?
30. To what extent may the hardness of a metal be related to its strength?
31. What factors should determine whether a tension or compression test is used to evaluate material properties?
32. In what ways can impact test data be used by designers and engineers?
33. What type of loading mode is most used in fatigue testing?
34. What is the difference between fatigue strength and endurance limit?
35. Why are proper manufacturing processes and good design often more important than material selection for components with fatigue applications?
36. What are some common misunderstandings related to progressive fracture?
37. What are the effects of temperature on the various strength properties of metals?
38. Why is it important to consider the rate of plastic deformation before using tabulated data for design or manufacture?
39. Why is creep an important property when metal must operate under tensile stress at high temperatures?
40. Why is it difficult to rate materials as to machinability?
41. What in general is the fracture mechanics approach?

Case study 1.
THE CASE OF THE MIXED-UP STEEL

You are an engineer employed by the ALO Company. Because of a shipping accident, 100 bars each of AISI 1020 and AISI 1040 hot-rolled steel have become mixed during shipment to the company's warehouse in Alaska. It is essential that the bars be correctly identified. You are being sent to the Alaska warehouse to identify each bar. The only equipment available at the warehouse that can be used in the identification is a 227-kN (60,000-lb) tensile testing machine, a Brinell hardness tester, and the equipment in a small machine shop. Determine the best procedure to use and justify your decision, making use of standard data-source references, such as Volume 1 of the *ASM Handbook.*

The nature of metals and alloys

The structure–property relationship. The fundamental engineering *properties* of materials presented in Chapter 2 are directly related to the *structure* of the particular material. Moreover, such properties as strength and ductility are often sensitive to minute variations of structure, some of which are macroscopic, others microscopic, and still others of which are on the atomic scale. For the engineer to control the properties of materials and intelligently use them to their optimum, he first must have a working knowledge of material structure.

The basic structure of materials. Since materials are all composed of the same basic components—*protons, neutrons,* and *electrons*—it is amazing that such a variety of materials exists with such widely varying properties. Variation is explained when one considers the many possible combinations of these units in macroscopic assembly. The subatomic components, listed above, combine in different arrangements to form the various elemental *atoms,* each having a

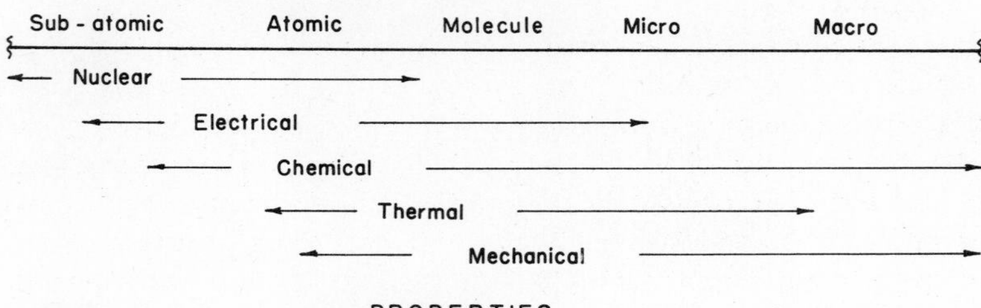

FIGURE 3-1. General relationship of structural level to engineering properties.

nucleus of protons and neutrons surrounded by the proper number of electrons to maintain charge neutrality. Atoms then combine in distinctive arrangements to form *molecules* or *crystals*. These units can then be assembled in differing amounts and configurations to form a microscopic scale structure, or *microstructure,* and ultimately an engineering component. The engineer, therefore, has at his disposal a wide variety of metals and nonmetals that possess an almost unlimited range of properties. Because the properties of materials depend upon all levels of structure, as shown schematically in Figure 3-1, it is important for the engineer to understand the entire structure spectrum, from atomic to macroscopic.

Atomic structure. Experiments have revealed that atoms consist of a relatively dense nucleus composed of positively charged protons and neutral particles of nearly identical mass, known as *neutrons.* Surrounding the nucleus are the negatively charged electrons, which have only $^1/_{1839}$ the mass of a neutron and appear in numbers equal to the protons to maintain a net charge balance. Distinct particle groupings produce the known elements, ranging from the relatively simple hydrogen atom to unstable transuranium atoms over 250 times as heavy. Except for density and specific heat, however, the weight of atoms has relatively little influence on engineering properties.

The light electrons that surround the nucleus play a far more significant role in determining properties. Again, experiments reveal that the electrons are arranged in a characteristic structure consisting of shells and subshells, each possessing a distinctive energy. Upon absorbing a small amount of energy, an electron can jump from a low-engergy shell near the nucleus to a higher-energy shell farther out. The reverse jump can occur with the release of a distinct amount or *quantum* of energy.

Each of the various shells and subshells contains only a limited number of electrons. The first shell, nearest the nucleus, can contain only two. The second shell can contain eight; the third, 32. Each shell and subshell is most stable when it is completely filled. For atoms containing electrons in the third shell and beyond, however, relative stability is associated with eight electrons in the outermost shell.

If, in its outer shell, a normal atom has slightly less than the number of electrons required for stability (for example, seven in its third shell), it will readily accept an electron from another source. It will then have one electron more than the number of protons and becomes a negatively charged atom or *negative ion.* Extra electrons may cause the formation of ions having negative charges of 1, 2, 3, and so on. If an atom has a slight excess of electrons beyond the number required for stability (such as sodium, with one electron in the third shell), it will readily give up the excess electron and become a *positive ion.* The remaining electrons become more strongly bound, making the removal of electrons progressively more difficult.

The number of electrons surrounding the nucleus of a neutral atom is called the *atomic number.* More important, however, are those electrons in the outermost shell (or subshell) known as *valence electrons.* These are influential in determining chemical properties, electrical conductivity, some mechanical properties, the nature of interatomic bonding, atom size, and optical characteristics. Elements with similar electron configurations in their outer shells will tend to have similar properties.

Atomic bonds. Atoms are rarely found as free and independent units, but usually are linked or *bonded* to other atoms in some manner as a result of interatomic forces. The electronic structure of the atoms influences the nature of the bond, which may be classed as *primary* (strong) or *secondary* (weak).

The simplest type of primary bond is the *ionic bond.* Electrons break free of atoms with excesses in their valence shell, producing positive ions, and unite with atoms having an incomplete outer shell to form negative ions. The positive and negative ions have a natural attraction for each other, producing a strong bonding force. Figure 3-2 illustrates the process for a bond between sodium and chlorine. In the ionic type of bonding, however, the atoms do not unite in simple pairs. All positively charged atoms attract all negatively charged atoms. Thus, for example, sodium ions surround themselves with

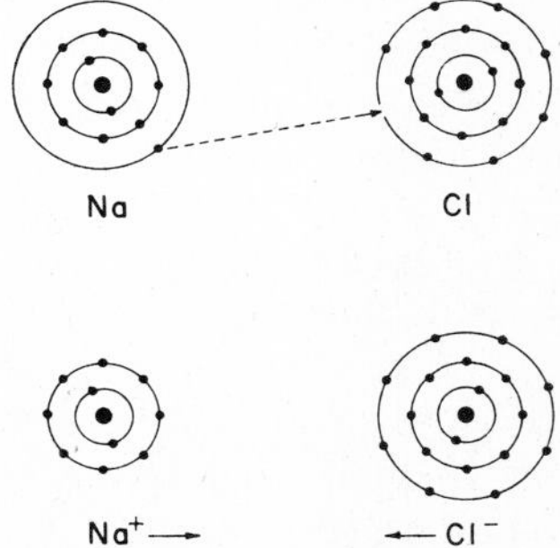

FIGURE 3-2. Mechanism of ionization of sodium and chlorine, producing stable outer shells by electron transfer.

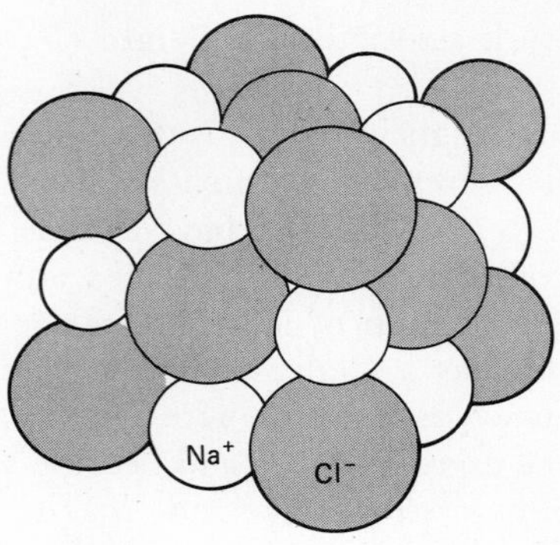

FIGURE 3-3. Three-dimensional structure of the sodium chloride molecule. Note how ions are surrounded by ions of opposite charge.

negative chlorine ions, and chlorine ions surround themselves with positive sodium ions. The attraction is equal in all directions and results in a three-dimensional structure, such as in Figure 3-3, rather than the simple link of the single bond. For stability in the structure, total charge neutrality must be maintained, thereby requiring equal numbers of positive and negative charges. General characteristics of materials joined by ionic bonds include moderate to high strength, high hardness, brittleness, high melting point, and electrically insulating properties (all charge transport must be through ion movement).

A second type of primary bond is the *covalent* type. Here the atoms being linked find it impossible to produce completed shells by electron transfer, but achieve the same goal by electron sharing. Adjacent atoms share outer-shell electrons so that each achieves a stable electron structure. Moreover, the shared negative electrons locate between the positive nuclei to form the bonding link. Figure 3-4 illustrates this type of bond for chlorine, where two atoms, each containing seven valence electrons, share a pair to form a stable *molecule.* Stable molecules can also form from the sharing of more than one electron from each atom, as in the case of nitrogen in Figure 3-5a. Atoms need not be identical (as in HF in Figure 3-5b), the sharing need not be equal, and an atom may share with more than one other atom. For elements such as carbon (four valence electrons), one atom may share valence electrons with each of four neighboring carbon atoms. The resulting structure becomes

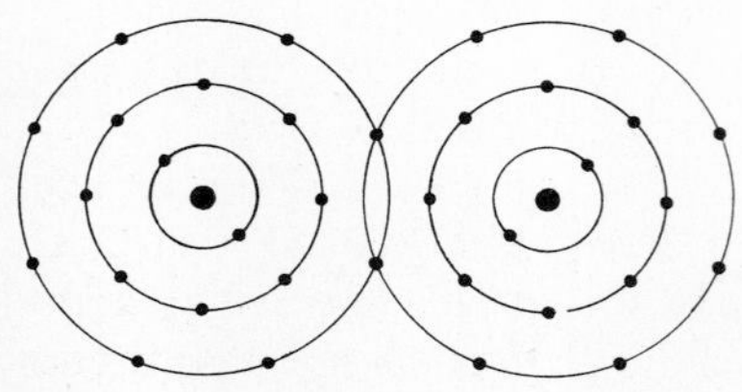

FIGURE 3-4. Formation of a chlorine molecule through a covalent bond.

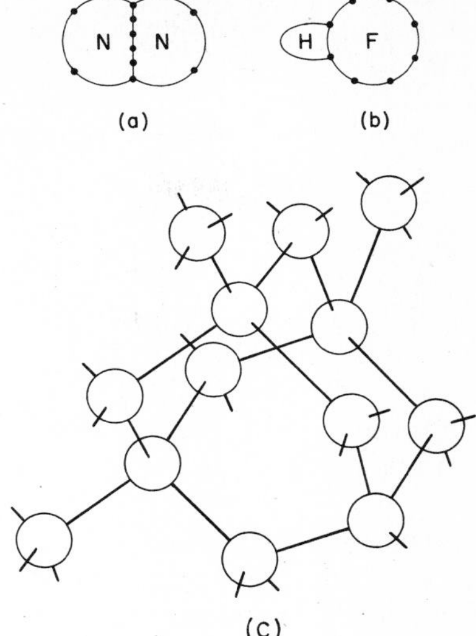

FIGURE 3-5. Examples of covalent bonding in N_2, HF, and diamond.

a network of bonded atoms, Figure 3-5c, instead of a finite, well-defined molecule. Like the ionic bond, the covalent bond tends to produce materials with high strength and high melting points. Atom movement within the material (deformation) requires the breaking of distinct bonds, thereby making the material characteristically brittle. Electrical conductivity depends upon bond strength, ranging from conductive tin (weak covalent bond) through semiconductive silicon and germanium to insulating diamond. Engineering materials possessing ionic and covalent bonds tend to be ceramic (refractories or abrasives) or polymeric in nature.

A third type of primary bond can result when a complete outer shell cannot be formed by either electron transfer or electron sharing, and is known as the *metallic bond.* If there are only a few valence electrons (1, 2, or 3) in each of a grouping of atoms, these electrons can be removed relatively easily while the remainder are held firmly to the nucleus. The result is a structure of positive ions (nucleus and nonvalence electrons) surrounded by a wandering assortment of universally shared electrons (electron cloud or gas), as in Figure 3-6. These highly mobile "free" electrons account for the observed high electrical and thermal conductivity as well as the opaque optical properties (electrons can absorb light radiation energies). Moreover, they provide the "cement" necessary to produce the positive–negative–positive attractions necessary for bonding. Bond strength, and therefore material strengh, vary over a wide range. More significantly, however, is the observation that the positive ions can move within the structure without the breaking of distinct bonds. Materials bonded by metallic bonds therefore can be deformed by atom movement mechanisms and produce a deformed material every bit as strong as

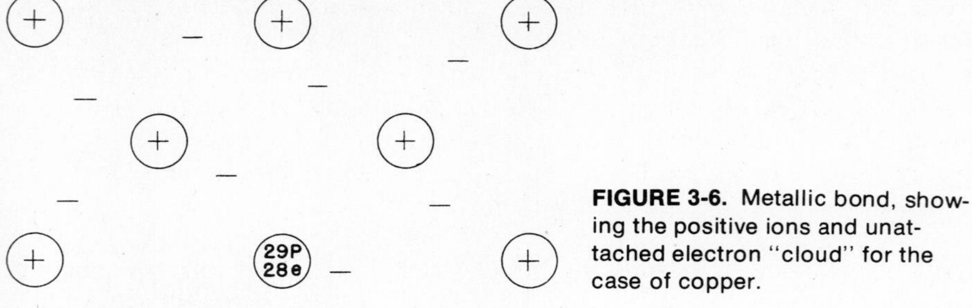

FIGURE 3-6. Metallic bond, showing the positive ions and unattached electron "cloud" for the case of copper.

the original. This is the basis of metal plasticity, ductility, and many of the shaping processes used in metal fabrication.

Secondary bonds. Weak or secondary bonds, known as *van der Waal's forces,* can link molecules that possess a nonsymmetric distribution of charge. Some molecules, such as hydrogen fluoride and water, can be viewed as electric dipoles, in that certain portions of the molecule tend to be more positive or negative than others (an effect referred to as *polarization*). The negative part of one molecule tends to attract the positive part of another to form a weak bond.

Another weak bond can result from momentary polarization caused by random movements of the electrons and the resulting momentary electrical unbalance. This random and momentary polarization leading to attractive forces is called the *dispersion effect.*

A third type of weak bond is the *hydrogen bridge,* where a small hydrogen nucleus is simultaneously attracted to the negative electrons of two different atoms, thereby forming a three-atom link. Such bonds play a significant role in biological systems but, as with all secondary bonds, are rarely of engineering significance.

Interatomic distances. Since the space occupied by the electron shells is very large compared to the size of the actual electrons, most of the total volume of an atom is vacant space, and the "size of an atom" becomes somewhat undefined. The interatomic or bonding forces tend to pull the atoms together; at the same time, there are repelling forces between the positive nuclei. There exists some equilibrium distance where the forces of attraction and repulsion are equal. The atoms will assume this separation, and to increase or decrease this distance will require energy, such as thermal energy or electrical or mechanical forces.

Thus atoms can be assigned a distinct size, the equilibrium distance between the centers of two neighboring atoms being considered to be the sum of the atomic radii. The atomic radius is not a constant, however. Temperature can change its value, producing an observable thermal expansion. Removal of electrons from the outer shell will decrease the radius and the addition of electrons will increase it. Consequently, a negative ion is larger than its base atom, and a positive ion is smaller. Atomic radius also changes with the number of adjacent or nearest-neighbor atoms. With more neighbors, there is less attraction to any single neighbor atom, and the interatomic distance is increased. Thus iron has a slightly different atomic size in its two different crystal forms, as will be discussed in a later chapter. In covalent bonds, the atomic radius will decrease as more atoms are shared.

Atom arrangements in materials. Continuing the development of material structure, we find that the arrangement of atoms in a material has a significant effect on its properties. Depending upon the manner of atomic grouping, materials are classified as having *molecular structures, crystal structures,* or *amorphous structures.*

Molecular structures have a distinct number of atoms that are held together by primary bonds, but they have only relatively weak bonds with other similar groups of atoms. Typical examples of molecules include O_2, H_2O, and C_2H_4 (ethylene). Each molecule is free to act more or less independently, giving these materials relatively low melting and boiling points. The materials tend to be soft, since the molecules can easily move past one another. Upon changes of state from solid to liquid or gas, the molecules remain intact as distinct entities.

Crystal structures are assumed by solid metals and most minerals. Here atoms are arranged in a regular geometric array known as a *space lattice.* These lattices are describable through a unit building block which is essentially repeated throughout space in a periodic manner. Such blocks are known as *unit cells.*

In amorphous structures, such as glass, the atoms have a certain degree of local order but, when viewed as an aggregate, have a more disorganized atom arrangement than the crystalline solids.

Crystal strucure of metals. From a manufacturing viewpoint, metals are the most important class of materials. Most often, they are the materials being processed and are used in the machines performing the processing. Consequently, in order to perform manufacturing operations intelligently, it is essential to have a basic knowledge of the fundamental nature of metals and their behavior when subjected to mechanical or thermal treatment.

More than 50 of the known chemical elements are classed as metals, and about 40 have commercial importance. These materials are characterized by a metallic bond and possess certain distinguishing characteristics: strength, good electrical and thermal conductivity, luster, the ability to be deformed permanently to a fair degree without fracturing, and relatively high specific gravities, as compared with nonmetals. The fact that some metals possess properties different from the general characteristics simply expands their engineering utility.

When metals solidify by cooling, they assume a crystalline structure;

TABLE 3-1. Types of lattices of common metals at room temperature

Metal	Lattice Type
Aluminum	Face-centered cubic
Copper	Face-centered cubic
Gold	Cubic (diamond)
Iron	Body-centered cubic
Lead	Face-centered cubic
Magnesium	Hexagonal
Silver	Face-centered cubic
Tin	Body-centered tetragonal
Titanium	Hexagonal

that is, the atoms arrange themselves in a space lattice. Most metals exist in only one lattice form. A few, however, can exist in the solid state in two or more lattice forms, the particular form depending on the conditions of temperature and pressure. These metals are said to be *allotropic,* and the change from one lattice form to another is called an *allotropic change.* The most notable example of such a metal is iron, where the property makes possible the use of heat-treating procedures to produce a wide range of characteristics. It is largely due to its allotropy that iron is the base of our most important alloys.

Metals are known to solidify into 14 different crystal structures. However, nearly all the important commercial metals solidify into one of three types of lattices, these being body-centered cubic, face-centered cubic, and hexagonal close-packed. Table 3-1 shows the lattice structure of a number of common metals at room temperature. Figure 3-7 compares these crystal structures to each other and to the easily visualized, but rarely observed, simple cubic structure.

The simple cubic structure of Figure 3-7a can be constructed by placing single atoms at all corners of a cube and subsequently linking identical cubes together. Assuming that the atoms are spheres of atomic radii touching each other, computation reveals that only 52 per cent of available space is occupied. Each atom has only six nearest neighbors. Both of these observations are unfavorable to the metallic bond, where atoms desire the greatest number of nearest neighbors and high-efficiency packing.

If the cube is expanded somewht to allow the insertion of an additional atom in the center, the *body-centered-cubic* (b.c.c.) structure results, as in Figure 3-7b. Each atom now has eight nearest neighbors and 68 per cent of the space is occupied. Such a structure is more favorable to metals and is observed in the elements Fe, Cr, Mn, and so on, as listed in Figure 3-7b.

If Ping-Pong balls, used to simulate atoms, were placed in a box and agitated until a stable arrangement was produced, we would find the structure to consist of layered *close-packed planes,* where each plane looks like Figure 3-8. Two different structures can result, depending upon the sequence in which the

	Lattice Structure	Unit Cell Schematic	Ping-Pong Ball Model	Number of Nearest Neighbors	Packing Efficiency	Typical Metals
a	Simple cubic	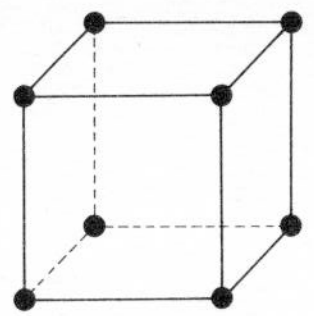	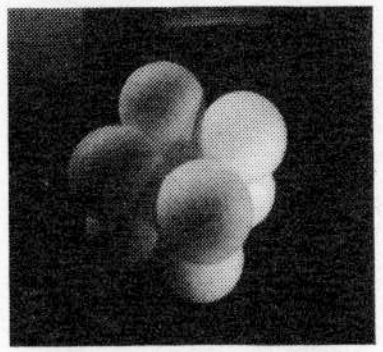	6	52%	None
b	Body-centered cubic	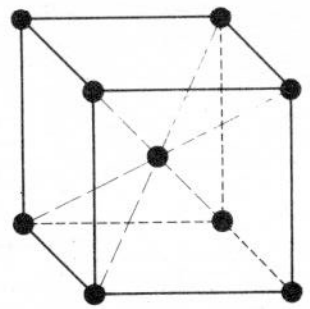		8	68%	Fe, Cr, Mn, Cb, W, Ta, Ti, V, Na, K
c	Face-centered cubic	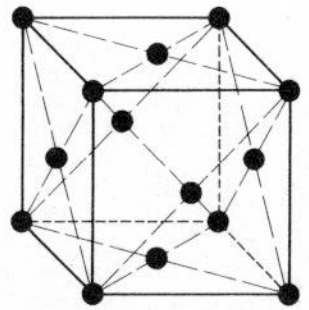		12	74%	Fe, Al, Cu, Ni, Ca, Au, Ag, Pb, Pt
d	Hexagonal close-packed	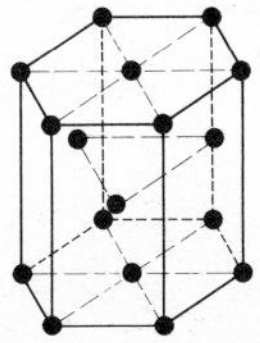		12	74%	Be, Cd, Mg, Zn, Zr

FIGURE 3-7. Comparison of crystal structures: simple cubic, body-centered cubic, face-centered cubic, and hexagonal close-packed.

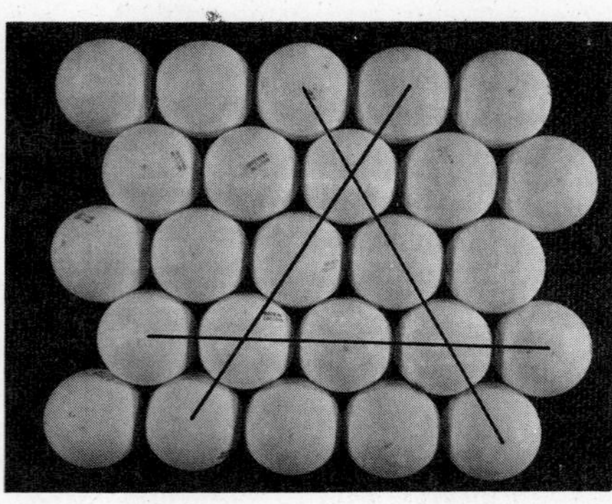

FIGURE 3-8. Close-packed atomic plane showing three directions of closest packing.

various planes are stacked, but both are identical in nearest neighbors (12) and efficiency of occupying space (74 per cent).

One of these sequences produces a structure that can be viewed as an expanded cube with an atom inserted in the center of each face, the *face-centered-cubic* (f.c.c.) structure of Figure 3-7c. Such a structure occurs in many of the most important engineering metals and tends to produce high formability (ability to be permanently deformed without fracture).

The second of the structures is known as the *hexagonal-close-packed* (h.c.p.), wherein the close-packed planes can be clearly identified (see Figure 3-7d). Metals having this structure tend to have poor formability and often require special processing procedures.

Development of metallic grains. As metals solidify, small particles of solid form in the liquid, each having the lattice structure characteristic of the given material. These particles then act as *seeds* or *nuclei* onto which other atoms in the vicinity tend to attach themselves, producing growth of the solid. The resulting arrangement is a crystal composed of repetitions of the same basic pattern throughout space, as illustrated in Figure 3-9.

In actual solidification it is expected that the seed or nuclei particles would form independently at various locations in the liquid mass and have random orientations. Each then grows until it begins to interfere with its neighbors, as illustrated in two dimensions in Figure 3-10. Since adjacent lattice structures have different alignments, growth cannot produce a single continuous structure. The small continuous segments of solid are known as *crystals* or *grains,* and the surfaces that divide them (i.e., the surfaces of crystalline discontinuity) are known as *grain boundaries.* The process through which the grain structure is produced is one of *nucleation and growth.*

Grains are the smallest structural units of metal that are observable with ordinary light microscopy. If a piece of metal is polished to a mirror finish with a series of abrasives and then exposed to an attacking chemical for a short time, the grain structure can be seen. The atoms on the grain boundaries are more loosely bonded and tend to react with the chemical more readily than

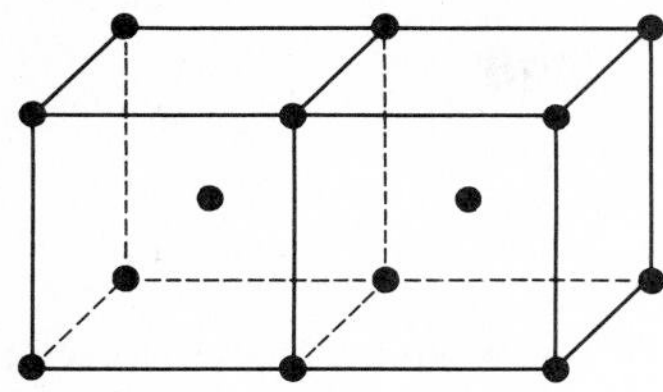

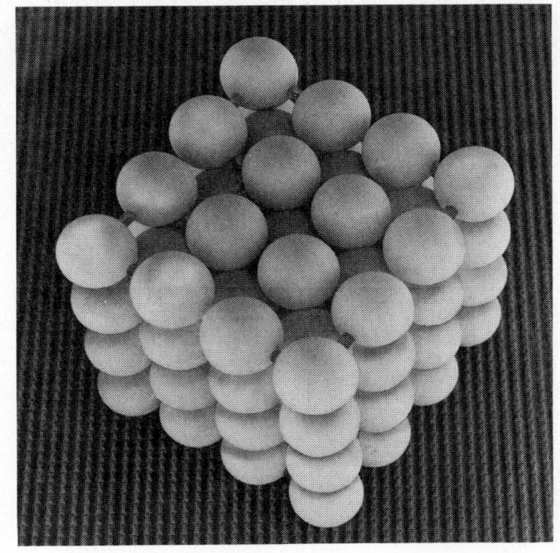

FIGURE 3-9. Growth of crystals to produce an extended lattice: (*left*) line schematic, (*right*) Ping-Pong ball model.

those that are part of the grain interior. When subsequently viewed under reflected light, the attacked boundaries appear dark compared to the relatively unaffected (still flat) grains, as in Figure 3-11. Occasionally, grains are large enough to be seen by the naked eye, as on some galvanized steel, but usually magnification is required.

The number and size of the grains in a metal are a function of two factors: the rate of nucleation and the rate of growth. The greater the rate of nucleation, the smaller the resulting grains. Similarly, the greater the rate of growth, the larger the grain size. Because the overall *grain structure* will influence certain mechanical and physical properties, it is an important property for the engineer to be able to both control and specify. One such specification scheme is the *ASTM grain-size number,* defined by

FIGURE 3-10. Schematic representation of the growth of crystals.

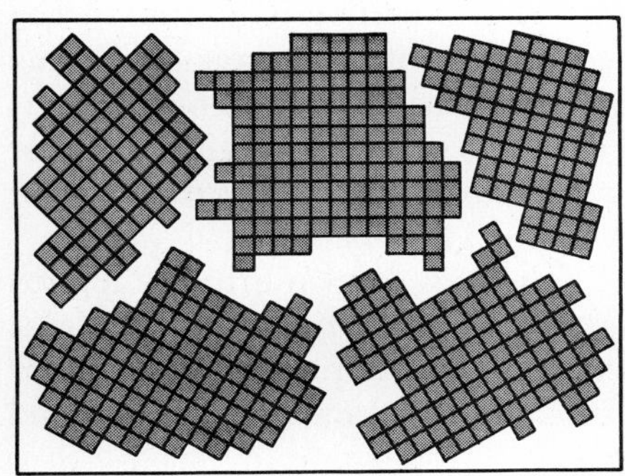

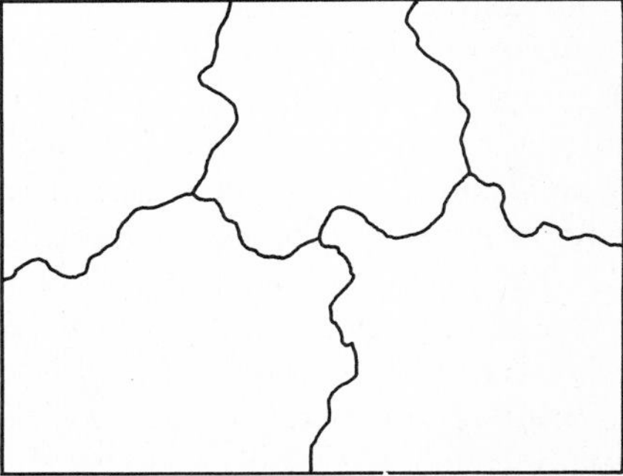

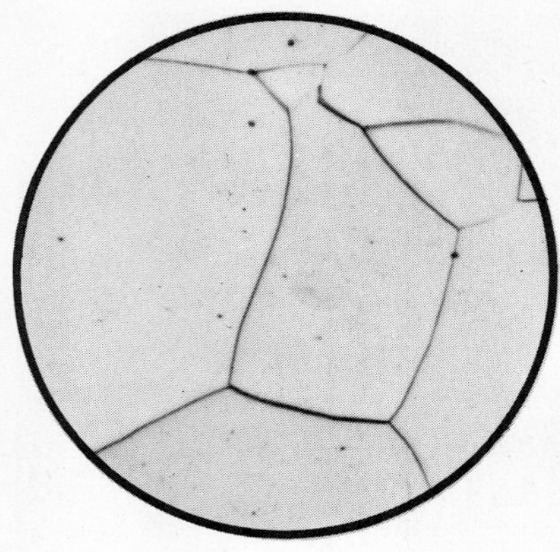

FIGURE 3-11. Photomicrograph of alpha ferrite (essentially pure iron), showing grain boundaries; 1000. *(Courtesy United States Steel Corporation.)*

$$N = 2^{n-1}$$

where N = number of grains per square inch visible in a prepared specimen at 100 ×

n = ASTM grain size number

Higher numbers relate to smaller grain sizes. Materials with ASTM grain size 7–9 are often desired when good formability is required.

Elastic deformation of single crystals. To a great extent, the mechanical behavior of materials is dependent upon their crystal structure. Therefore, to understand mechanical behavior, it is important for the engineer to have some understanding of the way crystals react when subjected to mechanical loading. Much of what is known about this subject has been obtained from the study of carefully prepared single crystals that may be several inches long. In general, observation reveals that the behavior of metal crystals depends upon (1) the lattice type, (2) the interatomic forces, (3) the spacing between planes of atoms, and (4) the density of atoms on various planes.

If the applied loads are relatively low, the crystal responds by simply stretching or compressing the distance between atoms, as in Figure 3-12. The lattice unit does not change, and the atoms retain their basic positions. The applied load serves only to disrupt the force balance of the atomic bonds in a manner so as to transmit the applied load through the body. When the load is removed, the balance is restored and the lattice resumes its original size and shape. The response to such loads is *elastic* in nature. The amount of stretch or compression (strain) is proportional to the applied load or stress.

Elongation or compression in one direction in response to an applied force also produces an opposite change in dimensions at right angles to that force. The ratio of lateral contraction to axial tensile strain under uniaxial tensile loading is known as *Poisson's ratio.* This ratio is always less than 0.5 and usually is about 0.3.

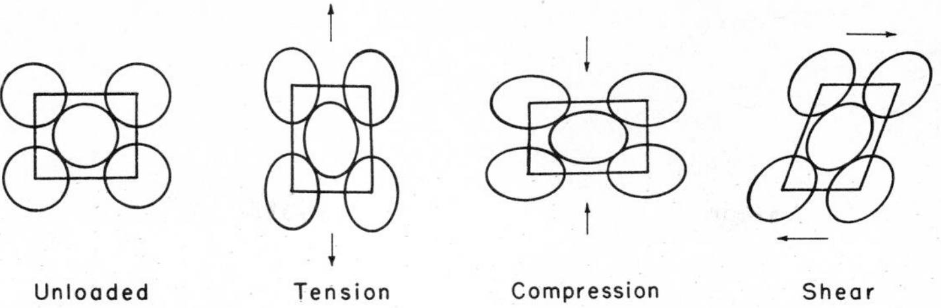

FIGURE 3-12. Distortion of the crystal lattice in response to elastic loadings.

Plastic deformation in a single crystal. As the magnitude of applied load is increased, the distortion increases to a point where the atoms must either (1) break bonds to produce a fracture, or (2) slide over one another to produce a permanent shift of atom positions. Fortunately for metallic materials, the second phenomenon requires lower loads and thus occurs preferentially in nature. The result is a plastic deformation, wherein a permanent change in shape occurs without a concurrent deterioration in properties.

Investigation reveals that the mechanism of plastic deformation is the shearing of atomic planes over one another to produce a net displacement. Conceptually, this is similar to the distortion of a deck of playing cards when one card slides over another. As we shall see, the actual mechanism, however, is a progressive one rather than all atoms in a plane shifting simultaneously.

Recalling that a crystal structure is a regular and periodic arrangement of atoms in space, it becomes possible to link atoms into flat planes in a nearly infinite number of ways. Planes having different orientations with respect to the basic lattice will have different atomic densities and different spacing between adjacent parallel planes, as illustrated in Figure 3-13. Given the choice of all possibilities, plastic deformation tends to take place along planes having the highest atomic density and greatest parallel separation. The reason for this may be seen in the simplified Figure 3-14. Planes A and A' have higher density and greater separation than planes B and B'. In visualizing relative motion, the atoms of B and B' would interfere significantly with one another, whereas planes A and A' do not experience this difficulty.

FIGURE 3-13. Schematic diagram showing crystalline planes with different atomic densities and interplanar spacings.

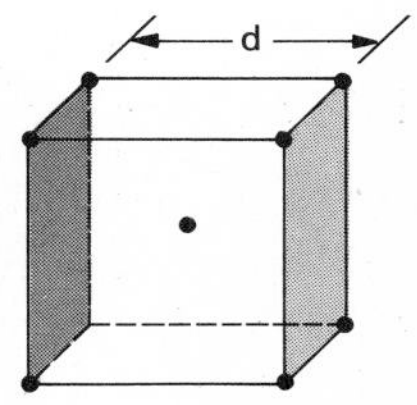

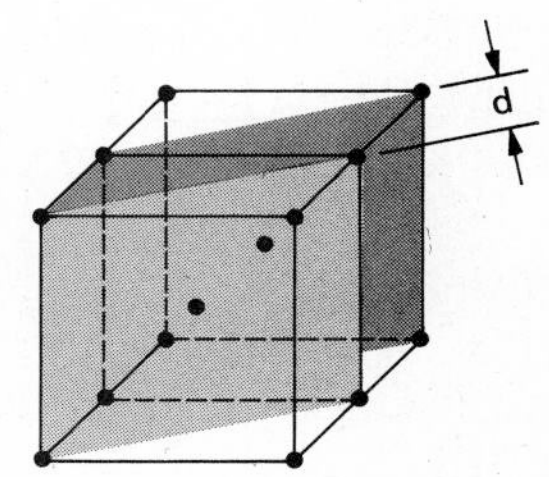

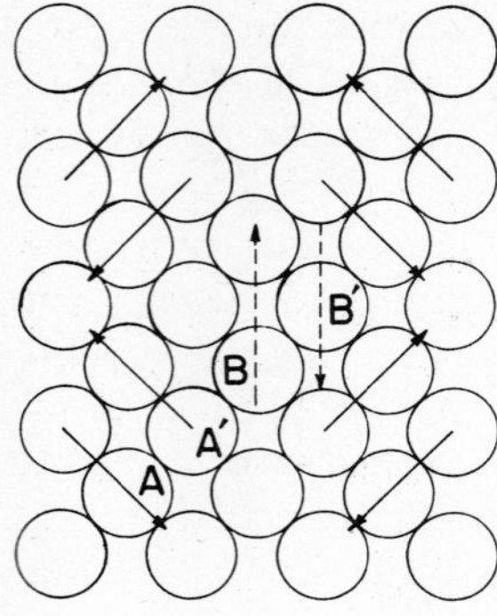

FIGURE 3-14. Planar schematic representing the greater deformation resistance of planes of lower atomic density and closer interplanar spacing.

Within the preferred planes are also preferred directions. If sliding occurs in a direction corresponding to close packing of atoms in the plane (as in Figure 3-8), atoms can simply follow one another rather than each having to negotiate its own path. Thus, plastic deformation occurs by the preferential sliding of maximum-density planes (close-packed planes if present) in directions of closest packing.

The ease with which a metal may be deformed depends on the ease of shearing one atomic plane over an adjacent one and the favorability with which the plane is oriented with respect to the load. For example, a deck of playing cards will not shear when laid flat on a table and pressed from the top, or stacked on edge and pressed uniformly. Only if the deck is skewed with respect to the applied load will shearing be produced.

With this understanding, let us now consider the properties of the various crystal structures:

Body-centered-cubic: In the body-centered-cubic structure, there are no close-packed planes. Slip therefore occurs on planes with large interplanar spacings (six of which are illustrated in Figure 3-15) in directions of closest packing (that is, the cube diagonals). If each combination of plane and direction is considered as a slip system, we find that 48 such systems exist. The

FIGURE 3-15. Slip planes in the various lattice types.

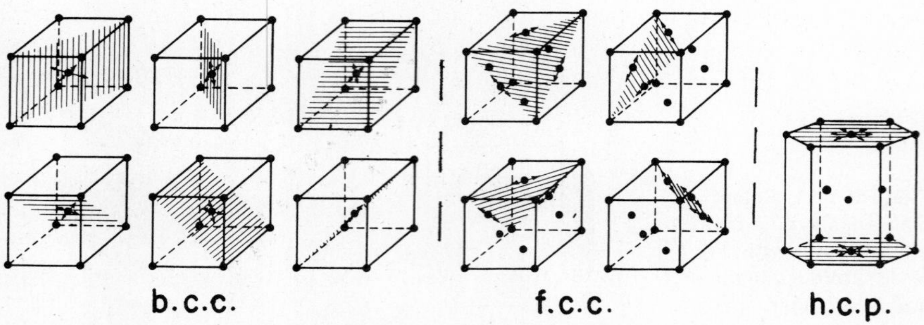

probability that one of these systems will be oriented for easy shear is great, but the force necessary to effect deformation is rather large. Materials with this structure possess only moderate deformation capabilities. (See the typical metals in Figure 3-7.)

Face-centered-cubic: In the face-centered-cubic structure, each unit cube possesses four close-packed planes, as illustrated in Figure 3-15. Each plane contains three close-packed directions (the face diagonals), giving 12 possible slip systems. Again the probability of one system being favorably oriented for shear is great, and the required force is relatively low. Face-centered metals possess excellent ductilities, as an inspection of Figure 3-7 will reveal.

Hexagonal-close-packed: The hexagonal lattice also contains close-packed planes, but only one such plane exists for the lattice. Although this plane contains three close-packed directions and the required force is rather low, the probability of favorable orientation to the applied load is rather small. Metals with the hcp structure tend to have quite low ductilities and often appear to be brittle.

Dislocation theory of slippage. A theoretical calculation of the strength of metals (resistance to plastic deformation) based on the sliding of atomic planes over one another gives yield strengths on the order of 20,000 MPa (3 million psi). Observed strengths are usually 100 to 150 times less than this value, indicating a discrepancy between theory and reality.

Explanation is provided by the fact that plastic deformation does not occur by all the atoms in one plane slipping simultaneously over all atoms in an adjacent plane. Instead, the motion takes place by the progressive slippage of a localized disruption, the disruption being known as a *dislocation.* Consider an analogy. An individual wants to move a carpet a short distance in a given direction. One approach would be to pull on one end and try to "shear the carpet on the floor." This would require a large force acting over a small distance. An alternative approach would be to form and work a wrinkle across the floor to produce a net shift of the whole carpet—a low-force-over-large-distance approach to the same task. In the region of the wrinkle, there was excess carpet with respect to the floor below, and motion of this excess was relatively easy.

It has been shown that metal crystals ordinarily do not have all their atoms in perfect arrangement, but contain certain localized imperfections. Two such imperfections are the *edge dislocation* and *screw dislocation,* as illustrated in Figure 3-16. Edge dislocations are the ends of extra half-planes of atoms, which produce localized disruptions such as illustrated in Figure 3-17. (Note the similarity to extra carpet in a wrinkle.) Screw dislocations correspond to a partial tearing of crystal planes. In each case, the dislocation is a disruption to the regular, symmetrical arrangement of atoms which can be moved about with rather low applied forces. It is the motion of these microscopic dislocations under applied loads that produces the observed microscopic deformation.

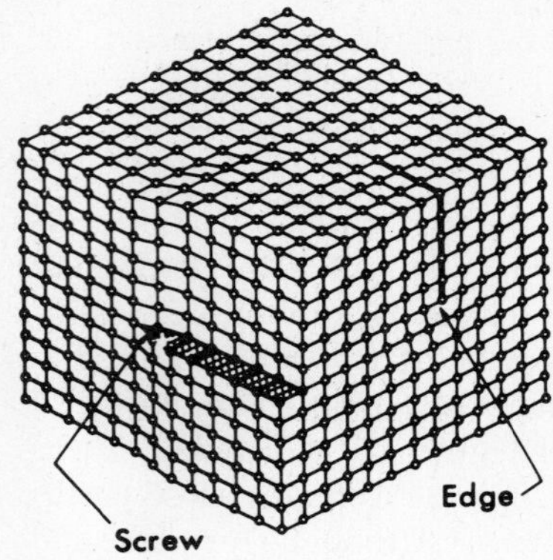

FIGURE 3-16. Schematic representation of screw and edge dislocations.

Strain hardening or work hardening. Many metals possess a unique property, in that after undergoing some deformation, the metal possesses greater resistance to further plastic flow. In essence, metals become stronger when plastically deformed, a phenomenon known as *strain hardening* or *work hardening.*

Understanding of this phenomenon can come from further consideration of the carpet analogy. Suppose this time that the goal is to move the carpet diagonally. The best way would be to move a wrinkle in one direction and then a second one 90° to the first. But suppose that both wrinkles were started simultaneously. We would find that wrinkle 1 would interfere with the motion of wrinkle 2, and vice versa. In essence, the device that made deformation easy can also serve to impede the motion of other, similar devices. Returning to metals, we find that plastic deformation is accomplished by the motion of dislocations. As dislocations move, they are more likely to encounter and interact with other such dislocations, thereby impeding further motion. Moreover, mechanisms exist to markedly increase the number of dislocations in a metal undergoing deformation, the effect being an increased probability of interaction.

FIGURE 3-17. Localized disruption of a crystal lattice at the end of an edge dislocation.

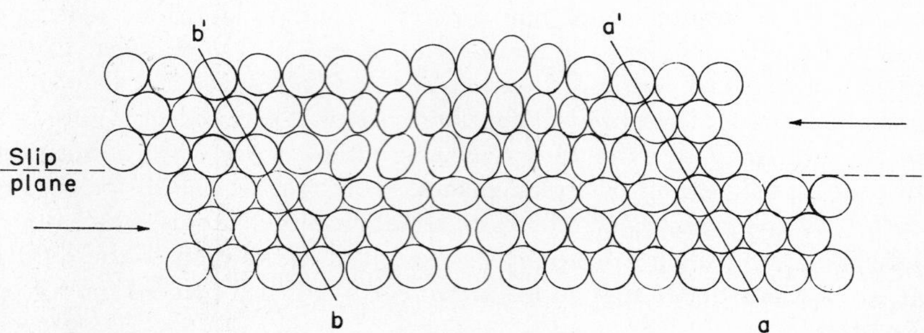

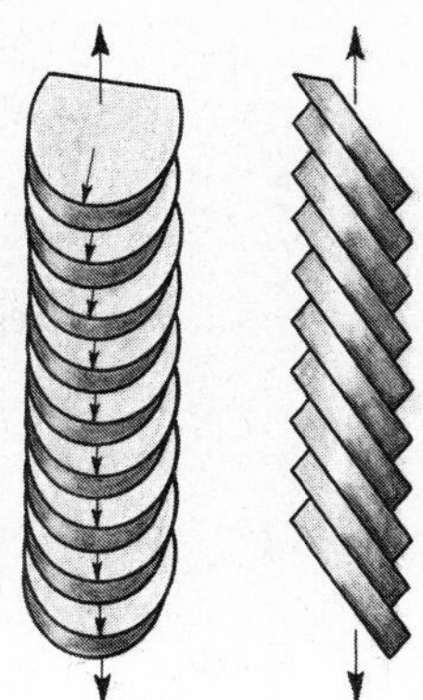

FIGURE 3-18. Schematic representation of slip and rotation resulting from deformation.

This phenomenon becomes significant when one considers mechanical working processes operating in the cold-working range. Strength properties of metals can be increased markedly, thereby enabling a deformed inexpensive metal to often substitute for a stronger, undeformed, but more costly one.

Experimental evidence strongly confirms the current theory. When a load is applied to a metal crystal, deformation will commence on the slip system most favorably oriented. The net result is often an observable slip and rotation in a macroscopic (but still single crystal) specimen, as illustrated in Figures 3-18 and 3-19. Often dislocation motion on one slip system becomes blocked as strain hardening produces increased resistance. Slip stops on the first system and, as load increases, further deformation will take place through alternative systems offering less resistance. This phenomenon, known as *cross slip,* also has been observed.

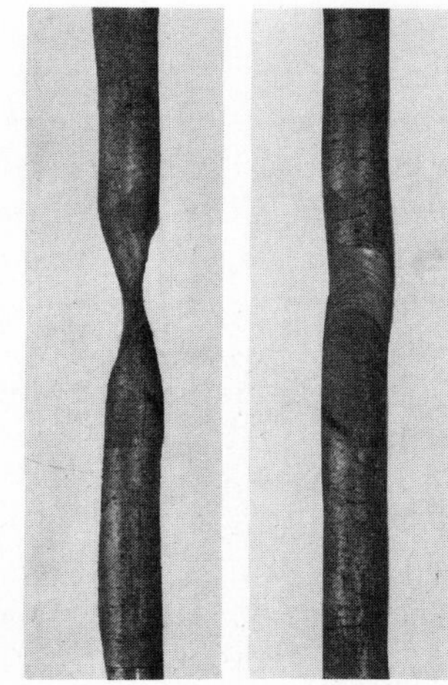

FIGURE 3-19. Front and side views of a single zinc crystal that has been elongated in uniaxial tension. (*Courtesy E. R. Parker.*)

FIGURE 3-20. Slip lines in a polycrystalline material. (*From Deformation and Fracture Mechanics of Engineering Materials, by Richard Hertzberg; courtesy John Wiley & Sons, Inc.*)

Plastic deformation in polycrystalline metals. Thus far, only the deformation of single crystals has been considered. Metals, as normally encountered, are polycrystalline. Within each crystal of a polycrystalline metal, deformation proceeds in the manner just described. However, since adjacent grains do not have their lattice structures aligned in the same orientation, an applied loading will cause different deformations within the various grains. This type of response is shown in Figure 3-20, where the slip lines in the various grains can be seen. It may be noted that the slip lines do not cross over from one grain to another. Thus the presence of neighboring grains tends to restrict the slip within a given grain. Finer grain structure, that is, more grains per unit area, generally tends to produce greater strength and hardness coupled with increased impact resistance. This "universal improvement of properties" is strong motivation for controlling grain size during processing.

Grain deformation and fiber structure. When a metal is deformed to a considerable degree, the grains become elongated in the direction of metal flow, as can be seen in Figure 3-21. At moderate magnification, a cross section of such a deformed metal may appear fibrous. Concurrent with this nonuniformity of structure is a nonuniformity of properties. Because of strain hardening and the fact that the intergranular boundaries are no longer randomly oriented, the

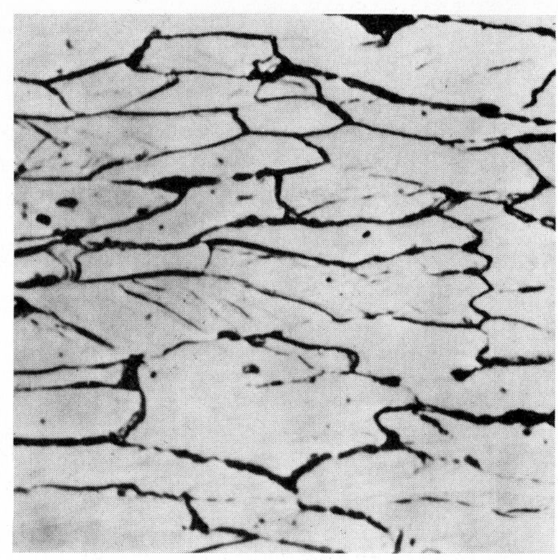

FIGURE 3-21. Deformed grains in cold-worked 1008 steel after 50 per cent reduction by rolling; 1000 ×. (*From Metals Handbook, 8th Edition; © 1972, American Society for Metals.*)

strength and other mechanical properties will not be the same in all directions. Electrical and magnetic properties may also show directional variation.

The possibility of employing this increase in strength in certain directions is important to both the designer and manufacturing engineer. Certain processes, such as forging, can be designed to utilize directional properties. Caution should be used, however, for improvement of properties in some directions usually is accompanied by a decline in some properties in other directions. Moreover, directional structures may impose serious difficulties in some operations, such as sheet metal drawing.

Fracture of metals. Under certain conditions of load, temperature, impact, and so on, metals may respond by fracture. *Brittle fracture* is the most catastrophic, for it occurs without the prior warning of plastic deformation and propagates rapidly through the metal. Such fractures usually are associated with metals having the bcc or hcp crystal structure. *Ductile fracture* generally occurs when plastic deformation is extended too far.

The actual mechanism and type of fracture will vary depending upon the material, temperature, state of stress, and rate of loading. Shear or slip fractures are due to extensive shear on slip planes. Intergranular fracture occurs by separation along the grain boundaries. Cleavage fracture results from the pulling apart of metal grains along distinct "cleavage planes" due to tensile loads within the crystal.

Recrystallization. If a polycrystalline metal is heated to a high enough temperature after being plastically deformed, new, equiaxed, unstrained crystals will form from the original distorted grains, as in Figure 3-22. This process is known as *recrystallization.* The temperature at which recrystallization takes place is different for each metal and varies with the amount of cold deformation. In general, the greater the amount of deformation, the lower is the recrystallization temperature. However, there is a practical lower limit below which recrystallization will not take place in a reasonable length of time. Table 3-2 gives the short-time recrystallization temperatures of several metals.

FIGURE 3-22. Recrystallization of 70–30 brass: as cold-worked 33 per cent (a); then heated at 580°C (1075°F)—(b) 3 seconds, (c) 4 seconds, (d) 8 seconds; 45×. (*Courtesy J. E. Burke, General Electric Co.*)

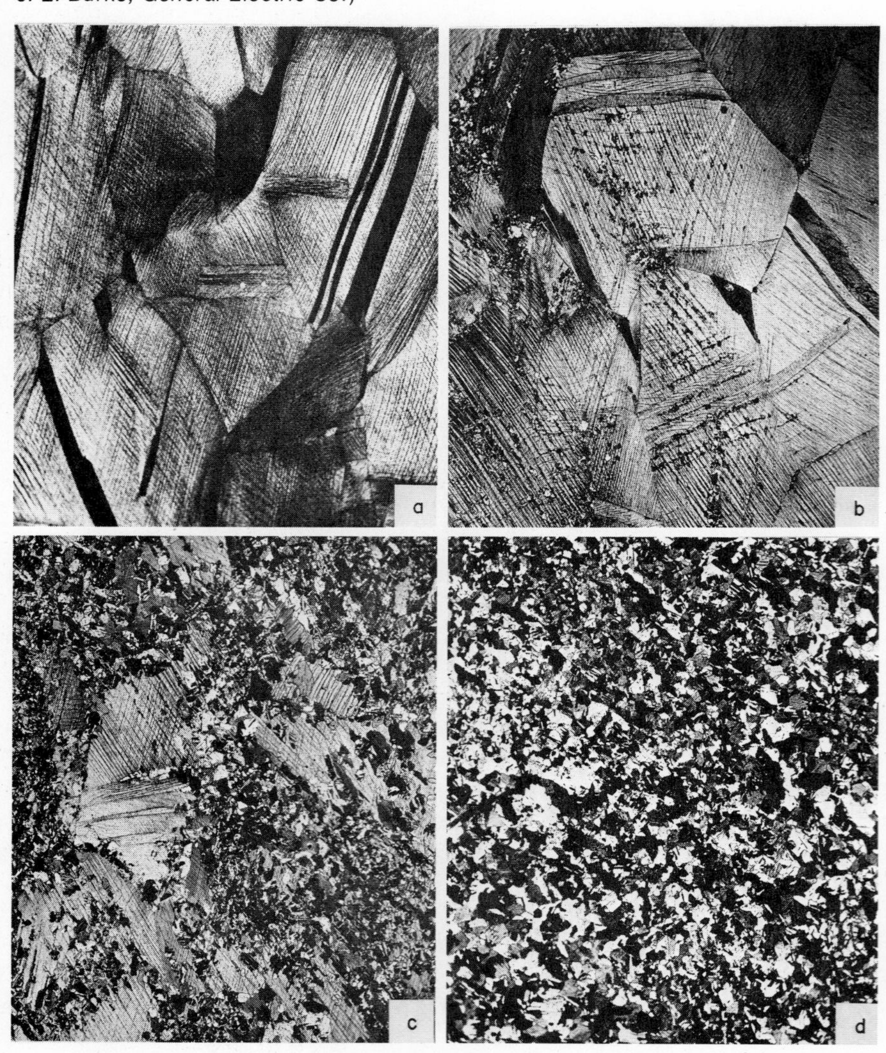

TABLE 3-2. Lowest recrystallization temperature of common metals

Metal	Temperature, °C (°F)
Aluminum	150 (300)
Copper	200 (390)
Gold	200 (390)
Iron	450 (840)
Lead	Below room temperature
Magnesium	150 (300)
Nickel	590 (1100)
Silver	200 (390)
Tin	Below room temperature
Zinc	Room temperature

Noting that metals may fracture if deformed too much, we find it common practice to recrystallize material after certain initial amounts of cold work. Ductility is restored and the material is ready for further deformation. This process, known as *recrystallization annealing,* enables deformation processes to be carried out to great lengths without danger of fracture and is important to many manufacturing processes. If metals are deformed above the recrystallization temperature, working and recrystallization take place simultaneously and large deformations are made possible.

Having already noted the desirability of fine grain size in improving properties, we find recrystallization as a means of grain-size control. In metals that do not undergo allotropic changes, a coarse grain structure can be converted to a fine grain structure through recrystallization. The material must first be plastically deformed to provide the driving force for recrystallization. Control of the recrystallization process establishes the final grain size.

Grain growth. The recrystallization process tends to produce uniform grains of comparatively small size. If the metal is held at or above the recrystallization temperature for any appreciable time, the new grains will start to "grow." In effect, some grains become larger at the expense of neighboring grains, the mechanisms being facilitated by the elevated temperature. Since properties tend to diminish with increased grain size, control is of prime importance here.

Hot and cold working. When metals are deformed plastically below the recrystallization temperature, the process is called *cold working.* The metal strain hardens and the structure consists of distorted grains. When deformation takes place above the recrystallization temperature, the process is called *hot working.* A recrystallized structure continually forms, and no strain hardening is apparent. The temperature above which hot forming can be performed depends on the material being worked, as shown in Table 3-2.

Alloys. Up to this point, the discussion in this chapter has been confined to the nature and behavior of pure metals. For most manufacturing applications, however, metals are not used in their pure form, but in the form of alloys. An *alloy* can be defined as a material composed of two or more elements, at least one of which is a metal, which possesses metallic properties. The addition of a second element to a first to form an alloy usually results in a change of properties. Knowledge of alloys is important to the intelligent selection of material for given applications.

Alloy types. Alloys can be formed by any of three mechanisms. The first, and probably the simplest type, is where *the two components are insoluble in each other in the solid state.* In this case, the base metal and the alloying element each maintain their individual identities, structures, and properties. The alloy, in effect, assumes a composite structure consisting of two types of building blocks in an intimate mechanical mixture.

The second mechanism occurs when *the two elements are soluble in each other in the solid state.* They thus form a *solid solution,* with the alloying element being dissolved in the base metal. These solid solutions may be of two types: (1) *substitutional* and (2) *interstitial.* In the substitutional type, some atoms of the alloy element occupy sites normally occupied by atoms of the host or base metal. Replacement is random in nature, the alloy atom being free to occupy any atom site in the base lattice. In the interstitial type, the alloy element atoms squeeze into the "unoccupied" spaces within the base metal lattice.

The third mechanism of alloying is where *the elements combine to form intermetallic compounds.* In this case, atoms of the alloying element combine with atoms of the base metal *in definite proportions and in definite relationships as to position.* Bonding is primarily of the nonmetallic variety (that is, ionic or covalent), and the lattice structures are more complex than in metallic materials. Such compounds tend to be hard, brittle, high-strength materials.

Even though alloys are composed of more than one type of atom, their structure is one of lattices and grains just as in pure metals. Their behavior when subjected to loading, therefore, should be similar to that of pure metals, with due provision for the structural modifications. Plastic flow along atomic planes may be impeded by the presence of unlike atoms. Grains in composite-type mixtures will show different responses to the same loading, reflecting the different properties of the component units.

Atomic structure and electrical properties. As with mechanical properties, the structure of materials strongly influences their electrical properties. Electrical conductivity involves the movement of valence electrons through the crystalline lattice so as to produce a net transport charge. The more perfect the atomic arrangement is in a metal, the higher is the conductivity. Conversely, the more lattice imperfections or irregularities, the higher the resistance to electrical conduction.

Electrical resistance depends largely upon two factors: (1) lattice imperfections, and (2) temperature. Vacant atomic sites, interstitial atoms, substitutional atoms, dislocations, and grain boundaries all act as disruptions to the regularity of a crystalline lattice. Temperature becomes important when one considers the associated atomic vibrations. We have already seen that mechanical energy can displace atoms from their equilibrium positions and stretch bonds. Thermal energy causes atoms to vibrate about their equilibrium positions with amplitudes as much as 10 per cent of the interatomic distances. This vibration interferes with electron transport, reducing conductivity at higher temperatures. At low temperatures, resistivity becomes primarily a function of crystal imperfections. Thus the best conductors are pure (defect-free) crystalline solids at low temperatures.

The conductivity of metals is primarily a result of the "free" electrons in the metallic bond. Materials with covalent bonds require bonds to be broken to provide electrons available for conduction. Thus the electrical properties of these materials depend on bond strength. Diamond, for instance, is a strong insulator. Silicon and germanium, however, have weak bonds that can easily be broken by thermal energy. These materials, when pure, are known as intrinsic *semiconductors,* since moderate amounts of applied energy can enable the materials to conduct small amounts of electricity.

The conductivity of the nonmetallic semiconductors can be substantially improved by a process known as *doping.* Both silicon and germanium have four valence electrons and four covalent bonds. If one of these elements is replaced with an atom containing five valence electrons, such as phosphorus, the four bonds form, leaving one excess electron, as in Figure 3-23. The extra electron is free to move about and conduct electricity. Such materials are known as *n*-type extrinsic semiconductors.

A similar effect can be created by substituting an atom with three valence electrons, such as aluminum. An electron is missing from a bond, creating an electron "hole." As an electron jumps into this hole, it creates a

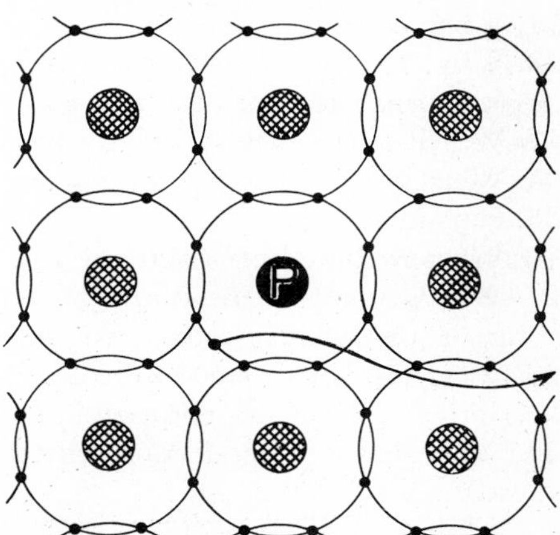

FIGURE 3-23. Schematic representation of an *n*-type semiconductor, with the excess electron of the phosphorus atom being free for conduction.

hole in the spot it vacated. Movement of electron holes is equivalent to a countermovement of electrons, and thereby produces improved conductivity. These materials are known as *p*-type semiconductors. The control of conductivity through semiconductor devices underlies much of the current advance in solid-state devices and circuitry.

Review questions

1. Why is it important for an engineer to have an understanding of structure?
2. Of what significance are the valence electrons in determining the properties of materials?
3. What are ions and why do they tend to form?
4. What is an ionic bond?
5. What properties are characteristic of ionically bonded materials?
6. What is a covalent bond? How does it differ from the ionic?
7. What properties are characteristic of covalently bonded materials?
8. How does the metallic bond differ from the ionic? From the covalent?
9. What properties are characteristic of materials bonded by metallic bonds?
10. Why do metals have high electrical and thermal conductivities?
11. Why is the "size of an atom" not a constant value for a particular element?
12. Contrast molecular, crystal, and amorphous structures.
13. What is a space lattice? A unit cell?
14. What is an allotropic material?
15. Sketch the three most common crystal lattices found in metals.
16. Why is the simple cubic structure not observed in nature?
17. What are the two controlling processes in solidification?
18. What is one way of characterizing grain structure?
19. What is the basic mechanism of elastic deformation?
20. How does plastic deformation differ from elastic deformation within a crystal?
21. What are the characteristics of the planes and directions preferred for slip?
22. Why do metals having different crystal lattice types exhibit different deformation behavior?
23. Which of the three metal crystal structures has the greatest ductility?
24. What is a dislocation? An edge dislocation? A screw dislocation?
25. Explain strain hardening.
26. What is the role of grain size and shape in plastic deformation?
27. Why do components formed by plastic deformation often show a directional variation in properties?
28. What conditions are required for recrystallization?
29. Why is control of grain size important during recrystallization?
30. What is the distinguishing difference between hot working and cold working?

31. What is an alloy?
32. What are the three classes of alloys presented in the text?
33. What properties tend to characterize intermetallic compounds?
34. What conditions favor high electrical conductivity in a metal?
35. What is an intrinsic semiconductor? An *n*-type extrinsic semiconductor? A *p*-type extrinsic semiconductor?

Case study 2.
THE MISUSED TEST DATA

Supreme Sheet Metal Company is a supplier of formed sheet metal panels for various applications. One of its customers has approached the company with a request to convert one such panel from 1008 steel sheet (a low carbon/high ductility steel) to a thinner-gage, high-strength material as a means of reducing the weight of its product. It was also its desire to retain the same design and, thereby, the same forming dies, if at all possible.

John Doakes, a young engineer with Supreme Sheet Metal, was assigned the task of determining the feasibility of the substitution. He first pulled a uniaxial test specimen from the new material and observed a 6 per cent uniform elongation. Next he placed a grid on one of the 1008 steel sheets, deformed it into the desired shape and observed a maximum tensile strain of only 4 per cent during forming. Concluding that the new material had adequate formability for the particular component, he reported favorably on the substitution.

When production began, numerous rupture-type failures occurred in the region of maximum strain, indicating insufficient ductility. What had John overlooked?

CHAPTER 4

Production and properties of common engineering metals

Engineering metals possess a wide range of usable properties. Some of these are inherent to the particular metal, but many can be varied by controlling the manner of production and processing. Metals, to a large extent, are history-dependent materials, with the final properties being affected by the specific details of the processing history. Thus it is helpful for the engineer to have a working knowledge of how metals are produced and processed and the effects on final properties and engineering utility.

IRON

Iron is the fourth most plentiful element in the earth's crust and, for centuries, has been the most important of the basic engineering metals. It is rarely found in the metallic state, but occurs in mineral compounds known as ores. The metallic iron or steel metal produced from these ores has played a central role in the development of civilization, and it appears that it will continue to do so in the foreseeable future. Iron-based tools have extended man's capabilities in numerous areas, including the necessities of food, clothing, and shelter. New advances in technology of iron and iron alloys continue to expand their utility in engineering application.

Pig iron. *Pig iron* is the first product in the process of converting iron ore into useful metal and is produced in a blast furnace. These furnaces, illustrated schematically in Figure 4-1, are large, round, costly steel structures from 30 to 40 meters (98 to 131 feet) tall, lined with refractory firebrick. The diameter varies with position on the furnace, but at the largest point, called the *bosh,* they usually are about 10 meters (33 feet) in diameter.

Iron ore, which basically is an oxide of iron with companion impurities, is processed in a manner so as to break the iron–oxygen bonds and produce metallic iron. This process is known as *reduction.* The ore may be of various forms and compositions, among them being (1) *hematite,* a mixture of Fe_2O_3 and earthy matter known as *gangue,* and (2) *magnetite,* the Fe_3O_4 form of iron oxide. Approximately two parts of iron ore, one part of coke, and one-half part of limestone are put into the blast furnace through the double bell hopper at the top. The process is then controlled so as to reduce the iron oxide to iron and simultaneously regulate the reactions involving the gangue, which may contain silica, alumina, calcium oxide, magnesium oxide, water, phosphorus, and sulfur. Control of these components is important because they may substantially affect the product material.

FIGURE 4-1. Schematic diagram of a blast furnace and its associated equipment.

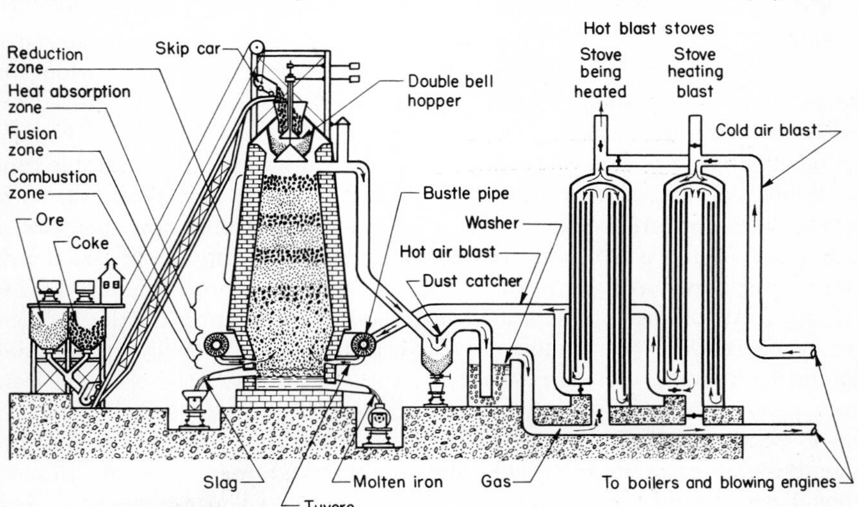

Ore is the source of the metallic iron. Limestone is added for several reasons, but primarily as a fluxing agent to enable the gangue material and coke ash to enter into a fusible liquid. A second purpose is to provide a material with which undesirable elements or compounds will combine in preference to the metallic iron being produced. The coke serves as a source of the heat necessary to cause the reducing processes to occur and produce molten iron. Air is the fourth raw material. It is preheated to 600 to 700°C (1100 to 1300°F) and is blown into the furnace through nozzles, called *tuyeres,* near the base of the furnace. As the air passes up through the incandescent coke, large volumes of carbon monoxide gas are formed. This gas, together with the carbon in the coke, enables the reduction of the iron oxide through several possible reactions:

$$Fe_2O_3 + 3CO \rightleftharpoons 2Fe + 3CO_2$$
$$Fe_2O_3 + 3C \rightleftharpoons 2Fe + 3CO$$

The actual reduction process does not take place in as simple a manner as indicated by these equations, but involves several sequential steps with the same net result. Although the reactions above are reversible, they are forced to go in the desired direction by regulating the charge, temperature, and amount of air.

As the input material settles near the bosh and the temperature increases, the iron is reduced to a spongy incandescent mass. In this state the iron absorbs more carbon, which lowers its melting point until it finally melts and drips down over the unburned incandescent coke into the base or hearth of the furnace.

Meanwhile, the limestone, which aids in reducing the iron oxides, combines with the calcium and magnesium oxides, alumina, and silica to form a fusible *slag.* Being lighter than the molten iron, the slag floats on top of it and is drained off (tapped) periodically through a hole (slag notch) located at a level above the iron tap hole. Although once discarded, now this material is processed in a variety of manners, among them being (1) solidification and crushing for road-bed aggregate, and (2) water spray atomizing to form a sandlike material for construction purposes.

Blast furnace operation is a continuous process with ore, limestone, and coke being fed in steadily and the molten pig iron being tapped off every 4 to 6 hours. Although most iron is now sent directly to steelmaking facilities in the form of molten metal, earlier processing involved the solidification into small molds known as pigs, from whence the name "pig iron." From 1000 to 4000 tons can be produced in a 24-hour period, depending on the furnace. Furnace operation is continuous, being terminated only during production lags and for relining or repair. Modern furnaces can usually be operated from 4 to 6 years without being shut down for repair.

For each ton of pig iron tapped from a blast furnace, about 3 tons of hot or combustible gases are obtained. These are used to preheat the air blast by means of stoves, part of the auxiliary equipment shown in Figure 4-1. Ordinarily, four stoves are associated with each blast furnace. These are rotated in operation so that three receive heat from the gases while the fourth is giving up its heat to the air blast. By-product gases may also be used to generate power for operation of additional equipment, such as compressors and blowers.

Recent advances in blast furnace technology tend to center on increased productivity, energy conservation, energy-form substitutions, and the use of lower-grade materials. *Taconite,* a low-grade ore that contains from 20 to 27 per cent recoverable iron, is now being used extensively. Beneficiation treatments are applied at or near the mine site to produce pellets that contain about 63 per cent iron and are suitable for use in the blast furnace.

Composition of pig iron. Along with the iron ore, other easily reducible oxides in the ore and coke also are reduced in the blast furnace. All the phosphorus and most of the manganese introduced to the furnace are present in the pig iron. Oxides of silicon and sulfur are only partially reduced. Calcium, magnesium, and aluminum oxides are fluxed by the calcium carbonate in the limestone, go into the fusible slag, and are removed. Consequently, the pig iron contains from 3 to 4.5 per cent carbon, all the phosphorus that was present, and most of the manganese. The silicon and sulfur content is determined to some extent by the raw materials, and also by controlling the chemistry of the slag and the temperature of the furnace. All of the reduced elements go into the melt, while those elements that are oxidized go into the slag. Pig iron therefore has about the following analysis:

Carbon	3.0–4.5%
Manganese	0.15–2.5%
Silicon	1.0–3.0%
Sulfur	0.05–0.1%
Phosphorus	0.1–2.0%

Types and uses of pig iron. Pig iron is important primarily as the raw material for other processes. Although a very small portion of the total output is cast into final shapes as it comes from the blast furnace, most iron is transfered in the molten state and fed into various types of furnaces to be made into steel. Several types of pig iron are made, each varying somewhat in composition and properties. These types and their uses are indicated in Table 4-1.

Direct reduction of iron ore. As a result of the desire to employ lower-grade ores and readily available fuels that are currently unsuitable for blast furnace operations, many attempts have been made to develop a process that would at least partially replace the blast furnace as a source of metallic iron. Such processes are referred to as *direct reduction processes* and have as their goal either (1) production of steel directly from iron ore, (2) manufacture of a product equivalent to blast-furnace pig iron for use in current steelmaking processes,

TABLE 4-1. Types and uses of pig iron

Type	Use
Basic pig	Steel castings Rolled shapes
Acid or Bessemer pig	Steel castings Rolled shapes Wrought iron products
Forging pig	Wrought iron products Crucible steels Alloys steels
Foundry pig	Gray iron castings Chilled gray iron castings Alloyed iron castings
Malleable pig	Malleable iron castings White iron castings Alloyed malleable iron castings

or (3) the production of a low-carbon iron to be used in a similar manner as scrap in the current steelmaking processes. Almost every type of apparatus suitable for the purpose has been adapted, including pot furnaces, reverberatory furnaces, regenerative furnaces, shaft furnaces, rotary and stationary kilns, retort furnaces, electric furnaces, and fluidized-bed reactors. Reducing agents include coke, coal, graphite, char, distillation residues, fuel oil, tar, combustible hydrocarbon gases, and hydrogen. Many processes have passed through the pilot-plant stage and several have become economically favorable as a result of local resource conditions. To date, however, no such process has shown sufficient promise that it would challenge the blast furnace as the chief source of iron for the steelmaking industry.

As an example, one process passes ore, coke, and limestone through a controlled reducing atmosphere in a rotary kiln at a temperature below the melting point of iron. The kiln output is then put through magnetic separators and ball mills and made into briquets. With high-grade ore, the briquets contain about 90 to 95 per cent iron and can be fed directly into a steelmaking furnace. When lower grade ore is used, only about 70 per cent of the briquets contain 90 to 95 per cent iron. The remaining 30 per cent contains about 80 to 90 per cent iron and can be used as input to a blast furnace.

Interest in direct reduction processes is high because they can be used for low-grade ores and with fuels other than coking-grade coal. Moreover, there is an economic advantage in that a direct reduction plant is probably less expensive than the traditional blast furnaces. Plants currently in successful operation, however, are competing largely due to favorable supplies of natural resources in their immediate vicinity. Many processes require extensive amounts of natural gas, a rapidly diminishing fuel material.

Cast iron. *Cast iron* is essentially the same as pig iron, but is the term applied to the metal when cast in product form. Ordinarily, pig iron and scrap are melted in a cupola and then cast into prepared molds. The metallurgical properties of cast iron are discussed in Chapter 5 and its melting and utilization in the casting process in Chapter 11.

Wrought iron. Although no longer produced in any substantial quantity, *wrought iron* was the most important structural metal prior to 1855. Pig iron was further refined to a very pure iron and was then combined with siliceous slag and processed to form an iron matrix with fibrous particles of slag. The product was strong, ductile, and corrosion-resistant, but has been almost entirely replaced by steel and other materials.

STEEL

The manufacture of *steel* is essentially an oxidation process that decreases the amount of carbon, silicon, manganese, phosphorus, and sulfur in a mixture of molten pig iron and steel scrap. In 1856, the *Kelly–Bessemer process* essentially opened up an industry by enabling the manufacture of commercial quantities of steel. The *open-hearth process* surpassed the Bessemer process in tonnage produced in 1908 and was producing over 90 per cent of all steel in 1960. *Oxygen furnaces* of a variety of types and *electric furnaces* now produce most of our commercial steels.

The Bessemer process. The Kelly–Bessemer process, or simply the *Bessemer process,* made use of the fact that air passed through molten pig iron enables exothermic reactions to occur that refine the metal into steel. Carbon oxidizes to produce gaseous CO or CO_2. Action of the air also forms large quantities of iron oxide, which further reacts to form oxides of silicon and manganese, taking these elements into the slag. In this process, however, the reactions do not affect the phosphorus and sulfur contents of the metal and leave these somewhat undesirable elements in the steel.

About 25 tons could be produced in a 15-minute blow, but chemical control of the product was difficult to achieve. High nitrogen contents were also introduced into the metal as the air passed through the melt.

The open-hearth process. The *open-hearth furnace,* developed commercially around 1870, is basically a shallow refractory-lined "dish" that may be up to 5 by 11 meters (16 by 36 feet) in surface and about ⅔ meter (2 feet) deep. Approximately 150 to 250 tons of selected material is heated to a molten state, raised to elevated temperatures, and processed to obtain a product of desired chemical composition. The furnaces are regenerating in nature, with heated air and fuel gas entering one end of the hearth, passing over the charge of pig iron, scrap steel, and limestone, and finally heating a heat reservoir

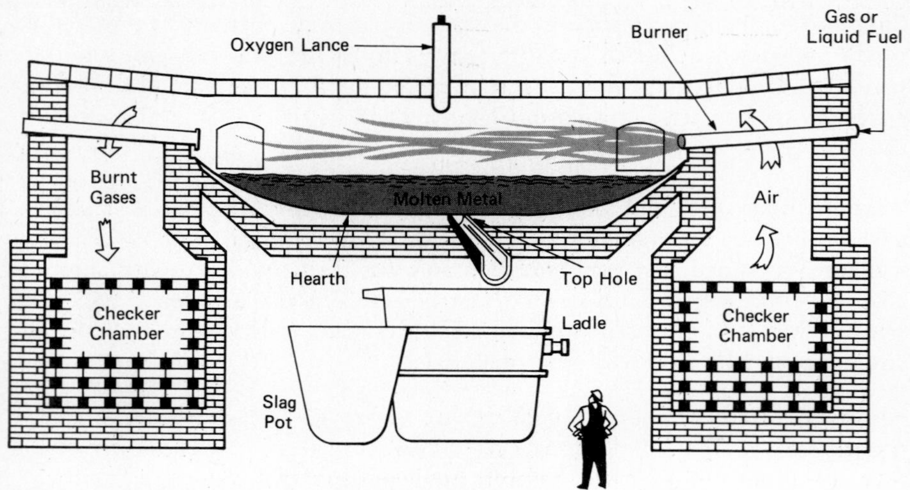

FIGURE 4-2. Schematic diagram of an open-hearth furnace, showing optional oxygen lancing. (*Courtesy Bethlehem Steel Corp.*)

(checker) on the other side of the furnace with the waste gases, as illustrated in Figure 4-2. Upon reversal of flow, the checkers yield their stored heat to the incoming air to continue the removal of manganese, phosphorus, silicon, and sulfur from the melt.

Approximately 7 to 10 hours are required for a melting and refining cycle, enabling periodic samplings and additions to the melt. The process is easy to control, produces uniform-quality material with low phosphorus and sulfur contents, has high flexibility, and can use up to 80 per cent scrap in its charge. More recent advances have speeded up the process either by blowing oxygen over the surface of the molten metal or by introducing it directly into the molten metal through a water-cooled pipe, a process known as *bottom lancing*.

Basic-oxygen process. In 1952, the *basic-oxygen process* was developed, taking advantage of the availability of large quantities of pure oxygen. Approximately 200 to 400 tons of steel scrap and molten pig iron are charged into a large, cylindrical, open-mouth vessel that has a basic refractory lining. The furnace is returned to its upright position and burned lime and flux are poured in through a chute. A water-cooled oxygen lance, roughly 18 meters (65 feet) long and 25 cm (10 inches) in diameter, is then lowered to within 2 meters of the bath and high-purity oxygen is blown onto the surface of the metal. The oxygen is ejected under considerable pressure at rates in excess of 1 ton per minute, blows the slag aside, and reacts violently with the exposed iron to form iron oxide. Part of the oxide reacts with the flux to form a basic slag,

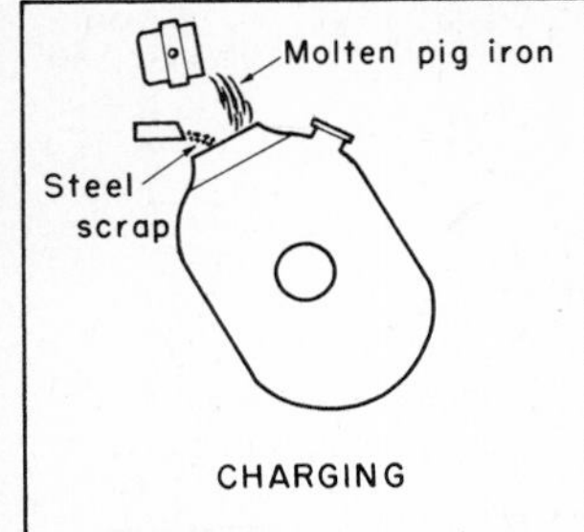

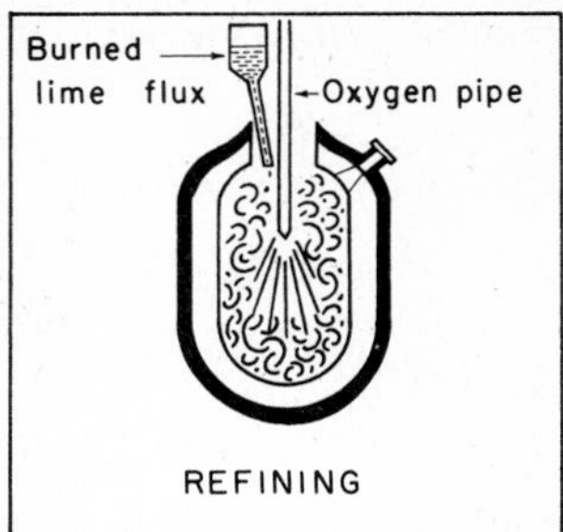

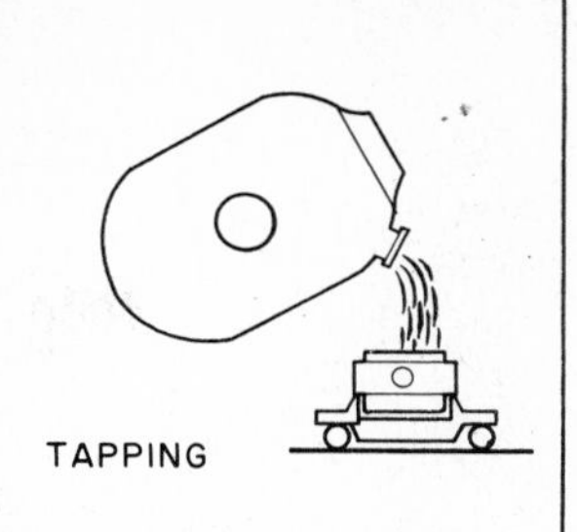

FIGURE 4-3. Schematic diagram of the basic-oxygen steelmaking process.

while the remainder is mixed with the bath, through turbulence, and oxidizes the impurities. The oxygen treatment requires about 15 to 20 minutes of the total 40- to 45-minute cycle. Samples are taken and analyzed and adjustments made to produce desired compositions prior to tapping. Figure 4-3 shows the process in schematic.

Production capacity for a single furnace shows a tenfold increase for the basic-oxygen furnace over the open-hearth furnace. Capital investment and operating expenses are lower for the oxygen furnace provided that a sufficient supply of hot metal is available. Improved ingot quality, lower energy requirements, and easier environmental control all favor the oxygen process over the open-hearth process. The major disadvantage to the process is the limited amount of scrap input (30 to 35 per cent) and large amount of hot metal required. Also, with the reliance on only a few furnaces, prolonged downtime can seriously affect an entire plant.

Current development efforts relate primarily to increasing the amount of scrap in the charge and improving the energy efficiency of the process. Bottom lancing modifications have recently been introduced.

Electric furnace processes. A substantial tonnage of steel is currently produced in electric furnaces. Although once associated with small quantities of specialty materials, such as tool steel, stainless steel, die steels, and aircraft-quality metal, these furnaces are now widely used when smaller quantities and high alloy contents are desired or where their ability to utilize up to 100 per cent scrap provides a distinct advantage.

In one type of furnace, an *electric arc* is maintained between electrodes and the metal being melted. Three electrodes are normally used in conjunction with a three-phase power source. High currents flowing through the conductive metal provide the heat necessary for the process. Various slag baths are used during the process to control the chemistry of the product. Commercial furnaces of the types illustrated in Figures 4-4 and 4-5 range in capacity from 5 tons to more than 150 tons.

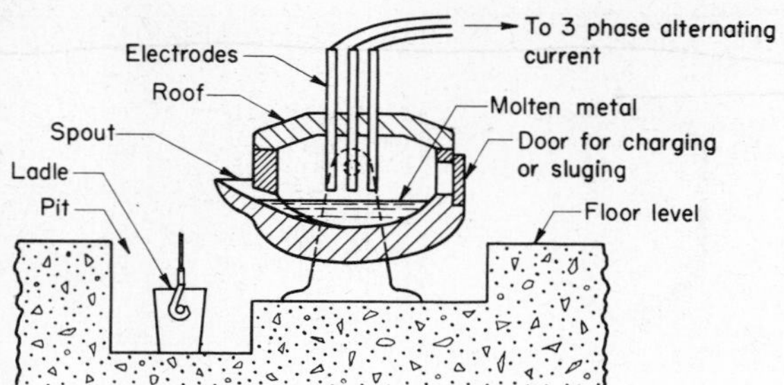

FIGURE 4-4. Schematic diagram of a three-phase electric arc furnace.

The *induction-type furnace,* illustrated in Figure 4-6, usually has a capacity of less than 5 tons. They are primarily used for making high-quality or high-alloy steels or for remelting metals such as cast iron in an iron foundry.

Major disadvantages of the electric furnace process relate to the inefficiency of the energy system once meltdown has been completed, and the pickup of undesired gases by the melt during processing.

The production of ingots. Regardless of the method by which steel is made, it must undergo a change of state from liquid to solid before it becomes a usable product. This solidification may be directed into a final desired shape by steel casting or into a form suitable for further processing. In most cases, the latter option is exercised through either *continuous casting* or the forming of *ingots* to produce the raw material for subsequent forging and rolling operations.

FIGURE 4-5. Electric arc furnace, tilted for pouring. (*Courtesy Pittsburgh Lectromelt Furnace Corporation.*)

FIGURE 4-6. Pouring molten metal from an electric induction furnace. Inset shows the principle of this type of furnace. (*Courtesy Ajax Magnethermic Corporation.*)

In casting ingots, the desire is to obtain metal as free of flaws as possible. The molten steel is tapped from the furnace into pouring ladles, nearly all of which are of the bottom-pouring type illustrated in Figure 4-7. By extracting the metal from the bottom of the ladle, slag and floating matter are not transferred to the ingots. When poured into the ingot molds, the metal may enter from either the top or the bottom. Although the majority of ingots are poured from the top, there are several disadvantages to the procedure. Hot

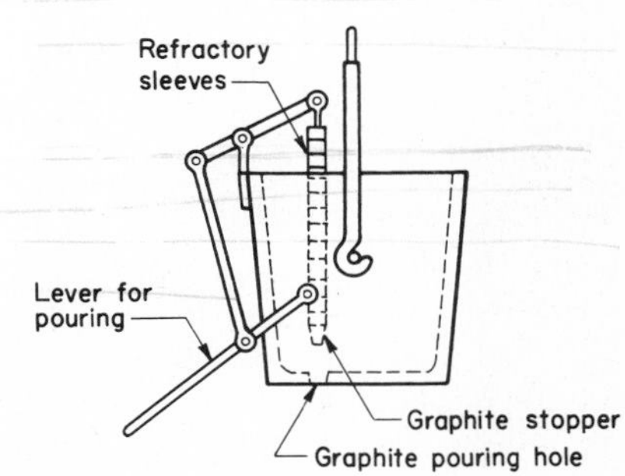

FIGURE 4-7. Schematic diagram of a bottom-pouring ladle.

FIGURE 4-8. Pouring steel ingots by the bottom-pouring process. The bottom of the center mold being poured is connected to the bottom of the remaining molds in the cluster by means of tile channels. (*Courtesy United States Steel Corporation.*)

metal may be splashed onto the side walls of the mold, solidify, and later become part of the ingot when the pouring level passes that point. This produces a disruption to the continuity of ingot's internal structure. Slag entrapment is also more likely in top poured ingots. To avoid these difficulties, ingots are sometimes poured from the bottom. In this procedure, illustrated in Figure 4-8, the bottom of several outer ingots is connected to the bottom of a central pouring ingot by ceramic tile tunnels to form a spiderlike arrangement. Hot metal poured into the center is conveyed through the tunnels to fill the outer molds from the bottom. Bottom pouring is more costly than top pouring so it is usually used only for high-quality steels.

Figure 4-9 schematically shows the contraction of a metal undergoing solidification, the jump at the melting point being known as the *solidification shrinkage.* When metals solidify, it is expected that a shrinkage will be observed in the region of the last material to solidify. In ingots, solidification proceeds inward from the mold walls and upward from the bottom. Shrinkage takes the form of a *pipe* coming in from the top as illustrated in Figure 4-10. Since the pipe surface has been exposed to the atmosphere at elevated temperature, oxides and surface contaminants form which prevent the metal from

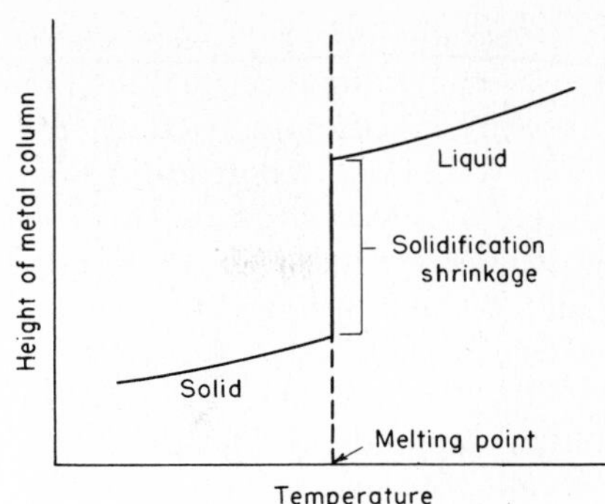

FIGURE 4-9. Height of a column of metal as a function of temperature, showing solidification shrinkage.

welding back together during subsequent processing. That portion of the ingot containing the pipe must be recycled as scrap, an amount that may be a substantial fraction of the ingot.

Although the amount of shrinkage cannot be changed, the shape and location can be greatly controlled. One procedure to reduce the amount of pipe is to use a ceramic *hot top* on the top of the ingot mold. By retaining heat at the top, the liquid reservoir at the end of solidification is more of a uniform layer on top of the ingot, thereby minimizing the depth of shrinkage. Variation of the shape of the mold controls not only the solidification shape, but also the type of ingot structure. The type having the big end up is commonly used for greatest soundness.

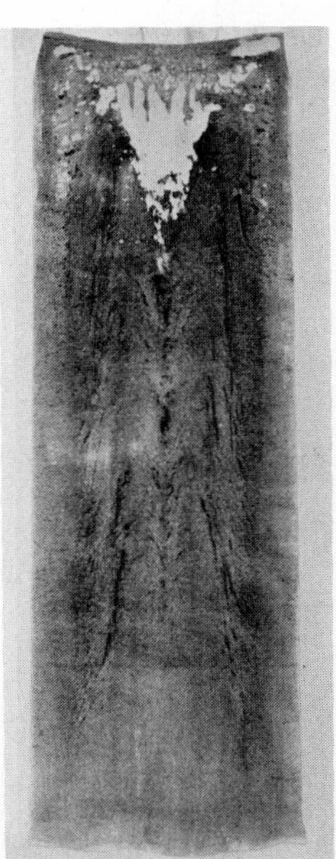

FIGURE 4-10. Section of an ingot, showing a "pipe" (top) and "segregation" (dark areas). (*Courtesy Bethlehem Steel Corporation.*)

Several new processes have been developed as a means of overcoming some of the ingot-related difficulties, such as piping, entrapped slag, and structure variation, as well as reducing or replacing the associated transport, mold stripping, and reheating operations required for further processing. A variety of *continuous-casting techniques* are being used extensively. Figure 4-11 illustrates the most common procedure, wherein liquid metal flows from a ladle, through a tundish, into a bottomless, water-cooled mold, usually made of copper. Cooling is controlled so that the outside has solidified before the metal exits the mold. The material is further cooled by direct water sprays to assure complete solidification. The cast solid is then either bent and fed horizontally through a reheat furnace for further processing or cut to desired lengths. Mold shape, and thus the shape of the cast product, may vary such that products may be cast with cross sections close to the desired final shape.

Continuous casting virtually eliminates the problems of piping and mold spatter, as well as the cost of ingot molds, handling, and stripping. Some small surface defects, such as small surface cracks, may occur and have to be removed where use involves critical applications. The steel, copper, and aluminum industries are all producing material by continuous casting.

FIGURE 4-11. Schematic representation of the continuous-casting process for producing billets, slabs, and bars. (*Courtesy Materials Engineering*.)

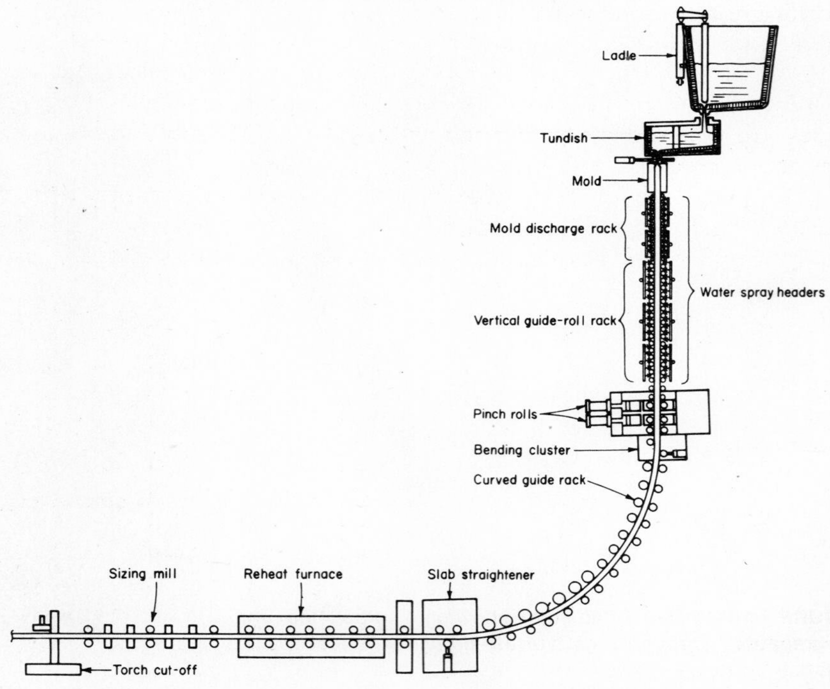

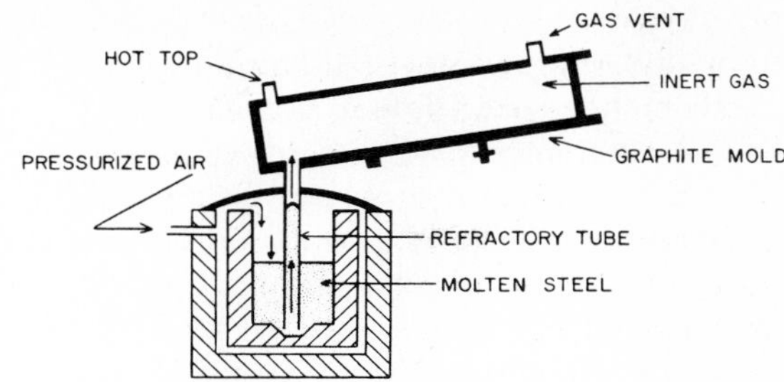

FIGURE 4-12. Schematic representation of the pressure-pouring process for casting ingots and slabs.

Molten steel may also be cast into slabs suitable for further processing by means of the pressure-pouring process depicted in Figure 4-12. Molten metal is forced up into a graphite mold by air pressure and solidifies. The mold is hinged for easy slab removal and can be used repeatedly. Because the metal is taken from the bottom of the ladle and is introduced through the bottom of the slab mold, excellent slab quality is obtained.

Degassification of ingots. During the oxidation that takes place in the making of steel, considerable amounts of oxygen can dissolve in the molten metal. When this molten metal is then cooled to produce solidification, oxygen and other gases are rejected from the solid as the saturation point decreases (see the schematic of Figure 4-13). The rejected oxygen then links with atomic carbon to produce a CO gas evolution. A porous structure results where the gas bubble-induced *porosity* has the form of either small and dispersed pores or large blowholes. Pores that are totally internal can be welded shut during subsequent hot forming, but if they are exposed to the air at elevated temperatures, the pore surfaces oxidize and will not weld. Cracks and internal defects may well appear in the finished product.

In many cases, it is desirable to avoid the porosity difficulties by removing the oxygen or rendering it nongaseous prior to solidification. Where high-quality steel is desired or where subsequent deformation may be inadequate to

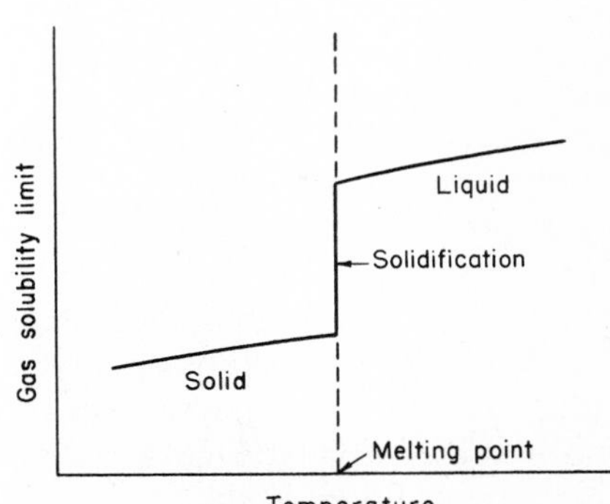

FIGURE 4-13. Gas solubility in a metal as a function of temperature.

produce welding of the pores, the metal is usually fully *deoxidized* (or *killed*), to produce a killed steel. Aluminum, ferromanganese, or ferrosilicon is added to the molten steel while it is in the ladle to introduce material with a higher affinity for oxygen than the carbon, a process particularly useful with higher-carbon steels. Rejected oxygen simply reacts to produce solid metallic oxides dispersed throughout the structure. Full shrinkage is observed as this material solidifies and a large portion of scrap may be generated when the exposed pipe region is cut off the solidified ingot. This may be somewhat offset by the use of a heated refractory "hot top" placed on the top of the ingot mold. The solidification is slowed at the top, giving a more uniform shrink instead of a pipe.

For steels with lower carbon contents, a partial deoxidation may be employed to produce a *semikilled steel.* Enough deoxidant is added to partially suppress bubble evolution, but not enough to completely eliminate the effect of oxygen. Some pores still form in the center of the ingot, their volume serving to cancel some of the solidification shrinkage and thereby reduce the extent of piping and scrap generation.

For steels of sufficiently low carbon (usually less than 0.2 per cent carbon), another process, known as *rimming,* may be employed. The steel is only partially deoxidized prior to solidification. After the material is poured into the ingot mold, the first metal to solidify is almost pure iron, being very low in carbon, oxygen, sulfur, and phosphorus. These elements are rejected from the solid into the liquid ahead of the solidification front. When the concentration of dissolved gases exceeds the saturation point, a layer of CO bubbles evolves and the effervescence tends to drive entrapped particles to the center of the ingot. Further solidification produces additional porosity in the inner portions of the ingot and, if properly controlled, the rimming action can produce a porosity level that will approximately compensate for solidification shrinkage.

The outside of the ingot is clean, defect-free metal which provides an excellent blemish-free surface when the ingot is rolled to flat strips or similar product. The holes on the inside of the rimmed ingots have bright, clean surfaces, because they have not been exposed to the air. When hot working is performed, the surfaces weld together to produce a sound product. Figure 4-14 presents a schematic comparison of killed, semikilled, and rimmed ingot structures.

In addition to the effects of oxygen, small amounts of other dissolved gases, particularly hydrogen and nitrogen, have deleterious effects on the performance of steels. This is particularly important in the case of alloy steels because several of the major alloying elements, particularly vanadium, columbium, and chromium, tend to increase the solubility of these gases. Consequently, several methods have been devised for *degassing* steels, and considerable quantities of degassed steel are now used in critical applications, such as turbine rotor shafts. Figure 4-15 illustrates a method that is widely used to produce degassed ingots, known as *vacuum degassing.* The ingot mold is en-

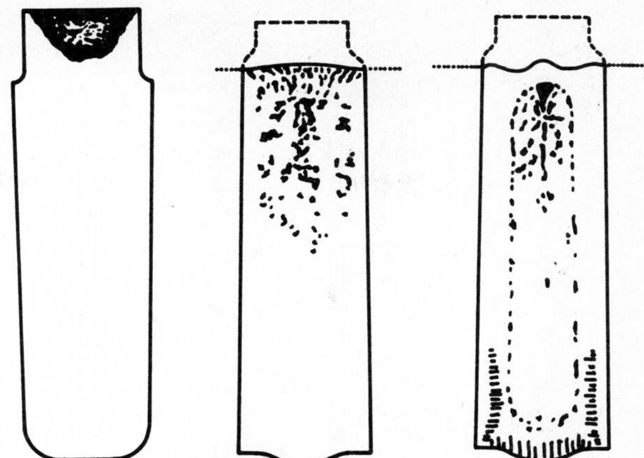

FIGURE 4-14. Schematic comparison of killed, semi-killed, and rimmed ingot structures. (*Courtesy American Iron and Steel Institute.*)

closed in an evacuated chamber and the metal stream passes through the vacuum during pouring, the vacuum serving to remove the dissolved gases. Another procedure is to perform the melting in an induction furnace and enclose both the furnace and the ingot mold in a vacuum chamber.

When exceptional purity is required, a *consumable-electrode remelting* process may be employed, as illustrated in Figure 4-16. A solidified metal electrode is remelted by electric arc in an evacuated chamber and resolidifies into a

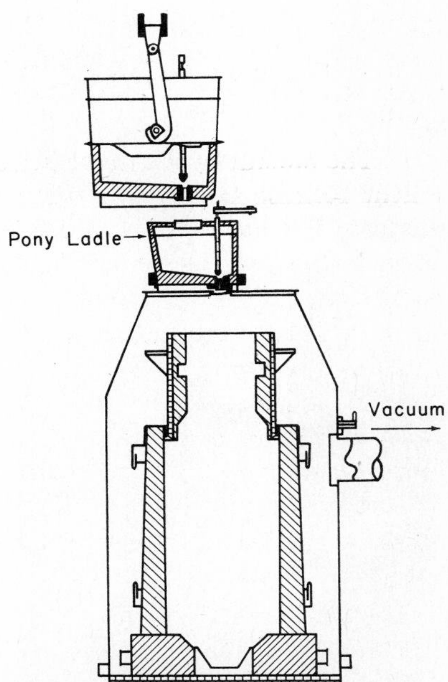

FIGURE 4-15. Method of vacuum degassing steel while pouring ingots.

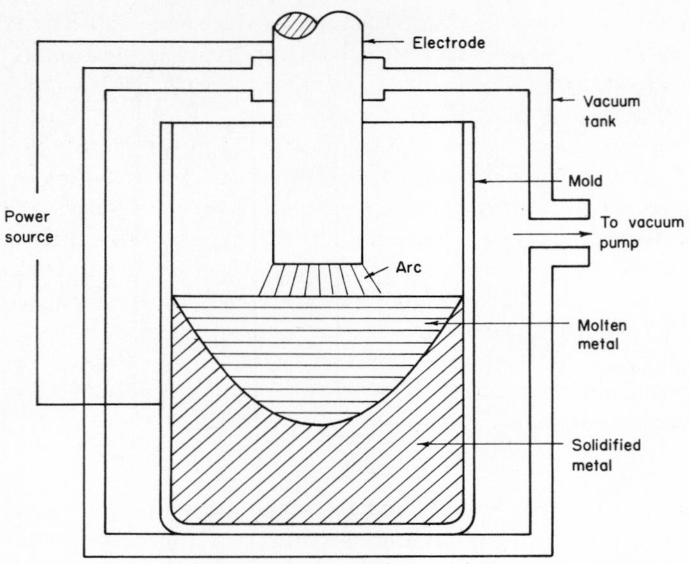

FIGURE 4-16. Method of producing degassed ingots by consumable-electrode vacuum remelting.

FIGURE 4-17. (*Left*) Production of an ingot by the electroslag remelting process. (*Right*) Schematic representation of this process. (*Courtesy Carpenter Technology Corporation.*)

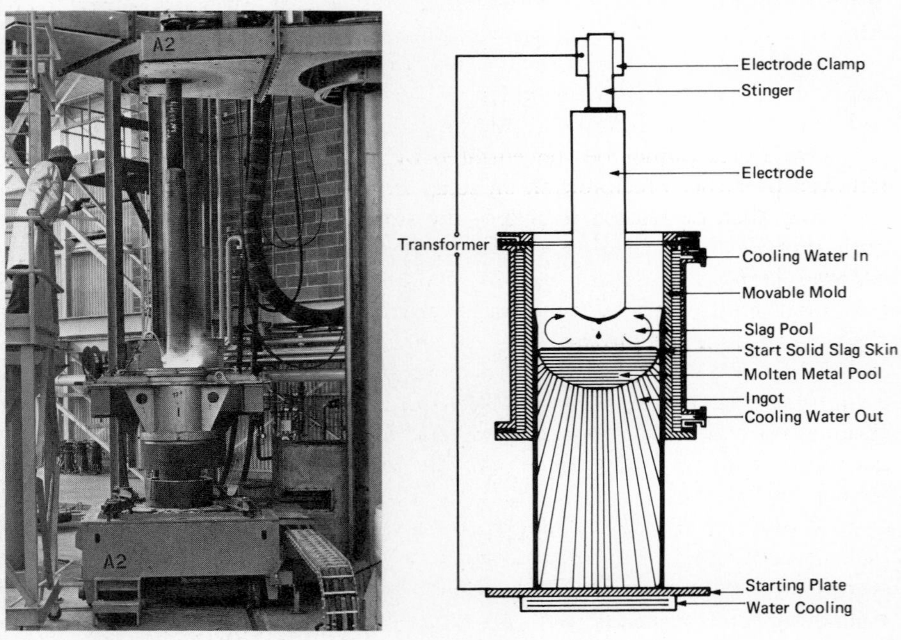

new form. This process, known as *vacuum arc remelting* (VAR) [or *vacuum induction melting* (VIM) if induction heating replaces the electric arc], is highly effective in removing all dissolved gases, but cannot remove any nonmetallic impurities from the original metal.

If extremely gas-free and clean metal is required, it may be obtained by using the *electroslag remelting process* (ESR), illustrated in Figure 4-17. In this process, an electrode made by the normal consumable-electrode process is remelted and cast by means of an arc with the surface of the melted metal now covered by a thick blanket of molten flux. Nonmetallic impurities are collected in the flux blanket and there is no danger of arc spatter collecting on the mold walls and then becoming part of the ingot.

In addition to being used to produce steel ingots of exceptional high purity, the consumable-electrode processes are also used to produce ingots of the highly reactive metals.

COPPER

Production. *Copper* is a very important engineering metal that has been used by man for over 6000 years. Used in its pure state, copper is the backbone of the electrical industry, although aluminum is now offering competition. It is also the major metal in several highly important engineering alloys.

Most copper ore is in the form of sulfides or oxides, with major U.S. deposits being found in Arizona, Utah, New Mexico, Montana, and Michigan. With the depletion of the best domestic ore, deposits in South America and Africa have recently assumed great importance.

Even the best ore contains a rather low percentage of copper, so the first step in production is a concentration process. Sulfide ores are crushed, ground, and concentrated by flotation to yield a product that is about 50 per cent copper. Oxide ores cannot be concentrated by flotation, so an acid leach is used followed by either precipitation on scrap iron or electrowinning. The concentrate may then be roasted to reduce the sulfur and arsenic. It is then melted with suitable fluxes in reverberatory or electric furnaces, a process known as *smelting.* Lighter impurities combine and float to the top as a slag while copper, iron, sulfur, and any precious metals form a product known as *matte* in the lower part of the furnace.

The molten matte (containing about 40 per cent copper) is transferred to a converter that is similar to those used in making Bessemer steel. Air blown through the matte oxidizes and blows out the sulfur and oxidizes the iron that goes into the slag. The product is known as *blister copper* and is approximately 99 per cent pure.

Blister copper may be upgraded to refined copper by melting in a furnace and removing the principal impurity, oxygen, in a reducing atmosphere formed by the use of green logs thrust into the melt (poling) or by injection of a reducing gas such as methane. Most blister copper, however, undergoes only a partial furnace refining and is then cast into copper anodes for electrolytic refining. These anodes and thin copper starting sheets, or cathodes, are suspended in tanks containing a solution of copper sulfate and sulfuric acid. An electric current passed through the solution dissolves the copper anodes and deposits refined copper on the cathode. Gold, silver, and other valuables are recovered from the sludge on the bottom of the tanks.

The resulting refined copper contains only about 0.07 per cent oxygen and is called *electrolytic tough-pitch* (ETP) *copper.* If the intended use of the copper is as the base for an alloy, refinement to low oxygen content may not be necessary and the poling process may be eliminated.

Copper with low oxygen content may also be obtained by two other methods. The first uses an inert-gas atmosphere in the reverberatory furnace and produces an *oxygen-free high-conductivity* (OFHC) *copper.* The second procedure is to use deoxidation with a strong reducing agent such as phosphorous or silicon. This approach has the disadvantage that the electrical conductivity is reduced by 10 to 20 per cent. The use of calcium, lithium, or boron as the reducing agent causes less reduction in the conductivity.

The most pressing problem currently facing the copper industry is environmental. Smelting produces large quantities of sulfur compounds, trace elements, and particulate emissions that must be contained before release to the atmosphere. Existing practices must be modified or new techniques developed as a solution to the problem. Similar problems also confront the lead industry, where the principal ore is also a sulfide. Processing techniques are in a state of dynamic change. Further modifications relate to the use of leaching techniques to recover copper from low-grade ore or previous waste material, economics now favoring a recovery process.

Properties. The wide use of copper is based, primarily, on three important properties: its high electrical *conductivity,* high *ductility,* and *corrosion resistance.* Obviously, its excellent conductivity accounts for its importance to the electrical industry. The better grades of copper conductor have a conductivity rating of about 102 per cent, reflecting the better oxidation procedures now available as opposed to 1913, when the standard was established.

While copper in the soft pure state has a tensile strength of only about 200 MPa (30,000 psi), its elongation in 2 inches is about 60 per cent. By cold working, it can be hardened and the tensile strength increased to above 450 MPa (65,000 psi) with a decrease in elongation to about 5 per cent. Its relatively low strength and high ductility make it a very desirable material where forming operations are necessary. Furthermore, the hardening effects of cold working may easily be removed because the recrystallization temperature is less than 260°C (500°F).

Copper, as a pure metal, is not used extensively in manufactured products except in electrical equipment. More often, it is the base material for some alloy to which it imparts its good ductility and corrosion resistance. In nonelectrical uses, it is usually one or both of these properties that accounts for its selection as the material to be used.

If copper is stressed at high temperatures over long periods of time, it is subject to intercrystalline failure at about half its normal room-temperature strength. Material containing more than 0.3 per cent oxygen is also subject to hydrogen embrittlement when exposed to reducing gases above 400°C (750°F).

ALUMINUM

Production. Although *aluminum* has been a commercial metal for less than 100 years, it now ranks second to steel in both worldwide quantity and expenditure. Besides replacing other metals in numerous uses, it has made possible many new advances that could not have been developed without it.

Aluminum is the most abundant metallic element in the earth's crust. Supply, therefore, is limited only by the economics of mining, extracting, and processing. Under current conditions, bauxite is the major source, bauxite being $Al_2O_3 \cdot nH_2O$ with impurity oxides of iron, silicon, and titanium. Most ore contains about 50 per cent aluminum oxide (alumina) and is sufficiently inexpensive that the major portion of the ore cost lies in transportation. Kaolin-type clays become an alternative source of aluminum if bauxite is limited.

The first stage of the processing is the separation of the alumina from the impurity oxides in the ore. If not performed, the reduction process that produces aluminum from its oxide would also reduce the other metallic oxides. Most industries use modifications of the Bayer process (developed in 1888) wherein the bauxite is digested in a caustic soda leach at elevated temperature and pressure. The alumina dissolves out as a solution of sodium aluminate, is separated, selectively precipitated as hydrated aluminum oxide, and finally converted to pure Al_2O_3 by calcination.

Further processing involves the reduction of the oxide to molten metal. Because alumina has a very high melting temperature (2045°C or 3720°F), it cannot be reduced by the usual furnace techniques used for iron. However, in 1886, Charles M. Hall and Paul Heroult discovered that if Al_2O_3 were dissolved in molten cryolite, the aluminum could be deposited at the cathode of an electrolytic cell.

Industry now carries out the electrolysis in cells made of steel shells lined with carbon, which acts as the cathode. The cells are filled with molten cryolite into which about 16 per cent of alumina is dissolved. Carbon anodes dip into the electrolyte and introduce the current. The separated aluminum is deposited on the bottom of the cell, being heavier than molten cryolite, and is drawn off periodically as it collects. Powdered alumina is added to the bath to replace the aluminum drawn off. Electrolytic refinement requires 15 to 18 kilowatt-hours of electricity per kilogram of aluminum. The significant energy saved by remelting scrap is a substantial motivation for recycling aluminum.

Properties and uses. The properties of aluminum that make it of engineering significance are its *workability, light weight, corrosion resistance,* and good electrical and thermal *conductivity.* Aluminum has a specific gravity of 2.7 as compared to 7.85 for steel. In addition, when exposed to air, aluminum forms an adherent surface oxide that possesses good corrosion resistance. Therefore, aluminum is an excellent material for those applications where relatively good corrosion resistance must be combined with light weight. Additional applications utilize the high electrical and thermal conductivities. Areas in which aluminum finds extensive use include transportation, building and construction, electrical products, containers, consumer durables, and machinery.

Probably the most serious weakness of aluminum from an engineering viewpoint is its relatively *low modulus of elasticity,* roughly one-third that of steel. (Under identical loading, aluminum will deflect three times as much as steel.) This factor, coupled with the higher cost of aluminum, makes it necessary to use sections that place the metal properly so as to obtain adequate stiffness while using minimal material. Fortunately, this often can be done with relative ease because of aluminum's good working qualities.

In its pure state aluminum is soft, ductile, and not very strong. As a result, it is not often used in this condition. Its electrical conductivity is about 60 per cent that of copper. For this reason it often is used for electrical transmission lines, but it is usually reinforced by a steel core so as to obtain adequate strength. In most other applications it is alloyed with copper, manganese, magnesium, or silicon to produce increased strength and hardness. Aluminum alloys are particularly well suited for use in various casting processes. Thus, aluminum serves as the base metal of a series of extremely useful alloys that are constantly increasing in importance. These alloys will be discussed in detail in a later chapter.

TITANIUM

Titanium is a strong, lightweight, corrosion-resistant metal that has been of commercial importance only since 1948. Because its properties are between those of steel and aluminum, its importance is increasing rapidly. Yield strength is about 415 MPa (60,000 psi) and can be raised to 1300 MPa (190,000 psi) by alloying—a strength comparable to many alloy steels. Density, on the other hand, is only 56 per cent that of steel, and the modulus of elasticity is about one-half that of the steel material. Properties are retained well up to temperatures of 480°C (900°F).

Aside from high cost, titanium and its alloys are also difficult to fabricate and possess a high reactivity at elevated temperatures (above 480°C).

Production. Titanium is difficult to produce. One method involves the reduction of titanium tetrachloride with magnesium in an inert atmosphere. The resulting magnesium chloride and free magnesium are then leached out with hydrochloric acid, leaving sponge or powdered titanium. The sponge or powder is then melted in a vacuum unit and chilled to form solid metal. Another method of production involves the electrolysis of titanium tetrachloride in a fused salt bath. Processing problems are numerous, but titanium is now being produced routinely and is readily available on the metals market.

Properties and uses. Commercially pure titanium has typical properties as follows:

	Annealed	Hard-Rolled
Yield strength	275 MPa	655 MPa
Ultimate strength	415 MPa	760 MPa
Elongation in 2 inches	25%	10%

Numerous alloys have been developed to improve these properties and are grouped into three classes on the basis of stable structure at room temperature. Some of these alloys have yield strengths of 1380 MPa (200,000 psi) and ultimate strengths of 1520 MPa (220,000 psi) at room temperature in the heat-treated condition.

Uses of titanium relate primarily to the *high strength-to-weight ratio* and *corrosion resistance* as well as *retention of these properties at elevated temperatures.* Applications are primarily in the area of aerospace but also include chemical and electrochemical processing equipment, marine implements, and ordnance equipment. The metal can be cast, forged, rolled, or extruded to desired shape with special process modifications or control. Some applications relating to bonding utilize the unique property that titanium wets glass and some ceramics. Titanium carbide is used where a very hard material is required, such as in cutting-tool tips.

MAGNESIUM

Production. *Magnesium* is the lightest of the commercially important metals, having a specific gravity of about 1.75. Two major processes are used to produce magnesium: electrolysis of molten anhydrous magnesium chloride and the thermic reduction of dolomite with ferrosilicon. The magnesium chloride for the first process may be obtained by processing various mineral deposits or by processing seawater, which contains about 1.07 kilograms of magnesium per cubic meter (1 pound per 15 cubic feet).

Properties and uses. Like aluminum, magnesium is relatively weak in the pure state and for engineering purposes is almost always used as an alloy. Its modulus of elasticity is even less than that of aluminum, being only about one-fifth that of steel. Therefore, it is usually necessary to require considerable thickness or deep sections to obtain adequate stiffness. In various types of castings, this restriction is easily accommodated. Exceedingly poor workability and ductility also limit the processing of most magnesium metal to the casting techniques.

Wear, creep, and fatigue properties are rather poor. Corrosion resistance is such that paint or some other type of surface protection is often required. Rapid oxidation or burning of the metal can also occur during elevated-temperature processing or operations that may generate high temperature, such as machining. A protective, oxygen-free, atmosphere may be required during these operations.

These restrictions, coupled with high cost, limit magnesium to applications where light weight is very important. Aside from the possibility of burning chips, the machining characteristics are sufficiently good that in many applications the savings in machining costs more than compensate for the increased material expense.

ZINC

Production. *Zinc* ore is most commonly a sulfide (zincblende), with zinc contents ranging from 2 to 15 per cent. For further processing, the ore must first be concentrated by a process of crushing and grinding followed by either "gravity" separation or flotation. The zinc sulfide concentrate, containing from 48 to 60 per cent zinc, is then put through a roasting process in which the sulfur is burned to SO_2 and the zinc converted to ZnO. The SO_2 is generally processed into sulfuric acid.

Metallic zinc is then produced from the oxide by either a carbon reduction process using a furnace or by an electrolytic process. In the furnace process, a mixture of zinc oxide and coal is briquetted or sintered and fired at a temperature in excess of 1000°C. Zinc is reduced and vaporized (since the temperature is in excess of the boiling point of zinc) and is then liquefied in a condenser. All grades of zinc, except special high purity (99.99 per cent Zn) can be produced by the furnace process. If further purity is desired, the impure zinc can be subjected to fractional distillation to produce 99.99+ per cent zinc.

In the electrolytic process, the roasted ore is dissolved in sulfuric acid and the zinc-bearing solution is filtered and purified. Electrolysis plates the zinc onto prepared aluminum cathodes, from which it is periodically stripped, melted, and cast into slabs. Purity exceeds the 99.99 per cent level.

Properties and uses. As a pure metal, zinc has only one important use: *galvanizing* iron and steel. This process, in which steel is acid-cleaned and then coated with a layer of zinc either by dipping in a bath of molten metal or by electrolytic plating, provides excellent corrosion resistance even when the coating is scratched or marred. Galvanizing accounts for about 35 per cent of all zinc used. *Sherardizing* is a similar process in which zinc is applied to the surface of steel components in the form of a diffusion coating. The goal again is corrosion resistance, but the form of the coating and process is different.

The second major use of zinc is in alloys used in the *die-casting* process. About 40 per cent of the annual consumption of zinc is for this process, which requires a fluid metal and low melting point. In its pure state, zinc softens at relatively low temperatures but has rather low strength and is brittle at low temperatures. The die-casting metals are alloys designed to retain the low melting point but have considerably improved strength properties. They will be discussed at greater length in Chapter 8. Most zinc die castings are subsequently plated or painted.

The third major use for zinc is as the principal alloying element in *brass* (copper–zinc alloys). Brass making utilizes about 15 per cent of the total zinc produced.

CHROMIUM, MOLYBDENUM, NICKEL, COBALT, TIN, AND LEAD

Chromium, molybdenum, nickel, cobalt, and tin are of importance as engineering metals but seldom are used in their pure forms. *Chromium, molybdenum,* and *nickel* have been used for years as alloying elements in steels. Nickel is used as the base metal of a number of corrosion-resistant nonferrous alloys. Nickel or *cobalt* are the basis for the *superalloys,* which can maintain useful strengths and properties at temperatures slightly in excess of 1100°C (2000°F). Cobalt is also used as a binder metal in various powder-based components and sintered carbides.

Tin is used primarily as a coating on steel to provide corrosion resistance, in combination with copper to produce bronze, or in certain alloys that are used as bearing materials.

Lead is often used in the pure state as a corrosion-resisting material. In most uses, however, it is used as an addition to provide corrosion resistance or machining characteristics to alloy metals. Principal uses, such as storage batteries, paint pigment, and cable covering, consume about 60 per cent of the annual output.

GRAPHITE

Properties and uses. *Graphite* is an old material that in improved forms has considerable importance as an engineering material. It possesses the unique property of having increased strength at higher temperatures. Recrystallized, polycrystalline graphites have mechanical strengths, as measured by modulus of rupture, up to 70 MPa (10,000 psi) at room temperature, which double at 2500°C (4500°F).

Large quantities of graphite are used as electrodes in arc furnaces, but other uses are developing rapidly. The addition of small amounts of borides, carbides, nitrides, and silicides greatly lowers the oxidation rate of graphite at high temperatures and also improves its mechanical strength. This makes it highly suitable for use in rocket-nozzle inserts and permanent molds for casting various products. It can be machined quite readily to excellent surface finishes.

"UNUSUAL METALS"

Several uncommon metals have achieved importance in modern technology as a result of their somewhat unique properties. *Hafnium, thorium,* and *beryllium* are used in nuclear reactors because of their low neutron-absorption characteristics. High-temperature applications often require metals with high melting points, such as *niobium* (2470°C), *molybdenum* (2610°C), *tantalum* (3000°C), *rhenium* (3170°C), and *tungsten* (3410°C). Depleted *uranium,* because of its very high density (19.1 grams per cubic centimeter), is useful in special applications where maximum weight must be put into a limited space, as in counterweights. *Zirconium* is used for its outstanding corrosion resistance to most acids, chlorides, and organic acids. When alloyed with a small percentage of hafnium, it has a yield strength of about 579 MPa (84,000 psi) and a tensile strength of about 620 MPa (90,000 psi). With extensive interest in space, nuclear, high-temperature, and electronic applications, a considerable amount of research and development work is being done on these uncommon metals.

Review questions

1. What are the several types of iron ore?
2. What are the four major raw materials that go into the blast furnace to produce pig iron?
3. What are the chemical processes involved in the production of pig iron?
4. What is the approximate percentage of carbon in pig iron? What is its primary source?
5. What would be the major advantage of the direct reduction of iron ore? What are the goals of the various processes?
6. How does steel differ from pig iron?
7. Describe the chronological evolution of the primary steelmaking processes.
8. What are the advantages of the basic-oxygen processes over the open-hearth process for making steel?
9. What are the major limitations of the basic-oxygen process?

10. What elements are usually present in steel beside iron and carbon?
11. Why are electric furnaces often used for making and melting high-alloy steels?
12. What is the advantage of bottom-pouring ladles? Of bottom pouring of ingots?
13. What is solidification shrinkage? Why is it important?
14. What are some of the benefits of continuous-casting techniques?
15. Why do dissolved gases cause problems during solidification?
16. Why is steel "killed"? How is this done?
17. What is a semikilled steel?
18. What is a rimmed steel?
19. What is the difference between the deoxidation and degassing of steel?
20. What is the difference between vacuum arc remelting and electroslag remelting?
21 What is the difference between ETP and OFHC copper?
22. What are the properties of copper that give it importance as an engineering material?
23. Why is aluminum an attractive metal for recycling?
24. What properties of aluminum make it attractive as an engineering material?
25. Why are aluminum alloys relatively corrosion resistant?
26. Why is the modulus of elasticity of aluminum often an engineering concern?
27. What are the relative electrical conductivities of copper and aluminum?
28. How do the properties of titanium compare to those of steel and aluminum?
29. Why is titanium often used for elevated-temperature applications?
30. What are the relative weights of steel, aluminum, and magnesium? Is there a correlation between weights and elastic moduli for these metals?
31. Why is casting a very common fabricating process for magnesium alloys?
32. What are the major uses of zinc?
33. Which metals form the basis of the superalloys?
34. What are some of the uses of graphite as an engineering material?
35. What are several metals used in ultra-high-temperature applications?

Case study 3. THE REPAIRED BICYCLE

The frame of a high-quality, ten-speed bicycle was made of cold-drawn alloy steel tubing, taking advantage of the cold-working characteristics to provide added strength. As a result of excessive abuse, the frame was broken, and a repair was made by conventional arc welding. The repair seemed adequate but, shortly thereafter, the frame again broke, this time adjacent to the repair weld. The break appeared to be ductile in nature, showing evidence of metal flow prior to fracture. What was the probable cause of the second failure?

CHAPTER 5

Equilibrium diagrams

As our study of engineering materials becomes more focused on specific metals and alloys, it is increasingly important that we know the natural characteristics and properties of the material under various environments. What condition is the material in? Is the composition uniform throughout? If not, how much of each component is present? Is something present that may give undesired properties? What will happen if temperature is increased or decreased; pressure is changed; or chemistry is varied? The answers to these and other important questions can be obtained through use of equilibrium phase diagrams.

Phases. Before moving to a discussion of these diagrams, it seems imperative that a working definition of the term *phase* be developed. As a first-order definition, a phase is simply a form of material possessing a single characteristic structure and associated characteristic properties. Uniformity of chemistry, structure, and properties is assumed throughout the phase. More rigorously, a phase is *any physically distinct, homogeneous, and mechanically separable portion of a substance.* In layman's terms, this requires a unique structure, uniform composition, and well-defined boundaries or interfaces.

A phase can be continuous, like the air in a room; or discontinuous, like grains of salt in a shaker. A phase can be solid, liquid, or gas. In addition, a phase can be a pure substance or a solution, provided that the structure and composition are uniform throughout. Alcohol and water mix in all proportions and will therefore form a single phase when placed in a beaker. Oil and water tend to form isolated regions with distinct boundaries and must be considered as two distinct phases.

Equilibrium phase diagram. The *equilibrium phase diagram* is a *graphical mapping* of the natural tendencies of a material system, assuming infinite time at the conditions specified. Areas of the diagram are assigned to the various phases, with the boundaries indicating the equilibrium conditions of transition.

With the background just developed, let us now consider the types of phase mappings that may be useful. Three *primary variables* are at our disposal: *temperature, pressure, and composition.* The simplest type of diagram is a pressure–temperature (*P–T*) diagram for a fixed composition material.

For simplicity, consider the *P–T* diagram for water presented in Figure 5-1. With composition fixed, the diagram enables the determination of the stable form of water at any condition of temperature and pressure. Holding pressure constant and varying temperature, the transition boundaries locate the melting and boiling points. Still other uses can be presented. Locate a temperature where the stable phase at atmospheric pressure is the solid (ice). Now maintain that temperature and begin to decrease the pressure. A transi-

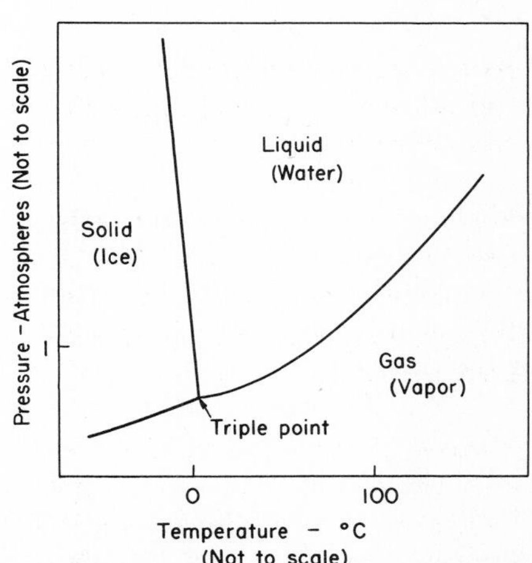

FIGURE 5-1. Schematic pressure–temperature diagram for water.

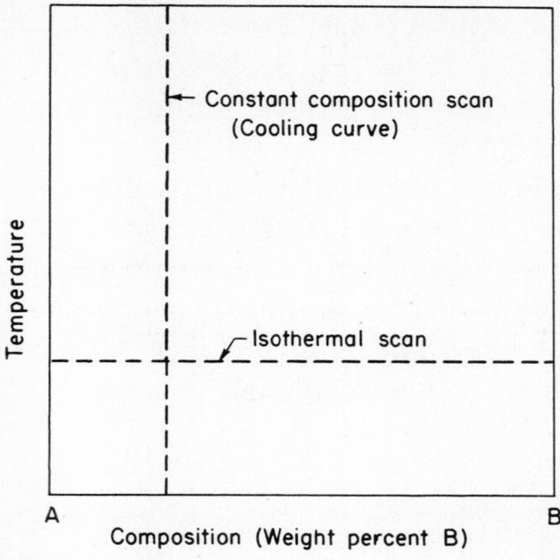

FIGURE 5-2. Temperature–composition equilibrium phase diagram mapping.

tion is encountered wherein solid goes directly to gas with no liquid intermediate. The process is that of freeze drying and is employed in the manufacture of numerous dehydrated products.

Temperature–composition diagrams. The water diagram serves as an example of the use of phase diagrams. In engineering applications, however, the *P–T* phase diagram is rarely used. Most processes are conducted at atmospheric pressure, with control variations coming primarily in temperature and composition. The most useful mapping, therefore, would be a *temperature–composition phase diagram* at atmospheric pressure. For the remainder of the chapter, it is this second form of diagram that will be considered.

For mapping purposes, therefore, temperature is placed on the vertical axis and composition on the horizontal. Figure 5-2 shows the form of such a mapping for the A–B system where the left-hand vertical corresponds to pure material A and the percentage of B increases as we move toward pure B at the right of the diagram. Experimental investigations to fill in the details of the diagram basically take the form of vertical or horizontal scans designed to locate transition points.

Cooling curves. Considerable information can be obtained from vertical scans through the diagram in which a fixed composition material is heated and subsequently slow-cooled by removing heat at a uniformly slow rate. Transitions in structure appear as characteristic points in a temperature as a function of the time plot of the cooling cycle, known as a *cooling curve.*

For the system composed of sodium chloride (common table salt) and water, five different cooling curves are presented in Figure 5-3. Curve (a) is for pure water being cooled from the liquid state. A smooth continuous line is observed for the liquid, the extraction of heat producing a concurrent drop in temperature. When the freezing point is reached (point *a*), the material

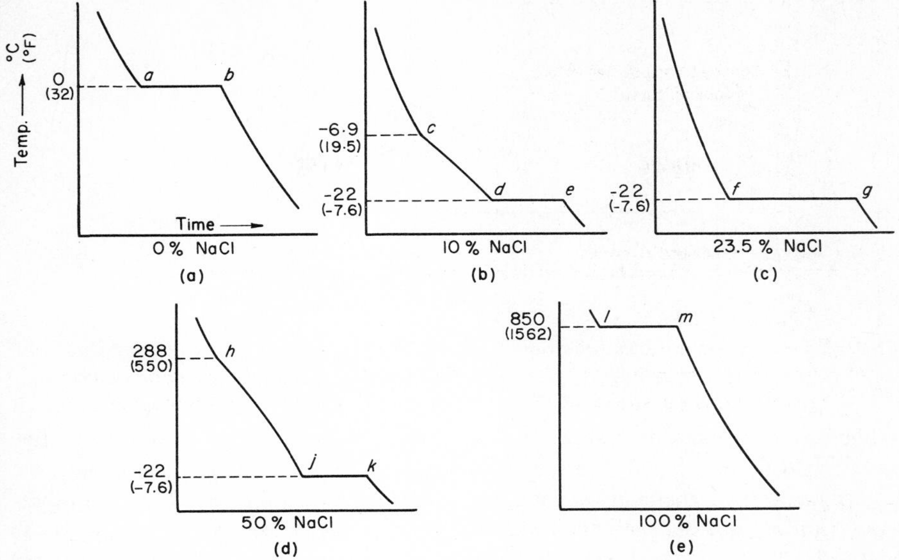

FIGURE 5-3. Cooling curves for various compositions of NaCl–H_2O solutions.

changes state and releases heat energy equivalent to the liquid-to-solid transition. Heat is still being extracted from the system, but its source is the change in state and not a decrease in temperature. Thus, an isothermal hold (*a–b*) is observed until solidification is completed. From this point, the newly formed solid experiences a smooth drop in temperature as heat extraction continues. Such a curve is characteristic of pure metals or substances with a distinct melting point.

Curve (b) of Figure 5-3 is the cooling curve for a 10 per cent solution of salt in water. The liquid region undergoes a continuous cooling down to point *c*, where the slope abruptly decreases. At this temperature small particles of ice begin to form and the slope change is due to the energy released in this transition. The formation of these ice particles leaves the remaining solution richer in salt and imparts a lower freezing temperature to it. Further cooling must take place for additional solid to form. The formation of more solid leaves the remaining liquid richer in salt and the freezing point is progressively lowered. Instead of possessing a distinct melting point or freezing point, the material is said to have a *freezing range.* When the temperature of point *d* is reached, the remaining liquid solidifies into an intimate mechanical mixture of solid salt and solid water (to be discussed later) and an isothermal hold is observed. Further heat extraction from the solid material produces a continuous drop in temperature.

For a solution of 23.5 per cent salt in water, a distinct freezing point is observed, as shown by curve (c) of Figure 5-3. Further compositions with

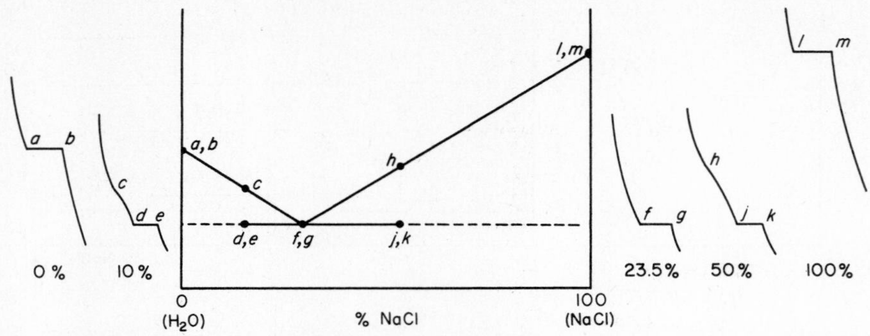

FIGURE 5-4. Derivation of equilibrium diagram for NaCl and H_2O from cooling curves.

richer salt concentrations show phenomena similar to those presented earlier but with solid salt being the first substance to form from the liquid melt.

The key transition points can now be transferred to a temperature–composition diagram with the designating letters corresponding to those on the cooling curves. Figure 5-4 presents such a map with several key lines being drawn. Line *a–f–l* denotes the lowest temperature at which the material is totally liquid and is known as the *liquidus line.* Line *d–f–j* is the highest temperature at which a material is completely solid and is called the *solidus line.* Between the liquidus and solidus, two phases coexist, one being liquid and the other being solid. Cooling-curve studies have thus enabled the determination of various bits of key information regarding the system being studied.

Solubility studies. The observant reader will note that the ends of the diagram still remain undetermined. Both pure materials have only one transition point below which they appear as a single-phase solid. Can ice retain some salt in solid solution? If so, how much? Can salt hold solid water and still remain a single phase? Completion of the diagram, therefore, requires several horizontal scans to determine any solubility limits or the condition of saturation at various temperatures.

These isothermal scans with variable composition will usually require analysis of the specimens by X-ray technique, microscopy, or other investigative approaches to determine the composition at which transitions occur. As the scan moves away from the pure metal, the first line encountered (provided that the temperature is in the all-solid region) denotes the solubility limit and is known as the *solvus line.* Figure 5-5 presents the equilibrium phase diagram for the lead–tin system, using the conventional notation wherein Greek letters are used to label the various single-phase solids. The upper portion closely resembles the salt–water diagram, but solubility of one metal in the other can be seen at both ends of the diagram.

Having now been exposed to the concepts of equilibrium diagrams, let

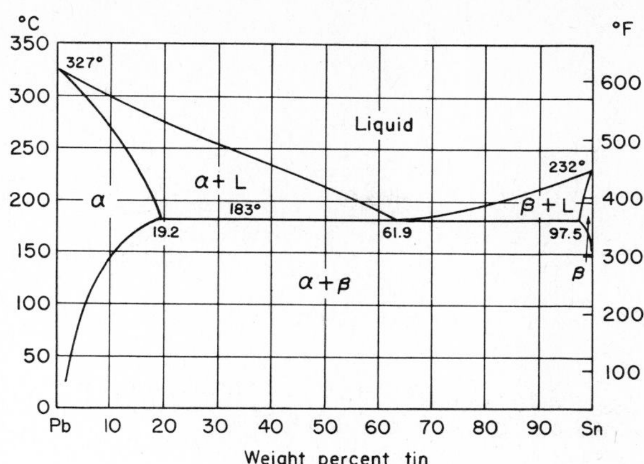

FIGURE 5-5. Lead–tin equilibrium diagram.

us move on to consider several specific examples. Presentation will move from the simple to the more complex.

Complete solubility in both liquid and solid states. If two metals are each completely soluble in the other in both the liquid and solid states, a rather simple diagram results, as illustrated in Figure 5-6 for the copper–nickel system. At temperatures above the liquidus line, the two materials are in liquid solution no matter what the composition. Similarly, below the solidus, the materials form a solid solution at all compositions. Between the liquidus and solidus is a two-phase region where liquid and solid solutions coexist.

Partial solid solubility. As might be expected, many materials exhibit neither complete solubility nor complete insolubility in the solid state. Each is soluble in the other up to a certain limit or saturation point, the value of this limit being a function of temperature. Such a diagram has already been observed in the lead–tin system of Figure 5-5.

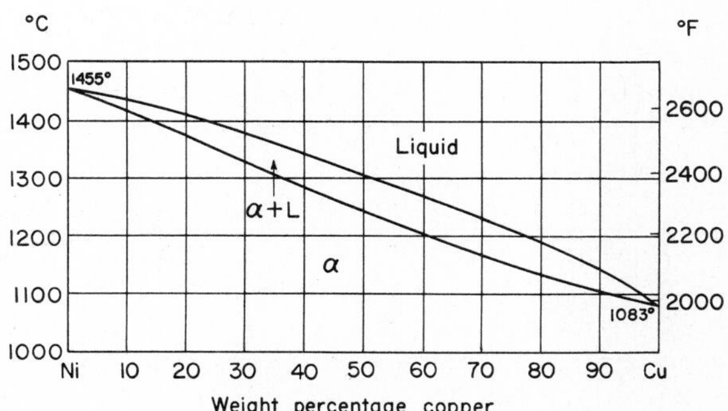

FIGURE 5-6. Copper–nickel equilibrium diagram, showing complete solubility.

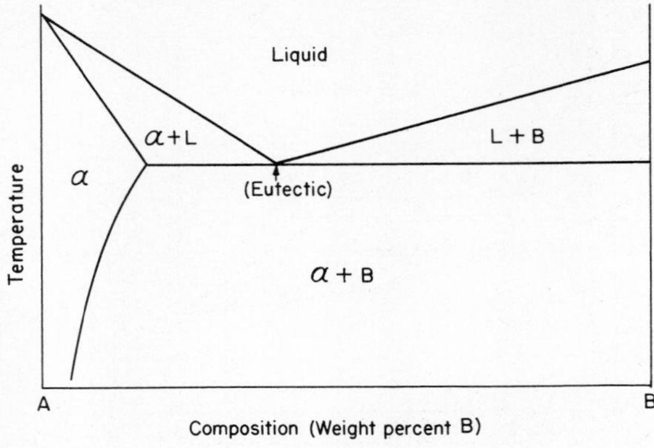

FIGURE 5-7. Equilibrium diagram of two metals, one of which is partially soluble in the other in the solid state.

Consideration of Figure 5-5 shows that the maximum solubility of tin in lead in the solid state is 19.2 per cent by weight. Similarly, tin will dissolve up to 2.5 weight per cent lead in solid solution. If the temperature is decreased from this point of maximum solubility, the amount of substance capable of being held in solution generally decreases. Thus, if a saturated solution of tin in lead is cooled from 183°C, the material moves from a single-phase region into a two-phase region. Some tin-rich second phase must precipitate from solution. This fact is used to control the properties of many engineering alloys.

Insolubility. If one or both of the components are insoluble in the other, the diagrams are modified to reflect this phenomenon. Figure 5-7 illustrates the case where component A is completely insoluble in component B. Figure 5-8 presents the extreme case where the metals are completely insoluble in each other in both the liquid and solid states.

Utilization of diagrams. Before considering some of the more complex components of phase diagrams, let us return to a complete solubility diagram (Figure 5-9) and pursue the utilization of these diagrams to obtain useful informa-

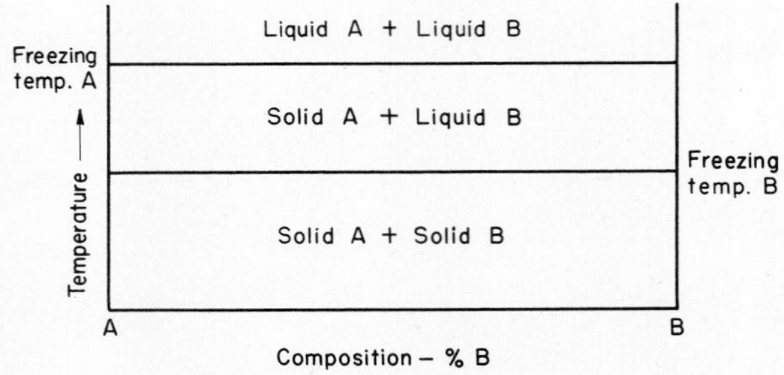

FIGURE 5-8. Equilibrium diagram of two metals that are completely insoluble in each other in both the liquid and solid states.

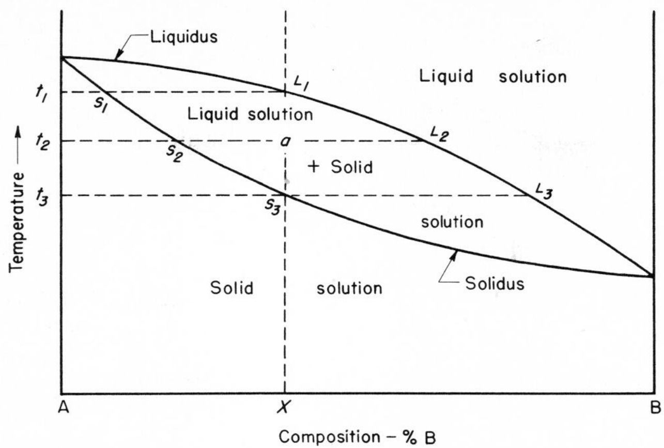

FIGURE 5-9. Equilibrium diagram of two metals that are completely soluble in each other in both the liquid and solid states.

tion. Three basic pieces of information can be obtained at each point on the diagram.

1. *The phases present.* By simply locating the point of consideration on the temperature–composition mapping and identifying the region of the diagram in which it appears, the stable phases can be determined.

2. *The composition of each phase.* If the point lies in a single-phase region, the composition of the phase is, by necessity, the overall composition of the material being considered. If the point lies in a two-phase region, a *tie-line* is drawn. A tie-line is simply an isothermal line drawn through the point of consideration, terminating at the boundaries of the single-phase regions on either side of the point in question. The composition at which the tie-line intersects the neighboring single-phase regions determines the compositions of these phases in the two-phase mixture. For example, consider point a in Figure 5-9. The tie-line for this point runs from S_2 to L_2. The point S_2 is the intersection of the tie-line with the solid phase, and thus the solid in the two-phase mixture at a will have the composition of point S_2. Similarly, the liquid will have the composition of point L_2.

3. *The amount of each phase present.* If the point lies in a single-phase region, the amount must be 100 per cent of that phase. If the point lies in a two-phase region, the relative amounts can be determined by a *lever-law* calculation, which considers a tie-line as a lever with one of the phases at each end and a fulcrum at the vertical composition line. Thus, in Figure 5-9,

$$\text{solid} \times (a - S_2) = \text{liquid} \times (L_2 - a)$$

From this the amount of liquid at t_2 is

$$\frac{a - S_2}{L_2 - S_2} \times 100\%$$

(This is an interesting concept but of virtually no concern to the engineer who selects or processes metals.)

Other applications relate to an overall view of the diagram or the location of the transition points for a specific alloy. For instance, the temperature necessary to put an alloy into a given phase field can easily be determined. Changes that may occur upon the slow heating or slow cooling of a given material can be predicted. In fact, most of the questions posed at the beginning of this chapter can now be answered.

Solidification of Alloy X. Using the tools discussed above, it now becomes relatively simple to follow the solidification of alloy X in Figure 5-9. At temperature t_1, the first minute amount of solid forms with the chemistry of point S_1. As temperature drops, more solid forms, but the chemistry of both solid and liquid phases is shifting in accordance with the tie-line end points. Finally, at t_3, solidification is complete, and the composition of the single phase solid is that of alloy X, as required.

The composition of the final solid solution is different from that of the first solid that was formed. If cooling is sufficiently slow (equilibrium conditions are approached) the composition of the entire mass of solid tends to become uniform at the value predicted by the tie-line. Compositional differences are removed by the phenomenon of diffusion, wherein atoms migrate from point to point in the crystal lattice under the energy impetus of elevated temperature. If the cooling rate is rapid, a nonuniform solid mass may result, with the initial solid that formed retaining a composition different from the latter portions of the solid. This structure is referred to as being *cored.*

Three-phase reactions. Several of the diagrams previously presented contain a distinct feature in which phase regions are separated by a horizontal line. These lines are further characterized by either a V intersecting from above or an inverted V intersecting from below, denoting the location of a *three-phase equilibrium reaction.*

One common type of three-phase reaction, known as a *eutectic,* has already been observed in Figures 5-4, 5-5, and 5-7. Understanding of such reactions is possible by the use of the tie-line and lever-law concepts and will be presented using the lead–tin diagram of Figure 5-5. Consider any alloy containing between 19.2 and 97.5 weight per cent tin at a temperature just above the 183°C horizontal line. Tie-line and lever-law computations show that the material contains either a lead-rich or tin-rich solid and remaining liquid, the liquid having a composition of 61.9 weight per cent tin. [Note that any liquid will always have a composition of 61.9 weight per cent at 183°C (361°F), regardless of the overall composition of the alloy.] If we now focus on the liquid and allow it to cool to just below 183°C (361°F), a transition occurs wherein liquid of composition 61.9 per cent tin goes to a mixture of lead-rich solid with 19.2 per cent tin and a tin-rich solid containing 97.5 per cent tin. The relative amounts of the two components maintain chemical uniformity. The form for this eutectic transition is similar to a chemical reaction:

$$\text{liquid} \rightarrow \text{solid}_1 + \text{solid}_2$$

Since the two solids have chemistries on either side of the intermediate liquid, a separation must have occurred within the system. Such a separation in a solidifying melt results from two metals that are soluble in the liquid state but only partially soluble in the solid state. Separation requires atom movement, so the distances between the two solids cannot be great. The resulting eutectic structure is an intimate mechanical mixture of two single-phase solids and assumes a characteristic set of physical and mechanical properties. Alloys of the eutectic composition have the lowest melting points of all alloys in a given system and are often used as casting alloys or as filler metal in soldering or brazing applications.

Figure 5-10 summarizes some other types of three-phase reactions that may occur in engineering systems. These include the *peritectic, monotectic,* and

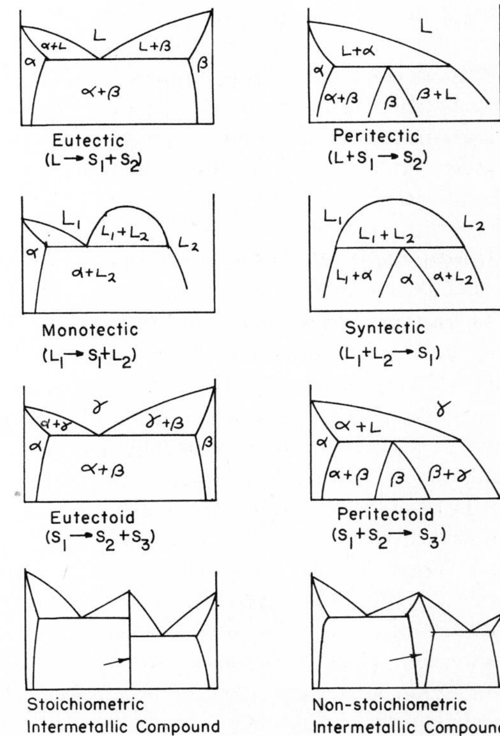

FIGURE 5-10. Summary schematic of three-phase reactions and intermetallic compounds.

syntectic, the suffix *-ic* denoting that at least one of the three phases in the reaction is a liquid. If the suffix *-oid* is used, it simply denotes that all phases involved are solids; the form of the reaction remains the same. Two all-solid reactions can occur: the *eutectoid* and *peritectoid.* These reactions tend to be a bit more sluggish since all changes must occur in the solid state.

Intermetallic compounds. A final feature occurs in systems wherein the bonding attractions of the component materials is sufficiently strong that compounds tend to form. These compounds are single-phase solids and tend to break the diagram into recognizable subareas. If components A and B form a compound A_xB_y and the compound cannot tolerate any deviation from that fixed ratio, the intermetallic is known as a *stoichiometric intermetallic* and appears as a single vertical line in the diagram. If some degree of deviation is tolerable, the vertical line expands into a region and the compound is a *nonstoichiometric intermetallic.* Figure 5-10 shows schematic representations for both types of intermetallic compounds.

In general, intermetallics tend to be hard, brittle materials, these properties being related to their ionic or covalent bonding. If they are present in large quantities or lie along grain boundaries, the overall alloy can be extremely brittle. If the same intermetallic can be uniformly distributed throughout the structure in small particles, the effect can be considerable strengthening of the alloy.

Complex diagrams. Most equilibrium diagrams of actual alloy systems will be one of the basic types just discussed or combinations thereof. Often, these diagrams appear to be quite complex and formidable to the manufacturing engineer. By focusing on the particular alloy in question and analyzing specific points using the tie-line and lever-law concept, even the most complex diagram can be simply dissected. Knowledge of the properties of the various components then enables predictions about the overall product.

The iron–carbon equilibrium diagram. Because steel, composed essentially of iron and carbon, is such an important engineering material, the iron–carbon equilibrium diagram of Figure 5-11 is by far the most important of those with which the average engineer must deal. Actually, the diagram most frequently used is an iron–iron carbide diagram, for a stoichiometric intermetallic carbide of the form Fe_3C can be used to terminate the useful range of the diagram at 6.67 weight per cent carbon. Names and notations have evolved historically and will be used in their generally accepted form.

Three distinct three-phase reactions can be identified in the diagram. At 1495°C (2723°F), a *peritectic* occurs for alloys with low weight per cent carbon. Because of its high temperature and the extensive single-phase gamma region immediately below it, the peritectic reaction rarely assumes any engineering significance. A *eutectic* is observed at 1148°C (2098°F), with the eutectic point at 4.3 weight per cent carbon. Alloys containing greater than

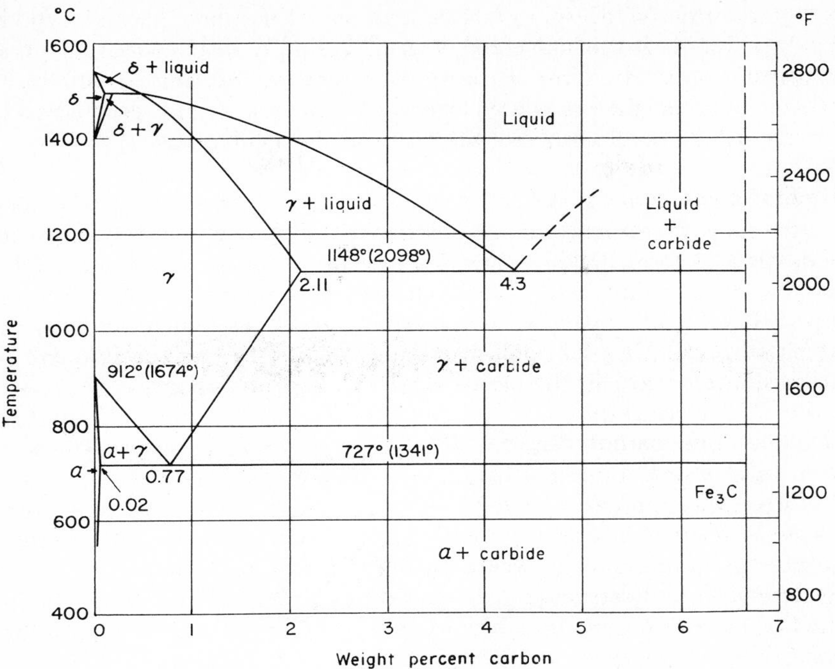

FIGURE 5-11. Iron–carbon equilibrium diagram.

2.11 per cent carbon will experience this eutectic and are classified by the general term *cast irons.* The final three-phase reaction is a *eutectoid* at 727°C (1341°F), with the eutectoid point at 0.77 weight per cent carbon. Alloys with less than 2.11 per cent carbon can undergo a transition from a single-phase solid solution (γ) through the eutectoid to a two-phase mixture and are known as steels. Thus the point of maximum carbon solubility in iron, 2.11 weight per cent, forms an arbitrary division between the steels and cast irons.

To further our understanding of the diagram, let us now consider the four single-phase components, three of which occur in pure iron, the fourth being the carbide at 6.67 per cent carbon. Upon first solidification, pure iron forms a body-centered-cubic solid that is stable down to 1394°C (2541°F). Known as *delta-ferrite,* this phase is only stable at extremely elevated temperature and has no significant engineering importance. From 1394°C (2541°F) to 912°C (1674°F), pure iron assumes a face-centered-cubic structure known as *austenite* (γ) (in honor of the famed metallurgist Sir Robert Austen of England). Key features of austenite are the high formability characteristic of the f.c.c. structure and its high solubility of carbon. Hot forming of steel benefits from these features of formability and compositional uniformity. Moreover, most heat treatment of steel begins in the single-phase austenite region. *Alpha-ferrite,* or more commonly just *ferrite,* is the stable form of iron at temperatures below 912°C (1674°F). This body-centered-cubic structure can hold only 0.02 weight per cent carbon in solid solution and forces a two-phase mixture in most steels. The only other change upon the further cooling of iron is the nonmagnetic-to-magnetic transition at the Curie point (770°C) (1418°F). Since this is not associated with any change in phase, it does not appear on the equilibrium phase diagram.

The fourth single-phase region is the brittle intermetallic, Fe_3C, which also goes by the name *cementite.* As with most intermetallics, it is quite hard and brittle and care should be exercised in controlling the structures in which it occurs. Alloys with excessive amounts of cementite or cementite in undesirable form tend to have brittle characteristics. Since cementite dissociates prior to melting, its exact melting point is unknown and the liquidus line remains undetermined in the high-carbon region of the diagram.

A simplified iron–carbon diagram. If we focus only on the materials normally known as steels, a simplified diagram is often used. Those portions of the iron–carbon diagram near the delta region and those above 2 per cent carbon content are of little importance to the engineer and are deleted. A simplified diagram, such as the one in Figure 5-12, focuses on the eutectoid region and is quite useful in understanding the properties and processing of steel.

The key transition described in this diagram is the decomposition of single-phase austenite (γ) to the two-phase ferrite and carbide structure as temperature drops. Control of this reaction, which arises due to the drastically

FIGURE 5-12. Simplified iron–carbon diagram for the processing of steels.

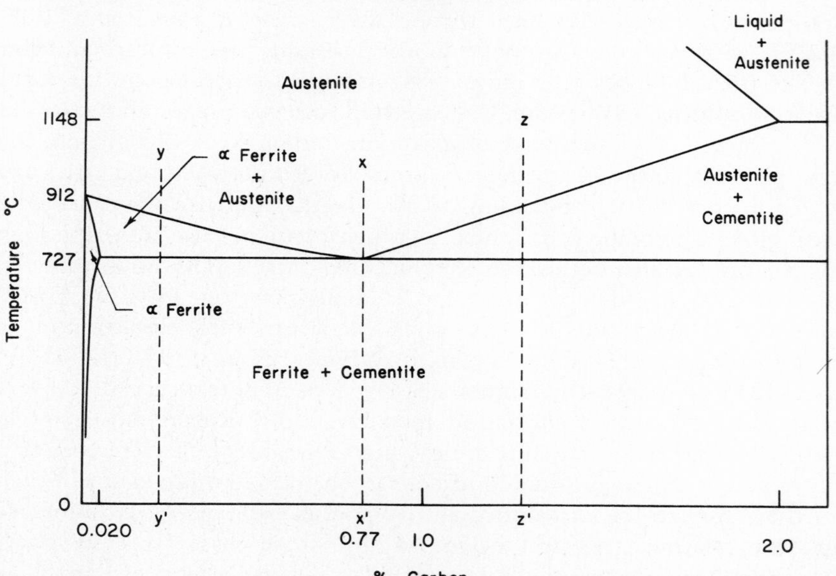

different carbon solubilities of austenite and ferrite, enables a wide range of properties to be achieved through heat treatment.

To begin to understand these processes, consider a steel of the eutectoid composition, 0.77 per cent carbon, being slow cooled along line *x–x′* in Figure 5-12. At the upper temperatures, only austenite is present, the 0.77 percent carbon being dissolved in solid solution with the iron. When the steel cools to 727°C (1341°F), several changes occur simultaneously. The iron wishes to change from the f.c.c. austenite structure to the b.c.c. ferrite structure, but the ferrite can only contain 0.02 per cent carbon in solid solution. The rejected carbon forms the carbon-rich cementite intermetallic with composition Fe_3C. In essence, the net reaction at the eutectoid is:

$$\text{austenite}_{0.77\%\ C} \rightarrow \text{ferrite}_{0.02\%\ C} + \text{cementite}_{6.67\%\ C}$$

Since this chemical separation of the carbon component occurs entirely in the solid state, the resulting structure is a fine mechanical mixture of ferrite and cementite. Specimens prepared by polishing and etching in a weak solution of nitric acid and alcohol reveal the lamellar structure of alternating plates that forms on slow cooling. This structure, seen in Figure 5-13, is composed of two distinct phases, but has its own set of characteristic properties and has the name *pearlite,* because of its resemblance to mother-of-pearl.

Steels having less than the eutectoid amount of carbon (less than 0.77 per cent) are known as *hypoeutectoid steels:* Consider now the transformation of such a material, represented by cooling along line *y–y′* in Figure 5-12. At high temperatures, the material is entirely austenite, but upon cooling enters a region where the stable phases are ferrite and austenite. Tie-line and lever-law calculations show that low-carbon ferrite nucleates and grows, leaving the remaining austenite richer in carbon. At 727°C (1341°F), the austenite is of eutectoid composition (0.77 per cent carbon) and further cooling transforms the remaining austenite to pearlite. The resulting structure is a mixture of primary or proeutectoid ferrite (ferrite that formed above the eutectoid reac-

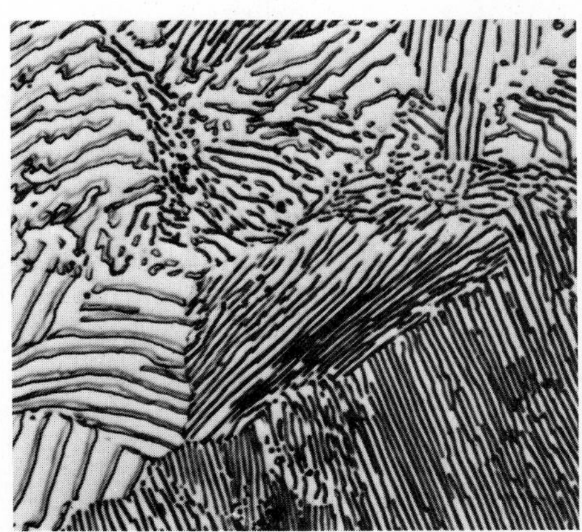

FIGURE 5-13. Pearlite; 1000×. (*Courtesy United States Steel Corporation.*)

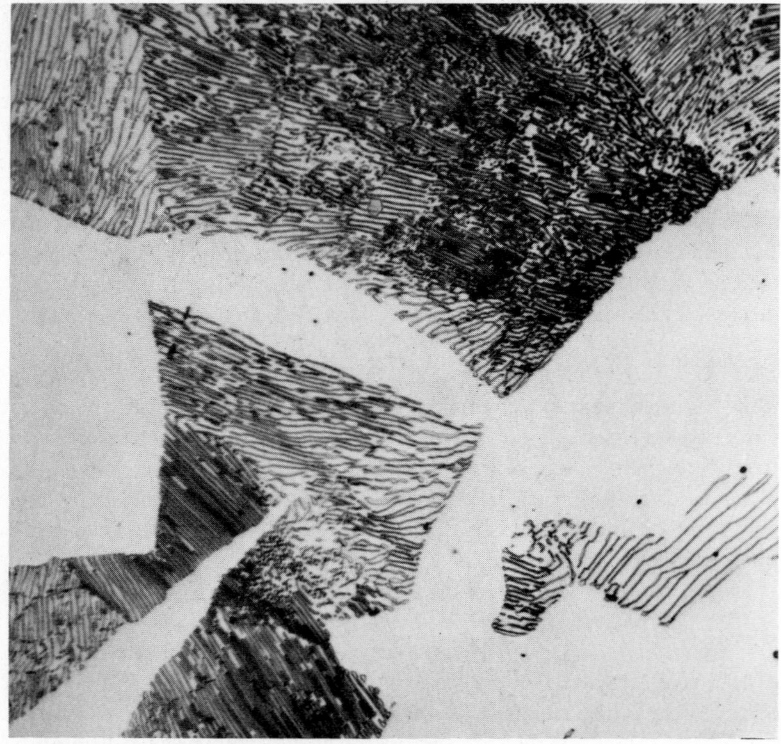

FIGURE 5-14. Photomicrograph of a hypoeutectoid steel; 500×. (*Courtesy United States Steel Corporation.*)

tion) and islands of pearlite. An example of this structure is shown in Figure 5-14.

Hypereutectoid steels are steels that contain greater than the eutectoid amount of carbon. When such a steel cools, as in line z–z' of Figure 5-12, the process is similar to the hypoeutectoid case, except that the primary or proeutectoid phase is now cementite instead of ferrite. As the carbon-rich phase forms, the remaining austenite decreases in carbon content, reaching the eutectoid composition at 727°C (1341°F). As before, any remaining austenite transforms to pearlite upon slow cooling through this temperature. Figure 5-15 is a photomicrograph of the resulting structure.

It should be remembered that the transitions that have been described in all phase diagrams are for equilibrium conditions that are approximated by slow cooling. With slow heating, these transitions occur in the reverse manner. However, when alloys are cooled rapidly, entirely different results may be obtained, because sufficient time is not provided for the normal phase reactions to occur. In such cases, the phase diagram is no longer a useful tool for engineering analysis. Since such processes are often important to the heat treatment of steels and other metals, their characteristics will be discussed in Chapter 6 and new tools will be developed to aid our understanding.

Cast irons. Alloys of iron and carbon that have more than 2.11 per cent carbon are called *cast irons.* In essence, these are alloys in which the carbon level exceeds the amount that can be retained as a solid solution in austenite at the eutectic temperature and therefore experience the eutectic transformation during cooling. Being relatively inexpensive, with good fluidity and rather low liquidus temperatures, these alloys occupy an important place in engineering applications.

Most commercial cast irons contain significant amounts of silicon in addition to iron and carbon, the general composition being 2.0 to 4.0 per cent carbon, 0.5 to 3.0 per cent silicon, less than 1.0 per cent manganese, and less than 0.2 per cent sulfur. Silicon has two major effects on the alloy. First, it partially substitutes for carbon, such that use of the phase diagram requires replacing the weight per cent carbon scale with a carbon equivalent. Although

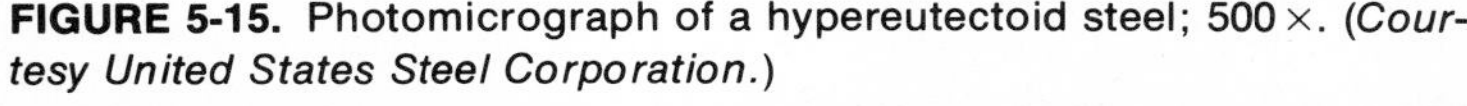

FIGURE 5-15. Photomicrograph of a hypereutectoid steel; 500×. (*Courtesy United States Steel Corporation.*)

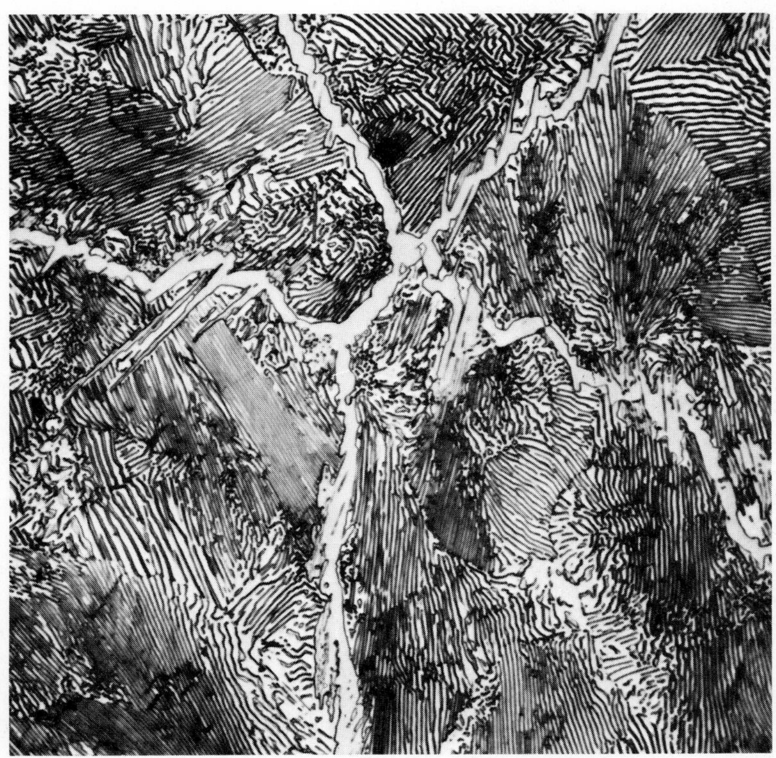

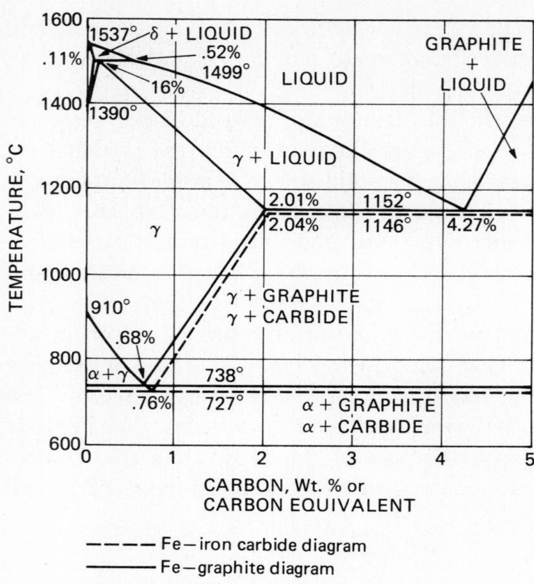

FIGURE 5-16. Iron–carbon diagram showing two possible eutectic reactions. Graphite is solid line; cementite is dashed. (*Courtesy United States Steel Corporation.*)

more complex formulations exist, carbon equivalent may be simply defined as per cent carbon plus one-third of the per cent silicon. Second, silicon tends to promote the formation of graphite as the carbon-rich single phase instead of the Fe_3C intermetallic. The eutectic reaction, therefore, has two distinct possibilities:

$$\text{liquid} \rightarrow \text{austenite} + Fe_3C$$

or

$$\text{liquid} \rightarrow \text{austenite} + \text{graphite}$$

Figure 5-16 shows a modified phase diagram reflecting the two possibilities.

Thus, the final microstructure has two possible extremes: (1) all the carbon-rich phase in the form of Fe_3C, and (2) all the carbon-rich phase in the form of graphite. In usual practice, these extremes can be approached in the various types of cast irons by control of the process variables. Graphite formation is promoted by slow cooling, high carbon and silicon contents, heavy section sizes, inoculation practices, and alloy additions of Ni and Cu. Cementite is favored by fast cooling, low carbon and silicon levels, thin sections, and alloy additions of Mn, Cr, and Mo.

Four basic types of cast irons are produced, the most common being *gray cast iron.* In this type, most of the carbon is in the form of graphite flakes formed at the eutectic reaction, although carbide may form at the lower eutectoid transformation. When fractured, the freshly exposed surface has a gray appearance (see Figure 5-17), and a graphite smudge can usually be obtained if one rubs a finger across a freshly fractured or machined surface.

FIGURE 5-17. (*Left to right*) Fractures of gray, white, and malleable iron. (*Courtesy Malleable Founders' Society.*)

Gray cast iron is the least expensive of the four types and is characterized by those features which promote the formation of graphite. Typical compositions range from 2.5 to 4.0 per cent carbon, 1.0 to 3.0 per cent silicon, and 0.4 to 1.0 per cent manganese. The microstructure consists of three-dimensional flakes of graphite in a matrix of either ferrite or pearlite, seen in cross section in Figure 5-18. Because the graphite flakes have no appreciable strength, they act essentially as voids in the structure. Moreover, the pointed

FIGURE 5-18. Photomicrographs of typical gray cast iron; 1000×. (*Left*) unetched; (*right*) etched. (*Courtesy Bethlehem Steel Corporation.*)

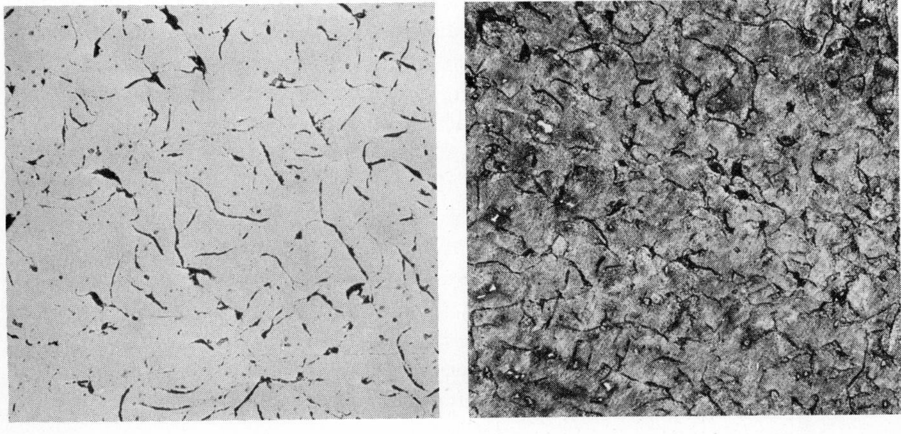

TABLE 5-1. Classification of gray irons according to ASTM Specification A48-64

Class Number	Minimum Tensile Strength	
	MPa	psi
20	138	20,000
25	172	25,000
30	207	30,000
35	241	35,000
40	276	40,000
45	310	45,000
50	345	50,000
55	379	55,000
60	414	60,000

ends of the flakes serve as preexisting notches or crack initiation sites, giving the material a somewhat brittle nature. Size and shape of the graphite flakes have considerable effect on the overall properties of gray cast iron. When maximum strength is desired, small, uniformly distributed flakes are desired with a minimum amount of mutual intersection.

Another means of controlling strength is through variation of the matrix structure. Several distinct classes of gray iron are identified on the basis of tensile strength and are presented in Table 5-1. Class 20 is composed of high-carbon equivalent metal and has a ferrite matrix. Higher strengths, up to class 40, are attainable by lower-carbon equivalents and a pearlite matrix. To go above class 40, alloying is required to provide added solid solution strengthening, or special heat-treatment practices are employed. In all gray irons, however, the presence of graphite flakes results in a near-zero ductility.

Gray cast irons possess excellent compressive strengths (compressive forces do not promote crack propagation), excellent machinability (graphite acts as a chip former and lubricant), good wear resistance (graphite flakes self-lubricate), and outstanding vibration damping characteristics (graphite flakes absorb vibration energy). High silicon contents promote good corrosion resistance and the fluidity desired for casting applications. For these reasons, along with its low cost, it is an excellent material for large machinery parts that are subjected to high compressive loads and vibration.

White cast iron has essentially all the carbon in the form of iron carbide. When fractured (see Figure 5-17) the surface is white in color, hence its name. Features promoting its formation are those which favor cementite over graphite: a low-carbon equivalent (1.8 to 3.6 per cent carbon, 0.5 to 1.9 per cent silicon, and 0.25 to 0.8 per cent manganese) and rapid cooling. By virtue of the primary carbide and pearlite microstructure, white iron is very hard and brittle and is seldom used as an engineering material. The rapid-cooling requirement also limits the thickness to which it can be produced.

Applications of white iron usually involve a surface layer with underlying substrate of other material. Mill rolls that require extreme wear resistance often have a white cast iron surface with steel backup. Variable cooling rates produced by tapered sections or the placing of metal chill bars in the molding sand can produce a surface chill of white cast iron on a gray iron casting. An example of this technique can be found in inexpensive scissors where white iron forms along the thin cutting edge and the somewhat more ductile gray iron forms in the heavier backup section. In this form the cast metal is known as *chilled cast iron.* A fracture surface will show white iron, a transition region of both white and gray iron known as the *mottled zone,* and gray iron.

If white cast iron is put through a controlled heat-treatment cycle, the cementite dissociates and some or all of the carbon is converted to irregular graphite spheroids. This product, known as *malleable cast iron,* has appreciably better ductility than that exhibited by gray cast iron (because the favorable graphite shape removes internal notches) but is limited by the fact that the casting must first be produced as white cast iron. The rapid cooling required for white iron production restricts the size and thickness of malleable iron components.

Two types of malleable iron can be produced, depending on the nature of the thermal cycle. If white iron is heated and held for a prolonged time just below the melting point, the carbon in cementite reverts to graphite. Subsequent slow cooling through the eutectoid reaction causes the austenite to go to ferrite and more graphite. The resulting product is known as *ferritic malleable iron* and has properties consistent with its structure of irregular graphite spheroids in a ferrite matrix: 10 per cent elongation, 240-MPa (35-ksi) yield strength, 345-MPa (50-ksi) tensile strength, and excellent impact strength, corrosion resistance, and machinability. Figure 5-19 shows a typical thermal cycle and Figure 5-20 shows the final structure.

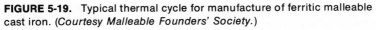

FIGURE 5-19. Typical thermal cycle for manufacture of ferritic malleable cast iron. (*Courtesy Malleable Founders' Society.*)

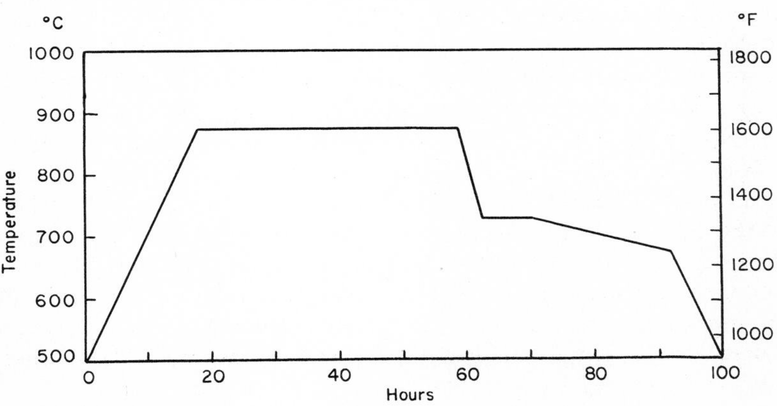

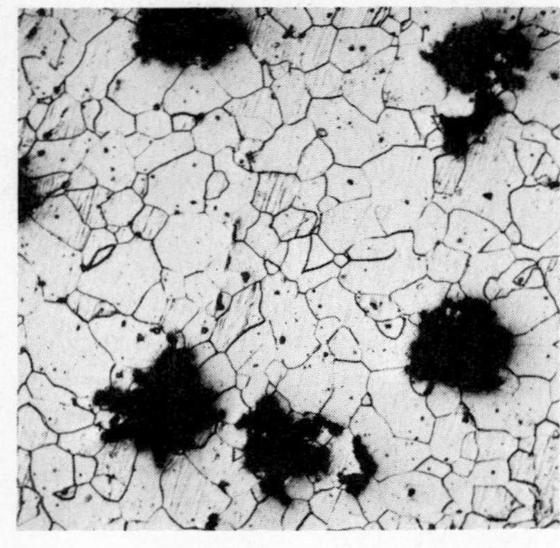

FIGURE 5-20. Photomicrograph of malleable iron. (*Courtesy Malleable Founders' Society.*)

If the casting is quenched to just below the eutectoid after the thermal hold, austenite transforms to pearlite. Generally, the matrix undergoes further thermal processing to increase ductility, but the resulting *pearlitic malleable iron* is characterized by higher strength and lower ductility than its ferritic counterpart. Typical properties range from 1 to 4 per cent elongation, 310 to 590 MPa (45 to 85 ksi) yield strength, and 450 to 725 MPa (65 to 105 ksi) tensile strength, with reduced machinability.

Most malleable iron (ferritic-type) has poorer wear resistance than gray cast iron but is as good or better in machinability. Because of its superior ductility, it is widely used for automotive parts, such as axle housings and brackets, and for pipe fittings and other applications where a uniform material of moderate ductility must be machined at rather low cost.

The spheroidal graphite structure of malleable iron provides quite an improvement in properties, but it would be even better if it could be obtained during solidification rather than by prolonged heat treatment at highly elevated temperatures. Certain materials have been found that promote graphite formation and change the morphology of the graphite product. If sufficient magnesium (added in the form of MgFeSi or MgNi alloy) or cerium is added to the liquid iron just prior to solidification, graphite tends to form as regular spheroids during solidification. The added material is known as a nodulizing agent and the product as *ductile or nodular cast iron.* Subsequent control of the thermal history can produce a wide range of matrix structures, with ferrite or pearlite being the most common (see Figure 5-21). Properties span a wide range from 2 to 18 per cent elongation, 275 to 620 MPa (40 to 90 ksi) yield strength, and 415 to 825 MPa (60 to 120 ksi) tensile strength.

The combination of good ductility, high strength, and castability makes nodular iron a rather desirable engineering material. Unfortunately, the cost of the nodulizer, higher-grade melting stock, better furnaces, and improved process control may make dutile iron almost as expensive as malleable iron. Nevertheless, it is replacing gray iron, malleable iron, and steel castings in numerous applications.

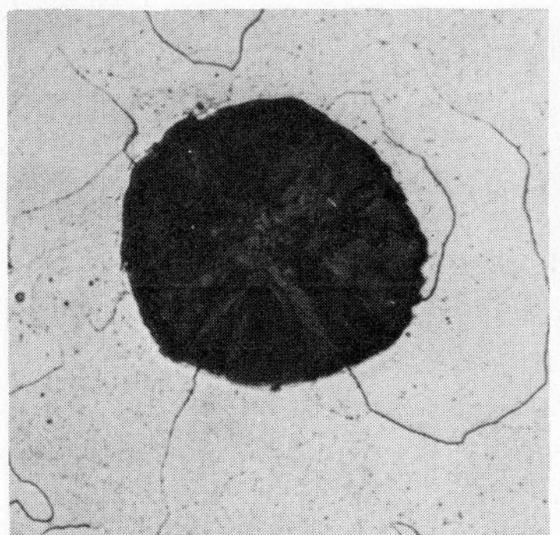 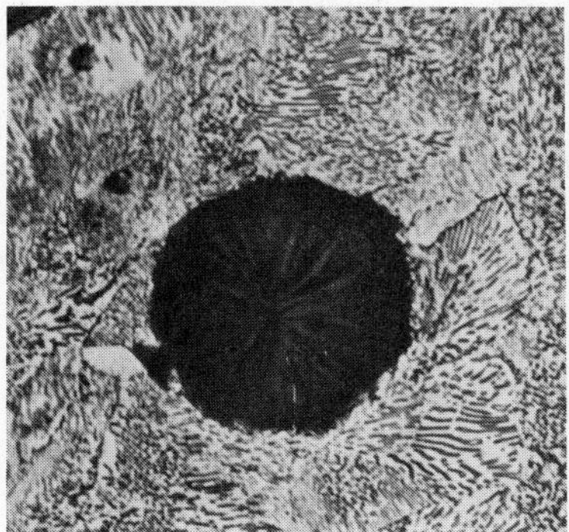

FIGURE 5-21. (*Left*) Ductile iron with ferrite matrix. (*Right*) Ductile iron with pearlite matrix; 500×. Note spheroidal graphite nodule.

Thus, by controlling the form of the carbon-rich phase, it is possible to produce a cast iron product with properties ranging from the hard, brittle, almost unmachinable white cast iron; to the soft, brittle, highly machinable gray cast iron; to the soft, ductile, and machinable, nodular or malleable irons.

Review questions

1. What is meant by the term "phase"?
2. What is an equilibrium phase diagram?
3. What is a cooling curve and how does it aid in determining phase diagrams?
4. How does the time–temperature cooling curve of an alloy of eutectic composition differ from that of a pure metal? That of a noneutectic composition alloy?
5. What is a liquidus line? A solidus line? A solvus line?
6. Why are isothermal scans of a temperature–composition diagram difficult to conduct?
7. What is a precipitation reaction and what causes it?
8. What pieces of information can be obtained for each point in a phase diagram? What kind of general information can be obtained?
9. What is a tie-line and what information can it provide?
10. What is a lever-law calculation and what information can it provide?
11. Explain what causes metal grains to be "cored."
12. What is a three-phase reaction? What key features in a diagram indicate such a reaction?
13. What is the form for a eutectic reaction? How does it differ from a eutectoid?
14. What properties are typical of most intermetallic compounds?

15. Why is the iron–carbon equilibrium diagram usually not shown beyond 6.67 per cent carbon?
16. What three-phase reactions occur in the iron–carbon diagram?
17. How do cast irons differ from steels in terms of carbon content and carbon solubility?
18. Define *austenite, ferrite, cementite,* and *pearlite.*
19. What is a hypoeutectoid steel? A hypereutectoid steel?
20. Describe the phase changes that occur when a molten 0.35 per cent carbon steel solidifies and cools slowly to room temperature.
21. Why may the results of various thermal processes not be exactly that indicated by the equilibrium diagram?
22. What additional element plays a significant role in cast irons?
23. What are the two possible eutectic reactions in cast iron metals?
24. What promotes graphite formation? What favors formation of cementite?
25. Why is the size, shape, and distribution of the graphite flakes in gray cast iron of such importance?
26. What is a class 40 cast iron?
27. What are the engineering assets of gray cast iron? Its liabilities?
28. How does white cast iron differ from gray cast iron?
29. What is malleable iron and how is it made?
30. What are the two common types of malleable iron?
31. What is nodular or ductile cast iron and how is it made?
32. Why do gray cast iron and nodular cast iron having the same carbon content differ so greatly in their ductility?
33. Why is the use of nodular cast iron increasing more rapidly than the use of malleable iron?

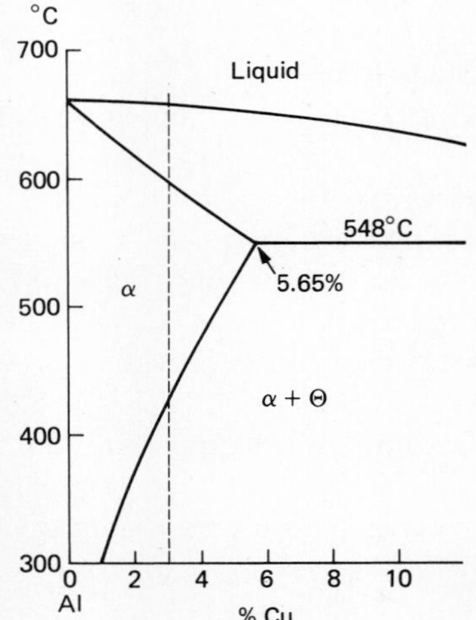

FIGURE CS-4. Equilibrium diagram for an aluminum–3 per cent copper alloy.

Case study 4.
IMPROPER UTILIZATION OF PHASE DIAGRAMS

Harry Simon, a production engineer with Missouri Machine Co., needed to reduce the thickness of a standard strip of aluminum–3 per cent copper alloy for use in a shimming application. Noting that the available rolling equipment had limited capacity, he decided to hot-roll the material in an effort to reduce the required forces. In selecting the heating temperature, he consulted the phase diagram in Figure CS-4 and selected 575°C (single-phase α region, 25°C below where liquid would form). Upon rolling at this temperature, however, the strip fragmented, breaking into pieces, rather than deforming uniformly. What had Harry overlooked?

CHAPTER 6

Heat treatment

THEORY AND PROCESSES OF HEAT TREATMENT

Heat treatment, by definition, is the *controlled heating and cooling of metals for the purpose of altering their properties.* Because the mechanical and physical properties can be altered so much by heat treatment, it is one of the most important and widely used manufacturing processes, providing a usually simple and low-cost means of obtaining desired properties. However, if performed improperly, more harm than good can result. Thus heat treatment must be understood and correlated with the other manufacturing processes in order to obtain effective results.

Although actual heat treatment applies only to processes where the heating and cooling are done for the specific purpose of altering properties, heating and cooling often occur as incidental phases of other manufacturing operations. (Obvious examples are hot-forming operations and welding.) But the effect on the properties of the metal—beneficial or harmful—is the same, even though no change in properties is desired. Thus the designer who selects the metal and the engineer who determines its processing must be aware of possible changes and take them into account.

Because proper application of heat treatment requires a thorough understanding of the material's response to various processes, both the theory of heat treatment and the various processes will be considered in this chapter.

Processing heat treatments. While heat treatment often is associated only with those thermal processes designed to increase strength, the definition permits inclusion of numerous processes which, for lack of a better term, we will call *processing heat treatments.* These are performed with a major goal of preparing the material for fabrication, including improving machining characteristics, reducing forming forces and energy consumption, and restoring ductility for further deformation. Thus heat treatment has tremendous capabilities, permitting the same metal to be softened for ease of fabrication and then, by another process, be given a totally different set of properties for service.

Equilibrium diagrams as aids to heat treatment. Most processing heat treatments involve rather slow cooling or extended times at elevated temperatures, thus tending to approximate equilibrium conditions. The resulting structures, therefore, can reasonably be predicted by the use of the equilibrium phase diagrams. The diagram indicates the temperatures that must be attained to achieve a desired product and the change that will occur upon subsequent cooling. It should be remembered, however, that the diagram is for truly equilibrium conditions, and departures from equilibrium may lead to substantially different results.

Processing heat treatments for steel. Because most processing heat treatments are applied to plain carbon and low alloy steels, they will be presented here with the simplified iron–carbon equilibrium diagram. Figure 6-1 shows this diagram with the key transition lines being labeled by standard notation. The eutectoid line is designated by the symbol A_1, and A_3 designates the boundary between austenite and ferrite + austenite for hypoeutectoid compositions. The transition from austenite to austenite + cementite is designed as the A_{cm} line.

A number of process heat-treating operations are classified under the general term *annealing.* These may be employed to reduce hardness, remove residual stresses, improve toughness, restore ductility, refine grain size, reduce segregation, or alter the mechanical, electrical, or magnetic properties of the material. By producing a certain desired structure, characteristics can be imparted that are favorable to the specific application. The temperature, cooling rate, and specific details of the process are determined by the material being treated and the objectives of the treatment.

In *full annealing,* hypoeutectoid steels are heated to 30 to 60°C (50 to

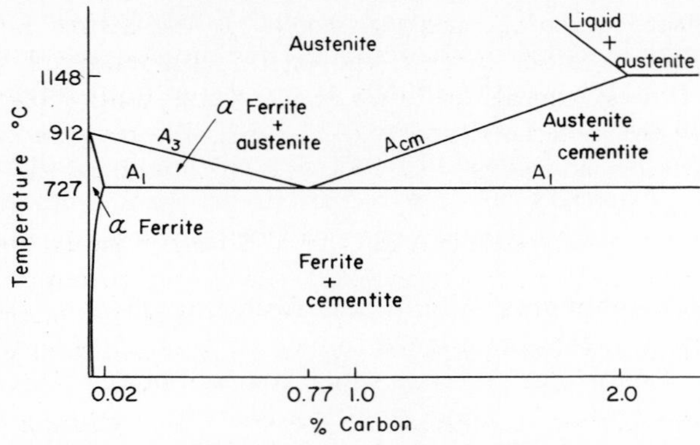

FIGURE 6-1. Simplified iron–carbon diagram for steels, with transition lines labeled in standard notation.

100°F) above the A_3 temperature to convert the structure to homogeneous single-phase austenite of uniform composition and temperature, held at this temperature for a period of time, and then slowly cooled at a controlled rate to below the A_1 temperature. A general rule is to provide 1 hour at temperature per inch of thickness of the largest section, but energy savings have motivated a reduction in this time. Cooling from the soaking temperature usually is done in the furnace, decreasing the temperature at a rate of 10 to 30°C (20 to 50°F) per hour to at least 30°C (50°F) below the A_1, followed by air cooling to room temperature. The resulting structure is coarse pearlite (widely spaced lamellae) with excess ferrite in amounts predicted by the phase diagram. The material is quite soft and ductile.

The procedure for hypereutectoid alloys is basically the same, except the original heating is only into the austenite plus cementite region (30 to 60°C above the A_1). If the material is slow-cooled from the pure austenite region, a network of cementite may form on the grain boundaries and make the material brittle. When properly annealed, the structure of a hypereutectoid steel will be coarse pearlite plus excess cementite in dispersed spheroidal form.

Full anneals are time-consuming and require considerable energy to maintain the elevated temperatures. When extreme softness is not required and cost savings are desired, *normalizing* may be employed. Here the metal is heated to 60°C (100°F) above the A_3 (hypoeutectoid) or A_{cm} (hypereutectoid), soaked to obtain uniform austenite, then removed from the furnace and allowed to cool in still air. Although a wide variety of structures is possible, depending on the size and geometry of the metal, fine pearlite with excess ferrite or cementite generally is produced.

Where cold working has severely strain-hardened a metal, it is often desirable to restore the ductility, either for service or to permit further processing without danger of fracture. A *process anneal* is often used for this purpose. The metal is heated to a temperature slightly below the A_1, held long enough to achieve softening, and then cooled at any desired rate (usually in air). Since austenite is not formed, the existing phases simply change their morphology. Process anneals are often used to recrystallize low-carbon steel sheets. Because the material is not heated to as high a temperature as in other processes, a process anneal is somewhat cheaper, more rapid, and tends to produce less scaling.

A *stress-relief anneal* often is employed to remove residual stresses in large steel castings and welded structures. Parts are heated to temperatures below the A_1 (550 to 650°C; 1000 to 1200°F), held for a period of time, and then slow-cooled. Times and temperatures vary with the conditions of the component.

When high-carbon steels must be prepared for machining or forming, a process known as *spheroidization* is employed. The goal is to produce a structure wherein all cementite is in the form of small, well-dispersed spheroids or globules in a ferrite matrix. This can be accomplished by a variety of techniques, including (1) prolonged heating at a temperature just below the A_1 followed by relatively slow cooling, (2) prolonged cycling between temperatures slightly above and slightly below the A_1, or (3) in the case of tool steel or high alloy steel, heating to 750 to 800°C (1400 to 1500°F) or higher and holding at this temperature for several hours, followed by slow cooling.

Although the selection of a processing heat treatment often depends on the desired objectives, steel composition also dominates the choice. Low-carbon steels (less than 0.3 per cent carbon) are most often normalized or process

FIGURE 6-2. Graphical summary of process heat treatments for steel on an equilibrium diagram.

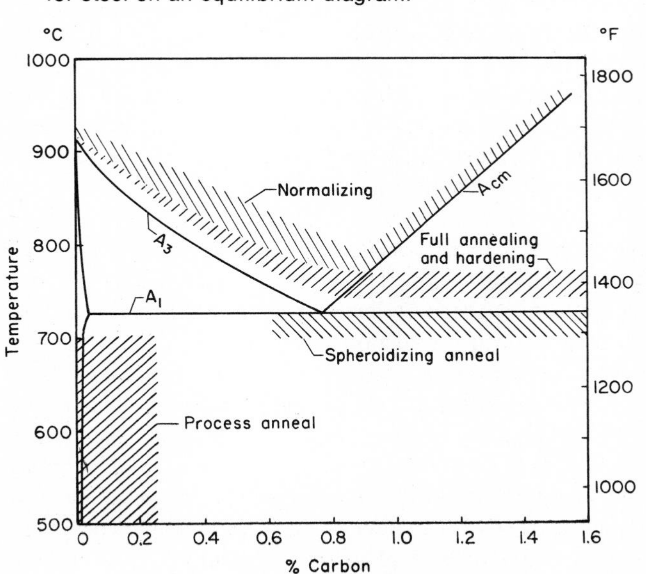

annealed. Steels of the medium (0.4 to 0.6 per cent) carbon range are usually full annealed. Above 0.6 per cent carbon a spheroidization treatment generally is required. Figure 6-2 provides a graphical summary of the process heat treatments.

Heat treatments for nonferrous metals. Most nonferrous metals do not have the significant phase transitions observed in the iron–carbon system and, for them, process heat treatments do not play such a significant role. Aside from precipitation hardening, which will be discussed later, nonferrous metals are ordinarily heat-treated for three purposes: (1) to obtain a uniform structure (such as to eliminate coring), (2) to provide stress relief, or (3) to bring about recrystallization. Coring, which may be present in castings that have cooled too rapidly, can be removed by heating to moderate temperatures and holding for a sufficient period to allow thorough diffusion to take place. Similarly, stresses that result from forming, welding, or brazing can be removed by heating for several hours at relatively low temperatures. Recrystallization, on the other hand, is a function of the particular metal, the degree of prior straining, and the time provided for completion. In general, the more a metal has been strained, the lower is the recrystallization temperature. If a nonferrous metal has been strained, it is relatively easy to bring about recrystallization and thus have new, equiaxed, and stress-free grains. Without straining, no recrystallization will occur and only grain growth will result. It is important to remember that only through recrystallization can new equiaxed grains (of the same or finer grain size as the original) be obtained in nonferrous metals in the solid state.

HEAT TREATMENTS TO INCREASE STRENGTH

Six major mechanisms are available to increase the strength of metals: (1) solid solution hardening, (2) strain hardening, (3) grain-size refinement, (4) precipitation hardening, (5) dispersion hardening, and (6) phase transformations. All but strain hardening involve heat treatment.

In *solid solution hardening,* a base metal dissolves other atoms in solid solution, either as *substitutional solutions,* where the new atoms occupy sites on the regular crystal lattice, or as *interstitial solutions,* where the new atoms squeeze into "holes" in the base lattice. The resulting distortions of the host lattice make dislocation motion more difficult.

Strain hardening, as discussed in Chapter 3, produces increased strength by plastic deformation under cold-working conditions.

Because grain boundaries act as barriers to dislocation motion, a metal with small grains will tend to be stronger than the same metal with larger grains. Thus *grain-size refinement* can be used to increase strength, except at elevated temperatures where failure is by creep, since this type of failure is related to grain boundary diffusion. Grain-size refinement is one of the few processes capable of increasing both strength and ductility.

All the mechanisms just described can be used to increase the strength of nonferrous metals: solid-solution strengthening for single-phase metals; strain hardening if sufficient ductility is present; dispersion hardening for eutectic-forming alloys. However, the most effective mechanism is precipitation hardening.

Precipitation or age hardening. Some alloy systems, mostly nonferrous, possess a sloping solvus line and can be heated into a single-phase solid solution and, owing to decreasing solubility, will form two distinct phases at lower temperatures. Consequently, if the heated single phase is rapidly cooled, a supersaturated solid solution is formed wherein the material required to form the second phase is trapped in the base lattice. An *aging* process then occurs, either at room or elevated temperature, wherein the excess solute atoms precipitate out of the supersaturated matrix and provide increased resistance to dislocation motion.

As an example, consider the silver-rich portion of the silver–copper system shown in Figures 6-3 and 6-4. If an alloy composed of 92.5 per cent silver and 7.5 per cent copper (sterling silver) is slowly cooled from 775°C (1430°F), upon crossing the solvus line A–B at 760°C (1400°F), β phase is precipitated out of the α solid solution because the solubility of copper in silver decreases from 8.4 per cent at 780°C (1435°F) to less than 1 per cent at room temperature. If this same alloy were cooled rapidly from 775°C (1430°F), there would not be sufficient time for the normal precipitation to take place, and the α phase would be retained in a highly supersaturated form. Because this supersaturated condition is not normal, the excess copper would like to precipitate out of solution and coalesce into β-phase particles.

FIGURE 6-3. Silver–copper equilibrium diagram.

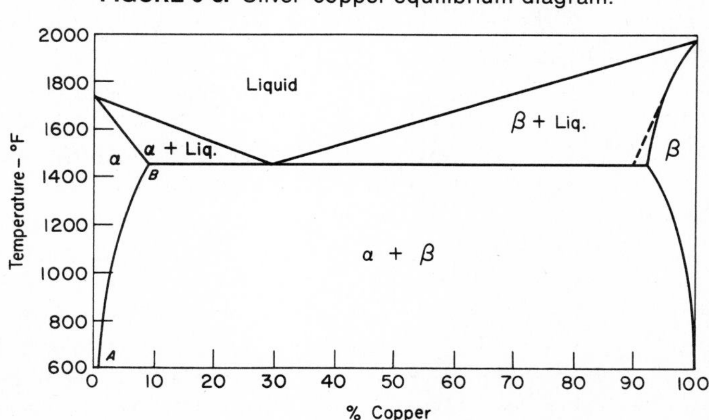

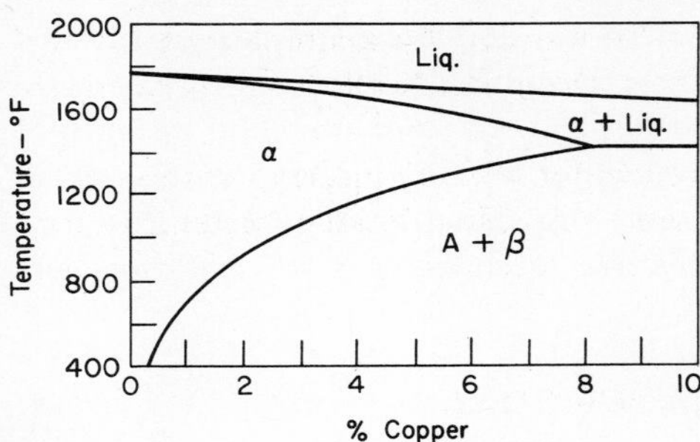

FIGURE 6-4. Enlargement of copper-rich section of the silver-copper equilibrium diagram.

Atom movement or diffusion is required, and therefore time at elevated temperature may have to be provided. If, in the case of sterling silver, the temperature is raised to around 260°C (500°F), precipitation occurs such that after 30 minutes the hardness will have increased from about 55 Brinell to about 110 Brinell.

Probably the most common application of precipitation hardening is in connection with aluminum–copper alloys. The first step is a heating (*solution treatment*) to put the material into a single-phase solid solution. This heating must not exceed the eutectic temperature so as to avoid melting in a cored structure. After soaking to achieve a uniform single phase, the material is *quenched* (cooled), usually in water, to prevent diffusion and to produce the supersaturated solid solution. In this state the material is rather soft, possibly softer than in the annealed condition. It can be straightened, formed, or machined while in this soft condition.

At this point, precipitation-hardening materials can be put into types: (1) *naturally aging,* in which the required diffusion from the unstable supersaturated to the stable two-phase structure can occur at room temperature, and (2) *artificially aging,* which require elevated temperatures. With the first type, the after-quench softness can be retained by refrigeration. (Aluminum alloy rivets are examples. Removed from the refrigeration and headed, they attain full strength in several days.) The properties of the second type can readily be controlled by means of temperature and time, the aging process being stopped at any intermediate condition by simple quenching.

Within the metal, the aging process begins with clustering of precipitated atoms on distinct planes in the base lattice. Transitions may then occur, leading to formation of a distinct phase with its own characteristic crystal structure. The key feature of precipitation hardening, however, is *coherency.* If the crystallographic planes are continuous in all directions, the resulting aggregates tend to distort the adjacent lattice for a sizable distance away, a very small aggregate acting like a much larger barrier to dislocation motion. Without coherency, interphase boundaries form, and the mechanism of strengthening is that of *dispersion hardening,* wherein the particles present only their physical dimensions as dislocation blocks.

Because artificial aging can readily be stopped by a simple quenching, thus avoiding decreased strength due to *overaging,* such alloys are used much more than those which age naturally.

Precipitation hardening is responsible for the engineering strengths of many aluminum, copper, and magnesium alloys. By special processing, some age-hardenable steels also have been produced.

STRENGTHENING HEAT TREATMENTS FOR STEEL

Iron-base metals have been heat-treated for centuries, and today over 90 per cent of all heat treating is performed on steel, utilizing phase transformations that occur in it. The striking changes that resulted from plunging red-hot steel into cold water or some other quenching medium were awe-inspiring to the ancients. Those who did such heat treatment in the making of swords or armor were looked upon as possessing unusual powers, and much superstition arose regarding the process. Because quality was directly related to the act of quenching, great importance was placed on the quenching medium that was used. For example, urine was thought to be a very superior quenching medium, and that from a red-haired boy was deemed particularly effective.

The isothermal transformation diagram. It has been only within the last hundred years that the art of heat treating has begun to turn into a science. One of the major barriers to understanding was the fact that the strengthening treatments were nonequilibrium in nature. One of the aids to understanding these nonequilibrium processes was the *isothermal transformation* (IT) or *time–temperature transformation* (T-T-T) *diagram,* obtained by heating thin specimens of a given metal to form uniform single-phase austenite, "instantaneously" quenching to a temperature where austenite was not the stable phase, and holding for variable periods of time.

For simplicity, consider a carbon steel of eutectoid composition and the resulting T-T-T diagram (Figure 6-5). Above 727°C (1341°F), austenite is the stable phase. Below this temperature, the face-centered austenite would like to transform to body-centered ferrite and carbon-rich cementite. Two factors control the rate of transition: (1) the motivation or driving force for the change, and (2) the ability to form the desired products (i.e., the ability to move atoms through diffusion). The result shown can be interpreted as follows. For any given temperature below 727°C (1341°F), time zero corresponds to a sample quenched "instantaneously." The structure usually is unstable austenite. As time passes (moving horizontally across the diagram), a line is encountered representing the start of transformation and a second line indicating completion of the phase change. At elevated temperatures (just below 727°C), diffusion is rapid, but the rather sluggish driving force dominates the kinetics. At lower temperature, the driving force is high but diffusion is quite limited. Kinetics are more rapid at a compromise intermediate temperature

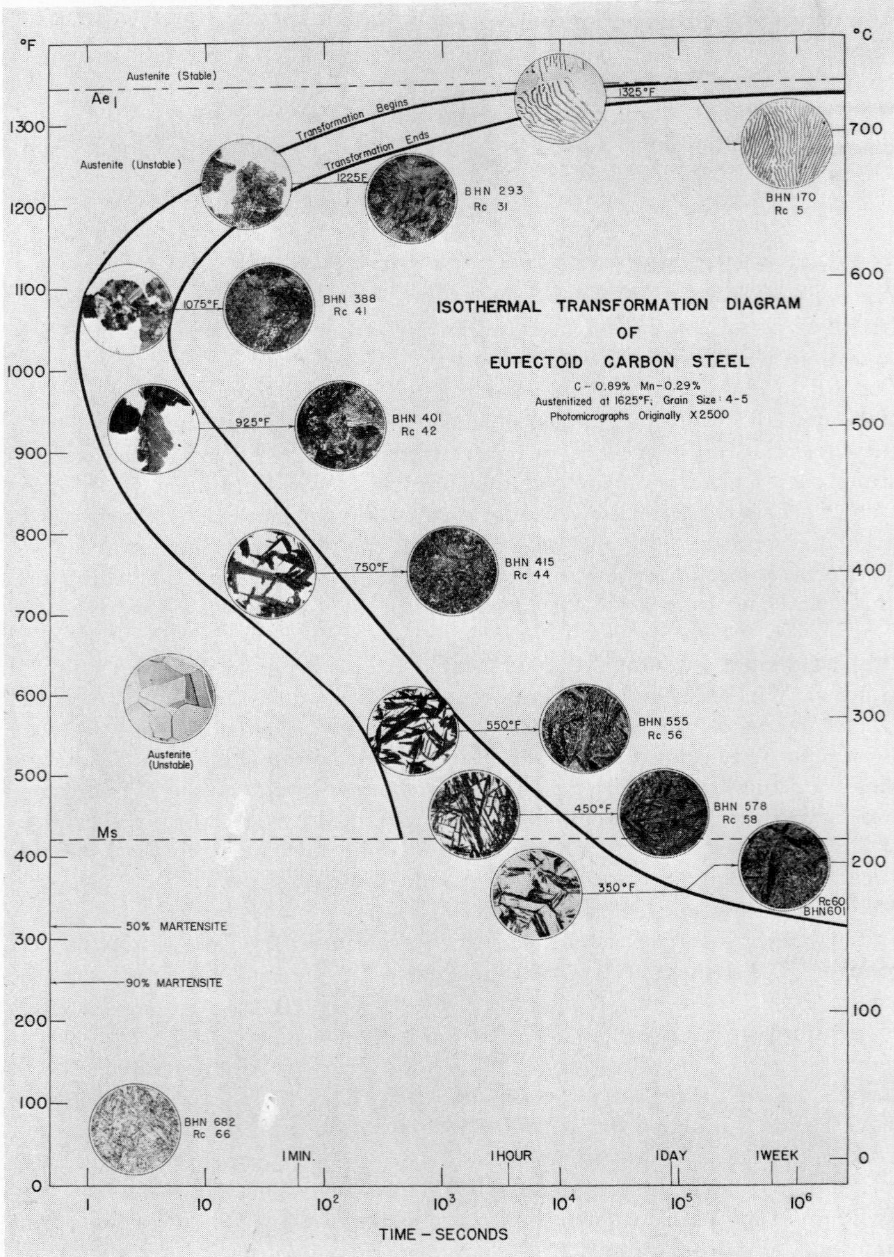

FIGURE 6-5. Isothermal transformation diagram (T-T-T diagram) for eutectoid composition steel. Structures resulting from transformation at various temperatures are shown as insets. (*Courtesy United States Steel Corporation.*)

than at either extreme, resulting in the often referred to *C-curve* terminology. That portion of the C which extends farthest to the left is known as the *nose* of the T-T-T diagram.

Now consider the products of the various transformations. If the transformation occurs between the nose of the curve and the A_1 temperature, the departure from equilibrium is not great. Austenite forms ferrite and cementite in the combined structure known as *pearlite.* Because of diffusion capabilities at higher temperatures, the lamellae spacing will be greater for the higher-temperature product than for that formed nearer the nose.

If quenched to a temperature between the nose and the temperature designated as M_s, another type of product must form. The process now is a significant departure from equilibrium, and the diffusion required to form lamellar pearlite is no longer available. The metal still has the same goal, however—to change its structure from austenite to a mixture of ferrite and cementite. The resulting structure is not that of alternating plates, but rather a dispersion of discrete cementite particles in either a lathlike or needlelike matrix of ferrite. Electron microscopy may be required to resolve the carbides in this structure, known as *bainite.* Because of the fine dispersion of carbide, its strength exceeds that of fine pearlite, and ductility is retained because soft ferrite is the continuous phase.

If the metal is quenched to below the M_s temperature, a different type of transformation occurs. The metal still desires to go from the face-centered to body-centered structure, but it cannot expel the required amount of carbon to form ferrite. As a response to the nonequilibrium conditions, the material undergoes an instantaneous change in crystal structure with no diffusion. The trapped carbon distorts the structure such that a body-centered tetragonal lattice results (a distorted body-centered cubic); the degree of distortion is a function of the amount of trapped carbon. The new structure, shown in Figure 6-6, is known as *martensite* and, with sufficient carbon, is exceptionally strong, hard, and brittle. Dislocation motion necessary for metal flow is effectively blocked by the highly distorted lattice.

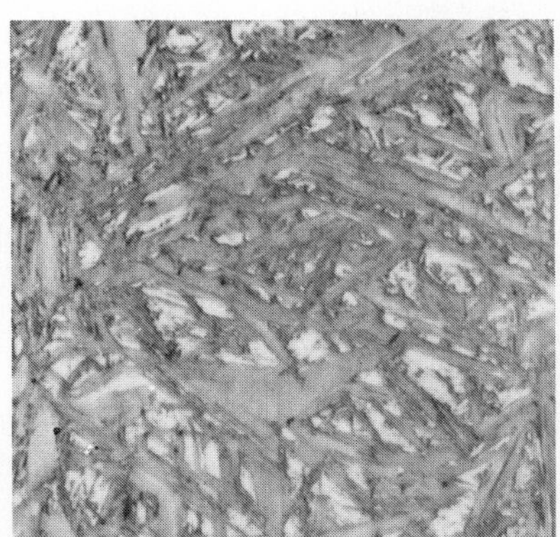

FIGURE 6-6. Photomicrograph of martensite; 1000×. (*Courtesy United States Steel Corporation.*)

Hardness and strength of steel in the martensitic form is a strong function of the carbon content. Below 0.10 per cent carbon, martensite is not very strong, although it is tough. Hardness typically is 30 to $35R_c$ at 0.1 per cent carbon and drops rapidly with decreasing carbon. Since no diffusion occurs in the transformation, higher-carbon-content material forms higher-carbon-containing martensite, with concurrent increase in strength and hardness and decrease in toughness and ductility. Thus from 0.3 to 0.7 per cent carbon, the maximum hardness increases rapidly; above 0.7 per cent carbon, the maximum hardness rises only slightly with increased carbon, a feature related to retained austenite.

As shown in Figure 6-7, the amount of martensite formed upon cooling is a function of the lowest temperature encountered and not the time at that temperature. Returning to Figure 6-5, we see below the M_s a temperature designated as M_{50} at which the structure will be 50 per cent martensite and 50 per cent untransformed austenite. At the lower M_{90} temperature, the structure is 90 per cent martensite. If no further cooling is undertaken, the remaining untransformed austenite can remain within the structure indefinitely. This *retained austenite* can cause loss of strength or hardness, dimensional instability or cracking, or brittleness. Since most quenches are to room temperature, retained austenite problems become significant when the martensite finish, or 100 per cent martensite, temperature lies below ambient. Higher carbon contents and alloy additions both decrease all martensite-related temperatures, and these materials may require refrigeration or a quench in liquid nitrogen to obtain full hardness.

Note that all the transformations that occur below the A_1 are one-way transitions, austenite-to-something. These are the only reactions possible, and it is impossible to convert one product to another without first reheating to above the A_1 to again form some stable austenite.

The T-T-T diagram is useful, the left-hand curve showing the elapsed time at constant temperature before transformation begins and the right-hand curve the time required for complete transformation at constant temperature.

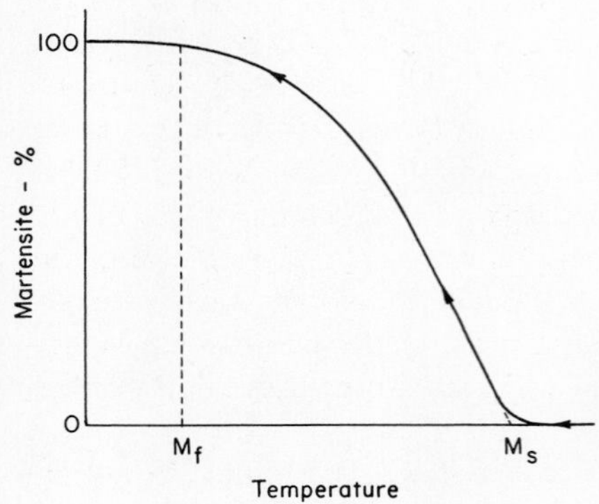

FIGURE 6-7. Schematic representation depicting the percentage of martensite resulting from quenching through various temperatures from M_s through M_f.

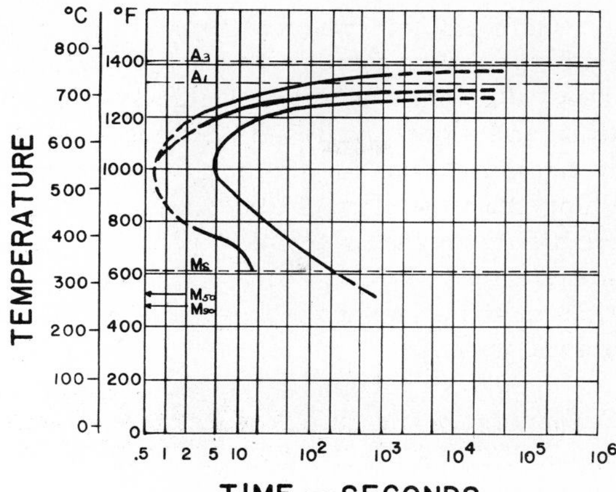

FIGURE 6-8. Isothermal transformation diagram for a hypoeutectoid steel (1050) showing additional region for primary ferrite.

If a hypo- or hypereutectoid steel were considered, additional regions would be added to correspond to the primary equilibrium phases. These regions would not extend below the nose, however, since the nonequilibrium bainite and martensite phases do not have to maintain fixed compositions as the phases in the equilibrium diagram do. Figure 6-8 is an *I–T* curve for a hypoeutectoid steel.

Tempering of martensite. Because it lacks good toughness and ductility, medium- or high-carbon martensite is not a useful engineering microstructure, despite its great strength. A subsequent heating, known as *tempering,* usually is required to restore some desired degree of toughness at the expense of a decrease in strength and hardness.

Martensite, in essence, is a supersaturated solid solution of carbon in alpha ferrite and therefore is metastable. By reheating in the range 100 to 700°C (200 to 1300°F), carbon atoms will be rejected from solution, and the structure will move to a mixture of the stable ferrite and cementite phases. This decomposition of martensite to ferrite and cementite is a time- and temperature-controlled diffusion phenomenon with a spectrum of intermediate and transitory conditions. An initial stage, which occurs at 100 to 200°C (200 to 400°F), is the precipitation of an intermediate carbide with the composition of $Fe_{2.4}C$, known as epsilon (ϵ) carbide. This allows the matrix to revert to the body-centered-cubic configuration. From 200 to 400°C (400 to 750°F), the structure becomes one of ferrite and cementite. Little is observable in the microscope, however, for the cementite particles are submicroscopic and the original martensite boundaries are retained. Figure 6-9 shows such a structure, which appears as a rather mottled mass with little well-defined structure. Electron microscope studies reveal the fine carbide structure responsible for the softening and improved ductility.

If tempering progresses into the 400 to 550°C (750 to 1000°F) range,

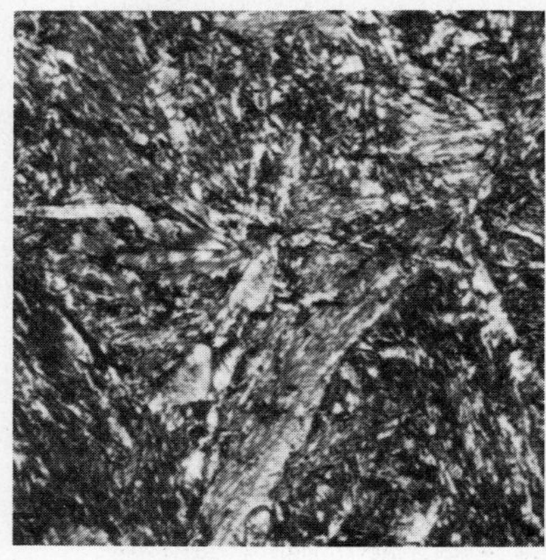

FIGURE 6-9. Eutectoid steel, hardened and tempered at 315°C (600°F); 1000×. (*Courtesy United States Steel Corporation.*)

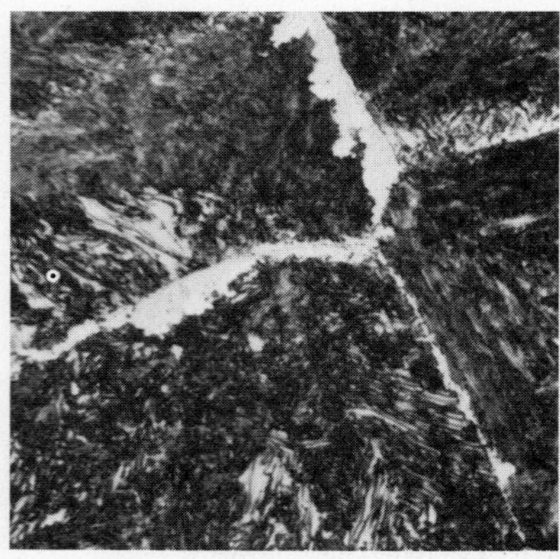

FIGURE 6-10. Eutectoid steel, hardened and tempered at 540°C (1000°F); 1000×. (*Courtesy United States Steel Corporation.*)

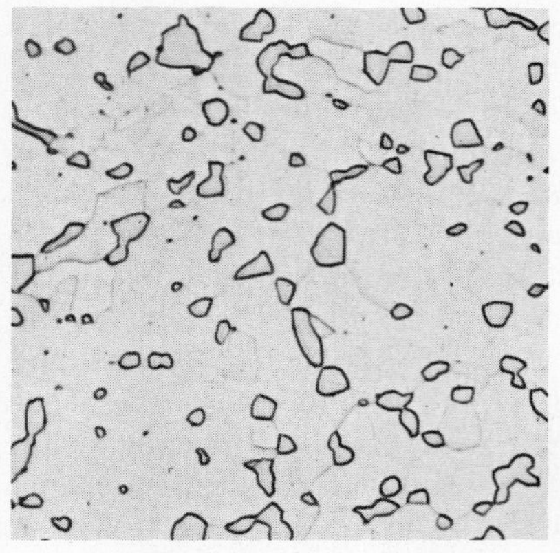

FIGURE 6-11. Photomicrograph of structure obtained by prolonged tempering above 540°C (1000°F), sometimes called spheroidite; 1000×. (*Courtesy United States Steel Corporation.*)

the martensite boundaries disappear and a ferrite structure nucleates and grows. As the precipitated carbides grow in size, the properties move farther in the direction of a weaker but more ductile material. Figure 6-10 shows some of the newly formed ferrite (white) in the tempered material.

Above 550°C (1000°F), the new ferrite grains totally consume the original structure, and the cementite particles become larger and more spheroidal. Figure 6-11 shows such a structure having the highest toughness and ductility and lowest strength of the tempered-martensite structures. Heating above the A_1 temperature will cause the structure to revert to stable austenite. Thus an infinite range of structures, and corresponding range of properties, can be obtained by quenching steel to obtain 100 per cent martensite and then tempering it to the desired state. This commonly is known as the *quench-and-temper* process.

Continuous-cooling transformations. While the T-T-T diagrams have provided significant information about the structures obtainable through nonequilibrium thermal processing, they usually are not applicable to direct engineering utilization because the assumptions of instantaneous cooling followed by complete isothermal transformation fail to match reality. More realistic would be a diagram showing the results of continuous cooling at various rates of temperature decrease. What will be the result if the temperature is dropped 300°C per second, 30°C per second, or 3°C per second?

A *continuous cooling transformation* (C-C-T) *diagram,* such as is shown schematically in Figure 6-12, can provide answers to these questions and several others. Critical cooling rates required to obtain various products can easily be determined. If cooled fast enough, the structure will be all martensite. A slow cool may produce coarse pearlite and some primary phase. Intermediate rates usually produce mixed structures, the time at any one temperature usually being insufficient for complete transformation. If each structure is considered as providing a companion set of properties, the wide range of possibilities obtainable through controlled heating and cooling of steel becomes evident.

Hardenability. When attempting to understand the heat treatment of steel, several key effects must be understood: the effect of carbon content, the effect of alloy additions, and the effect of various quenching conditions. The first two relate to the material and the third to the process.

Hardness is a mechanical property related to strength and is a strong function of the carbon content of a metal. *Hardenability,* on the other hand, is a measure of the depth to which full hardness can be attained under a normal hardening cycle and is related primarily to amounts and types of alloying elements. In Figure 6-13 all the steels have the same carbon content but differ in type and amounts of alloy elements. Maximum hardness is the same in all cases, but the depth of hardening varies considerably. Figure 6-14 shows the

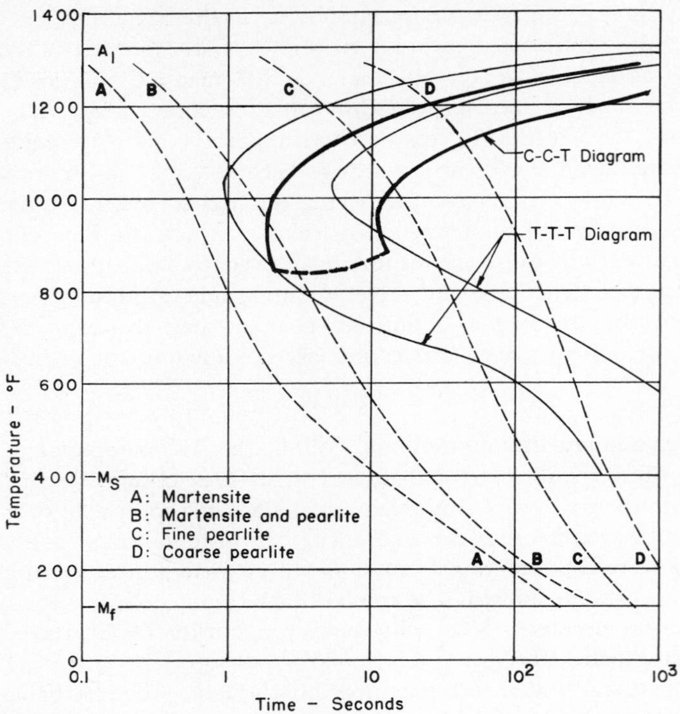

FIGURE 6-12. Schematic C-C-T diagram for a eutectoid composition steel, showing several superimposed cooling curves and the resultant structures. *(Courtesy United States Steel Corporation.)*

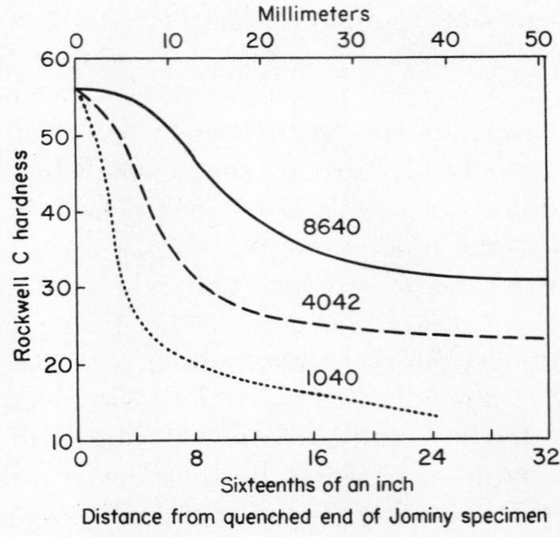

FIGURE 6-13. Jominy hardness curves for engineering steels with the same carbon content and varying types and amounts of alloy elements.

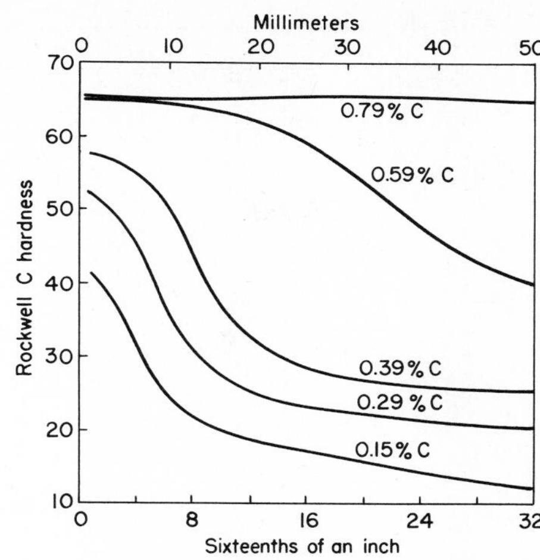

FIGURE 6-14. Jominy hardness curves for engineering steels with identical alloy conditions but variable carbon content.

results for steels containing the same alloying elements but variable amounts of carbon. Note the change in peak hardness.

The primary reason for adding alloy elements to commercial steels is to increase the hardenability, not to improve the strength properties. Steels with greater hardenability need not be cooled as rapidly to achieve a desired level of strength or hardness, and they can be completely hardened in thicker sections.

Materials selection for steels requires accurate determination of need. Strength and structure tend to be determined by carbon content, the general rule being to stay as low as possible and still meet specifications. Because heat can only be extracted from the surface of a metal, the depth of required hardening sets the conditions for hardenability and quench. For a given quenching condition, different alloys will produce different results. Because alloy elements increase the cost of a material, a general rule is to select only what is required to assure compliance with specifications. Money often is wasted by specifying an alloy steel for an application where a plain carbon steel, or steel with lower alloy content (and thus less costly), would be satisfactory. Another alternative when greater depth of hardness is required is to modify the quench conditions such that a faster cooling rate is achieved. Quench changes may be limited, however, by cracking or warping problems, depending on the shape, size, complexity, and precision of the part being treated.

The Jominy test for hardenability. A tool that is used to assist nonequilibrium heat treatment is the *Jominy end-quench hardenability test* and associated diagrams. In this test, depicted in Figure 6-15, an effort is made to reproduce the entire spectrum of cooling rates on a single specimen. The quench is standardized by specifying the water temperature (24°C or 75°F), internal nozzle

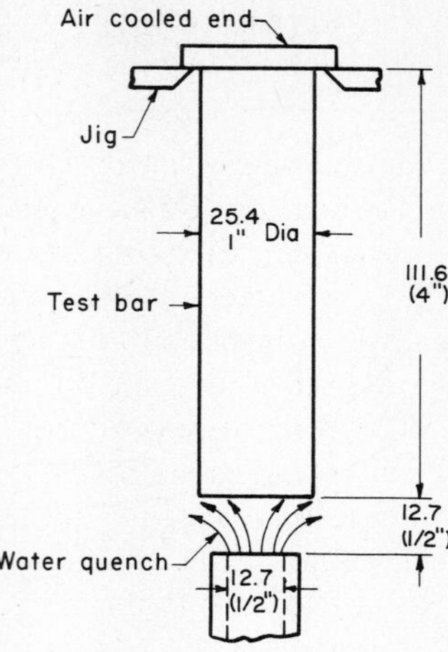

FIGURE 6-15. Schematic diagram of the Jominy hardenability test.

diameter, water pressure (63.5 mm or 2½-inch vertical travel unimpeded), and gap between the nozzle and specimen.

After the specimen has been cooled, a flat region is ground along one side, and R_c hardness readings are taken every 1.6 mm ($^1/_{16}$ inch) along the bar and plotted as shown in Figure 6-16. The hardness values then can be related to the cooling-rate data so that the cooling rate required to produce a given set of properties in the steel being considered can be determined rather precisely.

Applications of the test assume that identical results will be obtained if the same material undergoes identical cooling histories. If the cooling rate is known for a given location within a part (from experiment or theory), the properties at that location can be predicted as those at the equivalent cooling-rate location in a Jominy test bar. If specific properties are required, the necessary cooling rate for a given material can be determined, or, if the cooling rate

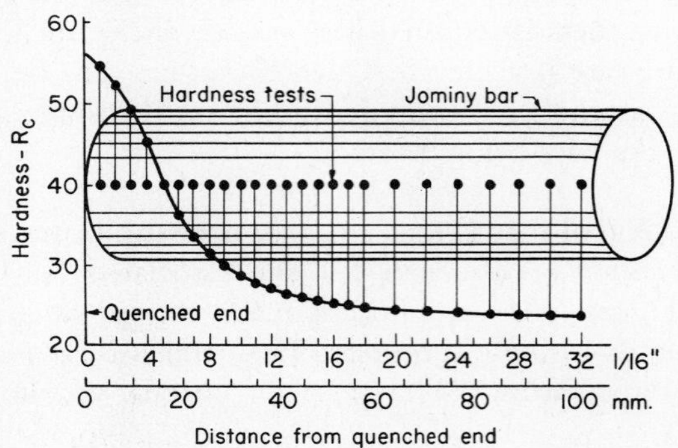

FIGURE 6-16. Typical hardness distribution in Jominy bars.

is restricted, an acceptable material can be selected. Figure 6-13 shows Jominy curves of several common engineering steels.

Quench media. Quench media vary in their effectiveness, the variation being best understood by considering the three stages of quenching. When a piece of hot metal is first inserted into a tank of liquid quenchant, that adjacent to the metal vaporizes and forms a gaseous layer separating the metal and liquid. Cooling is slow in this *vapor-jacket stage* (*first stage*) since all heat transport now must be through a gas. This stage occurs when the metal is above the boiling point of the quenchant. Soon bubbles nucleate and remove the gas, liquid contacts the metal, vaporizes (removing its heat of vaporization from the metal), forms a bubble, and the process continues. This *second stage of quenching* provides rapid cooling as a result of the large quantities of heat removed by the mechanism. When the metal cools below the boiling point of the quenchant, all heat transfer occurs through conduction across the solid–liquid interface, aided by convection or stirring within the quenchant; this is the *third stage.*

Water is a fairly good quenching medium because of its high heat of vaporization and the second stage of quenching extending down to 100°C (212°F), usually well into the martensite transition range or below. Water also is cheap, but the clinging tendency of the bubbles may cause soft spots in the metal. Agitation is recommended when using a water quench.

Brine is a more severe quenching medium than water because the salt nucleates bubbles, forcing a more rapid transition through the vapor-jacket stage. Unfortunately, brine tends to accentuate corrosion problems unless completely removed. Sodium or potassium hydroxide sometimes is used when very severe quenching is desired and one wishes to obtain good hardness in low-carbon steels. Various degrees of agitation or spraying of the quench can be used to increase the effectiveness of a given medium.

When a slower cooling rate is desired, oil quenches are often used. Various oils are available that have high flash points and different degrees of quenching effectiveness. Cooling often is entirely by conduction. The slower cooling through the M_s to M_f temperature range leads to a milder temperature gradient and less tendency to cracking. Molten salt baths also are used as quench mediums and, similarly, possess the property of going directly to the third stage.

Still slower cooling can be obtained by cooling in still air, packing the metal in sand, and a variety of other methods.

The role of design in the heat treatment of steel. Design details and material selection play important roles in the satisfactory and economical heat treatment of parts, and proper consideration of these factors usually does not impose serious limitations. In fact, proper consideration of them usually leads to more simple, more economical, and more reliable products. Failure to relate design and materials to heat-treatment procedure usually produces disappoint-

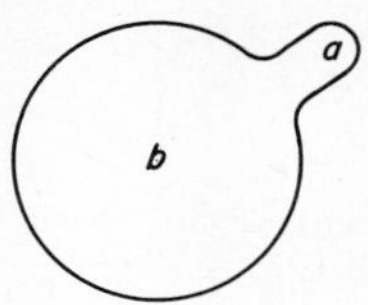

FIGURE 6-17. Nonuniform shape that will react unevenly in heat treatment.

ing or variable results and often service failure. Undesirable results may include nonuniform structure and properties, undesirable residual stresses, cracking, warping, and dimensional changes.

Undesirable design features are (1) nonuniform sections or thicknesses, (2) sharp interior corners, and (3) sharp exterior corners. Because these may easily find their way into the design of parts, the designer should be aware of their effect in heat-treating operations.

Heat can only be extracted from a piece of metal through its exposed surface. Thus, if a piece to be hardened has a nonuniform cross section, such as in Figure 6-17, the portion at *a* will cool rapidly and fully harden while that at *b* may not harden, except on the surface. The surface at *b* may even be tempered considerably by the heat retained in the center of the heavy section. A doughnut-shaped piece is probably closest to the ideal from the viewpoint of quenching, having a uniform cross section with maximum exposed surface. This is a good concept to keep in mind when designing parts that are to be heat-treated.

Residual stresses are the often-complex results of the various dimensional changes that occur during heat treatment. Thermal contraction during cooling is a well-understood phenomenon. In addition, the various phases and structures that may form are often characterized by different densities and, therefore, a volume expansion or contraction accompanies the phase transformations. When austenite transforms to martensite, there is a volume expansion of roughly 4 per cent. Austenite transforming to pearlite also experiences a volume expansion, but it is of smaller magnitude.

If all temperature changes occurred uniformly throughout the part, all changes in dimension would occur simultaneously and the result would be a component free of residual stresses. However, most parts being heat-treated experience nonuniform temperatures during the cooling or quenching process. Cross sections should be designed so that temperature differences are as low as possible and are not concentrated. If this is not possible, slower cooling and a material that will harden in an oil or air quench may be required. Because materials having greater hardenability invariably are more expensive, the design alternative clearly has advantages.

When temperature differences become severe or localized, additional problems, such as cracking or distortion, can result. Figure 6-18a shows an example where a sharp interior corner was placed at a change of cross section. Upon quenching, stresses will concentrate along line A–B and a crack is almost certain to result. When changes in cross section or other transition must

FIGURE 6-18. (*Left*) Shape containing nonuniform sections joined by a sharp interior corner that may crack in quenching. (*Right*) Improved design using a large radius to join sections and to avoid cracking in heat treatment.

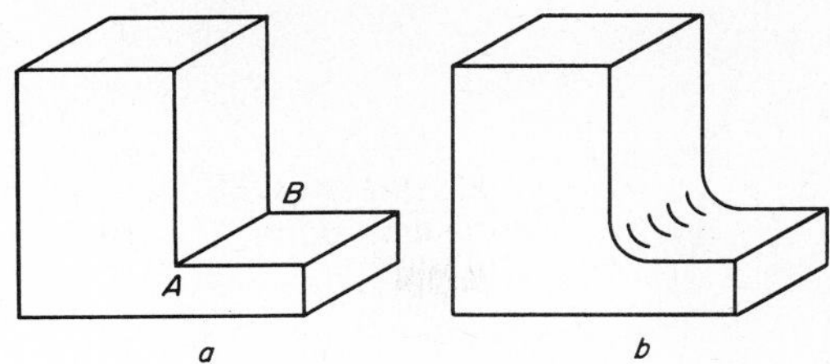

be made, they should be gradual, as in the redesigned Figure 6-18b. Generous fillets at interior corners, radiused external corners, and smooth transitions all reduce problems. Use of a more hardenable material or less severe quenching also will aid in preventing unnecessary problems.

Figure 6-19a shows the cross section of a die that consistently cracked during hardening. Eliminating the sharp corners and adding holes to provide a more uniform cross section during quenching, as in part (b) of the figure, eliminated the difficulty.

One of the ominous features about improperly designed heat-treated parts is the fact that the residual stresses may not produce immediate failure, but may contribute to failure at a later time. Applied stresses well within the "safe" designed limit may couple with residual stresses to produce loads sufficient to cause failure. Dimensions may change or warping result during subsequent machining or grinding operations. Corrosion reactions may be significantly accelerated in the presence of residual stresses. Time and money are lost that could be saved if proper design practices are employed. Heat treatment is an important manufacturing process that enables better results to be obtained with less costly materials, *if used properly*. The designer, however, must take into account all the facts and conditions related to it when he designs a part, selects a material, and specifies, directly or indirectly, the heat treatment for it.

Techniques to reduce cracking. Two variations on the quenching cycle have been designed to minimize the high temperature gradients, caused by a rapid quench, which often result in cracking. Rapid quenching still is required to prevent transformation to the weaker pearlite structure, but instead of quenching through the martensite transformation, the component is rapidly

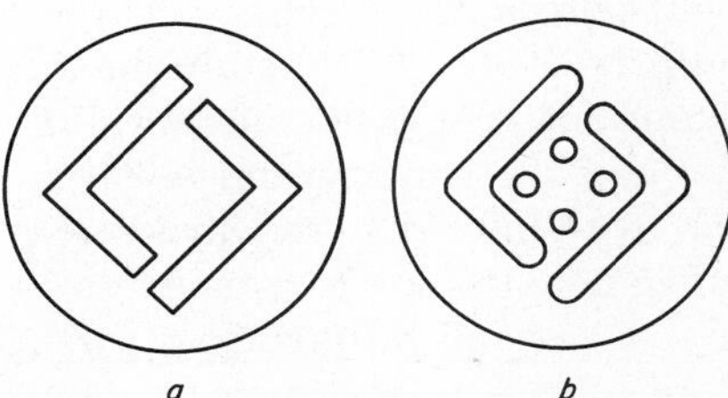

FIGURE 6-19. Improved design techniques to provide more uniform sections.

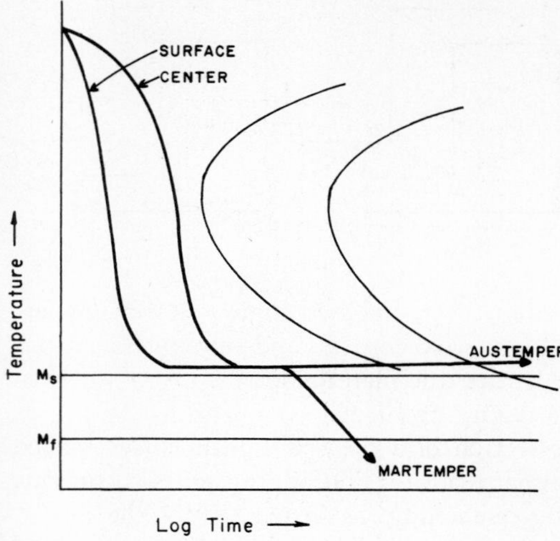

FIGURE 6-20. Schematic representation of the austempering and martempering processes.

quenched to a temperature several degrees above the M_S, usually in a bath of molten salt. Holding for a brief time at this temperature enables the piece to come to a nearly uniform temperature. If the piece is held at this temperature long enough, the austenite will transform to bainite, a process known as *austempering.* If the piece is stabilized and then slow cooled through the martensite transformation, the process is known as *martempering* or *marquenching.* Here the product is martensite, which must be tempered the same as martensite formed by rapid quenching. Figure 6-20 shows these processes schematically on a T-T-T diagram (a misuse of the diagram but good for visualization).

Pieces with complicated shapes, undesirable design features, or high precision can all benefit from these modified hardening techniques. Straightening can also be performed after stabilization in the quench bath, before final hardening occurs.

Ausforming. A process often confused with austempering is that of *ausforming.* Certain alloys tend to retard the pearlite transformation far more than the bainite reaction and produce a T-T-T curve such as that shown in Figure 6-21 for 4340 steel. If a metal is heated to form austenite and then quenched to the temperature of the "bay" between the pearlite and bainite regions, it can retain its austenitic structure for a useful period of time. Deformation can be performed here on an austenitic material, a structure that, technically, is not stable at the temperatures involved. Benefits include the increased ductility of the face-centered crystal structure, the finer grain size characteristic of recrystallization at a lower temperature, and some degree of possible strain hardening. Following deformation, the material can either be slowly cooled to produce a bainitic structure or quenched to form martensite, which then is

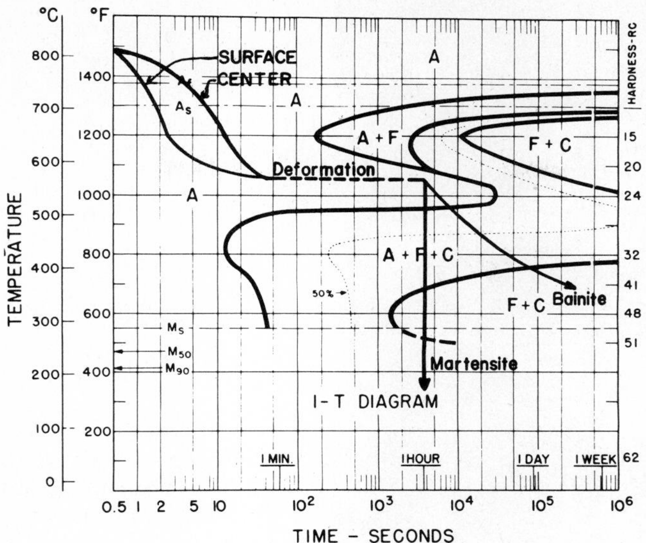

FIGURE 6-21. T-T-T diagram for 4340 steel, showing the "bay" and the ausforming process in schematic. (*Courtesy United States Steel Corporation.*)

tempered. The material that results shows extremely high strength, ductility, toughness, creep resistance, and fatigue life, far greater than if the processes of deformation and transformation were conducted in their normal sequence.

SURFACE HARDENING OF STEEL

Often it is desirable for parts to have a hard, wear-resistant surface coupled with a tough, impact-resistant core. Several methods have been developed to produce such properties by selectively hardening the surface regions.

Selective heating techniques. Two basic classes of processes exist. In the first, the steel has sufficient carbon content to attain the desired hardness. Different properties are obtained simply by varying the thermal histories of the various regions, a technique known as *selective heating.* In general, these processes require at least 0.3 per cent carbon, the basic rules of steel hardening still being applicable.

Flame hardening employs a high-intensity oxyacetylene flame to raise the surface temperature high enough to form the desired austenite. Heat input is quite rapid and is concentrated on the surface. The short time involved, and inadequate heat transfer, leave the interior at a much lower temperature and therefore free from any significant changes.

FIGURE 6-22. Surface hardening by the flame-hardening process. (*Courtesy Linde Division of Union Carbide Corporation.*)

Considerable flexibility is provided since the rate and depth of heating can be varied. Depth of hardening can range from thin skins to over 6 mm (¼ inch). Much flame hardening is confined to large objects, such as the teeth of large gears, where size or shape restrictions prohibit the use of alternative techniques. Equipment varies from crude hand-held torches to fully automated and computerized special units. Figure 6-22 shows a typical operation.

Induction heating is particularly well suited to surface hardening, the rate and depth of heating being controlled by the amperage and frequency of the current. The steel part is placed inside a wound coil which then is subjected to ac current. The changing magnetic field induces a current in the steel, flowing through the surface layers and heating by electrical resistance. Heating rates are extremely rapid and efficiency is high. In addition, the process lends itself to design and adaptation for special shapes, retaining close control and the benefits of automated processing. Figures 6-23 and 6-24 show the process and results for a large gear where the main body has been left soft and surface hardening has been applied to those areas subject to wear. Since nearly all the metal remains cool during the process, sufficient rigidity is provided to keep distortion to near-negligible amounts.

Other selective heating techniques use immersion in a *lead pot* or *salt bath* to heat the surface. *Laser beam hardening* and *electron beam hardening* are additional techniques showing high industrial potential.

Techniques using altered surface chemistry. When steel contains insufficient carbon to attain the desired surface properties by selective heating, techniques

FIGURE 6-23. (*Left*) Production setup for induction hardening teeth on a gear. (*Right*) Closeup of the head and teeth. (*Courtesy TOCCO Division, Park-Ohio Industries, Inc.*)

must be employed to alter the surface chemistry. The most common technique in this category is *carburizing.* In *pack carburizing,* the parts are packed in a high-carbon solid medium, enclosed in a gastight box, and heated in a furnace for 6 to 72 hours at roughly 900°C (1650°F). The carburizing compound releases CO gas which reacts with the metal and releases the desired carbon, which is absorbed into the austenite. The boxes are then taken from the furnace and the parts removed and processed further. Direct quenching from the carburizing operation produces different structures as a result of the different carbon contents of the metal. Slow cool and reheat or a duplex core-case treatment are alternative methods. The carbon content of the surface varies from 0.7 to 1.2 per cent, depending on the details of the process. Case depth may range from a few hundredths of a millimeter to several hundred millimeters, although cases over 1.5 mm (0.060 inch) are not often employed.

Several problems are encountered in pack carburizing. Heating is inefficient; temperature uniformity is questionable; handling is often difficult; and the process is not readily adaptable to continuous operation. *Gas carburizing*

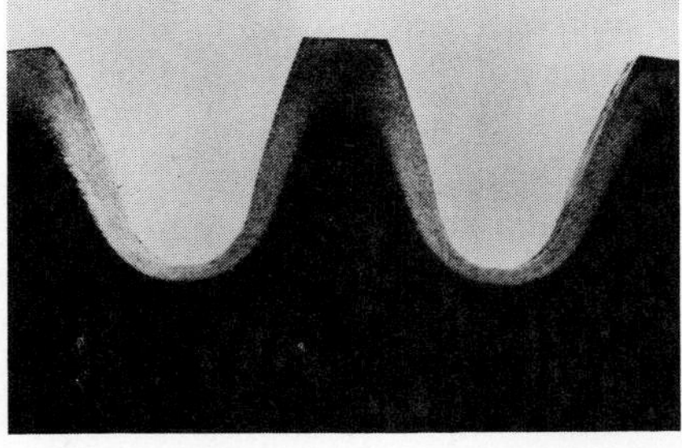

FIGURE 6-24. Section of gear teeth, showing localized surface hardening. (*Courtesy TOCCO Division, Park-Ohio Industries, Inc.*)

overcomes these difficulties by replacing the solid carburizing compound with a carbon-providing gas, usually containing an excess of CO. While the mechanism and processing are the same, the operation is faster and more easily controlled. Accuracy and uniformity are increased, and continuous operation is possible. Special types of furnaces are required, however.

In *liquid carburizing* or *cyaniding,* the carbon is supplied by immersing the part in a molten cyanide salt. Both carbon and nitrogen are added to the steel surface in this process, used primarily to put thin cases on small parts. Safety restrictions associated with the toxic cyanide fumes tend to favor the use of alternative techniques.

When only certain portions of a surface require hardening, *selective carburization* can be employed where a thin copper plating or special paints are used to prevent carbon diffusion. Slow cooling, followed by selective heating of the areas to be hardened, or machining away the unwanted carburized surfaces, are alternative techniques.

Nitriding produces surface hardening by formation of alloy nitrides in the surface layers of special steels (steels that contain nitride-forming elements, such as aluminum, chromium, molybdenum, or vanadium). The parts are heat-treated and tempered at 525 to 675°C (1000 to 1250°F) prior to nitriding. After cleaning and removal of any decarburized surface material, they are heated in a dissociated ammonia atmosphere (containing nitrogen and hydrogen) for 10 to 40 hours at 500 to 625°C (950 to 1150°F). Nitrogen diffusing into the steel will then form alloy nitrides, hardening the metal to a depth of about 0.65 mm (0.025 inch). Very hard cases are formed, and distortion is low (temperatures are low and slow cooling from nitriding is permissible). No subsequent thermal processing is required. Although the surface hardness is higher than for other hardening methods, the long times at elevated temperatures, coupled with the thin case and associated dimensional precision, restrict application to those cases requiring high quality.

HEAT-TREATMENT EQUIPMENT

Furnaces. Because heating always is involved in heat treating, numerous types of heating equipment have been developed in a range of sizes. A basic classification of furnaces is whether they are of the *batch* or *continuous* type. Batch-type furnaces are those in which the workpiece remains stationary during the time it is in the furnace. Batch furnaces may be either *horizontal* or *vertical.* Continuous furnaces are designed so that the parts go into and come out of the furnace one at a time at predetermined rates that match other continuous operations in the manufacturing sequence.

Horizontal batch-type furnaces are often called *box furnaces,* because they resemble a rectangular box. As shown in Figure 6-25, a door is provided in one end to permit the work to be put in and removed. Either gas or electricity is used as the heat source. In some furnaces, particularly those for lower tem-

FIGURE 6-25. Box-type electric heat treating furnace. *(Courtesy Lindberg Engineering Company.)*

peratures, a circulating fan is provided to obtain more uniform heating. For large or long workpieces, *car-bottom furnaces* are used, such as in Figure 6-26. The work is loaded onto a flatcar on wheels which is run into the furnace through large doors.

Horizontal-type furnaces are relatively easy to construct in any size, are easily insulated, and are thermally efficient. Nevertheless, it is difficult to heat-treat long, slender work in them because of the sagging or warping that is likely to occur. Vertical furnaces of the *pit type,* shown in Figure 6-27, have been designed for such work. These are cylindrical chambers sunk into the floor with a door on the top that can be swung aside to permit the work to be lowered and suspended in the furnace. Suspended in this manner, long workpieces are less likely to warp. This type of furnace also is used to heat batches of small parts, which are loaded into baskets and lowered into the furnace.

Another furnace type is the *bell furnace,* shown in Figure 6-28. The heating elements are contained within a bottomless bell that is lowered over the work. This type often is used for annealing batches of steel or other metal sheets. An airtight inner shell often is employed to contain a protective atmosphere during the heating and cooling cycles to prevent surface tarnish or scale. After the work has been heated, the furnace unit can be lifted off and transferred to another batch, the inner shell retaining the controlled atmosphere during cooling. Sometimes an insulated cover may be placed over the heated work if slower cooling is desired.

FIGURE 6-26. Car-bottom box-type furnace. (*Courtesy Hevi Duty Electric Company.*)

FIGURE 6-27. Pit-type heat-treating furnace. (*Courtesy Hevi Duty Electric Company.*)

FIGURE 6-28. Bell-type heat-treating furnace. (*Courtesy Surface Combustion Corporation.*)

The *elevator-type furnace* is a modification of the bell furnace, in which the bell is stationary and the work is raised up into it by means of a movable platform that forms the bottom of the furnace. An interesting variation of this furnace is one for which there are three vertical positions. In the middle position the work is loaded onto the platform elevator. In the upper position the work is in the bell furnace, and in the lower position it is in a quench tank. Such furnaces are used where work must be quenched as soon as possible after being removed from the furnace.

Continuous furnaces usually are equipped with some type of conveyor device onto which the work can be loaded one piece at a time and moved into, through, and out of the unit at a controlled rate. The conveyor system can be arranged so that the work falls off into a quench tank to complete the treatment. These are excellent heat-treating furnaces because complex heating, holding, and quenching or cooling cycles can be carried out in an exact and repeatable manner with very low labor cost.

Protective gas atmospheres can be introduced into nearly all types of furnaces, if properly designed, and are frequently used in production. Special atmospheres can be created to prevent decarburization or to eliminate the scaling or tarnishing that occurs in the presence of certain gases at elevated temperatures. Expensive and troublesome cleaning and descaling operations can be completely eliminated.

Salt-bath furnaces are often used for heat treating where a liquid heating medium is preferred over gas. In some types, an electrically conductive salt is used that is heated by a current passing between two electrodes suspended in the salt bath. Where salt baths of this type are used, the electrical currents also cause the bath to circulate and thus maintain uniform temperature. When nonconductive salts are used, they can be heated by some form of immersion heater. By proper selection, a salt bath also can provide a protective medium to prevent decarburization or scaling. A similar furnace is the *lead pot,* in which molten lead replaces the molten salt as the heating medium.

The use of *electrical induction heating* has simplified many heat-treating operations. Small parts can be through-heated and hardened as in other styles of furnace. Local or selective hardening, as well as surface hardening, is possible at a very rapid production rate. Furthermore, a standard induction unit can be adapted to a wide variety of products by simply changing the induction coil and adjusting the equipment settings.

Furnace controls. All heat-treating operations should be carried out under rigid controls if uniformity is desired. For this reason, most heat-treating furnaces are equipped with indicating and controlling pyrometers. Some furnaces are equipped with controllers that also regulate the rate of heating or cooling. It should be remembered that it is the temperature of the workpiece and not that of the furnace that controls the result. Since the temperature of the work may be many degrees different from the furnace, ample time must be allowed to bring the work to uniform temperature.

Review questions

1. What is heat treatment? Why is it an important manufacturing process?
2. To what extent, if any, do the effects resulting from incidental heating and cooling of a metal in conjunction with some processing operation differ from those resulting from a heat treatment that involves the same heating and cooling cycle?
3. What are some purposes of the processing heat treatments?
4. Why are equilibrium phase diagrams useful for understanding the processing heat treatments?
5. Why is the term "annealing" often confused or misunderstood?
6. Compare and contrast the processes of full annealing, normalizing, process annealing, stress-relief annealing, and spheroidizing.
7. What are the primary reasons for heat-treating nonferrous metals?
8. In what ways can the strength of a given metal be improved? Can all the techniques be applied to every alloy?

9. Why is precipitation hardening an important process for nonferrous metals?
10. What are the stages of a precipitation hardening treatment?
11. What are the two classes of precipitation hardening metals (based on aging characteristics)?
12. What are the structural differences among pearlite, bainite, and martensite?
13. What is retained austenite and what problems can it cause?
14. Why must martensite be tempered? What occurs during tempering?
15. Why are T-T-T diagrams usually not applicable to direct engineering utilization?
16. What is a continuous-cooling transformation (C-C-T) diagram?
17. What is the benefit of the Jominy end-quench hardenability test?
18. What is the difference between hardness and hardenability?
19. Explain the effect of carbon content and alloy content on maximum hardness and hardenability.
20. What are the three stages of quenching?
21. Rate and explain the order of effectiveness of the following quench media: oil, brine, water, and molten salt.
22. How are design and material selection linked to heat treatment?
23. Why are residual stresses so important?
24. What is austempering, martempering, and ausforming?
25. What are some of the selective heating techniques employed for surface hardening?
26. What are some surface-hardening techniques employing altered surface chemistry?
27. Contrast batch-type and continuous-type heat-treating furnaces.
28. Why are protective atmospheres often used?
29. What is the difference between a processing heat treatment and a hardening heat treatment when applied to steels?

Case study 5.
THE BROKEN MARINE ENGINE BEARINGS

The bearings on a small shipboard marine engine have been made of AISI 52100 grade steel that has been austenitized, quenched, and tempered to establish the desired final properties. Performance under normal operating conditions appears adequate. After exposure to a period of subzero temperatures, however, the engines failed. Tear-down revealed rather brittle cracks in the bearings, together with an observed expansion in the bearing dimensions. What would you suspect to be the cause of the failures?

CHAPTER 7

Alloy irons and steels

Without alloy irons and steels, the state of technology would be set back considerably. Many varieties of alloys have been developed to meet specific needs of an advancing civilization. However, the availability of many varieties has often resulted in poor selection and excess cost for an unnecessary and expensive alloy material. It is the responsibility of the design and manufacturing engineer to be knowledgeable in this area and to make the best selection from the available alternatives.

Plain-carbon steel. Steel theoretically is an alloy of iron and carbon. When produced commercially, however, certain other elements—notably manganese, phosphorus, sulfur, and silicon—are present in small quantities. When these four foreign elements are present in their normal percentages, the product is referred to as *plain-carbon steel.* Its strength is primarily a function of its carbon content. Unfortunately, the ductility of plain-carbon steel decreases as the carbon content is increased, and its hardenability is quite low. In addition, the properties of ordinary carbon steels are impaired by both high and low temperatures, and they are subject to corrosion in most environments.

Plain-carbon steels are generally classed into three subgroups, based on carbon content. *Low-carbon steels* have less than 0.30 per cent carbon, possess good formability and weldability, but not enough hardenability to be hardened to any significant depth. Their structures usually are ferrite and pearlite, and the material generally is used as it comes from the hot-forming or cold-forming process. *Medium-carbon steels* have between 0.30 and 0.80 per cent carbon, and they can be quenched to form martensite or bainite if section size is small and a severe water or brine quench is used. The best balance of properties is attained at these carbon levels, the high fatigue and toughness of the low-carbon material being in good compromise with the strength and hardness that comes with higher carbon content. These steels find numerous applications. *High-carbon steels* have more than 0.80 per cent carbon; toughness and formability are quite low, but hardness and wear resistance are high. Severe quenches can form martensite, but hardenability is still poor. Quench cracking is often a problem when the material is pushed to its limit.

Plain carbon steels are the lowest-cost steel material and should be considered for many applications. Often, however, their limitations become restrictive. When improved material is required, steels can be upgraded by the addition of one or more alloying elements.

Alloy steels. The differentiation between "plain carbon" and "alloy" steel is often somewhat arbitrary. Both contain carbon, manganese, and usually silicon. Copper and boron also are possible additions to both classes. Steels containing more than 1.65 per cent manganese, 0.60 per cent silicon, or 0.60 per cent copper are designated as alloy steels. Also, a steel is considered to be an alloy steel if a definite amount or minimum of other alloying element is specified or required. The most common alloy elements are chromium, nickel, molybdenum, vanadium, tungsten, cobalt, boron, and copper, as well as manganese, silicon, phosphorus, and sulfur in amounts greater than normally are present.

Effects of the alloying elements. In general, alloying elements are added to steel in small percentages—usually less than 5 per cent—to improve strength or hardenability, or in much larger amounts, often up to 20 per cent, to produce special properties such as corrosion resistance or stability at high or low temperatures. Certain additions may be made during the steelmaking process to remove dissolved oxygen from the melt; manganese, silicon, and aluminum are often used for this deoxidation. Aluminum and, to a lesser extent, vanadium, columbium, and titanium are used to control austenitic grain size. Machinability can be enhanced by additions of sulfur, lead, selenium, and tellurium. Other elements may be added to improve the strength or toughness properties of the product metal. Manganese, silicon, nickel, and copper add strength by forming solid solutions in ferrite. Chromium, vanadium, molybdenum, tungsten, and other elements increase strength by forming dispersed second-phase carbides. Columbium, vanadium, and zirconium can be used for ferrite grain-size control. Nickel and copper are added to low-alloy steels to provide improved corrosion resistance.

For constructional alloy steels, the principal effect of alloying elements is to increase hardenability. The commonly used elements, in order of decreasing effectiveness, are manganese, molybdenum, chromium, silicon, and nickel. Small quantities of vanadium are quite effective, but the response drops off as quantity is increased. Boron is also extremely significant in steels with less than 0.65 per cent carbon.

Still other characteristics of the individual elements are worthy of note.

Manganese is present in most plain-carbon steels to combine with sulfur and form soft manganese sulfides. This prevents formation of iron sulfide along grain boundaries, which would lead to brittleness of the metal. In alloy steels, it increases hardenability, lowers transformation temperatures, and causes the transformations to be more sluggish. Consequently, it often is added in amounts greater than 1 per cent. Manganese steel having about 1 to 1.5 per cent manganese and 0.9 to 1.0 per cent carbon has desirable nondistorting qualities when being hardened and is often used in die work. Sometimes 0.5 per cent chromium and vanadium are also added. When manganese is used in large percentages (11 to 14 per cent), an austenitic alloy known as *Hadfield steel* is produced. Its high hardness with good ductility, high strain hardening capability, and excellent wear resistance make it ideal for mining tools and similar applications.

As mentioned previously, *sulfur* usually is not desired in steel because of the embrittling effect of iron sulfide. In the form of manganese sulfide, however, sulfur is not harmful provided that the sulfides are not in large quantities and are well dispersed. If manganese sulfide is present in larger quantities and in proper form, it can impart desirable machinability properties. Therefore, some *free-machining steels,* which are to be machined automatically and are to be used for parts that will not be subjected to much impact, have 0.08 to 0.15 per cent sulfur added. The manganese content is usually increased to make certain that no iron sulfide is formed.

Nickel is added primarily for its increase in toughness and impact resistance, particularly at low temperature. It also lessens distortion in quenching, improves corrosion resistance, lowers the critical temperatures, and widens the temperature range for successful heat treatment. It is used in amounts of 2 to 5 per cent, often combined with other alloying elements to improve toughness. When 12 to 20 per cent nickel is used in steel with low carbon content, good corrosion resistance is provided. A steel with 36 per cent nickel has a thermal expansion coefficient of almost zero. Commonly known as *Invar,* this metal is used for measuring devices. Because of its high cost, nickel should only be used where it is uniquely effective, as in providing low-temperature impact resistance for cryogenic steels.

Although large percentages of *chromium* can impart corrosion resistance and heat resistance, in the amounts used in low-alloy steels, these effects are minor. Less than 2 per cent chromium usually is required, and often chromium and nickel are used together in a ratio of about 1 part chromium to 2 parts nickel. Chromium carbides often are desired for their superior wear resistance.

Molybdenum, as used in ordinary alloy steels, improves hardenability and increases strength properties, particularly under dynamic and high-temperature conditions. It tends to form stable carbides that persist at elevated temperatures, thereby retaining fine grain size and unusual toughness. Molybdenum steels are somewhat resistant to tempering and therefore maintain their strength and creep-resistant properties at elevated temperatures. Resistance to temper embrittlement is also noted.

Molybdenum is often used in conjunction with chromium, its amount seldom being in excess of 0.3 per cent. Molybdenum is used in larger amounts in tool steels because of its effect, similar to tungsten, of imparting hardness that persists at red heats. Within the temperature ranges where it is effective, it is about twice as potent as tungsten and much cheaper. It is commonly used in forging dies that must resist impact and abrasion at elevated temperature.

Vanadium is another alloying element that forms strong carbides that persist at elevated temperature. The carbides do not readily go into solution when the metal is heated prior to quenching, and therefore inhibit grain growth. The resulting effect of 0.03 to 0.25 per cent vanadium is an increase in strength properties, particularly the elastic limit, yield point, and impact strength, with almost no loss in ductility.

Tungsten is more effective than molybdenum in producing hardness at very high temperatures, again attributable to stable carbides. It thus is a primary alloying element in tool steels that must maintain their hardness at high operating temperatures. It also serves as a principal alloying element in some air-hardening steels.

Although the resistance of *copper* to atmospheric corrosion has been known for centuries, only recently has it been used as an addition to steel (in amounts from 0.10 to 0.50 per cent) to provide this property. It now is used extensively in low-carbon sheet steel, especially in thin gages, and in structural steels.

Silicon in small percentages has somewhat the same effect on steel as does nickel, increasing the strength properties, especially the elastic limit, with little loss in ductility. It is an important alloying element (usually 0.2 to 0.7 per cent) in certain high-yield-strength structural steels. It also is used in spring steels, which contain about 2 per cent silicon, 0.8 per cent manganese, and 0.6 per cent carbon. Another primary use of silicon is to promote the large grain size desirable for steels used for magnetic applications in electrical equipment.

Boron is a very powerful hardenability agent, being from 250 to 750 times as effective as nickel, 75 to 125 times as effective as molybdenum, and about 100 times as powerful as chromium. Only a few thousandths of a per cent are sufficient to produce the desired effect in low-carbon steels, but the results diminish rapidly with increasing carbon content. Since no carbide formation or ferrite strengthening is produced, improved machinability and cold-forming capability often result from the use of boron in place of other hardenability agents.

In addition to its use as a deoxidizer, *aluminum* may be added to steels in amounts of 0.95 to 1.30 per cent to produce a nitriding steel, as discussed in Chapter 6. *Titanium* and *niobium* (columbium) are other carbide formers. Steels with 0.15 to 0.35 per cent *lead* show substantially improved machinability.

Table 7-1 shows the basic effects of the common alloying elements. A working knowledge of the information contained in this table is useful to the design engineer in selecting an alloy steel for a given requirement. Of course, alloying elements are often used in combination, resulting in a large variety of

TABLE 7-1. Principal effects of major alloying elements in steel

Element	Percentage	Primary Function
Manganese	0.25–0.40	Combine with sulfur to prevent brittleness
	>1%	Increase hardenability, by lowering transformation points and causing transformations to be sluggish
Sulfur	0.08–0.15	Free-machining properties
Nickel	2–5	Toughener
	12–20	Corrosion resistance
Chromium	0.5–2	Increase hardenability
	4–18	Corrosion resistance
Molybdenum	0.2–5	Stable carbides; inhibits grain growth
Vanadium	0.15	Stable carbides; increases strength while retaining ductility; promotes fine grain size
Boron	0.001–0.003	Powerful hardenability agent
Tungsten		Hardness at high temperatures
Silicon	0.2–0.7	Increases strength
	2	Spring steels
	Higher percentages	Improve magnetic properties
Copper	0.1–0.4	Corrosion resistance
Aluminum	0.95–1.30	Alloying element in nitriding steels
Titanium	—	Fixes carbon in inert particles
		Reduces martensitic hardness in chromium steels
Lead	—	Improves machinability

alloy steels being available. To simplify matters, a classification system has been developed and is in general used in a variety of industries.

The AISI–SAE classification system. Undoubtedly the most important group of alloy steels, from the manufacturing viewpoint, is that designated by the AISI identification system. This system, which classifies alloys by chemistry, was started by the Society of Automotive Engineers (SAE) to provide some standardization of steels used in the automotive industry. It was later adopted and expanded by the American Iron and Steel Institute (AISI) and has become the most universal system in the United States. Both plain-carbon and low-alloy steels are identified by a four-digit number, the first number indicating the major alloying elements and the second number indicating a subgrouping of the major alloy system. Groupings by the first two digits is according to an arbitrary table. The last two digits indicate the approximate carbon content of the metal in "points" of carbon, where one point is equivalent to 0.01 per cent carbon. Table 7-2 presents the basic composition classification. As examples, a 1080 steel would be a plain-carbon steel with 0.80 per cent carbon. Similarly, a 4340 steel would be a Mo–Cr–Ni alloy with 0.40 per cent carbon.

A letter prefix may be used to indicate the process employed to produce the steel, such as basic open-hearth (C) or electric furnace (E). An X prefix is used to indicate permissible variations in the range of manganese, sulfur, or chromium. A letter B between the second and third digits indicates that the base metal has been supplemented by addition of boron.

The *H-grade AISI steels* are designated by the letter H as a suffix to the standard designation. These steels are for use where hardenability is a major requirement, slightly broader variations in the hardenability-producing elements being permitted. The steel is supplied to meet hardenability standards as specified by the customer in terms of hardness values at specific locations from the quenched end of a Jominy hardenability specimen.

Other systems of designation, such as the American Society for Testing and Materials (ASTM) and United States government (MIL and federal) specification systems focus more on specific applications. Acceptance for a given grade may be more on physical or mechanical properties than on chemistry of the metal. Many low-carbon structural and alloy steels are referred to by their ASTM designation.

Balanced alloy steels. From the previous discussion, it is apparent that two or more alloying elements may produce similar effects. Thus it is possible to obtain steels having almost identical properties although their chemical compositions are substantially different. This fact is strikingly demonstrated in Figures 7-1 and 7-2. Here the test data show that several steels of quite different compositions have almost identical property ratios *when heat-treated properly*. This fact should be kept in mind by all who select and specify the use of alloy steels. It is particularly important when one realizes that some alloying

TABLE 7-2. Some AISI–SAE standard steel designations

AISI Number	Type	Alloying Elements (%)					
		Mn	Ni	Cr	Mo	V	Other
1xxx	Carbon steels						
10xx	Plain carbon						
11xx	Free cutting (s)						
12xx	Free cutting (s and p)						
15xx	High manganese						
13xx	High manganese	1.60–1.90					
2xxx	Nickel steels		3.5–5.0				
3xxx	Nickel–chromium		1.0–3.5	0.5–1.75			
4xxx	Molybdenum						
40xx	Mo				0.15–0.30		
41xx	Mo, Cr			0.40–1.10	0.08–0.35		
43xx	Mo, Cr, Ni		1.65–2.00	0.40–0.90	0.20–0.30		
44xx	Mo				0.35–0.60		
46xx	Mo, Ni (low)		0.70–2.00		0.15–0.30		
47xx	Mo, Cr, Ni		0.90–1.20	0.35–0.55	0.15–0.40		
48xx	Mo, Ni (high)		3.25–3.75		0.20–0.30		
5xxx	Chromium						
50xx				0.20–0.60			
51xx				0.70–1.15			
6xxx	Chromium–vanadium						
61xx				0.50–1.10		0.10–0.15	
8xxx	Ni, Cr, Mo						
81xx			0.20–0.40	0.30–0.55	0.08–0.15		
86xx			0.40–0.70	0.40–0.60	0.15–0.25		
87xx			0.40–0.70	0.40–0.60	0.20–0.30		
88xx			0.40–0.70	0.40–0.60	0.30–0.40		
9xxx	Other						
92xx	High silicon						1.20–2.20 S:
93xx	Ni, Cr, Mo		3.00–3.50	1.00–1.40	0.08–0.15		
94xx	Ni, Cr, Mo		0.30–0.60	0.30–0.50	0.08–0.15		

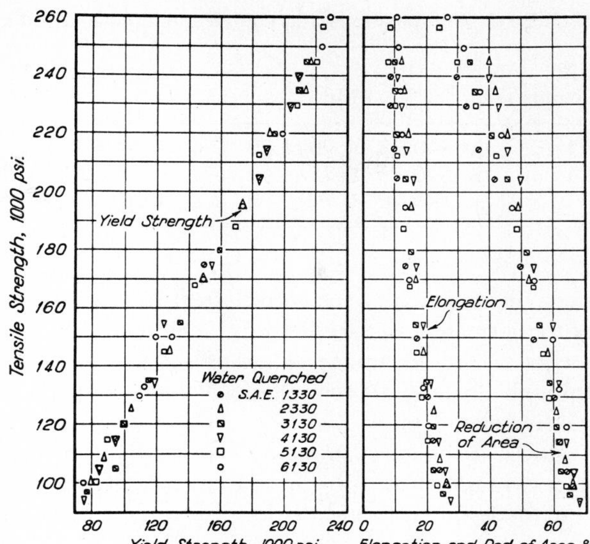

FIGURE 7-1. Straight-line relationship of mechanical properties of properly heat-treated SAE alloy steels. (*Courtesy American Society for Metals.*)

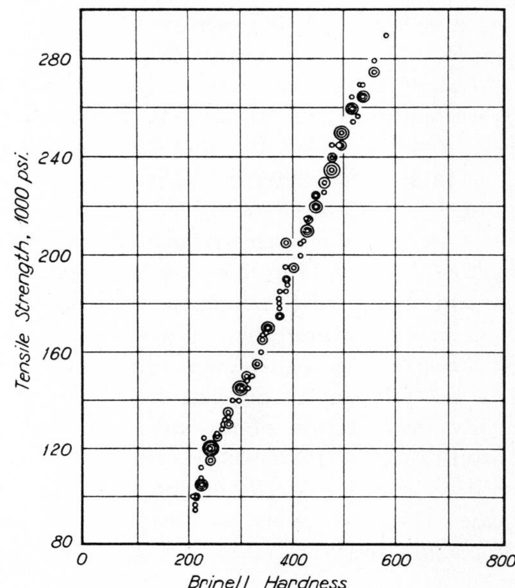

FIGURE 7-2. Relationship of hardness and tensile strength for a group of standard alloy steels. (*Courtesy American Society for Metals.*)

elements are much more costly than others, and that some may be in short supply in certain countries (the United States, for one) in times of emergency or as a result of political constraints. Overspecification is often employed to guarantee success in spite of sloppy manufacturing or heat-treatment prac-

TABLE 7-3. Compositions and equivalents of several EX steels

EX Number	C	Mn	Cr	Mo	Other	Equivalent AISI Grade
15	0.18–0.23	0.90–1.20	0.40–0.60	0.13–0.20	—	8620
24	0.18–0.23	0.75–1.00	0.45–0.65	0.20–0.30	—	8620
31	0.15–0.20	0.70–0.90	0.45–0.65	0.45–0.60	0.70-1.00Ni	4817

tices. In most cases, the correct steel to use is the least expensive one that can be heat-treated to satisfactorily achieve the desired properties. This usually means taking advantage of the effects provided by *all* the elements in a steel through "balanced" compositions that avoid needlessly large amounts of expensive elements.

An excellent example of what can be achieved is seen in the series of *EX steels* originated in 1963 by the Society of Automotive Engineers (SAE). The system designates new grades of wrought alloy steels on a temporary basis. Grades are removed from the list when they are either promoted to full SAE standard steel status or dropped for lack of interest. More than 50 of these EX steel compositions have been approved since 1963. Most have been designed to reduce the need for expensive alloying elements or those which are in short supply, notably nickel and chromium. Table 7-3 lists the compositions and equivalent standard grades (on the basis of hardenability) for several EX steels. By comparing the compositions of the EX steel and the equivalent composition from Table 7-2, the alloy savings become readily apparent.

In selecting alloy steels, it is important to keep usage in mind. For example, for one use it might be permissible to increase the carbon content in order to obtain greater strength. For a different usage, for example in welding, it would be better to keep the carbon content low and use a balanced amount of alloying elements to obtain the required strength without the cracking or fracture problems associated with possible higher-carbon martensitic structures. Steel selection often involves the defining of required properties, the selection of the best microstructure to provide those properties, and finally the selection of the steel with the proper carbon content and hardenability characterististics to achieve that goal. By not specifying exact composition, but by purchasing on the basis of properties, savings can often be made. The producer is free to supply any material that will meet the desired properties and can take advantage of the residual elements in increased amounts of recycled scrap. Reduced material costs result in lower steel prices.

High strength/low alloy (HSLA) structural steels. There are two general categories of alloy steels: the high strength/low alloy types, which rely largely on chemical composition to develop the desired mechanical properties in the as-rolled or normalized condition, and the constructional alloy steels, in which desired properties are developed by thermal treatment. Many manufactured products require steels with good hardenability, ductility, and fatigue strength—the constructional alloy steels. For structural applications, however, high yield strength, good weldability, and corrosion resistance are most desired, with only limited ductility and virtually no hardenability. Development of steels possessing these properties have made possible substantial weight savings in automobiles, trains, bridges, and buildings.

The low-alloy structural steels have about twice the yield strength of the plain-carbon structural steels. This increase in strength, coupled with resistance to martensite formation in a weld zone, is obtained by adding low percentages of several elements, notably manganese, silicon, niobium (columbium), and vanadium, as well as several others. About 0.2 per cent copper is usually added to improve corrosion resistance. Although numerous types of these steels have been obtained by the addition of alloying elements in various combinations and quantities, four of the more common ones are listed in Table 7-4.

These steels represent a significant contribution to the field of structural materials. Through their use, weight savings of 20 to 30 per cent have been achieved without any sacrifice in strength or safety. They currently are produced in sufficient tonnages that their cost is little more than that of ordinary grades of structural steel.

Quenched-and-tempered structural steel. The need for even stronger structural steels that can be welded, notably for use in submarines and pressure vessels, has led to the development of several alloys that are always used in the quenched-and-tempered condition. These steels have yield strengths in the

TABLE 7-4. Typical compositions and strength properties of several groups of low-alloy structural steels

	Chemical Composition[a] (%)					Strength Properties				
						Yield		Tensile		Elongation in 2 inches
Group	C	Mn	Si	Cb	V	MPa	ksi	MPa	ksi	(%)
Columbium or vanadium	0.20	1.25	0.30	0.01	0.01	379	55	483	70	20
Low manganese–vanadium	0.10	0.50	0.10		0.02	276	40	414	60	35
Manganese–copper	0.25	1.20	0.30			345	50	517	75	20
Manganese–vanadium–copper	0.22	1.25	0.30		0.02	345	50	483	70	22

[a] All have 0.04% P; 0.05% S; and 0.20% Cu.

range 550 to 1050 MPa (80 to 150 ksi); tensile strengths of 650 to 1400 MPa (95 to 200 ksi), and elongations in 2 inches of 13 to 20 per cent. The chemical compositions of most of these steels fall into two groups, one using nickel and low manganese and the other high manganese and silicon and some zirconium and boron but less chromium and no nickel. Typical compositions are:

Group A		Group B
0.18–0.20%	Carbon	0.15–0.21%
0.10–0.40%	Manganese	0.80–1.10%
0.025%	Phosphorus	0.035%
0.025%	Sulfur	0.04%
0.15–0.35%	Silicon	0.40–0.90%
2.00–3.50%	Nickel	
1.00–1.80%	Chromium	0.50–0.90%
0.06%	Molybdenum	0.28%
	Zirconium	0.05%
	Boron	0.0025%

Such steels are typically water-quenched from about 900°C (1650°F) and tempered at 625 to 650°C (1150 to 1200°F) to produce a tempered martensite structure. When welded they are still tough, even though not tempered after welding, because the resulting martensite has a very low carbon content. They have excellent impact resistance at low temperatures and good atmospheric corrosion resistance. Because of their superior strength properties, they usually permit considerable weight saving, which will offset the added material cost.

Free-machining steels. The increased use of high-speed machining, particularly on automatic machine tools, has spurred the development of several varieties of *free-machining steels.* These steels machine readily and form small chips so as to reduce the rubbing against the cutting tool and associated friction and heat. Formation of small chips also reduces the likelihood of chip entanglement in the machine and makes chip removal much easier.

These steels are basically ordinary carbon steels that have been modified in one of two ways. The first method is to increase the sulfur to a value of 0.08 to 0.33 per cent and the manganese to 0.7 to 1.6 per cent. The second modification utilizes 0.25 to 0.35 per cent sulfur and 0.15 to 0.35 per cent lead, with small amounts of tellurium, selenium, or bismuth. The additives cause the formation of inclusions, such as MnS, which act as discontinuities in the structure at which the chips will tend to break. It should be remembered, however, that these steels have somewhat reduced ductility and impact properties.

Alloy steels for electrical applications. There are two groups of alloy steels that are widely used in the electrical industry. The addition of small amounts of silicon, 0.5 to 5 per cent, in the *silicon steels* results in increased resistivity and permeability. Increased resistivity decreases eddy-current losses, while improved permeability decreases hysteresis losses. When such steel is used in electrical motors, generators, and transformers, the power losses and associated heat problems are reduced. For this reason, silicon steel frequently is used for the magnetic circuits of electrical equipment. For armature cores the silicon content is about 0.5 per cent, whereas in transformers about 5 per cent silicon is used. Silicon causes the steel to be brittle, so the amount used should be kept low where the steel is subjected to high rotative speeds.

Cobalt increases the magnetic saturation point in steel when it is used in amounts up to about 36 per cent. For this reason *cobalt alloy steels* are used in electrical equipment, where high magnetic densities must exist. For example, it is often used for the pole pieces of electromagnets. High-cobalt alloys are also used for most permanent magnets.

Corrosion-resistant or stainless steels. *Corrosion-resistant or stainless steels* contain sufficient amounts of chromium that they can no longer be considered low-alloy steels. The corrosion resistance is imparted by the formation of a strongly adherent chromium oxide on the surface. Good resistance to many corrosive media encountered in the chemical industry may be obtained by the addition of 4 to 6 per cent chromium to low-carbon steel. Usually, 0.4 to 0.8 per cent silicon and 0.5 per cent molybdenum are also added.

Where improved corrosion resistance and outstanding appearance are required, materials are designed to utilize a superior chromium oxide that forms when the amount of chromium in solution (excluding chromium carbides, etc.) exceeds 12 per cent. In this category, a variety of stainless steels exist, a major breakdown classification being on the basis of microstructural characteristics. The AISI designation scheme for these metals is a three-digit code that identifies family and particular alloy within the family.

Chromium is a ferrite stabilizer, the addition of chromium tending to decrease the temperature range over which austenite is stable. If sufficient chromium is added to iron, a material can be obtained that is ferritic at all temperatures below solidification. *Ferritic stainless steels* are therefore low carbon/high chromium alloys. They possess rather poor ductility or formability because of the b.c.c. crystal structure, but they are readily weldable. No martensite can form in these materials, because there is no possibility of austenite that can transform.

If the austenite region is not completely eliminated, stainless metals can be produced that are austenite at high temperature and ferrite at low. Such a metal can be quenched to form martensite and is known as a *martensitic stainless steel*. Carbon content can vary to produce the desired strength level, but chromium must be adjusted to assure more than 12 per cent in solution. In addition to the higher base cost, approximately 1½ times that of the ferritic material, martensitic stainlesses require more costly processing. They are usually annealed for fabrication and hardened by a full austenitize–quench–stress relieve–temper cycle.

Nickel is an austenite stabilizer, and with sufficient amounts of both chromium and nickel, it is possible to produce a metal in which austenite is the stable phase at room temperature, an *austenitic stainless steel.* From 3.5 to 22 per cent nickel is used and in some cases small amounts of molybdenum or titanium. The cost of this material is more than double that of the cheaper ferritic stainless.

Austenitic stainless steels are nonmagnetic and are highly corrosion resistant in almost all media except hydrochloric acid and other halide acids and salts. In addition, they may be polished to a mirror finish and thus combine attractive appearance with good corrosion resistance. Formability is outstanding (characteristic of the f.c.c. crystal structure), and they respond quite well to strengthening by cold work. The response of the popular 18–8 (18 per cent chromium and 8 per cent nickel) grade to a small amount of cold work is as follows:

	Water Quench	Cold-Rolled 15 per cent
Yield strength	260 MPa (38 ksi)	805 MPa (117 ksi)
Tensile strength	620 MPa (90 ksi)	965 MPa (140 ksi)
Elongation in 2 inches	68%	11%

One should note that these materials are often water-quenched to retain the alloys in solid solution—no transformation occurs since the stable phase is austenite for all temperatures involved.

Austenitic stainless steels are expensive metals and, although produced in large tonnages, they should not be specified where the less expensive ferritic or martensitic alloys would be adequate or where a true stainless steel is not required. Figure 7-3 lists several grades from each of the three major classifications and notes some key properties. Table 7-5 presents the typical compositions of the basic stainless classes.

Problems with the basic stainless grades generally relate to loss of corrosion resistance (sensitization) when the level of chromium in solution drops below 12 per cent. Since chromium depletion is usually caused by the formation of chromium carbides along grain boundaries, especially at elevated temperatures, one method of prevention is to keep the carbon content as low as possible, usually less than 0.10 per cent. Another approach is to tie up the carbon with small amounts of "stabilizing" elements such as titanium or columbium, alloys that have a stronger affinity to form carbides. Rapidly cooling these metals through the range 480 to 820°C (900 to 1500°F) also retards carbide formation.

Other stainless steels have been developed for special uses. Ordinary stainless steels are difficult to machine because of their work-hardening properties and the fact that they "seize" and thus prevent clean cutting. *Free-machining varieties,* produced by the addition of sulfur and molybdenum, are available that machine nearly as well as a medium-carbon steel.

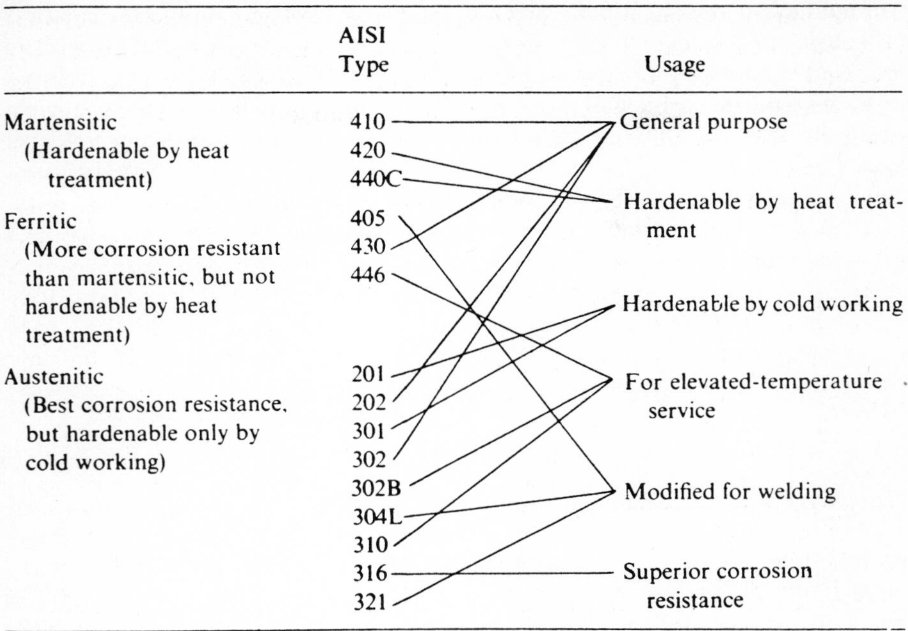

AISI
Type
Usage
Martensitic
(Hardenable by heat
treatment)
Ferritic
(More corrosion resistant
than martensitic, but not
hardenable by heat
treatment)
Austenitic
(Best corrosion resistance,
but hardenable only by
cold working)
410
420
440C
405
430
446
201
202
301
302
302B
304L
310
316
321
General purpose
Hardenable by heat treatment
Hardenable by cold working
For elevated-temperature service
Modified for welding
Superior corrosion resistance

FIGURE 7-3. Classification and uses of stainless steels.

Another class of stainless material is that of the *precipitation-hardening stainless steels.* These alloys are basically martensitic or austenitic types with low carbon, modified by the addition of alloying elements that permit age hardening at relatively low temperatures. Properties such as a 1790-MPa (260-ksi) yield strength and a 1825-MPa (265-ksi) tensile strength with 2 per cent elongation can be attained with none of the distortion associated with quenching. These grades, however, are among the most expensive stainlesses and should be used only when required.

TABLE 7-5. Typical compositions of the ferritic, martensitic, and austenitic stainless steels (values in parentheses are for one type)

Element	Ferritic	Martensitic	Austenitic
Carbon	0.08–0.20%	0.15–1.2%	0.03–0.25%
Manganese	1–1.5%	1%	2% (5.5–10%)
Silicon	1%	1%	1–2% (0%)
Chromium	11–27%	11.5–18%	16–26%
Nickel			3.5–22%
Phosphorus and sulfur			Normal (0%)
Molybdenum			Some cases
Titanium			Some cases

The maraging steels. When 18 to 25 per cent nickel and significant amounts of cobalt, molybdenum, and titanium are added to very low-carbon steel, a material results that air-quenches from about 815°C (1500°F) to form martensite, which, in turn, will age-harden at 480°C (900°F) to produce yield strengths in excess of 1725 MPa (250 ksi) with elongations in excess of 11 per cent. This material is very useful in such applications as rocket-motor cases, where high strength and good toughness are important.

A typical maraging steel in the 1725-MPa (250-ksi) strength class has a composition of:

0.03% C	0.10% Al
18.5% Ni	0.003% B
7.5% Co	0.10% Si maximum
4.8% Mo	0.10% Mn maximum
0.40% Ti	0.01% S maximum
0.01% Zr	0.01% P maximum

This steel can be hot-worked from 760 to 1260°C (1400 to 2300°F). Air cooled from 815°C (1500°F), it has a hardness of about 30 R_c and a structure of soft, tough martensite. It is easily machined and, because of its low work-hardening rate, can be cold-worked to a high degree. Aging at 480°C (900°F) for 3 to 6 hours, followed by air cooling, raises the hardness to about 52R_c and produces full strength. Maraging steels can be welded, if welding is followed by the full solution and aging treatment.

Steels for high-temperature service. Continued developments in rocketry, missiles, jet aircraft, and nuclear power have increased the need for metals that have good strength characteristics, corrosion resistance, and, particularly, creep resistance at high temperatures. Much work has been done to produce both ferrous and nonferrous alloys that have the desired properties at temperatures from 550 to 950°C (1000 to 1750°F). The nonferrous alloys will be discussed in a later chapter.

The ferrous alloys, which ordinarily are used below 760°C (1400°F), are low-carbon materials with less than 0.1 per cent carbon. One alloy is a modified 18–8 stainless steel stabilized with either columbium or titanium. A 1000-hour rupture stress of 40 to 50 MPa (6000 to 7000 psi) is observed at 760°C (1400°F), with considerably higher strengths at lower temperatures. Iron also is a major component of other high-temperature alloys, but when amounts become less than 50 per cent, the metal can hardly be classified as ferrous in nature. High strength at high temperature usually requires the more expensive nonferrous material.

Tool steels. *Tool steels* are metals designed to provide wear resistance and toughness combined with high strength. They are basically high-carbon alloys, where the chemistry provides the balance of toughness and wear desired.

Several classification or breakdown systems have been applied to tool steels, some using chemistry as a basis and another one employing hardening method and major mechanical property desired. The standard AISI–SAE designation system identifies letter grades by basic principles such as quenching method, primary application, special characteristic, or specific industry involved. Table 7-6 lists the seven basic types of tool steel and corresponding AISI–SAE grades. Individual alloys are then listed numerically within the grade to produce a letter–number identification system.

Water-hardening carbon tool steels (W grade) account for a large percentage of all tool steels used. They are the least expensive and are used for a wide variety of parts that are usually quite small and not subjected to severe usage or elevated temperatures. Because strength and hardness are functions of the carbon content, a wide range of these properties can be obtained through composition variation. These steels must be quenched in water to obtain high hardness and, because their hardenability is low, they can be used only for relatively light sections if full depth of hardness must be obtained. They also are rather brittle, particularly at higher hardness.

The uses of plain carbon steels, according to carbon content, are somewhat as follows:

0.60–0.75% carbon: machine parts, chisels, set screws, and the like, where medium hardness with considerable toughness and shock resistance is required.

0.75–0.90% carbon: forging dies, hammers, and sledges.

0.90–1.10% carbon: general-purpose tooling requirements, which

TABLE 7-6. Basic types of tool steel and corresponding AISI–SAE grades

Type	AISI–SAE Grade	
1. Water hardening	W	
2. Cold work	O	Oil-hardening
	A	Air-hardening medium-alloy
	D	High carbon/high chromium
3. Shock resisting	S	
4. High speed	T	Tungsten base
	M	Molybdenum base
5. Hot work	H	H1–H19: chromium base
		H20–H39: tungsten base
		H40–H59: molybdenum base
6. Plastic mold	P	
7. Special purpose	L	Low alloy
	F	Carbon–tungsten

require good balance of wear resistance and toughness, such as drills, cutters, shear blades, and other heavy-duty cutting edges.

1.10–1.30% carbon: small drills, lathe tools, razors, and similar light-duty applications, where extreme hardness is necessary without great toughness.

In applications where improved toughness is desired in conjunction with high strength and hardness, small amounts of manganese, silicon, and molybdenum are often added. Vanadium additions of about 0.20 per cent are often used to form strong stable carbides that retain fine grain size during heat treating. One of the main weaknesses of carbon tool steels is the fact that they do not hold their hardness at elevated temperatures. Prolonged exposure to temperatures over 150°C (300°F) results in undesired softening. This places a considerable limitation on their use.

When larger parts must be hardened or distortion must be minimized, *oil- or air-hardening grades* (O and A designations, respectively) are often employed. Because of the higher hardenability, they can be hardened by less severe quenches, maintaining close dimensional tolerances and minimizing the tendency to cracking. Manganese tool and die steels form one segment of this group. These contain 0.75 to 1.0 per cent carbon and 1.0 to 2.0 per cent manganese and are moderate in cost. In some cases, manganese is reduced and chromium, silicon, and nickel are added to give greater toughness. These steels are not as hard as the plain manganese types, but have less tendency to crack because of the greater hardenability and less severe quench requirement.

Chromium tool and die steels are also in this class. The low-chrome tool steels are much the same as plain carbon tool steels, with chromium added to produce the desired hardenability and toughness. Chromium levels are usually between 0.5 and 5.0 per cent. High-chromium tool steels are designated by a separate letter (D) and contain between 10 and 18 per cent chromium. These steels ordinarily must be annealed before they can be machined. After machining, they are hardened and tempered and will usually retain the final hardness at temperatures up to 425°C (800°F). They are used for tools and dies that must withstand hard usage over long periods, often at elevated temperatures. Forging dies, die-casting die blocks, and drawing dies are often made from such steel.

Shock-resisting tool steels (S designation) have been developed for both hot and cold impact applications. Low carbon contents (approximately 0.5 per cent) are used to provide toughness, with carbide-forming alloying elements being added to supply the desired abrasion resistance, hardenability, and hot-work characteristics.

High-speed tool steels are used primarily for cutting tools that machine metal and other special applications where retention of hardness at red heat is required. The most common type of high-speed steel is the tungsten-based T1 alloy, also known as 18-4-1 because the analysis contains 0.7 per cent carbon, 18 per cent tungsten, 4 per cent chromium, and 1 per cent vanadium. This alloy has a balance combination of shock resistance and abrasion resistance and is used for a wide variety of cutting applications. Other high-speed tool steels have cobalt added to improve hardness at elevated temperatures. Toughness diminishes and forming problems increase.

The *molybdenum high-speed steels* (M designation) were developed to reduce the amount of tungsten and chromium required to produce the desired properties. The M2 variety is now the most widely used high-speed steel, its higher carbon content and balanced analysis producing properties applicable to all general-purpose high-speed uses.

Hot-work tool steels (H designation) are designed to perform adequately under environments of prolonged high temperature. All employ additions of carbide-forming alloying elements: H1 to H19 are chromium-base types with about 5.0 per cent chromium; H20 to H39 are tungsten-base types with 9 to 18 per cent tungsten coupled with 3 to 4 per cent chromium; and H40 to H59 are molybdenum-based types.

Other types of tool steels include (1) the *plastic mold steels* (P designation) made specifically for the requirements of zinc die casting and plastic molding dies, (2) the *low-alloy special-purpose tool steels* (L designation), such as the L6 extreme toughness variety; and (3) the *carbon–tungsten type* of special-purpose tool steels (F designation), which are water hardening but substantially more wear-resistant than the plain-carbon tool steels.

Alloy cast steels and irons. The effects of alloying elements in steel are the same regardless of the method used to produce the final shape. To couple the attributes of the casting process and the benefits of alloy material, many alloy cast steels are produced. In the case of cast iron, however, alloys often perform functions in addition to those previously cited.

Whereas alloy steels are usually heat-treated, many alloy cast irons are not, except for stress relieving or annealing. If a cast iron is to be hardened, chromium, molybdenum, and nickel are frequently added to improve hardenability. In addition, chromium tends to offset the undesirable quenching effects of high carbon, and molybdenum and nickel offset the high silicon.

If alloy cast irons are not to be heat-treated, the alloy elements are often selected to alter properties through affecting the formation of graphite or cementite, modifying the morphology of the carbon-rich phase, or simply strengthening the matrix material. Alloys are often added in small amounts to improve strength properties or wear resistance. High-alloy cast irons are often designed to provide corrosion resistance, particularly at high temperatures such as are encountered in the chemical industry.

Nickel promotes graphite formation and tends to produce a finer graphite structure. Chromium, on the other hand, retards graphitization and stabilizes the cementite. They are frequently used together in the ratio of 2 to 3 parts of nickel to 1 part of chromium.

If the silicon content of gray cast iron is lowered, the strength is increased considerably. However, if this is carried very far, hard, white cast iron results. The addition of about 2 per cent nickel will minimize the formation of white cast iron. By adjusting the silicon content and adding a small amount of nickel, a cast iron of good strength is obtained without sacrificing machinability.

Molybdenum strengthens gray cast iron to some extent and often forms carbides. In addition, it is used to control the size of the graphite flakes. From 0.5 to 1.0 per cent is ordinarily added.

Among the high-alloy cast irons, *austenitic gray cast iron* is quite common. This contains about 14 per cent nickel, 5 per cent copper, and 2.5 per cent chromium. It possesses good corrosion resistance to many acids and alkalis at temperatures up to about 800°C (1500°F).

Review questions

1. What elements, other than iron and carbon, are normally found in plain-carbon steel?
2. What are the three primary groups of plain-carbon steels and the characteristics of each?
3. What are some of the common alloy elements in alloy steels?
4. What are some of the reasons for adding alloying elements to steel?
5. Why is sulfur more desirable in the form of manganese sulfide rather than as iron sulfide?
6. Which alloy elements are basically carbide formers?
7. What are the primary benefits of nickel, chromium, and copper as alloying elements in steel?
8. How is boron a unique alloying element?
9. What elements aid in strength retention at high temperatures?
10. What element is commonly used to improve the magnetic properties of steels used in electrical machinery?
11. Explain the AISI–SAE designation system for alloy steels.
12. What is an AISI 1040 steel?
13. What is the H-grade series of AISI steels?
14. How do ASTM, MIL and federal specifications differ from the AISI–SAE system?
15. What is the EX series of alloy steels?
16. What are some of the guidelines for selection of alloy steels?
17. What are HSLA steels? What are the property goals of these alloys?
18. How do the quenched-and-tempered structural steels maintain ductility and weldability with a martensitic structure?
19. What are free-machining steels? How are they produced?
20. Why do stainless steels lose their corrosion resistance when the chromium in solution drops below 12 per cent?
21. What is a ferritic stainless steel? A martensitic stainless? An austenitic stainless?

22. What properties are characteristic of each of the three basic types of stainless steels?

23. What are the relative costs of austenitic, ferritic, martensitic, and precipitation-hardening stainless steels?

24. What is sensitization of a stainless steel?

25. What is a maraging steel and how do they obtain their high strengths?

26. Explain the AISI–SAE designation system for tool steels.

27. What is the difference between water, oil, and air-hardening tool steels?

28. What alloying elements are used to provide high-speed or hot-work characteristics?

29. What are the primary alloy elements in alloy cast iron and their useful characteristics?

Case study 6.
THE SHORT-LIVED GEAR

A 250-mm (10-inch)-diameter gear was fabricated from AISI 1080 steel. The gear blank was hot-forged, air-cooled, and then full-annealed in preparation for machining. Following finish machining, the gear teeth were surface-hardened by flame hardening and quenching to a hardness of R_c 55. After a short period of service, the teeth began to deform in a manner consistent with subsurface flow, and the gear failed to mesh properly. What is a probable cause of the failure?

CHAPTER 8

Nonferrous alloys

Nonferrous metals and alloys are playing increasingly important roles in modern technology. Because of their number and the fact that the properties of the individual metals vary widely, either in their relatively pure form or as base metals for alloys, they provide an almost limitless range of properties for the design engineer. Even though they are not produced in as great tonnages and are more costly than iron and steel, they make available certain important properties or combinations of properties that cannot be obtained in steels, notably:

1. Resistance to corrosion.
2. Ease of fabrication.
3. High electrical and thermal conductivity.
4. Light weight.
5. Color.

While it is true that corrosion resistance can be obtained in certain ferrous alloys, several of the nonferrous alloys possess this property without requiring special and expensive alloying elements. Nearly all the nonferrous alloys possess at least two of the qualities listed above, and some possess all five. For many applications, certain combinations of these properties are highly desirable, and the availability of materials that provide them directly is a strong motivation for the use of nonferrous alloys.

In most cases, the nonferrous alloys are inferior to steel in respect to strength. Also, the modulus of elasticity may be considerably lower, a fact that places them at a disadvantage where stiffness is a necessary property. Fabrication, however, is usually easier then for steel. Those alloys with low melting points are often easy to cast, either in sand molds, permanent molds, or dies. Many alloys have high ductility coupled with low yield points, the ideal conditions for easy cold work and high formability. High machinability is characteristic of several nonferrous alloys. Fabrication savings can often overcome the higher cost of the nonferrous material and favor its use in place of steel. The one fabrication area in which the nonferrous alloys are somewhat inferior to steel is weldability. Recent developments, however, have made it possible to produce satisfactory weldments from the viewpoint of both quality and economy.

COPPER-BASE ALLOYS

As was discussed in Chapter 4, copper is seldom used in its pure state except in the electrical industry. For other engineering applications, it is almost always used in the form of an alloy with such elements as zinc, tin, nickel, aluminum, silicon, or beryllium. These alloys now are commonly identified through a system of copper alloy numbers developed by the Copper Development Association (CDA). Table 8-1 presents a breakdown of this system, which has been adopted by the ASTM, SAE, and the U.S. government.

Copper–zinc alloys. Zinc is by far the most popular alloying addition to copper, the alloys of copper and zinc being commonly known as *brasses.* If the zinc content is not over 36 per cent, brass is a single-phase solid solution of zinc in copper. Since this is identified as the alpha phase, these alloys are often called *alpha brasses.* They are quite ductile and formable, these characteristics increasing with the zinc content up to 36 per cent. Above 36 per cent zinc the alloys enter a two-phase region involving a brittle beta phase. Cold-working properties are rather poor for the high-zinc brasses, but deformation is easy when performed hot. Like copper, brass is hardenable by cold working and, in commercial grades, is available in various degrees of hardness.

Table 8-2 lists some of the most common copper–zinc alloys, their compositions, properties, and typical uses. Brasses range in color from copper to nearly white, the lower-zinc brasses being more coppery than those with more

TABLE 8-1. Standard designations for copper and copper alloys (CDA system)

Wrought Alloys		Cast Alloys	
100–155	Commercial coppers	833–838	Red brasses and leaded red brasses
162–199	High-copper alloys	842–848	Semi-red brasses and leaded semi-red brasses
200–299	Copper–zinc alloys (brasses)	852–858	Yellow brasses and leaded yellow brasses
300–399	Copper–zinc–lead alloys (leaded brasses)	861–868	Manganese and leaded manganese bronzes
400–499	Copper–zinc–tin alloys (tin brasses)	872–879	Silicon bronzes and silicon brasses
500–529	Copper–tin alloys (phosphor bronzes)	902–917	Tin bronzes
532–548	Copper–tin–lead alloys (leaded phosphor bronzes)	922–929	Leaded tin bronzes
600–642	Copper–aluminum alloys (aluminum bronzes)	932–945	High-leaded tin bronzes
647–661	Copper–silicon alloys (silicon bronzes)	947–949	Nickel–tin bronzes
667–699	Miscellaneous copper–zinc alloys	952–958	Aluminum bronzes
700–725	Copper–nickel alloys	962–966	Copper–nickels
732–799	Copper–nickel–zinc alloys (nickel silvers)	973–978	Leaded nickel bronzes

zinc. The addition of a third element, however, can change color considerably.

Most brasses have good corrosion resistance. In the 0 to 40 per cent zinc region, the addition of a small amount of tin produces excellent resistance to seawater corrosion. Cartridge brass with tin becomes admiralty brass; muntz metal with tin is known as naval brass. Brasses with 20 to 36 per cent zinc are subject to a selective corrosion, known as *dezincification,* when in acid or salt solutions. Another corrosion problem in brasses with more than 15 per cent zinc is season cracking or stress-corrosion cracking. Both stress and exposure to corrosive media are required for this failure to occur. Thus brasses often undergo a stress relief to remove the residual stresses induced by cold working prior to being put into service.

Where high machinability is needed, as in automatic screw-machine stock, 2 to 3 per cent lead is added to brass to assure production of free-breaking chips.

Many uses of brass relate to the high electrical and thermal conductivities coupled with adequate strength. Plating characteristics are outstanding and make brass an excellent base for decorative chrome or similar coatings. A unique property of alpha brass is its ability to have rubber vulcanized to it without any special treatment except thorough cleaning. It is widely used in mechanical rubber goods because of this property.

An alloy containing from 50 to 55 per cent copper and the remainder

TABLE 8-2. Composition, properties, and uses of some common copper–zinc alloys

CDA Number	Common Name	Composition (%) Cu	Zn	Sn	Pb	Mn	Condition	Tensile Strength MPa	(ksi)	Elongation in 2 in. (%)	Typical Uses
220	Commercial bronze	90	10				Soft sheet	262	38	45	Screen wire, hardware, screws, jewelry
							Hard sheet	441	64	4	
							Spring	503	73	3	
240	Low brass	80	20				Annealed sheet	324	47	47	Drawing, architectural work, ornamental
							Hard	517	75	7	
							Spring	627	91	3	
260	Cartridge brass	70	30				Annealed sheet	365	53	54	Munitions, hardware, musical instruments, tubing
							Hard	524	76	7	
							Spring	634	92	3	
270	Yellow brass	65	35				Annealed sheet	317	46	64	Cold forming, radiator cores, springs, screw
							Hard	524	76	7	
280	Muntz metal	60	40				Hot rolled	372	54	45	Architectural work, condenser tube
							Cold rolled	551	80	5	
443–445	Admiralty metal	71	28	1			Soft	310	45	60	Condenser tube (salt water), heat exchangers
							Hard	655	95	5	
360	Free-cutting brass	61.5	35.5		3		Soft	324	47	60	Screw-machine parts
							Hard	427	62	20	
675	Manganese bronze	58.5	39	1		0.1	Soft	448	65	33	Clutch disks, pump rods, valve stems, high-strength propellers
							Bars, Half Hard	579	84	19	

zinc is often used for filler metal in brazing. It is an effective agent for joining steel, cast iron, brasses, and copper, producing joints that are nearly as strong as those obtained by welding.

Copper–tin alloys. Alloys of copper and tin, commonly called tin *bronzes,* are considerably more expensive than the brasses because of the high price of tin. Consequently, they have now been replaced to a considerable degree by less expensive nonferrous alloys, but are used in certain applications because of special properties.

The term "bronze" is somewhat confusing, since some alloys that contain no tin are called bronzes because of their color. The true bronzes usually contain less than 12 per cent tin. Strength increases with tin content up to about 20 per cent, beyond which the alloys become brittle. Copper–tin alloys are characterized by good strength, toughness, wear resistance, and corrosion resistance. They are often used for bearings, gears, and fittings that are subjected to heavy compressive loads. When used as bearing material, up to 10 percent lead is often added.

The most popular wrought bronze is phosphor bronze, which usually contains from 1 to 11 per cent tin. Alloy 521 (CDA) is typical of this class and contains 92 per cent copper, 8 per cent tin, and 0.15 per cent phosphorus. Hard sheet of this material has a tensile strength of 758 MPa (110ksi) and an elongation in 2 inches of 3 per cent. Soft sheet has a 379-MPa (55-ksi) tensile strength and 65 per cent elongation. It is used for pump parts, gears, springs, and bearings.

Alloy 905 is a commonly used cast bronze containing 88 per cent copper, 10 per cent tin, and 2 per cent zinc. In the cast condition, the tensile strength is about 310 MPa (45 ksi), with an elongation of 25 per cent in 2 inches. It has very good resistance to seawater corrosion and is used on ships for pipe fittings, gears, pump parts, bushings, and bearings.

Copper–nickel alloys. Copper and nickel have complete solid solubility as seen in Figure 5-6, and a wide range of alloys have been developed. High thermal conductivity coupled with corrosion resistance make these materials a good choice for heat exchangers, cookware, and other heat-transfer applications. *Cupro-nickels* contain 2 to 30 per cent nickel. *Nickel silvers* have 10 to 30 per cent nickel and at least 5 per cent zinc. An alloy with 45 per cent nickel is known as *constantan,* and a 67 per cent nickel alloy is called *Monel.*

Other copper-base alloys. The alloys previously discussed acquire their strength primarily through solid-solution strengthening and cold work. In the copper alloy family, three alloying elements produce materials that are precipitation hardenable: aluminum, silicon, and beryllium.

Aluminum bronze alloys usually contain between 6 and 12 per cent aluminum and often 2 to 5 per cent iron. With aluminum contents below 8 per cent, the alloys are very ductile. With more than 9 per cent, hardness approaches that of steel, and higher aluminum contents result in brittle, but very wear resistant, materials. A cast aluminum bronze having 86.2 per cent copper, 10.2 per cent aluminum, and 3.3 per cent iron has a tensile strength of 480 to 550 MPa (70 to 80 ksi) and 18 to 22 per cent elongation in 2 inches. By varying the aluminum content and the heat treatment, the tensile strength may be varied from about 415 to 860 MPa (60 to 125 ksi). Large amounts of shrinkage occur in parts cast in aluminum bronze. Castings of this material should be designed with this in mind.

Silicon bronzes contain up to 4 per cent silicon and 1.5 per cent zinc (higher zinc contents are used when the material is cast). Strength, formability, machinability, and corrosion resistance are quite good. Tensile strengths can approach 900 MPa (130 ksi) with cold work, whereas the soft material has a tensile strength of about 380 MPa (55 ksi) and 65 per cent elongation. Uses include boiler, stove, and tank applications requiring high strength with corrosion resistance.

Copper–beryllium alloys can be age-hardened to produce the highest strengths of the copper-based metals. They ordinarily contain less than 2 per cent beryllium, but are quite expensive. When annealed the material has a yield strength of 172 MPa (25 ksi), tensile strength of 482 MPa (70 ksi), and elongation of 50 per cent. After heat treatment, these properties can rise to 760 MPa (110 ksi), 1345 MPa (195 ksi), and 3 per cent, respectively. The modulus of elasticity is about 125,000 MPa (18,000,000 ksi) and the endurance limit is around 275 MPa (40 ksi). These properties make the material excellent for springs, but cost limits application to small components requiring long life and high reliability. Other applications relate to the unique properties: strength of steel, nonsparking, nonmagnetic, and conductive.

ALUMINUM ALLOYS

Aluminum has become the most important of the nonferrous metals. Pure aluminum is outstanding for its light weight, high thermal and electrical conductivity, and corrosion resistance. In the annealed condition, however, pure aluminum has only about one-fifth the strength of ordinary structural steel, and its modulus of elasticity—a property that can be modified only slightly by alloying and not at all by heat treatment—is only one-third that of steel. Although it costs more per pound than steel, it is only a little over one-third as heavy, thereby making it cheaper per unit volume. Corrosion resistance is far superior to that of ordinary steel.

Electrical-conductor grade of aluminum is used in large quantities and has replaced copper in many electrical transmission lines and for bus bars. This grade, commonly designated by the letters EC, contains a minimum of 99.45 per cent aluminum and has an electrical conductivity 62 per cent that of copper for the same size wire and 200 per cent that of copper on an equal-weight basis.

Aside from its electrical uses, most aluminum is used in the form of alloys. These have much greater strength than pure aluminum, yet retain the advantages of light weight, good conductivity, and corrosion resistance. Some alloys are available that have tensile properties, except for ductility, superior to those of low alloy/high yield strength structural steel. On a strength-to-weight basis, most of the aluminum alloys are superior to steel, but wear, creep, and fatigue properties are usually somewhat poorer. The selection between steel and aluminum for any given application is largely a matter of cost, although in many cases the advantages of reduced weight or corrosion resistance may justify additional expense. In most cases, aluminum replaces steel or cast iron, where the need for lightness, corrosion resistance, low maintenance expense, or high thermal or electrical conductivity offsets the added cost.

Corrosion resistance of aluminum and its alloys. Pure aluminum is very reactive and forms a tight, adherent oxide coating on the surface as soon as it is exposed to air. This oxide is resistant to many corrosive media and serves as a corrosion-resistant barrier to protect the underlying aluminum. When alloying elements are added to aluminum, oxide formation is somewhat retarded, so the alloys, in general, do not have quite the superior corrosion resistance of pure aluminum.

The oxide coating on aluminum alloys causes some difficulty in relation to its weldability. In resistance welding, it usually is necessary to remove the oxide immediately before welding in order to obtain consistent results. In fusion welding processes, the aluminum oxidizes so readily that it is necessary to use special fluxes or protective inert-gas atmospheres. Suitable welding techniques have been developed to enable aluminum to be welded with complete success, from both quality and cost viewpoints.

Wrought aluminum alloys. *Wrought aluminum-base alloys* can be divided into two basic types: those that achieve strength by *solid solution alloying and work hardening* and those that can be *precipitation-hardened.* Table 8-3 lists some common wrought aluminum alloys, using the standard four-digit designation system for aluminums. The first digit indicates the major alloy group as follows:

Major Alloying Element	
Aluminum, 99.00% and greater	1xxx
Copper	2xxx
Manganese	3xxx
Silicon	4xxx
Magnesium	5xxx
Magnesium and silicon	6xxx
Zinc	7xxx
Other element	8xxx

The second digit indicates modifications of the original alloy or impurity limits, and the last two digits identify the particular alloy or indicate the aluminum purity.

Further specification of the condition of the alloy is done through the *temper* designation, a letter or letter–number suffix that follows the alloy-designation number code. Symbols and their meanings are as follows:

-F: as fabricated, as in casting
-H: strain-hardened
  -H1: strain-hardened by working to desired dimensions; a second digit, through 9, indicates the degree of hardness, 8 being commercially full-hard and 9 extra-hard
  -H2: strain-hardened by cold working, following by partial annealing; second-digit numbers 2 through 8, as above
  -H3: strain-hardened and stabilized
-O: annealed
-T: thermally treated (heat treated)
  -T1: cooled from hot working and naturally aged
  -T2: cooled from hot working, cold-worked, and naturally aged
  -T3: solution-heat-treated, cold-worked, and naturally aged
  -T4: solution-heat-treated and naturally aged
  -T5: cooled from hot working and artificially aged
  -T6: solution-heat-treated and artificially aged
  -T7: solution-heat-treated and stabilized
  -T8: solution-heat-treated, cold-worked, and artificially aged
  -T9: solution-heat-treated, artificially aged, and cold-worked
  -T10: cooled from hot working, cold-worked, and artificially aged
-W: solution-heat-treated only

Additional digits beyond those listed above indicate variations of the basic temper.

It can be noted from Table 8-3 that the work-hardenable alloys are primarily those in the 1000 (pure aluminum), 3000 (aluminum–manganese), and 5000 (aluminum–magnesium) series. Within these series the 1100, 3003, and 5052 alloys tend to be the most popular. Strength tends to increase with increasing alloy number (1100→3003→5052), but ductility decreases with increasing strength.

Because of their higher strengths, the precipitation-hardenable alloys are more numerous than those that are work-hardenable and are found primarily in the 2000, 6000, and 7000 series. Alloy 2017, the original *duralumin,* is probably the oldest hardenable aluminum. The 2024 aluminum alloy is stronger than 2017 and is widely used in aircraft application. Also, the ductility of the 2000 series alloys does not decrease significantly with the strength increases produced by heat treatment. The more recently developed precipitation-hardenable alloys are of the 7075 type. In the heat-treated condition, these alloys have yield strengths that approach or exceed those of the

TABLE 8-3. Compositions, typical properties, and designations of some wrought aluminum alloys

Designation[a]	Composition (%) Aluminum = Balance					Form Tested	Tensile Strength		Yield Strength[b]		Elongation in 2 in. (%)	Brinell Hardness	Uses and Characteristics
	Cu	Si	Mn	Mg	Others		ksi	MPa	ksi	MPa			
Work-hardening alloys—not heat-treatable													
1100–0	0.12				99 Al	1/16″ sheet	13	90	5	34	35	23	Commercial Al: good forming properties
1100–H14						1/16″ sheet	16	110	14	97	9	32	Good corrosion resistance, low yield strength
1100–H18						1/16″ sheet	24	165	21	145	5	44	Cooking utensils; sheet and tubing
3003–0	0.12		1.2			1/16″ sheet	16	110	6	41	30	28	Similar to 1100
3003–H14						1/16″ sheet	22	152	21	145	8	40	Slightly stronger and less ductile
3003–H18						1/16″ sheet	29	200	27	186	4	55	Cooking utensils; sheetmetal work
5052–0				2.5	0.25 Cr	1/16″ sheet	28	193	13	90	25	45	Strongest work-hardening alloy
5052–H32						1/16″ sheet	33	228	28	193	12	60	High yield strength and fatigue limit
5052–H36						1/16″ sheet	40	276	35	241	8	73	Highly stressed sheetmetal products
Precipitation-hardening alloys—heat-treatable													
2017–0	4.0	0.5	0.7	0.6		1/16″ sheet	26	179	10	69	20	45	Duralumin, original strong alloy
2017–T4						1/16″ sheet	62	428	40	276	20	105	Hardened by quenching and aging
2024–0	4.4		0.6	1.5		1/16″ sheet	27	186	11	76	20	42	Stronger than 2017
2024–T4						1/16″ sheet	64	441	42	290	19	120	Used widely in aircraft construction
2014–0	4.4	0.8	0.8	0.5		½″ extruded shapes	27	186	14	97	12	45	Strong alloy for extruded shapes

2014–T6						Forgings	65	448	55	379	10	125	Strong forging alloy
2014–T6						1/16" sheet	70	483	60	413	8		Higher yield strength than Alclad 2024
Alclad 2014–T6	4.5	1.0	0.8	0.4		1/16" sheet	63	434	56	386	7		Clad with heat-treatable alloy[c]
7075–0	1.6		0.2	2.5	0.3 Cr 5.6 Zn	1/16" sheet	33	228	15	103	17	60	Alloy of highest strength
7075–T6						1/16" sheet	76	524	67	462	11	150	Lower ductility than 2024
Alclad 7075–T6						1/16" sheet	76	524	67	462	11		Strongest Alclad product
7075–T6						½" extruded shapes	80	552	70	483	6		Strongest alloy for extrusions
6061–T6	0.28	0.6		1.0	0.20 Cr	½" extruded shapes	42	290	40	276	12	95	Strong, corrosion resistant
6063–T6		0.4		0.7		½" rod extruded	35	241	31	214	12	80	Good forming properties and corrosion resistance
6151–T6		0.9		0.6	0.25 Cr	Forgings	48	331	43	297	17	90	For intricate forgings
2025–T6	4.5	0.8	0.8			Forgings	55	379	30	207	18	100	Good forgeability, lower cost
2018–T6	4			0.7	2 Ni	Forgings	55	379	40	276	10	100	Strong at elevated temperatures; forged pistons
4032–T6	0.9	12.2		1.1	0.9 Ni	Forgings	55	379	46	317	9	115	Forged aircraft pistons
2011–T3	5.5			(0.5 Bi)	0.5 Pb	½" rod	55	379	43	297	15	95	Free cutting, screw machine products

[a] 0 = annealed; T = quenched and aged, H-cold rolled to hard temper.
[b] Yield strength taken at 0.2% permanent set.
[c] Cladding alloy: 1.0 Mg, 0.7 Si, 0.5 Mn.

high-yield-strength structural steel. Ductility, however, is less than that for steel, and fabrication is more difficult than for the 2024 alloy. Nevertheless, these alloys are widely used in aircraft.

The heat-treatable alloys tend to have poorer corrosion resistance than either pure aluminum or the work-hardenable alloys. Thus, where both high strength and superior corrosion resistance are needed, the wrought aluminum is often produced as *Alclad.* A thin layer of corrosion-resistant aluminum is bonded to one or both surfaces of the high-strength metal during rolling and the material is further processed as a composite. Galvanic protection retards corrosion even when the metal is severely scarred.

Because only moderate temperatures are required to lower the strength so that plastic flow occurs readily, aluminum-alloy extrusions and forgings are relatively easy to produce and are manufactured in large quantities. Deep drawing operations can also be carried out easily by using only moderately elevated temperatures. In general, the high ductility and low yield strength of the aluminum alloys make them appropriate for almost all forming operations. Good dimensional tolerances and fairly intricate shapes can be produced with relative ease.

The machinability of aluminum-base alloys varies greatly. Most cast alloys are machined easily. For the wrought alloys, with the exception of a few special types, special tools and techniques are desirable if larger-scale machining is to be done. Free-machining alloys, such as 2011, have been developed for screw-machine work. They can be machined at very high speeds and have replaced brass screw-machine stock in many cases.

Aluminum casting alloys. Although its low melting temperature tends to make it suitable for casting, pure aluminum is seldom cast. Its high shrinkage and susceptibility to hot cracking cause considerable difficulty and scrap is high. By adding small amounts of alloying elements, however, very suitable casting characteristics are obtained and strength is increased. Large amounts of aluminum alloys are cast, the principal alloying elements being copper, silicon, and zinc. Table 8-4 lists the common commercial aluminum casting alloys and employs the rather new designation system of the Aluminum Association. The first digit indicates the alloy group as follows:

Major Alloying Element	
Aluminum, 99.00% and greater	1xx.x
Copper	2xx.x
Silicon with Cu and/or Mg	3xx.x
Silicon	4xx.x
Magnesium	5xx.x
Zinc	7xx.x
Tin	8xx.x
Other elements	9xx.x

The second and third digits identify the particular alloy or aluminum purity, and the last digit, separated by a decimal point, indicates the product form (that is, casting, ingot, etc.). A modification of the original alloy is indicated by a letter before the numerical designation.

Alloys are designed for both properties and process. Where strength requirements are low, as-cast properties are employed. High-strength castings usually require the use of an alloy that can be subsequently heat-treated. Sand casting has the fewest process restrictions. The aluminum alloys used for permanent-mold castings are designed to have lower coefficients of thermal expansion (or contraction) because the molds offer restraint to the dimensional changes that occur upon cooling. Die-casting alloys require high degrees of fluidity and "castability," because they are often cast into thin sections. Moreover, since die castings ordinarily are not heat-treated, the alloys used are designed to produce rather high "as-cast" strength under rapid cooling conditions. Several of the permanent-mold and die-casting alloys have tensile strengths above 275 MPa (40 ksi).

MAGNESIUM-BASE ALLOYS

The primary reason for the extensive, but specialized, use of *magnesium-base alloys* is their light weight—about 1.74 grams per cubic centimeter, compared with 2.7 for aluminum and 7.8 for iron or steel. Thus, whereas aluminum alloys are most suitable for strength members of mechanically motivated structures, such as airplanes, trains, and trucks, magnesium alloys are best suited for those applications where lightness is the first consideration, with strength as a secondary requirement. Of course, the applications of aluminum and magnesium overlap to some degree.

The designation system for magnesium alloys is not as well standardized as in the case of steels or aluminums, but most producers follow a system using one or two prefix letters, two or three numerals, and a suffix letter. The prefix letters designate the two principal alloying metals according to the following format:

A	aluminum	H	thorium	Q	silver
B	bismuth	K	zirconium	R	chromium
C	copper	L	beryllium	S	silicon
D	cadmium	M	manganese	T	tin
E	rare earth	N	nickel	Z	zinc
F	iron	P	lead		

Aluminum, zinc, zirconium, and thorium promote precipitation hardening; manganese improves corrosion resistance; and tin improves castability. Aluminum is the most common alloying element. The numerals correspond to the rounded-off percentages of the two main alloy elements and are arranged in the same order as the letters. The suffix letter distinguishes between different

TABLE 8-4. Composition, designations, and properties of aluminum casting alloys

Alloy Designation[a]	Process[b]	Composition (%) (Major Alloys > 1%) Cu	Si	Mg	Zn	Fe	Other	Temper	Tensile Strength MPa	ksi[c]	Elongation in 2 in. (%)	Uses, characteristics, etc.
208	S	4.0	3.0		1.0	1.2		F	131	19	1.5	General-purpose sand castings, can be heat treated
213	P	7.0	2.0		2.5	1.2		F	131	19	—	General purpose alloy, not heat-treatable
222	S & P	10.0	2.0			1.5		T61	207	30	—	Withstands elevated temperatures
242	S & P	4.0		1.6		1.0	2.0 Ni	T61	276	40	—	Withstands elevated temperatures
295	S	4.5	1.0			1.0		T6	221	32	3.0	Structural castings, heat-treatable
B295	P	4.5	2.5			1.2		T6	241	35	2.0	Permanent-mold version of 295
308	P	4.5	5.5		1.0	1.0		F	166	24	—	General-purpose permanent mold
319	S & P	3.5	6.0		1.0	1.0		T6	214	31	1.5	Superior casting characteristics
328	S	1.5	8.0		1.5	1.0		T6	234	34	1.0	High strength and pressure tightness
A332	P	1.0	12.0	1.0		1.2	2.5 Ni	T65	276	40	—	Strength and wear resistance at elevated temperatures
F332	P	3.0	9.5	1.0	1.0	1.2		T5	214	31	—	Same as A332
333	P	3.5	9.0		1.0	1.0		T6	241	35	—	General-purpose permanent mold
354	P	1.8	9.0					—	—	—	—	High strength, aircraft
355	S & P	1.3	5.0					T6	221	32	2.0	Similar to 328
C355	S & P	1.3	5.0					T61	276	40	3.0	Stronger and more ductile than 355
356	S & P		7.0					T6	207	30	3.0	Excellent castability and impact strength
A356	S & P		7.0					T61	255	37	5.0	Stronger and more ductile than 356
357	S & P		7.0					T6	310	45	3.0	High strength-to-weight castings
359	S & P		9.0					—	—	—	—	High-strength aircraft usage
360	D		9.5			2.0		F	303	44[d]	2.5[d]	Good corrosion resistance and strength

A360	D		9.5			2.0			F	317	46[d]	3.5[d]	Similar to 360
380	D	3.5	8.5		3.0	2.0			F	317	46[d]	2.5[d]	High strength and hardness
A380	D	3.5	8.5		3.0	1.3			F	324	47[d]	3.5[d]	Similar to 380
383	D	1.5	10.5		3.0	1.3			F	310	45[d]	3.5[d]	High strength and hardness
384	D	3.75	11.3		1.0	1.3			F	331	48[d]	2.5	High strength and hardness
413	D	1.0	12.0			2.0			F	297	43[d]	2.5[d]	General purpose, good castability
A413	D	1.0	12.0			1.3			F	290	42[d]	3.5[d]	Similar to 413
443	D	5.25				2.0			F	228	33[d]	9.0[d]	General purpose, good castability
B443	S & P	5.25				2.0			F	117	17	3.0	General-purpose casting alloy
514	S			4.0					F	152	22	6.0	High corrosion resistance
A514	P			4.0	1.8				F	152	22	2.5	Similar to 514
B514	S		1.8	4.0					F	117	17	—	Similar to 514
518	D			8.0		1.8			F	310	45[d]	5.0[d]	Good corrosion resistance, strength, and toughness
520	S			10.0					T4	290	42	12.0	High strength with good ductility
535	S			6.9					F	241	35	9.0	Good corrosion resistance and machinability
705	S & P			1.6	3.0				F	207	30	5.0	High strength, good machinability
707	S & P			2.1	4.3				F	228	33	7.0	Similar to 705
A712	S				6.5				T5	221	32	2.0	Good properties without heat treatment
D712	S				5.8				F	234	34	4.0	Similar to A712
713	S & P				7.5	1.1			F	221	32	3.0	Similar to A712
771	S				7.0				T6	290	42	5.0	Aircraft and computer components
850	S & P	1.0					6.3 Sn	1.0 Ni	T5	110	16	5.0	Bearing alloy
A850	S & P	1.0	2.5				6.3 Sn		T5	117	17	3.0	Similar to 850
B850	S & P	2.0					6.3 Sn	1.2 Ni	T5	166	24	3.0	Similar to 850

[a] Aluminum Association.
[b] S, sand cast; P, permanent mold cast; D, die cast.
[c] Minimum figures unless noted.
[d] Typical values.

alloys with the percentage of the principal alloying elements, proceding alphabetically as compositions become standard. Temper designation is much the same as in the case of aluminum, using -F, -O, -H1, -H2, -T4, -T5, and -T6. Some of the more common magnesium alloys are listed in Table 8-5 together with their properties and uses.

Most of the magnesium alloys are rather easily cast and are particularly well suited for die casting. Their machinability is by far the best of any metal. However, it is necessary to keep tools sharp and provide ample space for the chips.

Most alloys can be cold-worked, but deformation is not very good (as one might expect from the hcp crystal structure). If the temperature is raised to the region betwen 160 and 400°C (325 to 750°F), forming and drawing characteristics improve measureably. Because these temperatures are relatively low and easily attained, many formed and drawn magnesium parts are produced.

Magnesium alloys can be spot-welded nearly as easily as aluminum. Scratch brushing or chemical cleaning is necessary before spot welding. Fusion welding is carried out most easily by using an inert shielding atmosphere of argon or helium gas.

From a use viewpoint, magnesium alloys can be characterized by low wear, creep, fatigue, and corrosion resistance. A low elastic modulus makes it necessary to use thick sections to provide adequate stiffness. Fortunately, the alloys are so light that it often is possible to use the thicker sections required for rigidity and still have a lighter structure than can be obtained with any other metal. Cost per unit volume is low, so the use of thick sections does not push the cost out of line. Moreover, since a large portion of magnesium components are cast, the thick sections actually become a desirable feature. Corrosion resistance is moderate unless exposed to salt water, salt air, or an unfavorable galvanic couple. Adequate corrosion resistance can usually be provided by enamel or lacquer finishes.

Another problem with magnesium alloys is the limited ductility. Here, designers should be aware of the brittle failures possible when components are loaded beyond the assumed conditions. Magnesium automobile wheels are a notable example.

Considerable misinformation has existed regarding the fire hazard that exists in processing magnesium alloys. It is true that magnesium alloys are highly combustible when they are in a finely divided form, such as powder or fine chips, and this hazard should not be ignored. Above 425°C (800°F) a noncombustible atmosphere is required to suppress burning. Castings require additional precautions due to the reactivity of magnesium with sand and water. In sheet, bar, extruded, or cast form, magnesium alloys present no fire hazard.

TABLE 8-5. Composition, properties, and uses of common magnesium alloys

Alloy	Temper	Composition (%)						Tensile Strength[a]		Yield Strength[a]		Elongation in 2 in. (%)	Uses and Characteristics
		Al	Rare Earths	Mn	Th	Zn	Zr	MPa	ksi	MPa	ksi		
AM60A	F	6.0		0.13				207	30	117	17	6	Die castings
AM100A	T4	10.0		0.1				234	34	69	10	6	Sand and permanent-mold castings
AZ31B	F	3.0				1.0		221	32	103	15	6	Sheet, plate, extrusions, forgings
AZ61A	F	6.5				1.0		248	36	110	16	7	Sheet, plate, extrusions, forgings
AZ63A	T4	6.0				3.0		234	34	76	11	7	Sand and permanent mold castings
AZ80A	T5	8.5				0.5		234	34	152	22	2	High-strength forgings, extrusions
AZ81A	T4	7.6				0.7		234	34	76	11	7	Sand and permanent-mold casting
AZ91A	F	9.0				0.7		234	34	159	23	3	Die castings
AZ92A	T4	9.0				2.0		234	34	76	11	6	High-strength sand and permanent mold castings
EZ33A	T5		3.2			2.6	0.7	138	20	97	14	2	Sand and permanent mold castings
HK31A	H24				3.2		0.7	228	33	166	24	4	Sheets and plates; castings in T6 temper
HM21A	T5			0.8	2.0			228	33	172	25	3	High-temperature (425°C) sheets, plates, forgings
HZ32A	T5				3.2	2.1		186	27	90	13	4	Sand and permanent-mold castings
ZH62A	T5				1.8	5.7	0.7	241	35	152	22	5	Sand and permanent-mold castings
ZK51A	T5					4.6	0.7	234	34	138	20	5	Sand and permanent-mold castings
ZK60A	T5					5.5	0.45	262	38	138	20	7	Extrusions, forgings

[a] Properties are minimums for the designated temper.

TABLE 8-6. Characteristics of Zamak 3 and Zamak 5

	Zamak 3 (ASTM AG40A) (SAE 903)	Zamak 5 (ASTM AC41A) (SAE 925)
Composition (%)		
Copper	0.25	0.75–1.25
Aluminum	3.5–4.3	3.5–4.3
Magnesium	0.02–0.05	0.03–0.08
Iron, maximum	0.1	0.1
Lead, maximum	0.005	0.005
Cadmium, maximum	0.004	0.004
Tin, maximum	0.003	0.003
Zinc	Remainder	Remainder
Properties		
Tensile strength, as cast (MPa (ksi))	283 (41)	328 (47.6)
Tensile strength, 10 years of aging	241 (35)	271 (39.3)
Elongation in 2 in., as cast (%)	10	7
Charpy impact strength, as cast (ft-lb)	43	48
Melting point (°F)	717	717

ZINC-BASE ALLOYS

Zinc-base alloys are of primary importance for their use in die castings. Zinc is low in cost, has a melting point of only 380°C (715°F), does not affect steel dies adversely, and can be made into alloys that have good strength properties and good dimensional stability. Two of the most widely used zinc alloys are characterized in Table 8-6. Zamak 3 is used widely because of its excellent dimensional stability. Zamak 5 offers higher strength and better corrosion resistance. A newer alloy, Zamak 7, provides better castability through a lower magnesium content. All alloys have good tensile strengths coupled with exceptional impact resistance. Their development has been responsible for the extensive use of zinc die castings.

Die-cast zinc has a strength greater than all other die-cast metals except the copper alloys. The alloys lend themselves to casting within close dimensional limits, permitting the thinnest sections yet produced, and are machinable with a minimum of cost. Resistance to surface corrosion is adequate for a wide range of applications. Prolonged contact with moisture results in the formation of white corrosion products, but surface treatments can be applied to prevent this corrosion.

NICKEL-BASE ALLOYS

Nickel-base alloys are noted for their outstanding strength and corrosion resistance, particularly at high temperatures. *Monel* metal, containing about 67 per cent nickel and 30 per cent copper, has been used for years in the chemical and food-processing industries because of its outstanding corrosion resistance. It probably has better overall corrosion resistance to more media than any other alloy. It is particularly resistant to saltwater corrosion, sulfuric acid, and it even resists high-velocity, high-temperature steam. For the latter reason, Monel has been used for steam turbine blades. It can be polished to have an excellent appearance, similar to stainless steel, and is often used for ornamental trim and household ware. In its common form, Monel has a tensile strength of from 480 to 1170 MPa (70 to 170 ksi), depending on the amount of cold working. The elongation in 2 inches varies from 50 to 2 per cent.

There are three special grades of Monel that contain small amounts of added alloying elements. K Monel contains about 3 per cent aluminum and can be precipitation-hardened to a tensile strength of 1100 to 1240 MPa (160 to 180 ksi). H Monel has 3 per cent silicon added and S Monel has 4 per cent silicon. They are used for castings and can be precipitation-hardened. To improve upon the machining charactristics of Monel, a special free-machining alloy known as R Monel is produced with about 0.35 per cent sulfur.

The nickel-base alloys that have been developed for extreme-high-temperature service will be discussed in the next section.

Another use for nickel-base alloys is as electrical resistors. Some, primary nickel–chromium alloys, are called *Nichromes.* One alloy contains 80 per cent nickel and 20 per cent chromium. Another has 60 per cent nickel, 16 per cent chromium, and 24 per cent iron. They have excellent resistance to oxidation while retaining their strength at red heats.

Most of the nickel alloys are somewhat difficult to cast, but they can be forged and hot-worked. The heating, however, usually must be done in controlled atmospheres to avoid intercrystalline embrittlement. Welding can be performed with little difficulty.

NONFERROUS ALLOYS FOR HIGH-TEMPERATURE SERVICE

The rapid developments in the jet engine, gas turbine, rocket, and nuclear fields have stimulated, and have been made possible by, the development of a number of nonferrous alloys that have high strength, creep resistance, and corrosion resistance at temperatures up to and in excess of 1100°C (2000°F). Several of the more common of these *superalloys* are listed in Table 8-7. It will be noted that nickel, iron and nickel, or cobalt forms the base metal in these alloys. Most are precipitation-hardenable, and yield strengths above 690 MPa (100 ksi) are readily attained. The nickel-base alloys tend to have higher strengths at room temperature with yield strengths up to 1200 MPa (175 ksi) and ultimate strengths up to 1450 MPa (210 ksi), as compared with 790 MPa (115 ksi) and 1170 MPa (170 ksi), respectively, for the cobalt-base alloys. The 1000-hour rupture strengths of the nickel-base alloys at 815°C (1500°F) are also higher than those of the cobalt-base alloys, up to 450 MPa (65 ksi)

TABLE 8-7. Some nonferrous alloys for high-temperature service

Alloy	Composition (%)													
	C	Mn	Si	Cr	Ni	Co	Mo	W	Cb	Ti	Al	B	Zr	Fe	Other
Nickel base															
Hastelloy X	0.1	1.0	1.0	21.8	Balance	2.5	9.0	0.6	—	—	—	—	—	18.5	—
IN-100	0.18	—	—	10.0	Balance	15.0	3.0	—	—	4.7	5.5	0.014	0.06	—	1.0 V
Inconel 601	0.05	0.5	0.25	23.0	Balance	—	—	—	—	—	1.4	—	—	14.1	0.2 Cu
Inconel 718	0.04	0.2	0.2	19.0	Balance	—	3.0	—	5.0	0.9	0.5	—	—	18.5	0.2 Cu
M-252	0.15	0.5	0.5	19.0	Balance	10.0	10.0	—	—	2.6	1.0	0.005	—	—	—
Rene 41	0.09	—	—	19.0	Balance	11.0	10.0	—	—	3.1	1.5	0.01	—	—	—
Rene 80	0.17	—	—	14.0	Balance	9.5	4.0	4.0	—	5.0	3.0	0.015	0.03	—	—
Rene 95	0.15	—	—	14.0	Balance	8.0	3.5	3.5	3.5	2.5	3.5	0.01	0.05	—	—
Udimet 500	0.08	—	—	19.0	Balance	18.0	4	—	—	3.0	3.0	0.005	—	0.5	—
Udimet 700	0.07	—	—	15.0	Balance	18.5	5.0	—	—	3.5	4.4	0.025	—	0.5	—
Waspaloy B	0.07	0.75	0.75	19.5	Balance	13.5	4.3	—	—	3.0	1.4	0.006	0.07	2.0	0.1 Cu
Iron–nickel base															
Illium P	0.20	—	—	28.0	8.0	—	2.0	—	—	—	—	—	—	Balance	3.0 Cu
Incoloy 825	0.03	0.5	0.2	21.5	42.0	—	3.0	—	—	0.9	0.1	—	—	30	2.2 Cu
Incoloy 901	0.05	0.4	0.4	13.5	42.7	—	6.2	—	—	2.5	0.2	—	—	34	—
16-25-6	0.08	1.35	0.7	16.0	25.0	—	6.0	—	—	—	—	—	—	Balance	0.15 N
Cobalt base															
Haynes 150	0.08	0.65	0.75	28.0	—	Balance	—	—	—	—	—	—	—	20.0	—
MAR-M322	1.00	0.10	0.1	21.5	—	Balance	—	9.0	—	0.75	—	—	2.25	—	4.5 Ta
S-816	0.38	1.20	0.4	20.0	20.0	Balance	4.0	4.0	4.0	—	—	—	—	4.0	—
WI-52	0.45	0.5	0.5	21.0	1.0	Balance	—	11.0	2.0	—	—	—	—	2.0	—

versus 228 (33 ksi). Materials such as TD-nickel (a nickel alloy containing 2 per cent dispersed thoria) and columbium give promise of operating at service temperatures above 1100°C (2000°F).

Many of the superalloys are virtually unmachinable except by electrodischarge, electrochemical, or ultrasonic methods. Consequently, they are often produced in the form of investment castings. Powder metallurgy techniques are also being used extensively in the manufacture of superalloy components. Because of their ingredients, all these alloys are expensive, and this limits their use to small or critical parts or applications where cost is not the determining factor.

LEAD–TIN ALLOYS

Lead and tin are nearly always used together in alloys of engineering importance, the two major uses being as bearing materials or as solders. One of the oldest and best bearing metals, composed of about 84 per cent tin, 8 per cent copper, and 8 per cent antimony, is called genuine or tin *babbitt.* Because of the high cost of tin, a more widely used babbitt metal is the lead babbitt composed of 85 per cent lead, 5 per cent tin, 10 per cent antimony, and 0.5 per cent copper. For high speeds and fairly heavy loads, the lead-base babbitts prove unsatisfactory; for slow speeds and moderate loads, they are quite adequate.

Figure 8-1 shows a photomicrograph of a lead babbitt metal. The antimony combines with the tin to form hard particles in the softer lead matrix, a structure typical of many bearing metals. The shaft rides on the harder particles with little friction while the softer matrix acts as a cushion that can distort sufficiently to take care of misalignment and assure a proper fit between the two surfaces.

Soft *solders* are basically lead–tin alloys near the eutectic composition (61.9 per cent tin). A variety of compositions exist, each with a characteristic melting range. Because of the higher price of tin, however, a 50–50 composition or alloys of the lead-rich variety are often used.

FIGURE 8-1. Photomicrograph of babbitt metal; 75×. Square or triangular white masses are SbSn; small white particles are CuSn.

Review questions

1. What properties do nonferrous alloys have that usually are not associated with ferrous metals and alloys?
2. What are some of the major limitations of nonferrous alloys compared with the ferrous materials?
3. Why is brass more extensively used than bronze?
4. Why should brass containing 25 per cent zinc not be used in sea water? What alternative material would you suggest?
5. What alloying element is used to impove the machining characteristics of brass?
6. Why is the term "bronze" often misleading?
7. Which alloying elements promote precipitation hardening when added to copper?
8. What are some of the outstanding properties of copper–beryllium alloys?
9. What are some typical applications of copper–beryllium alloys?
10. What properties have made aluminum and its alloys the most important nonferrous metal?
11. What are the major assets and liabilities of aluminum when compared to steel?
12. Why do aluminum alloys have good corrosion resistance?
13. What are the two types of wrought aluminum alloys?
14. What does the designation 2024-T6 tell about the material?
15. Why are the heat-treatable aluminum alloys often produced in the form of Alclad material?
16. Why are cast aluminum alloy compositions often different for sand, permanent-mold, and die-casting operations?
17. What is the primary reason for the use of magnesium alloys?
18. Why are magnesium alloys quite suitable for cast components but not so attractive for cold-worked or sheet structures?
19. Discuss the fire hazards and safety measures associated with magnesium alloys.
20. What is the principal use of the zinc-base alloys?
21. Why is Monel an important alloy?
22. What metals form the bases for most of the high-temperature nonferrous alloys?
23. What processing techniques are most often employed with the superalloy metals?
24. What are the functions of the hard and soft components in bearing materials?
25. What are the component metals in soft solders?

Case study 7.
THE SUBSTITUTE ALUMINUM CONNECTING RODS

Winning Racing, Inc., is a manufacturer of high-performance automotive components, specially designed for racing applications. One highly successful product line is a series of specially designed connecting rods, made of forged alloy steel.

Noting the successful use of aluminum alloys in certain racing applications, Team Rabbit has requested Winning Racing to produce a special set of lightweight aluminum connecting rods, using their highly successful existing design. The rods for three engines were made on a special run, using the existing dies, and were put into the engines for testing. During dynamometer testing, however, the engines failed in under 30 minutes, and the failures were attributed to the connecting rods, although none of the rods broke. What had been overlooked that caused the trouble?

CHAPTER 9

Nonmetallic materials: plastics, elastomers, ceramics

A number of nonmetallic materials have substantial importance in manufacturing. Consequently, it is imperative for the design engineer to have an understanding of their natures, properties, advantages, and limitations so he may know when and how they may be used advantageously in his designs. Except in furniture manufacturing, where wood is of prime importance, these materials are plastics, elastomers, and ceramics. Most of these are man-made, permitting a wide range of properties to be obtained, and entirely new materials, and variations of them, are being created almost continuously. As a result, it is difficult for one to keep abreast of all the individual materials that are available at a given time, and no attempt will be made in this chapter to give detailed information about all of them. Instead, the emphasis will be on the basic nature, properties, and processing of these materials, so that the reader may have a good idea as to whether they are potential materials for use in products. For more detailed information about specific materials of these types, texts, handbooks, and compilations that deal exclusively with these materials should be consulted.

PLASTICS

It is difficult to give a precise definition of the term *plastics*. Basically, it covers a group of materials characterized by large molecules that are built up by joining small molecules, usually artificially. Practically, it is sufficient to say that they are natural or synthetic resins, or their compounds, that can be molded, extruded, cast, or used as films or coatings. Most of them are organic substances, usually containing hydrogen, oxygen, carbon, and nitrogen.

The only natural plastic of engineering importance is shellac, and its importance has decreased greatly in recent years as the result of the development of synthetic resins.

The molecular structure of plastics. It is helpful to have an understanding of the basic molecular structure of plastics. Most are based on hydrocarbons, in which carbon and hydrogen combine in the relationship C_nH_{2n+2}, known as *paraffins*. Theoretically, these hydrocarbons can be linked together indefinitely to form very large molecules, as illustrated in Figure 9-1. The bonds between the atoms are single pairs of covalent electrons. Because there is no provision for additional atoms to be added to the chain, such molecules are said to be *saturated*. Such molecules have strong intramolecular bonds, but the intermolecular bonds are much weaker.

Carbon and hydrogen atoms also can form molecules by being held together by double pairs of covalent bonds. Ethylene and acetylene are examples, illustrated in Figure 9-2. Because such molecules do not have the maximum possible number of hydrogen atoms, they are said to be *unsaturated*. Such molecules are important in the *polymerization* (joining together) of small molecules into large ones having the same constituents.

In organic compounds, four electron pairs surround each carbon atom,

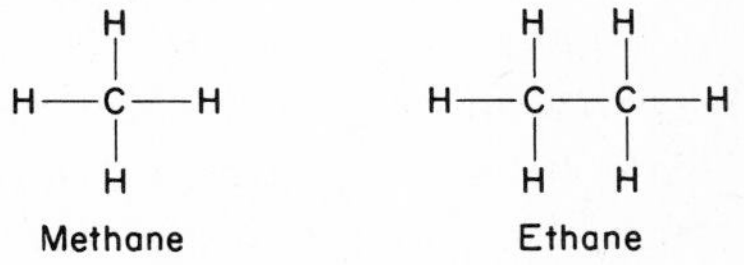

FIGURE 9-1. Linking of hydrogen and carbon in methane and ethane molecules.

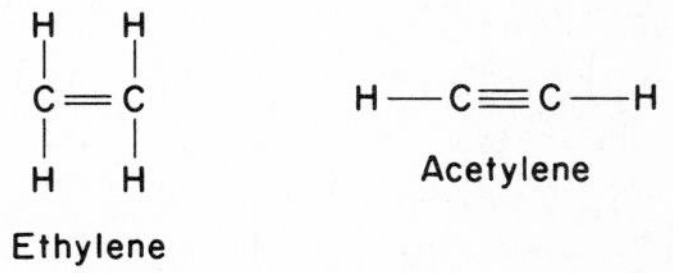

FIGURE 9-2. Covalent bonds in ethylene and acetylene molecules.

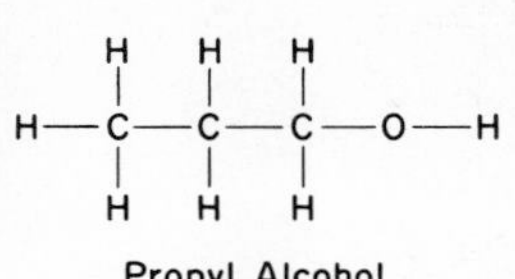

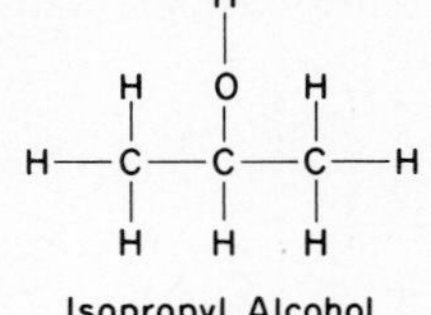

FIGURE 9-3. Linking of eight hydrogen, three carbon, and one oxygene atoms to form two isomers, propyl and isopropyl alcohol.

and one electron pair is shared jointly by each hydrogen atom. However, other kinds of atoms can be introduced, such as a benzene ring. Also, a carbon atom may be replaced by other elements, such as oxygen, sulfur, or nitrogen. Consequently, by these procedures, a wide range of organic materials can be created.

Isomers. The same kind and number of atoms can unite in different structural arrangements, thus forming different compounds which have completely different properties. Figure 9-3 shows such an example. Such compounds are called *isomers.* These are analogous to allotropism or polymorphism in the case of crystalline phases, such as body-centered-cubic and face-centered-cubic iron. A number of plastics are isomers.

Forming molecules by polymerization. Polymerization of large molecules in plastics takes place by either *addition* or *condensation.* Figure 9-4 illustrates polymerization by addition; a number of basic units (*monomers*) are added together to form a large molecule (*polymer*) in which there is a repeated unit (*mer*). All the components are utilized in the final product with this type of polymerization. Theoretically, addition polymerization could go on indefinitely, forming a single, large molecule. This does not occur, however, because as the molecules get larger it becomes more difficult for them to diffuse and be available at the needed place in the chain.

Polymerization also can take place utilizing two kinds of mers. This process, illustrated in Figure 9-5, is called *copolymerization.* It greatly expands the possibilities for creating additional types of plastics.

When polymerization occurs by condensation, as illustrated in Figure 9-6, a nonpolymerizable by-product is produced from each of the molecules that enter into the reaction.

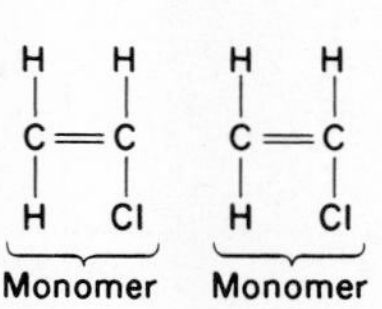

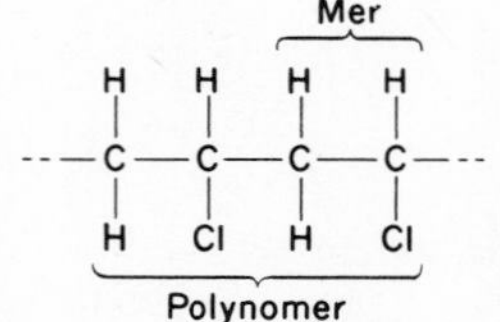

FIGURE 9-4. Polymerization by addition—the uniting of monomers.

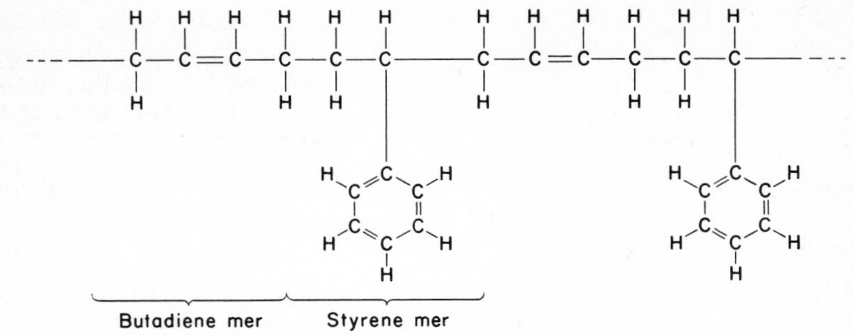

FIGURE 9-5. Polymerization by the addition of two kinds of mers—copolymerization.

Thermosetting and thermoplastic plastics. The molecules of plastics can be thought of as having backbones of carbon atoms with attached ribs, or pendants, of other atoms, such as hydrogen, fluorine, and chlorine. The characteristics of the plastics are affected by (1) the molecular bonds that join the backbone of the chain together, (2) the bonds that connect adjacent chains, and (3) the geometry of the chains. The molecules of the backbone and ribs within a polymer are joined by primary bonds. In some plastics, the bonds that hold adjacent chains together, and determine the resistance of the molecules to breakdown from chemical and thermal attack and weathering, also are strong, primary bonds. These plastics, having primary covalent bonds throughout, are known as *thermosetting plastics.* Their structure, in effect, is one large molecule. Their hardness is the result of chemical change, usually produced by elevated temperature and pressure but sometimes by the action of a catalyst at room temperature without pressure. Once hardened, they cannot be softened, and they maintain their hardness at elevated temperatures.

In another type of plastics the bonding between polymer chains is by much weaker, secondary (van der Waal's) forces. The mechanical and physical properties of such plastics are determined by these secondary bonds. Because these secondary bonds are weakened by temperature, plastics of this type soften with increasing temperature and become harder and stronger with

FIGURE 9-6. Formation of phenol-formaldehyde by condensation polymerization.

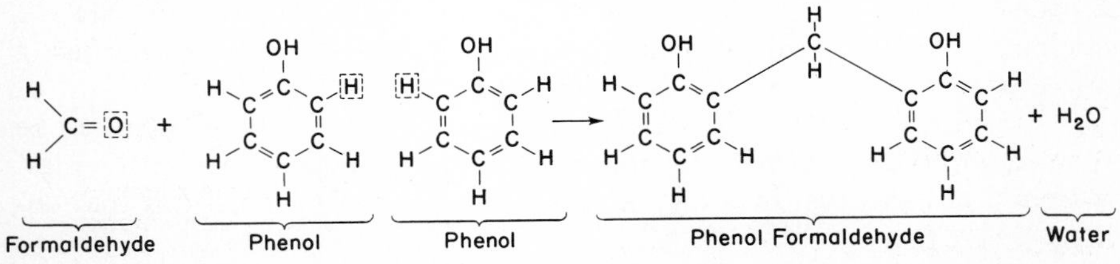

decreasing temperature. These are called *thermoplastic plastics*. The softening and hardening can be repeated as often as desired, and no chemical change is involved. Because they contain molecules of different sizes, thermoplastic materials do not have a definite melting temperature but, instead, soften over a temperature range.

Some thermosetting plastics have been developed that have partial ionic (primary) bonding between the chains, thus modifying their properties.

The bonding forces between plastic molecules are much weaker than those within the molecules. Consequently, deformation occurs by slippage between the molecules rather than by breakage of the molecular bonds. In thermoplastic materials this can occur very easily. However, because there are some linkages between the major molecules of thermosetting plastics, these materials do not deform or soften readily and tend to maintain their strengths up to the temperature at which they char.

Obviously, whether a plastic is thermosetting or thermoplastic is of great importance to the person who is selecting it for use, because considerable indication is given not only to its behavior in service but also as to how it must be processed. Thermoplastics, as a class, are easily molded. However, after the material is formed to shape in a mold at an elevated temperature (and ordinarily under considerable pressure), the mold must be cooled to cause the plastic to harden and thus retain its shape when removed. In producing products from thermosetting plastics, the mold remains at an elevated temperature throughout the molding cycle, and the material hardens as a result of the temperature and the pressure. It can then be removed without cooling the mold. Thus the molding procedure is controlled by the type of the plastic.

Types of plastics and their properties. Because there are so many plastics, with new ones becoming available almost continuously, it is helpful to have a knowledge of the general properties which they, as a whole, possess and the properties of the several basic types. First, all plastics are quite light in weight. Most have specific gravities between 1.1 and 1.6, as compared with about 1.75 for magnesium. Thus plastics are the lightest of the engineering materials.

Second, nearly all plastics have good electrical resistance; consequently, they are widely used as insulating materials. Third, they all have much lower thermal conductivity than metals and thus are relatively good heat insulators.

A fourth characteristic of plastics, as a group, is that they can be obtained in an almost unlimited range of colors, either transparent or opaque. Of course, not all of them have a color range, but many do; and not only is a wide range of colors available, but the color goes throughout and is not just a surface effect.

A fifth, important characteristic is that an excellent surface finish can be obtained without added operations beyond those required to convert the plastic from a raw material into the final shape. A sixth characteristic is that objects frequently can be produced from plastics in only one operation—from raw material to final shape—by several basic processes, such as casting, extrusion, and molding.

The general properties discussed thus far are desirable ones. The inferior properties of plastics have to do with their strength. None of them has strength properties that approach those of the engineering metals. The impact strength of most of them is low, but several—ABS, high-density polyethylene, polycarbonate, and cellulose propionate—have very good impact strengths. Because of their low weight, their strength-to-weight ratio is fair. However, as a class they are not suitable for applications that require high strength unless special strengthening filler materials are added. Another weakness is that the dimensional stability of most of them is greatly inferior to metals.

Table 9-1 lists properties of a number of common plastics. From a consideration of the properties just discussed, and as shown in this table, it is apparent that plastics are best suited for applications that require materials of only low or moderate strength, having low electrical and/or thermal conductivity, obtainable in a wide range of colors, and easily transformed from the raw to the finished state. In no other material but plastics can this combination of properties be obtained. Thus a large percentage of all plastics are used as "packaging" or container materials. This classification includes such items as radio cabinets, clock cases, and household appliance housings, which, primarily, serve as containers for the interior mechanisms.

Another important use of plastics results from their low electrical and/or thermal conductivity. Their use as insulators in electrical equipment and as handles for hot articles, such as electric irons, is due largely to these properties.

A third major use for plastics utilizes them in the form of foam. Soft, pliable foams are used extensively as cushioning material, while rigid foams are used inside thin sheet-metal structures, such as airplane and rocket stabilizers, to provide compression strength.

A fourth use is as adhesive and bonding agents in fastening together a multitude of metal and nonmetal objects. This use is discussed in Chapter 33.

A fifth use is as a substitute for costly metal dies in connection with tooling for various press-working operations for forming sheet metal.

Obviously there are many uses where the need for these groups of properties overlaps or where the use of plastics is determined by only one or two properties—in some cases others than those mentioned. Increasingly, fabric- or fiber-reinforced plastics are being used in applications where considerable tensile strength is required, but this strength is provided primarily by the reinforcing, not by the plastic. It is important to remember that to a large degree the use of plastics is due to the fact that they provide, *in combination,* several desirable properties that cannot be found in any other single material. A realization of this fact helps to clarify the position of plastics when assessing

TABLE 9-1. Properties and major characteristics of common types of plastics

Material	Specific Gravity	Tensile-Strength (1,000 lb/ in.²)	Impact Strength Izod ft-lb/in. of notch	Top Working Temper-ature °C (°F.)	Dielectric Strength volts/ mil.[b]
Thermoplastics					
ABS material	1.02–1.06	4–8	1.3–10.0		300–400
Acetal	1.4	10	1.5	121 (250)	1200
Acrylics	1.12–1.19	5.5–10	0.2–2.3	93 (200)	400–530
Cellulose acetate	1.25–1.50	3–8	0.75–4.0	127 (260)	300–600
Cellulose acetate butyrate	1.18–1.24	2–6	0.6–3.2	54 (130)	250–350
Cellulose propionate	1.19–1.24	1–5	0.8–9	60 (140)	300
Chlorinated polyether	1.4	6	3.3	149 (300)	400
Ethyl cellulose	1.16	3–6	1.8–4.0	66 (150)	350
TFE–fluorocarbon	2.1–2.3	1.5–3	2.5–4.0	260 (500)	450
CFE–fluorocarbon	2.1–2.15	4.5–6	3.5–3.6	199 (390)	550
Nylon	1.1–1.2	8–10	2	121 (250)	385–470
Polycarbonate	1.2	9.5	14	121 (250)	400
Polyethylene	0.96	4	10	93 (200)	440
Polypropylene	0.9–1.27	3.4–5.3	1.02	110 (230)	520–800
Polystyrene	1.05–1.15	5–9	0.3–0.6	88 (190)	400–600
Modified polystyrene	1.0–1.1	2.5–6	0.25–11.0	100 (212)	300–600
Vinyl	1.16–1.55	1–5.9	0.25–2.0	104 (220)	25–500
Thermosetting plastics					
Epoxy	1.1–1.7	4–13	0.4–1.5	163 (325)	500
Melamine	1.76–1.98	5–8		177 (350)	460
Phenolic	1.2–1.45	5–9	0.25–5	149 (300)	100–500
Polyester (other than molding compounds)	1.06–1.46	4–10	0.18–0.4	149 (300)	340–570
Polyester (alkyd, DAP)	1.6–1.75	3.2–8	3.6–8		
Silicone	2.0	3–5	0.2–3.0	288 (550)	250–350
Urea	1.41–1.80	4–8.5	0.2–0.5	85 (185)	300–600

[a] × denotes a principal reason for its use; 0 indicates a secondary reason.
[b] Short-time ASTM Test.

24-Hour Water Absorption (%)	Special Characteristics[a]: Weatherability	Colorability	Optical Clarity	Chemical Resistance	Common Forms: Injection Molding	Extrusions	Formable Sheet	Film	Fiber	Compression or Transfer Moldings	Castings	Reinforced Plastics Moldings	Industrial Thermosetting Laminates	Foam
0.2–0.3	0	×		0	√	√	√							
0.22		×		0	√	√								
0.2–0.4	×		×	0	√	√	√		√		√			
2.0–6.0		×	×		√	√	√	√	√					
1.8–2.1	×	×	×		√	√	√	√						
1.8–2.1		×	×		√	√	√	√						
0.01				×	√	√								
1.6–2.2		×			√	√								
0	×			×		√	√	√	√	√				
0	×			×	√	√	√	√						
0.4–5.5		0	0	0	√	√	√	√	√					
0.15	×	0		×	√	√								
0.003		0	×	×	√	√	√	√	√					√
0.03		0	×	×	√	√	√	√	√					
<0.2		×	×	0	√	√	√	√						
0.03–0.2		×		×	√	√	√	√						
0.2–1		×	×	×	√	√	√	√	√		√			√
0.1–0.5	×	×		×							√	√	√	√
0.1		×		0						√		√	√	
0.2–0.6				0						√	√	√	√	√
0.5		×	×	0				√	√		√	√	√	
0.16–0.67										√		√	√	
0.4–0.5										√	√	√	√	√
1–3		×		0										

their suitability for a particular application. Following are some comments about several types of plastics listed in Table 9-1 which may be helpful in selecting them for use:

Phenolics: oldest of the plastics, but still widely used; hard, relatively strong, low cost, and easily molded; opaque, but wide color range; wide variety of forms—sheets, rods, tubes, and laminates.

Urea formaldehyde: similar properties to phenolics, but available in lighter colors; useful for containers and housings, but not outdoors; used in lighting fixtures because of translucence in thin sections.

Melamines: excellent resistance to heat, water, and many chemicals; full range of translucent or opaque colors; excellent electrical-arc resistance; tableware, but stained by coffee; used extensively in treating paper and cloth to impart water-repellent properties.

Epoxides: good toughness, elasticity, chemical resistance, and dimensional stability; used as coatings, cements, and "potting" materials for electrical components; easily compounded to cure at room temperatures; widely used in tooling applications.

Silicones: semiorganic (spine molecules alternating silicon and oxygen atoms); heat resistant; low moisture absorption; high dielectric properties.

ABS: contain acrylonitrile, butadiene, or styrene; low weight, good strength, and very tough; good under severe service conditions.

Acrylics: highest optical clarity, transmitting over 90 per cent of light; common trade names are *Lucite* and *Plexiglas;* high impact, flexural, tensile, and dielectric strengths; wide range of colors; stretch rather easily.

Cellulose acetate: wide range of colors; good insulating qualities; easily molded; high moisture absorption in most grades and affected by alcohols and alkalies.

Cellulose acetate butyrate: higher impact strength and moisture resistance than cellulose acetate; will withstand rougher usage.

Ethyl cellulose: high electrical resistance and impact strength; retains toughness at low temperatures.

Fluorocarbons: inert to most chemicals; high temperature resistance; very low coefficients of friction (Teflon), used for nonlubricated bearings and nonstick coatings for cooking utensils and electric irons.

Nylon: good abrasion resistance and toughness; excellent dimensional stability; used as bearings with little lubrication; available as monofilaments for textiles, fishing lines, ropes, and so on; expensive (specialized applications only).

Polycarbonates: high strength and outstanding toughness.

Polyethylenes: tough; high electrical resistance; used for bottle caps, unbreakable kitchenware, and electrical wire insulation.

Polystyrenes: high dimensional stability and low water absorption; best all-around dielectric; burns readily and is adversely affected by citrus juices and cleaning fluids.

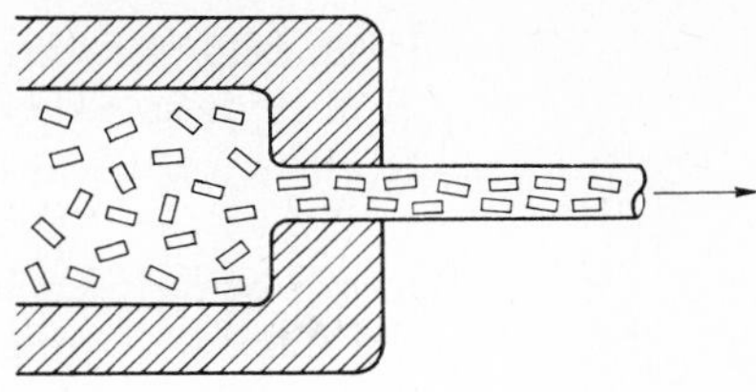

FIGURE 9-7. Schematic representation of the aligning of plastic molecules in the orienting process.

Vinyls: wide range of types, from thin, rubbery films to rigid forms; tear resistant; good aging properties; good dimensional stability and water resistance in rigid forms; used for floor and wall coverings, upholstery fabrics, and lightweight water hose.

Oriented plastics. Because the intermolecular bonds of thermoplastics are much weaker than the bonds between the atoms in their spines, these plastics can be given a much greater strength in one direction by rearranging the molecules so they are lined up in the direction of an applied load. This *orienting* process is accomplished either by stretching or by extrusion, as is illustrated in Figure 9-7. The material usually is heated somewhat during the orienting process to aid in overcoming the internal forces and is cooled immediately afterward to "freeze" the molecules in the desired orientation. Either uniaxial or biaxial orientation may be imparted.

Orienting may increase the tensile strength by more than 50 per cent, but a 25 per cent increase is more typical. The elongation may be increased several hundred per cent. One shortcoming is that if oriented plastics are reheated, they tend to return to their original shape, owing to the phenomenon of *elastic memory*.

Additive agents in plastics. For most uses, other materials usually are added to plastics to (1) improve their properties, (2) reduce the cost, (3) improve their moldability, and/or (4) impart color. Thus such additive constituents usually are classified as *fillers, plasticizers, coloring agents,* or *lubricants*.

Ordinarily fillers comprise a large percentage of the total volume of a molded plastic product, being added to improve the strength or to decrease cost. To a large degree they determine the general properties of a molded plastic. They also may aid in controlling shrinkage and in improving moldability, although they more commonly reduce the latter property. Whenever possible, fillers that are much less expensive than the plastic resin are used. The most common fillers, and the properties they impart, are:

1. Wood flour: the general-purpose filler; low cost with fair strength; good moldability.
2. Cloth fibers: improved impact strength; fair moldability.
3. Macerated cloth: high impact strength; limited moldability.

4. Glass fibers: high strength; dimensional stability; translucence.
5. Asbestos fiber: heat resistance; dimensional stability.
6. Mica: excellent electrical properties and low moisture absorption.

Other fillers are being used increasingly, particularly for imparting high strength, often at elevated temperatures. "Whiskers" of various metals and nonmetals, such as boron, stainless steel, columbium, tantalum, titanium, zirconium, and silicon carbide, are used. These are from 1 to 5 μm (39 to 197 microinches) in diameter and 30 to 1000 μm (0.0012 to 0.039 inch) in length and have high moduli of elasticity and tensile strengths up to 21 000 MPa (3,000,000 psi). More common is the use of filaments of glass, graphite, or boron, which usually are less than 0.1 mm in diameter but of any desired length. These can provide tensile strengths up to 2450 MPa (350,000 psi) with moduli of elasticity up to 420 000 MPa (60 million psi). Glass-fiber cloth is a very commonly used material.

When fillers are used, the resin acts as the binding material surrounding each filler particle and holding the mass together. Thus the surface of a molded plastic part is almost pure resin with no filler exposed.

Coloring agents may either be dyes, which actually alter the color of the resin, or colored pigments that through their presence impart a desired color. Most of the fillers do not in themselves produce attractive colors, so a dye is usually needed.

Plasticizers are added in small amounts to increase and control the flow of the plastic during molding. The amount needed for a given resin is governed by the intricacy of the mold. As a rule, the amount of plasticizer is held to a minimum, because it is likely to affect the stability of the finished product through gradual loss during aging.

Lubricants are added in small amounts to improve the moldability and to facilitate removal of parts from the molds. Wax, stearates, and occasionally soaps are used for this purpose. They also are held to a minimum because they affect the properties adversely.

Plastics as adhesives. The use of plastics as adhesives is highly developed, and their use for this purpose has expanded tremendously in recent years. This application of plastics is discussed in Chapter 33.

Ablative coatings. Plastics form the base of many ablative coating materials used on rockets and missile motors to provide short-time protection for low-melting-temperature alloys from intense heating conditions or the heating experienced by space vehicles in reentering the earth's atmosphere. Ablation involves thermal decomposition of high polymers into low-molecular-weight gaseous products and porous carbon char. Some of the heat absorption also may occur by sublimation and chemical decomposition.

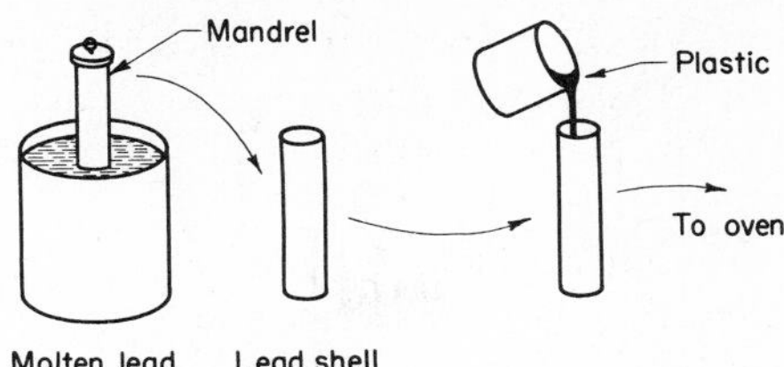

FIGURE 9-8. Steps in the casting of plastic parts.

Production processes for plastic products. Not only does the designer have a large number of plastics available from which he can select, there are a number of quite distinct processes by which a chosen plastic can be converted into a desired product. *Casting, hot-compression molding, injection molding, transfer molding, extrusion, laminating,* and *cold molding* are all used extensively. Each has certain advantages and limitations that bear on part design, material selection, and final cost, and not every plastic is suitable for each process. Because it usually is desirable to convert the material into the finished product in a single-process operation, it is important to have an understanding of the various processes so that the material-process selection will be optimal.

Casting is the simplest of the processes because no fillers are used and no pressure is involved. Of course, a mold is required. For certain simple shapes, a model of the product can be made, usually of steel, and then dipped into molten lead until a thin sheath of lead is formed over the model. The model, or mandrel, then is pulled out of the lead sheath, leaving a thin, lead mold into which the liquid plastic is poured. The resin then is cured, usually in an oven at low temperatures (65.6 to 93°C; 150 to 200°F), as indicated in Figure 9-8. After removal from the oven, the lead shell is stripped from the finished product. Some plastics, of course, can be cured at room temperatures.

Because cast plastics contain no fillers, they have a distinctive lustrous appearance. The process is inexpensive because no expensive dies or equipment are involved. However, it is limited to small objects of rather simple shape. Small radio cabinets, jewelry, and ornamental objects are commonly made by this process.

Blowmolding is used extensively for making hollow products, such as bottles and other containers. The steps in this process are illustrated in Figure 9-9.

In *hot-compression molding,* indicated schematically in Figure 9-10, granules or preformed tablets of the raw, mixed plastic material are loaded into the cavity of an open, heated mold. The plunger (male member) of the mold, usually attached to the upper portion of the press, descends, closing the mold and creating sufficient pressure to force the plastic, as it becomes fluid, into all portions of the cavity. After the material has set, or cured, the mold is opened and the part is removed. Usually, a number of cavities are contained within a single mold. The process is simple, but its use is restricted almost

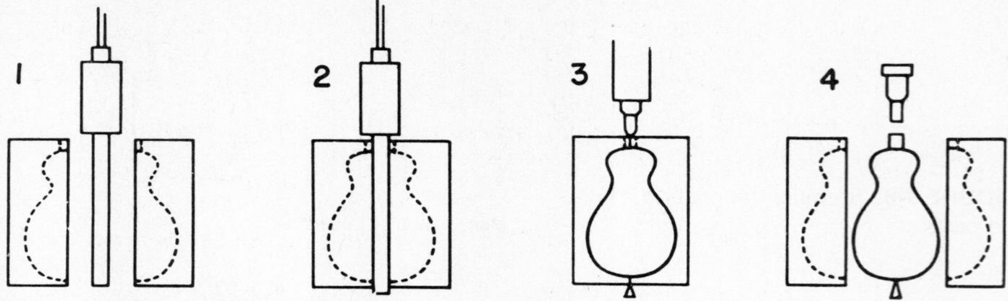

FIGURE 9-9. Steps in "blow molding" plastic parts: (1) Tube of heated plastic is placed in open mold. (2) Mold closes over extruded tube. (3) Air forces tube against mold sides. (4) Mold opens to release product.

exclusively to thermosetting materials. Alternate heating and cooling of the mold is required for thermoplastic materials and thus is uneconomic.

In order to contain the material within the die cavity and enable pressure to be built up, some type of seal is required on hot-compression molding dies. Three types of seals are employed, illustrated in Figure 9-11. In the *flash type,* during the final stage of the mold closing some of the excess material that is provided is squeezed out of the cavity. This resulting *flash* must be removed from the finished product, usually requiring an additional operation. This type of mold is relatively inexpensive to make, and it is not necessary to control the amount of raw material closely. However, the dimensions of the product across the mold-opening plane and the density will vary somewhat.

In the plunger or positive type, no material can escape from the mold. Thus to obtain close dimensional control, the raw material must be measured accurately. By controlling the pressure on the plunger, good density control is obtained.

The *landed-plunger type* of mold is often used. This provides partial pressure with positive final cutoff. Figure 9-12 shows a hot-compression mold in operation.

FIGURE 9-10. Schematic representation of the production of plastic parts by the hot-compression molding process.

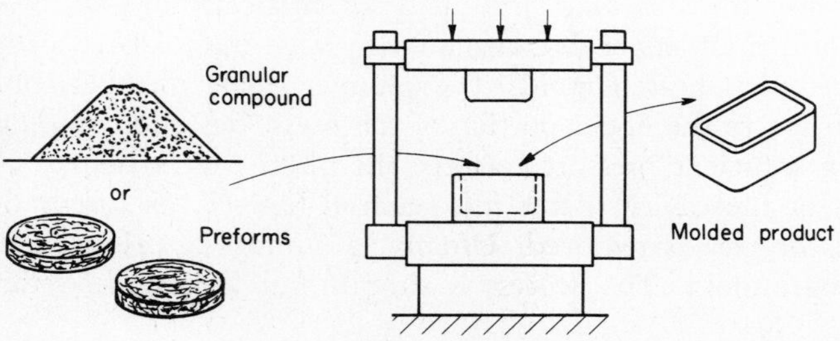

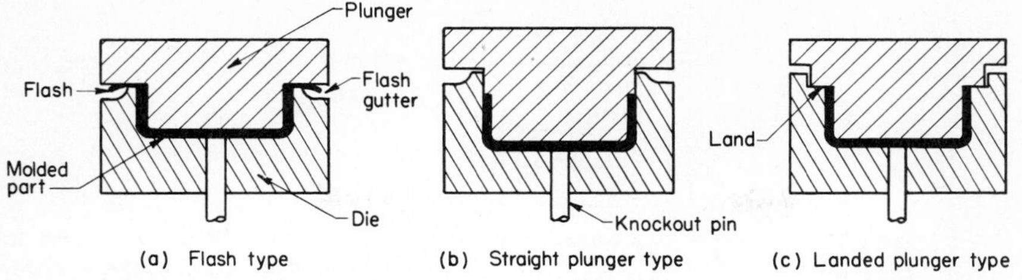

FIGURE 9-11. Three types of molds used in hot-compression molding.

In order to avoid the turbulence and uneven flow that often result from the nonuniform, high pressures in hot-compression molding, *transfer molding* sometimes is used. The raw material is placed in the plunger cavity, where it is heated until it is melted. The plunger then descends, forcing the molten plastic into the die cavities. Because the material enters the cavities as a liquid, there is little pressure until the cavity is completely filled. This makes it easier to obtain thin sections, excellent detail, and good tolerances and finish. The process is particularly useful when fragile inserts, which are to be incorporated into the product, must be inserted into and maintained in position in the cavity. An undercut on the plunger causes the sprue to be withdrawn with the plunger when the mold is opened. This procedure is illustrated in Figure 9-13.

Injection molding, illustrated schematically in Figure 9-14, is used to produce more thermoplastic products than any other process. Raw material is fed by gravity from a hopper into a pressure chamber ahead of a plunger. As the plunger advances, the plastic is forced into a heating chamber, where it is

FIGURE 9-12. Typical hot-compression molding press; mold being placed between platens. (*Courtesy Pennsalt Chemicals Corporation.*)

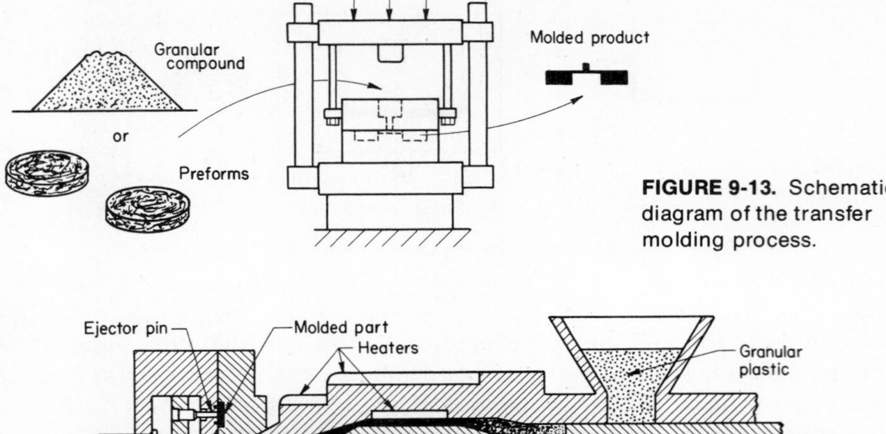

FIGURE 9-13. Schematic diagram of the transfer molding process.

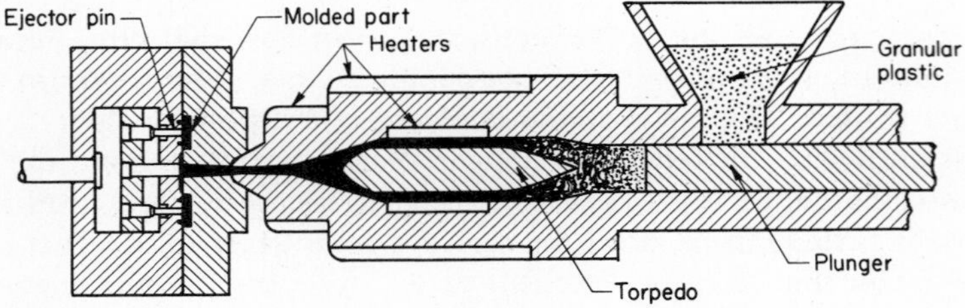

FIGURE 9-14. Schematic diagram of the injection-molding process.

FIGURE 9-15. Injection-molding machine. Inset shows plastic part being removed from mold. (*Courtesy Pennsalt Chemicals Corporation.*)

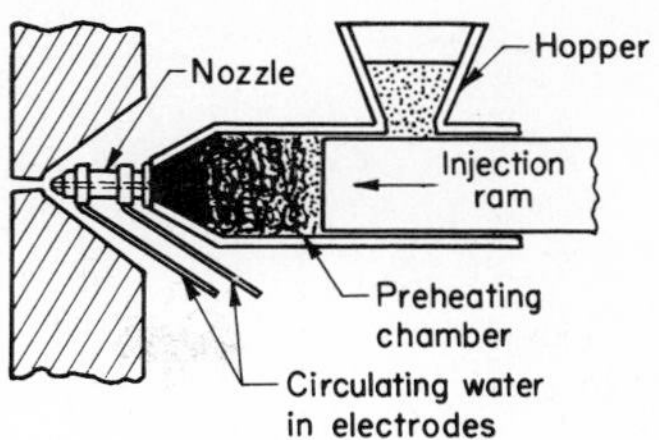

FIGURE 9-16. Jet molding process for injection-molding thermosetting plastics.

preheated. From the preheating chamber it is forced through the *torpedo* section, where it is melted and the flow regulated. It leaves the torpedo through a nozzle that seats against the mold and allows the molten plastic to enter the die cavities through suitable gates and runners. In this process the die remains cool, so the plastic solidifies almost as soon as the mold is filled. To ensure proper filling of the cavity, the material must be forced into the mold rapidly under considerable pressure; premature solidification would cause defective products. While the mold is being opened and closed and the part ejected, the material for the next part is being heated in the torpedo. The complete cycle requires only a few seconds. Figure 9-15 shows a typical injection molding machine.

Because thermosetting plastics must be held in the mold under temperature and pressure a sufficient time to permit the curing to be completed, a modification of the injection molding process must be used for this type of plastic. In the jet molding process shown in Figure 9-16, the plastic is preheated in the feed chamber to about 93°C (200°F) and then further heated to the polymerization temperature as it passes through the nozzle. The mold is held at an elevated temperature to complete the curing process. As soon as the charge for one cycle has nearly filled the die cavity, the nozzle is cooled to prevent the plastic in the nozzle from hardening and clogging the machine. Due to economic factors resulting from the relatively long cycle time, little injection molding of thermosetting plastics is done.

Plastic products with long, uniform cross sections are readily produced by *extrusion,* as depicted in Figure 9-17. The plastic material is fed from a hopper into a screw chamber from whence the rotating screw conveys it through a preheating section, where it is compressed, and then forces it

FIGURE 9-17. Extrusion process for producing plastic parts.

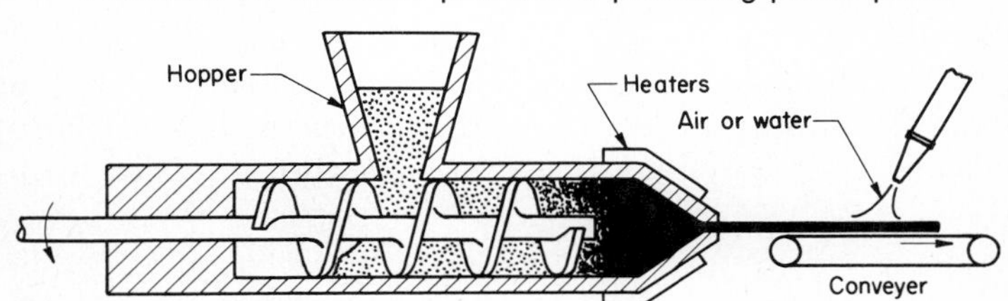

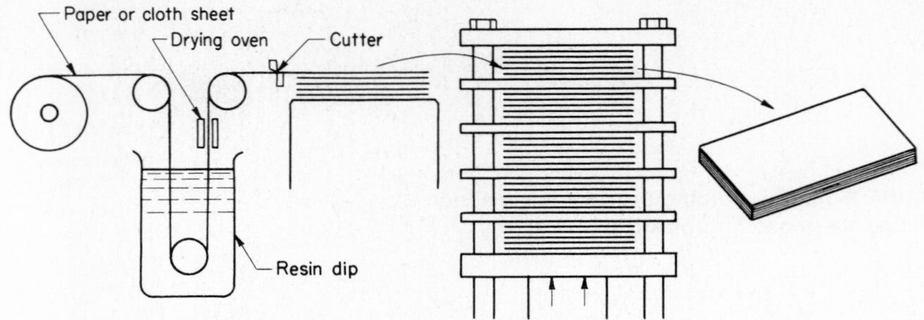

FIGURE 9-18. Method of producing flat laminated plastic sheets.

through a heated die and onto a conveyor belt. As the plastic passes onto the belt it is cooled by air or water sprays to harden it sufficiently to preserve the shape imparted to it by the die. It continues to cool as it passes along the belt. It is either cut to length, in the case of rigid plastics, or coiled, in the case of flexible plastics. In addition to providing a cheap and rapid method of molding, the extrusion process makes it possible to produce tubing and shapes having reentrant angles.

In the laminating process sheets of paper or cloth made from glass or other types of fibers are impregnated with thermosetting liquid resin and placed together to build up a desired thickness. The resulting "sandwich" is cured, usually under considerable pressure and at an elevated temperature. Such products can be produced to have unusual strength properties, which primarily are the result of the sheet filler that is used. Because the surface is a thin layer of pure resin, laminates usually possess a smooth, attractive appearance. By using transparent resins, the sheet filler can be made visible, so that various decorative effects can be obtained by using cloth, wood, or other suitable materials.

Laminated plastics are produced as sheets, tubes, and rods. Flat sheets are produced as indicated in Figure 9-18. Figure 9-19 shows a press used for curing laminated sheets.

Figure 9-20 illustrates the method that is used to produce laminated tubing. The impregnated sheet stock is wound on a mandrel of the proper diameter. It then is cured in a molding press, after which the mandrel is removed. Rods are produced in a similar manner by using a small mandrel that is removed prior to curing.

Because of their excellent strength qualities, plastic laminates find a wide variety of uses. Some laminated sheets can be blanked[1] and punched readily. Gears frequently are machined from thick laminated sheets; these have unusually quiet operating characteristics when matched with metal gears.

[1] See Chapter 15.

FIGURE 9-19. Laminating plastic sheets in a molding press. Polished metal sheets are placed on the top and bottom of each pack to give a smooth surface. (*Courtesy Bakelite Division, Union Carbide Corporation.*)

Many laminated-plastic products, which are not flat and contain relatively simple curved shapes, such as boats, automobile bodies, and safety helmets, are made using only moderate or no pressure and low temperatures, often supplied by heat lamps. Usually, a simple female mold is used, made from metal, hard wood, or particle board. The laminating material, often in the form of fabric or glass cloth dipped in the liquid plastic resin, is placed in the mold in layers until the desired thickness is obtained. The mold, containing the laminated material, is then placed in a bag from which the air is evacuated. The external air pressure holds the laminate against the mold dur-

FIGURE 9-20. Method of producing laminated plastic tubing.

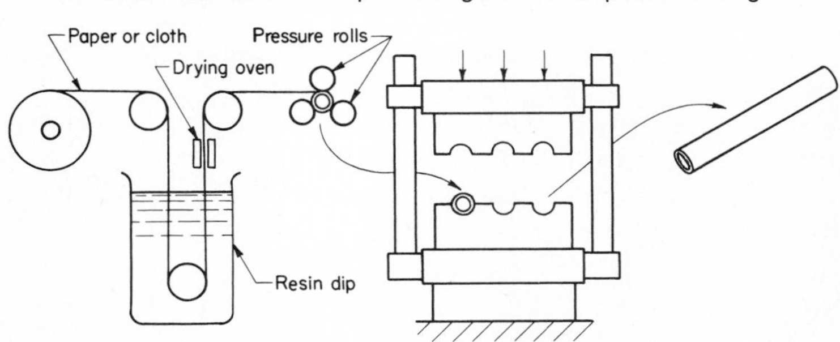

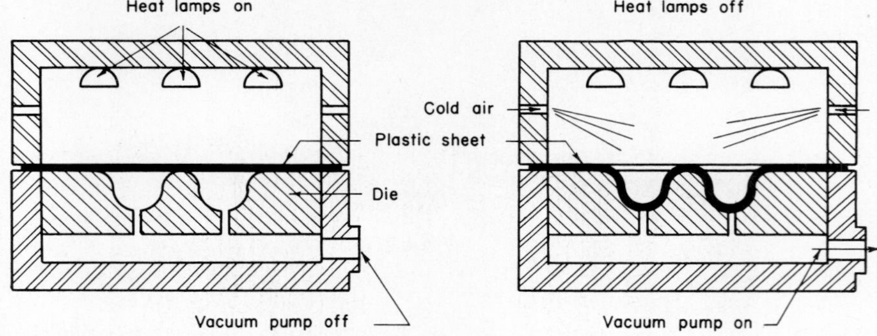

FIGURE 9-21. Method of molding plastic shapes by heat and vacuum (Thermoforming).

ing curing, in live steam or under heat lamps. The vacuum bag often can be eliminated when plastics are used that cure at room temperatures or at temperatures that can readily be obtained from a heat lamp; the pliable resin-dipped material is merely placed in the mold or over a form. Because of the low tooling costs, these processes make it possible to produce economically only one or a few large parts.

Vacuum forming is used extensively to form shapes from thermoplastic materials. As indicated in Figure 9-21 a sheet of plastic is placed over a die or form and heated until it becomes soft. A vacuum then is drawn between the sheet and the form so that the plastic takes the desired shape. It then is cooled, the vacuum dropped, and the part removed from the mold. The entire cycle usually requires only a few minutes. This process is quite economical and is used for producing a wide variety of products, ranging from panels for light fixtures to pages of Braille for the blind.

In the *cold-molding* process, depicted in Figure 9-22, the raw material is pressed to shape while cold, then removed from the mold and cured in an oven. The process is economical, but the resulting products do not have very good surface finish or dimensional tolerances. It is used, primarily, for compounds that are somewhat refractory.

FIGURE 9-22. Schematic diagram showing the steps in cold molding.

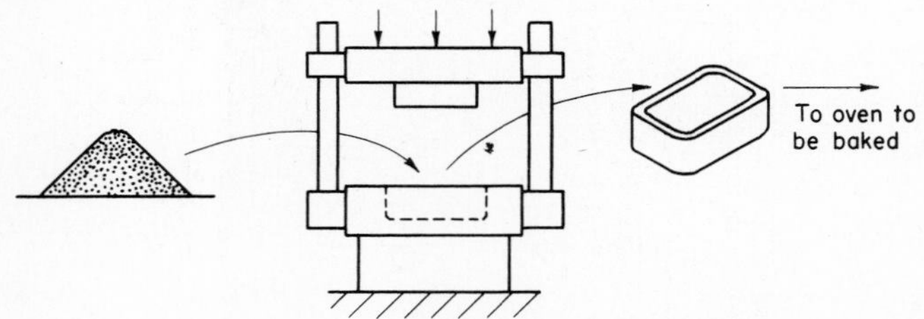

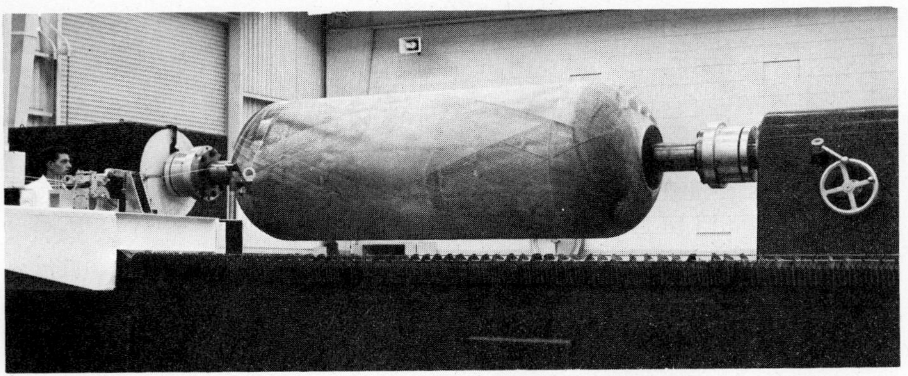

FIGURE 9-23. Large tank being made by filament winding. (*Courtesy Rohr Corporation.*)

Filament-wound products. The availability of plastic-coated, high-strength filaments of various materials, such as glass, graphite, and boron, has made it possible to produce containers of various shapes that have exceptional strength-to-weight ratios. The filaments are wound over a form, using longitudinal, circumferential, or helical patterns, or a combination of these, to take advantage of their highly directional strength properties and thus provide directional strength as needed in the product. Figure 9-23 shows a large tank being made by this process. Such tanks can be made in virtually any size, some as large as 4572 mm (15 feet) in diameter and 19 812 mm (65 feet) long, having been produced in fair quantities. The process has been highly mechanized so that uniform quality can be maintained. Because the special tooling for a new size or design is relatively inexpensive, the process is economical and flexible. A variety of plastics is used, with epoxies being very common.

Foamed plastics. *Foamed plastics* have become an important and widely used form. A foaming agent is mixed with the plastic resin and releases gas when the combination is heated during molding. The resulting products have very low densities, ranging from 32 to 641 kg/m^3 (2 to 40 lb/ft^3). Both rigid and flexible foamed plastics can easily be produced. The former type is useful for structural applications, packaging and shipping containers, as patterns in the full-mold casting process,[2] and for providing rigidity to thin-skinned metal components, such as aircraft fins and stabilizers, by being foamed in place in their interiors. Flexible foams are used primarily for cushioning. Quite a number of the basic plastics, both thermoplastic and thermosetting, are used for making foams.

It is possible to produce plastic products that have a solid, rigid outer skin and a rigid foam core. Such a product, made in a single molding process

[2] See Chapter 11.

FIGURE 9-24. Plastic gear having solid outer skin and rigid foam core. (*Courtesy American Machinist.*)

using two injection molding machines connected to a single mold, is shown in Figure 9-24.

Plastics for tooling. Because of their wide range of properties, their ease of conversion into desired shapes, and their excellent properties when loaded in compression, plastics are widely used in tooling to make jigs, fixtures and forming-die components. Both thermoplastics and thermosetting plastics, particularly cold-setting types, are used for drill and trim jigs, routing and assembly jigs and fixtures, and form blocks for forming. Thermoplastic materials are widely used for punch-press and drop-hammer punches, with the punches frequently formed in the female form block. Thermosetting and cold-setting plastics are used for stretch-press dies and tube-bending mandrels. The use of plastics in these tooling applications usually results in less costly tooling, permitting small quantities to be produced more economically. In addition, the tooling usually can be produced in a much shorter time so that production can get under way at an earlier date.

DESIGN AND SELECTION CONSIDERATIONS FOR PLASTICS

Plastics are not direct substitutes for metals. Also, in many cases, they are used closer to their design limits than are metals. In most cases, the conversion from raw material to finished product is a single step. But the range of properties is so great that if a proper selection is made, plastics can be used successfully in many more applications than is generally thought.

If one is to successfully design a plastic product, it is essential that four factors be considered: (1) the user requirements, particularly as to temperature, operating environment, and aging; (2) the material, with particular concern as to how well its properties are known; (3) the design; and (4) the production process required. Plastics offer the potential for substituting a single part for several, and various fabrication and fastening techniques usually are available. Obviously, the possibility of integral color, corrosion resistance, light weight, and thermal and/or electrical insulating properties offers unique advantages. But these seldom can be optimized unless one considers all four of the previously mentioned factors simultaneously. Thus the designer should keep abreast of the spectrum of plastics that are available and their properties, while also understanding how design details are related to the processing.

Design factors related to molding. In every casting process in which a fluid or semifluid material is introduced into a mold cavity and permitted to solidify in a desired shape, certain basic problems are encountered. First, the proper amount of material must be introduced and caused to completely fill the mold cavity. Second, any entrapped material within the cavity, usually air, must be removed. Third, any shrinkage of the material that occurs during solidification and/or cooling must be taken into account. Finally, it must be possible to remove the part from the mold after it has solidified; if the mold is to be reused, the mold must be opened. Unless the parts are designed properly, these requirements will not be met. The primary design consideration, of course, must be that the part satisfy its functional requirements. Thus a material must be selected that has the requisite properties in respect to tensile strength, impact strength, dimensional stability, color, and so on. This usually requires close cooperation between the designer and the molder. Often special attention must be given to appearance details, because plastics frequently are used for goods where consumer acceptance is of great importance.

Careful attention must be given to the problem of removing the part from the mold. Because metal molds are rigid, provision must be made so they can be opened. A small amount of unidirectional taper must be provided on each side of the mold parting plane to facilitate withdrawal of the part. Undercuts should be avoided whenever possible; they prevent removal unless special mold sections are provided that move at right angles to the opening motion of the major mold halves. Such mold construction is costly to make and to maintain.

As in all cast products, it is important to provide adequate fillets between adjacent sections to assure smooth flow of the plastic into all sections of the mold and also to eliminate stress concentrations at sharp interior corners. They also make the mold less expensive to produce and lessen the danger of mold breakage where thin, delicate mold sections are encountered. It is even desirable to round exterior edges slightly where permissible. A radius of 0.26 to 0.38 mm (0.010 to 0.015 inch) is scarcely noticeable but will do much to prevent an edge from chipping. Where plastics are used for electrical applications, sharp corners should be avoided because they increase voltage gradients and may lead to failure.

Wall section thickness is very important in plastic products. The curing time is determined by the thickest section. Thus it is desirable to keep sections as nearly uniform in thickness as possible. Primarily, the minimum wall thickness is determined by the size of the part and to some extent by the type of plastic used:

Minimum recommended	0.64 mm (0.025 in.)
Small parts	1.27 mm (0.050 in.)
Average-size parts	2.16 mm (0.085 in.)
Large parts	3.18 mm (0.125 in.)

Thick corners should be avoided because they are likely to lead to gas pockets, undercuring, or cracking. Where extra strength is desired at corners, this can usually best be accomplished by ribbing.

Economical production is greatly facilitated by adequate dimensional tolerances. A minimum tolerance of ±0.08 mm (0.003 inch) should be allowed in the direction parallel with the parting line of the mold. In the direction at right angles to the parting line a minimum tolerance of 0.26 mm (0.010 inch) is desirable. In both cases, increasing the tolerance by 50 per cent will reduce manufacturing difficulties and cost appreciably.

Design factors related to finishing. In designing plastic parts, a prime objective should be to eliminate any necessity for machining after molding. This is especially important where machined areas would be exposed; these are poor in appearance, and they also absorb moisture. Parting surfaces of molds are difficult to maintain in perfect condition so that they mate properly. Radii or curved surfaces where parting lines meet make it even more difficult to maintain perfect mating and should be avoided. The result of poor parting-line fit is a small fin or "flash," as illustrated in Figure 9-25. When fillers are used, as they are in most plastics, the exterior surface is a thin film of pure plastic without any filler, thus providing the smooth high luster that is characteristic of plastic parts. If the flash is trimmed off, a line of exposed filler remains, which may be objectionable. If parting lines are located at sharp corners, it not only is easier to maintain satisfactory mating of mold sections, but any exposed filler resulting from the removal of a fin will be confined to a corner, where it will be less noticeable.

Because plastics have low moduli of elasticity, large flat areas are not rigid and should be avoided whenever practicable. Ribbing or doming, as illustrated in Figure 9-26, are helpful in providing required stiffness. Flat surfaces also may reveal flow marks from molding and scratches that are bound to result from service. External ribbing can serve the dual function of increasing strength and rigidity and also of preventing scratches from showing. Dim-

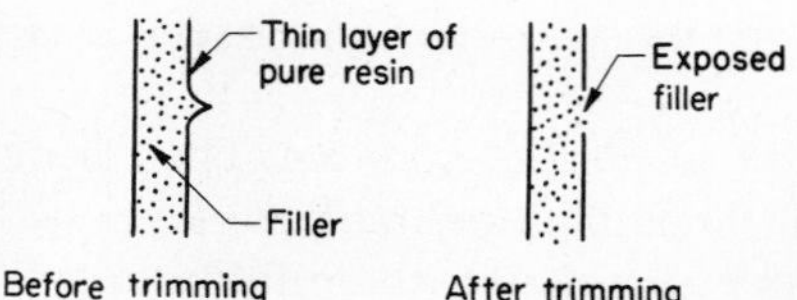

FIGURE 9-25. Effect of trimming the flash from a plastic part containing filler.

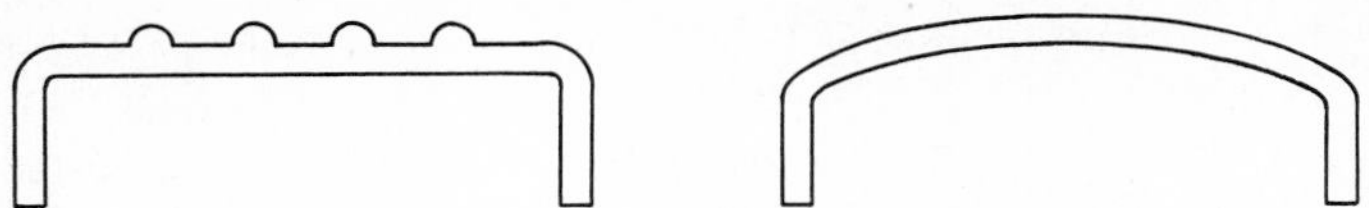

FIGURE 9-26. Method of providing stiffness in large surfaces of plastic parts by the use of ribbing and doming.

pled, or textured, surfaces often provide a pleasing appearance and do not reveal scratches.

Holes that must be formed by pins in the mold should be given special consideration. In compression-type molding, such pins are subjected to considerable bending influence during mold closing. Where these pins are supported at only one end, their length should not exceed twice the diameter. In transfer-type molds the length can be five times the diameter without maintenance becoming excessive.

Holes that are to be threaded after molding, or are to be used for self-tapping screws, should be countersunk slightly. This facilitates starting the tap or screw and avoids chipping at the outer edge of the hole. If the threaded hole is to be less than 6.35 mm (¼ inch) in diameter, it is best to cut the thread after molding by means of a thread tap. For diameters above 6.35 mm (¼ inch) it usually is better to mold the thread or to use an insert, which will be discussed later. If threads are molded, either a section of the mold must be removable, to permit later unscrewing from the part, or the part must be such that it can be unscrewed from the mold. Both procedures, particularly the latter, are not economical because of the mold delay that results.

Inserts. Because of the difficulty of molding threads in plastic parts and the fact that cut threads tend to chip, tapped or threaded inserts generally are used where considerable strength is required or frequent disassembly of the parts may occur. Several types of inserts are shown in Figure 9-27. The use of inserts requires attention to design details to obtain satisfactory results and economy. Inserts usually are made of brass or steel and are held in the plastic only by a mechanical bond. Therefore, it is necessary to provide suitable knurling or grooving so that the insert may be gripped firmly and not become loose in service. A medium or coarse knurl is quite satisfactory to resist torsional loads and moderate axial loads. A groove is excellent for axial loads but offers little resistance to torsional stresses.

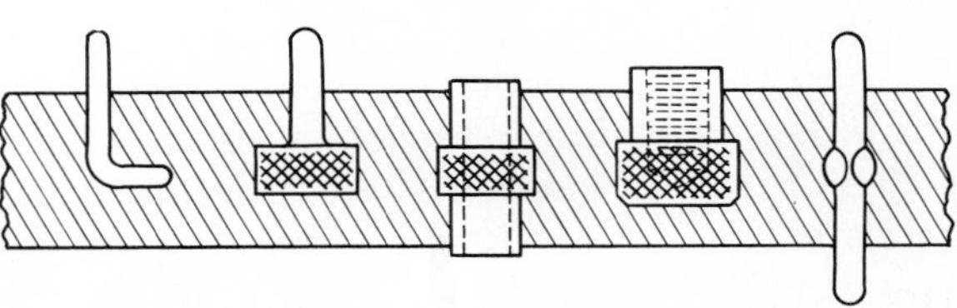

FIGURE 9-27. Typical metal inserts for use in plastic parts.

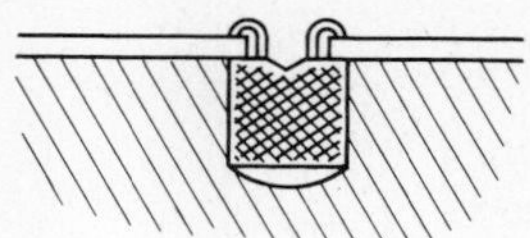

FIGURE 9-28. Type of spin-down insert for fastening plastic parts together permanently.

If an insert is to act as a boss for mounting or is an electrical terminal, it should protrude slightly above the surface of the plastic in which it is embedded. This permits a firm connection to be made without creating an axial load that would tend to pull the insert from the compound. On the other hand, if it is desired to use the insert to hold two mating parts closely together, the insert should be flush with the surface. In this way the parts can be held together snugly without danger of loosening the insert. Where it is necessary to keep the surface of an insert entirely free from any plastic, a shouldered design is most satisfactory. However, if an insert is used to fasten mating parts that must fit closely, a depression must be made in the mating part to provide clearance for the shoulder. Similarly, a depression has to be provided in the mold. Both operations add to the cost.

Where parts are to be fastened together permanently, the spun-down type of insert shown in Figure 9-28 is convenient and economical.

Inserts must have adequate support. The wall thickness of the surrounding plastic must be sufficient to support any load that may be transmitted by the insert. For small inserts the wall thickness should be at least half the diameter of the insert. Above 12.7 mm (½ inch) in diameter, the wall thickness should be at least 6.35 mm (¼ inch).

Machining plastics. Although most plastics are readily machined, because their properties vary so greatly, it is impossible to give instructions that are exactly correct for all. It is very important to remember some general characteristics that affect their machinability. First, all plastics are poor heat conductors. Consequently, little of the heat that results from chip formation will be conducted away through the material or be carried away in the chips. As a result, cutting tools run very hot and may fail more rapidly than when cutting metal. Carbide tools frequently are more economical to use than high-speed steel tools if cuts are of moderately long duration or if high-speed cutting is to be done.

Second, because considerable heat and high temperatures do develop at the point of cutting, thermoplastics tend to soften, swell, and bind or clog the cutting tool. Thermosetting plastics give less trouble in this regard.

Third, cutting tools should be kept very sharp at all times. Drilling is best done by means of straight-flute drills or by "dubbing" the cutting edge of a regular twist drill to produce a zero rake angle; these are shown in Figure 9-29. Rotary files and burrs, saws, and milling cutters should be run at high speeds so as to improve cooling, but with the feed carefully adjusted to avoid jamming the gullets. In some cases coolants can be used advantageously if

FIGURE 9-29. Straight-flute drill (*left*) and "dubbed" drill (*right*) used for drilling plastics.

they do not discolor the plastic or cause gumming. Water, soluble oil and water, and weak solutions of sodium silicate in water are used.

Fourth, filled and laminated plastics usually are quite abrasive and may produce a fine dust which may be a health hazard.

Finishing plastic parts. In the majority of cases plastic parts can be designed to require very little finishing or decorative treatment, thus promoting economy. In some cases fins and rough spots can be removed and smoothing and polishing done by barrel tumbling with suitable abrasives or polishing agents. Sometimes required decoration or lettering can be obtained by etching or engraving the mold. These procedures produce letters or designs that protrude from the surface of the plastic only about 0.1 mm (0.004 inch). When higher relief is desired, the mold must be engraved, which adds materially to mold cost.

Whenever possible, depressed letters that are to be filled with paint should be avoided. Such letters must be from 0.08 to 0.25 mm (0.003 to 0.010 inch) deep and must be raised above the surrounding surface of the mold, requiring the entire remaining mold surface to be cut away, at considerable expense. The cost frequently can be reduced by setting the letters in a small area raised above the main plastic surface. This requires only a small amount of die metal to be undercut.

Many plastic parts now are electroplated, as will be discussed in Chapter 37.

RUBBER AND ARTIFICIAL ELASTOMERS

Elastomers have unique characteristics in that, at room temperature, they can be stretched up to at least twice their original length and upon immediate release of the stress will return quickly to approximately their original length. Although they are elastic over a wide range, they do not obey Hooke's law. It is their structure, rather than their composition, that produces their elastic properties. They have remarkable capacity for storing energy, and they can be tailored to provide a wide range of stress–strain characteristics.

Natural rubber is the oldest, and still a widely used, elastomer. However, numerous artificial elatomers now are available that have been developed to meet specific needs and, in total, are used more than rubber.

The elastic characteristics of nonelastomers are usually due to the change in distance between adjacent atoms as the result of applied loads. The interatomic forces return the atoms to their normal positions when the load is removed. In elastomers, on the other hand, the elastic properties are due primarily to the fact that in the unstrained condition the basic molecule is in the form of a coil that, like a coil spring, can be stretched. When the load is removed, the stretched coil returns to its normal shape. Thus elastomers exhibit very large degrees of elasticity.

Rubber. Rubber is obtained from the *Hevea brasiliensis* tree, a native of Brazil but grown for commercial purposes primarily in the East Indies and Africa. The trees are tapped to obtain the sap, called *latex,* which consists of about 65 per cent water and 35 per cent rubber. The latex is coagulated with acetic acid, squeezed to remove the water, and the coagulate milled into sheets and dried. The dried sheet is known as *pale crepe,* or, if smoked after drying, as *smoked sheet.* The sheets are pressed into bales for shipment.

Some latex is imported in the liquid form for use in certain dipped products.

Rubber is used to only a limited extent in the crude form. It is an excellent adhesive and is thus used in many cements. These are made by dissolving the crude rubber in suitable solvents.

The modern extensive use of rubber dates from 1839, when Charles Goodyear discovered that it could be vulcanized by adding about 30 per cent sulfur and heating it at a suitable temperature. Sulfur causes sufficient cross-linking between the chains of molecules to restrict movement between them, and thus it imparts strength. Subsequently, it was found that the properties of vulcanized rubber could be greatly improved by adding certain pigments, notably carbon black, which would act as stiffeners, tougheners, and antioxidants. Certain *accelerators* have been found that greatly speed the vulcanization process with reduced amounts of sulfur, so most modern rubber compounds contain less than 3 per cent sulfur. As in the compounding of plastics, softeners and fillers are usually added to facilitate processing and to add bulk.

Rubber can be compounded over a wide range from soft and gummy to very hard, such as ebonite. Where high strength is required, textile cords or fabrics are coated with rubber to withstand the applied loads, the rubber largely serving to insulate the cords from each other and thus prevent chafing and friction. For severe service, steel wires may be coated with rubber and used as the load-carrying medium. This is done in some tires and heavy-duty conveyor belts.

Natural rubber compounds are outstanding for their high flexibility, good electrical insulation, low internal friction, and resistance to most inorganic acids and salts and to alkalies. However, their resistance to petroleum products, such as oil, gasoline, and naptha, is poor. In addition, they lose their strength at elevated temperatures, so it is not advisable to operate them at temperatures above 79 to 82°C (175 to 180°F). They also deteriorate fairly rapidly in direct sunlight unless specially compounded.

Artificial elastomers. The uncertainty of the supply and price of natural rubber to the highly industrialized countries in time of war led to the development of a number of artificial elastomers which have great commercial importance. One, polyisoprene, appears to have the same molecular structure as natural rubber and equal or superior properties. Some of the others are inferior to natural rubber, while others have distinctly different and, frequently, superior properties, thus extending their usefulness for specific applications. Table 9-2 lists the most widely used artificial elastomers, with natural rubber shown for comparison, and gives their typical properties and uses. However, it must be remembered that the properties can vary considerably, depending on how they are compounded and processed.

Processing of rubber and elastomers. Rubber products are made by several processes. The simplest is where they are formed from a liquid preparation or compound. These commonly are called *latex products.* Dipped products are made by immersing a form repeatedly into the latex compound, causing a certain amount of the liquid to adhere to the surface of the form each time. After each dipping, the film is allowed to dry, usually in air. Dipping is continued until the desired thickness is obtained. After vulcanization, usually in steam, the products are stripped from the forms.

Most latex products now are made by the *anode* process. This process utilizes the electrical charges on the latex particles. The charges are neutralized by being associated with a coagulant that has previously been deposited and that releases positively charged ions when dipped into the latex. The positive ions neutralize the charges on the adjacent latex particles and thus cause them to be deposited on the form. This process goes on continuously, so any desired thickness can be deposited.

When products are to be made from solid elastomers, the first step is the compounding of the elastomers, vulcanizers, fillers, autioxidants, accelerators, and other pigments. This usually is done in a *Banbury mixer,* which breaks down the elastomers and permits mixing in some of the other components to form a homogeneous mass.

Usually, the mix next is put on a *mill,* such as is shown in Figure 9-30, in which chilled iron rolls rotate toward each other at different speeds. They are cooled by the circulation of water through their interiors to remove the heat generated by the milling action, thus preventing the start of vulcanization. The sulfur and accelerators usually are added at this stage.

TABLE 9-2. Properties and uses of common elastomers

Elastomer	Specific Gravity	Durometer Hardness	Tensile Strength (psi) Pure Gum	Tensile Strength (psi) Black	Elongation (%) Pure Gum	Elongation (%) Black	Service Temp. °C (°F) Min.	Service Temp. °C (°F) Max.	Resistance to:* Oil	Resistance to:* Water Swell	Resistance to:* Tear	Typical Applications
Natural rubber	0.93	20–100	2500	4000	750	650	−54 (−65)	82 (180)	P	G	G	Tires, gaskets, hose
Polyacrylate	1.10	40–100	350	2500	600	400	−18 (0)	149 (300)	G	P	F	Oil hose, O-rings
EDPM (ethylene propylene)	0.85	30–100	1	3		500	−40 (−40)	149 (300)	P	G	G	Electric insulation, footware, hose, belts
Chlorosulfonated polyethylene	1.10	50–90	4	2		400	−54 (−65)	121 (250)	G	E	G	Tank linings, chemical hose, shoe soles and heels
Polychloroprene (neoprene)	1.23	20–90	3500	4000	800	550	−46 (−50)	107 (225)	G	G	G	Wire insulation, belts, hose, gaskets, seals, linings
Polybutadiene	1.93	30–100	1000	3000	800	550	−62 (−80)	100 (212)	P	P	G	Tires, soles and heels, gaskets, seals
Polyisoprene	0.94	20–100	3000	4000		600	−54 (−65)	82 (180)	P	G	G	Same as natural rubber
Polysulfide	1.34	20–80	350	1000	600	400	−54 (−65)	82 (180)	E	G	G	Seals, gaskets, diaphragms, valve disks
SBR (styrene butadiene)	0.94	40–100	2			1200	−54 (−65)	107 (225)	P	G	G	Molded mechanical goods, disposal pharmaceutical items
Silicone	1.1	25–90		1200		450	−84 (−120)	232 (450)	F	E	P	Electric insulation, seals, gaskets, O-rings
Epichlorohydrin	1.27	40–90		2		325	−46 (−50)	121 (250)	G	G	G	Diaphragms, seals, molded goods, low-temperature parts
Urethane	0.85	62–95	5000		700		−54 (−65)	100 (212)	E	F	E	Caster wheels, heels, foam padding
Fluoroelastomers	1.65	60–90	1	3		400	−40 (−40)	232 (450)	E	E	F	O-rings, seals, gaskets, roll coverings

* P = Poor; F = Fair; G = Good; E = Excellent

FIGURE 9-30. Twin rubber mill installation. Rubber is being fed to a tuber from the right-hand mill. (*Courtesy Adamson United Company.*)

Rubber compounds and plastics are made into sheet form on *calenders,* such as is shown in Figure 9-31 and illustrated schematically in Figure 9-32. The sheet coming from a calender is rolled into a fabric liner to prevent the adjacent layers from sticking together.

When cord or square-woven fabric is to be covered with rubber, this also

FIGURE 9-31. Three-roll calender used for producing rubber or plastics in sheet form. (*Courtesy Farrel-Birmingham Company.*)

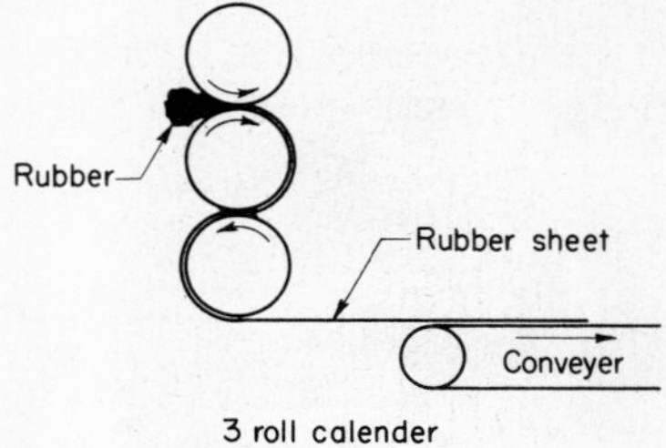

FIGURE 9-32. Schematic diagram showing the method of making sheets of rubber in a three-roll calender.

is done on a three- or four-roll calender. On a three-roll calender only one side of the fabric can be coated at each pass. A four-roll claneder, such as is indicated schematically in Figure 9-33, makes it possible to coat both sides of the fabric at a single pass.

Many rubber products, such as inner tubes, garden hose, tubing, and moldings, are produced by extrusion. The compound from a mill is forced through a die by a screw device similar to the meat grinder used by a butcher. Figure 9-34 shows such an extrusion machine.

Excellent adhesives have been developed which permit bonding rubber and artificial elastomers to metal, usually brass or steel. Tanks of all sizes are made by this procedure for transporting and storing a wide variety of corrosive liquids. Usually only moderate pressure and temperature are required to obtain excellent adhesion.

Elastomers for tooling. When an elastomer is confined, it will act as a fluid, transmitting force quite uniformly in all directions. This phenomenon often makes it possible to substitute an elastomer for one half of a metal die set in connection with metal forming. This procedure also makes it possible to do bulging and forming of reentrant sections, which would be impossible with steel dies except by very costly multipiece dies. Also, because elastomers can be compounded to range from very soft to very hard, hold up very well when subjected to compression loading, and can quickly and economically be made into a desired shape, they are being used increasingly as tooling materials.

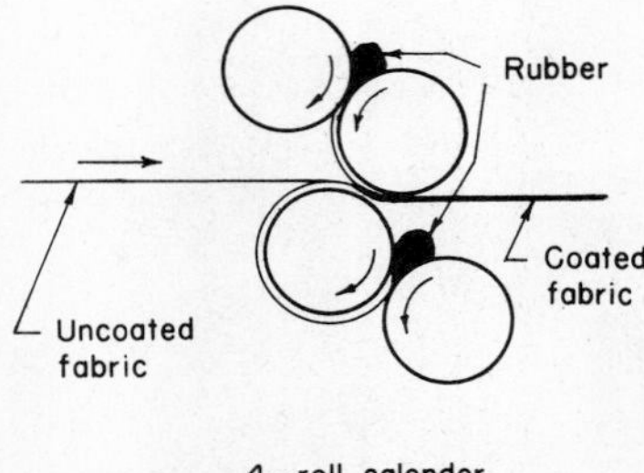

FIGURE 9-33. Arrangement of the rolls, fabric, and coating material for coating both sides of fabric in a Z-type four-roll calender.

FIGURE 9-34. Extrusion of automobile tire treads. Inset shows closeup of the tread emerging from the die. (*Courtesy Farrel-Birmingham Company.*)

CERAMICS

Long used in the electrial industry because of their high electrical resistance, in recent years ceramics have assumed considerable importance as a general engineering material because of their ability to withstand high temperatures. Generally, they are applied as coatings over metal to provide protection from hot and/or corrosive gases, as in jet and rocket engines.

Ceramics contain phases that are compounds of metallic and nonmetallic elements. Consequently, because there are many possible combinations of metals and nonmetals, there are a multitude of ceramic materials. Also, the same combination of metals and nonmetals may exist in more than one structural arrangement, thereby producing polymorphism. For example, depending on the temperature, silica can exist in three forms—*quartz, tridymite,* and *cristobalite.* The subject of ceramics is too large to be treated in detail here, but some details of their basic nature and properties will be presented so as to indicate their possible uses.

Molecular structure of ceramics. Most ceramics have crystal structures. However, unlike metals, they do not have large numbers of free electrons, the electrons being shared covalently or in ionic bonds. The absence of free electrons

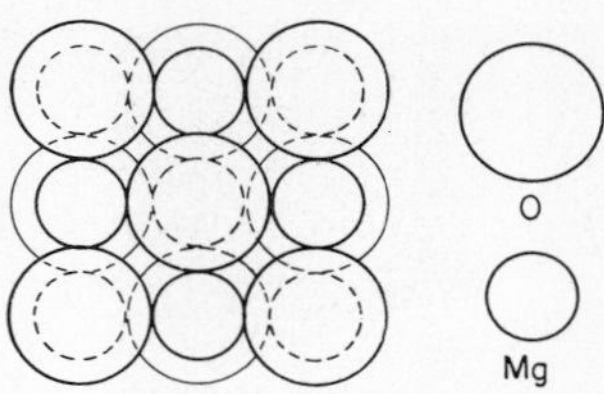

FIGURE 9-35. Arrangement of atoms of different sizes in a ceramic crystal.

makes ceramics poor electrical conductors and results in their being transparent in thin sections; because ionic bonds tend to produce high stability, ceramics have high melting temperatures.

The crystal structure of ceramics must accommodate atoms of different sizes. Figure 9-35 illustrates such an arrangement. Several basic crystal structures exist, the cubic and tetrahedral being very common.

In numerous cases ceramic crystals grow into chains, similar to plastic molecules. However, there is an important difference in that the chains are held together by ionic bonds instead of by weak van der Waal's forces. However, the bonds between chains are not so strong as those within the chains. Consequently, when forces are applied, cleavage occurs between the chains.

In some cases the molecules form *sheets* and result in *layered* structures; these have relatively weak bonds between the sheets.

Properties of ceramics. Most *mechanical ceramics* have specific gravities in the range 2.3 to 3.85. By comparing their molecular structure and bonding with those of metals, their behavior under load can be predicted. Metals have considerable ductility because they have lower shear resistance than tensile resistance. Ceramics, on the other hand, have stronger interatomic bonding and higher shear resistance, and thus they have low ductility. However, these same qualities impart high compressive strength. Theoretically, ceramics could also have high tensile strengths, but ordinarily they do not because of small cracks and pores that act as stress concentrators which are not reduced through ductility and plastic flow. Failure thus occurs at low average stress values, typical tensile strengths ranging from 21 to 210 MPa (3000 to 30,000 psi). By using special techniques to eliminate the cracks, very high strengths can be obtained; some glass fibers have strengths above 7000 MPa (1,000,000 psi). Some ceramics have melting points above 1649°C (3000°F).

Cermets. Cermets are combinations of metals and ceramics, bonded together in the same manner in which powder metallurgy parts are produced. They combine some of the high refractoriness of ceramics and the toughness and thermal-shock resistance of metals. Oxide cermets usually are chromium–alumina- or chromium–molybdenum–alumina–titania-based. Carbide cermets are based on tungsten, titanium, chromium, or tungsten–titanium carbides.

Parts are produced from cermet materials by pressing the powders in molds at pressures ranging from 70 to 280 MPa (10,000 to 40,000 psi) and then sintering them in controlled-atmosphere furnaces at about 1649°C (3000°F).

Cermets are used principally as cutting tools and crucibles and as nozzles for jet engines or other high-temperature devices.

Review questions

1. What is the distinguishing molecular characteristic of plastics?
2. What is meant by the term "unsaturated molecule"?
3. Explain what isomers are.
4. How does copolymerization differ from polymerization?
5. Basically, what causes the difference between thermosetting and thermoplastic materials?
6. Why is it important to know whether a plastic is thermoplastic or thermosetting?
7. Explain the mechanism by which plastics deform under load.
8. What are the characteristics of plastics which account for their wide use as engineering materials?
9. What is the most common use classification of plastics? What accounts for this?
10. Explain why the orienting process is applied to some plastics.
11. Why are plastic resins seldom used in their pure form?
12. Why should the addition of lubricants in compounding plastics be held to a minimum?
13. What properties account for the increased use of epoxies?
14. Name three common types of thermosetting plastics.
15. Since nylon has excellent properties, why is it not used more extensively?
16. What are the principal uses for fluorocarbon plastics?
17. What are six common processes for producing parts from plastics?
18. Why are thermoplastics not as suitable for hot-compression molding as are thermosetting materials?
19. Why is the transfer molding process used?
20. Why are thermosetting materials more difficult to injection mold than thermoplastics?
21. Explain how solid rods are produced by laminating.
22. Why are melamine plastics not used extensively for extrusions?
23. Why are large plastic parts seldom produced by casting?
24. Where is the best location for the mold parting line on a plastic part?
25. Why should the walls in plastic parts be kept as uniform as practicable?
26. Why can high strength-to-weight ratios be obtained in filament-wound plastic products?
27. Explain what results when the flash is trimmed from a plastic part.

28. When should threaded inserts for screws be used in plastic parts?
29. Why should an insert that acts as a mounting boss on a plastic part protrude slightly above the surface in which it is embedded?
30. Why is it desirable for lettering on a plastic part to be raised above the surface?
31. Why do elastomers exhibit elastic properties?
32. What is the purpose of adding an accelerator to an elastomer compound?
33. Which type of artificial rubber has the same molecular structure as natural rubber?
34. Explain how vulcanization increases the strength of elastomers.
35. What are the primary functions of rubber in a tubeless automobile tire?
36. What are two functions provided by calendering?
37. How do ceramic crystals differ basically from metal crystals?
38. Explain why there can be so many ceramic materials.
39. What is a basic difference between ceramic and plastic molecular chains?
40. Why do most ceramics not have high tensile strengths?
41. What are cermets?
42. What are two common uses for cermets?

CHAPTER 10

Material selection

During recent years the selection of the materials that are to be used in the design and manufacture has assumed great importance. For numerous reasons, it is certain that its importance is going to continue to increase. Although some new materials will become available, there will be decreasing availability of others. Problems associated with pollution and recycling will have to be considered. The necessity for weight reduction, to save energy, will require the use of different materials. Domestic, and particularly foreign, competition will force product reevaluation. Service and customer requirements will be more severe and critical. The great increase in product liability actions, too often growing out of improper use of materials, will have a marked impact. The interdependent relationship between materials and their processing will continue and, most likely, will be better recognized. Consequently, both design and manufacturing engineers will have to exercise knowledgeable care in selecting, specifying, and utilizing materials in order to achieve desired satisfactory results at reasonable cost and with assured quality.

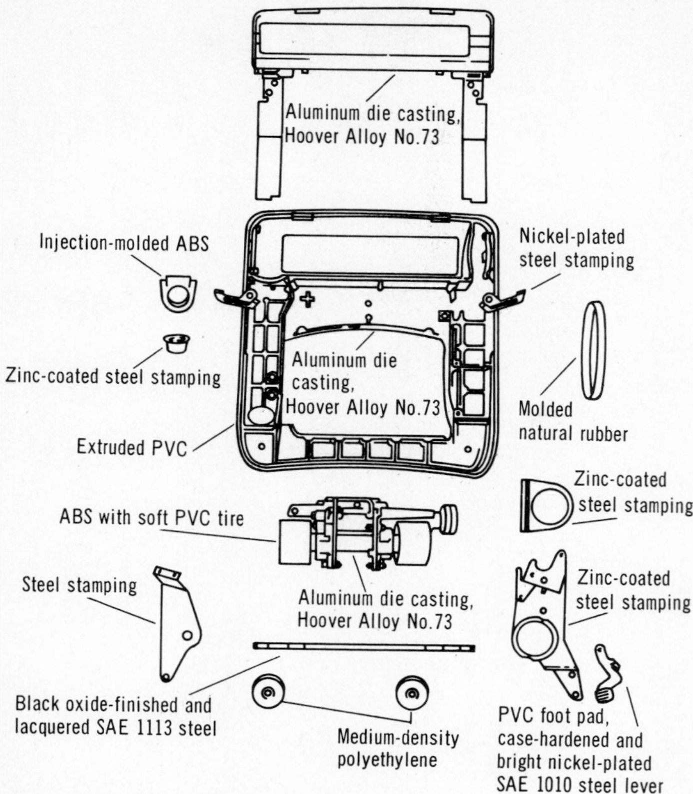

FIGURE 10-1. Materials used in various parts of a vacuum cleaner assembly. (*Courtesy Metal Progress.*)

Most modern products are relatively complex. To achieve a proper balance among functional fulfillment, pleasing appearance, and reasonable cost, it almost always is necessary to utilize a variety of materials. Further, with new materials almost constantly coming onto the market, manufacturers of existing products have a virtually continuous task of reevaluating the materials currently in use to assure that progress does not pass them by. The vacuum cleaner assembly shown in Figure 10-1 is typical, nine different materials being used in the assembly. As shown in Table 10-1, of the 13 components, the materials for 12 had been changed completely from those used originally, and one has been modified substantially. Eleven different reasons were given for making the changes. This example illustrates the fact that material selection is both a complex and a continuing process. It is not surprising that a single individual will have great difficulty in keeping abreast of, and making the necessary decisions concerning, the materials in even a single, fairly simple product and, as a consequence, many companies have established separate

TABLE 10-1. Examples of materials selection in modern vacuum cleaners

Part	Former Material	Present Material	Benefits
Bottom plate	Assembly of steel stampings	One-piece aluminum die casting	More convenient servicing
Wheels (carrier and caster)	Molded phenolic	Molded medium-density polyethylene	Reduced noise
Wheel mounting	Screw-machine parts	Preassembled with a cold-headed steel shaft	Simplified replacement, more economical
Agitator brush	Horsehair bristles in a die-cast zinc or aluminum brush back	Nylon bristles stapled to a polyethylene brush back	Nylon bristles last seven times longer and are now cheaper than horsehair
Switch toggle	Bakelite molding	Molded ABS	Breakage eliminated
Handle tube	AISI 1010 lock seam tubing	Electric seam-welded tubing	Less expensive, better dimensional control
Handle bail	Steel stamping	Die-cast aluminum	Better appearance, allowed lower profile for cleaning under furniture
Motor hood	Molded cellulose acetate (replaced Bakelite)	Molded ABS	Reasonable cost, equal impact strength, much improved heat and moisture resistance—eliminated warpage problems
Extension tube spring latch	Nickel-plated spring steel, extruded PVC cover	Molded acetal resin	More economical
Crevice tool	Wrapped fiber paper	Molded polyethylene	More flexibility
Rug nozzle	Molded ABS	High-impact styrene	Reduced costs
Hose	PVC-coated wire with a single-ply PVC extruded covering	PVC-coated wire with a two-ply PVC extruded covering separated by a nylon reinforcement	More durability, lower cost
Bellows, cleaning tool nozzles, cord insulation, bumper strips	Rubber	PVC	More economical, better aging and color, less marking

Source: Metal Progress magazine; by permission.

departments or groups to deal with this problem. But, obviously, the design engineer has a primary responsibility to know about, and understand, the properties and characteristics of the materials he or she selects for use in converting designs into reality.

Steps in the design process. Design usually takes place in several distinct steps: (1) conceptual, (2) functional, and (3) production design. Material selection enters into these steps in different ways. During the *conceptual-design* step, the designer is concerned primarily with the functions the product is to fulfill and the manner in which it will satisfy these requirements. Usually, several concepts are visualized and considered, and a decision is made either that the idea is not practicable or that the idea is sound and one or more of the conceptual designs will be developed further. At this stage, about the only thought given to material selection is whether materials are available that will provide the properties that are required, or, if no materials are available, whether there is a reasonable propspect that new ones can be developed within cost and time limitations.

At the *functional-* or *engineering-design* stage, a practical, workable design is developed. Usually, fairly complete drawings are made, and materials are selected and specified for the various components. Often a prototype or working model is made that can be tested to permit evaluation of the product as to function, reliability, appearance, serviceability, and so on. Although it is expected that such testing may show that some changes may have to be made in materials before the product is advanced to the production-design stage, this should not be taken as an excuse for not doing a thorough job of material selection in the engineering development stage. Consequently, appearance, cost, reliability, and producibility factors should be considered in detail, together with the functional factors. Although it is a somewhat unorthodox procedure, there is much merit to the practice of one very successful company which requires that all prototypes be built with the same materials that will be used in production and, insofar as possible, using the same manufacturing techniques. It is of little value to have a perfectly functioning prototype that cannot be manufactured economically in the expected sales volume, or that is substantially different from what the production units will be, particularly as to quality and reliability. Also, it is much better for design engineers to do a complete job of material analysis, selection, and specification at the development stage of design than to leave it to the production-design stage, where changes may be made by others that will adversely affect the functioning of the product they conceived and for which they hold high hopes.

At the *production-design* stage, the primary concern relative to materials should be that they are specified fully, that they are compatible with, and can be processed economically by, the existing equipment, and that they are readily available in the needed quantities.

As manufacturing progresses, it is inevitable that situations will arise that will require some modifications of the materials being used. Experience may reveal that some substitutions of cheaper materials can be made. But in most cases changes are much more costly to make after manufacturing is in progress than before it starts, and good selection during the production-design phase will eliminate the necessity for most of this type of change. The more common type of change that occurs after manufacturing starts results from the availability of new materials. These, of course, present possibilities for cost reduction and improved performance. However, *new materials must be evaluated very carefully to make sure that all their characteristics are well established.* One should always remember that it is indeed rare that as much is known about the properties and reliability of a new material as about those of an existing material. A large proportion of product failure and product-liability cases have resulted from new materials being substituted before their long-term properties were really known. There is no excuse for such poor material selection, and claimant attorneys thrive on such practices.

A procedure for material selection. Because there almost always are a number of factors to be considered, material selection is a complex process. Consequently, it is desirable to follow some routine procedure when selecting a material to assure that some important factor is not neglected. The obvious, but often overlooked, first step is to list all the functional requirements that the material must meet. This may be a very short list, but often it contains quite a few items. Such a list may include strength, hardness, ductility, hardenability, formability, impact resistance, fatigue properties, weldability, appearance, availability, machinability, cost, and so on. The important point is that *all* factors that are requisite be listed. It also is apparent that in most cases there will be several factors that are important, and a very simple selection and specification, such as "low-carbon steel," is totally inadequate.

In making a list of required material properties, it is very important that *all* service conditions and uses be considered. Many failures and product-liability claims have resulted from the designer not anticipating reasonable uses for a product or conditions that the product would have to encounter aside from the narrow, specific function for which he designed it. Figure 10-2 shows such an example. The designer of a large electrical transformer substituted a plastic laminate in place of maple for the members of the structure that supported the heavy, copper leads. In service, certain of these members would have been only in compression, and the laminate was excellent for this type of loading. However, the designer failed to consider that during *shipping,* some of the members might be subjected to bending loads, and the material was weak in this regard. As a consequence, several supporting members broke during shipping, with a domino effect, producing extensive failure and damage.[1] Admittedly, it is not easy to forsee all possible requirements, but obviously it is necessary to do so.

[1] Another example was the complete failure of a very large steel storage tank during its hydrostatic proof test. The tank would have been operated at a temperature of about 38°C (cont'd)

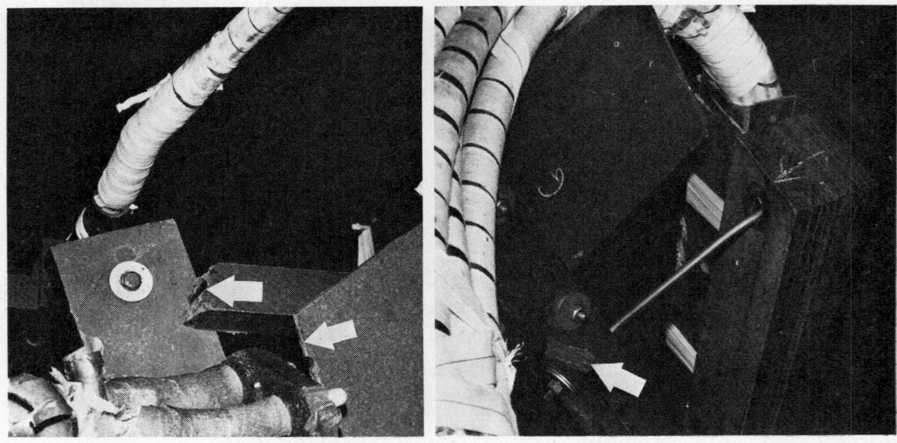

FIGURE 10-2. Failures in the lead-support structure of a large electrical transformer. Material selection failed to take into account stresses resulting from shipping.

After the complete list of required material properties is complete, the next step is to indicate which of the requirements are "absolutes" and which are "relative." Absolutes are those about which there can be no compromise. For example, if ductility is a "must," gray cast iron would be ruled out. If it must have electrical conductivity, plastics are out. On the basis of absolutes many materials are quickly eliminated from consideration, and attention is focused on certain possibilities. On the other hand, the relative requirements usually are not go–no go situations. If a requirement is weldability, this is not a precise-degree term; it can be met, with varying degrees of ease, in various ways, with various materials. One quickly recognizes that compromise and opinion will enter into the decisions regarding these factors: cost versus appearance, hardness versus formability. Decisions and compromises will not be avoided or made automatic by such a listing of the required properties, but it will go a long way in assuring that the person making the selection has given thought to *all* the factors that are required.

Additional factors to consider. There are several factors that frequently are overlooked in making material selections but that are worthy of special mention. One is possible misuse of the product by the user. If a product is to be used only by skilled, trained technicians, one may have some assurance that it will be used properly. But could it also be used by untrained people? If a product is to be used by the general public, one should anticipate the worst.

(100°F), but the proof test was to be conducted by filling it with water at ambient temperature. The water and ambient temperatures were about 7°C (45°F), and the tank failed with a brittle fracture, originating in a heat-affected zone adjacent to a weld, with the loss of one life. The designer neglected to consider the brittle fracture properties of the material at the test temperature, and the steel was found to have a ductile-to-brittle transition temperature of 21°C (70°F).

Very careful consideration should be given to the analysis of prior failures of a product or component, or of similar products or components. Quite often such failure-analysis information never seems to get into the hands of designers.

Another matter is the extent to which a particular material, or type of material, has contributed to excessive or unusual service requirements or difficulties. It is amazing how often such difficulties are carried over from year to year, or model to model, and it often appears that designers never consider that a product has to be serviced.

A fourth factor for special consideration is the extent to which materials can be standardized. Although one should not sacrifice function, reliability, and appearance for standardization and its possible attendant cost savings, neither should one overlook the possible savings and other benefits that often can be obtained without sacrificing any of the required properties.

Finally, it is desirable for anyone who selects and specifies materials to be somewhat familiar with the concepts of value analysis.[2] The type of thinking involved in this technique is also applicable to materials selection.

The effect of product liability on material selection. Product liability actions and awards in the courts, particularly in the United States, make it imperative that designers and companies employ the very best procedures in selecting materials. While most neutral persons would agree that the situation has gotten almost out of control, there unfortunately have been too many instances where designers and companies have not used sound procedures in selecting materials. No company designer can afford to do so, nor should do so, under today's conditions. Probably the five most common faults have been: (1) failure to know and use the latest and best information available about the materials utilized; (2) failure to foresee, and take into account, the *reasonable uses* for the product; (3) the use of materials about which there was insufficient data, particularly as to its long-term properties; (4) inadequate, and unverified, quality control procedures, and (5) material selection made by people who are completely unqualified to do so.

An examination of the five faults above will lead one to conclude that there is no good reason why they should exist, and a consideration of them provides guidance as to how they can be eliminated. The current situation regarding product liability claims is so serious that no designer or company can afford to ignore it. And the fact is that until some changes are made in our laws and legal procedures, following the very best methods in material selection is not going to eliminate all product-liability claims. There are too many plaintiff's attorneys who operate on the principle of "I don't care about

[2] See L. D. Miles, *Techniques of Value Analysis and Engineering* (New York: McGraw-Hill Book Company, 1961), or E. P. DeGarmo, J. R. Canada, and W. G. Sullivan, *Engineering Economy,* 6th ed. (New York: Macmillan Publishing Co., Inc., 1979), Chap. 4.

the facts, just let me get the case before a jury." However, proper procedures by designers and companies can greatly assist in reducing such claims.

Aids to material selection. From the previous discussion in this chapter, it is apparent that those who select materials should have a broad, basic understanding of the nature and properties of materials and their processing. This is, of course, a primary purpose of this book. However, the number of materials is so great, as is the mass of information that is available and useful in specific situations, that a single book of this type and size cannot be expected to furnish all the information required. Thus anyone who does much work in materials selection needs to have ready access to other sources of data. Fortunately, there are numerous sources of such data available.

One very useful source is the "Materials Selector" issue of *Materials Engineering.* This is a monthly magazine, and one issue each year, usually in November, contains only tabulated data and advertising about all the common, current engineering materials.

Another "must" is Volume 1 of the *Metals Handbook,* published by The American Society for Metals. This volume deals solely with the properties and selection of metals. It is voluminous and detailed—not as easy to use for quick reference as the "Materials Selector," but very complete and authoritative.

The two-volume *Source Book on Materials Selection,* also published by The American Society for Metals, is very useful. While it contains numerous verbatum sections from the *Metals Handbook,* it also contains reprints of numerous special papers and some original material relative to special products and applications.

The "Databook Issue" of *Metal Progress* magazine, published by The American Society for Metals, contains a vast amount of useful data regarding materials, process engineering, and fabrication technology. Unfortunately, the large amount of advertising that is closely interspersed with the data detracts considerably from its usefulness.

Anyone who selects materials should have available a number of the handbooks that are published by the various technical societies and trade associations, such as the *Cast Metals Handbook* of The American Foundrymen's Association. In addition, various supplier companies publish and make available, at low or no cost, data books relating to their particular products. For example, several aluminum and steel companies publish excellent books of this type.

In addition to books of the type described, it is helpful to compile and keep current lists or charts of cost indices, such as cost per kilogram or pound, cost per cubic millimeter or cubic inch, or cost per kilonewton or 1000 pounds of yield strength. These can be in whatever form is most applicable for the type of work most commonly done. However, it should be remembered that these can be only rough guides, because more than one variable usually must be considered.

The type of comparative rating chart shown in Figure 10-3 can be help-

RATING CHART FOR SELECTING MATERIALS

Material	Go-No-Go** Screening			Relative Rating Number (†Rating Number x *Weighting Factor)								Material Rating Number
	Corrosion	Weldability	Brazability	Strength (5)*	Toughness (5)	Stiffness (5)	Stability (5)	Fatigue (4)	As-Welded Strength (4)	Thermal Stresses (3)	Cost (1)	$\frac{\Sigma \text{ Rel Rating No.}}{\Sigma \text{ Rating Factors}}$

*Weighting Factor (Range = 1 Lowest to 5 Most Important)
†Range = 1 Poorest to 5 Best
**Code = S = Satisfactory
U = Unsatisfactory

FIGURE 10-3. Rating chart that may be used for comparing materials.

ful in selecting between several materials. Rating factors of whatever magnitude believed applicable, in terms of their importance, can be assigned to the various properties. Each material then is rated on a scale of 1 to 5 in each of these categories.

The availability of computers makes it possible to quickly compare materials, somewhat in the manner indicated in Figure 10-3, and also to store and print out detailed information about a given material. Although this procedure is quick and convenient, it should be remembered that the stored data must be kept up to date if it is to be useful.

Finally, it is apparent that material selection can be a very large and complex task when complex or many products are involved, as in many large organizations. Just keeping up with the new materials that appear is a very considerable task. Consequently, in such cases the organization of a materials group is important. Some specialists often are required, but it also is important that a variety of backgrounds is included so as to cover the chemical, metallurgical, mechanical, pollution, recycling, and production aspects.

Review questions

1. What are the three usual phases of product design, and how does the consideration of materials differ in each?

2. Name three factors which commonly bring about reevaluation of the materials that are being used in a product.
3. What factor must be considered with special care in considering the selection of an entirely new material?
4. Why is an established routine, or procedure, desirable in selecting materials?
5. What are four conditions (or failures) that often give rise to product liability claims?
6. Why is it important for a designer to consider the type of persons for which a product is designed?
7. What is meant by "absolute" and "relative" requirements of materials?
8. Why is it important to be able to show, by records, how a material-selection decision was made?
9. What are two excellent sources of data regarding materials?
10. Three materials, X, Y, and Z, are available for a certain usage. Any material selected must have good weldability. Tensile strength, stiffness, stability, and fatigue strength are required, with fatigue strength being considered most important and stiffness the least important of these factors. The three materials are rated as follows in these factors:

	X	Y	Z
Weldability	Excellent	Poor	Good
Tensile strength	Good	Excellent	Fair
Stiffness	Good	Good	Good
Stability	Good	Excellent	Good
Fatigue strength	Fair	Good	Excellent

Using the rating chart shown in Figure 10-3, which material should be selected?

Case study 8.
THE CASE OF THE CHIPPED HAMMER

The left-hand object in Figure CS-8 is an enlarged view of a chip that came off the face of a framing hammer at the point indicated by the arrow in the next two views, which show the condition of the face of the hammer. In the side view shown at the right, a duplicate, new hammer is shown for purposes of comparison. The chip lodged in the eye of the worker using the hammer, and he brought legal action against the manufacturer, alleging that the hammer was defective.

What factors would you check to determine whether the hammer was of proper quality as to material and manufacture, or whether the worker had misused the hammer in violation of any or all of the warning notice that was attached to the handle of the hammer when it was sold? The notice read: "WARNING—BE SAFE—WEAR SAFETY GOGGLES. This hammer is intended for driv-

FIGURE CS-8. (*Left*) Chip from framing hammer. (*Center*) View of end of hammer. (*Right*) Profiles of new and subject hammer.

ing and pulling common nails only. Hammer face may chip if struck against another hammer, hardened nails, or other hard objects, possibly resulting in eye or other bodily injury.''

Casting and forming processes

Casting processes

Basically, casting consists of pouring a liquid material—most frequently molten metal—into a previously prepared cavity, or mold, and permitting it to solidify and thereby acquire a desired shape. Thus, *in a single step,* simple or complex shapes can be made from any metal that can be melted, with the resulting product having virtually any configuration the designer desires for best resistance to working stresses, having minimal directional properties and, usually, a pleasing appearance.

Although some nonmetals are cast, the process is of primary importance in the production of metal products, and only metal casting will be considered in this chapter. The metals most frequently cast are iron, steel, aluminum, brass, bronze, magnesium, and certain zinc alloys. Of these, iron, because of its fluidity, low shrinkage, strength, rigidity, and ease of control, is outstanding for its suitability for casting and is used more than all others.

Cast parts range in size from a few millimeters and weighing a fraction

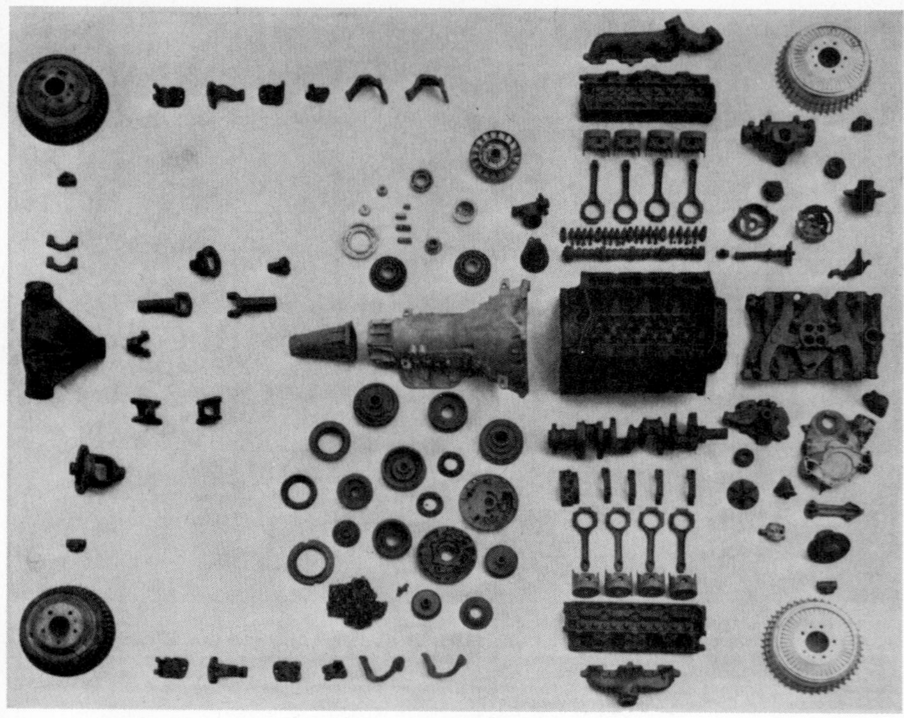

FIGURE 11-1. Cast metal parts in a typical American automobile.

of a gram, such as the individual teeth in a zipper, to 10 or more meters (32.8 feet) and weighing many tons, such as the huge propellors and stern frames of ocean liners. Although its use is not restricted to these types of parts, casting has marked advantages in the production of complex shapes, parts having hollow sections, parts that contain irregular curved surfaces (except those made from thin sheet metal), very large parts, and parts made from metals that are difficult to machine. Because of these obvious advantages, casting is one of the most important of the manufacturing processes. To illustrate its wide use, Figure 11-1 shows the cast metal parts in a typical eight-cylinder automobile—often as many as 135. In the United States, the cast metals industry is the sixth largest of the basic industries.

Today it is virtually impossible to design anything that cannot be cast by means of one or more of the available casting processes. However, as with other manufacturing processes, best results and economy can be achieved if the designer understands the various casting processes and adapts his designs so as to use the process most efficiently.

In all casting processes, six basic factors are involved. These are as follows:

1. A mold cavity, having the desired shape and size and with due allowance for shrinkage of the solidifying metal, must be produced. Any complexity of shape desired in the finished casting must also exist in the cavity. Consequently, the mold material must be such as to reproduce the desired detail and also have a refractory character so that it will not be too greatly affected by the molten metal that it must contain. Either a new mold must be prepared for each casting, or it must be made from a material that can withstand being used for repeated castings; the latter are called *permanent molds.* Inasmuch as permanent molds must be made of metal or graphite and are costly to make, much effort has been devoted to methods for economically producing single-usage molds that will enable castings to be made with good accuracy.

2. A suitable means must be available for melting the metal that is to be cast, providing not only adequate temperature but also satisfactory quality and low cost.

3. The molten metal must be introduced into the mold in such a manner that all air or gases in the mold, prior to pouring or generated by the action of the hot metal upon the mold, will escape, and the mold will be completely filled so that the resulting casting will be dense and free from defects, such as air holes.

4. Provision must be made so that the mold will not cause too much restraint to the shrinkage that accompanies cooling after the metal has solidified. Otherwise, the casting will crack while its strength is low. In addition, the design of the casting must be such that solidification and shrinkage can occur without producing cracks and internal porosity or voids.

5. It must be possible to remove the casting from the mold. Where the casting is done in molds made from materials such as sand, and the molds are broken up and destroyed after each casting is made, there is no serious difficulty. However, in certain processes where molds of a permanent nature are used, this is a major problem.

6. After removal from the mold, finishing operations may need to be performed to remove extraneous portions that are attached to the casting as the result of the method of introducing the metal into the cavity, or are picked up from the mold through contact of the metal with it.

Much of the development that has taken place in the foundary industry has been directed toward solving these six problems with greater economy. Seven major casting processes currently are used. These are:

1. Sand casting
2. Shell-mold casting
3. Permanent-mold casting
4. Die casting
5. Centrifugal casting
6. Plaster-mold casting
7. Investment casting

Sand casting accounts for by far the largest proportion of the total tonnage of castings produced. However, the use of permanent-mold, die, investment, and shell-mold castings has expanded very rapidly in recent years.

FIGURE 11-2. Essential steps in sand casting. (a) Bottom (drag) half of pattern in place on mold board between halves of flask ready to receive sand. (b) Drag half of mold completed, ready for turning over. (c) Top (cope) half of pattern and sprue and riser pins in place. (d) Cope half of mold packed with sand. (e) Mold opened, showing parting surface

SAND CASTING

Sand casting uses sand as the mold material. The sand grains, mixed with small amounts of other materials to improve the moldability and cohesive strength, are packed around a pattern that has the shape of the desired casting. Because the grains will pack into thin sections and can be used economically in large quantities, products covering a wide range of sizes and detail can be made by this method. A new mold must be made for each casting, and gravity usually is employed to cause the metal to flow into the mold. Except in the full-mold process, after the sand has been packed firmly around it, the pattern must be removed to leave a cavity of the desired shape. Consequently, the mold must be made in at least two pieces. An opening, called a *sprue hole,* is provided from the top of the mold through the sand and connected to the cavity through a system of channels, called *runners.* The molten metal is

(e)

(e′)

(f)

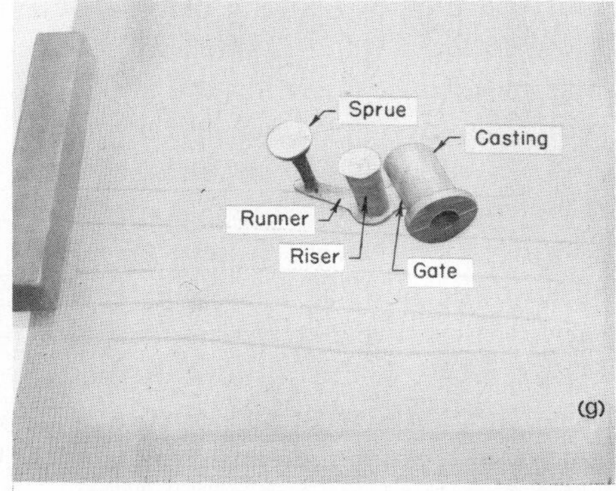

of drag half, with pattern drawn and runner and gate cut. (e[1]) Parting surface of cope half of mold, with pattern and pins removed. (f) Mold closed, ready for pouring metal. (g) Casting removed from mold.

poured into the sprue hole and enters the cavity through the runners and an opening called a *gate,* which controls the rate of flow. These essential steps and components are illustrated in Figure 11-2.

Patterns. The first requirement in sand casting is the design and making of a pattern. This is a duplicate of the part that is to be cast but modified in accordance with the basic requirements of the casting process and the particular molding technique that is to be employed. The pattern material is determined primarily by the number of castings to be made. Wood is most often used for small quantities, whereas for larger quantities aluminum, magnesium, or certain hard plastics are employed. In the full-mold process, discussed later, foamed polystyrene is used, but each pattern can be used only once.

The modifications that must be incorporated into a pattern are called *allowances. Shrinkage allowance* requires that the pattern be made slightly larger than the desired casting, to compensate for the shrinkage of the metal as it solidifies and cools. This allowance obviously is a function of the kind of metal that is to be cast. The following allowances commonly are used:

Cast iron	0.8–1.0% (1/10–1/8 in./ft)
Steel	1.5–2.0% (3/16–1/4 in./ft)
Aluminum	1.0–1.3% (1/8–5/32 in./ft)
Magnesium	1.0–1.3% (1/8–5/32 in./ft)
Brass	1.5% (3/16 in./ft)

The patternmaker incorporates these allowances into the pattern by using special *shrink rules,* which are longer than a standard rule by the desired shrink allowance. If a metal pattern is made by casting, a double shrink allowance must be included in the original wood pattern. Thus the total shrinkage allowance for a wood pattern to be used to make an aluminum pattern, which will in turn be used for the production of cast iron parts, would be 2.3 per cent (9/32 inch per foot).

In the casting processes wherein the pattern must be withdrawn from the mold, the mold must be made in two or more sections. This requires that consideration be given to the *parting line* or *surface* where one section fits against the sand of the other section, and to the *draft* or *taper* that must be provided on the pattern to facilitate its withdrawal. These are illustrated in Figure 11-3. If the surfaces of the pattern normal to the parting line were all exactly parallel with the direction in which the pattern had to be moved for withdrawal from the sand, the friction between the pattern and the sand, and any movement of the pattern perpendicular to this direction, would tend to cause sand particles to be broken away from the mold. This would be particularly severe at corners between the cavity and the parting surface. By providing a slight taper on all the surfaces parallel with the direction of withdrawal, this difficulty is minimized. Because of this draft, as soon as the pattern is withdrawn a slight amount, it is free from the sand on all surfaces, and it can be withdrawn without damaging the mold. Draft must be unidirectional with respect to the parting line. Some shapes, such as a hemisphere, provide their own draft.

The amount of draft is determined by the size and shape of the pattern, the depth of the draw, the method used to withdraw the pattern, and the molding procedure. Draft seldom is less than 1° or 1:100 (⅛ inch per foot),[1] with a minimum of about 1.6 mm (1/16 inch) on any surface. On interior surfaces where the opening is small, such as a hole in the center of a hub, the draft should be increased to about 1:24 (½ inch per foot). Because draft always increases the size of the pattern, and thus the size and weight of the casting, it is always desirable to keep it to the minimum that will permit sat-

[1] In this instance, and in similar cases that follow throughout the book, where the values are only typical of customary practice and thus are not specified precisely, the SI and English values given are not exact conversions.

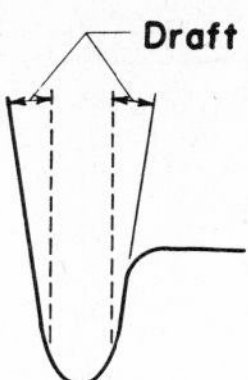

FIGURE 11-3. Relationship of draft to the mold parting line in castings.

isfactory pattern removal. Modern molding procedures, which provide higher strength to the molding sand before the pattern is withdrawn, and the use of molding machines, which substitute mechanical for manual pattern drawing, have permitted substantial reductions in draft allowances. These facilitate the production of lighter castings with thinner sections, thus saving weight and machining.

When machined surfaces must be provided on castings, it is necessary to provide a *finish allowance.* The amount of the finish allowance depends to a great extent on the casting process and the mold material. Ordinary sand castings have rougher surfaces than shell-mold castings. Die castings are sufficiently smooth that very little or no metal has to be removed, and investment castings frequently do not have to be machined. Consequently, the designer should relate the finish allowance to the casting process and also remember that draft may provide part or all of the extra metal needed for machining.

Some shapes require an allowance for *distortion.* For example, the arms of a **U**-shaped section may be restrained by the mold, while the base of the **U** is free to shrink, resulting in the arms sloping outward. Long, horizontal sections tend to sag in the center unless adequate support is provided by suitable ribbing. This type of distortion depends greatly on the particular configuration, and the designer must use experience and judgment in providing any required distortion allowance.

If patterns are withdrawn by hand, it may be necessary to consider a *rap allowance.* In order to facilitate removal of the pattern, it sometimes is rapped or struck lightly, usually perpendicular to the direction of withdrawal, to free it from the sand. This enlarges the mold slightly. Provision of the proper amount of draft reduces the amount of rapping necessary, and extensive use of mechanical vibrating and pattern-drawing equipment has almost eliminated the need for considering rap allowance.

The manner in which the various allowances are included in the pattern for a simple shape is illustrated in Figure 11-4. In general, allowances tend to increase the weight of the casting and the amount of metal that has to be removed by machining. It is evident, therefore, why the use of modern mechanical molding and pattern-drawing equipment, and processes that harden and strengthen the sand before the pattern is withdrawn, resulting in reduction in pattern allowances, have been adopted so widely.

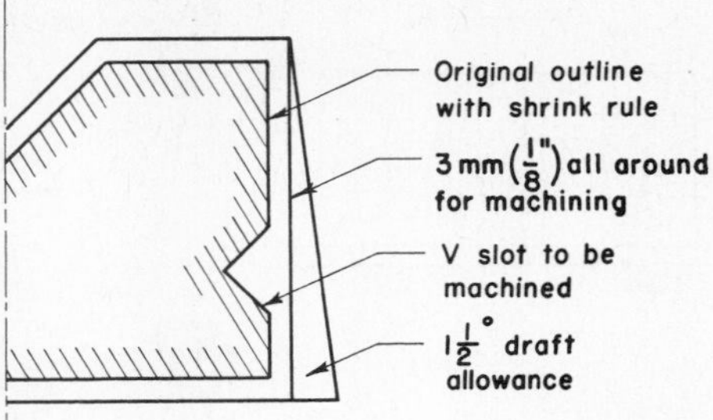

FIGURE 11-4. Various allowances provided on patterns.

Types of patterns. A number of basic types of patterns are in common use, the particular type being determined primarily by the number of duplicate castings required and the complexity of the part.

One-piece or *solid patterns,* such as are shown in Figure 11-5, are the simplest and cheapest type to make. Essentially, such a pattern is a duplicate of the part to be cast, except for the provision of the various allowances and the possible addition of core prints.[2] Although this type of pattern usually is inexpensive to construct, in most cases it slows molding and thus is used only where one or a few duplicate castings are to be made.

Unless a one-piece pattern is quite simple in shape and contains a flat surface that can be placed on the follow board to form a plane parting surface, it may be necessary for the molder to cut out an irregular parting surface by hand. This is time-consuming and costly and requires a skilled workman. It can be avoided by using a special follow board that is dug out so that the one-piece pattern fits down into the follow board up to the depth of the parting-line location. This type of follow-board pattern is illustrated in Figure 11-6. The follow board determines the parting surface, which may be either plane or curved, although it usually is the former.

Split patterns are used where moderate quantities of duplicate castings are

FIGURE 11-5. Single-piece pattern for a pinion gear.

[2] See page 284 for a discussion of core prints.

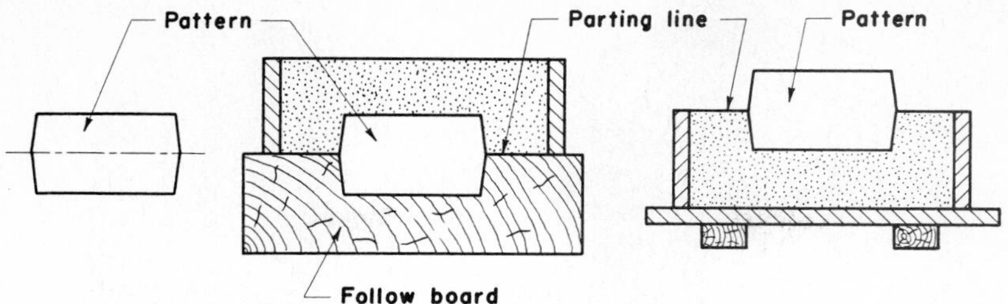

FIGURE 11-6. Method of using a follow-board pattern.

to be made and also to permit molding more complex shapes without resorting to forming the parting plane by hand or using cut-out follow boards. The pattern is split into two sections along a single plane, which will correspond to the parting plane of the mold, as shown in Figure 11-7. One half of the pattern forms the cavity in the lower, or drag, portion of the mold, and the other half serves a similar function in the upper, or cope, section. Tapered pins in the cope half of the pattern fit into corresponding holes in the drag half to hold the halves in proper position while the cope is being filled with sand.

Match-plate patterns, such as are shown in Figure 11-8, are widely used in modern foundry practice because of their suitability for use with several types of molding machines in making large quantities of duplicate castings. In these, the cope and drag sections are fastened to opposite sides of a wood or metal match plate that is equipped with holes or bushings that mate with

FIGURE 11-7. Split pattern, showing the two sections together and separated. Light-colored portions are core prints.

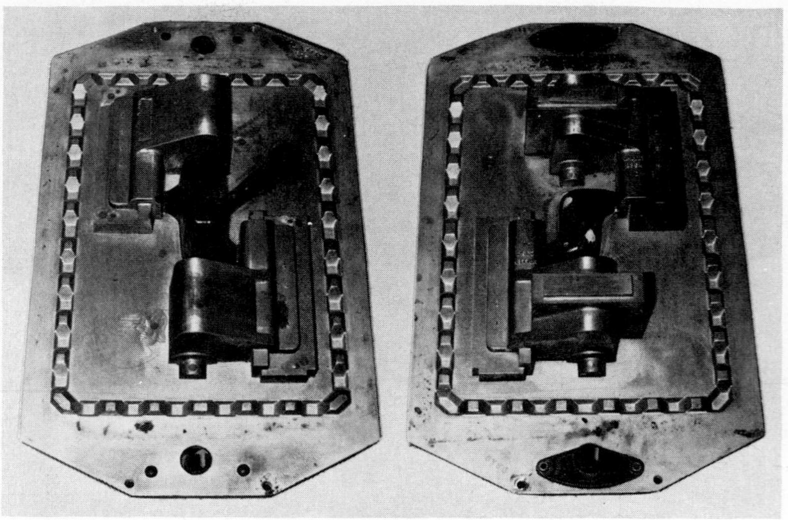

FIGURE 11-8. Match-plate pattern for molding two parts: (*left*) cope side; (*right*) drag side.

pins or guides on the halves of the mold flask. The match plate is fitted between the two flask sections, and the sand is packed into the flask to complete the cope and the drag. The entire match-plate pattern is then removed by separating the mold sections. Upon closing the mold, the cavities in the cope and drag will be in proper alignment with respect to each other, because the two halves of the pattern were correctly positioned on the two sides of the match plate, and the guide holes and pins assure identical alignment when the mold is closed. The holes and guide pins are arranged to prevent the cope and drag from being put together 180° out of proper position.

In most cases, the necessary gate and runner system is included on the match plate. This serves the double purpose of eliminating the necessity for the molder to cut the gates and runner by hand and also of assuring that they will be uniform and of the proper size in each mold, thus reducing the likelihood of defects. The gate and runner system can be seen on the match-plate pattern in Figure 11-8. This pattern also illustrates the common practice of having more than one pattern on one match plate. Core prints also are provided when required.

When large quantities are to be produced, or when the casting is quite large, it may be desirable to have the cope and drag halves of split patterns attached to two separate match plates instead of being attached to the opposite sides of a single plate. This permits large molds to be handled more easily, or for two workers, on two machines, to simultaneously produce the two portions of a mold. Such patterns are called cope-and-drag patterns. A pattern of this type is shown in Figure 11-9.

FIGURE 11-9. Cope-and-drag pattern for molding two heavy parts: cope section (*left*) drag section (*right*).

When an object to be cast has protruding sections arranged such that neither a one-piece pattern nor one split along a single parting plane can be removed from the molding sand, a *loose-piece pattern* sometimes can be used. Such a pattern is shown in Figure 11-10. Loose pieces are held to the remainder of the pattern by beveled grooves or by pins. This construction permits all the pattern except the loose pieces to be withdrawn directly from the sand, after which there is space within the cavity so that the loose pieces may be moved the necessary direction and amount to permit their removal. If the loose pieces cannot be held to the main portion of the pattern by grooves or stationary pins, a long, sliding pin may be used. After the sand is rammed, the pin is withdrawn, thereby freeing the loose pieces from the remainder of the pattern, permitting it to be removed. Obviously, loose-piece patterns are expensive to make, require careful maintenance, slow the molding process,

FIGURE 11-10. Loose-piece pattern for molding a large worm. After sufficient sand is packed around the pattern halves to hold the pieces in position, the wooden pins are withdrawn so the loose pieces can be removed from the mold. The remaining sand is then rammed around the pattern and the mold is opened.

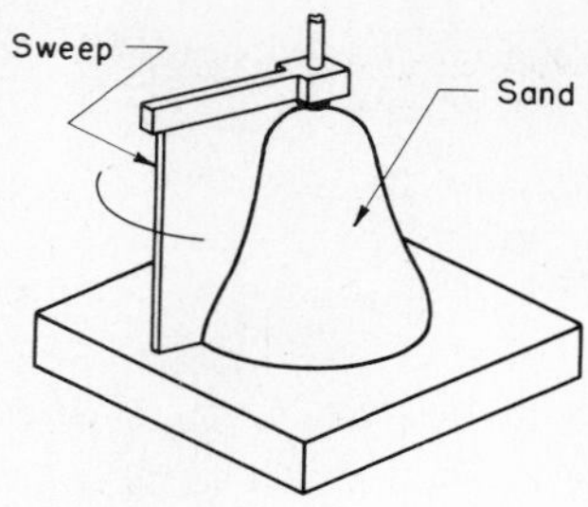

FIGURE 11-11. Method of making a mold using a "sweep."

and increase molding costs. They do make possible the casting of complex shapes that otherwise could not be cast, except by the full-mold or investment processes. However, it is desirable to eliminate their necessity by design changes when practicable.

Simple *sweeps* sometimes can be used in place of three-dimensional patterns. Ordinarily, they are used where the shape to be molded can be formed by the rotation of a curved-line element about an axis, as shown in Figure 11-11. The simple sweep eliminates the necessity for making a large and expensive pattern.

It is very important on all castings that intersecting surfaces be joined by a small radius, called a fillet, instead of permitting them to intersect in a line. Fillets avoid shrinkage cracks at such intersections and also eliminate stress concentrations. As a general rule, designers should make fillets generous in size, 6.35- and 3.18-mm (¼- and ⅛-inch) radii commonly being used. Fillets are added to wood patterns by means of wax, leather, or plastic strips of the desired radius, which are glued to the pattern or pressed into place with a heated fillet tool.

Sand conditioning and control. Sand used to make molds must be carefully conditioned and controlled in order to give satisfactory and uniform results. Ordinary silica (SiO_2), zircon, or olivine (forsterite and fayalite) are compounded with additives to meet four requirements:

1. *Refractoriness:* the ability to withstand high temperatures.
2. *Cohesiveness* (referred to as *bond*): the ability to retain a given shape when packed in a mold.
3. *Permeability:* the ability to permit gases to escape through it.
4. *Collapsibility:* the ability to permit the metal to shrink after it solidifies.

Refractoriness is provided by the basic nature of the sand. Cohesiveness, bond, or strength is obtained by coating the sand grains with clays, such as bentonite, kaolite, or illite, that become cohesive when moistened. Collapsibility is obtained by adding cereals or other organic materials that burn out when exposed to the hot metal, thereby reducing the volume of solid bulk and decreasing the strength of the restraining sand. Permeability is primarily a function of the size of the sand particles, the amount and types of clays or other bonding agents, and the moisture content.

Good molding sand always represents a compromise between conflicting factors, such as grain size of the sand particles, the amount of bonding agents such as clay, moisture content, and organic matter, combined so as to obtain a satisfactory combination of the four requirements listed previously. Consequently, the various factors must be carefully controlled to assure satisfactory and consistent results. In addition, because the sand is reclaimed and reused, control of its temperature also is important. Standard tests and procedures have been developed to permit consistent sand quality to be achieved.

Each grain of sand should be coated uniformly with the additive agents. To achieve this, the ingredients are put through a *muller,* which provides the necessary mixing. Figure 11-12 shows a modern, continuous-type muller. Frequently, the sand is discharged from the muller through an *aerator,* which fluffs the sand so that it does not tend to pack too hard in handling.

In modern foundry practice many molds and cores are made of sand that is given strength through the addition of about 4 per cent silicate of soda or various furfurals or furfural alcohols. Sand mixed with silicate of soda remains soft until exposed to CO_2 gas, after which it hardens in a few seconds. The setting is due to the reaction

$$Na_2SiO_3 + CO_2 \rightarrow Na_2CO_3 + SiO_2 \text{ (colloidal)}$$

Sands of this type can be mixed with the silicate of soda in a muller and handled in the normal manner, inasmuch as they are not gassed with CO_2 until after they have been packed around the pattern in the mold flask. Furfurals, with a catalyst, usually will cause hardening within a few minutes.

FIGURE 11-12. Continuous-type sand muller. (*Courtesy The Jeffrey Manufacturing Co.*)

Consequently, they are added and mixed with the sand by special equipment just prior to delivering the sand to the molding station.

Sand control. Although sand control is of little concern to the designer of castings, it is a matter of great concern to the foundry worker, who is expected to deliver castings of good and consistent quality. Standard tests are used to control *grain size, moisture content, clay content, mold hardness, permeability,* and *strength.*

Grain size is determined by shaking a known amount of dry silica grains downward through a set of 11 standard sieves having increasing fineness. After shaking for 15 minutes, the amount remaining in each sieve is weighed, and the weights are converted into an AFS[3] number.

Moisture content most commonly is determined by a special device which measures the electrical conductivity of a small sample of sand that is compressed between two prongs. Another device provides a continuous measure of the moisture content, by emission from a radioactive source, as the sand passes along a conveyor belt. A third method is to measure the direct weight loss from a 50-gram sample when it is subjected to a flow of air at about 110°C (230°F) for 3 minutes.

Clay content is determined by washing the clay from a 50-gram sample of molding sand in water that contains sufficient sodium hydroxide to make it alkaline. After several cycles of agitation and washing in such a solution, the clay will have been removed. The remaining sand is dried and then weighed to determine the proportion of the original sample that was clay.

Hardness, permeability, and strength tests are conducted on a standard specimen, 2 inches in diameter and 2 inches in height, that is prepared by use of the device shown in Figure 11-13. The sand is compacted inside a steel tube by means of a weight that is dropped from a predetermined height.

For the permeability test, the sample tube containing the rammed sand specimen is placed on the device shown in Figure 11-14 and subjected to air pressure of 10 grams per square centimeter. An electronic device measures the flow rate of the air through the sand, thus establishing its permeability.[4]

The strength of the rammed specimen is determined by removing it from the tube and placing it in the tester shown in Figure 11-15 in such a manner that the load on a beam scale is applied through the sand specimen.

[3] American Foundrymen's Society.

[4] An AFS permeability number is used as a measure. This is

$$\frac{V \times H}{P \times A \times T}$$

being the volume of air, in cm^3, that will pass through a specimen 1 cm^2 in area and 1 cm high under a pressure of 1 g/cm^2, where V is the volume of air (2000 cm^3), H the height of the specimen (5.08 cm), P the pressure (10 g/cm^2), A the cross section of specimen (20.268 cm^2), and T the time in seconds required for a flow of 2000 cm^3. Because of the constants, the permeability number $= 3007.2/T$.

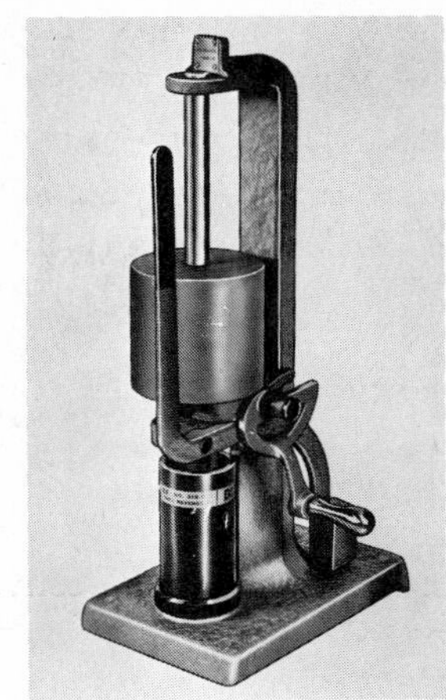

FIGURE 11-13. Sand rammer for preparing a standard rammed foundry sand specimen. (*Courtesy Harry W. Dietert Company.*)

The same tester also can be used to measure the transverse bend strength of baked or cured sand specimens.

The hardness to which sand is compacted in a mold is very important because it affects the strength–permeability relationship. It commonly is measured by means of the instrument shown in Figure 11-16, which measures the resistance of the sand to penetration by a 5.08-mm (0.2-inch)-diameter, spring-loaded steel ball.

The making of sand molds. Except in very small foundries or when only a very few castings of a given design are to be made, virtually all sand molds now are made with the use of various types of molding machines. These greatly reduce the labor and skill required and produce castings with better dimensional accuracy and consistency.

FIGURE 11-14. *Permeability tester for foundry sand.* Standard sample in sleeve is sealed by O-ring on top of unit. (*Courtesy Harry W. Dietert Company.*)

FIGURE 11-15. Universal strength-testing machine for foundry sand. (*Courtesy Harry W. Dietert Company.*)

Molding machines basically vary in the way the sand is packed within the molding flask, whether mechanical assistance is provided for turning and/or handling the mold, and whether a flask is required. The type of machine shown in Figure 11-17 is a common one. With such machines, a match-plate pattern usually is employed, being placed between the two halves of a flask, such as is shown in Figure 11-18. These usually are constructed of aluminum or magnesium and arranged so that they can be opened slightly to permit them to be slid off the mold after it is completed.

Packing of the sand is done by using one or a combination of several principles. One is to raise the table, on which the sand-filled flask is resting, a

FIGURE 11-16. Mold hardness tester. (*Courtesy Harry W. Dietert Company.*)

FIGURE 11-17. Jolt-squeeze, roll-over molding machine. (*Left*) in jolting and squeezing position and (*right*) in roll-over position. (*Courtesy The Osborn Manufacturing Company.*)

few inches by compressed air and permitting it to drop, producing a jolting action. This may be repeated several times. These *jolt-type machines* cause the sand to be compacted, but it is packed more firmly near the pattern than near the top of the flask. Consequently, simple jolt-type machines seldom are used.

Squeezing machines pack the sand by the squeezing action of either an air-operated squeezing head; a flexible diaphragm; or small, individually actuated squeeze heads. Squeezing packs the sand firmly near the squeezing head, but the density diminishes as the distance from the head increases. A high-pressure, flexible-diaphragm type of squeezing machine, commonly called a *Taccone machine,* tends to give a more uniform density around all parts of an irreg-

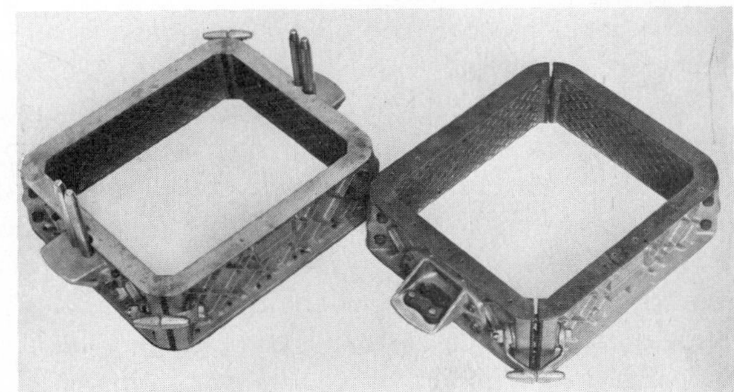

FIGURE 11-18. Halves of a tapered cam-latch snap flask, closed at left and open at right.

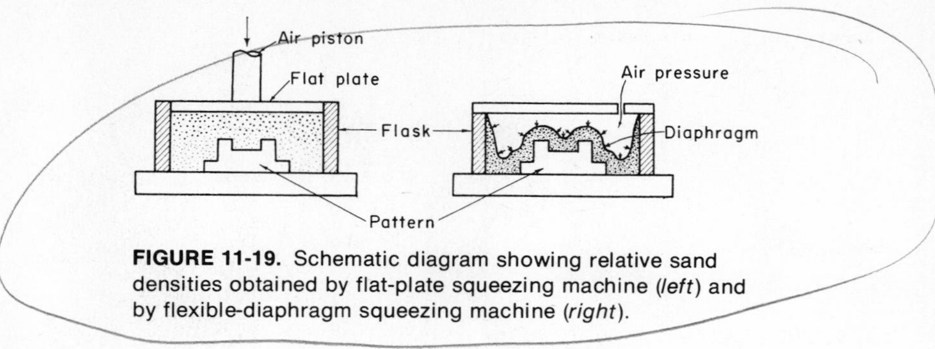

FIGURE 11-19. Schematic diagram showing relative sand densities obtained by flat-plate squeezing machine (*left*) and by flexible-diaphragm squeezing machine (*right*).

ular pattern than can be obtained with a regular flat-plate squeezing machine, as illustrated in Figure 11-19.

Commonly, a combination of jolting and squeezing is used to obtain more uniform density throughout the mold depth. On these the flask, with a match-plate pattern in place, is placed upside down on the table of the machine. After parting dust has been sprinkled on the pattern, the flask is filled with sand. The table and flask then are jolted the desired number of times to pack the sand around the pattern in the lower portion of the flask. The squeeze head then is lowered and pressure is applied to pack the sand in the upper portion of the flask. Both cope and drag may be made on the same machine by rolling the flask over and repeating the operations on the cope half, or the cope and drag portions may be made on separate machines, using cope-and-drag patterns.

Except for very small molds, jolt and/or squeeze machines usually provide mechanical assistance for turning over the heavy mold, as illustrated in Figure 11-17. Some patterns for use on molding machines form the sprue hole, but often it is cut by hand. Often a shallow pouring basin—a depression in the sand at the top of the sprue hole into which the metal is poured and then flows into the sprue hole—is formed by a protruding shape on the squeeze board. Usually, the pattern will form the runner and gate system, but sometimes these are cut by hand. Considerable effort is made in designing and constructing patterns to eliminate the necessity for such handwork.

After a mold is completed, the flask is removed so that it will not be damaged during the pouring of the mold. A *slip jacket*—an inexpensive metal band—may be slid down around the mold to hold the sand in place. Heavy metal weights often are placed on top of the molds to prevent the sections from being separated by the hydrostatic pressure of the molten metal. When used, slip jackets and pouring weights are needed only during pouring and for a few minutes afterward, while solidification occurs. They then can be removed and placed on other molds, so as to keep up with the pouring crew. Thus the amount of expensive equipment is kept to a minimum.

For mass-production molding, a number of machines have been devel-

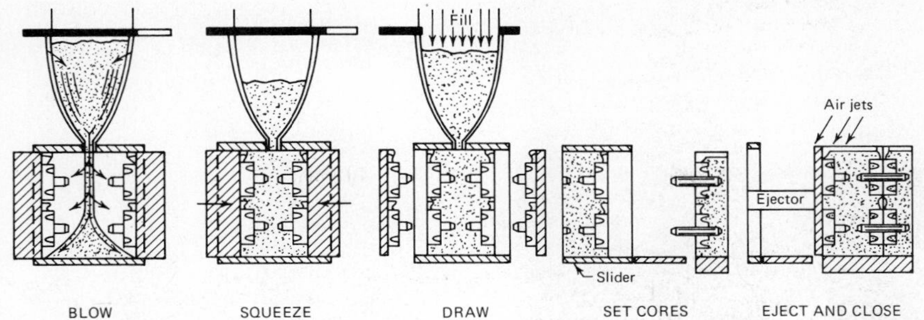

FIGURE 11-20. Method of making molds without the use of a flask. (*Courtesy Belens Corporation.*)

oped that produce a continuous flow of molds without the use of flasks. The principle utilized in one machine of this type is shown in Figure 11-20. The halves of two separate molds are made at one time, with a complete mold being formed subsequently by the joining of its two halves. One step in the cycle can be arranged so that cores can be set in place before the halves are closed.

A modification of the CO_2 process is used quite extensively in making certain portions of molds where better accuracy of detail and thinner sections are desired than can be achieved with ordinary molding sand. Silicate of soda-mixed sand is packed around the metal pattern to a depth of about 1 inch, followed by regular molding sand as backing material. After the mold is rammed, CO_2 is introduced through vents in the metal pattern, thereby setting the sand adjacent to the pattern. The pattern then can be withdrawn with reduced danger of damage to the mold. This procedure is particularly useful where deep draws are necessary. Sand mixed with various "cold-setting" additives, such as furfurals, is used extensively with molding machines of all types to permit thinner sections, deeper draws, and greater accuracy to be obtained.

For some purposes, molds of all sizes are either *skin-dried* or completely dried. This is done to strengthen the mold and also, sometimes, to reduce the amount of gases that will be evolved when hot metal comes in contact with the sand. Molds into which steel is to be cast are nearly always skin-dried just prior to pouring, because much higher temperatures are involved than in the case of cast iron. Such molds usually are also given a high-silica wash prior to drying to increase the refractoriness of the surface. Skin drying to the depth of about a half inch is accomplished by means of a gas torch. Frequently, additional binders, such as molasses, linseed oil, or corn flour, are added to the facing sand when skin drying is to be done.

For castings too large to be made by bench molding or in the type of molding machines that have been described, large flasks, which rest on the foundry floor, may be employed, and various mechanical aids are employed. Figure 11-21 shows the use of a *sand slinger,* which impels the sand into the

FIGURE 11-21. Sand slinger being used to fill a large mold. (*Courtesy Beardsley & Piper Division of Pettibone Mulliken Corporation.*)

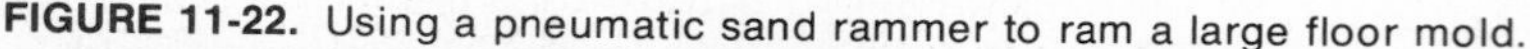

FIGURE 11-22. Using a pneumatic sand rammer to ram a large floor mold.

FIGURE 11-23. Using dry-sand sections in the construction of a large pit-type mold. (*Courtesy Pennsylvania Glass Sand Corporation.*)

mold with sufficient force to pack it to the desired hardness. Extra tamping may be done by means of a pneumatic rammer, as illustrated in Figure 11-22.

Very large molds usually are made in pits in the floor, as shown in Figure 11-23. Such a mold may be made in essentially the same manner as a floor mold. However, because such large castings usually are complex and portions of the mold cannot be turned over, as can a cope and drag, and because the preparation of pit molds is costly and it is desirable to reduce dimensional variations and to provide greater strength and stability to the mold, they frequently are built from baked or dried-sand sections. Sections made with the use of furan binders are being used more extensively.

Shell molding. Large numbers of molds now are made by the *shell-molding process.* This process has important advantages in that it provides better surface finish than can be obtained with ordinary sand molding, provides better dimensional accuracy, and less labor is required. In many cases, the process can be completely mechanized. These characteristics make it particularly suitable for mass production. The basic process involves six steps:

1. A mixture of sand and a thermosetting plastic binder is dumped onto a metal pattern that is heated to 150 to 232°C (300 to 450°F) and allowed to stand for a few minutes. During this interval the heat from the pattern partially cures a layer of the sand-plastic mixture, about 3.2 mm (⅛ inch) thick, adjacent to the pattern.
2. The pattern and the sand mixture are inverted to permit all the sand to drop off except the layer of partially cured material adhering to the pattern.
3. The pattern and the partially cured "shell" are then placed in an oven for a few minutes to complete the curing of the shell.
4. The hardened shell is stripped from the pattern.
5. Two shells are glued or clamped together to form a complete mold.
6. The glued pair of shells form a mold, ready for pouring. In some cases they are placed in a pouring jacket and backed up with shot or sand to provide extra support.

Because the plastic and sand shell is molded to a metal pattern and is compounded to undergo almost no shrinkage, it has virtually the same dimensional accuracy as the pattern. Tolerances of 0.08 to 0.13 mm (0.003 to 0.005) inch are readily obtained. The sand used is finer than ordinary foundry sand and, combined with the plastic resin, it results in a very smooth shell and casting surface. Also, the consistency between castings is superior to that obtainable by ordinary sand casting.

Figure 11-24 shows a set of patterns, the two shells before clamping, and the resulting casting. Machines for making shell molds vary from the very simple, but effective, type shown in Figure 11-25, suitable for a very small foundry, to large, completely automated types that are used in mass-production foundries. Even though the cost of the metal pattern is relatively high, and the gate and runner system must be included on the pattern, the high productivity, low labor cost, and saving in machining cost make the process economical for even moderate quantities. Often several relatively simple shells and suitable cores can be combined for producing complex castings.

The full-mold process. The *full-mold process* avoids two bothersome restrictions inherent in the making of ordinary sand molds. These are (1) the cost of making a relatively expensive wood pattern when only one or a few castings are required, and (2) the necessity for withdrawal of the pattern from the mold, often requiring either some modification in the design of the casting or a complex pattern and/or molding procedure. In the full-mold process, the pattern is made of foamed polystyrene, which remains in the mold *during* the pouring of the metal. When the molten metal is poured, the heat vaporizes the polystyrene pattern almost instantaneously, and the metal fills the space previously occupied by the pattern.

FIGURE 11-24. (*Bottom*) Two halves of a shell-mold pattern. (*Top*) Shell-mold shells and the resulting casting. (*Courtesy Shalco Systems, an Acme-Cleveland Co.*)

Foamed polystyrene is inexpensive, weighs only about 15.4 kg/m^3 (1.2 pounds per cubic foot), is very easily cut, and complex patterns, including the pouring basin, sprue hole, and runner system, can be made by gluing together several simple shapes. Because the pattern is not withdrawn, no draft need be provided. Molding sand, usually using a furan binder, is packed around the pattern in the usual manner to form the mold. Figure 11-26 shows a large, quite complex, styrene pattern, with the gate and runner system, and the resulting casting ready for machining.

The full-mold process can be used for castings of any size. Because of the lessened pattern cost, its application is most economical where only one or a

FIGURE 11-25. Machine for making shell-molding shells. Sand mix is placed on metal pattern at center station, and shells are baked under ovens at right and left. (*Courtesy Shalco Systems, an Acme-Cleveland Co.*)

few castings are to be made. However, it also is very useful for complex castings, which, because of reentrant sections, otherwise would require costly cores or loose-piece patterns. Also, the time required to obtain a needed single casting may be decreased substantially.

Cores and core making. One of the distinct advantages of castings is that hollow or reentrant sections can be included with relative ease. Often such configurations could not be made by any other process. However, to produce such castings it is necessary to use *cores.* An excellent example of the use of cores is shown in Figure 11-27, a product that could not be produced by any other process than casting. Of course, cores constitute an added cost, and in many cases the designer can do much to facilitate and simplify their use. Thus although cores are extremely useful, they must be made and present some problems. These aspects will be discussed next.

The use of cores can be illustrated by the example of the belt pulley shown in Figure 11-28. The shaft hole could be made in several ways. First, the pulley could be cast solid and the hole made by machining. However, this procedure would require a substantial amount of costly machining and quite a bit of metal would be wasted. Therefore, it is more economical to cast the pulley with a hole in it of the approximate size desired. One procedure would

FIGURE 11-26. Polystyrene pattern (*top*) used for making the casting (*bottom*) by the Full-Mold process. (*Courtesy Full Mold Process, Inc.*)

be to use a split pattern, as shown in Figure 11-28(b). Each half of the pattern contains a tapered hole into which green molding sand is packed, just as it is in the remainder of the mold. These sections of sand, which protrude into the hole in the pattern, are called *cores.* In this case, because they are made of green sand, they are called *green-sand cores.* It usually is desirable to make cores of green sand. However, it frequently is impracticable to do this because green-sand cores are weak; if they are narrow or long, the pattern cannot be drawn without breaking them, or they will have insufficient strength to support themselves. In addition, the amount of draft that must be provided on the pattern is such that if the core is very long, a considerable amount of machining may have to be done and the advantage of the core is lost. In other cases the shape is such that a green-sand core cannot be used. To overcome these difficulties, it usually is necessary to use *dry-sand cores.*

Dry-sand cores can be made in several ways. In each, the sand, mixed with some kind of binding material, is packed into a core box that contains a cavity of the desired shape. The core box, usually made of wood or metal, is analogous to the mold in which a casting is made. The *dump-type core* box,

FIGURE 11-27. Dry-sand cores used in making the V-8 engine block shown at center, bottom. (*Courtesy Central Foundry Division, General Motors Corp.*)

shown in Figure 11-29, is very common. The sand is packed into the box and struck off level with the top surface. Next, a metal plate is placed on top of the box, and the box is turned over and lifted upward, leaving the core resting on the plate. With the older and slower-setting binders, cores that did not have a flat surface on which to rest had to have special, metal core supports during drying. Modern, fast-setting binders have largely eliminated this.

Some cores can be made in a *split core box,* consisting of two halves that

FIGURE 11-28. Four methods of making a hole in a cast pulley.

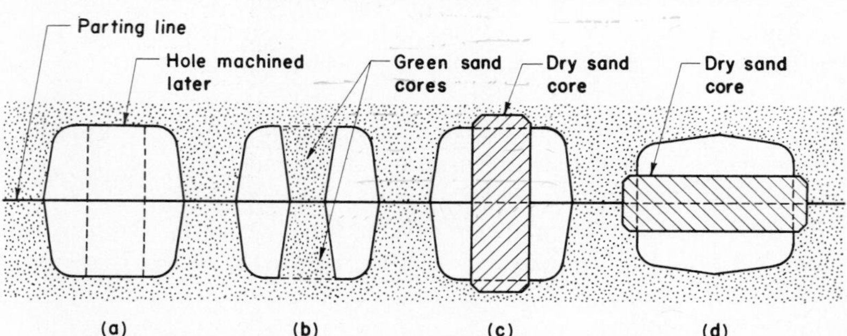

FIGURE 11-29. Core box, two core halves ready for baking, and completed core made by gluing two halves together.

are clamped together with an opening in one or both ends, through which the sand is rammed. The halves of the box are separated to permit removal of the core.

Most cores now are made with furan or resin binders, which harden with little or no heat, or by the silicate of soda–CO_2 process or by the shell-molding process. In many cases the sand mix is blown into the core box, using equipment similar to that depicted in Figure 11-30. Cores that have a uniform cross section throughout their length can be made by a core-extruding

FIGURE 11-30. Method of making cores on a core-blowing machine.

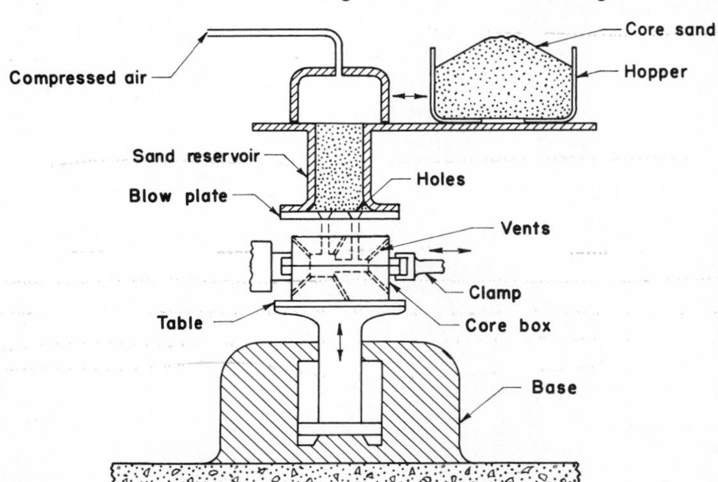

machine that is similar to the familiar meat grinder. The cores are cut to length as they come from the machine and placed in core supports for hardening.

Many cores are molded in halves and assembled after baking or hardening is completed. Rough spots are removed with coarse files or on sanding belts, and the halves are pasted together, often using hot-glue guns. Substantial numbers of cores, particularly those having a round cross section, are made by the shell process.

Quite often cores are given a thin coating to produce a smoother surface or greater refractoriness. Graphite, silica, or mica may be used, either sprayed on or brushed on.

To function properly, dry-sand cores must have the following characteristics:

1. Sufficient hardness and strength after baking or hardening to withstand handling and the forces of the molten metal.
2. Sufficient strength before hardening fully to permit any required handling.
3. Adequate permeability to permit the escape of gases.
4. Collapsibility to permit the shrinkage that accompanies cooling of the solidified metal, thereby preventing cracking of the casting and also permitting easy removal from the casting.
5. Adequate refractoriness.
6. A smooth surface.
7. Minimum generation of gases.

Cores occasionally are strengthened by means of wires or rods in order to have sufficient strength for handling and to resist the forces that act against them during pouring. In some cases, particularly in steel casting, where considerable shrinkage is encountered, cores are made hollow at their centers, or straw is put in the center, so they will collapse as the metal cools after solidifying. Failure to provide for free shrinkage may result in cracked castings.

All but very small cores must be vented to enable gases to escape. Vent holes may be produced by pushing small wires into the core or by using waxed strings that burn out and leave vent holes. Coke or cinders sometimes are placed in the center of large cores to provide venting.

When dry-sand cores are used, it usually is necessary to provide recesses in the mold into which the ends of the cores can be placed to provide support and/or to hold them in position. The recesses are made by *core prints*—extensions—on the pattern. The patterns shown in Figures 11-5 and 11-7 contain core prints.

In some cases the design of a casting does not permit the core to be supported from the sides of the mold. In such instances, the core can be supported, and can be prevented from being moved or floated by the molten metal, by means of small, metal supports called *chaplets*. Figure 11-31 shows

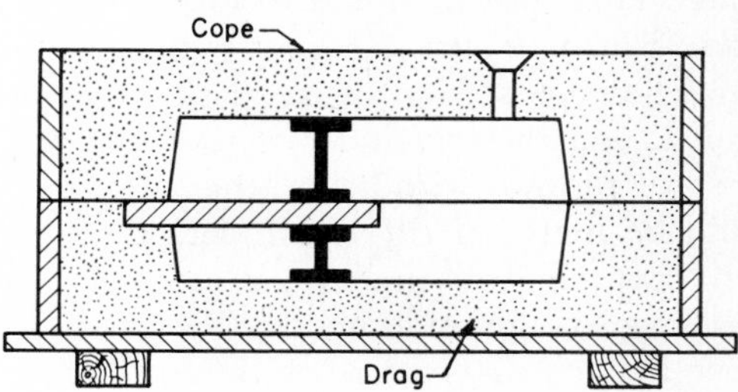

FIGURE 11-31. (*Top*) Typical chaplets. (*Bottom*) Method of supporting cores by use of chaplets (relative size of chaplets exaggerated.)

some chaplets and illustrates how they are used. The use of chaplets should be minimized because they become a part of the casting and may be a possible source of weakness.

As mentioned previously, dry-sand mold sections are used frequently to facilitate the molding of complex shapes, particularly those containing reentrant sections. Figures 11-32, 11-33, and 11-34 illustrate how such a procedure may be used advantageously. If the pulley were to be molded as shown in Figure 11-32, time-consuming handwork would be required by a skilled molder. He first would ram the drag section in the usual manner. After turning the drag over, he would cut the sand away, by hand, down to the lower

FIGURE 11-32. Method of molding a reentrant section by coping.

Cope
Drag
Upper parting surface
Lower parting surface

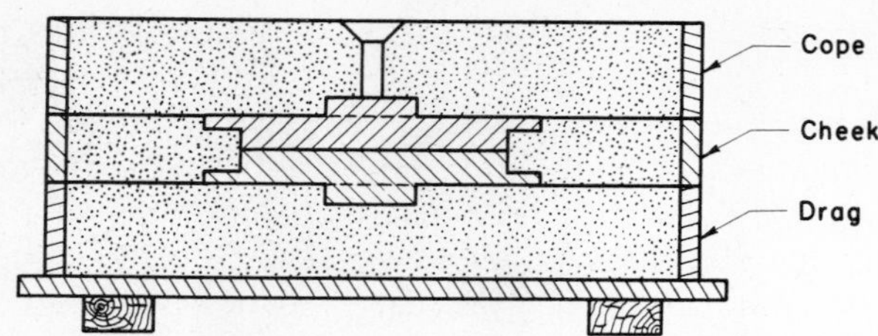

FIGURE 11-33. Method of molding a reentrant section by using a three-piece flask.

parting line. He next would place the top half of the split pattern in place, sprinkle parting dust on the pattern and lower parting surface, and riddle enough sand into the mold to make the green-sand core. After this sand was rammed firm, it would be cut away by hand to form the upper parting surface. The cope section of the flask would then be set in place, parting dust applied, and the remainder of the cope prepared in the usual manner. After the sprue pin was withdrawn, the entire mold would be turned over. The drag would be lifted off the remainder of the mold, the drag half of the pattern withdrawn, and the runner and gate cut. The drag would be replaced and the mold turned right side up. In this position the cope would be lifted and the upper half of the pattern withdrawn. The dry-sand core for the hub hole would be set in place and the cope put back on the drag. The mold then would be ready for pouring.

A similar procedure for using a green-sand core is shown in Figure 11-33. In this case an intermediate flask section, called a *cheek,* is used to help contain the green-sand core. In both these procedures using green-sand cores, considerable skilled handwork is required and the mold must be turned over additional times. Such molding procedures may be advantageous when only a few molds are to be made because they eliminate the necessity of making a special core box. However, if any substantial quantity of the pulley had to be made, the use of a dry-sand core, as illustrated in Figure 11-34, greatly simplifies and speeds the operation. In this case, the pattern is of a different

FIGURE 11-34. Molding a reentrant section by using a dry-sand core.

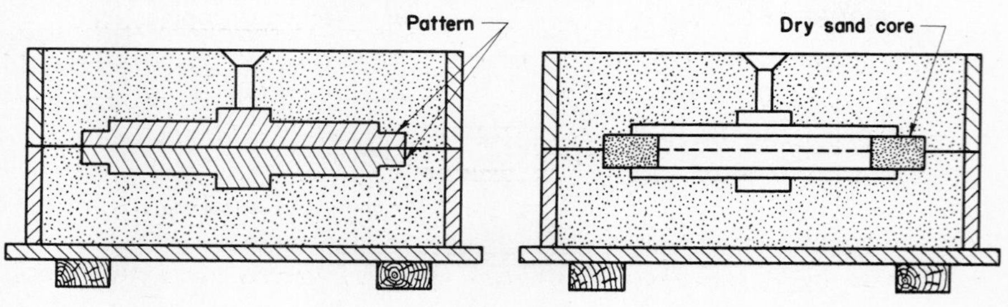

shape, to make a seat for the specially shaped dry-sand core that is set in place to form the reentrant section in the casting. The use of dry-sand cores is especially advantageous where rapid machine molding is used. The procedure, of course, requires less skilled molders. However, the cost of providing a core box and making the cores must be balanced against the saving in labor cost; quantity is a primary factor. If modifications can be made in the design of a casting that will eliminate the necessity for dry-sand cores or sections and still permit machine molding to be used, this obviously is the best solution.

PERMANENT-MOLD CASTING PROCESSES

Obviously, two major disadvantages of the sand-casting processes are the necessity for making a new mold for each casting and the possibility for some dimensional variations. Consequently, much effort has been devoted to developing *permanent-mold casting,* and very substantial success has been achieved. However, all these processes have important limitations which must be considered by designers. The molds are either metal or graphite and, consequently, most permanent-mold castings are restricted to nonferrous metals and alloys. However, a significant quantity of ferrous castings are now being made by permanent-mold processes.

Nonferrous permanent-mold casting. Of all permanent mold casting, a relatively small amount employs only gravity to introduce the metal. This type often is called *permanent-mold casting,* whereas the other types are given special names. In this process, the molds most commonly are made of fine-grain cast iron or steel, and aluminum-, magnesium-, or copper-base alloys are the metals cast. However, graphite molds are being used increasingly for casting iron and steel. In all cases, the mold halves or sections are hinged so that they can be opened and closed accurately and rapidly by mechanical means.

Permanent molds are heated at the beginning of a run and maintained at a fairly uniform temperature—usually by controlling the casting rate—to avoid too-rapid chilling of the metal. It usually is necessary to coat the cavity surfaces with a thin refractory wash to prevent the casting from sticking to the mold and to prolong the mold life. When used for cast iron, an additional coating of carbon black usually is added, often by means of an acetylene torch.

Obviously, a rigid mold offers great resistance to shrinkage of the casting. As a result, only relatively simple shapes are cast by the permanent-mold process. In addition, it is the practice to withdraw the cores, if any, and open the mold as soon as the casting has solidified to prevent tearing.

When cores must be used in making cast iron castings by the permanent-mold process, they usually are made of dry sand. Such a procedure is sometimes called *semipermanent-mold casting.*

Figure 11-35 shows a casting being removed from a permanent mold.

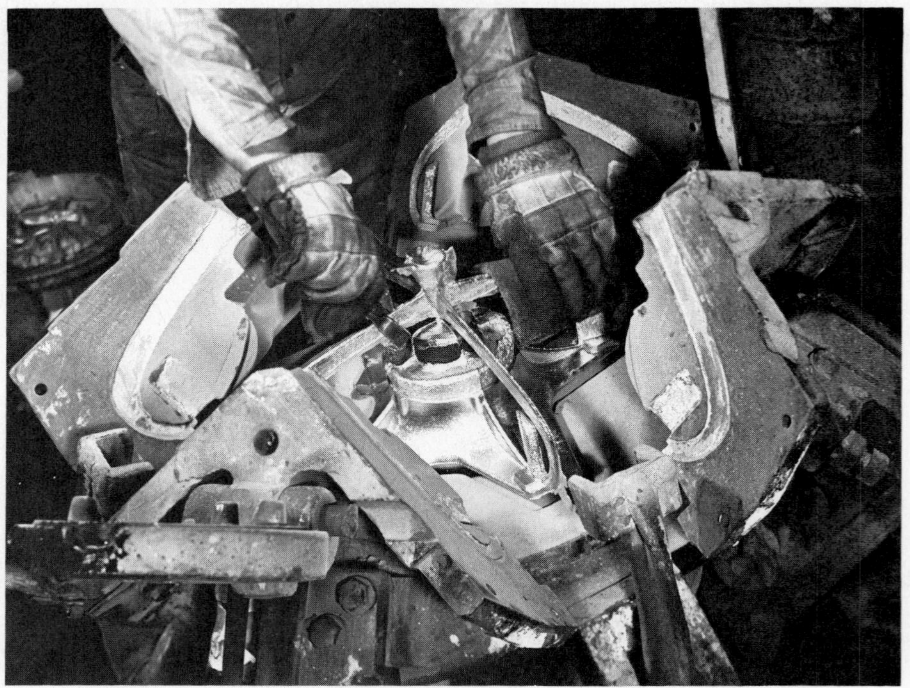

FIGURE 11-35. Removing a permanent-mold casting from an opened mold. (*Courtesy Aluminum Company of America.*)

Mass-production, permanent-molding machines often automatically coat the mold, pour the metal, and remove the casting.

Because of the nature of the mold, permanent-mold castings have better dimensional accuracy and smoother surfaces than can be obtained with sand casting. Accuracies of from 0.13 to 2.5 mm (0.005 to 0.010 inch) can readily be achieved. Special provision must be made for venting the molds, because they are not permeable. This usually is accomplished through the slight cracks between the mold halves, or by very small vent holes that will permit the flow of trapped air but not the metal.

Several variations of permanent-mold casting exist. *Slush casting* is a process wherein the metal is permitted to remain in the mold until a shell, adjacent to the mold cavity, has solidified to the desired thickness. The mold then is inverted, and the remaining molten metal is poured out. The resulting hollow casting has varying wall thickness, but the process is satisfactory for making ornamental objects, such as candlesticks and lamp bases.

Another variation, known as *Corthias casting,* utilizes a plunger that is pushed down into the mold cavity, sealing the sprue hole and displacing some of the molten metal into the outer portions of the mold cavity. The positive pressure produces good detail and permits thin sections to be cast successfully.

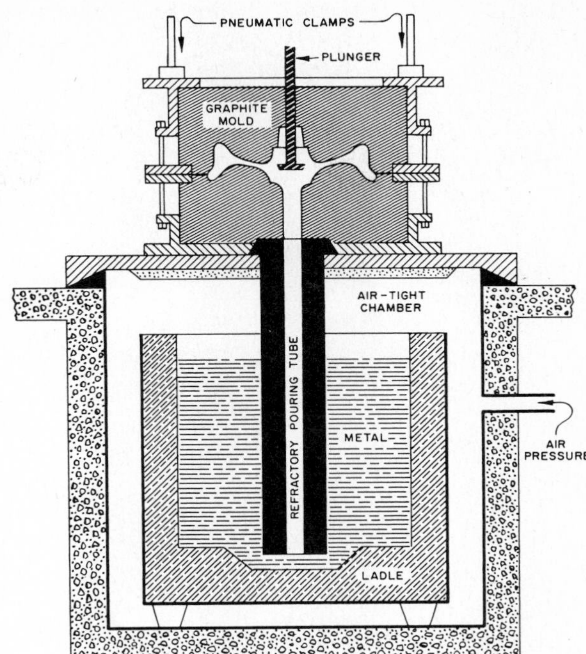

FIGURE 11-36. Schematic diagram of the pressure pouring process. (*Courtesy Amsted Industries.*)

Pressure pouring. The *pressure pouring process,* illustrated in Figure 11-36, is used extensively for making railroad-car wheels and steel ingots. A graphite mold is used, and the metal is forced upward into the mold by means of air pressure. When the mold is filled, a plunger is forced downward to seal the sprue hole so that the metal will not drain out of the mold when the pouring pressure is decreased. This requires that the expendable plunger must be at a location where it can be removed from the finished casting as scrap.

Pressure pouring protects the metal entering the mold from the atmosphere, and, by controlling the air pressure, the rate of flow can be regulated to avoid turbulence. It is used almost exclusively for casting steel and produces castings of excellent quality.

Die casting. *Die casting* differs from ordinary permanent-mold casting in that the molten metal is forced into the molds by pressure and held under pressure during solidification. Most die castings are made from nonferrous metals and alloys, but substantial quantities of ferrous die castings now are being produced routinely. Because of the combination of metal molds, dies, and pressure, fine sections and excellent detail can be achieved, together with long mold life. Zinc-. copper-. and aluminum-base alloys suitable for die casting have been developed which have excellent properties, thereby contributing to the very extensive use of the process.

FIGURE 11-37. Die-cast aluminum-alloy engine block for a modern compact car. (*Courtesy Chevrolet Motor Division, General Motors Corp.*)

Because die-casting dies usually are made from alloy steel, they are expensive to make. They may be relatively simple, containing only one or two mold cavities, but more often they are complex and may contain eight or more cavities. In order to be opened to remove the castings, the die must be made in at least two pieces. But very often they are much more complicated, containing sections that often move in several directions. In addition, the die sections must contain cooling-water passages and knock-out pins, which eject the casting. Consequently, such dies usually cost in excess of $3000, often over $10,000. It follows that the economical use of die casting is closely related to high production rates, the excellent surface qualities that are obtainable, and the almost complete elimination of machining. Thus die casting is not a suitable process for small quantities. The size and weight of die castings are increasing constantly, and parts weighing up to about 9 kg (20 pounds) and measuring up to 610 mm (24 inches) are made routinely. Figure 11-37 is an example of the complexity often found in die castings.

The die-casting cycle consists of the following steps:

1. Closing and locking the dies.
2. Forcing the metal into the die and maintaining the pressure.
3. Permitting the metal to solidify.
4. Opening the die.
5. Ejecting the casting.

Because the high injection pressures may cause turbulence and air entrapment, the amount and time of applying the pressure varies considerably. A recent trend is toward the use of larger gates and lower injection pressure, followed by much higher pressure after the mold has been completely filled and the metal has started to solidify. This tends to improve the density and reduce porosity. Where the shape of the casting permits, a very high semiforging pressure is applied when the casting has cooled to forging temperature.

Two basic types of die-casting machines are used. Figure 11-38 illustrates, schematically, the simple *gooseneck type.* A "gooseneck" is partially submerged in the molten metal, either continuously or periodically. Molten metal fills the gooseneck through a port that is opened by each cycle of the hydraulically or air-actuated injection plunger. The plunger thereafter forces the metal out of the gooseneck into the die. This type of machine is fast in operation but cannot be used with higher-melting-point metals such as brass and bronze. They also have a tendency to "pick up" some iron from the metal

FIGURE 11-38. Air-actuated type of gooseneck die-casting machine. (*Courtesy The New Jersey Zinc Company.*)

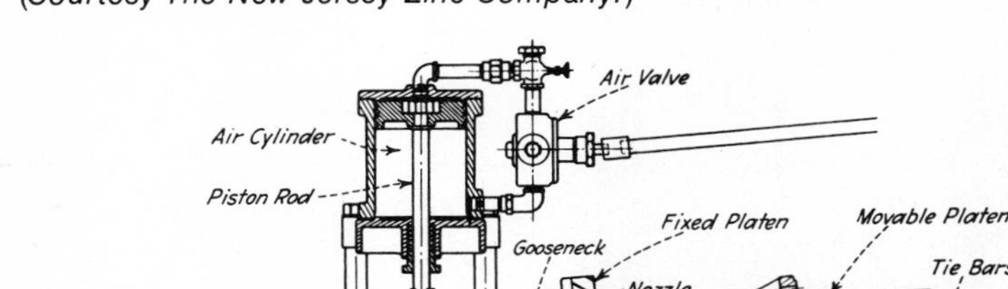

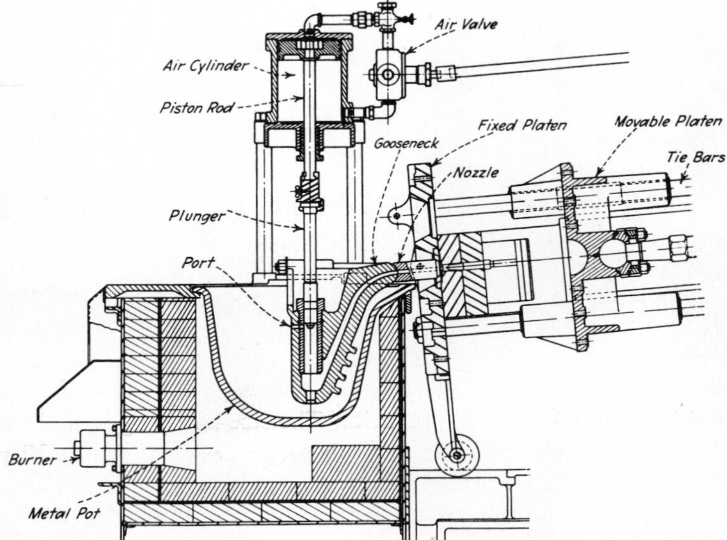

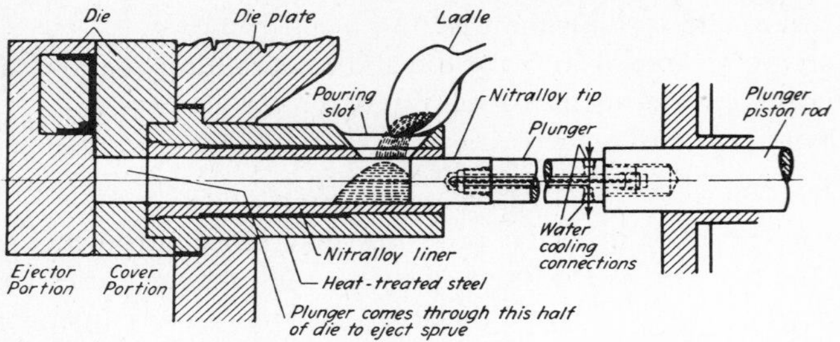

FIGURE 11-39. Schematic diagram of a cold-chamber die-casting machine. (*Courtesy The New Jersey Zinc Company.*)

container when used for casting aluminum. Consequently, they are used primarily with zinc- and tin-base alloys.

Cold-chamber machines, utilizing the principle illustrated in Figure 11-39 and shown in Figure 11-40, are used for most die casting. The metal for each casting is fed into the cold chamber and then is forced into the die by the plunger. Because the chamber remains relatively cool, there is much less tendency for the pickup of iron. In addition, the pressure cycle can be controlled as desired, permitting high-quality castings to be obtained. Although this

FIGURE 11-40. Cold-chamber die-casting machine. Inset shows a closeup of the die and the casting being removed. (*Courtesy Reed-Prentice Corporation.*)

type of machine has a longer operating cycle than do gooseneck machines, because the metal for each cast must be fed individually to the chamber, either manually or automatically, the productivity still is high, and up to 100 cycles per minute are not uncommon.

Metal cores are used extensively in die castings, and provision must be made for retracting them, usually before the die is opened for removal of the casting. Because a close fit must be maintained between the halves of the die and between any cores and the die sections to prevent the metal from flowing out of the die, both construction and maintenance costs are greatly increased when cores must be used. It is very important that the direction of the core-retracting motions be either a straight line or a circular arc. Otherwise, loose core pieces must be inserted into the die at the beginning of each cycle and then removed from the casting after it is ejected from the die. Such a procedure permits complex shapes to be cast, such as internal or external threads, but only with considerable reduction of the production rate and increased cost. Figure 11-41 is an example of a complex die casting that involves a number of cores.

Proper venting, to eliminate entrapped air, is a problem in die casting. Usually it is achieved by providing and maintaining proper fit between the die halves and cores. Also, very small vent holes or *overflows* may be used. Overflows result in excess metal that must be trimmed off after the casting is removed from the mold. This usually is done with trimming dies that also remove the sprues and runners.

Because of the method by which they are produced, die castings tend to have certain distinguishing characteristics. The surfaces tend to be harder than the interior as a result of the chilling action of the metal die on the molten metal. There is some tendency for the inner metal to be porous because of entrapped air. However, this can be eliminated by proper metal flow, venting, and higher pressures that are timed properly. Good casting design, die design

FIGURE 11-41. Die casting requiring complex cores. (*Courtesy The New Jersey Zinc Company.*)

and maintenance, proper procedure, and good equipment make excellent-quality die castings readily obtainable.

It is possible to make special bearing surfaces, or threaded studs or bosses, out of harder metals and cast them into die castings. These must be placed in position within the die before it is closed and the metal injected. Suitable recesses must be provided in the die for holding such parts, and the casting cycle is slowed down.

Excellent dimensional accuracy can be obtained with die casting. For aluminum-, magnesium-, zinc-, and copper-base alloys, linear tolerances of ±0.075 mm per 25 mm (±0.003 inch per inch) of length can be maintained. Minimum section thickness and draft depend on the kind of metal, as follows:

Metal	Minimum section	Minimum draft
Aluminum alloys	0.889 mm (0.035 in.)	1:100 (0.010 in./in.)
Brass and bronze	1.27 mm (0.050 in.)	1:80 (0.015 in./in.)
Magnesium alloys	1.27 mm (0.050 in.)	1:100 (0.010 in./in.)
Zinc alloys	0.635 mm (0.025 in.)	1:200 (0.005 in./in.)

As a result of the excellent dimensional accuracy and the smooth surfaces that can be obtained, most die castings require no machining except the removal of a small amount of fin, or flash, around the edge and possibly drilling or tapping of holes.

Comparison between permanent-mold and die casting. Obviously, ordinary permanent-mold and die casting are competitive processes. Figure 11-42 shows a comparison that will assist designers. It will be noted that, in general, permanent-mold casting provides greater flexibility and less lead time and cost where low quantities are involved. Die casting excels for large quantities where close tolerances, surface finish, and low labor cost are of great importance.

Centrifugal casting. In *centrifugal casting,* the molten metal conforms to the shape of the mold cavity as the result of the centrifugal force that is developed from the mold being rotated about its axial center line, at speeds of from 300 to 3000 rpm, while the molten metal is being introduced. In *true centrifugal casting,* the mold rotates about either a horizontal or vertical axis. Either a dry-sand or metal mold is used, and it determines the outer surface of the casting. Although a round shape is most common, hexagonal or other symmetrical but slightly out-of-round shapes can be cast. No mold or core is needed to form the inner surface of the casting. When a horizontal axis is used, as illustrated schematically in Figure 11-43, the inner surface always is cylindrical. If a vertical axis is used, the inner surface is a section of a parabola, as illustrated in Figure 11-44, the exact shape being a function of the speed of rotation.

In true centrifugal casting, the metal is forced against the walls of the

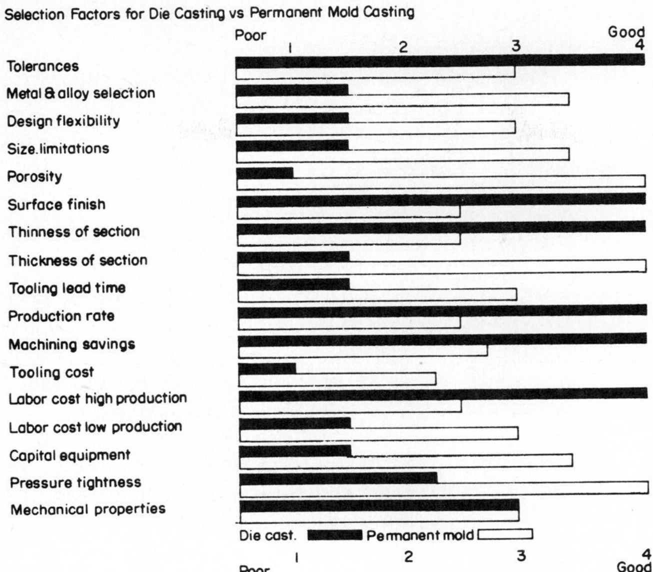

FIGURE 11-42. Factor chart for selecting between die casting and permanent-mold casting. (*Courtesy Materials Engineering.*)

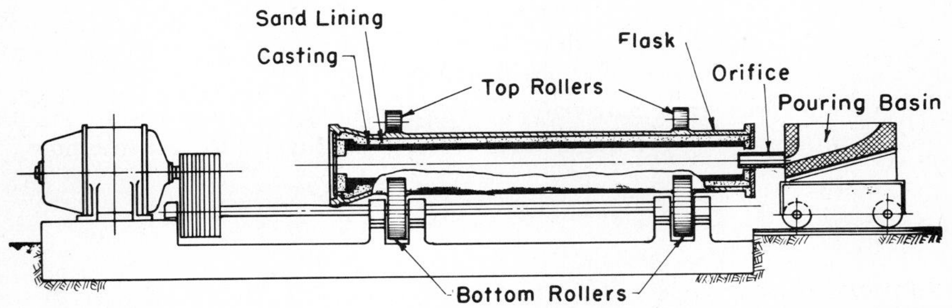

FIGURE 11-43. Schematic representation of the true horizontal centrifugal casting machine. (*Courtesy American Cast Iron Pipe Company.*)

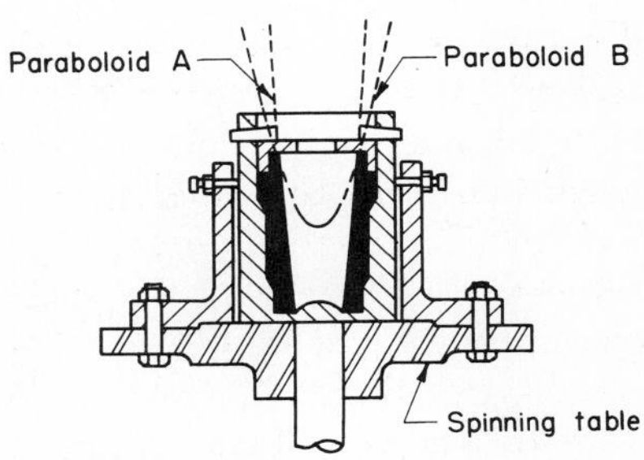

FIGURE 11-44. Vertical centrifugal casting, showing the effect of rotational speed on the shape of the inner surface.

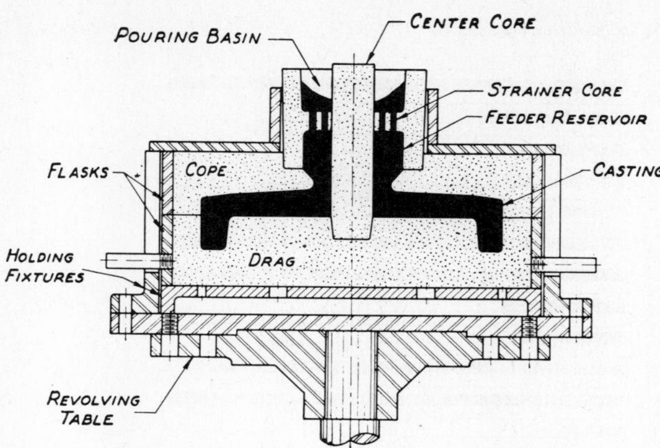

FIGURE 11-45. Semicentrifugal casting process. (*Courtesy American Cast Iron Pipe Company.*)

mold with considerable force, and it solidifies first at the outer surface. This results in a dense structure, with all the lighter impurities tending to be at the inner surface, thereby permitting them to be removed readily by a light machining cut, if required.

True centrifugal casting is widely used for the mass production of pipe, pressure vessels, cylinder liners, and brake drums. The equipment required is specialized and is expensive for large castings, but relatively simple and inexpensive equipment is available for small parts. The required permanent molds are relatively costly, but they have quite a long life. When ferrous metals are used, the molds are coated with some type of refractory dust or wash before the metal is introduced, to prolong their life.

In some cases the *semicentrifugal casting* principle, illustrated in Figure 11-45, can be used to advantage. In this case centrifugal force aids the flow of

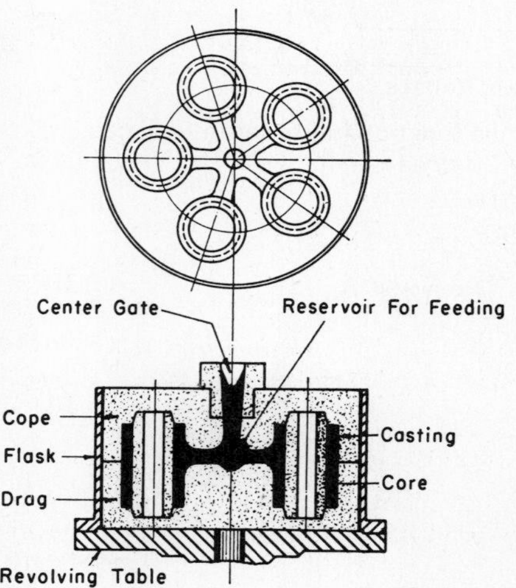

FIGURE 11-46. Method of casting by the centrifuging process. (*Courtesy American Cast Iron Pipe Company.*)

the molten metal from a central feeding reservoir and can thereby produce a somewhat more dense structure. The rotational speeds employed usually are considerably less than the case of true centrifugal casting. Frequently, several identical molds are stacked on top of each other and fed by a single, central reservoir. It is very important that the central feeding reservoir be sufficiently large to assure that the metal in it remains molten until the castings have solidified.

Centrifuging, illustrated in Figure 11-46, is another procedure that occasionally is used to provide forced metal flow from a central feeding reservoir to the molds in order to obtain a more dense grain structure. Note that the molds do not rotate about their axes. The rotational speeds are relatively low, and only simple shapes can be cast by this method.

PLASTER-MOLD CASTING

Two important casting processes utilize plaster molds. These are *investment casting* and the *Shaw process.* In most cases the molds basically are gypsum plaster with small additions of talc, terra alba, or magnesium oxide, to prevent cracking and to reduce the setting time. Lime or cement may be added to control expansion during baking, and about 25 per cent of fibers often are added to improve the strength. In one modification of investment casting, a slurry containing a high percentage of finely ground glass has been used.

Investment casting. *Investment casting* actually is a very old process. It existed in China for centuries, and Cellini employed a form of it in Italy in the sixteenth century. Dentists have utilized the process since 1897, but it was not until World War II that it attained industrial importance for making jet turbine blades from metals that were not readily machinable. Currently millions of castings are produced by the process each year, its unique characteristics permitting the designer almost unlimited freedom in the complexity and close tolerances he can utilize.

Investment casting involves the following steps:

1. *Produce a master pattern.* The pattern may be made from metal, wood, plastic, or some other easily worked material.

2. *From the master pattern, produce a master die.* This usually is made from low-melting-point metal or from steel, but sometimes is made from wood. When steel is used, the cavity ordinarily is engraved directly in the die. If low-melting-point metals are used, the dies can be cast, using the master pattern. In some cases the dies are machined directly without first making a master pattern, thereby skipping step 1.

3. *Produce the wax patterns.* These are made by pouring, or injecting under pressure, molten wax into the master die and allowing it to harden. A typical wax pattern, and the die in which it was molded, are shown in Figure 11-47.

FIGURE 11-47. Wax pattern for an investment casting being removed from the mold. (*Courtesy Haynes Stellite Company.*)

4. *Assemble the wax patterns to a common wax sprue.* The individual wax patterns are removed from the master die and several of them are attached to a central sprue and runners by means of heated tools and melted wax. In some cases, several pattern pieces may be united to form a complex, single pattern that if made in one piece could not be withdrawn from a master die. The result of this step is a cluster, such as is shown in Figure 11-48.

5. *Coat the cluster with a thin layer of investment material.* This step usually is accomplished by dipping the cluster into a thin slurry of finely ground refractory. Fine silicaceous material mixed with a special plaster is often used. This step produces a thin but very smooth layer of investment material adjacent to the wax patterns and ensures a smooth surface and good detail in the final product.

6. *Produce the final investment around the coated cluster.* The cluster can be dipped repeatedly in the investment material until the desired thickness is obtained, or it can be placed upside down in a flask and the investment material poured around it.

7. *Vibrate the flask to remove the entrapped air and settle the investment material around the cluster.* This step is necessary only when the investment material is poured around the cluster.

8. *Allow the investment to harden.*

9. *Melt or dissolve the wax pattern to permit it to run out of the mold.* This can be accomplished by placing the molds upside down in an oven or by passing them through a solvent, such as trichlorethylene. The wax is recovered for further use. This step is the most distinctive feature of the process, because it permits a complex pattern to be removed from a mold without the necessity of the mold being made in two or more sections so that it can be opened. Consequently, complicated shapes with reentrant sections can be cast by this

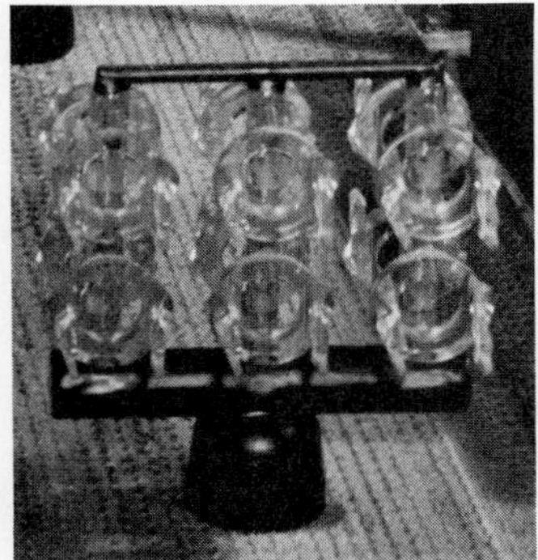

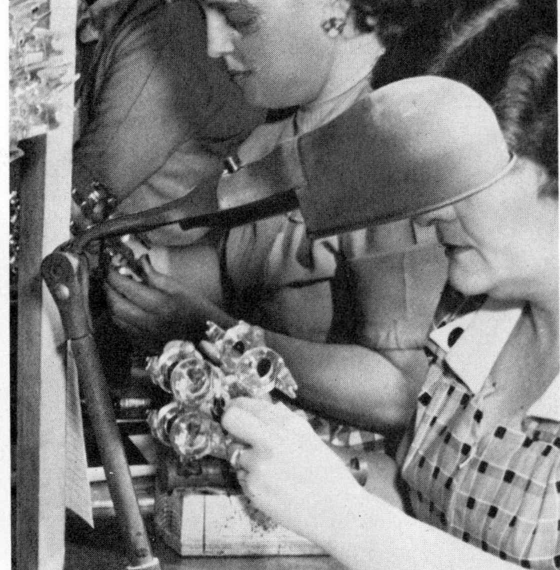

FIGURE 11-48. Wax patterns being assembled into clusters, shown in smaller view (*above*). (*Courtesy Haynes Stellite Company.*)

process, and better dimensional accuracy can be achieved. In the early history of the process, when only small parts were cast, the molds were placed in an oven and, as the wax was melted, it was absorbed into the porous investment. Because the wax disappeared from sight, the process was called the *lost-wax* process; this name still is used occasionally.

10. *Preheat the mold preparatory to pouring.* Heating to 538 to 1093°C (1000 to 2000°F) assures that the molten metal will flow more readily to all thin sections. It also gives better dimensional control because the mold and the metal shrink together during cooling.

11. *Pouring the molten metal.* Various methods, beyond simple pouring, are utilized to assure complete filling of the mold, especially where complex, thin sections are involved. Among these are the use of air pressure, evacuation of the air from the mold, and a centrifugal process. Figure 11-49 shows a sectioned investment mold with a casting in place.

12. *Remove the castings from the mold.* This is accomplished by breaking the mold away from the casting.

Investment casting obviously is a complex process and thus is expensive. Yet its unique advantages make it economically feasible in many cases. Not only can complex shapes be cast which could not be cast by any other process or be made by machining, but, because no mold material must be packed around the pattern, very thin sections—down to 0.38 mm (0.015 inch)—can be cast. The dimensional tolerances are excellent—0.1 to 1 per cent

FIGURE 11-49. Sectioned investment casting mold with casting in position.

(0.005 to 0.010 inch per inch) being routinely obtained. Combined with the very smooth surfaces, machining often can be completely eliminated or greatly reduced. Where machining is required, allowances of 0.38 to 1 mm (0.015 to 0.040 inch) usually are ample. These advantages are especially important where difficult-to-machine metals are involved.

Although most investment castings are less than 76.2 mm (3 inches) in

FIGURE 11-50. Group of typical parts produced by investment casting. (*Courtesy Haynes Stellite Company.*)

size and weigh less than 0.454 kg (1 pound), there are no specific size limitations, and castings up to 914 mm (36 inches) and weighing 36.3 kg (80 pounds) have been produced. Some typical investment castings are shown in Figure 11-50. It will be noted that complexity of shape is the most common characteristic. Although investment castings often cost from $0.91 to $3.63 per kg ($2 to $8 per pound), where proper use can be taken of their unique characteristics, they can be used advantageously and economically. However, designers should not use them indiscriminately.

RUBBER-MOLD CASTINGS

Several types of artificial rubbers—usually silicone varieties—are available which can be compounded in liquid form and then poured over patterns, forming semirigid molds upon hardening. The molds retain sufficient flexibility to permit them to be stripped from the pattern, thereby permitting quite intricate shapes with reentrant sections to be cast. They are suitable only for quite small castings, such as wax patterns and those of plastics and low-melting-point alloys which melt below about 260°C (500°F).

The Shaw process. Obtaining adequate permeability, while retaining the desired smooth mold-cavity surface, is a major problem in plaster-mold casting. The *Shaw process* provides an excellent solution.

In this process, a slurrylike mixture of a refractory aggregate—hydrolyzed ethyl silicate—and a jelling agent are poured over the pattern. This mixture sets in a rubbery jell that can be stripped from the pattern but having sufficient strength so that it will return to the exact shape it had while fitted to the pattern. The mold then is ignited to burn off the volatile elements in the mix. Next, it is brought to a red heat in a furnace. This firing makes the mold rigid and hard, but at the same time it causes a network of microscopic cracks to form. This *microcrazing* produces fissures that provide excellent permeability and good collapsibility, to accommodate the shrinkage of the solidifying metal.

The Shaw process can be used for castings of all sizes, and it produces excellent surface finish, detail, and dimensional accuracy. Dimensional tolerances between 0.051 mm and 0.25 mm in 25 mm (0.002 to 0.010 inch per inch) are readily obtainable. The molds may be one-piece or multipiece, depending on the type and complexity of the pattern. Figure 11-51 shows some of the wide variety of accurate and complex parts that are cast by this process, and Figure 11-52 shows how the Shaw process occasionally can be combined advantageously with the lost-wax process.

Cores can be used in Shaw-process and other plaster molds much the same as in sand casting. However, as a general rule, thinner cast sections can be obtained in plaster molds because the plaster retards the cooling. This possibility of obtaining thinner sections, better dimensional accuracy, fine de-

FIGURE 11-51. (*Top*) Ornamental casting and (*bottom*) group of cutters produced by the Shaw process. (*Courtesy Avnet Shaw Division of Avnet, Inc.*)

FIGURE 11-52. Method of combining the Shaw process and wax-pattern casting to produce the vanes of an impeller. A wax pattern is added to a metal pattern. After the metal pattern is withdrawn (*center*), the wax pattern is melted, leaving a cavity as shown at the right. (*Courtesy Avenet Shaw Division of Avnet, Inc.*)

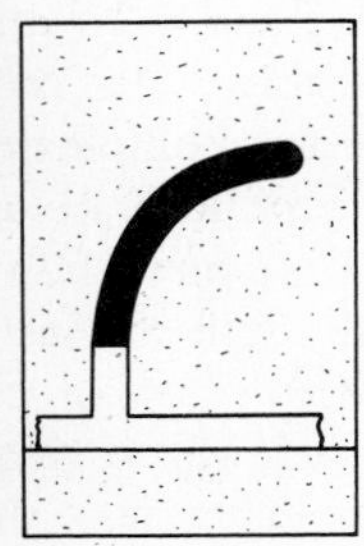

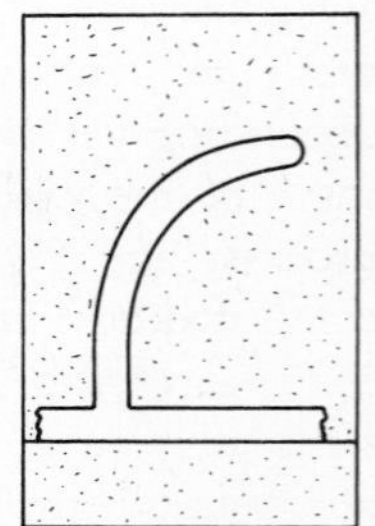

tails, elimination of machining, and exceptionally fine surface finish with plaster molds must be balanced against their greater cost.

MELTING AND POURING

All the casting processes require a furnace for melting the metal. Ideally, such a furnace should (1) provide adequate temperature, (2) minimize contamination, (3) make possible holding the metal at the required temperature without harmful effects when the chemical composition must be altered by alloy additions, (4) be economical, and (5) be capable of control to avoid atmospheric pollution. Except for experimental or very small operations, virtually all foundries use either a cupola, an air furnace, an electric-arc furnace, or an electric induction furnace. Occasionally, in fully integrated steel mills containing a foundry, steel is taken directly from an open-hearth or basic-oxygen furnace and poured into casting molds, but such a practice usually is not done regularly and is reserved for exceptionally large castings. For some experimental work, and in very small foundries, gas-fired crucible furnaces may be used, but these do not have sufficient capacity for most commercial operations.

Cupolas. By far the majority of gray, nodular, and white cast iron is melted in *cupolas,* although the use of electric induction furnaces is increasing. As shown in Figure 11-53, a cupola essentially is a refractory-lined, vertical, cylindrical, steel shell supported on legs, with its lower end 1 to 1.5 m (3 to 5 feet) above the floor. The lower end is closed by hinged doors that can be opened to permit cleaning the inside. A sand bottom, 102 to 152 mm (4 to 6 inches) deep and sloping downward to the pouring spout, is prepared on top of the closed bottom doors. The pouring spout has a taphole 25 to 50 mm (1 to 2 inches) in diameter.

After the sand bottom has been prepared, kindling wood is placed on the bottom and a bed of coke on top of this. The wood is ignited and the coke bed is increased until it is 457 to 1219 mm (18 to 48 inches) in depth, depending on the size of the cupola. Natural draft is supplied through the tuyeres, openings around the periphery of the cupola about 380 to 508 mm (15 to 20 inches) above the top of the sand bottom. When the coke bed has been built up to the desired depth and is thoroughly ignited, alternate charges of coke and iron are added through the charging door located about half the distance up the height of the cupola. The metal charge may be either pig iron or scrap. About one part of coke, by weight, is used to eight or 10 parts of iron. In addition, small amounts of limestone, or other materials, may be added to act as a flux and to increase the fluidity of the metal. It is possible to add certain alloy-rich materials to alter the chemical composition of the final product, although such additions often are added in the ladle at the time the cupola is tapped.

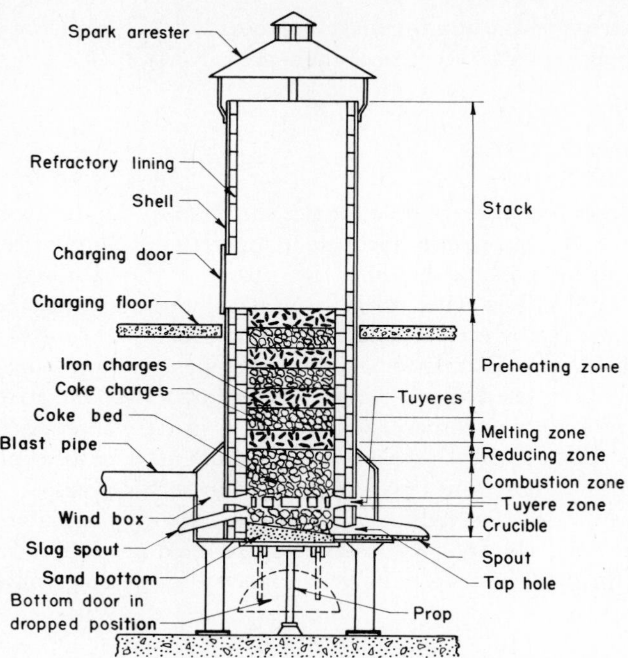

FIGURE 11-53. Sectional drawing of a cupola.

After the iron and coke charges have been built up to the level of the charging door, the charge is heated for about half an hour under natural draft. The natural draft openings then are closed and the blower is started. Under forced draft, the temperature in the cupola rises rapidly, and the iron charge starts to melt and run down through the coke. Final melting occurs in the zone just above the tuyeres. Molten metal should start to appear within 10 minutes after the air blast is turned on. Quartz-covered peepholes make it possible to see the metal flowing down through the incandescent coke. As soon as the first metal starts to flow, the taphole is plugged with a conical-shaped plug of fire clay.

The molten metal collects in the lower portion of the cupola between the sand bottom and the slag spout, located slightly below the tuyeres. Excess slag floats on top of the molten metal and can be drained off through the slag hole to obtain cleaner metal. When a sufficient amount of molten metal has collected, the air blast is turned off temporarily, the clay plug is punched out, and the metal is permitted to flow out into a pouring or holding ladle. When sufficient metal has been removed, the taphole is again plugged and the air blast is turned on.

From this point the operation of the cupola is a semicontinuous batch-type process. As metal is melted and removed, additional charges of coke and

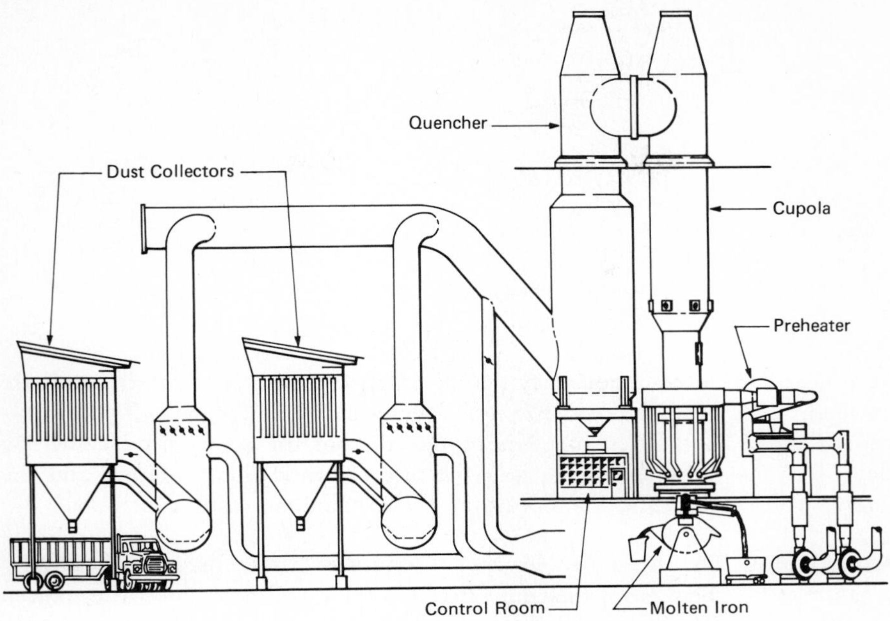

FIGURE 11-54. Modern cupola installation with associated antipollution equipment. *(Courtesy Foundry Management & Technology.)*

iron are added through the charging door. It is desirable to maintain the charge up to the level of the charging door at all times to obtain uniform combustion and melting conditions. The tapping process may continue up to as long as 14 to 16 hours.

Cupolas are simple and economical, can be obtained in a wide range of capacities, and can produce excellent quality cast iron if the proper raw materials and good control are used. Because the hot metal is in intimate contact with the coke, its chemical composition is influenced by what is in the coke. Coke with a too high sulfur content should be avoided.

In order to increase the melting rate and give greater economy, in the *hot-blast* type of cupola some of the stack gases are put through a heat exchanger to preheat the incoming air. In the *water-cooled* type, water is circulated around the shell to aid in preventing overheating of the lining in the melting zone and thus prolong its life. In all modern installations, the large volume of stack gases is passed through dust collectors of various types to avoid pollution of the atmosphere. Consequently, as indicated in Figure 11-54, a modern cupola installation is quite complex.

Air furnaces. Acid-hearth *air furnaces,* such as the one shown in Figure 11-55, are used only in malleable-iron foundries. They are somewhat similar to small

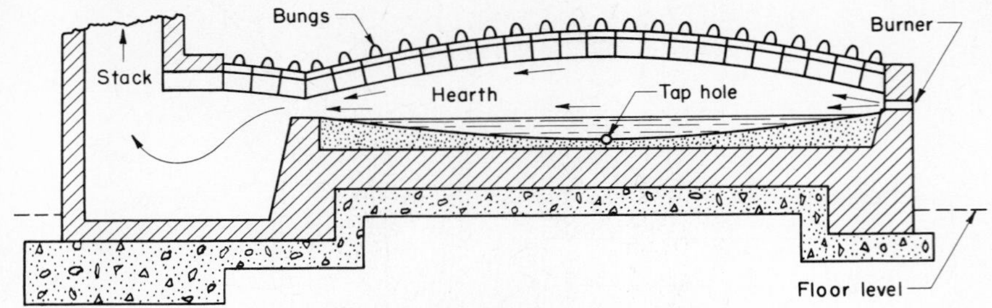

FIGURE 11-55. Section of an air furnace.

open-hearth furnaces, but they have no regenerative equipment and are charged through the removable *bungs.* They usually are fired by fuel oil or pulverized coal. The molten metal is protected from the combustion flames by slag. They have the advantage that the rate of heating and melting and the temperature and chemical composition of the metal are easily controlled.

Arc furnaces. The use of *arc furnaces* in foundries has increased substantially because of (1) their rapid melting rates, (2) their ability to hold the molten metal for any desired period of time to permit alloying, and (3) the greater ease of providing adequate pollution control.

The basic features of a *direct-arc* electric furnace are shown in Figure 4-4. In most types the top can be lifted or swung off to permit the charge to be introduced. Heating occurs through lowering the electrodes so that an arc is struck and maintained between the electrodes and the metal charge. The current path is from one electrode across the arc to the metal, through the metal, and back across the arc between the metal and another electrode.

Fluxing materials are added to provide a protective cover over the molten metal. Because the metal is thus protected and the metal maintained at a given temperature as long as desired, high-quality metal of any desired composition can be obtained. Furnaces of this type in capacities up to 100 tons are in use, but those below 25 tons are more common. Linings may be either acid or basic; the former is more common. A typical furnace of this type is shown in Figure 4-5.

An *indirect-arc* type of electric furnace also is used occasionally for melting small quantities of metal. In this type the arc is maintained between two horizontal, carbon electrodes that are above the metal, and melting is by radiation.

Induction furnaces. Because of their very rapid melting rates and relative ease of controlling pollution, the use of electric induction furnaces has increased extensively. Two types are used. The *high-frequency type,* shown in Figure 4-6, consists of a crucible surrounded by a water-cooled coil of copper tubing. A high-frequency electrical current passes through the coil, establishing an alter-

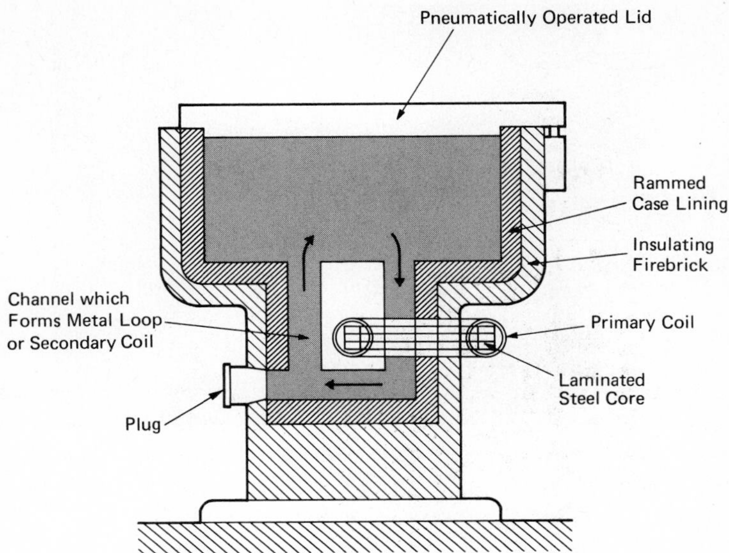

FIGURE 11-56. Principle of the low-frequency induction furnace.

nating magnetic field, which, in turn, induces secondary currents in the metal in the crucible. These secondary currents heat the metal very rapidly.

Low-frequency, or *channel-type,* induction furnaces are being used increasingly. As indicated in Figure 11-56, these have an ordinary alternating-current primary coil, but the secondary coil is formed by a loop, or channel, of the molten metal. Because the secondary coil has but one turn, a low voltage/high amperage current is induced in it, which provides the desired heating. Some molten metal, to form the secondary coil, must be used to start such furnaces, but their heating rate is very high, and the temperature is readily controlled. This makes them very useful, and widely used, as holding furnaces, where it is desired to maintain molten metal at a constant temperature for an extended period of time, as in die-casting machines or in mold-pouring systems.

Because there is no contamination from the heat source, induction furnaces produce very pure metal. Although the capacity of a given furnace usually is relatively small—from a few pounds up to 6 tons—the short melting cycle gives them high total productivity.

Pouring practice. In order to transfer the metal from the melting furnace into the molds, some type of pouring device, or ladle, must be used. Primary considerations are to maintain the metal at the proper temperature for pouring and to assure that only quality metal gets into the molds. The type of ladle used is determined largely by the size and number of castings to be poured. In small foundries, the hand-held, shank-type ladle shown in Figure 11-57 often

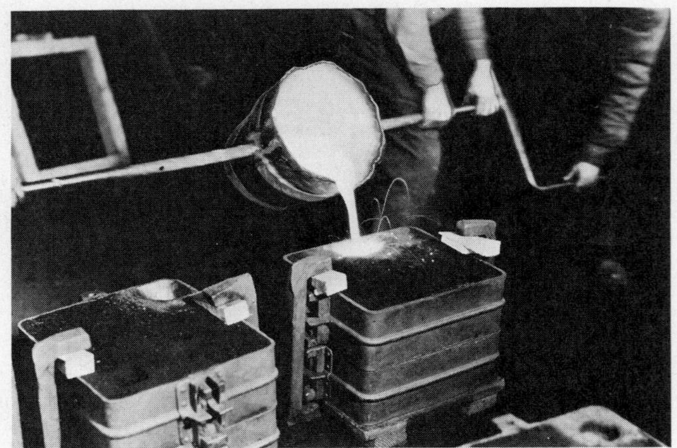

FIGURE 11-57. Pouring a mold from a shank-type ladle. (*Courtesy Steel Founders' Society of America.*)

is used. In larger foundries, either bottom-pour or teapot-type ladles, illustrated in Figure 11-58, are used, often in conjunction with a conveyor line on which the molds move past the pouring stations, as shown in Figure 11-59. These types of ladles help to assure that slag and oxidized metal do not enter the mold. In modern, mass-production foundries, automatic pouring systems, such as shown in Figure 11-60, are employed. The molten metal is brought from the main melting furnace to a holding furnace by overhead crane. A programmed amount of molten metal is poured into the individual pouring ladles from the holding furnace and, in turn, is poured automatically into the corresponding molds.

Because of its high combustibility at elevated temperatures, special precautions are required in melting and pouring magnesium. During melting, the metal must be kept covered by an adequate flux. Prior to pouring, the

FIGURE 11-58. Types of ladles for use in pouring castings.

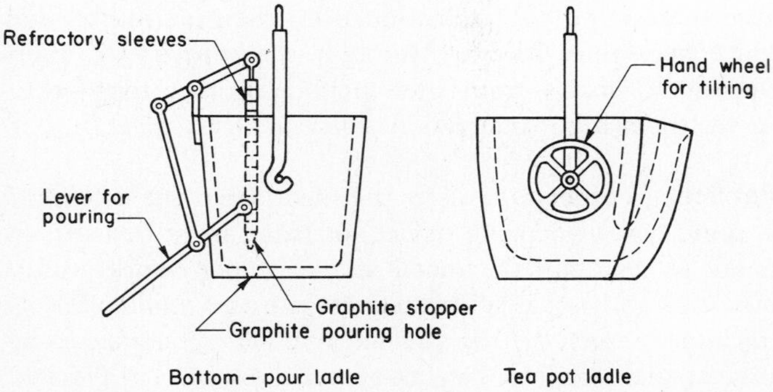

Refractory sleeves
Lever for pouring
Graphite stopper
Graphite pouring hole
Bottom – pour ladle
Hand wheel for tilting
Tea pot ladle

FIGURE 11-59. Pouring molds on a conveyor line. (*Courtesy The Link-Belt Company.*)

FIGURE 11-60. Machine for automatic pouring of molds on a conveyor line. (*Courtesy Roberts Corporation.*)

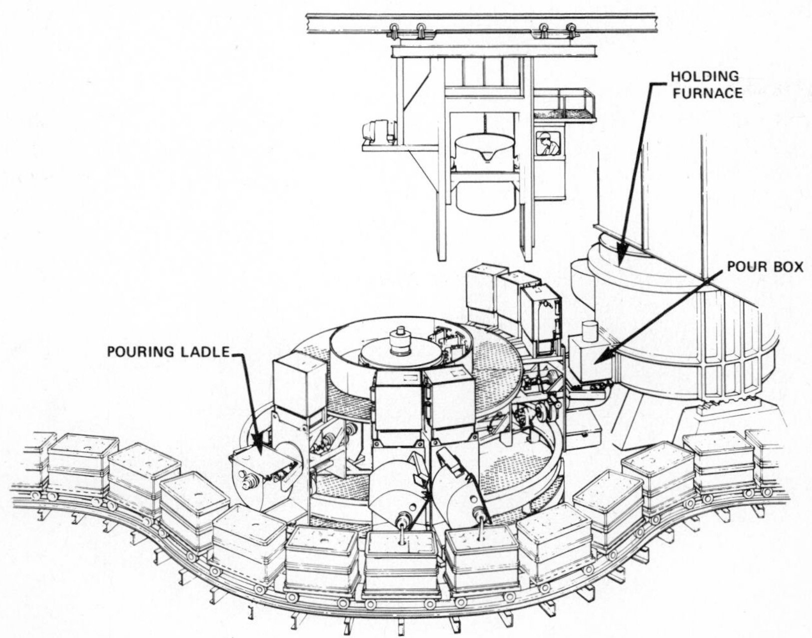

molten metal should be stirred with a steel rod to free the impurities, which often are not much lighter than the metal, thereby permitting them to be collected in the flux. The flux then should be skimmed from the metal and a layer of protective material, in the form of powder, immediately put over the surface of the magnesium. Such a protective layer should be maintained at all times, including the time the metal is in the pouring ladle and on the top of the risers.

Vacuum melting and pouring. Just as increasing amounts of metals are being melted in a vacuum to remove gases and assure higher purity, increasing numbers of castings are being poured in a vacuum to retain their purity in the cast form. This often is done by enclosing both an induction furnace and the mold in a chamber from which the air can be evacuated, and arranging for the metal to be poured directly from the furnace into the mold.

Another method of vacuum melting and pouring, used when extremely high-purity metal is necessary, is carried out by using an arc furnace in which a consumable metal electrode of the metal to be melted is used. This method is used to make titanium ingots. When castings are to be made, the electrode is melted into a crucible and then poured into the mold, all of the equipment being contained in an evacuated chamber. Obviously, these vacuum methods are expensive and should be employed only when the highest quality is essential.

Continuous casting. Long, continuous shapes, having special cross sections such as shown in Figure 11-61, can be produced readily by the *continuous cast-*

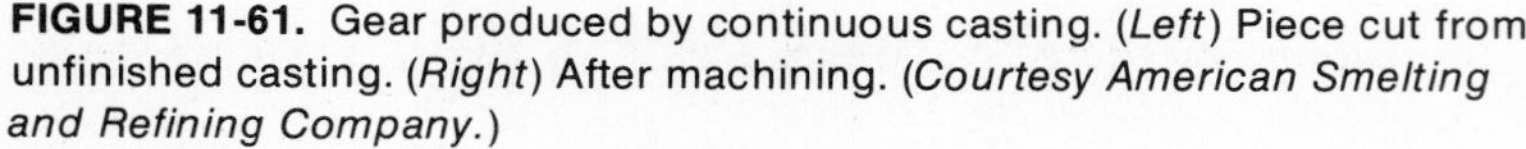

FIGURE 11-61. Gear produced by continuous casting. (*Left*) Piece cut from unfinished casting. (*Right*) After machining. (*Courtesy American Smelting and Refining Company.*)

ing process illustrated in Figure 4-12. Only a single mold is required, the individual units being cut from a single strand. High-quality metal can be obtained because it is protected from contamination while molten and being poured.

CLEANING, FINISHING, AND HEAT-TREATING CASTINGS

After solidification and removal from the molds, most castings require some cleaning and finishing operations. These may involve all or several of the following steps:

1. Removing cores.
2. Removing gates and risers.
3. Removing fins and rough spots from the surface.
4. Cleaning the surface.
5. Repairing any defects.

The required operations are not always done in the same order; the particular casting process used may eliminate some of them or reduce the amount of work required. Because these cleaning and finishing operations may involve considerable expense, some consideration should be given to them in designing castings and in selecting the casting method to be used. Often substantial savings can be effected. In recent years much attention has been given to mechanizing these operations.

Sand cores usually can be removed by shaking. Sometimes they must be removed by dissolving the core binder. On small castings, gates and risers often can be knocked off. However, on larger castings, and often on small castings, they usually must be cut off. On nonferrous and cast iron castings this usually is done by means of an abrasive cutoff wheel, as illustrated in Figure 11-62. Power hacksaws and bandsaws also are used. Gates and risers on steel castings, especially large ones, usually are removed by an oxyacetylene torch, as shown in Figure 11-63.

After the gates and risers are removed, small castings often are put through tumbling barrels, as shown in Figure 11-64, to remove fins, snags, and sand that adheres to the surface. Tumbling also may be used to remove cores and, in some cases, gates and risers. Frequently, some type of shot or slug material is added to the barrel to aid in the cleaning. Larger castings may be passed through a cleaning chamber on a conveyor, wherein they are subjected to blasts of abrasive or cleaning material, often using the principle illustrated in Figure 15-19. Large castings usually have to be finished manually, using pneumatic chisels, portable grinders, and manually directed blast hoses in separate cleaning rooms.

Although it is desirable that castings contain no defects, it is inevitable that some will occur, particularly in large castings where only one or a few of a particular design are made. Some types of defects can be repaired readily and

FIGURE 11-62. Removing a riser by means of an abrasive cutoff wheel. (*Courtesy The Norton Company.*)

satisfactorily by arc welding. However, it is imperative that the casting be of a material that can be welded satisfactorily, that *all* defective areas be removed down to sound metal by gouging, or chipping, and that a sound repair weld be made. When required, aluminum and magnesium castings can be given a treatment to make them watertight.

Heat treatment of castings. Because of the increased use of alloys, the heat treatment of castings has become more common, to obtain the full benefit of the alloy additions. Steel castings almost always are given a full anneal to reduce the hardness in rapidly cooled thin sections and to reduce internal stresses that result from uneven cooling. Nonferrous castings of some types are heat-treated to put them in a normalized condition.

FIGURE 11-63. Using an oxyacetylene torch to remove the riser from a steel casting. (*Courtesy Steel Founders' Society of America.*)

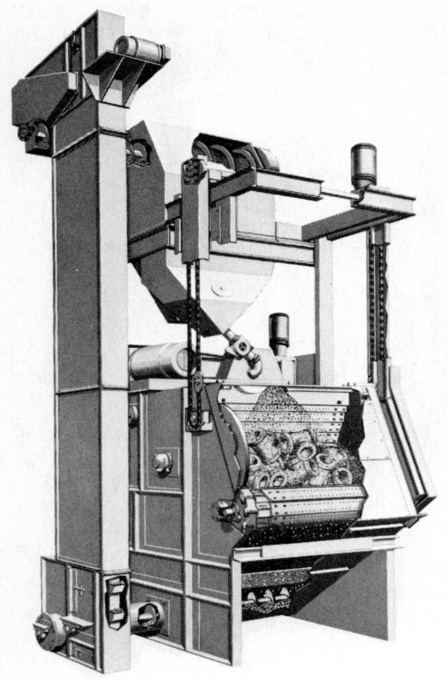

FIGURE 11-64. (*Left*) Tumbling machine for cleaning castings. (*Above*) Uncleaned castings in loading hopper at bottom while cleaned castings are being discharged at top. (*Courtesy Wheelabrator-Frye Inc.*)

DESIGN CONSIDERATIONS IN CASTINGS

It is very important that the designer of castings give careful attention to several requirements of the process and, if possible, cooperate closely with the foundry if economy and best results are to be obtained. Frequently, minor, and readily permissible, changes in design details will greatly facilitate and simplify the casting of a component and reduce the percentage of defects.

One of the first factors that must be considered by the designer is the location of the parting plane, except in the cases of castings to be produced by the full-mold or investment processes. The location of the parting plane affects the following:

1. The number of cores.
2. Effective and economical gating.
3. Casting weight.
4. Method of supporting cores.
5. Dimensional accuracy.
6. Ease of molding.

In general it is desirable to minimize the use of dry-sand cores. Often a change in the location of the parting plane will assist in this objective, as il-

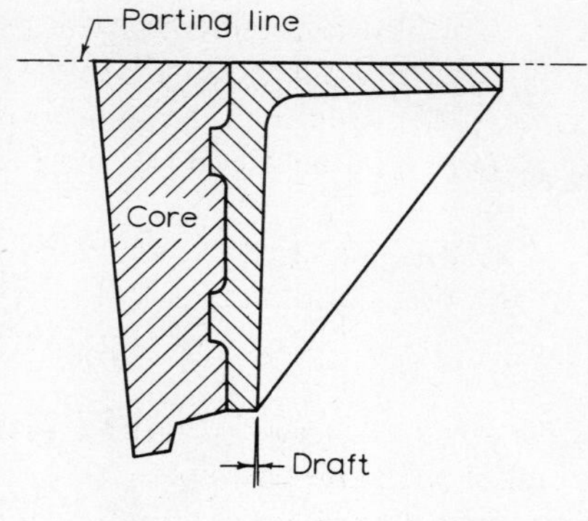

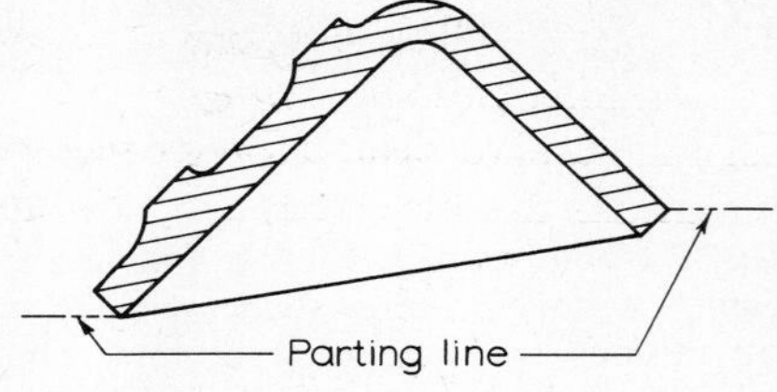

FIGURE 11-65. Showing how a core can be eliminated by changing the location of the parting line on a casting.

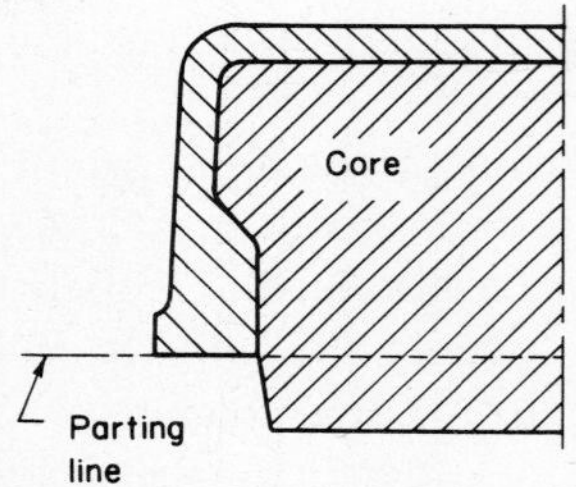

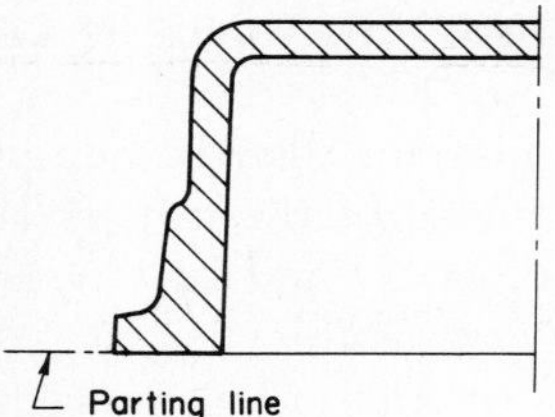

FIGURE 11-66. Elimination of a dry-sand core by change in design.

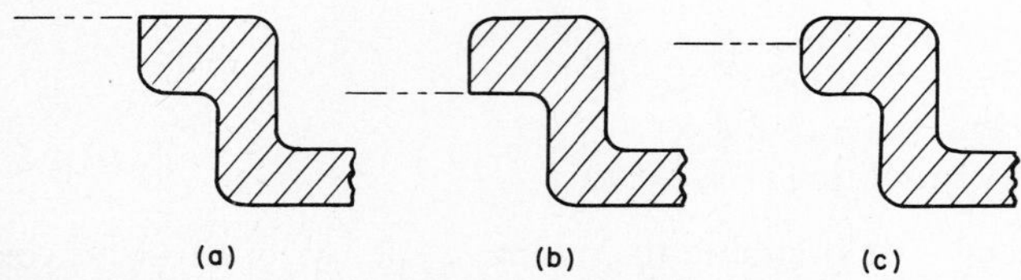

FIGURE 11-67. Effect of rounded edges on location of the parting line.

lustrated in Figure 11-65. Note that the change also reduced the weight of the casting by eliminating the need for draft. Figure 11-66 shows a typical example of how a simple design change eliminated the need for a dry-sand core. As shown in Figure 11-67, specification of round edges on a part often affects the location of the parting plane.

As illustrated in Figure 11-68, when draft is indicated on a part to be cast, the parting plane may be fixed, whereas more economical molding may be possible if provision for draft is indicated by a note or left to the option of the foundry.

The designer should remember that dimensions across a parting plane are subject to more variation than those parallel with the parting plane.

The solidification process is of prime importance in obtaining sound castings, and it is closely related to design. It should always be remembered that castings reflect the characteristics of cast metal. Those portions of a casting that have a high ratio of surface area to volume, and thus are subjected to rapid cooling rates, will tend to be hard or have a hard skin. Heavier sections cool more slowly and, unless special precautions are observed, may contain shrinkage cavities and porosity or have large grain structures.

Ideally, a given casting should have virtually uniform thickness in all sections. In most cases this is not possible. However, when the section thickness must change, such changes should be gradual. Figure 11-69 gives some guides regarding changes in sections.

When sections of castings intersect, two problems arise. The first is the matter of possible stress concentrations. This problem can be minimized by providing generous fillets at all changes of direction to avoid sharp, interior

FIGURE 11-68. Optional location of parting plane and results when drawing permits options.

As shown on drawing

As shown on drawing, with draft permitted by note

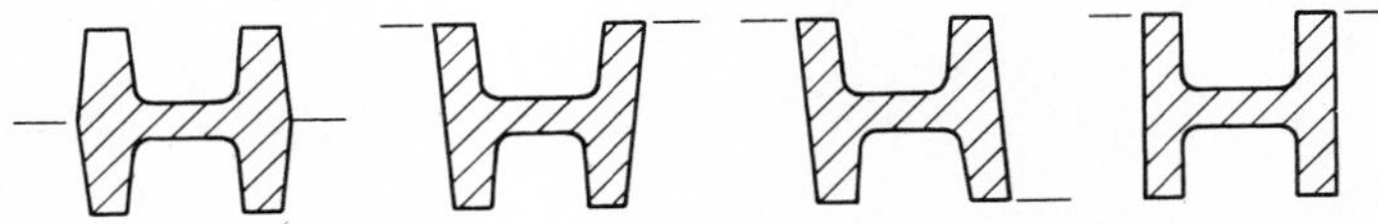

Optional results, with and without draft (exaggerated)

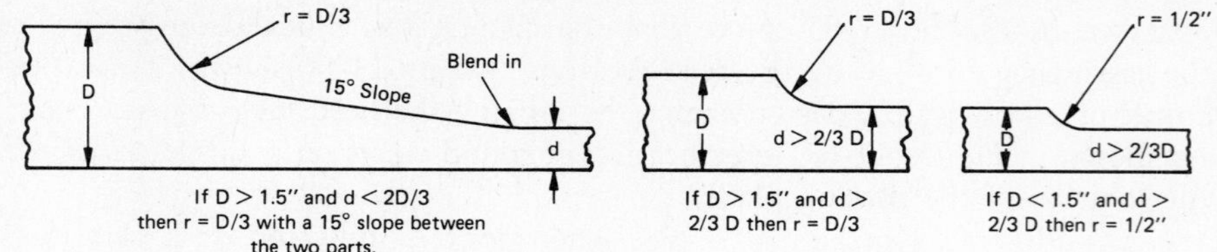

FIGURE 11-69. Guides for section changes in castings.

corners. However, too large fillets may cause difficulty because of the second problem—*hot spots.*

When sections of castings intersect, localized thick sections may result, as indicated in Figure 11-70. These thick sections cool more slowly than others and result in localized, abnormal shrinkage. When the differences in sections are large, as illustrated in Figure 11-71, the hot-spot areas are likely to result in serious defects in the form of porosity or shrinkage cavities.

Figure 11-72 shows three examples of the incorrect and correct use of radii in reducing hot spots, or shrinkage areas. When a condition exists as shown in Figure 11-72a and b, the metal at the heavy section remains liquid while that in the adjoining legs contracts and exerts tensile strains at the junction. The design in (c) greatly reduces this tendency, but it is not a desirable solution if the fillet is to be stressed in tension, due to the decreased section modulus.

Obviously, voids, porosity, and cracks in castings can result in serious failures. Sometimes cored holes, as illustrated in Figure 11-73, can be used to avoid hot spots. Where heavy sections must exist, an adjacent riser often can be provided to feed the section during shrinkage, as illustrated in Figure 11-74. The riser provides a reservoir of molten metal which, because of its mass, remains molten until after the adjacent portions of the casting have solidified and thus can *feed* the heavy section during shrinkage. A shrinkage cavity occurs only within the riser and is removed when the riser is cut off. Most risers are open to the atmosphere, but *blind risers,* such as illustrated in Figure 11-75, also are used. Both open and blind risers may be seen in Figure 11-26.

Risers add to the cost of castings because extra metal must be melted,

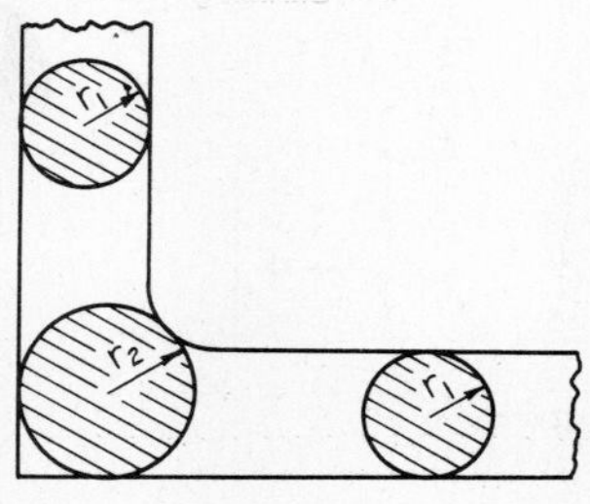

FIGURE 11-70. "Hot spot" at section r_2 caused by intersecting sections.

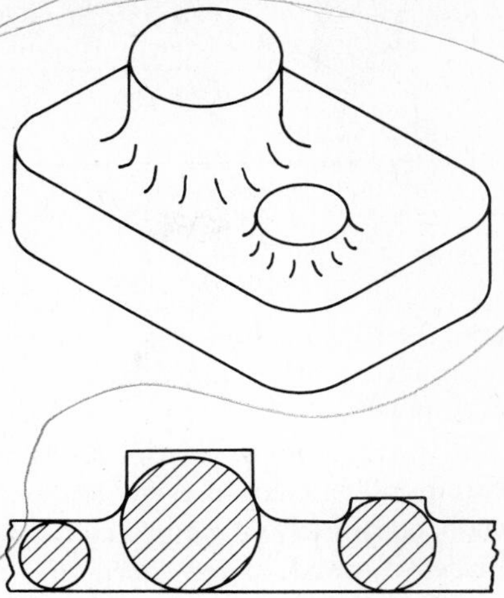

FIGURE 11-71. Hot spot resulting from intersecting sections of different thicknesses.

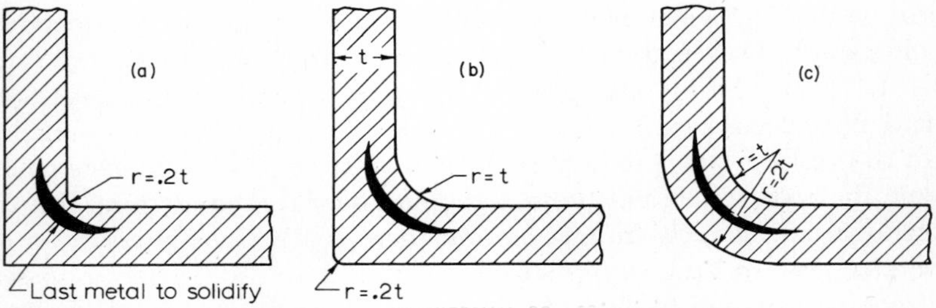

FIGURE 11-72. Last metal to solidify in sections resulting from incorrect (*left and center*) and correct (*right*) use of radii.

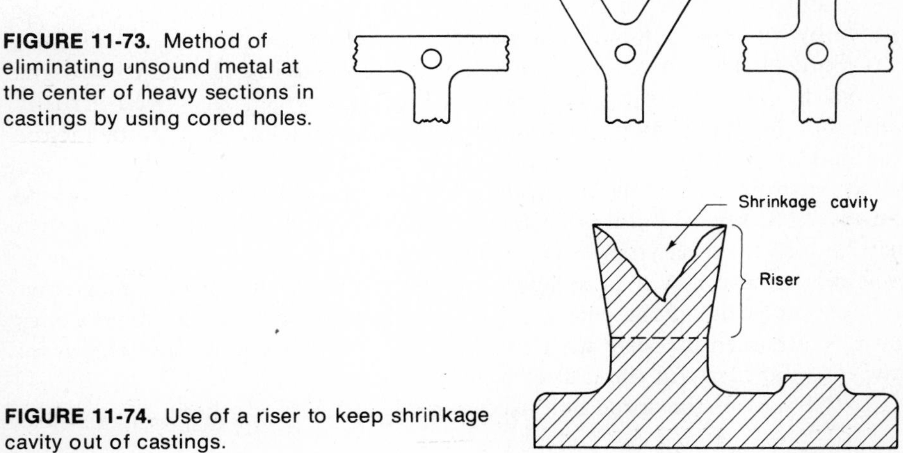

FIGURE 11-73. Method of eliminating unsound metal at the center of heavy sections in castings by using cored holes.

FIGURE 11-74. Use of a riser to keep shrinkage cavity out of castings.

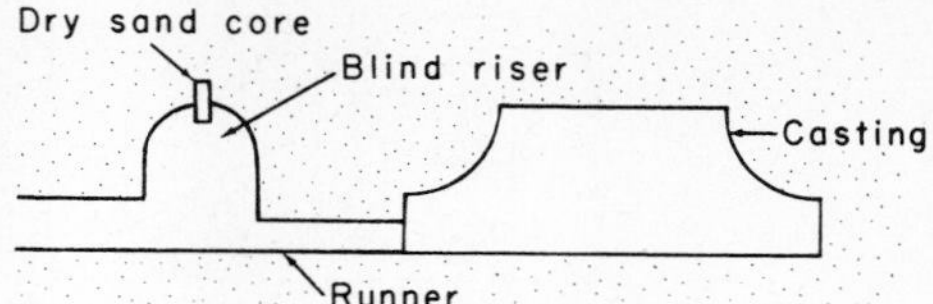

FIGURE 11-75. Use of a blind riser in casting with dry-sand core to prevent the creation of a vacuum.

and the solidified riser must be removed from the casting. Consequently, it is desirable to eliminate the necessity for them through design changes, whenever possible.

Other means besides risers may be used to help obtain proper solidification. One method is to increase the size of sections that otherwise might cool too quickly. This practice is called *padding.* Another procedure is to place pieces of metal, called *chills,* into the molding sand adjacent to heavy sections to accelerate the rate of solidification. External chills do not adhere to the casting or may be removed readily. Internal chills are pieces of metal, usually in the form of a coil of wire or like a large-headed nail, which are suspended in the mold cavity. They produce a chilling action, but they also become an integral part of the casting. As such they may be the source of internal weakness, and their use is restricted.

In die casting, the pressure on the metal during solidification tends to overcome the contraction and provide the necessary feeding action. Similarly, in centrifugal and semicentrifugal casting there is a positive pressure to force the metal to flow and thus offset shrinkage.

Proper gating and feeding of a mold can do much to control casting solidification. A satisfactory system will (1) permit the metal to flow rapidly and to fill the mold quickly; (2) allow the metal to flow with minimum turbulence, particularly within the cavity; (3) not cause aspiration of gases; (4) induce solidification at the center with progressive solidification toward the risers; and (5) not cause the surfaces of the cavity or runner system to be eroded by too rapid a flow of the metal.

Sprue holes should be adequate in size, decreasing in size toward the bottom to prevent aspiration. The *gate,* through which the metal actually flows into the mold cavity, is very important. Frequently, it is located at the parting line of the mold. Ordinarily, more than one gate is used, being connected with the sprue hole by means of a *runner.* In this way the metal can be introduced at several points to aid proper directional solidification. Gates also may be located at the top of a casting, but bottom or side gating usually is preferred because it will cause less turbulent metal flow within the mold cavity. Consequently, if the gate is located at the parting line, the mold cavity must be arranged properly with respect to the parting line to have the metal flow into the cavity in the desired manner.

The wall of the sprue hole is commonly made of exothermic materials,

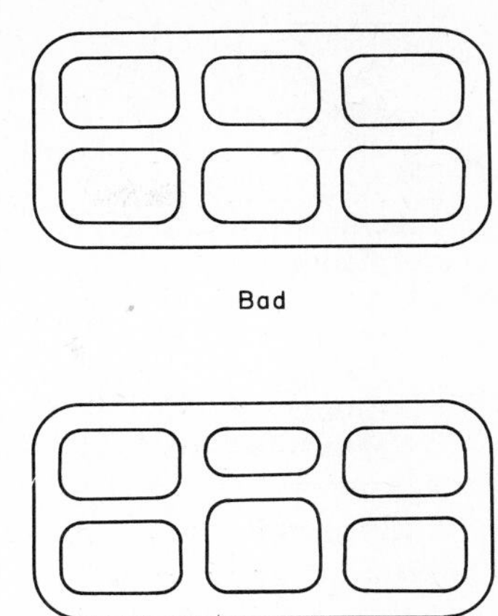

FIGURE 11-76. Method of using staggered ribs to prevent cracking during cooling.

or a stick of exothermic material is added to the sprue after the metal is poured, to keep the sprue molten longer and thus aid in feeding the casting.

Intersecting ribs can cause shrinkage problems and should be given special consideration by the designer. Where sections intersect, forming continuous ribs, contraction occurs in opposite directions in each rib. As a consequence, cracking frequently occurs during cooling. By staggering the ribs there is some opportunity for slight distortion to occur so that high stresses are not built up. Figure 11-76 illustrates this situation.

Large unsupported flat areas should be avoided in all types of castings. Such sections tend to warp during cooling. In die castings, good surface appearance of a flat area is difficult to maintain over an extended period of production because the molten metal reacts with the die and any deterioration of the die becomes apparent. Another consideration that is of particular importance in die castings is that the parting line of the die halves be at the corner of the casting whenever possible. Some small amount of fin, or flash, usually will be present at the parting line, particularly as the dies wear. When this flash is removed, or when it is so slight that it does not have to be removed, a small line of demarcation will remain. If this occurs in the middle of a flat surface it will be visible, whereas at a corner it usually will not be noticeable.

In designing all types of castings, minimum section thickness must be considered. Exact specifications as to economical and practicable section thicknesses cannot be given because the shape and size of the casting, the kind of metal, the method of casting, and the practice of the individual foundry are all factors that can affect the results obtainable. The following tabulation represents reasonable guidelines:

Material	Minimum mm	Minimum inches	Desirable mm	Desirable inches	Casting process
Steel	4.76	3/16	6.35	1/4	Sand
Gray iron	3.18	1/8	4.76	3/16	Sand
Malleable iron	3.18	1/8	4.76	3/16	Sand
Aluminum	3.18	1/8	4.76	3/16	Sand
Magnesium	4.76	3/16	6.35	1/4	Sand
Zinc alloys	0.51	0.020	0.76	0.030	Die
Aluminum alloys	1.27	0.050	1.52	0.060	Die
Magnesium alloys	1.27	0.050	1.52	0.060	Die

In conclusion, a final caveat: in casting design, probably more than in connection with any other manufacturing process, the designer can gain tremendous benefits by working closely with the supplying foundry.

Review questions

1. What are six basic factors that are involved in casting?
2. Explain the differences by which the six factors are met in sand molding and investment molding.
3. Which of the pattern allowances must *always* be provided?
4. Which of the pattern allowances can be omitted when the Full-Mold process is used?
5. What total shrinkage allowance would be made on a wood pattern that was to be used to cast an aluminum pattern, which, in turn, would be used in making gray iron castings?
6. What is the function of a core print?
7. What are four required characteristics of molding sand?
8. How is collapsibility imparted to molding sand?
9. What are three possible sources of gas in a mold?
10. Why must the amount of moisture in molding sand be limited?
11. What are two methods of imparting strength to molding sand other than by means of clay and water?
12. Give two reasons for the increased use of furan binders for molding sand.
13. What is a slip jacket, and what is its use?
14. What six factors must be controlled in green molding sand?
15. What primarily controls the rate of flow of molten metal into a mold?
16. Explain the difference in the packing action on the sand in using jolting and squeezing.
17. Why are molds for casting steel given a high-silica wash and thoroughly dried before pouring?
18. What are three advantages of shell molding?
19. Would you choose shell molding or the Full-Mold process for making only two duplicate castings? Why?

20. How is sand packed in a mold by means of a sand slinger?
21. For what conditions is the Full-Mold process best suited?
22. One of the requirements for casting nickel–aluminum bronze is that it be thoroughly degassed. Would the Full-Mold process be a suitable process for casting this alloy? Why?
23. Why should green-sand cores always replace dry-sand cores whenever possible?
24. Large dry-sand cores often are made hollow. Why?
25. What is a chaplet? For what are they used?
26. What are two factors that make it difficult to make permanent-mold castings of steel?
27. What are two advantages of the pressure-pouring process?
28. Why is die casting restricted primarily to nonferrous metals and alloys?
29. Why is die casting not suitable for small quantities?
30. What are the primary advantages of centrifugal casting?
31. What are two primary advantages of investment casting?
32. What advantages does the Shaw process have over ordinary plaster-mold casting?
33. Why do undercuts and reentrant sections not have to be considered in designing investment castings?
34. For what conditions can the more complicated and costly investment casting process often compete with less expensive casting processes?
35. What type of furnace is most commonly used for melting gray iron?
36. Give two reasons why electric induction furnaces are being used increasingly for melting iron and steel in foundries.
37. Why is an induction furnace advantageous for melting alloys?
38. Most automatic pouring units in foundries use what are in effect teapot-type ladles. Why?
39. What is the advantage of gating into the bottom of a large casting mold rather than directly into the top?
40. What is the function of a riser on a casting?
41. Explain the difference between a "chill" and a "pad."
42. Why are the dimensions across a parting line of a casting more likely to vary than those parallel with the parting surface?
43. What procedure is utilized on castings to avoid stress concentrations due to sharp interior corners?
44. Explain what a "hot spot" is in a casting.
45. What type of defect is likely to result from a hot spot in a casting?
46. What are three general rules that should be followed in designing castings?
47. What accounts for the increased use of vacuum melting and pouring?
48. What is likely to result if too-large fillets are incorporated in castings?
49. What are three basic advantages of the casting process?
50. In connection with casting design, why is it important to work closely with the foundry that will produce the part?

Case study 9.
THE DEFECTIVE PROPELLORS

The Propco Foundry Co. casts large ship propellors. A new-design propellor, 6.1 meters (20 feet) in diameter and weighing about 32 kN (35 tons), was cast of nickel–aluminum bronze, using the mold–gating–risering arrangement shown schematically in Figure CS-9. [The large end of the propellor hub was approximately 1.52 meters (5 feet) in diameter with a cored hole 0.46 meter (1.5 feet) in diameter.] While the ship on which one of the propellors was installed was at sea, an entire blade broke off, resulting in a damage claim of over $250,000. Investigation revealed excessive porosity and shrinkage cavities in the blades of this and two other duplicate propellors. What corrective measures would you recommend to eliminate the difficulty?

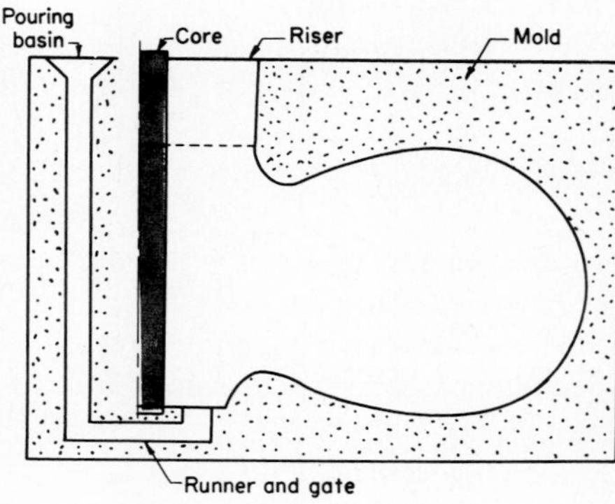

FIGURE CS-9. Schematic showing gating and risering of mold for a large marine propellor.

Powder metallurgy

Powder metallurgy is the name given to a process wherein fine metal powders are pressed into a desired shape, usually in a metal die and under high pressure, and the compacted powder is then heated (sintered), either concurrently or subsequently, for a period of time at a temperature below the melting point of the major constituent. The process (commonly designated as P/M) has a number of distinct advantages which account for its rapid growth in recent years, including: (1) no material is wasted, (2) usually no machining is required, (3) only semiskilled labor is required, and (4) some unique properties can be obtained, such as controlled degrees of porosity and built-in lubrication.

A crude form of powder metallurgy appears to have existed in Egypt as early as 3000 B.C., utilizing particles of sponge iron. In the nineteenth century it was utilized for producing platinum and tungsten wires. However, its first significant use related to general manufacturing was in Germany, following World War I, for making tungsten carbide cutting-tool tips. Since 1945 the process has been highly developed, and large quantities of a wide variety of P/M products are made annually, many of which could not be made by any other process. Most are under 50.8 mm (2 inches) in size, but many are larger, some weighing up to 22.7 kg (50 pounds) and measuring up to 508 mm (20 inches).

The powder metallurgy process normally consists of four basic steps:

1. Producing a fine metallic powder.
2. Mixing and preparing the powder for use.
3. Pressing the powder into the desired shape.
4. Heating (sintering) the shape at an elevated temperature.

Other operations can be added to obtain special results.

The pressing and sintering operations are of special importance. The pressing greatly affects the density of the product, which has a direct relationship to the strength properties. Sintering strips contaminants from the surfaces of the powder particles, permitting diffusion bonding to occur and resulting in a single piece of material. Sintering usually is done in a controlled, inert atmosphere, but sometimes it is done by the discharge of a spark through the powder while it is under compaction in the mold.

Metal powders. The properties of powder metallurgy products are highly dependent on the qualities of the metal powders that are used. Most important are *chemical qualities, particle size, size distribution, particle shape,* and *purity.* Several processes are used for making metal powders, and almost any metal or alloy can be obtained in powder form. For most uses, a very fine powder, of smooth particle shape and purity, is desired.

The most commonly used methods for making metal powders are *reduction of oxides or ores, electrolytic deposition* from solutions or fused salts, and *melt atomization.* The last process may take two forms. One, illustrated in Figure 12-1a, starts with molten metal, which is atomized by a stream of impinging gas or liquid as it emerges from an orifice. In the other, illustrated in Figure 12-1b, an electric arc impinges on a rapidly rotating electrode within an evacuated and helium-filled chamber. This process is particularly useful for obtaining prealloyed powders, since, by using an alloy electrode, each powder particle will have the same alloy composition. Other processes that are employed are *thermal decomposition of carbonyls, condensation of metallic vapors,* and *mill grinding.*

The *flow characteristics* of the powders also are very important. These determine the ability of the powder to flow under pressure, thus affecting the rapidity of the pressing cycle and, to a great extent, controlling the density of the compacted material. If a powder does not flow readily and has particles that are too large, the particles may resist the applied pressure by "bridging," like the stones in an arch, resulting in relatively large voids. Thus higher pressures will be required to obtain a dense, homogeneous compact.

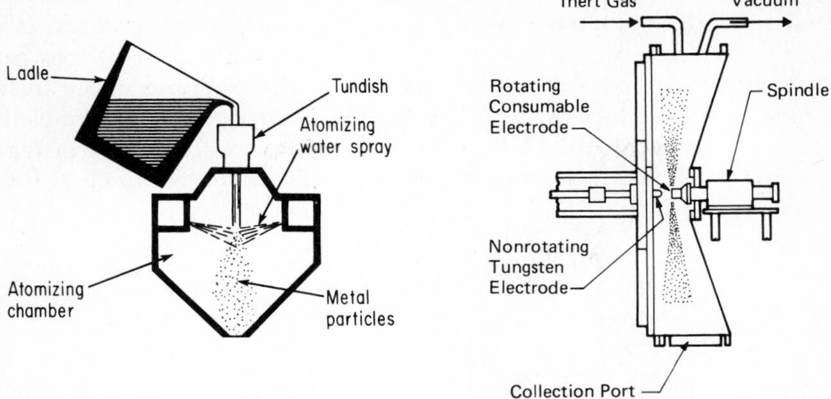

FIGURE 12-1. Two methods for producing metal powders. (*Left*) Melt atomization, and (*right*) from a rotating consumable electrode.

The strength obtained in a powder metallurgy product depends on the chemical composition of the powder(s) and the final density. For a given material, the greater the density, the greater the strength. While theoretically particles of varying sizes should result in greater density, *if they are properly distributed,* there is a tendency for the finer sizes to separate and segregate during the required handling and mixing. Consequently, many users prefer uniform-size particles, relying on pressing pressure to achieve the desired density.

Most commonly pure metal or nonmetal powders are used, powders of different materials being combined to obtain a variety of properties. However, in some cases superior properties can be obtained by using prealloyed powders, each powder particle being an alloy. Prealloyed powders of stainless steels, nickel-base alloys such as Monel, titanium alloys, cobalt-base alloys, and some low-alloy steels are available. Precoated powders also are available in which each particle is coated with another element. These permit obtaining a product that will have quite different properties from those of the base material—for instance, superior corrosion resistance much as is obtained by the use of stainless steel–clad plate.

Mixing and blending. Before the powders are pressed into shape, they usually are mixed or blended to (1) obtain uniform distribution of particle sizes, (2) mix powders of different materials, and, sometimes, (3) coat the particles with lubricants. The mixing can be done either wet or dry. Water or solvent may be used to obtain better mixing, reduce dusting, and lessen explosion hazards. Lubricants, such as graphite or stearic acid, improve the flow characteristics of the powder and also reduce die wear. Obtaining a uniform mix aids in the pressing operation and helps to assure uniformity throughout a run.

Pressing. Because pressing forms the powder into the desired shape and also determines the density, and the uniformity of the density, of the product, it is a very important operation. Pressures of 68.9 to 689 MPa (5 to 50 tons per square inch) are utilized, 138 to 414 MPa (10 to 30 tons per square inch) being most common. In most cases toggle-type mechanical presses are used, such as is shown in Figure 12-2. Most presses have capacities of less than 8.9×10^5 N (100 tons), but an increasing number have a capacity of 17.8 to

FIGURE 12-2. (*Top*) Toggle press for compacting metal powders. (*Bottom*) Compacts being automatically ejected from the dies and the press. (*Courtesy Stokes Equipment Division, Pennsalt Chemicals Corporation.*)

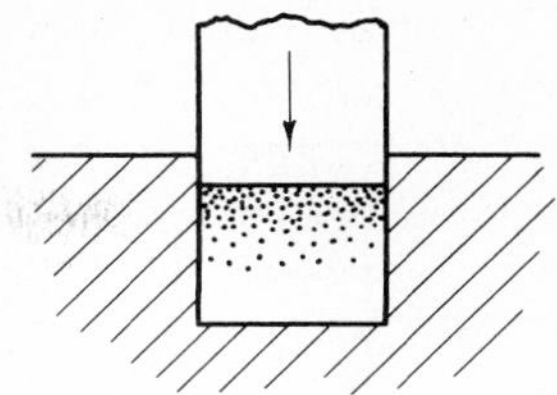

FIGURE 12-3. Compacting with a single plunger, showing resultant nonuniform density.

26.7×10^5 N (200 to 300 tons), and at least one with a capacity of 26.7 MN (3000 tons) is in operation. Consequently, it follows that most powder metallurgy products have cross sections less that 1935 mm^2 (3 in.^{2}), but sizes up the 6452 mm^2 (10 in.^{2}) now are quite common, and even larger ones are being produced.

In most cases, the prepared powder flows into the die by gravity until there is some excess. The excess is scraped off, providing measurement by volume, and the press closes, compressing the powder. Where the powder does not flow readily, the amount may be controlled by weighing. Another method is to preform the desired amount of powder into tablets in a tablet-making machine.

As stated previously, it is desirable to have uniform density throughout the compact. Because the powder does not flow like a liquid, it offers considerable internal resistance to flow, this resistance increasing as the pressure increases. As a result, it seldom is possible to obtain completely uniform density. As illustrated in Figure 12-3, when pressure is applied by only one plunger in the press, maximum density occurs near this plunger and decreases with the distance from the end of the plunger because of the internal friction in the powder mass and the friction between the powder and the die walls. A double-plunger press, as illustrated in Figure 12-4, enables a more uniform density to be obtained, or longer shapes to be formed, so that this type of press is used more frequently than the single-plunger type.

Because nonuniformity is a function of both the length and width of the part being pressed, the ratio l/w should be kept below 2 whenever possible. Ratios above 2 produce considerable variation in density.

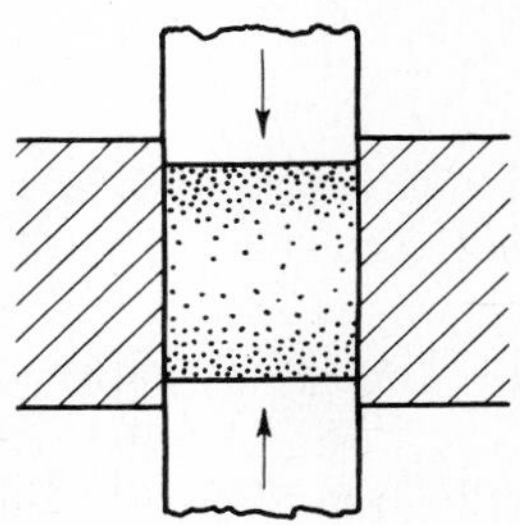

FIGURE 12-4. Density distribution obtained with a double-plunger press.

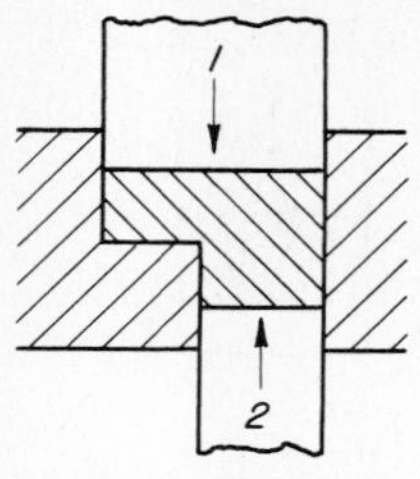

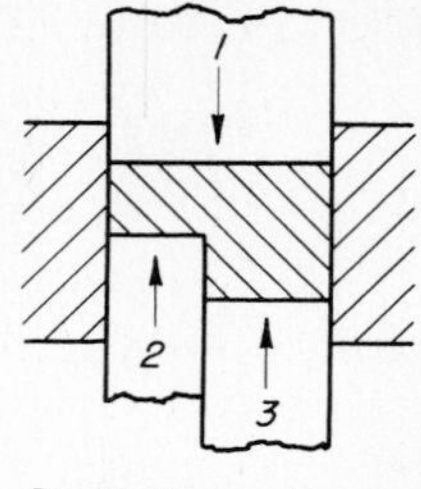

FIGURE 12-5. Two methods of pressing powders for powder metallurgy parts that have two cross sections.

Because density is a function of length, it is difficult to produce products when more than one length is involved. Although the ideal shape for a powder metallurgy product is a uniform cross section throughout the entire length, it is possible to produce other shapes by using more complicated presses. For example, Figure 12-5 illustrates how it is possible to press parts that have two cross sections by single and multiple lower punches in the die.

Another innovation is a hydraulic press that does the compacting by a series of rapid squeezes of controlled, increasing intensity. In some instances it is possible to add powder between the applications of pressure and thus obtain more uniform density. Obviously, the methods described in this paragraph require more complicated and expensive dies and/or presses.

Pressing rates vary widely, with about six pieces per minute being the minimum that might be expected, and rates of 100 pieces per minute are common. The pressed parts are ejected from the die mechanically. At this stage the compacts have sufficient strength to withstand a reasonable amount of handling.

Dies. Because the powder particles tend to be somewhat abrasive, coupled with the high pressures involved, there is considerable wear on the die walls. Consequently, they usually are made of hardened tool steel. For particularly abrasive powders, or for very high volume work, cemented carbide dies sometimes are employed. All die surfaces should be highly polished.

The dies must be very heavy to withstand the high pressing pressures, which adds materially to die cost.

Sintering. Sintering is carried out below the melting temperature of the major constituent. Most metals are sinterd at 70 to 80 per cent of the melting temperature. Certain refractory materials may be sintered at 90 per cent of the melting point. When the product is composed of more than one material, the sintering temperature may be above the melting temperature of some of the materials. Lower-melting-point materials tend to flow into the higher-melting-point material.

The sintering time varies from 30 minutes up to several hours. As the sintering proceeds, volatile materials that are present are driven off. Thus, if volatile materials are present in appreciable quantities, the final product will tend to be porous. Advantage is taken of the phenomenon in making certain types of bearings and fine filters by the powder metallurgy process.

In most cases the sintering atmosphere must be controlled carefully to prevent oxidation or combustion. This is critical because the fine powder particles have large surface areas that are exposed to the atmosphere at elevated temperatures—conditions that are ideal for rapid oxidation. It also is necessary to make provision for the rapid removal of any adsorbed or other gases that may be liberated during sintering. Hydrogen, dissociated ammonia, or cracked hydrocarbon atmospheres commonly are used. Sintering may be carried out in either batch or continuous-type ovens.

Spark-discharge sintering, wherein a high-energy electrical spark is discharged between the powder particles while they are under pressure in the compacting press, is being used increasingly. The energy discharge strips the contaminants from the surface of the particles, and because of the intimate contact resulting from the high pressure, diffusion bonding results almost instantaneously. After a 1- or 2-second spark discharge, the current is adjusted to heat the mass for about 10 seconds at a temperature considerably below its melting point. This procedure not only eliminates the separate sintering oven, but also provides easier size control by virtually eliminating the shrinkage that occurs during conventional sintering.

Presintering. Powder metallurgy frequently is employed to produce parts from materials that are very difficult to machine. When some machining is required on such parts, it often can be made much easier by employing a *presintering* operation, wherein the compacted parts are heated for a short time at a temperature considerably below the final sintering temperature. The presintering treatment imparts sufficient strength to the parts so they can be handled and machined without difficulty. Following machining, they are given the final sintering, during which very little dimensional change occurs. Thus machining after final sintering may be reduced to a minimum, or eliminated entirely. Presintering is particularly useful where holes must be drilled in hard-to-machine parts.

P/M forgings. One of the most rapidly developing uses for powder metallurgy is in the production of preforms for making forgings. A major limitation in forging is obtaining a uniform flow of the raw, solid material within a forging die. This often limits the complexity of the forging, or requires more steps in the forging process, with resulting multiple and costly dies, or substantial postforging machining. Consequently, the more nearly the raw-material shape conforms to the shape of the forging, the easier it is to obtain a desired product. The powder metallurgy process provides an easy and economical method for producing a block of solid metal in a simple shape, but one that will be easy to forge, often in a single forging step. At the same time, the forging of a P/M preform can increase its density—often up to 99 per cent of theoretical

FIGURE 12-6. Comparison of forging and the use of powder-metallurgy preforms to form a gear blank and gear. Top shows sheared stock, rough forging, forged flank, and scrap. By using powder-metallurgy preform (*bottom*), finished gear is formed with no scrap. (*Courtesy GKN Forging Limited.*)

density—and thus greatly improve its properties. Figure 12-6 shows an example of a P/M part and the preform from which it was made.

In Germany, powder preforms have been hot-forged at about 298°C (570°F) in mechanical presses prior to sintering at 1121°C (2000°F) with good results. Densities of 97 to 99 per cent are routinely obtained, with considerably lower pressures and forging temperatures being required.

Extrusions from powders. Rods, wires, and small, special billets are made from metal powders by an extrusion process illustrated in Figure 12-7. The metal powder is placed in a thin steel can that then is welded, evacuated, and sealed. The can protects the powder from contamination during heating and extrusion and the postextrusion cooling. After cooling, the can is removed. The process, often called *hot isostatic pressing* (HIP), is particularly useful in connection with some of the more exotic metals, such as uranium, zirconium, and high-strength titanium alloys, where high purity is required.

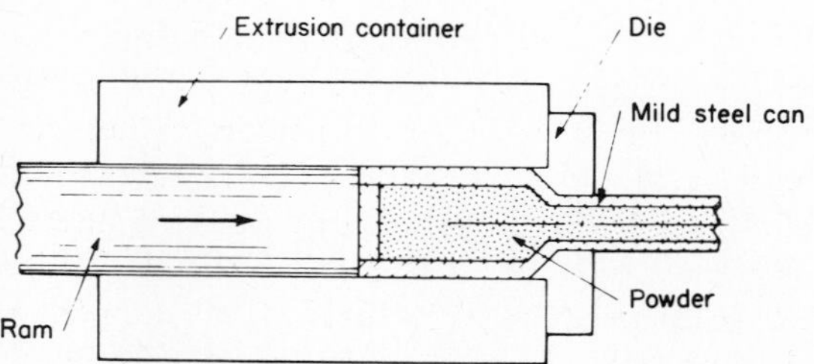

FIGURE 12-7. Method of extruding wire and rods from powdered metal.

Sizing and finishing. Many powder metallurgy products are ready for use in the form in which they come from the sintering oven. However, various sizing and finishing processes frequently are used. One of the most common is *coining* or *sizing.* Coining is an operation in which the part is put into a die and pressed under high pressure. The pressures used are equal to or greater than the intitial pressing pressures. A very small amount of plastic flow takes place, resulting in a very uniform product with respect to size and sharpness of detail. Coining also increases the strength from 25 to 50 per cent.

Another common finishing process is the introduction of oil into the pores of bearings made by the powder metallurgy process. Volatile materials that are mixed with the metal powders vaporize during sintering, leaving a fine, porus network throughout the bearing. Lubricants are forced into this porous network, either by immersing the bearing in oil and applying pressure or by a combination vacuum-pressure process. The finished bearing contains a considerable quantity of lubricant, which will be released slowly throughout its life as the applied load and resulting temperatures act upon it.

Another process used is *infiltration,* in which a molten metal of a lower melting point than that of the major constituent is forced into the product under pressure. The molten metal penetrates the pores and frequently will increase the strength by 70 to 100 per cent. Because the process is carried out at temperatures close to the sintering temperature, the part may be given a shortened sintering treatment to partially harden it and to drive out the volatile materials. The subsequent infiltration treatment serves the dual purpose of introducing the secondary metal and completing the sintering.

Properties of P/M products. Because the strength properties of powder metallurgy products depend on so many variables—type and size of powder, pressing pressure, sintering temperature, finishing treatments, and so on—it is difficult to give generalized information regarding them. In general, the strength properties of products that are made from pure metals (unalloyed) are about the same as those made from the same wrought metals. As alloying elements are added, the resulting strength properties of powder metallurgy products fall below those of wrought products by varying, but usually substantial, amounts. The ductility usually is markedly less, as might be expected because of the lower density. However, tensile strengths of 275.8 to 344.8 MPa (40,000 to 50,000 psi) are common, and strengths above 689.5 MPa (100,000 psi) can be obtained. Table 12-1 gives a few powder metallurgy materials and their strength properties, with some wrought metals shown for comparison. As larger presses and forging combined with P/M preforms are used, to provide greater density, the strength properties of powder metallurgy materials more nearly equal to those of wrought materials. Coining also can be used to increase the strength properties of powder metallurgy products and to improve their dimensional accuracy.

New developments continue to occur in the P/M field. For example, it recently has been found that the addition of pulverized ferrophosphorus to

Table 12-1. Typical properties of some powder metallurgy materials, with some similar wrought materials

Material	Composition	Theoretical Density (%)	Condition	Tensile Strength MPa	Tensile Strength 10^3 psi	Elongation in 2 in. (%)
Iron[a]			Wrought HR	331	48	30
Iron		84	As-sintered	214	31	2
		95	Repressed	283	41	25
Steel[a]	AISI 1025		HR	586	85	25
Steel	0.25%C; 99.75%Fe	84	As-sintered	234	34	2
Stainless steel[a]	Type 303		Annealed	621	90	50
Stainless steel	Type 303	75	As-sintered	241	35	2
Alloy steel[a]	AISI 4620		Annealed	552	80	30
Alloy steel	0.75% Mo; 0.35% Ni; 0.40% Mn; 0.25% Si; 0.5% C	85	As-sintered	352	51	Nil
Al. alloy[a]	7075		T6	572	83	11
Al. alloy	75	90	Heat-treated	310	45	3
Brass[a]	90% Cu; 10% Zn		Annealed	255	37	45
Brass	90% Cu; 10% Zn	89	As-sintered	234	34	29

[a] Shown for comparison.

iron powders greatly increases the toughness while requiring less sintering. Also, the soft magnetic properties are important in many electrical uses.

Design of metal-powder parts. In designing parts that are to be made by powder metallurgy it must be remembered that it is a special manufacturing process, and provision should be made for a number of factors that are related to this specific process. Because strength is so closely related to the density of the pressed powder, in order to obtain homogeneous strength in a product it is necessary that uniform pressures is obtained throughout the compact. This means that (1) an almost uniform cross section must exist throughout the length of the part, or (2) it be relatively short in length in comparison with its diameter, or (3) a relatively complex compacting press and die must be used. Two different cross sections can be achieved in a single part with a fair degree of success through the use of double-plunger dies. The ratio of unpressed length to pressed length should be kept below 2, if possible, and never exceed 3.

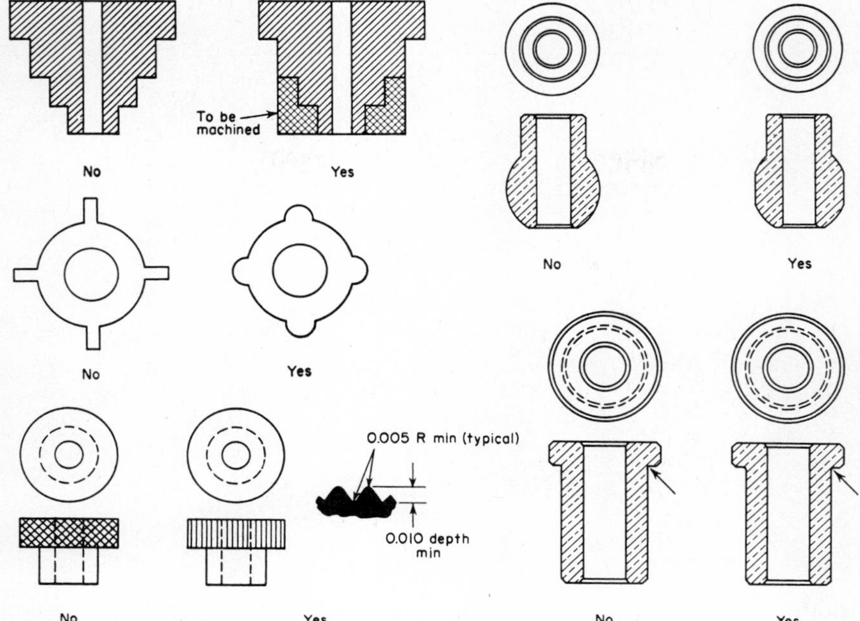

FIGURE 12-8. Some poor and good design details for use in powder metallurgy parts.

Designs should not contain holes whose axes are perpendicular to the direction of pressing. Multiple stepped diameters, reentrant holes, grooves, and undercuts should be eliminated. As in castings, abrupt changes in section and internal angles without generous fillets should be avoided. Straight serrations can be molded readily, but diamond knurls cannot. The meeting plane between mold punches should be on a cylindrical or flat surface, never on a spherical surface. Narrow, deep flutes should be avoided. Figure 12-8 illustrates some of these points.

Powder metallurgy products. The products that commonly are produced by the powder metallurgy process can be classified roughly into four groups:

1. Porus products; bearings and filters.
2. Products of complex shapes that require considerable machining when made by other processes; gears, and so forth.
3. Products made from materials that are very difficult to machine; tungsten carbide and other carbide products.
4. Products where the combined properties of two metals, or of metals and nonmetals, are desired; nonporous bearings, electrical contacts, electric motor brushes.

FIGURE 12-9. Typical parts produced by the powder metallurgy process. (*Courtesy Amplex Division, Chrysler Corporation*.)

Figure 12-9 shows some typical powder metallurgy products.

Porous bearings, made from either iron or copper alloys and containing from 10 to 40 per cent oil by volume, probably constitute the largest volume of powder metallurgy products. They are widely used in home appliance and automotive applications because they require no lubrication during their service life.

Filters can be made with pores of almost any size, some as small as 0.0025 mm (0.0001 inch). The pores may constitute as much as 85 per cent of the volume.

Large numbers of small gears are made by the powder metallurgy process. Because of the accuracy and fine finish obtainable, many require no further processing and others only a small amount of finish machining. Some gears are oil-impregnated to provide self-lubrication. Other complex shapes, such as pawls, cams, and small actuating levers, which ordinarily would involve very high machining costs, frequently can be made very economically by this process.

One of the earliest modern uses for powder metallurgy was in the production of tungsten-carbide tool tips, and this continues to be an important use, with other types of carbides also being utilized.

A unique characteristic of the powder metallurgy process is the fact that metals and nonmetals can readily be combined so as to utilize the characteristics of each material. In the electrical industry, copper and graphite frequently are combined in such applications as motor and generator brushes, the copper providing high current-carrying capacity while the graphite supplies lubrication. Similarly, bearings can be made of graphite combined with iron or copper mixtures.

Two metals also may be combined to provide unique characteristics. Electrical contacts may combine copper or silver with tungsten, nickel, or molybdenum. The copper or silver provides high conductivity while the tungsten, nickel, or molybdenum provides resistance to fusion, because of the high arcing temperature of each. In the bearing field, mixtures of two metals, such as tin and copper, are used to achieve a softer metal in a harder metal matrix. Such bearings are not self-lubricating.

Advantages and disadvantages. Like most other manufacturing processes, powder metallurgy has some distinct advantages and disadvantages with which the designer should be familiar in order to use it successfully and economically. Among the important advantages are

1. *Elimination of machining.* The dimensional accuracy and surface finish obtainable are such that for many applications all machining can be eliminated. If unusual dimensional accuracy is required, simple coining or sizing operations will give accuracies equal to those obtainable from most production machining.

2. *High production rates.* All the steps of the process are very simple and high rates of production can be achieved. Labor requirements are simple and labor costs are low. The process often can be automated, and the uniformity among pieces is excellent.

3. *Complex shapes can be produced.* Subject to the limitations discussed previously, quite complex shapes can be produced readily. Often it is possible to produce parts by powder metallurgy that cannot be machined or cast economically. Similarly, porous parts can be made that could not be made in any other way.

4. *Wide variations in compositions can be obtained.* Parts of very high purity can readily be produced, or entirely different materials, such as metals and nonmetals, can be combined.

5. *Scrap is eliminated.* Powder metallurgy is the only common production process wherein no material is wasted, whereas in casting, machining, or press forming the scrap often exceeds 50 per cent. This is important where expensive materials are involved and often makes it possible to use more costly materials without increasing the overall cost of the product.

The major disadvantages of the process are as follows:

1. *Inferior strength properties.* In most cases for a given material, powder metallurgy parts do not have as good physical properties as wrought parts or most cast structures. Consequently, their use is limited when high stresses are involved. However, the required strength frequently can be obtained by using

different and usually more costly materials, if the economics of the situation permit.

2. *Relatively high die cost.* Because of the high pressures and severe abrasion involved in the process, the dies must be made of expensive materials and be relatively massive. Thus a large number of parts must be made to spread the die costs over sufficient parts to give a low unit cost. It is almost never economic to use the process to produce fewer than 10,000 parts.

3. *High material cost.* Powdered metals are considerably more expensive than the cost of wrought forms. However, the absence of scrap and elimination of machining often will offset the higher material cost. Because powder metallurgy usually is employed for rather small parts, the actual material cost is not great.

4. *Design limitations.* The powder metallurgy process simply is not feasible for many shapes. Parts must have essentially uniform sections along the axis of compression. The length-to-diameter ratio is limited. Furthermore, the overall size must be within the capacity of available presses.

Review questions

1. What four properties are important relative to the quality of metal powders?
2. What are the steps that usually are involved in making products by powder metallurgy?
3. Why are the flow characteristics of the metal powders important?
4. For a given metal, which factor is most important in determining the strength of a P/M part?
5. Why are P/M parts usually relatively small?
6. Why are mechanical presses usually used for making P/M parts rather than hydraulic presses?
7. What is the maximum length-to-width ratio usually permissible on P/M parts?
8. Why is it desirable to have uniform cross sections in P/M parts?
9. Explain how fine-pore filters are made by powder metallurgy.
10. How does the sintering temperature used in making P/M parts compare with the melting point of the materials involved?
11. What occurs during the sintering process?
12. Why is it necessary to exercise careful control of the atmosphere during sintering P/M parts?
13. What are the advantages of high-energy spark sintering?
14. What is the purpose of presintering?
15. Why are P/M preforms advantageous for forging?
16. How are preoiled bearings made by powder metallurgy?
17. What is infiltration, and why is it used?

18. What are two important design limitations in connection with P/M products?
19. What are two methods of producing alloy parts by P/M?
20. List four advantages of the P/M process.
21. What are the major disadvantages of the P/M process?

CHAPTER 13

The theoretical basis for metal forming

Many manufacturing processes involve the plastic deformation of metals, converting crude forms—sheets, rods, and plates—directly into desired shapes. To a degree, it is because most metals can be deformed plastically without fracturing that they have such importance as manufacturing materials. But plastic deformation also is a coincident and important factor to be considered in the formation of chips in metal cutting. Consequently, to utilize these processes more effectively it is helpful to understand the stress, strain, and plasticity relationships that are involved when metals are deformed or cut. The acquisition and application of basic knowledge in this area has led to great improvement in, and development of new, metal-forming processes, and to a better understanding of metal cutting. In this chapter these stress, strain, and plasticity relationships will be discussed as they relate to metal forming; in a later chapter their relationship to metal cutting will be shown. Clearly these relationships are of primary importance to the manufacturing engineer, but they also have significance to the design engineer in providing an understanding of why certain design features are, or are not, easy to produce.

The states of stress in forming operations. It will be recalled[1] that in a work-hardening material, plastic, or permanent, deformation begins only after the yield condition is reached, and the true-stress true-strain curve can be considered as the locus of the yield stress for various amounts of plastic strain. If an isotropic material is tested in compression with no friction existing—with parallel and low-friction anvils—the true stress at yielding is nearly identical to that obtained in tension, except that the sign of the stress is negative. If the test metal remains reasonably isotropic and the Bridgeman correction is applied to the tensile stress data after necking, the stress–strain curves for tension and compression will, in effect, be images of each other. Consequently, if true stresses and true strains are converted to *effective stresses* and *effective strains,* which will be discussed later, for all practical purposes the resulting stress–strain curves for tension and compression are identical. Furthermore, each locus on the effective stress/effective strain curve represents identical deformation energy, stress, and strain, irrespective of whether the metal was pulled in tension or compressed between low-friction anvils.

One very important difference does occur when metals are tested in tension and compression. When pulled in tension, necking occurs; when it is compressed, it will buckle unless the length-to-radius ratio is about 3 or less or unless the buckling is prevented by external constraints. Consequently, when the stresses in a forming operation are tensile, necking, rather than fracturing, often is the limiting factor. Similarly, unless prevented, when the stresses are compressive, buckling usually is the limiting factor. If neither necking nor buckling can occur, as in certain shearing operations or where a solid bar is subjected only to torsion, fracture is the only limit to plastic deformation. It thus follows that it is the type of loading, or more precisely the state of stress within the metal, that is of prime importance in determining the extent of deformation that can be achieved. Unfortunately, most forming operations involve multiaxial stresses, and the forming limits due to buckling, necking, or fracturing are not readily determinable unless a detailed analysis of the state of stress throughout the plastically deforming metal can be made. Such a detailed analysis is beyond the scope of this book. However, some general remarks can be made as follows:

1. If the part is subjected to compressive loading and the slenderness ratio is large, forming is limited by buckling, and relatively small deformations can be obtained.
2. If the stresses are principally tension, the forming limit is determined by the condition for necking. This results in deformations that generally are greater than those achievable when the process is limited by buckling.

[1] See Chapter 2.

3. If the loading is carried out in such a way that buckling and necking will not occur, as in torsion of a solid bar, the forming limit is given by fracturing and the achievable deformation generally is appreciably greater than that of 1 and 2.
4. If loading is carried out under a condition such that neither buckling nor necking can occur, and if the operation is carried out under biaxial or even triaxial compressive loading, fracture is delayed and the extent of deformation that can be achieved becomes very large.
5. If the forming process for 4 is carried out above the recrystallization temperature, the strains can approach infinite proportions.

Table 13-1 classifies the various states of stress, and Table 13-2 shows a classification of forming operations in accordance with the approximate state of stress that prevails. Additional stresses imposed due to friction have been neglected.

Stress at a point acting on a given plane. Stress is the internal reactive force per unit area and is defined by

$$\sigma_{nr} = \lim_{\delta A \to 0} \frac{\delta F}{\delta A} = \frac{dF}{dA} \tag{13-1}$$

Figure 13-1a represents a stress σ_{nr}, which acts at point P on an arbitrary plane dA whose normal n makes an angle θ with the r direction of the stress. The stress σ_{nr} is a tensile stress when it acts away from the point P as shown and is considered positive. Compressive stresses are given by the symbol p and act toward the point P, from which $p_{nr} = -\sigma_{nr}$.

In using this designation for stresses, it is important to remember that the first subscript denotes the *normal* to the plane, and thus the plane on which the stress acts, whereas the second subscript denotes the *direction* in which it acts. Thus σ_{nr} is the stress that acts on the plane having the normal n and in the direction r. The same convention is used when decomposing a total stress σ_{nr} into its normal stress σ_{nn} and its shear stress τ_{nt} components, as shown in Figure 13-1b. If the stresses act in the same plane, they may be treated as vector quantities, and consequently,

$$\sigma_{nr}^2 = \sigma_{nm}^2 + \tau_{nt}^2 \tag{13-2}$$

If the stresses act in different planes, they cannot be treated as vectors, and it is necessary to specify both the direction of the normal to the plane and the direction in which the stress acts.

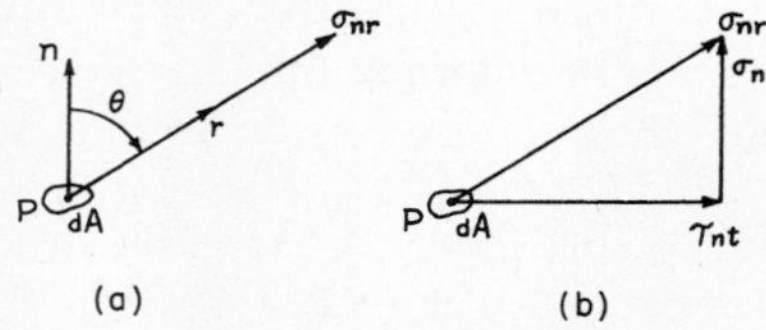

FIGURE 13-1. Stress σ acting at point P in direction r. Shown at (b) are the normal and shear components, σ_{nm} and τ_{nt}, which act normal and parallel to the plane dA.

TABLE 13-1. Classification of states of stress

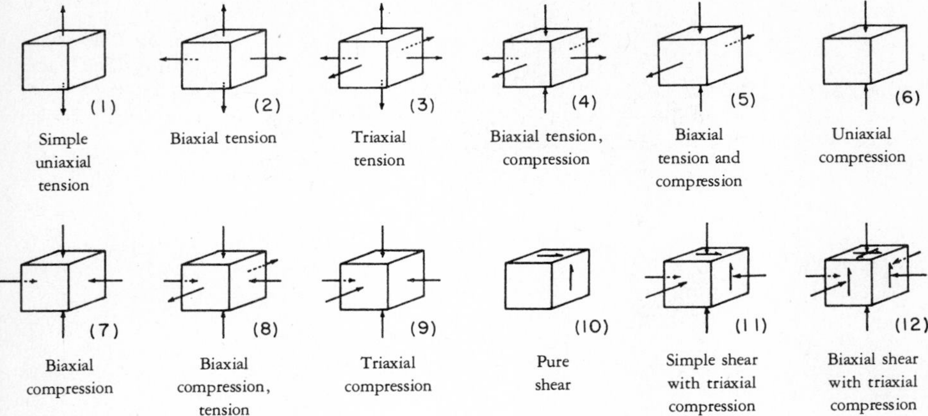

TABLE 13-2. Classification of some forming operations

Number	Process	Schematic Diagram	State of Stress in Main Part during Forming (See Table 13-1 for key)
1	Rolling		7
2	Forging		9
3	Extruding		9
4	Shear spinning		12

TABLE 13-2. Classification of some forming operations (*Continued*)

Number	Process	Schematic Diagram	State of Stress in Main Part during Forming (See Table 13-1 for key)
5	Tube spinning	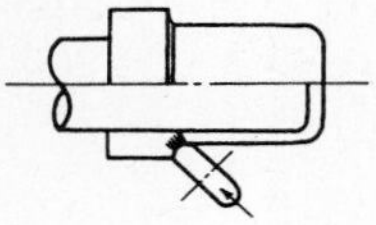	9
6	Swaging or kneading	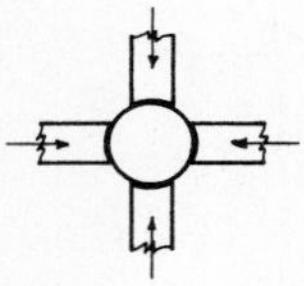	7
7	Deep drawing	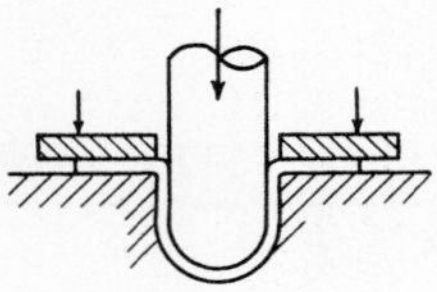	In flange of blank, 5 In wall of cup, 1
8	Wire and tube drawing	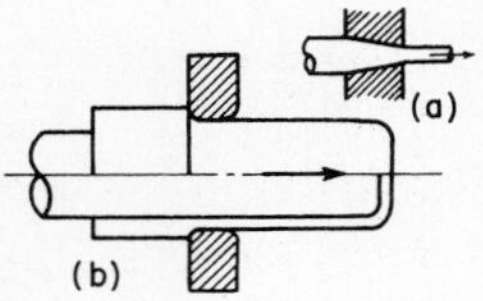 (a) (b)	8
9	Stretching	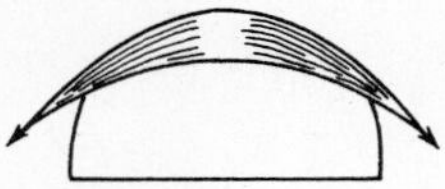	2
10	Straight bending	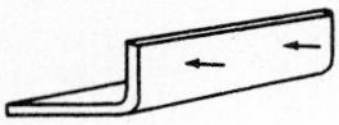	At bend, 2 and 7
11	Contoured flanging	(a) Convex 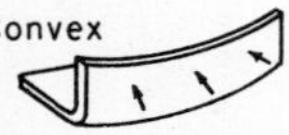	At outer flange, 6 At bend, 2 and 7
		(b) Concave 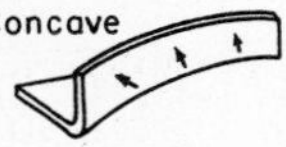	At outer flange, 1 At bend, 2 and 7

Stresses may be due to external loading or may be the result of nonuniform plastic deformation resulting from forming, machining, or welding operations. These latter stresses, called *residual stresses,* remain after external loading is removed and the mechanical or thermal operations have ceased. They are important because they may exist as the result of a number of important manufacturing processes. They usually vary from point to point throughout the workpiece. Similarly, the stresses in forming generally vary from point to point and cannot be expressed by a single numerical quantity as they can be in a case of simple loading, such as in a tensile test.

The general state of stress at a point. In working with stress problems, it usually is desirable to specify a general state of stress at a point by means of the stress components that act on the planes perpendicular to arbitrary X, Y, Z coordinate directions. Such a general state of stress is shown in Figure 13-2 and requires six independent components of the stress, σ_{xx}, σ_{yy}, σ_{zz}, τ_{xy}, τ_{yz}, *and* τ_{zx}, for its specification. As in the usual practice, the stress components that are normal to the planes have been identified by a single subscript that signifies both plane and direction, while the shearing stresses retain their mixed subscripts. The shearing stresses having the same mixed subscripts are equal in magnitude, that is,

$$\tau_{xy} = \tau_{yx}, \qquad \tau_{yz} = \tau_{zy}, \qquad \tau_{zx} = \tau_{xz} \tag{13-3}$$

and are considered positive if they act in positive coordinate directions.

Principal planes and principal stresses. Stress problems often can be reduced in complexity by selecting the coordinate directions such that the planes perpendicular to these directions contain no shear stresses. Such planes are known as *principal planes,* and the stress components that are normal to these planes are called *principal stresses.* Principal stresses commonly are given a numerical

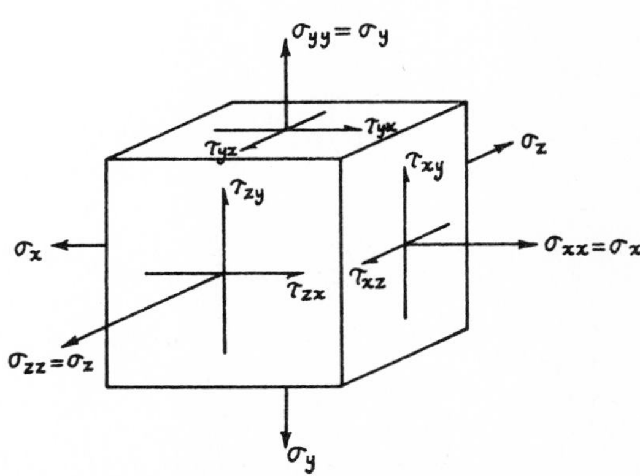

FIGURE 13-2. General state of stress at a point.

subscript, and the order of the number indicates the magnitudes of the stress, number 1 being largest. Thus in this discussion principal stresses will be indicated in the following manner:

$$\sigma_x = \sigma_1 \qquad \text{or} \qquad \sigma_x, \tau_{xy} = 0 \tag{13-4}$$

Other problems can be simplified by choosing coordinate axes that are perpendicular to planes that contain maximum shearing stresses. These planes are called the *maximum shear planes* and are identified by the condition that

$$\sigma_x = \sigma_y = \sigma_z = \sigma_m = -p_m \tag{13-5}$$

where σ_m is the mean stress and p_m the mean pressure. The mean stress can be calculated from *any* set of normal stresses, irrespective of the orientation of the coordinate axes, and always is given by the equation

$$\sigma_x + \sigma_y + \sigma_z = 3\sigma_m = -3p_m \tag{13-6}$$

Formulas for the principal stress magnitudes. The magnitudes of the principal stresses are given by the following formulas:

For *plane stress,* or *biaxial stress,*[2]

$$\left.\begin{aligned} \sigma_{\text{max}} &= \sigma_1 = \frac{\sigma_x + \sigma_y}{2} + \sqrt{\left(\frac{\sigma_x - \sigma_y}{2}\right)^2 + \tau_{xy}^2} \\ \sigma_{\text{intermediate}} &= \sigma_2 = \frac{\sigma_x + \sigma_y}{2} - \sqrt{\left(\frac{\sigma_x - \sigma_y}{2}\right)^2 + \tau_{xy}^2} \\ \sigma_{\text{min}} &= \sigma_3 = 0 \end{aligned}\right\} \tag{13-7}$$

$$\tau_{\text{max}} = \frac{\sigma_1 - \sigma_2}{2}. \tag{13-8}$$

For *plain strain,* or *triaxial stress,* resulting in zero strain in the 2-direction,

$$\left.\begin{aligned} \sigma_{\text{max}} &= \sigma_1 = \frac{\sigma_x + \sigma_y}{2} + \sqrt{\left(\frac{\sigma_x - \sigma_y}{2}\right)^2 + \tau_{xy}^2} \\ \sigma_{\text{intermediate}} &= \sigma_2 = \frac{\sigma_x + \sigma_y}{2} \\ \sigma_{\text{min}} &= \sigma_3 = \frac{\sigma_x + \sigma_y}{2} - \sqrt{\left(\frac{\sigma_x - \sigma_y}{2}\right)^2 + \tau_{xy}^2} \end{aligned}\right\} \tag{13-9}$$

$$\tau_{\text{max}} = \frac{\sigma_{\text{max}} - \sigma_{\text{min}}}{2} = \sqrt{\left(\frac{\sigma_x - \sigma_y}{2}\right)^2 + \tau_{xy}^2} \tag{13-10}$$

[2] For the derivation of these equations, and their explanation in terms of the Mohr's circle diagram, see any textbook on mechanics of materials.

Conditions for yielding under a general state of stress. In connection with forming and metal-cutting operations, one is much interested in the yielding of materials, beyond which permanent deformation occurs. In ductile materials, yielding starts when the yield-point stress is exceeded, or the yield strength (stress at 0.2 per cent permanent strain) is exceeded in materials having no definite yield point. For the analysis here, the stress at yielding under uniaxial tension, σ_0, will be idealized as being the uniaxial yield condition when the true plastic strain is zero. Thus σ_0 may be taken as either the yield-point stress or the yield strength in tension.

Although for many purposes design and structural engineers would consider the state of stress at which yielding occurs to be the condition for failure, in metal forming it is the beginning point of forming. No permanent change of shape occurs unless this stress condition has been exceeded, and this state of stress is referred to as the *initiation of yielding.* However, because most forming operations involve combined states of stress, it is necessary to consider the initiation of plastic deformation under these conditions.

Through experiment, it has been found that various materials appear to yield differently under the same states of stress. For isotropic *brittle* materials it has been found that the condition for yielding is fulfilled when the maximum shear stress exceeds a critical value. With reference to the yield stress σ_0 in simple tension, if τ_0 is the particular shearing stress at which yielding first is initiated,

$$\tau_0 = \frac{\sigma_0}{2} = \frac{\sigma_{\max} - \sigma_{\min}}{2} = \frac{\sigma_1 - \sigma_3}{2} \tag{13-11}$$

In words, because $(\sigma_1 - \sigma_3)/2$ is equal to the mean shearing stress, this means that isotropic brittle materials start to yield when the mean shearing stress exceeds one half the yield stress in uniaxial tension. This is known as the *Tresca yield condition.*

For isotropic ductile materials:

$$\sigma_0 = \bar{\sigma}_0 = \sqrt{\frac{(\sigma_x - \sigma_y)^2 + (\sigma_y - \sigma_z)^2 + (\sigma_z - \sigma_x)^2}{2} + 3(\tau_{xy}^2 + \tau_{yz}^2 + \tau_{zx}^2)} \tag{13-12}$$

$$= \sqrt{\frac{(\sigma_1 - \sigma_2)^2 + (\sigma_2 - \sigma_3)^2 + (\sigma_3 - \sigma_1)^2}{2}} \tag{13-12a}$$

where $\bar{\sigma}_0$ is the particular *effective stress* (referred to previously) at which yielding first is initiated. The yield condition, given by equation (13-12), also is known by several other names, such as *von Mises yield condition, flow stress,* and *condition for failure.*

Because forming deals almost entirely with ductile metals, the condition for yielding as given by $\bar{\sigma}_0$ is of special importance. Equation (13-12) can be shown to be that of a cylinder in three-dimensional stress space with the three principal stresses as the stress axes. This is illustrated in Figure 13-3. The cyl-

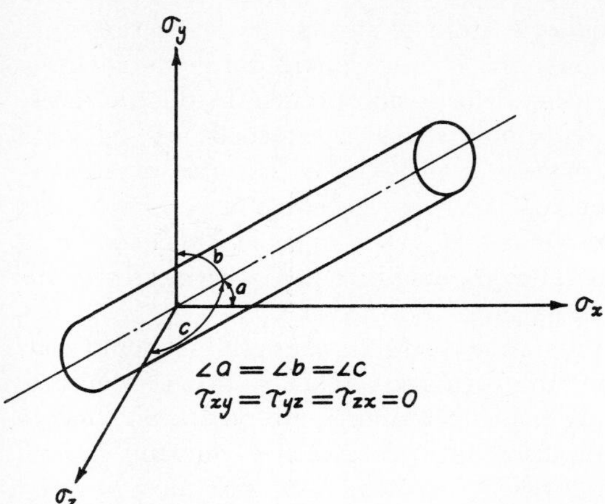

FIGURE 13-3. Yield cylinder in stress space.

inder is so inclined that its center line makes equal angles with the principal-stress axes, that is, $\angle a = \angle b = \angle c = 54°44'$. The significance of this cylinder is the fact that yielding will not occur unless the state of stress lies on its surface, which coincides with the condition that the effective stress has reached the value $\bar{\sigma}_0$. Any state of stress falling inside this cylindrical yield surface results only in elastic straining; plastic deformation has commenced when the state of stress falls beyond the cylindrical (initiation) surface.

The case of equal triaxial stressing is of special interest in that it follows the path given by the center line of the stress cylinder. Thus a body that is subjected to equal triaxial tension will not deform plastically but only elastically. Structures or parts made from metal that normally is ductile fail with a brittle fracture when subjected to this type of loading, indicating that plastic deformation essentially was absent. Similarly, a body subjected to hydrostatic pressure, such as on the deepest part of the ocean floor, will not deform plastically, only elastically. This fact has been proved by nature; rocks and other objects lifted from the ocean floor, or rocks brought from the near vacuum of the moon to earth, have not been deformed plastically.

For brittle materials, which yield according to the maximum-shear-stress criterion, the three-dimensional yield surface is an inclined hexagon.

Yielding under biaxial stress conditions. In a large percentage of service and processing cases, materials are subjected to loading that produces biaxial stresses. Consequently, when and how yielding occurs under these stress conditions is of considerable importance.

If a plane that is parallel with any two of the stress axes (thus containing biaxial stresses) is passed through the yield cylinder, the trace of the cylinder on this plane is an ellipse. Similarly, if such a plane is passed through a yield hexagon, the trace is an inscribed, elongated hexagon. These are illustrated in Figure 13-4 for an *x–y* stress plane. In this figure, it will be noted that yield-

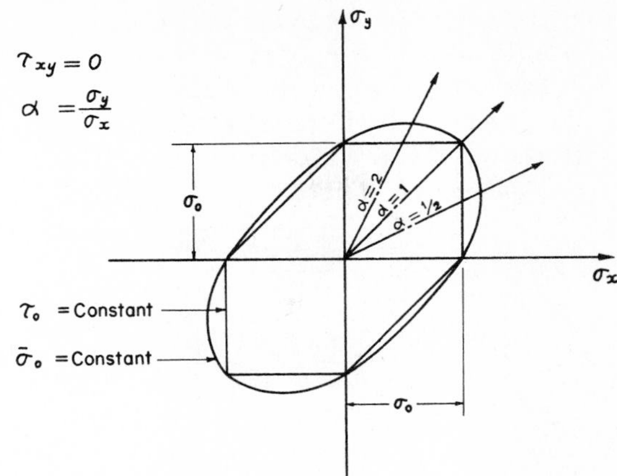

Figure 13-4. Yield ellipse and the inscribed maximum-shear-stress yield hexagon.

ing will occur in an isotropic material when any of the following stress relationships exist:

$$\sigma_x = \sigma_y = \sigma_0; \ \sigma_x = \sigma_0, \ \sigma_y = 0; \quad \text{or} \quad \sigma_y = \sigma_0, \ \sigma_x = 0 \tag{13-13}$$

These relationships also are obtained by substituting the condition of σ_x as the principal stress and $\sigma_y = \sigma_z = 0$ into equation (13-12). Note that if $\sigma_x = \sigma_y$, yielding occurs at the same stress magnitude as in the case of simple tension; that is, $\sigma_x = \sigma_y = \sigma_0$. But note that for the biaxial-tension case, where $\sigma_y/\sigma_x = 1/2 = \alpha$, the following results from equation (13-12):

$$\sigma_0 = \bar{\sigma}_0 = \sqrt{\frac{(\sigma_x - \sigma_y)^2 + (\sigma_y - 0)^2 + (0 - \sigma_x)^2}{2}}$$

$$= \sigma_x \sqrt{\frac{\left(\frac{\sigma_x - \sigma_y}{\sigma_x}\right)^2 + \left(\frac{\sigma_y}{\sigma_x}\right)^2 + \left(\frac{\sigma_x}{\sigma_x}\right)^2}{2}}$$

$$= \sigma_x \sqrt{(1/4 + 1/4 + 1) 1/2} = 0.866\sigma_x$$

$$\sigma_x = 1.155\sigma_0 = 1.155\sigma_0 \tag{13-14}$$

Thus, for some ratios of biaxial stresses, as in the case of ratios of $\alpha = 1/2$ or $\alpha = 2$, the maximum principal stress must be up to 15.5 per cent greater than the yield strength in simple tension before plastic deformation will be initiated. This can be seen clearly in Figure 13-4. This fact may be helpful in certain service conditions, or it may be detrimental in some forming operations.

Example of a yielding calculation. Inasmuch as yielding may not occur at the simple-tension yield stress when biaxial stress conditions exist, it is frequently necessary to determine whether it will occur under certain biaxial loads. This can readily be done by use of the yield equations. For example, the yield stress of a mild steel in simple tension is approximately 40,000 psi. In forming a part from this material the state of stress is $\sigma_x = \sigma_y = \tau_{xy} = 30{,}000$ psi and $\sigma_z = \tau_{yz} = \tau_{zx} = 0$. Determine whether the metal has yielded in accordance with the von Mises yield criterion.

Solution: Yielding has occurred if $40{,}000 = \bar{\sigma}_0 = \sigma_0 \leqslant \bar{\sigma}$.

$$\bar{\sigma} = \sqrt{\frac{(\sigma_x - \sigma_y)^2 + (\sigma_y - 0)^2 + (0 - \sigma_x)^2}{2} + 3\tau_{xy}^2}$$

$$= \sqrt{\frac{0 + (30{,}000)^2 + (-30{,}000)^2}{2} + 3(30{,}000)^2}$$

$$= 60{,}000 \text{ psi} \tag{13-15}$$

Consequently $\bar{\sigma}_0 < \bar{\sigma}$ and yielding has occurred. This example clearly illustrates the fact that in forming operations, owing to the multiaxial stress conditions that may exist, yielding and plastic flow may occur even though the individual stress components are below the tensile yield stress.

Loci of instantaneous yield conditions. If σ_x is the tensile stress in a tension test at any instantaneous condition of loading, with all other stresses equal to zero, then the effective stress, $\bar{\sigma}$ of equation (13-12), reduces to

$$\bar{\sigma} = \sqrt{\frac{(\sigma_x - 0)^2 + (0 - 0)^2 + (0 - \sigma_x)^2}{2} + 0} = \sigma_x \tag{13-16}$$

Consequently, although $\bar{\sigma} = \bar{\sigma}_0$ is the condition for initial yielding, $\bar{\sigma}$ now is the effective stress equal to the instantaneous true stress in tension. If the test were carried out in simple compression, the effective stress $\bar{\sigma} = p_x = -\sigma_x$, where p_x is the compression stress at the instantaneous condition of loading. Similarly, if a torsion test were performed the effective stress is related to the shear stress of the outer fiber of the test bar by $\bar{\sigma} = \sqrt{3}\tau_{xy}$, as can be seen from equation (13-12) for the condition when τ_{xy} is the only stress.

For each true strain there also exists in a tension test an *effective strain,* $\bar{\epsilon}$, being the strain at a point corresponding to the effective stress at that point. Thus in a tension test $\epsilon = \ln (l/l_0) = \bar{\epsilon}$, and similarly in a compression test $+\epsilon - \ln (l/l_0) = -\bar{\epsilon}$. Thus each point on the true stress/true strain curve not only represents the true stress and true strain, but also the effective stress $\bar{\sigma}$ and the effective strain $\bar{\epsilon}$. Hence the tensile true stress/true strain curve also is a $\bar{\sigma} \sim \bar{\epsilon}$ curve applicable for complex states of stress. Furthermore, because the area under the curve is given by

$$\int_0^{\epsilon_x} \sigma_x \, d\epsilon_x, \tag{13-17}$$

it also must be equal to the integral

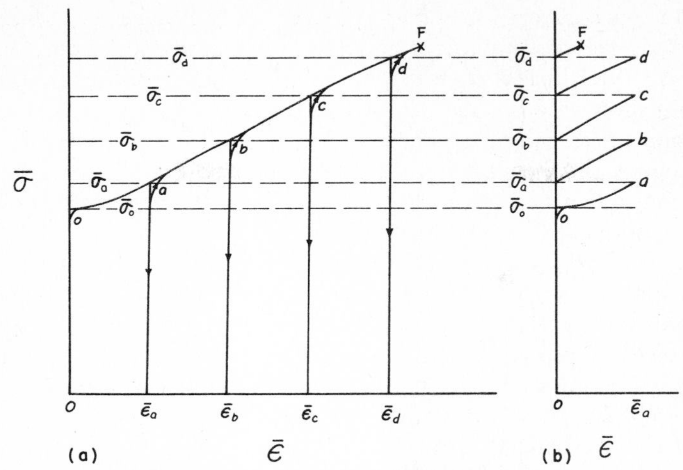

FIGURE 13-5. Interrupted stress–strain curve.

$$\int_0^{\bar{\epsilon}} \bar{\sigma}\, d\bar{\epsilon} \qquad (13\text{-}18)$$

where both represent specific energy or energy per unit volume in inch-pounds per cubic inch.

Now consider the stress–strain curve of Figure 13-5a to be composed of a series of interrupted tests. As already stated, yielding occurs when $\bar{\sigma} = \bar{\sigma}_0 = \sigma_0$ and $\bar{\epsilon} = 0$, and upon continued loading a work-hardening material has a rising stress–strain curve. At point *a* the effective stress or flow stress has risen to $\bar{\sigma}_a$, while the effective strain is $\bar{\epsilon}_a$. If the test is now stopped and the load is removed, the effective stress will drop to zero, while $\bar{\epsilon}_a$ remains at its value, except for a small negligible elastic recovery or springback equal to $-e_a$, which can be calculated by the equation $\bar{\sigma}_a/E = e_a$, where E is Young's modulus. Upon reloading of the test bar, plastic deformation will not start until the stress has again reached the former value, $\bar{\sigma}_a$. Other interruptions of the test at *b* and *c*, followed by reloading, would yield similar results.[3] If the finite strains $\bar{\epsilon}_a, \bar{\epsilon}_b$, and so forth, are transposed to the origin as shown in Figure 13-5, considering for simplicity that the deformation steps were of equal magnitude, that is, $\bar{\epsilon}_a = \bar{\epsilon}_b - \bar{\epsilon}_a$, and so forth, then it can be seen that $\bar{\sigma}_a, \bar{\sigma}_b, \bar{\sigma}_c$, and so forth, can be thought of as yield stresses at given states of deformation. Thus each point on the effective stress–strain curve represents, in effect, a yield stress locus, and the total curve may be thought of as loci of instantaneous yield conditions.

[3] Actually, there would be a slight departure from the idealized curve $\overline{od}$, which would be that of a continuous test.

Note from Figure 13-5 that if a metal already had been deformed to the condition represented by point d and then were tested in tension, it would have an initial yield stress $\bar{\sigma}_d = \sigma'_0$. Also, if F is the fracture condition, the metal would show very limited ductility. Thus a metal that is ductile to begin with becomes brittle after the potential ductility has been exhausted, or nearly exhausted, through the extensive plastic deformation that accompanies a forming operation and may have to be annealed in order to restore the ductility before being subjected to further forming.

Now consider the yield cylinder and its traces—yield ellipses—on certain stress planes, in the light of the interpretation of the yield loci of the $\bar{\sigma} \sim \bar{\epsilon}$ curve. Figure 13-6 shows the initial yield condition as the $\bar{\sigma}_0$ yield cylinder and yield ellipse. It is seen now that work hardening, which results in an increased effective stress and permanent effective strain, can be represented by a new, enlarged yield surface in stress space. Thus at any stage of deformation, such as $\bar{\epsilon}_c$, plastic deformation can occur only if the loading is maintained in such a way that the effective stress remains at some point on this larger yield surface. If the load should decrease and follow an arbitrary stress path as shown in Figure 13-7 by arrows, then yielding will cease as the state of stress at c' falls inside the new yield ellipse. Only elastic deformation will occur now, and plastic deformation will not commence again until the stress path once more crosses the yield ellipse, as, for example, at point c''.

In many cases where metal components are deformed, particularly in uncontrolled conditions—as in accidents—the plastic deformation does not occur in a single, continuous step or direction, and the stress path may cross back and forth from plastic to elastic deformation, and vice versa. As a consequence of not knowing the exact manner in which the total deformation occurred, a

FIGURE 13-6. Yield circles and ellipses corresponding to various instantaneous yield conditions.

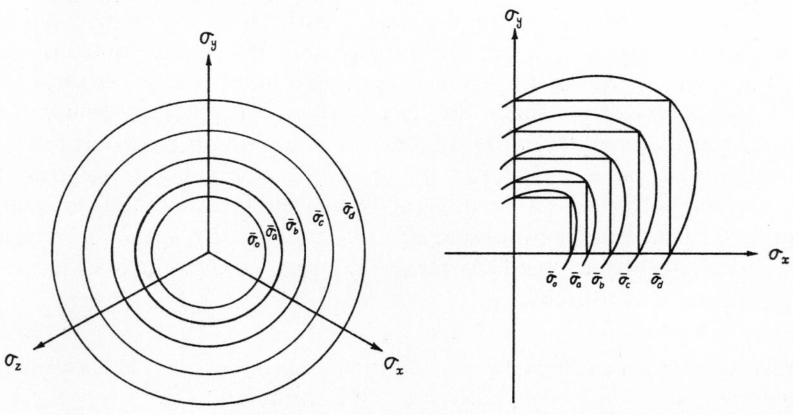

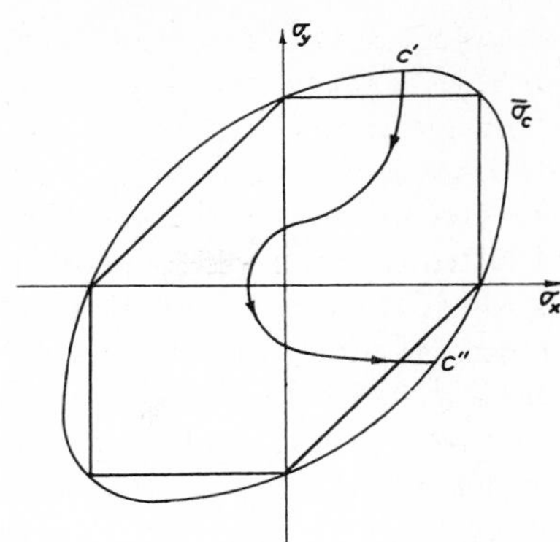

FIGURE 13-7. Arbitrary stress path within the yield surface.

valid determination of the forces and energy that were involved cannot be made on the basis of the final deformation.

Anisotropy developed during plastic deformation may warp or distort the yield cylinder. This complicates the theoretical treatment of the instantaneous yield condition and plastic deformation and somewhat limits the use of the resulting equations. Fortunately, the errors resulting from neglect of the anisotropic behavior of metals and their alloys in forming problems is usually of minor consequence, and it will be assumed here that the static tensile stress–strain curve can be applied to all types of forming problems carried out under complex states of stress.

Equations have been developed that express the relationships among plastic deformation, required energy, and power. These are useful in designing metal-forming equipment but seldom are of concern to the ordinary design engineer. Thus they will not be considered in this book.

Review questions

1. How is the effective stress–strain curve for a given isotropic metal in compression related to the curve in tension?
2. In forming operations where the stresses are compressive, what factor limits the amount of deformation that can usefully be achieved?
3. In tensile forming operations, why is the fracture point not the limiting point?
4. What would occur if forming could be carried out so that neither necking nor buckling could occur?
5. Explain the meaning of σ_{nr}.
6. What is a principal plane?
7. What is a principal stress?
8. How does a residual stress differ from an ordinary stress?

9. What are some causes of residual stresses?
10. Why is the Tresca yield condition of no great importance in metal-forming operations?
11. What is the condition for yielding for isotropic ductile materials?
12. How is "effective stress" related to the yield stress in simple tension?
13. What is the significance of the three-dimensional yield cylinder?
14. What would occur if three equally increasing and mutually perpendicular tension forces were applied to a ductile material?
15. How does a ductile material behave if subjected to equal triaxial pressures?
16. A piece of AISI 1020 steel was found to have failed without exhibiting any measurable change in dimensions. What valid assumption could be made regarding the cause of the failure?
17. What is the relationship of the maximum principal stress to the yield strength in simple tension of a metal that is subjected to biaxial stresses having a 2: 1 ratio?
18. An aluminum alloy sheet 2 feet wide, ¼ inch thick, and 6 feet long, with a tensile yield strength of 50,000 psi, is to be stretcher-straightened by applying a tensile load in the length direction. Assuming the von Mises criterion for yielding applies and that the sheet is straight as soon as yielding occurs, calculate the required axial load that must be applied to the ends of the sheet. (Assume that the induced stresses in the width and thickness directions are equal to one-half of the axial stress and zero, respectively.)
19. In forming a sheet of brass, which has a tensile yield stress of 18,000 psi, the state of stress is $\sigma_x = \sigma_y = \tau_{xy} = 10{,}000$ psi and $\sigma_z = \tau_{yz} = \tau_{zx} = 0$. Has yielding occurred?
20. An alleged "expert" claimed that he could calculate the speed and forces involved when an automobile struck a concrete wall, based upon the deformation sustained by the various components of the vehicle. Was his claim valid? Why?

Case study 10.
MATERIAL FOR A CORROSION-RESISTANT PROCESSING VESSEL

You are an engineer employed by the Petro Refining Company and have been assigned to evaluate and make a recommendation regarding two bids which the company has received for a processing vessel. The request for bids stated: "This vessel will be used in refining certain petroleum products that contain weak H_2SO_4. It must operate at approximately 427°C (800°F) for a period of at least 10 years. The corrosion-resistant steel used in its construction must be suitable for these conditions and must have a minimum tensile strength of 483 MPa (70,000 psi) and ductility of at least 15 per cent as measured with a 50.8-mm (2-inch) gage-length specimen."

The Weld-Fab Company's bid is based on AISI type 446 corrosion-resistant steel, whereas that of the Best-Weld Company is for the use of AISI type 302B corrosion-resistant steel. The bid from the Best-Weld Company is approximately 20 per cent higher than that of the Weld-Fab Company, and the company justifies its higher price by stating that type 302B steel is much superior to type 446. Investigate and give your recommendations and reasons.

CHAPTER 14

Hot-working processes

In the manufacture of metal components, four basic alternatives are available for the production of a desired shape: (1) casting, (2) machining, (3) consolidating smaller pieces (welding, powder metallurgy, mechanical fasteners, epoxy, etc.), and (4) deformation processes. Metal deformation processes exploit a remarkable property of metals—their ability to flow plastically in the solid state without concurrent deterioration of properties. Moreover, by simply moving the metal to the desired shape, there is little or no waste.

A large portion of all metal products is subjected to metal deformation at some stage of manufacture. Processes span a wide spectrum from breakdown rolling of an ingot to decorative embossing of fine silver. Deformation may be bulk flow in three dimensions, simple shearing, simple or compound bending, or any combination of these and others. Speeds, temperatures, tolerances, surface finishes, and deformation amounts span a wide spectrum. Any method of classification, therefore, is an arbitrary one and, by necessity, will have large areas that are somewhat indefinite or possibly overlap. For this presentation, division will be on the basis of temperature. Processes that are usually performed "hot" will be presented here and processes normally performed "cold" will be deferred to Chapter 15. One should be aware that the emphasis on energy conservation, the growth of "warm working," and advances in technology have considerably blurred any distinct temperature classification. Processes discussed as hot-working processes are often performed cold. Similarly, cold-forming processes can often be aided by some degree of heating.

Definition of hot working. Hot working is defined as the plastic deformation of metals above their recrystallization temperature. Here it is important to note that the recrystallization temperature varies greatly with different materials. Lead and tin are hot-worked at room temperature, steels require temperatures near 1100°C (2000°F), and tungsten is still in a cold or warm working state at 1100°C (2000°F). Hot working does not necessarily imply high absolute temperatures.

As was discussed in Chapter 3, plastic deformation above the recrystallization temperature does not produce strain hardening. Therefore, hot working does not cause any increase in yield strength or hardness, or corresponding decrease in ductility. In addition, Figures 2-30 and 2-31 showed that the yield strength of metals decreases as temperature increases, and the ductility improves. The result is a true stress/true strain curve that is essentially horizontal for strains above the yield point, rather than the rising shape observed below the recrystallization temperature. Thus it is possible to alter the shape of metals drastically by hot working without causing them to fracture and without the necessity for using excessively large forces. In addition, the elevated temperatures promote diffusion that can remove chemical inhomogeneities, pores can be welded shut or reduced in size during deformation, and the metallurgical structure can be altered to improve the final properties. Consequently, the hot-working processes become extremely important, not only for the benefits they can provide, but also because their products often form the starting material for further processing. Figure 14-1 depicts just a small segment of the hot-deformation processes.

From a negative viewpoint, the high temperatures may promote undesirable reactions between the metal and its surroundings. Tolerances are poorer due to thermal contractions and possible nonuniform cooling. Metallurgical structure may also be nonuniform, the final grain size depending on reduction, temperature at last deformation, cooling history after deformation, and other factors.

Grain alteration through hot working. When metals solidify, particularly in larger sections as in ingots or continuous-cast strands, coarse dendritic grains form and a certain amount of segregation of impurities occurs. Moreover, undesirable grain shapes, such as the columnar grains so common in ingots, can

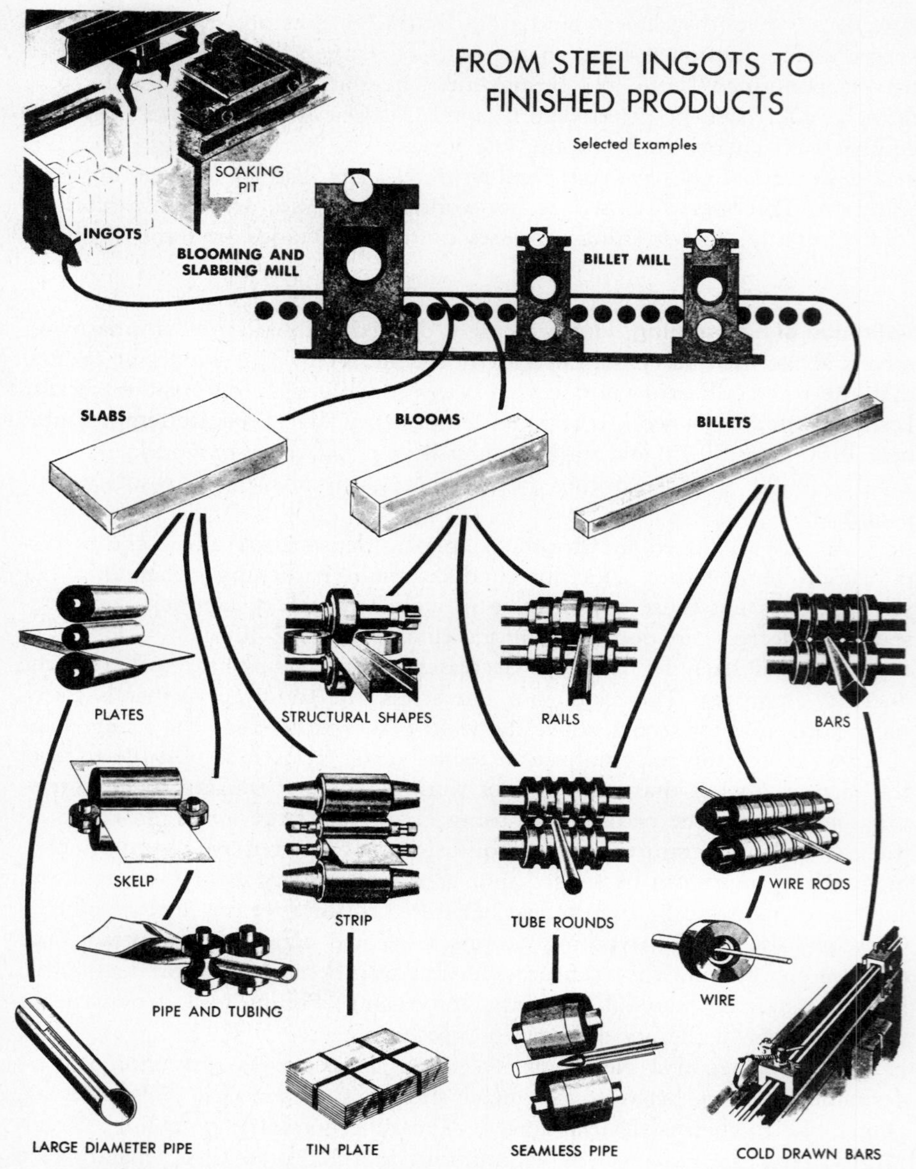

FIGURE 14-1. Schematic flow chart for the production of a variety of finished and semifinished steel shapes. (*Courtesy American Iron and Steel Institute.*)

FIGURE 14-2. Cross section of a cast copper bar (100 mm in diameter) showing as-cast grain structure.

form as seen in Figure 14-2. Small gas cavities or shrinkage porosity can also form. An as-solidified metal typically has a nonuniform grain structure with rather large grain size.

Reheating of the metal without prior deformation will simply promote grain growth and a concurrent decrease in properties. However, if the metal is sufficiently above the recrystallization temperature, the distorted structure is rapidly eliminated by the formation of new strain-free grains. The metal can enter into a state of grain growth or be further deformed and recrystallized. The final structure will be that formed by the last recrystallization and subsequent thermal history and will depend on the factors listed previously. Production of a fine, randomly oriented, spherical-shaped grain structure can result in a net increase not only in strength but also in ductility and toughness.

Another improvement that can be obtained from hot working is the

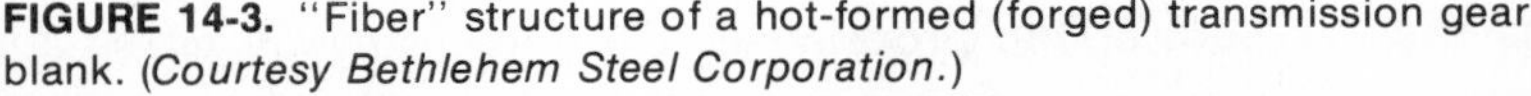

FIGURE 14-3. "Fiber" structure of a hot-formed (forged) transmission gear blank. (*Courtesy Bethlehem Steel Corporation.*)

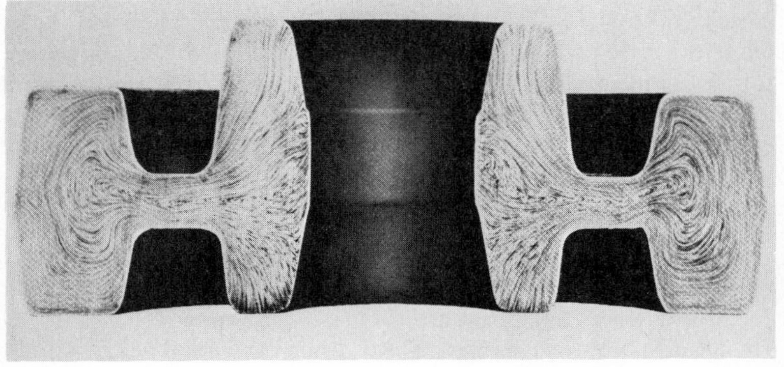

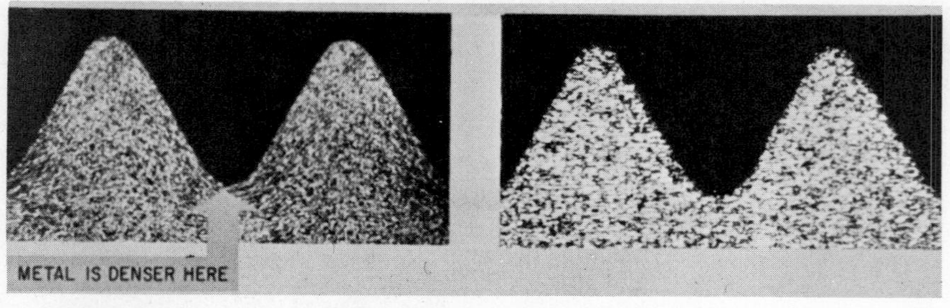

FIGURE 14-4. Comparison of a machined thread (*right*) and a rolled thread (*left*), showing the benefit of a hot-formed fiber texture. (*Courtesy Standard Pressed Steel Co.*)

reorientation of inclusions or impurity material in the metal. With normal melting and cooling, many impurities tend to locate along grain boundary interfaces and, if favorably oriented, can assist a crack in its propagation through a metal. When a piece of metal is plastically deformed, the impurity material often distorts and flows along with the metal. This material, however, does not recrystallize with the base metal and often produces a *fiber structure,* as seen in Figure 14-3. Such a structure clearly has directional properties, being stronger in one direction than in another. Moreover, an impurity originally oriented so as to aid crack movement through the metal is often reoriented into a "crack-arrestor" configuration, perpendicular to crack propagation. Through proper design, extensive use can be made of these results. Figure 14-4 compares a machined thread and a rolled thread in a threaded fastener. By removing potential failure sites where defects intersect the surface, the rolled thread with its fibered structure possesses improved strength characteristics. Improper design or deformation can significantly increase the likelihood of failure.

Classification of hot-working processes. The most obvious reason for the popularity of hot working is that it provides an attractive means of forming a desired shape. *Forging* is the oldest metalworking process known. From the days when prehistoric peoples discovered that they could heat sponge iron and beat it into a useful implement by hammering with a stone, until the present time, hot working has been an effective method of producing many useful shapes.

Some of the hot-working processes that are of major importance in modern manufacturing are:

1. Rolling
2. Forging
   a. Smith
   b. Drop
   c. Press
   d. Upset
   e. Roll
   f. Swaging
3. Extrusion
4. Drawing
5. Spinning
6. Pipe welding
   a. Butt
   b. Lap
7. Piercing

ROLLING

Rolling usually is the first step in converting ingots and billets into finished parts. Many finished parts, such as hot-rolled structural shapes, are completed entirely by hot rolling. More often, however, hot-rolled products, such as sheets, plates, bars, and strips, serve as input material for other processing, such as cold forming or machining. From a tonnage viewpoint, hot rolling is predominant among all manufacturing processes, and modern hot-rolling equipment and practices are sufficiently advanced that standardized, uniform quality products can be produced at low cost. Because the equipment is so massive and costly, however, hot-rolled products normally can be obtained only in standard sizes and shapes for which there is sufficient demand to permit economical production.

The basic rolling processes. Basically hot rolling consists of passing heated metal between two rolls that revolve in opposite directions, the space between the rolls being somewhat less than the thickness of the entering metal, as depicted in Figure 14-5. Because the rolls rotate with a surface velocity exceeding the speed of the incoming metal, friction along the contact interface acts to propel the metal forward. The metal is squeezed and elongated and usually changed in cross section. The amount of deformation that can be achieved in a single pass between a given pair of rolls depends on the friction conditions along the interface. If too much is demanded, the rolls will simply skid over stationary metal. Too little deformation per pass results in excessive cost.

Rolling temperatures. In hot rolling, as in all hot working, it is very important that the metal be heated uniformly throughout to the proper temperature before processing. This usually requires prolonged heating at the desired temperature, a procedure known as *soaking.* If the temperature is not uniform, the subsequent deformation will also be nonuniform, the hotter exterior flowing in preference to the cooler, and therefore stronger, interior. Cracking, tearing, and associated problems may result.

Today, much hot rolling is done in integrated mills where the flow of

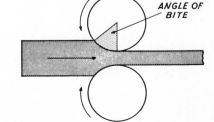

FIGURE 14-5. Schematic representation of the hot-rolling process. (*Courtesy American Iron and Steel Institute.*)

cast ingots or continuous cast material is directed toward certain specific products. Cast ingots are placed in gas- or oil-fired soaking pits as soon as they have solidified and the molds have been stripped. Heat is retained and less energy is required to attain the uniform 1200°C (2200°F) temperature often used for the rolling of carbon steel material. Continuous cast material often goes directly to the rolling stands without additional heating being required.

Hot rolling usually is completed about 50 to 100°C (100 to 200°F) above the recrystallization temperature. Maintenance of such a *finishing temperature* assures the production of a uniform fine grain size and prevents the possibility of unwanted stain hardening.

Rolling mills. Hot rolling is usually done in stages by a series of rolling mill stands. Cast stock is first rolled into large bars, called *blooms,* usually having a minimum thickness greater than 150 mm (6 in.) and often being square in cross section; or *slabs,* with a distinctly rectangular shape. Blooms, in turn, are reduced in size to form *billets,* and slabs are further rolled into *plate* or *strip.* These products then become the raw material for further hot-working or other forming processes.

FIGURE 14-6. Roll configurations used in rolling stands.

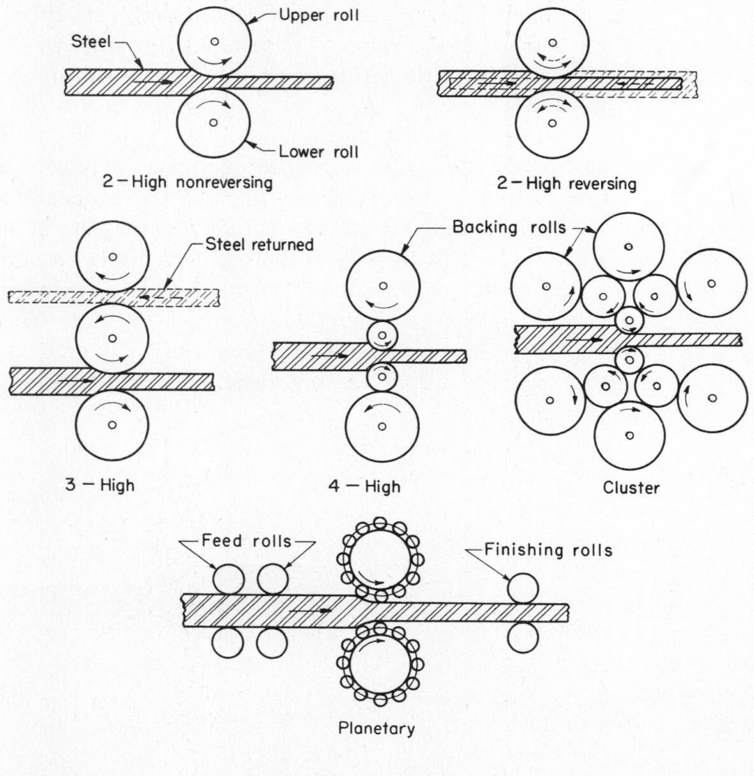

FIGURE 14-7. Bloom entering a three-high blooming mill. (*Courtesy Mesta Machine Company.*)

Rolling mill stands are available in a variety of roll configurations, as illustrated in Figure 14-6. Early reductions, often called *primary roughing* or *breakdown* passes, usually employ a two-high or three-high configuration with 600 to 1400-mm (24- to 55-inch)-diameter rolls. The three-high mill shown in Figure 14-7 is equipped with an elevator on each side of the stand for raising or lowering the bloom and mechanical manipulators for turning the bloom and shifting it for the various passes as it is rolled back and forth.

Smaller-diameter rolls produce less length of contact for a given reduction and therefore require lower loads and less energy for a given change in shape. The smaller cross section, however, provides reduced stiffness, and the rolls are prone to flex elastically under load. Four-high and cluster-roll arrangements use backup rolls to provide the necessary work-roll support. These configurations are used in the hot rolling of wide plate and sheets and in cold rolling, where the deflection in the center of the roll would result in a variation in thickness. Continuous hot-strip mills often involve a roughing train of approximately four four-high mills and a finishing train of six or seven four-high mills.

The planetary mill configuration enables much larger reductions to be performed in a single pass. Each roll consists of a set of planetary rolls carried about a backup or support roll, much like a roller bearing. External drives are required to push the material through the mill and speeds are quite slow. In a typical case, 57.1-mm (2¼-inch) slabs are reduced to 2.5-mm (0.10-in.) sheet, with the planetary mill providing 85 per cent of the reduction. Entering speed is about 1.8 meters per minute (6 feet per minute) and the metal exits at approximately 30.5 meters per minute (100 feet per minute). Production savings relate to the reduction in the required capital investment, floor space, and operating personnel. Speed and productivity are compromised and a small surface reduction usually is required to remove a slight scalloping effect produced by the planetary mill.

In the rolling of nonflat or shaped products, the sets of rolls are grooved to provide a series of openings designed to produce the desired cross section and also to provide metal flow in all directions. Figure 14-8 shows an example of the roll shapes and pass sequence used in the rolling of an I-beam. In some cases the operator adjusts the roll clearance between the passes, whereas in others the rolls are designed so that the various grooves cut into the rolls provide the proper, decreasing clearance for successive passes without any intermediate adjustment of the rolls.

When the volume of product justifies the investment, finished shapes, such as sheets, bars, plates, and some structurals, may be rolled from billets,

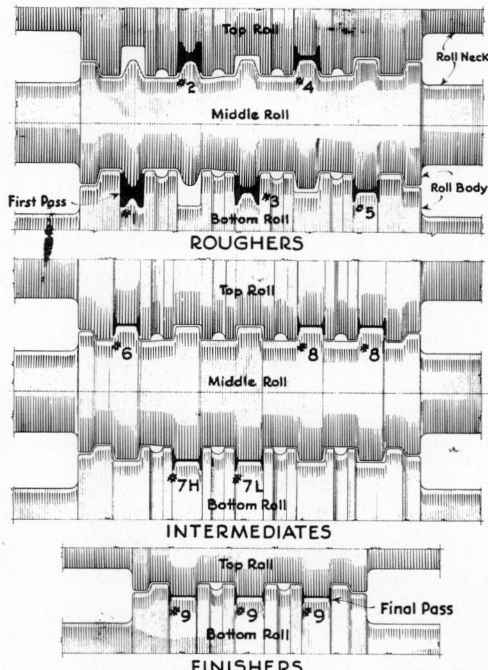

FIGURE 14-8. Rolling sequence used in producing a structural beam. (*Courtesy American Iron and Steel Institute.*)

FIGURE 14-9. Take-off end of a continuous hot-finishing mill. (*Courtesy Mesta Machine Company.*)

blooms, or slabs in a continuous operation on continuous rolling mills. These mills, as shown in Figure 14-9, consist of a series of nonreversing stands through which the metal passes in a continuous piece. The rolls of each successive stand must turn faster than those of the preceding one by a precise amount to accommodate the increased length produced by the reduction in thickness. The metal may leave the final stand at speeds in excess of 110 kilometers per hour (70 miles per hour). In some modern plants continuous rolling mills are fed from continuous casting units so that only a few seconds elapse from the time the metal is solidified until it is rolled into its final shape.

Isothermal rolling. The ordinary rolling of some high-strength metals, such as titanium and stainless steels, particularly in thicknesses below about 3.8 mm (0.150 inch), is difficult because the heat in the sheet is transferred rapidly to the cold, and much more massive, rolls. This has been overcome by *isothermal rolling,* depicted in Figure 14-10. Localized heating is accomplished in the area of deformation by the passage of a large electrical current between the

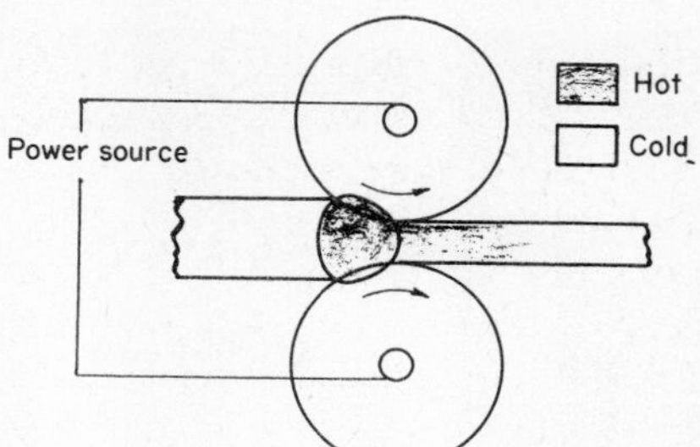

FIGURE 14-10. Schematic representation of the isothermal rolling process.

rolls, through the sheet. Reductions up to 90 per cent per roll pass have been achieved. The process usually is restricted to widths below 50 mm (2 inches).

Characteristics, quality, and tolerances of hot-rolled products. Because they are rolled and finished above the recrystallization temperature, hot-rolled products have minimum directional properties and are relatively free of residual stresses. These characteristics, however, often depend on the thickness of the product and the existence of complex sections that may prevent uniform working in all directions or cause nonuniform cooling. Thin sheets show some definite directional characteristics, whereas thicker plate, for example above 20 mm (0.8 inch), usually will have very little. A complex shape, such as an I- or H-beam, will warp substantially if a portion of one flange is cut away, because of the state of residual tension in the edges.

As a result of the hot plastic working, and the good control that is maintained during processing, hot-rolled products normally are of uniform and dependable quality and considerable reliance can be placed on them. It is quite unusual to find any voids, seams, or laminations when these products are produced by reliable manufacturers.

The surfaces of hot-rolled products are, of course, slightly rough and covered with a tenacious high-temperature oxide known as mill scale. With modern procedures, however, surprisingly smooth surfaces can be obtained.

Dimensional tolerances of hot-rolled products vary with the kind of metal and the size of the product. For most products produced in reasonably large tonnages, the tolerance is from 2 to 5 per cent.

The variety of hot-rolled products is considerable despite the fact that only standard sizes and shapes generally are available. Special sizes and shapes can be obtained, but not at reasonable cost unless ordered in considerable volume.

FORGING

Forging is the plastic working of metal by means of localized compressive forces exerted by manual or power hammers, presses, or special forging machines. It may be done either hot or cold. However, when it is done cold, special names usually are given to the processes. Consequently, the term "forging" usually implies hot forging done above the recrystallization temperature.

Modern forging is a development from the ancient art practiced by the armor makers and the immortalized village blacksmith. High-powered hammers and mechanical presses have replaced the strong arm, the hammer, and the anvil, and modern metallurgical knowledge supplements the art and skill of the craftsman in controlling the heating and handling of the metal.

Various types of forging have been developed to provide great flexibility, making it economically possible to forge a single piece or to mass produce thousands of identical parts. The metal may be (1) drawn out, increasing its length and decreasing its cross section; (2) upset, increasing the cross section and decreasing the length; or (3) squeezed in closed impression dies to produce multidirectional flow. As indicated in Table 13-2, the state of stress in the work is primarily uniaxial or multiaxial compression.

The common forging processes are:

1. Open-die hammer or smith forging.
2. Impression-die drop forging.
3. Press forging.
4. Upset forging.
5. Roll forging.
6. Swaging.

Open-die hammer forging. Basically, *open-die hammer forging* is the same type of forging done by the blacksmith of old, but now massive mechanical equipment is used to impart the repeated blows. The metal to be formed is heated throughout to the proper temperature before being placed on the anvil. Gas, oil, or electric furnaces are usually employed, although induction heating has become attractive for many applications. The impact is then delivered by some type of mechanical hammer, the simplest type being the gravity drop or *board hammer.* Here the hammer is attached to the lower end of a hardwood board, which is raised by being gripped between two driven rollers and then released for free-fall. Although some of these are still in use, steam or air hammers, which use pressure to both raise and propel the hammer, are far more common. These give higher striking velocities, more control of striking force, easier automation, and are capable of shaping pieces up to several tons. Figure 14-11 shows a large double-frame steam hammer. Another style is the open-frame design, which allows more room to manipulate the work and therefore more flexibility. The open-frame design is not as strong as the double frame, however. Open-frame hammers usually range in capacity up to about 2500 kilograms (5000 lb) and the double-frame type up to about 12,000 kilograms (25,000 lb).

Open-die forging does not confine the flow of metal, the hammer and

FIGURE 14-11. Double-frame steam forging hammer. (*Courtesy Erie Foundry Company.*)

anvil often being completely flat. The operator obtains the desired shape by manipulating the workpiece between blows. He may use specially shaped tools or a slightly shaped die between the workpiece and the hammer or anvil to aid in shaping sections (round, concave, or convex), making holes, or performing cutoff operations. Mechanical manipulators are used to hold and manipulate large workpieces, sometimes weighing many tons. Although some finished parts can be made by this technique, it is most often used to preshape metal for some further operation, as in the case of massive parts such as turbine rotors, where it is used to preshape the metal to minimize subsequent machining.

Impression-die drop forging. The open-die hammer or smith forging is a simple flexible process, but it is not practical for large-scale production because it is slow and the resulting size and shape of the workpiece are dependent on the skill of the operator. *Impression-die* or *closed-die forging* overcomes these difficulties by using shaped dies to control the flow of metal. Figure 14-12 shows a typical set of such dies, one-half of which attaches to the hammer and the other half to the anvil. The heated metal is placed in the lower cavity

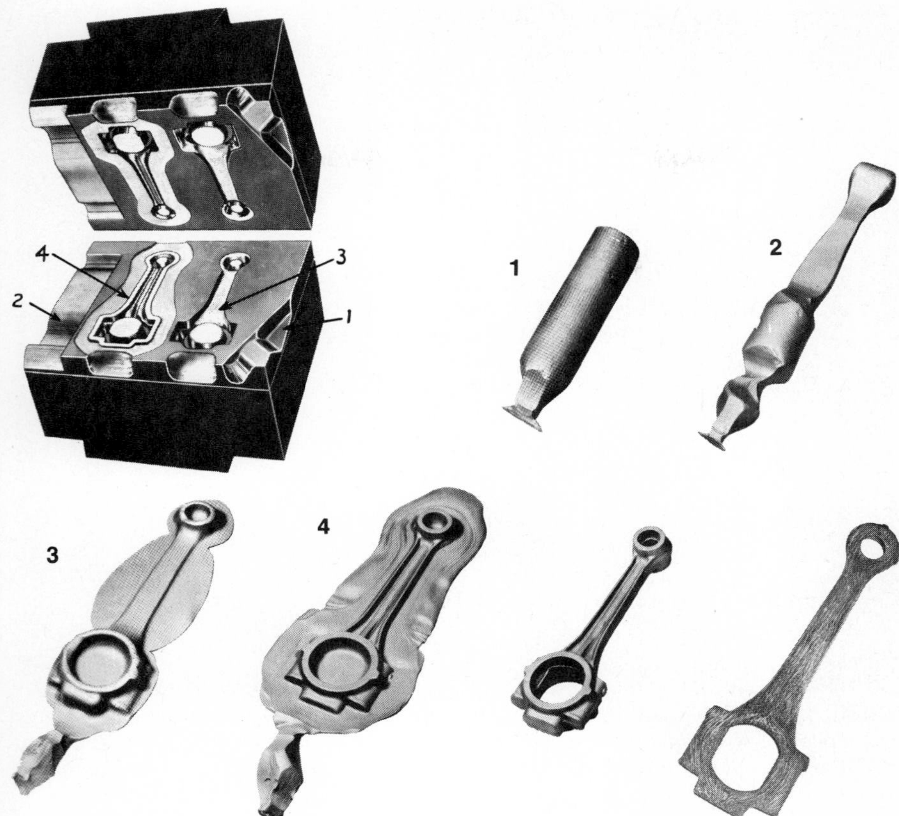

FIGURE 14-12. Impression drop-forging dies and the product resulting from each impression. The flash is trimmed from the finished connecting rod in a separate trimming die. The sectional view shows the grain fiber resulting from the forging process. (*Courtesy Drop Forging Association.*)

and struck one or more blows with the upper die. This hammering causes the metal to flow so as to fill the die cavity. Excess metal is squeezed out between the die faces along the periphery of the cavity to form a *flash*. When forging is completed, the flash is trimmed off by means of a trimming die.

Most drop-forging dies contain several cavities. The first impression usually is an *edging, fullering,* or *bending* impression for roughly distributing the metal in accordance with the requirements of the later impressions. The intermediate impressions are for *blocking* the metal to approximately its final shape. Final shape and size are imparted in the *final impression.* These steps and the shape of a part at the conclusion of each step are shown in Figure 14-12. Because each part produced in a set of dies is shaped in the same die cavities, each is very closely a duplicate of all others, subject to slight die wear.

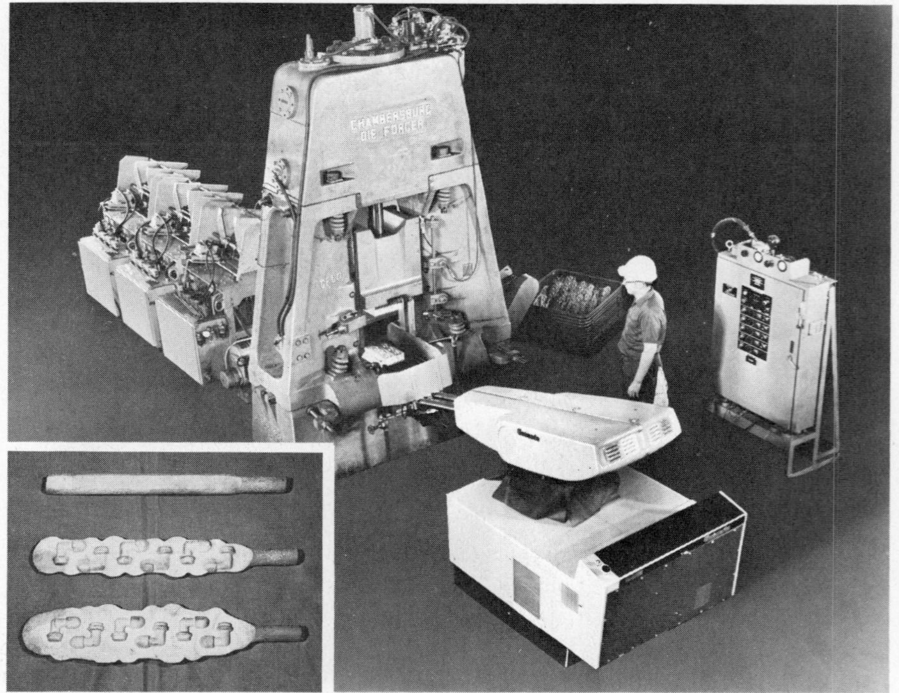

FIGURE 14-13. Fully automated drop forging system, combining a heating and feeding unit, a pneumatic hammer, and an automatic positioning and handling unit. Inset shows the three states of the forging process. (*Courtesy Chambersburg Engineering Company.*)

The restriction to flow that is imposed in certain directions by the shape of the cavity causes the metal to flow in desired directions, and a favorable fiber structure may thus be obtained. In addition, the metal may be placed where it is needed to provide the most favorable section modulus to resist the load stresses. These factors, together with the fine-grain structure and surety of the absence of voids, make it possible to obtain higher strength-to-weight ratios with forgings than with cast or machined parts of the same material.

Board hammers, steam hammers, and air hammers, such as shown in Figure 14-13, are all used in impression die forging. An alternative to the hammer and anvil arrangement is the counterblow or impact machine, such as shown in Figure 14-14. These machines have two horizontal hammers that move together simultaneously and forge the workpiece between them. Because the work is not supported on an anvil, no energy is lost to the machine foundation and the necessity for a heavy base is eliminated. In addition, the machine operates more quietly and with less vibration. In many installations, the operation of the machine can be almost completely automated. The work

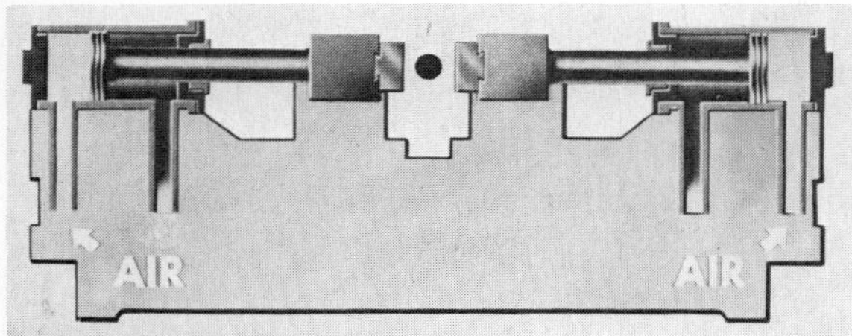

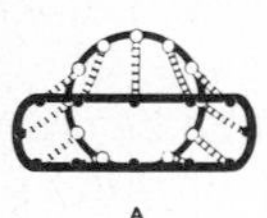

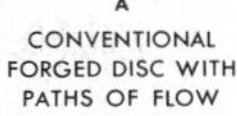

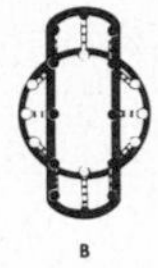

FIGURE 14-14. Automatic impact forging of aluminum rocket fin blades. Parts are forged between two rams, as shown in the upper left-hand diagram, at the rate of 40 per minute. Upper right-hand diagram compares the metal flow in impact and conventional forging. (*Courtesy Chambersburg Engineering Company.*)

can be heated by induction heating and mechanically fed into the machine, forged, and removed.

A recently developed modification of impression die forging utilizes a cast preform that is removed from the mold while hot and forged in a die. After forging, the flash is trimmed in the usual manner. In some cases the four-step process—casting, transfer from the mold to the forging die, forging, and trimming—is completely mechanized. This process is used mostly for nonferrous metals. Most recently, preforms have been made by gas-atomizing a stream of molten metal and directing the resulting spray of hot metal particles onto a shaped collector or mold. Final shape and properties are then imparted by forging.

Die design factors. There are several important factors that must be kept in mind when designing parts that are to be made by closed-die forging. Several of these relate to the design and maintenance of the forming dies, which are usually made of high alloy or tool steel and are quite costly to construct. Impact resistance, wear resistance, strength at elevated temperature, and the ability to withstand alternating rapid heating and cooling all must be outstanding. Considerable maintenance is often required to assure a smooth and accurate cavity and parting-plane surface. Better and more economical results can be obtained if the following rules are observed in the design of the forging:

1. The parting line of the die should be in a single plane if possible.
2. The parting line should lie in a plane through the center of the forging and not near an upper or lower edge.
3. Adequate draft should be provided—at least 3° for aluminum and 5 to 7° for steel.
4. Generous fillets and radii should be provided.
5. Ribs should be low and wide.
6. The various sections should be balanced to avoid extreme differences in metal flow.
7. Full advantage should be taken of fiber flow lines.
8. Dimensional tolerances should not be closer than necessary.

Design details, such as the number of intermediate steps and the shape of each, the amount of excess metal required to assure die filling, and the dimensions of the flash at each step, are often a matter of experience. Each component is a new design entity, and although computer-aided design has made notable advances, good die design is still largely an art.

Good dimensional accuracy is a major reason for using closed-die forging. It must be remembered, however, that dimensions across the parting line are dependent on die wear and maintenance of the parting surfaces. With reasonable precautions, the tolerances shown in Table 14-1 can readily be maintained perpendicular to the parting line. These can be improved by careful practice and die maintenance. Draft angles approaching zero can also be used with some designs, but these are not recommended for general use.

Press forging. In hammer or impact forging, metal flow is a response to the energy in the hammer–workpiece collision. If all the energy can be dissipated through flow of the surface layers of metal and absorption by the press foundation, the interior regions of the workpiece can go undeformed. Therefore, when the forging of large sections is required, *press forging* must be employed. Here the slow squeezing action penetrates throughout the metal and produces a more uniform metal flow. Problems arise, however, because of the long time

TABLE 14-1. Thickness tolerances for steel drop forgings

Mass of Forging		Minus		Plus	
kg	pounds	mm	inches	mm	inches
0.45	1	0.15	0.006	0.48	0.018
0.91	2	0.20	0.008	0.61	0.024
2.27	5	0.25	0.010	0.76	0.030
4.54	10	0.28	0.011	0.84	0.033
0.07	20	0.33	0.013	0.99	0.039
22.68	50	0.48	0.019	1.45	0.057
45.36	100	0.74	0.029	2.21	0.087

FIGURE 14-15. A 311-mN (35,000-ton) forging press. Foreground shows a 3.1-meter (121-inch) aluminum part, weighing 119 kilograms (262 pounds), that was forged on this press. (*Courtesy Wyman-Gordon Company.*)

of die contact with the hot workpiece. If the workpiece surface cools, it becomes stronger and less ductile, and it may crack during forming. Heated dies are often used during press forging operations to minimize this problem.

Forging presses are of two basic types, mechanical and hydraulic, and are usually quite massive. Presses with capacities up to 445MN (50,000 tons) are currently in operation in the United States. Figure 14-15 shows one of these presses designed and manufactured to enable very large sections of aircraft structures to be forged in a single piece.

Press forgings usually require somewhat less draft than drop forgings and are therefore more accurate dimensionally. In addition, press forgings often can be completed in a single closing of the dies.

Upset forging. *Upset forging* involves increasing the diameter of the end or central portion of a bar of metal by compressing its length. In terms of the number of pieces produced, it is the most widely used of all forging processes, and its use has increased greatly in recent years. Parts are upset-forged both hot and cold on special machines such as the one shown in Figure 14-16. Some machines can forge bars up to 250 mm (10 inches) in diameter.

In this type of forging, split dies having several positions or cavities, such as the set shown in Figure 14-17, are commonly used. The split dies move apart slightly for the heated bar to move through them into position. They are then forced together and a heading tool or ram moves longitudinally against the bar, upsetting it into the die cavity. Separation of the die permits transfer to the next position or removal of the product.

In a modification of the upset-forging process, segments of heated metal can be sheared from the bar as it moves into position in the dies. This permits continuous production of a number of pieces from a single coil or length of feedstock.

Upset-forging machines are used to forge heads on bolts and other fasteners, valves, couplings, and many other small components. The following three rules, illustrated in Figure 14-18, should be followed in designing parts that are to be upset-forged:

1. The limiting length of unsupported metal that can be gathered or upset in one blow without injurious buckling is three times the diameter of the bar.
2. Lengths of stock greater than three times the diameter may be upset successfully provided that the diameter of the die cavity is not more than 1½ times the diameter of the bar.
3. In an upset requiring stock with length more than three times the diameter of the bar and where the diameter of the upset is less than 1½ times the diameter of the bar, the length of unsupported metal beyond the face of the die must not exceed the diameter of the bar.

FIGURE 14-16. Modern upset-forging machine. Inset shows forged part being removed. (*Courtesy The Ajax Manufacturing Company.*)

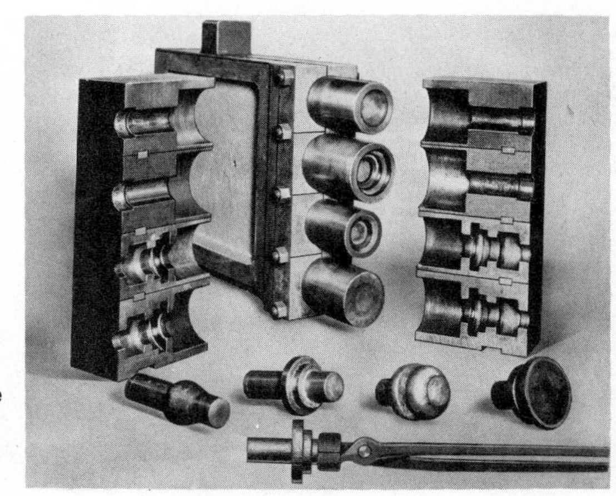

FIGURE 14-17. Set of upset forging dies having four positions. The product resulting from each position also is shown. (*Courtesy The Ajax Manufacturing Company.*)

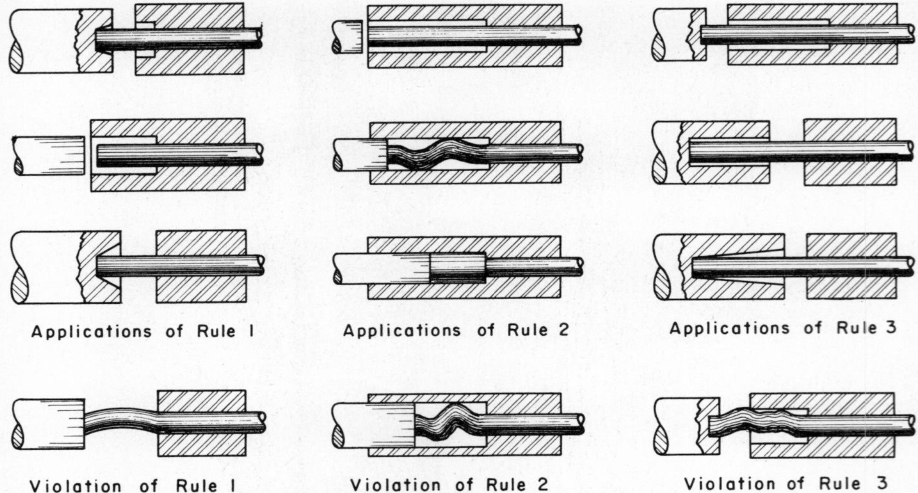

FIGURE 14-18. Rules governing upset forging. (*Courtesy National Machinery Company.*)

Figure 14-19 illustrates the variety of parts that can be produced by upsetting and subsequent piercing, trimming, and machining operations.

Roll forging. *Roll forging,* in which round or flat bar stock is reduced in thickness and increased in length, is used to produce such components as axles, tapered levers, and leaf springs. As shown in Figure 14-20, roll forging

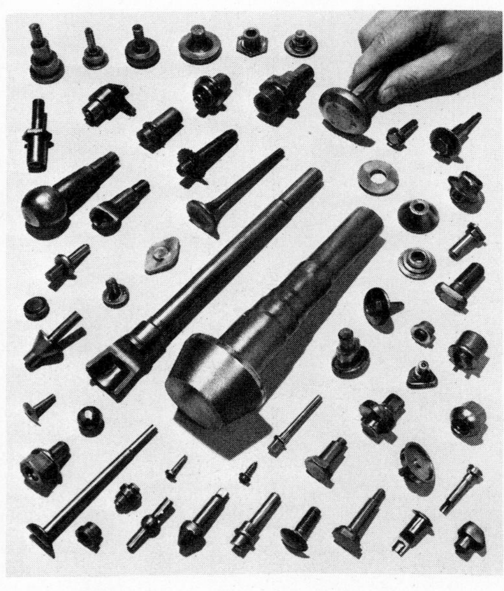

FIGURE 14-19. Typical parts made by upsetting and related operations. (*Courtesy National Machinery Company.*)

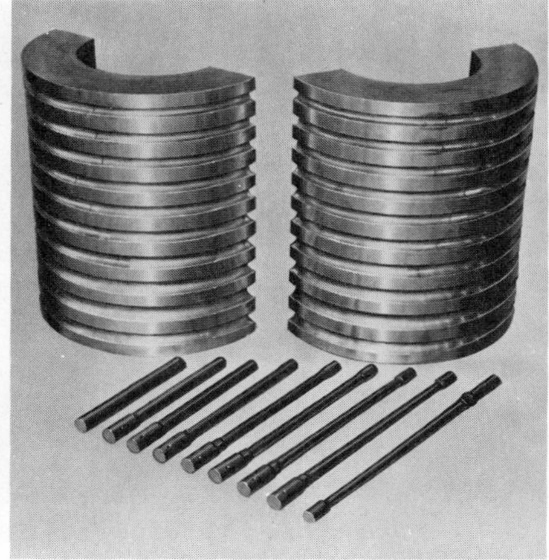

FIGURE 14-20. (*Left*) Roll forging machine. (*Right*) Rolls from a roll forging machine and stages in roll forging a part. (*Courtesy The Ajax Manufacturing Company.*)

is done on machines that have two semicylindrical rolls, containing shaped grooves that are slightly eccentric with the axis of rotation. The heated bar is placed between the rolls while they are in the open position. As the rolls turn one half-revolution, the bar is progressively squeezed and rolled out from between them toward the operator. The operator then inserts the forging between another set of smaller grooves and the process is repeated until the desired size and shape are obtained.

Swaging. *Swaging* involves hammering or forcing a tube or rod into a confining die to reduce its diameter, the die often playing the role of the hammer. Repeated blows cause the metal to flow inward and take the internal form of the die. Figure 14-21 illustrates the application of swaging to close and form the open end of a gas cylinder.

FIGURE 14-21. Steps in swaging a tube to form the neck of a cylinder. (*Courtesy United States Steel Corporation.*)

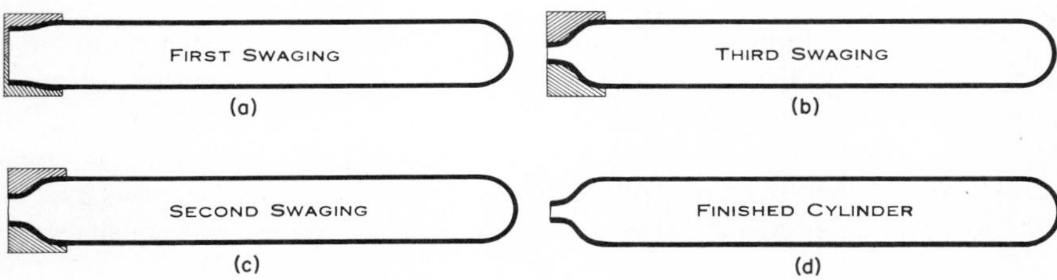

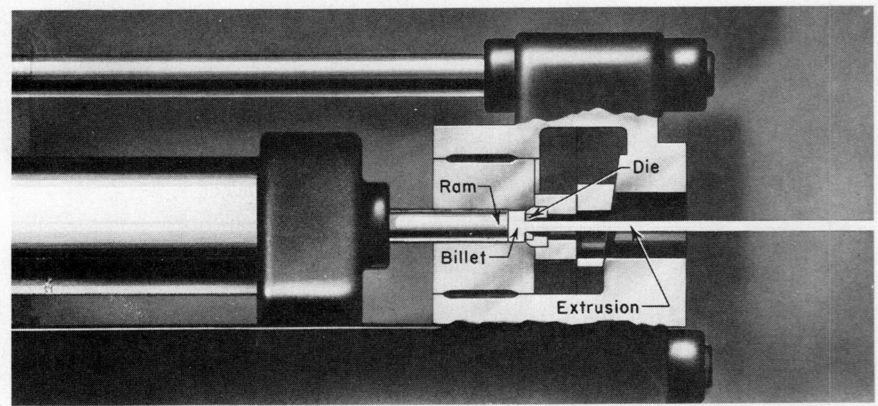

FIGURE 14-22. Basic components in direct extrusion. (*Courtesy Aluminum Company of America.*)

EXTRUSION

In the *extrusion process,* metal is compressively forced to flow through a suitably shaped die to form a product with reduced cross section. Although it may be performed either hot or cold, hot extrusion is employed for many metals to reduce the forces required, eliminate cold-working effects, and reduce directional properties. Basically, the extrusion process is like squeezing toothpaste out of a tube. In the case of metals, a common arrangement is to have a heated billet placed inside a confining chamber. A ram advances from one end causing the material to flow plastically through the die at the other, as illustrated in Figure 14-22. The stress state within the material is triaxial compression.

Lead, copper, aluminum, magnesium, and alloys of these metals are commonly extruded, taking advantage of the relatively low yield strengths and extrusion temperatures. Steel is more difficult to extrude. Yield strengths are high and the metal has a tendency to weld to the walls of the die and confining chamber under the conditions of high temperature and pressure. With the development and use of phosphate-based and molten glass lubricants, substantial quantities of hot steel extrusions are now produced. These lubricants adhere to the billet and prevent metal-to-metal contact throughout the process.

As shown in Figure 14-23, almost any cross-sectional shape can be extruded from the nonferrous metals. Size limitations are few because presses are now available that can extrude any shape that can be enclosed within a 750-mm (30-inch) circle. Shapes and sizes are much more limited in the case of steel and the higher-strength metals, but advances are rapidly being made.

Extrusion is an attractive process for numerous reasons. Many shapes can

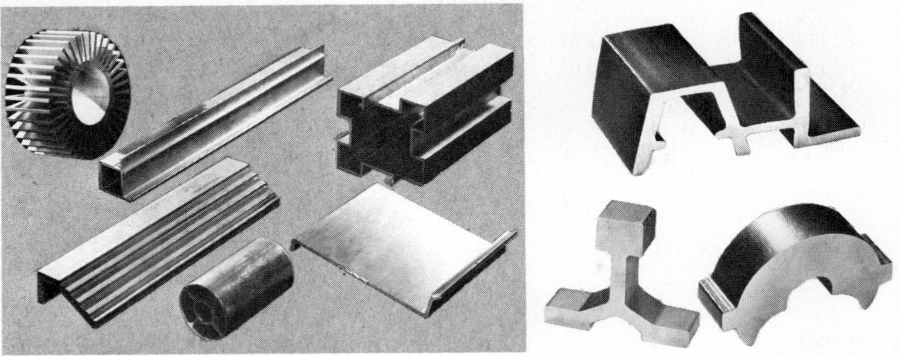

FIGURE 14-23. Typical shapes obtainable by extrusion: (*Left*) aluminum, (*right*) steel. (*Courtesy Aluminum Company of America and Allegheny Ludlum Steel Corporation.*)

be produced as extrusions that are not possible by rolling, such as ones containing reentrant angles or that are hollow. No draft is required, thereby enabling the saving of metal and weight. Being compressive in nature, the amount of reduction in a single step is limited only by the capacity of the equipment. Billet-to-product area ratios in excess of 100 have become quite common. In addition, extrusion dies are relatively inexpensive—often less than \$500—and often only one die is required for a given product. Product changes require only a die change, so small quantities of a desired shape can often be produced economically by extrusion. The major limitation of the process is the requirement that the cross section usually must be the same for the length of the product being extruded; some extrusions with two cross sections have been made, but at a very high cost.

The dimensional tolerances of extrusions are very good, for most shapes ± 0.003 mm/mm or a minimum of ± 0.07 mm is easily attainable. Grain structure is typical of other hot-worked metals, but strong directional properties usually accompany extrusions. Standard product lengths are about 7 to 8 meters (20 to 24 feet). Lengths in excess of 13 meters (40 feet) have been produced.

Extrusion methods. Three basic methods are employed to produce extrusions. Hot extrusion is usually done by either *direct extrusion* or *indirect extrusion,* illustrated in Figure 14-24. Although the indirect configuration reduces friction between the billet and chamber wall, added equipment complexity and restricted length of product favors the direct method. Figure 14-25 shows a typical large extrusion press with the extrusion emerging from the die. The third method, *impact extrusion,* is usually performed cold and will be discussed in Chapter 15. Extrusion speeds are often rather fast to minimize the cooling of the billet within the metal chamber.

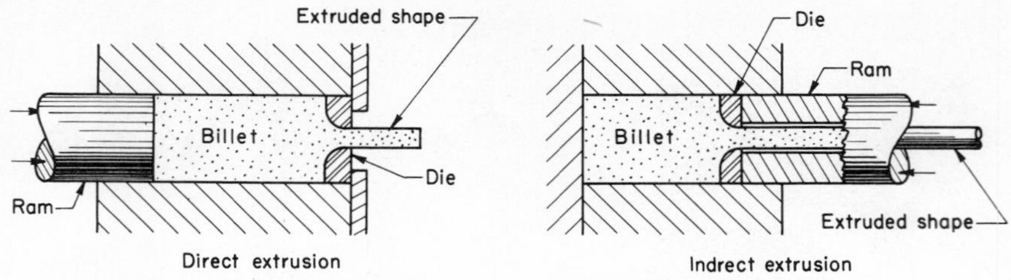

FIGURE 14-24. Direct and indirect extrusion.

Extrusion of hollow shapes. Hollow shapes can be extruded by several methods. For tubular products, the stationary or moving mandrel processes of Figure 14-26 are often employed. For more complex internal cavities, a spider mandrel or torpedo die, such as illustrated in Figure 14-27, is used. As the hot metal flows beyond the spider, further reduction between the die and mandrel forces the seams to close and weld back together. Since the metal has never been exposed to contamination, perfect welds result, and the location of the spider seams cannot be found in the product. Obviously, the cost for hollow extrusions is considerably greater than for solid ones, but a wide variety of shapes can be produced that cannot be made by any other process.

Metal flow in extrusion. The flow of metal during extrusion is rather complex, and some care must be exercised to prevent cracks and defects from forming. In direct extrusion, friction restricts motion between the billet and the chamber and forming die. Metal near the center can often pass through the die with little distortion. Metal near the surface undergoes considerable shear-

FIGURE 14-25. A 125-mN (14,000-ton) extrusion press. Inset shows an extrusion coming from the die. (*Courtesy Aluminum Company of America.*)

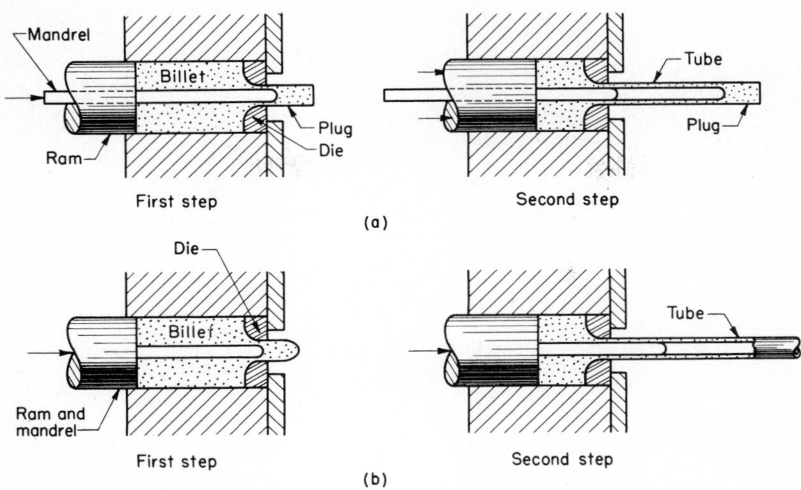

FIGURE 14-26. Two methods of extruding hollow shapes using internal mandrels.

ing to produce a deformation pattern such as is shown in Figure 14-28. Often flow of metal on the surface cannot keep up with the flow in the center, particularly if the surface regions of the billet have cooled, and surface cracks result. Process control is critical in the areas of design, lubrication, extrusion speed, and temperature uniformity.

HOT DRAWING OF SHEET AND PLATE

Drawing is a plastic forming process in which a flat sheet or plate of metal is formed into a recessed, three-dimensional part with a depth several times the

FIGURE 14-27. Extrusion of a hollow shape using a spider mandrel.

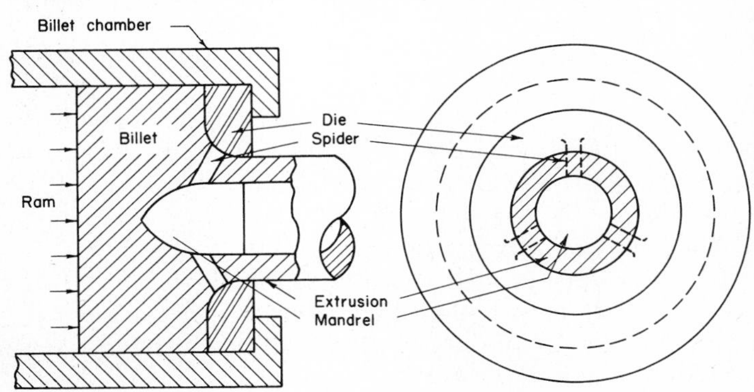

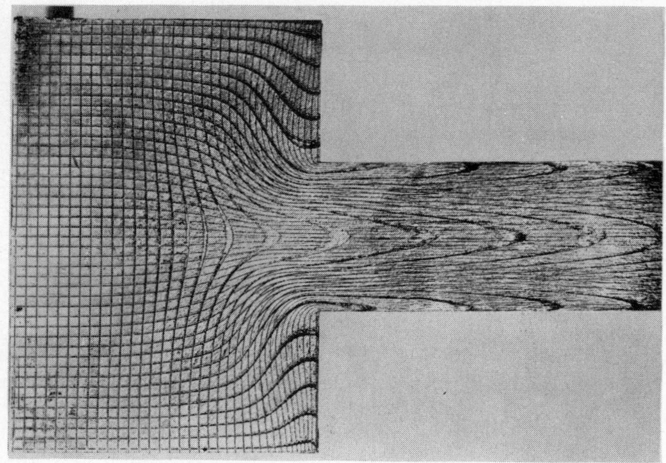

FIGURE 14-28. Grid pattern showing metal flow in direct extrusion. Billet was sectioned and grid pattern engraved prior to extrusion.

thickness of the metal. As the punch descends into the die (or the die moves over the punch), the metal assumes the configuration of the mating punch and die set. Hot drawing is used for forming relatively thick-walled parts of simple geometry, usually cylindrical. There is often considerable thinning of the metal as it passes through the dies. Cold drawing, on the other hand, utilizes relatively thin metal, changes the thickness very little, or not at all, and produces parts in quite a variety of shapes.

Hot drawing is illustrated schematically in Figure 14-29. As shown, a heated, flat dish of metal is placed on top of a female die. The punch then descends, pushing the metal through the die, converting the flat blank into a cup. If the difference between punch diameter and die opening is less than twice the thickness of the metal being drawn, the cup wall is thinned and elongated during drawing (a process often called *ironing*). The stress condition during deformation, as illustrated in Table 13-2, is primarily biaxial compression and uniaxial tension.

Further reduction in diameter can be obtained by redrawing through a smaller punch and die set as shown in Figure 14-29. This figure also illustrates another possibility wherein the cup is pushed through several dies by a single punch. Figure 14-30 shows a thick-walled cylinder emerging from a hot-draw bench after being drawn through a series of dies.

Hot drawing is used primarily for forming thick-walled cylindrical components such as oxygen tanks and artillery shells. It can also be used for shaping other parts, where the female die is closed. The male die or punch de-

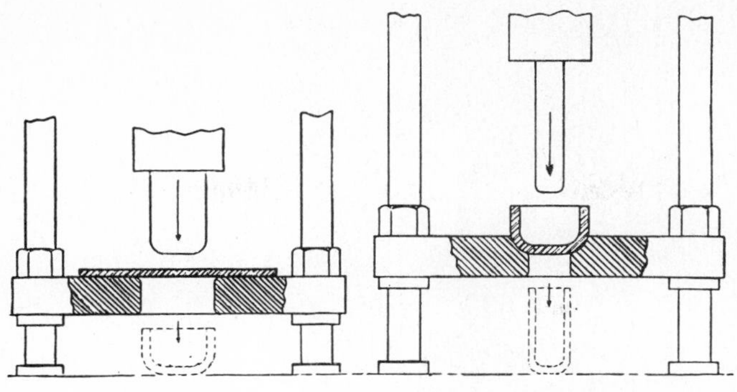

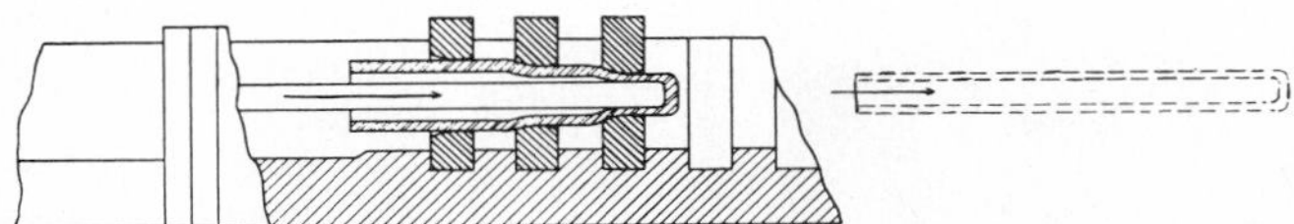

FIGURE 14-29. Methods of cupping or hot drawing by the use of single and multiple dies. (*Courtesy United States Steel Corporation.*)

FIGURE 14-30. Hot-draw bench, showing a cylinder emerging from the last die. (*Courtesy United States Steel Corporation.*)

FIGURE 14-31. Hot drawing a tank half from 15-mm (½-inch) plate on a 17.8 mN (2,000-ton) press. The dies for this operation weigh 63 500 kg (70 tons). (*Courtesy Lukens Steel Company.*)

scends to form the shape and then retracts to permit removal of the part. Figure 14-31 shows an example of such an operation.

HOT SPINNING

Spinning is the plastic forming of metal parts from a flat rotating disk, the application of localized pressure to one side of the disk forcing the metal to flow against a rotating male form that is held against the opposite side. The basic process is illustrated in Figure 15-57. Although most spinning is done cold using thin sheets of metal, hot spinning is used to form thicker plates of metal, usually steel, into axisymmetric shapes. Metal up to 150 mm (6 inches) in thickness is routinely spun into dished pressure vessel and tank heads. Thinner plates of hard-to-form metals, such as titanium, are also shaped by hot spinning. The basic theory of spinning is the same whether hot or cold, but, as with other hot processes, hot spinning enables the forming of greater thicknesses of metal with no strain hardening. Simple, hand-held wooden or metal tools are usually employed for cold spinning. In hot spinning, however, heavy rollers are employed with mechanical holding and control mechanisms, as shown in Figure 14-32.

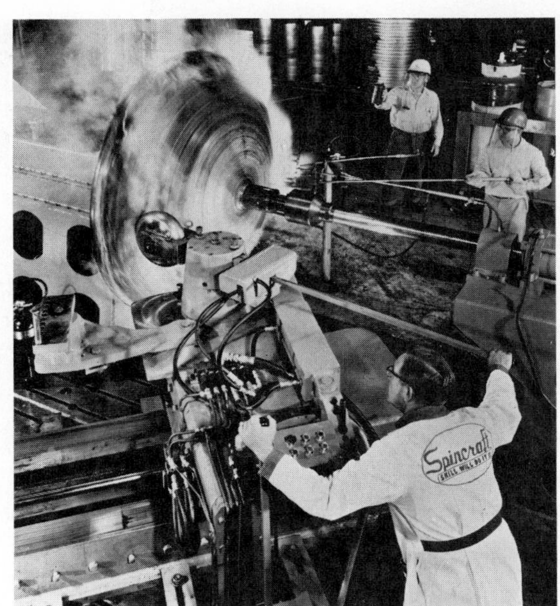

FIGURE 14-32. Hot-spinning a large workpiece using a machine equipped with power-assist controls. (*Courtesy Spincraft, Inc.*)

PIPE WELDING

Large quantities of small-diameter steel pipe are produced by two processes that involve hot forming of metal strip and welding of its edges through utilization of the heat contained in the metal. Both of these processes, *butt welding* and *lap welding* of pipe, utilize steel in the form of *skelp*—long, narrow strips of the desired thickness. Because the skelp has been previously hot rolled and the welding process produces further compressive working and recrystallization, pipe welded by these processes is uniform in quality.

Butt-welded pipe. Figure 14-33 illustrates schematically the *butt-welding process* for making pipe. The skelp is unwound from a continuous coil and is

FIGURE 14-33. Method of making butt-welded pipe from continuous skelp. (*Courtesy American Iron and Steel Institute.*)

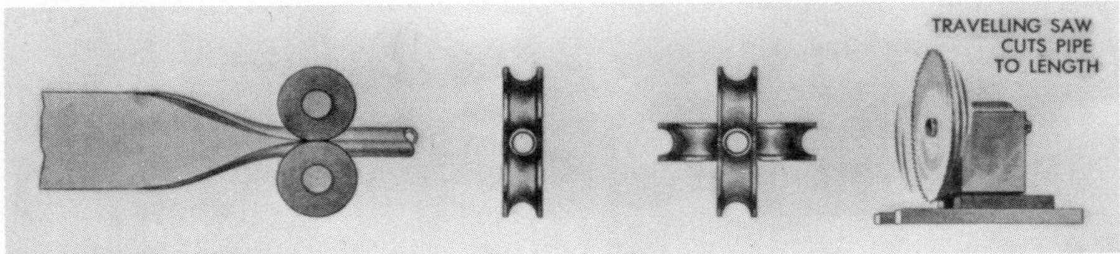

FIGURE 14-34. Continuous butt-welded pipe emerging from the rollers and being cut by the flying shears. (*Courtesy Bethlehem Steel Corporation.*)

heated to forging temperature as it passes through a furnace. Upon leaving the furnace, it is pulled through forming rolls that shape it into a cylinder. The pressure exerted between the edges of the skelp as it passes through the rolls is sufficient to upset the metal and weld the edges together. Additional sets of rollers size and shape the pipe, after which it is cut to length by "flying shears" as it moves. Figure 14-34 illustrates the series of continuous processes used to produce this pipe. Normal pipe diameters range from 3 mm (⅛ inch) to 75 mm (3 inches). Production rates of a single unit often exceed 10,000 meters per hour (30,000 feet per hour).

Lap-welded pipe. The *lap-welding process* for making pipe, illustrated in Figure 14-35, differs from butt welding in that the skelp has beveled edges and a mandrel is used in conjunction with a set of rollers to make the weld. The process is used primarily for larger sizes of pipe, from about 50 mm (2 in.) to 400 mm (14 in.) in diameter. Because of the necessity for supporting and removing the mandrel, lengths are limited to about 7 meters (20 feet).

FIGURE 14-35. Method of making lap-welded pipe from skelp. (*Courtesy American Iron and Steel Institute.*)

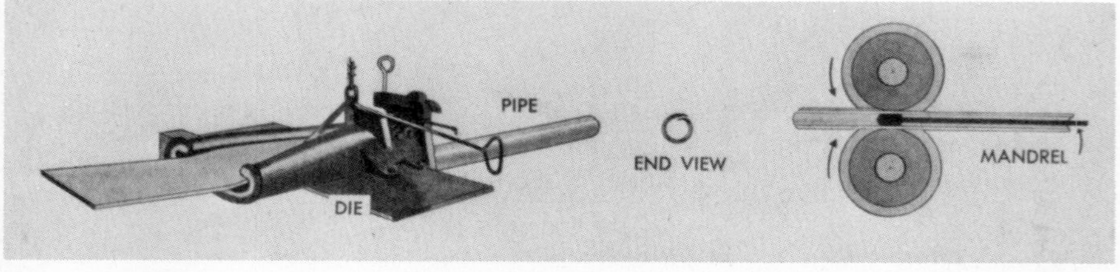

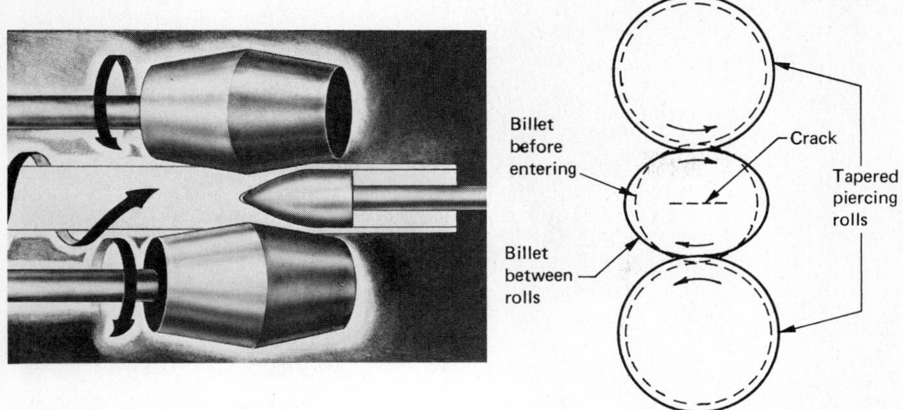

FIGURE 14-36. (*Left*) Principle of the Mannesmann process of producing seamless tubing. (*Courtesy The American Brass Company.*) (*Right*) Mechanism of crack formation in the Mannesmann process.

PIERCING

Thick-walled and seamless tubing is made by the *piercing process,* the basic part of which is illustrated in Figure 14-36. A heated, round billet, with its leading end center-punched, is pushed longitudinally in between two large, convex-tapered rolls that revolve in the same direction, their axes being inclined at opposite angles of about 6° from the axis of the billet. The clearance between the rolls is somewhat less than the diameter of the billet. As the billet is caught by the rolls and rotated, their inclination causes the billet to be drawn forward into them. The reduced clearance between the rolls forces the rotating billet to deform into an elliptical shape. To rotate with an elliptical cross section, the metal must undergo shear about the major axis, which causes a crack to open. As the crack opens, the billet is forced over a pointed mandrel that enlarges and shapes the opening, forming a seamless tube. After the complete tube has been formed, the mandrel drops off and the mandrel

FIGURE 14-37. Schematic diagram of the steps in the production of seamless tubing by the Mannesmann process. (*Courtesy American Iron and Steel Institute.*)

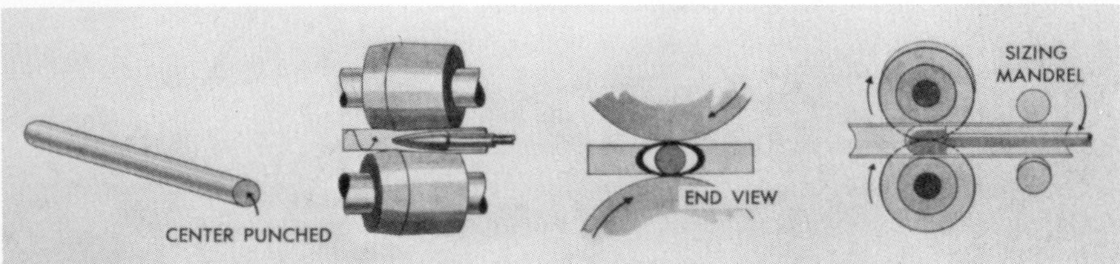

FIGURE 14-38. Pierced tube emerging from a Mannesmann mill. (*Courtesy The Timken Roller Bearing Company.*)

backup bar is removed. The tube is then passed between reeler and sizing rolls, as shown in Figure 14-37, to straighten it and bring it to the desired size. Figure 14-38 shows a thick-walled seamless tube emerging from a Mannesmann mill. When required, the tube may be put through a plug rolling mill, where it is rolled over a larger mandrel to increase its diameter and reduce its wall thickness before it is reeled and sized.

The *Mannesmann-type mill,* depicted in Figures 14-37 and 14-38, is used to produce tubing up to about 300 mm (12 inches) in diameter. Larger sizes sometimes are manufactured on a *Stiefel piercing mill,* which involves the same principle of piercing but uses larger-diameter conical disks instead of convex rolls.

Review questions

1. What are the four basic alternatives for the manufacture of a desired metal shape?
2. What is hot working and how is it related to temperature and material?
3. What are some of the benefits of hot working as opposed to cold working?
4. What are some of the disadvantages of hot forming?

5. Why are the conditions of the last recrystallization so important in hot working?
6. Why is a fiber structure often desired in a hot-formed component?
7. What is the purpose of obtaining temperature uniformity in a metal prior to rolling?
8. Why is hot rolling seldom a feasible method of producing small lengths of a special shape of metal?
9. What are the advantages and disadvantages of small-diameter work rolls? How are the disadvantages overcome?
10. Compare the relative dimensional tolerances and surface finishes of hot- and cold-rolled products.
11. Contrast the open-die and closed-die types of forging.
12. Why is it necessary to have the metal more than fill the cavity in impression-die forging?
13. Why do forgings usually have better strength-to-weight ratios than cast or machined parts made from the same composition of metal?
14. What are the advantages of impact forging as compared with ordinary forging?
15. Why is a draft allowance necessary for closed-die forged components?
16. Why are forging presses usually employed for forging thick sections rather than forging hammers?
17. What are the three rules that should be followed when designing parts that are to be upset-forged?
18. Why can extrusion produce a shape containing reentrant angles while rolling cannot?
19. Why is extrusion often economically practicable for making relatively small quantities of special shapes?
20. Since indirect extrusion involves less frictional resistance and requires less energy, why is it not used as extensively as direct extrusion?
21. How can hollow shapes be made by extrusion?
22. What types of products are normally produced by hot drawing?
23. What is the primary factor in determining whether a part is to be made by hot spinning or cold spinning?
24. Why is the continuous butt-welding process not suitable for making large-diameter pipe?
25. Explain how the hole is created in a bar by a Mannesmann mill.
26. Why is the piercing process of a Mannesmann mill not suitable for making long lengths of seamless tubing?

CHAPTER 15

Cold-working processes

Cold working is the plastic deformation of metals below the recrystallization temperature. In most cases of manufacturing, such cold forming is done at room temperature. In some cases, however, the working may be done at elevated temperatures that will provide increased ductility and reduced strength, but will be below the recrystallization temperature. From a manufacturing viewpoint, cold working has a number of distinct advantages, and the various cold-working processes have become extremely important. Significant advances in recent years have extended the use of cold forming and the trend appears likely to continue.

When compared to hot working, cold-working processes have certain distinct advantages:

1. No heating is required.
2. Better surface finish is obtained.
3. Superior dimension control.

4. Better reproducibility and interchangeability of parts.
5. Improved strength properties.
6. Directional properties can be imparted.
7. Contamination problems are minimized.

Some disadvantages associated with cold-working processes include:

1. Higher forces are required for deformation.
2. Heavier and more powerful equipment is required.
3. Less ductility is available.
4. Metal surfaces must be clean and scale-free.
5. Strain hardening occurs (may require intermediate anneals).
6. Imparted directional properties may be detrimental.
7. May produce undesirable residual stresses.

If one examines the advantages and disadvantages, it becomes evident that the cold-working processes are particularly suited for large-volume production where the quantity involved can readily justify the cost of the required equipment and tooling. Considerable effort has been devoted to developing and improving cold-forming equipment. In addition, better and more ductile metals and an improved understanding of basic plastic flow have done much to reduce the difficulties experienced in earlier years. To a very large extent, modern mass production has paralleled, and been made possible by, the development of cold-forming processes. Automated, high-quality production enables the manufacture of low-cost metal products. In addition, most cold-working processes eliminate or minimize the production of waste material and the need for subsequent machining. With increasing efforts in conservation and materials recycling, these benefits become quite significant.

Although the cold-forming processes tend to be better suited to large-scale manufacturing, much effort has been devoted to developing methods that enable these processes and their associated equipment to be used economically for quite modest production quantities. A substantial amount of the equipment required is well standardized and not excessively costly.

Classification of cold-working operations. The major cold-working operations can be classified basically under the headings of *squeezing, bending, shearing,* and *drawing* as follows:

Squeezing		Bending	
1. Rolling	7. Staking	1. Angle	4. Seaming
2. Swaging	8. Coining	2. Roll	5. Flanging
3. Cold forging	9. Peening	3. Roll forming	6. Straightening
4. Sizing	10. Burnishing		
5. Extrusion	11. Die hobbing (hubbing)		
6. Riveting	12. Thread rolling (see Chapter 27)		

Shearing		Drawing	
1. Shearing Slitting 2. Blanking 3. Piercing Lancing Perforating	4. Notching Nibbling 5. Shaving 6. Trimming 7. Cutoff 8. Dinking	1. Bar and tube drawing 2. Wire drawing 3. Spinning 4. Embossing	5. Stretch forming 6. Shell drawing 7. Ironing

Relationship of metal properties to cold working. The suitability of a metal for cold working is determined primarily by its tensile properties, these being directly influenced by its metallurgical structure. Similarly, the cold working of a metal has a direct relationship to the resulting tensile properties. Both of these relationships should be considered by the designer when selecting metals that are to be processed by cold working.

No plastic deformation of a metal can occur until the elastic limit is exceeded. Thus, in Figure 15-1, in order to obtain permanent deformation, the strain must exceed $0–X_1$, the strain associated with the elastic limit, *a,* on each stress–strain curve. If the strain exceeds $0–X_4$, the metal will rupture. Consequently, from the viewpoint of cold working, two factors are of prime importance: (1) the magnitude of the yield-point stress, which determines the force required to initiate permanent deformation, and (2) the extent of the strain region from 0 to X_4, which indicates the amount of plastic deformation, or ductility, that is available to be utilized. If considerable deformation must be imparted to a metal without rupture, one having tensile properties similar to those depicted in the left-hand diagram of Figure 15-1 is more desirable than one having properties like those shown in the right-hand diagram. Greater ductility would be available and less force would be required to initiate and continue plastic deformation. A metal having the characteristics of the right-hand diagram would be work-hardened to a greater extent by a

FIGURE 15-1. Relationship of the tensile properties of a metal to its suitability for cold working, as shown by its stress–strain diagram.

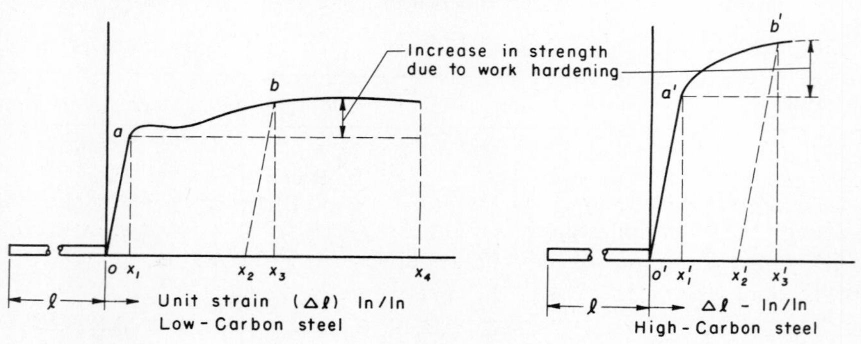

given amount of cold working and thus would not be as suitable for most forming operations. On the other hand, it might be more satisfactory for shearing operations and would be easier to machine, as will be discussed in Chapter 17. Cold-working properties are also affected by grain size, too large or too small, both producing undesirable results.

Springback is an ever-present phenomenon in cold-working operations that also can be explained with the aid of a stress–strain diagram. When a metal is deformed through the application of a load, part of the resulting total deformation is elastic. For example, if a metal is strained to point X_1 in Figure 15-1 and the load is removed, it will return to its original shape and size because all the deformation is elastic. If, on the other hand, the same metal is strained to X_3, corresponding to point b on the stress–strain curve, the total strain $0–X_3$ is made up of two parts, a portion that is elastic and another that is plastic. If the deforming load is removed, the stress reduction will follow line bX_2, and the residual strain will be only $0–X_2$. The decrease in strain, $X_3 - X_2$, is known as *elastic springback.* Quite clearly, springback is an important phenomenon in cold working because the deformation must always be carried beyond the desired point by an amount equal to the springback. Moreover, since different materials have different elastic moduli, the amount of springback from a given load will differ from one material to another. Change in material will therefore require changes in the forming process. Springback is a design consideration, and most difficulties can be overcome by proper design procedures.

Preparation of metals for cold working. In order to obtain several of the benefits of cold working, the metal often must receive special treatment prior to processing. First, if better surface finish and dimensional accuracy are to be obtained than those produced by hot working, the metal must be free of existing scale to avoid abrasion and damage to the dies or rolls that are used. Scale is removed by *pickling,* in which the metal is dipped in acid and then washed. Second, to assure good dimensional tolerances in cold-worked parts, it often is necessary to start with metal that is uniform in thickness and has a smooth surface. For this reason, sheet metal sometimes is given a light cold rolling prior to the major cold working. This pass also serves to remove the yield-point phenomenon and associated problems of nonuniform deformation and surface irregularities in the product.

A third treatment that may be given to metal prior to cold working is annealing. If the cold working is to involve considerable deformation, it is desirable to have as much ductility available as possible. In many cases, annealing is performed after the workpiece has been partially shaped by cold working. Here annealing restores sufficient ductility to permit the final stages of forming to be done without danger of fracture. The desired grain size can also be obtained by proper control of the annealing process.

Sometimes metal must be straightened prior to cold working. Bars and wires are straightened by passing through straightening rolls (see Figure 15-33), and sheets are straightened by stretcher leveling (see Figure 15-34).

SQUEEZING PROCESSES

Most of the cold-working squeezing processes have identical hot-working counterparts or are extensions of them. The primary reasons for deforming cold rather than hot are to obtain better dimensional accuracy and surface finish. In many cases the equipment is basically the same, except that it must be more powerful.

Cold rolling. *Cold rolling* accounts for by far the greatest tonnage of cold-worked products. Sheets, strip, bars, and rods are cold-rolled to obtain products that have smooth surfaces and accurate dimensions. Most cold rolling is done on four-high, cluster, or planetary rolling mills.

Cold-rolled sheet and strip is obtainable in four conditions: *skin-rolled, quarter-hard, half-hard,* and *full-hard.* Skin-rolled metal is given only a ½ to 1 per cent reduction to produce a smooth surface and uniform thickness and to remove the yield-point phenomenon. It is well suited for further cold-working operations, where good ductility is needed. Quarter-hard, half-hard, and full-hard sheet and strip have undergone increasing cold reductions, up to 50 per cent. Consequently, their yield points have been increased, they have definite directional properties, and they have correspondingly decreased ductility. Quarter-hard steel can be bent back on itself across the grain without breaking. Half-hard and full-hard can be bent back 90° and 45°, respectively, about a radius equal to the material thickness.

Swaging. *Swaging,* basically, is a process for reducing the diameter, tapering, or pointing round bars or tubes by external hammering. A useful extension of the process involves the formation of internal cavities. A shaped mandrel is inserted inside a tube, and the tube is then collapsed around it by swaging.

Cold swaging is done by means of a rotary machine, as shown in Figures 15-2 and 15-3. Rotation of the spindle within the cage causes the backer blocks to alternately move apart and then move inward to pass beneath the rollers. The dies therefore open and close from various angles around the workpiece. The operator inserts the bar or tube between the dies and gradually pushes it inward until the desired length of material has been swaged. As the diameter is reduced, the product is elongated.

An important modification of rotary swaging utilizes the principle depicted in Figure 15-4. An open- or closed-end tubular workpiece is placed over a shaped mandrel and inserted between the rotating swaging dies. As the swaging dies reciprocate and rotate, they force the interior of the workpiece to conform to the shape of the mandrel, thereby imparting an accurate internal shape. The operation requires only a few seconds, and the accuracy and surface

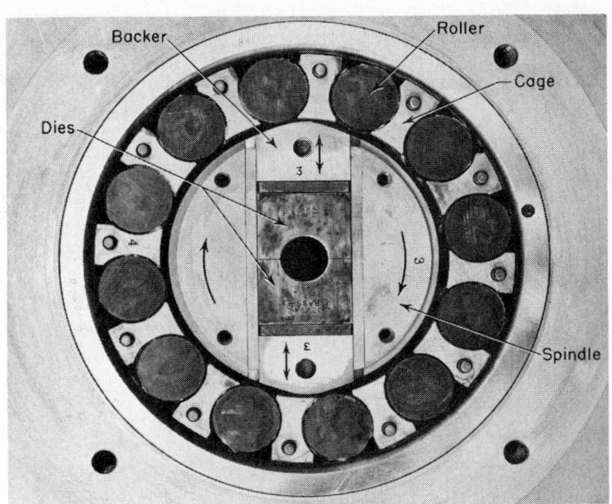

FIGURE 15-2. Basic components and motions in a rotary swaging machine. (*Courtesy The Torrington Company.*)

finish are excellent. When a long product is desired, the workpiece can be fed over a stationary mandrel, enabling a short, relatively inexpensive mandrel to be used. The process is particularly well suited for making internal gears and splines, recesses, and sockets. Figure 15-5 shows some examples of parts formed by this process.

Cold forging. Extremely large quantities of products are made by *cold forging,* in which the metal is squeezed into a die cavity that imparts the desired

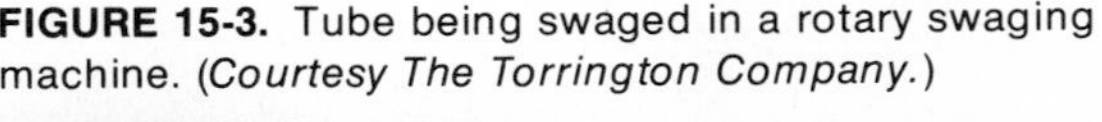

FIGURE 15-3. Tube being swaged in a rotary swaging machine. (*Courtesy The Torrington Company.*)

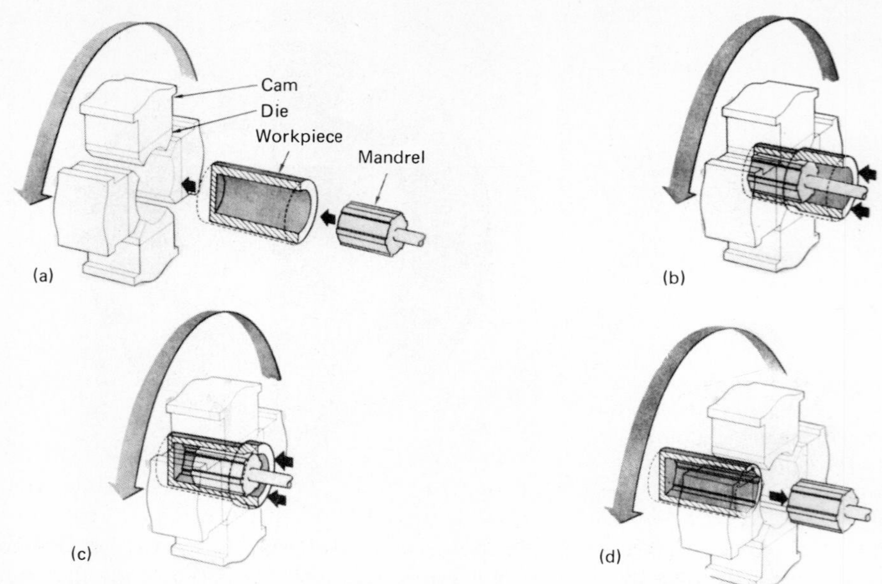

FIGURE 15-4. Principle of the Intraform process. (*Courtesy Cincinnati Milacron, Inc.*)

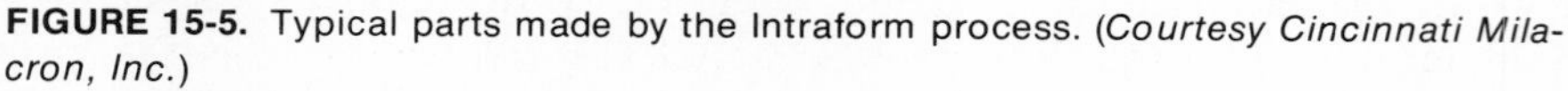

FIGURE 15-5. Typical parts made by the Intraform process. (*Courtesy Cincinnati Milacron, Inc.*)

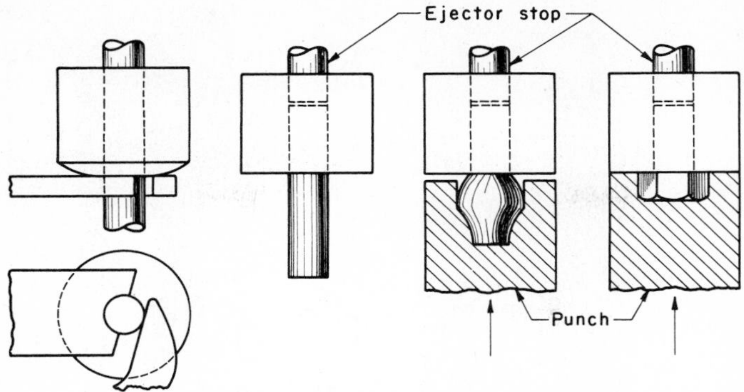

FIGURE 15-6. Steps in a cold heading operation.

shape. *Cold heading,* illustrated schematically in Figure 15-6, is used for making enlarged sections on the ends of a piece of rod or wire, such as the heads on bolts, nails, rivets, and other fasteners. Two variations of the process are used. In the sequence of Figure 15-6, a piece of rod is cut off in the first step and is then transferred to a holder–ejector assembly for subsequent operations. Upsetting is then done in one or more strokes of the heading punches. If two or more blows are required, the heading punches rotate into position between strokes. When heading is completed, the ejector stop advances to remove the work from the holding die.

In the other variation, a continuous rod is fed forward, clamped, and the head formed. The rod is further advanced, cut to length, and the cycle repeats. This procedure is used for making nails where the point is formed by the cutoff operation.

Enlarged sections at locations other than the ends of rods can be made by the *upsetting* process illustrated in Figure 15-7. In this procedure, ejector pins must be provided in both the punch and the die.

Through the use of various types of dies, and by combining cold heading and upsetting, relatively complex parts can be made to close tolerances very rapidly by cold forging. Usually, there is almost no wastage of material. As an

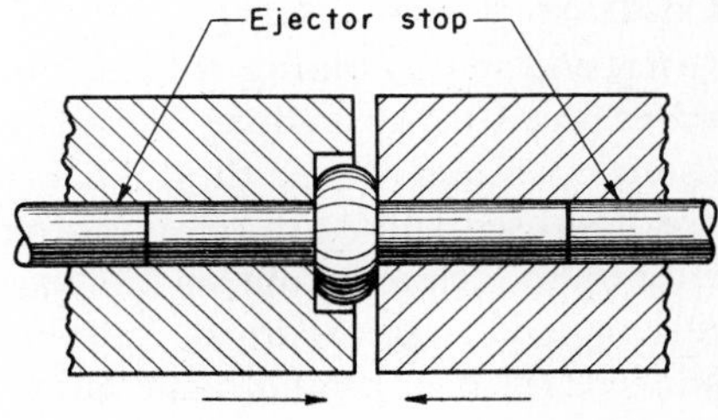

FIGURE 15-7. Method for upsetting the center portion of a rod. Both dies grip the stock during upsetting.

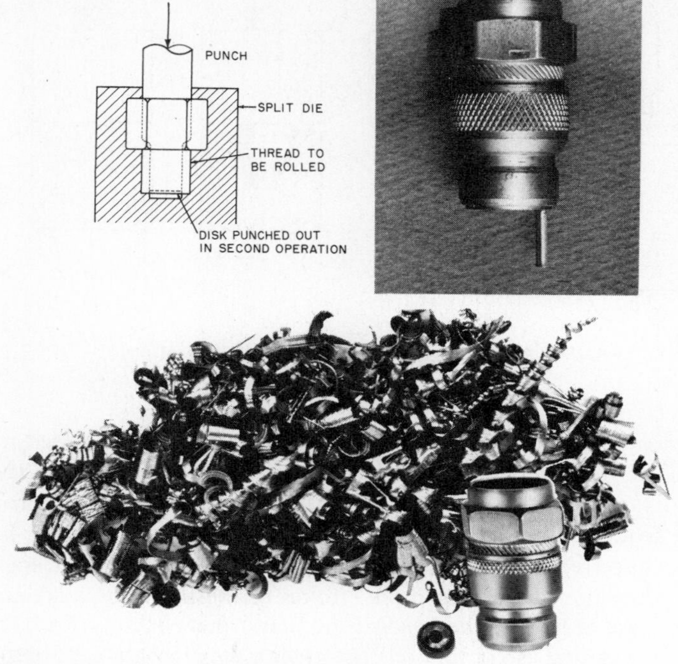

FIGURE 15-8. Manufacture of a spark-plug body by cold forming. (*Top left*) Method of forming body. (*Top right*) Body before rolling of threads. (*Bottom*) Chips produced when spark-plug body was made by machining. Small disk at the bottom is the only scrap from the cold-forming process. (*Courtesy American Iron and Steel Institute.*)

example, Figure 15-8 illustrates how a spark-plug body is formed from a cylindrical slug by a single forging stroke. Not only is considerable machining time and cost eliminated, but the saving in material often pays for the cost of the forging equipment. Although cold forging is particularly suitable for the ductile nonferrous metals, it is also used quite extensively for steel products.

Extrusion. Great advances have been made in cold *extrusion* during recent years, and in combining cold extrusion with cold heading. Figure 15-9 illustrates the basic principles of *forward* and *backward* types of cold extrusion using open and closed dies. This process is often called *impact extrusion* and was first used only with the low-strength, ductile metals such as lead, tin, and aluminum for producing such items as: collapsible tubes for toothpaste, medications, and so forth; small "cans" such as are used for shielding in electronics and electrical apparatus; and larger cans for food and beverages. In recent years, cold extrusion has been used for forming mild steel parts, often being

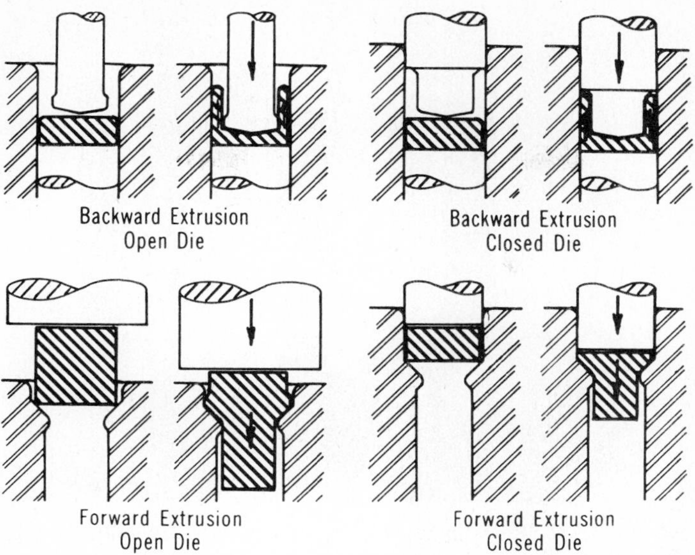

FIGURE 15-9. Methods of cold extrusion.

combined with cold heading. When cold heading alone is used, there is a definite limit to the ratio of the head and stock diameters, as discussed in connection with Figure 14-17. By combining extrusion and cold heading, this difficulty can be easily overcome, as is illustrated by the examples shown in Figure 15-10 and 15-11. Not only is considerable metal saved by not machining the parts from large-diameter stock but, in addition to overcoming the limitations of cold heading, the intermediate-size rod used is cheaper than a smaller-diameter rod that might be employed with multiple-step heading operations.

Another type of cold extrusion, known as *hydrostatic extrusion,* utilizes high fluid pressure to extrude a billet through a die, either into atmospheric pressure or into a lower-pressure chamber. The pressure-to-pressure process, illustrated in Figure 15-12, makes possible the extrusion of relatively brittle materials, such as molybdenum, beryllium, and tungsten. Billet-chamber friction is eliminated, billet-die lubrication is enhanced by the pressure, and the surrounding pressurized atmosphere suppresses crack initiation and growth.

Roll extrusion. Thin-wall cylinders can be produced from thicker-wall material by the *roll-extrusion process* depicted in Figure 15-13. The squeezing action of a rotating roller forces the metal to flow forward between the roller and the external confining ring. Although cylinders from 19 mm (0.75 inch) to 4000 mm (156 inches) in diameter have been made by this procedure, it is most commonly used for those in the range 75 mm (3 inches) to 500 mm (20 inch).

FIGURE 15-10. Steps in making a shaft and rotor for a refrigerator compressor by cold heading and cold extrusion. (*Left to right*) Blank, extruded, headed, machined. (*Courtesy Whirlpool Corporation.*)

Sizing. *Sizing* involves squeezing areas of forgings or ductile castings to a desired thickness. It is used principally on bosses and flats, with only enough deformation occurring to bring the region to a desired dimension. By this procedure, designers may make the general dimensional tolerances of a part more liberal, enabling the use of less costly production methods. The few close dimensions are then obtained by one or two simple and inexpensive siz-

FIGURE 15-11. Steps in cold-forging a bolt by extrusion, cold heading, and thread rolling. (*Courtesy The National Machinery Company.*)

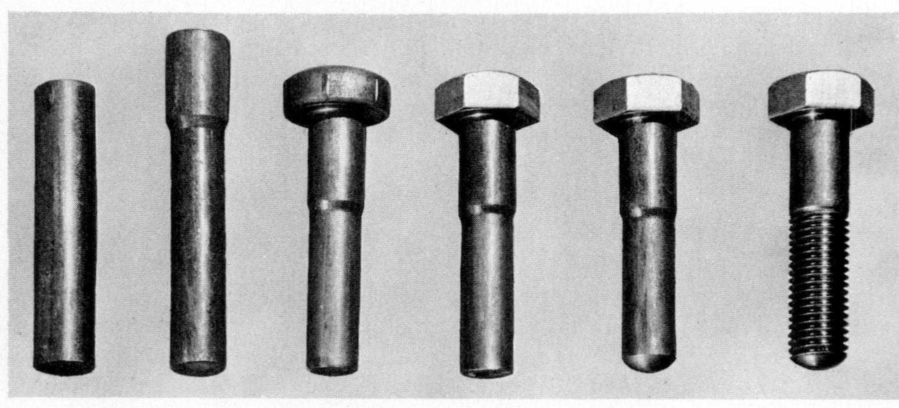

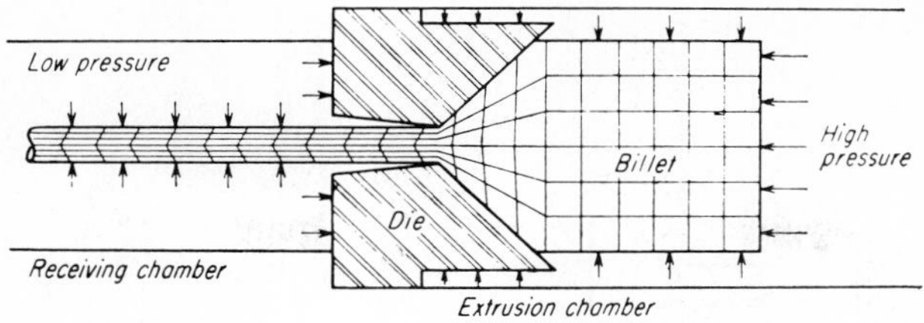

FIGURE 15-12. Hydrostatic extrusion method for extrusion of relatively brittle materials, utilizing differential hydrostatic pressures. (*Courtesy American Machinist.*)

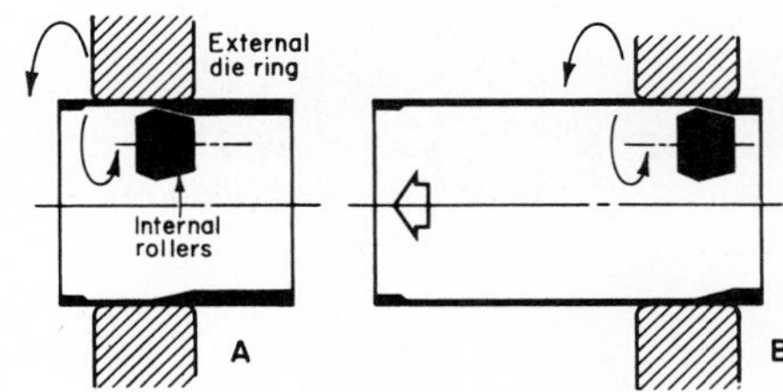

FIGURE 15-13. Roll extrusion process. (*Courtesy Materials Engineering.*)

ing operations. Sizing usually is done between simple dies in a mechanical-driven press, thereby assuring positive dimensional control.

Riveting. In *riveting,* a head is formed on the shank end of a fastener to provide a permanent method of joining sheets or plates of metal together. Although riveting usually is done hot in structural work, in manufacturing it almost always is done cold. Quite commonly, where there is access to both sides of the work, the method illustrated in Figure 15-14 is used. The shaped punch may be held and advanced by a press or contained in a hand-held pneumatic riveting hammer. When a press is used, the rivet usually is headed by a single squeezing action. Sometimes the heading punch also rotates. Special machines, such as those used in aircraft manufacturing, punch the hole for the

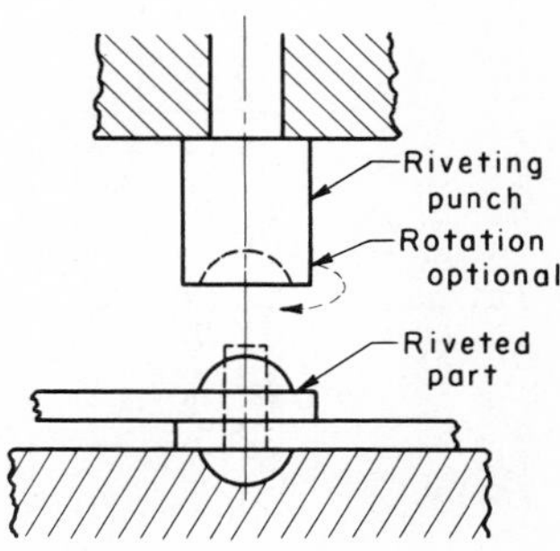

FIGURE 15-14. Method of fastening by riveting.

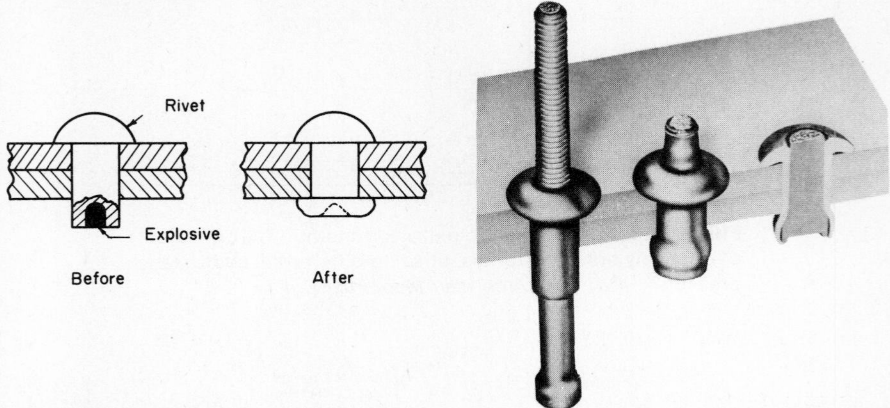

FIGURE 15-15. Rivets for use in "blind" riveting: (*left*) explosive type; (*right*) shank-type pull-up. (*Courtesy Huck Manufacturing Company.*)

rivet, place the rivet in position, and perform the riveting operation in about 1 second.

It often is desirable to use riveting in situations where there is access to only one side of the assembly. Several types of rivets are available for these applications, two of which are illustrated in Figure 15-15. Both involve cold working. The explosive type is activated by the application of a heated tool to the rivet head, causing the charge to explode and expand the shank into a retaining head. The pull-type or pop rivet mechanically expands the shank, after which the pull pin is cut off flush with the head.

Staking. *Staking* is a commonly used cold-working method for permanently fastening two parts together where one protrudes through a hole in the other. It is so simple in method and in appearance in the final product that it is often overlooked by designers. As shown in Figure 15-16, a shaped punch is driven into one of the pieces, deforming the metal sufficiently to squeeze it outward

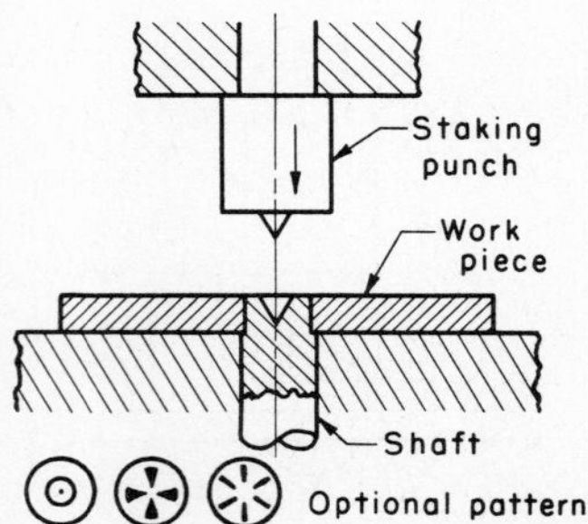

FIGURE 15-16. Fastening by staking.

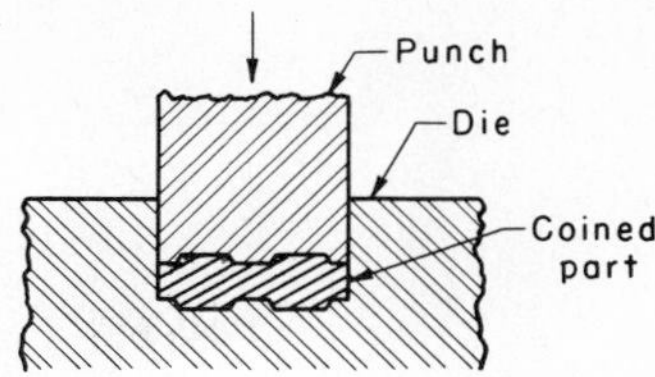

FIGURE 15-17. Coining process.

against the second piece so that they are tightly locked together. As illustrated, various types of punch patterns may be used. Because the staking punch is simple and the operation is completed with a single stroke of the press, it is a convenient and economical fastening method where permanence is desired and the appearance of the punch mark is not objectionable.

Coining. *Coining* involves cold working by means of a positive displacement punch while the metal is completely confined within a set of dies. The process, illustrated schematically in Figure 15-17, is used to produce coins, medals, and other products where exact size and fine detail are required in a variable thickness product. Because of the confinement of the metal and the positive displacement of the punch, there is no possibility for excess metal to flow from the die, and very high pressures are required. Pressures as high as 1400 MPa (200 ksi) are often used. Accurate volumetric measurement of the metal put into the die is essential to avoid breakage of the dies or press.

Hobbing. *Hobbing* is a cold-working process that is used to form cavities in various types of dies, such as those used for molding plastics. As shown in Figure 15-18, a male *hob* is made with the contour of the part that ultimately will be formed by the die. After the hob is hardened, it is slowly pressed into an annealed die block by means of a hydraulic press until the desired impression is produced. Flow of metal in the die block can be aided and controlled by machining away some of the metal in the block where large amounts of plastic flow would occur. The die block usually is round during hobbing and is reinforced by a heavy steel ring. When hobbing is completed, the die block is removed from the reinforcing ring, the excess metal is machined away, and the piece is hardened by heat treatment.

Because one hob may be used to form a number of identical cavities in a mold, hobbing is frequently more economical than producing dies by conventional die sinking by machining. The process is also referred to as *hubbing*.

Surface improvement by cold working. Two cold-working methods are used extensively for improving or altering the surface of metal products. *Peening* involves striking the surface repeated blows by impelled shot or a round-nose tool. The highly localized blows deform and tend to stretch the metal surface. Because the surface deformation is resisted by the metal underneath, the result

FIGURE 15-18. Hobbing of a die block in a hydraulic press. (*Inset*) Close-up of the hardened hob and the impression in the die block. Die block is contained in a reinforcing ring. Outer surface of the die block is machined flat to remove bulged metal.

is a surface layer under residual compression. This condition is highly favorable to resist cracking under fatigue conditions, such as repeated bending, because the compressive stresses are subtractive from the applied tensile loads. For this reason, shafting, crankshafts, gear teeth, and other cyclic-loaded components are frequently peened.

In most manufacturing, except in welding, peening is done by means of shot which is impelled by the type of mechanism shown in Figure 15-19.

FIGURE 15-19. Wheelapeening mechanism for impelling shot for peening. (*Inset*) Surface of steel peened with steel shot; 25X. (*Courtesy Wheelabrator Corporation.*)

When used after welding, peening usually is done by means of manual or pneumatic hammers to avoid distortion and prevent contraction cracking.

Burnishing involves rubbing a smooth hard object under considerable pressure over the minute surface protrusions that are formed on a metal surface during machining or shearing, thereby reducing their depth and sharpness through plastic flow. In one of two major techniques, the edge surfaces of sheet metal stampings are burnished by pushing the stamped parts through a slightly tapered die having its entrance end a little larger than the workpiece and its exit end a slight amount smaller than the workpiece. The rubbing against the sides of the die occurs under sufficient pressure to smooth the slightly rough edges that are produced by the blanking of the part (see Figure 15-36).

Roll burnishing is used to improve the size and finish of internal and external cylindrical and conical surfaces after machining by metal cutting. The process is illustrated in Figure 15-20. The hardened rolls of the tool press against the surface and roll the protrusions down into a more nearly flat surface. Being cold-worked, the surface also has better wearing properties.

FIGURE 15-20. Tool for roller burnishing. Burnishing rolls are moved outward by means of a taper. (*Top left*) Section of surface before burnishing. (*Top right*) Surface after burnishing. (*Courtesy Madison Industries, Incorporated.*)

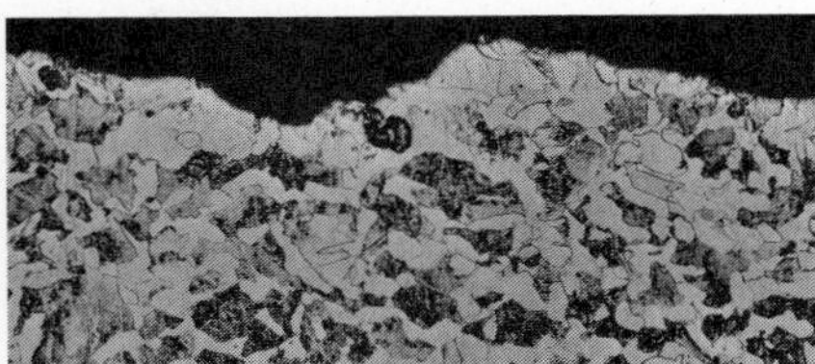

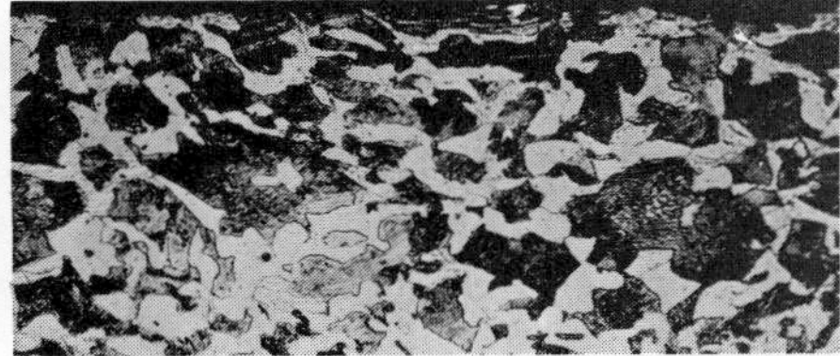

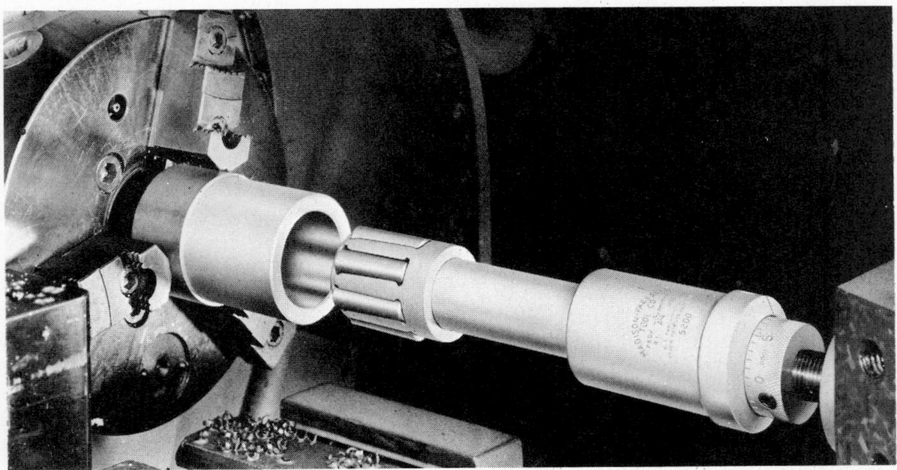

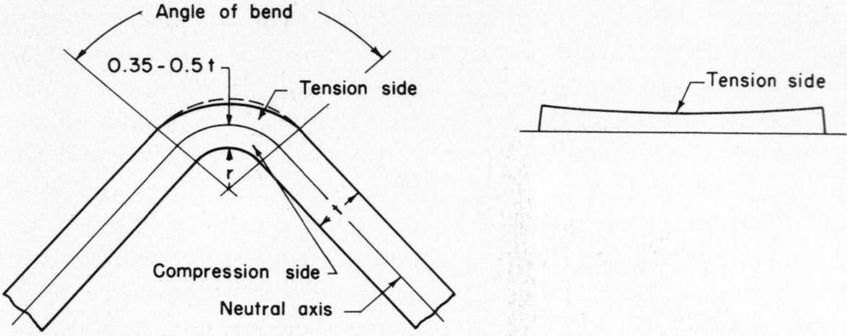

FIGURE 15-21. (*Left*) Nature of a bend in sheet metal. (*Right*) Cross section (exaggerated) of tension side of bent sheet metal or bar, showing variation in thickness due to restraint at edges.

BENDING

Bending is the plastic deformation of metals about a linear axis with little or no change in the surface area. When two or more bends are made simultaneously with the use of a die, the process is sometimes called *forming.* In some cases, the two axes involved in forming may be at an angle to each other, but each axis must be linear and independent of the other to be classed as a bending operation. In these cases, only simple bending theory is involved. If the axes about which deformation occurs are not linear, or are not independent, the process is one of *drawing* and not bending.

As shown in Figure 15-21, bending causes the metal on the outside of the neutral axis to be stretched, while that on the inside is compressed. Because the yield strength of metals in compression is somewhat higher than the yield strength in tension, the metal on the outer side of the bend yields first, and the neutral axis is not located equidistant between the two surfaces. Instead, the neutral axis tends to be from one-third to one-half the thickness of the metal from the inner surface, depending on the bend radius. Because of the preferred plastic flow in the metal outside the neutral axis, it is thinned somewhat at the bend, being more pronounced in the center of a sheet or bar than at the edges, as shown in Figure 15-21. This can cause difficulties in some applications. There is a tendency for the metal on the inner side of the neutral axis to be upset plastically as a result of the compressive forces, and to become somewhat wider in the direction parallel with the bend axis. This can be quite substantial and noticeable when thick metals are bent. A further consequence of the condition of combined tension and compression is the tendency of the metal to unbend somewhat after forming (that is, springback).

Angle bending. Angle bends up to 150° in sheet metal under about 1.5 mm (1/16 inch) in thickness may be made in a *bar folder,* as shown in Figure 15-22. These machines are manually operated and usually are less than 2½ meters (8

FIGURE 15-22. Phantom section of a bar folder. (*Courtesy Niagara Machine and Tool Works.*)

feet) long. After the sheet of metal is inserted under the folding leaf at the desired position, raising the handle first actuates a cam that causes the blade to clamp the sheet and then further motion of the handle bends the metal to the desired angle.

Bends in heavier sheet metal, and more complex bends in thinner sheets, are made on a *press brake,* such as shown in Figure 15-23. These are mechanically or hydraulically driven presses having a long, narrow bed and relatively slow, short, adjustable strokes. The metal is bent between interchangeable dies that are attached to the bed and the ram. As illustrated in Figure 15-24, different dies can be used to produce many types of bends. The metal can be fed inward between successive strokes to produce various types of repeated bends, such as corrugations. Figure 15-25 illustrates how a complex bend can be formed progressively by repeated strokes and the use of more than one die. Seaming, embossing, punching, and other operations can also be done on press brakes by using suitable dies, but they can usually be done more efficiently on other types of equipment when the volume is sufficient to justify their use.

Design for bending. Several factors must be considered when designing parts that are to be made by bending. Of primary importance is the minimum radius that can be bent successfully without metal cracking. This, of course, is related to the ductility of the metal. It has been shown that the ratio of the

FIGURE 15-23. (*Left*) Bending sheet metal on a large press brake: (*right*) close-up view of the dies. (*Courtesy Cincinnati Incorporated.*)

minimum bend radius R to the thickness of the metal t, for a wide range of metals, can be related to the per cent reduction in area by the curve shown in Figure 15-26. As can be noted, it is seldom feasible to call for a minimum radius of less than the thickness of the metal. Moreover, if the metal has been cold-worked previously, or has marked directional properties, this will have considerable effect on its bending properties. Whenever possible, it is wise to make the bend axis normal to the grain of the metal. If two perpendicular bend axes are involved, the metal should be oriented so they are at 45° to the direction of the grain, if at all possible.

A second matter that is of concern to the designer is that of determining the length of a flat blank that will produce a bent part of given dimensions. The fact that the neutral axis is not at the center line of the metal makes it

FIGURE 15-24. Several types of dies used on press brakes for forming angles and rounds. (*Courtesy Cincinnati Incorporated.*)

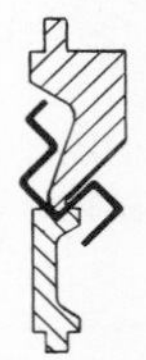
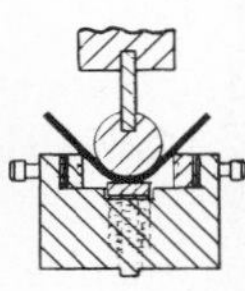
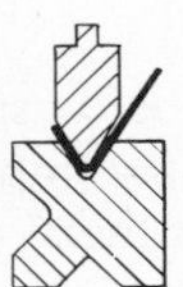
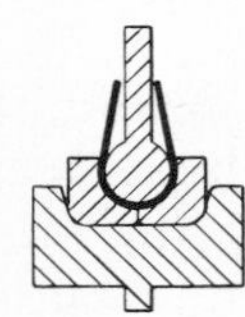
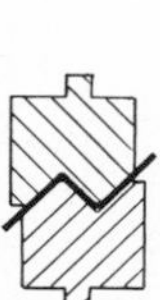

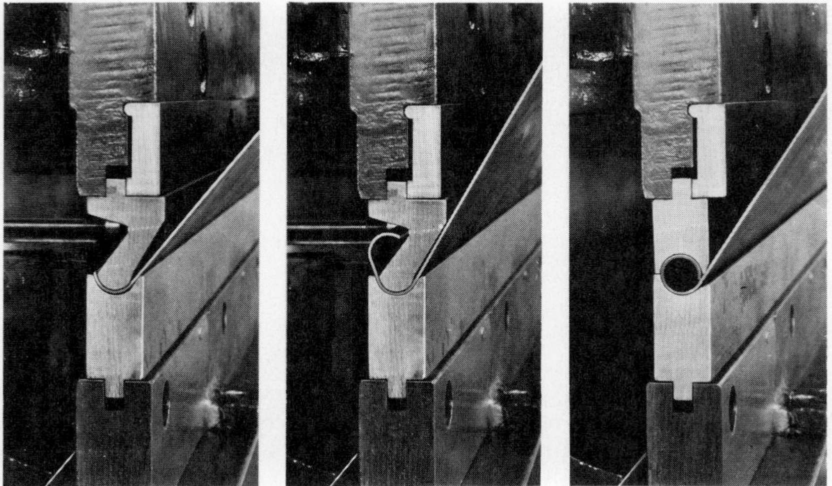

FIGURE 15-25. Dies and steps used in forming a roll bead on a press brake. (*Courtesy Cincinnati Incorporated.*)

necessary to make some adjustments that are functions of the stock thickness and bend radius. The method shown in Figure 15-27 has been found to give satisfactory results for determining blank length.

A third important design factor is the length of the minimum leg that can be bent successfully. In most cases the leg should be at least 1½ times the metal thickness plus the bend radius.

Whenever possible, the tolerance on bent parts should not be less than 0.8 mm (1/32 inch). Ninety-degree bends should not be specified without first determining whether the bending method will permit a full right angle to be obtained.

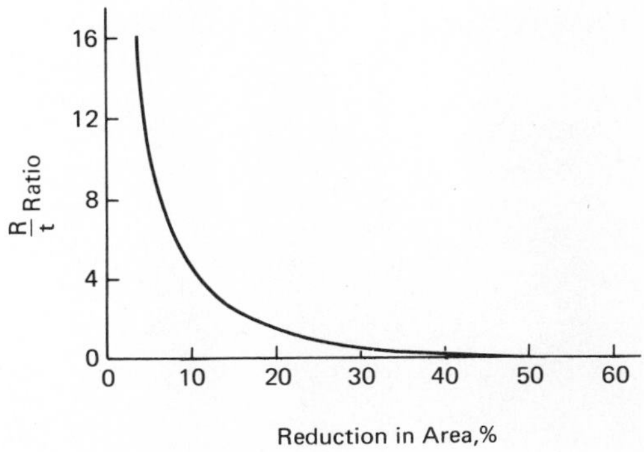

$\frac{R}{t}$ Ratio

16
12
8
4
0

0 10 20 30 40 50 60

Reduction in Area,%

FIGURE 15-26. Curve for relating minimum bend radius (relative to thickness) to metal ductility, measured by reduction in area.

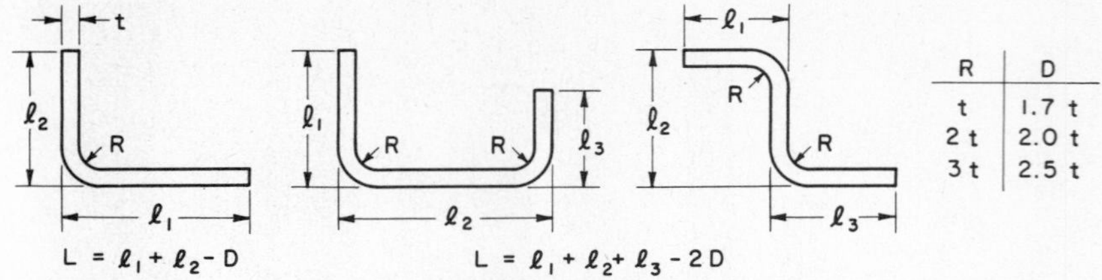

R	D
t	1.7 t
2 t	2.0 t
3 t	2.5 t

FIGURE 15-27. Method for determining blank length for bending operations.

Roll bending. Plates, heavy sheets, and rolled shapes can be bent to a desired curvature on forming rolls of the type shown in Figures 15-28 and 15-29. These usually have three rolls in the form of a pyramid, with the two lower rolls being driven and the upper roll adjustable to control the degree of curvature. Where the rolls are supported by a frame on each end, one of these supports can be swung clear to permit removal of a closed shape from the rolls. Bending rolls are available in a wide range of sizes, some being capable of bending plate up to 150 mm (6 inches) thick.

Cold-roll forming. *Cold-roll forming* of flat strip metal into complex sections has been highly developed in recent years. This process, depicted in Figures 15-30 and 15-31, involves the progressive bending of metal strip as it passes through a series of forming rolls. A wide variety of moldings, channeling, and other shapes can be formed on machines that produce up to 3000 meters (10,000 feet) of product per day. By changing the rolls, a single machine can be adapted to the production of many different shapes. Since changeover,

Figure 15-28. Cold-roll bending of structural shapes. (*Courtesy Buffalo Forge Company.*)

FIGURE 15-29. Bending a heavy steel ring on a set of smooth bending rolls. The ring is 152 mm (6 inches) thick, 0.4 meter (16 inches) wide, and 2.2 meters (88 inches) outside diameter. (*Courtesy Lukens Steel Company.*)

FIGURE 15-30. Schematic representation of the cold-roll forming process. (*Inset*) Typical shapes formed by this process. (*Courtesy Van Huffel Tube Corporation.*)

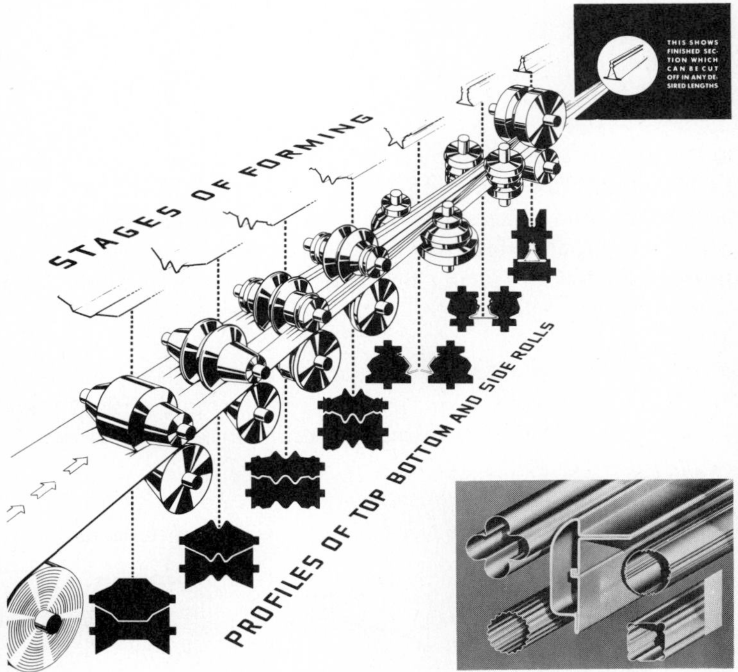

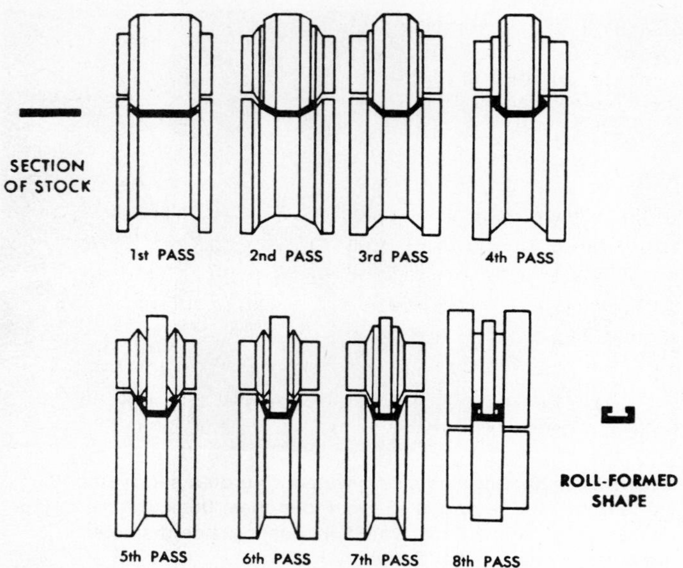

FIGURE 15-31. Eight-roll sequence for forming a box channel. (*Courtesy The Aluminum Association.*)

setup, and adjustment time may take several hours, it usually is not economical to use the process for less than about 3000 meters (10,000 feet) of product. When tubes or pipe are desired, a resistance welding unit is often combined with roll-forming equipment.

Seaming. *Seaming* is used to join ends of sheet metal to form containers such as cans, pails and drums. Figure 15-32 shows some of the most common types of seams that are used. The seams are formed by a series of small rollers on seaming machines that range from small hand-operated types to large automatic units capable of producing hundreds of seams per minute in the mass production of cans.

Flanging. *Flanges* can be rolled on sheet metal in essentially the same manner as seaming is done. In many cases, however, the forming of flanges and seams involves drawing, since localized bending occurs on a curved axis.

FIGURE 15-32. Various types of seams used on sheet metal.

FIGURE 15-33. Method of straightening rod or sheet by passing through a set of straightening rolls. For rod, another set of rolls provides straightening in the third dimension.

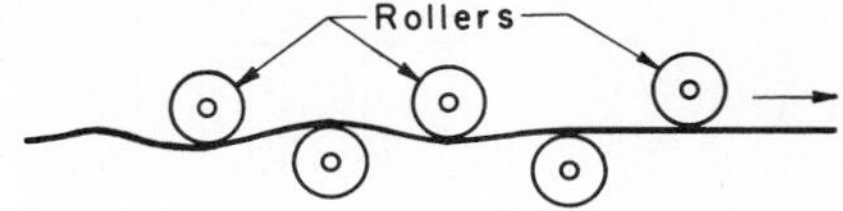

Straightening. *Straightening* or *flattening* has as its objective the opposite of bending and often is done before other cold-forming operations to assure that flat or straight material is available. Two different techniques are quite common. *Roll straightening* or *roller leveling,* illustrated in Figure 15-33, involves a series of reverse bends. The rod, sheet, or wire is passed through a series of rolls having decreased offsets from a straight line. These bend the metal back and forth in all directions, stressing it slightly beyond its previous elastic limit and thereby removing all previous permanent set.

Sheet may also be straightened by a process called *stretcher leveling.* As shown in Figure 15-34, the sheets are grabbed mechanically at each end and stretched slightly beyond the elastic limit to remove previous stresses and thus produce the desired flatness.

SHEARING

Shearing is the mechanical cutting of materials in sheet or plate form without the formation of chips or use of burning or melting. When the two cutting

FIGURE 15-34. Straightening sheets of brass by stretching, a process sometimes called stretcher leveling. (*Courtesy Scovill Manufacturing Company.*)

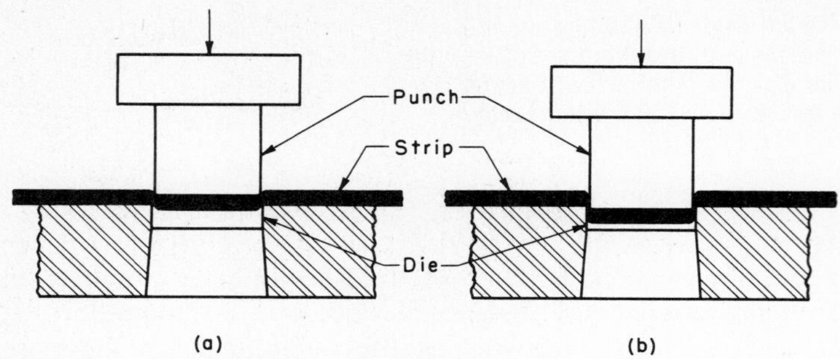

FIGURE 15-35. Basic mechanism of shearing by means of a punch and die.

blades are straight, the process is called shearing. Other processes, in which the shearing blades are in the form of the curved edges of punches and dies, are called by other names, such as *blanking, piercing, notching, shaving,* and *trimming.* These all are basically shearing operations, however.

The basic shearing process is illustrated in Figure 15-35. As the punch (upper blade) descends against the workpiece, as shown in the left-hand diagram, the metal is deformed plastically into the die (lower blade). Because the clearance between the punch and the die is only 5 to 10 per cent of the thickness of the metal being cut, the deformation is severely localized. The punch penetrates into the metal, and correspondingly on the opposite surface the metal bulges slightly and flows into the die. When the localized penetration reaches about 15 to 60 per cent of the thickness of the metal, depending on the ductility and strength, the applied stress exceeds the shear strength and the metal suddenly shears or ruptures through the remainder of its thickness. These two stages of the shearing process can often be seen on the edges of sheared parts and are clearly visible in Figure 15-36.

Because of the normal inhomogeneities in a metal and the lack of uniform clearance between the shearing blades, the final shearing does not occur uniformly. Fracture and tearing start at the weakest points and proceed progressively and intermittently to the next-stronger points, producing a somewhat rough sheared edge. If the punch and die or shearing blades have proper clearance and are maintained in good condition, sheared edges may be produced that are often sufficiently smooth for use without further finishing. The quality of sheared edges can be further improved if the strip stock is clamped firmly against the die from above, the clearance between the punch and die is reduced to a minimum, and the movement of the piece through the die (and thus the shearing) is controlled by resistance from a plunger acting upward against the workpiece under pressure, such as exerted by a rubber die cushion. These controls cause shearing to take place uniformly around the entire edge rather than randomly at the weakest points.

Research into the effect of superimposed pressure on the shearing process showed that as pressure increased, the relative amount of smooth edge in-

FIGURE 15-36. Conventionally sheared surface showing the two stages of fracture. (*Courtesy American Feintool Inc.*).

creased. Above a certain pressure, a 100 per cent smooth edge could be obtained. Figure 15-37 schematically presents the production extension of this work, designed to economically produce smooth and square sheared edges, a process known as *fineblanking*. A V-shaped protrusion is incorporated into the holddown or pressure plate lying slightly external to the contour of the cut. The holddown is first pressed into the plate being sheared and the protrusion places the region to be cut into a localized state of compression. When the punch starts its action, the compressed metal is held tightly against it throughout shearing and a virtually smooth and square edge results, as in Figure 15-38. Usually, less clearance between the punch and die is used, and a controlled upward-acting plunger, described previously, may be employed.

FIGURE 15-37. (*Left*) Method of obtaining a smooth edge in shearing by using a shaped pressure plate to put the metal into localized compression. (*Courtesy Metal Progress.*) (*Right*) Stock skeleton after shearing, showing the compression indentation. (*Courtesy Clark Metal Products Company.*)

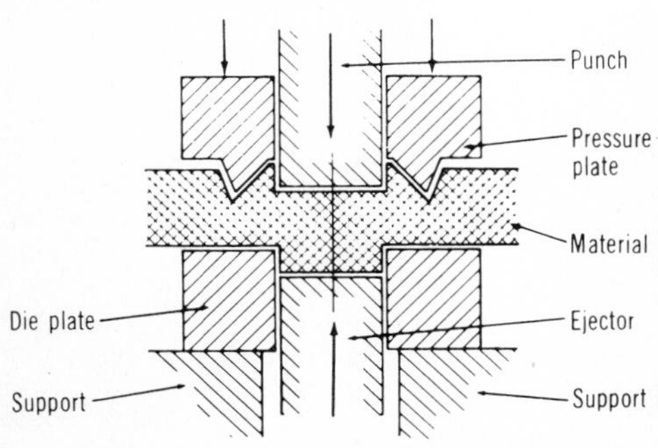

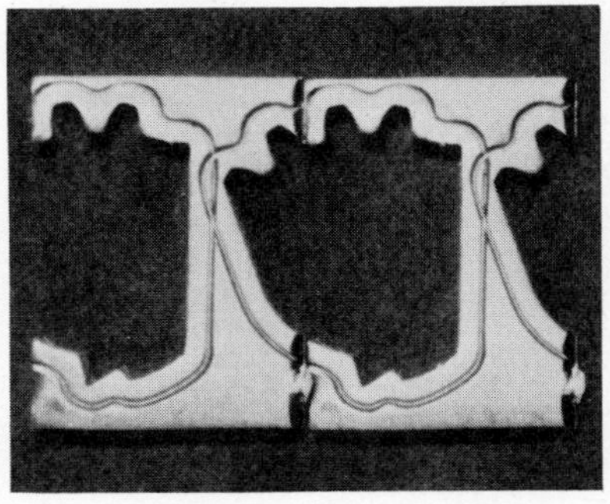

FIGURE 15-38. Fineblanked surface for the same component shown in Figure 15-36. (*Courtesy American Feintool Inc.*)

A similar, but somewhat simpler, technique using the same principle is illustrated in Figure 15-39. Incoming bar stock is pressed against the closed end of a feed hole, putting the stock in compression and thereby permitting the shearing of burr-free slugs for use in further processing.

When sheets of metal are to be sheared on a straight line, foot or power-operated *squaring shears* may be used, such as those illustrated in Figure 15-40. As the ram descends, a clamping bar or a set of clamping fingers presses the sheet of metal against the machine table to hold it in position. The moving blade then comes down across the fixed blade and shears the metal. On larger shears the moving blade is usually set at an angle or "rocks" as it descends, to make the cut progressively from one end to the other. This action reduces the amount of cutting force required, although the total energy expended is the same.

Slitting is the shearing process used to cut rolls of sheet metal into several rolls of narrower width. As shown in Figure 15-41, the shearing blades are in the form of circumferential mating grooves on cylindrical rolls, the ribs on one roll mating with the grooves in the other. The process is continuous and can be done rapidly and economically. Moreover, because the distance between adjacent sets of shearing edges is fixed, the width of the slit strips is very accurate and constant, more so than can be obtained by alternative procedures.

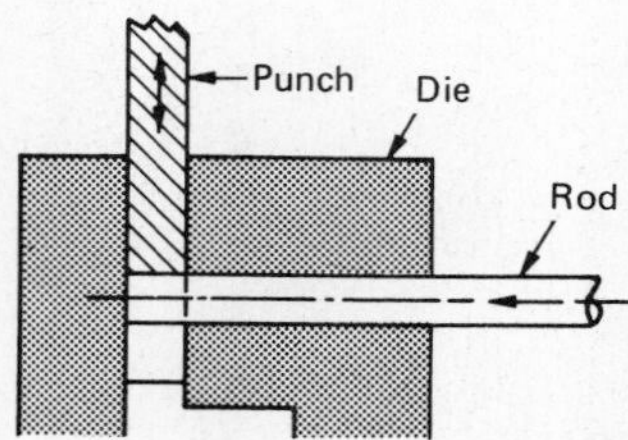

FIGURE 15-39. Method for smooth shearing rod by putting it into compression during shearing.

FIGURE 15-40. A 3-meter (10-foot) power shear for 6.4-mm (¼-inch) steel. (*Courtesy Cincinnati Incorporated.*)

FIGURE 15-41. Slitting a roll of sheet metal by means of slitting rolls. (*Courtesy The Yoder Company.*)

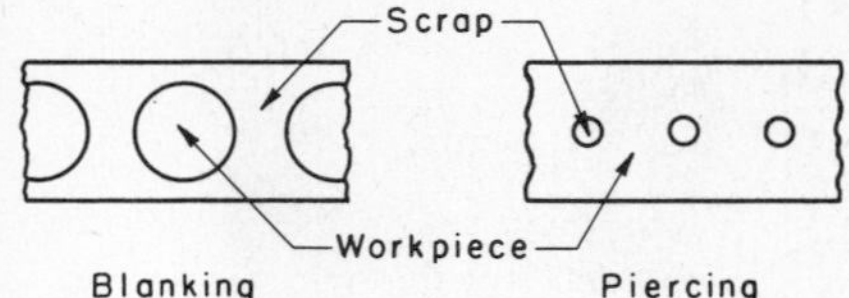

FIGURE 15-42. Difference between piercing and blanking.

Piercing and blanking. *Piercing* and *blanking* are shearing operations wherein the shearing blades take the form of closed, curved lines on the edges of a punch and die. They are basically the same cutting action, the difference being primarily one of definition, as shown in Figure 15-42. In blanking, the piece punched out is the desired workpiece, and major burrs or undesirable features should be left on the strip. In piercing, the piece punched out is the scrap and the remainder of the strip becomes the desired workpiece. Piercing and blanking are usually done by some type of mechanical press.

Several variations of piercing and blanking are used and have come to acquire specific names. *Lancing* is a piercing operation that may take the form of a slit in the metal or an actual hole as shown in Figure 15-43. The purpose of lancing is to permit adjacent metal to flow more readily in subsequent forming operations. In the case illustrated, the lancing makes it easier to form the grooves, which were shaped before the ashtray was blanked from the strip of stock and drawn to final shape.

Perforating consists of piercing a large number of closely spaced holes.

Notching is essentially the same as piercing except that the edge of the sheet of metal forms a portion of the periphery of the piece that is punched out. It is used to form notches of any desired shape along the edge of a sheet.

Nibbling is a variation of notching in which a special machine makes a series of overlapping notches, each further into the sheet of metal. As can be seen in Figure 15-44, the already sheared edge forms one end of the notch being cut. By repeating the procedure, any desired shape can be cut from sheets of metal up to about 6.5 mm (¼ inch) in thickness. Nibbling machines

FIGURE 15-43. Steps in making an ashtray, showing (*left to right*) piercing and lancing, blanking, and the final formed ashtray.

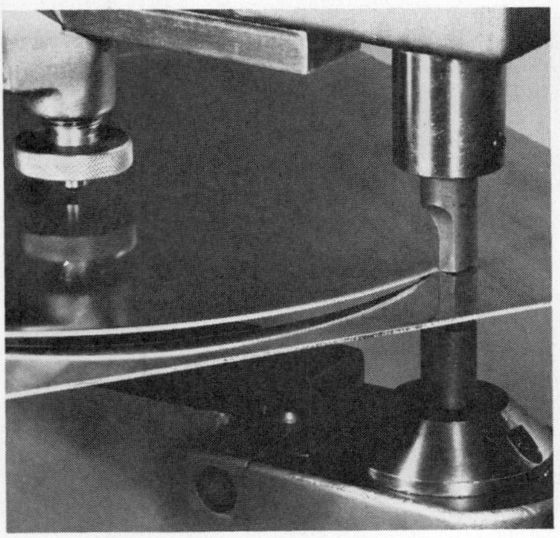

FIGURE 15-44. Operations being performed on a nibbling machine. (*Courtesy Tech-Pacific.*)

are simple, inexpensive, and versatile. Moreover, by starting the operation from a punched or drilled hole, the interior of a sheet can be cut away.

Shaving is a finishing operation in which a very small amount of metal is sheared away around the edge of a blanked part. Its primary use is to obtain greater dimensional accuracy, but it also may be used to obtain a square or smoother edge. Because only a very small amount of metal is removed, the punches and dies may be made with very little clearance. Parts, such as small gears, can be shaved to tolerances of 0.025mm (0.001 inch) after blanking.

Trimming is used to remove the excess metal that remains after a drawing, forging, or casting operation. It is essentially the same as blanking.

A *cutoff* operation is one in which a stamping is removed from a strip of stock by means of a punch and die. The cutoff punch and die cut across the entire width of the strip. Frequently, an irregularly shaped cutoff operation may simultaneously give the workpiece all or part of the desired shape.

Dinking is a modified shearing operation that is used to blank shapes

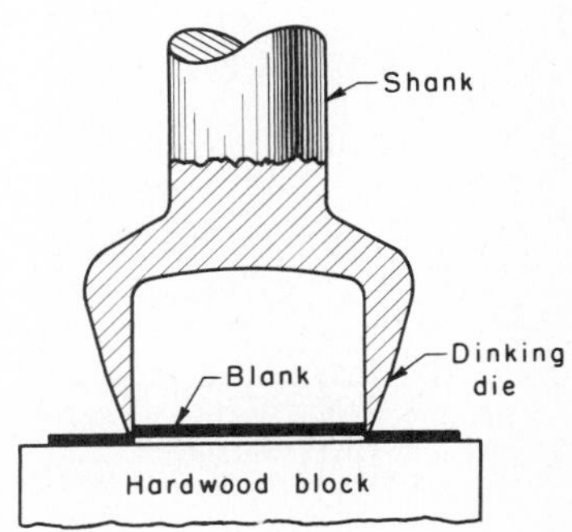

FIGURE 15-45. Dinking process.

from low-strength materials, primarily rubber, fiber, and cloth. The procedure is illustrated in Figure 15-45. The shank of the die can be struck with a hammer or mallet, or it can be operated by a press of some type.

Piercing and blanking dies. The basic components of piercing and blanking die sets, shown in Figure 15-46, are a *punch,* a *die,* and a *stripper plate.* Theoretically, the punch should be of dimensions such that it would just fit within the die with a uniform clearance approaching zero and, on its downward stroke, it should not enter into the die. In most practice, however, the clearance is from 5 to 12 per cent of the stock thickness, with 5 to 7 per cent being common, and the punch enters the die by a small amount. In operation, the punch and die should be maintained in alignment so that a uniform clearance is obtained around the entire perphery. The die is usually attached to the bolster plate of the press, which, in turn, is attached to the main press frame. The punch is attached to the movable ram, moving in and out of the die with each stroke of the press. Frequently, the punch and die are mounted on a *punch holder* and *die shoe* to form a die set such as is shown in Figure 15-47. The holder and shoe are permanently aligned and guided by two or more guide pins. Once a punch and die are correctly aligned and fastened to the die set, the unit can be inserted directly into a press without having to check the alignment, thereby reducing setup time. When a given punch and die are no longer needed, they may be removed and new ones attached to the shoe and holder assembly.

The punch face may be ground normal to the axis of motion, or it may have a slight angle, referred to by the term *shear.* Shear reduces the maximum cutting force required because all the periphery of the cut is not made at the same time. The amount of shear is limited to the thickness of the metal and frequently is less. One-half shear reduces the cutting load by about 25 per cent and full shear by 50 per cent. The total energy required is not changed, however.

FIGURE 15-46. Basic components in piercing and blanking dies.

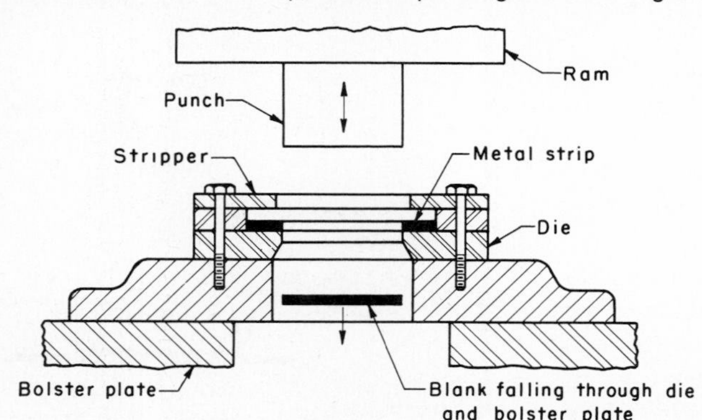

FIGURE 15-47. Typical die set having two guideposts. (*Courtesy Danly Machine Specialties, Inc.*)

The function of the stripper plate is to prevent the material from riding upward on the punch as it moves upward. It is located a sufficient distance above the die so that the sheet metal can easily slide between it and the die. The hole in the stripper plate is larger than the punch so there will be no friction between them.

In most cases, the punch holder of the die set is attached to the ram of the press so that the punch is raised, as well as lowered, by the press ram. On smaller die sets, a modern practice is for the punch and punch holder to be raised by springs in the die set. These small die-set units are known as *subpress dies*. The press ram simply contacts the top of the punch holder and forces it downward, the springs providing the return. This construction makes the set self-contained so that it can be put into and removed from a press quite rapidly and thereby reduce setup time. Numerous varieties of standardized, self-contained die sets have been developed that can be combined in various patterns on the bed of a press to produce large parts that would otherwise require large and costly die sets. One such setup is shown in Figure 15-48.

Punches and dies usually are made of nondeforming, or air-hardening, tool steel so they can be hardened after machining without danger of warpage. Beyond a depth of about 3 mm (⅛ inch) from the face, the die is usually provided with angular clearance or back relief to reduce friction between the part and the die and to permit the part to fall freely from the die after being sheared. The 3-mm land provides adequate strength and sufficient metal so that the die can be resharpened by grinding a few hundreths of a millimeter from its face.

Dies may be made in a single piece, which results in a basically simple, but costly, die, or made in sections that are fastened together on the die shoe. The latter procedure usually simplifies making the die and repairing it in case of damage, because only the broken piece must be replaced. Many standardized punch and die parts are available from which complex die sets can often

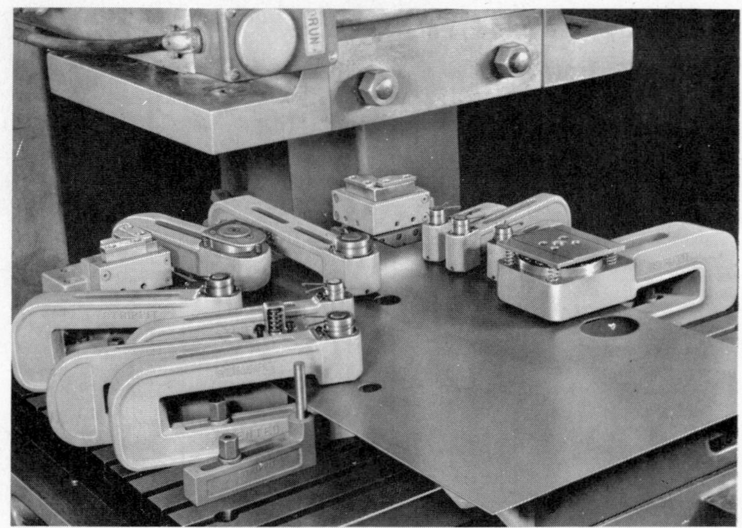

FIGURE 15-48. Typical setup for piercing and blanking using self-contained subpress tooling units. (*Courtesy Strippit Division of Houdaille Industries, Inc.*)

FIGURE 15-49. Progressive piercing, forming, and cutoff die set built up mostly from standard components. Part produced is shown below. (*Courtesy Oak Manufacturing Company.*)

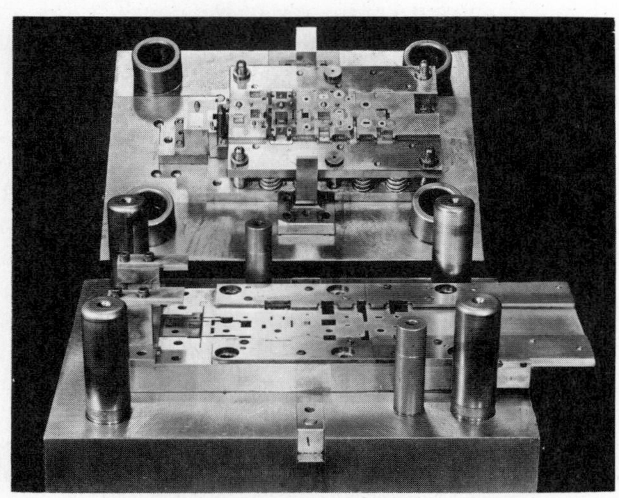

be constructed at greatly reduced cost. Figure 15-49 shows such a die set. Substantial savings can often be obtained if designers would determine what standard die components are available and modify the design of parts that are to be pierced and blanked so that such components can be utilized. An added advantage is that when the die set is no longer needed, the parts can be disassembled and used to construct another die set.

Another procedure that can be used to cut metal up to about 13 mm (½ inch) thick is to use "steel-rule" dies, as illustrated in Figure 15-50. Here the die is made from hardened, relatively thin steel strips that are mounted and supported on edge. A die plate takes the place of the conventional punch and holds the strips. Neoprene rubber pads take the place of the usual stripper plate, pushing the blank out from between the die strips. Such die sets are usually much less expensive to construct than solid dies and are thereby more feasible for small quantities of parts.

Only piercing or blanking can be done in the simple type of punch and die illustrated in Figure 15-46. Many parts require both piercing and blanking, and these operations can be combined in the two types of dies shown in Figures 15-51 and 15-52. Their operation may be understood by considering the steps required to form a simple, flat washer by piercing and blanking.

The *progressive die set,* such as depicted in Figure 15-51, is the simpler of the two types. Basically it consists of two or more sets of punches and dies mounted in tandem. The strip stock is fed into the first die where a hole is pierced when the ram closes on the first stroke. When the ram raises, the stock is moved over into position under the blanking punch. Positioning is

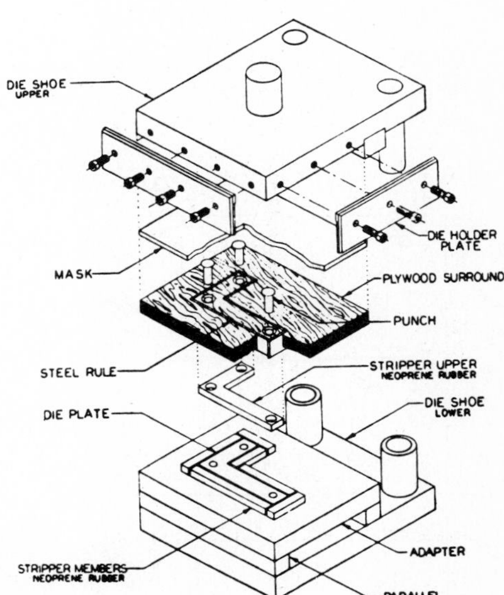

FIGURE 15-50. Construction of a steel-rule die set. (*Courtesy J. J. Raphael.*)

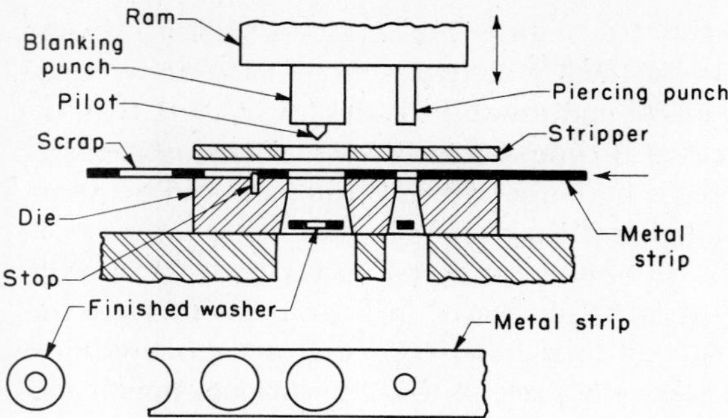

FIGURE 15-51. Progressive piercing and blanking die for making a washer.

accomplished automatically by a stop mechanism that engages a previously punched hole. As the ram descends on the second stroke, the pilot on the bottom of the blanking punch enters the hole that was pierced on the previous stroke to ensure accurate alignment. Further descent of the punch blanks the completed washer from the strip of stock and, at the same time, the first punch pierces the hole for the next washer. Another part is completed with each successive stroke of the press.

Progressive dies can be used for many variations of piercing, blanking, forming, lancing, drawing, and so forth. (An example is shown in Figure 15-43.) They have the advantage of being fairly simple to construct and are economical to repair because a broken punch or die does not necessitate the replacement of the entire set. However, if highly accurate alignment of the various operations is required, they are not as satisfactory as compound dies.

FIGURE 15-52. Method of making a simple washer in a compound piercing and blanking die. Part is blanked (a) and subsequently pierced (b). Blanking punch contains the die for piercing.

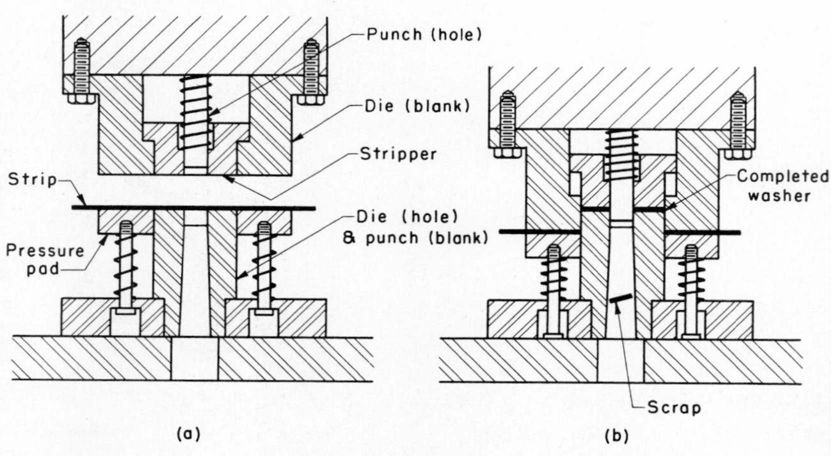

In *Compound dies,* such as shown in Figure 15-52, piercing and blanking, or other combinations, occur progressively within a single stroke of the ram while the strip of stock remains in one position. Dies of this type are more accurate, but they usually are more expensive to construct and are more subject to breakage.

Numerically controlled, turret-type punch presses, such as the one shown in Figure 39-30, are widely used for punching large numbers of holes in sheet metal components. In these, a variety, often up to 60, piercing punches and dies are contained in coordinated turrets and can quickly be rotated into operation as desired. The workpiece is moved into position through X–Y movements of the worktable.

Designing for piercing and blanking. The construction, operation, and maintenance of piercing and blanking dies can be greatly facilitated if designers of the parts to be fabricated keep a few simple rules in mind:

1. If solid dies are to be used, blank corners should be true radii whenever possible. Square corners are preferred for sectional dies.
2. The width of any projection or slot should be at least 1½ times the metal thickness and never less than 2½ mm (3/32 inch).
3. Diameters of pierced holes should not be less than the thickness of the metal or a minimum of 0.65 mm (0.025 inch). Smaller holes can be made, but with difficulty.
4. The minimum distance between holes, or between a hole and the edge of stock, should be at least equal to the metal thickness.
5. Keep tolerances as great as possible. Tolerances below about ±0.08 mm (±0.003 inch) mean that shaving will be required.
6. Arrange the pattern of parts on the strip stock to minimize scrap.

DRAWING

Cold drawing is a term that can refer to two somewhat different operations. If the stock is in the form of sheet metal, cold drawing is the forming of parts wherein plastic flow occurs over a curved axis. This is one of the most important of all cold-working operations because a wide range of parts, from small cups to large automobile body tops and fenders, can be drawn in a few seconds each. Cold drawing is similar to hot drawing, but the higher deformation forces, thinner metal, limited ductility, and closer dimensional tolerances create some distinctive problems.

If the stock is wire, rod, or tubing, "cold drawing" refers to the process of reducing the cross section of the material by pulling it through a die, a sort of tensile equivalent to extrusion.

Bar and tube drawing. One of the simplest cold-drawing operations, *bar* or *rod drawing,* is illustrated in Figure 15-53. One end of a bar is reduced or

FIGURE 15-53. Cold drawing of rods on a multiple-die bench; dies at left, grips in center, and draw chain at right. (*Courtesy Scovill Manufacturing Company.*)

pointed, inserted through a die of somewhat smaller cross section than the original bar, grasped by grips and pulled in tension, drawing the remainder of the bar through the die. The bars reduce in section, elongate, and strain-harden. The reduction in area per pass is usually 20 to 50 per cent, to avoid wire fracture, with several steps often being required to obtain a desired product. Intermediate annealing is often necessary to restore ductility and enable further working.

Tube drawing, depicted in Figure 15-54 is used to produce seamless tubing and is essentially the same as the hot drawing of tubes discussed in Chapter 14. Because it is drawn cold from descaled tubing, the product has smoother surfaces, thinner walls, and more accurate dimensions and than can be obtained by hot drawing. Mandrels are used for tubes from about 12.5 mm (½ inch) to 250 mm (10 inches) in diameter. Heavy-walled tubes and those less than 12.5 mm (½ inch) in diameter are drawn without a mandrel in a process known as *tube sinking.* This is the procedure used for drawing hypoder-

FIGURE 15-54. Method of cold drawing smaller tubing from large seamless tubing. (*Courtesy American Iron and Steel Institute.*)

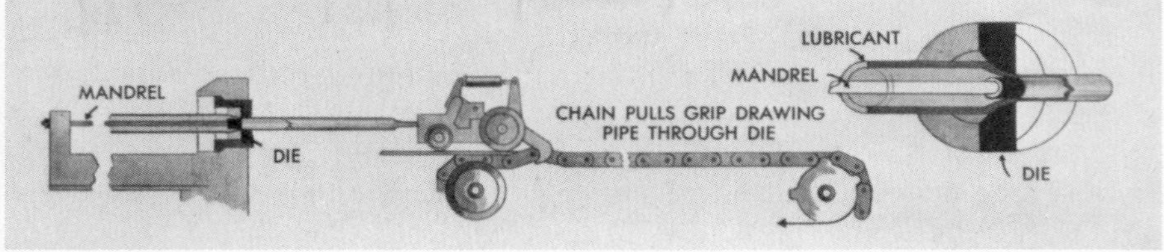

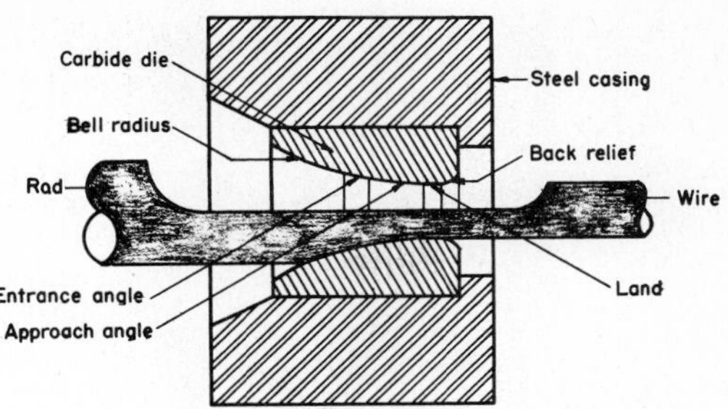

FIGURE 15-55. Cross section through a typical carbide drawing die, showing characteristic regions in the contour.

mic needles that may have an outside diameter as small as 0.2 mm (0.008 inch) and an inside diameter about half as large.

Wire drawing. *Wire drawing* is essentially the same as bar drawing except that it involves smaller diameters and is done as a continuous process through a succession of drawing dies. Large coils of hot-rolled material roughly 6.5 mm (¼ inch) in diameter are first descaled either by mechanical flexing or acid pickling and rinsing. The cleaned product is then further processed by immersion in a lime bath or other procedure to provide neutralization of remaining acid, a corrosion protection, and a carrier for surface lubricant. One end of the coil is pointed, fed through a die, and the drawing process begins. Dies have the configuration shown in Figure 15-55 and usually are made of tungsten carbide. Diamond dies are often used for drawing very fine wire. Lubrication boxes often precede the individual dies to assure coating of the material.

Small-diameter wire usually is drawn on tandem machines of the type shown in Figure 15-56, which contain 3 to 12 dies, each held in water-cooled die blocks. The reduction in each die is controlled so that each station uses about the same power. Speed is controlled at each station to avoid bunch-ups or excessive tension in the wire. For finer wire, intermediate anneals must be performed between various stages of drawing. Wires of a variety of tempers can be produced by controlling the placement of the last anneal in the process cycle. Full-soft wire is annealed in a controlled-atmosphere furnace after the final drawing.

Spinning. *Spinning* is a rather fascinating cold-working operation in which a rotating disk of sheet metal is drawn over a male form by applying localized pressure to the outside of the disk with a simple, round-ended wooden or metal tool or a small roller. The basic process is illustrated in Figure 15-57.

The form, or *chuck,* is rotated on a rapidly rotating spindle, often in a simple type of lathe. The disk of metal is centered and then held against the small end of the form by a follower attached to the tailstock of the lathe. As the disk and form rotate, the operator applies localized pressure against the metal, causing it to flow against the form, as shown in Figure 15-58. Because the final diameter of the formed part is less than that of the initial disk, thus

FIGURE 15-56. Wire being drawn on a machine having eight die blocks. (*Courtesy The Vaughn Machinery Company.*)

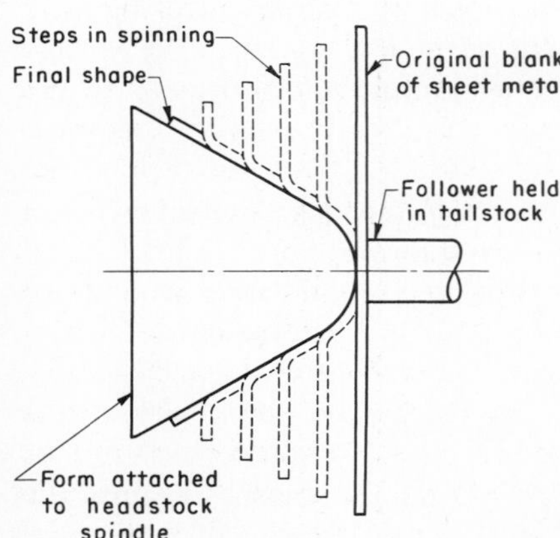

FIGURE 15-57. Progressive stages of forming a sheet of metal by spinning.

FIGURE 15-58. Two stages in spinning a metal reflector. (*Courtesy Spincraft Inc.*)

shortening the circumference, the operator must stretch the metal radially a corresponding amount to avoid circumferential buckling. (This is a shrink-forming operation, as explained later and illustrated in Figure 15-68.) This requires considerable skill on the part of the operator. Usually there is some thinning of the metal, but it can be kept to a very small percentage by a skilled worker.

Inasmuch as the metal is not pulled across it under pressure, the form block can often be made of hardwood; it is only essential that it have a smooth surface because any roughness will show in the finished part. Thus, the tooling cost is low, making the process economical for small quantities. The process also is used in many continuous-production applications, however, for making such parts as lamp reflectors, cooking utensils, bowls, and the bells of some musical instruments. If considerable numbers of identical parts are to be spun, a metal form block can be used.

Spinning is best suited for shapes that can be withdrawn directly from a one-piece chuck, but other shapes can be spun successfully by using multisection or offset chucks, such as is shown in Figure 15-59. Such shapes have also been spun successfully on form blocks made by freezing water, which is melted out after spinning is completed.

To retain the economic advantages of spinning but not require so much skill from the operator, spinning machines such as that in Figure 15-60 have been developed. The action of the spinning rollers is controlled automatically, being programmed for each particular part. Each successive part undergoes the same deformation sequence, resulting in better consistency and fewer rejects. Such machines also make possible the spinning of thicker metal. In some cases numerical control is used (see Chapter 39). The type of machine shown in Fig-

FIGURE 15-59. Method of spinning reentrant shapes using an offset chuck. (*Courtesy Aluminum Company of America.*)

ure 14-31 can also be used. On it, the spinning rolls are manipulated by hand through a power-assist system similar to that used in power steering on automobiles.

Shear forming. Shear forming or *flow turning,* illustrated in Figure 15-61, is a simplified variation of the spinning process in which the distance of each element in the blank from the axis of rotation remains constant. Because of this

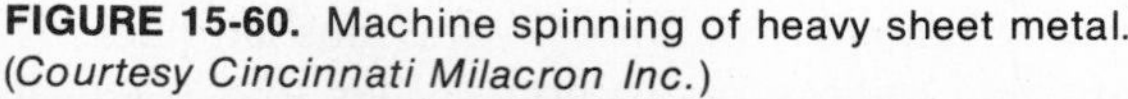

FIGURE 15-60. Machine spinning of heavy sheet metal. (*Courtesy Cincinnati Milacron Inc.*)

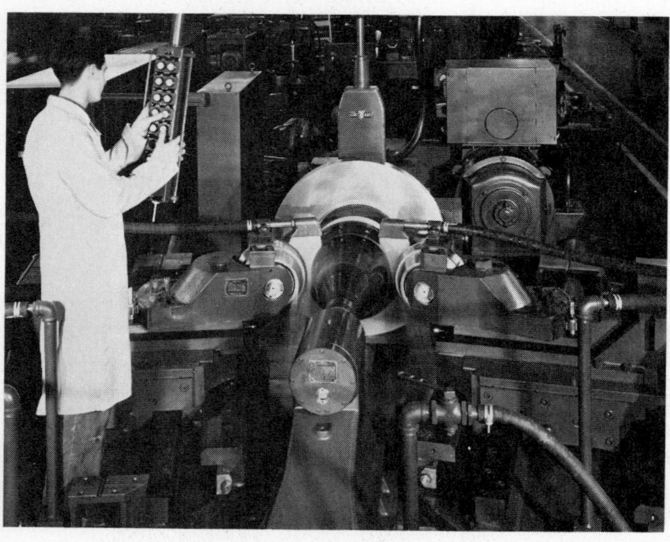

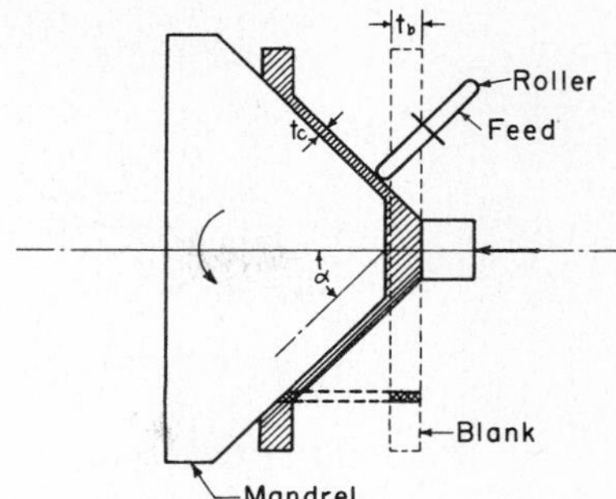

FIGURE 15-61. Basic shear-forming process.

fact, the metal flow is entirely in shear (except in the case of cylinders where it is essentially so), and no radial stretch has to take place to compensate for the circumferential shrinkage that occurs in ordinary spinning. As a consequence, conical, hemispherical, and cylindrical shapes, and modifications of them, can be spun more readily by this process, particularly by mechanized means, than by the normal spinning procedure. Wall thickness will vary with the angle of the region. As shown in Figure 15-61, for conical parts, the relationship between the wall thickness of the final part and that of the blank is $t_c = t_b \sin \alpha$. If α is less than 30°, it may be necessary to complete the forming in two stages with an intermediate anneal between. Reductions in wall thickness as high as 8:1 are possible, but the limit more generally is about 5:1, or an 80 per cent reduction.

In shear-forming straight cones, the wall thickness of the finished part will be uniform if a blank of uniform thickness is used. By using a blank of varying thickness, a straight cone of varying wall thickness can be obtained, or a wall of uniform thickness can be produced in shapes that do not have straight conical sides.

Conical shapes usually are shear-formed by the *direct process* depicted in Figure 15-61. They can also be made, however, by the *reverse process,* illustrated in Figure 15-62. By varying the direction of feed of the forming rollers

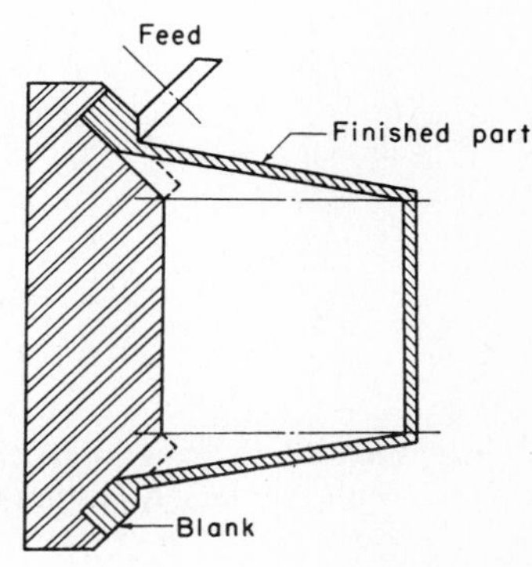

FIGURE 15-62. Forming a conical part by reverse shear spinning.

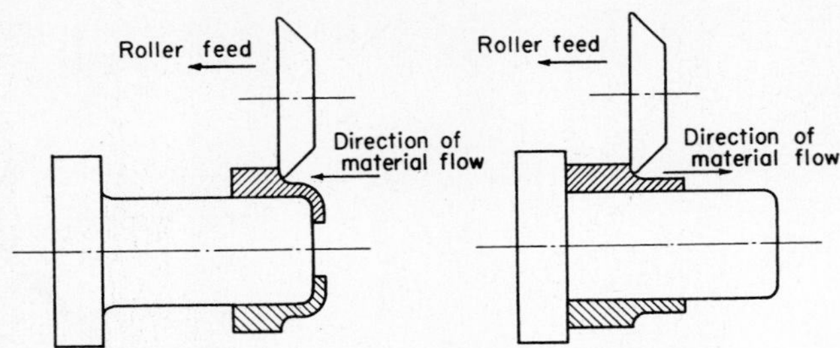

FIGURE 15-63. Shear forming a cylinder by the direct process (*left*) and the reverse process (*right*).

in the reverse process, it is possible to form convex or concave parts without the necessity of having a matching mandrel.

Shear-formed cylinders can be made by either the direct or indirect processes, as shown in Figure 15-63. The reverse process has the advantage that a cylinder can be formed that is longer than the mandrel. When long, thin-walled cylinders are desired, a "flo-reform" process, illustrated in Figure 15-64, can be used. The process is a combination of shear forming and conventional spinning, the first steps involving no change in the diameter of portions of the workpiece, while such a change does occur in the last step. Figure 15-65 shows the type of equipment commonly used in shear forming and a blank being formed into a cone.

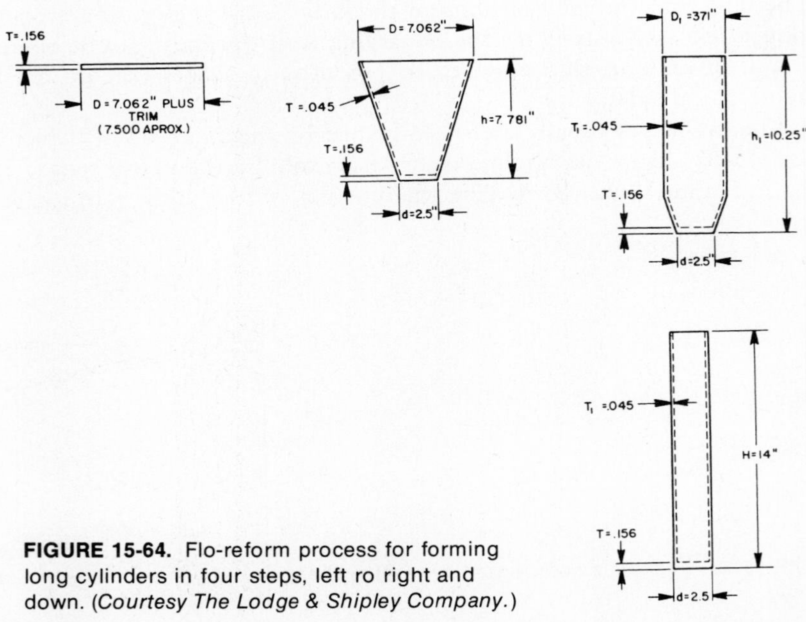

FIGURE 15-64. Flo-reform process for forming long cylinders in four steps, left ro right and down. (*Courtesy The Lodge & Shipley Company.*)

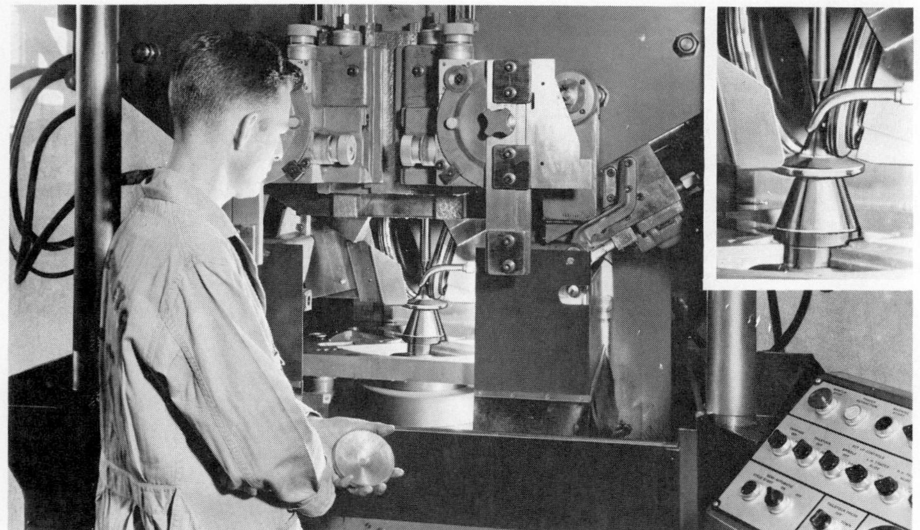

FIGURE 15-65. Cone being shear formed in a Floturn machine. Note blank in operator's hand and action of rollers, shown in inset. (*Courtesy The Lodge & Shipley Company.*)

Stretch forming. *Stretch forming,* the principle of which is illustrated in Figure 15-66, was developed by the aircraft industry to permit certain sheet-metal parts, particularly large ones, to be formed economically in small quantities. As shown, only a single male form block is required. The sheet of metal is gripped by two or more sets of jaws that stretch it and wrap it around the form block as the latter raises upward. Various combinations of stretching, wrapping, and upward motion of the block are employed, depending on the shape of the part. Figure 15-67 shows a typical part being formed by this process.

Through proper control of the stretching, most or all of the compressive stresses that accompany normal bending and forming are eliminated. Consequently, there is very little springback, and the workpiece conforms closely to the shape of the form block. Because the form block is almost completely in compression, it can be made of wood, Kirksite, or sprayed or laminated

FIGURE 15-66. Schematic representation of the motions and steps involved in stretch-wrap forming.

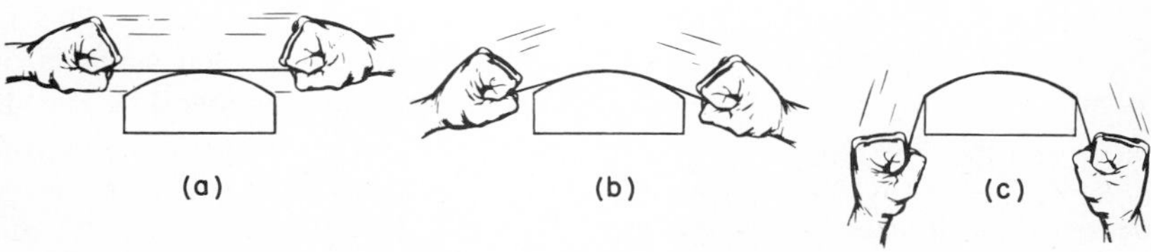

FIGURE 15-67. Large aircraft part being formed by stretch-wrap forming. (*Courtesy Hufford Machine Works.*)

plastics. The process is used in making cowlings, wing tips, scoops, and large aircraft panels out of aluminum or stainless steel, and large automobile and truck panels from low-carbon steel. *Stretch-wrap forming* is another name for the process.

Stretch-draw forming is essentially the same process except that mating male and female dies shape the stretch metal.

Shell or deep drawing. The drawing of closed cylindrical or rectangular containers, or a variation of these shapes, with a depth frequently greater than the narrower dimension of their opening, is one of the most important and widely used manufacturing processes. Because the process had its earliest uses in manufacturing artillery shells and cartridge cases, it is sometimes called *shell drawing.* When the depth of the drawn part is less than the diameter, or minimum surface dimension, of the blank, the process is considered to be *shallow drawing.* If the depth is greater than the diameter, it is considered to be *deep drawing.*

Basically there are two types of drawing, as illustrated in Figure 15-68, both involving multiaxial or curved-axis bending. As noted in Table 13-2, the stress conditions are complex. In *shrink forming,* there is circumferential compression during drawing resulting from the decrease of diameter d_1 to d_1', and the metal tends to thicken. Because the metal is thin, there is a tendency to relieve the circumferential compression by buckling or wrinkling. This tendency must be prevented by compressing the sheet between flat surfaces during forming and causing a controlled compensating flow of metal as the shell is drawn. In *stretch forming,* on the other hand, there is a thinning of the metal as a result of the circumferential stretching that must occur in order for the diameter d_2 to increase to d_2'. This can lead to tearing of the metal. In many

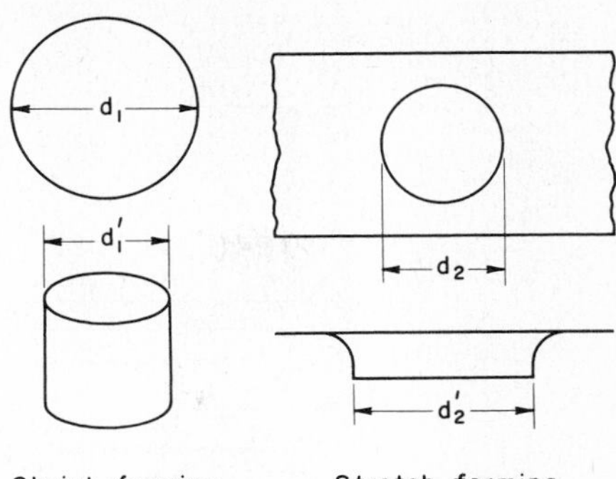

FIGURE 15-68. Two basic types of drawing operations.

cases drawn parts contain regions of both shrink and stretch forming. This, of course, presents complex problems, and designers should recognize that drawn parts with regions of large shrink or stretch will cost more than those that avoid the associated difficulties.

The design of complex parts that are to be drawn has been aided considerably by computer techniques, but is far from being completely and successfully solved. Consequently, such design still involves a mix of science, experience, empirical data, and actual experimentation. The body of known information is quite substantial, however, and is being used with outstanding results.

To avoid wrinkling and variations in thickness, the flow of metal must be controlled. This is usually accomplished by some type of pressure ring or pad. In single-action presses, as shown in Figure 15-69, where there is only one movement of the slide, spring or air pressure is used to control the flow of

FIGURE 15-69. Method of deep drawing on a single-action press.

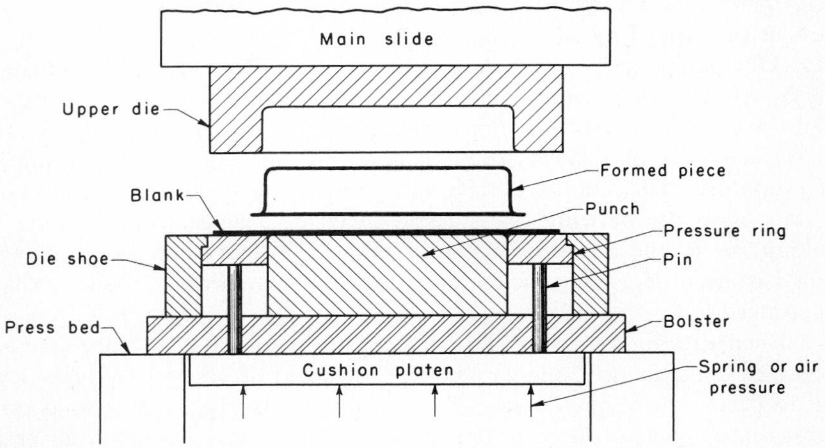

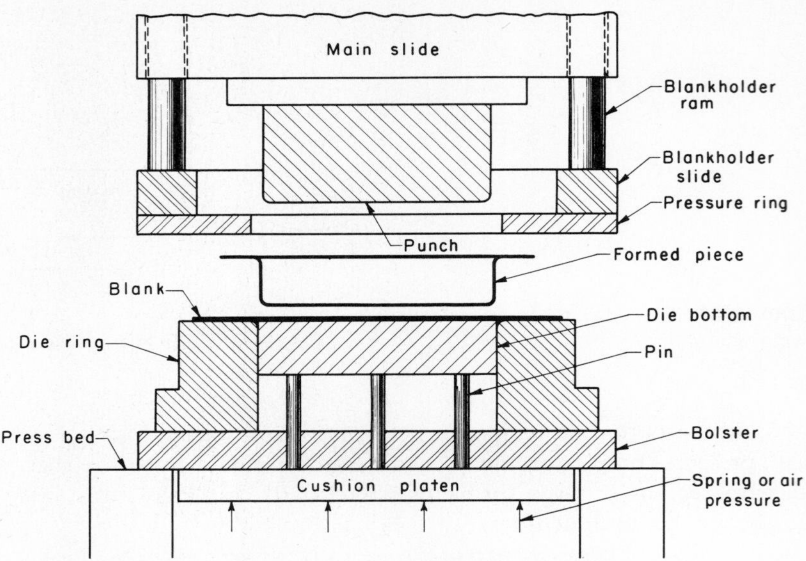

FIGURE 15-70. Drawing dies for use on a double-action press.

metal between the upper die and the pressure ring. When double-action presses are used, with two or more independent rams, as illustrated in Figure 15-70, the force applied to the pressure ring can be controlled independent of the position of the main slide. This permits the pressure to be varied as needed during the drawing operation. For this reason, double-action presses are usually used for drawing more complex parts, whereas single-action presses are often satisfactory for the more simple types of operations. Figure 15-71 shows a large, double-action press in operation.

Because the flow of metal is often not uniform throughout the workpiece, many drawn parts have to be trimmed, after forming, to remove the excess or undersired metal. Figure 15-72 shows a part before and after trimming. Obviously, such trimming adds to the cost, because it must either be done by hand-guided operations or by use of a separate and special trimming die. In many cases trimming after drawing can be avoided by using a blank that has been cut to a special shape before drawing. Figure 15-73 illustrates this procedure. This approach, of course, requires a special-shaped blanking die, but often the drawing blank is produced by blanking and making the blanking die to the desired shape may involve no extra cost. The choice of method is often dependent on the complexity of the part and the quantity to be produced.

When dies for blanking and drawing are made of steel, they are quite expensive. Consequently, numerous procedures and processes have been developed to permit these operations to be done with less expensive tooling, or to permit more extensive deformation to be done with fewer sets of dies or less

FIGURE 15-71. Double-action hydraulic press being used for drawing sheet metal. Inset shows part being removed. (*Courtesy The Hydraulic Press Manufacturing Company.*)

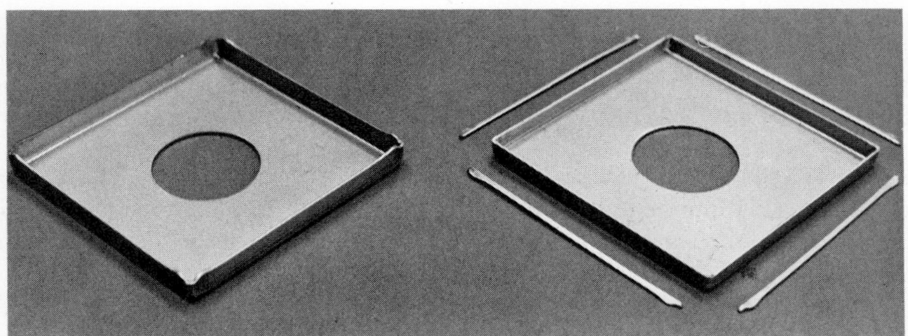

FIGURE 15-72. Pierced, blanked, and formed part before and after trimming.

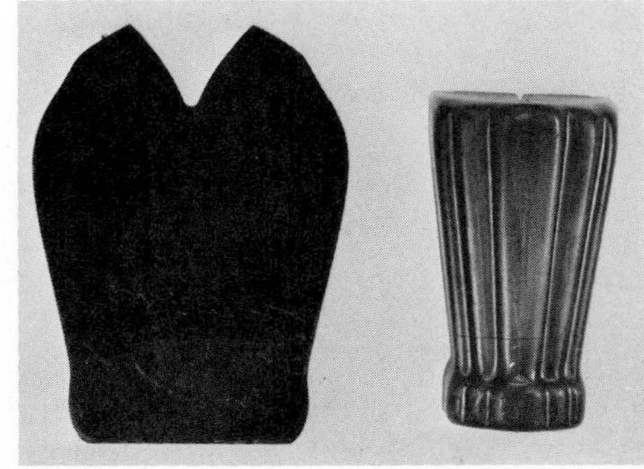

FIGURE 15-73. Irregularly shaped blank and finished drawn stove leg. No trimming is necessary.

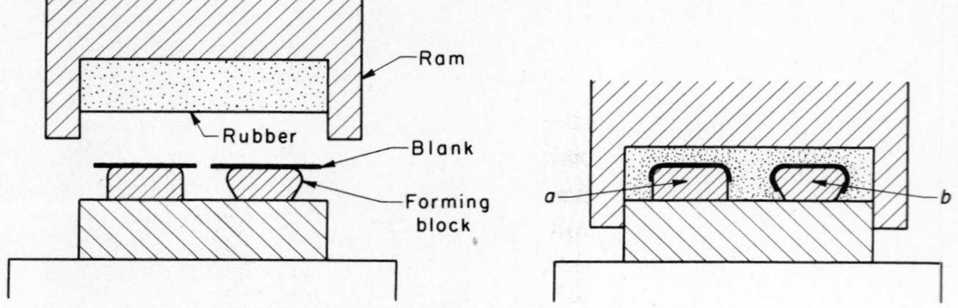

FIGURE 15-74. Guerin process for forming sheet metal.

in-process heat treating. Even though most of these processes have certain limitations as to the shapes or types of metal that can be formed, they are very useful where they can be applied. Some of them will now be discussed.

Forming with rubber or fluid pressure. Several methods of forming employ rubber or fluid pressure to obtain the desired deformation, and thereby eliminate either the male or female member of the die set. The *Guerin process,* depicted in Figure 15-74, utilizes the phenomenon that rubber of the proper consistency, *when totally confined,* acts essentially as a fluid and will transmit pressure uniformly in all directions. Blanks of sheet metal are placed on top of form blocks, which usually are made of wood. The upper ram, which contains a pad of rubber 200 to 250 mm (8 to 10 inches) thick in a steel container, then descends. The rubber pad is confined and transmits force to the metal, causing it to bend to the desired shape. Since no female die is used and form blocks replace the male die, die cost is quite low. Process flexibility is quite high (different shapes can even be formed at the same time), wear on material and tooling is low, and workpiece surface quality is easily maintained. When reentrant sections are formed, as in *b* in Figure 15-74, it must be possible to slide the parts lengthwise from the form blocks or disassemble the form block from within the product.

Guerin forming was developed in the aircraft industry, where small numbers of duplicate parts often must be formed, thus favoring low tooling cost. It can be used for aluminum up to about 3 mm (⅛ inch) thick and

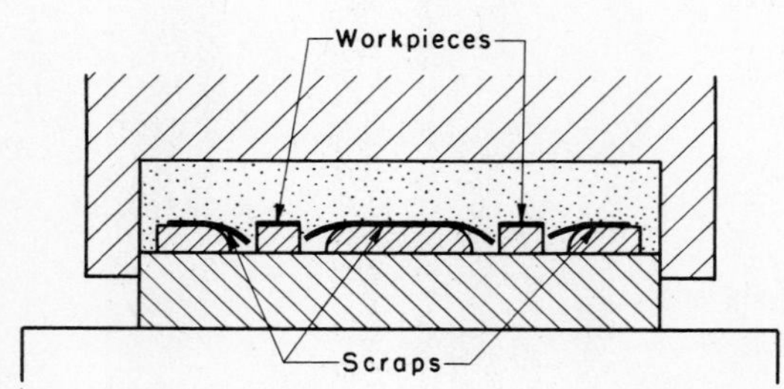

FIGURE 15-75. Method of blanking sheet metal by the Guerin process.

1.6 mm ($^1/_{16}$-inch) stainless steel. Magnesium can also be formed if it is heated and heated form blocks are used.

Most forming done by the Guerin process is multiple-axis bending, but some shallow drawing can be done. It is also used extensively for piercing and blanking thin gages of aluminum, as illustrated in Figure 15-75. For this purpose, the blanking blocks, shaped the same as the desired workpiece, have a face, or edge, made of hardened steel. Round-edge supporting blocks are spaced a short distance from the shearing (blanking) blocks to support the scrap skeleton and permit the metal to bend away from the shearing edges.

The *Hydroform process* or "rubber bag forming" replaces the rubber pad with a flexible diaphragm backed by controlled hydraulic pressure. Deeper parts can be formed with truly uniform fluid pressure, as illustrated in Figure 15-76.

In *bulging,* oil or rubber is used for applying an internal bulging force to

FIGURE 15-76. Hydroform process showing (1) blank in place, no pressure in cavity, (2) press closed and cavity pressurized, (3) ram advanced with cavity maintaining fluid pressure, and (4) pressure released and ram retracted. (*Courtesy The Aluminum Association.*)

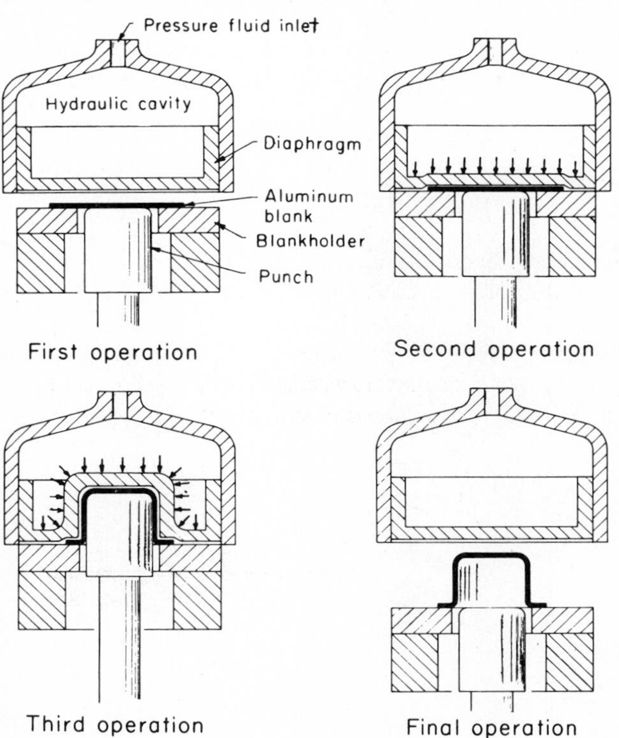

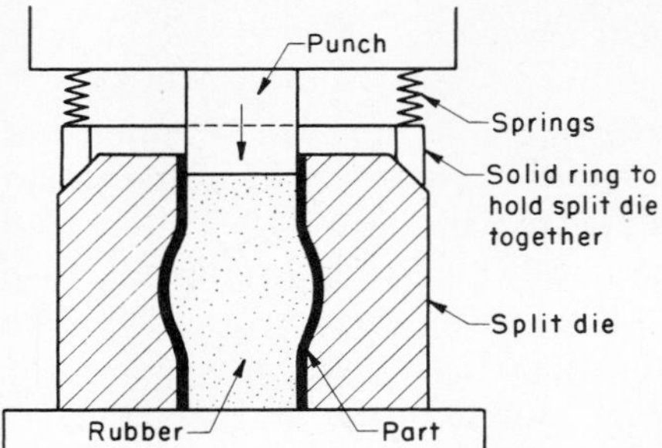

FIGURE 15-77. Method of bulging using rubber tooling.

expand a metal blank or tube outward against a female mold or die, and thereby eliminates the necessity for a complicated, multiple-piece male die member. Split female dies are often used to facilitate product removal. Complex equipment is used in the fluid technique to provide the necessary seals, yet enable easy input and removal. For less complicated shapes, rubber can replace the internal fluid, as illustrated in Figure 15-77.

Drawing on a drop hammer. Where small quantities of shallow-drawn parts are required, they can often be made most economically through the use of Kirksite dies and a rope-type drop hammer, as illustrated in Figure 15-78. The Kirksite dies can be cast, thus avoiding the expense of costly machined steel dies, and they can be melted down and cast into other shapes when no

FIGURE 15-78. Drawing of aluminum on a drop hammer using Kirksite dies. (*Courtesy Boeing Company.*)

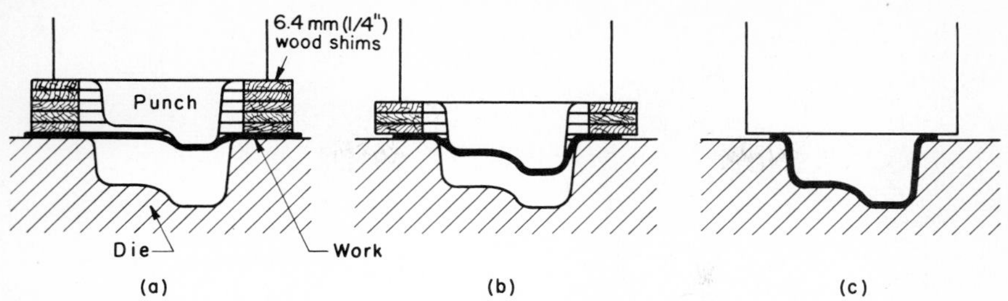

FIGURE 15-79. Method of deep drawing on a drop hammer using Kirksite dies and a book of shims.

longer needed. Because the raising and lowering of the ram is obtained through a rope passing several times around a rotating shaft, very good control can be obtained by manually applying and releasing tension on the end of the rope. The dies are permitted to close only partially on the early strokes, thereby forming the metal gradually. Wrinkles that form can be hammered out between strokes by means of a hand mallet. In many cases, the modification illustrated in Figure 15-79 is employed. A stack of shims, made of thin plywood, is placed on top of the sheet of metal to permit the ram to descend only partway. At the same time, they act as a pressure pad to restrict and control the flow of metal and thus inhibit wrinkling. One shim is withdrawn after each stroke of the ram, permitting the ram to fall farther and deepen the draw on each successive stroke.

Drawing on a drop hammer is considerably cruder than drawing with steel dies, but it often is the most economical method for small quantities. It is most suitable for aluminum alloys, but thin carbon and stainless sheets can also be drawn successfully.

High-energy-rate forming. A number of methods have been developed for forming metals through the release and application of large amounts of energy in a very short interval. These processes are called *high-energy-rate forming processes,* commonly designated HERF. Many metals tend to deform more readily under the ultrarapid rates of load application used in these processes, a phenomenon apparently related to the relative rates of load application and the movement of dislocations through the metal. As a consequence, HERF makes it possible to form large workpieces and difficult-to-form metals with less-expensive equipment and tooling than would otherwise be required.

Another advantage of HERF is that there is less difficulty due to springback. This is probably associated with two factors: (1) high compressive stresses are set up in the metal when it is forced against the die, and (2) some slight elastic deformation of the die occurs under the high pressure, resulting in the workpiece being slightly overformed and thereby giving the appearance of no springback.

The high energy-release rates are obtained by five methods: (1) un-

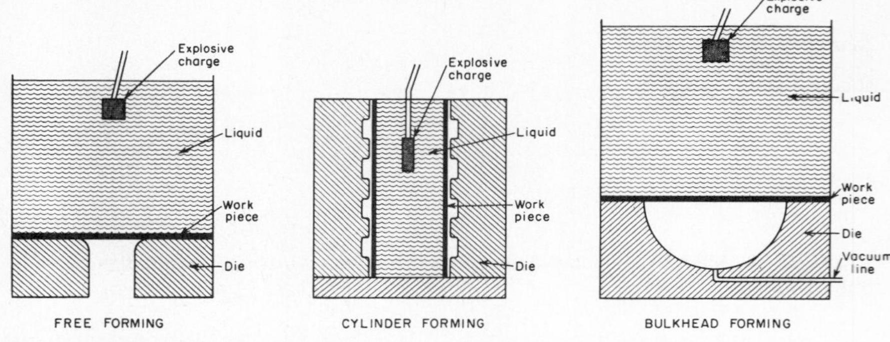

FIGURE 15-80. Three methods of high-energy-rate forming with explosive charges. (*Courtesy Materials Engineering.*)

derwater explosions, (2) underwater spark discharge (electrohydraulic techniques), (3) pneumatic-mechanical means, (4) internal combustion of gaseous mixtures, and (5) the use of rapidly formed magnetic fields (electromagnetic techniques).

Figure 15-80 illustrates three commonly used procedures involving the use of *explosive charges*. Although these procedures can be used for a wide range

FIGURE 15-81. Tailpipe section formed from 1.6 mm (0.063 inch) high-temperature alloy by high-energy-rate forming and the ductile iron die in which it was formed. The diameter tolerance is 0.05 mm (0.002 inch). (*Courtesy Ryan Aeronautical Company.*)

FIGURE 15-82. Elliptical dome 3 meters (10 feet) in diameter being removed from explosive forming die. (*Courtesy NASA.*)

of parts, as seen by the example of Figure 15-81, they are particularly suitable for large parts or thick material as in the 3-meter (10-foot)-diameter elliptical dome of Figure 15-82. Only a tank of water in the ground with about 2 meters (6 feet) of water above the workpiece is required, along with a female die that can be made of inexpensive materials such as wood, plastic, or Kirksite.

The *spark-discharge method,* shown in Figure 15-83, uses the energy of an electrical discharge to shape the metal. Electrical energy is stored in large capacitor banks and then released in a controlled discharge, either between two electrodes or across an exploding bridgewire submerged in a medium. High-energy shockwaves propagate from the discharge and deform the metal. The initiating wire can be preshaped, and shock-wave reflectors can be used to adapt the process to a variety of components. The space between the blank and the die is usually evacuated before the discharge occurs, removing the possibility of puckering due to inescapable air.

The spark-discharge methods are most often used for bulging operations in small parts, such as those in Figure 15-84, but parts up to 1.3 meters (50 inches) in diameter can be formed. Compared to explosive forming, the dis-

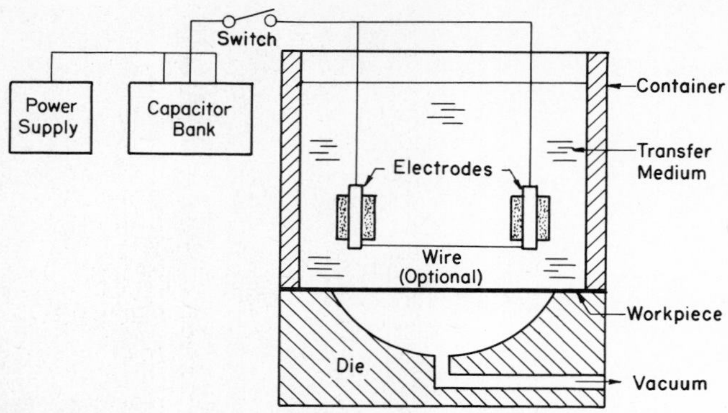

FIGURE 15-83. Components in the spark-discharge method of forming.

charge techniques are much easier and safer, do not require as large tanks, and do not have to be performed in remote areas.

To make the HERF techniques adaptable for rapid use within a plant, the *pneumatic-mechanical* and *internal-combustion* presses were developed. In the pneumatic-mechanical presses, such as shown in Figure 15-85, one portion of the forming die is attached to the stationary bolster of the press bed and the other to a piston rod. Low-pressure gas acts on the entire area of the piston in a cylinder, holding it up against a seal such that only a small area on the other side of the piston is exposed to high-pressure gas. The small and large areas are balanced so that when the pressure of the high-pressure gas is increased above a certain value, the seal is broken, exposing the toral area of the piston to high pressure. The piston is driven downward very rapidly, bringing the dies together.

Basically, the internal combustion presses operate on the same principle as an automobile piston and cylinder. A gaseous mixture is exploded within a cylinder, causing the piston to be driven downward very rapidly, one of the die members being attached to its lower end. This type of press can produce ram velocities up to 15 meters per second (50 feet per second) and up to 60 strokes per minute. Either single or repeated strokes can be obtained.

Electromagnetic forming is based on the principle that the electromagnetic field of an induced current always opposes the electromagnetic field of the inducing current. A capacitor is discharged through a coiled conductor that is either within or surrounding a cylinder or adjacent to a flat sheet of metal that is to be formed. This induces a current in the workpiece, causing it to be repelled from the coil and to be deformed against a die or mating workpiece. The process is very rapid and is useful for expanding or contracting tubing to various shapes, coining, forming, swaging, and permanently assembling component parts. Figure 15-86 shows some typical components.

FIGURE 15-84. Stainless steel parts formed by spark-discharge techniques. (*Courtesy General Dynamics, Fort Worth Division.*)

FIGURE 15-85. Pneumatic-mechanical press for high-energy-rate forming. (*Courtesy Clearing Division of U.S. Industries, Inc.*)

FIGURE 15-86. (*Top*) Blower rotor assembled by the Magneform process. (*Bottom*) Two parts formed by the Magneform process. Both are approximately 125 mm (5 inches) in diameter. (*Courtesy General Atomic Division of General Dynamics Corporation.*)

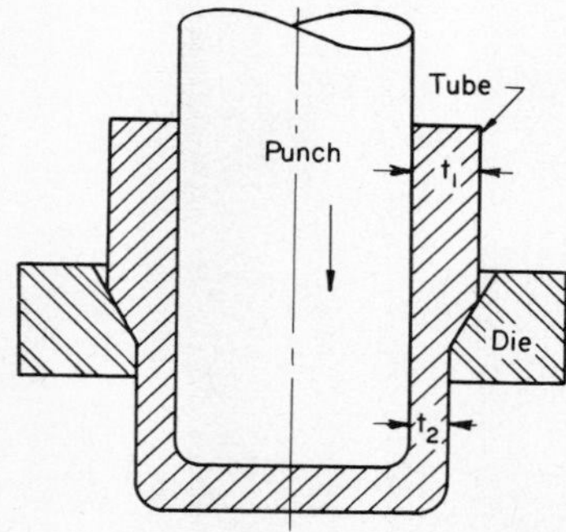

FIGURE 15-87. Ironing process in schematic.

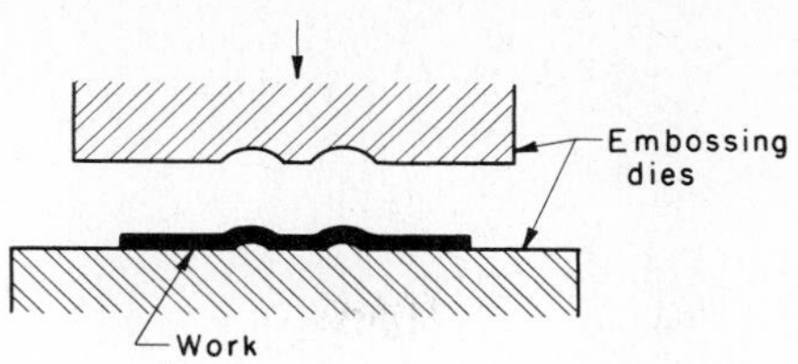

FIGURE 15-88. Embossing.

Ironing. *Ironing* is the name given to the process of thinning the walls of a drawn cylinder by passing it between a punch and a die where the separation is less than the original wall thickness. The walls are elongated and thinned while the base remains unchanged, as shown in the schematic of Figure 15-87. The most common example of an ironed product is the thin-walled all-aluminum beverage can.

Embossing. *Embossing,* shown in Figure 15-88, is a method for producing lettering or other designs in thin sheet metal. Basically, it is a very shallow drawing operation, usually in open dies, with the depth of the draw being from one to three times the thickness of the metal.

PRESSES

Classification of presses. Many types of presses have been developed to perform the variety of cold-working operations. When selecting a press for a given application, consideration should be given to the capacity required, whether the power source should be *hydraulic* or *mechanical*, and the method of transmitting power to the ram (the type of drive). The table below indicates the various types available and Figure 15-89 illustrates the more important drive mechanisms. In general, mechanical drives provide faster action and more positive displacements than do hydraulic, whereas greater forces can be obtained more readily with hydraulic drives, with more flexibility.

Foot	Mechanical	Hydraulic

Kick presses	Crank
	  Single
	  Double
	Eccentric
	Cam
	Knuckle joint
	Toggle
	Screw
	Rack and pinion
		Single-slide
		Multiple-slide

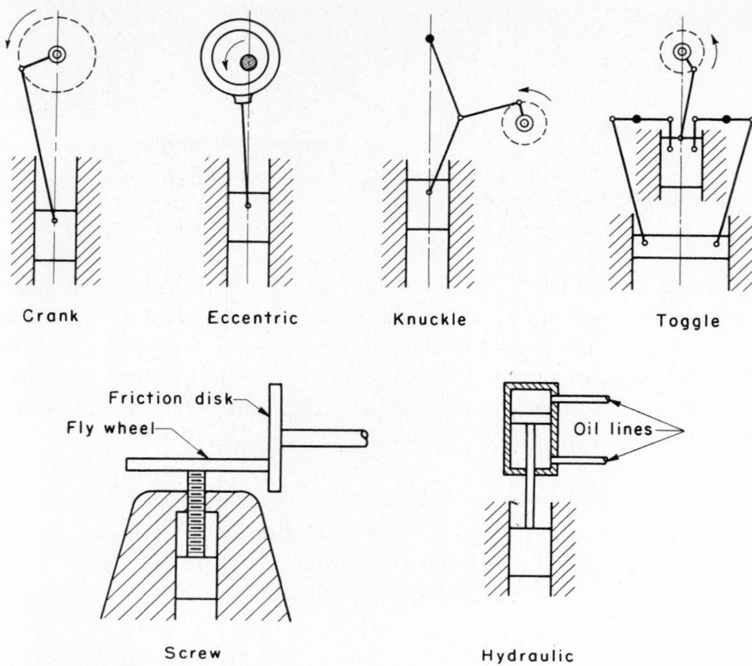

FIGURE 15-89. Schematic representation of the various drive mechanisms for presses.

Foot-operated presses, commonly called *kick presses,* are used only for very light work. *Crank-driven presses* are the most common type because of their simplicity. They are used for most piercing and blanking operations and for simple drawing. Double-crank presses provide a method of actuating blank holders or operating multiple-action dies. *Eccentric* or *cam drives* are used where only a short ram stroke is required. Cam action can provide a dwell at the bottom of the stroke and are sometimes used to actuate the blank holding ring in deep-drawing processes. *Knuckle-joint drives* provide a very high mechanical advantage along with fast action. They are often used in coining, sizing, and Guerin forming. *Toggle mechanisms* are used principally in drawing presses to actuate the blank holder, because they provide two sets of motions. The press shown in Figure 15-90 is of this type.

Hydraulic presses are available in many varieties and sizes. Because almost unlimited capacities can be provided, most large drawing presses are of this type. By using several hydraulic cylinders, programmed loads can be applied to the ram and any desired force and timing can be applied independently to the blank holder. Figure 15-71 shows a typical hydraulic press. Although most hydraulic presses tend to be relatively slow, types are available that provide up to 600 strokes per minute for high-speed blanking operations.

FIGURE 15-90. Toggle press. (*Courtesy E. W. Bliss Company.*)

Press frame types. Another matter of importance in selecting a press is the type of frame, because this may impose limitations on the size and type of work that can be done. The following is a classification according to frame type:

Arch	Gap	Straight-sided

Crank or eccentric
Percussion
	Foot
	Bench
	  Vertical
	  Inclinable
	Inclinable
	Open back
	Horn
	Turret
		Many variations,
		  but all with
		  straight-sided frames

FIGURE 15-91. Inclinable gap-frame press with sliding bolster to accommodate two die sets for rapid change of setup. (*Courtesy Niagara Machine & Tool Works.*)

Arch-frame presses, having their frames in the shape of an arch, are seldom used today, except with screw drives for coining operations.

Gap-frame presses, having a C-shaped frame as shown in Figure 15-91, provide good clearance for the dies and permit large stock to be fed into the press. They are made in a wide variety of sizes, capacity ranging from 8.9 to over 890 KN (1 ton to 100 tons). Often, they can be inclined to permit the completed work to drop out of the back side of the press, as in Figure 15-91. Also included in that press is a sliding bolster plate, a common feature that permits a second die to be set up on the press while another is in operation. Die changeover then requires only a few minutes to unclamp the punch segment of one die set from the ram, move the second die set into position, clamp the upper segment of the new set of the press ram, and start a new operation.

Bench presses are small, inclinable, gap-frame presses, 8.9 to 71.2 kN (1 to 8 tons) in capacity, that are made to be mounted on a bench. *Open-back presses* are gap-frame presses that are not inclinable. *Horn presses* are upright, open-back presses that have a heavy cylindrical shaft or "horn" in place of the usual bed. This permits curved or cylindrical workpieces to be placed over the horn for such operations as seaming, punching, riveting, and so forth, as shown in Figure 15-92. On some types both a horn and a bed are provided, with provision for swinging the horn aside when not needed.

Turret presses utilize a modified gap-frame construction but have upper and lower turrets that carry a number of punches and dies. The two turrets are

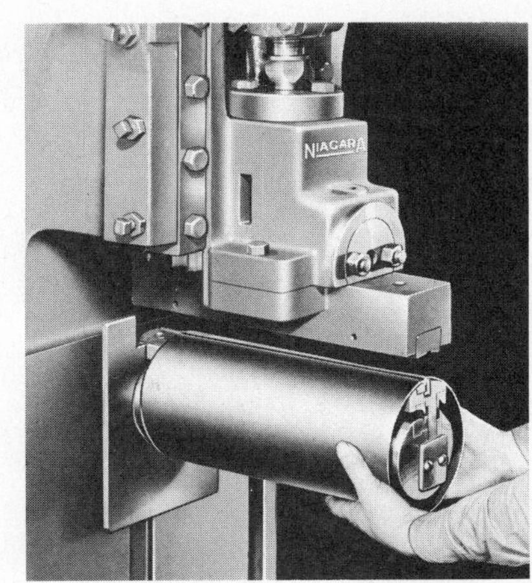

FIGURE 15-92. Making a seam on a horn press. (*Courtesy Niagara Machine & Tool Works.*)

geared together so any desired set can quickly be rotated into position. Another type uses a single turret on which subpress die sets are mounted.

Straight-sided presses are available in a wide variety of sizes and designs. This type of frame is used for most hydraulic presses and for larger-sized and specialized mechanical-driven presses. Typical examples are shown in Figures 15-71, 15-93, and 15-94.

FIGURE 15-93. Two straight-sided presses being operated in tandem. Strip stock is being fed automatically by a feeder at the right. (*Courtesy E. W. Bliss Company.*)

FIGURE 15-94. Transfer-type press. Inset shows parts at several of the eight stations. (*Courtesy Verson All-steel Press Company.*)

Special press types. A number of special types of presses are available, designed for doing special types of operations. Two interesting examples are the *transfer press* and the *multislide press.* The transfer press, illustrated in Figure 15-94, has a single, long slide with the provision for mounting a number of die sets side by side. Each die set can be adjusted individually. Stock is fed automatically to the first die station. After the completion of each ram stroke, the part is automatically and progressively transferred to the next die station by the mechanism shown in Figure 15-95. Such presses have high production rates, up to 1500 parts per hour, but are also very flexible.

The multislide machine, shown in Figures 15-96 and 15-97, contains a series of bending or forming slides that move horizontally. As indicated on

FIGURE 15-95. Transfer mechanism used in transfer presses. (*Courtesy Verson Allsteel Press Company.*)

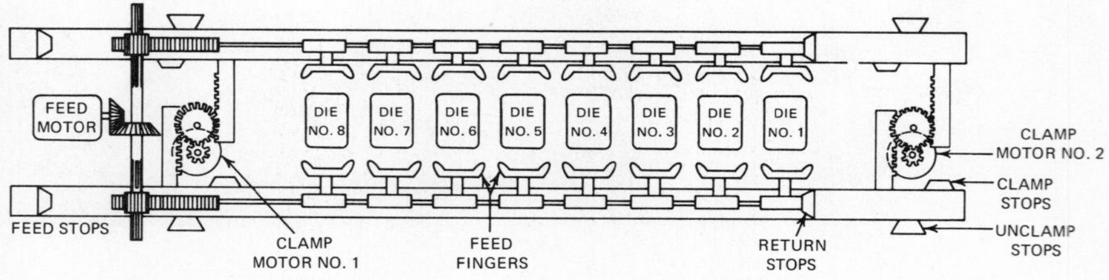

FIGURE 15-96. Multislide machine viewed from "back" side. (*Courtesy The U. S. Baird Corporation.*)

FIGURE 15-97. Schematic diagram of the operating mechanism of a multislide machine. (*Courtesy The U. S. Baird Corporation.*)

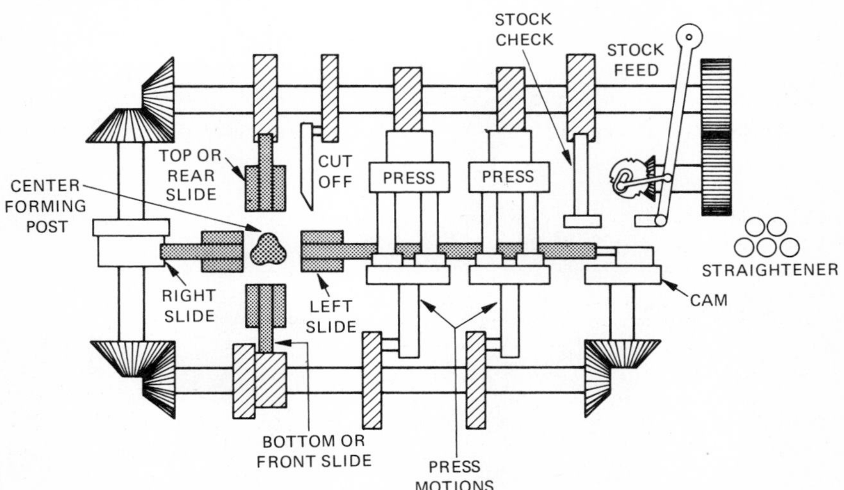

FIGURE 15-98. Example of the piercing, blanking, and forming of a part in a multislide machine. (*Courtesy The U. S. Baird Corporation.*)

the schematic diagram, these slides are driven by cams on four shafts that are located on the four sides of the machine, the shafts being driven by means of miter gears. Coiled strip stock is fed into the machine automatically and is progressively pierced, notched, bent, and cut off at the various slide stations in the manner indicated in Figure 15-98. Strip stock up to about 75 mm (3 inches) wide and 2.5 mm ($^3/_{32}$ inch) thick and wire up to about 3 mm ($\frac{1}{8}$ inch) in diameter are commonly processed. Parts such as hinges, links, clips, razor blades, and the like are processed at very high rates on such machines.

Press feeding devices. Although hand feeding is still used in many press operations, improved operator safety and increased productivity have motivated a strong shift to feeding by some type of mechanical device. *Dial-feed mechanisms,* such as the one shown in Figure 15-99, are often employed. The operator places the workpieces in the front holes of the dial, and the dial then indexes with each stroke of the ram to feed parts into proper position between the punch and die. When continuous strip stock is used, it can be fed automatically into a press by the type of mechanism illustrated in Figure 15-93. Lightweight parts can be fed into presses by suction-cup mechanisms, vibratory-bed feeders, or similar mechanisms. Automated mechanisms such as the

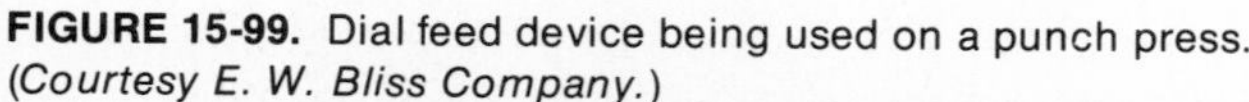

FIGURE 15-99. Dial feed device being used on a punch press. (*Courtesy E. W. Bliss Company.*)

one shown in Figure 38-15 are being used increasingly to place large parts into presses and remove them after processing. The technology and equipment exist to replace manual feeding in most cases if such a transition is desired.

Review questions

1. What distinguishes cold forming from hot forming?
2. What are some of the advantages of cold-working processes? Some disadvantages?
3. What are the four basic cold-working operations?
4. Why is springback of importance in cold-forming operations?
5. Explain the relationship of the stress–strain diagram of a metal to its suitability for use in cold-forming operations.
6. How would you expect the properties of cold-rolled steel to differ from those of hot-rolled steel of the same composition and shape?
7. Why might too-large or too-small grain size be objectionable in cold-forming operations?
8. What types of preparations may be required before a metal is cold-rolled?
9. What are some of the benefits of an intermediate anneal?
10. What are the four conditions listed for cold-rolled sheet and strip?
11. What is swaging?
12. How may shaped internal cavities be produced by swaging?
13. Why is cold heading used so extensively?
14. What are the advantages of combining cold extrusion with cold heading?
15. What is the primary advantage of the roll extrusion process?
16. What is the economic motivation for sizing operations?
17. Describe two techniques for riveting where access is only provided to one side of the work.
18. Why is staking not a suitable process for fastening parts that must be disassembled frequently?
19. How does sizing differ from coining?
20. Explain how peening may aid in preventing surface cracking.
21. How is peening usually accomplished?
22. Explain why burnishing produces a surface that has improved wearing properties.
23. What is the hobbing process used for?
24. What is the difference between bending and drawing?
25. Why does metal thin somewhat when bent?
26. What is the minimum bend radius and how is it tied to the material properties?
27. For what two processes might cold-roll forming be a substitute?
28. Why is cold-roll forming not economical for lots of 100 meters?
29. Explain the theoretical basis for roll straightening.
30. Why are sheared edges of metal usually not smooth?

31. What are two methods by which parts can be blanked to have smooth edges?
32. What is the difference between piercing and blanking?
33. Why is lancing often done prior to a drawing operation?
34. Why is the punch holder and die shoe arrangement an aid to reducing setup time?
35. Why are subpress die sets being used increasingly?
36. What is the basic difference between progressive and compound dies?
37. What is an advantage of a steel-rule die?
38. What will usually result if holes are to be pierced that are less than the stock thickness from the edge of the metal?
39. Why must wire drawing usually be done by means of a series of dies rather than by a single die?
40. Explain why the form block used in ordinary spinning can be made of inexpensive materials such as wood.
41. Why is it easier to mechanize shear forming than ordinary spinning?
42. How might a conical part, which is to have walls of nonuniform thickness, be formed by the direct shear-forming process?
43. Why is stretch-wrap forming particularly suitable for making large sheet-metal parts?
44. What is the distinction between shallow drawing and deep drawing?
45. What is the function of the pressure pad in deep drawing?
46. What conditions are required in order to use rubber tooling in forming operations?
47. What are some advantages of high-energy-rate forming?
48. It has been proposed that an indentation in a closed sheet-metal container be removed by means of electromagnetic forming. Will this work? Why?
49. How does embossing differ from coining?
50. Why are automatic feeding devices being used increasingly with blanking and forming presses?

Case study 11.
THE THREADED BALL STUDS

Forward Motors Corporation makes about 2,000,000 threaded ball studs each year, as shown in Figure CS-11. They currently are machined from hot-rolled

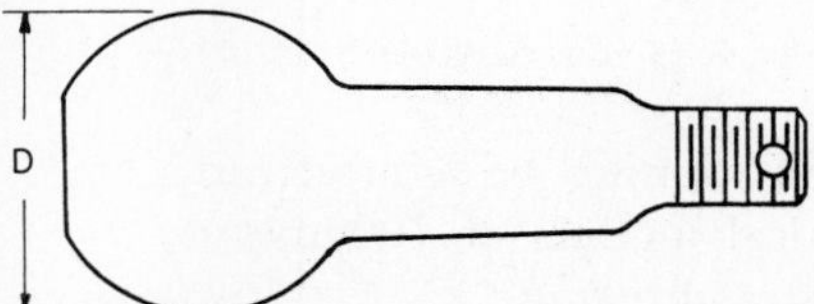

FIGURE CS-11. Threaded ball stud.

AISI 1030 bar stock of diameter *D*, heat-treated to 490 MPa (70,000 psi) ultimate strength, and the ball surface then ground to the required finish. Tolerances on the ball and threaded portion are ±0.05 mm (0.002 inch). Other dimensions are to ±0.64 mm (0.025 inch). Dependable quality is important.

It is desired to reduce the cost of this part and, if possible, reduce the weight. How would you bring about the desired improvement? [*D* is approximately 38 mm (1½ inches)].

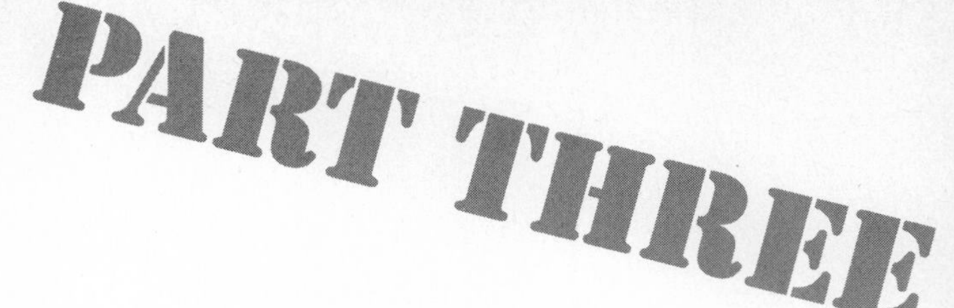

PART THREE

Machining processes

CHAPTER 16

Measurement, gaging, and quality control

The economical manufacture of products of satisfactory quality, from ballpoint pens to space vehicles, depends on precise specification, measurement, and control of dimensions. Shapes are wanted in specific sizes. This involves control of the processes that are utilized to make them and measurement of the resulting shapes. Such control and measurement are not easy and, as shown in Figure 16-1, the difficulty and cost increase geometrically as the preciseness of dimensions is increased.

In order to have the enormous output of reliable, low-cost products that are produced routinely today, utilizing the principle of interchangeable manufacture, a wide range of manufacturing and measuring equipment has had to be developed. Although the possibilities of interchangeable manufacture were recognized after the work of Eli Whitney in 1794, large-scale manufacturing based on this principle did not become well established until about 1915. It is difficult for us today to realize that until about 1920, if one broke an axle in

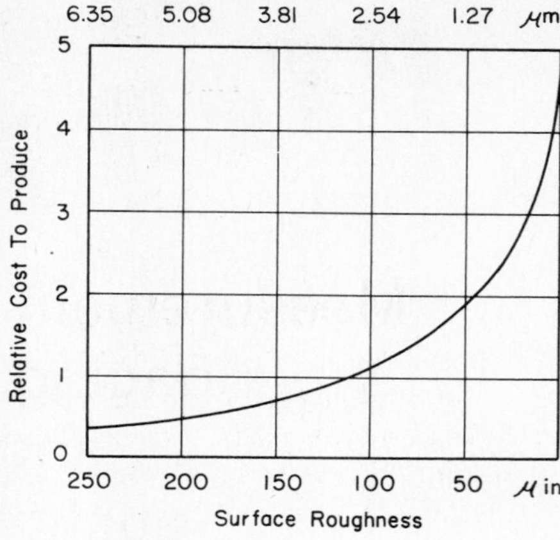

FIGURE 16-1. Relationship of surface finish to cost of producing the surface.

his automobile (a not uncommon occurrence in those days), he usually waited at least a week to obtain delivery of a new one from the factory and then had to take it to the local garage or machine shop to have it machined to fit.

Whitney achieved interchangeability through careful handwork, with the dimensions being controlled by his filing jigs. Such a procedure obviously is not practicable in modern, large-scale manufacturing. Size control must be built into machine tools and tooling through precise manufacture and be checked by suitable measurement and control of the quality of the resulting products. When a designer specifies the dimensions and tolerances of a part, he not only determines the functioning of the manufactured product, but also indirectly determines the machines, processes, and labor skills required to produce the part. Consequently, considerable attention will be given to the accuracy that can be achieved by the machines and processes that are discussed in this book, inasmuch as this has an important bearing on the selection of the proper tools or processes for manufacturing a given part.

Unfortunately, one cannot just specify the desired and proper dimensions for a part and assume that they will automatically be achieved in manufacture. This results only from the proper exercise of measurement and control at most of the steps in the production process. *Measurement* is the determination of a dimension through the use of some type of calibrated device that permits the magnitude of the dimension to be determined, directly or indirectly, in scalar units. An example is determining the width of a desk by means of an ordinary measuring tape. *Gaging*, on the other hand, is determining whether a dimension of a part is larger or smaller than an established standard, or standards, by direct or indirect comparison. A simple example of this is determining whether a desk will go through a door opening. A simple stick, without any graduated scale, will enable the determination to be made. Or, the door itself may serve as the gage, the desk being pushed against it to determine whether it is smaller or larger than the opening.

Linear standards. Satisfactory measurement and gaging must be based on a reliable, and preferably universal, standard or standards. These have not always existed. For example, although the musket parts made in Eli Whitney's shop were interchangeable, they were not interchangeable with parts made by another contemporary gun maker from the same drawings, because the two gunsmiths had different foot rulers. In another instance, it is reported that when Joseph Whitworth, who developed the Whitworth screw thread and had an unusually advanced set of standards, put his 1¼- and ¾-inch plug gages together, they would not go through his 2-inch ring gage. Today the entire industrialized world has adopted the *international meter* as the standard of linear measurement. From 1880 until 1960 it was officially defined as the distance between two fine scribed lines on a platinum–iridium bar maintained near Paris. In 1960, it was officially redefined as being 1,650,763.73 wavelengths of the orange-red light given off by electrically excited krypton 86, a rare gas from the atmosphere. Both the United States and English inches have been officially defined as 2.54 centimeters. Thus the United States inch is 41,929.399 wavelengths of the orange-red light from krypton 86.

Although *officially* the United States is committed to conversion to the metric (SI) system of measurement, which utilizes millimeters for virtually all linear measurements in manufacturing, the English system of feet and inches still is being utilized by most manufacturing plants and its use probably will continue for some time.[1]

Gage blocks. *Gage blocks* provide industry with linear standards of high accuracy that are necessary for everyday use in manufacturing plants. These are small, steel blocks, usually rectangular in cross section, having two very flat and parallel surfaces that are certain specified distances apart. Such gage blocks were first conceived by Carl E. Johansson in Sweden just prior to 1900. By 1911, he was able to produce sets of such blocks on a very limited scale, and they came into limited but significant use during World War I. Shortly after the war, Henry Ford recognized the importance of having such gage blocks generally available. He arranged for Johansson to come to the United States and, through facilities provided by the Ford Motor Company, methods were devised for large-scale production of gage block sets. As a result, since about 1940, gage block sets of excellent quality have been produced by a number of companies in this country and abroad.

Gage blocks usually are made of alloy steel, hardened and carefully heat-treated to produce "seasoning" to relieve internal stresses and minimize subsequent dimensional change. Some are made entirely of carbides, such as chromium or tungsten carbide, to provide extra wear resistance. The measuring surfaces of each block are ground to approximately the required dimension and

[1] One obvious objection to the exclusive use of millimeters is the magnitude of the numbers that result—for example, 60 inches being 1524 mm. In the aircraft industry numbers above 25,000 are not uncommon.

then lapped to reduce the block to the final dimension and to produce a very flat and smooth surface.

Gage blocks commonly are made to meet the U.S. Federal Specifications GGG-G-15a and 15b, having Grades 1, 2, and 3 with accuracies as follows:

	15a (inches)	15b (millimeters)
Grade 1 (Laboratory)	±0.000,002	±0.000 05
Grade 2 (Precision)	+0.000,004 −0.000,002	+0.000 10 −0.000 05
Grade 3 (Working)	+0.000,008 −0.000,004	+0.000 20 −0.000 10

Blocks up to 1 inch in length have absolute accuracies as stated, whereas the tolerances are per inch of length for blocks larger than 1 inch. Some companies supply blocks in AA quality, which corresponds to Grade 1, and A+ quality, to meet Grade 2.

Grade 1 (*Laboratory-Grade*) blocks are used for checking and calibrating other grades of gage blocks. Grade 2 (*Precision-Grade*) blocks are used for checking master gages and Grade 3 blocks. Grade 3 (B or *Working-Grade*) blocks are used to check routine measuring devices, such as micrometers, or in actual gaging operations.

All gage blocks should be checked periodically to ensure their accuracy. Federal Specification GGG-G-15a recommends that AA blocks should be checked annually; A+ blocks should be checked semiannually; A blocks should be checked quarterly; and B blocks should be checked semiannually or as often as is warranted by the conditions of use.

Gage blocks usually come in sets containing various numbers of blocks of various sizes, such as those shown in Figure 16-2. By wringing the blocks together in various combinations, as shown in Figure 16-3, any desired dimension can be obtained. As examples, one manufacturer's 81-block English set and 88-block metric set have the following blocks:

English			Metric		
Series	Number of Blocks	Range (inches)	Series	Number of Blocks	Range (mm)
0.0001 in.	9	0.1001–0.1009		2	0.5 and 1.0005
0.001 in.	49	0.101–0.149	0.001 mm	9	1.001–1.009
0.050 in.	19	0.050–0.950	0.01 mm	49	1.01–1.49
1.000 in.	4	1.000–4.000	0.5 mm	18	1.0–9.5
			10 mm	10	10–120

The 81 blocks in the English system can be combined in over 120,000 combinations in increments of 0.0001 inch from 0.1001 inch to over 25 inches. For

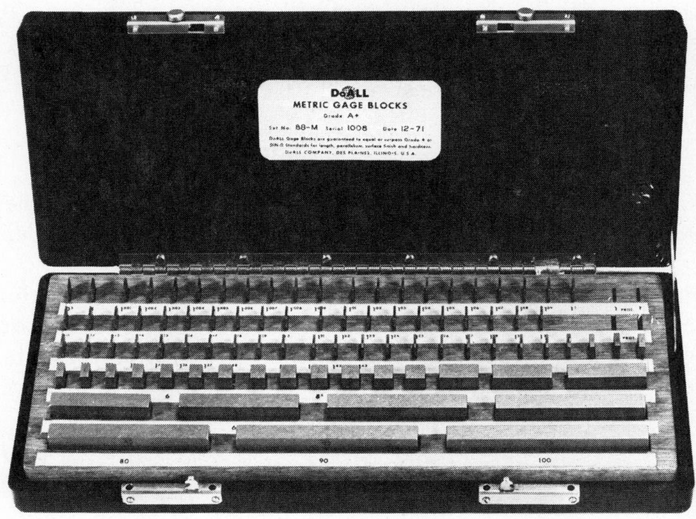

FIGURE 16-2. Metric gage block set, containing 88 blocks. (*Courtesy The DoALL Company.*)

example, if a dimension of 2.6243 inches is desired, it may be obtained by combining the following blocks:

From the first series	0.1003 in.
From the second series	0.124
From the third series	0.400
From the fourth series	2.000
	2.6243 in.

FIGURE 16-3. Seven gage blocks wrung together to build up a desired dimension. (*Courtesy The DoALL Company.*)

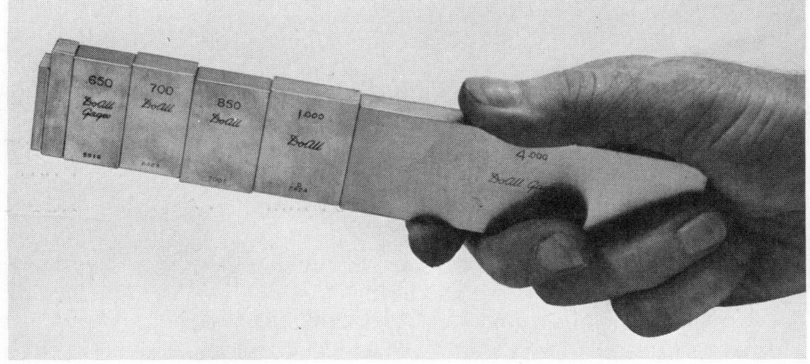

For the corresponding 66.66 mm, one would select:

From the fifth metric series	60.00 mm
From the fourth metric series	5.50
From the third metric series	1.16
	66.66 mm

When gage blocks are wrung together (slid together firmly), they adhere with considerable force and should not be left in contact for extended periods of time.

Several types of gage blocks are available: fractional series, thin series, angle series, and so on, so that standards of high accuracy can be obtained to fill almost any need. In addition, various auxiliary clamping, scribing, and base block attachments are available that make it possible to form very accurate gaging devices, such as shown in Figure 16-4.

A very useful variation of gage blocks is the device shown in Figure 16-5. It consists of a column of permanently wrung, 1-inch gage blocks arranged in a staggered pattern so that the entire column can be raised and lowered by means of an accurate, calibrated micrometer screw. Direct digital reading of the height of the block steps above the base is shown to 0.001-inch increments and on the micrometer dial to 0.0001-inch increments. Models also are available having direct electronic digital readout to 0.0001-inch increments. Such units are very useful in conjunction with deviation-type gages, which will be discussed later, in making accurate vertical measurements.

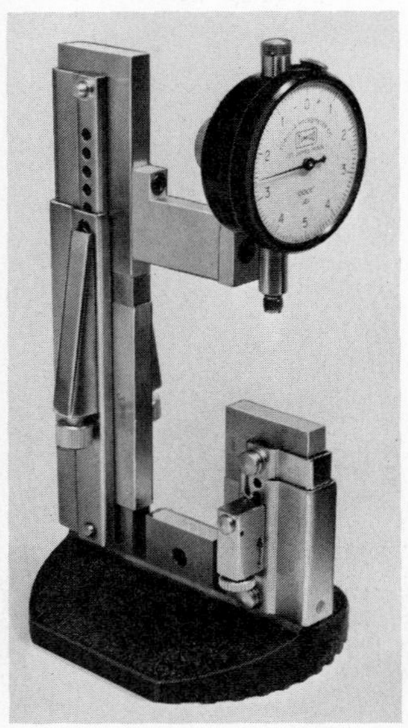

FIGURE 16-4. Wrung-together gage blocks in a special holder and used with a dial gage to form an accurate comparator. (*Courtesy The DoALL Company.*)

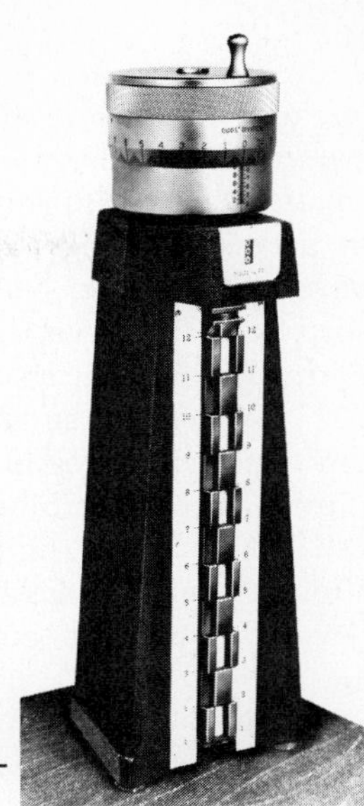

FIGURE 16-5. Digi-check height gage. (*Courtesy The L. S. Starrett Company.*)

Standard measuring temperature. Because all the commonly used metals are affected dimensionally by temperature, a standard measuring temperature of 20°C (68°F) has been adopted for precision-measuring work. All gage blocks, gages, and other precision-measuring instruments are calibrated at this temperature. Consequently, when measurements are to be made to accuracies greater than 0.0025 mm (0.0001 inch), the work should be done in a room in which the temperature is controlled at standard. Although it is true that to some extent both the workpiece and the measuring or gaging device *may* be affected to about the same extent by temperature variations, one should not rely on this. Measurements to even 0.0025 mm (0.0001 inch) should not be relied on if the temperature is very far from 20°C (68°F).

The fit of mating parts. The accuracy that must be specified and achieved in the manufacture of a given part very often is determined by the manner in which it must function with respect to other parts. If it is necessary only that one part always fit inside a second, it makes little difference if the dimension of the smaller part varies considerably as long as its maximum dimension is always smaller than the inside dimension of the part into which it must go. However, if the smaller part is to rotate smoothly within the larger part at high speeds with a minimum of vibration, as in a high-speed ball bearing, considerably more attention must be given to the specification and control of the dimensions of the parts. Thus *function* controls the specification of dimensions that determine the manner in which parts fit.

Two factors, allowance and tolerance, must be specified to obtain the desired fit between mating parts. *Allowance* is the intentional, desired difference between the dimensions of two mating parts. It is the difference between the dimension of the largest interior-fitting part (shaft) and that of the smallest exterior-fitting part (hole). It thus determines the condition of *tightest* fit between mating parts. Allowance may be specified so that either *clearance* or *interference* exists between the mating parts. With clearance fits, the largest shaft is smaller than the smallest hole, whereas with interference fits, the hole is smaller than the shaft.

Tolerance is an undesirable, but permissible, deviation from a desired dimension in recognition that no part can be made *exactly* to a specified dimension, except by chance, and that such is neither necessary nor economical. Consequently, it is necessary to permit the actual dimension to deviate slightly from the desired, theoretical dimension and to control the degree of deviation such that satisfactory functioning of the mating parts still will be assured.

Tolerance can be specified in two ways. Bilateral tolerance is specified as a *plus* or *minus* deviation from the basic dimension, such as 2.000 in. ±0.002 in. More modern practice uses the unilateral system, where the deviation is in one direction from the basic size, such as

$$2.000 \text{ in.} \begin{matrix} +0.004 \text{ in.} \\ -0.000 \text{ in.} \end{matrix} \qquad \text{or} \qquad 50.8 \text{ mm} \begin{matrix} +0.1 \text{ mm} \\ -0.0 \text{ mm} \end{matrix}$$

In the first case, that of bilateral tolerance, the dimension of the part could vary between 1.998 and 2.002 inches, a total tolerance of 0.004 inch. For the example of unilateral tolerance, the dimension could vary between 2.000 and 2.004 inches, again a tolerance of 0.004 inch. Obviously, in order to obtain the same maximum and minimum dimensions with the two systems, different basic sizes must be used. The maximum and minimum dimensions that result from the application of the designated tolerance are called *limit dimensions.*

There can be no rigid rules as to the amount of clearance that should be provided between mating parts; the decision must be made by the designer in consideration of how he wants them to function. The American National Standards Institute, Inc., (ANSI) has established eight classes of fits that serve as a useful guide in specifying the allowance and tolerance for typical applications, and that permit the amount of allowance and tolerance to be determined merely by specifying a particular class of fit. These classes are as follows:

Class 1. Loose fit; large allowance. For applications where accuracy is not essential.

Class 2. Free fit; liberal allowance. For running fits where speeds are over 600 rpm and pressures are 4.1 MPa (600 psi) or over.

Class 3. Medium fit; medium allowance. For running fits under 600 rpm and pressures less than 4.1 MPa (600 psi) and for sliding fits.

Class 4. Snug fit; zero allowance. For applications where no movement under load is intended and no shaking is wanted. The tightest fit that can be assembled by hand.

Class 5. Wringing fit; zero to negative allowance. Assemblies are selective and not interchangeable.

Class 6. Tight fit; slight negative allowance. An interference fit for parts that must not come apart in service and are not to be disassembled, or disassembled only seldom. Light pressure is required for assembly. Not to be used to withstand other than very light loads.

Class 7. Medium force fit; an interference fit requiring considerable pressure to assemble; ordinarily assembled by heating the external member or cooling the internal member to provide expansion or shrinkage. Used for fastening wheels, crank disks, and the like to shafting. The tightest fit that should be used on cast iron external members.

Class 8. Heavy force and shrink fits; considerable negative allowance. Used for permanent shrink fits on steel members.

The allowances and tolerances that are associated with the ANSI classes of fits are determined according to the theoretical relationship shown in Table 16-1. The actual resulting dimensional values for a wide range of basic sizes can be found in tabulations in drafting and machine-design books.

In the ANSI system the hole size is always considered basic, because the majority of holes are produced through the use of standard-size drills and reamers. The internal, or shaft, member can be made to any one dimension as readily as to another. The allowance and tolerances are applied to the basic hole size to determine the limit dimensions of the mating parts. For example, for a basic hole size of 50.8 mm (2 inches) and a Class 3 fit, the dimensions would be:

Allowance:		0.036 mm (0.0014 in.)
Tolerance:		0.025 mm (0.0010 in.)
Hole:	Maximum	50.825 mm (2.0010 in.)
	Minimum	50.800 mm (2.0000 in.)
Shaft:	Maximum	50.764 mm (1.9986 in.)
	Minimum	50.739 mm (1.9976 in.)

TABLE 16-1. ANSI recommended allowances and tolerations

Class of Fit	Allowance	Average Interference	Hole Tolerance	Shaft Tolerance
1	$0.0025\sqrt[3]{d^2}$		$+0.0025\sqrt[3]{d}$	$-0.0025\sqrt[3]{d}$
2	$0.0014\sqrt[3]{d^2}$		$+0.0013\sqrt[3]{d}$	$-0.0013\sqrt[3]{d}$
3	$0.0009\sqrt[3]{d^2}$		$+0.0008\sqrt[3]{d}$	$-0.0008\sqrt[3]{d}$
4	0		$+0.0006\sqrt[3]{d}$	$-0.0004\sqrt[3]{d}$
5		0	$+0.0006\sqrt[3]{d}$	$+0.0004\sqrt[3]{d}$
6		$0.00025d$	$+0.0006\sqrt[3]{d}$	$+0.0006\sqrt[3]{d}$
7		$0.0005d$	$+0.0006\sqrt[3]{d}$	$+0.0006\sqrt[3]{d}$
8		$0.001d$	$+0.0006\sqrt[3]{d}$	$+0.0006\sqrt[3]{d}$

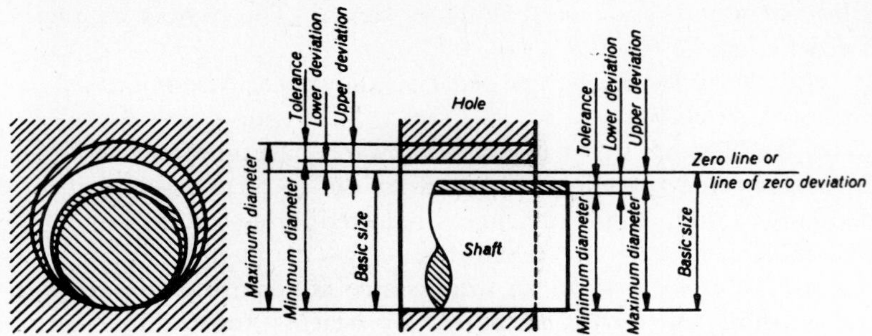

FIGURE 16-6. Basic size, deviation, and tolerance in the ISO system. (*By permission from ISO Recommendation R286-1962, System of Limits and Fits, copyright 1962 by American National Standards Institute.*)

It should be noted that for both clearance and interference fits, the permissible tolerances tend to result in a looser fit.

The *ISO System of Limits and Fits,*[2] widely used in a number of leading metric countries, is considerably more complex than the ANSI system just discussed. In this system, each part has a *basic size.* Each limit of size of a part, high and low, is defined by its *deviation* from the basic size, the magnitude and sign being obtained by subtracting the basic size from the limit in question. The difference between the two limits of size of a part is called the *tolerance,* an absolute amount without sign. Figure 16-6 illustrates these definitions.

There are three classes of fits: (1) *clearance fits,* (2) *transition fits* (the assembly may have either clearance or interference), and (3) *interference fits.*

Either a *shaft-basis system* or a *hole-basis system* may be used, as illustrated in Figure 16-7. For any given basic size, a range of tolerances and deviations may be specified with respect to the line of zero deviation, called the *zero line.*

FIGURE 16-7. "Shaft-basis" and "hole-basis" systems for specifying fits in the ISO system. (*By permission from ISO Recommendation R286-1962, System of Limits and Fits, copyright 1962 by American National Standards Institute.*)

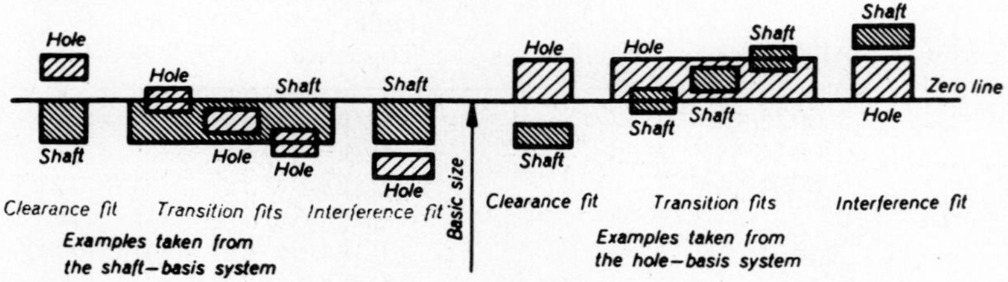

[2] This system and the necessary tables for its application are contained in ISO publication R286-1962, *System of Limits and Fits.*

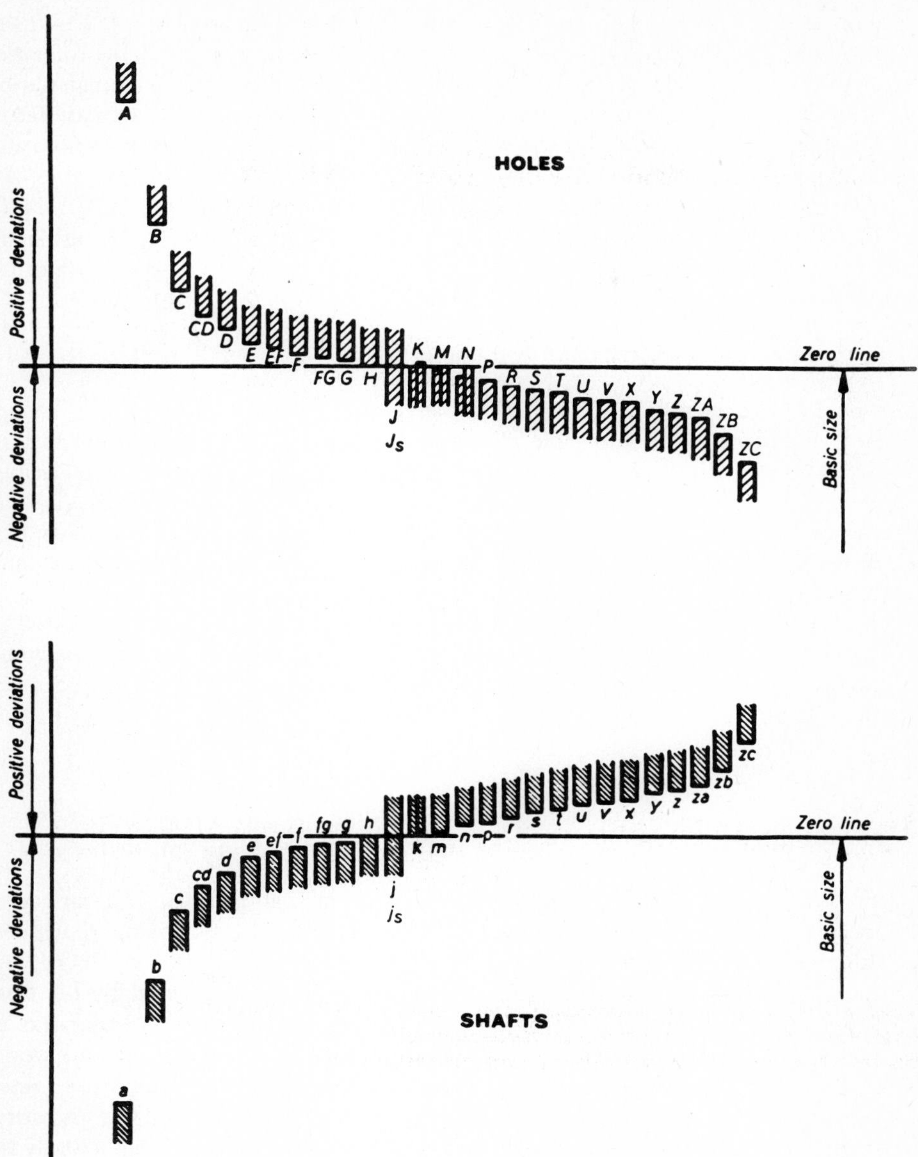

FIGURE 16-8. Positions of the various tolerance zones for a given diameter in the ISO system. (*By permission from ISO Recommendation R286-1962, System of Limits and Fits, copyright 1962 by American National Standards Institute.*)

The tolerance is a function of the basic size and is designated by a number symbol, called the *grade*—thus the *tolerance grade.* The *position* of the tolerance with respect to the zero line—also a function of the basic size—is indicated by a letter symbol (or two letters), a capital letter for holes and a lowercase letter for shafts, as illustrated in Figure 16-8.[3] Thus the specification for a hole and shaft having a basic size of 45 mm might be 45 H8/g7.

Eighteen standard grades of tolerances are provided, called IT 01, IT 0, IT 1–16, providing numerical values for each nominal diameter, in arbitrary steps up to 500 mm (for example 0–3, 3–6, 6–10, . . . , 400–500 mm). The value of the tolerance unit, i, for grades 5–16 is

$$i = 0.45\sqrt[3]{D} + 0.001D$$

where i is in microns and D in millimeters.

Standard shaft and hole deviations similarly are provided by sets of formulas. However, for practical application, both tolerances and deviations are provided in three sets of rather complex tables. Additional tables give the values for basic sizes above 500 mm and for "Commonly Used Shafts and Holes" in two categories—"General Purpose" and "Fine Mechanisms and Horology."

Inasmuch as the ANSI recommended allowances and tolerances, listed in Table 16-1, may be expressed either in inches or millimeters, one may use this relatively simple system or the much more complex ISO system in dimensioning metric drawings.

MEASURING INSTRUMENTS

Because of the great importance of measuring in manufacturing, a variety of instruments is available that permits measurements to be made routinely, ranging in accuracy from 0.5 to 0.0003 mm and $^{1}/_{64}$ to 0.000,01 inch. The ease and accuracy of making such measurements can be affected by (1) least count of the subdivisions on the instrument, (2) line matching, (3) parallax in reading the instrument, (4) elastic deformation of the instrument and workpiece, and (5) temperature effects. Some instruments are more subject to these factors than others. In addition, the skill of the person making the measurement, and in overcoming these factors, is very important, particularly in using some instruments. The inclusion of digital readout systems in measuring instruments, as will be discussed later, can lessen or eliminate the effect of some of these factors.

[3] It will be recognized that the "position" in the ISO system essentially provides "allowance" of the ANSI system.

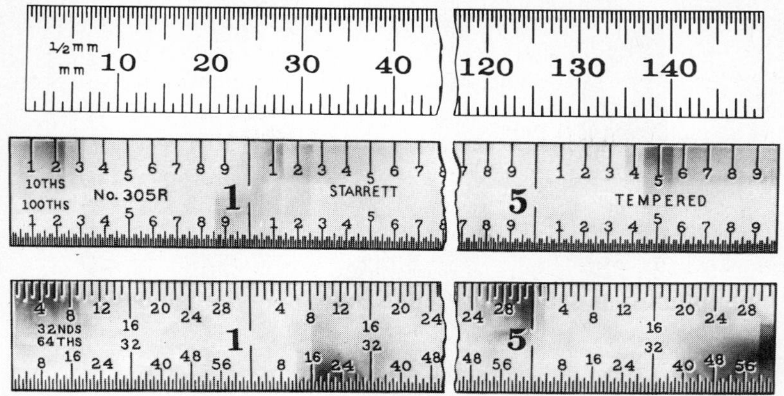

FIGURE 16-9. Machinist's rules. (*Top*) Metric. (*Center and bottom*) Inch graduations; 10ths and 100ths on one side, 32nd and 64ths on opposite side. (*Courtesy The L. S. Starrett Company.*)

Linear measuring instruments. Linear measuring instruments are of two types: direct reading and indirect reading. *Direct-reading instruments* contain a line-graduated scale so that the size of the object being measured can be read directly on this scale. *Indirect-reading instruments* do not contain line graduations and are used to transfer the size of the dimension being measured to a direct-reading scale, thus obtaining the desired size information in an indirect manner.

The simplest, and most common, direct-reading linear measuring instrument is the machinist's rule, shown in Figure 16-9. Metric rules usually have two sets of line graduations on each side—½- and 1-mm divisions—whereas English rules have four sets—$^1/_{16}$-, $^1/_{32}$-, $^1/_{64}$-, and $^1/_{100}$-inch divisions. Other combinations can be obtained in each type.

The machinist's rule is an end- or line-matching device; an end and a line, or two lines, must be aligned with the extremities of the object or distance being measured in order to obtain the desired reading. Thus the accuracy of the resulting reading is a function of the alignment and the magnitude of the smallest scale division. Such scales ordinarily are not used for accuracies greater than ½ mm, $^1/_{64}$ inch, or 0.01 inch.

Several attachments can be added to a machinist's rule to extend its usefulness. Shown in Figure 16-10, the *square head* can be used as a miter or try square or to hold the rule in an upright position on a flat surface for making height measurements. It also contains a small bubble-type level so that it can be used by itself as a level. The *bevel protractor* permits the measurement or layout of angles. The *center head* permits the center of cylindrical work to be determined.

The *rule depth gage,* shown in Figure 16-11, consists of a special head

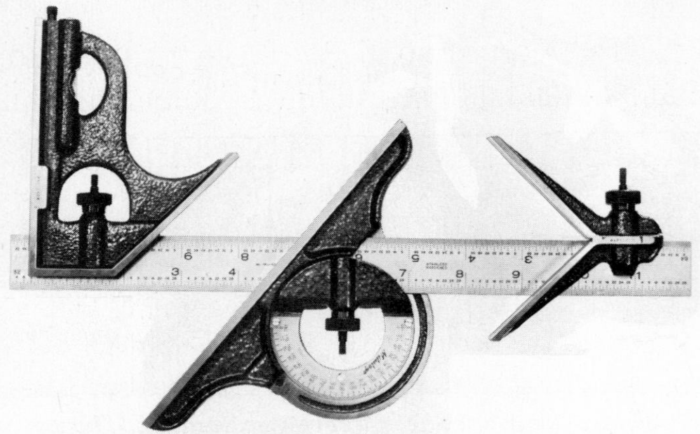

FIGURE 16-10. Combination set. (*Courtesy MTI Corporation.*)

that slides on a small rule and thus permits the depth of holes or shoulders below a given surface to be measured. The slide can be clamped at any desired position on the rule.

The *vernier caliper,* illustrated in Figure 16-12, is an end-measuring instrument, available in various sizes, that can be used to make both outside and inside measurements to theoretical accuracies of 0.01 mm or 0.001 inch. End-measuring instruments are more accurate and somewhat easier to use than line-matching types because their jaws are placed against either end of the ob-

FIGURE 16-11. Rule depth gage being used to measure the depth of a groove from a surface. (*Courtesy Brown & Sharpe Manufacturing Company.*)

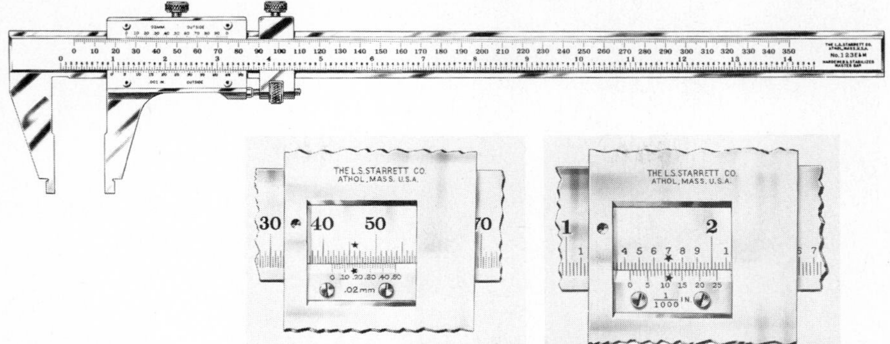

FIGURE 16-12. Vernier caliper. Insets show enlarged diagrams of the vernier scale: metric (*left*) and English (*right*). (*Courtesy The L. S. Starrett Company.*)

ject being measured, so that any difficulty in aligning edges or lines is avoided. However, the difficulty in obtaining uniform contact pressure, or "feel," between the legs or the instrument and the object being measured remains.

A major feature of the vernier caliper is the auxiliary scale, shown in Figure 16-12. The main English scale is divided into inches and tenths, with each tenth being subdivided into four divisions of 0.025 inch each. The 25 divisions on the vernier plate are equal to the length of 24 divisions on the main scale. Thus each division on the vernier scale is $^1/_{25}$ of $(24 \times 0.025$ inch$) = 0.024$ inch. This is $^1/_{1000}$ inch less than each division on the main scale. Thus if the zero readings in each scale were in line, the first lines in each scale would be $^1/_{1000}$ inch apart, the second lines $^2/_{1000}$ inch apart, and so on. The main English scale, as shown in the enlarged portion of Figure 16-12, is first read in inches and thousandths—1.425 inches. To this is added the reading of the vernier scale—the point at which a line on this scale coincides exactly with a line of the main scale—in this case 11, indicating that the zero line on the vernier scale and the next line to the left of it on the main scale are 0.011 inch apart. Thus the total reading is 1.425 inches plus 0.011 inch, or 1.436 inches. Similarly, the metric reading is 41.5 mm plus 0.18 mm, or 41.68 mm.

Vernier calipers are versatile but are rather slow to use. Consequently, they are not used as extensively as other instruments that have a more limited range but are easier to use.

The *vernier height gage,* shown in Figure 16-13, consists of a rule attached to a special base at one end and having a sliding head containing a vernier scale so that readings can be made to 0.01 mm or 0.001 inch. A beveled point, or other types of contact device, can be attached to the sliding head for making height measurements. It also is used in layout work to mark or transfer a desired height to a workpiece.

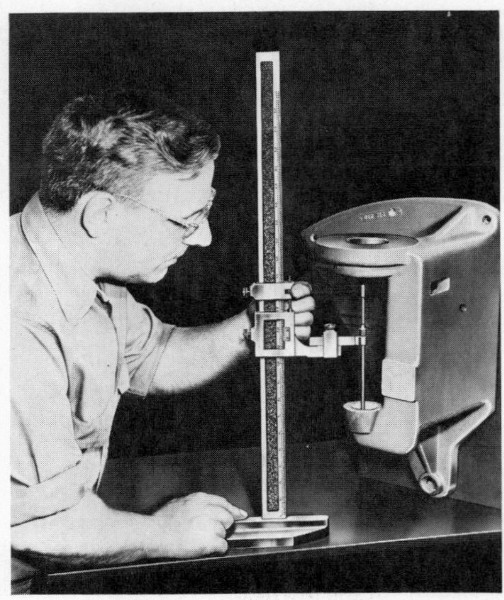

FIGURE 16-13. Measuring the height of the bottom of a cup above the base by means of a vernier height gage and special attachment. (*Courtesy The L. S. Starrett Company.*)

The *micrometer caliper,* more commonly called a *micrometer,* is one of the most widely used measuring devices. Until recently the type shown in Figure 16-14 was virtually standard. It consists of a fixed anvil and a movable spindle. By rotating the thimble on the end of the caliper, the spindle is moved away from the anvil by means of an accurate screw thread. On English types, this thread has a lead of 0.025 inch, and one revolution of the thimble moves the spindle this distance. The barrel, or sleeve, is calibrated in 0.025-inch divisions, with each tenth of an inch being numbered. The circumference

FIGURE 16-14. Cutaway view of a micrometer caliper, measuring in inches. Inset shows the method of reading the scales. (*Courtesy Brown & Sharpe Manufacturing Company.*)

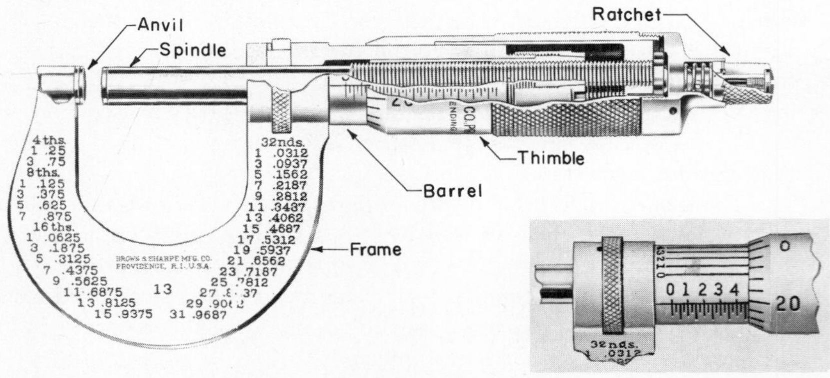

at the edge of the thimble is graduated into 25 divisions, each representing 0.001 inch.

A major difficulty of this type of micrometer is in making the reading of the dimension shown on the instrument. To read the instrument, the division on the thimble that coincides with the longitudinal line in the barrel is added to the largest reading exposed on the barrel. For example, on the English-type shown in Figure 16-14, the largest exposed number on the barrel is 4, representing 0.400 inch. Two additional lines exposed on the barrel represent 0.050 inch. The 20th line on the thimble nearly coincides with the longitudinal line on the barrel, representing 0.020 inch. Thus the total reading is: 0.400 inch + 0.050 inch + 0.020 inch = 0.470 inch. The micrometer shown also has vernier graduations on the barrel that enable it to be read to 0.0001 inch. Thus a more exact reading is 0.4697 inch. However, owing to the lack of pressure control, micrometers can seldom be relied on for accuracy beyond 0.0005 inch, and such vernier scales are not used extensively.

On metric micrometers the graduations on the barrel and thimble usually are 0.5 mm and 0.01 mm, respectively. As shown in Figure 16-15, this permits measuring to 0.01 mm, the total measurement in the example shown being 5.78 mm.

Many errors have resulted from the ordinary micrometer being misread, the error being ±0.025 inch or ±0.5 mm. Consequently, several direct-reading types have been developed. Figure 16-16 shows one of the latest developments, an electronic type that provides constant spindle force, metric or English units at the flick of a switch, and an LED-crystal digital display of the measurement. Such micrometers undoubtedly will come into wide use.

Although micrometer calipers are quite easy to use and can be read quickly, the range of each instrument is limited to about 1 inch. Thus a number of micrometers of various sizes are required to cover a wide range of dimensions. Being an end-measuring device, it has the advantages of this type of instrument. Its greatest limitation to accuracy is control of the pressure be-

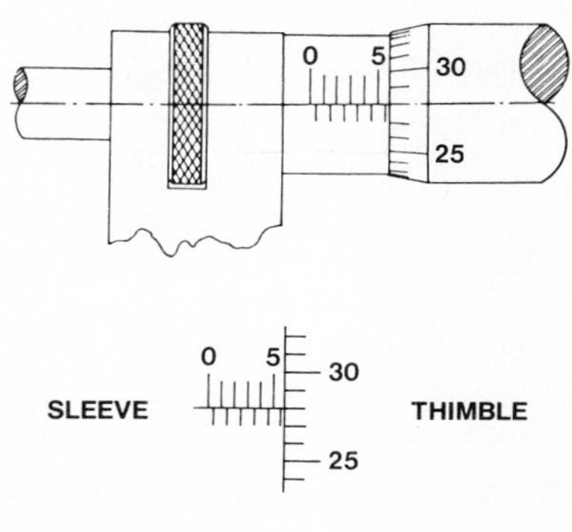

FIGURE 16-15. Method of reading scales of a metric micrometer caliper. (*Courtesy The L. S. Starrett Company.*)

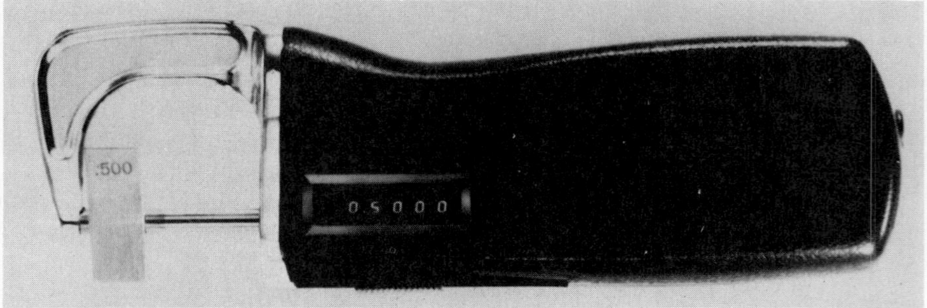

FIGURE 16-16. Electronic micrometer caliper having LED digital display and which can read in either English or metric units. (*Courtesy Quality Measurement Systems, Inc.*)

tween the anvil, spindle, and the piece being measured. Most micrometers are equipped with a ratchet or friction device, as shown in Figure 16-14, by means of which the thimble can be turned. This controls the pressure to a fair degree if used carefully. Calipers that do not have this device may be sprung several thousandths by applying excess torque to the thimble. Micrometer calipers usually should not be relied on for measurements of greater accuracy than 0.01 mm or 0.001 inch, except the electronic type shown in Figure 16-16.

Micrometer calipers are available with a variety of specially shaped anvils and/or spindles, such as points, balls, and disks, for measuring special shapes.

Two types of micrometers are available for inside measurements. The caliper type is shown in Figure 16-17 and the plain-rod type in Figure 16-18.

The micrometer principle also is incorporated in the *micrometer depth gage,* shown in Figure 16-19.

The *supermicrometer,* shown in Figure 16-20, is typical of a number of instruments that are designed to make linear measurements directly to 0.0001 inch by incorporating special features that overcome several of the factors mentioned at the beginning of this section. First, as can be observed, it is made with very heavy members to avoid any possibility of distortion. Second, although it is an end-measuring instrument, the pressure between the anvil and spindle and the workpieces is maintained constant by a pressure spring against which the anvil pushes. The spindle is tightened against the workpiece until an indicator on the tailstock shows that a selected pressure has been obtained. Third, the spindle screw is very accurate and is actuated by a large dial that, in conjunction with a vernier, reads directly to 0.0001 inch. The electronic meter permits the instrument to be used as a comparator and to measure to 0.00002 inch. Such instruments are used both for inspection and toolroom checking and for production inspection work. Some are available with the anvil connected to an electronic pressure-indicating device for even more precise control of the measuring pressure, and some of these, called stan-

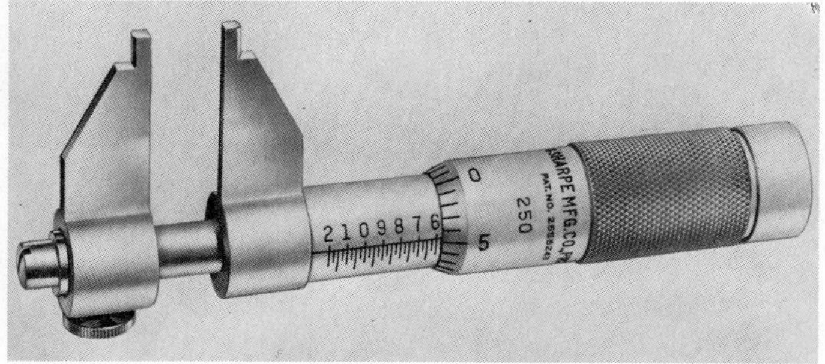

FIGURE 16-17. Caliper-type inside micrometer. (*Courtesy Brown & Sharpe Manufacturing Company.*)

FIGURE 16-18. Inside micrometer being used to measure an inside dimension. This type uses interchangeable rods. (*Courtesy Brown & Sharpe Manufacturing Company.*)

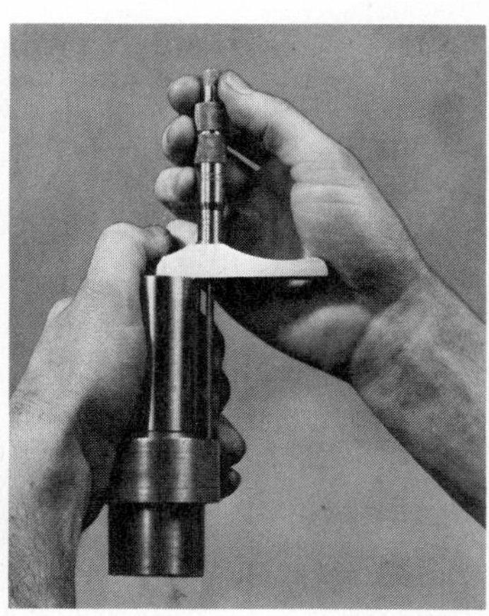

FIGURE 16-19. Measuring the depth to a shoulder using a micrometer depth gage. (*Courtesy Brown & Sharpe Manufacturing Company.*)

FIGURE 16-20. Supermicrometer. (*Courtesy Colt Industries, Pratt & Whitney Small Tool Division.*)

dard measuring machines, are capable of measuring directly to 0.000,01 inch when used in a controlled-temperature room.

The toolmakers' microscope, shown in Figure 16-21, is a versatile instrument that measures by optical means with no pressure being involved. It thus is very useful for making accurate measurements on small or delicate

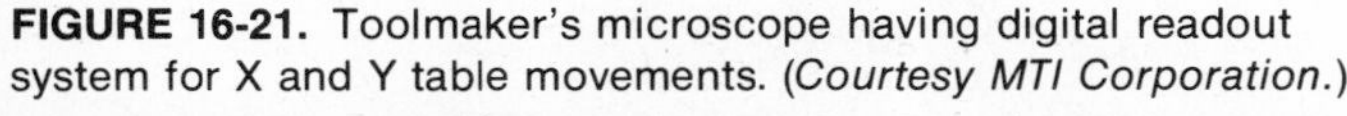

FIGURE 16-21. Toolmaker's microscope having digital readout system for X and Y table movements. (*Courtesy MTI Corporation.*)

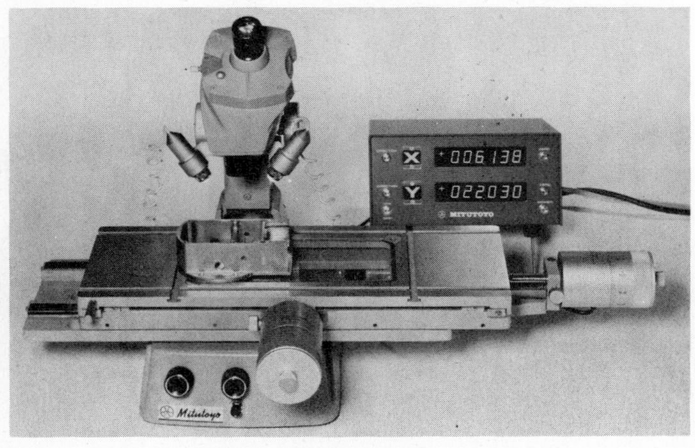

parts. The base, on which the special microscope is mounted, has a table that can be moved in two mutually perpendicular, horizontal directions by means of accurate micrometer screws that can be read to 0.0001 inch, or, if so equipped, by means of the digital readout. Parts to be measured are mounted on the table, the microscope is focused, and one end of the desired part brought into coincidence with the crossline in the microscope. The reading is then noted and the table moved until the other extremity of the part is in coincidence with the crossline. From the final reading, the desired measurement can be determined. In addition to a wide variety of linear measurements, accurate angular measurements can also be made by means of a special protractor eyepiece that can be put into the microscope.

The *optical projector* or *comparator,* shown in Figure 16-22, is a larger optical device on which both linear and angular measurements can be made. As with the toolmakers' microscope, the part to be measured is mounted on a table that can be moved in two directions by accurate micrometer screws. By means of an optical system the image of the part is projected on a screen, magnified from 5 to more than 100 times. Measurements can be made directly, either by means of the micrometer dials or on the magnified image on the screen by means of an accurate rule. A very common use for this type of instrument is the checking of parts, such as dies, screws, and so forth, against a template that is drawn to an enlarged scale, placed on the screen, and compared with the projected image of the part.

FIGURE 16-22. Optical contour comparator. Note workpiece on table. (*Courtesy Jones & Lamson Division of Waterbury Farrel.*)

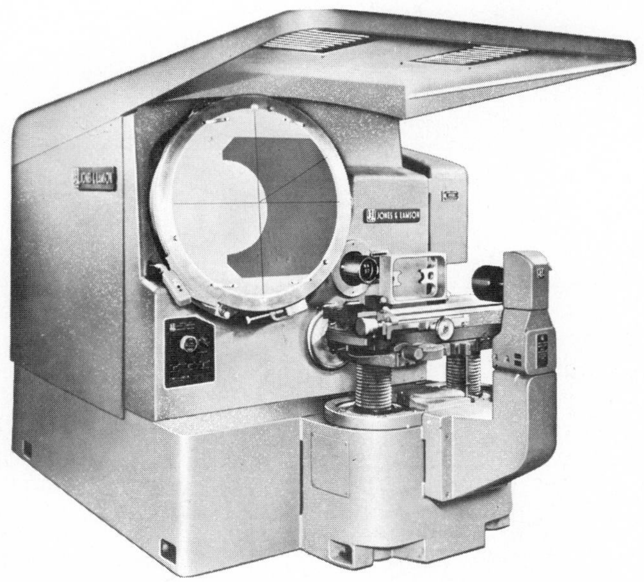

Accurate measurement of distances greater than a few inches was very difficult until the development of laser interferometry. Instruments based on this principle permit accuracies of ±0.000 25 mm (±0.000,01 inch) to be achieved routinely, and substantially greater accuracy is possible. Such equipment is particularly useful in checking the movement of machine tool tables, aligning and checking large assembly jigs, and making measurements of intricate machined parts, such as tire-tread molds. Figure 16-23 shows one type which employs the Doppler effect of a split laser beam, with the reflecting target attached to one end of the distance to be measured. Another type utilizes the principle depicted in Figure 16-24, wherein the laser beam acts as a probe and is reflected by the surface that is a variable distance from the beam

FIGURE 16-23. (*Top*) Calibrating the table movement of a machine tool by means of a laser interferometer. (*Bottom*) Schematic diagram of the components of a laser interferometer. (*Courtesy Hewlett-Packard.*)

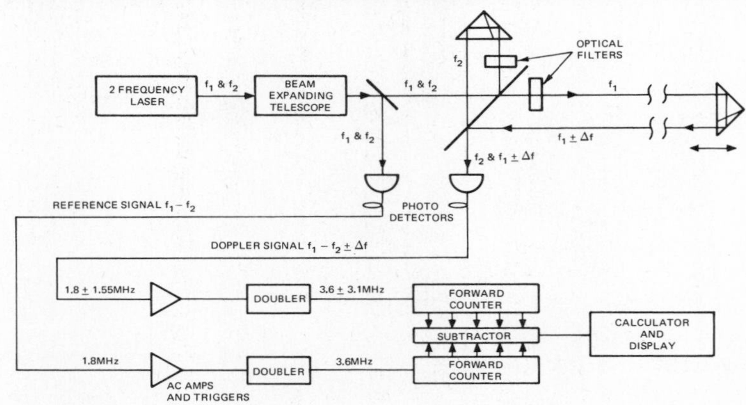

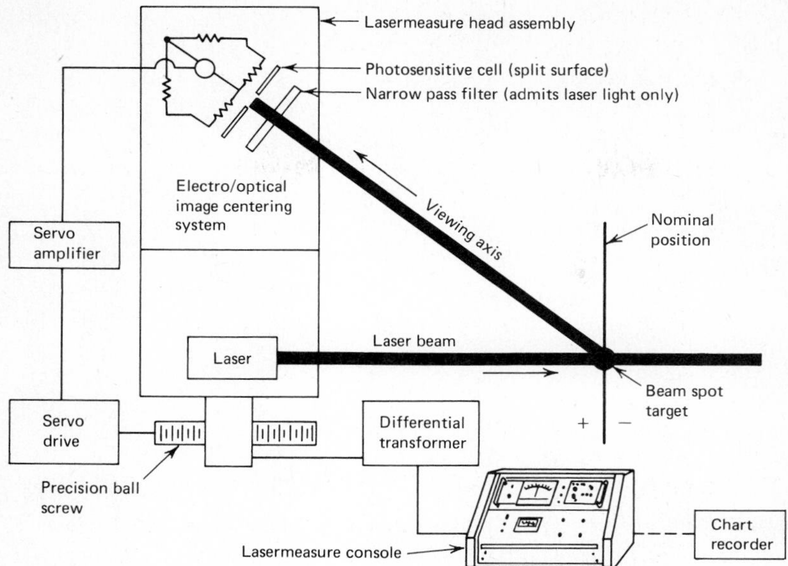

FIGURE 16-24. Arrangement for direct linear measurement by means of a laser beam.

source and the electro-optical image-reading system. An accurate measuring screw moves the source-reading head so as to center the reflected image and thus measure the variation in distance. Both types of laser systems incorporate a digital readout.

A number of the commonly used measuring instruments are of the indirect or transfer-measuring type. The most common are inside and outside calipers and dividers. These are shown in Figure 16-25. To obtain a measurement with these instruments they first are adjusted so that their legs just contact the desired portions of the object to be measured. Consequently, a considerable amount of care and experience is required to get the proper "feel" of the caliper legs against the work—a matter of contact pressure. The calipers or dividers are then held against a rule and the distance between the points of the instrument is read from the scale. Although such instruments are versatile, their accuracy of measurement is limited.

Telescoping gages are used for making indirect, internal measurements. *Plain telescoping gages,* such as shown in Figure 16-26, are the most common type. These consist of a knurled handle and a head composed of three sections. The two outer sections telescope inside each other against spring pressure and can be locked in any position by means of a knurled screw on the end of the handle. Their outer ends are rounded to a radius that is less than half the minimum length to which the head will telescope. To use such a gage, the plungers are pushed inward until the telescoping head will fit within the in-

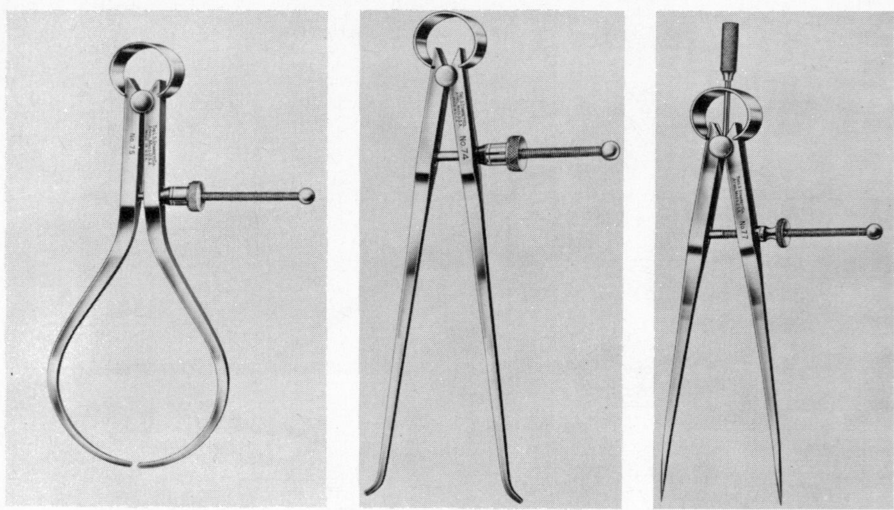

FIGURE 16-25. (*Left to right*) Outside calipers, inside calipers, and dividers.

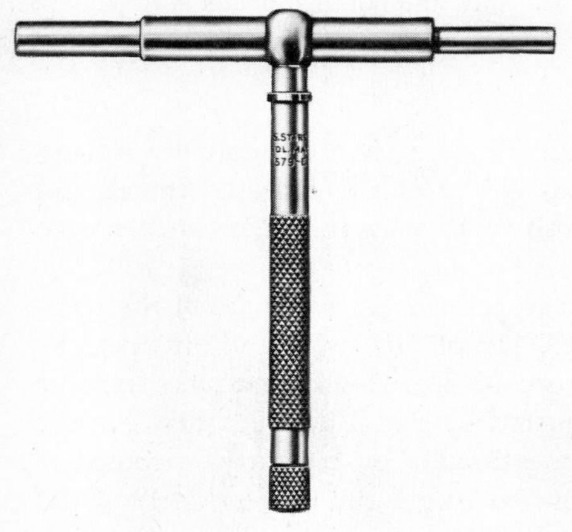

FIGURE 16-26. Telescoping gage used for making internal measurements. (*Courtesy The L. S. Starrett Company.*)

ternal dimension that is to be measured. The plungers are then locked and the gage withdrawn. A micrometer caliper commonly is used to measure the distance between the extremities of the gage head.

Small-hole gages, shown in Figure 16-27, are a special type of telescoping gage.

Angle-measuring instruments. Accurate angle measurements usually are more difficult to make than linear measurements. Angles are measured in *degrees*—

FIGURE 16-27. Small-hole gage. (*Courtesy The L. S. Starrett Company.*)

$^1/_{360}$ part of a circle—and decimal subdivisions thereof (or in minutes and seconds of arc).[4]

The *bevel protractor,* illustrated in Figure 16-28, is the most general angle-measuring instrument. The two movable blades are brought into contact with the sides of the angular part, and the angle can be read on the vernier scale to five minutes of arc. A clamping device is provided to lock the

FIGURE 16-28. Measuring an angle with a bevel protractor. (*Courtesy Brown & Sharpe Manufacturing Company.*)

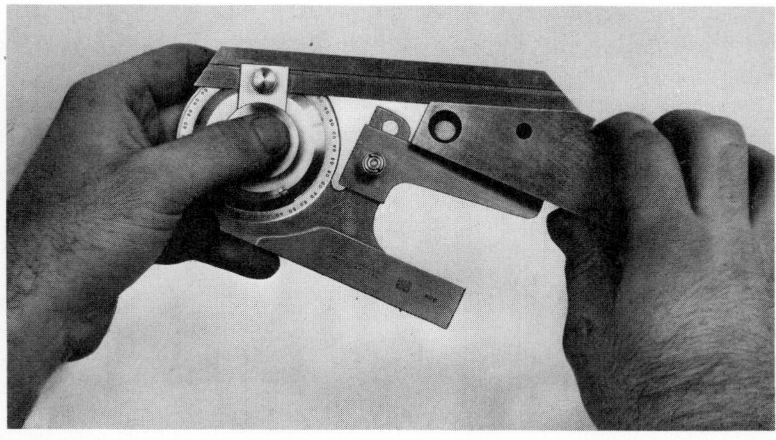

[4]The SI system calls for measurement of plane angles in radians, but degrees are permissible and, for several reasons, it appears likely that the use of degrees will continue to predominate in manufacturing, but with minutes and seconds of arc possibly being replaced by decimal portions of a degree.

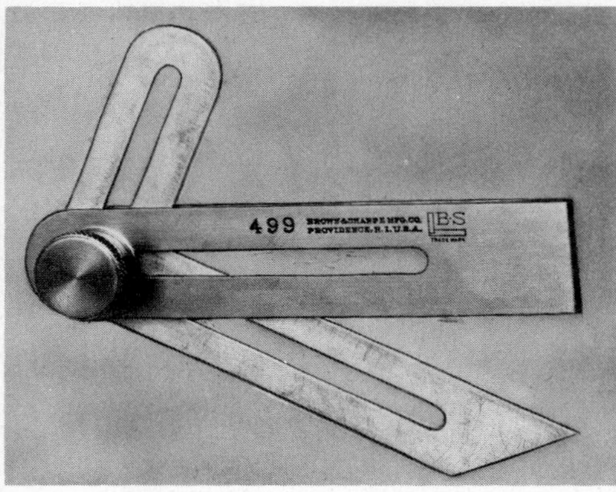

FIGURE 16-29. Universal bevel. (*Courtesy Browne & Sharpe Manufacturing Company.*)

blades in any desired position so that the instrument can be used for both direct measurement and layout work.

As indicated previously, the angle attachment on the combination set also can be used to measure angles in a manner similar to the bevel protractor, but usually with somewhat less accuracy.

The toolmakers' microscope is very satisfactory for making angle measurements, but its use is restricted to small parts. The accuracy obtainable is 5 minutes of arc. Similarly, angles can be measured on the contour projector.

The *universal bevel,* shown in Figure 16-29, is an indirect-measuring instrument for angles.

When very accurate angle measurements are required, a *sine bar* may be employed if the physical conditions will permit. This device, as illustrated in Figure 16-30, consists of an accurately ground bar on which two accurately ground pins, of the same diameter, are mounted an exact distance apart. The distances used are usually either 5 or 10 inches, and the resulting instrument is called a 5- or 10-inch sine bar.[5] Measurements are made by using the prin-

FIGURE 16-30. Method of measuring an angle by means of a sine bar.

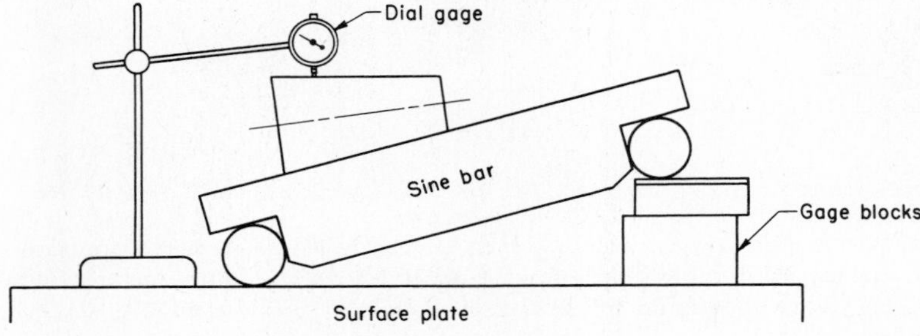

[5] Sine bars also are available with millimeter dimensions.

FIGURE 16-31. Measuring an angle by the use of angle gage blocks. (*Courtesy Webber Gage Company.*)

ciple that the sine of a given angle is the ratio of the opposite side of the right triangle to the hypotenuse. The object being measured is attached to the sine bar and the inclination of the assembly raised until the top surface is exactly parallel with the surface plate. If a stack of gage blocks is used to elevate one end of the sine bar, as shown, the height of the stack directly determines the difference in height of the two pins. The difference in height of the pins also can be determined by a dial gage or some other type of gage. The difference in elevation is then equal to either five or ten times the sine of the angle being measured, depending on whether a 5- or 10-inch bar is being used. Tabulated values of the angles corresponding to any measured elevation difference for 5- or 10-inch sine bars are available in various handbooks. Several types of sine bars are available to suit various requirements.

Accurate measurements of angles to 1 second of arc can be made by means of *angle gage blocks,* as shown in Figure 16-31. These come in sets of 16 blocks that can be assembled in desired combinations. Some type of indicator must be used, as shown. Angle measurements also can be made to $\pm 0.001°$ on rotary indexing tables having suitable numerical control.[6]

GAGES

In manufacturing, particularly in mass production, it usually is not necessary to know the exact dimensions of a part, only that it is within previously es-

[6] See Chapter 39.

tablished limits. This can be determined much more easily than specific dimensions by the use of *comparison-type* instruments called gages, either manually or mechanically. They may be either *fixed type* or *deviation type* and are used for both linear and angular dimensions.

Fixed-type gages. *Fixed-type gages* are designed to gage only one dimension and indicate whether it is larger or smaller than a previously established standard. They do not determine how much larger or smaller the measured dimension is than the standard. Because such gages fulfill a simple and limited function, they are relatively inexpensive and are easy to use.

Gages of this type ordinarily are made of hardened steel of proper composition and heat treatment to produce dimensional stability. Hardness is essential so that wear will be minimized and accuracy maintained. Because steels of high hardness tend to be dimensionally unstable, some fixed gages are made of softer steel with a hard chrome plating on the surface to provide surface hardness. Chrome plating also can be used for reclaiming some worn gages. Where gages are to be subjected to great use, they may be made of tungsten carbide at the wear points.

Plug gages are one of the most common types of fixed gages. As shown in Figure 16-32, these are accurately ground cylinders, held in a handle, that are used to gage internal dimensions such as holes. The gaging element of a *plain* plug gage has a single diameter. To control the minimum and maximum limits of a given hole, two plug gages are required. The smaller, or "go," gage controls the minimum because it must go (slide) into any hole that is large enough to meet the required minimum. The larger, "not-go," gage controls the maximum dimension because it must not go into any hole that is not larger than the maximum permissible size. As shown in Figure 16-32, the go and not-go plugs often are fastened into the two ends of a single handle for convenience in use. The not-go plug usually is much shorter than the go plug; it is subjected to little wear because it seldom slides into any holes. *Step-type* go, not-go gages, also shown in Figure 16-32, have the go and not-go diameters on a single plug, the go portion being the outer end. In gaging an acceptable part, the go portion should slide into the hole but the not-go section

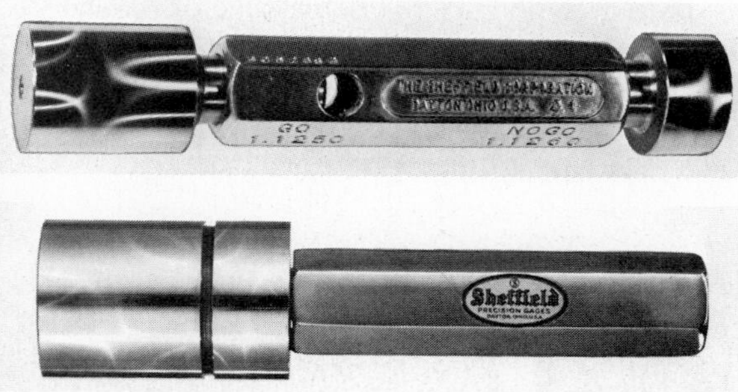

FIGURE 16-32. (*Top*) Plain plug gage having go member on one end and not-go member of the other. (*Bottom*) Plug gage with stepped go and not-go member. (*Courtesy The Bendix Corporation, Automation and Measurement Division.*)

FIGURE 16-33. Go and not-go ring gages. (*Courtesy The Bendix Corporation, Automation and Measurement Division.*)

should not enter. In using a plug or any other type of fixed gage, the gage should never be forced into, or onto, the part being measured.

Plug-type gages also are made for gaging shapes other than cylindrical holes. Three common types are *taper plug gages, thread plug gages,* and *spline gages.* Taper plug gages gage both the angle of the taper and its size. Any deviation from the correct angle is indicated by looseness between the plug and the tapered hole. Size is indicated by the depth to which the plug fits into the hole—the correct depth being denoted by a mark on the plug. Thread plug gages come in go and not-go types. The go gage must screw into the threaded holes while the not-go gage must not enter.

Ring gages are used to gage shafts or other external round members. These also are made in go and not-go types as shown in Figure 16-33. Go ring gages have plain knurled exteriors, whereas not-go ring gages have a circumferential groove in the knurling so that they can easily be distinguished.

Ring thread gages, shown in Figure 16-34, are made to be slightly adjust-

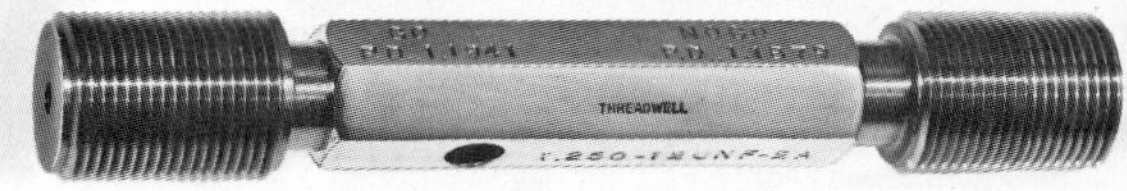

FIGURE 16-34. Plug and ring thread gages. (*Courtesy The Bendix Corporation, Automation and Measurement Division.*)

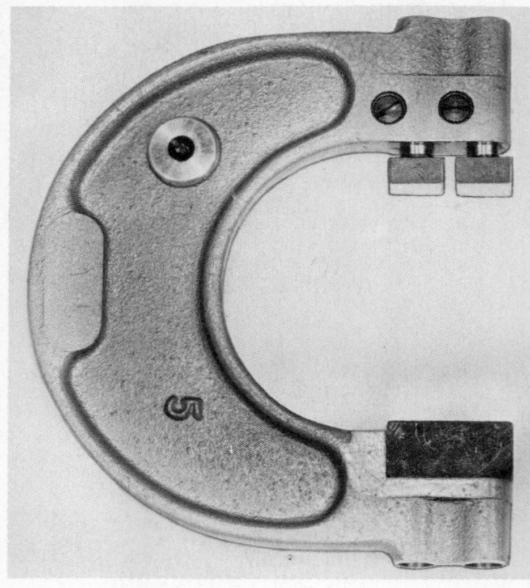

FIGURE 16-35. Adjustable go and not-go snap gage. (*Courtesy The Bendix Corporation, Automation and Measurement Division.*)

able because it is almost impossible to make them exactly to the desired size. Thus they are adjusted to exact, final size after the final grinding and polishing have been completed.

Snap gages are the most common type of fixed gage for measuring external dimensions. As shown in Figure 16-35, they have a rigid, U-shaped frame on which are two or three gaging surfaces, usually made of hardened steel or tungsten carbide. In the adjustable type shown, one gaging surface is fixed while the other(s) may be adjusted over a small range and locked at the desired position(s). Because, in most cases, one wishes to control both the maximum and minimum dimensions, the *progressive* or *step-type* snap gage, shown in Figure 16-35, is used most frequently. These have one fixed anvil and two adjustable surfaces to form the outer go and inner not-go openings, thus eliminating the use of separate go and not-go gages.

Snap gages are available in several types and a wide range of sizes. The gaging surfaces may be round or rectangular. They are set to the desired dimensions with the aid of gage blocks.

Many types of special gages are available or can be constructed for special applications. The *flush-pin gage,* illustrated in Figure 16-36, is an example. Used for gaging the depth of shoulders or holes, the main section is placed on

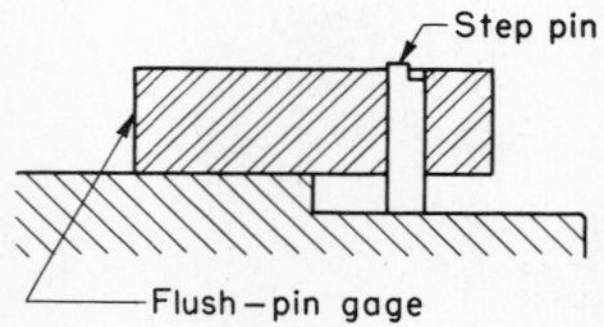

FIGURE 16-36. Principle of the flush-pin gage.

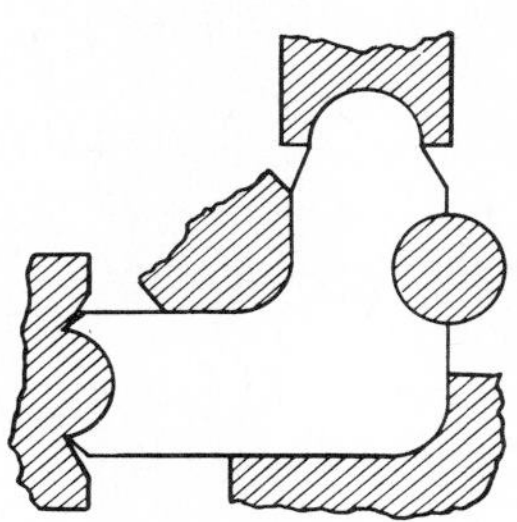

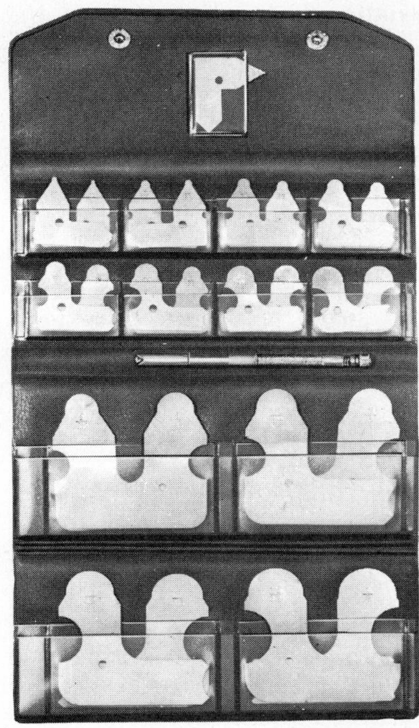

FIGURE 16-37. Set of radius gages, showing how they are used. (*Courtesy MTI Corporation.*)

the higher of the two surfaces with the movable step pin resting on the lower surface. If the depth between the two surfaces is sufficient but not too great, the top of the pin, but not the lower step, will be slightly above the top surface of the gage body. If the depth is too great, the top of the pin will be below the surface. Similarly, if the depth is not great enough, the lower step on the top of the pin will be above the surface of the gage body. By running a finger, or fingernail, across the top of the pin, its position with respect to the surface of the gage body can readily be determined.

Several types of *form gages* are available for use in checking the *profile* of various objects. Two of the most common types are *radius gages,* shown in Figure 16-37, and *screw-thread pitch gages,* shown in Figure 16-38.

FIGURE 16-38. Thread pitch gages. (*Courtesy The L. S. Starrett Company.*)

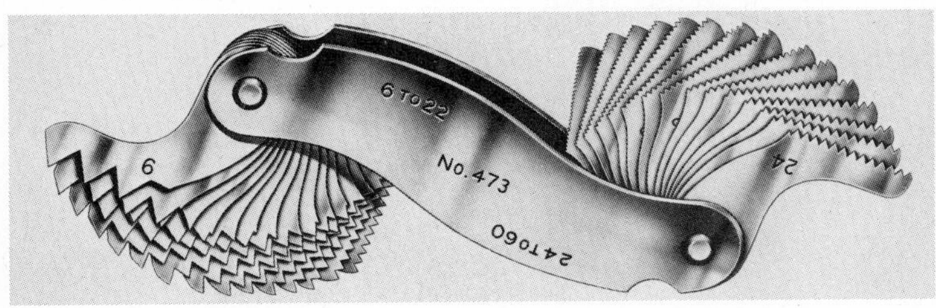

Deviation-type gages. A large amount of gaging, and some measurement, is done through the use of *deviation-type gages,* which determine the amount by which a measured part deviates, plus or minus, from a standard dimension to which the instrument has been set. In most cases the deviation is indicated directly in units of measurement, but in some cases, notably in production inspection, the gage shows only whether the deviation is within a permissible range. Such gages employ mechanical, electrical, or fluidic amplification techniques so that very small linear deviations can be detected. Most are quite rugged, and they are available in a variety of amplifications and sizes. As a consequence, it is quite easy to obtain a suitable gaging device of this type for use in almost any kind of gaging or measuring situation.

Dial indicators, such as shown in Figure 16-4, are a simple and widely used form of deviation-type gage. Movement of the gaging spindle is amplified mechanically through a rack and pinion and a gear train and indicated by a pointer on a graduated dial. Most dial indicators have a spindle travel equal to about 2½ revolutions of the indicating pointer and read in either 0.02- or 0.002-mm or 0.001- or 0.0001-inch units. The dial can be rotated by means of the knurled bezel ring to align the zero point with any position of the pointer. The indicator often is mounted on an adjustable arm to permit its being brought into proper relationship with the work. It is important that the axis of the spindle be aligned exactly with the dimension being gaged if accuracy is to be achieved.

Dial indicators should be checked occasionally to assure that accuracy has not been lost through wear in the gear train. Also, it should be remembered that the pressure of the spindle on the work varies due to spring pressure as the spindle moves into the gage. This normally will cause no difficulty unless the spindles are used on soft or flexible parts.

Three types of *deviation comparators* are widely used for routine gaging

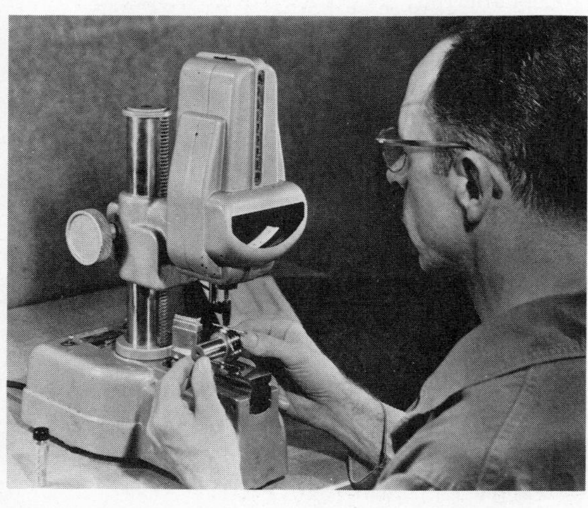

FIGURE 16-39. Visual comparator.

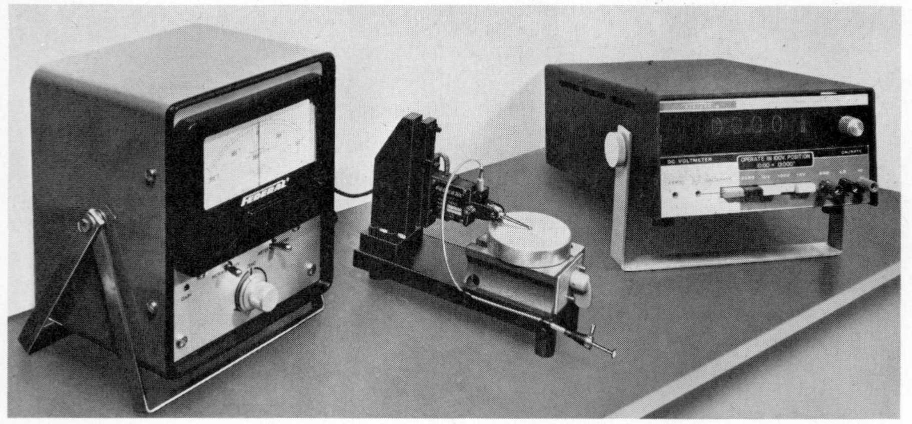

FIGURE 16-40. Electronic-magnification gage being used to gage a computer memory core 0.76 mm (0.030 inch) in diameter and 0.15 mm (0.006 inch) thick. (*Courtesy Federal Products Corporation.*)

and inspection. All are set to zero by means of standard gages or gage blocks and indicate the plus or minus deviation of the part being gaged. The mechanical type, shown in Figure 16-39, is quite rugged and readily portable. Depending on the magnification, each division on the dial may be from 0.002 to 0.0002 mm (0.001 to 0.0001 inch), with a total range of about ±0.25 mm (±0.010 inch).

Comparators employing electronic magnification are being used increasingly. As shown in Figure 16-40, the gaging head is small, readily portable, and can be mounted in many ways. The end of the sensing lever is shaped so as to automatically compensate for misalignment in the measuring plane up to about ±15°. The indicator may employ either a pointer and graduated scale or a digital readout. Accuracies up to ±0.0003 mm (0.000,01 inch) are available, and several ranges usually can be selected by merely turning a knob. Figure 16-41 shows an example of the versatility of this type of gage.

Another type of electronic gage is based on a linear variable differential transformer size-sensing element that produces a high-resolution, solid-state diode display. As shown in Figure 16-42, these gages frequently are combined into multiple units for simultaneous gaging of several dimensions. Ranges and resolutions down to 0.013 and 0.000 25 mm (0.000,5 and 0.000,01 inch), respectively, are available.

Air gages have special characteristics that make them especially suitable for gaging holes or internal dimensions of various shapes. A typical gage of this type is shown in Figure 16-43. These gages indicate the clearance between the gaging head and the hole by measuring either the volume of air that escapes or the pressure drop resulting from the air flow. The gage is calibrated directly in 0.02-mm or 0.001-inch divisions. Air gages have an ad-

FIGURE 16-41. Electronic comparator-type gage with gaging head being used with a vertical stand for measuring a large workpiece. (*Courtesy The Bendix Corporation, Automation and Measurement Division.*)

FIGURE 16-42. Three electronic-column gages being used as a unit to check a piston. A single indicating light shows if any of the three dimensions is out of tolerance. (*Courtesy The Bendix Corporation, Automation and Measurement Division.*)

FIGURE 16-43. Gaging the diameter of a hole with an air gage. (*Courtesy The Bendix Corporation, Automation and Measurement Division.*)

vantage over mechanical or electronic gages for this purpose in that they not only detect linear size deviations but also out-of-round conditions. Also they are subject to very little wear because the gaging member always is slightly smaller than the hole and the air flow minimizes rubbing. Special types of air gages can be used for external gaging.

The *wiggler indicator,* shown in Figure 16-44, is a simple, rugged, deviation-type gage that is widely used by machinists in making workpiece setups in machine tools. Through a lever system, 0.001-inch movements of the gaging point are indicated by the pointer.

FIGURE 16-44. Wiggler indicator. (*Courtesy Ideal Tool Company.*)

Measurement by lightwave interference. The phenomenon of lightwave interference can be used for making very precise measurements, such as in calibrating gage blocks. Three pieces of equipment are required. The first is an *optical flat.* These are quartz or special glass disks, from 50 to 250 mm (2 to 10 inches) in diameter and about 12 to 25 mm (½ to 1 inch) thick, whose surfaces are very nearly true planes and nearly parallel. These flats can be obtained with the surfaces within 0.000 03 mm (0.000,001 inch) of true flatness. It is not essential that both surfaces be accurate or that they be exactly parallel, but one must be certain that only the accurate surface is used in making measurements.

The second item of equipment required is a *toolmaker's flat.* These are similar to optical flats but are made of steel and usually have only one surface that is accurate.

The third requirement is a *monochromatic light source,* emitting light of a single wavelength. Selenium, helium, or cadmium sources are commonly used.

Lightwave interference bands are created by the phenomenon illustrated in Figure 16-45. A portion of the light rays coming from the source are reflected from the bottom surface of the optical flat while the remaining portions pass through the flat and are reflected from the surface being tested. If the optical flat and the work surface are separated by a very small angle, at certain intervals along the surface, the distance between the flat and the work surface will be such that the extra distance the light reflected from the work surface must travel will cause it to be 180° out of phase with the portion reflected from the lower surface of the optical flat. As a result, the two portions cancel each other and cause a dark line. At *A,* in Figure 16-45, the light ray striking at this point is thrown out of phase 180° upon being reflected from the work surface, and it also travels a distance of ½ wavelength in going each direction across the gap. It thus is 180° out of phase with the portion of the ray that is reflected from the lower surface of the optical flat. This produces a dark band at *A.* At *B* the distance between the flat and the work surface is ¾ wavelength, so the 1½ wavelengths of extra distance, plus the 180° phase change upon reflection, puts two reflected rays in phase and produces a

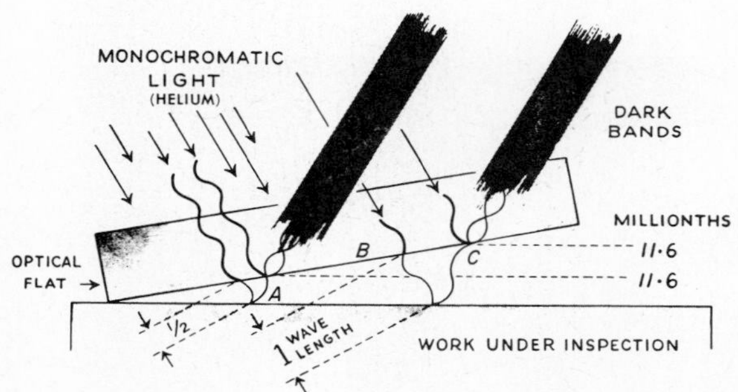

FIGURE 16-45. Explanation of the method of accurate measurement by light-wave interference phenomenon. (*Courtesy The DoALL Company.*)

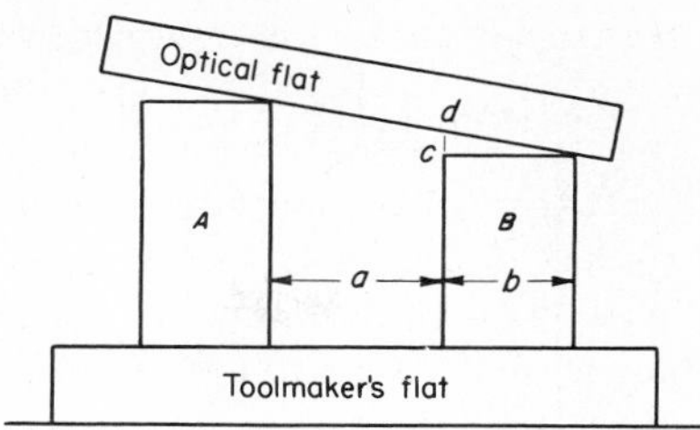

FIGURE 16-46. Method of calibrating gage blocks by light-wave interference.

bright band. Again at *C* a dark band is produced. Because the difference in the distances between the optical flat and the work surface at *A* and *C* is ½ wavelength, each dark band indicates a change of ½ wavelength in the elevation of the work surface. If a monochromatic light source having a wavelength of 23.2 microinches (0.589 μm) is used, each interference band represents 11.6 microinches (0.295 μm).

Figure 16-46 illustrates how lightwave interference is used for calibrating gage blocks. The block to be calibrated is placed at *B* and a calibrated block at *A*. Distances *a* and *b* must be known but do not have to be measured with great accuracy. By counting the number of interference bands showing on the surface of gage block *B*, the distance *c–d* can be determined and then, by simple geometry, the difference in the heights of the two blocks can be computed. The same method is applicable for making precise measurements of other objects by comparing them with a known gage block.

Light interference also makes it possible to determine easily whether a surface is exactly flat. This is illustrated in Figure 16-47.

FIGURE 16-47. Using a helium light source and an optical flat to determine the flatness of a surface. Each light band indicates a deviation of 0.295 μm (11.6 microinches). (*Courtesy The DoALL Company.*)

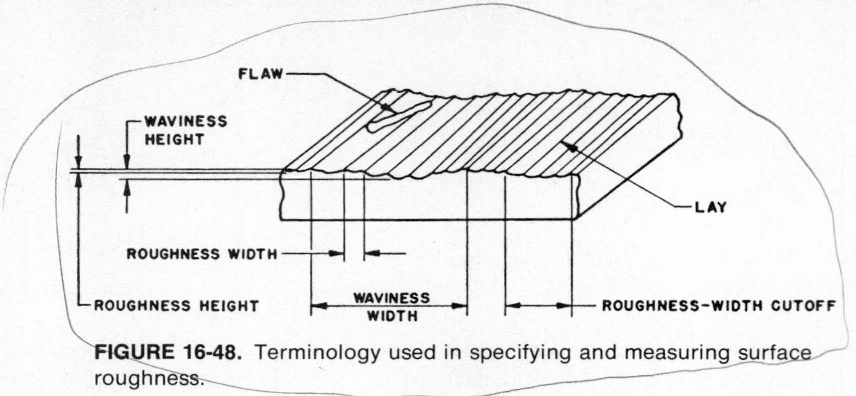

FIGURE 16-48. Terminology used in specifying and measuring surface roughness.

Surface roughness measurement. Increasingly, designers specify closer fits between mating parts in order to improve the performance and useful life of products. However, the quality of the mating surfaces also plays an important role in the performance and wear of mating parts. Consequently, an understanding of the nature of the surfaces that are obtained by various manufacturing processes, and the determination of surface quality, have become increasingly important.

Certain standard terms and symbols have been developed for use in specifying and measuring surface roughness. The most important terms are roughness, waviness, and lay, which are illustrated in Figure 16-48. *Roughness* refers to the finely spaced surface irregularities. It results from machining operations in the case of machined surfaces. *Waviness* is surface irregularity of greater spacing than occurs in roughness. It may be the result of warping, vibration, or the work deflecting during machining. *Lay* is the direction of the predominant surface pattern produced by the roughness or waviness.

Roughness is measured by the heights of the irregularities with respect to an average line, as illustrated in Figure 16-49. These measurements usually are expressed in micrometers or microinches. In most cases the arithmetical average (AA) is used. In terms of the measurements indicated in Figure 16-49, this would be as follows:

$$\frac{y_1 + y_2 + y_3 + y_4 + y_5 + y_6 + y_7 + \cdots + y_n}{n}$$

where n is the total number of vertical measurements. Occasionally the *root-mean-square* (rms) value is used. This is defined as

$$\text{rms} = \sqrt{\frac{\Sigma\, y^2}{n}}$$

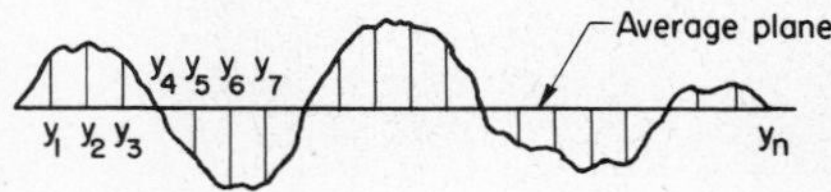

FIGURE 16-49. Method of designating surface roughness by reference to a plane.

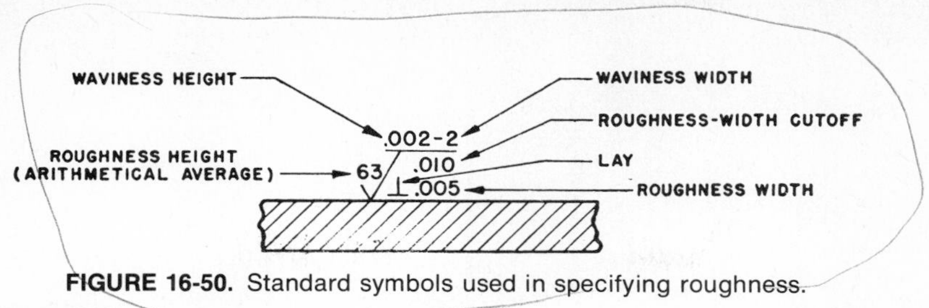

FIGURE 16-50. Standard symbols used in specifying roughness.

In specifying surface finish, the symbols established by the ANSI and shown in Figure 16-50 are commonly used.

Several instruments are available for measuring surface roughness. In each, a sharp diamond stylus is moved at a constant rate across the surface to be measured. The rise and fall of the stylus are amplified electrically and indicated on a meter that indicates the average surface roughness. The unit that contains the stylus and driving motor may be held in the hand or supported on the workpiece or other supporting surface. One of these instruments is shown in Figure 16-51. The output from the stylus also may be connected to a recording device so as to record the surface quality on a moving paper chart.

The instrument shown in Figure 16-52 provides a means for gaining an unusually good understanding of the nature and quality of a surface by providing a series of offset and parallel traces of the surface profile, as shown in the

FIGURE 16-51. Using Brush Surfindicator with portable head to measure surface roughness of a workpiece while mounted in a lathe. (*Courtesy Brush Instrument Division, Clevite Corporation.*)

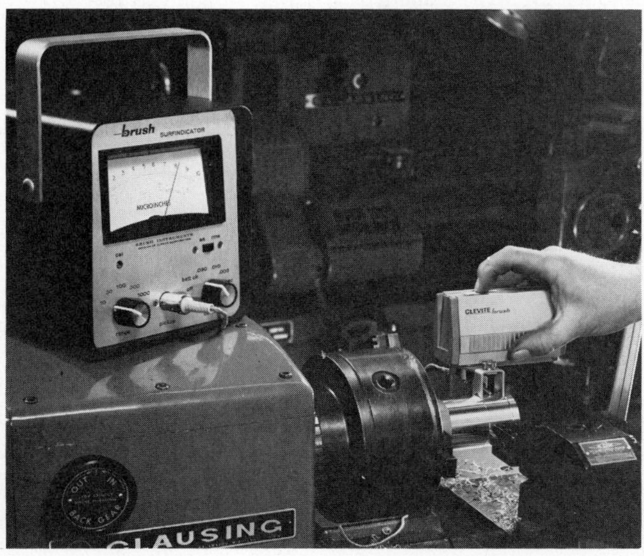

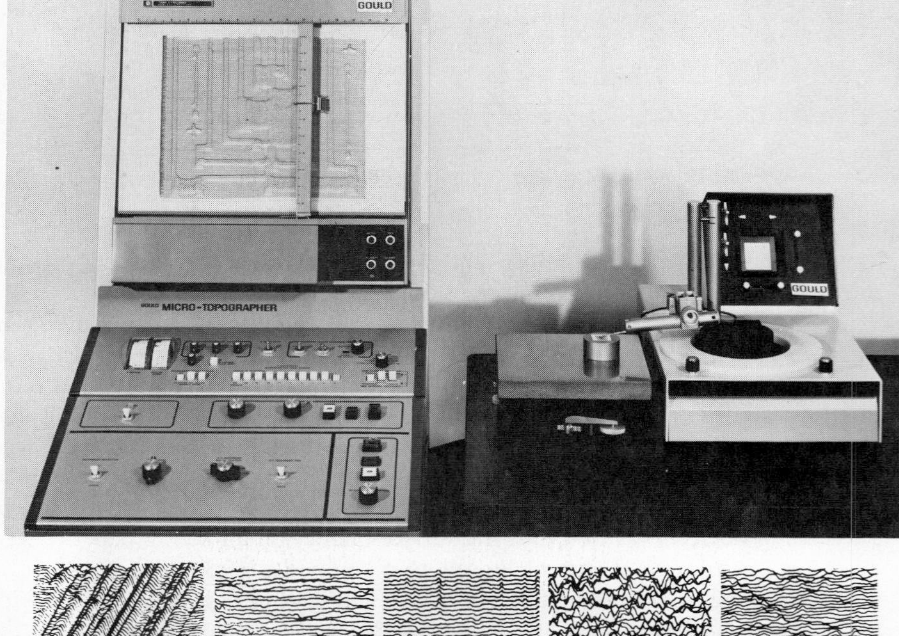

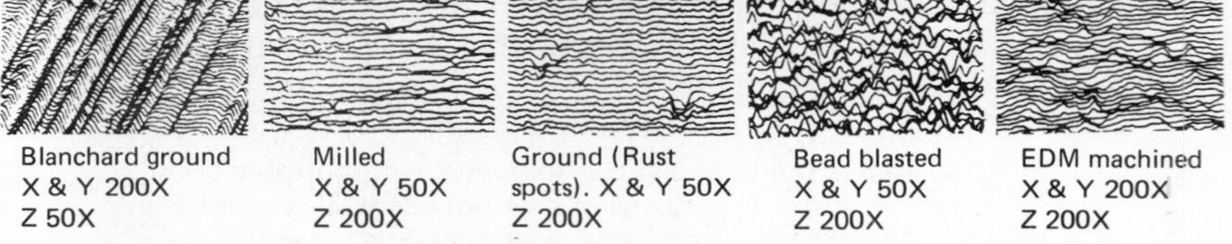

FIGURE 16-52. (*Top*) Micro-Topographer, used to measure and depict surface roughness and character. (*Bottom*) Some typical surface-roughness profiles. (*Courtesy Gould Inc., Measurement Systems Division.*)

lower portion of the illustration. It measures an area of from 0.13 × 0.13 mm (0.005 × 0.005 inch) up to 50.8 × 50.8 mm (2 × 2 inches), depending on the magnification selected.

Surface roughnesses that typically are produced by various manufacturing processes are indicated in Figure 16-53. For the designer's use, sets of *surface-finish blocks,* such as are shown in Figure 16-54, are very useful. It is difficult to visualize what a surface having a given microinch roughness looks like, inasmuch as the same value of roughness may reflect different surface characteristics when produced by different processes.

Gage tolerances. Like other machined products, gages cannot be made to exact sizes and must have specified tolerances. Obviously, the tolerances permitted in a gage must be less than the permissible tolerance for the part that it is to measure. In general, the tolerance that is permitted on a gage is about

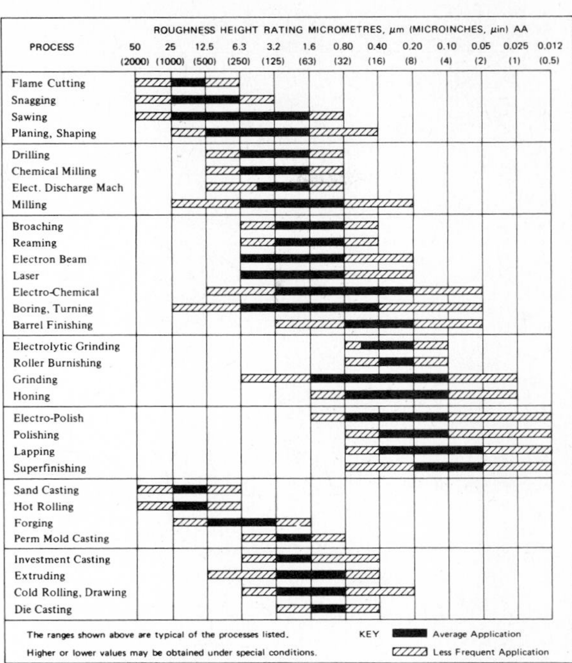

Extracted from General Motors Drafting Standards, June 1973 revision

FIGURE 16-53. Comparison of surface roughness produced by common production processes. (*Courtesy American Machinist.*)

FIGURE 16-54. Set of surface-roughness standards being used in a drafting room. (*Courtesy Surface Checking Gage Co.*)

TABLE 16-2. Gagemakers' tolerances

Nominal Size (inches)		Class			
Above	To and Including	XX	X	Y	Z
0.029	0.825	0.00002	0.00004	0.00007	0.00010
0.825	1.510	0.00003	0.00006	0.00009	0.00012
1.510	2.510	0.00004	0.00008	0.00012	0.00016
2.510	4.510	0.00005	0.00010	0.00015	0.00020
4.510	6.510	0.000065	0.00013	0.00019	0.00025
6.510	9.010	0.00008	0.00016	0.00024	0.00032
9.010	12.010	0.00010	0.00020	0.00030	0.00040

10 per cent of the tolerance on the part that is to be gaged. Table 16-2 is an example of the tolerances that are recommended for gages of various classes in the English system. Class XX gages are of the highest precision and are used primarily for master or reference gages. Class Y gages are those that are used for the most accurate commercial work.

In making gages, provision must be made for wear. The wear allowance is applied so that as some wear occurs, the gage will continue to assure that the inspected products are within the desired dimensions. Thus a new gage may reject some parts that are within the specified limits.

Selective assembly. In mass production, it often is possible to achieve a desired fit between mating parts of specific assemblies while not requiring the parts to be made to extremely close tolerances. The parts are inspected for size, often automatically, and are grouped in several classes according to actual size. Mating parts then are selected from the classes that will provide the desired allowance. Such a procedure can lower manufacturing costs appreciably, while maintaining desired fits, by permitting greater tolerances, or better fits can be obtained from the use of given tolerances. However, such a procedure may cause some difficulty if parts must be replaced later.

INSPECTION AND QUALITY CONTROL

In mass production, it is extremely important that the dimensions and quality of individual parts be known and maintained. This is of particular importance where large quantities of parts, often made in widely separated plants, must be capable of interchangeable assembly. Otherwise, difficulty may be experienced in subsequent assembly or in service, and costly delays and failures may result and, too often, extremely expensive litigation and damage awards may ensue. Whether dimensions are controlled by the operators of machines or automatically by the machines, a check on their performance is required.

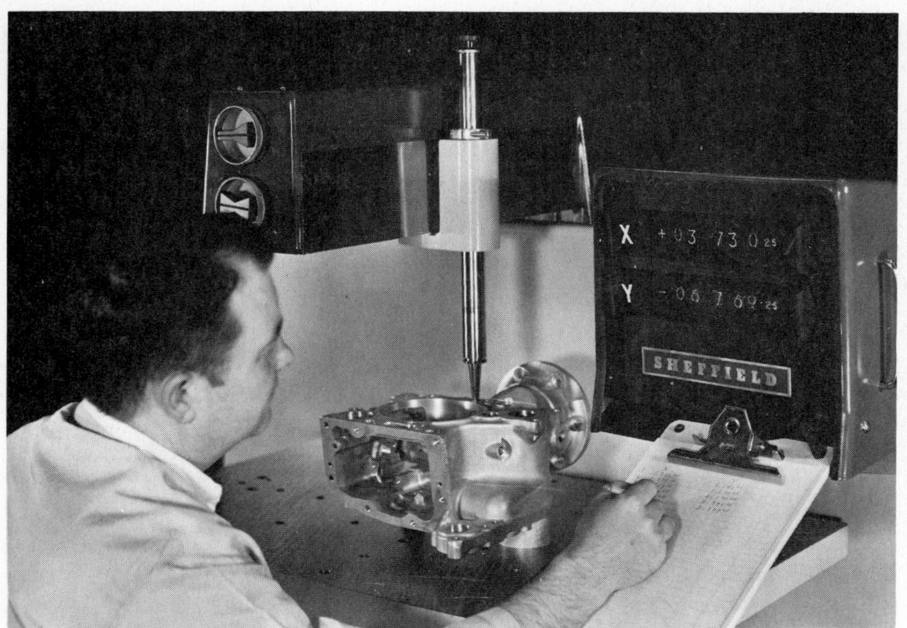

FIGURE 16-55. Measuring machine used for inspection; digital readout is provided in two axes. (*Courtesy The Bendix Corporation, Automation and Measurement Division.*)

Consequently, various means of inspection are used, and special equipment has been developed to assist in such measuring.

Modern measuring machines permit the making of direct linear measurements with ease and great accuracy. Such machines, as shown in Figure 16-55, can make measurements in two or three mutually perpendicular directions, reading directly in 0.002-mm or 0.0001-inch units. A transducer translates the motion of the table or a measuring head, depending on which moves, into electronic signals that actuate a digital readout display. Usually a floating-point zero is provided so that distances can be measured from any desired point. The gaging point may be optical cross hairs or a very sensitive probe that operates at almost zero pressure. Such machines are available in a wide range of sizes and have greatly facilitated the dimensional inspection of complex and large parts. Some are numerically controlled and will automatically inspect and record the results for a given workpiece after it is set up on the table.

Statistical quality control. Inspection of every part produced is costly, particularly if it must be done by people. In most cases, owing to the fallibility of inspection personnel and inspection machines, such a procedure will not assure that no defective parts are passed. In some cases, where the inspection process destroys or damages the product, 100 per cent inspection obviously is not feasible. It is also very important that when a process or operation is going out of control that this be determined before defective units are produced. *Statistical quality control procedures,* which are based on well-established statistical concepts, have been widely adopted to aid in determining and controlling quality.

In any continuing manufacturing process, variations from established standards are of two types: (1) *assignable cause variations,* such as those due to malfunctioning equipment or personnel, or to defective material, or due to a worn or broken tool; and (2) normal *chance variations,* resulting from the inherent nonuniformities that exist in materials and in machine motions and operations. Deviations due to assignable causes may vary greatly; their magnitude and occurrence are unpredictable, and one thus wishes to prevent their occurring. On the other hand, if the assignable causes of variation are removed from a given operation, the magnitude and frequency of the chance variations can be predicted with great accuracy. Thus, if one can be assured that only chance variations will occur, he knows what the quality of the product will be and manufacturing can proceed with assurance as to the results. By using statistical quality control procedures, one may detect the presence of an assignable-cause variation and remove the cause, often before it causes quality to become unacceptable.

Figure 16-56 shows the frequency distribution curve of the diameters of 100 ground pins having a nominal diameter of 12.700 mm (0.5000 inch). This is the type of *normal* or *bell-shaped* frequency distribution *curve* that results when only chance variations occur in a process. The *arithmetic average,* $\overline{X}$, obtained by the usual calculation, is

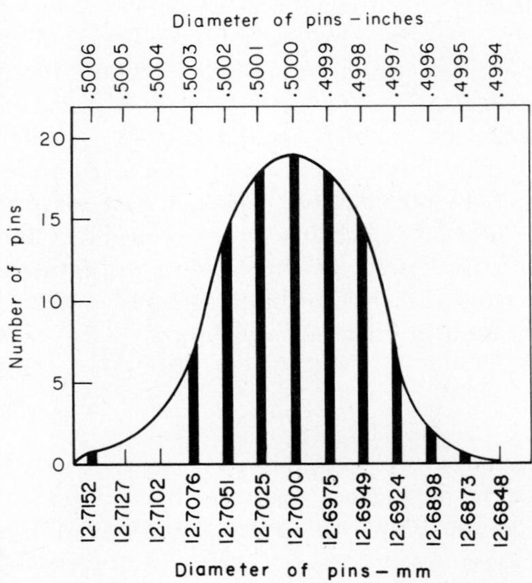

FIGURE 16-56. Frequency distribution of 100 pins having a nominal diameter of 12.700 mm (0.5000 inch).

$$\overline{X} = \frac{\sum_{1}^{n} X}{n} \tag{16-1}$$

where X is an individual measurement and *n* the number of items or measurements. It is a measure of the central tendency about which the individual measurements tend to group. The pattern of the way in which the individual measurements tend to group about the average may be indicated by the *standard deviation*, σ, where

$$\sigma = \sqrt{\frac{\sum_{1}^{n}(X - \overline{X})^2}{n}} \tag{16-2}$$

The standard deviation is of particular value in that for normal, or chance, distributions 68.26 per cent of all measurement values will lie with $\pm 1\sigma$ range from the average, 95.46 per cent within $\pm 2\sigma$ range, and 99.73 per cent within $\pm 3\sigma$ range. Thus, if one knows what the standard deviation is for a population of items, he knows that not over 0.3 per cent of them will fall outside the limits $X \pm 3\sigma$ *as long as only chance variations occur.* For the pin-grinding operation the average diameter, X, was 12.700 mm (0.5000 inch), and the standard deviation, σ, was 0.0048 mm (0.00019 inch), making $3\sigma = 0.014$ mm (0.0006 inch). The producer could thus set his manufacturing limits at 12.714 mm (0.5006 inch) / 12.682 mm (0.4994 inch) with considerable assurance that less than 3 parts in 1000 would fall outside those limits as long as no assignable causes of variation affected the process.

Quality control charts. Quality control charts are widely used as aids in maintaining quality and in achieving the objective of detecting trends in quality variation before defective parts are actually produced. These charts are based on the previously discussed concept that if only chance causes of variation are present, the deviations from the specified dimension or attribute will fall within predetermined limits.

In most cases sampling inspection is used—sample sizes of from three to six units being employed. This permits more rapid and more current inspection and also is less costly. Figure 16-57 shows a portion of the control charts for the 12.7-mm pins discussed previously. These charts are based on a sample size of 5. The control line of the $\overline{X}$ chart is the grand average of the averages of the diameters of each sample of five pins, computed as

$$\overline{X} = \frac{\sum_{1}^{k} \overline{X}}{k} \tag{16-3}$$

where *k* is the number of sample groups. The upper and lower control limits are set at $\overline{X} \pm 3\sigma_x$, respectively, where

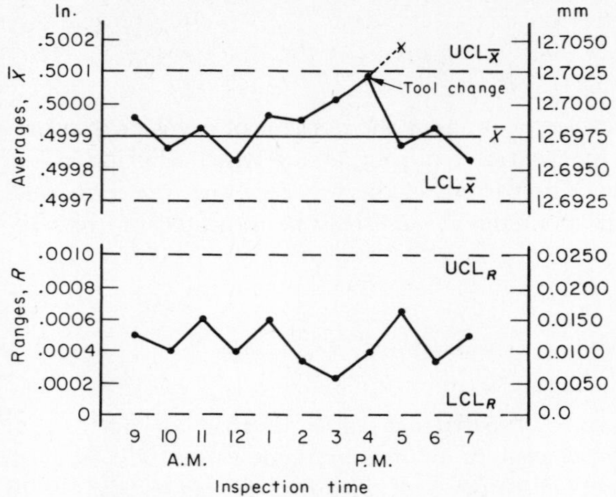

FIGURE 16-57. Statistical quality control charts for 12 700 mm (0.5000-inch) diameter pins. Note trend in *X* curve before tool change and probable point if tool had not been changed.

$$\sigma_x = \sqrt{\frac{\sum_{1}^{k}(\bar{X} - \bar{\bar{X}})^2}{k}} \tag{16-4}$$

In most cases a *range chart* is used as a second control. The value for $\bar{R}$ is the grand average of the ranges of the k subgroups of five pins each. However, to obtain the upper and lower control limits for the R-chart, the following relationships must be used:

upper control limit $= D_4\bar{R} = 2.115 \times \bar{R}$
lower control limit $= D_3\bar{R} = 0 \times \bar{R}$

D_4 and D_3 are constants that vary with the size of the sample and can be obtained from tables in books on quality control or from suitable handbooks. (The values of 2.115 and 0, given here, are for a sample size of 5.)

After the control charts have been established, the average and range for each sample group are plotted, as shown. If *both* values fall within the control limits, one may safely assume that only chance variations exist and that the quality is in control. However, if either the $\bar{X}$ or the $\bar{R}$ value falls outside the limits, some assignable cause is assumed to exist, and corrective action should be taken.

As shown in Figure 16-57, quality control charts often will indicate the existence and effect of an assignable-cause factor before a process actually produces units that are outside the control limits. In Figure 16-57, the $\bar{X}$ values for samples 6, 7, and 8 show a definite trend toward an out-of-control condition, thus signaling that corrective action should be taken, as was done in this case.[7] This warning feature of control charts and sampling inspection is of great value when parts are being produced at high-volume rates.

Review questions

1. What two factors did Eli Whitney accomplish with his filing jigs so as to produce interchangeable parts?
2. Explain how size specification may determine what process must be used to produce a part.
3. What is the difference between measurement and gaging?
4. What is the standard of linear measurement in the United States?
5. What is the official conversion factor between millimeters and inches under the SI system?
6. What gage blocks would you use from an 88-block metric set to obtain a dimension of 72.634 mm, using the minimum number of blocks?
7. What gage blocks would you select from an 81-block set to obtain a dimension of 3.4673 inches?
8. What is the standard measuring temperature?
9. It is has been said that "allowance" is desirable, whereas "tolerance" is necessary. Explain.
10. Why is hole size considered the basic dimension in applying allowances and tolerances?
11. Using a steel device for making the measurement, an aluminum part was passed as being 14.2503 inches; the specified dimensions were 14.2500 ± 0.0005 inch. Later it was learned when the inspection was made that both the part and the instrument were at 91°F. Using the linear coefficients for aluminum and steel as 0.000,013 and 0.000,006 inch per inch per degree F, respectively, determine whether the part actually was within tolerance.
12. Upon what factor should the allowance between mating parts be based?
13. What will be the limiting dimensions of a hole and mating shaft if the hole has a basic size of 52 mm and a Class 3 fit is desired?
14. What will be the limiting sizes of mating parts that are to have a Class 7 fit with a basic hole size of 2 inches?
15. What are four factors that can affect the accuracy obtainable with measuring instruments?
16. Give an example of a direct-reading and an indirect-reading linear measuring instrument.

[7] Under SQC theory a defective part can be produced while the system still is "in control," because with $\pm 3\sigma$ limits, 0.27 per cent of the product can be outside the limits. However, this is unlikely to occur if both the $\overline{X}$ and R values are within the control limits and no obvious trend has developed.

17. Why is a gage more often used in mass production to control dimensions than a measuring instrument?
18. What are three uses for a combination set?
19. What are the two most likely sources of error in using micrometer calipers?
20. What is the major disadvantage of a micrometer caliper as compared with a vernier caliper?
21. What would be the major difficulty in obtaining an accurate measurement with a micrometer depth gage (Figure 16-16) if it were not equipped with a ratchet or friction device for turning the thimble?
22. Explain how possible distortion of the frame is avoided in the super-micrometer.
23. Why is the toolmaker's microscope particularly useful for making measurements on delicate parts?
24. In what two ways can linear measurements be made using a contour projector?
25. What type of instrument would you select for checking the accuracy of linear movement of a machine tool table through a distance of 50 inches?
26. Upon what principle is a sine bar based?
27. How can the not-go member of a plug gage be easily distinguished from the go member?
28. What is the advantage of a progressive-type snap gage?
29. How does a taper plug gage check both the angle of taper and the size?
30. What is the primary precaution that should be observed in using a dial gage?
31. Why are air gages particularly well suited for gaging the diameter of a hole?
32. Explain the principle of measurement by lightwave interference.
33. In checking a 1-inch-square gage block by means of a helium light source, five dark bands were observed. There was a 2-inch distance between the front edges of the two blocks. What was the difference in height between the two blocks?
34. How does a toolmaker's flat differ from an optical flat?
35. Why may two surfaces that have the same microinch roughness be quite different in appearance?
36. Why are surface-finish blocks often used for specifying surface finish rather than just microinch values?
37. Why must the gagemaker's tolerance be less than the tolerance permitted on the part to be gaged?
38. Why are statistical quality control charts and sampling of particular value when parts are being made on automatic machines?
39. Explain how selective assembly can have an economic advantage.
40. What is one difficulty that may be experienced when selective assembly is utilized?

41. Does the use of sampling and statistical quality control rule out the possibility of any out-of-tolerance parts being accepted?

42. For a given part the specified basic dimension is 4.000 mm. A group of 10 samples of five each is measured and the sum of the averages of these groups is 40.0005 mm. The value for σ_x for these same samples is 0.0015, and $\bar{R}$ is 0.0025 mm. The measurements of five sample groups of five each from a production run are:

Sample				
1	2	3	4	5
3.996	4.001	4.002	3.999	4.002
3.998	4.002	4.003	3.998	4.001
3.997	4.002	4.002	3.999	4.002
3.998	4.003	4.004	3.997	4.002
3.999	4.002	4.003	3.998	3.999

Draw the $\bar{X}$ and $\bar{R}$ charts, plot the average and range values for each sample, and determine whether the operation is in control.

Case study 12.
THE UNDERGROUND STEAM LINE

An underground steam line for a military base in Alaska, approximately 1.6 km (1 mile) in length, utilized the units depicted in Figure CS-12, each unit being 9.14 meters (30 feet) in length. As shown, the steam line for each unit was enclosed in a cylindrical conduit made of sheet steel, and weighed approximately 356 kg (785 pounds). The conduit was to serve as the return drain line and to provide insulation. Each length of steam line was supported within the

FIGURE CS-12. Schematic diagram showing method of supporting steam line in drainage conduit for an underground steam line.

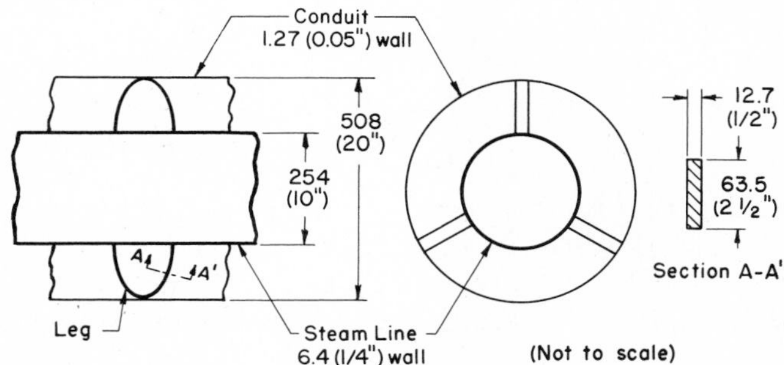

conduit by means of two sets of three U-shaped legs, made by cold-bending 12.7 × 63.5 mm (1/2 × 2 1/2 inch) hot-rolled steel bar stock, which were welded to the pipe about 2 meters (6.5 feet) from each end.

The units were fabricated in California and were transported to Alaska by a sequence of truck, barge, and railroad. Because of an early winter, only one-half of the line was installed the first summer. Before work was resumed the following spring, someone decided to test the line and found that there were numerous holes in the conduit. Consequently, a great amount of costly delay and litigation ensued.

(1) Where do you think the holes occurred? (2) What design error was the primary cause of the failures? (3) How would you modify the design to avoid the difficulty?

Metal cutting

Metal cutting—commonly called *machining*—is the removal of unwanted portions from a block of metal, in the form of chips, so as to leave a shape having the desired size and finish. The vast majority of manufactured products require machining at some stage in their production, ranging from relatively rough or nonprecision work, such as cleanup of castings or forgings, to high-precision work involving tolerances of 0.002 mm (0.0001 inch) or less. Thus it undoubtedly is the most important of the basic manufacturing processes.

Metal cutting basically involves the formation of chips, which may constitute from a few to more than 50 per cent of the initial workpiece and are thrown away, being only a means to the desired end. Chip formation is an inefficient process often producing some undesired side effects, is wasteful of material, and even the disposal of the chips frequently presents some difficulties. Yet it is not likely that the importance and use of metal cutting will decrease significantly in the foreseeable future. Because of its importance, a large amount of basic research and experimentation on metal cutting has been carried out during the past 30 years, which has led to better understanding of the chip-forming process and to significant improvements to machining operations. This work also has been influential in the development of processes wherein metal is removed without the formation of chips—called the *chipless machining processes.*[1] It is the purpose of this chapter to provide a fundamental understanding of the basic metal-cutting process and its requirements, advantages, and limitations. Subsequent chapters will discuss its application in individual machining processes and operations.

Chip formation. Metal cutting involves several interrelated factors, the most important being (1) the properties of the work material, (2) the properties and geometry of the cutting tool, and (3) the interaction between the tool and the workpiece. The last factor is most basic in chip formation.

Basically, chip formation is a mechanism of localized shear deformation resulting in the failure of the work material immediately ahead of the cutting edge of the tool. The relative motion between the tool and the workpiece causes compression of the work material near the tool and induces shear-type deformation within it. This *shear deformation,* or *plastic strain,* occurs in a narrow zone, of the order of 0.025 mm (0.001 inch) or less and approximating a plane, extending from the cutting edge of the tool to the surface of the work.

It is easiest to describe the chip-forming mechanism for the case of *orthogonal* (two-dimensional) *cutting.* This is the type of cutting that is encountered by the end cutting of a tube in a lathe when the width of the cutting tool exceeds the thickness of the tube wall. However, only rather simple transformations are required to apply the results obtained from the relatively simple orthogonal case to more complex cutting conditions.

Figure 17-1a depicts the basic relationship between the tool and the workpiece for orthogonal cutting. As the tool advances at velocity v, it exerts a compressive force (in the direction of its motion) against the workpiece, and

FIGURE 17-1. (a) Relationship of tool and work in chip formation. (b) Schematic representation of the mechanism of chip formation.

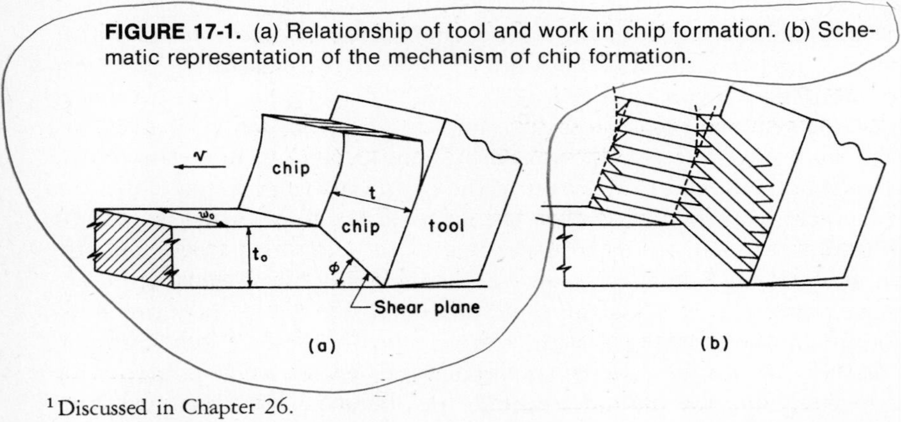

[1] Discussed in Chapter 26.

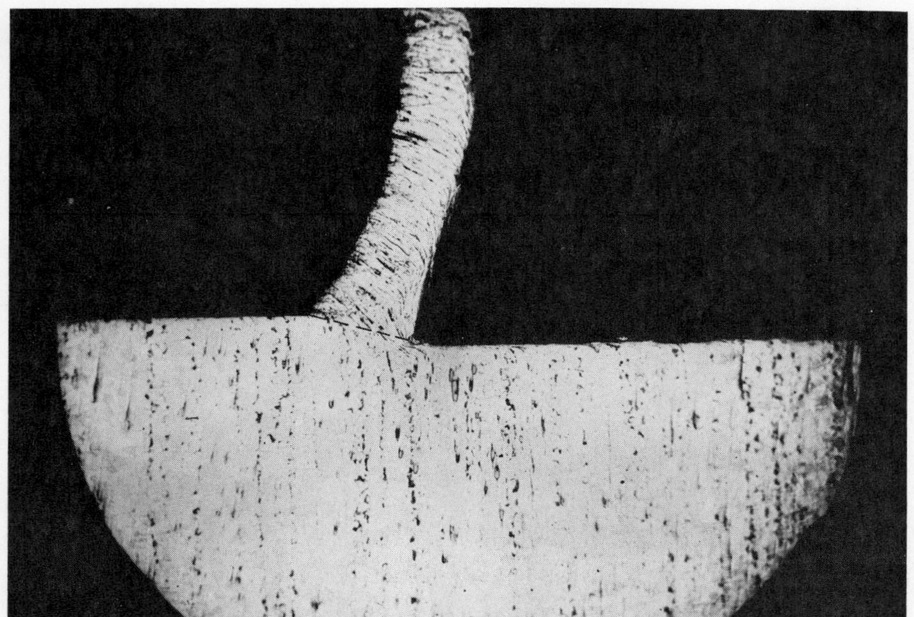

FIGURE 17-2. Magnified view of a metal chip, showing shear plane and deformed grain structure ahead of the cutting tool. (*Courtesy Cincinnati Milacron Inc.*)

the compressed and deformed metal shears along the shear angle ϕ. This action thus transforms metal of thickness t_0 (also known as the *undeformed chip thickness*) and width w_0 into a chip of thickness t, which slides along the face of the tool. This deformation by shear occurs by a process analogous to the displacement of a stack of cards, with each card sliding slightly over the adjacent card. This process of chip formation is illustrated schematically in Figure 17-1b.

Both the compression and shear deformation in an actual chip can be seen clearly in Figure 17-2, a magnified view of an interrupted cut in which the tool was stopped suddenly during its cutting action. The severely deformed grain structure immediately ahead of the cutting tool not only delineates the location and direction of the zone of shear deformation, but also illustrates the extensive shear strain that may accompany chip formation.

Several important conditions are readily recognized to be associated with this generalized mechanism of chip formation. First, shear and compression stress levels within the material ahead of the tool are sufficient to cause plastic strain throughout the shear zone and to exceed the material strength locally near the cutting edge. Such stress is associated with extreme pressure and with forces—which can be large—acting on the cutting tool and the workpiece.

Second, the action involved in chip formation also causes distortion in the workpiece, and the machined surface is not smooth, but torn and frag-

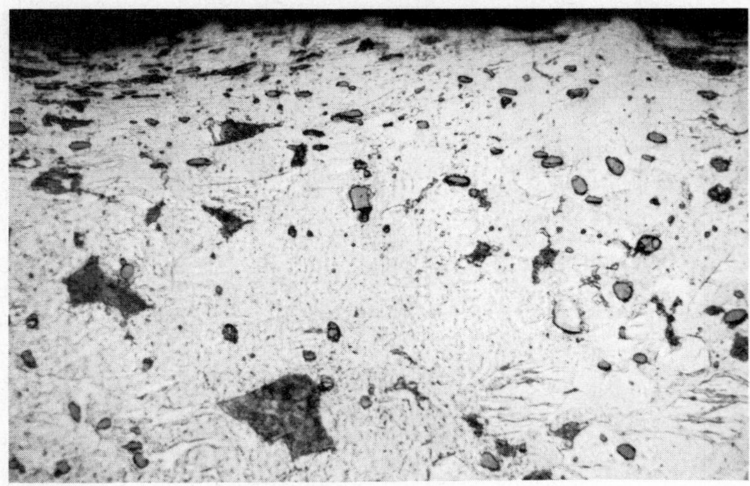

FIGURE 17-3. Photomicrograph of section of machined metal showing deformation below the machined surface (top).

mented. These effects can be seen clearly in Figure 17-3, which shows a section of a workpiece at and below a machined surface. These effects can be harmful and can adversely affect service performance unless eliminated.

Third, as a consequence of the work required to deform the workpiece and the chip, and also because of the friction between the chip and the face of the cutting tool, considerable heat is generated in the cutting zone. Studies indicate that in a typical cutting operation about 75 per cent of the developed heat is caused by metal deformation and shearing and about 25 per cent by friction. This heat generation results in elevated temperatures in the cutting zone, especially in the chips and the cutting tool, which is in essentially continuous contact with the work during cutting. Although it varies considerably, typically about 60 to 80 per cent of this heat is carried away in the chips, about 15 per cent is dissipated by the tool, and about 5 to 25 per cent by the workpiece.

Effects of work material properties. As noted previously, the properties of the work material are important in chip formation. High-strength materials require larger forces than do materials of lower strength, causing greater tool and work deflection, increased friction and heat generation and operating temperatures, and requiring greater work input. The structure and composition also influence metal cutting. Hard or abrasive constituents, such as carbides in steel, accelerate tool wear.

Work material ductility is an important factor. Highly ductile materials not only permit extensive plastic deformation of the chip during cutting, which increases work, heat generation, and temperature, but it also results in

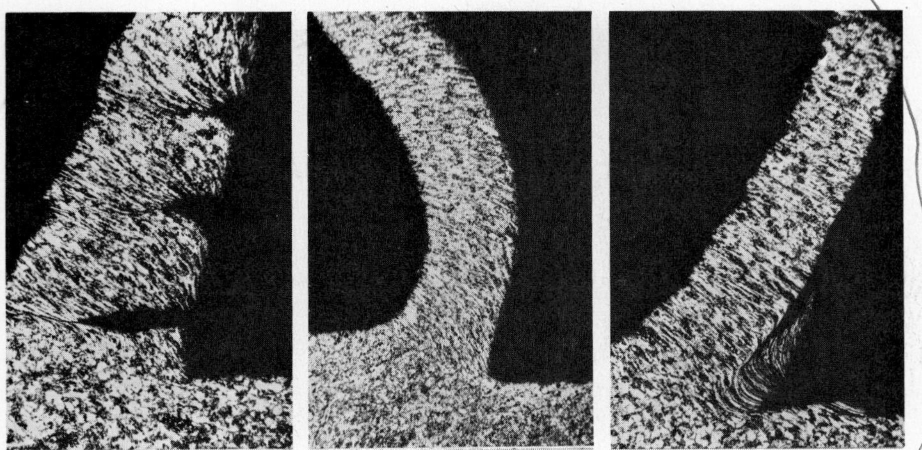

FIGURE 17-4. Three characteristic types of chips (*left to right*): discontinuous, continuous, and continuous with built-up edge. (*Courtesy Cincinnati Milacron Inc.*)

longer, "continuous" chips that remain in contact longer with the tool face, thus causing more frictional heat. Chips of this type are severely deformed and characteristically coil in a helix. On the other hand, brittle materials, such as gray cast iron, lack the ductility necessary for appreciable plastic chip formation. Consequently, the compressed material ahead of the tool fails in a brittle manner along the shear zone, producing small fragments. Such chips are termed *discontinuous* or *segmented.* The addition of sulfur or other constituents to "free-machining" steels reduces their ductility and causes their chips to break up and thus reduce the friction between the chips and the tool.

A variation of the continuous chip, often encountered in machining ductile materials, is associated with a "built-up" edge on the cutting tool. The local high temperature and extreme pressure in the cutting zone cause the work material to adhere or weld to the cutting edge of the tool forming the built-up edge. Although this material protects the cutting edge from wear, it modifies the geometry of the tool and also sloughs off periodically by adhering to the chip or passing under the tool and remaining on the machined surface. The periodic loss and replacement of the built-up edge is undesirable because it is a source of vibration and poor surface finish. A built-up edge often can be eliminated or minimized by reducing the depth of cut and increasing the cutting speed. Figure 17-4 illustrates the three characteristic types of chips.

Tool geometry. The shape and position of the tool relative to the workpiece have a very important effect in metal cutting. The most important geometry elements, relative to chip formation, are the location of the cutting edge and the orientation of the tool face with respect to the workpiece and the direction of cut. Other shape considerations are concerned primarily with relief or

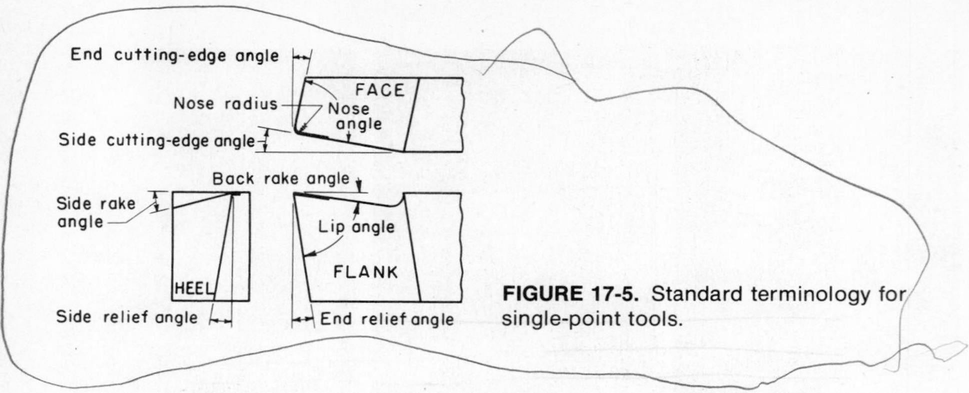

FIGURE 17-5. Standard terminology for single-point tools.

clearance—taper applied to tool surfaces to prevent rubbing or dragging against the work.

Figure 17-5 details the shape and descriptive terminology for a typical, single-point cutting tool used for lathe, shaper, and planer operations. Figure 17-6 shows a similar tool mounted in a shaper and making a surfacing cut. The portion of the tool cutting edge effective in the pictured operation is indicated by the heavy outline in Figure 17-5.

The face of the tool over which the chip flows during cutting is a plane of compound slope established by the back rake and side rake angles. For a specific cut, the *true rake* is defined as the inclination of the tool face at the cutting edge as measured in the direction of actual chip flow. In the example shown in Figure 17-6, the true rake very nearly coincides with the side rake angle of the tool; the back rake serves primarily to direct the chip upward from the machined surface in the form of a helix.

True rake inclination of a cutting tool has a major effect in determining the degree of compression and the inclination of the shear zone in the work ahead of the cutting edge. A small rake angle causes high compression, tool forces, and friction, resulting in a thick, highly deformed, hot chip. Increased rake angle reduces the compression, the forces, and the friction, yielding a thinner, less-deformed, and cooler chip. Unfortunately, it is difficult, and usually not possible, to take much advantage of these desirable effects of larger positive rake angles, since they are offset by the reduced strength of the cutting tool, due to the reduced tool section, and by its greatly reduced capacity to conduct heat away from the cutting edge. An interesting development along these lines will be discussed later.

In order to provide greater strength at the cutting edge and better heat conductivity, negative rake angles commonly are employed on sintered carbide and ceramic cutting tools. These materials tend to be brittle, but their ability to hold their superior hardness at high temperatures makes them preferred for high-speed and continuous machining operations. With a negative rake angle the face of the tool in the path of chip flow is inclined toward the direction of cutting rather than away from it, as in the case of positive rake angle tools. A negative rake angle increases tool forces to some extent,

FIGURE 17-6. Four simultaneous views of a cutting tool during cutting. The center view showing the side of the tool is magnified. It should be noted that the other views, taken with the aid of mirrors, are *mirror images* and thus are not third-angle projections.

but this minor disadvantage is offset by the added support to the cutting edge. This is particularly important in making intermittent cuts and in absorbing the impact during the initial engagement of the tool and work.

A shallow *chip-breaker* groove frequently is ground in the tool face a short distance back from the cutting edge and parallel with it. Such a groove is shown in Figure 17-7. Its effect is to deflect the chip at a sharp angle and cause it to break into short pieces that are easier to remove and are not so likely to become tangled in the machine and possibly cause damage to personnel. This is particularly important on high-speed, mass-production machines. Another means for breaking up chips is to use an adjustable chip breaker such as is shown in Figure 17-8. Another important reason for breaking up chips is to reduce the amount of friction and heat along the face of the tool. This will be discussed later.

The shapes of cutting tools as used for various operations and materials are compromises, resulting from experience and research so as to provide good overall performance. Table 17-1 gives representative rake angles and suggested cutting speeds.

In addition to side and back rake, other angles are provided on cutting

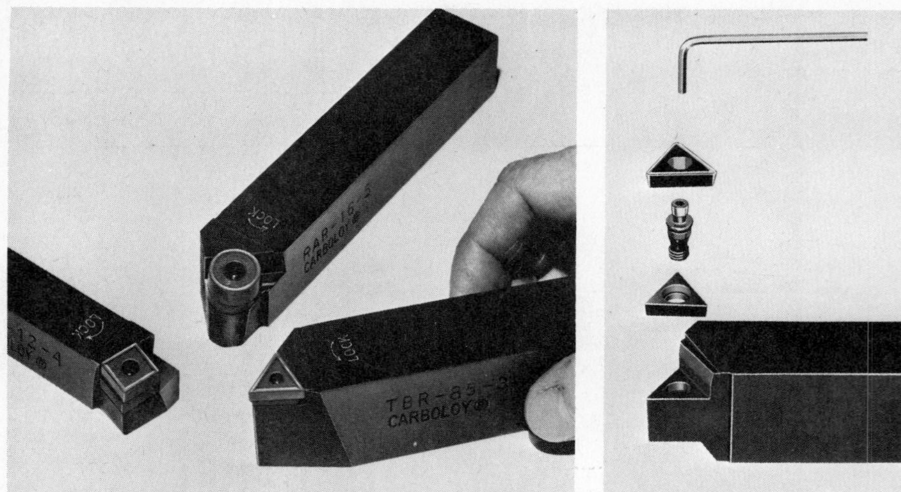

FIGURE 17-7. (*Left*) Examples of throwaway cutting tool tips. (*Right*) Components of mounting holder. (*Courtesy Metallurgical Products Division, General Electric Company.*)

tools to provide clearance between the tool and the work, thus avoiding rubbing and resulting friction. These are shown in Figure 17-5 and also can be seen in Figure 17-6.

Cutting tools used in various machining operations often appear to be very different from the one depicted in Figure 17-5. Often they have several cutting edges, as in the case of drills, broaches, saws, and milling cutters. Essentially, such tools are comprised of a number of single-point cutting edges arranged so as to cut simultaneously or sequentially. Simple analysis will disclose the same basic geometrical relationship between each cutting edge and the work, as is shown in Figure 17-9 for a drill having two cutting edges.

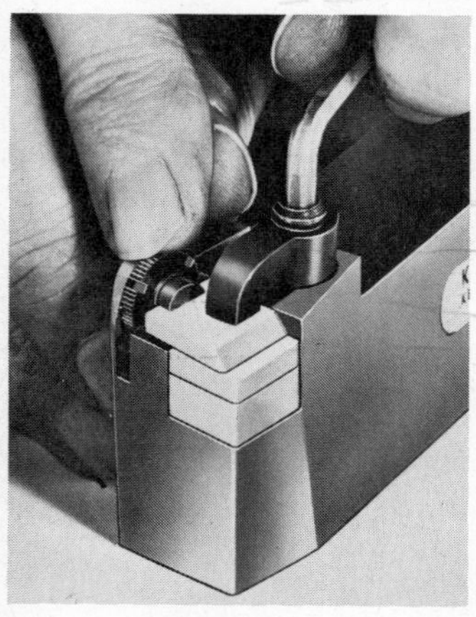

FIGURE 17-8. Adjustable chip breaker used with disposable tool tips. (*Courtesy Kennametal, Inc.*)

TABLE 17-1. Representative machining conditions for various work and tool–material combinations

Lathe Turning Operation Single-Point Tool				Feed: 0.38 mm/rev (0.015 ipr) Depth: 3.18 mm (0.125 in.)	
		Rake Angles (degrees)		Cutting Speed	
Work Material	Tool	Back	Side	m/min	fpm
B1112 steel	HSS	16	22	69	225
	WC	0	3	168	550
	Ceramic	− 5	− 5	427	1400
1020 steel	HSS	16	14	55	180
	WC	0	3	152	500
	Ceramic	− 5	− 5	366	1200
4140 steel	HSS	12	14	40	130
	WC	0	3	91	300
	Ceramic	− 5	− 5	274	900
18-8 steel (stainless)	HSS	8	14	27	90
	WC	4	8	84	275
	Ceramic	− 5	− 5	152	500
Gray cast iron (medium)	HSS	5	12	34	110
	WC	0–4	2–4	69	225
	Ceramic	− 5	− 5	244	800
Brass (free-machining)	HSS	0	0	76	250
	WC	0	4	221	725
Aluminum alloys	HSS	35	15	91	300 plus
	WC	10–20	10–20	122	400 plus
Magnesium alloys	HSS	0	10	91	300 plus
	WC	10	10	213	700 plus
Titanium (turning)	WC	0	5	46	150

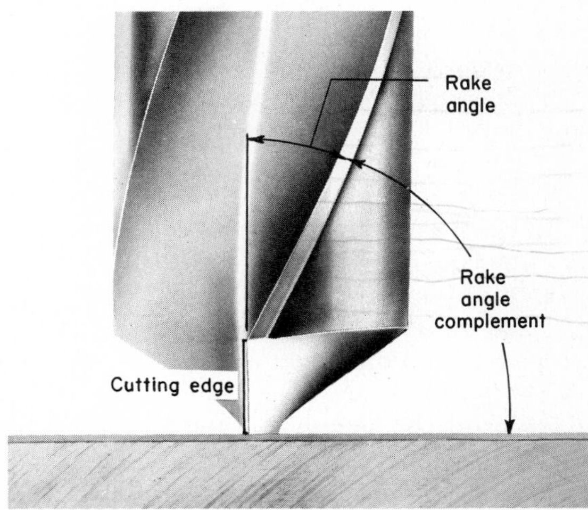

FIGURE 17-9. Drill, showing one cutting edge and the rake angles on this two-edged cutting tool. (*Courtesy The Cleveland Twist Drill Company.*)

Recognition of such individual cutting action aids and simplifies understanding of the action and use of such tools.

Cutting-tool materials. Because chip formation involves high local stresses, friction, abrasion, and the generation of considerable heat, materials for use as cutting tools must combine strength, toughness, and hardness or wear resistance at elevated temperatures. A number of materials having these properties in varying degrees have been developed for use in cutting tools.

Tool steel: Plain carbon steel of 0.90 to 1.30 per cent carbon when hardened and tempered has good hardness and strength, adequate toughness, and can be given a keen cutting edge. However, it loses its hardness at temperatures above 204°C (400°F) because of tempering, and it has largely been replaced by other materials for metal cutting.

High-speed steel: First introduced in 1900, this high-alloy steel is remarkably superior to tool steel in that it retains its cutting ability at temperatures up to 593°C (1100°F), exhibiting "red hardness." Compared with tool steel, it can operate at about double the cutting speed with equal life, resulting in its name *high-speed steel,* often abbreviated HSS. Although several formulations are used, a typical composition is that of the 18-4-1 type (tungsten 18 per cent, chromium 4 per cent, vanadium 1 per cent). Comparable performance can also be obtained by the substitution of approximately 8 per cent molybdenum for the tungsten. High-speed steel still is widely used for drills and many types of general-purpose milling cutters and in single-point tools used in general machining, but it has been almost completely replaced for high-production machining.

Sintered carbides: There are a variety of these nonferrous alloys, produced by powder metallurgy. The earliest, and still widely used, has tungsten carbide as the major constituent, with cobalt used as a binder in amounts from 3 to 13 per cent. Recent types utilize very fine microparticles dispersed in the carbide structure, improving their toughness and tool life, particularly when subjected to impact, as in making interrupted cuts. Various other carbides, especially titanium, tantalum, and columbium, can be added or substituted for the tungsten. Another type is composed of titanium carbonitrides, titanium–molybdenum transition phases, and a nickel alloy binder, thus requiring no scarce or imported materials.

Carbide tools are extremely hard (90 to $93R_c$) and can be operated at cutting speeds 200 to 500 per cent greater than those used for high-speed steel. Consequently, they have virtually replaced high-speed steel in most manufacturing machining.

Many tungsten carbide tools, particularly those in the form of inserts, are coated with a thin layer of titanium carbide or titanium nitride, or with multiple layers—for instance, titanium carbide, titanium carbonitride, and titanium nitride, in sequence. These retain the toughness of tungsten carbide but have improved resistance to cratering and edge wear. At the same time, they have the advantages of titanium carbide plus reduced face friction.

Many carbide tools are made in the form of "throwaway" inserts, such as shown in Figure 17-7. They contain from three to eight cutting edges and are held mechanically in a tool holder. When one cutting edge becomes dull, the insert is rotated or turned over to a new edge; when all the edges are dull, the insert is thrown away. It is very important that the carbide inserts be held firmly and backed up well. Some very significant improvements in tool life have been obtained in milling cutters where the inserts are held to present the length of the insert in the direction of the cut. This suggests that in trying to obtain economy by using a small amount of the expensive carbide, many inserts may have been made too thin.

Ceramics: These are the latest very important addition to the list of cutting-tool materials. Most are made of pure aluminum oxide. Very fine particles are formed into cutting tips under a pressure of 267 to 386 MPa (20 to 28 tons/in.^{2}) and sintered at about 982°C (1800°F). Unlike the case with ordinary ceramics, sintering occurs without a vitreous phase.

Ceramics usually are in the form of disposable tips. They can be operated at from two to three times the cutting speeds of tungsten carbide, almost completely resist cratering, usually require no coolant, and have about the same tool life at their higher speeds as tungsten carbide does at lower speeds. However, in order to take full advantage of their capabilities, special and more rigid machine tools are required.

Diamond: Diamond is the hardest material known. Industrial diamonds, either natural or artifical, have limited but important application in machining operations, such as in boring, to produce accurate, smooth bearing surfaces. When used for finishing operations, extremely high cutting speeds with fine feed are employed. Other applications for diamond are for truing grinding wheels and, in their crushed form ("bort"), as an abrasive in wheels for grinding or in lapping.

Cutting-tool operating conditions. A variety of machine tools have been developed for doing specific types of machining operations. Often a given machine can perform several of the basic operations; this currently is an important trend. However, despite differences in construction, motions, and terminology, one must deal with the basic tool–work relationship during cutting. This generally is adequately described by the factors tool shape and position, cutting speed, depth of cut, and feed.

Cutting speed refers to the relative surface speed between the tool and the work, expressed in meters per minute[2] or feet per minute (CS, fpm). It is a relative term, since either the work, the tool, or both may move during cutting.

[2] These units for cutting speed are not in strict accord with SI recommendations but are those commonly used in metric countries and by most builders of machine tools. A similar situation exists relative to some expressions for feed (mm/min rather than m/sec) and for certain other measures commonly encountered in metal cutting, where strict use of SI units results in quite meaningless quantities.

Depth of cut is the depth the cutting edge engages the work in millimeters or inches (thousandths). It determines one linear dimension of the area of cut (chip cross section).

Feed is more complex in that it is expressed differently for various operations. It is the second linear dimension of the area of cut, being associated with the spacing of tool or feed marks on the machined surface. For the shaper and planer, feed is the lateral offset between the tool and work for each stroke, expressed as millimeters or inches per stroke (mmps or ips). In turning or

TABLE 17-2. Operations and machines for machining flat surfaces.

Operation	Block Diagram	Most Commonly Used Machines	Machines Less Frequently Used	Machines Seldom Used
Shaping	Tool, Work	Horizontal shaper	Vertical shaper	
Planing	Tool, Work	Planer		
Milling	slab milling: Tool, Work; face milling: Work, Tool	Milling machine		Lathe (with special attachment)
Facing	Work, Tool	Lathe	Boring mill	
Broaching	Tool, Work	Broaching machine		
Grinding	Work, Tool	Surface grinder		Lathe (with special attachment)
Sawing	Tool, Work	Cutoff saw	Contour saw	

⟵⟶ Tool and work motion
⟵ - - ⟶ Feed only

TABLE 17-3. Operations and machines for machining external cylindrical surfaces

Operation	Block Diagram	Most Commonly Used Machines	Machines Less Frequently Used	Machines Seldom Used
Turning	Work, Tool	Lathe	Boring mill	Vertical shaper Milling machine
Grinding	Tool, Work	Cylindrical grinder		Lathe (with special attachment)
Sawing	Tool, Work	Contour saw		

drilling, feed is the axial advance of the tool along or through the work during each revolution of the work or tool expressed as millimeters or inches per revolution (mmpr or ipr). For multitooth milling cutters, feed is the advance of the work or cutter between the cutting action of two successive teeth, expressed basically as millimeters or inches per tooth. The diagrams in Tables 17-2, 17-3, and 17-4 depict schematically the tool–work relationship for the common machine operations and indicate the machine tools that perform them.

Feed and depth of cut determine the area of cut and relate directly to cutting-force requirements. For a single-point tool, the area of cut multiplied by the cutting speed in meters per minute (CS × 1000) or inches per minute (CS × 12) yields the rate of metal removal in cubic millimeters or cubic inches per minute, respectively. The metal removal rate determines machine-tool productivity during actual machining intervals and also establishes power requirements, because it determines the rate at which work is done.

For a given area of cut, a large ratio of depth of cut to feed usually offers the most efficient performance as well as a better surface finish, because feed marks are more closely spaced. Cutting speed has little effect on machining forces but has a marked effect on tool wear and is treated at greater length in this connection.

Tool life and tool failure. As a consequence of continued use, cutting tools cease to perform satisfactorily. Unsatisfactory performance may involve loss of dimensional accuracy, excessive surface roughness, increased power require-

TABLE 17-4. Operations and machines for machining internal cylindrical surfaces

Operation	Block Diagram	Most Commonly Used Machines	Machines Less Frequently Used	Machines Seldom Used
Drilling	Work; Tool	Drill press	Lathe	Milling machine Boring mill Horizontal boring machine
Boring	Work; Tool	Lathe Boring mill Horizontal boring machine		Milling machine Drill press
Reaming	Work; Tool	Lathe Drill press Boring mill Horizontal boring machine	Milling machine	
Grinding	Work; Tool	Cylindrical grinder		Lathe (with special attachment)
Sawing	Tool; Work	Contour saw		
Broaching	Tool; Work	Broaching machine		

ments, physical loss of the cutting edge, or combinations of these. The actual cutting time accumulated before failure is termed *tool life* and is expressed in minutes. When failure occurs, a tool must be removed and either thrown away or reconditioned, usually by grinding. This causes loss in production from machine downtime, in addition to the cost of replacing or reconditioning the tool (with loss of tool material).

Although tools commonly are said to become dull, their failure to perform properly usually is the result of several factors. Wear, in several forms, is

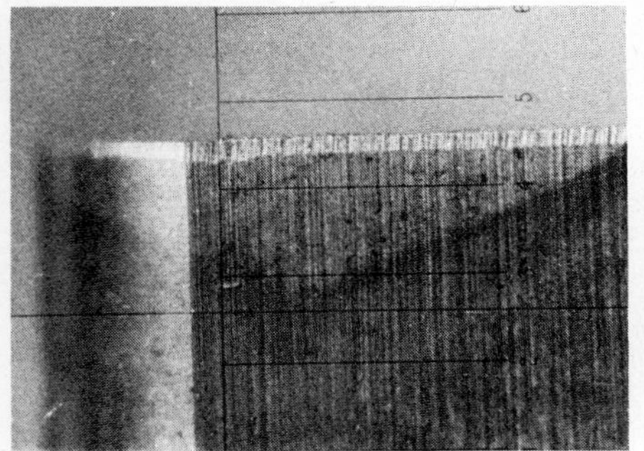

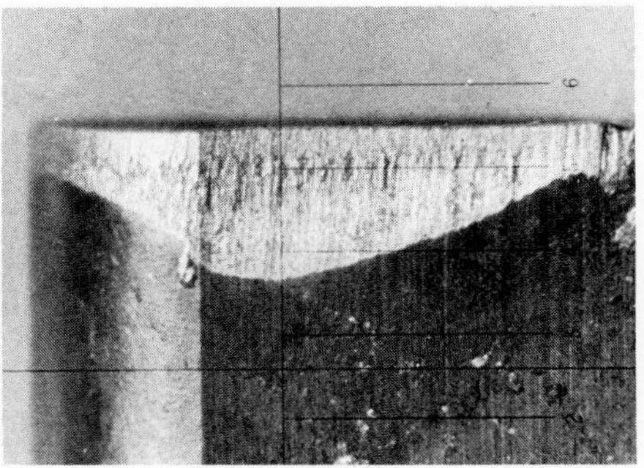

FIGURE 17-10. (*Left*) Heat marks on a cutting tool during early service and (*right*) subsequent wear pattern. (*Courtesy K. J. Trigger.*)

the primary, evident factor. However, heat also is a major factor, because sufficiently high temperatures can cause the tool material to lose its hardness and thus make it more subject to wear. That there is a very close correlation between temperature and wear is shown in Figure 17-10.

Tool wear occurs in two areas. The major tendency for wear is on the tool face a short distance from the cutting edge, the result of the severe abrasion between the chip and the tool face. This results in a crater being formed in the tool face, being more common on HSS tools than on carbide or ceramic tools, which have much greater hot-hardness. Cratering also is most pronounced in machining ductile materials, which form continuous chips. The second area of wear is on the flank below the cutting edge, resulting from contact with the abrasive machined surface. Flank wear results in a rough machined surface. On carbide and ceramic tools, which virtually do not crater, tool life is taken arbitrarily as corresponding to 0.76 mm (0.030 inch) of wear land on the flank.

The rate of tool wear is influenced by the abrasive properties of the work materials, the magnitude of the cutting forces, and the operating temperature of the tool. Although cutting forces can be controlled by proper selection of feed and depth of cut, tool temperature is closely related to cutting speed. With increased cutting speed, the rate of energy input into the cutting zone increases more than does the output, which is largely heat flow. This results in temperature rise, loss of tool properties, and decreased tool life. Whereas tool temperature is seldom measured, much investigation has been made of the effect of cutting speed on tool life.

In tool-life testing, tools are operated to failure at different cutting speeds and the test results are plotted. In general, a parabolic decrease in tool life with increased cutting speed is obtained. Such a relationship plots as a straight line on a log-log graph, as shown in Figure 17-11. The analytical expression for the relationship is of the form

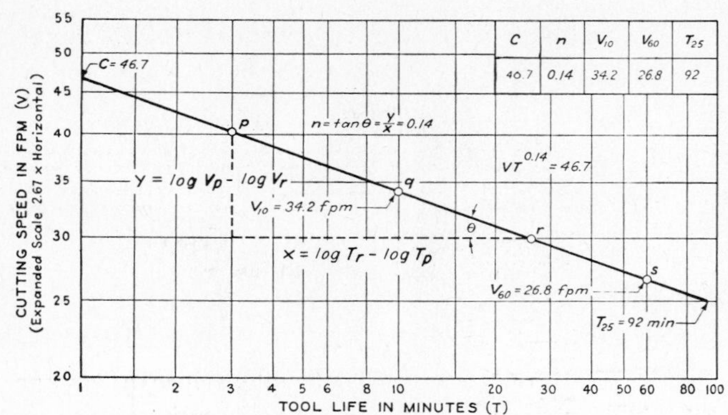

FIGURE 17-11. Relationship between cutting speed and tool life (*From ANSI Standard B*5.19-1946).

$$VT^n = c \tag{17-1}$$

where V = cutting speed, fpm
T = tool life, minutes
n = slope of the curve
c = intercept value, 1 minute of tool life

Obviously, the results are valid only for the particular test conditions employed. However, their usefulness can be extended by intelligent extrapolations.

Tool-life tests, made on specific materials, make it possible to select a cutting speed to obtain a desired tool life. This will establish both the rate of metal removal or productivity and also the frequency of tool reconditioning. There remains the difficulty that in the absence of such a test on the particular material to be machined it is difficult to select the optimum cutting speed. In general, cutting speeds are inversely proportional to the strength and hardness of a material. However, there are great exceptions to this simple rule. Figure 17-12 shows a concept that relates practical cutting speeds to material properties. The thermal dispersion factor takes into account some of the thermal characteristics of the work material, thus showing the importance of heat in affecting tool life and other cutting results.

In practice, the cutting speed used is a compromise that will provide economical results. Typical values of cutting speeds for various combinations of tool and work materials representative of general practice are given in Table 17-1.

Cutting fluids. Various fluids, usually liquids, are used to improve machining performance. In most cases, their primary effect is as a coolant to reduce the tool-operating temperature. The reduction in temperature aids in retaining the hardness of the tool, thereby extending tool life or permitting increased

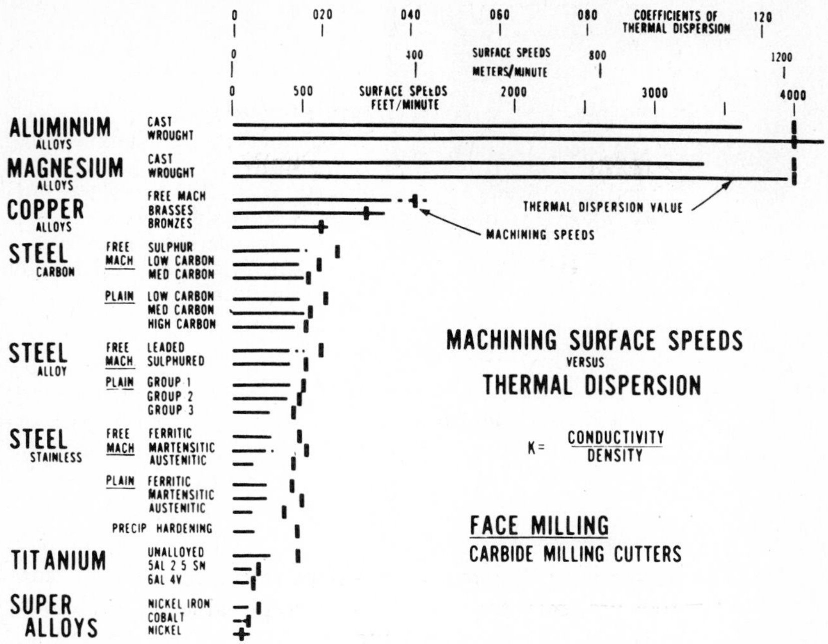

FIGURE 17-12. Correlation between machining surface speeds and thermal dispersion of the metal being machined. (*From A New Machinability Index, by J. R. Ewell, Courtesy American Machinist.*)

cutting speed with equal tool life. In addition, the removal of heat from the cutting zone reduces thermal distortion of the work and permits better dimensional control. Coolant effectiveness is closely related to the thermal capacity and conductivity of the fluid used. Water is very effective in this respect but presents a rust hazard to both the work and tools and also is ineffective as a lubricant. Oils offer less effective coolant capacity but do not cause rust and have some lubricant value. In practice, straight cutting oils or emulsion combinations of oil and water or wax and water are frequently used. Various chemicals also can be added to serve as wetting agents or detergents, rust inhibitors, or polarizing agents to promote formation of a protective oil film on the work. The extent to which the flow of a cutting fluid washes the very hot chips away from the cutting area is an important factor in heat removal. Thus the application of a coolant should be copious and with some velocity.

The possibility of a cutting fluid providing lubrication between the chip and the tool face is an attractive one. An effective lubricant can modify the geometry of chip formation so as to yield a thinner, less-deformed, and cooler chip. This is shown in Figure 17-13, where the cut was started dry and a cutting fluid was then added. As soon as the cutting fluid was added, thus reducing the friction, the shear angle increased, and the chip thickness was

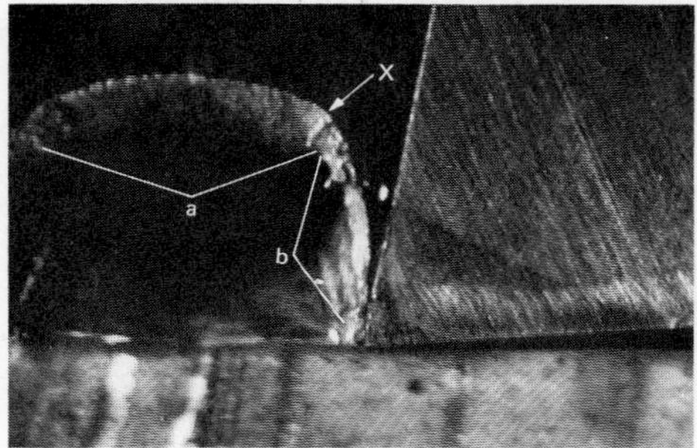

FIGURE 17-13. Effect of reducing friction between the chip and the tool face by means of a cutting fluid. Portion of chip at (a) was formed by dry cutting; Thinner portion at (b) was formed by wet cutting. Note abrupt change in thickness and nature of chip at (x) where cutting fluid was added.

decreased. Such action further discourages the formation of a built-up edge on the tool and thus promotes improved surface finish. However, the extreme pressure existing between the somewhat rough chip and the tool, the very high pressures involved, and the rapid flow of the chip away from the cutting edge make it virtually impossible to maintain a conventional hydrodynamic lubricating film where it would be needed to obtain reduced friction by this method. Consequently, any lubrication that can be provided is associated primarily with the formation of solid chemical compounds of low shear strength on the freshly cut chip face, thereby reducing chip-tool shear forces or friction. For example, carbon tetrachloride is very effective in reducing friction in machining several different metals and yet would hardly be classed as a good lubricant in the usual sense. Chemically active compounds, such as chlorinated or sulfurized oils, can be added to cutting fluids to achieve such a lubrication effect. Extreme-pressure lubricants of this type are especially valuable in severe operations, such as internal threading by a tap where the extensive tool–work contact results in much friction with limited access for a fluid.

In addition to functional effectiveness as coolant and lubricant, cutting fluids should be stable in use and storage, noncorrosive to work and machines, and nontoxic to operating personnel.

Forces in chip formation. In order to better understand metal cutting and the possibilities and limitations for improving the effectiveness of the process, it is helpful to consider the forces involved in chip formation. Figure 17-14 depicts these forces with the chip treated as a free body that is in equilibrium under the collinear forces R and R'. This condition is approximated in practice, although there also are other forces, such as the "ploughing" force at the cutting edge, that may require a small correction of the force R. It is assumed that all the energy appears as shear energy and friction energy (on the tool face) and that the kinetic energy associated with the creation of a new surface and the flank-friction energy are negligible. These assumptions are substantially true at conventional cutting speeds up to 305 to 610 meters (1000 to 2000 feet) per minute, but a built-up edge on the tool may require modification in the analysis at low speeds.

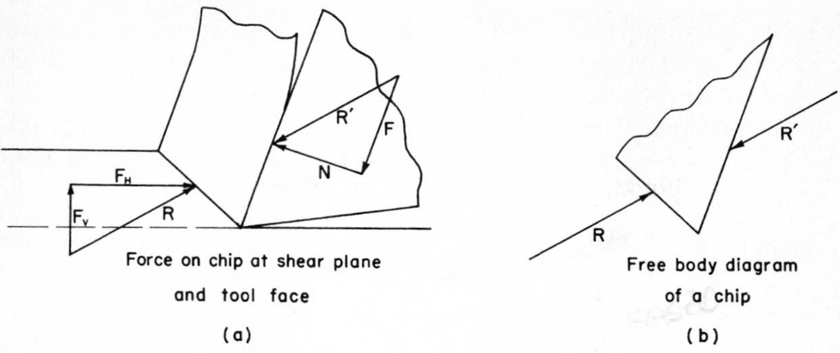

FIGURE 17-14. Forces acting in chip formation.

The forces shown in Figure 17-14 can conveniently be represented on a *force circle,* as shown in Figure 17-15. In this diagram, for convenience, the resultant forces have been moved to the point of the tool; such notation avoids moments in the force system at equilibrium. As in Figure 17-14, the force components of R are shown and also the shear force F_s and the normal force on the shear plane, N_s. Angle α is the *rake angle* of the tool, the angle between the tool face and the normal to the direction of motion of the tool. It is to be noted that angle β, which is known as the *friction angle* on the tool face and is given by $\tan \beta = F/N$, also appears in the angle λ, which is the angle between the resultant force R and the shear force F_s. Thus

$$\phi + \beta - \alpha = \lambda \tag{17-2}$$

as can be proved from the figure.

Research has shown that the angle λ appears to vary little with variations in w_0, t_0, and α. Consequently, as a first approximation in many prob-

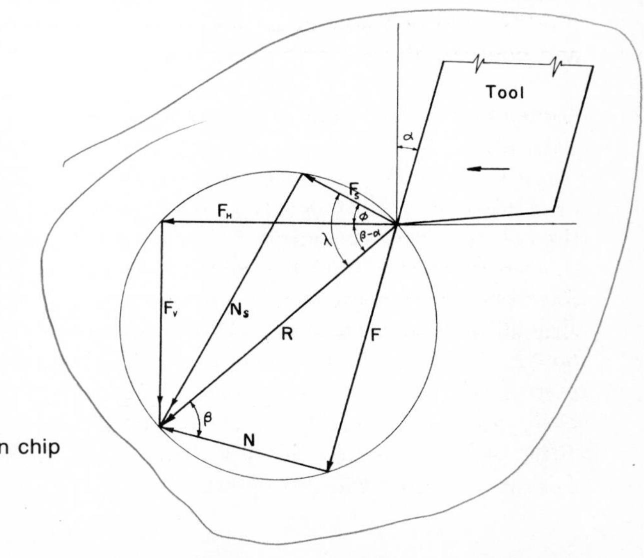

FIGURE 17-15. Force circle in chip formation.

lems, λ can be assumed to be constant. The angle λ is, however, a function of the combination of tool geometry, work material, and cutting conditions—that is, use of cutting fluid and cutting speed.

In an orthogonal cutting operation, the tool forces F_H and F_V can be measured with a tool dynamometer, and the shear angle ϕ can be obtained from the equation

$$\tan \phi = \frac{r \cos \alpha}{1 - r \sin \alpha} \tag{17-3}$$

where r is the chip-thickness ratio t_0/t. The chip-thickness ratio in conventional cutting is always less than unity, because of the compression that takes place before shearing occurs. Thus, the actual chip always is thicker than the undeformed chip. The chip-thickness ratio can be obtained in several ways, such as sectioning frozen chips, length measurements (assuming a constant-volume relationship), and weight–length measurements.

Having determined F_H, F_V, α and r, as described, the stresses on the shear plane can be obtained from the following relationships:

$$\left.\begin{aligned} R &= \sqrt{F_V^2 + F_H^2} = \frac{F_H}{\cos(\beta - \alpha)} \\ F_s &= R \cos \lambda \\ N_s &= R \sin \lambda \\ \tau_s &= \frac{R \cos \lambda \sin \phi}{t_0 w_0} \\ \sigma_s &= \frac{R \sin \lambda \sin \phi}{t_0 w_0} \end{aligned}\right\} \tag{17-4}$$

On the tool face,

$$\left.\begin{aligned} F &= R \sin \beta = F_H \sin \alpha + F_V \cos \alpha \\ N &= R \cos \beta = F_H \cos \alpha - F_V \sin \alpha \\ \beta &= \tan^{-1} \frac{F}{N} \end{aligned}\right\} \tag{17-5}$$

While the effect of changing certain cutting conditions can be derived through equations 17-4 and 17-5, the effect can be shown more vividly by means of force–circle diagrams. For example, the effect of decreasing the friction between the chip and the tool face is illustrated in Figure 17-16. In (b) all conditions are assumed to be the same as in (a), except that the friction force F is decreased. Inasmuch as R and F_s are unchanged, the force circle is rotated clockwise about the tool tip so that the shear angle ϕ is increased. In actuality, if the friction force were reduced, R and F_s would not remain unchanged. Because $\tau_s = \text{constant} \times \sin \phi$, and the shear strength of the workpiece is constant, an increase in the shear angle results in the shearing

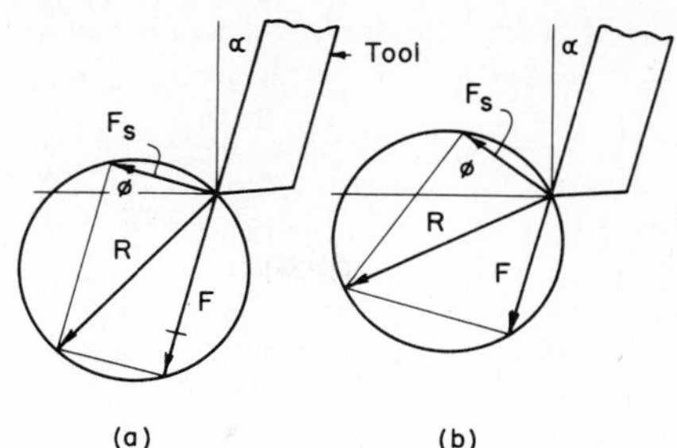

FIGURE 17-16. Effect of reducing friction on the tool face, shown by means of force circles.

taking place with a lower resultant force. Consequently, the shearing would occur at a lower resultant force and a somewhat smaller shear angle than is depicted at (b). This example emphasizes the twofold effect that results from reducing the tool-face friction in metal cutting—a reduction in friction increases the shear angle, which reduces the cutting force, which in turn results in less friction. It thus is apparent that reducing the friction between the chip and the tool face is extremely desirable—increasing the shear angle (with its accompanying reduction in friction) and reduction in chip thickness with accompanying decrease in metal deformation (Figure 17-13)—even though its achievement by use of a lubricant is limited.

Figure 17-17 illustrates the effect of increasing the rake angle of a cutting tool. In (b) of this illustration R, F, and F_s are the same as in (a). However, the increased positive rake angle in (b) causes the force circle to be rotated clockwise about the point of the cutting tool to increase the shear angle. From the previous discussion it is clear that this increased shear angle is beneficial. Figure 17-18 shows proof of this theory. As shown in the diagram, the carbide tool used had a high rake angle and greater than normal depth. Although it had a small negative-rake-angle land at its tip, a built-up edge formed on it and provided a positive rake angle where the chip formed. As can be seen, there is greatly reduced compression in the chip, and the chip actually does not contact the tool face. These two phenomena result in much less heat being generated. Although this procedure has some interesting possibilities, the chip is completely continuous, causing some disposal problems, and it appears to be suitable only for certain turning operations.

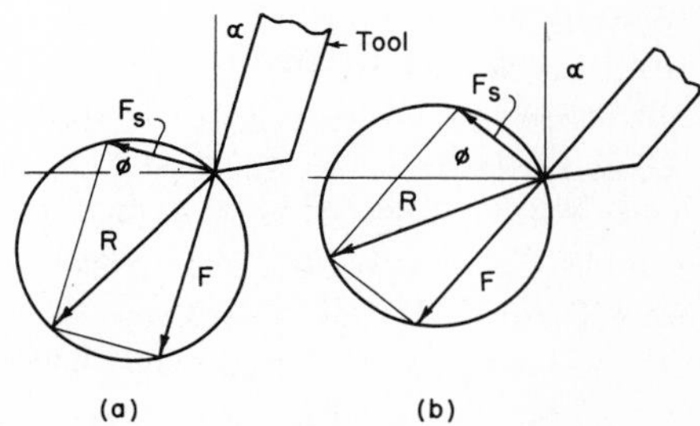

FIGURE 17-17. Effect on cutting forces and shear angle of an increase in positive rake angle on cutting tool.

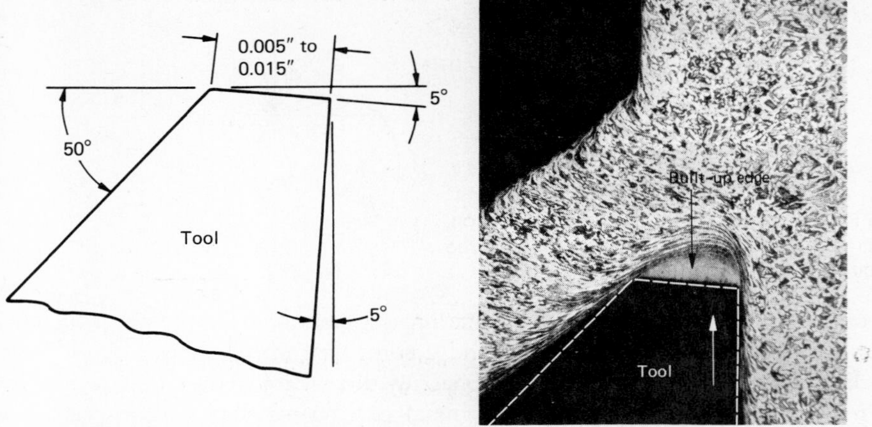

FIGURE 17-18. (*Left*) High-positive-rake-angle tool with very small negative-rake-angle land. (*Right*) Resulting chip and built-up edge. Note that chip does not rub on tool face. (*Courtesy Carroll M. Gordon.*)

From a practical viewpoint it thus is difficult to obtain the benefits that theoretically are possible by increasing the rake angle. This is due largely to the decreased strength and lowered heat capacity and conductivity of the tool.

By using a procedure similar to that illustrated in Figure 17-17, it can be shown that using negative rake angles causes the cutting forces to increase. However, negative rake angles often are advantageous because of the greater strength of the cutting tool and the consequent heavier cuts that can be taken when carbide or ceramic tools are used, which can withstand the higher temperatures that result. However, full advantage usually cannot be taken of using negative rake angles with these tool materials unless the machines on which they are employed have adequate strength and power to compensate for the higher cutting forces.

Stress–strain relationships in metal cutting. As might be expected, comparisons of metal-cutting stress–strain data do not correlate well with ordinary static test data, such as those obtained from the usual tensile test. In metal cutting, neither necking nor buckling can occur, and strain thus is limited only by fracture. Consequently, the strains often exceed 1, whereas in a tension test a reduction in area of 50 per cent ($\epsilon = \ln 2 = 0.69$) is considered excellent ductility. Just as fracture is delayed when a metal is deformed plastically while subjected to high hydrostatic pressure, fracture normally will not occur in metal cutting except when brittle materials are machined, as in the case of gray cast iron. However, if static tensile-test data are extrapolated to higher strains, or if compression tests are conducted by special techniques so

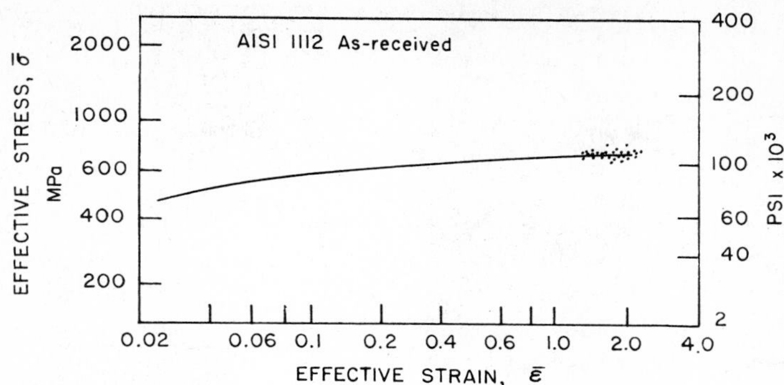

FIGURE 17-19. Correlation of metal-cutting data with compression-test data for AISI 1112 steel, as received, based on distortion-energy theory.

as to achieve strains larger than unity, there is good correlation with metal-cutting data, as is shown in Figure 17-19.

On the other hand, the dynamic shear stress in metal cutting essentially is not influenced by the strain rate. Figure 17-20 illustrates the theory that the energy in metal cutting is given by the area at constant shear stress. It is not entirely clear why the dynamic shear stress in metal cutting should have the same magnitude as in a slow tensile or compression test at identical strains. One theory is that the shearing process in metal cutting is at such a high rate and the shear zone moves so rapidly that the process is essentially isothermal. Under these conditions the plastic-flow process may be initiated at a stress level that is sufficiently high to overcome the internal stress resisting the movement of dislocations. As a result, after the initiation of plastic deformation the process can go to completion without further increase in the shear stress and unaided by a rise in temperature.

Predicting forces and power in metal cutting. One often needs to determine the forces and power that will be required for metal-cutting operations in order to design jigs and fixtures and select feed and depth of cut in accordance with the power available on a particular machine tool. Sometimes this can be

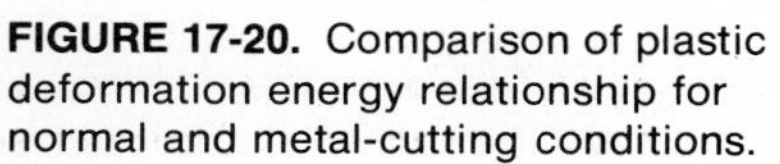
FIGURE 17-20. Comparison of plastic deformation energy relationship for normal and metal-cutting conditions.

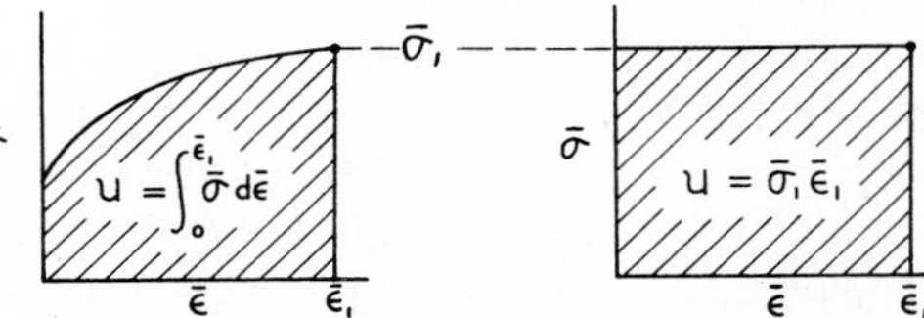

done by using empirical data that have been collected for a variety of cutting conditions. For example, in lathe turning,

$$F_H = Cf^m d$$
$$F_v = C_1 f^n d^q \qquad (17\text{-}6)$$
$$F_r = C_2 f^o d^p$$

where m, n, o, p, and q are exponents that must be evaluated from experiments, and F_r is the radial force in a three-dimensional force system. Unfortunately, the conditions for which such equations hold are limited, because the shape of the tool cannot be varied without affecting the results. More commonly and conveniently, power and force are determined through the use of *specific power*—the power required to remove one unit volume of material in 1 minute in the form of chips, expressed in watts per cubic millimeter or horsepower per cubic inch. Such values can be determined by cutting tests. Tables are available giving tabulated values of specific power as a function of some variable. For example, one such table gives specific horsepower as a function of the Brinell hardness number of the material for various feeds. Such tables are fairly useful, but they cannot be expected to give accurate results in many cases because they attempt to combine the effects of a number of variables. For example, AISI 1112 and AISI 1020 may, under certain conditions, have identical Brinell hardness numbers, yet they will machine entirely differently and require differing amounts of power. Tables that give specific power for various materials, such as is shown in abbreviated form in Table 17-5, usually produce more accurate results.

Example: What power is required to face-mill a block of gray cast iron, 127 mm (5 inches) wide, using a tungsten carbide cutter that is 152 mm (6 inches) in diameter and has eight teeth? The depth of cut is 3.18 mm (⅛ inch) and a chip thickness of 0.254 mm (0.010 inch) is to be used.

From Table 17-1, a suitable cutting speed would be 69 m/min (225 ft/min). The rpm would be 144. For this cutter and chip thickness, the feed

TABLE 17-5. Specific power of metal removal in machining various metals

Material	Specific Power		Material	Specific Power	
	W/mm³	hp/in.³		W/mm³	hp/in.³
Magnesium alloys	0.018	0.39	Malleable iron	0.025	0.55
Aluminum alloys			Steel		
Soft	0.018	0.40	AISI 1112	0.027	0.59
24-ST	0.020	0.43	AISI 1120	0.028	0.61
Brass			AISI 4140	0.029	0.64
Red	0.019	0.42	AISI 1020	0.030	0.67
Free-cutting	0.019	0.41	AISI 2345	0.032	0.70
Copper, annealed	0.033	0.72	Nickel	0.041	0.89
Cast iron	0.021	0.47			

of the table would be $144 \times 8 \times 0.254 = 293$ millimeters per minute (11.5 in./min). The metal-removal rate is $127 \times 3.18 \times 293 = 118{,}331$ cubic millimeters per minute (7.2 in.^{3}/min). From Table 17-5 the specific power for cast iron is 0.021 watt per cubic millimeter, so the power required would be $118{,}331 \times 0.021 = 2485$ watts (3.3 horsepower).

Procedures have also been developed to determine power and force requirements by applying plastic-strain theory, using ordinary tensile-test data for the material involved, a simple cutting test to determine the chip-thickness ratio, and the physical geometry of the cutting tool.[3]

Improving cutting efficiency. Chip formation obviously is an inefficient process because of the large amount of useless compression, the considerable friction along the tool face, and the consequent heat and high temperatures. What steps, if any, can be taken to improve the efficiency?

Unfortunately, *for a given work material,* little can be done to reduce the extensive compression and the heat resulting from it except as the shear angle can be increased. This is achieved primarily by the relatively limited changes that can be made in rake angles or by the reduction of friction. Thus reduction of tool-face friction is an attractive avenue to improved cutting efficiency but, as has been pointed out, here, too, there are major difficulties that limit what can be achieved. However, several procedures are helpful.

One means for reducing friction is to reduce the chip length by means of chip breakers or by metallurgical modifications of the metal—as in the free-machining alloys. Unfortunately, these modified alloys usually have somewhat decreased ductility and strength. However, there is much evidence that those who specify and select materials do not give sufficient consideration to the free-machining grades in many cases where the slightly decreased properties would not disqualify them and where their use would significantly increase machining rates, with accompanying decrease in overall cost.

Another effective procedure for reducing friction is to improve the surface characteristics of the tool face by honing, chrome plating, or other procedures. This is an increasing practice with carbide inserts. Still another procedure that is effective in some cases with negative-rake-angle tools is to use a double rake angle, as illustrated in Figure 17-21. The shortened negative-rake-angle face forms the chip but limits the area over which the chip can slide.

Increasing cutting speeds, if they are below 91 meters (300 feet) per minute, produces some increase in efficiency. Above about 122 meters (400 feet) per minute, increases in speed have little effect. Obviously increased speed affects tool life adversely. Increased depth of cut increases cutting efficiency a small amount.

In view of these facts, it must be concluded that, given our present state

[3] For a discussion of this method, see E. G. Thomsen et al., *Mechanics of Plastic Deformation in Metal Processing* (New York: Macmillan Publishing Co., Inc., 1965).

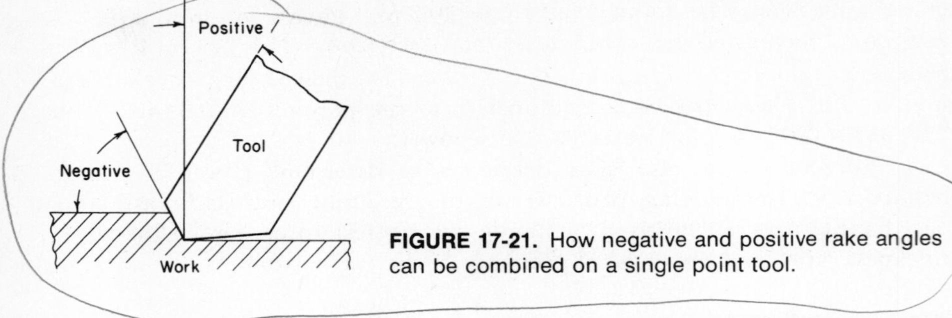

FIGURE 17-21. How negative and positive rake angles can be combined on a single point tool.

of knowledge, metal removal by chip formation is an inefficient method for obtaining desired shapes and surfaces. It is fortunate that superior cutting-tool materials—the various carbides and ceramics—have been developed. It also is clear why much attention, with considerable success, has been, and is being, given to the development of metal removal and placement processes that do not require the production of chips. But a very large proportion of our products are, and for the foreseeable future will continue to be, made by machining.

Modern trends in chip-producing machines. The early machine tools were developed to do only one or a few basic machining operations: for example, the early lathes for turning and Wilkinson's boring mill. This situation existed for many years. With the development of carbide tools and the accompanying, much-greater metal-removal rates, a large portion of the total time a machine tool was available was utilized in making setups of the workpiece and tooling, and often a given machine tool was not being utilized because required operations could not be performed on it. Further, the labor cost for making repeated setups was excessive. Consequently, in recent years there has been a trend toward machine tools that can do a variety of chip-producing operations. Some can do as many as five distinctly different basic machining operations, but in nearly all, commonly called *machining centers,* the workpiece is either stationary (except for feed), utilizing rotating tools and thus permitting milling, drilling, limited boring, and so on, or the tools are stationary and the workpiece rotates, thus permitting turning, drilling, boring, and so on, but almost no milling; these commonly are called *turning centers.*

There is a recent trend toward bridging this gap, as is demonstrated by the machine tool shown in Figure 17-22. It appears inevitable that more machines of this general type will be developed in the future.

The trend toward making machine tools stronger and more powerful, so that carbide and ceramic cutting tools can be used with optimum effectiveness, now is firmly established. A fourth trend has been the building of machining units composed of a number of powered and essentially self-contained units, each of which can do a single machining operation, assembled onto a common base. A fifth, and very significant, trend has been the development of machines that are numerically controlled through the use of punched tape or computers. These provide much greater versatility, accuracy, and productivity.

In following chapters the basic machining operations and the machines that can accomplish these operations will be examined.

FIGURE 17-22. Modern machining center which provides six tool–work motions and two rotating spindles, one mounted on the main tool turret, so as to permit both turning and milling to be done. (*Courtesy Okuma Machinery Works, Ltd.*)

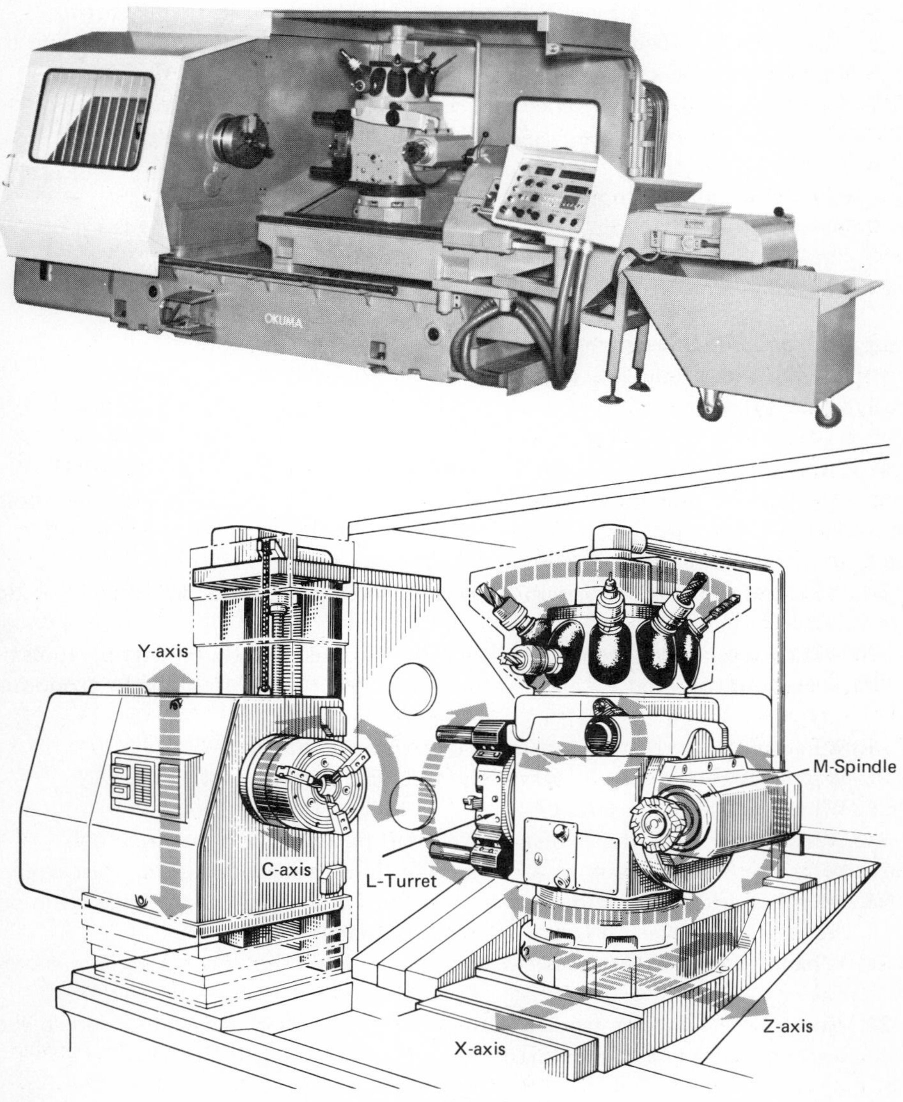

Review questions

1. Since chips are a waste product, why is the mechanism of chip formation of such great importance?
2. Basically, what is the mechanism of chip formation?
3. What are three important interrelated variables in metal cutting?
4. If chip formation is thought of as a process involving compression, shear, and friction, for a given material, which of these factors offer possibilities for improving the efficiency of cutting?
5. What is the approximate thickness of the shear zone in metal cutting?
6. About what percentages of the heat generated in metal cutting are due to metal deformation and friction?
7. To what extent can the efficiency of metal cutting be improved by more effective cooling?
8. About what percentage of the heat generated in metal cutting can be dissipated through the tool?
9. Why are discontinuous-type chips preferred over the continuous type?
10. Explain why a built-up edge on a cutting tool is undesirable.
11. Why may the true rake angle in a single-point cutting tool be different than either the back or side rake?
12. Basically, why does a chip breaker break up the chips?
13. What are two benefits derived from the use of chip breakers?
14. Why do carbide tools employ negative rake angles more often than HSS tools?
15. How does the wear of a carbide tool differ from that of an HSS tool?
16. What is the advantage gained by coating the face of tungsten carbide tools with a thin coating of titanium carbide?
17. What is the purpose of clearance angles on cutting tools?
18. What advantages do ceramic cutting tools have over tungsten carbide tools?
19. What prevents one from using large positive rake angles on cutting tools?
20. Since their material is costly, why are "throw away" inserts often economical?
21. What are the principal functions of a cutting fluid?
22. Why is it difficult to lubricate the face of a cutting tool?
23. What does feed determine?
24. What is the basic mechanism of tool failure of an HSS cutting tool?
25. Why is it not necessary to use a cutting fluid in machining gray cast iron?
26. By using a force-circle diagram, show the effect of reducing the friction on the tool face. (Keep all other factors constant.)
27. What is the relationship between λ, ϕ, β, and α in cutting a given material?
28. How does changing from a positive to a negative rake angle affect the power required in metal cutting?

29. How do the strains involved in metal cutting compare with those in tension testing?

30. What is meant by specific horsepower in connection with metal cutting?

31. Aluminum alloy 24-ST is machined dry at a cutting speed of 152 m (500 feet) per minute in a turning operation, using a depth of cut of 3.2 mm (⅛ inch) and a feed of 0.635 mm (0.025 inch) per revolution. Using Table 17-5, determine the power required.

32. In turning a mill roll that is 1829 mm (72 inches) long and 508 mm (20 inches) in diameter, a cut 3.18 mm (0.125 inch) deep must be made across the entire length. If an HSS tool bit, costing $2.25, is used with a feed of 0.51 mm (0.020 inch) per revolution, it will have to be ground originally and then reground six times during the cut. Grinding or regrinding the tool costs $1.20, and removing and resetting the tool requires 6 minutes. Each tool can be ground 40 times before being discarded.

A triangular, throwaway ceramic tool, having six cutting edges, costs $4.50 but can make the same cut with the use of only two edges. Resetting this tool requires 3 minutes.

If labor and machine costs are $5.50 and $4.50 per hour, respectively, which type of tool should be used? (Use cutting speeds for 4140 steel as shown in Table 17-1.)

33. If the time required for placing the roll in the lathe and removing it is a total of 15 minutes, what is the net effect of using the better tool on the total cycle time and the total cost in question 32?

CHAPTER 18

Shaping and planing

From a consideration of the relative motions between the tool and the workpiece, shaping and planing are the simplest of a machining operations, and the machines that do the operations are among the simplest of all machine tools. As indicated in Table 17-2 and Figure 18-1, each involves straight-line relative motion between a single-point cutting tool and a workpiece. A flat surface is generated by the work or the tool being fed at right angles to the cutting motion between successive strokes. Although it is possible to machine curved surfaces by these processes, it is not done frequently, because either a form-shaped tool must be used, with no feed, which severely limits the area machined, or difficult simultaneous movements must be made to control the relative positions of the work and tool. Consequently, in addition to plain flat surfaces, the shapes most commonly produced on the shaper and planer are those illustrated in Figure 18-2. Relatively skilled workers are required to operate shapers and planers, and most of the shapes that can be produced on

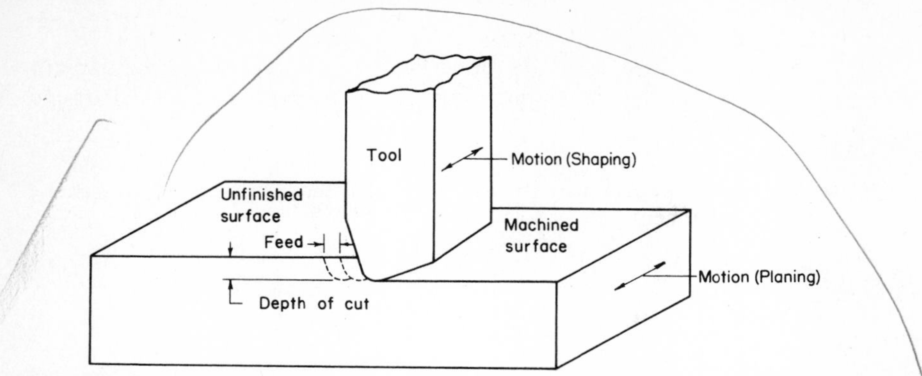

FIGURE 18-1. Basic relationships of tool motion, feed, and depth of cut in shaping and planing.

them also can be made by much more productive processes, such as milling, broaching, or grinding. Consequently, except for certain special types, planers that will do only planing have become obsolete, and shapers are used very little in manufacturing except in tool and die work or in very low volume production. Their basic simplicity and flexibility make them very useful in these situations.

SHAPING AND SHAPERS

Shaping is done on a *shaper,* utilizing a reciprocating, single-point cutting tool that moves in a straight line across the workpiece as illustrated in Figure 18-1. To produce a flat horizontal surface, the work is fed across the line of motion of the tool, between strokes. For vertical and angled flat surfaces, the tool usually is fed across the workpiece. Because the tool cuts on only one direction of the stroke, shaping is a relatively slow process. However, setups on shapers can be made easily and quickly, and they are flexible, thus making them very useful where only one or a few identical parts are required.

Shaping usually is confined to machining flat surfaces that do not exceed 610 × 610 mm (24 × 24 inches) in size. External and internal surfaces, ei-

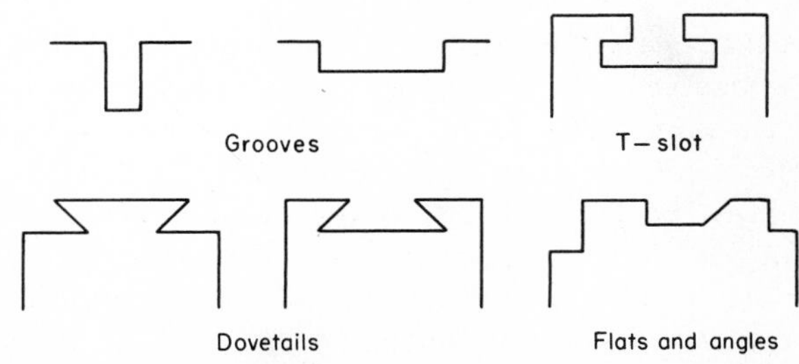

FIGURE 18-2. Types of surfaces commonly machined by shaping and planing.

ther horizontal or inclined, can be produced readily; curved and irregular surfaces also can be made, but with much less ease or by the use of special attachments.

Types of shapers. Shapers usually are classified according to their general design features as follow:

1. Horizontal
   a. Push-cut
   b. Pull-cut
2. Vertical
   a. Regular or slotters
   b. Keyseaters
3. Special

They are also designated as to the type of drive employed—*mechanical drive* or *hydraulic drive.*

Horizontal push-cut shapers. Most shapers are of the *horizontal push-cut* type. As indicated in the block diagram of Figure 18-3, cutting occurs as the ram *pushes* the tool across the work. Figure 18-4 shows a typical shaper of this type.

The main components of a horizontal push-cut shaper are the *base* and *column,* the *ram,* the *toolhead assembly,* the *driving mechanism,* the *table,* and the *cross-rail* and *feeding mechanism.* The base and column support all the other components, the column containing the ram-driving mechanism. Dovetail ways on the top of the column support the ram, and the cross rail is mounted on vertical ways on the front of the column.

Two types of mechanisms—mechanical and hydraulic—are employed for reciprocating the ram, on which the toolhead assembly is mounted at the forward end. The driving mechanism should provide ample force during the forward, cutting stroke, provide a more rapid return-stroke speed, and permit

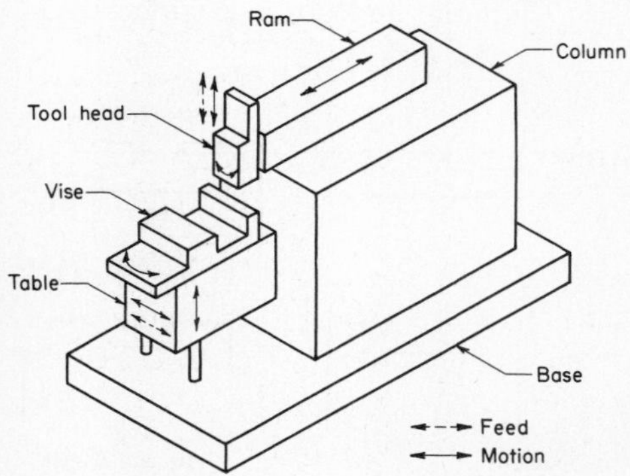

FIGURE 18-3. Block diagram showing the basic components of a shaper.

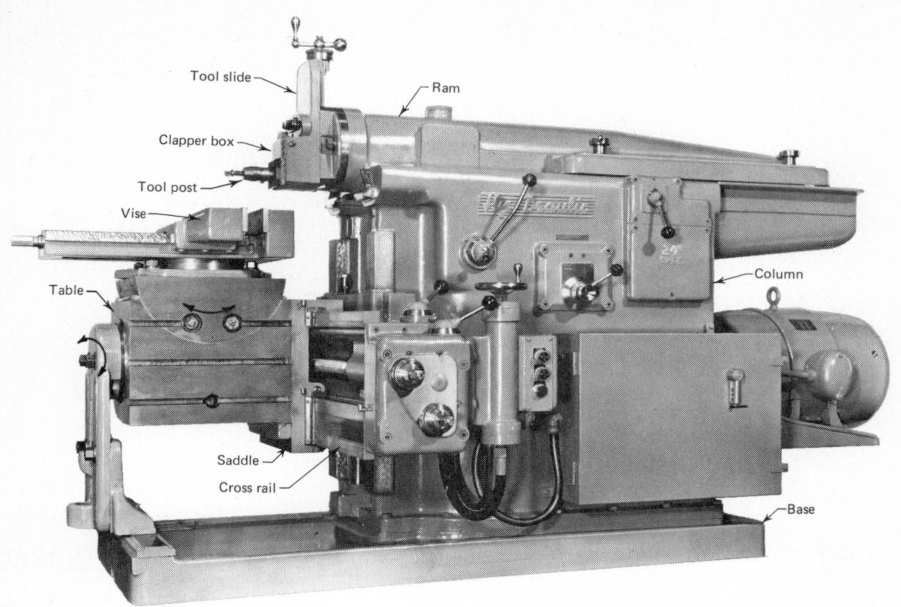

FIGURE 18-4. Hydraulic-drive, horizontal push-cut shaper with a universal table. (*Courtesy Rockford Machine Tool Company.*)

adjustment of the stroke length and stroke speeds. Figure 18-5 shows the sliding-block driving mechanism that is used in mechanical-drive shapers. Hydraulic-drive shapers employ the type of system shown in Figure 18-6. Hydraulic drives provide more uniform velocity during the cutting stroke and more rapid return-stroke speeds. Mechanical drives are somewhat cheaper and provide absolute limits to the stroke length.

The *ram and toolhead assembly* provides means for adjusting the horizontal position of the ram stroke with respect to the column and for mounting the tool on its forward end. As shown in Figure 18-7, the toolhead assembly has a swivel mounting, which permits the toolslide to be turned about a horizontal axis so that cuts can be made at any desired angle. A graduated scale is provided to facilitate making angular settings. The toolslide is carried on the swivel mounting in dovetail ways so that it can be moved a few inches by means of the manually operated lead screw to provide feed or depth of cut, as required, for the tool. A calibrated dial provides a means for controlling this movement.

A two-piece clapper box is mounted on the toolslide. The outer piece is pivoted at its upper end so that the tool, which is mounted on this section of the clapper box by means of a tool post, can lift upward. This permits the tool to clear the work on the reverse ram stroke. On most shapers the tool slides over the work surface on the backward stroke of the ram, but in some the tool

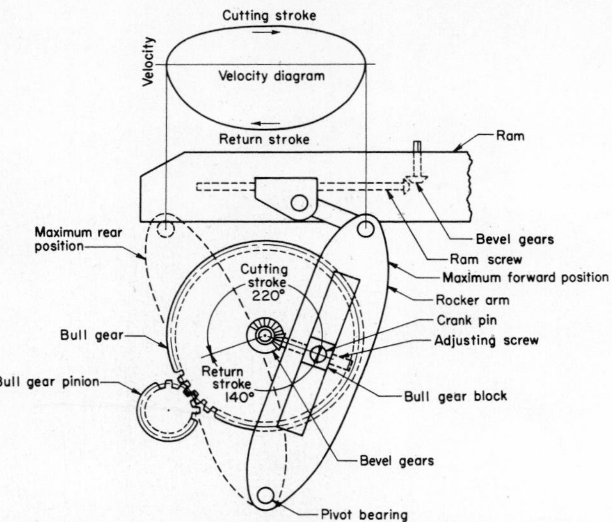

FIGURE 18-5. Sliding-block driving mechanism and resulting velocity diagram used on mechanical-drive shapers.

is lifted mechanically so that it does not slide on the work. The bottom piece of the clapper box is pivoted about a horizontal axis so that it can be tilted about 20° in either direction with respect to the toolslide. This angular displacement, shown in Figure 18-7, is necessary to permit the tool to swing away from, instead of into, the work surface as the tool lifts in cutting vertical or inclined surfaces. The clapper box should be swiveled so that it inclines away from the surface that is to be machined.

As shown in Figure 18-4, the shaper table is carried on a horizontal

FIGURE 18-6. Hydraulic drive used on shapers and planers. (*Courtesy Rockford Machine Tool Company..*

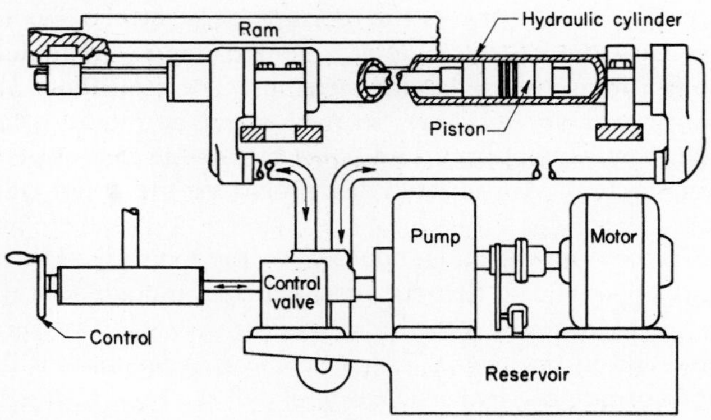

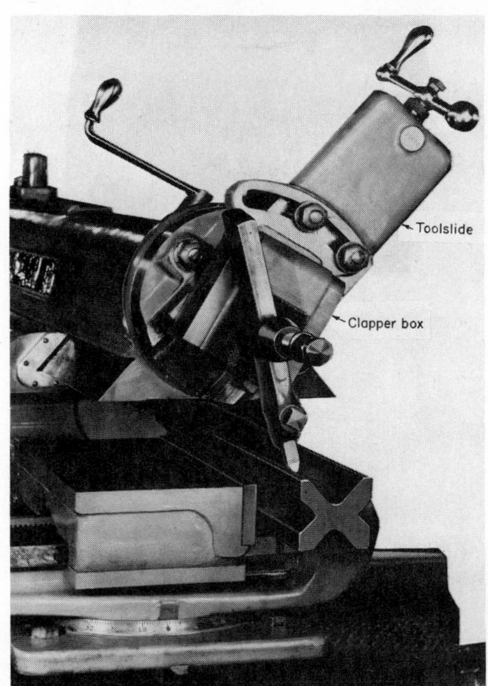

FIGURE 18-7. Toolslide and clapper box tilted for shaping an inclined surface. (*Courtesy The Cincinnati Shaper Company.*)

cross-rail, which, in turn, is mounted on vertical ways on the front of the shaper column. The cross rail can be raised and lowered by means of an elevating screw, or screws, and hand crank. On some larger shapers the cross rail can be raised and lowered by power, and in some machines vertical power feed is provided. The crossrail can be clamped to the column in any desired position to give added rigidity during machining.

A saddle slides on upper and lower ways on the cross rail, and the table is attached to the saddle. A *plain table* is nonadjustable and has T-slots and/or a heavy vise by means of which the work can be held. A *universal table,* shown in Figure 18-4, is provided with a means of rotating the entire table about a horizontal axis parallel with the rail. Such a table is particularly convenient in toolroom work, where inclined flat surfaces must frequently by machined.

In machining horizontal surfaces on a shaper, the depth of cut is set by means of the toolslide, and feed is provided by moving the table along the rail prior to the beginning of each stroke, usually by means of a power mechanism and lead screw. The table remains stationary throughout the cutting stroke. Feed is expressed in *millimeters or inches per stroke.*

To machine a vertical or inclined surface, feed usually is accomplished through manual movement of the tool slide before each cutting stroke. Some larger shapers have vertical power feed for the table so that vertical surfaces can be machined by using this vertical power feed. Also, special attachments

FIGURE 18-8. Draw-cut shaper machining a large die block. (*Courtesy Morton Manufacturing Company.*)

can be obtained for providing powered feed to the toolslide, but these are seldom used.

Horizontal draw-cut shapers. In order to eliminate the long, unsupported overhang of the ram that would result when shaping larger surfaces with push-cut shapers, *draw-cut shapers* are available. As illustrated in Figure 18-8, a large overarm supports the outer end of the ram, and cutting is done during the return stroke. Because the cutting forces act inward against the column, it is easier to support the work and avoid vibration and chatter. Such shapers are used primarily in tool and die shops.

Vertical shapers and slotters. A vertical shaper, shown in Figure 18-9, has a vertical ram and a round table that can be rotated in a horizontal plane, by either manual or power feed. These machine tools are sometimes called *slotters.* Usually, the ram is pivoted near the top so that it can be swung outward from the column through an arc of about 10°.

Because one circular and two straight-line motions and feeds are available, vertical shapers are very versatile tools and thus find considerable use in die shops. Not only can vertical and inclined flat surfaces be machined, but external and internal cylindrical surfaces can be generated by circular feeding of the table between strokes, or they can be made by turning or boring, respectively, by using a stationary tool and rotating the workpiece. A vertical shaper is particularly useful in machining curved surfaces, interior surfaces, and arcs by using a stationary tool and rotating the workpiece.

Vertical shapers are particularly useful in machining curved exterior and interior surfaces and arcs that have projections at each end so they cannot be made by turning. The inset in Figure 18-9 shows a typical job of this type.

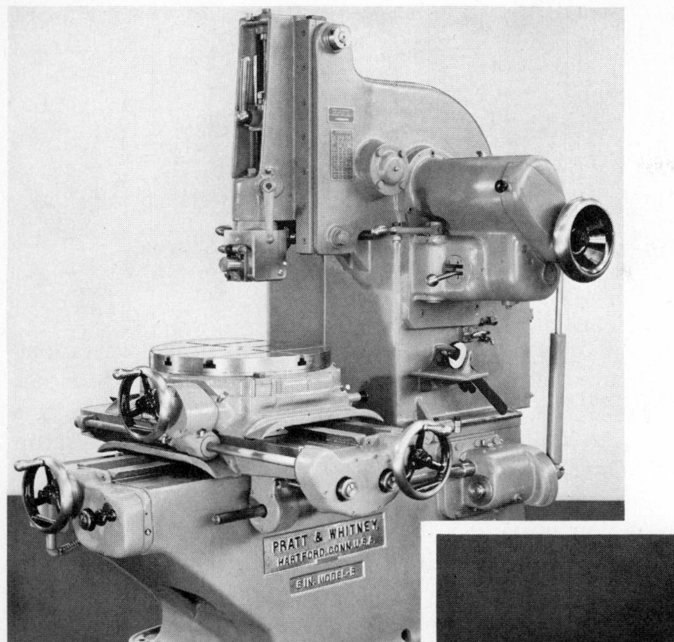

FIGURE 18-9. A 6-inch vertical shaper, equipped with a rotary table. Inset shows a typical job that can be done on this type of machine. (*Courtesy Colt Industries, Pratt & Whitney Machine Tool Division.*)

A *keyseater* is a special type of vertical shaper designed and used exclusively for machining keyways on the inside of wheel and gear hubs.

Shaper size. Horizontal-shaper size is designated by the maximum length of the stroke of the ram in millimeters or inches. Push-cut shapers are available in sizes from 100 to 915 mm (4 to 36 inches), whereas draw-cut shapers are made in sizes up to 1830 mm (72 inches). The table and crossrail usually are designed to provide table movement equal to the maximum ram stroke so that a 406-mm (16-inch) shaper, for example, will machine a horizontal surface 406 × 406 mm (16 × 16 inches). Vertical shapers also are designated by the length of the ram stroke, with the diameter of the worktable also given.

Work setup in shapers. The mounting and holding of the workpiece is of prime importance in all machining operations. The setup not only requires

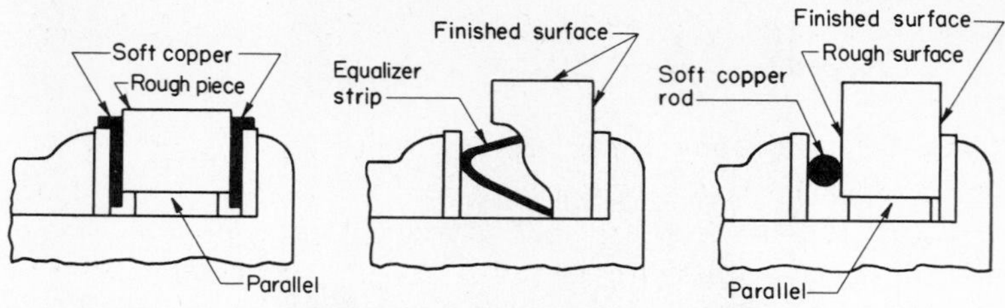

FIGURE 18-10. Methods of clamping workpieces in a shaper vise.

labor, but the machine usually is idle during setup time. If several operations are done on a given machine, or if the workpiece is of an odd shape or is not strong, this usually complicates the setup problem. One of the virtues of a shaper is that only a single, simple cutting action is involved, and the setup usually is very simple. On horizontal push-cut shapers, the work usually is held in a heavy vise mounted on the top side of the table. As shown in Figure 18-4, shaper vises have a very heavy movable jaw, because the vise must often be turned so that the cutting forces are directed against this jaw. Shaper vises can be rotated and clamped about a vertical axis.

In clamping work in a shaper vise, care must be exercised to make sure that it rests solidly against the bottom of the vise, or on parallel bars, so that it will not be deflected by the cutting force, and that it is held securely yet not distorted by the clamping pressure. Figure 18-10 illustrates the use of parallel bars for raising work to the proper height in the vise jaws, and several methods of clamping rough and irregularly shaped work. These latter procedures are required when the workpiece surfaces are rough, or the sides are not parallel, to prevent the clamping action from tilting the work in the vise.

Work that cannot be held conveneiently in a vise can be clamped directly to the top or sides of the table, using the T-slots.

Shaper tools. Most shaping is done with simple high-speed or carbide-tipped tool bits held in a heavy, forged tool holder, as shown in Figure 18-7. The toolholder is held in a slotted tool post by means of a setscrew. In some cases where a tool of irregular shape is required, as in cutting a T-slot, or where very heavy cuts are to be made, a large, forged tool bit is held directly in the tool post.

Special attachments for shapers. Several special attachments are available for shapers that permit special work to be done more conveniently. *Index centers* provide a means for holding work and spacing cuts around it. Irregular surfaces can be machined by the use of *profiling attachments,* in which a follower mechanism moves in contact with a master surface and causes the tool to be raised or lowered hydraulically to reproduce the master surface. *Automatic tool lifters* are available that lift the tool to clear the work on the return stroke, thus avoiding dragging carbide tools over rough work surfaces and causing them to be chipped.

Computing shaper speeds and machining time. Ram speeds on shapers are indicated in *strokes per minute.* Because cutting occurs only in one direction and the cutting-stroke speed is less than the return-stroke speed, these factors must be taken into account in order to determine the number of strokes per minute necessary to give a desired cutting speed.

On mechanical-drive shapers, the ratio of cutting time to return-stroke time averages about 1.6: 1; thus cutting takes place approximately 61 per cent of the total stroke time. Thus the number of strokes per minute, N, to produce a desired cutting speed, C.S., with a stroke length of l is given by the following equations:

$$\begin{array}{ll} \text{C.S.} = \text{m/min} & \text{C.S.} = \text{ft/min} \\ l = \text{mm} & l = \text{inches} \\ N = 608.4\dfrac{\text{C.S.}}{l} & N = 7.3\dfrac{\text{C.S.}}{l} \end{array} \qquad (18\text{-}1)$$

Because hydraulic shapers have a ratio of 2: 1, for a given number of strokes per minute, a smaller portion of the total stroke time is required for the return stroke. Consquently, the equations become

$$\begin{array}{ll} \text{Metric} & \text{English} \\ N = 666.7\dfrac{\text{C.S.}}{l} & N = 8\dfrac{\text{C.S.}}{l} \end{array} \qquad (18\text{-}2)$$

This means that on a hydraulic-drive shaper, a given number of strokes per minute produces a slightly lower cutting speed than on a mechanical-drive shaper. Because shaper feed is expressed in *millimeters* (or *inches*) *per stroke,* the time required to machine a given surface is given by the equation

$$T = \frac{W}{N \times f} \qquad (18\text{-}3)$$

where T is the time in minutes, W the width of the surface in millimeters (or inches), and f the feed in millimeters (or inches) per stroke. By substituting the equivalent values of N from equation (18-1) or (18-2) in equation (18-3), the approximate machining time can be determined directly from the cutting speed, width, and length of the surface to be machined.[1]

Shaper accuracy. Although shapers are versatile tools, the accuracy of the work done on them is greatly dependent on the operator. Feed dials on shapers nearly always are graduated in 0.02-mm or 0.001-inch divisions, and work seldom is done to greater accuracy than this. A tolerance of 0.04 to 0.06

[1]The stroke length always is slightly longer than that of the workpiece, to provide for some overrun of the tool on both ends of the stroke.

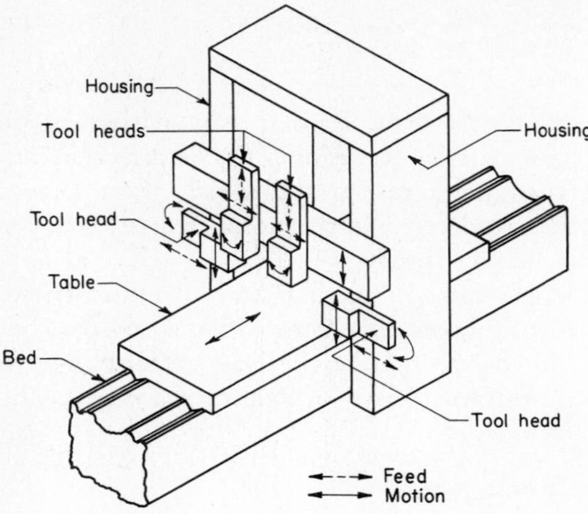

FIGURE 18-11. Block diagram showing the basic components of a double-housing planer.

mm (0.002 to 0.003 inch) is desirable on parts that are to be machined on a shaper, because this gives some provision for errors due to clamping and possible looseness or deflection of the table.

PLANING AND PLANERS

Planing can be used to produce horizontal, vertical, or inclined flat surfaces on workpieces that are too large to be accommodated on shapers. However, as will be pointed out, planing is much less efficient than other basic machining processes that will produce such surfaces; consequently, planing and planers have largely been replaced. However, some understanding of planing and planers is desirable so as to be aware of the possibilities and limitations.

Figure 18-11 shows the basic components and motions of a planer. In most planing, the action is opposite to that of shaping, in that the work is moved past one or more stationary, single-point cutting tools. Because a large and heavy workpiece and table must be reciprocated at relatively low speeds, several tool heads are provided, often with multiple tools in each head, as illustrated in Figure 18-12. In addition, many planers are provided with tool heads arranged so that cuts occur on both directions of the table movement. However, beause only single-point cutting tools are used and the cutting speeds are quite low, planers are low in productivity as compared with some other types of machine tools. As a result, they have been replaced, almost entirely, by planer-type milling machines or by similar combination machines which can do both milling and planing. These will be discussed in Chapter 22.

FIGURE 18-12. Interchangeable multiple tool holder for use on planers. (*Courtesy Gebr. Boehringer GMBH.*)

Types of planers. Planers have been made in four basic types. Figure 18-11 depicts the most-common, double-housing type, and Figure 18-17 shows a special planer of this type. It has a closed housing structure, spanning the reciprocating worktable, with a crossrail supported at each end on a vertical column and carrying two toolheads. An additional toolhead usually is mounted on each column, so that four tools (or four sets) can cut during each stroke of the table. Obviously, the closed-frame structure of this type of planer limits the size of the work that can be machined. Consequently, open-side planers are available, as indicted in Figure 18-13, which have the crossrail supported on a single column. This provides unrestricted access to one side of the table to permit wider workpieces to be accommodated. Some open-side planers are convertible, in that a second column can be attached to the bed when desired so as to provide added support for the crossrail.

As indicated previously, double-housing and open-side planers seldom are made currently, but two special types of planers probably will continue to

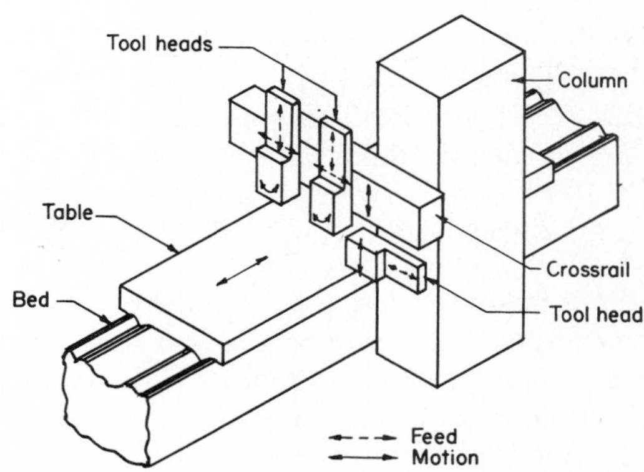

FIGURE 18-13. Block diagram of an open-side planer.

FIGURE 18-14. Plate or edge planer. Inset shows a close-up of the tool. (*Courtesy Baldwin-Lima-Hamilton Corporation.*)

have significant, special use. One is the *plate* or *edge planer,* illustrated in Figure 18-14, designed specifically for planing the edges of plates. The work is stationary, and the tool reciprocates on a side-mounted carriage that often also carries the operator. The cutting tool can be adjusted vertically and horizontally on its support and rotated to cut during both directions of travel. The other type is the *pit planer,* shown in Figure 18-15, on which both the work and table are stationary, but the columns and crossrail move longitudinally on massive rails. These are used only for very large work, where the weight of the workpiece and the required table would make reciprocation difficult and severely limit cutting speeds. The toolheads usually revolve to permit cutting in both directions. They are not common but are very useful for very large work.

Planer size. The size of double-housing planers is designated by the distance between the vertical housings in inches, the distance from the table to the rail in its uppermost position in inches, and the maximum length of the table travel in feet. These dimensions indicate the size of the largest workpiece that can be planed on the machine tool.

For open-side planers, a "planing width" is substituted for the distance between housings. This width is the distance from the table side of the housing to the maximum outward position of the tool in the outer toolhead.

Plate planers are designated by the length of the maximum plate that can be machined.

Work setup on planers. Two factors must be given special consideration in making a work setup on planers. First, because the workpieces are unusually heavy and must be reciprocated, they must be fastened to not only resist the cutting forces but also the high inertia forces that result from the rapid acceleration changes at the ends of the strokes. Special stops often are provided at each end of the workpiece to prevent it from shifting. Figure 18-16 shows a

FIGURE 18-15. Pit planer. (*Courtesy Consolidated Machine Tool Division, Farrel Birmingham Company, Inc.*)

FIGURE 18-16. Heavy workpiece mounted on planer table, using standard supporting and clamping devices. (*Courtesy Armstrong Bros. Tool Co.*)

FIGURE 18-17. Divided-table planer. One workpiece is being set up on a table while another is being machined on a second table. (*Courtesy The Cincinnati Shaper Company.*)

typical setup on a planer table, utilizing standard clamping and supporting devices.

The second factor is that considerable time usually is required to set up a workpiece, thus reducing the time the costly machine is available for producing chips. Sometimes special setup plates are used, to which the workpiece is fastened. These plates are designed for quick attachment to the planer table. Another procedure involves a planer having two tables, as shown in Figure 18-17. Work is set up on one table while another workpiece is being machined on the other. Both tables can be fastened together for machining long workpieces. This same table arrangement is available on planer-type milling machines, which will be discussed in Chapter 22.

Planer tools. Because a large percentage of available time is used for setup, and the large workpieces involved usually will permit heavy cutting forces, when planing is used the cuts should be as heavy as practicable. Consequently, planer tools usually are quite massive. Usually, the main shank of the tool is made of plain-carbon steel, and a high-speed or carbide tip is clamped or brazed to it. Because the cuts usually are quite long, in planing steel or other ductile materials it is very important that chip breakers be provided to avoid long and dangerous chips.

Many modern planers have toolholders arranged so that cutting occurs during both directions of the table movement, thereby further increasing the productivity. Others are equipped with hydraulic tool lifters to raise the tool from the work on the return stroke. This is important when using carbide tool tips.

Planer accuracy and use. Theoretically, planers have about the same accuracy as shapers. The feed and other dimension-controlling dials usually are graduated in 0.02-mm or 0.001-inch divisions. However, because larger and heavier workpieces usually are involved, and much longer beds and tables, the working tolerances for planer work should be somewhat greater than for shaping.

Review questions

1. What is the basic difference between a shaper and a planer?
2. Why are shapers not well suited for producing curved surfaces?
3. For what type of work are shapers best suited?
4. What is the function of the clapper box on a shaper?
5. What are the advantages and disadvantages of mechanical-drive and hydraulic-drive shapers?
6. Why is a pull-cut shaper more suitable for large work than a push-cut shaper?
7. How is shaper feed expressed?
8. How is the size of a shaper designated?
9. What are two ways in which angular surfaces can be machined on a shaper?
10. Why is it seldom necessary to use carbide cutting tools on a shaper?
11. How many strokes per minute would be required to obtain a cutting speed of 36.6 m (120 feet) per minute on a typical mechanical-drive shaper if a 254-mm (10-inch) stroke is used?
12. How much time would be required to shape a flat surface 254 mm (10 inches) wide and 203 mm (8 inches) long on a hydraulic-drive shaper, using a cutting speed of 45.7 m (150 feet) per minute, a feed of 0.51 mm (0.020 inch) per stroke, and an overrun of 12.7 mm (½ inch) at each end of the cut?
13. What is the metal-removal rate in question 13 if the depth of cut is 6.35 mm (¼ inch)?
14. If the metal being removed in question 12 is gray cast iron, what will be the power required? (Use Table 17-5.)
15. A hydraulic-drive shaper has a 5.6-kW (7½-hp) motor, and 75 per cent of the motor output is available at the cutting tool. The specific power for cutting a certain metal is 0.03 W/mm^3 (0.67 hp/in.^{3}). What is the maximum depth of cut that can be taken in shaping a surface in this material if the surface is 305 × 305 mm (12 × 12 inches), the feed is 0.64 mm (0.025 inch) per stroke, and the cutting speed is 54.9 m (180 feet) per minute?
16. Why is it difficult to use high cutting speeds on a planer?
17. How does the designation of planer size differ from that of a shaper?
18. What are four basic types of planers?
19. How does a plate-edge planer differ from other types?
20. Why is it so desirable to make more than one cut simultaneously on a planer?

21. What is the basic advantage of an open-side planer? What is an inherent disadvantage?

22. Why is a vertical shaper more versatile than an ordinary shaper?

23. Assume that the bearing block shown in figure 36-4 is to have all the flat surfaces machined on a horizontal shaper. List the sequence in which the operations would be done.

Case study 13.
ALUMINUM RETAINER RINGS

The Owlco Corporation has to make 5000 retainer rings, as shown in Figure CS-13. It is essential that the surfaces be smooth with no sharp corners on the circumferential edges. Determine the most economical method for manufacturing these rings.

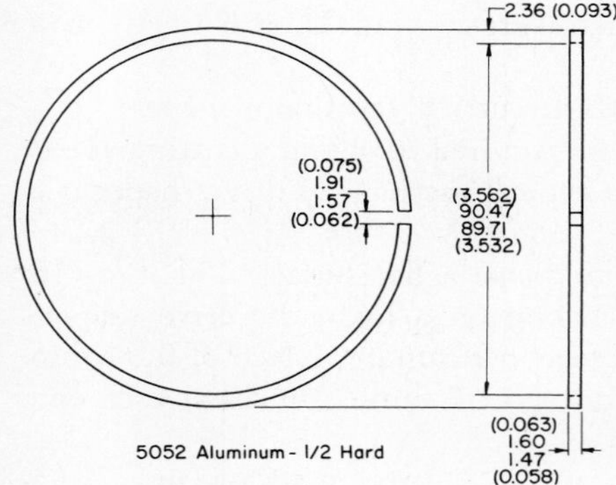

FIGURE CS-13. Aluminum snap-in retainer ring.

CHAPTER 19

Drilling and reaming

DRILLING

In manufacturing, it is probable that more internal cylindrical surfaces—holes—are produced than any other shape, and a large proportion of these are made by drilling. Consequently, drilling is a very important process. Although drilling, basically, is a relatively simple process, certain aspects of it can cause considerable difficulty. Most drilling is done with a tool having two cutting edges. However, these edges are at the end of a relatively flexible tool, and the cutting action takes place within the workpiece, so that the chips must come out of the hole while the drill is filling a large portion of it. Also, there is friction between the body of the drill and the wall of the workpiece, resulting in heat in addition to that due to chip formation. As a result of these conditions, substantial difficulty can be experienced due to poor heat removal, the counterflow of the chips makes lubrication and cooling difficult, and obtaining desired accuracy often is not easy.

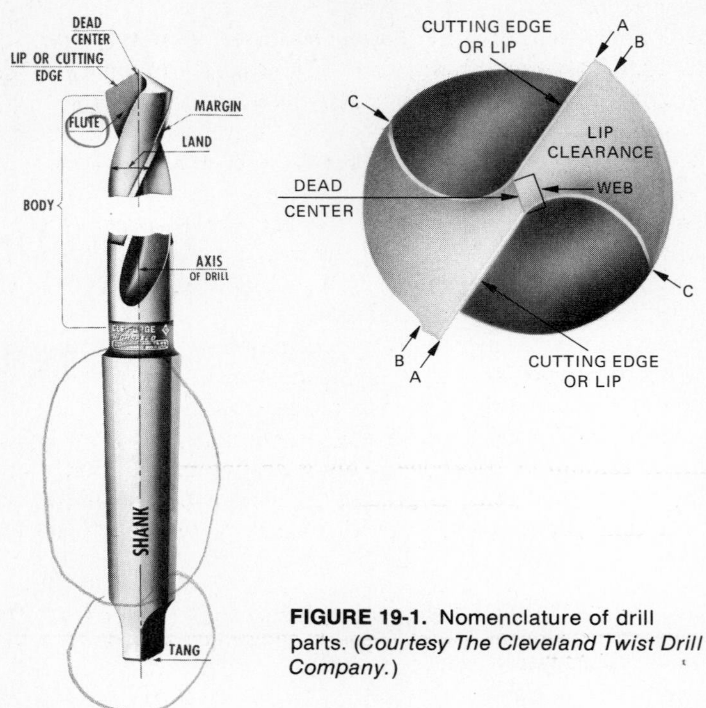

FIGURE 19-1. Nomenclature of drill parts. (*Courtesy The Cleveland Twist Drill Company.*)

Types of drills. The most common types of drills are *twist drills.* These have three basic parts: the *body,* the *point,* and the *shank,* shown in Figure 19-1. The body contains two or more spiral grooves, called *flutes,* in the form of a helix along opposite sides. To reduce the friction between the drill and the hole, each land is reduced in diameter except at the leading edge, leaving a narrow *margin* of full diameter to aid in supporting and guiding the drill and thus aiding in obtaining an accurate hole. The lands terminate in the point, with the leading edge of each land forming a cutting edge.

As shown in Figure 17-9, the principal rake angles behind the cutting edges are formed by the relation of the flute helix angle to the work. Because the helix angle is built into the drill, the primary rake angle cannot be varied by normal grinding. The helix angle of most drills is 24°, but drills with other helix angles are available. Larger helix angles—often above 30°—are used for materials that can be drilled very rapidly, resulting in a large volume of chips. Helix angles ranging from 0 to 20° are used for soft materials, such as plastics and copper. Straight-flute drills (zero helix and rake angles) also are used for drilling thin sheets of soft materials. It is possible to change the rake angle adjacent to the cutting edge by a special grinding procedure, called *dubbing* (see Figure 9-29).

The cone-shaped point on a drill contains the cutting edges and the various clearance angles. This angle affects the direction of flow of the chips across the tool face and into the flute. Obviously, it is very important. The 118° cone angle that is used most often is an arbitrary one that has been found to provide good cutting conditions and reasonable tool life for mild steel, thus making it suitable for much general-purpose drilling. Smaller cone angles—from 90 to 118°—sometimes are used for drilling more brittle materials, such as gray cast iron and magnesium alloys. Cone angles from 118 to 135° often are used for the more ductile materials, such as alloy steels and aluminum alloys. Cone angles less than 90° frequently are used for drilling plastics. As will be discussed later, several methods of grinding drills have been developed that produce points that are not plain cones.

The flutes serve as channels through which the chips come out of the hole and also to permit coolant to get to the cutting edges. Although most drills have two flutes, some, as shown in Figure 19-2, have three.

As shown in Figure 19-1, the relatively thin *web* between the flutes forms a metal column or backbone. This is an unfortunate feature of a twist drill. If a plain conical point is ground on the drill, the intersection of the web and the cone produces a straight-line *chisel center,* which can be seen in

FIGURE 19-2. Types of twist drills and shanks: (*left to right*) straight-shank, three-flute core drill; taper-shank; straight-shank; bit-shank; straight-shank, high-helix-angle; straight-shank; straight-flute.

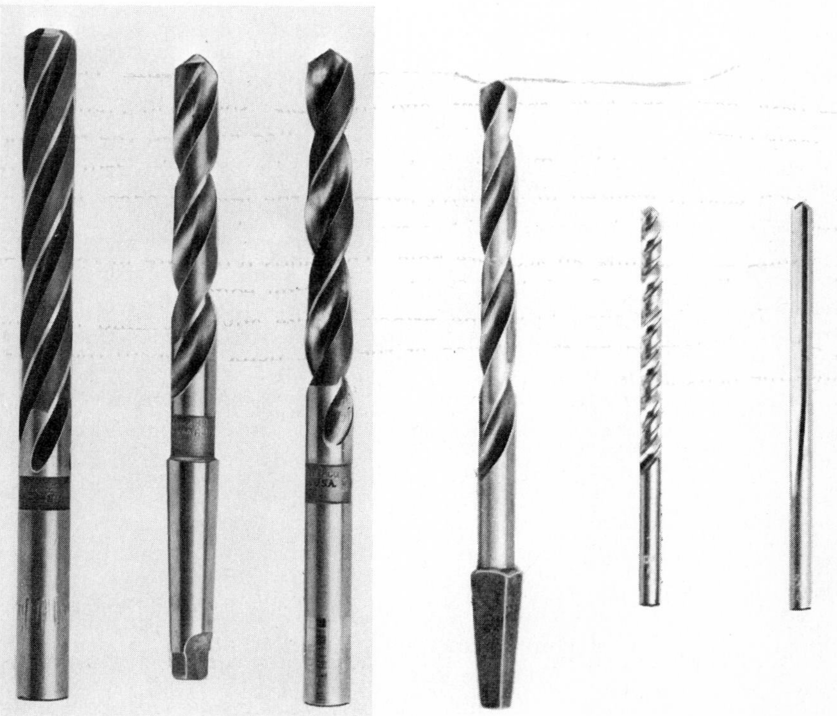

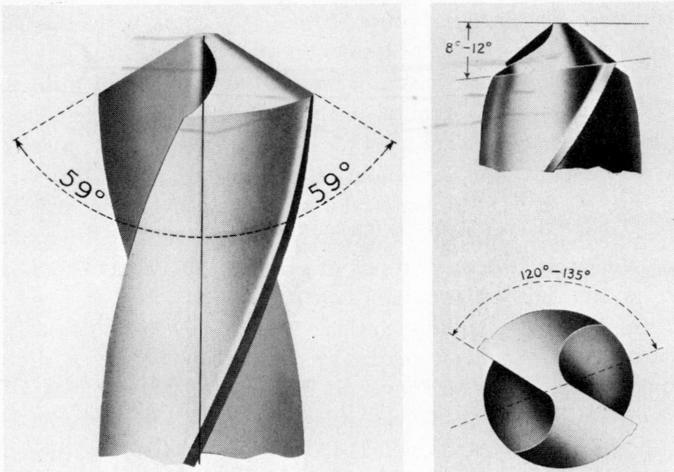

FIGURE 19-3. Cone, lip, and clearance angles for twist drills. (*Courtesy Cleveland Twist Drill, an Acme-Cleveland Company.*)

the three views of Figure 19-3. Unfortunately, the chisel point, which also must act as a cutting edge, forms a 56° negative rake angle with the conical surface. Such a large negative rake angle does not cut efficiently, causing excessive deformation of the metal. This results in high axial forces being required and excessive heat being developed at the point. This condition is further complicated by the obvious fact that the cutting speed at the drill center is low, approaching zero. As a consequence, drill failure on a normally ground drill occurs both at the center, where the cutting speed is lowest, and at the outer tips of the cutting edges, where the speed is highest.

Another difficulty resulting from a chisel point is that when this rotating, straight-line point comes in contact with the workpiece, it has a distinct tendency to "walk" along the surface, thus moving the drill away from the desired location, unless positive means are provided to prevent this action, usually at extra cost.

Several special methods of grinding drill points have been developed to eliminate or minimize the difficulties caused by the chisel point and to obtain better cutting action and tool life. Two such methods are shown in Figure 19-4. Such methods have had varying degrees of success, and they require special drill-grinding equipment. Another procedure is *web thinning,* in which a narrow grinding wheel is used to remove a portion of the web near the point of the drill.

Because of the use of higher cutting speeds, particularly in automatic machines, some type of chip breaker often is incorporated in drills. One procedure is to grind a small groove in the tool face, parallel with and a short distance from the cutting edge. This, of course, requires an additional grind-

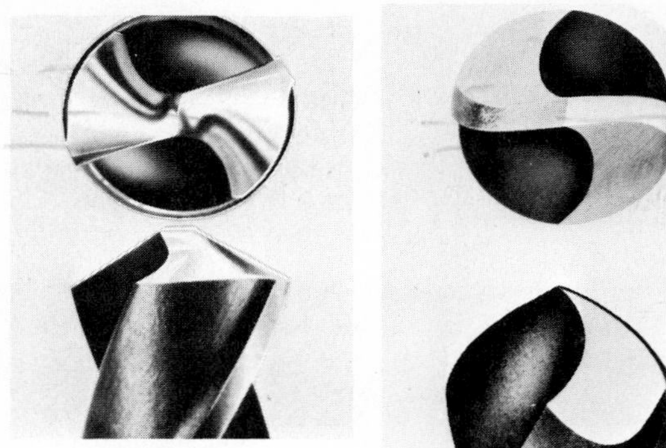

FIGURE 19-4. (*Left*) Spiral-point drill. (*Courtesy Omark Industries Machine Tool Division.*) (*Right*) Radial-lip-point drill. (*Courtesy Radial Lip Machine, Inc.*)

ing operation. A more effective procedure is to employ drills that have a special chip-breaker rib as an integral part of the flute, as shown in Figure 19-5. The rib interrupts the flow of the chip, causing an abrupt change in direction, and breaks it into short lengths. Such a drill requires no special grinding.

Some success has been achieved in improving the effectiveness of drilling some hard-to-drill materials, such as titanium, by adding a reciprocating motion having an amplitude of about 0.02 mm (0.001 inch) and an ultrasonic frequency of 20 kHz to a rotating drill.

Drill shanks are made in several types. Figure 19-2 shows several shank styles. The two most common types are the straight and the taper. *Straight-shank* drills are used only for sizes up to 12.7 mm (½ inch) and must be held in some type of drill chuck. *Taper shanks* are available on drills from 3.18 to 12.7 mm (⅛ to ½ inch) and are standard on drills above 12.7 mm (½ inch). Morse tapers, having a taper of approximately 1:19 (⅝ inch per foot), are used on taper-shank drills, ranging from a number 1 taper on ⅛-inch drills to a number 6 on a 3½-inch drill.

Taper-shank drills are held in a female taper in the end of the machine tool spindle. If the taper on the drill is smaller than the spindle taper, one or

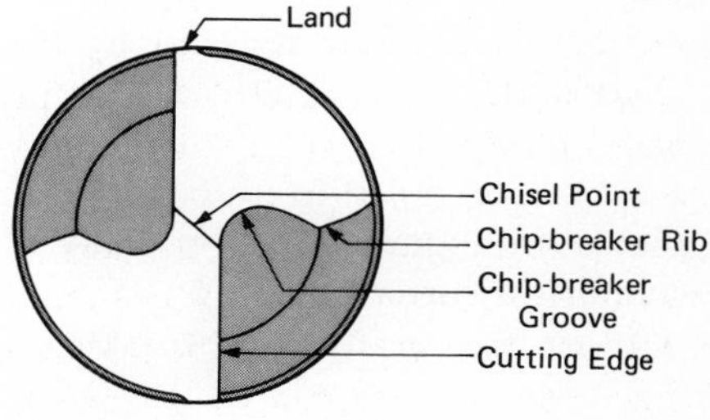

FIGURE 19-5. End view of drill containing a chip-breaker rib.

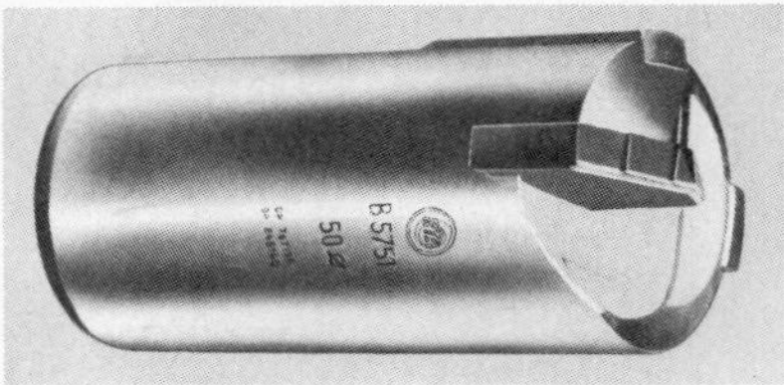

FIGURE 19-6. Drills for drilling deep holes. (*Top*) central chip-hole type; (*bottom*) single-flute type. (*Courtesy Colt Industries, Pratt & Whitney Machine Tool Division, and American Heller Corporation, respectively.*)

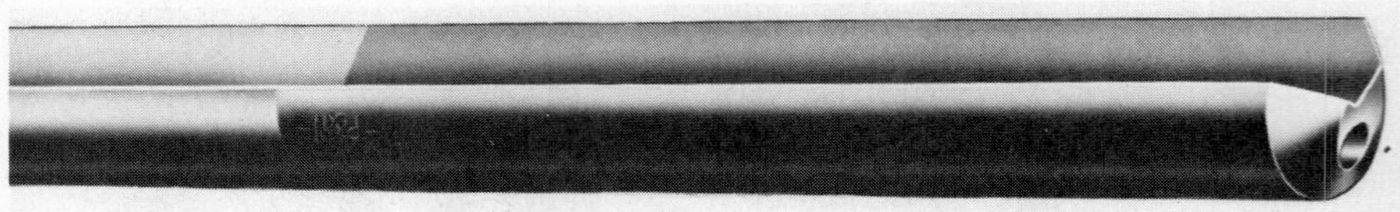

more adapter sleeves are placed in the hole to reduce it to the same size as the taper on the drill. Taper-shank drills have the advantage that the taper assures the drill being accurately centered in the spindle. The tang at the end of the taper shank fits loosely in a slot at the end of the tapered hole in the spindle. Its primary function is to provide a means by which the drill may be loosened for removal by driving a tapered *drift* through a hole in the side of the spindle and against the end of the tang. It also acts as a safety device to prevent the drill from rotating in the spindle hole under heavy loads. However, if the tapers on the drill and in the spindle are in proper condition, no slipping should occur; all driving of the drill is through the friction between the two tapered members.

Deep-hole drills, which contain passages through which coolant is forced to the cutting edges, and which also aids in pushing chips back out of the hole, are used when deep holes are to be drilled. Their special form reduces the tendency of the drill to drift, thus producing a more accurately aligned hole. The one shown in the bottom portion of Figure 19-6 is designed so that the chips are forced back along the central, straight flute, whereas with the type shown in the upper portion of this figure, the chips flow through a hole in the center of the drill. This construction provides better centering support to the drill and prevents chips from abrading the walls of the hole.

Two-flute drills are available that have holes extending throughout the length of each land to permit coolant to be supplied, under pressure, to the point adjacent to each cutting edge. These are helpful in providing cooling and also in promoting chip removal from the hole when drilling to moderate depths. They require special fittings through which the coolant can be supplied to the rotating drill and are used primarily on automatic and semiautomatic machines.

Larger holes in thin material may be made with a *trepanning cutter,* shown in Figure 19-7, whereby the main hole is produced by the thin, cylindrical cutter or saw.

Because drills are relatively slender, and also because of the "walking"

FIGURE 19-7. Hole cutter. (*Courtesy Armstrong-Blum Manufacturing Company.*)

action of the chisel point, they can be deflected rather easily when starting to drill a hole. Consequently, to assure that a hole is started accurately, a *center drill and countersink,* illustrated in Figure 19-8, is used prior to a regular chisel-point twist drill. The center drill and countersink has a short, straight drill section extending beyond a 60° taper portion. The heavy, short body provides rigidity so that a hole can be started with little possibility of the center drill being deflected. The hole should only be drilled partway up on the tapered section of the center drill. The conical portion of the hole that is left in the workpiece by the center drill serves to guide the drill that is subsequently used to make the main hole. Combination center drills are made in four sizes to provide the proper-size starting hole for any drill. If the drill is sufficiently large in diameter, or if it is sufficiently short, satisfactory accuracy often may be obtained without center drilling. Special drill holders are available that permit drills to be held with only a very short length protruding (see Figure 19-13).

FIGURE 19-8. Combination center drill and countersink. (*Courtesy Chicago-Latrobe Twist Drill Works.*)

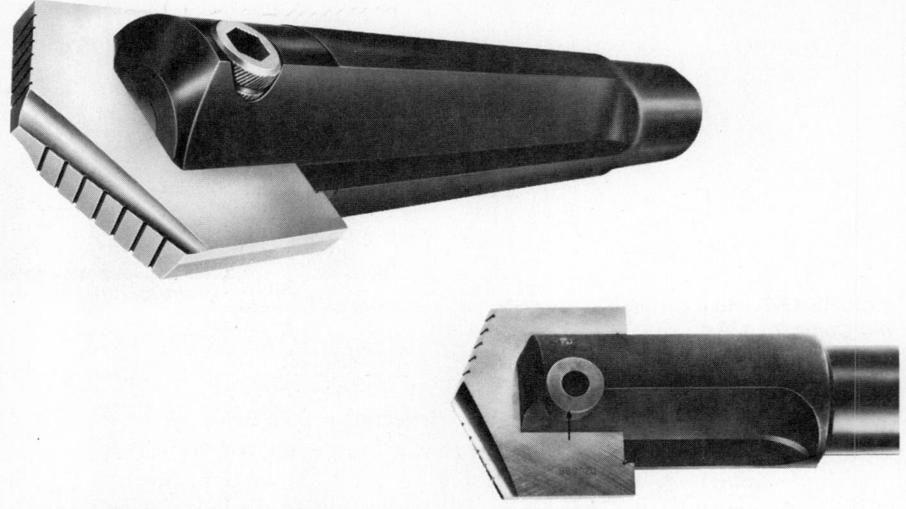

FIGURE 19-9. (*Top*) Typical spade drill. (*Bottom*) Spade drill with carbide inserts (arrow) to provide lateral guidance. (*Courtesy Erickson Tool Company.*)

Spade drills, such as is shown in Figure 19-9, are widely used for making holes 25.4 mm (1 inch) or larger in diameter. Such drills have several advantages: (1) less of the costly cutting-tool material is required because the long supporting bar can be made of ordinary steel; (2) the drill point can be ground with a minimum chisel point; (3) the main body can be made more rigid because no flutes are required; (4) the cutting blade is easier to sharpen; and (5) the main body can be provided with a central hole through which a fluid can be circulated to aid in cooling and in chip removal.

Spade drills are being used increasingly to machine a shallow locating cone for a subsequent smaller drill and at the same time to provide a small bevel around the hole to facilitate later tapping or assembly operations. Such a bevel also frequently eliminates the need for deburring. This practice is particularly useful on mass-production and numerically controlled machines.

Although the use of a center drill aids materially in assuring that a drill starts drilling at the desired location, because of its flexibility the drill may drift off the center line during drilling as the result of nonhomogeneity in the workpiece. Such nonhomogeneities and imperfect drill sharpening also may cause the hole to be oversize. Thus there is little assurance that a drilled hole is accurate as to alignment and size. If accuracy in these respects is desired, it is necessary to follow center drilling and drilling by boring and reaming, as illustrated in Figure 19-10. Boring trues the hole alignment, whereas reaming brings the hole to accurate size and improves the surface finish.

Special *combination drills* are made that can drill two or more diameters, or drill and countersink and/or counterbore, in a single operation. Some of

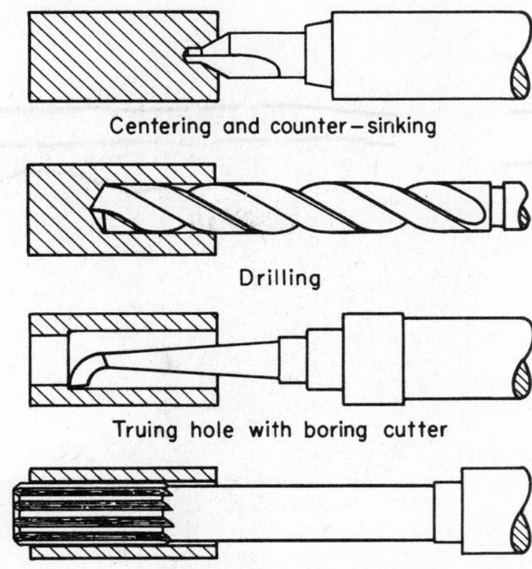

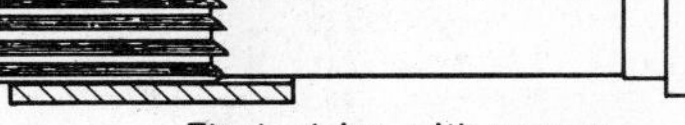

FIGURE 19-10. (*Top to bottom*) Steps required to obtain a hole that is accurate as to size and alignment.

these are illustrated in Figure 19-11. A *step drill* has a single set of flutes and is ground to two or more diameters. *Subland drills* have a separate set of flutes, on a single body, for each diameter or operation; they provide better chip flow, and the cutting edges can be ground to give proper cutting conditions for each operation. Combination drills are expensive but are economical for production-type operations because they reduce work handling and separate machines and operations.

FIGURE 19-11. Special-purpose subland or multicut drills, and some of the operations possible with such drills. (*Courtesy Chicago-Latrobe Twist Drill Works.*)

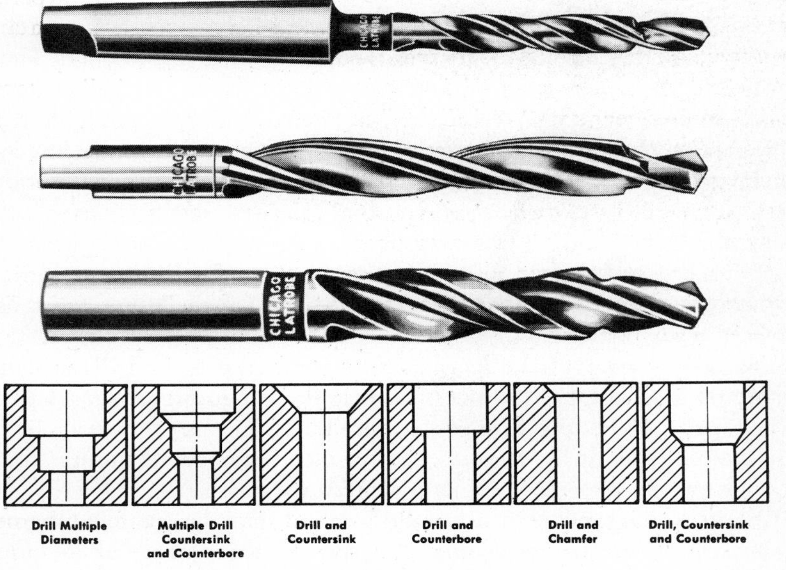

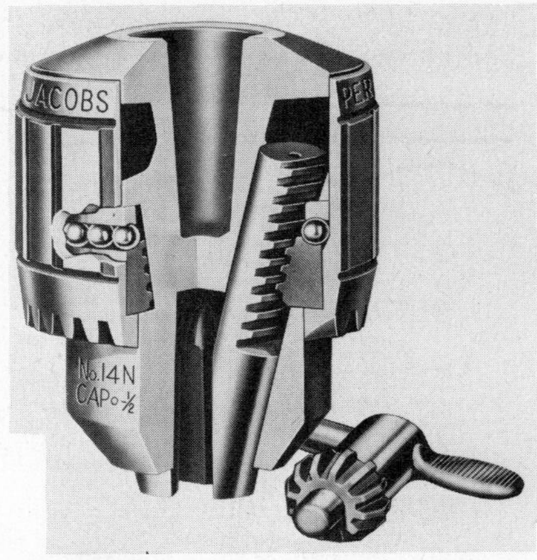

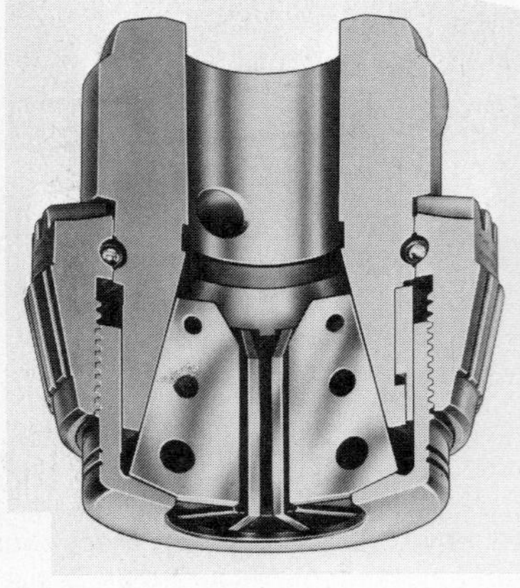

FIGURE 19-12. Two types of drill chucks. (*Courtesy The Jacobs Manufacturing Company.*)

Drill sizes. Standard drills are available in four size series—the size indicating the diameter of the drill body.

Millimeter series: 0.01- to 0.50-mm increments, according to size, in diameters from 0.015 mm
Numerical series: No. 80 to No. 1 (0.0135 to 0.228 inch)
Lettered series: A to *Z* (0.234 to 0.413 inch)
Fractional series: 1/64 to 4 inches (and over) by 64ths

Drill chucks. Straight-shank drills must be held in some type of drill chuck. Those shown in Figure 19-12 are the types used most commonly. The one shown in Figure 19-12a is adjustable over a considerable size range. The one in Figure 19-12b has radial steel fingers that are attached to, and held in position by, synthetic rubber. When the chuck is tightened, these fingers are forced inward against the drill—the rubber permitting sufficient motion. It has less size range than the other type but provides excellent centering and holding of the drill. On smaller drill presses, the chuck often is permanently attached to the machine spindle, whereas on larger drilling machines the chucks have a tapered member that fits into the female Morse taper of the machine spindle.

Two special types of chucks frequently are used to hold drills in semi- or fully automatic machines. The one shown at the left in Figure 19-13 permits quite a range of drills to be held in a single chuck. The one shown at the right permits holding the drill with only a controlled amount protruding, so as to reduce flexing and give more accurate centering.

Chucks of the types shown in Figure 19-12 must be tightened by means of a chuck key, requiring the machine spindle to be stopped in order to

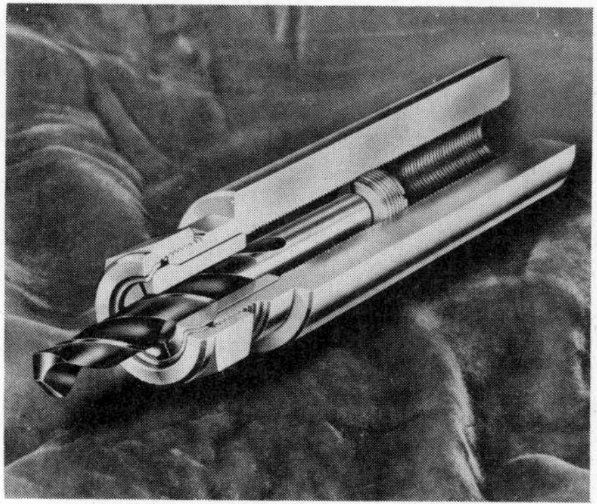

FIGURE 19-13. Two types of drill chucks used on automatic equipment. (*Left*) Universal type. (*Courtesy Brookfield Tool Company.*) (*Right*) Collet type. (*Courtesy Erickson Tool Company.*)

change a drill. To reduce the downtime where drills must be changed frequently, *quick-change chucks,* such as shown in Figure 19-14, are used. Each drill is fastened in a simple, round collet that can be inserted into, and removed from, the chuck hole while it is turning by merely raising and lowering a ring on the chuck body. By using this type of chuck, center drills, drills, counterbores, reamers, and so on, can be used in quick succession.

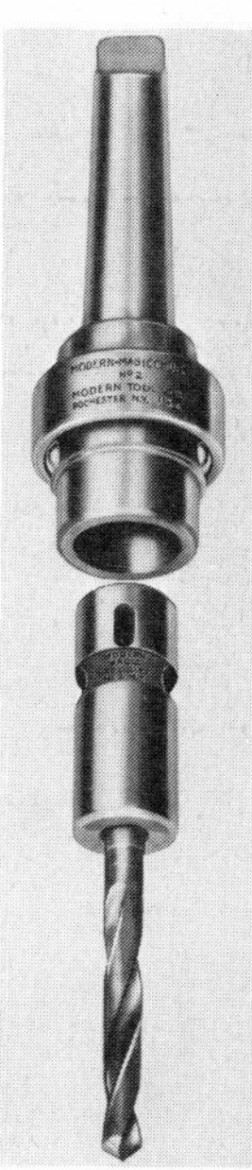

FIGURE 19-14. Quick-change drill chuck. (*Courtesy Consolidated Machine Tool Division, Farrel Birmingham Company, Inc.*)

Speeds and feeds in drilling. The usual cutting speeds, as given in Chapter 17, are used in drilling; these are the surface speeds at the outside of the drill. These surface speeds are used to compute the rotational speed of the drill. Also, one should consider whether heat will be conducted away from the cutting edges by conduction through the workpiece and rapid chip flow. In drilling deep holes, or in drilling holes in material that does not conduct heat readily, cutting speeds may have to be reduced unless an ample supply of coolant can be provided at the cutting edges.

Drilling feeds are expressed as *millimeters* or *inches per revolution* (mmpr or ipr). The following are representative:

Diameter of Drill	mmpr	ipr
Less than 3 mm (⅛ in.)	0.03–0.05	0.001–0.002
3 to 6.4 mm (⅛ to ¼ in.)	0.05–0.10	0.002–0.004
6.4 to 12.7 mm (¼ to ½ in.)	0.10–0.18	0.004–0.007
12.7 to 25.4 mm (½ to 1 in.)	0.18–0.38	0.007–0.015
Over 25.4 mm (1 in.)	0.38–0.64	0.015–0.025

If the conditions are unusual, the feed may have to be varied from these suggested values.

Cutting fluids in drilling. For shallow holes, the general rules relating to cutting fluids, as given in Chapter 17, are applicable. Where the hole depth exceeds one diameter, it is desirable to increase the lubricating quality of the fluid because of the rubbing between the drill margins and the wall of the hole. The effectiveness of a cutting fluid as a coolant is quite variable in drilling. While the rapid egress of the chips is a primary factor in heat removal, this action also tends to restrict entry of the cutting fluid. This is of particular importance in drilling materials that have poor heat conductivity.

If the hole depth exceeds two or three diameters, it usually is advantageous to withdraw the drill each time it has drilled about one diameter of depth in order to clear chips from the hole. Some machines are equipped to provide this "pecking" action automatically.

Where cooling is desired, the fluid should be applied copiously. For severe conditions, drills containing coolant holes have considerable advantage; not only is the fluid supplied near the cutting edges, but the flow of the fluid aids in chip removal from the hole. Where feasible, drilling horizontally has distinct advantages over drilling vertically downward.

Drill grinding. Proper grinding of a drill is a more complex and important operation than often is assumed. If satisfactory cutting and hole size are to be achieved, it is essential that the point angle, lip clearance, and lip length be correct. As illustrated in Figure 19-15, incorrect sharpening often results in

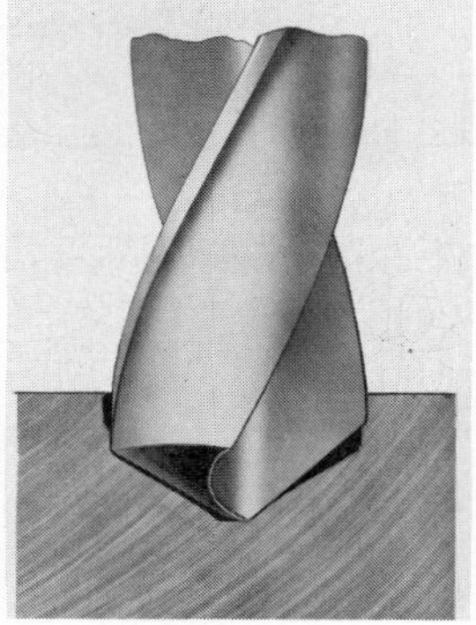
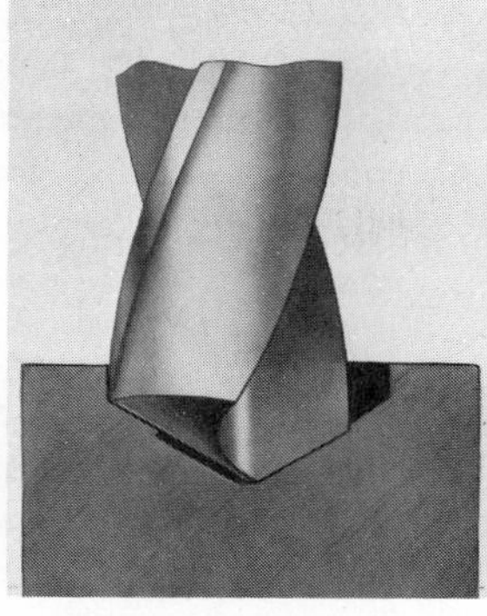

FIGURE 19-15. Effects of improper drill grinding: (*left*) angles of the two lips are different; (*right*) lengths of the lips are not equal. (*Courtesy The Cleveland Twist Drill Company.*)

oversize holes. Relatively few machinists can sharpen drills correctly and consistently by offhand grinding, even small drills. To obtain correct hole sizes, good drill life, and consistent results, special drill grinders, such as is shown in Figure 19-16, should be used. Such grinders have built-in mechanical features that assure the desired lip clearance, cone angle, and balanced lip lengths. This is extremely important where drills are used on mass-production or numerically controlled machines.

FIGURE 19-16. Drill grinder, showing the relationship of the drill to the grinding wheel. (*Courtesy Covel Manufacturing Company.*)

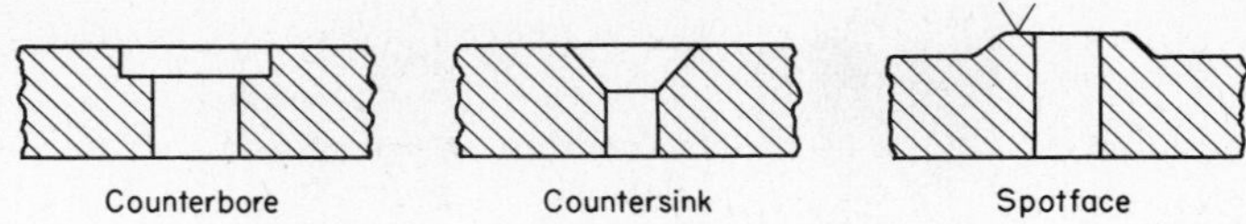

FIGURE 19-17. Surfaces produced by counterboring, countersinking, and spot facing.

Counterboring, countersinking, and spot facing. Drilling often is followed by counterboring, countersinking, or spot facing. As shown in Figure 19-17, each provides a bearing surface at one end of a drilled hole. They usually are done with a special tool having from three to six cutting edges.

Counterboring provides an enlarged, cylindrical hole with a flat bottom so that a bolt head, or a nut, will have a smooth bearing surface that is normal to the axis of the hole; the depth may be sufficient so that the entire bolt head or nut will be below the surface of the part. Counterboring usually is done with tools similar to those shown in Figure 19-18. The pilot on the end of the tool fits into the drilled hole and helps to assure concentricity with the original hole. Two or more diameters may be produced in a single counterboring operation. Counterboring also can be done with a single-point tool, although

FIGURE 19-18. Counterboring tools (*bottom to top*): interchangeable counterbore; solid, taper-shank counterbore with integral pilot; replaceable counterbore and pilot; replaceable counterbore, disassembled. (*Courtesy Ex-Cell-O Corporation* [*bottom*] *and Chicago-Latrobe Twist Drill Works.*)

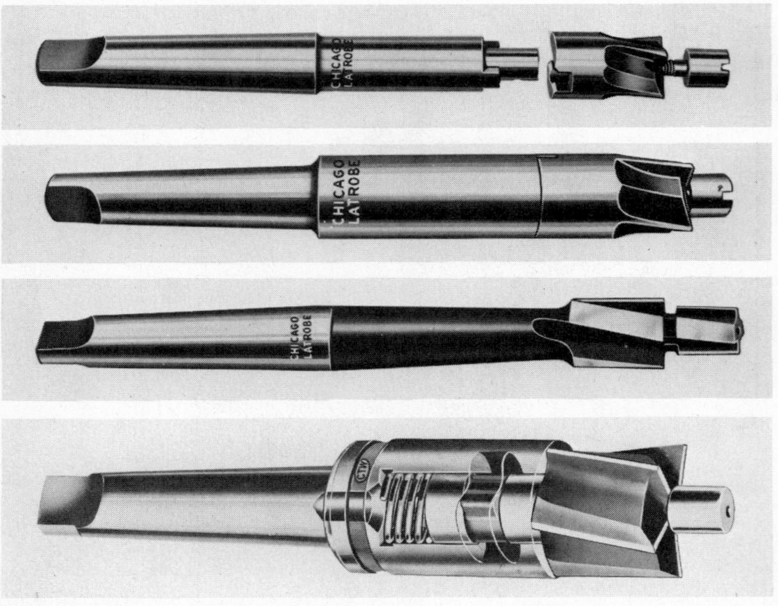

this method ordinarily is used only on large holes and essentially is a boring operation.

Countersinking makes a beveled section at the end of a drilled hole to provide a proper seat for a flat-head screw or rivet. The most common angles are 60, 82, and 90°. Countersinking tools are similar to counterboring tools except that the cutting edges are elements of a cone and they usually do not have a pilot because the bevel of the tool causes them to be self-centering.

Spot facing is done to provide a smooth bearing area on an otherwise rough surface at the opening of a hole and normal to its axis. Machining is limited to the minimum depth that will provide a smooth, uniform surface. Spot faces thus are somewhat easier and more economical to produce than counterbores. They usually are made with a multiedged end-cutting tool that does not have a pilot, although counterboring tools frequently are used.

REAMING

Reaming is done for two purposes—to bring holes to a more exact size, and to improve the finish of an existing hole by machining a small amount from its surface. Multiedged cutting tools are used, and no special machines are built especially for reaming; it usually is done on the same machine that was employed for drilling the hole that is to be reamed.

In order to obtain proper results, only a minimum amount of material should be left for removal by reaming. As little as 0.13 mm (0.005 inch) is desirable, and in no case should the amount exceed 0.38 mm (0.015 inch). A properly reamed hole should be within 0.03 mm (0.001 inch) of correct size and have a fine finish.

Types of reamers. The principal types of reamers, shown in Figure 19-19, are:

1. Hand reamers
   a. Straight
   b. Taper
2. Machine or chucking reamers
   a. Rose
   b. Fluted
3. Shell reamers
4. Expansion reamers
5. Adjustable reamers

Hand reamers are intended to be turned and fed by hand and to remove only a few thousandths of metal. They have a straight shank with a square tang for a wrench. They can have straight or spiral flutes and be solid or expansible. The teeth have relief along their edges and thus may cut along their entire length. However the reamer is tapered from 0.13 to 0.25 mm (0.005

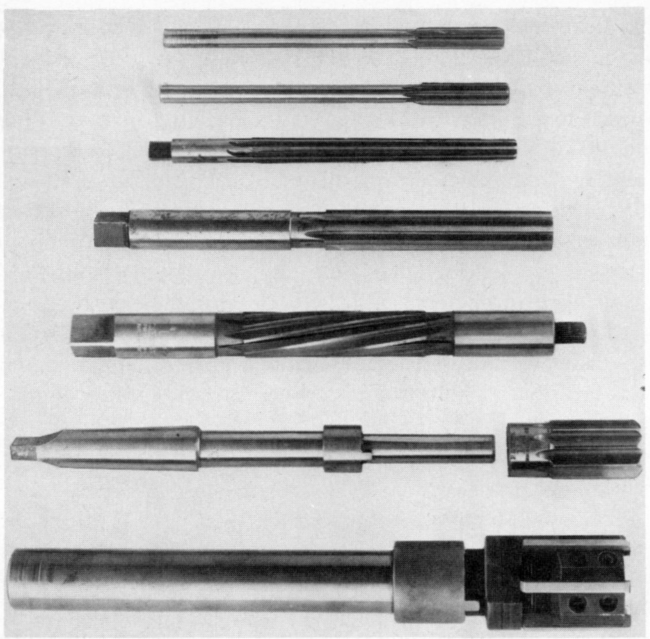

FIGURE 19-19. Types of reamers: (*top to bottom*) straight-fluted rose reamer; straight-fluted chucking reamer; straight-fluted taper reamer; straight-fluted hand reamer; expansion reamer; shell reamer; adjustable reamer.

to 0.010 inch) in the first third of its length to assist in starting it in the hole, and most of the cutting therefore takes place in this portion.

Machine or *chucking reamers* are for use with various machine tools at slow speeds. They have straight or tapered shanks and either straight or spiral flutes. *Rose chucking reamers* are ground cylindrical and have no relief behind the outer edges of the teeth. All cutting is done on the beveled ends of the teeth. *Fluted chucking reamers,* on the other hand, have relief behind the edges of the teeth as well as beveled ends. They thus can cut on all portions of the teeth. Their flutes are relatively short and they are intended for light finishing cuts. For best results they should not be held rigidly but permitted to float and be aligned by the hole.

Shell reamers often are used for sizes over 19 mm (¾ inch) in order to save cutting-tool material. The shell, made of tool steel for smaller sizes and with carbide edges for larger sizes or for mass-production work, is held on an arbor that is made of ordinary steel. One arbor may be used with any number of shells. Only the shell is subject to wear and need be replaced when worn. They may be ground as rose or fluted reamers.

Expansion reamers can be adjusted over a few thousandths of an inch to compensate for wear, or to permit some variation in hole size to be obtained. They are available in both hand and machine types.

Adjustable reamers have cutting edges in the form of blades that are locked in a body. The blades can be adjusted over a considerably greater range than in the case of expansion reamers. This permits adjustment for size and to compensate for regrinding. When the blades become too small from regrinding, they can be replaced. Both tool steel and carbide blades are used.

Taper reamers are used for finishing holes to an exact taper. They may have up to eight straight or spiral flutes. Standard tapers, such as Morse, Jarno, or Brown and Sharpe, come in sets of two. The *roughing reamer* has nicks along the cutting edges to break up the heavy chips that result as a cylindrical hole is cut to a taper. The *finishing reamer* has smooth cutting edges.

MACHINES FOR DRILLING

The basic work—tool motions that are required for drilling—relative rotation between the workpiece and the tool, with relative longitudinal feeding—also occur in a number of other machining operations. Consequently, as indicated in Table 17-4, drilling can be done on a variety of machine tools, such as lathes, milling machines, and boring machines, which will be discussed in later chapters. This chapter will consider only machines that are designed, constructed, and used primarily for drilling.

Basically, as illustrated in Figure 19-20, drilling machines consist of a *base,* a *column* that supports a *powerhead,* a *spindle,* and a *worktable.* On small machines the base rests on a bench, whereas on larger machines it rests on the floor. The column may be either round or of box-type construction, the latter being used on larger, heavy-duty machines, except in radial types. The powerhead contains an electric motor and means for driving the spindle in rotation at several speeds. On small drilling machines this may be accomplished by shifting a belt on a step-cone pulley, but on larger machines a geared transmission is used.

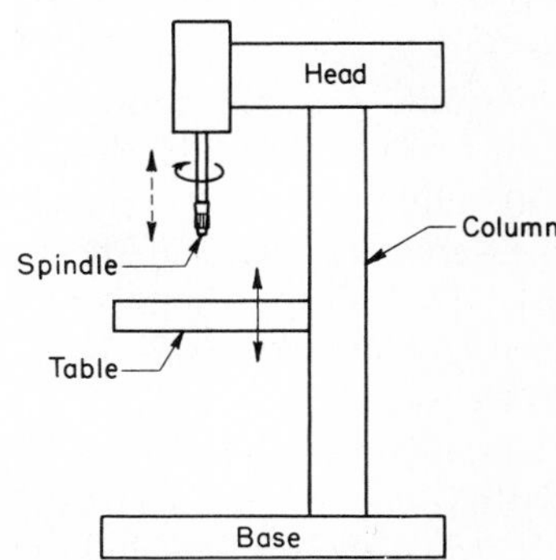

FIGURE 19-20. Block diagram showing the basic components of a drill press.

The heart of any drilling machine is its spindle. In order to drill satisfactorily, the spindle must rotate accurately and also resist whatever side forces result from the drilling. In virtually all machines the spindle rotates in preloaded ball or tapered-roller bearings. In addition to powered rotation, provision is made so that the spindle can be moved axially to feed the drill into the work. On small machines the spindle is fed by hand, whereas on larger machines power feed is provided. Except on some small bench types, the spindle contains a hole with a Morse taper in its lower end into which taper-shank drills or drill chucks can be inserted.

The worktables on drilling machines are mounted so they may be moved up and down on the column to accommodate work of various sizes. On round-column machines the table usually can also be swung sideways so that workpieces can be supported directly on the base. On some box-column machines the table is mounted on a subbase so that it can be moved in two directions in a horizontal plane by means of feed screws.

Types of drilling machines. Drilling machines usually are classified in the following manner:

1. Bench
   a. Plain
   b. Sensitive
2. Upright
   a. Single-spindle
   b. Turret
3. Radial
   a. Plain
   b. Semiuniversal
   c. Universal
4. Gang
5. Multiple-spindle
6. Deep-hole
   a. Vertical
   b. Horizontal
7. Transfer

Bench-type drilling machines. Figure 19-21 shows a typical bench-type drilling machine and details of the spindle construction. The spindle rotates on ball bearings within a nonrotating *quill* that can be moved up and down in the machine head to provide feed to the drill. The vertical motion is imparted by a hand-operated capstan wheel through a pinion that meshes with a rack on the quill. A spring raises the quill-and-spindle assembly to the highest position when the hand lever is released. The spindle is driven by means of a step-cone pulley that rides on a splined shaft, thus imparting rotation regardless of the vertical position of the spindle.

Drilling machines of this type—commonly called *drill presses*—usually have eight spindle speeds, from about 600 to 3000 rpm, and accommodate drills to 13 mm (½ inch). Some worktables contain holes and slots for use in clamping work; others do not. The same type of machine can be obtained with a long column so that it can stand on the floor instead of on a bench.

The size of bench and upright drilling machines is designated by *twice*

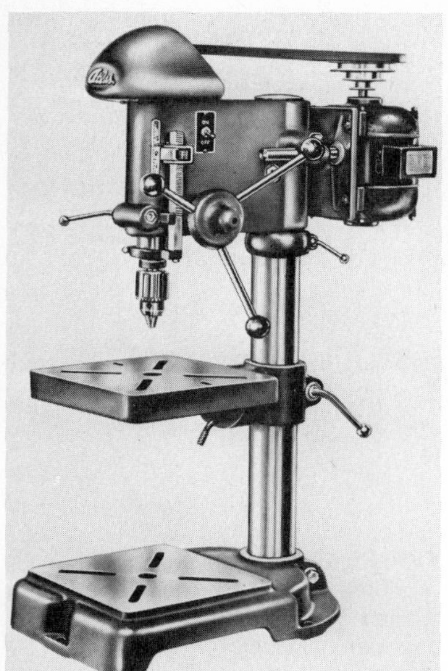

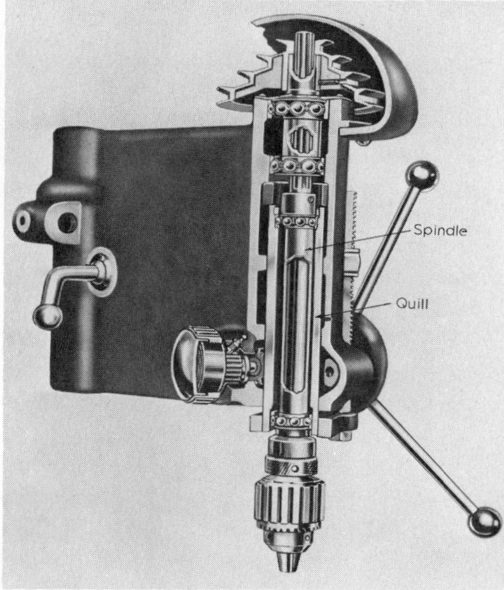

FIGURE 19-21. Fifteen-inch bench-type drill press and sectional view of the spindle construction. (*Courtesy Atlas Press Company.*)

the distance from the center line of the spindle to the nearest point on the column, thus being an indication of the maximum size of the work that can be drilled in the machine. For example, a 380-mm (15-inch) drill press will permit a hole to be drilled at the center of a workpiece 380 mm (15 inches) in diameter.

Sensitive drilling machines are essentially the same as plain bench-type machines except that they usually are smaller, are provided with more accurate spindles and bearings, and operate at higher speeds—up to 30,000 rpm. Very sensitive, hand-operated feeding mechanisms are provided for use in drilling small holes. Such machines are used for tool and die work and for drilling very small holes, often less than 1 millimeter (a few thousandths of an inch) in diameter, where high spindle speeds are necessary to obtain proper cutting speed and sensitive feel in order to provide delicate feeding to avoid the breakage of very small drills.

Upright drilling machines. The term *upright* is applied to drilling machines that stand on the floor, have single vertical columns, and have power spindle feed. These essentially are the same as bench-type machines in respect to spindle design but are heavier. Figure 19-22 shows such a machine. Upright drilling machines are widely used for heavy-duty drilling. Both round and box-column designs are manufactured, but box-column machines are more common because of their rigidity.

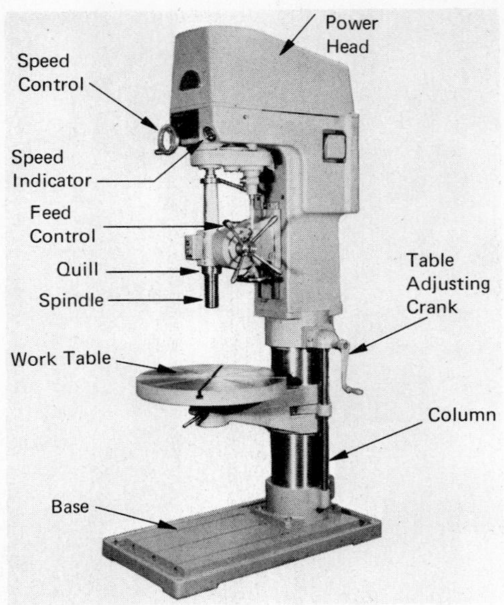

FIGURE 19-22. Upright drilling machine, with the principal parts named. (*Courtesy Buffalo Forge Company.*)

Upright drilling machines usually have spindle speed ranges from 60 to 3500 rpm and power feed rates, in from four to 12 steps, from about 0.10 to 0.60 mm (0.004 to 0.025 inch) per revolution. Most modern machines use a single-speed motor and a geared transmission to provide the range of speeds and feeds, but some utilize a multispeed motor to obtain some of the spindle speeds. The feed clutch usually is designed so that it disengages automatically when the spindle reaches a preset depth or when it reaches the limits of its travel.

Worktables on most upright drilling machines contain holes and slots for use in clamping work and nearly always have a channel around the edges to collect cutting fluid, when it is used. On box-column machines, the table is mounted on vertical ways on the front of the column and can be raised or lowered by means of a crank-operated elevating screw.

Gang-drilling machines. In mass production, *gang-drilling machines,* shown in Figure 19-23, often are used where several related operations, such as holes of different sizes, reaming, or counterboring, must be done on a single part. These consist essentially of several independent columns, heads, and spindles mounted on a common base and having a single table, on which the work can be slid into position for the operation at each spindle. They are available with or without power feed. One or several operators may be used.

Turret-type drilling machines. *Turret-type, upright drilling machines,* such as is shown in Figure 19-24, are used where a series of holes of different size, or a

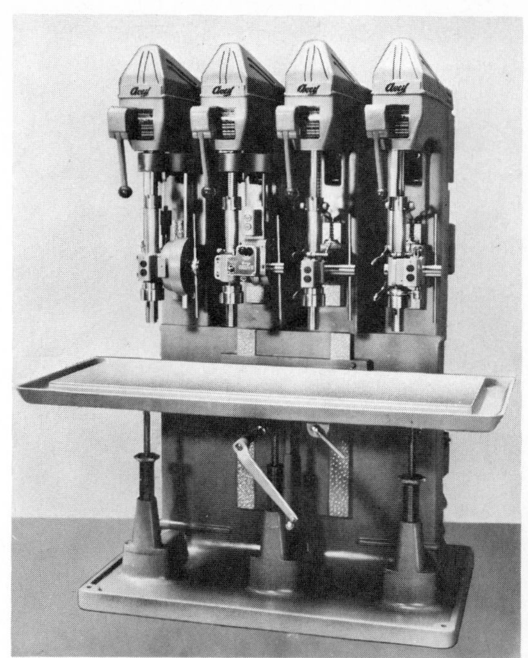

FIGURE 19-23. Four-spindle gang-drilling machine. The spindles on this machine are equipped (*left to right*) with automatic feed, power feed, hand feed, and hand-feed tapper. (*Courtesy Avey Machine Tool Company.*)

series of operations (such as center drilling, drilling, reaming, and spot facing), must be done repeatedly in succession. After the selected tools are set in the turret, each can quickly be brought into position to be driven by the power spindle merely by rotating the turret, rather than requiring moving and positioning of the workpiece, as with a gang-drilling machine. These machines also can be equipped with an automatic device, as shown in Figure 19-24, which will automatically provide individual feed rates for each spindle. Such machines are particularly adaptable for numerical and tape control.

Radial drilling machines. When holes must be drilled at different locations on large workpieces which cannot readily be moved and clamped on an upright drilling machine, *radial drilling machines* are employed. As shown in Figure 19-25, these have a large, heavy, round, vertical column supported on a large base. The column supports a radial arm that can be raised and lowered by power, and the entire column can rotate on the base. The spindle head, with its speed- and feed-changing mechanism, is mounted on the radial arm so that it can be moved horizontally to any desired position on the arm. Thus, by the combined movements of raising or lowering and swinging the radial arm, and horizontal movement of the spindle assembly, the spindle can quickly be brought into proper position for drilling holes at any desired point on a large workpiece mounted either on the base of the machine or on the floor.

Plain radial drilling machines provide only a vertical spindle motion. On

FIGURE 19-24. Turret-type upright drilling machine. Inset shows control panel for automatic operation. (*Courtesy Burgmaster Division of Houdaille Industries, Inc.*)

semiuniversal machines, the spindle head can be swung about a horizontal axis normal to the arm to permit the drilling of holes at an angle in a vertical plane. On *universal machines,* an additional angular adjustment is provided by rotation of the radial arm about a horizontal axis. This permits holes to be drilled at any desired angle.

The size of radial drilling machines is designated by the radius, in millimeters or feet, of the largest disk in which a center hole can be drilled when the spindle head is at its outermost position. Sizes from 900 to 3650 mm (3 to 12 feet) are available. Usually the diameter of the column also is given; these range from about 225 to 660 mm (9 to 26 inches). Most radial drilling machines have a wide range of speeds and feeds. For example, one machine has 32 spindle speeds from 20 to 1600 rpm and 16 feeds from 0.076 to 3.18 mm (0.003 to 0.125 inch) per revolution. These also include provision for tapping threads having pitches from 1.4 to 3 mm (8 to 18 threads per inch).

Large workpieces usually are fastened directly on the base of radial drill-

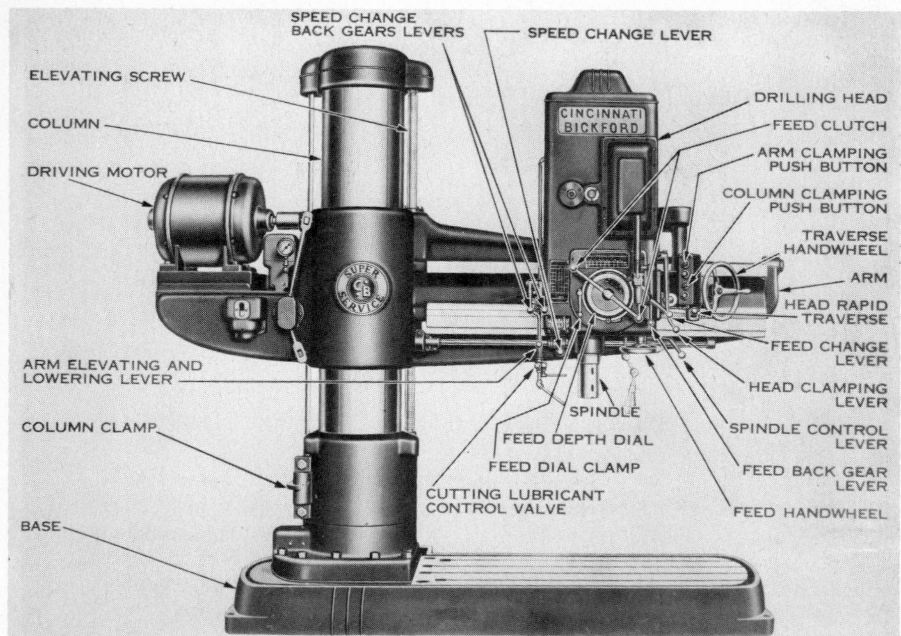

FIGURE 19-25. Radial drilling machine. (*Courtesy The Cincinnati Bickford Tool Company.*)

ing machines; small pieces can be mounted on a worktable that is attached to the base. Special jigs or fixtures also can be attached to the base to hold the work when multiple pieces are to be drilled.

Several special types of radial drilling machines are available. A track-type machine rests on rails so that it can be moved and clamped in any desired position along the rails. On *sliding-base machines,* the column can be moved along the base on ways to permit drilling over a wider area. A *portable* modification of the sliding-base type can readily be picked up and moved to a desired location by an overhead crane.

Most radial drilling machines are equipped with adequately heavy spindle bearings so they can also be used to do boring (see Chapter 21).

Multiple-spindle drilling machines. Where a number of parallel holes must be drilled in a part, *multiple-spindle drilling machines* are used. As shown in Figure 19-26, these are mass-production machines with the several spindles driven by a single powerhead and fed simultaneously into the work. Figure 19-27 shows the method of driving and positioning the spindles, which permits them to be adjusted over limited, but overlapping, areas so that holes can be drilled at any location within the overall capacity of the head. For example, one machine having 20 spindles can drill holes at any location within a 762-mm (30-

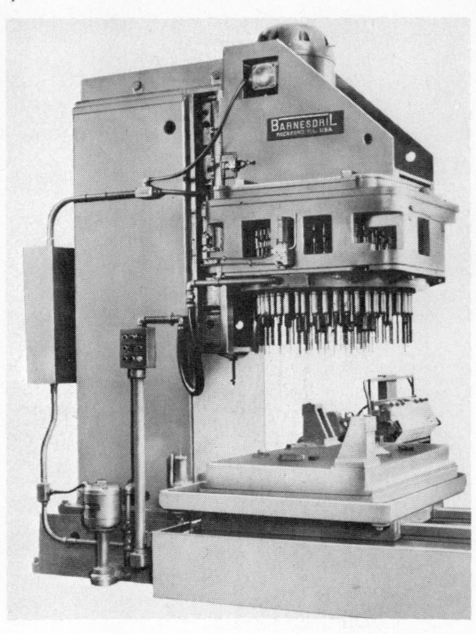

FIGURE 19-26. Multiple-spindle drilling machine equipped with 50 spindles. *(Courtesy Barnes Drill Company.)*

inch)-diameter circle. A special drill jig is made for each job to provide accurate guidance for each drill. Although such machines are quite costly, they can readily be converted for use on different jobs where the quantity to be produced will justify the small setup cost and the cost of the jig.

Multiple-spindle drilling machines are available with a wide range of numbers of spindles in a single head, and two or more heads frequently are combined in a single machine, as shown in Figure 19-28. Often drilling operations are performed simultaneously on two or more sides of a workpiece.

Deep-hole drilling machines. Special machines are used for drilling long (deep) holes, such as are found in rifle barrels, connecting rods, and long spindles. High cutting speeds, very light feeds, a positive and copious flow of cutting fluid to assure rapid chip removal, and adequate support for the long, slender drill are required. In most cases horizontal machines of the type shown in Figure 19-29 are used, wherein the work is rotated in a chuck with steady rests providing support along its length, as required. The drill does not rotate and is fed into the work. Vertical machines also are available for work that is not very long.

Work holding in drilling. Work that is to be drilled ordinarily is held in a vise or a special jig or fixture. Even in light drilling the work should not be held on the table by hand unless very adequate leverage is available. This is a dangerous practice and can lead to serious accidents, because the drill has a tendency to catch on the workpiece and cause it to rotate. Drilling vises and jigs

FIGURE 19-27. Multiple-spindle drill head, showing method of driving and positioning the spindles. (*Courtesy Thrift-master Products Incorporated.*)

FIGURE 19-28. Two-station multiple-spindle drilling machine, having two power heads and a rotary indexing table. (*Courtesy Baker Brothers, Inc.*)

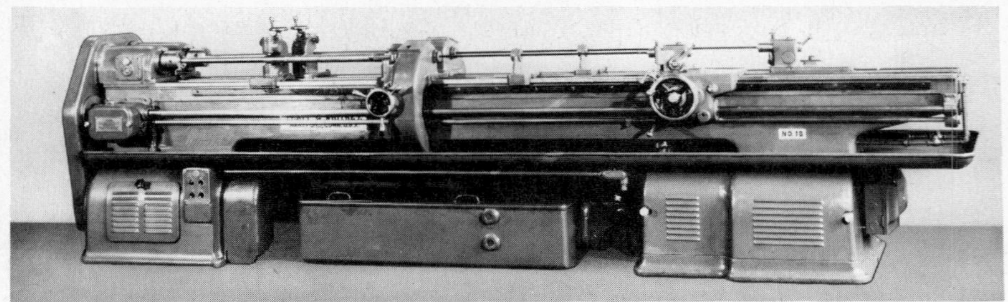

FIGURE 19-29. Two-spindle deep-hole drilling machine. (*Courtesy Colt Industries, Pratt & Whitney Machine Tool Division.*)

frequently are made so they can be turned on two faces to permit drilling to be done on two faces of the work with a single clamping.

Work that is too large to be held in a vise can be clamped directly to the machine table, using suitable bolts and clamps and the slots or holes in the table.

Estimating drilling time. The time required to drill a hole of a given depth is easily computed. However, one must allow for the length of the drill point and the *overtravel,* or distance, the drill goes beyond the far surface of the work, if the hole is clear through. An amount equal to half the hole diameter is sufficient to provide for these allowances. Then, because the drill advances the amount of the feed each revolution, the required time, t (in minutes), is

$$t = \frac{\text{hole depth} + \frac{1}{2}\ \text{hole diameter}}{\text{feed} \times \text{rpm}}$$

In terms of the cutting speed, this expression is

Metric (C.S. = m/min)

$$t = \frac{\text{hole depth} + \frac{1}{2}\ \text{hole diameter}}{\dfrac{\text{feed} \times (1000 \times \text{C.S.})}{\pi \times \text{hole diameter}}}$$

English (C.S. = ft/min)

$$t = \frac{\text{hole depth} + \frac{1}{2}\ \text{hole diameter}}{\dfrac{\text{feed} \times (12 \times \text{C.S.})}{\pi \times \text{hole diameter}}}$$

with all hole dimensions in millimeters or inches.

Review questions

1. What functions are performed by the flutes on a drill?
2. What determines the rake angle of a drill?
3. Basically, what determines what helix angle a drill should have?
4. If an ordinary two-flute twist drill were made to have negative rake angles,

describe its appearance as compared with an ordinary positive-rake-angle drill.

5. When a large-diameter hole is to be drilled, why is a small-diameter hole often drilled first?
6. How is the rotating torque applied to a taper-shank drill?
7. What results when a drill is improperly ground?
8. Why are very few drilled holes encountered that are below the specified diameter?
9. What are the two primary functions of a combination center drill?
10. What is the function of the margins on a twist drill?
11. What factors tend to cause a drill to "drift" off the center line of a hole?
12. For what types of holes are drills having coolant passages in the flutes advantageous?
13. How does a subland drill differ from a step drill?
14. Why do cutting fluids for drilling usually have more lubricating qualities than those for most other machining operations?
15. How is feed expressed in drilling?
16. How does a gang-drilling machine differ from a multiple-spindle drilling machine?
17. For what type of work is a quick-change drill chuck advantageous?
18. What may result from holding the workpiece by hand when drilling?
19. What is meant by a 305-mm (12-inch) upright drilling machine?
20. How much time will be required to drill a 25.4-mm (1-inch)-diameter hole through a piece of gray cast iron that is 38 mm (1½ inches) thick, using a high-speed drill? (Use values for feed and cutting speed from Table 17-1.)
21. What is the purpose of spot facing?
22. How does the purpose of counterboring differ from that of spot facing?
23. What are the primary purposes of reaming?
24. Explain why it is not desirable to hold a fluted chucking reamer rigidly?
25. What are the advantages of shell reamers?
26. A drill that operated satisfactorily for drilling cast iron gave very short life when used for drilling a plastic. Why?
27. What precautionary procedures should be used in drilling a fairly deep, vertical hole in mild steel when using an ordinary twist drill?
28. What is the metal-removal rate when a 1½-inch-diameter hole, 2 inches deep, is drilled in 1020 steel at a cutting speed of 120 fpm with a feed of 0.020 ipr?
29. If the specific horsepower for the steel in question 28 is 0.7, what horsepower would be required?
30. What is the advantage of a spade drill?
31. If the specific power of AISI 1020 steel is 0.03 W/mm^3, and 75 per cent of the output of the 1.5-kW motor of a drilling machine is available at the tool, what is the maximum feed that can be used in drilling a 51-mm-diameter hole with a HSS drill? (Use the cutting speed suggested in Table 17-1.)

Case study 14.
BOLTING LEG ON A CASTING

Figure CS-14 shows the design of one of four attachment legs on a component manufactured by the Hardhat Company and the type of service loading to which it is subjected. Difficulty has been experienced in machining these legs, and a substantial number of in-service failures have occurred. (1) What machining difficulties would you expect to occur with this design? (2) Where would you expect the failures to occur? (3) Redesign this portion of the component to eliminate both difficulties.

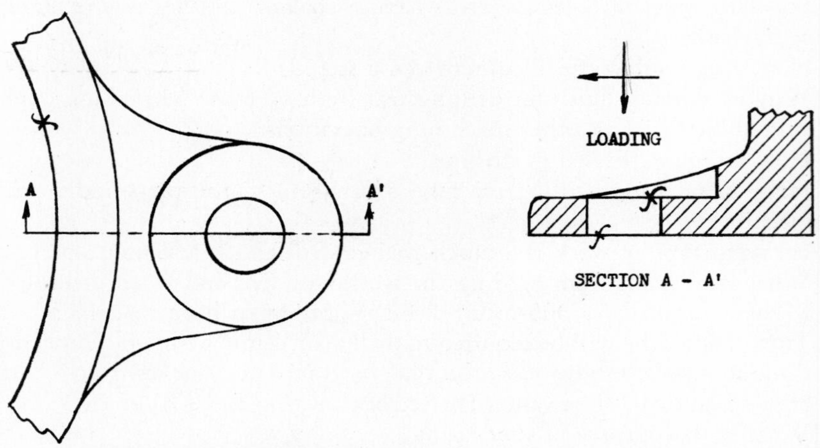

FIGURE CS-14. Bolting leg on cast iron casting.

CHAPTER 20

Turning and related operations

Turning provides a widely used means for machining external cylindrical and conical surfaces. As indicated in Figure 20-1, relatively simple work and tool movements are involved in turning a cylindrical surface—the workpiece rotates and a longitudinally fed, single-point tool does the cutting. If the tool is fed at an angle to the axis of rotation, an external conical surface results. If the tool is fed at 90° to the axis of rotation, using a tool that is wider than the width of the cut as shown in Figure 20-2d, the operation is called *facing*, and a flat surface is produced. Such a surface can be thought of as a conical surface having a 180° apex angle.

External cylindrical, conical, and irregular surfaces of limited length can also be turned by using a tool having a specific shape and feeding it inward against the work, as indicated in Figure 20-2b and c. The shape of the resulting surface is determined by the shape and size of the cutting tool. Such machining is called *form turning*. Obviously, if feeding of the tool continues to

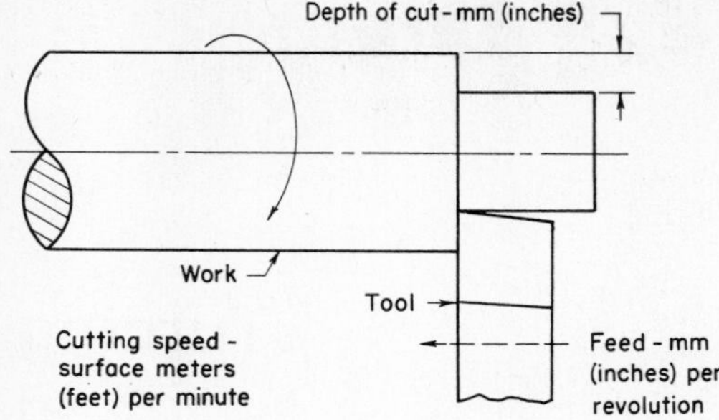

FIGURE 20-1. Tool–work relationships in turning.

the axis of the workpiece, it will be cut in two; this is called a *cutoff* operation and, when done for this purpose, a simple, thin tool is used, as portrayed in Figure 20-2a.

Boring is a variation of turning. Essentially it is internal turning, in that a single-point cutting tool produces internal cylindrical or conical surfaces. Consequently, boring can be done on most machine tools that can do turning. However, boring also can be done using a rotating tool with the workpiece remaining stationary. Also, specialized machine tools have been developed that will do boring, drilling, and reaming but will not do turning. Therefore, boring will be discussed more fully in Chapter 21.

Because of the rotational and longitudinal-feed relationships between the workpiece and the tool in turning, it is apparent that most machines that can be used for turning can also be used for drilling and other operations that require these work–tool relationships.

LATHES

Lathes are machine tools designed primarily to do turning, facing, and boring. As indicated in Tables 17-2, 17-3, and 17-4, very little turning is done on other types of machine tools, and none can do it with equal facility. Because lathes also can do facing, drilling, and reaming, their versatility permits sev-

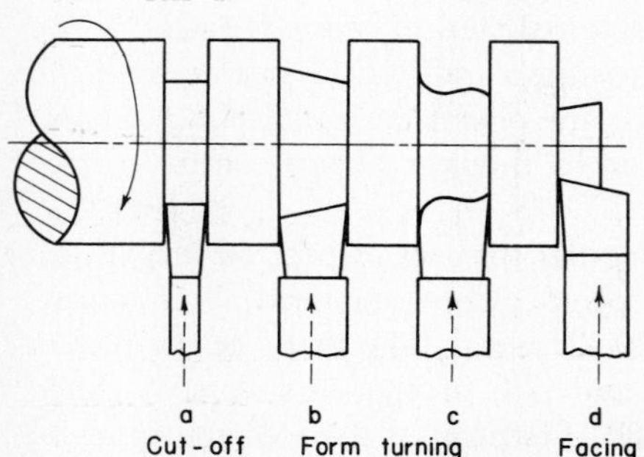

FIGURE 20-2. Tool–work relationships for cutoff, form turning, and facing operations.

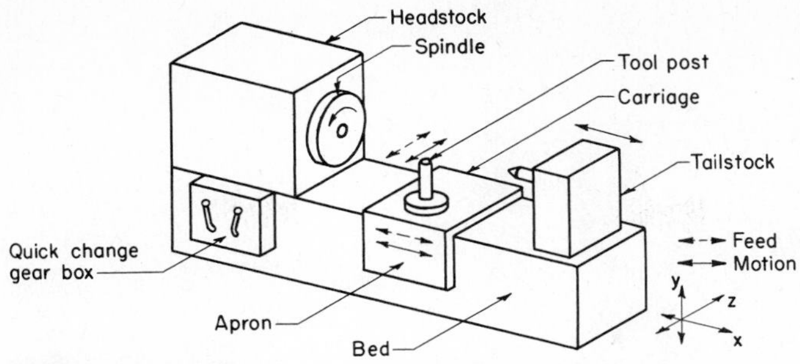

FIGURE 20-3. Block diagram of the basic components of a lathe.

eral operations to be done with a single setup of the workpiece. Consequently, more lathes of various types are used in manufacturing than any other machine tool.

Lathes in various forms have existed for more than 2000 years, but modern lathes date from about 1797, when Henry Maudsley developed one with a leadscrew (see Figure 1-12), providing controlled, mechanical feed of the tool. This ingenious Englishman also developed a change-gear system that could connect the motions of the spindle and leadscrew and thus enable threads to be cut.

Lathe construction. The essential components of a lathe are depicted in Figure 20-3. These are the *bed, headstock assembly, tailstock assembly, carriage assembly, quick-change gearbox,* and the *leadscrew* and *feed rod.* A typical modern engine lathe is shown in Figure 20-4.

The *bed* is the backbone of a lathe. It usually is made of well-normalized or aged gray or nodular cast iron and provides a heavy, rigid frame on which all the other basic components are mounted. Two sets of parallel, longitudinal *ways,* inner and outer, are contained on the bed, usually on the upper side. Some makers use an inverted **V**-shape for all four ways, whereas others utilize one inverted **V** and one flat way in one or both sets. Because several other components are mounted and/or move on the ways, they are precision-machined to assure accuracy of alignment. On most modern lathes the ways are surface-hardened to resist wear and abrasion, but precaution should be taken in operating a lathe to assure that the ways are not damaged. Any inaccuracy in them usually means that the accuracy of the entire lathe is destroyed.

The *headstock* is mounted in a fixed position on the inner ways, usually at the left end of the bed. It provides a powered means of rotating the work at various speeds. Essentially, it consists of a hollow spindle, mounted in accurate bearings, and a set of transmission gears—similar to a truck transmission—through which the spindle can be rotated at a number of speeds. Figure 20-5 shows the arrangement of the gears and spindle in a typical lathe. Most

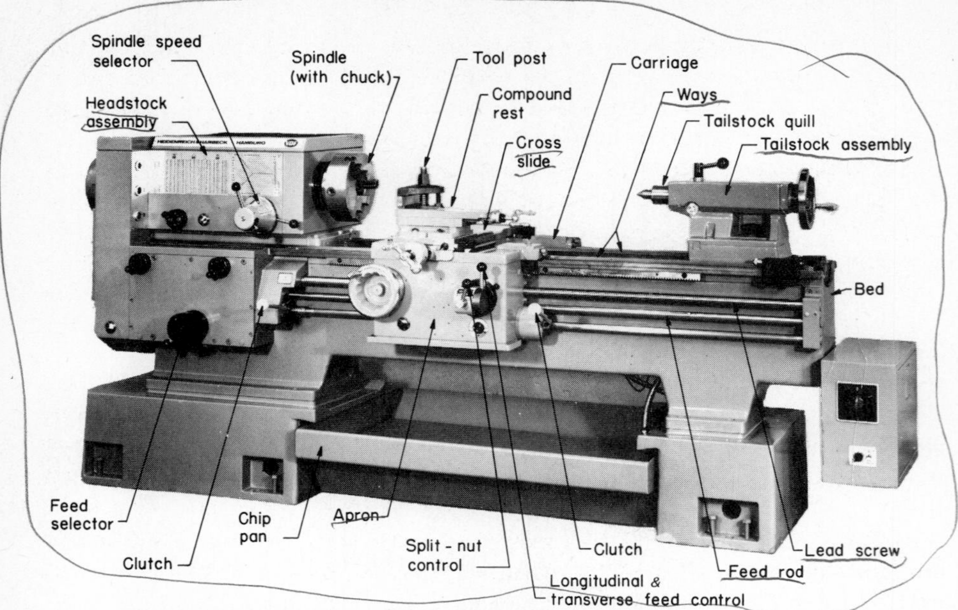

FIGURE 20-4. Modern engine lathe, with the principal parts named. (*Courtesy Heidenreich & Harbeck.*)

FIGURE 20-5. Gears and spindle in the headstock of a modern lathe. (*Courtesy The American Tool Works Company.*)

lathes provide from eight to 18 speeds, usually in a geometric ratio, and on modern lathes all the speeds can be obtained merely by moving from two to four levers. An increasing trend is to provide a continuously variable speed range through electrical or mechanical drives.

Because the accuracy of a lathe is greatly dependent on the spindle, it is of heavy construction and mounted in heavy bearings, usually preloaded tapered roller or ball types. The spindle has a hole extending through its length, through which long bar stock can be fed. The size of this hole is an important dimension of a lathe because it determines the maximum size of bar stock that can be machined when the material must be fed through the spindle.

The inner end of the spindle protrudes from the gearbox and contains a means for mounting various types of chucks, face plates, and dog plates on it. Whereas small lathes often employ a threaded section to which the chucks are screwed, most large lathes utilize either *cam-lock* or *key-drive taper* noses, shown in Figure 20-16. These provide a large-diameter taper that assures the accurate alignment of the chuck, and a mechanism that permits the chuck or face plate to be locked or unlocked in position without the necessity of having to rotate these heavy attachments.

Power is supplied to the spindle by means of an electric motor through a V-belt or silent-chain drive. Most modern lathes have motors of from 5 to 15 horsepower to provide adequate power for carbide and ceramic tools at their high cutting speeds.

As indicated in Figure 20-4, the tailstock assembly consists, essentially, of three parts. A lower casting fits on the inner ways of the bed and can slide longitudinally thereon, with a means for clamping the entire assembly in any desired location. An upper casting fits on the lower one and can be moved transversely upon it, on some type of keyed ways, to permit aligning the tailstock and headstock spindles. It also provides a method of turning tapers, as will be discussed later. The third major component of the assembly is the *tailstock quill.* This is a hollow steel cylinder, usually about 51 to 76 mm (2 to 3 inches) in diameter, that can be moved several inches longitudinally in and out of the upper casting by means of a handwheel and screw. The open end of the quill hole terminates in a Morse taper in which a lathe center, or various tools such as drills, can be held. A graduated scale usually is engraved on the outside of the quill to aid in controlling its motion in and out of the upper casting. A locking device permits clamping the quill in any desired position.

The *carriage assembly,* shown in Figure 20-6, provides the means for mounting and moving cutting tools. The *carriage,* a relatively flat H-shaped casting, rests and moves on the outer set of ways on the bed. The *cross slide* is mounted on ways on the transverse bar of the carriage and can be moved by means of a feed screw that is controlled by a small handwheel and a graduated dial. The cross slide thus provides a means for moving the lathe tool in the direction normal to the axis of rotation of the workpiece.

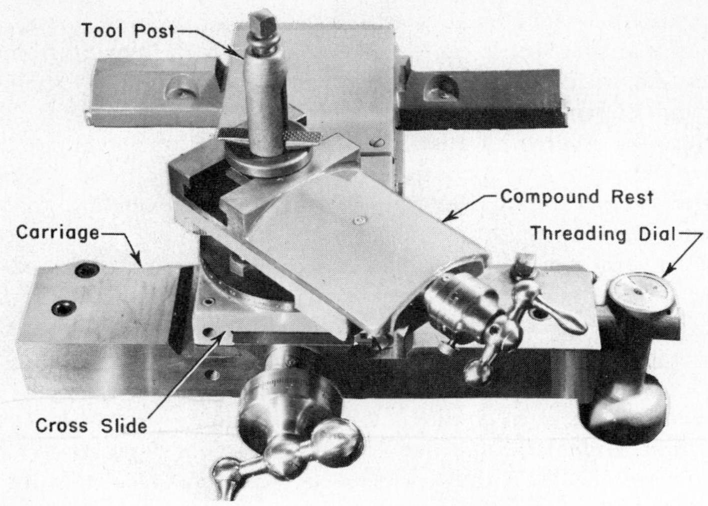

FIGURE 20-6. Lathe carriage assembly. (*Courtesy Sheldon Machine Company, Inc.*)

On most lathes the tool post actually is mounted on a *compound rest,* also shown in Figure 20-6. This consists of a base, which is mounted on the cross slide so that it can be pivoted about a vertical axis, and an upper casting. The upper casting is mounted on ways on this base so that it can be moved back and forth and controlled by means of a short lead screw operated by a handwheel and a calibrated dial.

The *apron,* attached to the front of the carriage, contains the mechanism and controls for providing manual and powered motion for the carriage and powered motion for the cross slide. Figures 20-7 and 20-8 show front and rear views of a typical apron. Manual movement of the carriage along the bed is effected by turning a handwheel on the front of the apron, which is geared to a pinion on the back side. This pinion engages a rack that is attached beneath the upper front edge of the bed in an inverted position.

Powered movement of the carriage and cross slide is provided by a rotating *feed rod,* shown in Figure 20-4. (The driving of this feed rod will be discussed later.) The feed rod, which contains a keyway throughout most of its length, passes through the two reversing bevel pinions shown in Figure 20-8, and is keyed to them. Either pinion can be brought into mesh with a mating bevel gear by means of the reversing lever on the front of the apron and thus provide "forward" or "reverse" power to the carriage. Suitable clutches connect either the rack pinion or the cross-slide screw to provide longitudinal motion of the carriage or transverse motion of the cross slide.

For cutting threads, a second means of longitudinal drive is provided by a *lead screw,* shown in Figure 20-4. Whereas motion of the carriage when driven by the feed-rod mechanism takes place through a friction clutch in

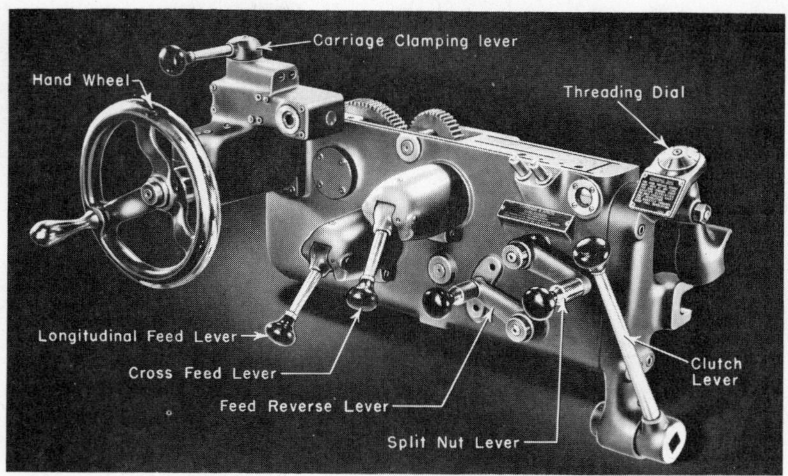

FIGURE 20-7. Front side of a lathe apron. (*Courtesy The American Tool Works Company.*)

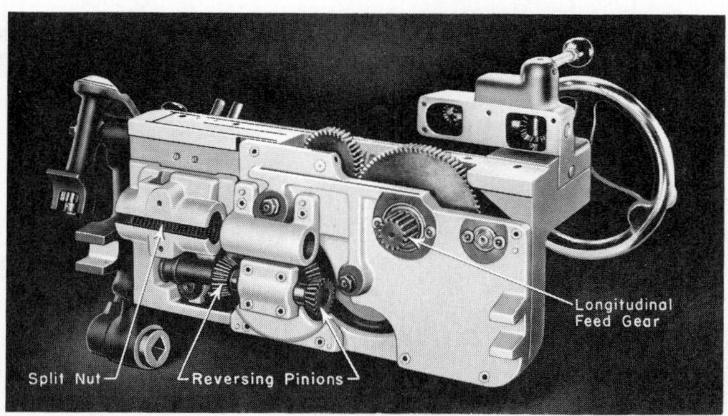

FIGURE 20-8. Back side of a lathe apron. (*Courtesy The American Tool Works Company.*)

which slippage is possible, motion through the lead screw is by a direct, mechanical connection between the apron and the lead screw. This is achieved by a *split nut,* shown in Figure 20-8, which can be closed around the lead screw by means of a lever on the front of the apron. With the split nut closed, the carriage is moved along the lead screw by direct drive without possibility of slippage.

On modern lathes, the input end of a *quick-change gear box,* such as shown in Figure 20-9, is driven from the spindle by means of suitable gearing, as shown in Figure 20-10, and the output end is connected to the feed rod and lead screw. Thus, through the quick-change gear box, the associated

FIGURE 20-9. Exterior and interior views of a quick-change gear box. (*Courtesy The American Tool Works Company.*)

gearing, and the lead screw and feed rod, the carriage is connected to the spindle, and the cutting tool can be made to move a specific distance, either longitudinally or transversely, for each revolution of the spindle. Typical lathes may provide, through the feed rod, as many as 48 feeds, ranging from 0.05 to 3 mm (0.002 to 0.118 inch) per revolution of the spindle, and, through the lead screw, leads from 0.28 to 17 mm (1½ to 92 threads per inch). On some small or inexpensive lathes, one or two gears in the gear train

FIGURE 20-10. Phantom view of an engine lathe showing the gear train. (*Courtesy The Monarch Machine Tool Company.*)

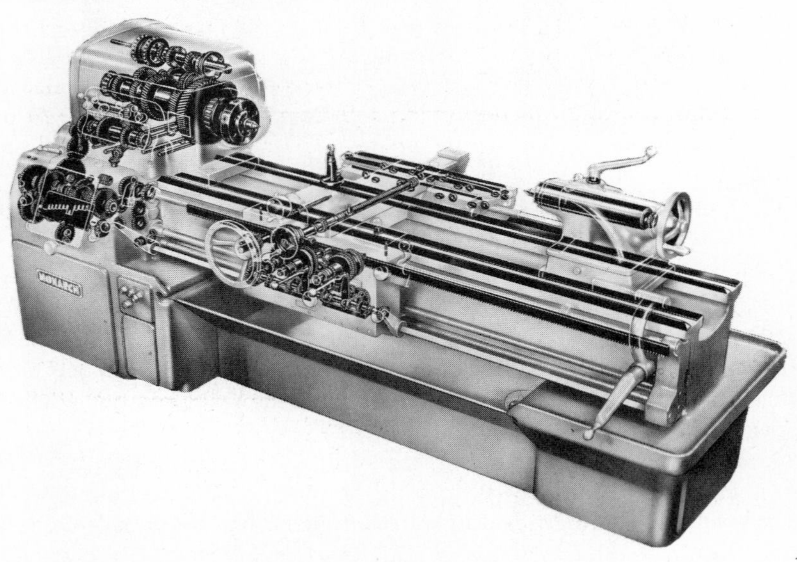

between the gear box and the spindle must be changed to obtain a full range of threads and feeds.

Size designation of lathes. The size of a lathe is designated by two dimensions. The first is known as the *swing.* This is the maximum diameter of work that can be rotated on a lathe. It is approximately twice the distance between the line connecting the lathe centers and the nearest point on the ways. The second size dimension is the *maximum distance between centers.* The swing thus indicates the maximum workpiece diameter that can be turned in the lathe, while the distance between centers indicates the maximum length of workpiece that can be mounted between centers. The maximum diameter of a workpiece that can be mounted between centers is somewhat less than the swing diameter—usually 127 to 152 mm (5 to 6 inches)—because such a workpiece must clear the carriage assembly, which is above the ways.[1] Thus these two dimensions indicate approximately the diameter and length of the largest workpiece that can be machined in a lathe. For example, one 356 × 762 mm (14 × 30 inches) lathe will handle work between centers that is 203 mm (8 inches) in diameter and 762 mm (30 inches) long.

Types of lathes. Lathes used in manufacturing can be classified as speed, engine, toolroom, and special types.

Speed lathes usually have only a headstock, tailstock, and a simple tool post mounted on a light bed. They ordinarily have only three or four speeds and are used primarily for wood turning, polishing, or metal spinning. Spindle speeds up to about 4000 rpm are common.

Engine lathes are the type most frequently used in manufacturing. Figure 20-4 is an example of this type. They are heavy-duty machine tools with all the components described previously and have power drive for all tool movements except on the compound rest. In most cases the bed is mounted on two pedestal legs, as shown in Figure 20-4. They commonly range in size from 305 to 610 mm (12 to 24 inches) swing and from 610 to 1219 mm (24 to 48 inches) center distances, but swings up to 1270 mm (50 inches) and center distances up to 3658 mm (12 feet) are not uncommon. Most have chip pans and a built-in coolant circulating system. Smaller engine lathes—with swings usually not over 330 mm (13 inches)—also are available in *bench type,* designed for the bed to be mounted on a bench or cabinet.

Toolroom lathes have somewhat greater accuracy and, usually, a wider range of speeds and feeds than ordinary engine lathes. Designed to have greater versatility to meet the requirements of tool and die work, they often have a continuously variable spindle speed range and shorter beds than ordinary engine lathes of comparable swing, since they are generally used for machining relatively small parts. They may be either bench or pedestal type.

[1] Lathe manufacturers usually supply the maximum diameter that can be machined between centers.

Several types of special-purpose lathes are made to accommodate specific types of work. On a *gap-bed lathe,* for example, a section of the bed, adjacent to the headstock, can be removed to permit work of unusually large diameter to be swung. Another example is the *wheel lathe,* which is designed to permit the turning of the journals and wheel treads of railroad-car wheel-and-axle assemblies; a special headstock drives the assembly at a point between the two wheels.

Supporting work in lathes. Five methods commonly are used for supporting workpieces in lathes:

1. Held between centers.
2. Held in a chuck.
3. Held in a collet.
4. Mounted on a face plate.
5. Mounted on the carriage.

In the first four of these methods the workpiece is rotated during machining. In the fifth method, which is not used extensively, the tool rotates while the workpiece remains stationary, except for being fed into the tool.

Lathe centers. Workpieces that are relatively long with respect to their diameters usually are machined between centers. Two *lathe centers* are used, one in the spindle hole and the other in the hole in the tailstock quill. Two types are used. The *plain* or *solid* type, shown in Figure 20-11, is made of hardened steel with a Morse taper on one end so that it will fit into the spindle hole. The other end is ground to a 60° taper; sometimes the tip of this taper is made of tungsten carbide to provide better wear resistance. Before a center is placed in position, the spindle hole should be wiped carefully to make sure that no dirt or chips are in the hole. The presence of foreign material will prevent the center from seating properly and it will not be aligned accurately.

Before a workpiece can be mounted between lathe centers, a 60° center hole must be drilled in each end. This can be done in a drill press, or in a lathe by holding the work in a chuck. A combination center drill and countersink ordinarily is used, taking care that the center hole is deep enough so that it will not be machined away in any facing operation, and yet is not drilled to the full depth of the tapered portion of the center drill. Table 20-1 gives recommended center-hole sizes.

The work and the center at the headstock end rotate together, so no lubricant is needed in the center hole at this end. However, because the center in the tailstock quill is dead with respect to the rotating workpiece, adequate lubrication must be provided. This usually is accomplished by putting a mix-

FIGURE 20-11. Solid lathe center. (*Courtesy Chicago-Latrobe Twist Drill Works.*)

TABLE 20-1. Size of center holes and combination countersink

Diameter of Work		Diameter of Large End of Hole		Diameter of Body of Countersink	
mm	inches	mm	inches	mm	inches
5.0–8.0	$^3/_{16}$–$^5/_{16}$	3.2	$^1/_8$	5.15	$^{13}/_{64}$
9.5–25.5	$^3/_8$–1	4.75	$^3/_{16}$	7.95	$^5/_{16}$
27.0–51.0	$1^1/_{16}$–2	6.35	$^1/_4$	7.95	$^5/_{16}$
52.5–102.0	$2^1/_{16}$–4	7.95	$^5/_{16}$	11.10	$^7/_{16}$

ture of white lead and oil in the center hole before the dead center is tightened in the hole. Failure to provide proper lubrication at all times will result in scoring of the workpiece center hole and the center, and inaccuracy and serious damage may occur.

Proper tightness must be maintained between the centers and the workpiece. The workpiece must rotate freely, yet no looseness should exist. Looseness usually will be manifested in "chattering" of the workpiece during cutting. Tightness of the centers should be checked after cutting has been done for a short time; the resulting heating and thermal expansion of the workpiece will increase the tightness.

Live centers often are used in the tailstock quill. In this type, shown in Figure 20-12, the end that fits into the workpiece is mounted on ball or roller bearings so that it is free to rotate; thus no lubrication of the center hole is required. However, they may not be as accurate as the plain type, so they often are not used for precision work.

A mechanical connection must be provided between the spindle and the workpiece to cause it to rotate. This is accomplished by some type of lathe *dog* and a *dog plate,* as shown in Figure 20-13. The dog is a forging that fits over the end of the workpiece and is clamped to it by means of a setscrew. The tail of the dog enters a slot in the dog plate, which is rigidly attached to the lathe spindle in the same manner as a lathe chuck. If the dog must be attached to work that has a finished surface, a piece of soft metal, such as copper or aluminum, can be placed between the work and the setscrew to avoid marring.

Mandrels. Workpieces that must be machined on both ends, and those that are disklike in shape, are often mounted on mandrels for turning between cen-

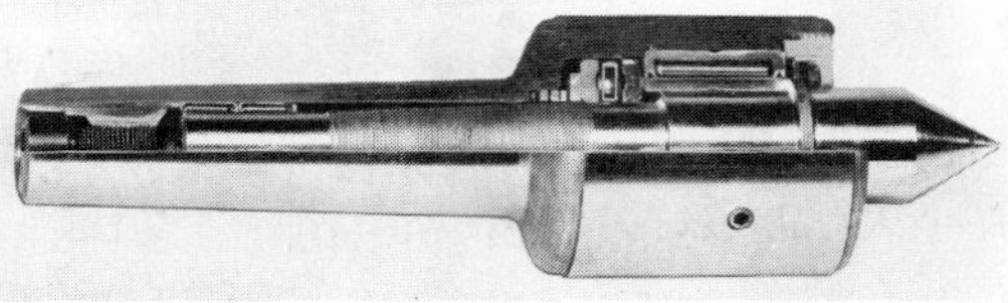

FIGURE 20-12. "Live" type of lathe center. (*Courtesy Motor Tool Manufacturing Company.*)

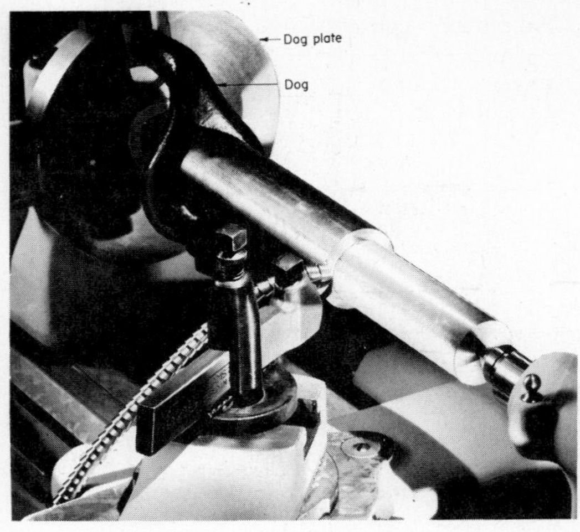

FIGURE 20-13. Work being turned between centers in a lathe, showing the use of a dog and dog plate. (*Courtesy South Bend Lathe.*)

ters. Three common types of mandrels are shown in Figure 20-14. *Solid mandrels* usually vary from 102 to 305 mm (4 to 12 inches) in length and are accurately ground with a 1:2000 taper (0.006 inch per foot). After the workpiece is drilled and/or bored to fit, it is pressed on the mandrel. The mandrel should be mounted between centers so that the cutting force tends to tighten the work on the mandrel taper. Solid mandrels permit the work to be machined on both ends as well as on the cylindrical surface. They are available in stock sizes but can be made to any desired size.

Gang or *disk mandrels* are used for production-type work, because the workpieces do not have to be pressed on and thus can be put in position and removed more rapidly. However, only the cylindrical surface of the workpiece can be machined when this type of mandrel is used.

Cone mandrels have the advantage that they can be used to center workpieces having a range of hole sizes.

Lathe chucks. Lathe chucks are used to support a wider variety of workpiece shapes and to permit more operations to be performed than can be accomplished when the work is held between centers. Two basic types of chucks are used. These are illustrated in Figure 20-15.

FIGURE 20-14. Three types of mandrels.

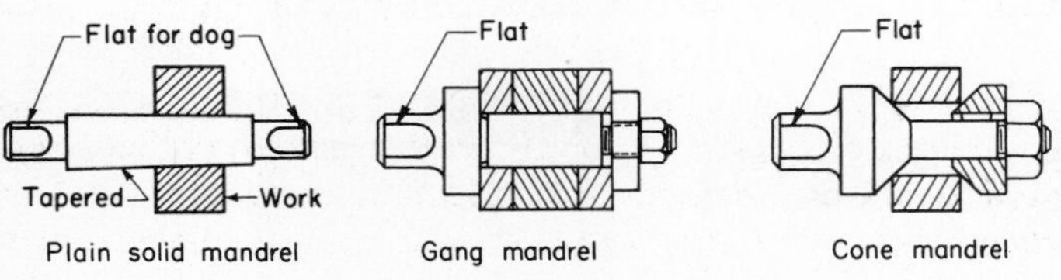

FIGURE 20-15. Lathe chucks: (*left*) four-jaw independent; (*right*) three-jaw, self-centering. (*Courtesy Cushman Industries, Inc.*)

Three-jaw, self-centering chucks are used for work that has a round or hexagonal cross section. The three jaws are moved inward or outward simultaneously by the rotation of a spiral cam, which is operated by means of a special wrench through a bevel gear. If they are not abused, these chucks will provide automatic centering to within about 0.025 mm (0.001 inch). However, they can be damaged through use and will then be considerably less accurate.

Each jaw in a *four-jaw independent chuck* can be moved inward and outward independent of the others by means of a chuck wrench. Thus they can be used to support a wide variety of work shapes. A series of concentric circles, engraved on the chuck face, aid in adjusting the jaws to fit a given workpiece. Four-jaw chucks are heavier and more rugged than the three-jaw type and, because undue pressure on one jaw does not destroy the accuracy of the chuck, they should be used for all heavy work. The jaws on both three- and four-jaw chucks can be reversed to facilitate gripping either the inside or the outside of workpieces.

Combination four-jaw chucks are available in which each jaw can be moved independently or all can be moved simultaneously by means of a spiral cam. Two-jaw chucks also are available. For mass-production work, special chucks often are used in which the jaws are actuated by air or hydraulic pressure, permitting very rapid clamping of the work.

Lathe chucks are bolted to a special chuck plate, which, in turn, can be attached to the spindle nose. Except on smaller and more inexpensive types, most modern lathes use methods that assure accurate and rapid attachment without the necessity for rotating the chuck, such as in Figure 20-16.

FIGURE 20-16. Methods for attaching chucks to lathe spindles. (*Left*) Cam-lock spindle nose and mating dog plate. (*Right*) Key-drive, taper-nose lathe spindle and dog plate. (*Courtesy The American Tool Works Company.*)

Collets. By the use of *collets*, smooth bar stock or workpieces that have been machined to a given diameter can be held more accurately than normally can be achieved in a regular three- or four-jaw chuck. As shown in Figure 20-17. collets are relatively thin tubular steel bushings that are split into three longitudinal segments over about two-thirds of their length. At the split end, the smooth internal surface is shaped to fit the piece of stock that is to be held, and the external surface is a taper that fits within an internal taper of a collet sleeve placed in the spindle hole, as shown in Figure 20-18. When the collet

FIGURE 20-17. Several types of lathe collets. (*Courtesy South Bend Lathe.*)

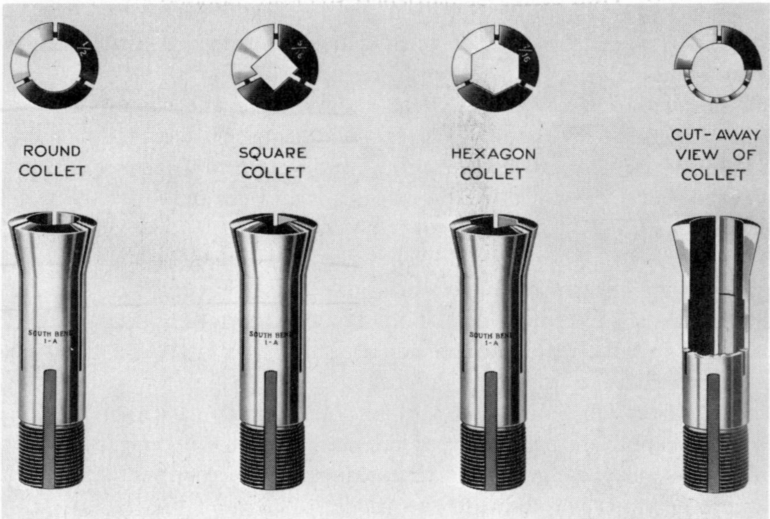

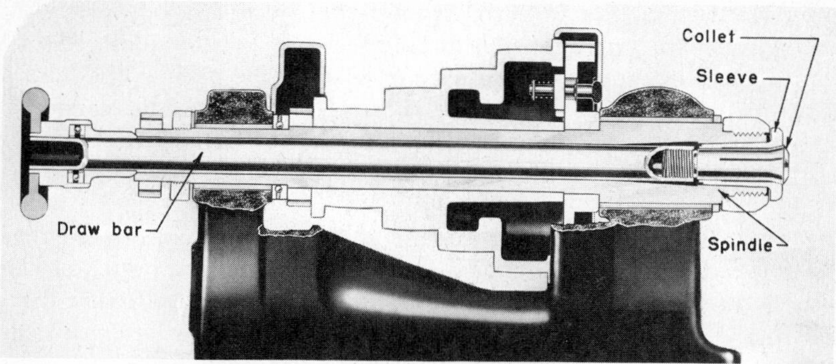

FIGURE 20-18. Method of using a draw-in collet in a lathe spindle. (*Courtesy South Bend Lathe.*)

is pulled inward into the spindle, by means of the draw bar that engages threads in its inner end, the action of the two mating tapers squeezes the collet segments together, causing them to grip the workpiece.

As shown in Figure 20-17, collets are made to fit a variety of symmetrical shapes. If the stock surface is smooth and accurate, good collets will provide very accurate centering; maximum runout should be less than 0.013 mm (0.0005 inch). However, the work should be no more than 0.05 mm (0.002 inch) larger or 0.13 mm (0.005 inch) smaller than the nominal size of the collet. Consequently, collets of the type shown in Figure 20-17 are used only on drill-rod, cold-rolled, extruded, or previously machined stock.

Another type of collet, shown in Figure 20-19, has a size range of about 3.2 mm (⅛ inch). These are similar in construction to the drill chuck shown in Figure 19-12b. Thin strips of hardened steel are bonded together on their

FIGURE 20-19. Jacobs lathe chuck. (*Courtesy The Jacobs Manufacturing Company.*)

sides by synthetic rubber to form a truncated cone with a central, cylindrical hole. The collet fits into a tapered spindle sleeve so that the outer edges of the metal strips are in contact with the inner taper of the sleeve. The inner edges bear against the workpiece. Pulling the collet into the adapter sleeve thus causes the strips to grip the work. Because of their greater size range, fewer collets are required than with the ordinary type.

Face plates. *Face plates* are used to support irregularly shaped work that cannot be gripped easily in chucks or collets. The work can be bolted or clamped directly on the face plate or can be supported on an auxiliary fixture that is attached to the face plate, as shown in Figure 20-20. The latter procedure is time-saving when several identical pieces are to be machined.

Steady and follow rests. If one attempts to turn a long, slender piece between centers, the radial force exerted by the cutting tool, or the weight of the workpiece itself, may cause it to be deflected out of line. *Steady rests* and *follow rests,* shown in Figure 20-21, provide means for supporting such work between the headstock and the tailstock. The steady rest is clamped to the lathe ways and has three movable fingers that are adjusted to contact the work and align it. A light cut should be taken before adjusting the fingers to provide a smooth contact-surface area.

A steady rest also can be used in place of the tailstock as a means of supporting the end of long pieces, pieces having too large an internal hole to permit using a regular dead center, or work where the end must be open for boring. In such cases the headstock end of the work must he held in a chuck to prevent its moving longitudinally, and tool feed should be toward the headstock.

The follow rest is bolted to the lathe carriage. It has two contact fingers that are adjusted to bear against the workpiece, opposite the cutting tool, so as to prevent the work from being deflected away from the cutting tool by the cutting forces.

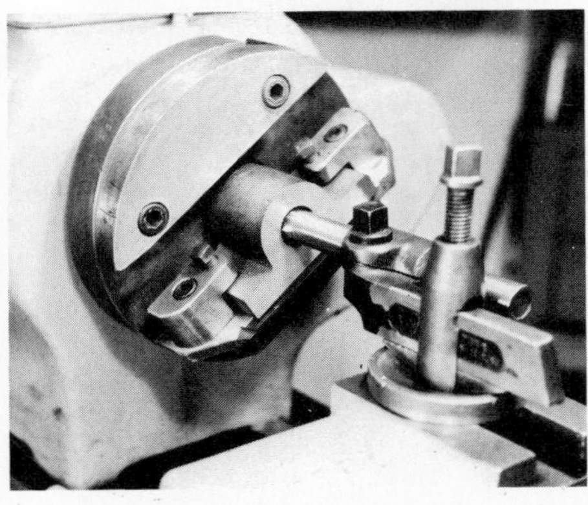

FIGURE 20-20. Boring in a lathe with the work held in a special fixture mounted on a face plate.

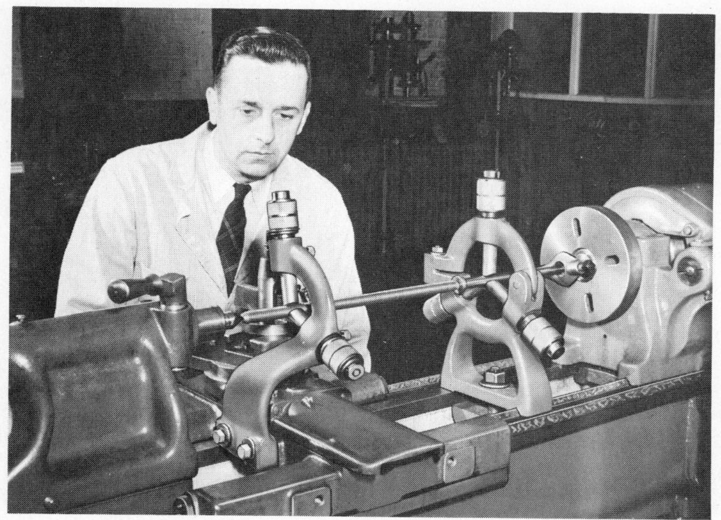

FIGURE 20-21. Cutting a thread on a long, slender workpiece, using a follow rest (*left*) and a steady rest (*right*). (*Courtesy South Bend Lathe.*)

Mounting work on the carriage. When no other means is available, boring occasionally is done on a lathe by mounting the work on the carriage, with the boring bar mounted between centers and driven by means of a dog. This procedure is illustrated in Figure 20-22.

Lathe tools. Most lathe operations are done with relatively simple, single-point cutting tools, such as illustrated in Figure 20-23. On right-hand and left-hand turning and facing tools, the cutting takes place on the side of the tool so that the side rake angle is of primary importance and deep cuts can be made. On the round-nose turning tools, cutoff tools, finishing tools, and some threading tools, cutting takes place on or near the end of the tool, so that the back rake is of importance. Such tools are used with relatively light depths of cut.

FIGURE 20-22. Method of boring on a lathe with the work mounted on the carriage.

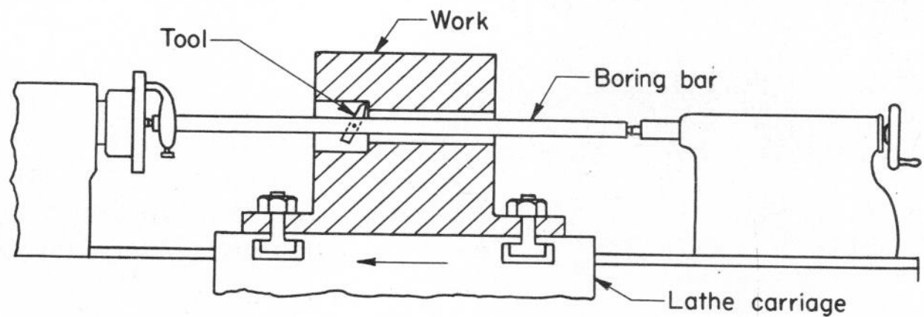

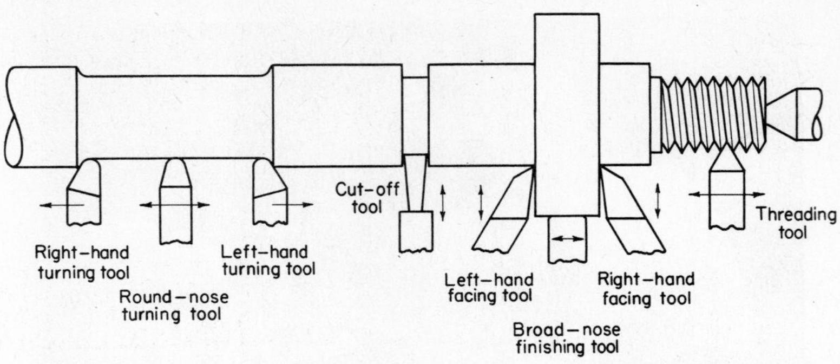

FIGURE 20-23. Shapes and uses of common single-point lathe tools.

Most lathe work is done with high-speed steel, carbide, or ceramic tools. For mass-production work, the throwaway types of carbide or ceramic tips are used, with either integral chip breakers or the adjustable type shown in Figure 20-24.

Toolholders. Because cutting-tool materials are expensive, it is desirable to use as small amounts as possible. At the same time, it is essential that the cutting tool be supported in a strong, rigid manner to minimize deflection and possible vibration. Consequently, lathe tools are supported in various types of heavy, forged toolholders, such as shown in Figure 20-25. The tool bit should be clamped in the tool post with minimum overhang. Otherwise, tool chatter and poor surface finish may result.

Where large tool bits are required, the type of forged toolholder shown in Figure 20-26 may be used. It provides a more adequate method of clamp-

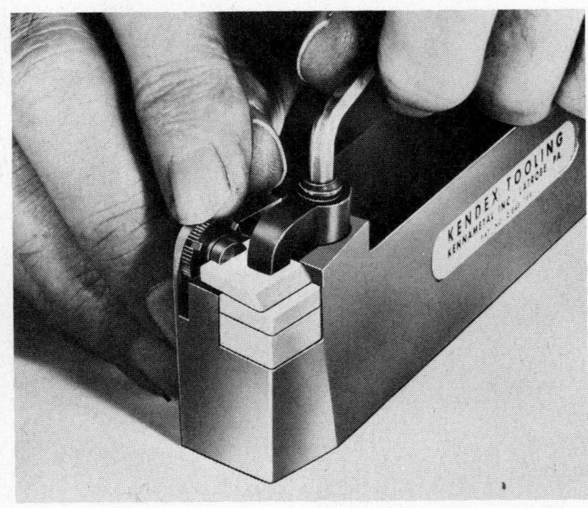

FIGURE 20-24. Tool holder for ceramic and carbide tips, having an adjustable chip breaker. (*Courtesy Kennametal, Inc.*)

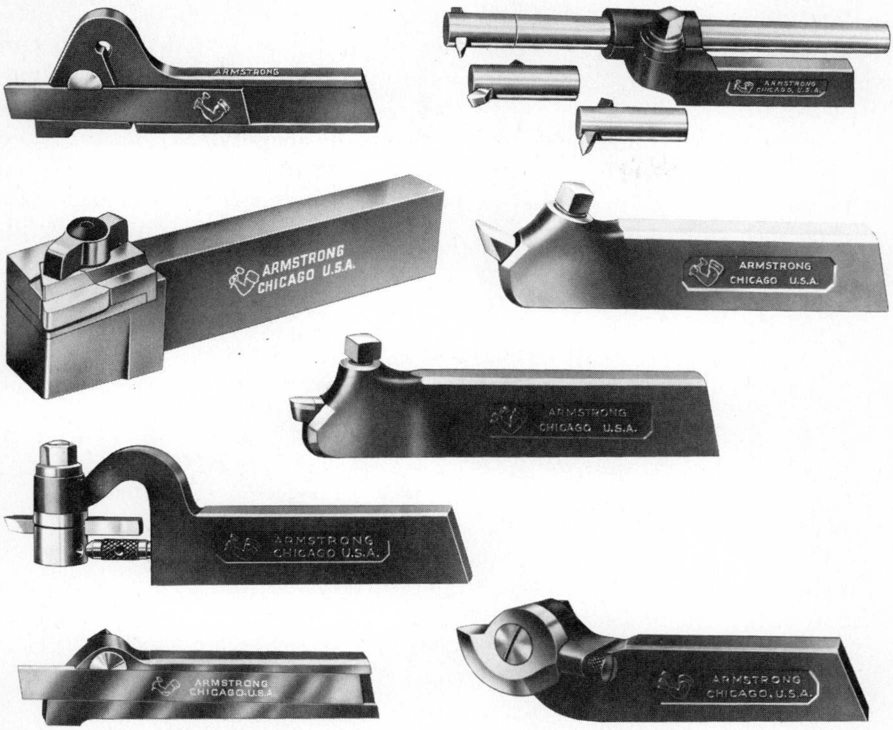

FIGURE 20-25. Common types of forged toolholders. (*Courtesy Armstrong Bros. Tool Company.*)

FIGURE 20-26. Heavy forged toolholder for holding large, carbide-tipped turning tools. (*Courtesy South Bend Lathe.*)

ing and supporting the tool than is provided by an ordinary tool post. The tools used in such cases have a heavy shank of forged or hot-rolled bar stock in which a carbide tip is brazed, thus reducing the amount of expensive tool material required.

Where several different operations on a lathe are performed repeatedly in sequence, the time required for changing and setting tools may constitute as much as 50 per cent of the total time. As a consequence, quick-change toolholders, such as shown in Figure 20-27, are being used increasingly. The in-

FIGURE 20-27. Quick-change tool post and accompanying toolholders. (*Courtesy Armstrong Bros. Tool Co.*)

FIGURE 20-28. Circular and block types of form tools. (*Courtesy Speedi Tool Company, Inc.*)

dividual tools, preset in their holders, can be interchanged in the special tool post in a few seconds. With some systems a second tool may be set in the tool post while a cut is being made with the first tool, and then be brought into proper position by rotating the post.

In lathe work the nose of the tool should be set exactly at the same height as the axis of rotation of the work. However, because any setting below the axis causes the work to tend to "climb" up on the tool, most machinists set their tools a few thousandths of an inch above the axis, except for cutoff, threading, and some facing operations.

Form tools. Form tools, made by grinding the inverse of the desired work contour on a piece of tool steel, are used to a considerable extent on lathes. For example, a threading tool often is a form tool. Although form tools are relatively expensive to make, they make it possible to machine a fairly complex surface with a single inward feeding of one tool. For mass-production work, adjustable form tools of either flat or rotary types, such as are shown in Figure 20-28, are used. These, of course, are expensive to make but can be resharpened by merely grinding a small amount off the face and then raising or rotating the cutting edge to the correct position.

The use of form tools is limited by the difficulty of grinding adequate rake angles for all points along the cutting edge.

Drill chucks. As mentioned in Chapter 19, drill chucks are used on lathes for holding drills, center drills, and reamers mounted in the tailstock quill. Large drills are mounted directly in the quill hole by means of their taper shanks.

Speeds, feeds, and coolants for turning. The discussion of speeds, feeds, and coolants given in Chapter 17 is applicable for all lathe operations. It should be remembered that when two or more diameters are involved on a single workpiece, the rotational speed of the work should be changed for each diameter in order to maintain maximum cutting efficiency and tool life. A less effective alternative is to base the rotational speed on the largest diameter involved.

Lathe feed, in millimeters or inches per revolution of the work, determines the chip thickness.

Cutting speeds and feeds in lathe work are very high when using carbide and ceramic tools. Consequently, the proper use of coolants has become very important. When used, they should be applied in ample quantities in order to achieve their objective.

LATHE OPERATIONS

Turning. *Turning* constitutes the majority of lathe work. The work usually is held between centers or in a chuck, and a right-hand turning tool is used, so that the cutting forces, resulting from feeding the tool from right to left, tend to force the workpiece against the headstock and thus provide better work support.

If good finish and accurate size are desired, one or more roughing cuts usually are followed by one or more finish cuts. Roughing cuts may be as heavy as proper chip thickness, tool life, and lathe capacity permit. Large depths of cut and smaller feeds are preferred to the reverse procedure, because fewer cuts are required and less time is lost in reversing the carriage and resetting the tool for the following cut.

On workpieces that have a hard surface, such as castings or hot-rolled materials containing mill scale, the initial roughing cut should be deep enough to penetrate the hard material. Otherwise, the entire cutting edge operates in hard, abrasive material throughout the cut, and the tool will dull rapidly. If the surface is unusually hard, the cutting speed on the first roughing cut should be reduced accordingly.

Finishing cuts are light, usually being less than 0.38 mm (0.015 inch) in depth, with the feed as fine as necessary to give the desired finish. Sometimes a special finishing tool is used, but often the same tool is used for both roughing and finishing cuts. In most cases one finishing cut is all that is required. However, where exceptional accuracy is required, two finishing cuts may be made. If the diameter is controlled manually, it usually is desirable to make a short finishing cut (about 6.4. mm or ¼ inch long) and check the diameter before completing the cut. Because the previous micrometer measurements were made on a rougher surface, some readjustment of the tool setting may be necessary in order to have the final measurement, made on a smoother surface, check exactly.

In turning operations, diameters usually are measured with micrometer calipers, although spring calipers may be used to check roughing cuts or where close accuracy is not required. The method of making length measurements is controlled, primarily, by the shape and accessibility of the surfaces over which measurement must be made. Spring, hermaphrodite, vernier, or micrometer calipers or micrometer depth gages can be used.

FIGURE 20-29. Making a facing cut in a lathe, the tool being fed outward.

Facing. *Facing* is the producing of a flat surface as the result of the tool being fed across the end of the rotating workpiece. The work may be held in a chuck, on a face plate, or between centers. Unless the work is held on a mandrel, if both ends of the work are to be faced, it must be turned end for end after the first end is completed and the facing operation repeated.

Because most facing operations are performed on surfaces that are away from the headstock, a right-hand tool is used most frequently. The spindle speed should be determined from the largest diameter of the surface to be faced. Facing may be done either from the outside inward or from the center outward. In either case, the point of the tool must be set exactly at the height of the center of rotation. Because the cutting force tends to push the tool away from the work, it usually is desirable to clamp the carriage to the lathe bed during each facing cut to prevent it from moving slightly and thus producing a surface that is not flat.

When facing castings or other materials that have a hard surface, the depth of the first cut should be sufficient to penetrate the hard material to avoid excessive tool wear.

Figure 20-29 shows a typical facing operation being done on a lathe.

Boring. *Boring* always involves the enlarging of an existing hole, which may have been made by a drill or be the result of a core in a casting. An equally important, and concurrent, purpose of boring may be to make the hole concentric with the axis of rotation of the workpiece and thus correct any eccentricity that may have resulted from the drill having drifted off the center line. Concentricity is an important attribute of bored holes.

When boring is done in a lathe, the work usually is held in a chuck or on a face plate. Holes may be bored straight, tapered, or to irregular contours. The boring of tapered holes will be discussed later.

Figure 20-20 shows the relationship of the tool and the workpiece for boring. Two types of boring tools are used. The one shown in Figure 20-20 is used more frequently. It consists of a conventional left-hand lathe tool held in the end of a round bar that, in turn, is mounted in cantilever fashion in a special forged toolholder. The other type, shown in Figure 20-37, is forged from a single piece of tool steel and is held either in the tool post or in a forged toolholder. It is used for boring holes that are too small to permit the entry of the other type of boring bar.

In most respects the same principles are used for boring as for turning. However, the tool should be set exactly at the same height as the axis of rotation. Slightly larger end clearance angles sometimes have to be used to prevent the heel of the tool from rubbing on the inner surface of the hole. Because the tool overhangs its support a considerable amount, feeds and depths of cut may have to be somewhat less than for turning to prevent tool vibration and chatter. In some cases the boring bar may be made of tungsten carbide because of its greater stiffness.

There always is a tendency for bored holes to be slightly bell-mouthed because of the tool springing away from the work as it progresses into the hole. This usually can be corrected by repeating the cut with the same tool setting.

Drilling. Most *drilling* on lathes is done with the drill held in the tailstock quill and fed against a workpiece that is rotated in a chuck. Drills with taper shanks are mounted directly in the quill hole, as shown in Figure 20-30, whereas those with straight shanks are held in a drill chuck that is mounted in

FIGURE 20-30. Drilling in a lathe, using a taper-shank drill in the tailstock quill.

FIGURE 20-31. Drilling in a lathe, using a drill chuck.

the quill hole, as illustrated in Figure 20-31. Feeding is by hand by means of the handwheel on the outer end of the tailstock assembly.

It also is possible to do drilling on a lathe with the drill mounted and rotated in the spindle while the work remains stationary, supported by a special pad mounted in the tailstock quill. This procedure, illustrated in Figure 20-32, is seldom used but is useful in special cases.

Usual speeds are used for drilling in a lathe. Because feeding is by hand, care must be exercised, particularly in drilling small holes. Coolants should be used where required. In drilling deep holes, the drill should be withdrawn occasionally to clear chips from the hole and to aid in getting coolant to the cutting edges.

Reaming. *Reaming* in a lathe involves no special precautions. Reamers are held in the tailstock quill, taper-shank types being mounted directly and straight-shank types by means of a drill chuck. Rose-chucking reamers usually are used. Fluted-chucking reamers also may be used, but these should be held in

FIGURE 20-32. Drilling in a lathe with the drill held in a drill chuck in the spindle and with the work supported by a drill pad in the tailstock quill. (*Courtesy South Bend Lathe.*)

FIGURE 20-33. Using a floating reamer holder for reaming in a lathe. (*Courtesy South Bend Lathe.*)

some type of holder that will permit the reamer to float, such as is illustrated in Figure 20-33.

Parting. *Parting* is the operation by which one section of a workpiece is severed from the remainder by means of a cutoff tool. Because parting tools are quite thin and must have considerable overhang from the more rigid toolholder, it is a somewhat difficult operation. The tool, such as shown in Figures 20-23 and 20-25, should be set exactly at the height of the axis of rotation, be kept sharp, have proper clearance angles, and be fed into the workpiece at a proper and uniform rate.

Knurling. *Knurling* produces a regularly shaped, roughened surface on a workpiece. Although knurling also can be done on other machine tools, even on flat surfaces, in most cases it is done on external cylindrical surfaces in some

FIGURE 20-34. Knurling tool with forming rolls. (*Courtesy Armstrong Bros. Tool Company.*)

FIGURE 20-35. Knurling in a lathe, using a forming type tool, and showing the resulting pattern on the workpiece.

FIGURE 20-36. Knurling with a chip-type knurling tool. (*Courtesy Gebrüder Miller, Ges.m.b.H.*)

type of lathe. In most cases, knurling is a chipless, cold-forming process, using a tool of the type shown in Figures 20-34 and 20-35. The two hardened rolls are pressed against the rotating workpiece with sufficient force to cause a slight outward and lateral displacement of the metal so as to form the knurling in a raised, diamond pattern. Another type of knurling tool, shown in Figure 20-36, produces the knurled pattern by cutting chips. Because it involves less pressure and thus does not tend to bend the workpiece, it often is preferred for workpieces of small diameter and for use on automatic or semiautomatic machines.

Turning and boring tapers. The turning and boring of uniform tapers are common lathe operations. Such tapers can be specified either in degrees of included angle between the sides or as the change in diameter per unit of length—millimeters per millimeter or inches per foot.

Three methods are available for turning external tapers on a lathe, and two for boring internal tapers. The simplest is to use the compound rest; this method is suitable for both external and internal tapers. However, because the length of travel of the compound rest is quite limited—seldom over 150 mm (a few inches)—only short tapers can be turned or bored by this method. It is particularly useful for steep tapers. The compound rest is swiveled to the desired angle and locked in position, as shown in Figure 20-37. The compound slide then is fed manually to produce the desired taper. The tool should be set at exactly the height of the axis of rotation of the workpiece in all taper turning and boring.

Because the graduated scale on the base of the compound rest usually is calibrated only to 1° divisions, it is difficult to make the angle setting with accuracy. If accuracy is required, tapers made by this method are checked by means of plug or ring gages, readjusting the setting of the compound rest until the gage fits perfectly. Also, the compound rest cannot be set directly to the correct angle if the taper is dimensioned in millimeters per millimeter or inches per foot.

FIGURE 20-37. Boring a short, internal taper, using the compound rest. (*Courtesy South Bend Lathe.*)

FIGURE 20-38. Taper-turning attachment. (*Courtesy The American Tool Works Company.*)

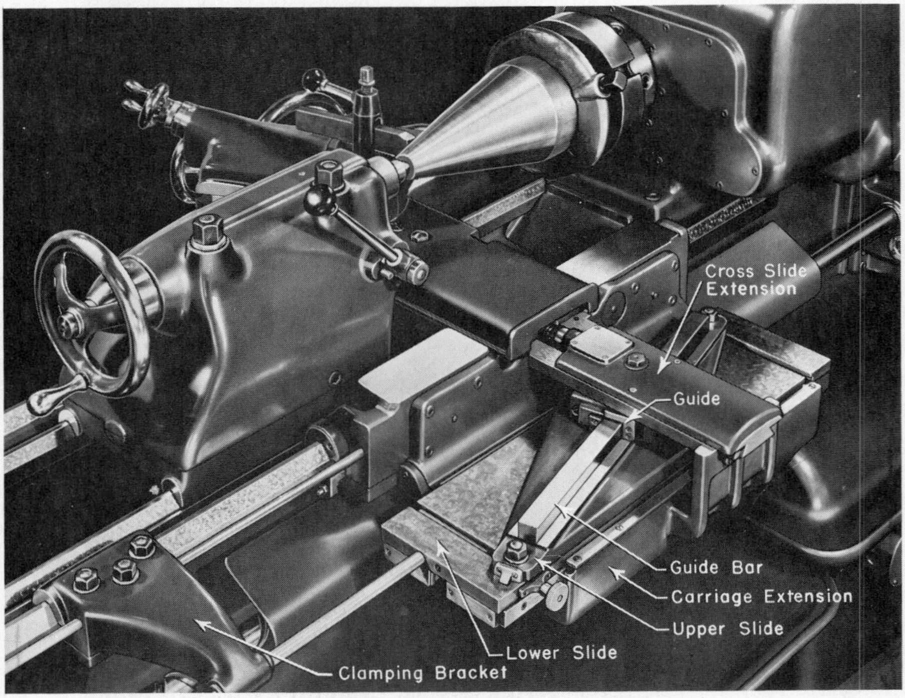

Both external and internal tapers can be made on a lathe by using a *taper attachment,* such as is shown in Figure 20-38. In this device, an *extension* is bolted to the rear of the carriage and supports a *lower slide,* which, in turn, is connected to a clamping bracket so that when this bracket is clamped to the lathe bed, the lower slide is prevented from moving. When the carriage is moved, the lower slide moves in the carriage extension.

An *upper slide,* containing a raised *guide bar,* is pivoted to the lower slide so that it can be swiveled to any desired angle (within its limits). A *guide,* on an extension of the cross slide, slides on the guide bar, so that when the carriage is moved longitudinally, the guide follows the guide bar and moves the cross slide and tool post transversely according to the angle at which the upper slide is set.

Most lathes have a telescoping cross-slide screw that permits the cross slide to be moved by the taper attachment. On some lathes the screw must be disconnected in order to use the taper attachment.

Graduations of taper in mm/mm or in./ft are provided at one end of the lower slide and in degrees at the other end so that the upper slide can be set to the desired taper. While taper attachments provide an excellent and convenient method of cutting tapers, they ordinarily can be used only for tapers of less than 1:2 (0.5 mm/mm or 6 in./ft).

External tapers also can be turned on workpieces that are mounted between centers by *setting over the tailstock.* The theory of this method is illustrated in Figure 20-39. The tailstock is moved out of line with the headstock spindle by means of the mechanism illustrated in Figure 20-40. The distance the tailstock must be set off from the center line is given by the following formulas:

Metric:

$$\text{set off (mm)} = \frac{\text{length of workpiece (mm)} \times \text{taper (mm/mm)}}{2}$$

FIGURE 20-39. Theory of turning tapers by the tailstock setover method.

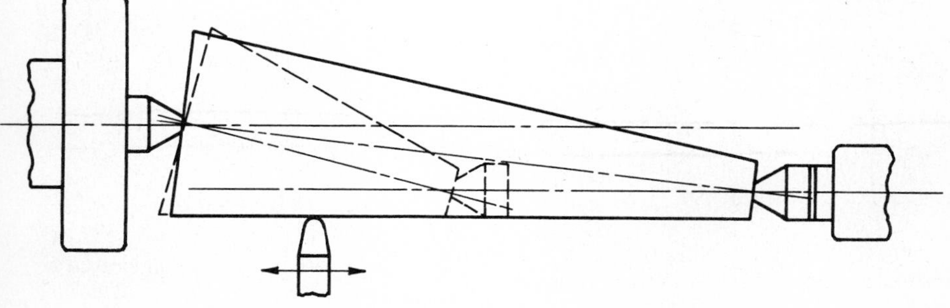

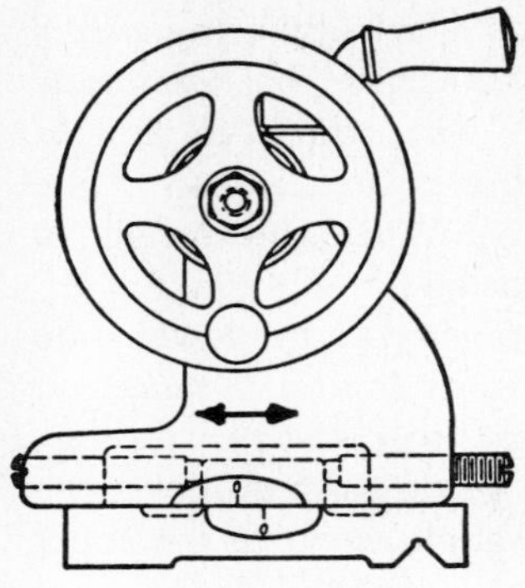

FIGURE 20-40. Method of setting the tailstock off center for turning tapers. (*Courtesy South Bend Lathe.*)

English:

$$\text{set off (in.)} = \frac{\text{length of workpiece (ft)} \times \text{taper (in./ft)}}{2}$$

This method is limited to small tapers and is seldom used.

When specifying tapers on drawings, the designer or draftsman should remember that it is difficult for the machinist to measure the smaller diameter of a taper accurately if it is the end of a workpiece.

Special lathe accessories. Several attachments are available that facilitate doing special types of work on lathes. The *milling attachment,* shown in Figure 20-41, is a special vise that attaches to the cross slide to hold work while milling is being done by a cutter that is rotated by the spindle. The work is fed by means of the cross-slide screw.

A *turret attachment,* as shown in Figure 20-42, replaces the usual tail-

FIGURE 20-41. Milling a dovetail slot on a lathe, using a milling attachment. (*Courtesy South Bend Lathe.*)

FIGURE 20-42. Conversion of a regular engine lathe into a turret lathe through the use of a turret attachment, which has been substituted for the tailstock, and a special cross slide. (*Courtesy South Bend Lathe.*)

stock assembly and holds six tools that can be brought into operating position by rotating the turret head.

The *relieving attachment,* shown in Figure 20-43, moves the cutting tool back and forth, either transversely or longitudinally, in synchronization with the rotation of the work. It thus makes possible the machining of cams or the provision of relief on rotary cutters, such as end mills. Movement of the cross

FIGURE 20-43. Relieving attachment on a lathe (the covers of the two cam boxes are open). The right-hand shaft controls transverse movement of the cross slide, while the left-hand shaft moves the carriage for side relief. (*Courtesy Colt Industries, Pratt & Whitney Machine Tool Division.*)

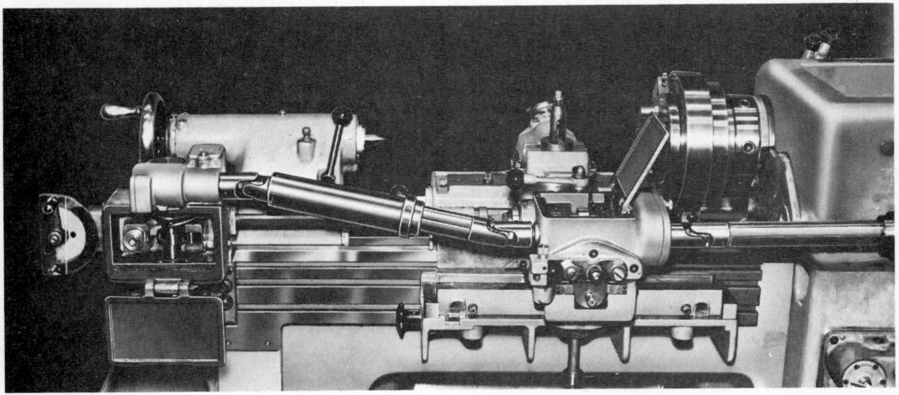

slide is provided by a cam, driven from the headstock. Another cam actuates the carriage and moves it longitudinally.

Tool-post grinders are often used to permit grinding to be done on a lathe. Such an attachment is discussed in Chapter 25 and illustrated in Figure 25-18.

Duplicating attachments are available that, guided by a template, will automatically control the tool movements for turning irregularly shaped parts. In some cases the first piece, produced in the normal manner, may serve as the template for duplicate parts. To a large extent, duplicating lathes using templates have been replaced by tape- or computer-controlled lathes. These will be discussed in Chapter 39.

TURRET LATHES

Although engine lathes are versatile and very useful, because of the time required for changing and setting tools and for making measurements on the workpiece, they are not suitable for quantity production. Often the actual chip-production time is less than 50 per cent of the total cycle time. In addition, a skilled machinist is required for all the operations, and such persons are costly and often in short supply. However, much of the operator's time is consumed by simple, repetitious adjustments and in watching chips being made. Consequently, to reduce or eliminate the amount of skilled labor that is required, turret lathes, screw machines, and other types of semiautomatic and automatic lathes have been highly developed and are widely used in manufacturing.

Turret lathe construction. The basic components of a *turret lathe* are depicted in Figure 20-44. Basically, a longitudinally feedable, multisided turret replaces the compound rest and tool post on the cross slide. The main turret usually has six sides on which tools can be mounted. This turret can be rotated about a vertical axis to bring each tool into operating position, and the entire unit can be moved longitudinally, either manually or by power, to provide feed for the tools. When the turret assembly is moved away from the

FIGURE 20-44. Block diagram showing the basic components of a turret lathe.

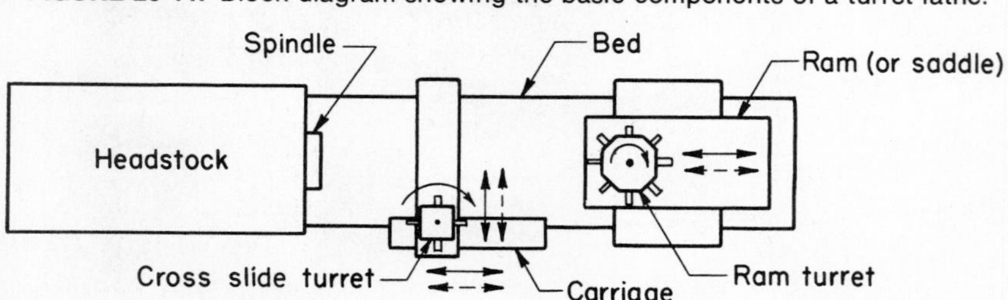

spindle by means of a capstan wheel, the turret indexes automatically at the end of its movement, thus bringing the next tool into operating position.

The turret on the cross slide can be rotated manually about a vertical axis to bring each of the four tools into operating position. On most machines, the turret can be moved transversely, either manually or by power, by means of the cross slide, and longitudinally through power or manual operation of the carriage. In most cases, a fixed toolholder also is added to the back end of the cross slide; this often carries a parting tool.

Through these basic features of a turret lathe, a number of tools can be set up on the machine and then quickly be brought successively into working position so that a complete part can be machined without the necessity for further adjusting, changing tools, or making measurements. A skilled machinist is required for making the setup, but a relatively low-skilled operator thereafter can operate a turret lathe and produce parts with as good accuracy and with as much speed as though the operation were performed by a skilled machinist.

Two basic types of turret lathes are made, differing in the manner in which the main turret is mounted. Figure 20-45 shows a *ram-type turret lathe,* on which the main turret is pivoted and carried on a ram that slides back and forth in a saddle. When making a setup, the saddle is moved on the bed ways and clamped in the desired position. Because only the ram and turret must be moved during operation, this type of mounting provides easy and rapid motion.

FIGURE 20-45. Ram-type turret lathe. (*Courtesy Sheldon Machine Co., Inc.*)

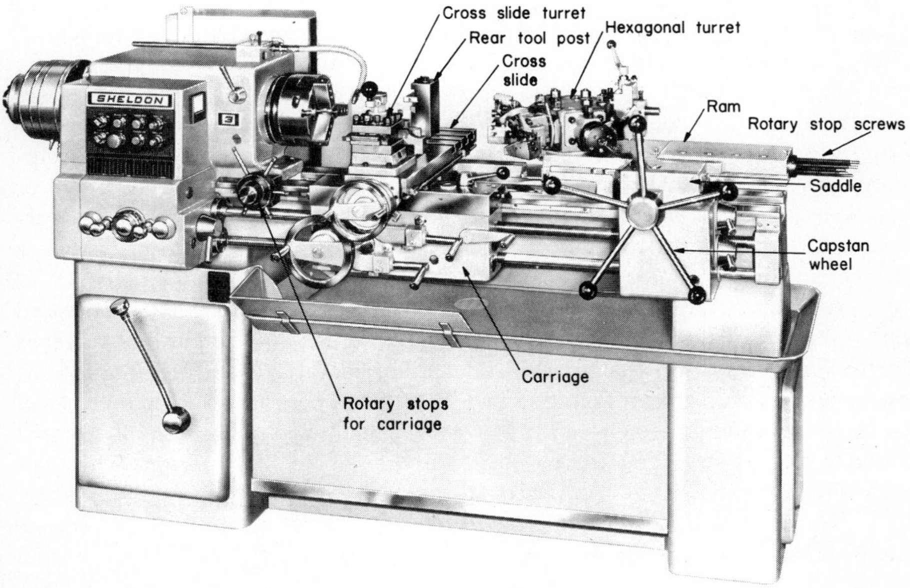

FIGURE 20-46. Saddle-type turret lathe, having a side-hung carriage. (*Courtesy The Warner & Swasey Company.*)

Ordinarily, the ram and turret are moved up to the cutting position by means of the capstan wheel, and the power feed then is engaged. As the ram is moved toward the headstock, the turret is automatically clamped in its rotational position so that rigid tool support is obtained. A set of rotary stopscrews, such as those shown in Figure 20-45, can be set to control the inward travel of the ram, one stop being provided and set for each face on the turret. The proper stop is brought into operating position automatically when the turret is indexed. A similar set usually is provided to limit movement of the cross slide.

Saddle-type turret lathes provide a more rugged mounting for the main turret than can be obtained by the ram-type mounting. On these lathes, as shown in Figure 20-46, the main turret is mounted directly on the saddle, and the entire saddle and turret assembly reciprocates. Larger turret lathes usually have this type of mounting. However, because the saddle-turret assembly is rather heavy, this type of mounting provides less rapid turret reciprocation. When such lathes are used with heavy tooling for making heavy or multiple cuts, a *pilot arm* attached to the headstock engages a pilot hole attached to one or more faces of the turret to give additional rigidity. Such a device is shown in Figure 20-46.

A different arrangement of the turret and turret mounting is provided on a German-made turret lathe. As shown in Figure 20-47, the tool turret is in the form of a drum that rotates about a horizontal axis and is mounted directly on, and reciprocates on, the bed ways. This permits a larger number of tools to be mounted on the turret. Also, the drum turret can be rotated slightly during an operation to provide an additional tool motion. This permits such operations as boring internal grooves.

Turret lathe headstocks have two features not found on ordinary engine lathes. One permits rapid shifting between at least two spindle speeds, with a

FIGURE 20-47. Drum-type turret lathe. Inset shows a close-up view of the tools in the drum turret. (*Courtesy Werkzeugmaschinen-Fabrik Gildemeister & Comp. Akt.-Ges.*)

brake to stop the spindle very rapidly. The second feature is an automatic stock-feeding device for feeding bar stock through the spindle hole. As shown in Figure 20-48, a spring collet is used so that when a lever is thrown, the collet is released and the feed mechanism moves the bar stock forward until it reaches a stop on the main turret. Moving the control lever in the opposite direction clamps the stock in the collet. The entire operation can be done without stopping the spindle.

FIGURE 20-48. Stationary-type collet chuck mechanism. (*Courtesy The Warner & Swasey Company.*)

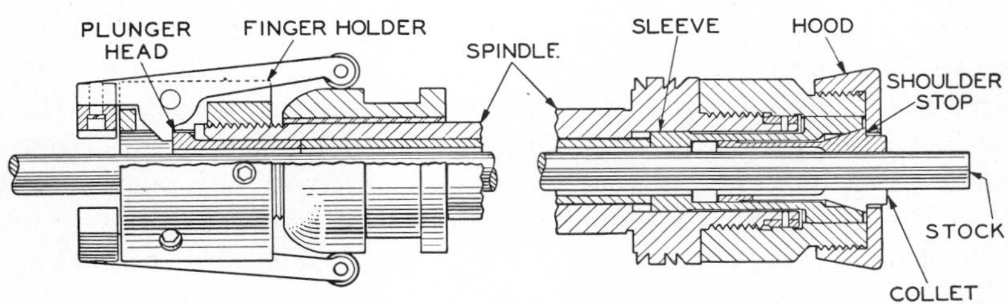

On smaller lathes the stock feed usually is actuated by a weight and the collet by hand lever. On larger lathes the stock feeding and collet clamping are done by either pneumatic or hydraulic means.

If the work is to be held in a chuck, some type of air-operated chuck, or special clamping fixture, frequently is employed to reduce the work setup time to a minimum.

Two types of carriage assemblies are used on turret lathes. The *reach-over* type is essentially the same as on the ordinary engine lathe and permits both a front and rear tool post. On the *side-hung* type, shown in Figure 20-47, the carriage does not extend across the top of the lathe, being supported entirely by front and side ways. This construction provides greater swing but does not permit a rear tool post.

Turret-lathe tooling. Eight factors should be kept in mind when considering turret-lathe tooling:

1. Setup time.
2. Work-handling time.
3. Machine-controlling time.
4. Cutting time.
5. Tool cost.
6. Setup labor cost.
7. Lathe operator labor cost.
8. Number of pieces to be produced.

The first four of these factors are basic in determining the arrangement of the tools in the turret and tool holders; the last four relate to the cost of producing parts on a turret lathe and thus determine whether such a lathe should be used for producing a given lot of parts.

Setup time is that required for a skilled machinist or tool setter to set the various cutting tools and work holders in the turret lathe, adjust them to produce the desired dimensions on the workpiece, and set the various stops. Because machinists who are capable of doing this type of work are highly paid and often scarce, it is desirable to keep the setup time as short as practicable. Another reason for minimizing setup time is the fact that this time reduces the amount of cutting time that is available. One common method of accomplishing this objective is to make use of standardized toolholders and tools whenever possible.

Work-handling time is that required for putting work into and removing it from the lathe. Where bar stock is fed through the spindle, automatic feeding devices provide a solution for this problem. For chucking work, power-actuated chucks or fixtures reduce this time.

Machine-controlling time is that required to manipulate the controls that reverse and rotate the turrets, change speeds, and so on. It is determined primarily by the design of the machine and may become a considerable part of the total machining cycle when high-speed cutting is employed. However, machine-controlling time also can be reduced by combining operations in a single-turret position or in a single tool, such as parting and chamfering with the same tool, as illustrated in Figure 20-49.

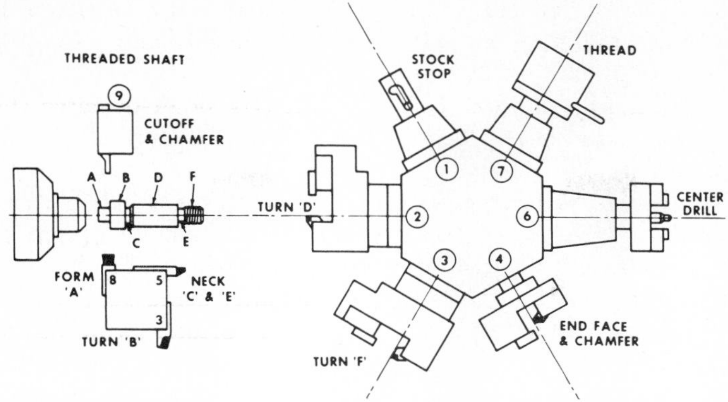

FIGURE 20-49. Turret-lathe tooling for producing the part shown.

Cutting time is that during which chips are produced. Although this time should be as large a part of the total machining cycle as possible, it should be reduced to a minimum in absolute amount. This can be done only by increasing cutting speeds and feeds and by using simultaneous cuts.

Figure 20-49 illustrates how turret lathes are tooled in accordance with these principles.

The hexagonal turret on most turret lathes is designed to permit tools to be mounted on it in two ways. Tools such as drills, reamers, and taps may be inserted in the central hole on each turret face and clamped by means of a nut on the top of the turret. Other tools, such as single-point turning tools, can be held in various types of standard holders that can, in turn, be quickly bolted to the faces of the turret. By the latter method, tools may be preset in the holders and mounted and changed very rapidly, thereby reducing setup time. Consequently, even drills, reamers, and so on, now are often mounted in standard holders for attaching to the turret faces.

Because heavy and/or simultaneous cuts frequently are made in doing turning on a turret lathe, it usually is necessary to hold the cutting tools in a type of holder that will also provide support for the work to prevent its being deflected from the tool. A tool with pressure rollers, such as illustrated in Figure 20-50, can be used for this purpose. If the surface being machined must be concentric with the existing surface, the rolls are set ahead of the cutting tool on the original surface. Otherwise, they are set a little behind the cutting tool on the surface being formed. When two or more cuts are taken, the same result sometimes can be achieved by spacing the tools around the work periphery.

Vertical turret lathes. Where chucking-type work is too large and heavy to permit holding it in a vertical chuck and rotating it about a horizontal axis,

FIGURE 20-50. Method of using backing rollers to prevent the work from being deflected by the cutting-tool forces. (*Courtesy The Warner & Swasey Company.*)

vertical turret lathes are used. As shown in Figure 20-51, these essentially are regular turret lathes turned on end. These machines resemble certain types of vertical boring mills and frequently are thought of as special types of the latter (see Chapter 21). Their rotary work tables commonly range from about 610 to 1200 mm (24 to 48 inches) in diameter and are equipped with both removable chuck jaws and T-slots for clamping the work. The cross rail carries one or two five- or six-sided turrets and, often, another smaller, four-sided turret is mounted on one side of the machine on an independent cross slide. Usually, each motion of successive tools can be controlled by means of stops so that duplicate workpieces can be machined with one tooling setup.

FIGURE 20-51. Vertical turret lathe: (*left*) normal front view; (*right*) side view, machine turned on its back. (*Courtesy The Bullard Company.*)

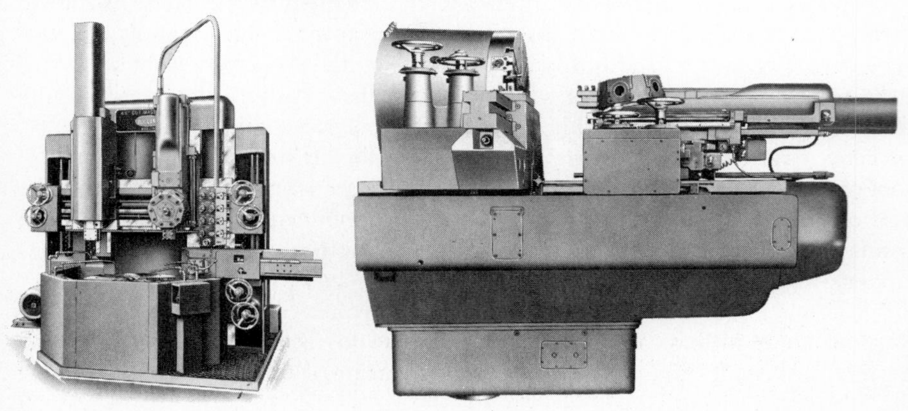

Example: How many units of a given part would have to be made to justify tooling and using a turret lathe under the following conditions:

Time per piece on turret lathe	3½ minutes
Turret lathe operator	\$3.50 per hour
Turret lathe overhead, including interest on capital	\$4.50 per hour
Setup time	2 hours
Setup man	\$6.50 per hour
Time per piece on engine lathe	12 minutes
Lathe machinist	\$4.50 per hour
Lathe overhead, including interest on capital	\$3.50 per hour

For the cost per piece to be equal by the two methods,

$$\frac{3.5}{60}(\$3.50 + \$4.50) + \frac{2 \times \$6.50}{N} = \frac{12}{60}(\$4.50 + \$3.50)$$

$$N = 11.5 \quad \text{or} \quad 12 \text{ pieces}$$

Therefore, if 12 or more pieces were to be produced, it would be more economical to tool and use the turret lathe.[2]

Automatic turret lathes. After a turret lathe is tooled, the skill required of the operator is very low, and the motions are simple and repetitive. As a result, several types of automatic turret lathes have been developed which require no operator. In addition, a number of machines of this type permit setting the controls for all the machine motions very quickly by means of buttons and knobs on a control panel. One type of automatic turret lathe, shown in Figure 20-52, has an ordinary ram-type turret and a cross slide.[3] A second type, shown in Figure 20-53, has a turret that indexes about a horizontal axis and reciprocates horizontally to provide feed. It has up to three cross slides, and the turret also can be rotated a small distance while making cuts to cut internal grooves. On this machine the operations are controlled by setting trip blocks and pins. A third type of automatic turret lathe for heavier chucking work, shown in Figure 20-54, has a vertical spindle.

SCREW MACHINES

Ordinary turret lathes eliminate the necessity for making tooling setups for each piece machined and minimize machine-controlling time. However, an operator is required to control the machine and to feed the work into machin-

[2] This solution assumes that no overhead is charged on the turret lathe while it is being tooled.
[3] This particular lathe also has a vertical cutoff slide and a slant bed. This latter feature, which permits the chips to fall out of the cutting zone more readily, is incorporated in many modern, high-productivity machine tools.

ing position. Automatic turret lathes can eliminate these last two functions, but because they usually have provision for manual operation, they are not as productive as *screw machines,* which are lathes designed for completely automatic operation. They originally were designed for machining small parts,

FIGURE 20-52. Automatic turret lathe and close-up view of control panel, on which all operations can be programmed by means of push-buttons and switches. (*Courtesy Bardons & Oliver, Inc.*)

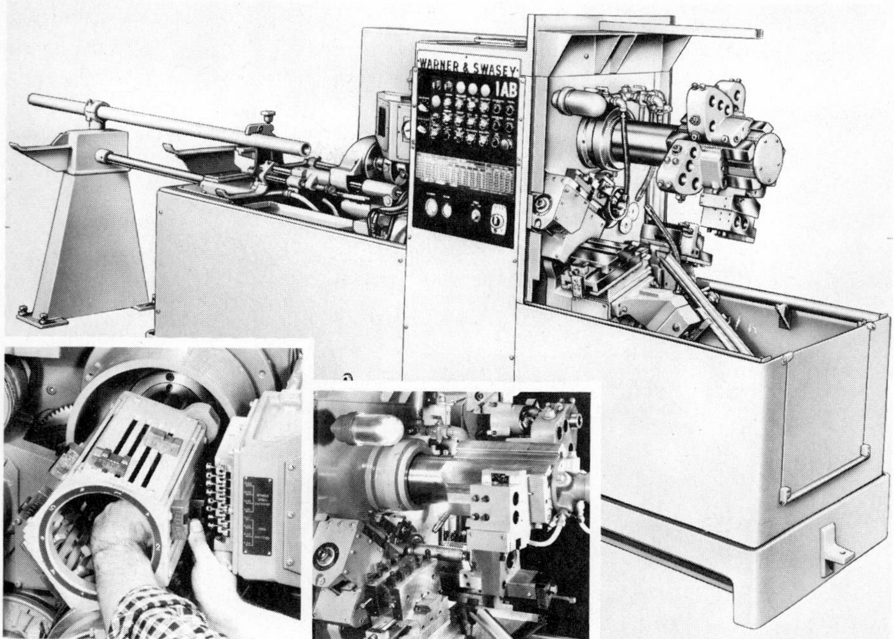

FIGURE 20-53. Automatic single-spindle turret lathe having turret that revolves about a horizontal axis. Insets show setting of control trip blocks and pins and a close-up view of the tooling. (*Courtesy The Warner & Swasey Company.*)

FIGURE 20-54. Vertical turret lathe with a Man-Au-Trol automatic control. (*Courtesy The Bullard Company.*)

FIGURE 20-55. Brown & Sharpe single-spindle screw machine. (*Courtesy Brown & Sharpe Manufacturing Company.*)

such as screws, bolts, bushings, and so on, from bar stock—hence the name "screw machines." Now they are used for producing a wide variety of parts, covering a considerable range of sizes, and are even used for some chucking-type work.

Single-spindle screw machines. There are two common types of *single-spindle screw machines.* One, an American development and commonly called the Brown and Sharpe type, is shown in Figure 20-55. The other, shown in Figure 20-57, is of Swiss origin and is referred to as the Swiss type.

As can be seen in Figures 20-55 and 20-56, the *Brown and Sharpe screw machine* is essentially a small automatic turret lathe, designed for bar stock, with the main turret mounted in a vertical plane on a ram. Front and rear toolholders can be mounted on the cross slide. All motions of the turret, cross slide, spindle, chuck, and stock-feed mechanism are controlled by disk cams. These machines usually are equipped with an automatic rod-feeding magazine that feeds a new length of bar stock into the collet as soon as one rod is completely used.

FIGURE 20-56. Close-up view showing the cams and tooling on a Brown & Sharpe screw machine. (*Courtesy Brown & Sharpe Manufacturing Company.*)

FIGURE 20-57. Swiss-type screw machine. (*Courtesy George Gorton Machine Corporation.*)

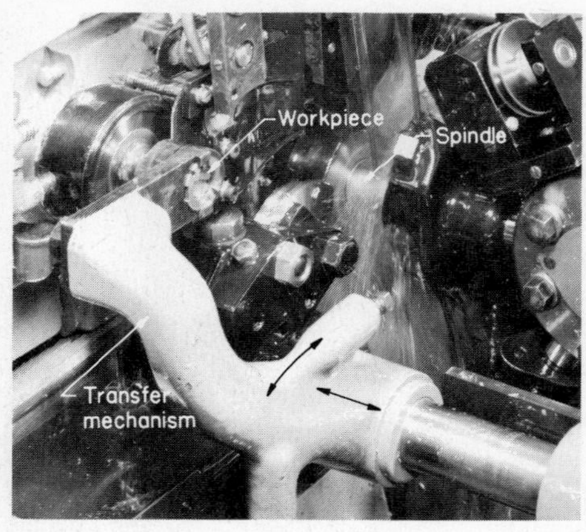

FIGURE 20-58. Transfer mechanism (sometimes called a "picking attachment") on a Brown & Sharpe screw machine. (*Courtesy Brown & Sharpe Manufacturing Company.*)

Often Brown and Sharpe-type screw machines are equipped with a transfer, or "picking," attachment, shown in Figure 20-58. This device swings over and picks up the workpiece from the spindle as it is cut off and carries it to, and holds it in, the position shown while a secondary operation is performed by a small, auxiliary power head. In this manner screwdriver slots are put in screw heads, small flats are milled parallel with the axis of the workpiece, or holes are drilled normal to the axis.

On the *Swiss-type screw machine,* shown in Figures 20-57 and 20-59, the cutting tools are held and moved in radial slides. Disk cams move the tools into cutting position and provide feed into the work in a radial direction only; they provide any required longitudinal feed by reciprocating the headstock.

Most machining on Swiss-type screw machines is done with single-point cutting tools. Because they are located close to the spindle collet, the workpiece is not subjected to much deflection. Consequently, these machines are particularly well suited for machining very small parts and are used primarily for such work.

Both types of single-spindle screw machines are capable of producing work to close tolerances, the Swiss-type probably being somewhat superior for very small work. Tolerances of 0.005 to 0.013 mm (0.0002 to 0.0005 inch) are not uncommon. The time required for setting the tooling usually is only an hour or two, and one person can tend several machines, once they have been properly tooled. They are highly productive; the entire cycle time frequently is less than ½ minute per piece.

Multiple-spindle screw machines. Although single-spindle screw machines eliminate the need for constant operator attendance, only one or two of the tooling positions are utilized at any given time. Thus the total cycle time per workpiece is the sum of the individual machining and tool-positioning times of the several cutting tools. On *multiple-spindle screw machines,* sufficient spind-

FIGURE 20-59. Close-up view of a Swiss-type screw machine, showing the tooling and radial tool slides, actuated by rocker arms. (*Courtesy George Gorton Machine Corporation.*)

les, usually four, six, or eight, are provided so that all tools cut simultaneously. Thus the cycle time per piece is equal to the maximum cutting time of a single tool position plus the time required to index the spindles from one position to the next.

The two distinctive features of multiple-spindle screw machines are shown in Figures 20-60 and 20-61. First, the multiple spindles are carried in a rotatable drum that indexes in order to bring each spindle into a different working position. Second, a nonrotating tool slide contains the same number of tool holders as there are spindles and thus provides and positions a cutting tool (or tools) for each spindle and imparts feed to these tools by longitudinal, reciprocating motion. In addition, most machines have a cross slide at each spindle position so that an additional tool can be fed from the side for facing, grooving, knurling, beveling, and cutoff operations. These slides also are shown in Figure 20-61. All motions are controlled automatically.

With a tool position available on the end tool slide for each spindle (except for a stock-feed stop at one position), when the slide moves forward, these tools cut essentially simultaneously. At the same time, the tools in the cross slides move inward and make their cuts. When the forward cutting mo-

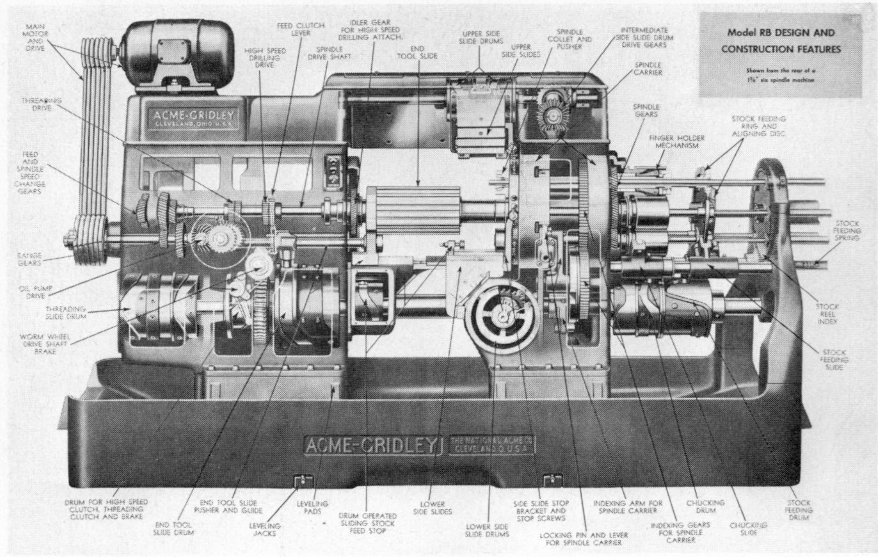

FIGURE 20-60. Mechanism in a six-spindle screw machine. (*Courtesy The National Acme Company.*)

FIGURE 20-61. Close-up view of the spindle carrier, spindles, tooling in the end and cross slides, and parts being produced in a six-spindle screw machine. (*Courtesy The National Acme Company.*)

FIGURE 20-62. (*Top*) Large six-spindle screw machine with the sound-control housing (*shown in bottom view*) removed. (*Courtesy The National Acme Company.*)

tion of the end tool slide is completed, it moves away from the work, accompanied by the outward movement of the radial slides, and the spindles are indexed one position, by rotation of the spindle carrier, to position each part for the next operation to be performed. At one spindle position finished pieces are cut off, and the bar stock is fed to correct length for the beginning of the next operation. Thus a piece is completed each time the tool slide moves forward and back.

Multiple-spindle screw machines are made in a considerable range of sizes, determined by the diameter of the stock that can be accommodated in the spindles. Figure 20-62 shows a rather large one.

In tooling a multiple-spindle screw machine, it should be remembered that because all cutting operations occur simultaneously, the operating cycle of the main tool slide is determined by the operation that requires the longest time. Consequently, one attempts to balance all the operations so that each requires the same amount of time. Although this ideal frequently cannot be achieved, careful planning permits it to be approached. Proper sequencing is important, and long operations may be broken up so as to be completed using two or more positions. For example, in the tooling program shown in Figure 20-63, if the ½-inch-diameter hole were drilled before the one having a $^{47}/_{64}$-inch diameter, this operation would require much more time than any of the others. By reversing the sequence, better balance is achieved, a total feed of 0.720 inch being required in one case and 0.750 inch for the other.

Where a simple part is to be made, it sometimes is possible to arrange

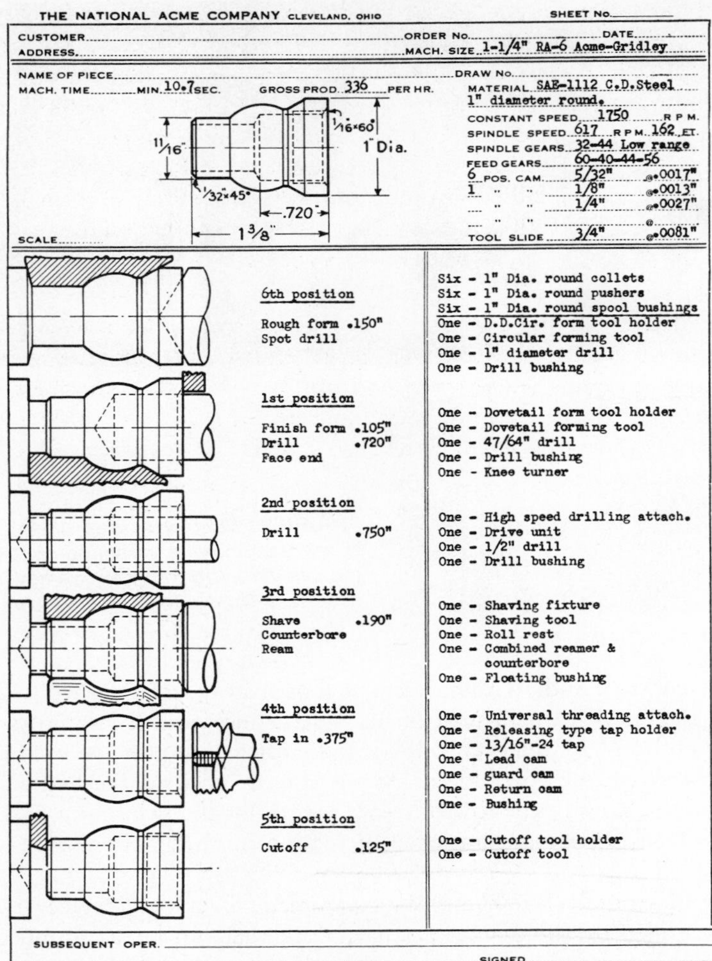

THE NATIONAL ACME COMPANY CLEVELAND, OHIO — SHEET No.

CUSTOMER — ORDER No. — DATE

ADDRESS — MACH. SIZE 1-1/4" RA-6 Acme-Gridley

NAME OF PIECE — DRAW No.

MACH. TIME — MIN. 10.7 SEC. — GROSS PROD. 336 PER HR.

MATERIAL SAE-1112 C.D.Steel 1" diameter round.

CONSTANT SPEED 1750 R.P.M.

SPINDLE SPEED 617 R.P.M. 162 E.T.

SPINDLE GEARS 32-44 Low range

FEED GEARS 60-40-44-56

6 POS. CAM 5/32" @.0017"

1 " " 1/8" @.0013"

" " 1/4" @.0027"

" " @

TOOL SLIDE 3/4" @.0081"

SCALE

Operation	Tools
	Six - 1" Dia. round collets Six - 1" Dia. round pushers Six - 1" Dia. round spool bushings
6th position Rough form .150" Spot drill	One - D.D.Cir. form tool holder One - Circular forming tool One - 1" diameter drill One - Drill bushing
1st position Finish form .105" Drill .720" Face end	One - Dovetail form tool holder One - Dovetail forming tool One - 47/64" drill One - Drill bushing One - Knee turner
2nd position Drill .750"	One - High speed drilling attach. One - Drive unit One - 1/2" drill One - Drill bushing
3rd position Shave .190" Counterbore Ream	One - Shaving fixture One - Shaving tool One - Roll rest One - Combined reamer & counterbore One - Floating bushing
4th position Tap in .375"	One - Universal threading attach. One - Releasing type tap holder One - 13/16"-24 tap One - Lead cam One - guard cam One - Return cam One - Bushing
5th position Cutoff .125"	One - Cutoff tool holder One - Cutoff tool

SUBSEQUENT OPER.

SIGNED

FIGURE 20-63. Tooling sheet for making a part on a six-spindle screw machine. (*Courtesy The National Acme Company.*)

the tooling on a multiple-spindle automatic so that two pieces are produced in each revolution of the spindle carrier.

The only attention a multiple-spindle screw machine requires is to keep the bar stock feed rack supplied and to check the finished products occasionally to make sure they are within the desired tolerances. One operator usually services several machines.

Most multiple-spindle screw machines utilize cams that are composed of specially shaped segments that are bolted onto a drum to control the motions, thereby reducing the need for special cams. Setting these cams and the tooling for a given job may require from 2 to 20 hours. However, once such a machine is placed in operation, the productivity is very great. Often a piece may be completed each 10 seconds. Typically, from 2000 to 5000 parts are required in a lot to justify tooling a multiple-spindle automatic.

The accuracy of multiple-spindle screw machines is good, but seldom as good as that of single-spindle machines. However, tolerances of from 0.013 to 0.025 mm (0.0005 to 0.001 inch) on the diameter are common.

Vertical multistation lathes. *Multiple-station lathes* are available for work that must be held in chucks. As shown in Figure 20-64, these essentially are vertical, multiple-spindle screw machines. A number of chuck-equipped spindles are mounted in a rotary indexing table and are indexed successively under a series of vertical rams in which the cutting tools are mounted. One chuck position remains at rest and has no tool ram, so that workpieces can be loaded and unloaded at this position while machining takes place at all the others.

FIGURE 20-64. Vertical multistation lathe having 12 spindles. (*Courtesy The Bullard Company.*)

AUTOMATIC LATHES

Although screw machines and automatic turret lathes are automatic *types* of lathes, the term *automatic lathe* generally is applied to a group of lathes that are semiautomatic and make simultaneous cuts, but that do not involve the use of the turret or screw-machine principles. The tools are fed to the work and retracted automatically by means of cam-controlled mechanisms. In most cases an operator is required to place the work in and remove it from the machine, so they are not truly automatic. However, in some instances a magazine type of workpiece feeder is employed so that no operator is required. The majority of automatic lathes have only a single spindle, but some specialized multispindle machines are also used.

Single-spindle automatics. A typical *single-spindle automatic lathe* is shown in Figure 20-65. Figure 20-66 shows typical tooling and several special features of this type of lathe. First the cutting tools are held in *tool blocks,* or *slides,* which are power-actuated and controlled to move the tools into and feed them along the work. Sometimes the front block provides only radial motion for facing, form cutting, and cutoff operations. The rear block has both radial and longitudinal motions, which are controlled by a plate cam, as illustrated in Figure

FIGURE 20-65. Single-spindle automatic lathe without tooling. (*Courtesy Gisholt Corporation.*)

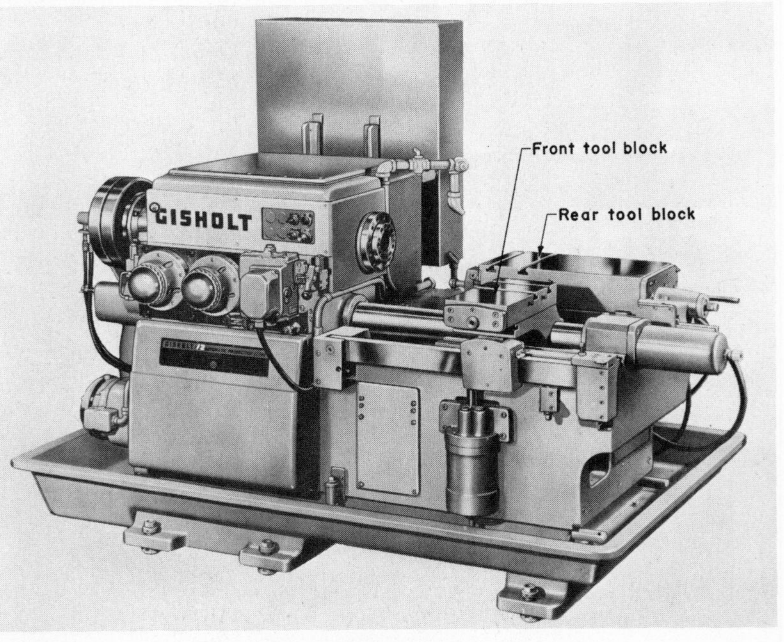

FIGURE 20-66. Multidiameter cylinder being machined in an automatic lathe, showing the tool blocks and the cutting tools. (*Courtesy Gisholt Corporation.*)

20-67, which shows the motions of the tool blocks and the portion of the metal removed by each tool. On some machines a third, overhead tool block also is provided.

In most cases the work is held between centers, utilizing various types of power-actuated chucks, collets, and tailstocks so that the work-handling time

FIGURE 20-67. Movements of the tool blocks for the machining process shown in Figure 20-66, and the metal to be removed by each tool. (*Courtesy Gisholt Corporation.*)

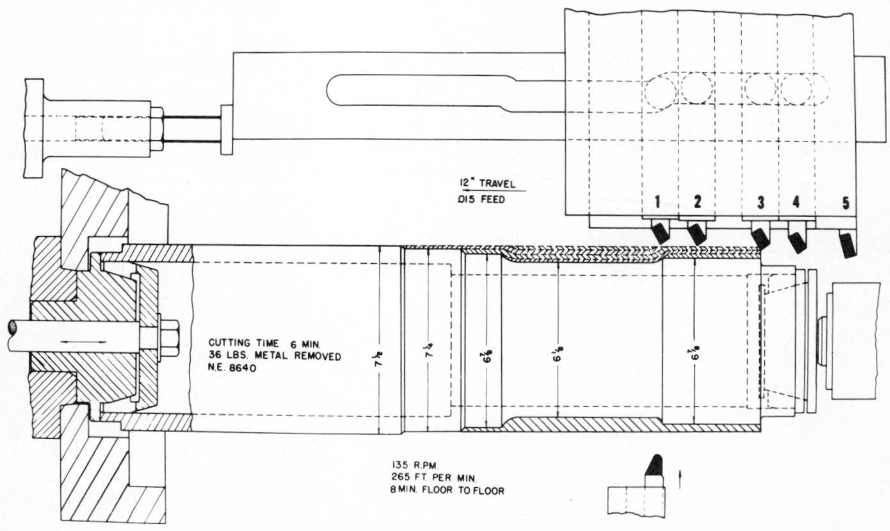

is minimized. In some cases the work is fed into the machine from a hopper-type feeding device and clamped and discharged automatically.

The total machining cycle on an automatic lathe usually is very short—often less than 1 minute. Sometimes a part is put successively into from two to four automatic lathes to complete its machining. They are fairly flexible, so that quite a variety of shapes and sizes can be handled in one lathe by changing the tooling setup.

Review questions

1. What is the tool–work relationship in turning?
2. What different kinds of surfaces can be produced by turning?
3. How does form turning differ from ordinary turning?
4. How does facing differ basically from a cutoff operation?
5. Name six different machining operations that can be done on a lathe.
6. Why is it difficult to make heavy cuts if a form-turning tool is complex in shape?
7. Why is the bed of a lathe of such great importance?
8. How is the size of a lathe designated?
9. What is the "swing" of a lathe?
10. Why is a lathe spindle hollow?
11. What functions does a lathe carriage have?
12. By what two methods is powered motion imparted to a lathe carriage?
13. Why would it not be desirable to drive the lathe carriage by the feed rod for cutting threads?
14. How is feed specified on a lathe?
15. What function is provided by the lead screw on a lathe that is not provided by the feed rod?
16. How is rotation provided to a workpiece that is mounted between centers on a lathe?
17. What are four ways for supporting work in a lathe?
18. What will result if work is mounted between centers in a lathe and the centers are not exactly in line?
19. Why is it not advisable to hold hot-rolled steel stock in a collet?
20. What is the unique feature of a gap-bed lathe?
21. What is a lathe "dog"?
22. What precautions are necessary in mounting work between centers?
23. How does a steady rest differ from a follow rest?
24. What are the advantages and disadvantages of a four-jaw independent chuck?
25. Why should the distance a lathe tool projects from the toolholder be minimized?
26. What occurs if a lathe tool is set below the center line of the workpiece in turning?

27. In what ways may tapers be turned on a lathe?
28. What should be the amount of tailstock setover to turn a taper of 2 per cent on a workpiece that is 254 mm long?
29. Why is it desirable to use a heavy depth and a light feed in turning rather than the opposite?
30. On what is the rpm based for a facing cut, assuming given work and tool materials?
31. Why is it usually necessary to take relatively light cuts when boring on a lathe?
32. How is a taper-shank drill held in a lathe?
33. What are two basic ways in which knurling is made?
34. Why are saddle-type turret lathes used much less than ram-type lathes?
35. What are two advantages of drum-type turret lathes?
36. What important factor must be kept in mind in tooling multiple-spindle screw machines that does not have to be considered in single-spindle machines?
37. What is the major difference between Brown and Sharpe-type and Swiss-type screw machines?
38. What would be the effect on the number of pieces required to justify the use of the turret lathe if all labor rates in the example on page 621 were increased 15 per cent?
39. At what speed should a 76.2-mm (3-inch)-diameter bar be rotated to provide a cutting speed of 61 meters (2000 feet) per minute?
40. Assume that the workpiece in question 39 is 203.2 mm (8 inches) long and a feed of 0.51 mm (0.020 inch) per revolution is used. How long will a cut across its entire length require?
41. If the depth of cut in question 40 is 4.76 mm (3/16 inch), what is the metal removal rate?
42. How long will be required to cut off a bar of B1112 steel that is 152.4 mm (6 inches) in diameter using a HSS tool and a feed of 0.254 mm (0.010 inch)? Use Table 17-1 to select the proper cutting speed.
43. Why is it sometimes necessary to split one operation between two positions on a multiple-spindle screw machine?
44. The following data apply for machining a part on a turret lathe and on an engine lathe:

	Lathe	Turret Lathe
Cycle time	20 min	5 min
Labor rate	\$5.50/hr	\$4.00/hr
Machine rate	\$3.50/hr	\$5.00/hr

The setup cost and cost for special tooling on the turret lathe would be \$30. How many pieces would have to be made to justify using the turret lathe?

CHAPTER 21

Boring

Boring is a machining process by which a hole of desired size and contour is obtained by enlarging an existing hole that has resulted from prior processes, such as drilling or casting. It is accomplished in two ways. In one, depicted in Figure 21-1a, a tool, offset from the axis of rotation, rotates within a stationary workpiece. In the second—Figure 21b—the work rotates past a stationary tool, thus being essentially interior turning. In both cases the tool is fed axially.

Although the primary purpose usually is to enlarge an existing hole, boring also accomplishes an important secondary result in that the bored hole always is concentric with the axis or rotation of the work or tool. This is an important attribute of bored holes, and often it is the primary reason for using boring for enlarging a hole rather than some other method, such as drilling.

Because the rotational relationship between the work and the tool is a simple one and is employed on several types of machine tools, such as lathes,

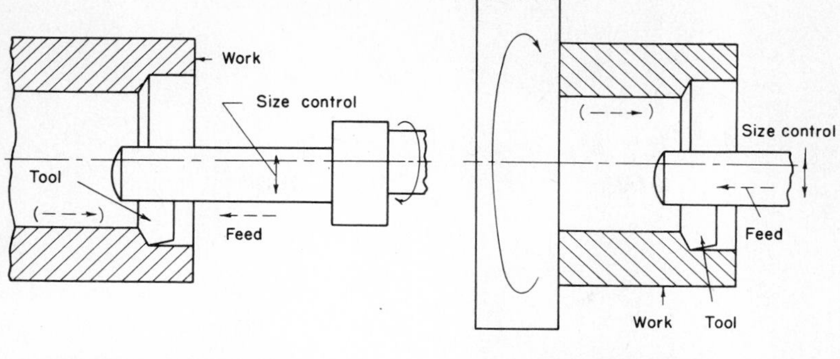

FIGURE 21-1. Tool–work relationships in boring.

drilling machines, and milling machines, boring very frequently is done on such machines. However, several machine tools have been developed primarily for boring, especially where large workpieces are involved or for large-volume boring of smaller parts. Some of these are also capable of performing other operations, such as milling and turning. Because boring frequently follows drilling, many boring machines also can do drilling, permitting both operations to be done with a single setup of the work.

Vertical boring and turning machines. Figure 21-2 shows the basic elements of a vertical boring and turning machine, and an actual machine is shown in Figure 21-3. It will be noted that these structurally are similar to double-housing planers, except that the table rotates instead of reciprocating. Functionally, a vertical boring machine essentially is the same as a vertical turret lathe, but it usually has two main tool heads instead of a turret. Thus turning, facing, and usually drilling also can be done on these machines, which often are called *vertical boring mills.*

Vertical boring machines customarily are used when holes larger than about 305 mm (12 inches) are to be bored or where the workpiece is so large,

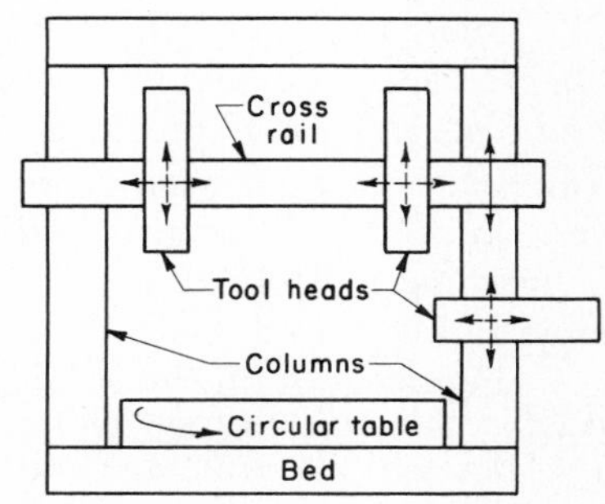

FIGURE 21-2. Block diagram of the basic components of a vertical boring machine.

FIGURE 21-3. Large vertical boring machine. (*Courtesy The Farrel Corporation.*)

or of such shape, that it would be difficult to hold and rotate about a horizontal axis on a lathe.

Vertical boring machines come with tables ranging from about 900 to 12200 mm (3 to 40 feet) in diameter. Usually, two toolheads are mounted on an elevatable cross rail, and these are provided with both horizontal and vertical feed, so they can be used for boring and for facing cuts. Usually, one or both can also be swiveled about a horizontal axis to permit boring at an angle. Most machines also have a side toolhead, sometimes provided with a four-sided turret. This toolhead has vertical and horizontal feed and is used primarily for turning. Although single-point tools customarily are used, three or more cuts can be made simultaneously. Thus turning, facing, and boring, or roughing and finishing cuts can be done at the same time.

Many modern boring machines are equipped with numerical control (see Chapter 39). This permits the operator to make tool settings merely by setting dials and also to preset the adjustment for a cut while one is being made. The pressing of a button at the conclusion of a cut causes the tool to move very quickly to the proper position for the next cut. This reduces the amount of machine-controlling time and thereby increases the productivity of such large and costly machines.

Figure 21-4 shows an important variation in vertical boring machine design, in which the rotary table can be moved horizontally with respect to

FIGURE 21-4. Vertical boring machine having provision for moving the table horizontally to accommodate workpieces of widely different sizes. (*Courtesy Société Anonyme Des Anciens Etablissements Charles Berthiez.*)

the single, massive column and cross rail. This permits a large range of sizes of workpieces to be machined while minimizing the amount of overhang of the toolhead on the cross rail.

The size of vertical boring machines is designated by the diameter of the worktable, expressed in millimeters, feet, or inches.

Automatic vertical boring machines. *Automatic vertical boring machines* are available for use in quantity machining of relatively small parts. One is shown in Figure 21-5. After the machine is set up during the making of the first workpiece, it will automatically make the required settings and tool adjustments for succeeding duplicate parts.

On another type the workpiece remains stationary while one or more boring tools are rotated by sliding power heads.[1] Once the machine is set in motion, the boring is carried out automatically, roughing and finishing cuts often being done in sequence. On such machines, several holes can be bored simultaneously, as in an automobile engine cylinder block.

[1] See Chapter 38.

FIGURE 21-5. Automatic vertical boring machine. (*Courtesy Ex-Cell-O Corporation.*)

Horizontal boring, drilling, and milling machines. *Horizontal boring, drilling, and milling machines* are very versatile and thus particularly useful in machining large parts. The basic components of these machines are indicated in Figure 21-6, and an actual machine in use is shown in Figure 21-7.[2] The essential features are as follows:

1. A table that can be moved and fed in two directions in a horizontal plane.
2. A headstock that can be moved vertically.
3. A rotating spindle that can be fed horizontally.
4. An outboard bearing support for a long boring bar.

These features provide accurately controlled, relative motion between the work and a rotating tool in three mutually perpendicular directions. Thus in addition to boring, drilling, and reaming, longitudinal, transverse, and vertical milling cuts can be made. In addition, accurate layout can be accomplished in conjunction with these several machining operations with a single setup and clamping of large, heavy workpieces.

[2] The reader will find it interesting to compare Figure 21-7 with Figure 1-11.

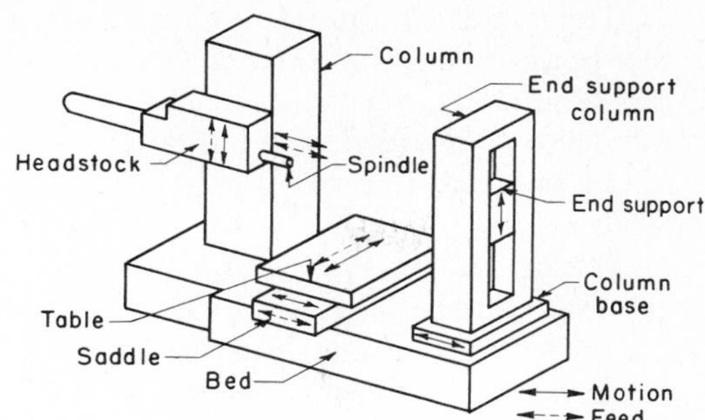

FIGURE 21-6. Block diagram showing the basic components and motions of a horizontal boring, drilling, and milling machine.

FIGURE 21-7. Boring a weldment on a horizontal boring, drilling, and milling machine. A line-type boring bar is being used with an outboard bearing support. (*Courtesy Lucas Machine Division, The New Britain Machine Company.*)

The spindle is similar to an oversized drilling-machine spindle and will accept both drills and milling cutters. A wide range of speeds is provided, and heavy bearings are incorporated that will absorb thrust in all directions. The spindle also is provided with longitudinal power feed so that drilling and boring can be done through a considerable distance without the table being moved.

Boring on this type of machine is done by means of a rotating single-point tool. The tool can be mounted in either a stub-type bar, held only in the spindle, or in a long line-type bar that has its outer end supported in a bearing on the outboard column, as shown in Figure 21-7. The outboard bearing provides rigid support for the boring bar and permits very accurate work to be done. However, because of the flexibility inherent in a long boring bar and offset tool holder, horizontal boring machines are used primarily for boring holes less than 305 mm (12 inches) in diameter, or for long holes, or for a series of in-line holes. Unless they are very long or unless the shape of the workpiece does not permit, larger holes usually are bored more readily on a vertical boring mill.

Although slab milling can be done, face milling is more commonly done on these machines.

Because a sequence of tools frequently is used in the spindle, special devices often are used to reduce the time for changing tools. The turret head,

FIGURE 21-8. Turret head for horizontal boring, drilling, and milling machine. This head will hold eight tools. (*Courtesy Innocenti.*)

shown in Figure 21-8, is an example. Eight tools can quickly be brought into operation by indexing the turret head.

Several special types of horizontal boring, drilling, and milling machines are built. On *floor-type machines* the work is mounted on a fixed base, and the two columns move along runways that parallel the base. *Planer-type machines* have a reciprocating table. *Multiple-head machines* are essentially the same as a double-housing planer, but with boring-machine headstocks substituted for the regular planer toolheads.

The size of horizontal boring machines is designated by the diameter of the spindle in millimeters or inches. This obviously tells very little about the size of the work that can be done on the machine, and supplementary dimensions, such as table size and the distance between the spindle and the end-support column, usually are given so as to indicate the size of the workpiece that can be machined.

Mass-production boring machines. Special boring machines are built for machining specific parts in mass production. In these the workpiece usually remains stationary, and boring is done by one or more rotating boring tools, typically carried in a reciprocating powerhead, such as is shown in Figure 21-9. In most cases the operation is automatic once the workpiece is placed in the fixture. Such machines usually are very accurate and often are equipped with automatic gaging and sizing controls.

FIGURE 21-9. (*Left*) Production-type boring machine, having multiple heads, that completes a part in 51 seconds. (*Above*) Close-up view of one multiple-spindle boring head on a production-type machine. (*Courtesy The Heald Machine Company.*)

FIGURE 21-10. Adjustable boring bar, using the offset-radius principle.

Boring tools. Several types of tools are used for boring. Two simple, general-purpose types were discussed in Chapter 20 and illustrated in Figures 20-20 and 20-37. These are used with rotating workpieces, and the size of the hole is controlled by transverse movement of the toolholder. When boring is done with a rotating tool, size is controlled by changing the offset radius of the cutting-tool tip with respect to the axis of rotation. A general-purpose type of adjustable boring bar is shown in Figure 21-10. The type shown in Figure 21-11 has more precise control and is used in larger-scale manufacturing. It also can be obtained with two or more adjustable cutting tools on a single bar,

FIGURE 21-11. Adjustable boring tool. Extension of the single-point tool from the bar is adjustable, as shown in sectional view. (*Courtesy DeVlieg Machine Company.*)

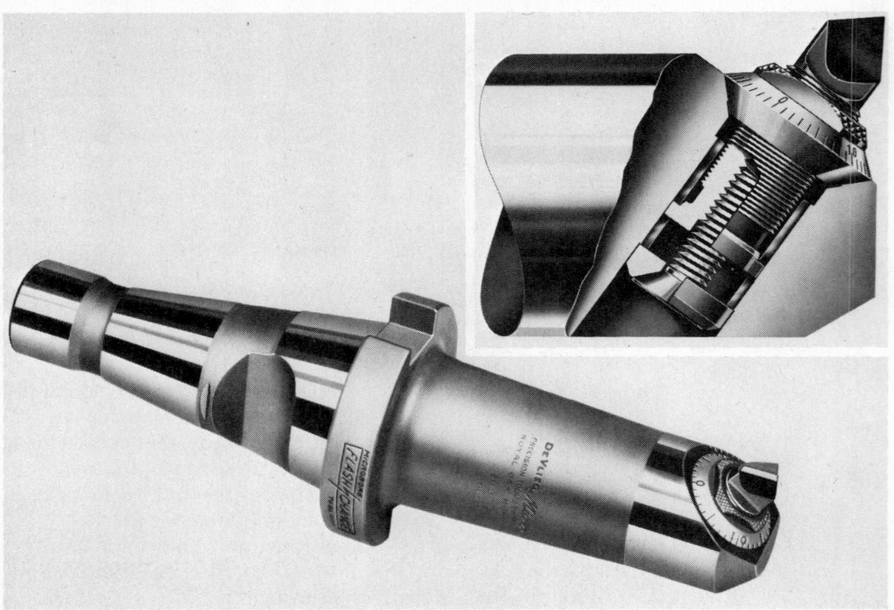

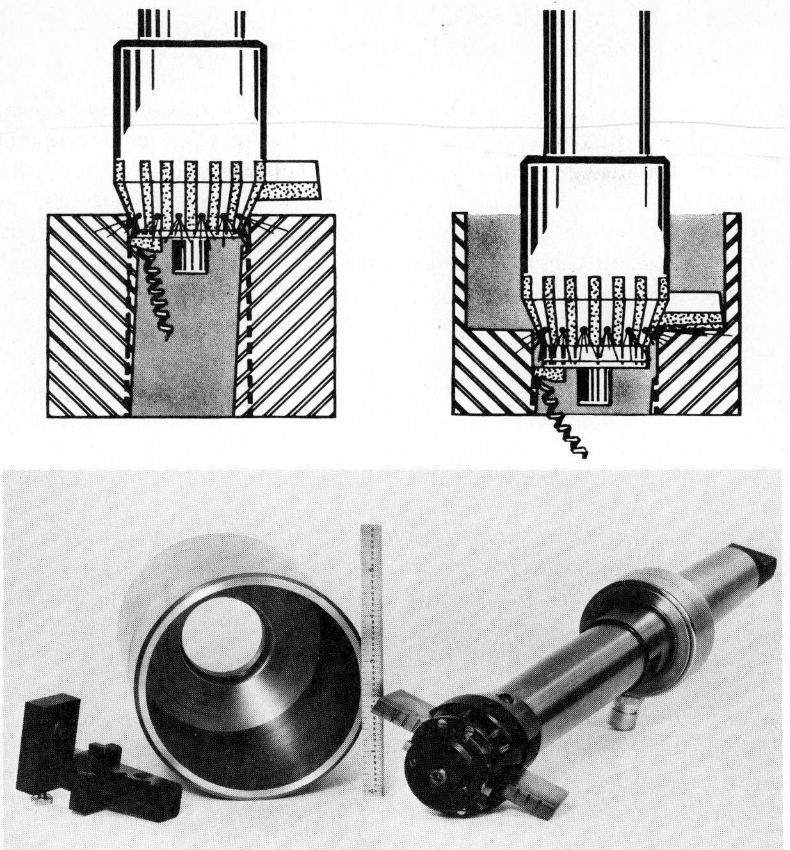

FIGURE 21-12. Boring tool employing a centering tool and conical guide, for boring large holes in a single operation. (*Courtesy Vernon Devices, Inc.*)

thus permitting more than one diameter to be bored simultaneously. For boring relatively long holes, the type of boring bar shown in Figure 21-12 has a special advantage. As shown, the smaller, forward bit corrects misalignment of the original hole and provides a guide hole for the nose cone. The nose cone then provides good alignment and support for the rear bit, which bores the final hole to size.

Boring machine accuracy. Because boring is essentially the same as turning, the accuracy obtainable is similar except for the fact that the tool support may be less rigid. Thus the accuracy depends considerably on the rigidity of the tool support. On specialized, production-type boring machines, tolerances are readily held to within 0.013 mm (0.0005 inch) on small diameters, whereas on general-purpose machines tolerances of 0.025 mm (0.001 inch) are typical unless the boring bar overhang becomes excessive.

Jig borers. *Jig borers* are very accurate vertical-type boring machines designed for use in making jigs and fixtures. From the viewpoint of boring operations they contain no unusual features, except that the spindle and spindle bearings are constructed with very high accuracy. Their unique features are in the design of the worktable controls, which permits very precise movement and control, thus making them especially useful in layout work. These machines will be discussed in Chapter 35. However, the accuracy of many modern machining centers (see Chapter 39) is such that these, to a considerable extent, have taken the place of jig borers.

Review questions

1. What are the two objectives of boring?
2. Why does boring assure concentricity of the hole axis and the axis of rotation of the workpiece (or boring tool) whereas drilling does not?
3. What operations can be done on a horizontal boring machine without the use of special equipment?
4. Why are vertical boring mills better suited for machining large workpieces than a lathe?
5. What feature do horizontal boring, drilling, and milling machines and Wilkinson's boring machine have in common?
6. What is the principal advantage of a horizontal boring machine over a vertical boring machine for large workpieces?
7. What accuracy ordinarily can be obtained on horizontal boring machines and vertical boring mills?
8. What is the deficiency of the simple method of size designation that commonly is given for horizontal boring, drilling, and milling machines?
9. How does a double-column vertical boring machine differ basically from a planer?
10. Explain why a horizontal boring, drilling, and milling machine is such an important tool for machining very large workpieces.
11. A hole 89 mm in diameter is to be drilled and bored through a piece of gray cast iron that is 200 mm long, using a horizontal boring, drilling, and milling machine. High-speed tools will be used. The job will be done by center drilling, drilling with a 18-mm drill, followed by a 76-mm drill, then bored to size in one cut, using a feed of 0.50 mm/rev. Drilling feeds will be 0.25 mm/rev for the smaller drill and 0.64 mm/rev for the larger drill. The center drilling operation requires ½ minute. To set or change any given tool and set the proper machine speed and feed requires 1 minute. Use Table 17-1 to select cutting speeds, and compute the total time required for doing the job. (Neglect setup time for the workpiece.)

CHAPTER 22

Milling

Milling is a basic machining process by which a surface is generated progressively by the removal of chips from a workpiece as it is fed to a rotating cutter in a direction perpendicular to the axis of the cutter. In some cases the workpiece remains stationary, and the cutter is fed to the work. In nearly all cases, a multiple-tooth cutter is used so that the material removal rate is high. Often the desired surface is obtained in a single pass of the cutter or work and, because very good surface finish can be obtained, milling is particularly well suited, and widely used, for mass-production work. Several types of milling machines are used, ranging from relatively simple and versatile machines that are used for general-purpose machining in job shops and tool-and-die work to highly specialized machines for mass production. Unquestionably, more flat surfaces are produced by milling than by any other machining process.

The tool used in milling is known as a *milling cutter.* It usually consists of a cylindrical body which rotates on its axis and contains equally spaced pe-

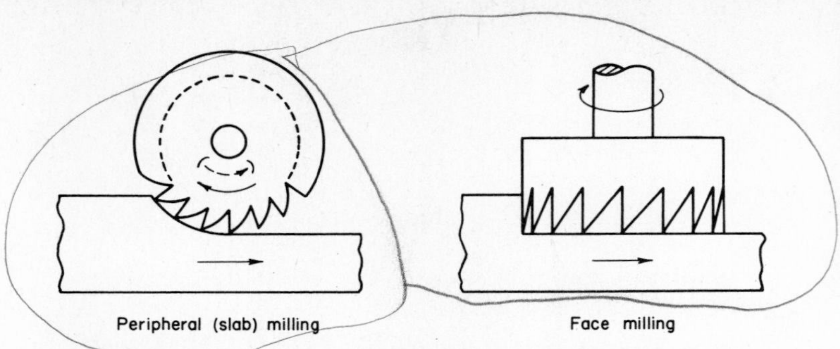

FIGURE 22-1. Tool–work relationships in peripheral and face milling.

ripheral teeth that intermittently engage and cut the workpiece (see Figure 22-4). In some cases the teeth extend partway across one or both ends of the cylinder.

Types of milling operations. Milling operations can be classified into two broad categories, each having some variations. The basic concepts of these two types are indicated in Figure 22-1.

In *peripheral milling* a surface is generated by teeth located on the periphery of the cutter body. The surface is parallel with the axis of rotation of the cutter. Both flat and formed surfaces can be produced by this method, the cross section of the resulting surface corresponding to the axial contour of the cutter. This process often is called *slab milling*.

In face milling the generated surface is at right angles to the cutter axis and is the combined result of the actions of the portions of the teeth located on both the periphery and the face of the cutter. Most of the cutting is done by the peripheral portions of the teeth, with the face portions providing some finishing action. Peripheral milling operations usually are performed on machines having horizontal spindles, whereas face milling is done on both horizontal- and vertical-spindle machines.

The generation of surfaces in milling. In milling, surfaces can be generated by two distinctly different methods, illustrated in Figure 22-2. In *up milling,* the cutter rotates against the direction of feed of the workpiece, whereas in *down milling* the rotation is in the same direction as the feed. As shown in Figures 22-2 and 22-3, the method of chip formation is completely different in the two cases. In up milling the chip is very thin at the beginning, where the tooth first contacts the work, and increases in thickness, becoming a maximum where the tooth leaves the work. The cutter tends to push the work along and lift it upward from the table. This action tends to eliminate any effect of looseness in the feed screw and nut of the milling machine table and results in a smooth cut. However, the action also tends to loosen the work from the clamping device so that greater clamping forces must be employed.

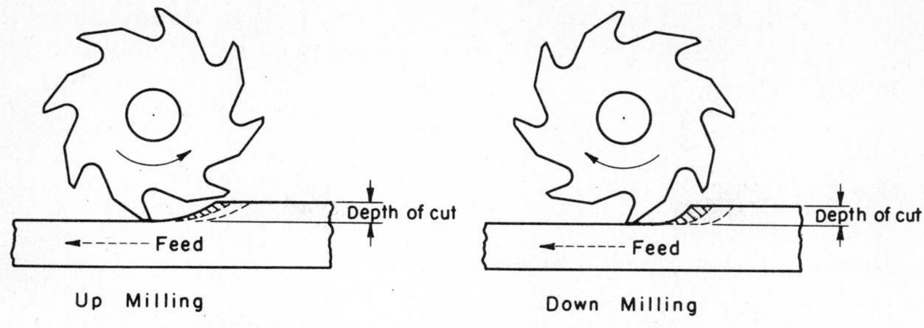

FIGURE 22-2. Cutting action in up milling and down milling.

In addition, the smoothness of the generated surface depends greatly on the sharpness of the cutting edges.

In down milling, maximum chip thickness occurs close to the point at which the tooth contacts the work. Because the relative motion tends to pull the workpiece into the cutter, all possibility of looseness in the table feed screw must be eliminated if down milling is to be used. It should never be attempted on machines that are not designed for this type of milling; virtually all modern milling machines are so equipped. Because the material yields in approximately a tangential direction at the end of the tooth engagement, there is less tendency for the machined surface to show toothmarks than when up milling is used. Another advantage of down milling is that the cutting force tends to hold the work against the machine table, permitting lower clamping forces. However, the fact that the cutter teeth strike against the sur-

FIGURE 22-3. Chip formation in up milling and down milling. Photos were taken at high magnification. (*Courtesy Cincinnati Milacron Inc.*)

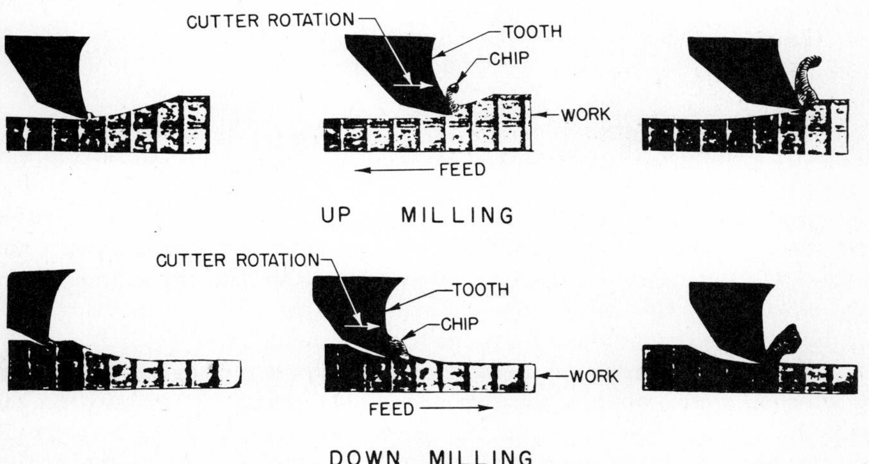

face of the work at the beginning of each chip can be a disadvantage if the workpiece has a hard surface, as castings sometimes do. This may cause the teeth to dull rapidly.

MILLING CUTTERS

Milling cutters can be classified in several ways. One is to group them into two broad classes, based on tooth relief, as follows:

1. On *profile cutters,* relief is provided on each tooth by grinding a small land back of the cutting edge. The cutting edge may be straight or curved.

2. On *form* or *cam-relieved cutters,* the cross section of each tooth is an eccentric curve behind the cutting edge, thus providing relief. All sections of the eccentric relief, parallel with the cutting edge, have the same contour as the cutting edge. Cutters of this type are sharpened by grinding only the face of the teeth; the contour of the cutting edge thus remains unchanged.

Another useful method of classification is according to the manner of mounting the cutter. *Arbor cutters* have a center hole so they can be mounted on an arbor. *Shank cutters* have either a tapered or straight integral shank. Those with tapered shanks can be mounted directly in the milling machine spindle, whereas straight-shank cutters are held in a chuck. *Facing cutters* usually are bolted to the end of a stub arbor. Common types of milling cutters, classified in this manner, are as follows:

Arbor Cutters	Shank Cutters
Plain	End mills
Side	Solid
Staggered-tooth	Inserted-tooth
Slitting saws	Shell
Angle	Hollow
Inserted-tooth	T-slot
Form	Woodruff key seat
Fly	Fly

Figures 22-4 and 22-5 show several types of arbor-type and shank-type milling cutters, respectively.

Another method of classification applies only to face and end-mill cutters and relates to the direction of rotation. A *right-hand cutter* must rotate counterclockwise when viewed from the front end of the machine spindle. Similarly, a *left-hand cutter* must rotate clockwise. All other cutters can be reversed on the arbor to change them from one hand to the other.

Positive rake angles are used on general-purpose milling cutters, such as in job-shop work, whereas negative rake angles commonly are used on carbide- and ceramic-tipped cutters that are employed in mass-production milling, in order to obtain the greater strength and cooling capacity which they

FIGURE 22-4. Arbor-type milling cutters: (*top*) side, plain, staggered-tooth; (*center*) angle, fly; (*bottom*) metal-slitting saw, inserted-tooth, form.

FIGURE 22-5. Shank-type milling cutters: (*top*) T-slot, shell end mill, Woodruff key seat, hollow end mill, solid end mill; (*bottom*) fly cutter.

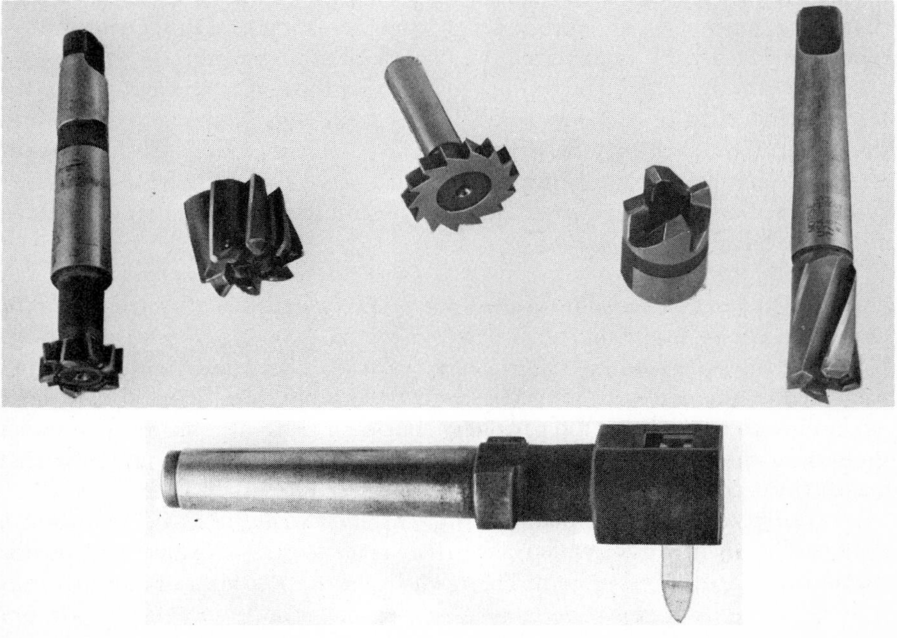

FIGURE 22-6. Manner in which chips are formed pregressively by the teeth of a plain helical-tooth milling cutter in up milling. (*Courtesy Cincinnati Milacron Inc.*)

provide. Occasionally dual rake angles are used—a short negative rake face adjacent to the cutting edge, followed by a longer face having a positive rake angle.

Types of milling cutters. *Plain milling cutters* are cylindrical or disk-shaped, have straight or helical teeth on the periphery, and are used for milling flat surfaces. This type of operation is called *plain,* or *slab milling.* As shown in Figure 22-6, each tooth in a helical cutter engages the work gradually, and usually more than one tooth cuts at a given time. This reduces shock and chattering tendencies and promotes a smoother surface. Consequently, this type of cutter usually is preferred over one with straight teeth.

Side milling cutters are similar to plain milling cutters except that the teeth extend radially part way across one or both ends of the cylinder toward the center. The teeth may be either straight or helical. Frequently, these cutters are relatively narrow, being disklike in shape. Two or more side milling cutters often are spaced on an arbor to make simultaneous, parallel cuts, in an operation called *straddle milling.*

Interlocking slotting cutters consist of two cutters similar to side mills, but made to operate as a unit for milling slots. The two cutters are adjusted to the desired width by inserting shims between them.

Staggered-tooth milling cutters are narrow cylindrical cutters having staggered teeth, and with alternate teeth having opposite helix angles. They are ground to cut only on the periphery, but each tooth also has chip clearance ground on the protruding side. These cutters have a free cutting action that makes them particularly effective in milling deep slots.

Slitting saws are thin, plain milling cutters, usually from 0.8 to 4.8 mm ($^1/_{32}$ to $^3/_{16}$ inch) thick, which have their sides slightly "dished" to provide clearance and prevent binding. They usually have more teeth per unit of diameter than ordinary plain milling cutters and are used for milling deep, narrow slots and for cutting-off operations.

Angle milling cutters are made in two types—single-angle and double-angle. *Single-angle cutters* have teeth on the conical surface, usually at an angle of 45 to 60° to the plane face. The teeth may also extend radially on the larger plain face. *Double-angle cutters* have **V**-shaped teeth, with both conical surfaces at an angle to the end faces, but not necessarily at the same angle. The **V**-angle usually is 45, 60, or 90°. Angle cutters are used for milling slots of various angles or for milling the edges of workpieces to a desired angle.

Most larger-sized milling cutters are of the *inserted-tooth type.* The cutter body is made of ordinary steel, with the teeth made of high-speed steel, cemented carbide, or ceramic, fastened to the body by various methods. Most commonly, the teeth are throwaway, indexable carbide or ceramic inserts, as shown in Figure 22-7. This type of construction reduces the amount of costly material that is required and can be used for any type of cutter but most often is used with face mills.

Form milling cutters have the teeth ground to a special shape—usually an irregular contour—to produce a surface having a desired transverse contour. They are cam-relieved and are sharpened by grinding only the tooth face, thereby retaining the original contour as long as the plane of the face remains unchanged with respect to the axis of rotation. Convex, concave, corner-rounding, and gear-tooth cutters are common examples.

End mills are shank-type cutters having teeth on the circumferential surface and one end. They thus can be used for facing, profiling, and end milling. The teeth may be either straight or helical, but the latter is more common. Small end mills have straight shanks, whereas taper shanks are used on larger sizes.

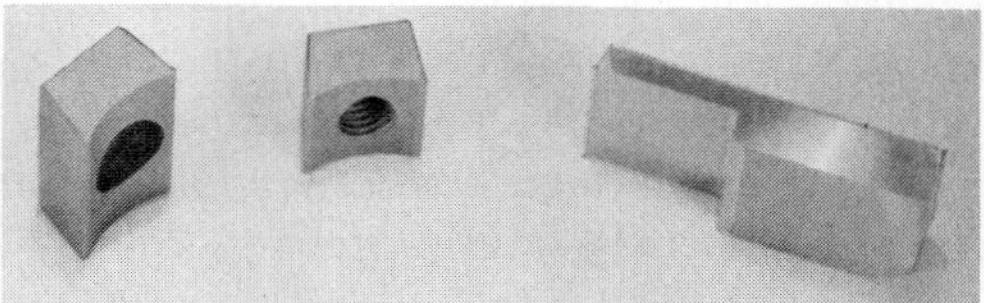

FIGURE 22-7. (*Left*) Inserted-tooth milling cutter, using throwaway tungsten-carbide cutting inserts and locking wedges. (*Above*) Locking wedges and insert support. (*Courtesy Lovejoy Tool Company, Inc.*)

Plain end mills have multiple teeth that extend only about halfway toward the center on the end. They are used in milling slots, profiling, and facing narrow surfaces. *Two-lip end mills* have two straight or helical teeth that extend to the center. Thus they may be sunk into material, like a drill, and then fed lengthwise to form a groove.

Shell end mills are solid-type, multiple-tooth cutters, similar to plain end mills but without a shank. The center of the face is recessed to receive a screw head or nut for mounting the cutter on a separate shank or a stub arbor. The back of the cutter contains driving slots that engage collar keys on the shank. This design enables one shank to hold any of several cutters and thus provides great economy for larger-sized end mills. The cutter shown in Figure 22-7 is of this type.

Hollow end mills are tubular in cross section, with teeth only on the end but having internal clearance. They are used primarily on automatic screw machines for sizing cylindrical stock, producing a short, cylindrical surface of accurate diameter.

T-*slot cutters* are integral-shank cutters with teeth on the periphery and *both* sides. They are used for milling the wide groove of a T-slot. In order to use them, the vertical groove must first be made with a slotting mill or an end mill to provide clearance for the shank. Because the T-slot cutter cuts on five surfaces simultaneously, it must be fed with care.

Woodruff keyseat cutters are made for the single purpose of milling the semicylindrical seats required in shafts for Woodruff keys. They come in standard sizes corresponding to Woodruff key sizes. Those below 50.8 mm (2 inches) in diameter have integral shanks; the larger sizes may be arbor-mounting.

Occasionally, *fly cutters* may be used as shown in Figure 22-5; these have a single-point cutting tool attached to a special shank, usually with provision for adjusting the effective radius of the cutting tool with respect to the axis of rotation. The cutting edge can be made in any desired shape and, because it is a single-point tool, is very easy to grind. These cutters can be used for face milling and also for boring, and frequently are used where both of these operations need to be done with a single tool at one setup of the work. They are used primarily in experimental and toolroom work.

MILLING MACHINES

Because the milling process is versatile and highly productive, a variety of machines have been developed to employ the milling principle. One type, commonly called *milling machines,* are basic, general-purpose machines that provide a high degree of flexibility. Another type, used exclusively for reproducing parts from templates or patterns, often are called *duplicators.* A third type encompasses *special-purpose machines* that are used in mass-production manufacturing. These will be discussed in Chapter 38. Machines of the fourth type are highly versatile but are designed to do other basic machining operations as well as milling. These commonly are called *machining centers* and will be discussed in Chapter 39.

Inasmuch as basic milling machines provide an accurate, rugged, rotating spindle, they can also be used for other machining operations, such as drilling and boring. Consequently, because they can do several types of operations and produce several types of surfaces with a single setup of the workpiece, they are among the most important of machine tools. The common types may be classified according to their general design characteristics as follows:

1. Column-and-knee type (general purpose)
   a. Plain
      (1) Power table feed
      (2) Hand table feed
   b. Universal
   c. Vertical
   d. Ram-type universal
2. Bed type (manufacturing)
   a. Simplex
   b. Duplex
   c. Triplex
3. Planer type (large work only)
4. Special
   a. Rotary table
   b. Drum type
   c. Profilers
   d. Duplicators

Basic milling-machine construction. Most basic milling machines are of column-and-knee construction, employing the components and motions shown in Figure 22-8. The column, mounted on the base, is the main supporting frame for all the other parts and contains the spindle with its driving mechanism. As indicated, this construction provides controlled motion of the worktable in three mutually perpendicular directions: (1) through the *knee* moving vertically on ways on the front of the column, (2) through the *saddle* moving transversely on ways on the knee, and (3) through the *table* moving longitudinally on ways on the saddle. All these motions can be imparted either by manual or powered means. In most cases, a powered rapid traverse is provided in addition to the regular feed rates for use in setting up work and in returning the table at the end of a cut.

Milling machines having only the three mutually perpendicular table motions just described are called *plain column-and-knee type*. These are available with both horizontal and vertical spindles, shown in Figures 22-9 and 22-10, respectively. On the horizontal type, an adjustable over-arm is mounted on

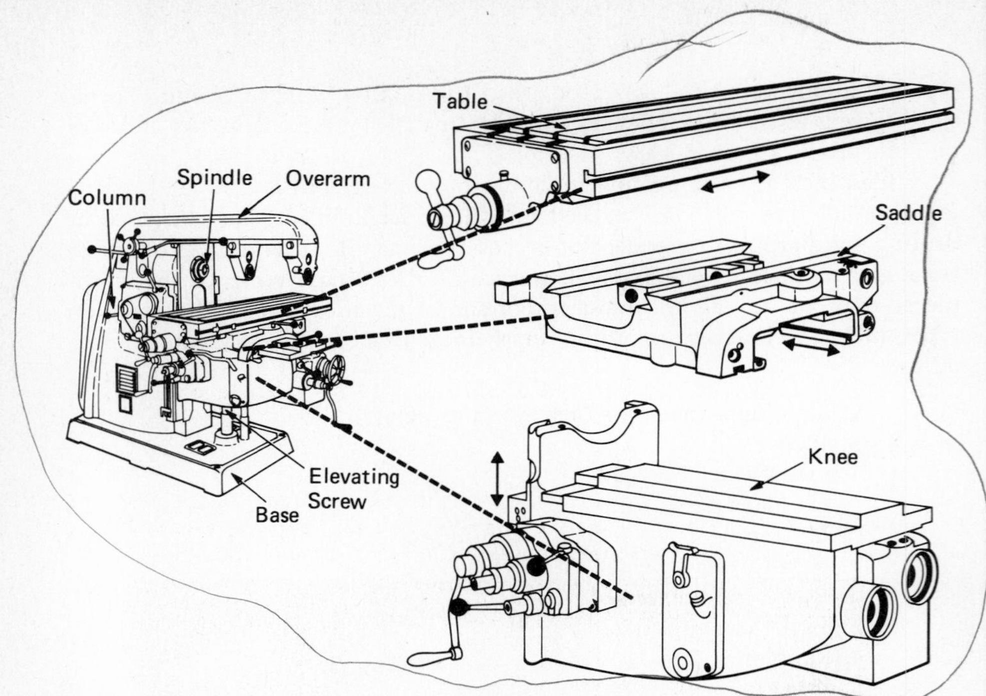

FIGURE 22-8. Major components of a plain column-and-knee-type milling machine. (*Courtesy Cincinnati Milacron Inc.*)

the top of the column to provide an outboard bearing support for the end of the cutter arbor, when required. In some vertical-spindle machines the spindle is mounted in a sliding head that can be fed up and down either by power or by hand. Vertical-spindle machines are especially well suited for face- and end-milling operations. They also are very useful for drilling and boring, particularly where holes must be spaced accurately in a horizontal plane, because of the controlled table motion.

Universal column-and-knee milling machines. *Universal column-and-knee milling machines* differ from plain column-and-knee machines in that the table is mounted on a housing that can be swiveled in a horizontal plane, thereby increasing its flexibility so as to permit the milling of helices, as found in twist drills, milling cutters, and helical gear teeth. This type of machine is shown in Figure 22-11.

Ram-type milling machines. *Ram-type column-and-knee milling machines,* as shown in Figure 22-12, have a spindle head that can be swiveled about a horizontal axis near the end of a horizontally adjustable ram. This permits milling to be done horizontally, vertically, or at any angle. This added flexibility is advantageous where a variety of work has to be done, as in tool and die or experimental shops. They are available with either plain or universal tables.

Milling machine size. Milling machine size is designated by numbers from 1 through 6, which are approximate indicators of the longitudinal table travel as follows:

Size		1	2	3	4	5	6
Table travel	(in.)	22	28	32	40	50	60
	(mm)	559	711	812	1016	1270	1524

These relationships are not standardized, and a particular machine may vary considerably from these values.

FIGURE 22-9. Plain column-and-knee-type horizontal milling machine performing a straddle or gang milling operation. (*Courtesy Cincinnati Milacron Inc.*)

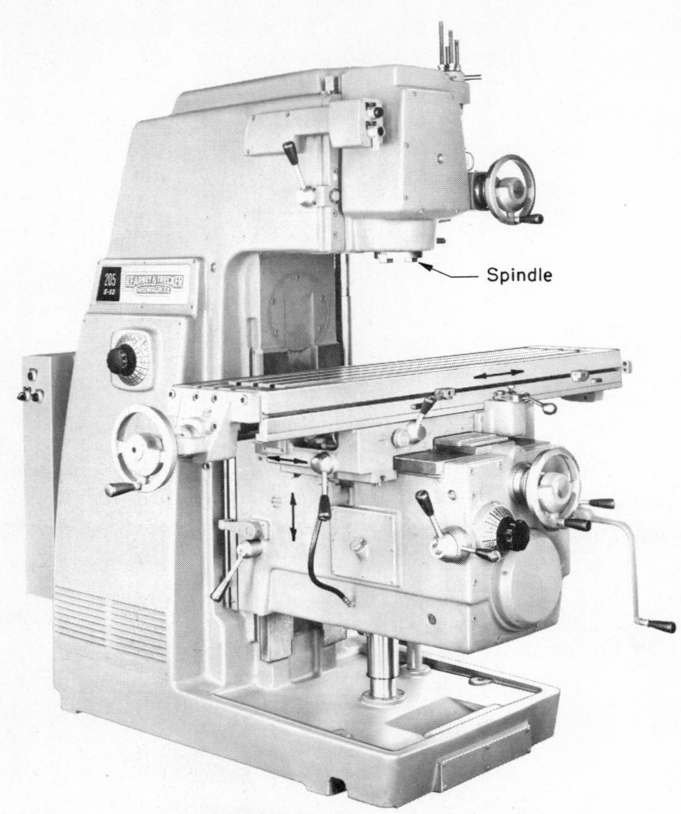

FIGURE 22-10. High-speed, vertical column-and-knee-type milling machine (*Courtesy Kearney & Trecker Corporation.*)

Hand-feed milling machines. *Hand-feed milling machines are* small, simple machines on which the table is fed longitudinally by means of a hand lever which rotates a pinion that engages a rack on the bottom of the table. Such machines are used for light milling of short slots, grooves, and so forth on small parts.

Bed-type milling machines. In production manufacturing operations, ruggedness and the capability of making heavy cuts are of more importance than versatility. *Bed-type milling machines,* such as shown in Figure 22-13, are made for these conditions. The table is mounted directly on the bed and has only longitudinal motion. The spindle head can be moved vertically in order to set up the machine for a given operation. Normally, once the setup is completed, the spindle head is clamped in position and no further motion of it occurs during machining. However, on some machines vertical motion of the spindle occurs during each cycle.

FIGURE 22-11. Milling the flutes in a helical milling cutter on a universal column-and-knee-type milling machine. A universal dividing head is mounted on the right-hand end of the table to support and turn the workpiece. (*Courtesy Cincinnati Milacron Inc.*)

FIGURE 22-12. Ram-type milling machine with the head positioned for vertical milling. (*Courtesy Van Norman Machine Company.*)

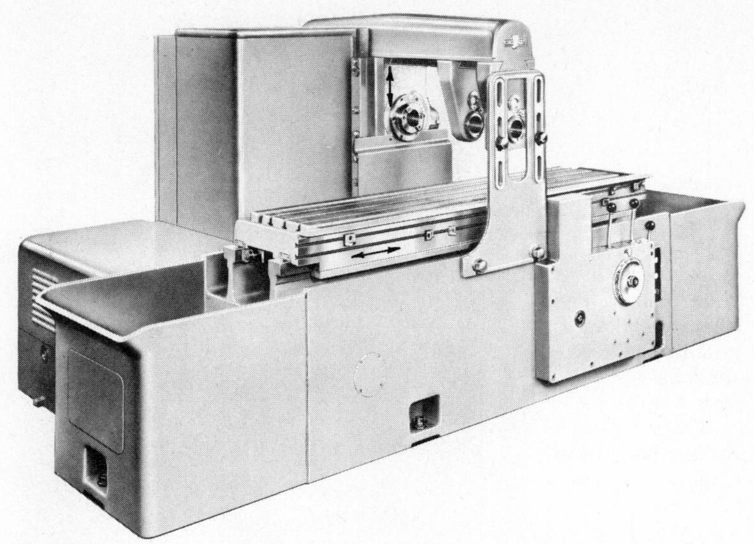

FIGURE 22-13. Bed-type milling machine. (*Courtesy Cincinnati Milacron Inc.*)

FIGURE 22-14. Triplex milling machine for face milling three surfaces simultaneously. (*Courtesy Cincinnati Milacron Inc.*)

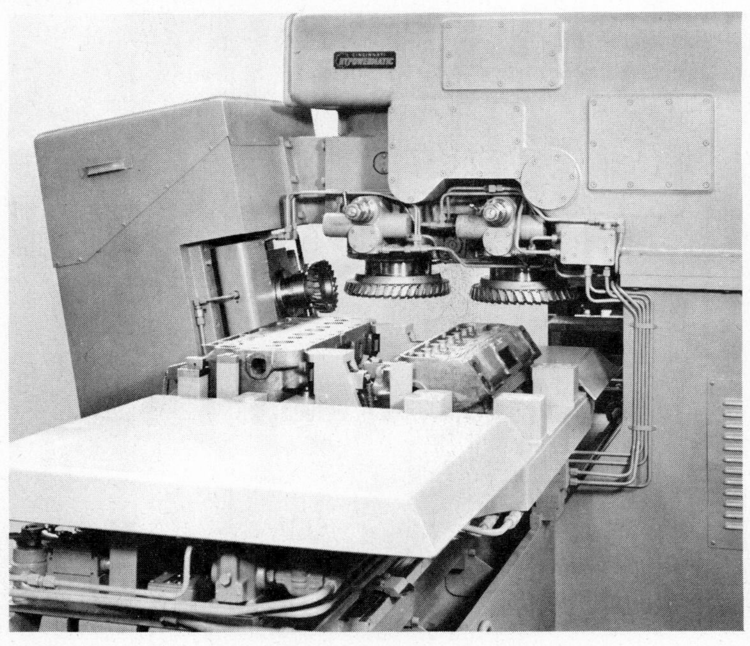

After such milling machines are set up, little skill is required to operate them. This permits the use of semiskilled operators. Some machines of this type are equipped with automatic controls so that the only activity required of the operator is to put the workpiece into a fixture and set the machine into operation. Often a fixture is provided at each end of the table so that one workpiece can be unloaded while another is being machined.

Bed-type milling machines with single spindles sometimes are called *simplex milling machines;* they are made with both horizontal and vertical spindles. Bed-type machines also are made in *duplex* and *triplex* types, having two and three spindles, respectively. These permit the simultaneous milling of two or three surfaces at a single pass. Figure 22-14 shows a setup of a triplex milling machine.

Figure 22-15 illustrates the trend to provide both flexibility and the rigidity required for production-type operations. By adding transverse table motion to the table of a bed-type milling machine, some of the flexibility of a plain column-and-knee-type machine is obtained.

FIGURE 22-15. VerciPower milling machine, which combines the rigidity of the bed type with most of the flexibility of the column-and-knee type. (*Courtesy Cincinnati Milacron Inc.*)

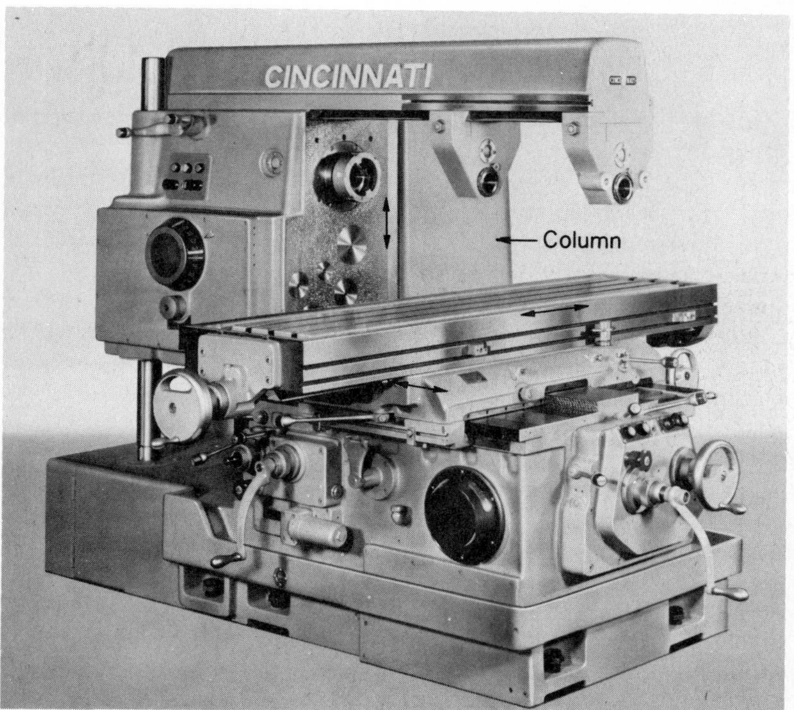

FIGURE 22-16. Large planer-type milling machine. (*Inset*) 90° head being used. (*Courtesy Cosa Corporation.*)

Planer-type milling machines. As was pointed out in Chapter 18, planers have two serious disadvantages in that they utilize only single-point cutting tools, and a large table and heavy workpiece cannot be reciprocated rapidly. Consequently, they have largely been replaced by *planer-type milling machines,* which, as illustrated in Figure 22-16, utilize several milling heads, which can remove large amounts of metal while permitting the table and workpiece to move quite slowly. Often only a single pass of the workpiece past the cutters is required. Through the use of different types of milling heads and cutters, a wide variety of surfaces can be machined with a single setup of the workpiece. This is a great advantage where heavy workpieces are involved.

Rotary-table milling machines. Some types of face milling in mass-production manufacturing are often done on *rotary-table milling machines,* such as is shown in Figure 22-17. Roughing and finishing cuts can be made in succession as the workpieces are moved past the several milling cutters while held in fixtures on the rotating table. The operator can load and unload the work without stopping the machine.

Profilers and duplicators. Milling machines that can duplicate external or internal profiles in two dimensions are called *profilers.* As shown in Figure 22-

FIGURE 22-17. Two-spindle 48-inch rotary-table milling machine being used to rough and finish the surface of automobile timing-gear covers. Eighty covers per hour are machined. (*Courtesy The Ingersoll Milling Machine Company.*)

18, a tracing probe follows a two-dimensional template and, through electronic or air-actuated mechanisms, controls the cutting spindles in two mutually perpendicular directions. The spindles—usually more than one—are set manually in the third dimension.

Duplicators reproduce forms in three dimensions, as illustrated in Figures 22-19 and 22-20. A tracing probe follows a three-dimensional master. Often the probe does not actually contact the master, a variation in the length of a spark between the probe and the master controlling the drives to the quill and the table, thereby avoiding wear on the master or possible deflection of the probe. On some machines, the ratio between the movements of the probe and cutter can be varied.

Duplicators are widely used to machine molds and dies and sometimes are called *die-sinking machines*. They also are used extensively in the aerospace industry to machine parts from wrought plate or bar stock as substitutes for forgings, where the small number required would make the cost of forging dies uneconomical.

To a great extent, ordinary duplicators and profilers are being replaced by numerically controlled machines in which a punched tape or computer input eliminates the necessity of making a template or master mold.

FIGURE 22-18. Four-spindle profile milling machine. Note the two-dimensional template at the right end of the table and a completed part leaning against one of the fixtures and shown in the inset. (*Courtesy Cincinnati Milacron Inc.*)

FIGURE 22-19. Hydro-tel duplicating milling machine. The three-dimensional templates are mounted in the vise, three typical templates being shown below the vise. In this case the cutter is duplicating the design on a cylindrical surface. (*Courtesy Cincinnati Milacron Inc.*)

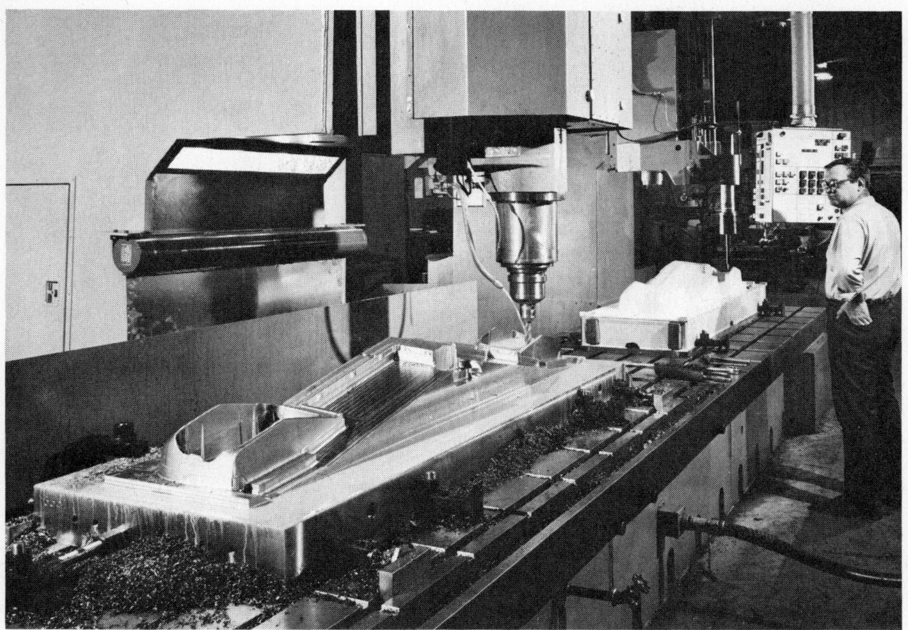

FIGURE 22-20. Machining a die block on a three-dimensional duplicating milling machine, using a plster model shown at the right. (*Courtesy Pratt & Whitney Machine Tool Division, Colt Industries.*)

Accessories for milling machines. The usefulness of ordinary milling machines is greatly extended by employing various accessories.

The *vertical milling attachment,* shown in Figure 22-21, is used on a horizontal milling machine to permit vertical milling to be done. Ordinarily, only moderately heavy work can be done with such an attachment.

The *universal milling attachment,* shown in Figure 22-22, is similar to the vertical attachment but can be swiveled about both the axis of the milling machine spindle and a second, perpendicular axis to permit milling to be done at any angle.

The *slotting attachment,* illustrated in Figure 22-23, permits adapting a horizontal milling machine to the work of a vertical shaper. Although not used extensively, it is useful for cutting small keyways.

The *universal dividing head* is by far the most widely used milling machine accessory, providing a means for holding and indexing work through any desired arc of rotation. The work may be mounted between centers, as shown in Figure 22-24, or held in a chuck that is mounted in the spindle hole

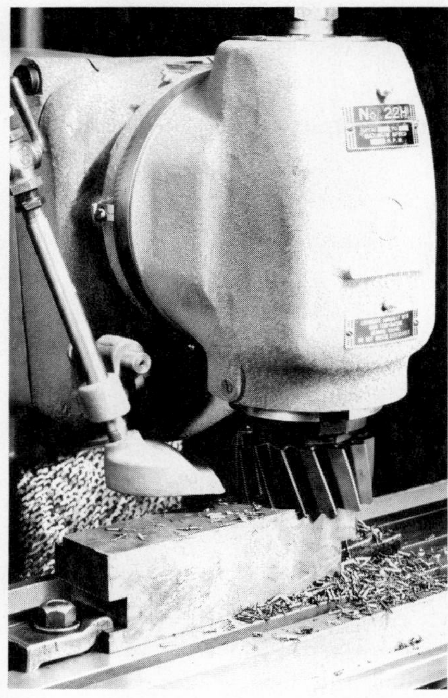

FIGURE 22-21. Vertical milling attachment. (*Courtesy Brown & Sharpe Manufacturing Company.*)

FIGURE 22-22. Milling a bevel surface with a universal milling attachment. (*Courtesy Brown & Sharpe Manufacturing Company.*)

FIGURE 22-23. Machining an internal keyway in a milling machine, using a slotting attachment. (*Courtesy Cincinnati Milacron Inc.*)

FIGURE 22-24. Milling a helical groove using a universal dividing head and a universal milling attachment. (*Courtesy Cincinnati Milacron Inc.*)

FIGURE 22-25. Milling a cam, with the work held in a chuck which is mounted in the spindle of a universal dividing head, having the swivel block turned to a vertical position. The cutter is mounted in a universal milling attachment. (*Courtesy Cincinnati Milacron Inc.*)

of the dividing head, as illustrated in Figure 22-25.

Basically, a dividing head is a rugged, accurate, 40:1 worm-gear reduction unit. The spindle is carried in a swivel block so that it can be tilted from about 5° below horizontal to beyond the vertical position illustrated in Figure 22-25.

The spindle of the dividing head is revolved one revolution by turning the input crank 40 turns. An index plate, mounted beneath the crank, contains a number of holes, arranged in concentric circles and equally spaced, with each circle having a different number of holes. A plunger pin on the crank handle can be adjusted to engage the holes of any circle. This permits the crank to be turned an accurate, fractional part of a complete circle as represented by the increment between any two holes of a given circle on the index plate. Utilizing the 40:1 gear ratio and the proper hole circle on the index plate, the spindle can be rotated a precise amount by the application of either of the following equations:

$$\text{(1) number of turns of crank} = \frac{40}{\text{cuts per revolution of the workpiece}}$$

$$\text{(2) holes to be indexed} = \frac{40 \times \text{holes in index circle}}{\text{cuts per revolution of the workpiece}}$$

If the first rule is used, an index circle must be selected that has the proper number of holes to be divisible by the denominator of any resulting fractional portion of a turn of the crank. In using the second rule, the number of holes in the index circle must be such that the numerator of the fraction is an even multiple of the denominator. For example, if 24 cuts are to be taken about the circumference of a workpiece, the number of turns of the crank required would be 1⅔. An index circle having 12, 15, 18, and so on, holes could be used with one full turn plus 8, 10, or 12 additional holes, respectively. Obviously, use of the second rule would give the same answer. Adjustable *sector arms* are provided on the index plate that can be set to a desired number of holes, less than a full turn, so that fractional turns can be made readily without the necessity for counting holes each time.

Dividing heads are made having ratios other than 40:1, so the ratio should be checked before using.

Because each full turn of the crank on a standard dividing head represents 360/40, or 9° of rotation of the spindle, indexing to a fraction of a degree can be obtained. For example, the space between two adjacent holes on a 36-hole circle represents ¼°.

Indexing can be done in three ways. *Plain indexing* is done solely by the use of the 40:1 ratio in the dividing head. In *compound indexing,* the index plate is moved forward or backward a number of hole spaces each time the crank handle is advanced. For *differential indexing* the spindle and the index plate are connected by suitable gearing so that as the spindle is turned, by means of the crank, the index plate is rotated a proportionate amount.

The dividing head also can be connected to the feed screw of the milling-machine table by means of gearing. This procedure is used to provide a definite rotation of the workpiece with respect to the longitudinal movement of the table, as in cutting helical gears. This procedure is illustrated in Figures 22-11 and 28-9, the connecting gearing being shown in the latter.

Although T-slots are provided on milling machine tables so that workpieces can be clamped directly to the table, more often various work-holding accessories are utilized. Smaller workpieces usually are held in a vise mounted on the table, as shown in Figure 22-23. A universal vise, shown in Figure 22-26, is particularly useful in tool-and-die work. In mass-production work,

FIGURE 22-26. Universal vise for use on a milling machine. (*Courtesy Cincinnati Milacron Inc.*)

FIGURE 22-27. Milling a circular slot, using a circular-milling attachment. (*Courtesy Cincinnati Milacron Inc.*)

special fixtures usually are employed, such as those shown in Figure 22-17; these reduce machine-loading time and assure proper clamping. The circular-milling attachment, shown in Figure 22-27, imparts rotary motion to the work, either by manual or power feed, and thus permits cylindrical surfaces to be milled.

ESTIMATING MILLING TIME

Cutting speeds in milling are selected as indicated in Chapter 17, with consideration given to machine capacity and workpiece strength and rigidity. Table 22-1 gives suggested values of feed, in chip thickness per tooth. It must

TABLE 22-1. Suggested feed per tooth for HSS milling cutters, in millimeters and (inches)

Material	Face Mills	Helical Mills	Slotting or Side Mills	End Mills	Saws
Magnesium and aluminum alloys	0.56 (0.022)	0.46 (0.018)	0.33 (0.013)	0.28 (0.011)	0.13 (0.005)
Medium brass	0.36 (0.014)	0.28 (0.011)	0.20 (0.008)	0.18 (0.007)	0.08 (0.003)
Medium cast iron Cast steel Carbon steel Free-machining steel	0.30 (0.012)	0.25 (0.010)	0.18 (0.007)	0.15 (0.006)	0.08 (0.003)

be remembered that it is the chip thickness per tooth that must be used in calculating milling machine feed, and thus milling time. The feed F per minute is

$$F = t \times N \times \text{rpm}$$

where

t = chip thickness, in millimeters or inches
N = number of teeth in the cutter

In calculating the time required for milling a given length of surface, allowance must be made for the approach and overtravel of the cutter. For *rough* face milling cuts, the approach distance is approximately equal to half the cutter diameter. Therefore, for rough facing cuts, the time T required to mill a piece of length L with a cutter of diameter D is

$$T = \frac{L + D/2}{F}$$

all dimensions being in either millimeters or inches. For *finish face milling* cuts, the entire cutter usually is permitted to travel beyond the end of the workpiece so that the trailing edges give the same wiping action to the entire surface. In this case, the time required is

$$T = \frac{L + D}{F}$$

For *plain* or *slab milling,* the approach distance is somewhat more complex. However, for practical purposes it can be assumed to be equal to half the cutter diameter. Thus the same formula can be used as for rough face milling.

Review questions

1. Why is milling better suited than shaping for producing flat surfaces in mass-production machining?
2. How does face milling differ basically from peripheral milling?
3. Why does down milling tend to produce better surface finish than up milling?
4. Why may down milling dull the cutter more rapidly than up milling when machining castings?
5. What are two common ways of classifying milling cutters?
6. Why do arbor-type slab milling cutters not have to be designated as "right-hand" or "left-hand"?
7. What is the advantage of a helical-tooth cutter over a straight-tooth cutter for slab milling?
8. For what is a form cutter used?
9. Why is a narrow milling cutter not well suited for milling deep slots?

10. Why are helical-tooth cutters preferred over straight-tooth cutters for most slab milling?
11. Explain what steps are required to produce a T-slot by milling.
12. What is the distinctive feature of a fly cutter?
13. Why would a plain column-and-knee milling machine not be suitable for milling the flutes on a large twist drill?
14. Why is a ram-type milling machine often preferred for tool-and-die work?
15. What is the distinctive feature of a universal milling machine?
16. Explain how controlled movements of the work in three mutually perpendicular directions are obtained in column-and-knee-type milling machines.
17. What is a triplex milling machine?
18. Why are bed-type milling machines preferred over column-and-knee types for production milling?
19. How does a duplicator differ from a profiler?
20. Why have planer-type milling machines replaced ordinary planers?
21. What are four common milling machine accessories?
22. What is the basic principle of a universal dividing head?
23. What is the purpose of connecting the input end of a universal dividing head to the feed screw of the milling machine table?
24. What is the purpose of the hole-circle plate on a universal dividing head?
25. Explain how a standard universal dividing head, having hole circles of 21, 24, 27, 30, and 32 holes, would be operated to cut an 18-tooth gear.
26. Explain what action occurs when a dividing head is connected for differential indexing.
27. Explain how feed is specified and computed on a milling machine.
28. Why must the number of teeth on the cutter be known when calculating milling machine feed and setting it on the machine controls?
29. How long will be required to face mill an AISI 1020 steel surface that is 300 mm long and 127 mm wide, using a 152-mm-diameter, eight-tooth tungsten carbide inserted-tooth cutter? Select values of chip thickness from Table 22-1 and cutting speed from Table 17-1.
30. If the depth of cut is 8 mm, what is the metal-removal rate in question 29?
31. Using data from Table 17-5, determine the power required for the operation of question 30.
32. If all the flat surfaces on the bearing block shown in Figure 36-4 must be machined, would you machine them in the same sequence on a shaper as on a milling machine? Explain why.
33. A gray cast iron surface 152 mm wide and 457 mm long may be machined on either a vertical milling machine, using a 203-mm-diameter cutter having 10 inserted HSS teeth, or on a hydraulic shaper with a HSS tool. If milling is used, a chip thickness of 0.25 mm would be employed, and in shaping a feed of 0.38 mm would be used. Setup time for the milling machine would be 20 minutes and for the shaper would be 10 minutes. The time required to put each piece in either machine and to remove it

would be 4 minutes. Machine-hour charges for the milling machine and shaper would be $4.50 and $3.50, respectively. Labor cost would be $4.75 per hour in each case. Which machine would be more economical for this job? (Use Table 17-1 for cutting speeds.)

Case study 15.
THE TITANIUM STIFFENERS

Figure C5-15 shows the design for 500 fittings that have to be milled from a titanium alloy at the Brandon Aircraft Company. The manufacturing manager has turned back the design, stating that it will be unnecessarily costly to produce. Analyze the design, state what features make it costly to manufacture, and make design modifications that will eliminate the difficulty.

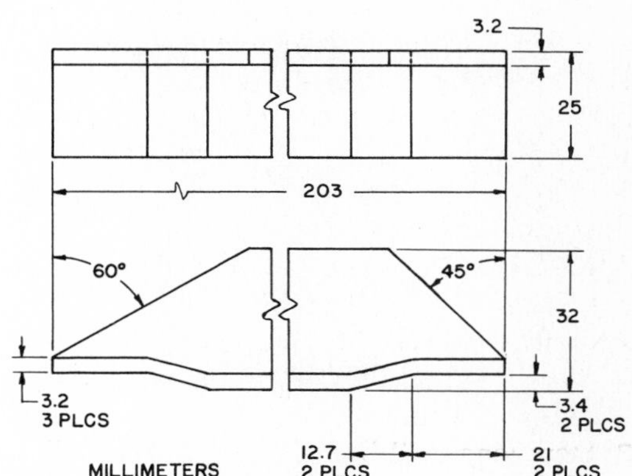

FIGURE CS-15. Titanium stiffener.

CHAPTER 23

Sawing and filing

SAWING

Sawing is a basic machining process in which chips are produced by a succession of small cutting edges, or *teeth,* arranged in a narrow line on a saw "blade." As shown in Figure 23-1, each tooth forms a chip progressively as it passes through the workpiece, and the chip is contained within the space between two successive teeth until these teeth pass from the work. Because sections of considerable size can be severed from the workpiece with the removal of only a small amount of the material in the form of chips, sawing is probably the most economical of the basic machining processes with respect to the wastage of material and power consumption, and in many instances with respect to labor. Its principal disadvantages are its limited accuracy and applicability.

While sawing is an old machining process, during recent years vast im-

FIGURE 23-1. Formation of chips in sawing. (*Courtesy The DoALL Company.*)

provements have been made in saw blades and the machines for using them and, as a result, the usefulness of the process has been greatly expanded. Although it still is true that most sawing is done to sever bar stock and shapes into desired lengths for use in other operations, there are many cases where it is used to produce desired shapes. Frequently, and especially for producing only a few parts, contour sawing may be more economical than any other machining process.

Saw blades. Saw blades are made in three basic configurations. The first, commonly called a *hacksaw* blade, is straight, relatively rigid, and of limited length with teeth on one edge. The second is sufficiently flexible so that a long length can be formed into a continuous band with teeth on one edge; these are known as *band-saw* blades. The third form is a rigid disk, having teeth on the periphery; these are *circular saws*.

All saw blades have certain common and basic features. These are (1) material, (2) tooth form, (3) tooth spacing, (4) tooth set, and (5) blade thickness. Small hacksaw blades usually are made entirely of tungsten or moloybdenum high-speed steel. Blades for power-operated hacksaws often are made with teeth cut from a strip of high-speed steel that has been electron-beam-welded to the heavy, main portion of the blade, which is made from a tougher and cheaper alloy steel. Band-saw blades frequently are made with this same type of construction, as shown in Figure 23-2, but with the main portion of the blade made of relatively thin, high-tensile-strength alloy steel to provide the required flexibility. Band-saw blades also are available with tungsten carbide teeth.

The three most common tooth forms are shown in Figure 23-3.

Tooth spacing is very important in all sawing because it determines three factors. First, it controls the size of the teeth. From a strength viewpoint, large teeth are desirable. Second, tooth spacing determines the space (*gullet*) available to contain the chip that is formed. As shown in Figure 23-1, the chip cannot drop from this space until it emerges from the slot, or *kerf*. The space must be such that there is no crowding of the chip and no tendency for chips to become wedged between the teeth and not drop out. Third, tooth

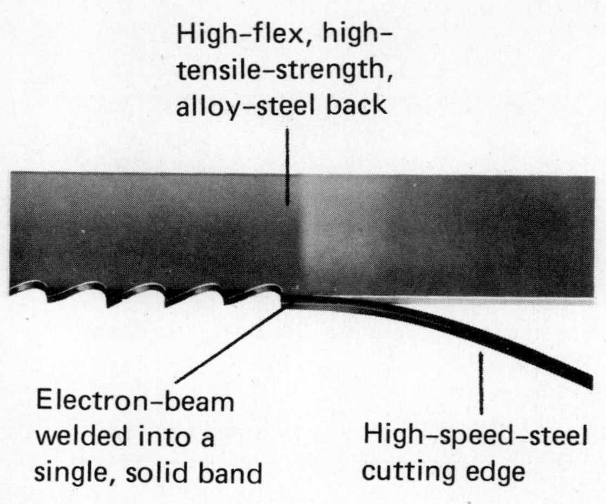

FIGURE 23-2. Method of providing HSS teeth on a softer steel band. (*Courtesy The DoALL Company.*)

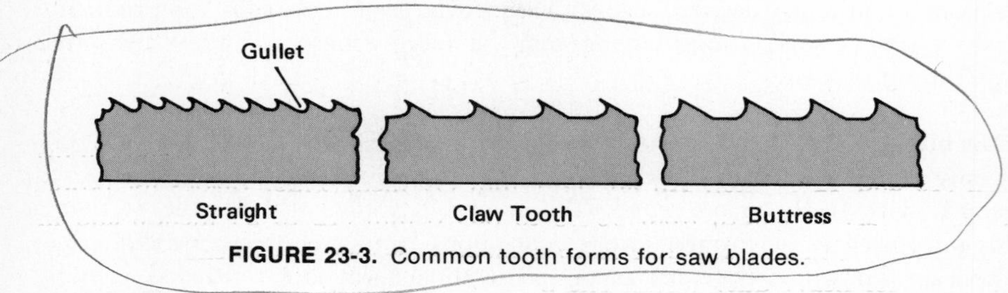

FIGURE 23-3. Common tooth forms for saw blades.

spacing determines how many teeth will bear against the work. This is very important in cutting thin material, such as tubing, as illustrated in Figure 23-4. At least two teeth should be in contact with the work at all times. If the teeth are too coarse, only one tooth rests on the work at a given time, permitting the saw to rock, and the teeth may be stripped from the saw. Hand hacksaw blades can be obtained with tooth spacings from 1.8 to 0.78 mm (14 to 32 teeth per inch). In order to make it easier to start a cut, some hand hacksaw blades are made with a short section at the forward end having teeth of a special form with negative rake angles, as shown in Figure 23-5. Tooth

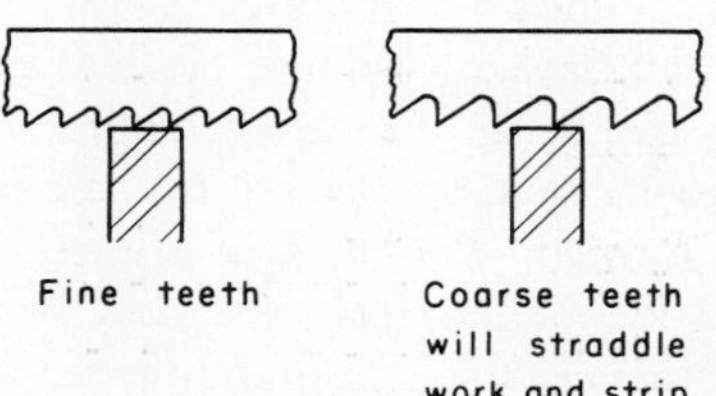

FIGURE 23-4. Relationship of tooth size to material thickness in sawing.

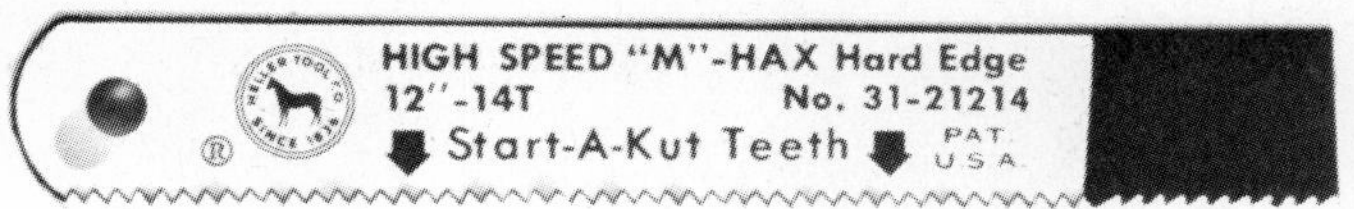

FIGURE 23-5. Hacksaw blade having specially shaped starting teeth. (*Courtesy Heller Tool Company.*)

spacings for power hacksaw blades range from 6.4 to 1.4 mm (4 to 18 teeth per inch).

Tooth set, illustrated in Figure 23-6, refers to the manner in which the teeth are offset from the center line in order to make a cut that is wider than the thickness of the back portions of the blade. The resulting wider cut permits the saw to move more freely in the kerf and reduces friction and heating. *Raker-tooth saws* are used in cutting most steel and iron. *Straight-set teeth* are used for sawing brass, copper, and plastics. Saws with *wave-set teeth* are used primarily for cutting thin sheets and thin-walled tubing.

The *blade thickness* of nearly all hand hacksaw blades is 0.64 mm (0.025 inch). Saw blades for power hacksaws vary in thickness from 1.27 to 2.54 mm (0.050 to 0.100 inch).

Hand hacksaw blades come in two standard lengths—254 and 304.8 mm (10 and 12 inches). All are 12.7 mm (½ inch) wide. Blades for power hacksaws vary in length from 304.8 to 609.6 mm (12 to 24 inches) and in width from 25.4 to 50.8 mm (1 to 2 inches). Wider and thicker blades are desirable for heavy-duty work. The blade should be at least twice as long as the maximum length of cut that is to be made.

Band-saw blades are available in long rolls in straight, raker, wave, or combination sets. In order to reduce the noise from high-speed bandsawing, it is becoming increasingly common to use blades that have more than one pitch, size of teeth, and type of set. One example is shown in Figure 23-7. The most common widths are from 1.6 to 12.7 mm ($^1/_{16}$ to ½ inch), although wider blades can be obtained. Blade width is very important in band

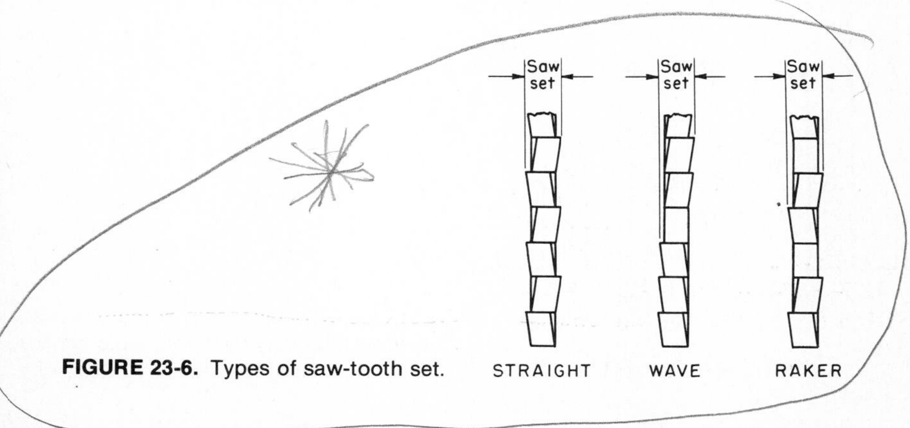

FIGURE 23-6. Types of saw-tooth set.

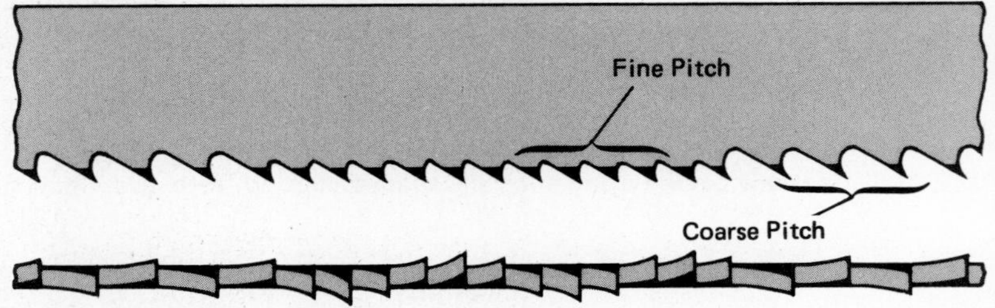

FIGURE 23-7. Use of alternating groups of saw teeth, having either coarse pitch with alternating set or fine pitch with wavy set, in order to reduce noise.

sawing because it determines the minimum radius that can be cut. This relationship is illustrated in Figure 23-8. Because wider blades are stronger, as wide a blade as possible should be used. Consequently, cutting small radii requires a narrower and weaker blade and is more time-consuming.

Band-saw blades come in tooth spacings from 12.7 to 0.79 mm (2 to 32 teeth per inch). The *buttress* form, illustrated in Figure 23-3, is often used for tooth spacings of 4.23 mm or greater (6 teeth or less per inch). Other tooth forms are available for special purposes, such as a scallop edge for cutting fabric. Another type is in the form of a tightly wound spiral of metal, forming a blade about 3.2 mm (1/8 inch) in diameter. This provides a continuous

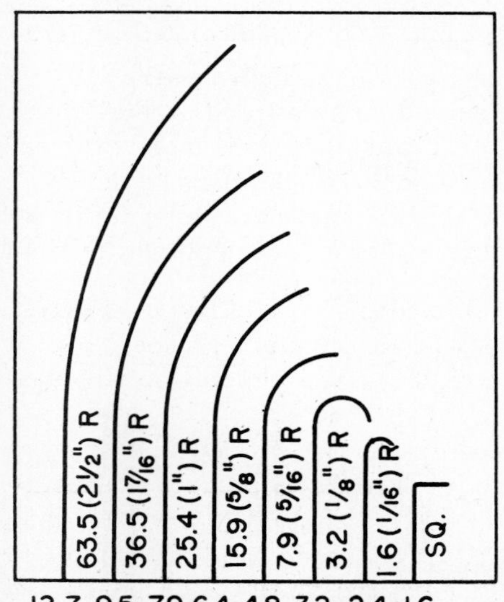

FIGURE 23-8. Relationship of band-saw width to the minimum radius that can be cut.

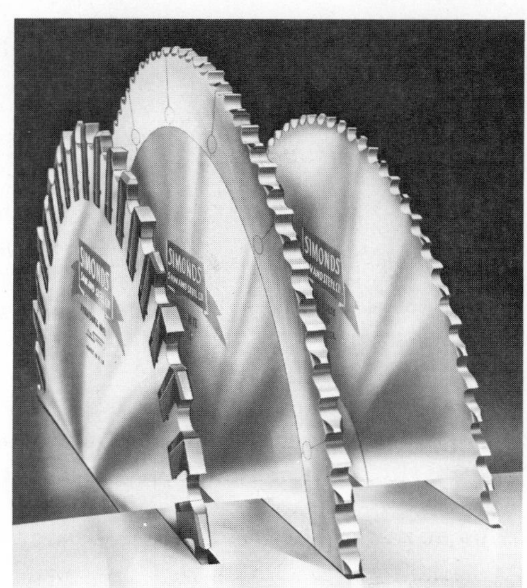

FIGURE 23-9. (*Left to right*) Inserted-tooth, segmental-tooth, and integral-tooth forms of saw construction. (*Courtesy Simonds Saw and Steel Co.*)

cutting edge along the entire spiral so that work can be fed against the blade in all directions.

Circular saws necessarily differ somewhat from straight blade forms. Because they must be relatively large in comparison with the work, only the sizes up to about 457 mm (18 inches) in diameter have teeth that are cut into the disk. Larger saws use either *segment* or *inserted* teeth, illustrated in Figure 23-9. Only the teeth are made of high-speed steel, or tungsten carbide; the remainder of the disk is made of ordinary, less expensive, and tougher steel. Segmental blades are composed of segments mounted around the periphery of the disk, usually fitted with a tongue and groove and fastened by means of screws or rivets. Each segment contains several teeth. If a single tooth is broken, only one segment need be replaced to restore the saw to operating condition.

Large circular saws, some as large as 1829 mm (72 inches) in diameter, often have inserted teeth. Each tooth is a separate piece of high-speed steel or tungsten carbide and is attached to the central disk by means of wedges or screws, or by brazing.

Figure 23-10 shows a common tooth form used in circular saws, in which successive teeth are beveled on opposite sides to produce a smoother cut. Another method is to bevel both sides of every other tooth. A third procedure has the first tooth beveled on the left side; the second tooth on both sides; the third tooth on the right side; the fourth tooth on the left side, and so forth.

Circular saws for cutting metal are often called *cold saws* to distinguish them from friction-type disk saws that heat the metal to the melting tempera-

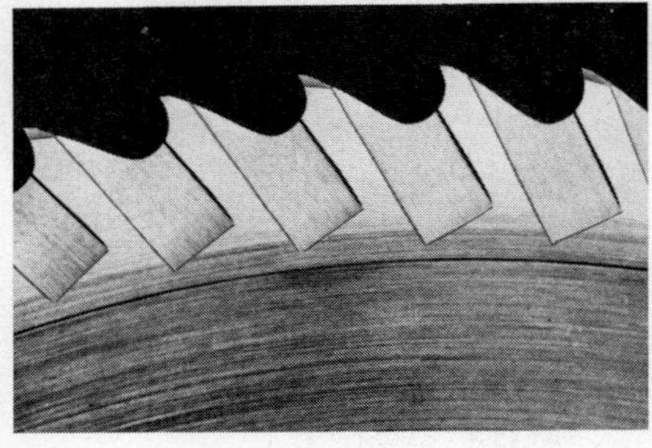

FIGURE 23-10. Method of beveling teeth on alternate sides to produce smoother cutting. (*Courtesy Simonds Saw and Steel Co.*)

ture at the point of metal removal. Cold saws cut very rapidly and can produce cuts that are comparable in smoothness and accuracy with surfaces made by slitting saws in a milling machine or by a cutoff tool in a lathe.

Types of sawing machines. Metal-sawing machines may be classified as follows:

1. Reciprocating saw
   a. Manual hacksaw
   b. Power hacksaw
2. Band saw
   a. Vertical cutoff
   b. Horizontal cutoff
   c. Combination cutoff and contour
   d. Friction
3. Circular saw
   a. Cold saw
   b. Steel friction disk
   c. Abrasive disk

Power hacksaws. As the name implies, power hacksaws are machines that mechanically reciprocate a large hacksaw blade. As shown in Figure 23-11, they consist of a bed, a work-holding frame, a power mechanism for reciprocating the saw frame, and some type of feeding mechanism. Because of the inherent inefficiency of cutting on only one stroke direction, they have been virtually replaced by more efficient, horizontal band-sawing machines.

Band-sawing machines. While the earliest metal-cutting band-sawing machines were direct adaptations from wood-cutting band saws, modern machines of this type are much more sophisticated and versatile and have been developed specifically for metal cutting. To a large degree, they were made possible by the development of vastly better and more flexible band-saw blades and simple flash-welding equipment, which can weld the two ends of a

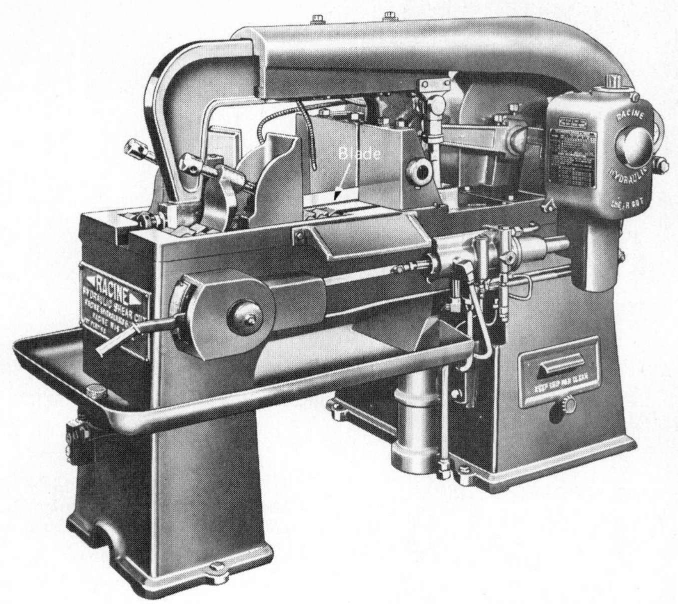

FIGURE 23-11. Heavy-duty power hacksaw having hydraulic feed mechanism. (*Courtesy Racine Hydraulics & Machinery, Inc.*)

strip of band-saw blade together to form a band of any desired length. Three basic types of band-sawing machines are in common use.

Upright, cutoff, band-sawing machines, such as shown in Figure 23-12, are designed primarily for cutoff work on single, stationary workpieces that can be held on a table. On most machines the blade mechanism can be tilted to about 45°, as shown, to permit cutting at an angle. They usually have automatic power feed of the blade into the work, automatic stops, and provision for supplying coolant.

Horizontal, metal-cutting band-sawing machines were developed to combine the flexibility of reciprocating power hacksaws and the continuous cutting action of vertical band saws. A typical machine of this type is shown in Figure 23-13. This is a heavy-duty automatic type with the saw fed vertically by a hydraulic mechanism. It also has automatic stock feed that can be set to feed the stock laterally any desired distance after a cut is completed and automatically clamp it for the next cut. Such machines can be arranged to hold, clamp, and cut several bars of material simultaneously. Smaller and less expensive types have swing-frame construction, with the band-saw head mounted in a pivot on the rear of the machine. Feed is accomplished by gravity through rotation of the head about the pivot point. Because of their continuous cutting action, horizontal band-sawing machines are very efficient.

FIGURE 23-12. Cutting a pipe at a 45° angle on an upright cutoff band-sawing machine. (*Courtesy Armstrong-Blum Mfg. Co.*)

FIGURE 23-13. Horizontal band-sawing machine, having automatic stock-feeding mechanism. (*Courtesy Armstrong-Blum Mfg. Co.*)

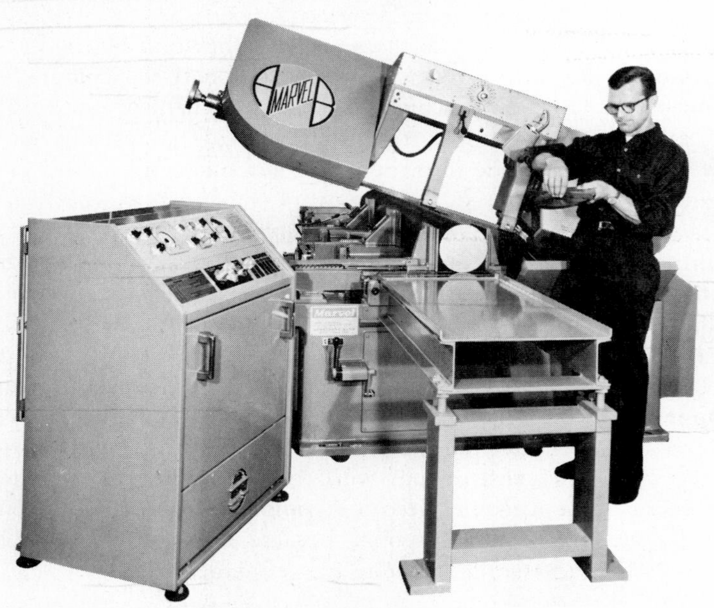

FIGURE 23-14. Combination cutoff and contour band-sawing machine. (*Courtesy The DoALL Company.*)

Combination cutoff and contour band-sawing machines, such as shown in Figure 23-14, can be used not only for cutoff work but also for contour sawing. They are widely used for cutting irregular shapes in connection with making dies and the production of small numbers of parts.

Several features distinguish these machines from regular vertical cutoff band saws. First, the table is pivoted so that it can be tilted to any angle up to 45°. Second, a small flash welder is provided, on the vertical column, so that a straight length of band-saw blade can be welded quickly into a continuous band. A small grinding wheel is located beneath the welder so that the flash can be ground from the weld to provide a smooth joint that will pass through the saw guides. This welding and grinding unit makes it possible to cut internal openings by drilling a hole, inserting one end of the saw blade through the hole, and then butt-welding the two ends together. When the cut is finished, the band is cut apart and removed from the opening. Third,

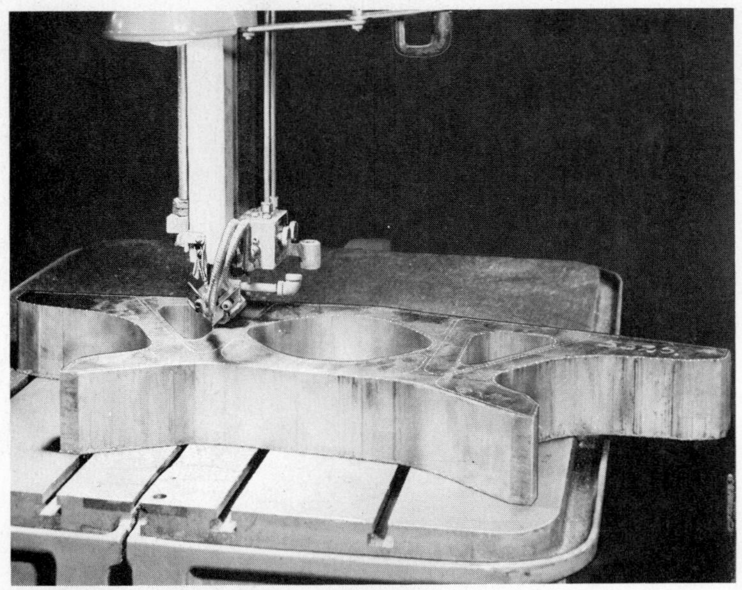

FIGURE 23-15. Contour sawing a part from a piece of steel 76 mm (3 inches) thick. (*Courtesy The DoALL Company.*)

the speed of the saw blade can be varied continuously over a wide range to provide correct operating conditions for any material. Fourth, a method of power feeding the work is provided, sometimes gravity-actuated.

Contour-sawing machines are made in a wide range of sizes, the principal size dimension being the throat depth. Sizes from 305 to 1829 mm (12 to 72 inches) are available. The speeds available on most machines range from about 15 to 610 meters (50 to 2000 feet) per minute. Figure 23-15 shows a typical job being cut on a contour-sawing machine.

Special band-sawing machines are available with very high speed ranges, up to 4572 meters (15,000 feet) per minute. These are known as *friction band-sawing machines.* Material is not cut by chip formation. Instead, the friction between the rapidly moving saw and the work is sufficient to raise the temperature of the material at the end of the kerf to or just below the melting point where its strength is very low. The saw blade then pulls the molten, or weakened, material out of the kerf. Consequently, the blades do not need to be sharp; they frequently have no teeth, only occasional notches in the blade to aid in removing the metal.

Almost any material, including ceramics, can be cut by friction sawing. Because only a small portion of the blade is in contact with the work for an instant and then is cooled by its passage through the air, it remains cool. Usually the major portion of the work, away from contact with the saw blade, also remains quite cool. Figure 23-16 shows a hardened steel cutter being cut

FIGURE 23-16. Cutting a hardened steel cutter by friction sawing. (*Courtesy The DoALL Company.*)

by this method. It also is a very rapid method for trimming the flash from pressed sheet-metal parts.

Circular-blade sawing machines. Machines employing rotating, circular saw blades are used exclusively for cutoff work. These range from small, simple types, such as shown in Figure 23-17, in which the saw is fed manually, to very large ones having power feed and built-in coolant systems. A common use for the large ones is to cut off hot-rolled shapes as they come from a rolling mill. In some cases friction saws are used for this purpose, having

FIGURE 23-17. Small rotary cutoff saw. (*Courtesy Rockwell Manufacturing Company.*)

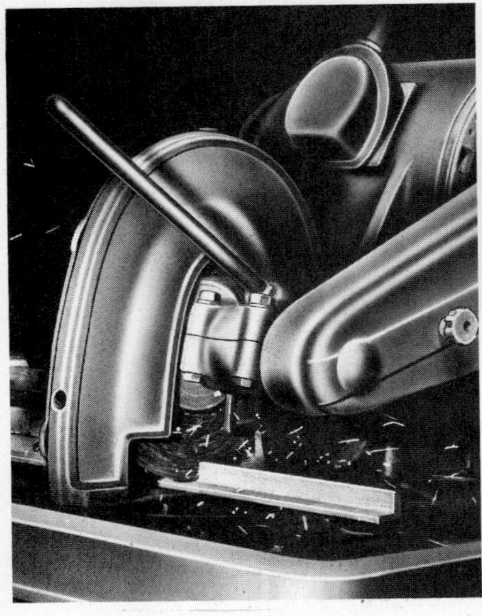

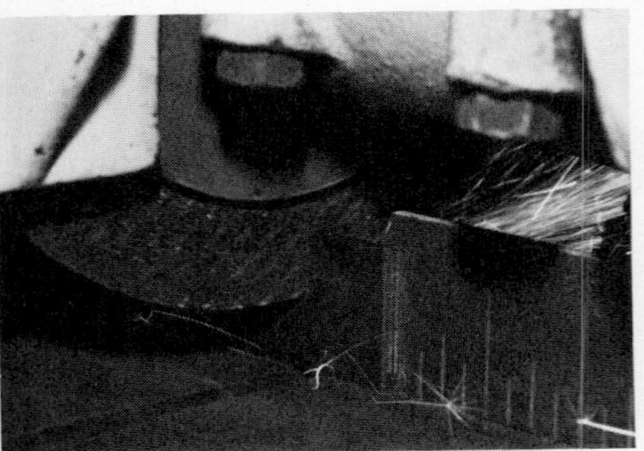

FIGURE 23-18. (*Left*) Cutting a steel angle by means of an abrasive disk. (*Courtesy The Carborundum Company.*) (*Above*) Close-up view of abrasive disk cutting flat bar stock.

disks up to 1.83 meters (6 feet) in diameter and operating at surface speeds up to 7620 meters (25,000 feet) per minute. Steel sections up to 610 mm (24 inches) can be cut in less than 1 minute by this technique.

Although technically not a sawing operation, cutoff work up to about 152 mm (6 inches) often is done utilizing thin *abrasive* disks. The equipment used is the same as for sawing. It has the advantage that very hard materials, which would be very difficult to saw, can be cut readily. A thin rubber- or resinoid-bonded abrasive wheel is used. Usually, a somewhat smoother surface is produced. Figure 23-18 shows an example of this process in use.

FILING

Basically, the metal-removing action in filing is the same as in sawing, in that chips are removed by cutting teeth that are arranged in succession along the same plane on the surface of a tool, called a *file*. There are two differences in that (1) the chips are very small, so that the cutting action is slow and easily controlled; and (2) the cutting teeth are much wider. Consequently, fine and accurate work can be done.

Types of files. Files are classified according to the following:

1. The type, or *cut*, of the teeth
2. The degree of coarseness of the teeth

3. Construction
   a. Single, solid units for hand use or in die-filing machines
   b. Band segments, for use in band-filing machines
   c. Disks, for use in disk-filing machines

Four types of *cuts* are available. *Single-cut files* have rows of parallel teeth that extend across the entire width of the file at the angle of from 65 to 85°. *Double-cut files* have two series of parallel teeth that extend across the width of the file. One series is cut at an angle of 40 to 45°. The other series is coarser and is cut at an opposite angle that varies from about 10° to 80°. A *vixen-cut file* has a series of parallel, curved teeth, each extending across the file face. On a *rasp-cut* file, each tooth is short and is raised out of the surface by means of a punch. These four types of cuts are shown in Figure 23-19.

The coarseness of files is designated by the following terms, arranged in order of increasing coarseness: *dead smooth, smooth, second cut, bastard, coarse,* and *rough.* There also is a series of finer Swiss pattern files, designated by numbers from 00 to 8.

Files are available in a number of cross-sectional shapes—*flat, round, square, triangular,* and *half-round.* Flat files can be obtained with no teeth on one or both narrow edges, known as *safe edges,* so as to prevent material from being removed from a surface that is normal to the one being filed.

Most files for hand filing are from 254 to 355 mm (10 to 14 inches) in length and have a pointed *tang* on one end on which a wood or metal handle can be fitted for easy grasping.

Filing machines. Although an experienced machinist can do very accurate work by hand filing, it is a slow and tiresome task. Consequently, three types of filing machines have been developed that permit quite accurate results to be

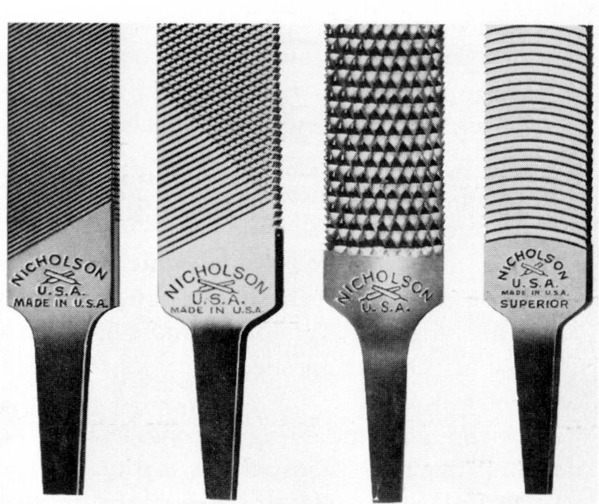

FIGURE 23-19. Four cuts of files: (*left to right*) single, double, rasp, and curved (*Vixen*). (*Courtesy Nicholson File Company.*)

FIGURE 23-20. Die-filing machine.

obtained much more rapidly and with much less effort. *Die filing machines,* such as is shown in Figure 23-20, hold and reciprocate a file that extends upward through the worktable. The file rides against a roller guide at its upper end, and cutting occurs on the downward stroke, so the cutting force tends to hold the work against the table. The table can be tilted to any desired angle. Such machines operate at from 300 to 500 strokes per minute, and the resulting surface tends to be at a uniform angle with respect to the table. Quite accurate work can be done but, because of the reciprocating action, approximately 50 per cent of the operating time is nonproductive.

Band-filing machines provide continuous cutting action. Most band filing is done on contour band-sawing machines by means of a special band file that is substituted for the usual band-saw blade.

The principle of a band file is shown in Figure 23-21. Rigid, straight file segments, about 3 inches long, are riveted to a flexible steel band near their leading ends. One end of the steel band contains a slot that can be hooked over a pin in the other end to form a continuous band. As the band passes over the drive and idler wheels of the machine, it flexes so that the ends of adjacent file segments move apart. When the band becomes straight, the ends of adjacent segments move together and interlock to form a continuous straight file. Where the file passes through the worktable, it is guided and supported by a grooved guide, which provides the necessary support to resist the pressure of the work against the file.

Band files are available in most of the standard cuts and in several widths and shapes. Operating speeds reange from about 15 to 76 meters (50 to 250 feet) per minute.

Although band filing is considerably more rapid than can be done on a die-filing machine, it usually is not quite as accurate. Frequently, band filing may be followed by some finish filing on a die-filing machine. Figure 23-22 shows a typical job being done on a band-filing machine.

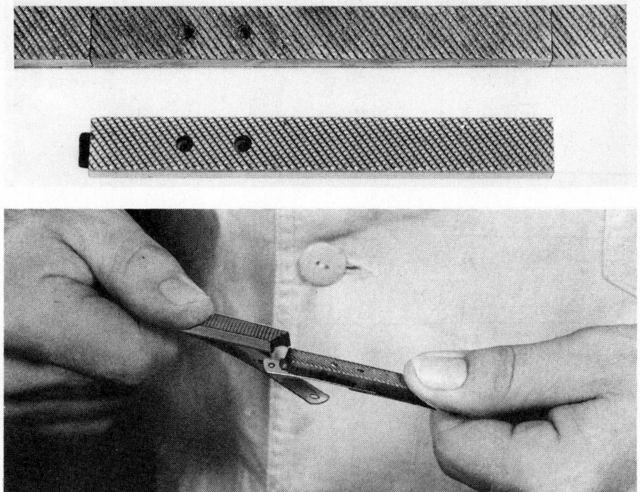

FIGURE 23-21. Band-file segments, and the method of joining the band ends to form a continuous band. (*Courtesy The DoALL Company.*)

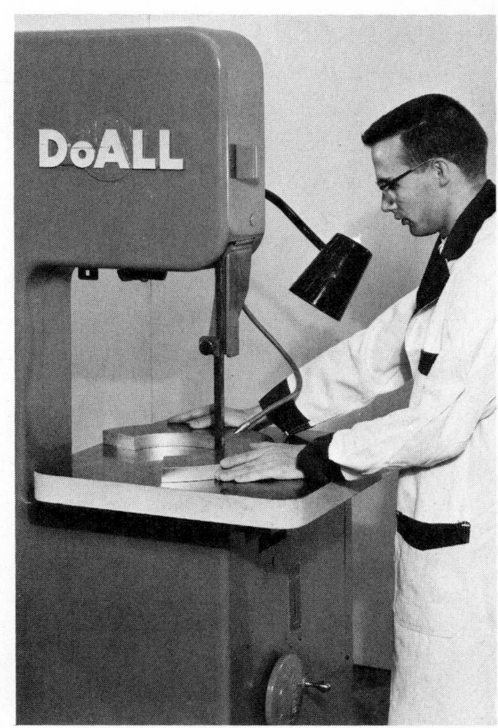

FIGURE 23-22. Typical operation on a band-filing machine. (*Courtesy The DoALL Company.*)

FIGURE 23-23. (*Above*) Disc-type filing machine. (*Left*) Some of the available types of disc files. (*Courtesy Jersey Manufacturing Company.*)

Some *disk-filing machines* are used, having files in the form of disks, as shown in Figure 23-23. These are even more simple than die-filing machines and provide continuous cutting action. However, it is difficult to obtain accurate results by their use.

Review questions

1. Why is sawing one of the most efficient of the chip-forming processes?
2. Explain why tooth spacing is important in sawing.
3. Why do the specially shaped teeth on the hacksaw blade shown in Figure 23-5 make it easier to start a cut?
4. Why are not all the teeth on a hacksaw blade shaped like the starting teeth shown in Figure 23-5?
5. Explain what is meant by the "set" of the teeth on a saw blade.
6. Why can a band-saw blade not be hardened throughout the entire width of the band?
7. What is the general relationship between the width of a band-saw blade and the radius of a cut that can be made with it?
8. What is the advantage of a spiral-type band-saw blade?
9. What are the advantages and disadvantages of circular saws?
10. Why have band-sawing machines largely replaced those using reciprocating saws?

11. Explain how a square hole can be made on a band-sawing machine.
12. How does friction sawing differ from ordinary band sawing?
13. What is the disadvantage of using gravity to feed a saw in cutting round bar stock?
14. Why does the blade of a friction band saw not melt?
15. To what extent is filing different from sawing?
16. What is a safe edge on a file?
17. Why is a band-filing machine more efficient than a die-filing machine?
18. How does a rasp-cut file differ from other types?
19. In making a 152.4-mm (6-inch) cut in a piece of AISI 1020 CR steel, 25.4 mm (1 inch) thick, the material is fed to a bandsaw blade with teeth having a pitch of 1.27 mm (20 pitch) at the rate of 0.0025 mm (0.0001 inch) per tooth. How long should be required for the cut?

Case study 16. THE GASOLINE TANK FASTENER

The gasoline tank on a vehicle was held in place by means of two steel straps, secured at one end by being hooked into a rectangular hole in the top flange of the frame cross-member, as indicated schematically in Figure CS-16a. The frame member was a relatively light U-shaped steel channel, and the rectangular hole was made by a piercing operation in a punch press. It was known that in normal usage the channel could be subjected to substantial, concentrated impact loads, as indicated by the arrow.

As a result of a severe impact, the channel was deformed, and the flange was torn open, as indicated in (b). This permitted the gasoline tank to come free, catch fire, and an occupant of the vehicle was killed.

How could the design have been altered, without any significant increase in manufacturing cost, so as to retain the ease of attaching the strap end but so as to greatly reduce or eliminate the possibility of this type of failure?

FIGURE CS-16. (a) Method of attaching end of gasoline tank securing strap. (b) Mode of failure of slot.

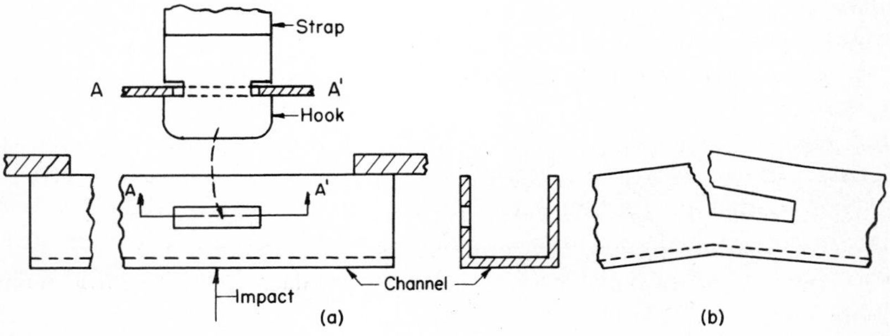

CHAPTER 24

Broaching

Broaching is unique in that it is the only one of the basic machining processes in which the feed of the cutting edges into the workpiece, determining the chip thickness, is built into the tool, called a *broach.* The machined surface always is the inverse of the profile of the broach and, in most cases, it is produced with a single, linear stroke of the tool across the workpiece (or the workpiece across the broach).

As illustrated in Figure 24-1, a broach is composed of a series of single-point cutting edges projecting from a rigid bar, with successive edges protruding farther from the axis of the bar than previous ones. This increased projection, known as *step,* determines the depth of cut by each tooth (chip thickness), so that no feeding of the broaching tool is required. The frontal contour of the teeth determines the shape of the resulting machined surface. As the result of these conditions built into the tool, no complex motion of the tool relative to the workpiece is required.

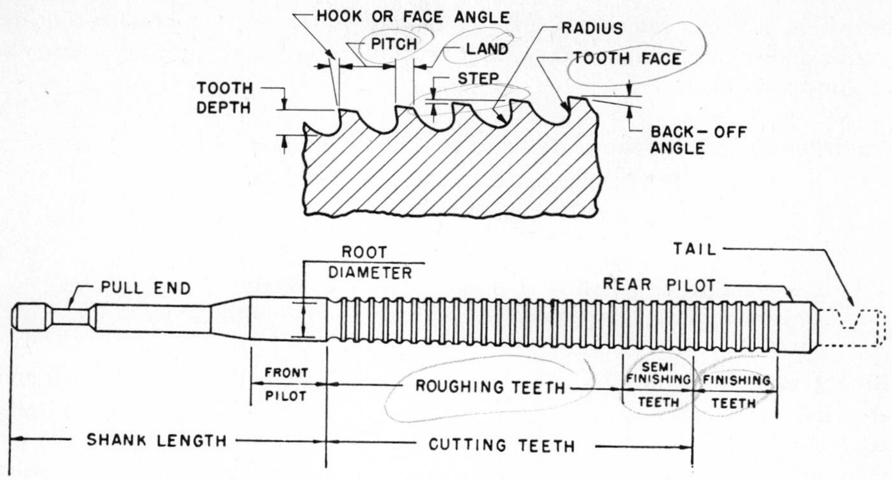

FIGURE 24-1. Standard broach parts and nomenclature.

Obviously, because of the features built into a broach, it is a simple and rapid method of machining. But it is equally evident that there is a close relationship between the contour of the surface to be produced, the amount of material that must be removed, and the design of the broach. For example, the total depth of the material to be removed cannot exceed the total step provided in the broach, and the step of each tooth must be such as to provide proper chip thickness for the type of material to be machined. Consequently, either a special broach must be made for each job, or the workpiece must be designed so that a standard broach can be used. This means that broaching is particularly well suited for, and is widely used in, mass production, where the volume can easily justify the cost of a rather expensive tool, and for certain simple and standardized shapes, such as keyways, where stock broaches can be used.

Broaching originally was developed for machining internal keyways. However, its obvious advantages quickly led to its development for mass-production machining of various surfaces, such as flat, interior and exterior cylindrical and semicylindrical, and many irregular surfaces. Because there are few limitations as to the contour form that broach teeth may have, there is almost no limitation in the shape of surfaces that can be produced by broaching. The only physical limitations are that there must be no obstruction to interfere with the passage of the entire tool over the surface to be machined and that the workpiece must be strong enough to withstand the forces involved. In internal broaching, a hole must exist in the workpiece into which the broach may enter. Such a hole can be made by drilling, boring, or coring.

Broaching usually produces better accuracy and finish than can be obtained by milling or reaming. Although the relative motion between the broaching tool and the work usually is a single linear one, a rotational motion can be added to permit the broaching of spiral grooves, as in spiral splines or in gun-barrel rifling.

Classification of broaches. Broaches commonly are classified as follows:

Purpose	Motion	Construction	Function
Single	Push	Solid	Roughing
Combination	Pull	Built-up	Sizing
	Stationary		Burnishing

Broach design. Figure 24-1 shows the principal components of a broach and the shape and arrangement of the teeth. Each tooth is an extensive, single-edge cutting tool, arranged essentially like the teeth on a saw except for the step, which determines the depth cut by each tooth. The depth of cut varies from about 0.15 mm (0.006 inch) for roughing teeth in machining free-cutting steel to a minimum of 0.025 mm (0.001 inch) for finishing teeth. The exact amount depends on several factors. Too-large cuts impose undue stresses on the teeth and the work; too-small result in rubbing rather than cutting action. The strength and ductility of the metal being cut are the primary factors.

Where it is desirable for each tooth to take a deep cut, as in broaching castings or forgings that have a hard, abrasive surface layer, *rotor-* or *jump-cut* tooth design, shown in Figure 24-2, may be used. In this design, two or three teeth in succession have the same diameter, or height, but each tooth of the group is notched or cut away so that it cuts only a portion of the circumference or width. This permits deeper, but narrower, cuts by each tooth without increasing the total load per tooth, and also reduces the forces and the power requirements.

FIGURE 24-2. Special types of broach teeth: (*top*) rotor-cut; (*bottom*) double-cut with chip-breaker grooves on alternate teeth except at the finishing end. (*Courtesy Colonial Broach & Machine Company.*)

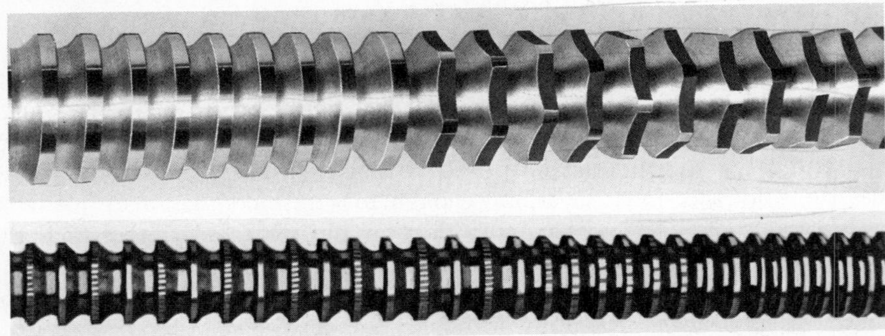

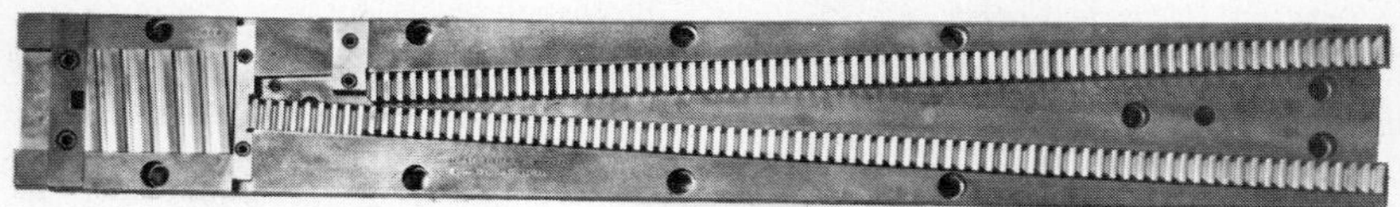

FIGURE 24-3. Progressive surface broach. (*Courtesy Detroit Broach & Machine Company.*)

Tooth loads and cutting forces also can be reduced by using the *double-cut* construction shown in Figure 24-2. Pairs of teeth have the same size, but the first has extra-wide chip-breaker notches and removes metal over only a part of its width, while the smooth second tooth completes the cut.

A third construction for reducing tooth loads utilizes the principle illustrated in Figure 24-3. Employed primarily for broaching wide, flat surfaces, the first few teeth in *progressive* broaches completely machine the center, while succeeding teeth are offset in two groups to complete the remainder of the surface. Rotor, double-cut, and progressive designs require the broach to be made longer than if normal teeth were used, and they thus can be used only on a machine having adequate stroke length.

The faces of the teeth on surface broaches may be either normal to the direction of motion or at an angle of from 5 to 20 degrees. The latter, *shear-cut,* broaches provide smoother cutting action with less tendency to vibrate. Although the majority of surface broaching is done on flat surfaces, other shapes can be broached, as shown in Figure 24-4.

As in saws, the pitch of the teeth and the gullet between them must be

FIGURE 24-4. Broaching the teeth of a segment gear. (*Courtesy Colonial Broach & Machine Company.*)

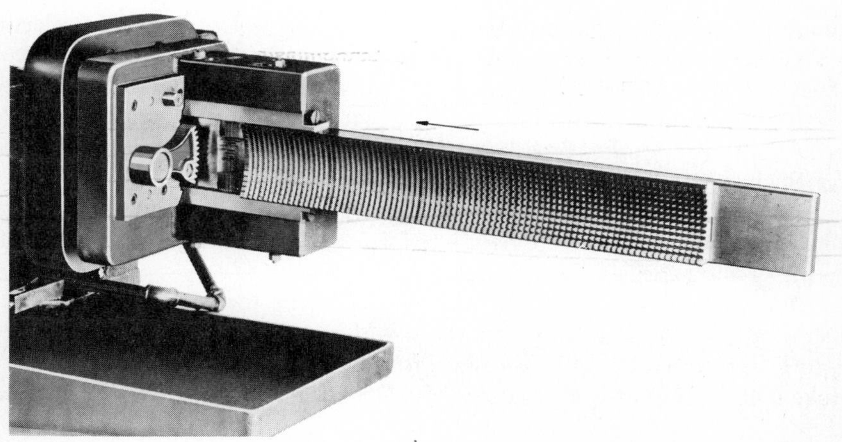

sufficient to provide ample room for chip clearance. All chips produced by a given tooth during its engagement throughout the length of the workpiece must be contained in the space between successive teeth. At the same time, it is desirable to have the pitch sufficiently small so that at least two or three teeth are cutting at all times.

The *hook* determines the primary rake angle and is a function of the material being cut, being 15 to 20° for steel and 6 to 8° for cast iron. *Back-off* or end clearance angles are from 1 to 3° to prevent rubbing.

Most of the metal removal is done by the *roughing teeth. Semifinishing teeth* provide surface smoothness, whereas *finishing teeth* produce exact size. On a new broach all the finishing teeth usually are made the same size. As the first finishing teeth become worn, those behind continue the sizing function. On some round broaches, *burnishing teeth* are provided for finishing. These have no cutting edges but are button-shaped and from 0.025 to 0.076 mm (0.001 to 0.003 inch) larger than the size of the hole. The resulting rubbing action smooths and sizes the hole. They are used primarily on cast iron and nonferrous metals.

The *pull end* of a broach (Figure 24-1) occurs only on pull broaches and is to provide a means of quickly attaching the broach to the pulling mechanism. The *front pilot* aligns the broach in the hole before it begins to cut, and the *rear pilot* keeps the tool square with the finished hole as it leaves the workpiece. *Shank length* must be sufficient to permit the broach to pass through the workpiece and be attached to the puller before the roughing teeth engage the work. If a broach is to be used on a vertical machine that has a tool-handling mechanism, a *tail* is necessary.

A broach should not be used to remove a greater depth of metal than that for which it is designed—the sum of the steps of all the teeth. In designing workpieces, a minimum of 0.51 mm (0.020 inch) should be provided on surfaces that are to be broached, and about 6.35 mm (¼ inch) is the practical maximum.

Broaching speeds. Broaching speeds are relatively low, seldom exceeding 15.24 meters (50 feet) per minute. However, because a surface usually is completed in a single stroke, the productivity is high. A complete cycle usually requires only from five to thirty seconds, with most of that time being taken up by the return stroke, broach-handling, and workpiece loading and unloading. Such cutting conditions facilitate cooling and lubrication and result in very slow tool wear. This is an advantage of broaching as a mass-production process because it reduces the necessity for frequent resharpening and prolongs the life of the expensive broaching tool.

For a given cutting speed and material, the force required to pull or push a broach is a function of the tooth width, step, and the number of teeth cutting. Consequently, it is necessary to design or specify a broach within the stroke length and power limitations of the machine on which it is to be used.

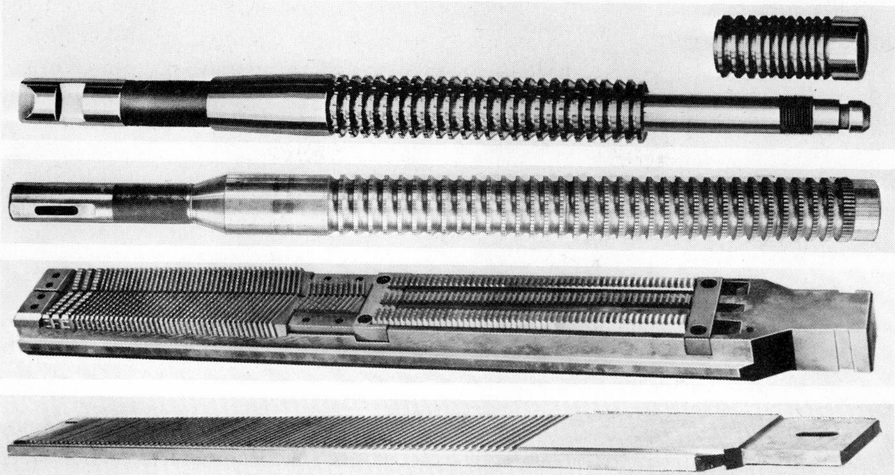

FIGURE 24-5. Types of broach construction: (*Top to bottom*) Round, shell-type, pull broach; round, solid-type, pull broach; sectional surface broach; solid surface broach. (*Courtesy Ex-Cell-O Corporation and Colonial Broach & Machine Company.*)

Broach materials and construction. Because of the low cutting speeds employed, most broaches are made of alloy or high-speed tool steel, even for some mass-production work. Where they are used in continuous mass-production lines, particularly in surface broaching, tungsten carbide teeth may be used, permitting them to be used for long periods of time without resharpening.

Most internal broaches are of solid construction. Quite often, however, they are made of *shells* mounted on a bar. When the broach, or a section of it, is subject to rapid wear, a single shell can be replaced and will be much cheaper than an entire solid broach. Shell construction, however, is more expensive than a solid broach of comparable size. Both types are shown in Figure 24-5.

Small surface broaches may be of solid construction, but larger ones usually are built up from sections, as shown in Figure 24-5. Sectional construction makes the broach easier and cheaper to construct and sharpen. It also often provides some degree of interchangeability of the sections.

Sharpening broaches. Most broaches are resharpened by grinding the faces of the teeth. The lands of internal broaches must not be reground because this would change the size of the broach. Lands of flat surface broaches sometimes are ground, in which case all of them must be ground to maintain their proper relationship.

FIGURE 24-6. Automatic broach puller: (*left*) open; (*right*) closed. (*Courtesy Detroit Broach & Machine Company.*)

Broach pullers. When pull broaches are used, means must be provided for connecting the broach to the pulling head of the machine. If the broach does not have to be disconnected each cycle, a simple threaded connection can be used. On machines where the broach is connected by hand, a slot is provided in the pull end of the broach and a key is inserted through this slot and corresponding ones in the puller.

Most broaching machines automatically return the broach to its initial position. On such machines, automatic, round pullers of the type shown in Figure 24-6 are used. At the beginning of the cutting stroke the puller automatically grips the shank of the broach. As the pulling ram reaches the end of its return stroke, the outer sleeve is forced back by a stop, thus releasing the jaw and permitting the shank to be withdrawn.

BROACHING MACHINES

Because all the factors that determine the shape of the machined surface and which determine all cutting conditions, except speed, are built into the broaching tool, broaching machines are relatively simple. Their primary functions are to impart plain reciprocating motion to the broach and to provide a means for automatically handling the broach.

Most broaching machines are driven hydraulically, although mechanical drive is used in a few special types. The major classification relates to whether the motion of the broach is vertical or horizontal as follows:

Vertical	Horizontal	Rotary
Broaching presses (push broaching)	Pull	Special types
Pull down	Surface	
Pull up	Continuous	
Surface		

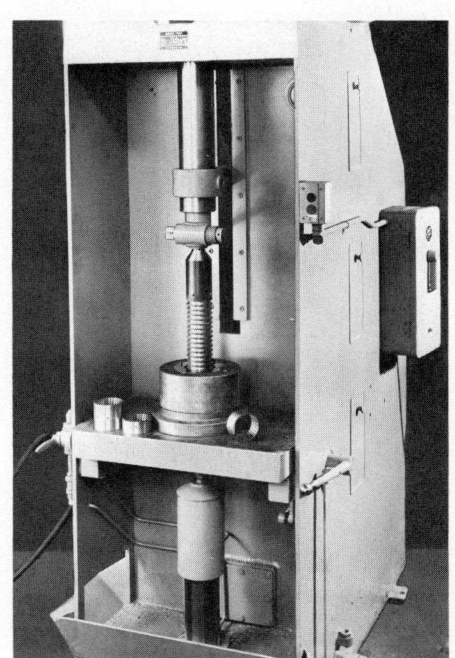

FIGURE 24-7. Broaching internal helical grooves in a broaching press, using a guided ram. (*Courtesy The Oilgear Company.*)

The choice between vertical and horizontal machines is determined primarily by the length of the stroke required and the available floor space. Vertical machines seldom have greater than 1524 mm (60 inch) strokes, because of height limitations. Horizontal machines can have almost any length of stroke, but they require greater floor space.

Broaching presses. As shown in Figure 24-7, *broaching presses* essentially are arbor presses with a guided ram. They are used with push broaches, have a capacity of from 44.5 to 445 kN (5 to 50 tons), and are used only for internal broaching. The forward guide of the broach is inserted through the hole in the workpiece as it rests on the press table, often in a fixture. As the ram descends, it engages the upper end of the broach and pushes it through the work.

Broaching presses are relatively slow, in comparison with other broaching machines, but they are inexpensive, flexible, and can be used for other types of operations, such as bending and staking.

Vertical pull-down machines. The major components of vertical pull-down machines are a worktable, usually having a spherical-seated workholder, a broach elevator above the table, and a pulling mechanism below the table. As shown in Figure 24-8, when the elevator raises the broach above the table, the work can be placed into position. The elevator then lowers the pilot end of the

FIGURE 24-8. Broaching a round hole on a vertical pull-down machine having two broaches. (*Courtesy The Oilgear Company.*)

broach through the hole in the workpiece, where it is engaged by the puller. The elevator then releases the upper end of the broach, and it is pulled through the workpiece. The workpieces are removed from the table, and the broach is raised upward to be engaged by the elevator mechanism. In some cases where machines have two rams, they are arranged so that one broach is being pulled down while the work is being unloaded and the broach raised at the other station.

Vertical pull-up machines. In *vertical pull-up machines,* shown in Figure 24-9, the pulling ram is above the worktable and the broach-handling mechanism below it. The work is placed in position, above the pilot, while the broach is lowered. The handling mechanism then raises the broach until it engages the puller head. As the broach is pulled upward, the work comes to rest against the underside of the table, where it is held until the broach has been pulled through. The work then falls free, often sliding down a chute into a tote bin.

Pull-up machines may have up to eight rams. Because the workpieces need only be placed in the machines, and the broach handling and work removal are automatic, they are highly productive. For certain types of work, automatic feeding can be provided.

FIGURE 24-9. Broaching 19.91-mm (0.784-inch) holes in valve rocker-arm supports on a vertical pull-up machine having four broaches. Parts are fed automatically from the magazines and fall from the machine upon completion. (*Courtesy American Broach & Machine Company.*)

Vertical surface-broaching machines. On *vertical surface-broaching machines* the broaches usually are mounted on guided slides, as shown in Figure 24-10, so as to provide support against lateral thrust. Because there is no need for handling the broach, they are simpler but much heavier than pull- or push-broaching machines. Many have two or more slides so that work can be loaded at one while another part is machined at the other. Inasmuch as there is no handling of the broach, the operating cycle is very short. Usually, slide or rotary-indexing fixtures are used to hold the work, reducing the work-handling time and minimizing the total cycle time.

Horizontal broachng machines. The primary reason for employing a *horizontal* configuration for pull- and surface-broaching machines is to make possible longer strokes and the use of longer broaches than can conveniently be accommodated in vertical machines. As shown in Figure 24-11, horizontal pull-broaching machines essentially are vertical machines turned on their sides. Where internal surfaces are to be broached, such as holes, the broach must

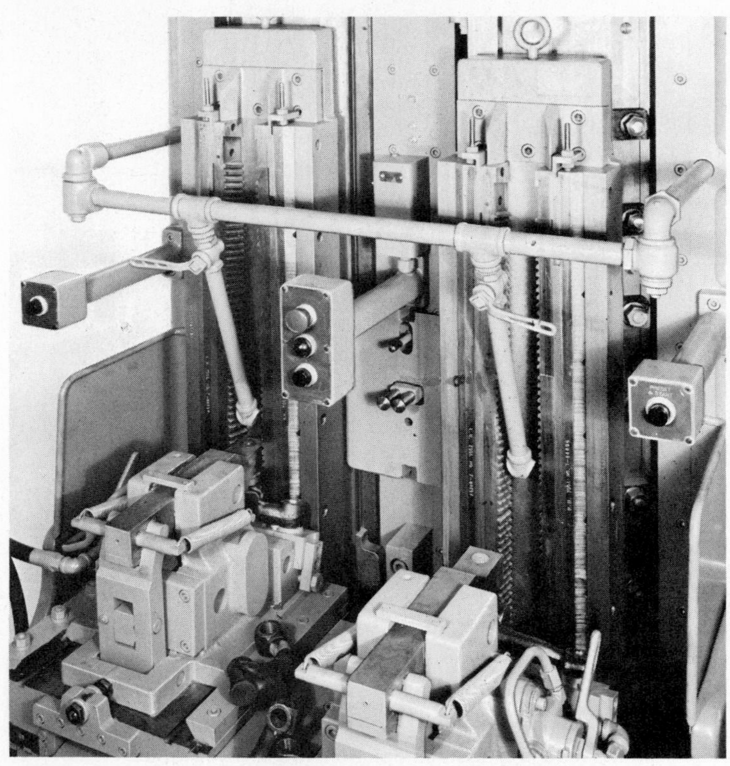

FIGURE 24-10. Close-up view of the fixtures and broaching inserts on a vertical duplex hydrobroaching machine. Work is being broached in the left-hand fixture while a workpiece has just been placed in the right-hand fixture. One fixture moves forward while work is being loaded in the second, retracted fixture. (*Courtesy Cincinnati Milacron Inc.*)

have a diameter-to-length ratio large enough to make it self-supporting without appreciable deflection. Consequently, horizontal machines seldom are used for small holes. In surface broaching, the broach always is supported in guides, so no such limitation is encountered. However, most horizontal broaching machines are quite large. Broaching that requires rotation of the broach, as in rifling and spiral splines, usually is done on horizontal machines.

Horizontal surface-broaching machines vary widely in size and design. As in the vertical type, the broaches are mounted on heavy, ram-driven slides. Some cut in only one direction of the ram stroke, whereas others are provided with two sets of broaches, arranged so that cutting occurs during both motions of the ram. Figure 24-12 shows a typical horizontal surface-broaching machine, such as often is incorporated into production lines.

Continuous surface-broaching machines. In *continuous surface-broaching machines,* the broaches usually are stationary, and the work is pulled past them

FIGURE 24-11. Broaching an internal spline on a horizontal, pull-type machine. (*Courtesy Colonial Broach & Machine Company.*)

FIGURE 24-12. Simultaneously machining semicircular and flat surfaces for bearings in an engine crankcase on a horizontal surface-broaching machine. (*Courtesy Cincinnati Milacron Inc.*)

on an endless conveyor. Fixtures usually are attached to the conveyor chain so that the workpieces can be placed in them at one end of the machine and removed at the other, sometimes automatically. Such machines are being used increasingly in mass-production lines.

Rotary broaching machines. In *rotary broaching machines,* occasionally used in mass production, the broaches are stationary, and the work is passed beneath, or between, them while held in fixtures on a rotary table. They have the advantage that there is no lost time due to noncutting, reciprocating strokes.

Review questions

1. What is unique about broaching, as compared with the other basic machining processes?
2. Why can a thick saw blade not be used as a broach?
3. Why are broaching machines more simple in basic design than most other machine tools?
4. Why is broaching particularly well suited for mass production?
5. Explain how internal spiral grooves are produced by broaching.
6. Why is it necessary to relate the design of a broach to the specific workpiece that is to be machined?
7. What two methods can be utilized to reduce the force and power requirements for a particular broaching cut?
8. For a given job, how would a broach having rotor-tooth design compare in length with one having regular, full-width teeth?
9. Why are the pitch and radius between teeth on a broach of importance?
10. Why are broaching speeds usually relatively low, as compared with other machining operations?
11. Why are some broaches made with shell-type construction?
12. Why are most broaches made from alloy or high-speed steel rather than from tungsten carbide?
13. How does a simple broaching press differ from an arbor press?
14. For mass-production operations, why are pull-up broaching machines usually preferred over pull-down machines?
15. Why can continuous broaching machines not be used for broaching holes?
16. The sides of a square, dead-end hole must be machined clear to the bottom. Would it be possible to do this by broaching? Why?
17. The interior, flat surfaces of socket wrenches, which have one "closed" end, often are finished to size by broaching. By examining one of these, what design modification was incorporated to make this operation possible?
18. A surface 304.8 mm long is to be machined with a flat, solid broach that has a tooth step of 0.12 mm. What is the minimum cross-sectional area that must be provided in the chip gullet between adjacent teeth?

19. Why is a continuous surface-broaching machine basically more productive than an ordinary horizontal surface-broaching machine?

20. The pitch of the teeth on a simple broach can be determined by the equation $p = \frac{1}{8}\sqrt{\text{length of surface}}$. If a broach is to remove 6.35 mm of material from a nodular iron casting that is 76 mm wide and 450 mm long, and if each tooth has a step of 0.10 mm, what will be the minimum length of the broach?

Case study 17.
THE JO-KO COMPANY'S COLLARS

The Jo-Ko Company requires 25,000 of the collars shown in Figure CS-17. They are to be made of 18-8-type stainless steel, and will be purchased from suitable suppliers. Determine two practicable methods for manufacturing these collars, give an estimate of the relative cost by the two methods, and state the reasons for the difference in cost, if any. (If convenient, contact two companies that would make the collars by the methods you have selected and verify your relative-cost estimate.)

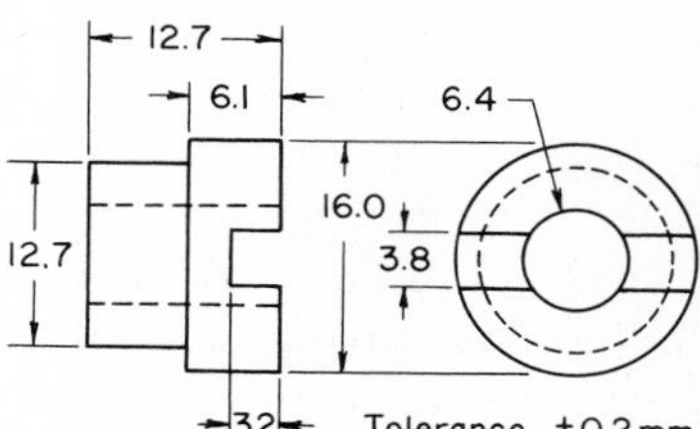

FIGURE CS-17. Connecting collar.

CHAPTER 25

Abrasive machining processes

Abrasive machining is the basic process in which chips are formed by very small cutting edges that are integral parts of abrasive particles, usually manmade. Unquestionably, abrasive machining is the oldest of the basic machining processes. Museums abound with examples of utensils, tools, and weapons that ancient man produced by rubbing hard stones against softer materials to abrade away unwanted portions, leaving desired shapes. For centuries, only natural, nonuniform, and relatively ineffective abrasives were available, and abrasive machining was far surpassed in importance and use by more modern, basic machining processes, which were developed around superior cutting materials. However, during this century two developments have changed this situation. First was the development of man-made abrasives. Second, in recent years basic research provided a fundamental understanding of the abrasive machining process. As a consequence, several variations of abrasive machining are among the most important of all the basic machining processes.

The results that can be obtained by abrasive machining range from the finest and smoothest surfaces produced by any machining process, in which very little material is removed, to rough, coarse surfaces that accompany high material-removal rates. In many cases the abrasive particles are bonded into "wheels" of various shapes, which are rotated against the workpiece. Such processes commonly are called *grinding* when used primarily for accurate sizing and smoothing by the removal of small amounts of material; they sometimes are called *abrasive machining* when used primarily for rapid metal removal to obtain a desired shape and approximate size.[1] In another abrasive machining process, fine abrasive particles are bonded into "stones," which are moved against the workpiece, usually in reciprocating or oscillatory motions; this process commonly is called *honing.* In another process very fine abrasive particles are used to impregnate, or "charge," a soft metal or cloth "lap," which then is moved in contact with the workpiece; this process is called *lapping.* In still another process, fine abrasive particles are hurled against the workpiece at ultrasonic velocities; this is called *ultrasonic machining.* Thus machining processes that use abrasive particles as the cutting tools vary widely in their application and results.

The abrasive machining processes have two unique characteristics. First, because each cutting edge is very small and a number of these edges can cut simultaneously when suitable machines are employed, very fine cuts are possible, and fine surfaces and close dimensional control can be obtained. Second, because extremely hard abrasive particles can be produced, very hard materials, such as hardened steel, glass, carbides, and ceramics, can be machined very readily. As a result, the abrasive machining processes are not only important as manufacturing processes, they are essential. Many of our modern products, such as automobiles, space vehicles, and aircraft, would not be possible without them. To a very large degree they have made possible longer-lived machines and mechanisms that will operate efficiently and dependably at sustained high speeds.

ABRASIVES

An *abrasive* is a hard material that can cut or abrade other substances. As noted previously, certain abrasives have existed from the earliest times. Sandstone was used by ancient people to sharpen tools and weapons. Early grinding wheels were cut from slabs of sandstone but, because they were not uniform throughout, they wore unevenly and did not produce consistent results. Thus it was not until artificial abrasives, having uniform and known qualities, were available that abrasive machining became a true manufacturing process.

[1] The application of the term "abrasive machining" to this one particular use of the grinding process is unfortunate, because all the processes that utilize abrasives to form chips obviously are abrasive machining.

Today, the only natural abrasives that have commercial importance are quartz sand, garnets, and diamonds. Quartz sand is used primarily in coated abrasives and in air blasting (which will be discussed later), but artificial abrasives are also making rapid inroads in these applications.

Diamonds are the hardest of all materials. Those that are used for abrasives are either natural, off-color stones that are not suitable for gems or small, man-made stones that are produced specifically for abrasive purposes. Neither natural nor synthetic diamonds have clear superiority over each other; man-made stones appear to be somewhat more friable and thus tend to cut faster and cooler. They do not perform as satisfactorily in metal-bonded wheels. Diamond abrasive wheels are used extensively for sharpening carbide and ceramic cutting tools. Diamonds also are used for truing and dressing other types of abrasive wheels. Obviously, because of their cost, diamonds are used only when cheaper abrasives will not produce the desired results.

Garnets are used primarily in the form of very finely crushed and graded powders for fine polishing.

Artificial abrasives date from 1891, when E. G. Acheson, while attempting to produce precious gems, discovered how to make *silicon carbide,* SiC. Silicon carbide is made by charging an electric furnace with silica sand, petroleum coke, salt, and sawdust. A temperature of over 2200°C (4000°F) is maintained for several hours, by passing large amounts of current through the charge, and a solid mass of silicon carbide crystals results. After the furnace has cooled, the mass of crystals is removed, crushed, and graded to various desired sizes. As can be seen in Figure 25-1, the resulting crystals, or grains, are irregular in shape, and most of the resulting cutting edges have negative rake.

Silicon carbide crystals are very hard, being about 9.5 on the Moh's scale, where diamond has a hardness of 10. Unfortunately, the crystals are rather brittle, and this limits their use. Silicon carbide is sold under the trade names *Carborundum* and *Crystolon.*

FIGURE 25-1. Abrasive grains at high magnification, showing their irregular, sharp cutting edges. (*Courtesy Norton Company.*)

Aluminum oxide, Al_2O_3, is the most widely used artificial abrasive. Also produced in an arc furnace, from bauxite, iron filings, and small amounts of coke, it contains aluminum hydroxide, ferric oxide, silica, and some other impurities. The mass of aluminum oxide that is formed is crushed, and the particles are graded to size. Common trade names for aluminum oxide abrasives are *Alundum* and *Aloxite*.

Although aluminum oxide is softer than silicon carbide, it is considerably tougher. Consequently, it is a more general-purpose abrasive.

A modification of aluminum oxide, containing vanadium, is somewhat harder and cuts more freely. It appears to be superior for grinding very hard steels.

Another relatively new abrasive is *boron nitride*. It is harder than either silicon carbide or aluminum oxide and is advantageous for grinding certain hardened tool-and-die steels. It often gives grinding ratios of from 150 to 700 in these applications. For some applications it is superior to diamonds, even though it is substantially softer.

Abrasive grain size. To assure uniformity of cutting action, abrasive grains are graded into various sizes, specified by numbers 4 through 600. These numbers originally indicated the number of openings per linear inch in a standard screen through which most of the particles of a particular size would pass. A 24 grit would pass through a standard screen having 24 openings per inch but would not pass through one having 30 openings per inch. These numbers have since been specified in terms of millimeters and micrometers.[2] Commercial practice commonly designates grain sizes from 4 to 24 inclusive, as *coarse;* 30 to 60, inclusive, as *medium;* and 70 to 600, inclusive, as *fine.* Silicon carbide is obtainable in grit sizes ranging from 2 to 240 and aluminum oxide in sizes from 4 to 240. Sizes from 240 to 600 are designated as *flour* sizes. These are used primarily for lapping, or in fine honing stones.

GRINDING

Grinding, wherein the abrasives are bonded together into a wheel of some shape, is the most common abrasive machining process. The cutting action of such wheels is greatly affected by the bonding material and the spatial arrangement of the abrasive particles, known as the structure.

Bonding materials for abrasive wheels. Inasmuch as the bonding material affects the cutting action of an abrasive wheel, it is a very important factor to be considered in selecting a grinding wheel. First, it determines the strength of the wheel, thus establishing the maximum speed at which it can safely be operated. Second, it determines whether the wheel is rigid or flexible. Third,

[2] See ANSI Standard B74. 12 for details.

it determines the force that is required to dislodge an abrasive particle from the wheel and thus plays a major role in the cutting action. Five types of bonding materials are in common use.

Vitrified bonds are used most extensively. They are composed of clays and other ceramic substances. The abrasive particles are mixed with the wet clays so that each grain is coated. Wheels are formed from the mix, usually by pressing, and then dried. They then are fired in a kiln, which results in the bonding material becoming hard and strong, having properties similar to glass. Vitrified wheels are porous and unaffected by oils, water, or temperature over the ranges usually encountered. The operating speed range in most cases is from 1676 to 1981 meters (5500 to 6500 feet) per minute, but some wheels now operate at surface speeds up to 4877 meters (16,000 feet) per minute.

Silicate wheels use silicate of soda (waterglass) as the bond, the wheels formed from the mixture being baked at about 260°C (500°F) for a day or more. They are not as strong as vitrified wheels, and the abrasive grains are released more readily. Consequently, they operate somewhat cooler than vitrified wheels and are useful in grinding tools where the temperature must be kept to a minimum.

Shellac-bonded wheels are made by mixing the abrasive grains with shellac in a heated mixer, pressing or rolling into the desired shapes, and baking for several hours at about 150°C (300°F). This type of bond is used primarily for strong, thin wheels having some elasticity. They tend to produce a high polish and thus are used in grinding such parts as camshafts and mill rolls.

Rubber bonding is used to produce wheels that can operate at high speeds but must have a considerable degree of flexibility so as to resist side thrust. Rubber, sulfur, and other vulcanizing agents are mixed with the abrasive grains. The mixture then is rolled out into sheets of the desired thickness, and the wheels are cut from these sheets and vulcanized. Rubber-bonded wheels can be operated at speeds up to 4877 meters (16,000 feet) per minute. They commonly are used for snagging work in foundries and for thin cutoff wheels.

Resinoid, or plastic, bonding now is widely used. Because plastics can be compounded to have a wide range of properties, such wheels can be obtained to cover a variety of work conditions. They have, to a considerable extent, replaced shellac and rubber wheels.

Some type of reinforcing frequently is added in rubber- or resinoid-bonded wheels that are to have some degree of flexibility, or are to be used in applications where they are apt to receive considerable abuse and side loading. Various natural and synthetic fabrics and fibers, glass fibers, and nonferrous wire mesh are used for this purpose.

Grinding-wheel cutting action. To select and use grinding wheels effectively, it is necessary to understand the manner in which the cutting action occurs. Because each exposed grain constitutes a small cutting edge, thus providing a multiplicity of cutting edges, the spacing of the abrasive particles and the ef-

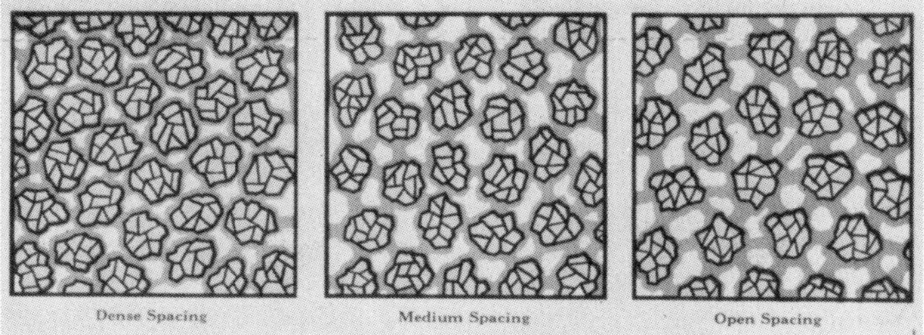

FIGURE 25-2. Meaning of grinding wheel "structure." (*Courtesy The Carborundum Company.*)

fect of the bonding material, which holds them in the wheel, are very important. Obviously, the number of cutting edges exposed per unit of area at any time is determined by the spacing of the abrasive grains; this is referred to as the **structure** of the wheel. The meaning of this term is illustrated in Figure 25-2.

As illustrated in Figure 25-3, each cutting edge acts the same as any other cutting edge under similar conditions. Because the cutting edges are very small, the resulting chips are small, as shown in Figure 25-4, but the same basic mechanism of compression and shear exists. In addition, the depth of cut and feed also must be small. However, the combination of a large number of cutting edges, small depth of cut, and small feed permits close tolerances and an excellent surface finish to be obtained.

As with any other cutting edge, as cutting progresses the edge dulls. Inasmuch as the cutting edges cannot be resharpened in the usual manner, it is desirable to have dulled grains removed from the wheel to permit unused, sharp grains to do the cutting. This can occur automatically in a grinding

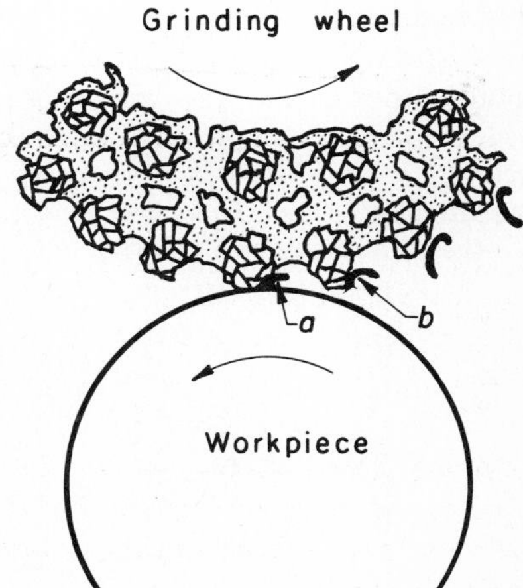

FIGURE 25-3. Relationship between the abrasive grains in a grinding wheel and the workpiece; chip being formed at (a) and completely formed at (b).

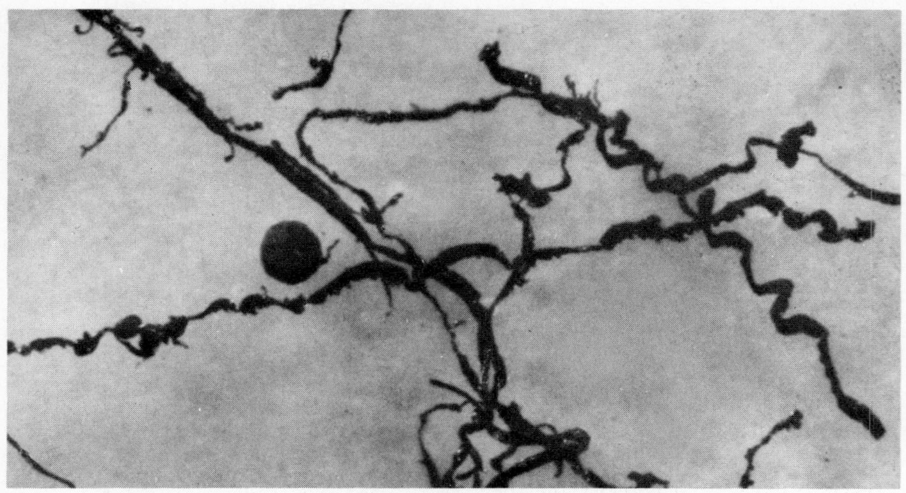

FIGURE 25-4. Chips produced by grinding; 100X. Black ball is a chip that is rolled up and fused together.

wheel through the fact that as the grains become dull the cutting force increases and there is an increased tendency for the grains to be pulled free from the bonding material. This action can be controlled by varying the strength of the bond, known as the *grade.* Grade thus is a measure of how strongly the grains are held in the wheel. It really is dependent on two factors: the strength of the bonding material, and the amount of the bonding agent connecting the grains. This latter factor is illustrated in Figure 25-5. Because abrasive wheels are usually porous, the grains are connected and held together as "posts" of bonding material. If these posts are large in cross section, the force required to break a grain free from the wheel is greater than when the posts are small. If a high dislodging force is required, the bond is said to be *hard.* If only a small force is required, the bond is said to be *soft.* Wheels commonly are referred to as hard or soft, referring to the net strength of the bond

FIGURE 25-5. Effect of the amount of bonding agent on the holding of abrasive grains in a wheel. (*Courtesy The Carborundum Company.*)

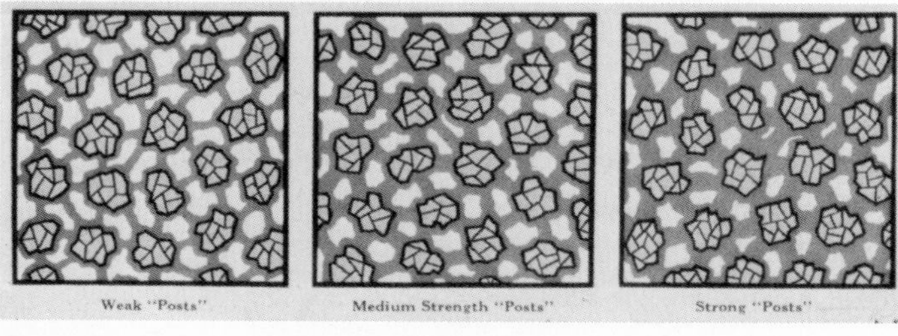

resulting from both the strength of the bonding material and its disposition between the grains.

Obviously, the shedding of grains from a grinding wheel depletes the available abrasive remaining and is a cost factor. In most grinding work the *grinding ratio*—the ratio of metal removal to abrasive used—is in the range of 40:1 to 80:1.

Another procedure for eliminating dulled abrasive edges is to fracture the grains by *crush dressing*. This procedure will be discussed later.

As a grinding wheel is used, there is a tendency for the wheel to become "loaded" with metal and dirt particles that become lodged between the exposed and dulling abrasive grains. Unless the wheel surface is cleaned and sharpened, by removing the dulled edges, the wheel does not cut properly. Ideally, as the cutting forces increase, the dulled grains should be pulled from the bond, thus providing a continuous exposure of sharp cutting edges. Such a continuous action ordinarily will not occur when light feeds and depths of cut are used, as commonly must be employed when grinding to obtain accurate dimensions or fine finishes. However, if considerably heavier cuts are taken, grinding wheels do become self-sharpening and large amounts of metal can be removed very efficiently. The net effects are shown in Figure 25-6. As the downfeed is increased, the required power increases rapidly up to the point at which the wheel becomes completely self-sharpening. After this the required power decreases sharply and later increases more slowly than during the stages of light feed. The metal-removal rate increases very markedly, accompanied by increased wheel wear. By operating with the conditions represented by the shaded area, much faster metal removal can be obtained with only slightly greater wheel wear and at much lower cost.

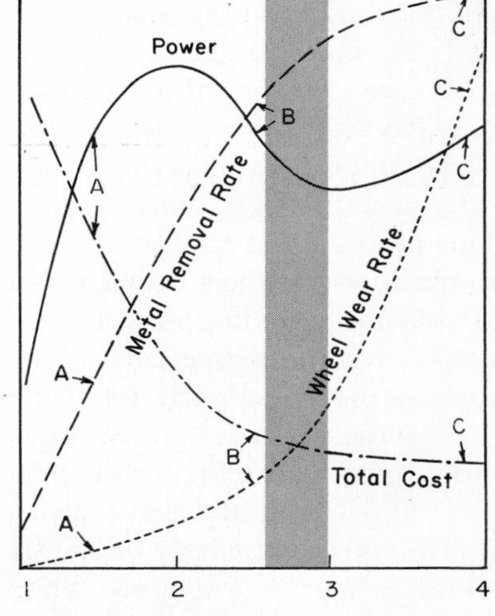

FIGURE 25-6. Relationship between power required, metal removal rate, wheel wear, and total cost as functions of downfeed. (*Courtesy Norton Company.*)

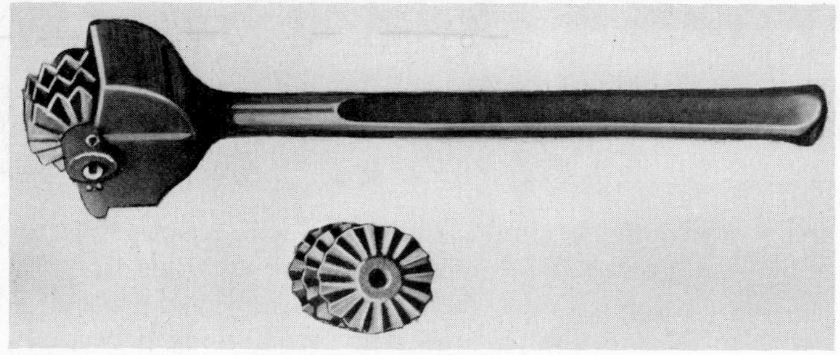

FIGURE 25-7. Tool for dressing grinding wheels. (*Courtesy Norton Company.*)

The condition where very rapid metal removal can be achieved by grinding is the one to which some have applied the term *abrasive machining,* undoubtedly due to the fact that the metal-removal rates are comparable with, or exceed, those obtainable by conventional cutter machining, and the size tolerances are comparable. But it obviously is just a special type of grinding, using abrasive grains as cutting tools, as do all other types of abrasive machining. When this procedure is used, it is followed by a few passes of conventional grinding, using small feeds, if better tolerances and finish are required.

Two procedures are employed to expose fresh, sharp cutting edges and to "unload" grinding wheels. One is to "dress" the wheel by using a dressing tool. A steel dressing tool, such as is shown in Figure 25-7, consists of irregularly shaped, hardened steel disks that are free to rotate on an axle. The tool is held at an angle against the rotating grinding wheel and moved across its face. The revolving points of the disks remove the foreign particles that have become lodged between the grains and also remove or fracture dulled grains, thus exposing sharp edges. A diamond dressing tool primarily fractures the abrasive grains and thus exposes new cutting edges.

Crush dressing, illustrated in Figure 25-8, consists of forcing a hard steel roll, having the same contour as the part to be ground, against the grinding wheel while it is revolving—usually quite slowly. The crushing action fractures and dislodges some of the abrasive grains, exposing fresh, sharp edges. This procedure usually is employed to produce and maintain a special contour to the abrasive wheel in form grinding. Crush dressing is a very rapid method of dressing grinding wheels and, because it fractures more of the abrasive grains, results in free cutting and somewhat cooler grinding. The resulting surfaces may be slightly rougher than when diamond dressing is used.

Some machines are equipped so that the wheel can be dressed continuously or intermittently while grinding continues.

Obviously, the self-sharpening tendency of a grinding wheel, determined by the strength or hardness of the bond, is important in grinding-wheel action. A soft-grade wheel loses its grains readily and thus tends to remain sharp at all times. However, it wears rapidly. Conversely, a hard

FIGURE 25-8. Machine equipped for crush dressing the grinding wheel. The crush-dressing roll is above the abrasive wheel and denoted by the arrow. (*Courtesy Bendix Automation & Measurement Division.*)

wheel wears less rapidly but is likely to become dull, particularly if used on hard materials. In grinding hard materials, the abrasive grains dull more rapidly than in grinding soft materials. Therefore, a wheel for grinding hard materials should lose its grains rapidly. Thus softer-grade wheels are used for grinding hard materials and harder-grade wheels for soft materials. However, several other factors also affect the apparent hardness or softness of a wheel.

If the speed of the work is held constant, increasing the speed of a grinding wheel makes it act harder; the higher speed tends to dull the grains more rapidly. However, because the path of the grain through the work is decreased, the dislodging force is less. Similarly, increasing the work speed, while maintaining a constant wheel speed, results in a softer action. A larger wheel will act harder than a smaller wheel of the same type and bond because of the larger contact area between the wheel and the work.

The spacing between the abrasive grains also is important in the cutting action. As shown in Figure 25-3, this spacing must provide room for the chips that are formed. If there is insufficient space, the chips will become wedged between the grains, protrude beyond the periphery of the wheel, and scratch the work surface. Thus, if heavy cuts are to be taken, or if the path of contact is long, greater spacing between the grains is necessary.

Grinding-wheel identification. Most grinding wheels are identified by a standard marking system that has been established by the American National Standards Institute, Inc. This system is illustrated and explained in Figure

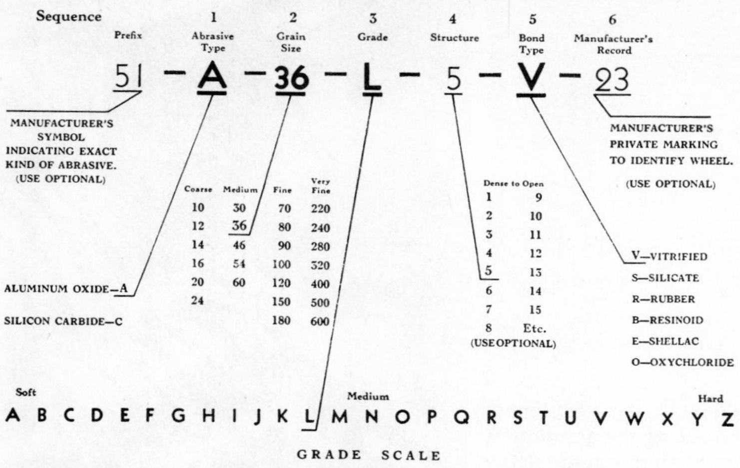

FIGURE 25-9. Standard marking system for grinding wheels.

25-9. It should be remembered that the first and last symbols in the marking are left to the discretion of the manufacturer.

Grinding-wheel selection. If optimum results are to be obtained, it is very important that the proper grinding wheel be selected for a given job. Several factors must be considered. Probably the first is the shape and size of the wheel. Obviously, the shape must permit proper contact between the wheel and all of the surface that must be ground. Grinding-wheel shapes have been standardized by the Grinding Wheel Manufacturers' Association. Eight of the most commonly used types are shown in Figure 25-10. Types 1, 2, and 5 are used primarily for grinding external or internal cylindrical surfaces and for plain surface grinding. Type 2 can be mounted for grinding either on the periphery or the side of the wheel. Type 4 is used with tapered safety flanges so that if the wheel is broken in doing rough grinding, such as snagging, these flanges will prevent the pieces of the wheel from flying and causing damage. Type 6, the straight cup, is used primarily for surface grinding but can also be used for certain types of offhand grinding. The flaring-cup type of wheel is used for tool grinding. Dish-type wheels are used for grinding tools and saws.

Straight grinding wheels can be obtained with a variety of standard faces. Some of these are shown in Figure 25-11.

The size of the wheel to be used is determined, primarily, by the spindle speeds available on the grinding machine and the proper cutting speed for the wheel, as dictated by the type of bond. For most grinding operations the cutting speed is about 1981 meters (6500 feet) per minute, but different types

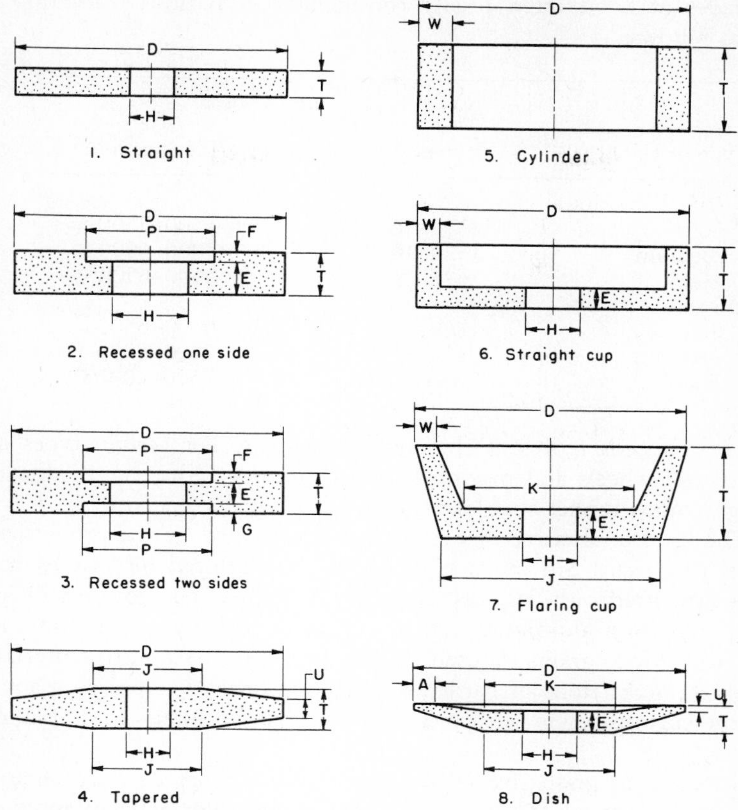

FIGURE 25-10. Common standard grinding-wheel shapes.

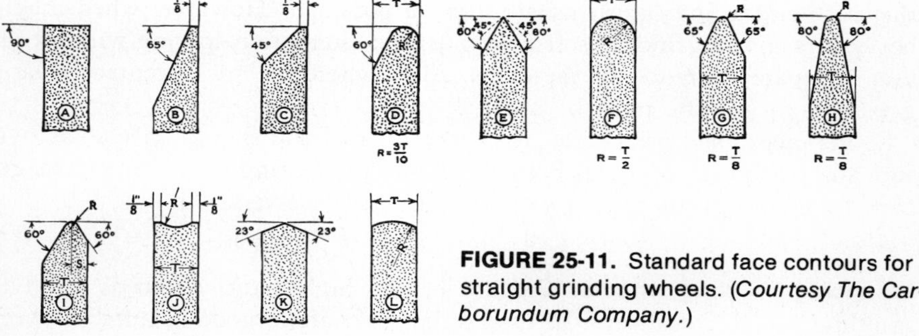

FIGURE 25-11. Standard face contours for straight grinding wheels. (*Courtesy The Carborundum Company.*)

and grades of bond often justify considerable deviations from this average speed as follows:

Type of Bond	Surface Speed	
	Meters per minute	Feet per minute
Vitrified or silicate		
Soft	1372–1676	4500–5500
Medium	1524–1829	5000–6000
Hard	1676–1981	5500–6500
Organic		
Soft	1829–1981	6000–6500
Medium	2134–2438	7000–8000
Hard	2286–3048	7500–10,000

These speeds represent the common ranges. For certain types of work using special wheels and machines, as in thread grinding and "abrasive machining," much higher speeds are used.

The kind of abrasive that is used is determined by the properties of the material to be cut. Silicon carbide usually is employed for brittle materials, such as cemented carbides, cast iron, and ceramics, and low-tensile-strength metals, including aluminum, brass, copper, and bronze. Aluminum oxide, having less brittle grains, is used for tougher, higher-strength materials, such as steel, wrought iron, or hardened steel. However, because so many factors affect the cutting action, either type of abrasive can be used for certain applications.

Selection of grain size is determined by whether coarse or fine cutting and finish are desired. Coarse grains take larger bites and cut more rapidly; fine grains leave smaller scratches and thus usually are selected for finishing cuts. Because of their cleaner cutting action, slightly large grains may be preferred for finishing work if there is a tendency for the work material to load the wheel.

The spacing of the grains (structure) is determined by the problem of chip clearance. With closer spacing, more cutting edges are in contact with the work and a finer finish usually can be obtained. However, when making heavy cuts and in grinding soft materials, too-close grain spacing will not provide adequate clearance for the chips. Also, where the arc of contact is large, greater grain spacing must be used.

As mentioned previously, the grade of the bond should be such as to assure sharp abrasive particles being available for cutting at all times. In general, harder materials call for softer bonds, and vice versa.

Balancing grinding wheels. Because of the high rotative speeds involved, grinding wheels must never be used unless they are in good balance. Not only will slight unbalance produce vibrations that will cause waviness in the work surface, but it may cause a wheel to break, with the probability of serious damage and injury.

The wheel should be properly bushed so it fits snugly on the spindle of the machine. Rings of blotting paper should be placed between the wheel and the flanges to assure that the clamping pressure is evenly distributed. Most grinding wheels will run in good balance if they are mounted properly and trued. Most machines have provision for compensating for a small amount of wheel unbalance by attaching weights to one mounting flange. Some have provision for semiautomatic balancing with weights that are permanently attached to the machine spindle.

Safety in grinding. Because the rotational speeds are quite high, and the strength of grinding wheels usually is much less than the materials on which they may be used or against which they may accidentally be struck, serious accidents occur much too frequently in connection with the use of grinding wheels. Virtually all such accidents could be avoided and are due to one or more of four causes. First, grinding wheels occasionally are operated at unsafe, and improper, speeds. All grinding wheels are clearly marked with the maximum rpm at which they should be rotated. They all are tested to considerably above the designated rpm and are safe at the specified speed *unless abused. They should never, under any condition, be operated above the rated speed.* Second, a most common form of abuse, frequently accidental, is dropping the wheel or striking it against a hard object. This can result in a crack, which may not be readily visible, and subsequent failure of the wheel while rotating at high speed. If a wheel is dropped or struck against a hard object, it should be discarded and never used unless tested at above the rated speed in a properly designed test stand. A third common cause of grinding wheel failure is improper use, such as grinding against the side of a wheel that was designed for grinding only on its periphery. The fourth and most common cause of injury from grinding is the absence of a proper safety guard over the wheel and/or over the eyes or face of the operator. The frequency with which people will remove safety guards from grinding equipment, or fail to use safety goggles or face shields, is amazing and inexcusable.

The use of fluids in grinding. Because grinding is a true cutting process, the selection and use of a cutting fluid is governed by the basic principles discussed in Chapter 17. If a fluid is used, it should be applied in sufficient quantities and in a manner that will assure that the chips are washed away and not trapped between the wheel and the work. This is of particular importance in grinding horizontal surfaces. The use of a fluid can help to prevent fine microcracks that may result from highly localized heating when hardened steels are ground.

Much grinding is done dry—for example, most snagging and offhand grinding. On some types of material, dry grinding produces a better finish than can be obtained by wet grinding.

Electrolytic grinding. Actually, in electrolytic grinding about 99.5 per cent of the material is removed by electrochemical action without the formation of chips. Consequently, it will be discussed in a later chapter on chipless machining processes. The name derives from the fact that it is a substitute for grinding and that about 0.5 per cent of the material unavoidably is removed by grinding. It is used primarily for sharpening tungsten carbide cutting tools.

Grinding machines. Grinding machines commonly are classified according to the type of surface they are used to produce. The following is such a classification, with further subdivision to indicate characteristic features of different types of machines within each classification:

Type of Surface	Characteristic Features
Cylindrical	Work rotated between centers Centerless Chucking Tool post Crankshaft, cam, etc.
Internal	Chucking Planetary (work stationary) Centerless

Surface	Reciprocating table
	  Horizontal spindle
	  Vertical spindle
	Rotating table
	  Horizontal spindle
	  Vertical spindle

Tool	Universal Special
Special	Disk Flexible shaft Swing frame Pedestal, etc.

Basically, grinding on all machines is done in two ways. In the first the depth of cut is obtained by *infeed*—moving the wheel into the work, or the work into the wheel. The desired surface is then produced by traversing the wheel across the workpiece, or vice versa. In the second method, known as *plunge-cut* grinding, the basic movement is of the wheel being fed radially into the work while the latter revolves on centers. It is similar to form cutting on a lathe; usually a formed grinding wheel is used.

Grinding machines that are used for precision work have certain important characteristics which permit them to produce parts having close dimensional tolerances. They are constructed very accurately, with heavy, rigid frames to assure permanency of alignment. Rotating parts are accurately bal-

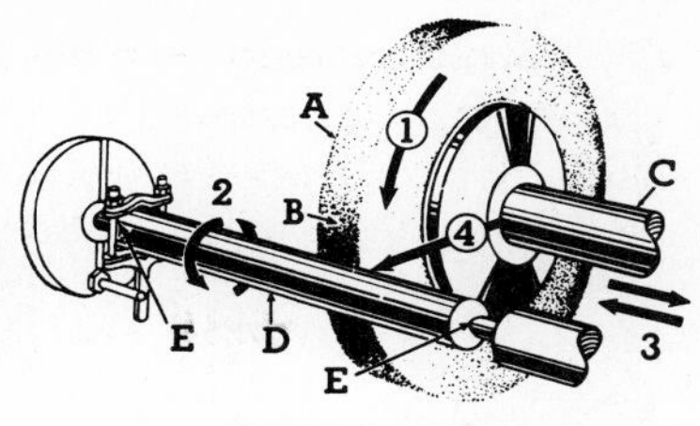

A. Grinding Wheel
B. Grinding Face
C. Wheel Spindle
D. Work Piece
E. Work Centers

MOVEMENTS
1. Wheel 2. Work
3. Traverse 4. Infeed

FIGURE 25-12. Basic motions in center-type cylindrical grinding. (*Courtesy The Carborundum Company.*)

anced to avoid vibration. Spindles are mounted in very accurate bearings, usually of the preloaded ball-bearing type. Controls are provided so that all movements that determine dimensions of the workpiece can be made with accuracy—usually to 0.0025 mm (0.0001 inch).

Another requisite is provision to prevent abrasive dust, resulting from grinding, from entering between moving parts. All ways and bearings must be fully covered or protected by seals. If this were not done, the abrasive dust between moving parts would become embedded in the softer of the two, causing it to become a charged lap. Subsequent movement would cause the harder of the two parts to be worn, resulting in permanent loss of accuracy of the machine.[3]

These special characteristics add considerably to the cost of these machines. Also, they often must be operated by skilled personnel. Such machines can be operated economically only if they have relatively high use factors. However, production-type grinders are designed to be used by less-skilled labor. Their metal-removal rates are high, and excellent dimensional accuracy and fine surface finish can be obtained very economically on them.

Center-type cylindrical grinders. *Center-type grinders* are the type used most commonly for producing external cylindrical surfaces. Figure 25-12 shows the basic principles and motions of these machines. The grinding wheel revolves at an ordinary cutting speed, and the workpiece rotates on centers at a much slower speed, usually from 23 to 38 meters (75 to 125 feet) per minute. The grinding wheel and the workpiece move in opposite directions at their point of contact. The depth of cut is determined by infeed of the wheel or workpiece. Because this motion also determines the finished diameter of the workpiece, accurate control of this movement is required. Provision is made to traverse the workpiece with respect to the wheel, usually by reciprocating the

[3] It is desirable to locate grinding machines apart from other machine tools, preferably in a separate room, to prevent this type of damage from occurring to machines that do not have such built-in protection.

FIGURE 25-13. Plain, center-type cylindrical grinding machine. (*Courtesy Cincinnati Milacron Inc.*)

work. However, in very large grinders the wheel is reciprocated due to the massiveness of the work.

A *plain center-type clyindrical grinder* is shown in Figure 25-13. On this type the work is mounted between headstock and tailstock centers. The tailstock center always is dead, and provision usually is made so that the headstock center can be operated either dead or live. High-precision work usually is ground with a dead headstock center, because this eliminates any possibility of the workpiece running out of round due to any eccentricity in the headstock.

The headstock spindle, mounted in accurate bearings, usually is driven through a belt drive. The workpiece is driven in rotation by means of a face plate and dog.

On this type of machine, the headstock and tailstock are mounted on a table, which can be swiveled approximately 10° about a vertical axis with respect to the table carrier on which it is mounted. This permits either straight cylinders or tapered cylinders, up to about 10° of taper, to be ground.

The table assembly can be reciprocated along the ways on the main frame, either manually or by power. In most cases hydraulic drive is used. The speed of the table movement can be varied, and the length of the movement can be controlled by means of adjustable tripping dogs.

Infeed is provided by movement of the wheelhead at right angles to the longitudinal axis of the table. The spindle is driven by an electric motor that also is mounted on the wheelhead. A flat-belt drive usually is employed to connect the driving motor and the spindle because this type of drive minimizes vibrations that might produce chatter marks on the work surface.

The infeed movement usually is controlled manually by some type of vernier drive to provide control to 0.025 mm (0.001 inch) or less. Some machines are equipped with digital readout equipment to show the exact size being produced. Some production-type grinders have devices that automatically infeed the wheel and then retract it when the desired size has been obtained. Such machines usually are equipped with an automatic diamond wheel-truing device that dresses the wheel and resets the measuring element before grinding is started on each piece.

The longitudinal traverse should be about one-fourth to three-fourths of the wheel width for each revolution of the work. For light machines and fine finishes, it should be held to the smaller end of this range.

The depth of cut (infeed) varies with the purpose of the grinding operation and the finish desired. When grinding is done to obtain accurate size, feeds of 0.051 to 0.010 mm (0.002 to 0.004 inch) commonly are used for roughing cuts. For finishing, the feed is reduced to 0.006 to 0.013 mm (0.00025 to 0.0005 inch). The design allowance for grinding should be from 0.13 to 0.25 mm (0.005 to 0.010 inch) on short parts and on parts that are not to be hardened. On long or large parts and on work that is to be hardened, a grinding allowance of from 0.38 to 0.76 mm (0.015 to 0.030 inch) is desirable. When grinding is used primarily for metal removal (so-called abrasive machining), feeds are much higher, 0.51 to 1.02 mm (0.020 to 0.040 inch) being common. Continuous downfeed often is used, rates up to 2.54 mm (0.100 inch) per minute being common.

Plain center-type cylindrical grinders contain systems for storing, filtering, and circulating adequate amounts of grinding fluid. Heavy wheel guards and adequate protection for the ways always are included.

The size of center-type grinders is designated by the *maximum diameter* and *length of work* that can be ground between centers. Plain grinders are obtainable in sizes ranging from 76.2 × 305 mm (3 × 12 inches) up to 610 × 4572 mm (24 × 180 inches) and larger.

Universal center-type grinders. *Universal center-type grinders* are basically the same as plain center-type grinders, except for two features. First, both the headstock and the wheelhead can be swiveled about vertical axes. This permits the grinding of tapers of all angles and certain other types of work that cannot be done on plain cylindrical grinders. The second feature is that most machines have dual spindles on the swiveling wheelhead, one for external grinding and the other for internal grinding. Either spindle can be brought into use by swiveling or tilting the wheelhead. Figure 25-14 shows a taper being ground on a universal center-type cylindrical grinder.

Roll grinders. *Roll grinders* basically are plain center-type machines designed for grinding large, cylindrical mill and calender rolls, which may be up to 1524 mm (60 inches) in diameter. Because of the weight of such workpieces, the wheelhead, instead of the work, reciprocates. Such a machine is shown in

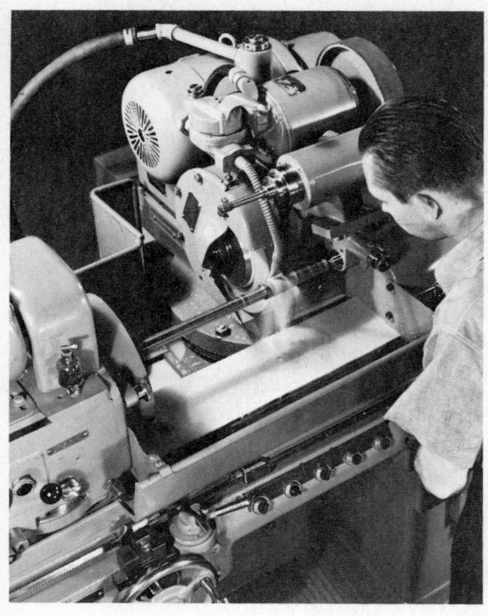

FIGURE 25-14. Grinding a taper on a universal center-type grinding machine. Wheels for both external and internal grinding are visible. (*Courtesy Cincinnati Milacron Inc.*)

Figure 25-15. Often there is provision for grinding a crown, or camber, on the roll to compensate for the deflection that occurs when they are subjected to heavy loads.

Chucking-type grinders. Grinding machines are available in which the workpiece is held in a chuck for grinding both external and internal cylindrical surfaces. *Chucking-type external grinders* are production-type machines, for use in

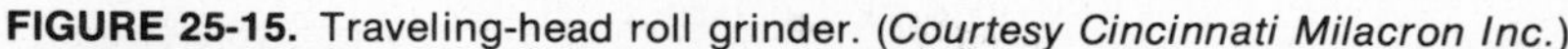

FIGURE 25-15. Traveling-head roll grinder. (*Courtesy Cincinnati Milacron Inc.*)

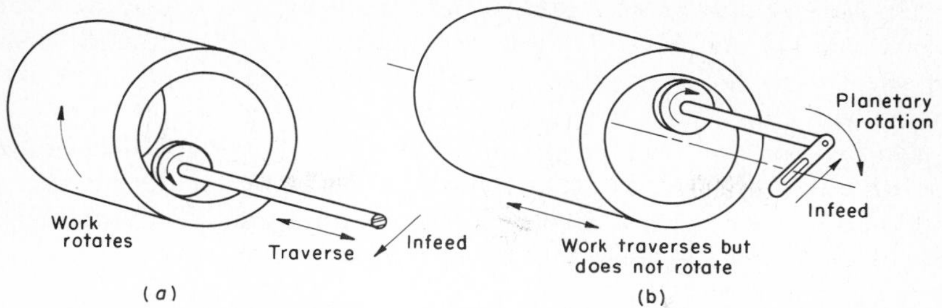

FIGURE 25-16. Two basic types of chucking-type internal grinders.

rapidly grinding relatively short parts, such as ball-bearing races. Both chucks and collets are used for holding the work, dictated by the shape of the workpiece and rapid loading and removal. Frequently, such machines have two head spindles so that work can be removed from one while another piece is being ground in the other.

Chucking-type internal grinding machines employ two basic principles, illustrated in Figure 25-16. The type illustrated in Figure 25-16a is more common. The chuck-held workpiece revolves, and a relatively small, high-speed grinding wheel is rotated on a spindle arranged so that it can be reciprocated in and out of the workpiece. Infeed movement of the wheelhead is normal to the axis of rotation of the work. A machine of this type is shown in Figure 25-17.

FIGURE 25-17. Size-matic chucking-type internal grinder. The machine is set up to grind two inside diameters on a gear. Two truing diamonds are mounted between the work and the grinding wheels. (*Courtesy The Heald Machine Company.*)

On *plain internal grinders* of this type, the workhead can be swiveled so that both straight internal cylinders and beveled holes can be ground. On *universal internal grinders* the workhead not only can be swiveled, but it also is mounted on a cross slide.

On some production-type machines the control of the infeed and traverse of the wheelhead is automatic. The control can be set for the desired size, and the grinding wheel is withdrawn and the machine is shut off automatically when the desired size is achieved. On one type of machine the hole size is automatically gaged after each reciprocation.

Planetary-type internal grinders, illustrated schematically in Figure 25-16b, are used for work that is too large to be rotated conveniently. The revolving grinding wheel also has planetary rotation about an axis that is coincident with the axis of the finished cylinder. The diameter of the ground surface is controlled by adjusting the radius of the planetary rotation. On this type of machine, the work is reciprocated past the wheel.

Tool-post grinders. *Tool-post grinders,* illustrated in Figure 25-18, are used to permit occasional grinding to be done on a lathe. The wheelhead is either a high-speed electric or air motor with the grinding wheel often mounted directly on the motor shaft. The entire mechanism is mounted either on the tool post or on the compound rest. The lathe spindle provides rotation for the workpiece, and the lathe carriage is used to reciprocate the wheelhead.

Although tool-post grinders are versatile and useful, care should be taken to cover the ways of the lathe with a closely woven cloth to provide pro-

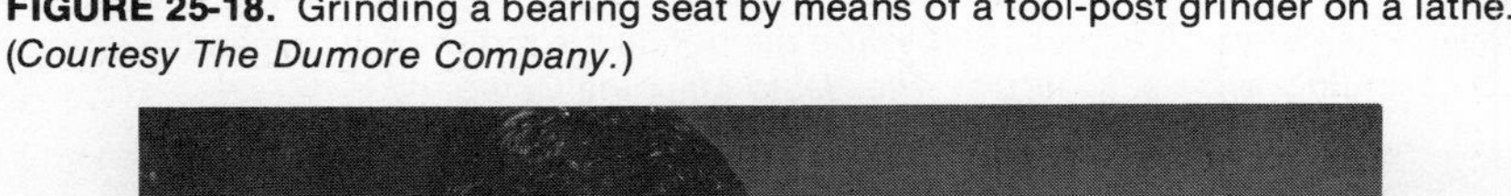

FIGURE 25-18. Grinding a bearing seat by means of a tool-post grinder on a lathe. (*Courtesy The Dumore Company.*)

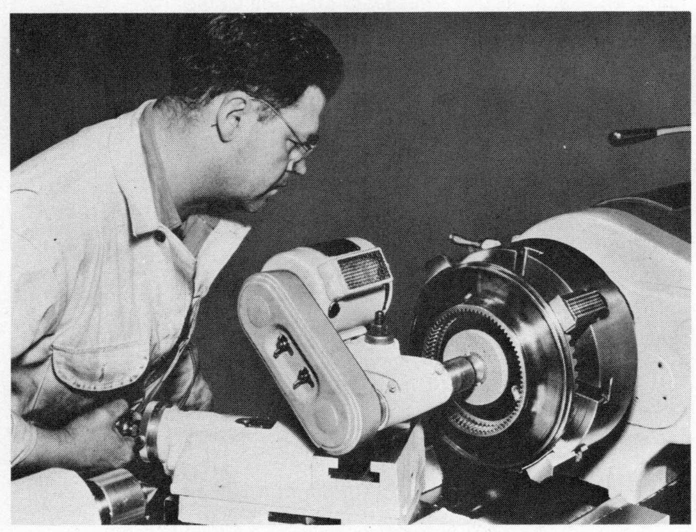

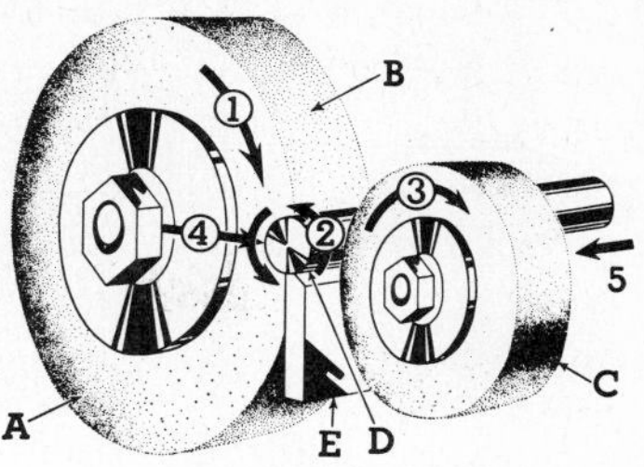

FIGURE 25-19. Principle of centerless grinding. (*Courtesy The Carborundum Company.*)

tection from the abrasive dust that can become entrapped between the moving parts.

Centerless grinding. *Centerless grinding* makes it possible to grind both external and internal cylindrical surfaces without the necessity of the workpiece being mounted between centers or in a chuck. This eliminates the requirement of center holes in some workpieces and the necessity for mounting the workpiece, thereby reducing the cycle time.

The principle of *centerless external grinding* is illustrated in Figure 25-19. Two abrasive wheels are used. The larger one operates at regular grinding speeds and does the actual grinding. The smaller wheel is the *regulating* wheel and is mounted at an angle to the plane of the grinding wheel, as shown in Figure 25-20. Revolving at a much slower surface speed—usually 15 to 61 meters (50 to 200 feet) per minute—it controls the rotation and longitudinal motion of the workpiece. It usually is a plastic- or rubber-bonded wheel with a fairly wide face.

The workpiece is held against the work-rest blade by the cutting force exerted by the grinding wheel and rotates at approximately the same surface

FIGURE 25-20. Relationship of the grinding wheels and the workpiece in centerless grinding.

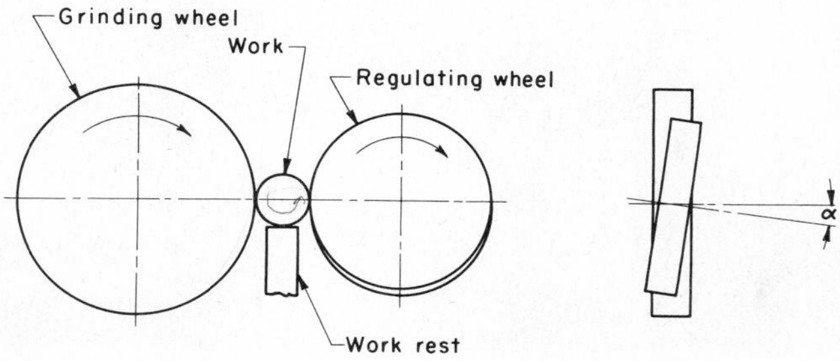

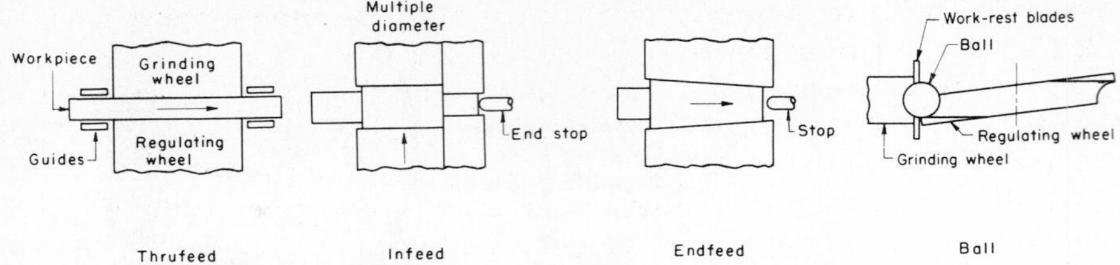

FIGURE 25-21. Three types of external centerless grinding.

speed as that of the regulating wheel. This axial feed is calculated approximately by the equation

$$F = dN \sin \alpha$$

where F = feed, millimeters or inches per minute
d = diameter of the regulating wheel, millimeters or inches
N = revolutions per mimute of the regulating wheel
α = angle of inclination of the regulating wheel

Three types of external centerless grinding operations can be done. These are illustrated in Figure 25-21. In *thrufeed grinding,* the workpiece is of constant diameter and is fed completely through between the rolls, starting at

FIGURE 25-22. External centerless grinding operation of the thrufeed type. (*Courtesy Cincinnati Milacron Inc.*)

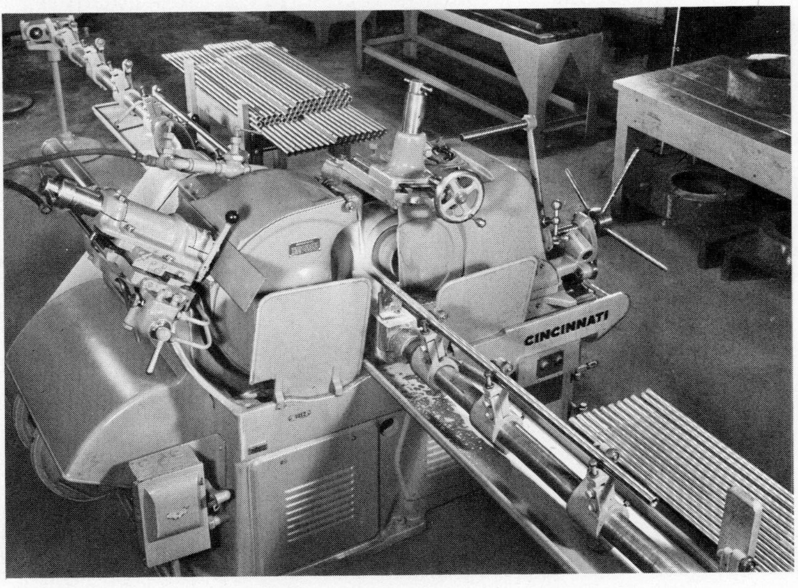

one end. This is the simplest type and can easily be made automatic. Figure 25-22 shows a typical grinding operation of this type.

In *infeed centerless grinding,* the work rest and the regulating wheel are retracted so that the work can be put in position and removed when grinding is completed. When the work is in position, the regulating wheel is fed inward until the desired diameter is obtained. This arrangement permits multiple diameters and curved parts to be ground. When multiple diameters are to be ground, a slight inclination of the regulating wheel holds the work against the end stop. For curved parts, a formed grinding wheel is used in conjunction with a plain regulating wheel to rotate the work about a single axis. For grinding balls, the regulating wheel is grooved and inclined to impart random rotation.

In *endfeed centerless grinding,* both the grinding and regulating wheels are tapered and thus produce tapered workpieces. The stock is fed from one side until it reaches the stop. This arrangement results in exact tapers and size.

Centerless grinding has several important advantages:

1. It is very rapid; infeed centerless grinding is almost continuous.
2. Very little skill is required of the operator.
3. It often can be made automatic.
4. Where the cutting occurs, the work is fully supported by the work rest and the regulating wheel. This permits heavy cuts to be made.
5. Because there is no distortion of the workpiece, accurate size control is easily achieved.
6. Large grinding wheels can be used, thereby minimizing wheel wear.

The major disadvantages are as follows:

1. Special machines are required that can do no other type of work.
2. The work must be round—no flats, such as keyways, can be present.
3. Its use on work having more than one diameter is limited.

Thus centerless grinding is ideally suited to certain types of mass-production operations.

Centerless internal grinding utilizes the principle illustrated in Figure 25-23. Three rolls support the workpiece on its outer surface and impart rotation

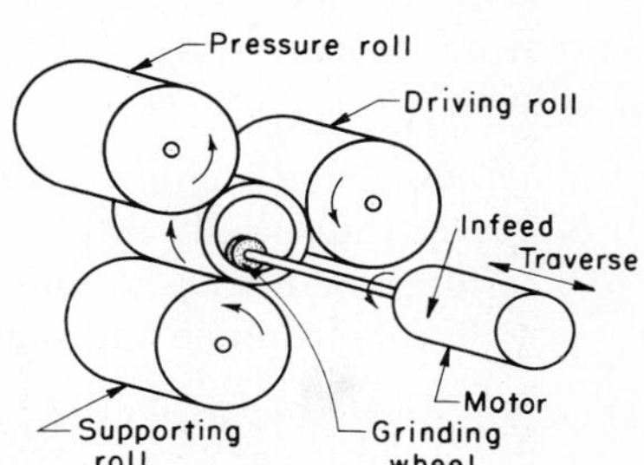

FIGURE 25-23. Principle of centerline internal grinding.

FIGURE 25-24. Automatic centerless internal grinding. The size is controlled continuously and automatically. The truing diamond is visible just below the raised cover. (*Courtesy The Heald Machine Company.*)

to it; one retracts to permit the work to be placed in position and removed. The grinding wheel traverses into the workpiece. This type of grinding, of course, requires that the external surface of the cylinder be finished accurately before the operation is started, but it assures that the internal and external surfaces will be concentric. The operation is easily mechanized for many applications, Figure 25-24 showing such an example. Usually, some type of automatic truing device is included on such machines.

Surface grinding. *Surface grinding* is used primarily to grind flat surfaces. However, formed, irregular surfaces can be produced on some types of surface grinders by using a formed wheel.

There are four basic types of surface grinding machines, differing in the movement of their tables and the orientation of the grinding wheel spindles, as follows:

1. Horizontal spindle and reciprocating table.
2. Vertical spindle and reciprocating table.
3. Horizontal spindle and rotary table.
4. Vertical spindle and rotary table.

These are illustrated in Figure 25-25.

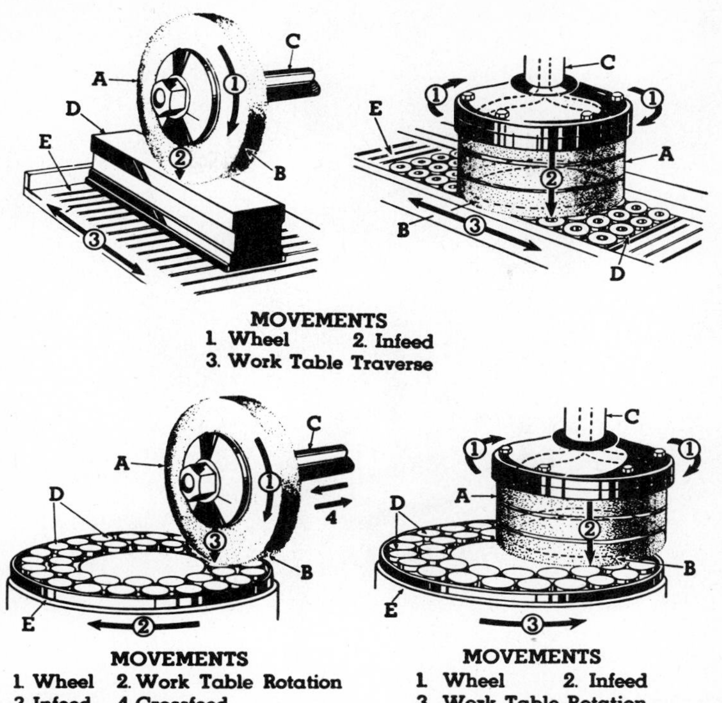

FIGURE 25-25. Motions in four types of surface grinding. *(Courtesy The Carborundum Company.)*

Surface grinding machines. The most common type of surface grinding machine has a reciprocating table and horizontal spindle, such as shown in Figure 25-26. Mounted on horizontal ways, the table can be reciprocated longitudinally either by handwheel or by hydraulic power. The wheelhead is given transverse motion at the end of each table motion, again either by handwheel or by hydraulic power feed. Both the longitudinal and transverse motions can be controlled by limit dogs that are easily set. Infeed on such grinders is controlled by a handwheel that lowers the grinding wheel toward the work. Most modern horizontal-spindle surface grinders have the grinding wheel mounted directly on the motor spindle.

The size of such machines is designated by the size of the surface that can be ground. Thus a 203 × 610 mm (8 × 24 inch) surface grinder has a sufficiently large table, and corresponding table and wheel motions, to permit grinding a plane surface 203 mm (8 inches) wide and 610 mm (24 inches) long. A wide range of sizes is available.

In using such machines, the wheel should overtravel the work at both

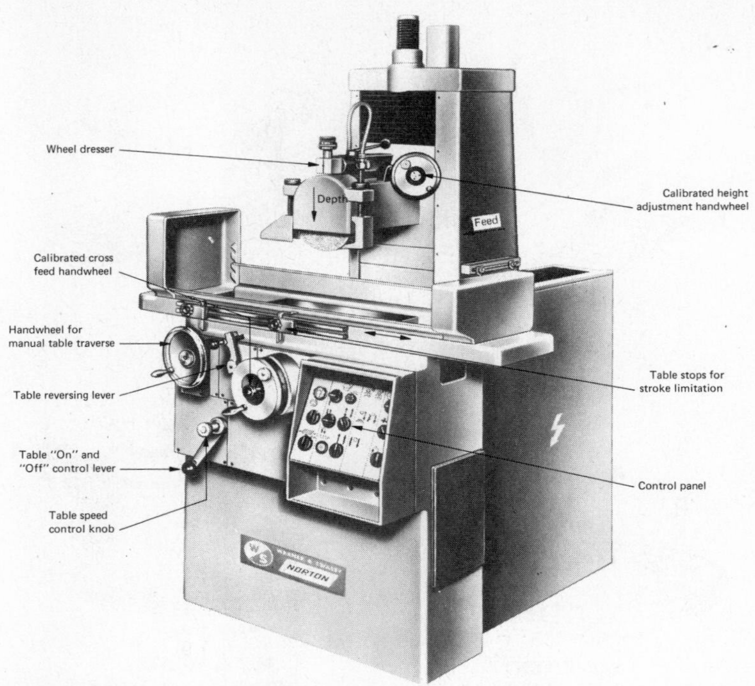

FIGURE 25-26. Reciprocating-table, horizontal-spindle surface grinder. (*Courtesy The Warner & Swasey Co., Grinding Machine Division.*)

ends of the table reciprocation, so as to prevent the wheel from grinding in one spot while the table is being reversed. The transverse motion should be one-fourth to three-fourths of the wheel width between each stroke.

As illustrated in Figure 25-27, *vertical-spindle reciprocating-table surface grinders* differ basically from those with horizontal spindles only in that their spindles are vertical and that the wheel diameter must exceed the width of the surface to be ground. Usually, no traverse motion of either the table or the wheelhead is provided. Such machines can produce very flat surfaces and are used primarily for production-type work.

Rotary-table surface grinders are of two types. Those with horizontal spindles will produce very flat surfaces but, because they are limited in the type of work they will accommodate, they are not used to a great extent. They usually are made in rather small sizes.

Vertical-spindle rotary-table surface grinders are primarily production-type machines. They frequently have two or more grinding heads, as illustrated in Figure 25-28, so both rough and finish grinding is accomplished in one rotation of the worktable. The work can be held either on a magnetic chuck or in special fixtures attached to the table.

FIGURE 25-27. Reciprocating-table, vertical-spindle surface grinder. (*Courtesy The Thompson Grinder Company.*)

By using special rotary feeding mechanisms, machines of this type often are made automatic. Parts are dumped on the rotary feeding table and fed automatically onto work-holding devices and moved past the grinding wheels. After they pass the last grinding head, they are automatically removed from the machine.

FIGURE 25-28. Rotary-table surface grinder having two vertical-spindle heads. The workpieces are automatically fed into the fixture and discharged when the grinding is completed. (*Courtesy The Blanchard Machine Company.*)

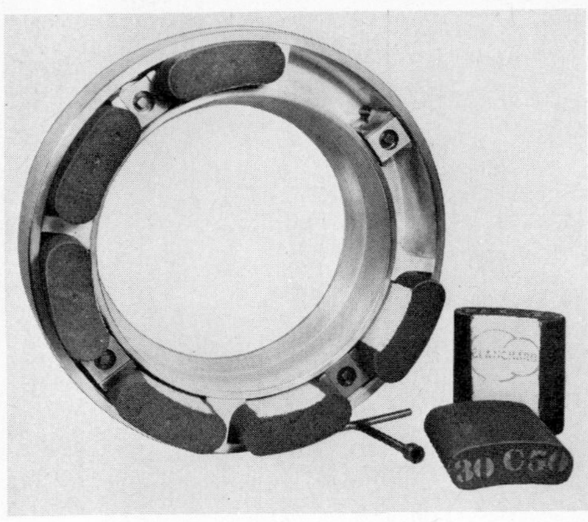

FIGURE 25-29. Segment-type grinding wheel. (*Courtesy The Blanchard Machine Company.*)

Large machines of this type usually employ a segment-type wheel, such as is shown in Figure 25-29.

Holding work on surface grinders. Workpieces usually are held in a different manner on surface grinders than on other machine tools. To obtain high accuracy, it is desirable to reduce clamping forces and distribute them over the entire area of the workpiece. Also, grinding very frequently is done on quite thin or relatively delicate workpieces, which would be difficult to clamp by normal methods. In addition, there often is the problem of grinding a number of small, duplicate workpieces. Magnetic, electrostatic, and vacuum chucks solve all these problems very satisfactorily.

Magnetic chucks are used most frequently. Two shapes of electric-powered chucks are shown in Figure 25-30. These use dry-disk rectifiers to provide the necessary direct-current power. Another type of magnetic chuck

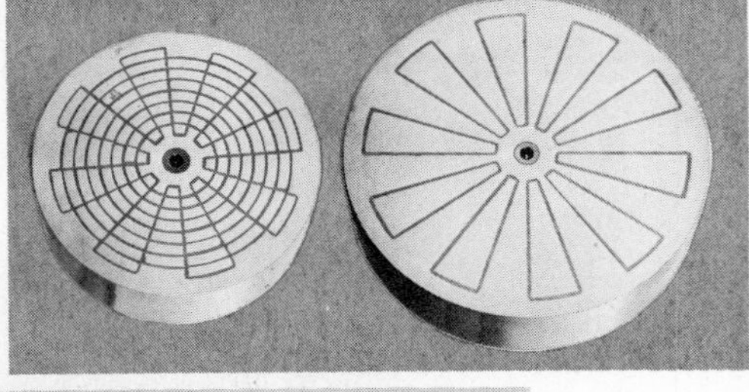

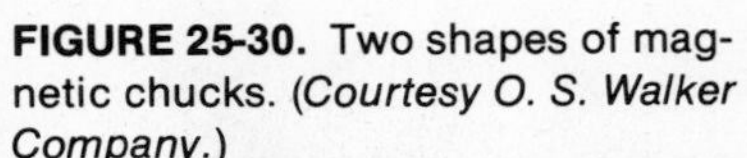

FIGURE 25-30. Two shapes of magnetic chucks. (*Courtesy O. S. Walker Company.*)

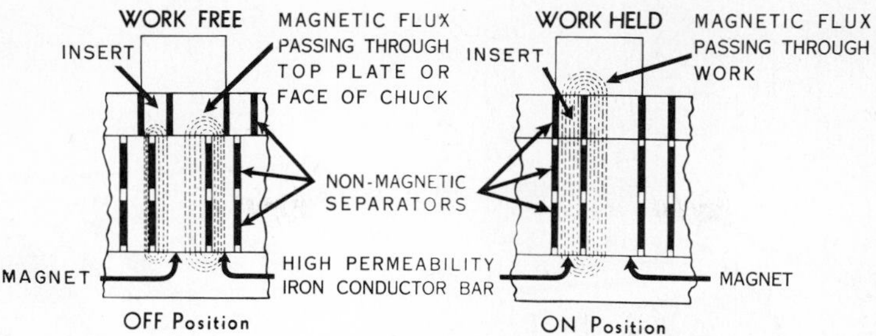

FIGURE 25-31. Principle of operation of permanent-magnet chucks. (*Courtesy Brown & Sharpe Manufacturing Company.*)

utilizes permanent magnets. The operation of this type is shown in Figure 25-31. In the operating position the magnetic flux lines pass through the work and thus hold it to the chuck. If the "off" position the top plate short-circuits the flux lines, and no holding force is created between the work and the chuck.

When the cutting forces are not too high, magnetic chucks provide an excellent means of holding workpieces. The holding force is distributed over the entire contact surface of the work, the clamping stresses are low, and there thus is little tendency for the work to be distorted. Consequently, pieces can be held and ground accurately. Also, a number of small pieces can be mounted on a chuck and ground at the same time.

It often is necessary to demagnetize work that has been held on a magnetic chuck. Some electrically powered chucks provide satisfactory demagnetization by reversing the direct current briefly when the power is shut off.

Obviously, magnetic chucks can be used only with ferromagnetic materials. Electrostatic chucks, on the other hand, can be used with any electrically conductive material. Their principle, indicated in Figure 25-32, involves the work being held by mutually attracting electrostatic fields in the chuck and the workpiece. These provide a holding force of up to 21 000 Pa (30 psi). Nonmetal parts usually can be held if they are flashed with a thin

FIGURE 25-32. Principle of the electrostatic chuck.

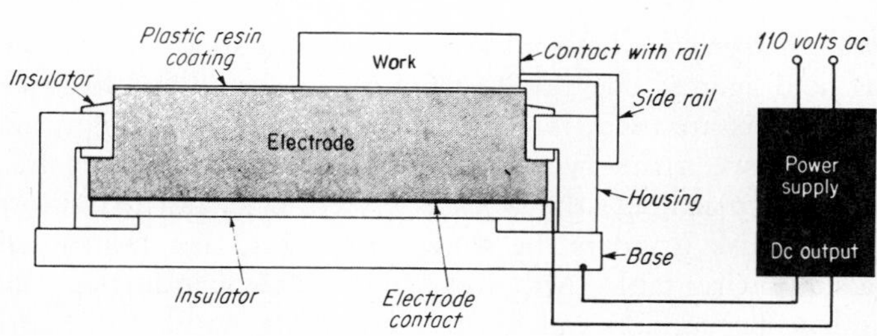

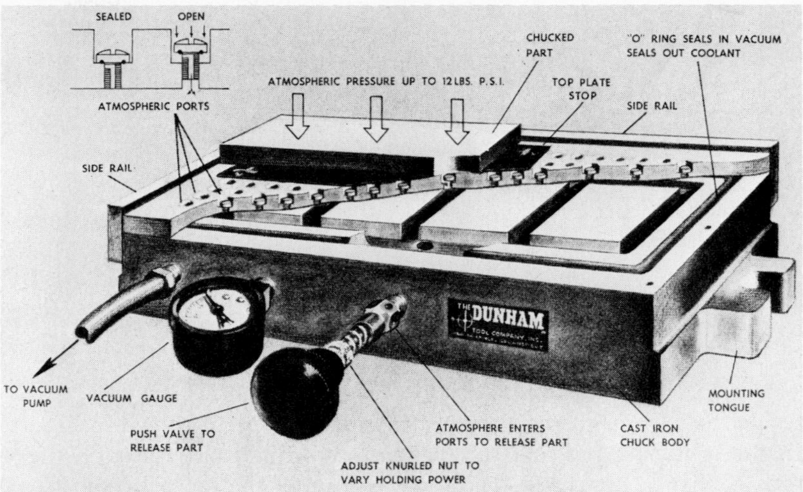

FIGURE 25-33. Cutaway view of a vacuum chuck. (*Courtesy The Dunham Tool Company Inc.*)

metal coating. These chucks have the added advantage of not inducing residual magnetism in the work.

Two types of vacuum chucks are available. In one, illustrated in Figure 25-33, the holes in the work plate are connected to a vacuum pump and can be opened or closed by means of valve screws. The valves are opened in the area on which the work is to rest. The other type has a porous plate on which the work rests. After the workpiece, or pieces, are placed in position on the porous plate, they are covered with a piece of special polyethylene film, which then is sealed around the edges of the chuck by a metal frame. When the vacuum is turned on, the film forms around the workpiece, covering and sealing the holes not covered by the workpiece and thus producing a seal. The first cut removes the film covering the workpiece.

Vacuum chucks have the advantage that they can be used on both nonmetals and metals.

Magnetic, electrostatic, and vacuum chucks have been so satisfactory that they also are used for some milling and turning operations.

Disk grinders. *Disk grinders* have relatively large, side-mounted abrasive disks. The work is held against one side of the disk for grinding. Both single and double disk grinders are used; in the latter type, shown in Figure 25-34, the work is passed between the two disks and is ground on both sides simultaneously. On these machines, the work is always held and fed automatically. On small, single-disk grinders the work can be held and fed by hand while resting on a supporting table. Although manual disk grinding is nonprecision in nature, fairly flat surfaces can be obtained quite rapidly with little or no

FIGURE 25-34. Automatic, double horizontal-spindle disk grinder. The parts are carried between the two grinding wheels by the notched carrier wheel. (*Courtesy Gardner Machine Company.*)

tooling cost. On specialized, production-type machines, excellent accuracy can be obtained very economically.

Tool and cutter grinders. Simple, single-point tools often are sharpened by hand on bench or pedestal grinders (*offhand grinding*). More complex tools, such as milling cutters, reamers, and hobs, and single-point tools for production-type operations, require more complex grinding machines, commonly called *universal tool and cutter grinders.* As can be noted in Figure 25-35, these are similar to small universal cylindrical center-type grinders, but they differ in four important respects:

1. The headstock is not motorized.
2. The headstock can be swiveled about a horizontal as well as a vertical axis.
3. The wheelhead can be raised and lowered and can be swiveled through a 360° rotation about a vertical axis.
4. All table motions are manual, no power feeds being provided.

Specific rake and clearance angles must be created, often repeatedly, on a given tool or on duplicate tools. Tool and cutter grinders have a high degree of flexibility built into them so that the required relationships between the tool and the grinding wheel can be established for almost any type of tool. Although setting up such a grinder is quite complicated and requires a highly skilled worker, after the setup is made for a particular job, the actual grinding

FIGURE 25-35. Universal tool and cutter grinder. (*Courtesy Cincinnati Milacron Inc.*)

is accomplished rather easily. Figure 25-36 shows several typical setups on a tool and cutter grinder.

There also are several specialized types of tool grinders for grinding specific tools, such as the *drill grinder* shown in Figure 19-16. Figure 25-37 shows another type that is equipped with a projection-type comparator that projects a magnified image of the grinding wheel and work contour on a screen, making it especially useful in grinding form tools.

Snagging. *Snagging* is a type of rough grinding that is done in a foundry to remove fins, gates, risers, and rough spots from castings, preparatory to further machining. The primary objective is to remove substantial amounts of metal rapidly without much regard for accuracy. Pedestal-type or *swing grinders,* shown in Figures 25-38 and 25-39, respectively, ordinarily are used. Portable electric or air grinders also are used for this purpose and for miscellaneous grinding in connection with welding.

Mounted wheels and points. *Mounted wheels and points* are small grinding wheels of various shapes that are permanently attached to metal shanks that

FIGURE 25-36. Typical setups for grinding single- and multiple-point tools on a universal tool and cutter grinder. (*Courtesy Cincinnati Milacron Inc.*)

FIGURE 25-37. Micromaster grinder, which combines a projection comparator and a horizontal-spindle grinder to permit grinding work to an enlarged template image. (*Courtesy Brown & Sharpe Manufacturing Company.*)

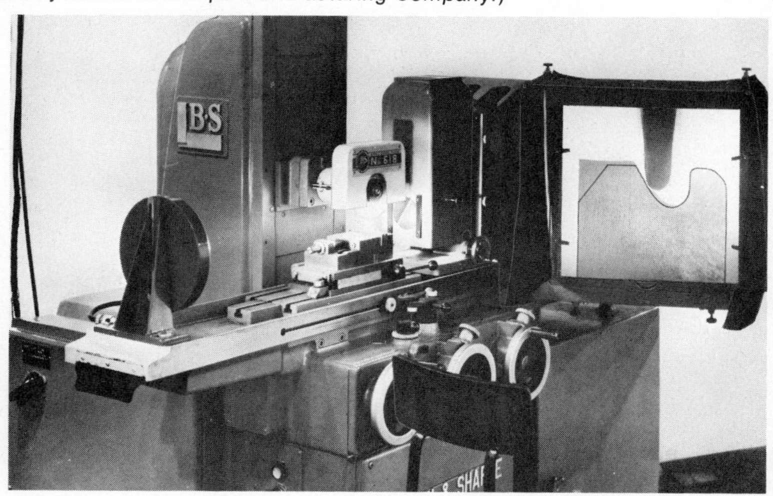

FIGURE 25-38. Rough grinding castings on a pedestal grinder. (*Courtesy Norton Company.*)

FIGURE 25-39. Grinding castings with a swing grinder. (*Courtesy Norton Company.*)

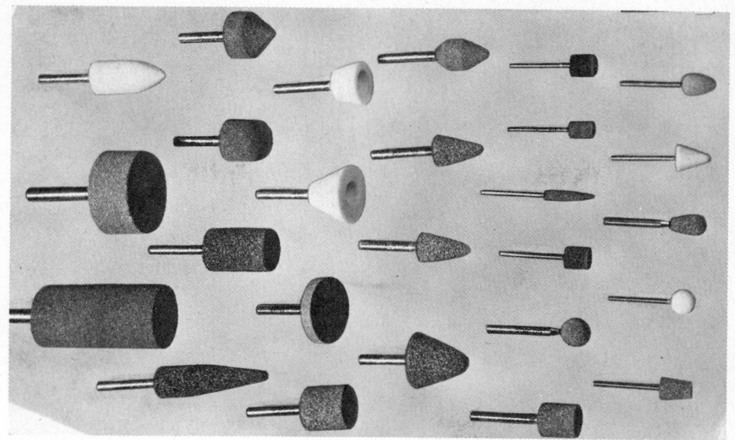

FIGURE 25-40. Mounted abrasive wheels and points. (*Courtesy Norton Company.*)

can be inserted in the chucks of portable, high-speed electric or air motors. They are operated at speeds up to 100,000 rpm, depending on their diameters, and are used primarily for deburring and finishing in mold and die work. Several types are shown in Figure 25-40.

Coated abrasives. *Coated abrasives* are being used increasingly in finishing both metal and nonmetal products. These are made by gluing abrasive grains onto a cloth or paper backing. Synthetic abrasives—Carborundum and Alundum—are used most commonly, but some natural abrasives—sea sand, flint, garnet, and emery—also are employed. Various types of glues are uti-

FIGURE 25-41. Various forms of coated abrasives. (*Courtesy Carborundum Company.*)

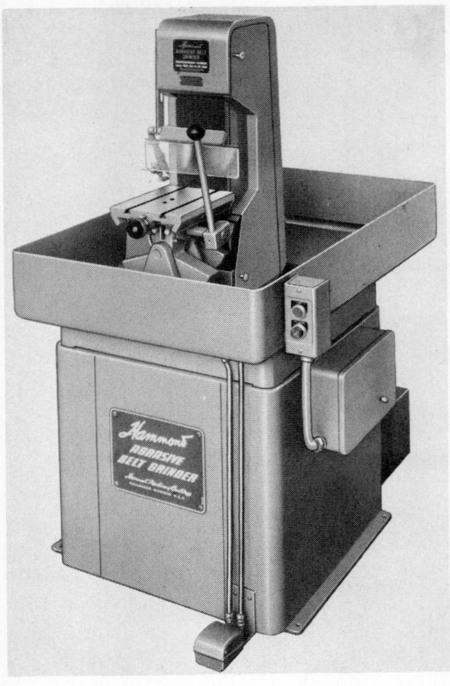

FIGURE 25-42. Production-type abrasive belt grinder with semiautomatic work table. *(Courtesy Hammond Machinery Builders, Inc.)*

lized to attach the abrasive grains to the backing, usually compounded to allow the finished product to have some flexibility.

Coated abrasives are available in sheets, rolls, endless belts, and disks of various sizes. Some of the available forms are shown in Figure 25-41. Although the cutting action of coated abrasives basically is the same as with grinding wheels, there is one major difference: they have little tendency to be self-sharpening through dull grains being pulled from the backing to expose sharp particles. Consequently, when the abrasive particles become dull, the belt or other article must be discarded.

Figure 25-42 shows a typical belt grinder such as is used in metal finishing.

HONING

Honing uses fine abrasive stones to remove very small amounts of metal. It is used as both a sizing and finishing operation, but the primary purpose usually is to remove scratches that remain from grinding. The amount of metal removed usually is less than 0.13 mm (0.005 inch).

Honing stones. Virtually all honing is done with stones made by bonding together various fine artificial abrasives. *Honing stones* differ from grinding wheels in that additional materials, such as sulfur, resin, or wax, often are added to the bonding agent to modify the cutting action. The abrasive grains range in size from 80 to 600 grit.

Honing equipment. Although honing occasionally is done by hand, as in finishing the face of a cutting tool, it usually is done with special equipment. Either flat or round surfaces may be honed, but by far the majority of honing is done on internal, cylindrical surfaces, such as automobile cylinder walls. The honing stones usually are held in a honing head, such as is shown in Figure 25-43, with the stones being held against the work with controlled, light pressure. The honing head is not guided externally but, instead, *floats* in the hole, being guided by the work surface.

Surface speeds in honing vary from 15 to 91 meters (50 to 300 feet) per minute. The stones are given a complex motion so as to prevent a single grit from repeating its path over the work surface. For example, in honing internal cylinders a slow rotation is combined with an oscillatory axial motion. For external and flat surfaces, varying oscillatory motions are used. The length of the motions should be such that the stones extend beyond the work surface at the end of each stroke. A cutting fluid is used in virtually all honing operations.

Single- and multiple-spindle honing machines are available in both horizontal and vertical types; Figure 25-44 shows an example of the latter. Some are equipped with special, sensitive measuring devices that collapse the honing head when the desired size has been reached.

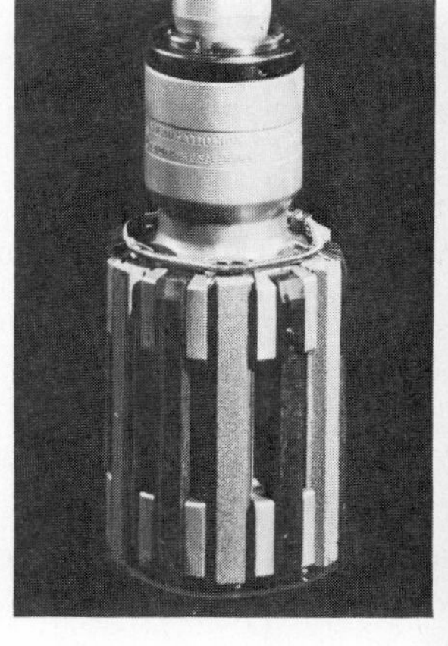

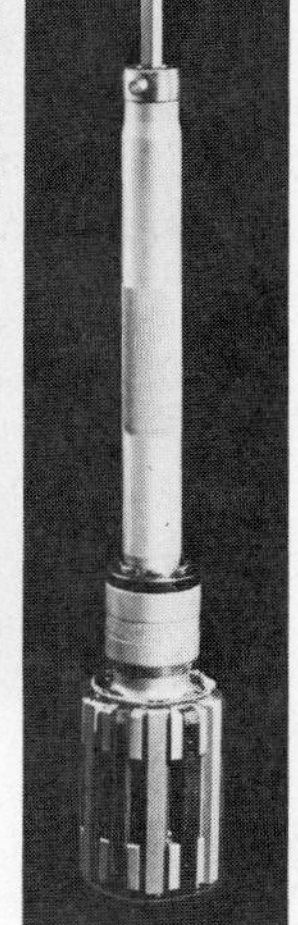

FIGURE 25-43. Typical honing head, with an enlarged view showing the manner in which the stones are held. (*Courtesy Micromatic Hone Corporation.*)

FIGURE 25-44. Multiple-spindle honing machine for honing cylinders of engine blocks. (*Courtesy Barnes Drill Company.*)

For honing single, small, internal cylindrical surfaces a procedure is often used wherein the workpiece is manually held and reciprocated over a rotating hone.

If the volume of work is sufficient so that the equipment can be fully utilized, honing is a fairly inexpensive process. A complete honing cycle, including loading and unloading the work, often is less than 1 minute. Size control within 0.0076 mm (0.0003 inch) is achieved routinely.

Superfinishing is a variation of honing that employs:

1. Very light, controlled pressure 0.07 to 0.28 MPa (10 to 40 psi).
2. Rapid (over 400 per minute), short strokes—less than 6.35 mm (¼ inch).
3. Stroke paths controlled so that a single grit never traverses the same path twice.
4. Copious amounts of low-viscosity lubricant–coolant flooded over the work surface.

This procedure, illustrated in Figures 25-45 and 25-46, results in surfaces of very uniform, repeatable smoothness.

Superfinishing is based on the phenomenon that a lubricant of a given viscosity will establish and maintain a separating, lubricating film between

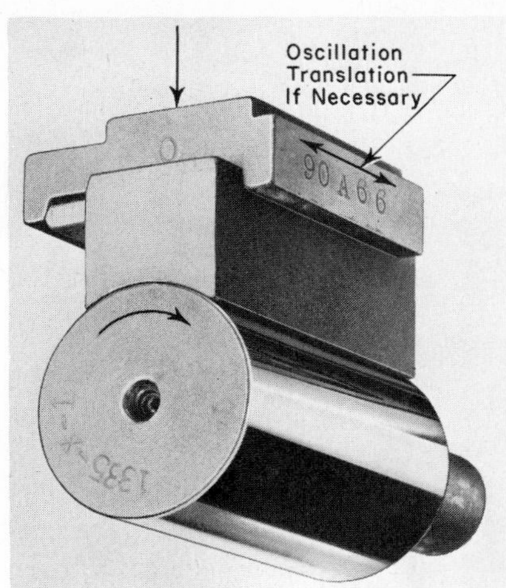

FIGURE 25-45. Basic motions in Superfinishing. (*Courtesy Gisholt Machine Company.*)

two mating surfaces if their roughness does not exceed a certain value and if a certain critical pressure, holding them together, is not exceeded. Consequently, as the minute peaks on a surface being superfinished are cut away by the honing stone, applied with a controlled pressure, when a certain degree of smoothness is achieved, the controlled-viscosity lubricant establishes a continuous lubricating film between the stone and the workpiece and separates them so that no further cutting action occurs. Thus with a given pressure, lubricant, and honing stone, each workpiece is honed to the same degree of smoothness.

Superfinishing is applied to both cylindrical and plane surfaces. The amount of metal removed usually is less than 0.05 mm (0.002 inch), most of it being the peaks of the surface roughness. The copious amounts of lubricant–coolant maintain the work at a uniform temperature and wash away all abraded metal particles so as to prevent scratching. Although usually used to

FIGURE 25-46. Manner in which a film of lubricant is established between the work and the abrasive stone in Superfinishing as the work becomes smoother.

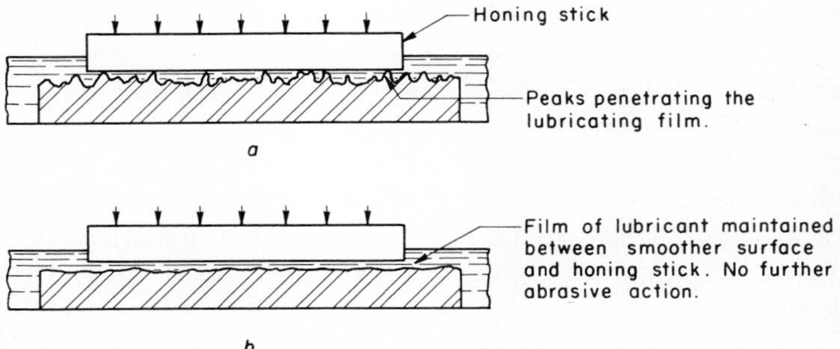

FIGURE 25-47. Superfinishing machine for finishing the bearing surfaces on a crankshaft. (*Courtesy Gisholt Machine Company.*)

produce smooth surfaces, superfinishing also can be employed to produce definite crosshatch patterns on surfaces to aid in lubrication.

Figure 25-47 shows a typical superfinishing machine for use on external cylindrical surfaces.

LAPPING

Lapping is an abrasive surface-finishing process wherein fine abrasive particles are *charged* (caused to become embedded) into a soft material, called a lap. The material of the lap may range from cloth to cast iron or copper, but it always is softer than the material to be finished, being only a holder for the hard abrasive particles. Lapping is applied to both metals and nonmetals.

As the charged lap is rubbed against a surface, the abrasive particles in the surface of the lap remove small amounts of material from the harder surface. Thus it is the abrasive that does the cutting, and the soft lap is not worn away because the abrasive particles become embedded in its surface instead of moving across it. This action always occurs when two materials rub together in the presence of a fine abrasive—the softer one forms a lap, and the harder one is abraded away.

In lapping, the abrasive usually is carried between the lap and the work surface in some sort of a vehicle, such as grease, oil, or water. The abrasive particles are from 120 grit up to the finest powder sizes. As a result, only very small amounts of metal are removed—usually considerably less than 0.025 mm (0.001 inch). Because it is such a slow metal-removing process, it is used only to remove scratch marks left by grinding or honing, or to obtain very flat or smooth surfaces, such as are required on gage blocks or for liquid-tight seals where high pressures are involved.

Materials of almost any hardness can be lapped. However, it is difficult to lap soft materials because the abrasive tends to become embedded. The most common lap material is fine-grained cast iron. Copper is used quite often and is the common material for lapping diamonds. For lapping hardened metals for metallographic examination, cloth laps are used.

Lapping can be done either by hand or by special machines. In hand lapping the lap is flat, similar to a surface plate. Grooves usually are cut across the surface of a lap to collect the excess abrasive and chips. The work is moved across the surface of the lap, using an irregular, rotary motion, and is turned frequently to obtain a uniform cutting action.

In lapping machines for obtaining flat surfaces, such as is shown in Figure 25-48, the workpieces are placed loosely in holders and are held against the rotating lap by means of floating heads. The holders, rotating slowly, move the workpieces in an irregular path. When two parallel surfaces are to be produced, two laps may be employed, one rotating below and the other above the workpieces.

FIGURE 25-48. Rotary lapping machine. The descending heads hold the workpieces against the lap as they are carried around by the holders. (*Courtesy Norton Company.*)

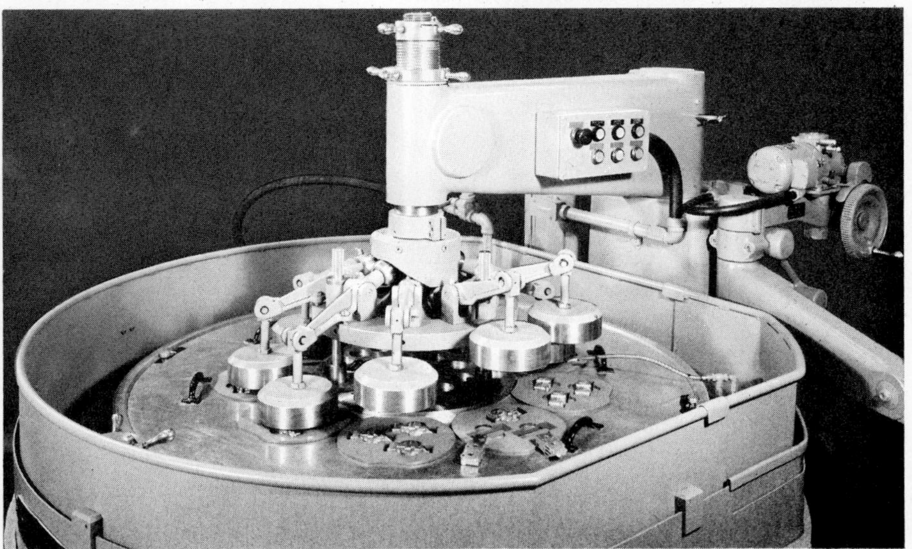

Various types of lapping machines are available for lapping round surfaces. A special type of centerless lapping machine is used for lapping small, cylindrical parts, such as piston pins and ball-bearing races.

Because the demand for surfaces having only a few micrometers of roughness on hardened materials has become quite common, the use of lapping has increased greatly. However, because it is a very slow method of removing metal, it obviously is costly, compared with other methods, and it should not be specified unless such a surface is absolutely necessary.

ULTRASONIC MACHINING

In *ultransonic machining,* material is removed from the workpiece through high-velocity bombardment by abrasive particles. The fine abrasive particles—usually boron carbide, silicon carbide, or aluminum oxide in the form of a slurry—are impelled against the workpiece through the action of an ultrasonic transducer.

As depicted in Figure 25-49, a high-frequency current drives a magnetostrictive transducer, converting electrical energy into mechanical energy. The tool oscillates linearly through about 0.015 to 0.10 mm (0.0006 to 0.004 inch) at ultransonic frequencies of from 15 to 25 kHz while immersed in a liquid slurry containing the abrasive particles. The resulting cavitation action within the liquid drives the abrasive particles against the workpiece with an impact-grinding action, thus machining away very minute chips of the work material. The turbulence within the liquid often provides sufficient pumping action to keep the abrasive slurry in circulation between the face of the tool and the workpiece, but more often a pump is used to assure adequate flow of the slurry.

Because all the cutting action occurs between the face of the tool and the

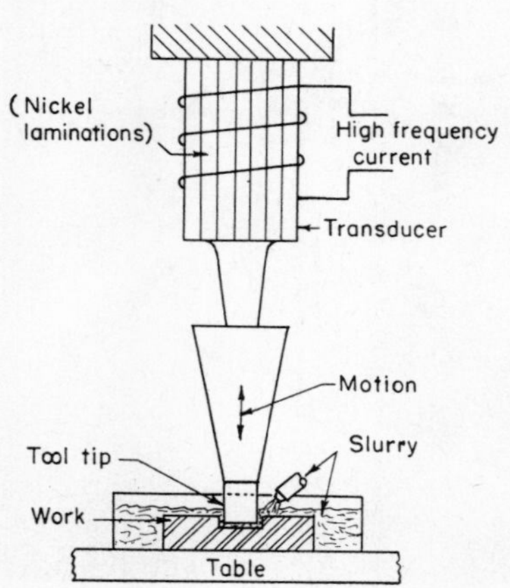

FIGURE 25-49. Principle of ultrasonic machining.

workpiece, holes of any desired shape can be obtained by making the face of the tool of the proper shape. Also, dead-end cavities of virtually any desired contour can be made.

Ultrasonic machining can be used to make holes or cavities in material of virtually any kind and hardness. Holes from 0.025 to 127 mm (0.001 to 5 inches) in diameter and up to 38 mm (1½ inches) in depth are common. Tolerances of 0.025 mm (0.001 inch) and surface finishes up to 0.15 μm (6 microinches) can be obtained. The cutting rate is a function of the hardness and density of the workpiece and the coarseness of the abrasive grains. The time per hole may vary from 1 minute to 1 hour.

The tool usually is made of a material that is easily machined, such as brass or soft steel. Although there is some wearing of the tool, it is quite slow, so total tool cost is not high.

Because the minute cutting action produces almost no thermal stresses, ultrasonic machining can be used on very hard and brittle materials. It is used quite frequently in tool-and-die work, particularly on nonmetals. However, it should be remembered that some of the "chipless" processes have much faster metal-removal rates and are to be preferred over ultrasonic machining, where applicable.

Review questions

1. What are four machining processes in which abrasive particles are the cutting tools?
2. What are the abrasives commonly used in machining?
3. Why are artificial abrasives superior to natural abrasives?
4. What are the two most-used artificial abrasives?
5. Why is aluminum oxide used more frequently than silicon carbide?
6. Why is boron nitride superior to silicon carbide as an abrasive in some applications?
7. What materials commonly are used as bonding agents in grinding wheels?
8. Explain what is meant by the grade of a bond in a grinding wheel. Why is it important?
9. What is a hard grinding wheel?
10. Why is a soft abrasive wheel used for grinding hard materials?
11. What is meant by loading in connection with a grinding wheel?
12. How does the required power vary with feed in grinding?
13. What is accomplished in dressing a grinding wheel?
14. How does abrasive machining differ from ordinary grinding?
15. What is a grinding ratio?
16. Basically, what determines the kind of abrasive to use in grinding?
17. Why is grain spacing important in grinding wheels?
18. What are some safety precautions that should be observed in using grinding wheels?

19. Why should a cutting fluid be used in copious quantities when doing wet grinding?
20. What is plunge-cut grinding?
21. Why is it not good practice to locate grinding machines among other machine tools?
22. Why does the headstock center on a center-type grinder seldom revolve?
23. How are tapers ground on a plain center-type cylindrical grinder?
24. List four advantages of centerless external cylindrical grinders.
25. What determines the through-feed rate on a centerless external cylindrical grinder?
26. How do roll grinders differ basically from ordinary cylindrical grinders?
27. What are the advantages of vacuum chucks?
28. How does a tool-and-cutter grinder differ from a center-type cylindrical grinder?
29. What basic principle enables Superfinishing to produce highly repeatable surfaces?
30. Why does a lap not wear, since it is softer than the material being lapped?
31. How does cutting occur in ultrasonic machining?
32. How do honing stones differ from grinding wheels?
33. How would you classify Superfinishing as an abrasive machining process?
34. Why is a honing head permitted to float in a hole that is being honed?
35. In what respect does the cutting action of a coated abrasive differ from that of an abrasive wheel?
36. What is crush dressing? Why is it often used in form grinding?

Case study 18.
THE CASE OF HSS VERSUS TUNGSTEN CARBIDE

The Quality Machine Works, which does job-shop machining, has received an order to make 40 duplicate pieces, made of AISI 4140 steel, which will require one hour per piece for actual cutting time if an ordinary HSS milling cutter is used. John Young, a new machinist, says the cutting time can be reduced to not over 25 minutes per piece if the company will purchase a suitable tungsten carbide milling cutter. Hans Oldman, the shop foreman, says he does not believe that John's estimate is realistic, and he is not going to spend $450 of the company's money on a carbide cutter that probably would not be used again. The machine-hour rate, including labor, is $30 per hour. Who is right?

Chipless machining processes: chemical, electrochemical, electrodischarge

Machining processes that involve compression-shear chip formation have a number of inherently adverse characteristics and limitations. As has been discussed previously, the formation of chips, basically, is an inefficient process. Large amounts of energy are utilized in producing an unwanted product—chips. Further expenditure of energy and money is required to remove these chips and to dispose of, or recycle, them. A large amount of energy ends up as undesirable heat that often produces problems of distortion and cooling. The high forces involved create problems in regard to holding the work, and sometimes cause distortion. Undesirable cold working often results, sometimes requiring further processing to remove the effects. Finally, there are definite limitations in regard to the delicacy of the work that can be done. For example, the production of such parts as the semiconductor "chip" shown in Figure 26-1, which now play such an important role in our economy (and even in many chip-making machine tools, as will be discussed in Chapter 39),

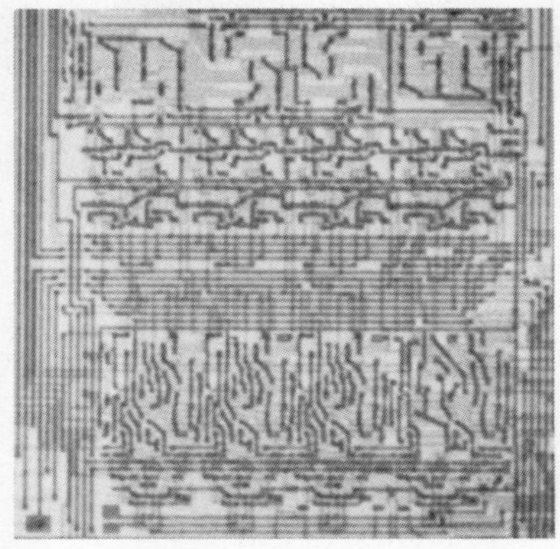

FIGURE 26-1. (*Left*) Enlarged view of one portion of a microprocessor chip. This chip, shown in full size above, measures only 5 millimeters on a side and contains over 3000 transistors. (*Courtesy Bell Laboratories.*)

would not be possible with any of the chip-making processes. In view of these adverse and limiting characteristics, it is not surprising that in recent years substantial effort has been devoted to developing and perfecting several material-removal processes that do not involve the formation of chips. These often are called *chipless machining processes.*

The chipless machining processes are of three basic types: (1) chemical, (2) electrochemical, and (3) electrodischarge. Material also can be removed by a fourth chipless procedure—*laser machining.* However, this process basically involves the melting and vaporization of the material, and it is quite limited in its applications. Consequently, it usually is not considered as one of the chipless machining processes. It will be discussed in Chapter 31.

CHEMICAL MACHINING

Basically, *chemical machining* is the simplest and oldest of the chipless machining processes. It has been used for many years in the production of engraved plates for printing and in making small name plates. However, in its use as a machining process, it is applied to parts ranging from very small "chips," such as shown in Figure 26-1, to very large parts, up to 15 meters (50 feet) long.

In chemical machining, material is removed from selected areas of a workpiece by exposing it to a chemical reagent. The material is removed by microscopic electrochemical cell action, as occurs in corrosion or chemical dissolution of a metal; no external circuit is involved. The process is called by several names: *chemical milling,* because in its earliest uses it replaced milling; and *photofabrication,* because photographic processes have come to be used to produce the necessary mask that shields from the chemical the portions of the workpiece from which no material is to be removed. *Chemical blanking* is another term that is applied where the resulting parts otherwise might have been produced by the use of blanking dies; in most such instances the size and details of the parts are such that they could not, in fact, be made by ordinary blanking.

Essentially, the metal-removal portion of the processes is very simple. The part is either immersed in, or exposed to a spray of, a chemical reagent, either acidic or basic, depending on the material. It is permitted to remain until the desired amount of material has been removed by the chemical action. If the immersion technique is employed, the bath is agitated or circulated so as to sweep away the waste products and thereby keep the metal exposed to the reagent. With the spray technique this is not required, so this procedure usually is preferred when the size and shape of the workpiece permit. The major complexity in the process involves providing a maskant on the surface of the workpiece so that removal will occur only on desired areas. This usually is accomplished by photographic means through the use of *photosensitive resists*. However, in some cases it is achieved by completely covering the surface with a maskant and then manually peeling away the maskant from those areas where metal removal is desired.

Chemical machining with photosensitive resists. Figure 26-2 shows the steps that are involved when chemical machining is done through the use of photosensitive resists. These are as follows:

1. Prepare the "artwork." An accurate drawing of the workpiece is made, usually on polyester drafting film or glass and up to 50 or more times the size of the final part. With such magnifications, an accuracy of 0.64 mm (0.025 inch) in the original drawing will permit 0.013 mm (0.0005 inch) to be achieved in the workpiece. By special procedures, lines only 8 nm have been made.

2. Reduce the original drawing by photographic means to obtain a negative, master pattern that is exactly the size of the finished part. This reduction may require several steps, using regular industrial photographic equipment.

3. Coat the workpiece with a light-sensitive emulsion, usually by dipping or by spraying. The emulsion, or resist, then is dried, usually in an oven.

4. After the sensitized workpiece has been placed against the negative, usually in a vacuum frame to ensure good contact, it is exposed to blue light, passing through the negative. Mercury-vapor lamps commonly are employed as the light source. Exposure to the light hardens the selected areas of the resist so that it will not be washed away in the subsequent developing.

5. Develop the workpiece. This removes, or dissolves away, the unexposed areas of the resist, thereby exposing the areas of the workpiece that are to be acted upon by the chemical reagent. The final developing step is to rinse away all residual material.

6. Spray the workpiece with (or immerse it in) the reagent.

FIGURE 26-2. Basic steps in chemical machining with the use of photoresists. (*Left, top*) Preparing the artwork; (*left, center*) making reduced-size template from the artwork by photography; (*left, bottom*) dipping and draining metal sheets in photoresist; (*right, top*) drying photoresist coating in oven; (*right, center*) placing sensitized metal behind template; (*right, bottom*) chemically machined parts emerging from etching machine. (*Courtesy Chemcut Corporation.*)

Chemical machining with the aid of photosensitive resists has revolutionized the production of small, complex parts, such as electronic circuit boards, and very thin parts, such as are shown in Figure 26-3, that are too small to permit their being blanked by ordinary blanking dies. Supplemented by plating, sputtering, and vacuum deposition, which make it possible to de-

FIGURE 26-3. Typical parts produced by chemical machining. (*Courtesy Chemcut Corporation.*)

posit metallic films of controlled thickness, it has made possible the tremendous developments of solid-state circuitry, much of it in miniature size.

The use of scribed maskants. Although photosensitive resists are used in the majority of chemical machining operations, there still are some cases where the older scribed-and-peeled maskants are employed. These usually are (1) where the workpiece is not flat, (2) where it is very large, and (3) for low-volume work where the several steps required in using photosensitive resists are not economically justified. In this procedure the maskant is applied to the entire surface of the workpiece, either by dipping or spraying. It then is removed from those areas where metal removal is desired, by scribing through the maskant with a knife and peeling away the desired portions. Where volume permits, scribing templates can be used to assist in the scribing. Figure 26-4 shows the use of a scribing template and the stripping of the maskant from the workpiece.

Chemical machining to multiple depths. If all areas are to be machined to the same depth, only a single masking, or resist application sequence, and immersing are required. Machining to two or more depths, called *step machining,* can be accomplished by removing the maskant from additional areas after the original immersion. Figure 26-5 illustrates the steps required for stepped chemical machining.

Parts having either uniformly or variably tapered cross sections can be produced by chemical machining through the relatively simple procedure of withdrawing them vertically from the bath at controlled rates. In this way, different areas are exposed to the chemical action for differing amounts of time.

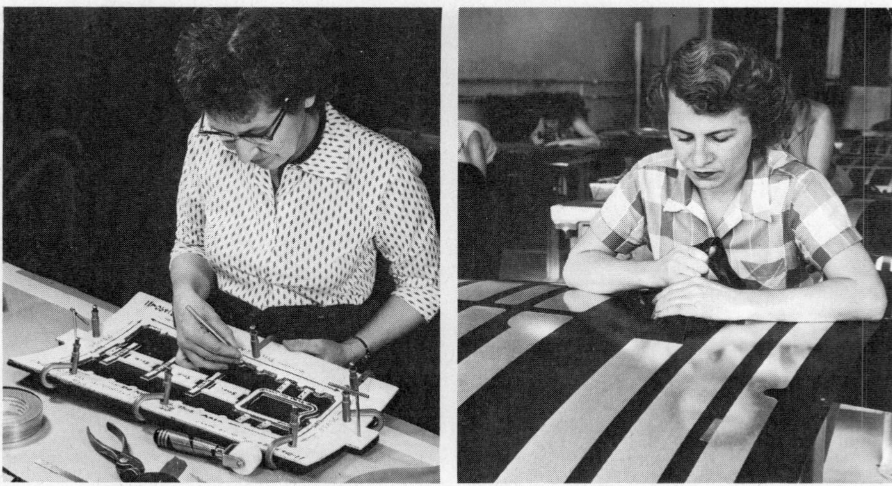

FIGURE 26-4. (*Left*) Scribing a maskant on a part to be chemically machined, using a scribing template. (*Right*) Stripping the maskant from a part after it has been scribed. (*Courtesy United States Chemical Milling Corporation.*)

FIGURE 26-5. Steps required to produce a stepped contour by chemical machining.

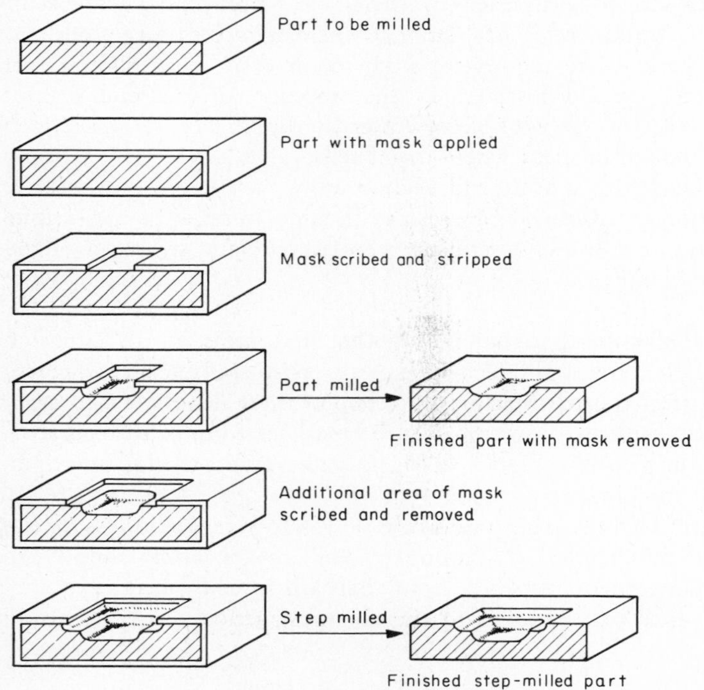

Design factors in chemical machining. When designing parts that are to be made by chemical machining, several unique factors related to the process must be kept in mind. First, dimensional variations can occur through size changes in the art work due to temperature and humidity changes. These usually can be eliminated or controlled by drawing the art work on thicker polyester films or on glass. If very accurate dimensions must be held, the room temperature and humidity should be controlled. The photographic film used in making the master negative also can be affected to some degree by temperature and humidity, but control of handling and processing conditions can eliminate this difficulty.

The second item that must be considered is the "etch factor" or "etch radius." The etchant acts on whatever surface is exposed. Areas that are exposed longer will have more metal removed from them. Consequently, as the depth of etch increases, there is a tendency to undercut or etch under the maskant, as illustrated in Figure 26-6. When the etch depth is only a few hundredths of a millimeter, as often is the case, this causes little or no difficulty. But when the depth is substantial, whether etching from only one or both sides, and when doing chemical blanking, the conditions shown in Figure 26-6 result. In making grooves, the width of the opening in the maskant must be reduced by an amount sufficient to compensate for the etch radius. This radius varies from about one-fourth to three-fourths of the *depth* of the etch (for

FIGURE 26-6. Effect of the "etch factor" in chemical machining.

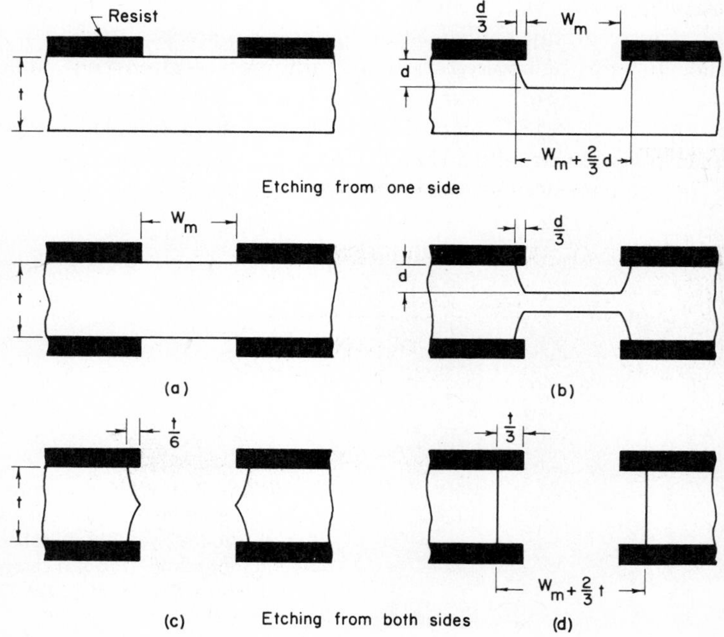

breakthrough in chemical blanking, $t/2$), depending on the type of material and to some extent on the depth of the etch. Consequently, it is difficult to produce narrow grooves except when the etch depth is quite small.

All allowances for the etch factor must be taken into account in designing the part and the original artwork or scribing template. The values indicated in Figure 26-6 are *minimum* values; it has been found that results will vary between etching machines, and actual etch allowances will have to be somewhat greater and adapted to the specific conditions.

In chemical blanking, with etching occurring from both sides, a sharp edge remains along the line at which breakthrough occurs, as in Figure 26-6c. Such an edge usually is objectionable, so etching ordinarily is continued to produce the straight-wall condition shown in (d). An example of the straight edge that can be obtained is shown in Figure 26-7. Obviously, if this procedure is used, allowance must be provided in the width of the opening in the maskant.

Etching from both sides, of course, requires the preparation of two maskant patterns and careful registration of them on the two sides of the workpiece.

If the bath is not agitated properly the "overhang" condition depicted in Figure 26-8 may result, particularly on deep cuts. Not only is the resulting dimension of the opening incorrect, but a very sharp edge may be produced.

Advantages and disadvantages of chemical machining. Chemical machining has a number of distinct advantages. Except for the preparation of the artwork and master negative, or a scribing template, the process is relatively simple, does not require highly skilled labor, induces no stresses or cold working in the metal, and can be applied to almost any metal—aluminum, magnesium, titanium, and steel being most common. Large areas can be machined—tanks for parts up to 3.7 × 15.2 meters (12 × 50 feet) are available. Machining can be done on parts of virtually any shape, and thin sections, such as honeycomb, can be machined because there are no mechanical forces involved. Con-

FIGURE 26-7. (*Left*) Sharp edge left by chemical machining from both sides. (*Right*) Square edge obtainable by use of special technique. (*From a copyrighted publication of Eastman Kodak Company, by permission.*)

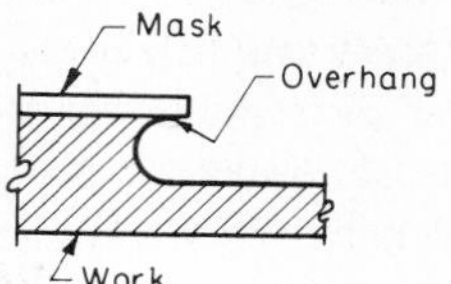

FIGURE 26-8. Overhang condition that can result from deep cuts in chemical machining.

sequently, chemical machining is very useful and economical for weight reduction. Figure 26-9 shows a typical large, thin part that has been chemically milled.

The tolerances obtainable with chemical machining are good—from ±0.013 mm (±0.0005 inch) with care on small etch depths to ±0.076 mm (±0.003 inch) in routine production involving substantial depths. The surface finish is good, seldom having a roughness greater than 0.0025 mm (0.0001 inch).

In using chemical machining, some disadvantages and limitations should be kept in mind. The metal-removal rate is slow *in terms of unit area exposed,* being about 0.1 to 0.2 kg per minute per square meter exposed (0.02 to 0.04 pound per minute per square foot exposed) in the case of steel. However, where large areas can be exposed at a time, or where the metal is quite thin, the overall removal rate may compare favorably with other metal-removal processes.

FIGURE 26-9. Inspecting curved parts that have been chemically machined. (*Courtesy North American Aviation, Inc.*)

The soundness and homogeneity of the metal are very important. Wrought materials should be uniformly heat-treated and stress-relieved prior to processing. Although chemical machining induces no stresses, it may release existing residual stresses in the metal and thus cause warping. Castings can be chemically machined provided they are sound and have uniform grain size. Lack of the latter can cause difficulty. Because of the different grain structures that exist near welds, weldments usually are not suitable for chemical machining.

Chemical milling and conventional machining often can be combined advantageously for producing parts, a fact that often is overlooked.

Chemical deburring. As shown in Figure 26-6c and d, as etching progresses in chemical machining, a sharp edge will be removed; greater amounts of metal are removed where larger surface areas are exposed. *Chemical deburring* utilizes this phenomenon to remove unwanted burrs that result from other machining operations. By exposing the parts to a suitable chemical spray, fine burrs can be removed quickly and at much less cost than if done by hand. *Of course, some smaller amount of metal is removed from all exposed surfaces, and this must be permissible if the process is to be used.* Consequently, the procedure usually can be used only for removing very small burrs.

Electrochemical Machining

Electrochemical machining, commonly designated as ECM, is a deplating process. The workpiece is made the anode in an electrical circuit and the tool, the face of which has the inverse shape of the desired workpiece, is the cathode. A film of the electrolyte, 0.05 to 0.08 mm (0.002 to 0.003 inch) thick, is forced between the anode and the cathode with sufficient velocity to sweep away the waste products. The tool must be properly shaped, and often provided with passages for circulating the electrolyte, to make this possible. The electrolytes commonly used are water-base saline solutions, such as sodium chloride or sodium chlorate solutions. A servomechanism automatically advances the tool and controls the gap between the electrodes.

As shown in Figure 26-10, the metal-removal rate, in terms of penetration of the tool into the workpiece, is primarily a function of the current density. Consequently, current densities from 2.33 to 3.10 A/mm^2 (1500 to 2000 amperes per square inch) are used and, in suitable applications, ECM provides faster metal-removal rates than any other machining process. Because the cutting rate is solely a function of the ion-exchange rate, it is not affected by the hardness or toughness of the work material. Cutting rates up to 2.54 mm (0.1 inch) of depth per minute are obtained routinely in very hard materials such as Waspalloy.

Holes of any uniform cross section can be produced and, by using two

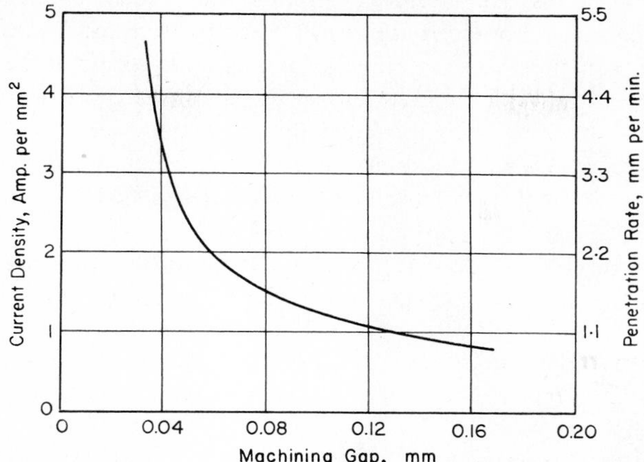

FIGURE 26-10. Relationship of current density, penetration rate, and machining gap in electrochemical machining.

shaped electrodes on opposite sides of a workpiece, as illustrated in Figure 26-11, rough shapes, such as forgings, can be machined to final shape with excellent surface finish in a few minutes. Figure 26-12 shows an example of an ECM machine producing the part that is displayed in the inset; this part would be quite difficult and much more costly to make by any other method. ECM also provides a rapid means of cutting off bars of hard metals by using a thin tool as the cathode.

Advantages and disadvantages of ECM. Eclectrochemical machining is simple and very fast. Because the action is entirely by metallic ion exchange, there are no cutting forces, no stresses are introduced into the work, and very thin sections can be made. Accuracies of 0.051 mm (0.002 inch) are readily

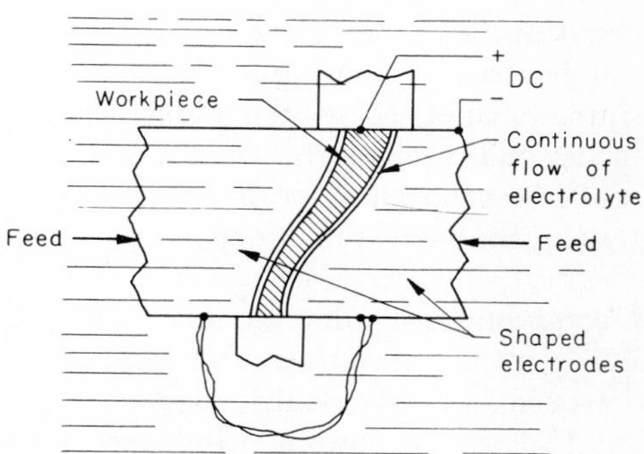

FIGURE 26-11. Schematic diagram of the method of forming contours by electrochemical machining.

FIGURE 26-12. Machine for electrochemical machining, showing workpiece and electrode. Inset shows completed workpiece. (*Courtesy Ex-Cell-O Corporation.*)

obtainable, and 0.0127 mm (0.0005 inch) is possible. For most applications, the surface finish is sufficiently good so that no other finishing is required.

For individual jobs, the principal tooling cost is for the preparation of the tool electrode. There is no wear on the tool during the actual cutting, and it is protected cathodically while the current is flowing, but it should resist any chemical reaction of the corrosive electrolyte while the current is not flowing. Consequently, if the tool is to be used repeatedly, it usually is made of stainless steel. As a result, ECM may not be economical except for (1) very simple shapes, (2) where a substantial number of duplicate parts are to be made, or (3) where the work material is so hard that it cannot be machined readily by conventional processes. Obviously, ECM can be used only on electrically conductive materials.

Electrochemical grinding. *Electrochemical grinding,* commonly designated as ECG, is a variant of electrochemical machining in which the tool electrode is a rotating, metal-bonded, diamond grit grinding wheel. The arrangement of the tool and workpiece is indicated in Figure 26-13. The metal bond of the wheel is the cathode. The diamond particles serve three functions: (1) as

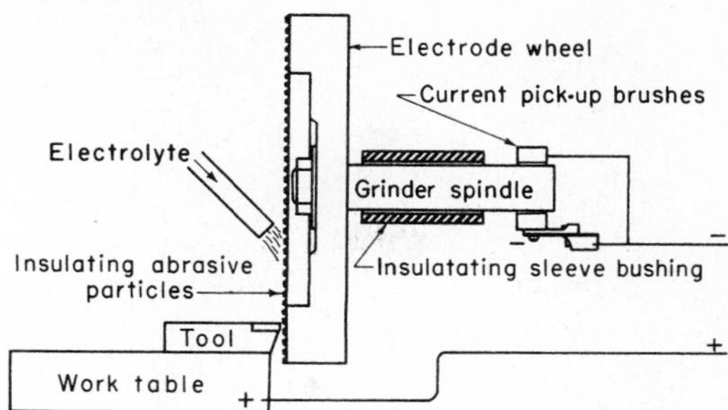

FIGURE 26-13. Equipment setup and electrical circuit for electrochemical grinding.

insulators to preserve a small gap between the cathode and the work, (2) to wipe away the residue, and (3) to cut chips if the wheel should contact the workpiece, particularly in the event of a power failure. When operated properly, less than 0.5 per cent of the material is removed by normal chip forming. The process is used almost exclusively for shaping and sharpening carbide cutting tools, which cause high wear rates on expensive diamond wheels in normal grinding. Electrochemical grinding greatly reduces this tool wear.

Electrochemical grinding not only is much more rapid than ordinary grinding, but it also produces a smoother surface and does not induce surface stresses.

Electrochemical deburring. Limited use is made of the ECM principle for removing burrs from parts. The work is put into a rotating, electrically insulated drum that contains two current-carrying electrodes that are insulated from the drum. Small graphite spheres, added to the electrolyte, receive an inductive charge from the electrodes and thus have sufficient potential gradient across the sphere-to-workpiece gap to cause electrochemical machining to occur as they move randomly over all areas of the workpiece. Because the current density is higher at the protrusions of the burrs than at smooth areas on the workpiece, they are preferentially removed. As in chemical deburring, there is a slight dimensional change throughout the workpiece, in this case due to the general ECM action and to the natural abrasive character of the graphite spheres.

ELECTRODISCHARGE MACHINING

Electrodischarge machining, designated EDM, removes metal through the action of high-energy electric sparks on the surface of the workpiece. In earlier years, it often was called *electrospark machining.* The mechanical setup and the electrical circuit involved are shown in Figure 26-14. The tool and the workpiece are submerged in a fluid having poor electrical conductivity—usually a light oil. A very small gap, of approximately 0.025 mm (0.001 inch), is main-

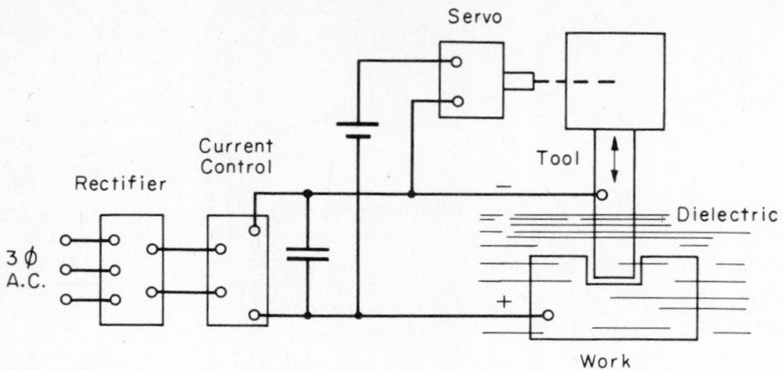

FIGURE 26-14. Mechanical setup and electrical circuit for electrodischarge machining.

tained between the tool and the workpiece by means of a servosystem. When the voltage across the gap becomes sufficiently high, condensers discharge current across the gap, in the form of a spark, in an interval of from 10 to 30 microseconds and with a current density of the order of 15 500 A/mm^2 (10^6 A/in.^{2}). When the voltage has dropped to about 12 volts, the spark discharge is extinguished, and the condensers start to recharge. This cycle is repeated thousands of times per second, and thousands of spark-discharge paths occur over all the area of the tool and its mating work surface. Each discharge removes minute amounts of material from both the tool and the workpiece. The resulting workpiece surface is composed of extremely small craters, as shown in Figure 26-15, so small that a surface finish of about 3.81 μm (150 microinches) is obtained on roughing cuts and about 0.762 μm (30 microinches) on finishing cuts.

The mechanism of metal removal by electrodischarge machining is not completely clear and appears to involve more than one phenomenon. Most of the removal is by fusion of minute particles of metal that are thrown from the surface by the thermal action accompanying the highly localized heating and cooling of the metal surface as a consequence of the concentrated release of

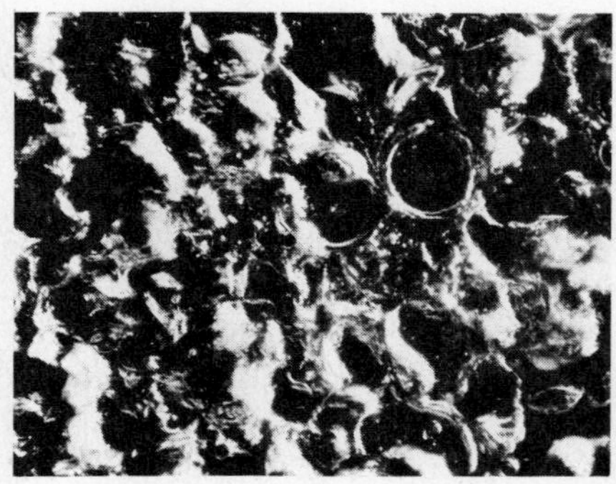

FIGURE 26-15. Surface formed by electrodischarge machining; 400X. (*Courtesy Cincinnati Milacron Inc.*)

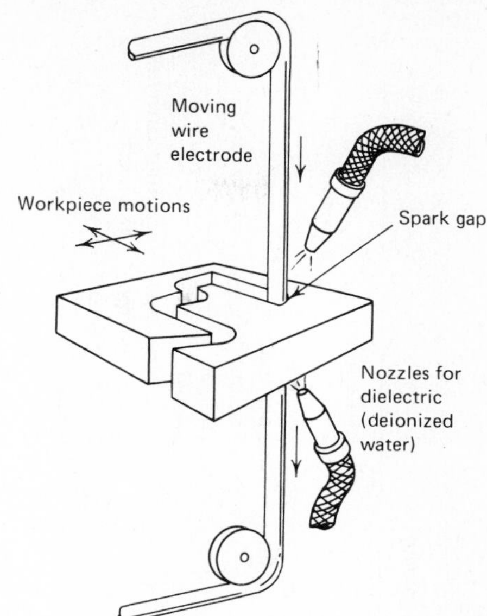

FIGURE 26-16. Schematic diagram of equipment for electrodischarge machining using a moving wire electrode.

energy in the spark. However, not all the particles released are fused, particularly in the case of some metals. Thus it appears that highly localized thermal stresses also play a role in the metal-removal process.

Figure 26-16 shows schematically a modification of EDM in which the tool electrode is a soft copper wire, about 0.20 mm (0.008 inch) in diameter. This wire moves slowly from a take-off spool to a take-up spool in order to compensate for its erosion and to remove the dislodged particles from the workpiece that may adhere to it. Deionized water usually is used as the dielectric. This arrangement permits cutting intricate openings and tight-radius contours, both internal and external, without a specially shaped tool. Machines are available that provide the careful two-dimensional guidance and control of the workpiece in order to maintain the proper gap length and at the same time obtain the desired shape.

Advantages and disadvantages of EDM. Electrodischarge machining is applicable to all materials that are fairly good electrical conductors, thus including metals, alloys, and most carbides. The melting point, hardness, toughness, or brittleness of the material impose no limitations. Thus it provides a relatively simple method for making holes of any desired cross section in materials that are too hard or brittle to be machined by most other methods. There are virtually zero forces between the tool and the workpiece, so that very delicate work can be done. The process leaves no burrs on the edges. Its use has expanded very rapidly, and it not only is used to produce the type of work

TABLE 26-1. Comparison of several chipless machining processes and ultrasonic machining

Process	Tooling Requirements	Material Limitations	Nominal Accuracy	Metal-Removal Rate	Remarks
Chemical machining	Photo negative of template	All homogeneous metals except Monel and stainless steel	±0.076 mm (±0.003 in.)	Low per unit of area, but large areas can be exposed	Etch radius is important
ECM	Surface must be electrically conductive, tool must not be attacked by electrolyte	Any electrically conductive material	±0.051 mm (±0.002 in.)	High penetration up to 2.54 mm (0.1 in.) per minute	Very high currents required
Electro-chemical grinding	Metal-bonded diamond wheel	Same as for ECM	Same as grinding	Similar to ECM, but less area exposed	Primarily for carbides
EDM	Removal ratio 4:1 to 20:1; brass most common	Any conductive metal	Rough; ±0.127 mm (±0.005 in.) Finish; up to ±0.025 mm (±0.001 in.)	Low; 0.41 mm (0.016 in.) penetration per minute	May produce thin, hard surface and possible cracks on some metals
Ultrasonic	Brass or soft steel	No restrictions if hard	Shallow; ±0.051 mm (±0.002 in.) Deep; ±0.10 mm (±0.004 in.)	Faster than EDM	Useful for very small holes and thin sections: 0.30 mm (0.012 in.)

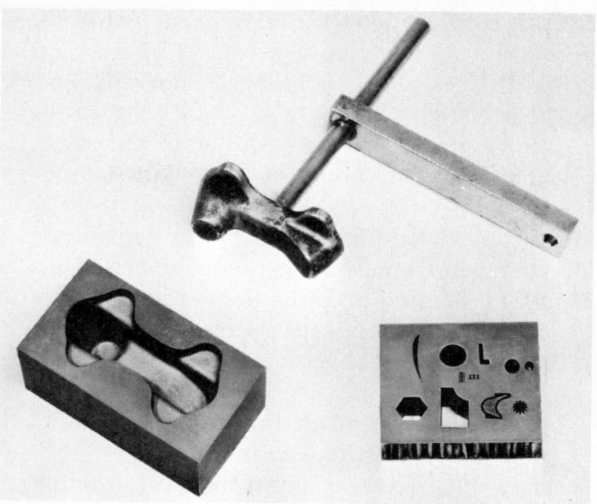

FIGURE 26-17. (*Left*) Die cavity formed by electrodischarge machining. (*Top*) Electrode used for forming the die cavity. (*Right*) Holes formed in honeycomb panel by electrodischarge machining. (*Courtesy Cincinnati Milacron Inc.*)

shown in Figure 26-17, but now it is widely used to produce large body-forming dies in the automotive industry.

On some materials the process produces a thin, hard surface, which may be an advantage or a disadvantage, depending on the use. When the work-piece material is one that tends to be brittle at room temperatures, the surface may contain fine cracks caused by the thermally induced stresses. Consequently, some other finishing process often is used subsequent to EDM to remove a thin surface layer.

Comparison of chipless and ultrasonic machining methods. Table 26-1 shows a useful comparison of the requirements and the results usually obtainable by the several chipless and ultrasonic machining processes, thus indicating their applicability.

Review questions

1. What are three basic types of chipless machining processes?
2. Give three reasons why chipless machining processes are likely to have greater importance in the future.
3. What is the purpose of a photosensitive resist?
4. What are the six steps in chemical machining using photosensitive resists?
5. Why is it perferable in chemical machining to apply the etchant by spraying instead of by immersion?

6. For what types of parts are photosensitive resists not suitable?
7. Explain how multiple depths can be produced by chemical machining.
8. Would it be feasible to produce a groove 2 mm wide and 3 mm deep by chemical machining?
9. A drawing calls for making a groove 23 mm wide and 3 mm deep by chemical machining. What should be the width of the opening in the maskant?
10. Could an ordinary steel weldment be chemically machined? Why?
11. How would you produce a tapered section by chemical machining?
12. What distinguishes chemical blanking from ordinary chemical machining?
13. How is ECM related to chemical machining?
14. What effect does work-material hardness have on the metal-removal rate in ECM?
15. Explain the basic principle involved in electrochemical deburring.
16. What is the principal cause of tool wear in ECM?
17. Basically, what type of process is electrochemical grinding?
18. Would electrochemical grinding be a suitable process for sharpening ceramic tools? Why?
19. Upon what factor does the metal-removal rate primarily depend in ECM?
20. What is the nature of the surface obtained by electrodischarge machining?
21. What is the principal advantage of using a moving-wire electrode in electrodischarge machining?
22. What effect would increasing the voltage have on the metal-removal rate in electrodischarge machining? Why?
23. If the metal from which a part is to be made is quite brittle and the part will be subjected to repeated tensile loads, would you select ECM or electrodischarge machining for making it? Why?
24. If you had to make several holes in a large number of duplicate parts, would you prefer ECM or EDM? Why?

Case study 19. HOW DID THE TUBING GET FLARED?

Figure CS-19 shows an enlarged view of a portion of the inside of a piece of 19.05-mm (0.750-inch)-O.D. soft steel tubing, which had a wall thickness of 24.67 mm (0.097 inch). This tubing had been assembled into a self-flare hydraulic fitting. If the tubing was assembled properly in the fitting, the inside would be forced against a smooth surface in the fitting, and a bevel would be formed on the inside of the end of the tubing. The joint failed in service, resulting in a fire and very substantial damage to a die-casting plant.

In subsequent litigation, one of the parties alleged that the tubing and fitting had been properly assembled. The other party claimed that someone had beveled the inside of the tubing by some means—probably by some hand-machining process—prior to assembly into the fitting. Using your knowledge of how such tubing is made, of how metals react when cold-worked and/or machined, and how the resulting metal surface would appear, which claim do you believe was correct? State your reasons.

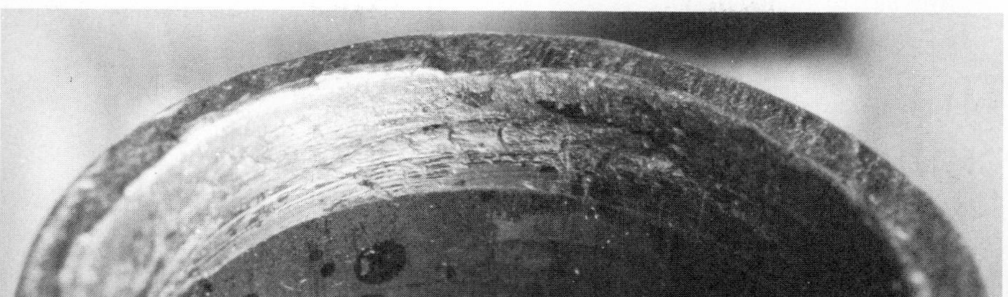

FIGURE CS-19. View of end of piece of hydraulic tubing showing marks on bevel on inside of tubing and variation of remaining wall thickness on end of tubing.

CHAPTER 27

Thread cutting and forming

Screw threads probably are the most important of all the machine elements. Without them, our present technological society would come to a grinding halt. More are made each year than any other machine element. They range in size from those used in the balance wheels of small watches to more than 254 mm (10 inches) in diameter, used to withstand or transmit tremendous forces. They are made in quantities ranging from one to several million duplicate threads. Their accuracy varies from that of cheap dime-store screws to that of micrometer calipers and lead screws on the most accurate machine tools. Consequently, it is not surprising that several very different procedures have been developed for making screw threads and that the production cost by the various methods varies greatly. Fortunately, some of the most economical methods can provide very accurate results. However, as in the design of most products, the designer can greatly affect the ease and cost of producing specified screw threads; thus reasonable understanding of the various processes is most helpful in permitting the designer to specify and incorporate screw threads into his designs while avoiding needless and excessive cost.

Technically, a screw thread is a ridge of uniform section in the form of a helix on the external or internal surface of a cylinder, or in the form of a conical spiral on the external or internal surface of a frustrum of a cone. These are called *straight* or *tapered* threads, respectively. Tapered threads are used on pipe joints or other applications where liquid-tight joints are required. Straight threads, on the other hand, are used in a wide variety of applications, most commonly on fastening devices, such as bolts, screws, and nuts, and as integral elements on parts that are to be fastened together. But, as mentioned previously, they find very important applications in transmitting controlled motion, as in lead screws and precision measuring equipment.

The basic problem in manufacturing screw threads is how to produce the desired ridge on the workpiece. Three basic methods are used: *cutting, rolling,* and *casting.* Although both external and internal threads can be cast, relatively few are made in this manner, primarily in connection with die casting or the molding of plastics. Today by far the largest number of threads are made by rolling. Both external and internal threads can be made by rolling, but the material must be ductile, and it is a less flexible process than thread cutting, thus it essentially is restricted to standardized and simple parts. Consequently, large numbers of threads still are, and will continue to be, made by cutting.

Screw-thread standardization. Starting with Sir Joseph Whitworth in England, in 1841, and William Sellers in the United States, in 1864, a great amount of effort has been devoted to screw-thread standardization. In 1948, representatives of the United States, Canada, and Great Britain adopted the Unified and American Screw Thread Standards, based on the form shown in Figure 27-1. In 1968 the International Organization for Standards (ISO) recommended the adoption of a set of metric standards, based on the basic thread profile shown in Figure 27-2. It appears likely that both types of threads will continue to be used for some time to come.

Screw-thread nomenclature. The standard nomenclature for screw-thread components is illustrated in Figure 27-3. In both the *Unifed* and the *ISO systems,* the crests of external threads may be flat or rounded. The root usually is made rounded to minimize stress concentration at this critical area. The internal thread has a flat crest in order to mate with either a rounded or V-root of the external thread. A small round is used at the root to provide clearance for the flat crest of the external thread.

In the metric system, the pitch always is expressed in millimeters, whereas in the American (Unified) system it is a fraction having as the numerator 1 and as the denominator the number of threads per inch—thus $^1/_{16}$ pitch being $^1/_{16}$ of an inch. Consequently, in the Unified system, threads

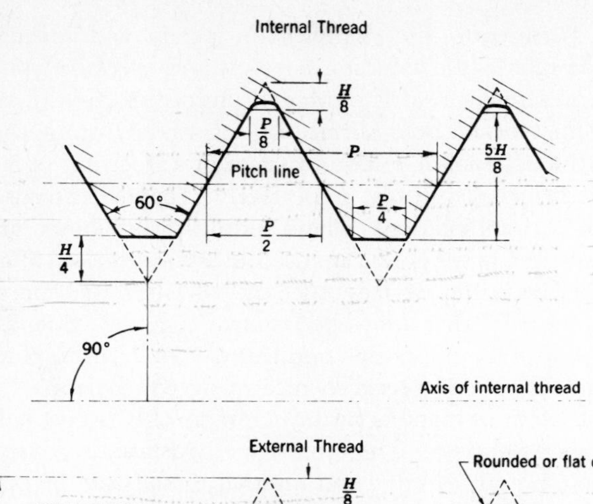

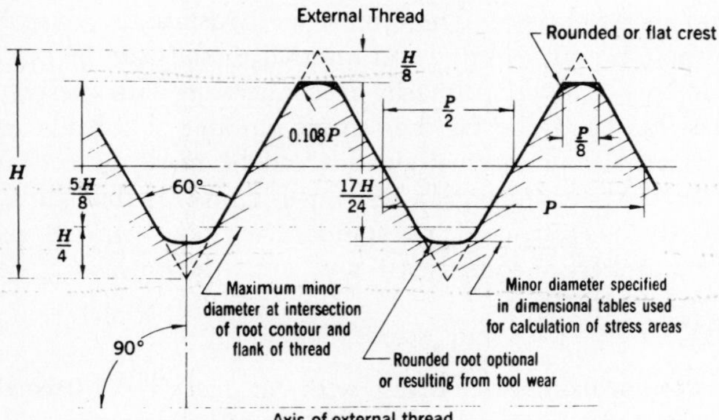

FIGURE 27-1. Unified and American screw-thread form.

FIGURE 27-2. ISO metric screw-thread forms: (*left*) basic profile; (*right*) design dimensions. (*Courtesy ASME, Report B1, ISO Metric Screw Threads.*)

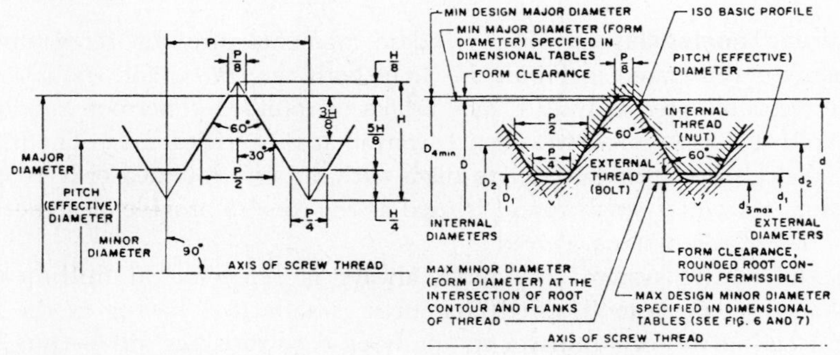

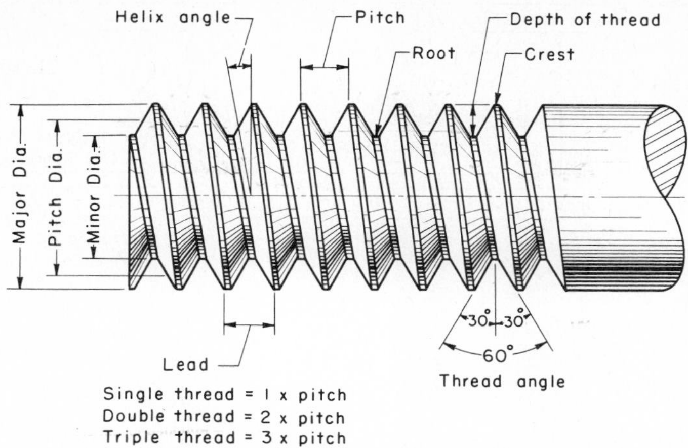

FIGURE 27-3. Standard screw-thread nomenclature.

more commonly are expressed in terms of threads per inch rather than by the pitch.

While all elements of the thread form are based on the *pitch diameter,* screw-thread sizes are expressed in terms of the *outside,* or *major,* diameter and the *pitch* or *number of threads per inch.*

Types of screw threads. Eleven types, or series, of threads are of commercial importance, several having equivalent series in the metric system and Unified systems:

1. *Coarse-thread series* (UNC and NC). For general use where not subjected to vibration.
2. *Fine-thread series* (UNF and NF). For most automotive and aircraft work.
3. *Extra-fine-thread series* (UNEF and NEF). For use with thin-walled material or where a maximum number of threads are required in a given length.
4. *Eight-thread series* (8UN and 8N). Eight threads per inch for all diameters from 1 through 6 inches. It is used primarily for bolts on pipe flanges and cylinder-head studs where an initial tension must be set up to resist steam or air pressures.
5. *Twelve-thread series* (12 UN and 12N). Twelve threads per inch for diameters from $^1/_2$ through 6 inches. It is not used extensively.
6. *Sixteen-thread series* (16 UN and 16N). Sixteen threads per inch for diameters from $^3/_4$ through 6 inches. It is used for a wide variety of applications that require a fine thread.

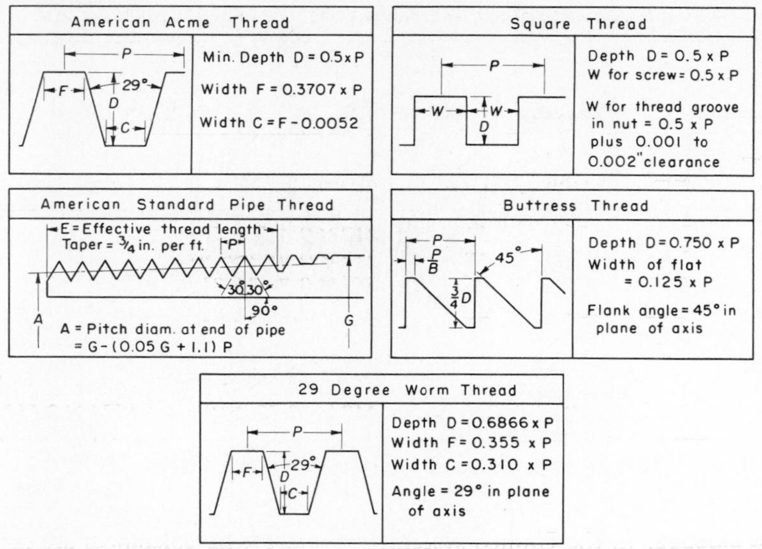

FIGURE 27-4. Special thread forms.

7. *American Acme Thread.*
8. *Buttress Thread.*
9. *Square Thread.*
10. *29° Worm Thread.* These last four of the threads, shown in Figure 27-4, are used primarily in transmitting power and motion.
11. *American Standard Pipe Thread.* This thread, shown in Figure 27-4, is the standard tapered thread used on pipe joints in this country. The taper on all pipe threads is $^{3}/_{4}$ inch per foot.

As has been indicated, the Unified threads are available in a coarse (UNC and NC), fine (UNF and NF), extra-fine (UNEF and NEF), and three "pitch" (8, 12, and 16) series, the number of threads per inch being according to an arbitrary determination based on the major diameter. Similarly, the ISO threads have a "Coarse" and a "Fine" series, with the pitch varying with the major diameter. In addition, they have a "Constant Pitch" series, wherein the pitch (from 6 mm to 0.2 mm) is the same for all diameters. Table 27-1 gives a comparison between some Unified and Metric threads.

The availability of fasteners, particularly nuts, containing plastic inserts to make them self-locking and thus able to resist loosening due to vibration, and the use of special coatings that serve the same purpose, have resulted in less use of finer-thread-series fasteners in mass production. Coarser-thread fasteners are easier to assemble and less subject to cross threading.

TABLE 27-1. Comparison between selected Unified and ISO threads

Unified				ISO	
Diameter		Threads per Inch		[Threads per Inch]	
Number; inches	mm	UNC	UNF	Coarse	Fine
#2	2.18	56	64	M2 × 0.4 [63.5]	
#4	2.84	40	48	M2 × 0.45 [56.4]	
#8	4.17	32	36	M4 × 0.7 [36.3]	
#10	4.82	24	28	M5 × 0.8 [31.8]	
¼ in.	6.35	20	28	M6 × 1.0 [25.4]	
½ in.	12.7	13	20	M12 × 1.75 [14.5]	M12 × 1.25 [20.3]
¾ in.	10.05	10	16	M20 × 2.5 [10.2]	M20 × 1.5 [16.9]
1 in.	25.4	8	14	M24 × 3 [8.47]	M24 × 2 [12.7]

Thread classes. In the Unified system, manufacturing tolerances are specified by three classes. Class 1 is for ordnance and other special applications. Class 2 threads are the normal production grade, and Class 3 threads have minimum tolerances where tight fits are required. The letters A and B are added after the class numerals to indicate external and internal threads, respectively.

In the ISO system, tolerances are applied to "positions" and "grades." Tolerance positions denote the limits of pitch and crest diameters, using "e" (large), "g" (small), and "h" (no allowance) for external threads and "G" (small) and "H" (no allowance) for internal threads. The grade is expressed by numerals 3 through 9. Grade 6 is medium quality and for normal length of engagement. Below 6 is fine quality and/or short engagement. Above 6 is coarse quality and/or long length of engagement.

Thread designation. In the Unified system, screw threads are designated by symbols as follows:

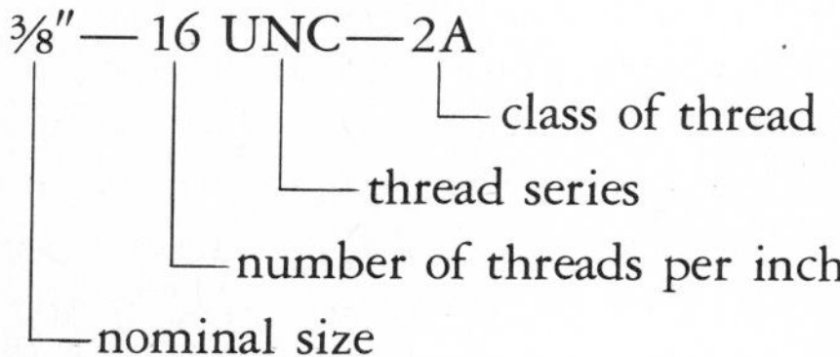

This type of designation applies to right-hand threads. For left-hand threads, the letters LH are added after the thread class symbol.

In the ISO system, threads are designated as follows:

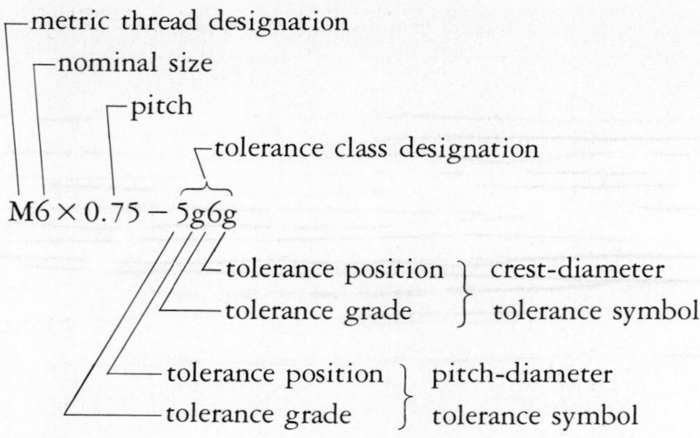

THREAD CUTTING

Threads can be cut by the following methods:

External	Internal
On an engine lathe	On an engine lathe
With a stock and die (manual)	With a tap and holder
With an automatic die	(manual, semiautomatic, or automatic)
(turret lathe or screw machine)	With a collapsible tap
By milling	(turret lathe, screw machine, or
By grinding	special threading machine)
	By milling

Cutting threads on a lathe. Lathes provided the first method for cutting threads by machine. Although most threads now are produced by other methods, lathes still provide the most versatile and fundamentally simple method. Consequently, they often are used for cutting threads on special workpieces where the configuration or nonstandard size does not permit them to be made by less costly methods.

There are two basic requirements for cutting a thread on a lathe. The first is an accurately shaped and mounted tool, because thread cutting is a form-cutting operation; the resulting thread profile is determined by the shape of the tool and its position relative to the workpiece. The second requirement is that the tool must move longitudinally in a specific relationship to the rotation of the workpiece, because this determines the lead of the thread. This requirement is met through the use of the lead screw and the split nut, which provide positive motion of the carriage relative to the rotation of the spindle.

External threads can be cut with the work mounted either between centers, as shown in Figure 27-5, or held in a chuck. For internal threads, the work must be held in a chuck. The cutting tool usually is checked for shape

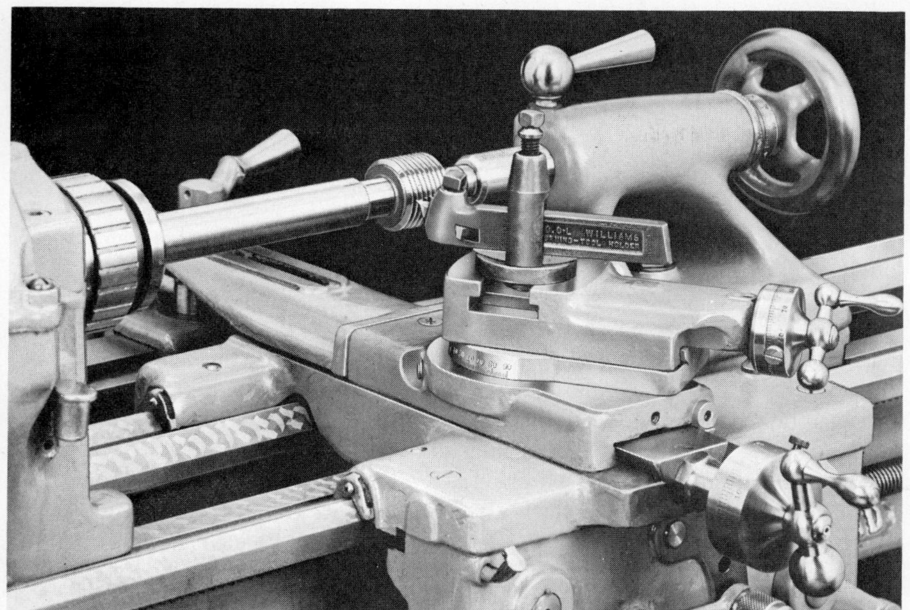

FIGURE 27-5. Cutting a screw thread on a lathe, showing the method of supporting the work and the relationship of the tool to the work. (*Courtesy South Bend Lathe.*)

and alignment by means of a thread template, as indicated in Figure 27-6. Figure 27-7 illustrates two methods of feeding the tool into the work. If the tool is fed radially, cutting takes place simultaneously on both sides of the tool. With this true form-cutting procedure, no rake should be ground on the tool, and the top of the tool must be horizontal and be set exactly in line with the axis of rotation of the work, as shown in Figure 27-8; otherwise, the resulting thread profile will not be correct. An obvious disadvantage of this

FIGURE 27-6. Methods of checking the form and setting of the cutting tool for thread cutting by means of a template. (*Courtesy South Bend Lathe.*)

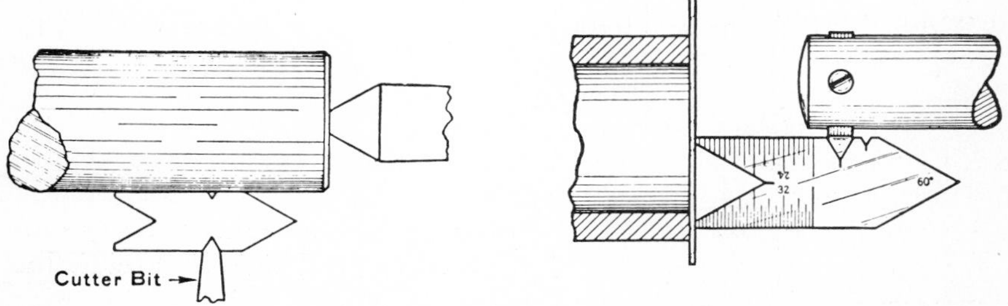

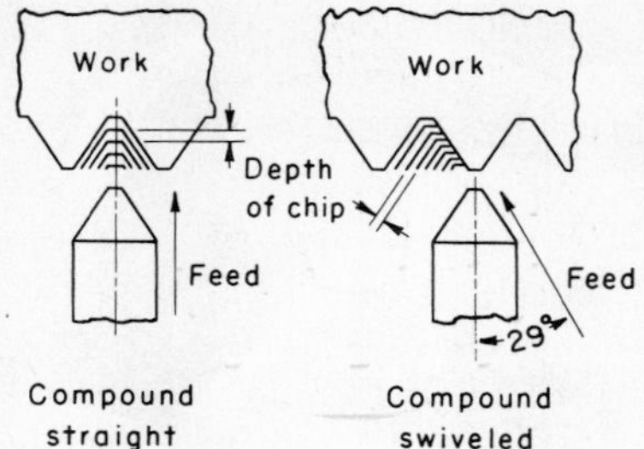

FIGURE 27-7. Two methods of feeding the tool into the work in cutting threads on a lathe.

method is that the absence of side and back rake will not produce proper cutting except on cast iron or brass. On steel the surface usually will be rough. Consequently, the second method commonly is used, with the compound swiveled 29°. The cutting then occurs primarily on the left-hand edge of the tool, and some side rake can be provided.

Proper speed ratio between the spindle and the lead screw is set by means of the gear-change box. Modern industrial lathes have ranges of ratios available so that nearly all standard threads can be cut merely by setting the proper levers on the quick-change gear box.

To cut a thread, it also is essential that a constant positional relationship be maintained between the workpiece, the cutting tool, and the lead screw. If this is not done, on successive cuts the tool will not be positioned correctly in the thread space. Correct relationship is obtained by means of a *threading dial,* shown in Figure 27-9, which is attached to the carriage and driven directly by the lead screw through a worm gear. Because the workpiece and the lead screw are directly connected, the threading dial provides a means for establishing the desired positional relationship between the workpiece and the cutting tool.

As shown in Figure 27-9, the threading dial is graduated into an even number of major and half divisions. If the split nut is closed in accordance with the following rules, correct positioning of the tool will result:

1. For even-number threads: at any line on the dial.
2. For odd-number threads: at any numbered line on the dial.
3. For threads involving $^1/_2$ numbers: at any odd-numbered line on the dial.
4. For $^1/_4$ or $^1/_8$ threads: return to the original starting line on the dial.

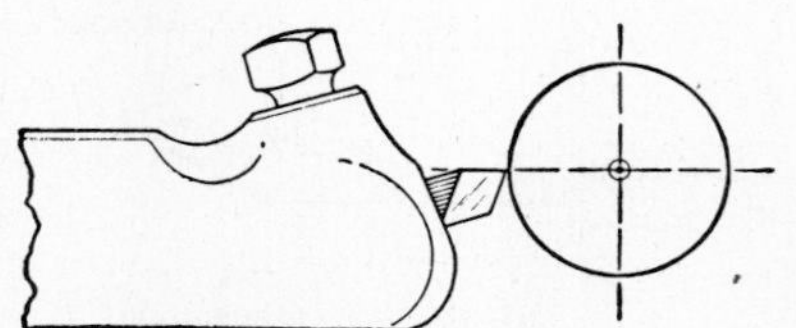

FIGURE 27-8. Proper relationship of the thread-cutting tool to the workpiece center line. (*Courtesy South Bend Lathe.*)

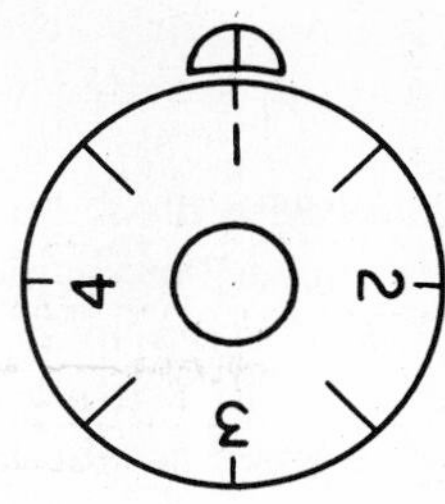

FIGURE 27-9. Face of a threading dial.

To start cutting a thread, the tool usually is fed inward until it just scratches the work, and the cross-slide dial reading is the noted, or set at zero. The split nut is engaged and the tool permitted to run over the desired thread length. When the tool reaches the end of the thread, it is quickly withdrawn by means of the cross-slide control. The split nut is then disengaged and the carriage returned to the starting position, where the tool is clear of the workpiece. At this point the future thread will be indicated by a fine scratch line. This permits the operator to check the thread lead, by means of a scale, or thread gage, to assure that all settings have been made correctly.

Next, the tool is returned to its initial zero depth position by returning the cross slide to the zero setting. By using the compound rest, the tool can be fed inward the proper amount for the first cut. A depth of 0.25 to 0.38 mm (0.010 to 0.015 inch) usually is used for the first cut, and smaller amounts on each successive cut, until the final cut is made with a depth of only 0.025 to 0.076 mm (0.001 to 0.003 inch) to produce a good finish.

Each successive cut is repeated in the manner described. At the end of each cut, the tool is quickly backed away from the workpiece by means of the cross-slide screw, the split nut disengaged, and the carriage returned to the starting position. The tool is fed inward to its previous position by means of the cross slide, and additional depth for the next cut provided by the compound rest. When the thread has been cut nearly to its full depth, it is checked for size by means of a mating nut or thread gage. Cutting is continued until a proper fit is obtained.

To cut right-hand threads, the tool is moved from right to left. For left-hand threads the tool must be moved from left to right. Otherwise, the procedure is essentially the same. Internal threads are cut in the same basic manner except that the tool is held in a boring bar. Tapered threads can be cut either with a taper attachment or by setting the tailstock off center. It should be remembered that in cutting a tapered thread the tool must be set normal to the axis of rotation of the workpiece.

It is apparent that cutting screw threads on a lathe is a slow, repetitious process and requires considerable skill on the part of the operator. The cutting speeds usually employed are from one-third to one-half of regular speeds to enable the operator to have time to manipulate the controls and to ensure better cutting. Consequently, the total process is quite costly, and it is evident why other methods are used whenever possible.

Cutting threads with dies. Straight and tapered external threads up to about 38 mm ($1^1/_2$ inches) in diameter can be cut quickly by means of threading dies, such as are shown in Figure 27-10. Basically, these are similar to hardened, threaded nuts with several longitudinal grooves that expose multiple cutting edges formed by the intersection of each groove and each thread. The cutting edges at the starting end are beveled to aid in starting the dies on the workpiece. As a consequence, a few threads at the inner end of the workpiece are not cut to full depth. Such threading dies are made of carbon or high-speed tool steel.

The solid-type dies, shown in Figure 27-10a, seldom are used in manufacturing because they have no provision for compensating for wear. The solid-adjustable type, shown in (b), is split and can be adjusted over a small range by means of a screw. This permits some adjustment to compensate for wear or to provide a variation in the fit of the resulting screw thread.

These types of threading dies usually are held in a *stock* and are rotated by hand. In using them, a suitable lubricant is desirable to produce a smoother thread and to prolong the life of the die, since there is a large amount of friction because the die is fed onto the workpiece by the screw action of the die on the threads being, or already, cut.

Self-opening die heads. A major disadvantage of solid-type threading dies is that they must be unscrewed from the workpiece to remove them. They thus are not suitable for use on high-speed, production-type machines. Therefore, *self-opening die heads* are used on turret lathes, screw machines, and special threading machines for cutting external threads.

FIGURE 27-10. (*a*) Solid threading die; (*b*) solid-adjustable threading die; (*c*) threading-die stock. (*Courtesy Greenfield Tap and Die Corporation.*)

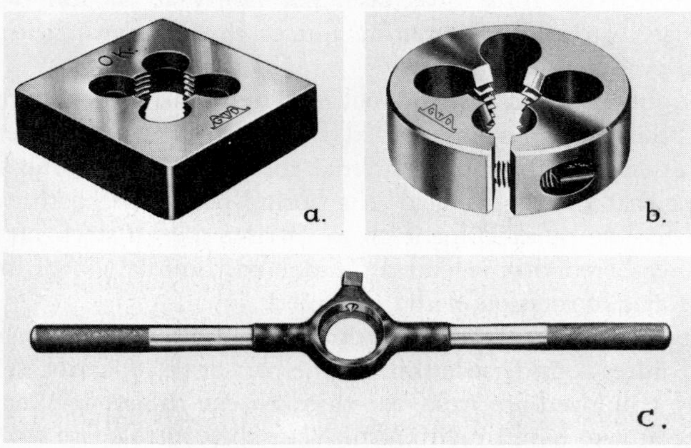

FIGURE 27-11. Self-opening die heads, with radial cutters (*left*) and tangential cutters (*right*). (*Courtesy Geometric Tool Company and The Warner & Swasey Company, respectively.*)

There are three types of self-opening die heads, all having four sets of adjustable, multiple-point cutters that can be removed for sharpening or for interchanging for different thread sizes. This permits one head to be used for a range of thread sizes. The two types shown in Figure 27-11 differ in the positioning of the cutters; those in one are positioned radially, whereas in the others they are positioned tangentially so as to result in somewhat less contact area and friction. In the third type, shown in Figure 27-12, the cutters are circular, with an interruption in the circular form to provide an easily sharpened cutting face. The cutters are mounted on the holder at an angle equal to the helix angle of the thread.

As the name implies, the cutters in self-opening die heads are arranged to open automatically when the thread has been cut to the desired length, thereby permitting the die head to quickly be withdrawn from the workpiece. On die heads used on turret lathes, the operator usually must reset the cutters in the closed position before making the next thread. The die heads used on

FIGURE 27-12. Self-opening die head in which circular cutters are used. (*Courtesy The National Acme Company.*)

screw machines and automatic threading machines are provided with a mechanism that automatically closes the cutters after the heads are withdrawn.

Cutting threads by means of self-opening die heads frequently is called *thread chasing.* However, some people apply this term to other methods of thread cutting—even to cutting a thread in a lathe.

Thread tapping. The cutting of an internal thread by means of a multiple-point tool is called *thread tapping,* and the tool is called a *tap.* A hole of diameter slightly larger than the minor diameter of the thread must already exist, made by drilling, boring, or casting.

For small holes, solid taps, such as shown in Figure 27-13 usually are used. These are similar to threaded bolts, with from one to four flutes to provide cutting edges. Such taps are made of either carbon or high-speed steel. The flutes can be either straight, helical, spiral, or spiral-pointed.

Hand taps, shown in Figure 27-13, have square shanks and are made in three types, usually in sets. The tapered end of *taper taps* will enter the hole a sufficient distance to help align the tap. In addition, the threads increase gradually to full depth, so this type of tap requires less effort to use. However, only a through hole can be threaded completely with a taper tap, because it cuts to full depth only behind the tapered portion. If a blind, or dead-end, hole must be threaded to the bottom, all three types of taps should be used in succession. After the taper tap has the thread started in proper alignment, a *plug tap,* which has only a few tapered threads, to provide gradual cutting of the threads to depth, is used to cut the threads as deep into the hole as its shape will permit. A *bottoming tap,* having no tapered threads, is used to finish the few remaining threads at the bottom of the hole to full depth. Obviously, producing threads to the full depth of a blind hole is time-consuming, and it also frequently results in broken taps and spoiled work-

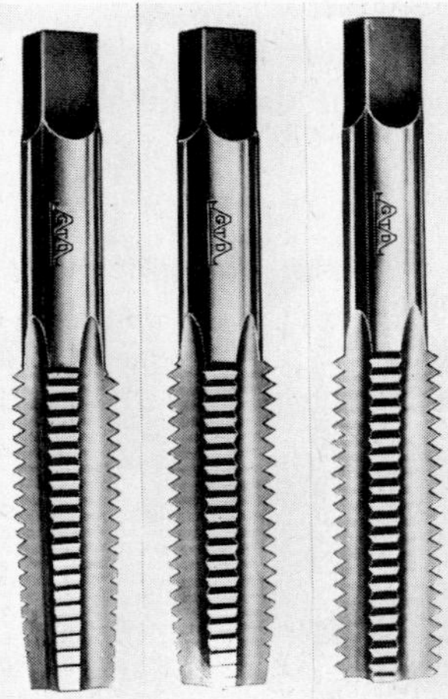

FIGURE 27-13. Solid thread taps: (*left to right*) taper, plug, and bottoming. (*Courtesy Greenfield Tap and Die Corporation.*)

FIGURE 27-14. Spiral-flute tap. (*Courtesy Greenfield Tap and Die Corporation.*)

pieces. Such configurations usually can be avoided if designers will give reasonable thought to the matter.

Taps operate under very severe conditions, both because of the severe friction involved and the difficulty of chip removal. Also, taps are relatively fragile. Spiral-fluted taps, illustrated in Figure 27-14, provide better removal of chips from a hole—particularly in tapping materials which produce long, curling chips. They also are helpful in tapping holes where the cutting action is interrupted by slots or keyways. In tapping through holes or blind holes that are not tapped to the bottom, the type of tap shown in Figure 27-15 is very useful. These have a spiral point that projects the chips ahead of the tap so that they do not interfere with the cutting action and the flow of cutting fluid into the hole.

Care must be exercised in using taps, particularly in tapping by hand. Proper cutting fluid should be used, and the tap should be reversed a partial turn after each two or three forward turns to assist in clearing the chips.

Solid taps also are used in tapping operations on machine tools, such as lathes, drill presses, and special tapping machines. In tapping on a drill press, a tapping attachment often is used. These devices rotate the tap slowly when the drill press spindle is fed downward against the work. When the tapping is completed and the spindle raised, the tap is automatically driven in the reverse direction at a higher speed to reduce the time required to back the tap out of the hole. Some modern machine tools provide for extremely fast spindle reverse for backing taps out of holes.

When solid taps are used on a screw machine or turret lathe, a special holder is employed in which a pin prevents the tap from turning while it is being fed into the work. As the tap reaches the end of the hole, it pulls the pin away from its stop so that the tap is free to rotate with the work. The rotation of the work is then reversed and the pin again prevents the tap from turning while it is backed out of the hole.

Collapsing taps. *Collapsing taps* are similar to self-opening die heads in that the cutting elements collapse inward automatically when the thread is completed. This permits withdrawing the tap from the workpiece without the

FIGURE 27-15. Spiral-point, straight-flute tap. (*Courtesy Greenfield Tap and Die Corporation.*)

FIGURE 27-16. Collapsing taps. (*Left*) Tap using circular cutters. (*Courtesy Landis Machine Company.*) (*Right*) Tap using radial cutters. (*Courtesy Geometric Tool Company Division, Greenfield Tap and Die Corporation.*)

necessity of unscrewing it from the thread. They can be either self-setting, for use on automatic machines, or require manual setting for each cycle. As shown in Figure 27-16, two types are available. The one with radial cutters generally is used for smaller size threads, whereas the type having circular cutters is used for larger-sized threads.

Tap-drill sizes. In most cases the hole that must be made before an internal thread is tapped is produced by drilling. Consequently, the drill size that is used is very important because it determines the depth of the thread contour and the force required in the tapping operation. In most applications a drill size is selected that will result in the thread having about 75 per cent of full thread depth. This practice makes tapping much easier than if full thread depth were attempted, and the resulting thread is only slightly less strong. Table 27-2 gives the drill sizes used to provide 75 per cent thread depth for several sizes of UNC threads. Full tables of tap-drill sizes for both Unified and Metric threads are readily available in machinists' handbooks and textbooks on drafting practice.

Threading and tapping machines. Special machines are available for production threading and tapping. Threading machines usually have one or more spindles on which a self-opening die head is mounted, with suitable means for clamping and feeding the workpiece. A typical machine of this type is shown in Figure 27-17.

Some special tapping machines are similar in construction, with self-collapsing taps substituted for the threading dies. More commonly, tapping machines resemble drill presses, modified to provide spindle feeds both up-

TABLE 27-2. Recommended tap-drill sizes for standard screw-thread pitches

American National Coarse-Thread Series

Number or Diameter	Threads per Inch	Outside Diameter of Screw	Tap Drill Sizes	Decimal Equivalent of Drill
6	32	0.138	36	0.1065
8	32	0.164	29	0.1360
10	24	0.190	25	0.1495
12	24	0.216	16	0.1770
1/4	20	0.250	7	0.2010
3/8	16	0.375	5/16	0.3125
1/2	13	0.500	27/64	0.4219
3/4	10	0.750	21/32	0.6562
1	8	1.000	7/8	0.875

FIGURE 27-17. Two-spindle, automatic threading machine. (*Courtesy Landis Machine Company.*)

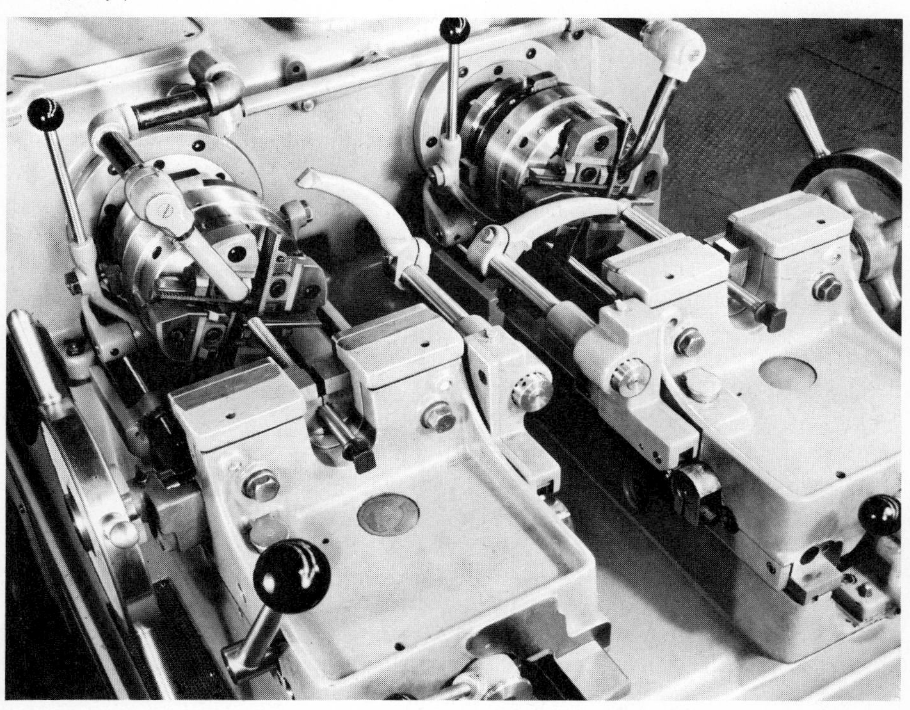

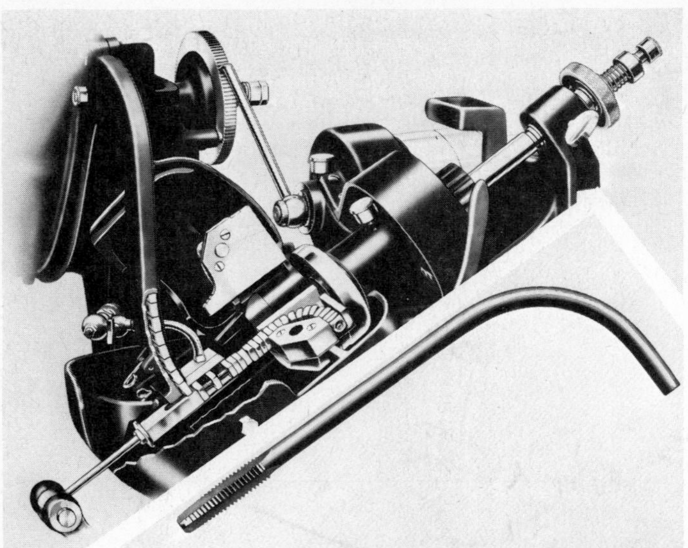

FIGURE 27-18. Automatic nut-tapping machine, using a bent tap. Inset shows a bent tap. (*Courtesy The National Machinery Company.*)

ward and downward, with the speed and feed more rapid on the upward motion.

For threading nuts, the type of machine shown in section in Figure 27-18 is used, in conjunction with a bent tap as pictured. The bent shank of the tap prevents it from rotating, yet permits the nuts to be threaded and slide continuously up and off the shank.

Thread milling. High-accuracy threads, particularly in larger sizes, are often cut by milling. Either a single- or a multiple-form cutter may be used, but the procedures are quite different.

A single-form cutter has a single, annular row of teeth. As shown in Figure 27-19, with the cutter tilted through an angle equal to the helix angle of the thread, it is fed inward radially to full depth while the work is stationary. The workpiece then is rotated slowly, and the cutter simultaneously is moved longitudinally, parallel with the axis of the work (or vice versa), by means of a lead screw, until the thread is completed. The thread can be completed in a single cut, or roughing and finish cuts can be used. This process is used primarily for large-lead or multiple-lead threads.

Some threads can be milled more quickly by using a multiple-form cutter, having several annular rows of teeth that are perpendicular to the cutter axis (the rows having no lead). The cutter must be slightly longer than the thread to be cut. It is set parallel with the axis of the workpiece and fed inward to full-thread depth while the work is stationary. The work then is ro-

FIGURE 27-19. Milling a large thread with a single-form cutter. The cutter can be seen behind the thread. (*Courtesy The Lees-Bradner Company.*)

tated slowly for a little over one revolution, and the rotating cutter is simultaneously moved longitudinally with respect to the workpiece (or vice versa) according to the thread lead. When the work has revolved one revolution, the thread is complete.

This process, illustrated in Figure 27-20, cannot be used on threads having a helix angle greater than about 3°, because clearance between the sides of the threads and the cutter depends on the cutter diameter being substantially less than that of the workpiece. Thus, although the process is rapid, its use is restricted to threads of substantial diameter and not more than about 51 mm (2 inches) long.

Thread grinding. Grinding can produce very accurate threads, and it also permits threads to be made on hardened materials. Three basic methods are used. *Center-type grinding with axial feed* is the most common method, being similar to cutting a thread on a lathe; a shaped grinding wheel replaces the single-point tool. Usually, a single-ribbed grinding wheel is employed, but multiple-ribbed wheels are used occasionally. The grinding wheels are shaped by special diamond dressers or by crush dressing and must be inclined to the

FIGURE 27-20. Milling a large thread with a multiple-form cutter. (*Courtesy The Lees-Bradner Company.*)

FIGURE 27-21. (*Left*) Grinding a worm, using a form wheel. Threads are ground in the same manner. (*Above*) Closeup view of the abrasive wheel and thread. (*Courtesy Ex-Cell-O Corporation.*)

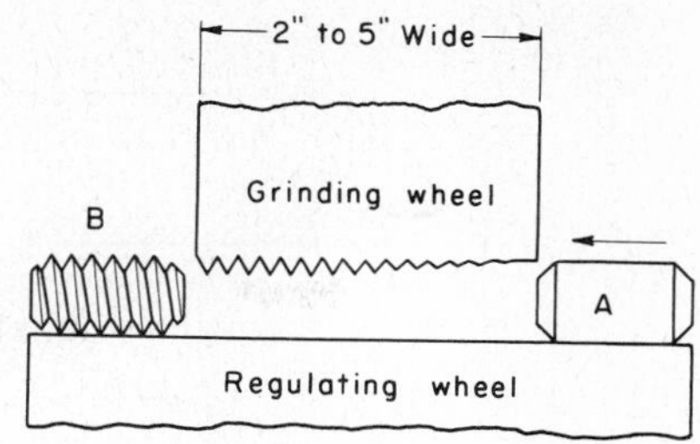

FIGURE 27-22. Principle of centerless thread grinding.

helix angle of the thread. Wheel speeds are in the high range. Several passes usually are required to complete the thread. Figure 27-21 shows a thread being ground by this procedure.

Center-type infeed thread grinding is similar to multiple-form milling in that a multiple-ribbed wheel, as wide as the length of the desired thread, is used. The wheel is fed inward radially to full thread depth, and the thread blank is then turned through about $1^1/_2$ turns as the grinding wheel is fed axially a little more than the width of one thread.

Centerless thread grinding, illustrated in Figure 27-22, is used for making headless set screws. The blanks are hopperfed to position *A*. The regulating wheel causes them to traverse the grinding wheel face, from which they emerge at position *B* in completed form. A production rate of 60 to 70 screws of 12.7-mm ($^1/_2$-inch) length per minute is possible.

THREAD ROLLING

As stated previously, virtually all threads that are produced in substantial quantities are made by *rolling*. This is a simple cold-forming operation in which the threads are formed by rolling a thread blank between hardened dies that cause the metal to flow radially into the desired shape. Because no metal is removed in the form of chips, less material is required, resulting in substantial savings. In addition, because of cold working, the threads have greater strength than cut threads, and a smoother, harder, and more wear-resistant surface is obtained. In addition, the process is almost unbelievably fast; the action on an actual machine cannot be observed without slowing the machine or using a stroboscopic light. On large threads, hot rolling is used occasionally.

Thread rolling is done by two basic methods. The simpler of these employs one fixed and one movable flat rolling die, as illustrated in Figure 27-23. For Class 2 threads, the normal commercial grade, the diameter of the thread blank is equal to the pitch diameter of the thread. After the blank is placed in position on the stationary die, movement of the moving die causes the blank to be rolled between the two dies and the metal in the blank is displaced to form the threads. As the blank rolls, it moves across the die par-

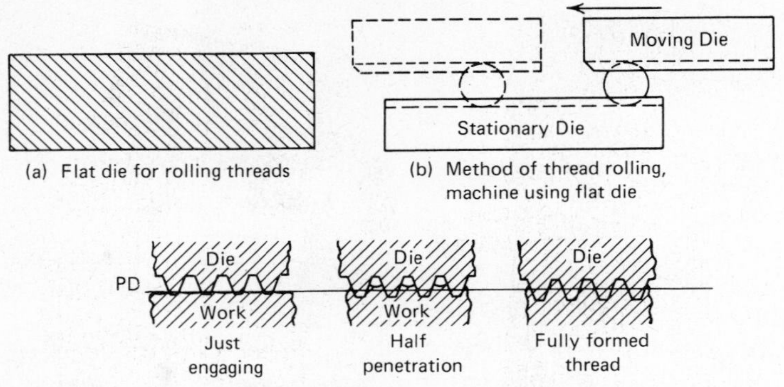

FIGURE 27-23. Schematic diagrams, showing the method of rolling threads with a flat die and the action of the die in forming the threads.

allel with its longitudinal axis. Prior to the end of the stroke of the moving die, the blank rolls off the end of the stationary die, its thread being completed.

One obvious characteristic of a rolled thread is that its major diameter always is greater than the diameter of the blank. When an accurate class of fit is desired, the diameter of the blank is made about 0.05 mm (0.002 inch) larger than the thread-pitch diameter. If it is desired to have the body of a bolt larger than the outside diameter of the rolled thread, the blank for the thread is made smaller than the body (see Figure 15-11).

The second method of thread rolling uses two or three roller-type dies. Figure 27-24 illustrates the three-roll method and shows the type of dies commonly employed on turret lathes and screw machines. Two variations are used. In one, the rolls are retracted while the blank is placed in position. They then move inward radially, while rotating, to form the thread. In the more common procedure, used on turret lathes and screw machines, the three rolls are contained in a self-opening die head, similar to the conventional type used for cutting external threads. The die head is fed onto the blank longitudinally and forms the thread progressively as the blank rotates. With this procedure, as in the case of cut threads, the innermost $1^1/_2$ to 2 threads are not formed to full depth because of the progressive action of the rollers.

A two-roll method, illustrated in Figure 27-25, is commonly employed for automatically producing large quantities of externally threaded parts.

Not only is thread rolling very economical, the threads are excellent as to form and strength. The cold working contributes to increased strength, particularly at the critical root areas. There is less likelihood of scratches and tears that can result from machining, which can act as stress raisers.

Whether a thread has been rolled or cut usually can be determined by examining the extreme inner end of the thread and the crests. As shown in

FIGURE 27-24. (*Left*) Method of rolling threads with three-roller-type dies. (*Right*) Three-roller-type die used on automatic machines. (*Courtesy Reed Rolled Thread Die Co. and TRW Geometric Tool Division, respectively.*)

Figure 27-26, the inner end usually will show evidence of the metal flow where the crest is not completely formed, and a fine line often will show along the center of the crest, where the metal met in flowing from the two sides. These effects are more noticeable on large threads than on small threads and usually are of no consequence.

Large numbers of threads are rolled on thin, tubular products. In this

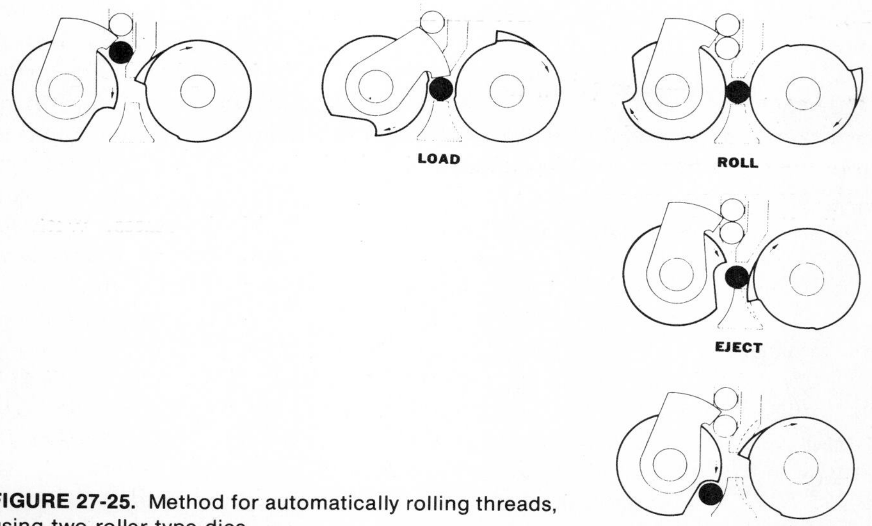

FIGURE 27-25. Method for automatically rolling threads, using two-roller-type dies.

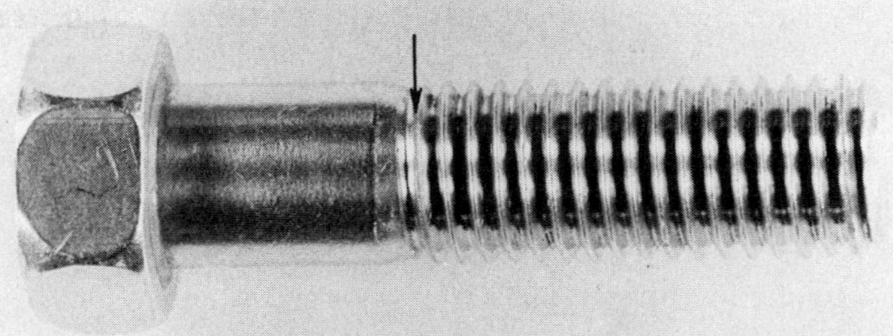

FIGURE 27-26. Bolt having rolled thread. Note that inner thread (arrow) is not completely formed.

case external and internal rolls are used. The threads on electric lamp bases and sockets are examples of this type of thread.

Cold forming internal threads. Unfortunately, most internal threads cannot be made by rolling; there is insufficient space within the hole to permit the required rolls to be arranged and supported, and the required forces are too high. However, many internal threads, up to about 12.7 mm (½ inch) in diameter, are formed in holes in ductile metals by means of *fluteless taps*. Such a tap and its special cross section are shown in Figure 27-27. As illustrated in Figure 27-28, the forming action is essentially the same as in rolling external threads. Because of the forming involved and the high friction, the torque required is about double that for cutting taps. Also, the hole diameter must be controlled carefully to obtain full thread depth without excessive torque.

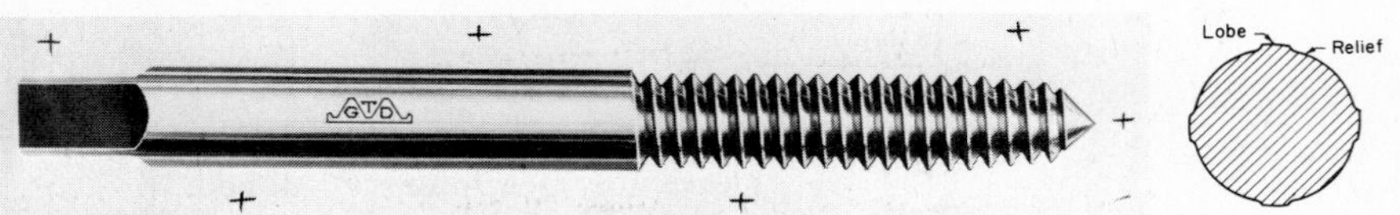

FIGURE 27-27. (*Left*) Fluteless tap for forming internal threads. (*Right*) Cross section of fluteless tap. (*Courtesy Besley-Welles Corporation.*)

FIGURE 27-28. Action of a fluteless tap in forming an internal thread.

However, fluteless taps produce somewhat better accuracy than cutting taps. A lubricating fluid should be used—water-soluble oils being quite effective.

Fluteless taps are especially suitable for forming threads in dead-end holes because no chips are produced. They come in both plug and bottoming types.

Review questions

1. How does the pitch diameter differ from the major diameter?
2. For what types of threads are the pitch and the lead the same?
3. What is the purpose of pipe threads being on a frustrum of a cone?
4. By what three basic methods can external threads be produced?
5. Explain the meaning of ¼″-20 UNC-3A.
6. What is meant by the designation M20 × 2.5-6g6g?
7. What are two reasons why fine-series threads are being used less now than in former years?
8. In cutting a thread on a lathe, how is the pitch controlled?
9. Why, when possible, should parts be designed so that any required threads can be made by methods other than cutting on a lathe with a single-point tool?
10. What is the function of a threading dial on a lathe?
11. What controls the lead of a thread when it is cut by a threading die?
12. What is the basic purpose of a self-opening die head?
13. Why is it essential that if threads are to be cut on a turret lathe, they should be designed to a standard diameter, whereas it is not so if they are to be cut on a lathe?
14. What is the reason for using a taper tap before a plug tap in tapping a hole?
15. What difficulties are encountered if full threads are specified to the bottom of a dead-end hole?
16. Why can a fluteless tap not be used for threading a hole in gray cast iron?
17. What provisions should a designer make if she or he desires a dead-end hole to be threaded?
18. What is the major advantage of a spiral-point tap?
19. Why can a fluteless tap not be used for threading to the bottom of a dead-end hole?
20. Is it desirable for a tapping fluid to have lubricating qualities? Why?
21. How does thread milling differ when using single- and double-form cutters?
22. What are the advantages of making threads by grinding?
23. Why has thread rolling become the most commonly used method for making threads?
24. How may one determine whether a thread has been produced by rolling rather than by cutting?

Case study 20.
THE STEERING SHAFT CONNECTION

Figure CS-20 shows the design of a connection that was incorporated in the steering shaft of an internationally known sports car. A failure occurred, resulting in a death and a large damage claim and award. (1) Analyze the design and determine its inherent deficiencies. (2) What manufacturing difficulties would you anticipate with this design, and which might lead to failure? (3) How would you modify the design to provide proper performance?

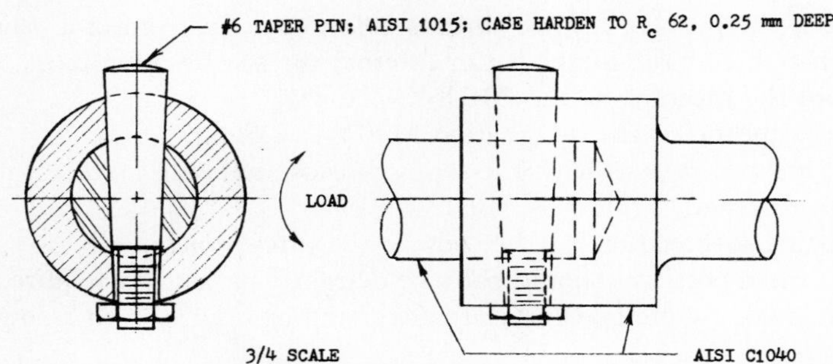

FIGURE CS-20. Method of coupling two sections of steering column shaft in an automobile.

CHAPTER 28

Gear manufacturing

Gears transmit power or motion mechnically between parallel, intersecting, or nonintersecting shafts. Although unsung and usually hidden from sight, they are one of the most important mechanical elements in our civilization, possibly even surpassing the wheel, since most wheels would not be turning without power being applied to them through gears. They operate at almost unlimited speeds under a wide variety of conditions. Millions are produced each year in sizes from a few millimeters up to more than 6 meters (20 feet) in diameter. Often the requirements that must be, and routinely are, met in their manufacture are amazingly precise. Consequently, the machines and processes that have been developed for producing them are among the most ingenious we have. In order to understand the functional requirements of these machines and processes, it is helpful first to consider the basic theory of gears and their operation.

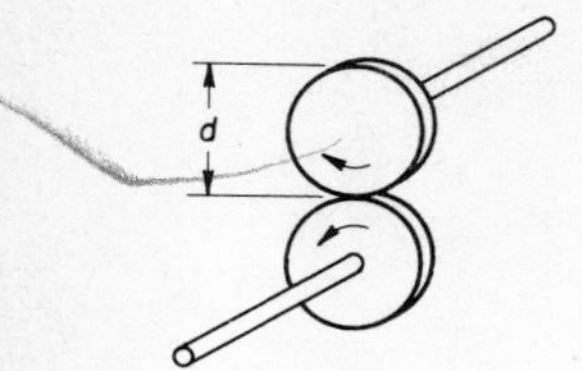

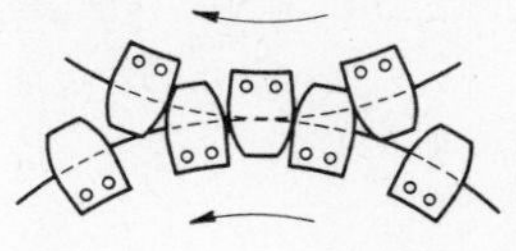

FIGURE 28-1. Relationship between the transmission of rotation through friction disks and by gear teeth.

Gear theory and terminology. Gears, basically, are modifications of friction disks, as illustrated in Figure 28-1, teeth being added to prevent slipping and to assure that their relative motions are constant. However, it should be noted that the addition of teeth does not change the relative velocities of the disks and shafts; the velocity ratio is determined by the diameters of the disks.

Although wooden teeth or pegs were attached to disks to make gears in ancient times, the teeth of modern gears are produced by making cuts into disks that are sufficiently large to contain the outer portions of the teeth, or by forming processes that cause the metal in the teeth to plastically flow outward from a disk. But the basic concept of a disk remains, with the *pitch circle,* shown in Figure 28-2, corresponding to the diameter of the friction disk. Thus the angular velocity of a gear is determined by the all-important diameter of this imaginary pitch circle, and all design calculations relating to gear performance are based on the pitch-circle diameter or, more simply, the *pitch diameter* (*PD*).

For two gears to operate properly, their pitch circles must be tangent to each other. The point at which the two pitch circles are tangent, and at which they intersect the center line connecting their centers of rotation, is called the

FIGURE 28-2. Gear-tooth nomenclature.

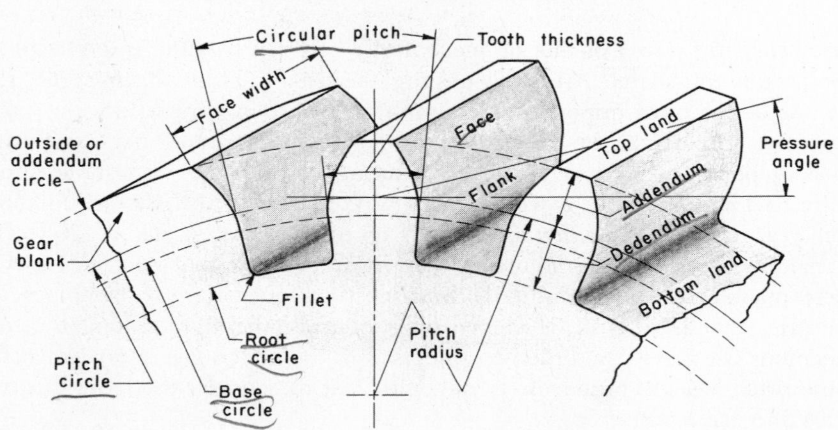

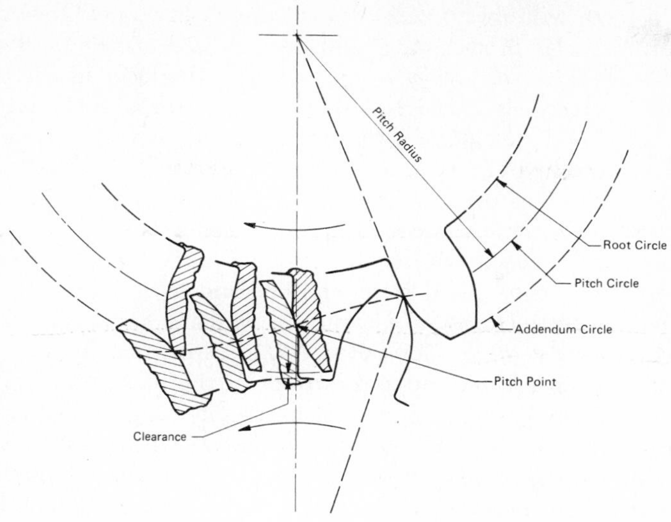

FIGURE 28-3. Action between mating gear teeth.

pitch point. The common normal at the point of contact of mating teeth must pass through the pitch point. This condition is illustrated in Figure 28-3.

To minimize friction and wear, and thus increase their life and efficiency, gears are designed to have rolling motion between mating teeth, rather than sliding motion. To achieve this condition, most gears utilize a tooth form that is based on an *involute curve.* This is the curve that is generated by a *point* on a straight line when the line rolls around a *base circle.* A somewhat simpler method of developing an involute curve is that shown in Figure 28-4, by unwinding a tautly held string from a base circle; point *A* generates an involute curve.

There are four reasons for using the involute form for gear teeth. First, such a tooth form provides the desired pure rolling action. Second, even if a pair of involute gears is operated with the distance between the centers slightly too large or too small, the common normal at the point of contact be-

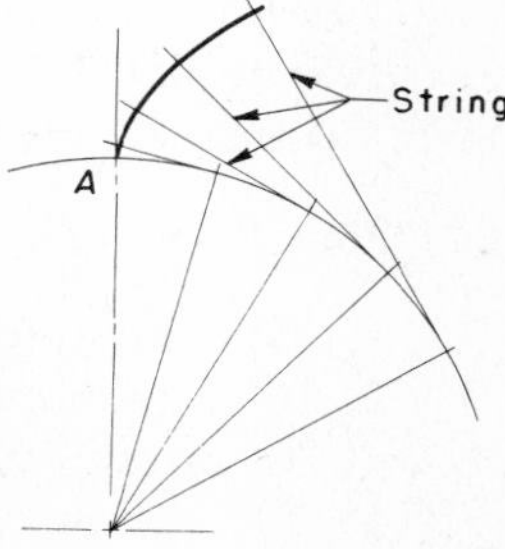

FIGURE 28-4. Method of generating an involute curve by unwinding a string from a cylinder.

tween mating teeth will always pass through the pitch point. Obviously, the theoretical pitch circles in such cases will be increased or decreased slightly. Third, the *line of action,* or *path of contact,* that is, the locus of the points of contact of mating teeth, is a straight line that passes through the pitch point and is tangent to the base circles of the two gears. The fourth very important reason is that a true involute tooth form can be generated by a rack that has straight-sided teeth. This permits a very accurate tooth profile to be obtained through the use of a simple and easily made cutting tool.

The basic size of gear teeth may be expressed in two ways. The common practice, especially in the United States and England, is to express the dimensions as a function of the *diametral pitch* (DP), which is *the number of teeth* (N) *per unit of pitch diameter* (PD); thus $DP = N/PD$. Dimensionally, DP involves inches in the English system and millimeters in the SI system, and it is a measure of tooth size. The second method for specifying gear tooth size is by means of the *module* (M), defined as *the pitch diameter divided by the number of teeth,* thus $M = PD/N$. It thus is the reciprocal of diametral pitch and is expressed in inches or millimeters. Any two gears having the same diametral pitch or module will mesh properly if they are mounted so as to have the correct distances and relationship.

The important tooth elements, shown in Figure 28-2, can be specified in terms of the diametral pitch or the module and are as follows:

1. *Addendum:* The radial distance from the pitch circle to the outside diameter.
2. *Dedendum:* The radial distance from the pitch circle to the root circle. It is equal to the addendum plus the *clearance,* which is provided to prevent the outer corner of a tooth from touching against the bottom of the tooth space.
3. *Circular pitch:* The distance between corresponding points of adjacent teeth, measured along the pitch circle. It is numerically equal to π/diametral pitch.
4. *Tooth thickness:* The thickness of a tooth, measured along the pitch circle. When tooth thickness and the corresponding *tooth space* are equal, no *backlash* exists in a pair of mating gears.
5. *Face width:* The length of the gear teeth in an axial plane.
6. *Tooth face:* The mating surface between the pitch circle and the addendum circle.
7. *Tooth flank:* The mating surface between the pitch circle and the root circle.

Four shapes of involute gear teeth are used in this country:

1. 14½° pressure angle, full-depth (used most frequently).
2. 14½° pressure angle, composite (seldom used).
3. 20° pressure angle, full-depth (seldom used).
4. 20° pressure angle, stub-tooth (next most common).

TABLE 28-1. Standard dimensions for involute gear teeth

	14½° Full Depth	20° Stub Tooth
Pitch diameter	$\frac{N}{DP}$	$\frac{N}{DP}$
Addendum	$\frac{1}{DP}$	$\frac{0.8}{DP}$
Dedendum	$\frac{1.157}{DP}$	$\frac{1}{DP}$
Outside diameter	$\frac{N+2}{DP}$	$\frac{N+1.6}{DP}$
Clearance	$\frac{0.157}{DP}$	$\frac{0.2}{DP}$
Tooth thickness	$\frac{1.5708}{DP}$	$\frac{1.5708}{DP}$

In the 14½° full-depth system the tooth profile outside the base circle is an involute curve. Inward from the base circle the profile is a straight, radial line that is joined with the bottom land by a small fillet. With this system the teeth of the basic rack have straight sides.

The 14½° composite system and the 20° full-depth system provide somewhat stronger teeth. However, with the 20° full-depth system considerable undercutting occurs in the dedendum area, so stub teeth often are used; in these the addendum is shortened by 20 per cent, thus permitting the dedendum to be shortened a similar amount. This results in very strong teeth without undercutting.

Table 28-1 gives the formulas for computing the dimensions of gear teeth in the 14½° full-depth and 20° stub-tooth systems.

Physical requirements of gears. A consideration of gear theory leads to five requirements that must be met in order for gears to operate satisfactorily:

1. The actual tooth profile must be identical to the theoretical profile.
2. Tooth spacing must be uniform and correct.
3. The *actual* and theoretical pitch circles must be coincident and be concentric with the axis of rotation of the gear.
4. The face and flank surfaces must be smooth and sufficiently hard to resist wear and prevent noisy operation.
5. Adequate shafts and bearings must be provided so that desired center-to-center distances are retained under operational loads.

The first four of these requirements are determined by the material selection and manufacturing process. The various methods of manufacture that are used represent attempts to meet these requirements to varying degrees with minimum cost, and their effectiveness must be measured in terms of the extent to which the resulting gears embody these requirements.

The more common types of gears are shown in Figure 28-5. *Spur gears* have straight teeth and are used to connect parallel shafts. They are the most easily made and the cheapest of all types.

The teeth on *helical gears* lie along a helix, the angle of the helix being the angle between the helix and a pitch cylinder element parallel with the gear shaft. Helical gears can connect either parallel or nonparallel, nonintersecting shafts. Such gears are stronger and quieter than spur gears because the contact between mating teeth increases more gradually and more teeth are in contact at a given time. Although they usually are slightly more expensive to make than spur gears, they can be manufactured in several ways and are produced in large numbers.

FIGURE 28-5. Several types of gears. (*Top row*) Spur gear and rack; worm and worm gear; continuous herringbone gears. (*Center row*) Spiral bevel gear; helical gears; crown gear. (*Bottom row*) Straight, zerol, and hypoid bevel gears. (*Courtesy Gleason Works.*)

Helical gears have one disadvantage in that a side thrust is created when they are loaded, which must be absorbed in the bearings. *Herringbone gears* neutralize this side thrust by having, in effect, two helical-gear halves, one having a right-hand and the other a left-hand helix. The *continuous* herringbone type, shown in Figure 28-5, is rather difficult to machine but is very strong. A modified herringbone type is made by machining a groove, or gap, around the gear blank where the two sets of teeth would come together. This provides a runout space for the cutting tool in making each set of teeth.

A *rack* is a gear with infinite radius, having teeth that lie on a straight line on a plane. The teeth may be normal to the axis of the rack or helical, so as to mate with spur or helical gears, respectively.

A *worm* is similar to a screw. It may have one or more threads, the multiple-thread type being very common. Worms usually are used in conjunction with a *worm gear.* High gear ratios are easily obtainable with this combination. The axes of the worm and worm gear are nonintersecting and usually are at right angles. If the worm has a small helix angle, it cannot be driven by the mating worm gear. This principle frequently is employed to obtain nonreversible drives. As is shown in Figure 28-5, worm gears usually are made with the top land concave, to permit greater area of contact between the worm and the gear. A similar effect can be achieved by using a *conical worm,* in which the helical teeth are cut on a double-conical blank, thus producing a worm that has an hourglass shape.

Bevel gears are used to transmit motion between intersecting shafts. They are conical in form, the teeth being cut on the surface of a truncated cone. Several types of bevel gears are made—the types varying as to whether the teeth are straight or curved, and whether the axes of the mating gears intersect. On *straight-tooth* bevel gears the teeth are straight, and if extended all would pass through a common apex. *Spiral-tooth* bevel gears have teeth that are segments of spirals. Like helical gears, this design provides tooth overlap so that more teeth are engaged at a given time and the engagement is progressive. *Hypoid* bevel gears also have a curved-tooth shape but are designed to operate with nonintersecting axes. They are used in the rear axles of most automobiles so that the drive shaft axis can be below the axis of the axle and thus permit a lower floor height. *Zerol* bevel gears have teeth that are circular arcs, providing somewhat stronger teeth than can be obtained in a comparable straight-tooth gear. They are not used extensively. When a pair of bevel gears are the same size and have their shafts at right angles, they are termed *miter gears.*

A *crown gear* is a special form of bevel gear having a 180° cone apex angle. In effect, it is a disk with the teeth on the side of the disk. It also may be thought of as a rack that has been bent into a circle so that its teeth lie in a plane. The teeth may be straight or curved. On straight-tooth crown gears the teeth are radial. Crown gears seldom are used, but they have the important quality that they will mesh properly with a bevel gear of any cone angle, provided that the bevel gear has the same tooth form and diametral pitch.

This important principle is incorporated in the design and operation of two very important types of gear generating machines that will be discussed later.

Most gears are of the *external* type, the teeth forming the outer periphery of the gear. Internal gears have the teeth on the inside of a solid ring, pointing toward the center of the gear.

GEAR MANUFACTURING

Gears are made in very large numbers by both machining and by cold-roll forming. In addition, significant quantities are made by extrusion, by blanking, and some by powder metallurgy and by a forging process. However, it is only by machining that all types of gears can be made in all sizes, and although roll-formed gears can be made with accuracy sufficient for most applications—even for automobile transmissions—machining still is unsurpassed for gears that must have very high accuracy. Also, roll forming can be used only on ductile metals.

Basic methods for machining gears. Three basic methods are employed for machining gears, each having certain advantages and limitations as to quality, flexibility, and cost.

Form cutting utilizes the principle illustrated in Figure 28-6, the cutter having the same form as the *space* between adjacent teeth. Usually a multiple-point cutter is used, either rotating about an axis as shown in Figure 28-6 or, occasionally, rotating about an axis that is normal to the axis of the gear blank. However, it is possible to use a single-point cutter, and the principle can be employed with a reciprocating tool. The tool is fed radially toward the center of the gear blank, to obtain the desired tooth depth; it then is moved across the tooth face to obtain the required tooth width. When one tooth has been completed, the tool is withdrawn, the gear blank is indexed, and the cutting of the next tooth space is started. It is possible to cut all the tooth

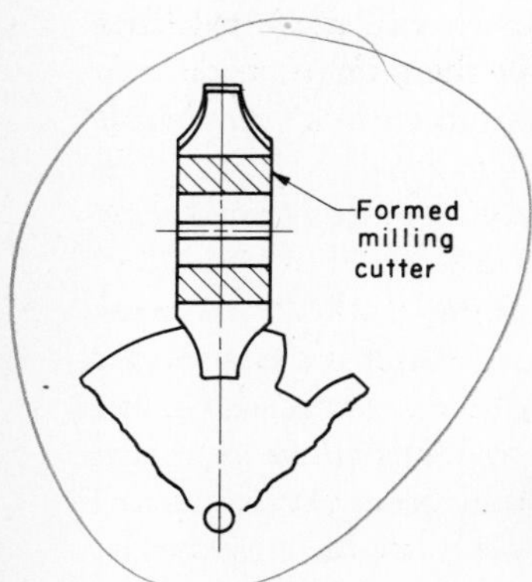

FIGURE 28-6. Basic method of machining a gear by form cutting.

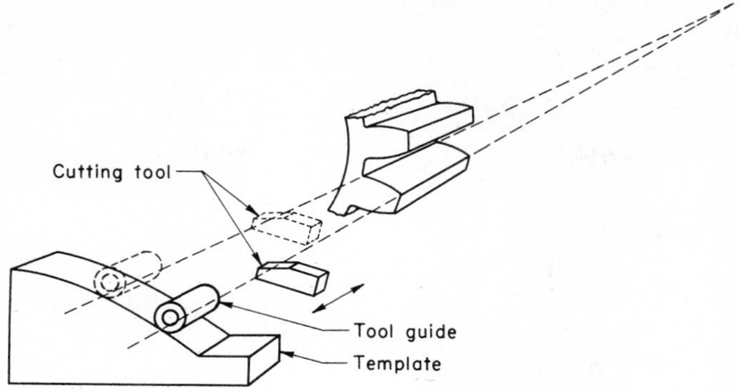

FIGURE 28-7. Method of machining gear teeth by means of a tool that is guided by a template.

spaces simultaneously by using a number of cutting tools equal to the number of teeth in the gear; this is done in one type of gear broaching.

Basically, form cutting is a simple and flexible method of machining gears. The equipment and cutters required are relatively simple, and standard machine tools (milling machines) often are used. However, in most cases the procedure is quite slow, and considerable care is required on the part of the operator, so it usually is employed where only one or a few gears are to be made.

Template machining utilizes a simple, single-point cutting tool that is guided by a template. This principle is illustrated in Figure 28-7. By using a template that is several times larger than the gear tooth that is to be cut, good accuracy can be achieved. However, the equipment is specialized, and the method is seldom used except for making large bevel gears.

Most high-quality gears that are made by machining are made by the *generating process*. This process is based on the principle that any two involute gears, or any gear and a rack, of the same diametral pitch will mesh together properly. Utilizing this principle, if one of the gears (or the rack) is made into a cutter by proper sharpening, it can be used to cut into a mating gear blank and thus generate teeth on the blank.

To carry out the process, the cutter gear and the gear blank must be attached rigidly to their corresponding shafts, and the two shafts must be interconnected by suitable gearing so that the cutter and the blank rotate positively with respect to each other and with the same pitch-line velocities. To start cutting a gear, the cutter gear is reciprocated and is fed radially into the blank between successive strokes. When the desired tooth depth has been obtained, the cutter and blank are then rotated slightly after each cutting stroke. The resulting generating action is indicated schematically in the upper dia-

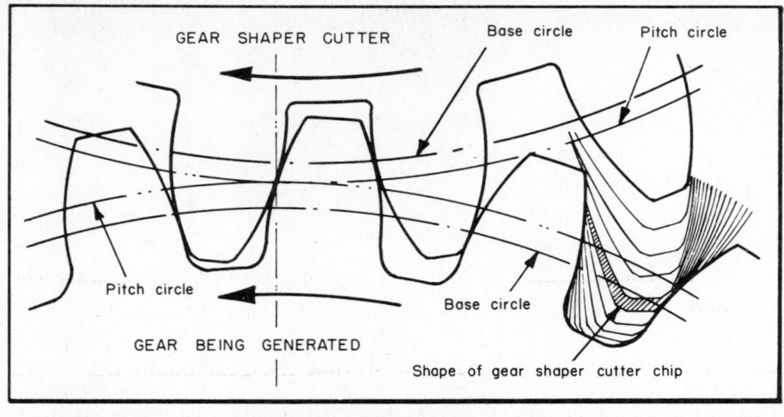

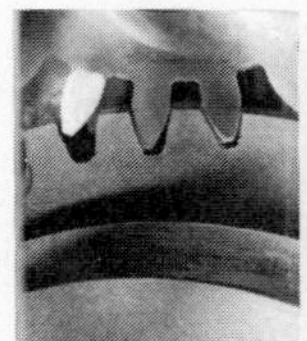
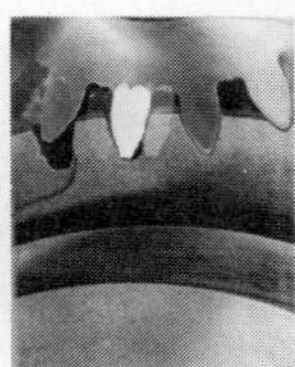
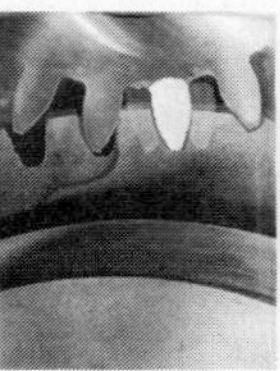
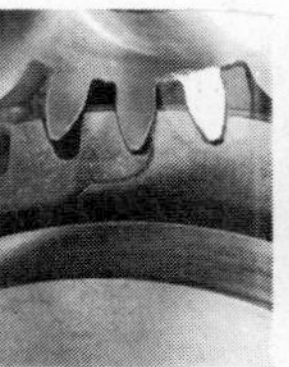
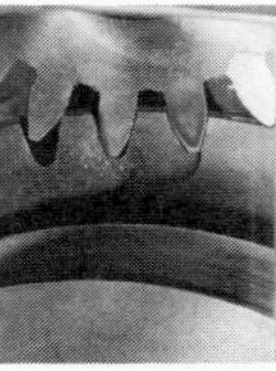

FIGURE 28-8. (*Top*) Generating action of a Fellows gear-shaper cutter. (*Courtesy The Fellows Gear Shaper Company.*) (*Bottom*) Series of photographs showing various stages in generating one tooth in a gear by means of a gear shaper, action taking place from right to left, corresponding to diagram above. One tooth of the cutter was painted white.

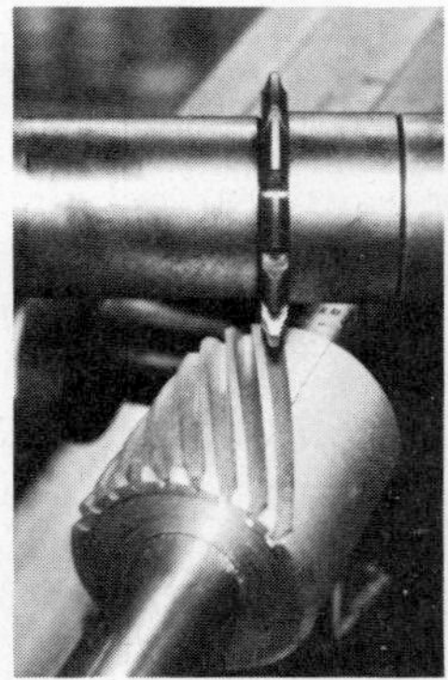

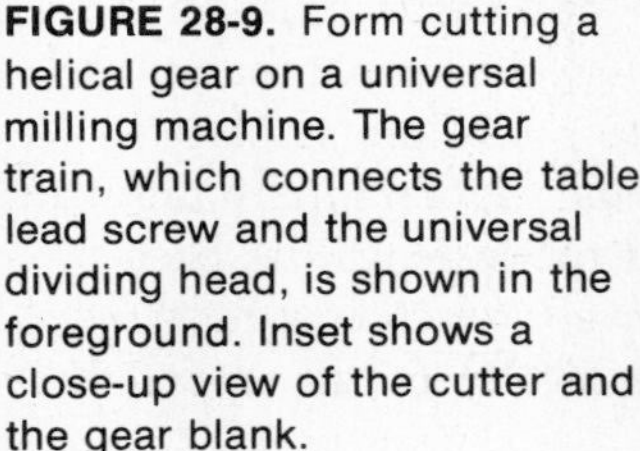

FIGURE 28-9. Form cutting a helical gear on a universal milling machine. The gear train, which connects the table lead screw and the universal dividing head, is shown in the foreground. Inset shows a close-up view of the cutter and the gear blank.

gram of Figure 28-8 and shown in the cutting of an actual gear tooth in the photographs in the lower portion of the same figure.

Machines for form-cutting gears. In machining gears by the form-cutting process, the form cutter is mounted on the machine spindle, and the gear blank is mounted on an arbor held between the centers of some type of indexing device. Figure 28-9 shows the arrangement that is employed when, as often is the case, the work is done on a universal milling machine; the cutter is mounted on the spindle, and a dividing head is used to index the gear blank. When a helical gear is to be cut, as in the case shown, the table must be set at an angle equal to the helix angle, and the dividing head is geared to the longitudinal feed screw of the table so that the gear blank will rotate as it moves longitudinally.

Standard cutters usually are employed in form-cutting gears. In the United States, these come in eight sizes for each diametral pitch and will cut gears having the number of teeth indicated in the following tabulation:

Cutter Number	Gear Tooth Range
1	135 teeth to rack
2	55–134
3	35–54
4	26–34
5	21–25
6	17–20
7	14–16
8	12–13

A single cutter will not produce a theoretically perfect tooth profile for all sizes of gears in the range for which it is intended. However, the change in tooth profile over the range covered by each cutter is very slight and for most purposes satisfactory results are achieved. Where greater accuracy is required, half-number cutters (such as 3½) can be obtained. A typical cutter is shown in Figure 28-10. Cutters are available for all common diametral pitches and 14½ and 20° pressure angles.

To cut a gear on a milling machine, the geometric center of the cutter must be exactly aligned vertically with the center line of the index-head spindle. A gear blank of the proper outside diameter is placed on an arbor, which, in turn, is mounted between a foot-stock dead center and the live center in the index head and is connected positively to the latter by a dog. The table of the machine is raised until the cutter just makes contact with the periphery of the gear blank. The vertical feed dial then is set to zero and the table moved horizontally until the cutter clears the blank. The table is then fed upward an amount equal to the desired tooth depth, or a lesser amount if two or more cuts are to be made. The longitudinal power feed of the table then is engaged, and the tooth space is cut in the blank. After one tooth space

FIGURE 28-10. Typical form cutter for machining gear teeth. (*Courtesy Brown & Sharpe Manufacturing Company.*)

is cut, the table movement is reversed until the cutter is again clear of the blank. The blank is then indexed to the proper position for cutting the next tooth space. This cutting procedure is repeated until all the teeth have been formed.

If the amount of metal that must be removed to form a tooth space is large, roughing cuts may be taken with a *stocking cutter,* shown in Figure 28-11. The stepped sides of the stocking cutter remove most of the metal and leave only a small amount to be removed subsequently by the regular form cutter in a finish cut.

Straight-tooth bevel gears can be form cut on a milling machine, but this seldom is done. Because the tooth profile varies from one end of the tooth to the other, after one cut is taken to form the correct tooth profile at the

FIGURE 28-11. Stocking cutter for making roughing cuts in machining gear teeth. (*Courtesy Brown & Sharpe Manufacturing Company.*)

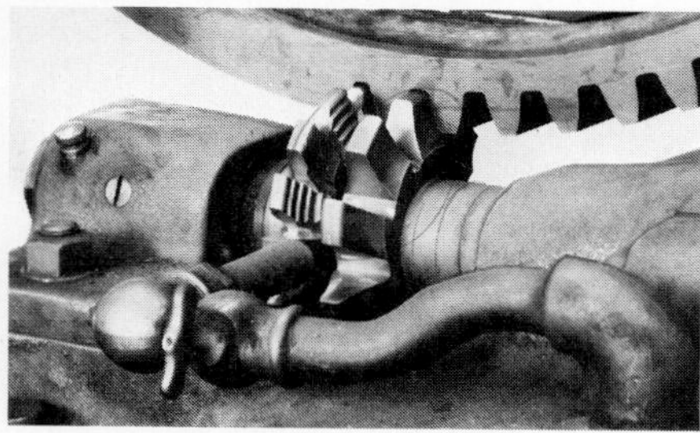

FIGURE 28-12. Cutting a gear on a semiautomatic gear-cutting machine. (*Above*) Cutters making simultaneous roughing and finish cuts. (*Courtesy Brown & Sharpe Manufacturing Company.*)

smaller end, the relationship between the cutter and the blank must be altered. Shaving cuts then are taken on the side of each tooth to form the correct profile throughout the entire tooth length.

Although the form cutting of gears on a milling machine is a flexible process and is suitable for gears that are not to be operated at high speeds, or that need not operate with extreme quietness, the process is slow and requires skilled labor. It obviously is not suitable for quantity production.

Semiautomatic machines are available for making gears by the form-cutting process. Such a machine is shown in Figure 28-12. The procedure utilized is essentially the same as on a milling machine, except that after it is set for depth of cut, indexing, and so on, the various operations are completed automatically. Gears made on such machines are no more accurate than those produced on a milling machine, but the possibility of error is less, and they are much cheaper because of reduced labor requirements. But for large quantities, form cutting is not used.

Template-type machines. A gear-making machine employing the template principle is shown in Figure 28-13. However, these are very seldom used.

Cutter-gear generating machines. A machine that generates gears by means of a gear-type cutter is shown in Figure 28-14. Figure 28-15 shows details of the cutter design. The gear blank is mounted on a vertical spindle and the cutter on the end of a second, vertical, reciprocating spindle. The two spin-

FIGURE 28-13. Template-type bevel-gear planing machine. The template and the follower that guide the cutter are shown in the inset. (*Courtesy Gleason Works.*)

dles are connected by means of change gears so that the cutter and gear blank revolve with the same pitch-line velocity. Cutting occurs on the down stroke on some machines and on the up stroke on others. At the end of each cutting stroke the spindle carrying the blank retracts slightly to provide clearance between the work and the tool on the return stroke. Because of the reciprocating action of the cutter, these machines commonly are called *gear shapers.*

To start cutting a gear, the cutter starts to feed inward before each cutting stroke as it and the blank rotate. When the proper depth is reached, the inward feed stops and the cutter and blank continue their rotation until all the teeth have been formed by the generation process.

Either straight- or helical-tooth gears can be cut on gear shapers. To cut helical teeth, both the cutter and the blank are given an oscillating rotational motion during each stroke of the cutter, turning in one direction during the cutting stroke and in the opposite direction during the return stroke. Because the cutting stroke can be adjusted to end at any desired point, gear shapers are particularly useful for cutting cluster gears. Some machines can be equipped with two cutters to simultaneously cut two gears, often of different diameters. Gear shapers also can be adapted for cutting internal gears, as shown in Figure 28-16.

Two special types of gear shapers have been developed for mass-production purposes. The *rotary gear shaper* essentially is 10 shaper units mounted on a rotating base and having a single drive mechanism. Nine gears are cut si-

FIGURE 28-14. Fellows gear shaper, showing the motions of the gear blank and the cutter. Inset shows a close-up view of the cutter and teeth. (*Courtesy The Fellows Gear Shaper Company.*)

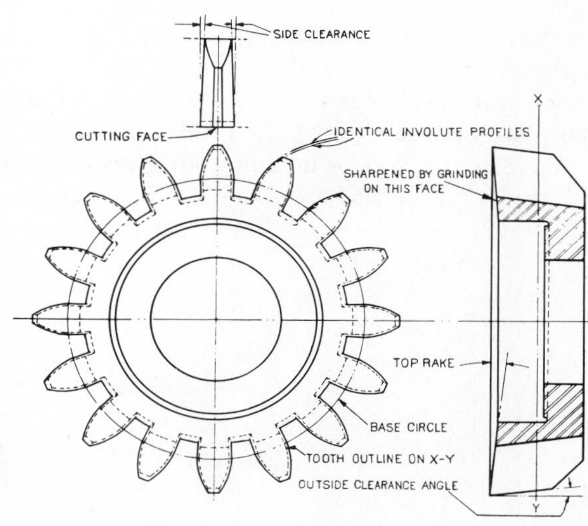

FIGURE 28-15. Details of the cutter used on the Fellows gear shaper. (*Courtesy The Fellows Gear Shaper Company.*)

FIGURE 28-16. Cutting an internal gear on a gear shaper. (*Courtesy The Fellows Gear Shaper Company.*)

multaneously while a finished gear is removed and a new blank is put in place on the tenth unit. In the *planetary gear shaper,* shown in Figure 28-17, six gear blanks move in planetary motion about a large, central gear cutter. The cutter has no teeth in one portion to provide a space where the gear can be removed and a new blank placed on the empty spindle.

FIGURE 28-17. Cutter and six gear blanks in a planetary gear shaper. (*Courtesy The Fellows Gear Shaper Company.*)

FIGURE 28-18. Cutting a continuous herringbone gear on a Sykes gear-cutting machine; (*top*) showing the gear blank at left end of the stroke and (*bottom*) at the right end. Cutters, behind gear, are denoted by arrows. (*Courtesy Western Gear Works.*)

The *Sykes gear-generating machine* is employed primarily for cutting continuous herringbone gears; it is the only machine that will cut gears of this type. Its unique feature is that it employs two cutter gears, mounted a fixed distance apart, and that either the cutters or the gear blank are reciprocated with respect to each other, parallel with the gear-blank axis, as shown in Figure 28-18. One cutter cuts one half of the gear as the reciprocating motion takes place in one direction, and the other cutter cuts the other half on the return stroke. Cutting takes place as each cutter moves toward the center line of the gear. Change gears connect the gear blank and the cutters to provide the necessary continuous rotation to bring about the generating process. In addition, the cutters and blank are given an oscillatory, twisting motion during each reciprocation cycle to provide the desired helix angle. These machines produce gears of excellent quality.

Gear-hobbing machines. As mentioned previously, involute gear teeth can be generated by a cutter that has the form of a rack, having teeth with straight sides. Although the use of a rack as a cutter would have a great advantage—in that the cutter would be simple to make—such a use has two major disadvantages. First, the cutter or the blank would have to reciprocate, with cutting occurring during only one stroke direction. Second, the rack would have to move longitudinally as the blank rotated (or the blank roll, in mesh, on the rack). Unless the rack were very long, or the gear very small, the two would not be in mesh after a few teeth were cut.

A *hob* overcomes the preceding two difficulties. As shown in Figures 28-

FIGURE 28-19. Typical gear hob. (*Courtesy Barber-Colman Company.*)

19 and 28-20, a hob can be thought of, basically, as one long rack tooth that has been wrapped around a cylinder in the form of a helix and gashed at intervals to provide a number of cutting edges. Relief is provided behind each of the helically arranged cutting faces. As shown in Figure 28-20, the cross section of each tooth, normal to the helix, is the same as that of a rack tooth. A hob also can be thought of as a gashed worm.

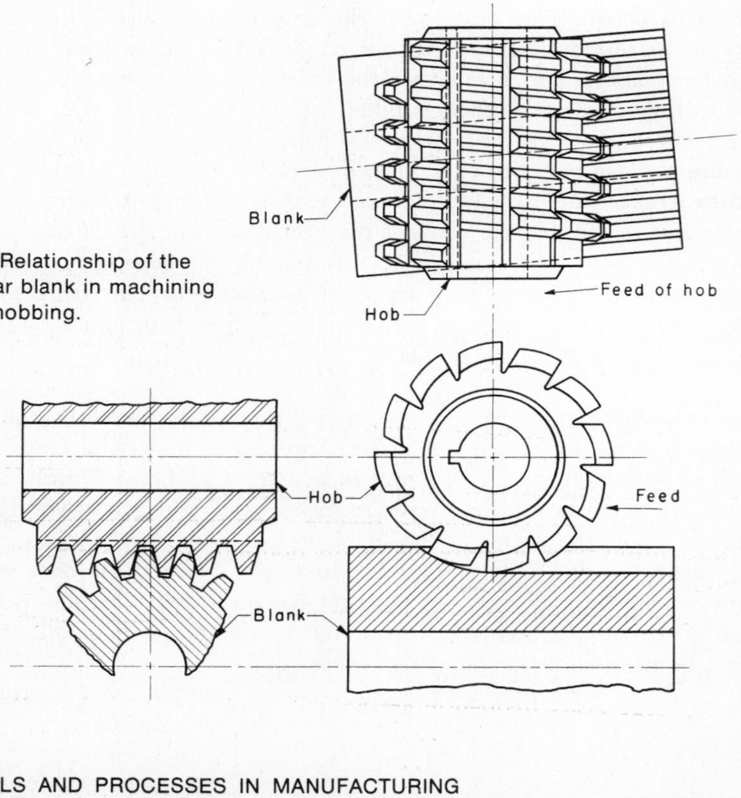

FIGURE 28-20. Relationship of the hob and the gear blank in machining a spur gear by hobbing.

FIGURE 28-21. Close-up view, showing the hobbing of a spur gear. (*Courtesy Barber-Colman Company.*)

The action of a hob in cutting a gear is illustrated in Figure 28-20, with the actual hobbing of a small spur gear shown in Figure 28-21. To cut a spur gear, the axis of the hob must be set off from the normal to the rotational axis of the blank by the helix angle of the hob. In cutting helical gears, the hob must be set over an additional amount equal to the helix angle of the gear.

The cutting of a gear by means of a hob is a continuous action. The hob and the blank are connected by means of proper change gearing so they rotate in mesh. To start cutting a gear, the hob is located so as to clear the blank and then moved inward until the proper setting for tooth depth is obtained. The hob is then fed in a direction parallel with the axis of rotation of the blank. As the gear blank rotates, the teeth are generated and the feed of the hob across the face of the blank extends the teeth to the desired tooth face width.

Because hobbing is a continuous action involving a multipoint cutting tool, it is rapid and economical. More gears are cut by this process than by any other. The process produces excellent gears and can also be used for splines and sprockets. Single-, double-, and triple-thread hobs are used. Multiple-thread types increase the production rate but do not produce accuracy as high as single-thread hobs.

Gear-hobbing machines are made in a wide range of sizes. Figure 28-21 shows a small machine while Figure 28-22 shows a hobbing machine cutting a gear over 2.44 meters (8 feet) in diameter. Such machines for cutting accurate, large gears frequently are housed in temperature-controlled rooms, and

FIGURE 28-22. (*Left*) Close-up showing the hobbing of a large gear. (*Right*) Completed gear. This gear will transmit 22.4 MW (30,000 hp). Allowable error in tooth spacing is 0.0025 mm (0.0001 inch). (*Courtesy Westinghouse Electric and Manufacturing Company.*)

the temperature of the cutting fluid is controlled to avoid dimensional changes due to variations in temperature.

Bevel-gear generating machines. The machines used to generate the teeth on various types of bevel gears are among the most ingenious of all machine tools. Two basic types utilize the principle that a crown gear will mate properly with any bevel gear having the same diametral pitch and tooth form. One is used for cutting straight-tooth bevel gears, and the other for cutting curved-tooth bevel gears.

The basic principle of these machines is indicated in Figure 28-23. The blank and a connecting gear, having the same cone angle and diametral pitch, are mounted on a common shaft so that the connecting gear meshes with the regular teeth on the crown gear, which also has a reciprocating cutter tooth that meshes with the teeth that it will cut in the gear blank. As the connecting gear rolls on the crown gear, the reciprocating cutter generates a tooth in the gear blank. Obviously, with only one cutter tooth, only a single tooth space would be cut in the gear blank. However, this limitation is overcome in actual machines by two modifications. First, by indexing the connecting gear and blank unit, the process is repeated as often as required to cut all the teeth in the blank. Second, instead of a single, reciprocating cutter, two half-tooth cutters are used—approximating the inner sides of two adjacent teeth, so that they cut both faces of one tooth on the blank.

FIGURE 28-23. Principle of bevel-gear generating machines.

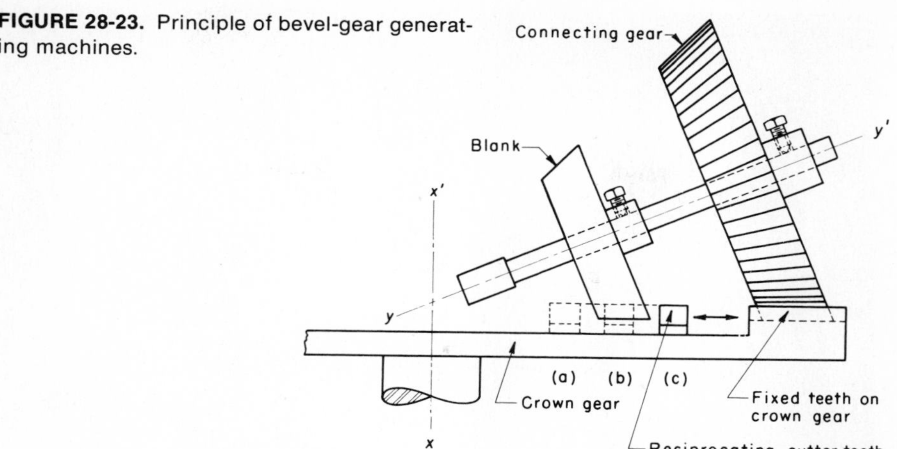

The application of this modified basic principle in actual machines is shown in Figures 28-24 and 28-25. The cradle acts as a crown gear and carries the two cutters, which reciprocate in slides simultaneously in opposite directions. It can be noted in Figure 28-25 that straight lines through the faces of the cutters, the axis of rotation of the gear blank, and elements of the pitch cone of the gear blank all meet at a common point that is in the plane of the imaginary crown gear and on the axis of the cradle, thus fulfilling the requirements shown in Figure 28-23.

In operation, while the cradle is at the extreme upward position, the gear blank is fed inward toward the cutter until the position for nearly full

FIGURE 28-24. (*Top*) Schematic diagram showing the roll of the gear blank and cutters during the cutting of one tooth on a bevel-gear generating machine, as seen from the front. (*Bottom*) Photographs showing the same roll action, as seen from the back of the machine. (*Courtesy Heidenreich & Harbeck.*)

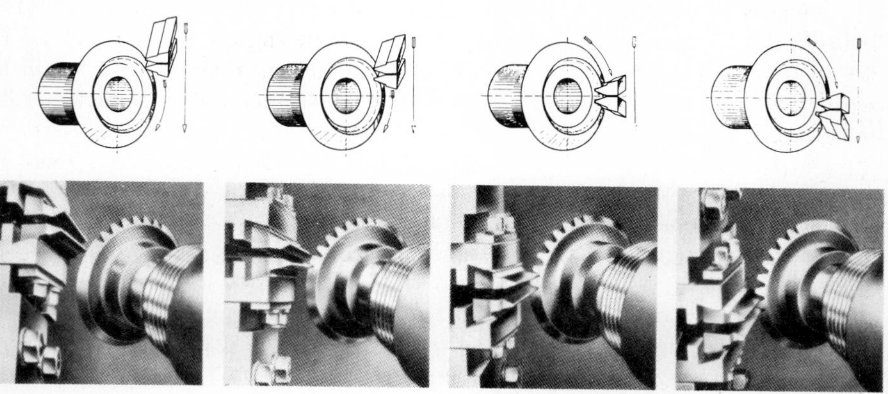

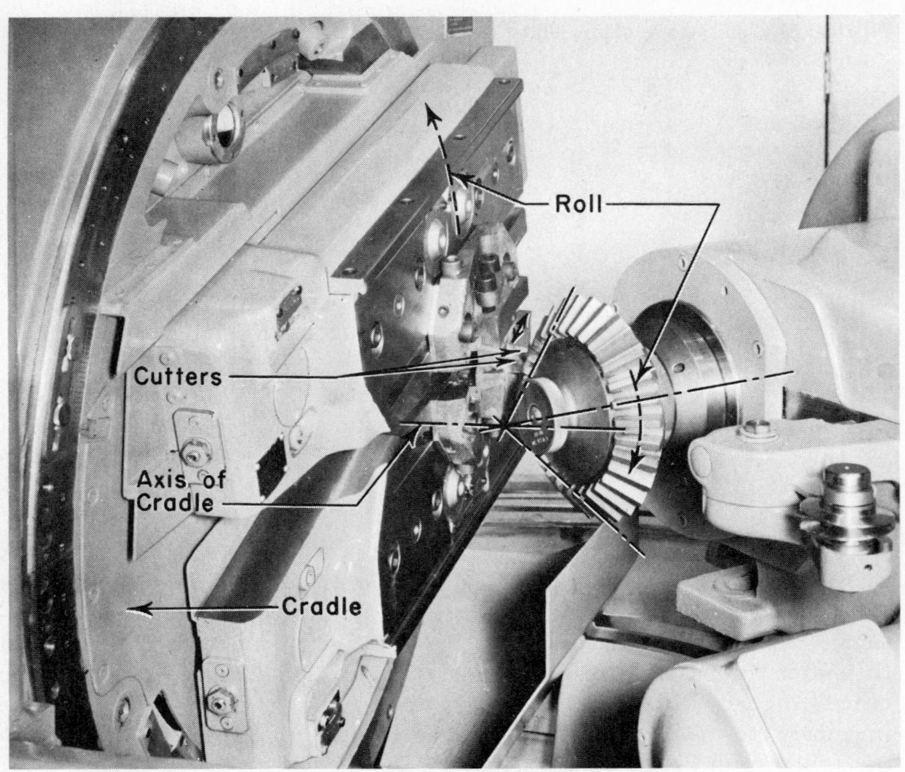

FIGURE 28-25. Cutters and gear blank on a Gleason straight-tooth gear-generating machine, showing the relationship between the axis of rotation of the blank and the axis of the cradle, which with the cutter simulates a crown gear. (*Courtesy Gleason Works.*)

tooth depth is reached. The cradle then starts to roll downward, and a tooth is generated during the downward roll. When the cradle reaches the full down position, the blank is automatically fed inward a small amount—usually 0.25 to 0.38 mm (0.010 to 0.015 inch)—and a finish cut is made during the upward roll of the cradle. At the completion of the upward roll, the blank is withdrawn, indexed, and moved inward automatically, ready to start the next tooth.

Provision is made so that the spindle on which the gear blank is mounted can be swung through an angle to accommodate bevel gears of different cone angles. This spindle is connected to the cradle by means of suitable internal change gears to provide the required positive rolling motion between the two.

For cutting small straight-tooth bevel gears, the type of machine shown in Figure 28-26 often is used. These employ two revolving, disk-type cutters having interlocking teeth. These cutters reciprocate on the cradle slides as

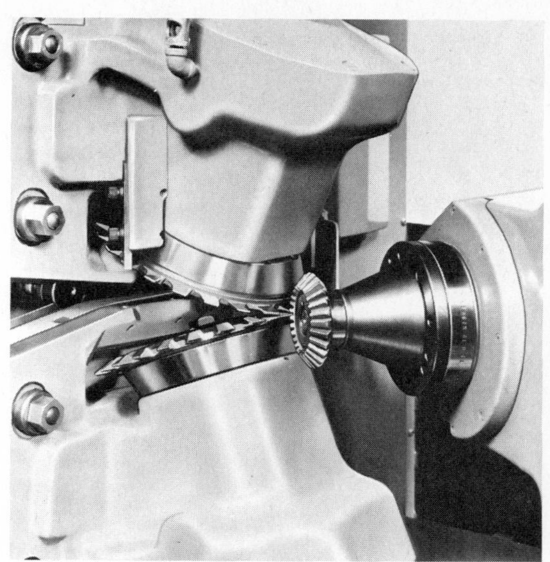

FIGURE 28-26. Bevel-gear generating machine, utilizing two disk-type cutters. (*Courtesy Gleason Works.*)

they rotate. Because of their multiple cutting edges and continuous rotary, rather than reciprocating, cutting action, they are much more productive than the other type.

Generating machines for straight-tooth bevel gears often are provided with a mechanism that produces a slight crown on the teeth. Such teeth are slightly thicker at the middle than at the ends, to avoid having applied loads concentrated at the tooth ends where they are weakest.

Machines used for generating spiral, Zerol, and hypoid bevel gears employ the same basic crown-gear principle, but a multitooth rotating cutter is used. This cutter has its axis of rotation parallel with the axis of roll of the cradle, as is shown in Figure 28-27. Figure 28-28 shows the relationship of the gear blank to the axis of the theoretical crown gear (the cradle) and of the cutter teeth to the theoretical mating bevel gear.

In cutting a spiral bevel gear, the blank is fed inward toward the cutter to the full tooth depth while the cradle is in the downward position. In this position the cutter clears the blank. As the cradle and the gear blank roll upward together, the cutter engages the blank, starting at the smaller end of the gear, and the teeth are generated. At the same time the cutter progresses across the width of the tooth face. In some cases both sides of a gear tooth are generated simultaneously, whereas in others only one side is cut in a single roll of the gear and cradle.

A special process is sometimes employed to make curved-tooth pairs of bevel gears somewhat more cheaply than can be done by true generation. It produces fairly satisfactory gears where the gear ratio is greater than 3 or 4 : 1. In this process the larger gear is cut on a special machine, not by generation.

FIGURE 28-27. Cutting a hypoid bevel gear on a Gleason curved-tooth generating machine. (*Courtesy Gleason Works.*)

FIGURE 28-28. Relationship of the gear blank to the axis of the theoretical crown gear (the cradle) and of the cutter teeth to the theoretical mating bevel gear. (*Courtesy Gleason Works.*)

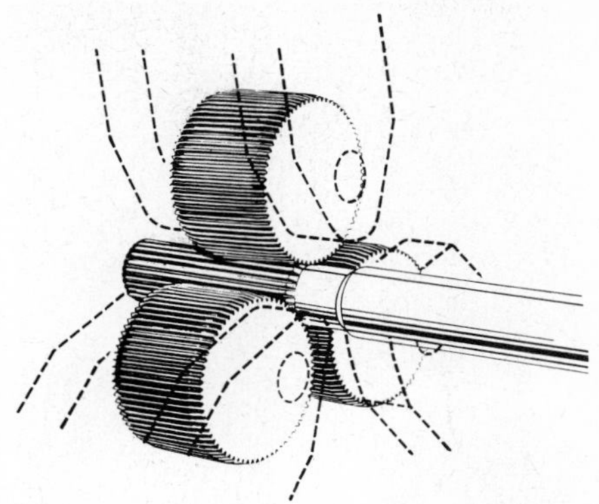

FIGURE 28-29. Method for forming gear teeth and splines by cold forming.

The smaller, mating gear is then generated to mate properly with the tooth profile of the larger gear. Such gears are known as *Formate gears*.

Cold-roll forming of gears. The manufacture of gears by cold-roll forming has been highly developed and widely adopted in recent years. Currently, millions of high-quality gears are produced annually by this process; many of the gears in automobile transmissions are made this way. As indicated in Figure 28-29, the process is basically the same as that by which screw threads are roll-formed, except that in most cases the teeth cannot be formed in a single rotation of the forming rolls; the rolls are fed inward gradually during several revolutions.

Because of the metal flow that occurs, the top lands of roll-formed teeth are not smooth and perfect in shape—a depressed line between two slight protrusions often can be seen, as shown encircled in Figure 28-30. However, because the top land plays no part in gear-tooth action, if there is sufficient clearance in the mating gear, this causes no difficulty. Where desired, a light turning cut is used to provide a smooth top land and correct addendum diameter.

The hardened forming rolls are very accurately made, and the roll-formed gear teeth usually have excellent accuracy. In addition, because the severe cold working produces tooth faces that are much smoother and harder than those on ordinary machined gears, they seldom require hardening or further finishing, and they have excellent wear characteristics.

Owing to the rapidity of the process, its ease of being mechanized, the fact that no chips are made and thus less material is needed, and because skilled labor is not required, roll-formed gears are rapidly replacing machined gears whenever the process can be used. Small gears often are made by rolling a length of shaft and then slicing off the individual gear units.

Miscellaneous gear-making processes. Gears can be made by the various casting processes. *Sand-cast gears* have rough surfaces and are not accurate

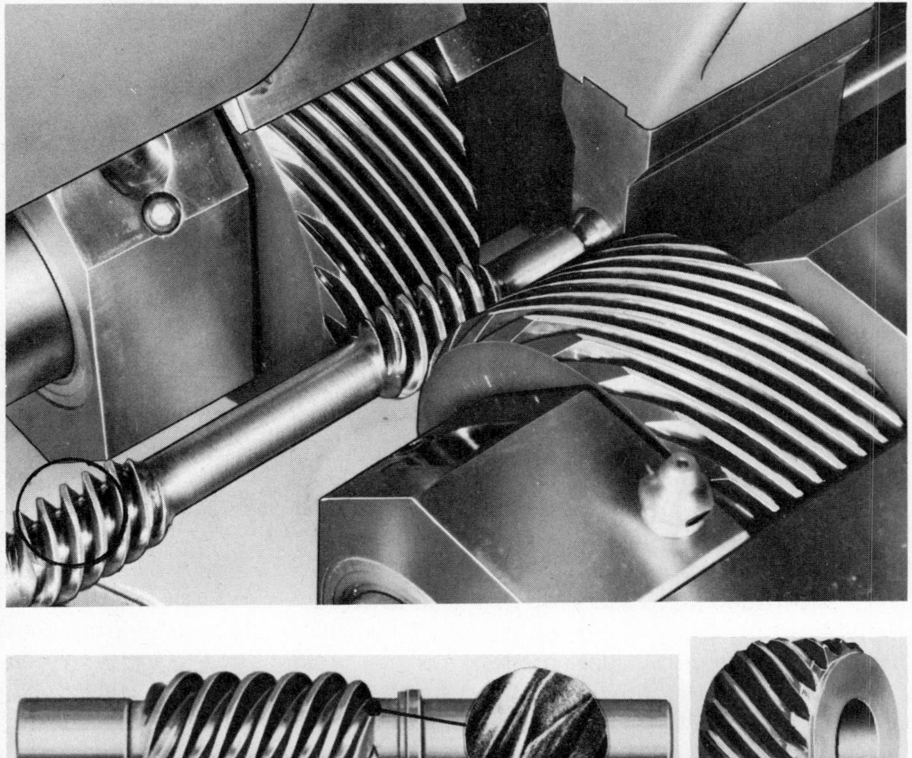

FIGURE 28-30. (*Top*) Worm being rolled by means of rotating rolling tools. (*Bottom left*) Typical worm made by rolling, with enlarged view of end of one tooth. (*Right*) Gear made by rolling. (*Courtesy Landis Machine Company.*)

dimensionally. They are used only for services where the gear moves slowly and where noise and inaccuracy of motion can be tolerated. Gears made by *die casting* are fairly accurate and have fair surface finish. They can be used to transmit light loads at moderate speeds. Gears made by *investment casting* may be accurate and have good surface characteristics. They can be made of strong materials to permit their use in transmitting heavy loads. In many instances gears that are to be finished by machining are made from cast blanks, and in some larger gears the teeth can be cast to approximate shape to reduce the amount of machining.

Large quantities of gears are produced by *blanking* in a punch press. The thickness of such gears usually does not exceed about 1.6 mm ($^1/_{16}$ inch). By shaving the gears after they are blanked, excellent accuracy can be achieved. Such gears are used in clocks, watches, meters, and calculating machines.

Excellent gears can be made by *broaching*. However, because of the very high cost of the broach, and the fact that a separate broach must be provided for each size of gear, this method is used relatively little.

High-quality gears, both as to dimensional accuracy and surface quality, can be made by the powder metallurgy process. Usually, this process is employed only for small sizes, ordinarily less than 25 mm (1 inch) in diameter. However, larger and excellent gears are made by forging powder metallurgy preforms. An example is shown in Figure 28-31. As discussed in Chapter 12, this results in a product of much greater density and strength than usually can be obtained by ordinary powder metallurgy methods, and the resulting gears give excellent service at reduced cost. Gears made by this process often require little or no finishing.

Large quantities of plastic gears are made by *plastic molding*. The quality of such gears is only fair, and they are suitable only for light loads. Accurate gears suitable for heavy loads frequently are machined out of laminated plastic materials. When such gears are mated with metal gears, they have the quality of reducing noise.

Quite accurate small-sized gears can be made by the *extrusion* process. Long lengths of rod, having the cross section of the desired gear, are extruded. The individual gears are then sliced from this rod. The only materials suitable for this process are brass, bronze, and aluminum and magnesium alloys.

Flame machining (oxyacetylene cutting) can be used to produce gears that are to be used for slow-moving applications, where accuracy is not required.

A few gears are made by the hot roll-forming process. In this process a cold master gear is pressed into a hot blank as the two are rolled together.

FIGURE 28-31. Powder metallurgy preform and finished forged gear. (*Courtesy Cincinnati Incorporated.*)

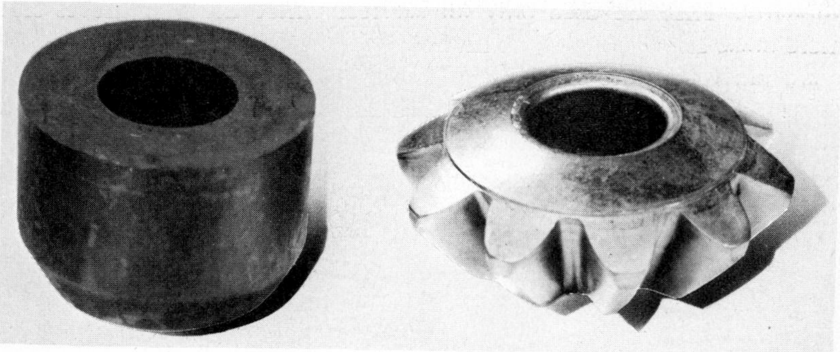

Gear finishing. In order to operate efficiently and have satisfactory life, gears must have accurate tooth profiles, and the faces of the teeth must be smooth and hard. These qualities are particularly important when gears must operate quietly at high speeds. When they are produced rapidly and economically by most of the processes except cold-roll forming, the tooth profiles may not be as accurate as desired, and the surfaces are somewhat rough and subject to rapid wear. Also, it is difficult to cut gear teeth in a hardened gear blank, so economy dictates that the gear be cut in a relatively soft blank and subsequently be heat treated to obtain greater hardness, if this is required. Such heat treatment usually results in some slight distortion and surface roughness. Although most roll-formed gears have sufficiently accurate profiles, and the tooth faces are adequately smooth and frequently have sufficient hardness, this process is feasible only for relatively small gears. Consequently, a large proportion of high-quality gears are given some type of finishing operation after they have received primary machining or heat treatment. Most of these finishing operations can be done quite economically, because only minute amounts of metal are removed.

Gear shaving is the most commonly used method for gear finishing. The gear is run, at high speed, in contact with a shaving tool, usually of the type shown in Figure 28-32. Such a tool is a very accurate, hardened, and ground gear that contains a number of peripheral gashes, or grooves, thus forming a series of sharp cutting edges on each tooth. The gear and shaving cutter are run in mesh with their axes crossed at a small angle, usually about 10°. As they rotate, the gear is reciprocated longitudinally across the shaving tool (or vice versa). During this action, which usually requires less than a minute,

FIGURE 28-32. (*Left*) Gear being shaved by a rotary shaving cutter. (*Right*) Cutter for rotary gear shaving. (*Courtesy National Broach and Machine Company.*)

very fine chips are removed from the gear-tooth faces, thus eliminating any high spots and producing a very accurate tooth profile.

Rack-type shaving cutters sometimes are used for shaving small gears—the cutter reciprocating lengthwise, causing the gear to roll along it, as it is moved sideways across the cutter and fed inward.

Although shaving cutters are costly, they have a relatively long life because only a very small amount of metal is removed—usually 0.025 to 0.10 mm (0.001 to 0.004 inch). Most gears are not hardened prior to shaving, although it is possible to remove very small amounts of metal from hardened gears, if they are not too hard. However, modern heat-treating equipment makes it possible to harden gears after shaving without harmful effects, so this practice usually is followed.

Some gear-shaving machines produce a slight crown on the gear teeth during shaving.

Gear burnishing is a finishing process that is used to a limited extent on unhardened gears. The unhardened gear is rolled under pressure contact with three hardened, accurately formed, burnishing gears. As a result of this action, any high points on the unhardened gear are plastically deformed so that a smoother surface and more accurate tooth form are achieved. Because the operation is one of localized cold working, some undesirable effects may accrue, such as localized residual stresses and nonuniform surface characteristics.

Grinding is used to obtain very accurate teeth on hardened gears. Two methods are used. One employs a formed grinding wheel that is trued to the exact form of a tooth by means of diamonds mounted on a special holder and guided by a large template. The other method utilizes a straight-sided grinding wheel that simulates one side of a rack tooth. The surface of the gear tooth is ground by the generating principle as the gear rolls past the grinding wheel, in the same manner that it would roll on a rack. This method has the advantage of using a simpler shape of grinding wheel, but it is much slower than form grinding because only one side of a tooth is ground at a time. Grinding produces very accurate gears, but because it is slow and expensive, it is used only on the highest quality, hardened gears.

Lapping also can be used for finishing hardened gears. The gear to be finished is run in contact with one or more cast-iron lapping gears under a flow of very fine abrasive in oil. Because lapping removes only a very small amount of metal, it usually is employed on gears that previously have been shaved and hardened. This combination of processes produces gears that are nearly equal to ground gears in quality, but at considerably lower cost.

Gear inspection. As with all manufactured products, gears must be checked to determine whether the resulting product meets the design specifications and requirements. Because of their irregular shape and the number of factors that must be measured, such inspection of gears is somewhat difficult. Among

FIGURE 28-33. Using gear-tooth vernier calipers to check the tooth thickness at the pitch circle.

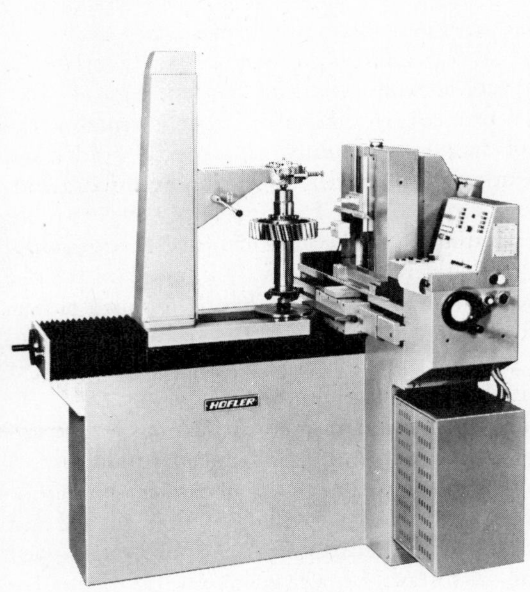

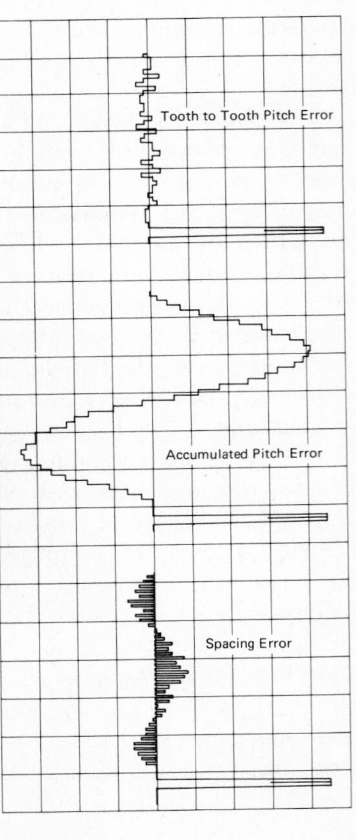

FIGURE 28-34. (*Above*) Gear being checked on a special gear-checking machine. (*Right*) Resulting charts. (*Courtesy American Pfauter Corporation.*)

the factors to be checked are the linear tooth dimensions—thickness, spacing, depth, and so on—tooth profile, surface roughness, and noise. Several special devices, most of them automatic or semiautomatic, are used for such inspection.

Gear-tooth vernier calipers can be used to measure the thickness of gear teeth on the pitch circle, as shown in Figure 28-33. However, most inspection is made by special machines, such as shown in Figure 28-34, which in one or a series of operations check several factors, including eccentricity, variations in circular pitch, variations in pressure angle, fillet interference, and lack of continuous action. The gear usually is mounted and moved in contact with a master gear. The movement of the latter is amplified and recorded on moving charts, as shown in Figure 28-34.

Because noise level is important in many applications, not only from the viewpoint of noise pollution but also as an indicator of probable gear life, special equipment for its measurement is quite widely used, sometimes integrated into mass-production assembly lines.

Review questions

1. Why can the relative angular velocities of two mating spur gears not be determined by their outside diameters alone?
2. Why is the involute form used for gear teeth?
3. What is the diametral pitch of a gear?
4. What is the relationship between the diametral pitch and the module of a gear?
5. On a sketch of a gear, indicate the pitch circle, addendum circle, dedendum circle, and the circular pitch.
6. What five requirements must be met in order for gears to operate satisfactorily? Which of these are determined by the manufacturing process?
7. What are the advantages of helical gears, compared with spur gears?
8. What is the principal disadvantage of helical gears?
9. A gear that has a pitch diameter of 152.4 mm (6 inches) has a diametral pitch of 4. What number of form cutter would be used in cutting it?
10. What difficulty would be encountered in hobbing a herringbone gear?
11. What is the only type of machine on which full-herringbone gears can be cut?
12. What modification in design is made to enable the major advantage of herringbone gears to be obtained and still permit them to be cut by hobbing?
13. Why are not more gears made by broaching?
14. What is the most important property of a crown gear?
15. What are three basic processes for machining gears?
16. Which basic gear-machining process is utilized in a Fellows gear shaper?
17. Could a helical gear be machined on a plain milling machine? Why?

18. Explain how the gear blank and the machine table are interconnected when a helical gear is machined on a milling machine.
19. What is the relationship between a crown gear and Gleason gear generators?
20. Why is a gear-hobbing machine much more productive than a gear shaper?
21. Why is a gear shaper more likely to be used for machining cluster gears than a hobbing machine?
22. What are the advantages of cold-roll forming for making gears?
23. Assume that 10,000 spur gears, 28.6 mm (1⅛ inch) in diameter and 9.53 mm (⅜ inch) thick are to be made of 70–30 brass. What manufacturing methods would you consider?
24. If only 3 gears described in question 23 were to be made, what process would you select?
25. Why is cold-roll forming not suitable for making gray cast iron gears?
26. Under what conditions can shaving not be used for finishing gears?
27. What inherent property accrues from cold-roll forming of gears that may result in improved gear life?
28. Why can lapping not be used to finish cast iron gears?
29. What factors usually are checked in inspecting gears?
30. A single thread hob that has a pitch diameter of 76.2 mm is used to cut a gear having 36 teeth. If a cutting speed of 27.4 meters per minute is used, what will be the rpm of the gear blank?
31. If the gear in question 30 has a face width of 76.2 mm, a feed of 1.9 mm per revolution of the workpiece is used, and the approach and overtravel distances of the hob are 38 mm, how much time will be required to hob the gear?

Case study 21.
THE VENTED CAP SCREWS

The machine shop at the Hi-Fly Space Laboratories received an order from their engineering department for 10 of the vented cap screws shown in Figure CS-21. The machine shop foreman returned the order, stating that there was

FIGURE CS-21. Design for venting threads of a cap screw.

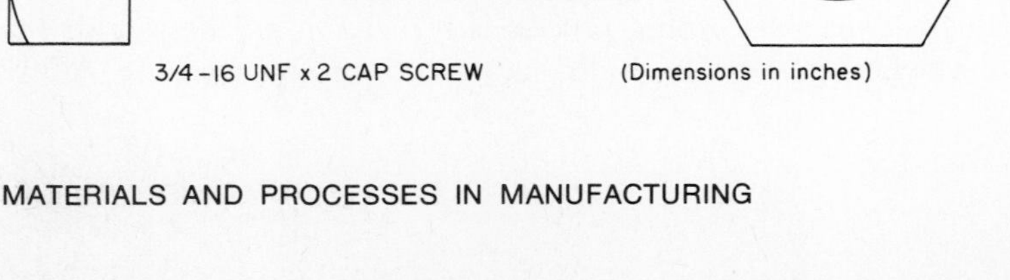

no practical way to make these other than by hand. The design engineer insisted that the vent slot, as designed, was essential to assure no pressure buildup around the threads of the screw body in the intended application, and that he knew they could be made because he had seen such cap screws. (1) Who was correct? (2) If there is a practicable way to make the slot, how should it be done?

PART FOUR

Welding and allied processes

CHAPTER 29

Forge, oxyfuel gas, and arc welding

Welding is a process in which two materials, usually metals, are permanently joined together through localized coalescence, resulting from a suitable combination of temperature, pressure, and metallurgical conditions. Because the combination of temperature and pressure can range from high temperature with no pressure to high pressure with no increased temperature, welding can be accomplished under a very wide variety of conditions, and numerous welding processes have been developed and are used routinely in manufacturing. The average person usually has little concept of the importance of welding as a manufacturing process, yet a large proportion of our metal products would have to be drastically modified, and would be considerably more costly or could not perform as efficiently, if it were not for the use of welding.

To obtain coalescence between two metals, there must be a combination of sufficient proximity and activity between the molecules of the pieces being joined to cause the formation of common metallic crystals. Such proximity and activity are impeded by (1) the usual asperities on the contacting surfaces, (2) any oxide layer or foreign "dirt" such as usually coats metal surfaces, or (3) a thin layer of adsorbed gas on the oxide surface. These deterrents must be overcome in some manner. Surface roughness is overcome either by force, causing plastic deformation of the asperities and thus bringing the workpieces into closer contact, or by melting the two surfaces, so that fusion occurs. In solid-state welding, the contaiminant layers can be removed by mechanical or chemical cleaning prior to welding or by causing sufficient metal flow along the interface so that they are squeezed out of the weld. In fusion welding, where a pool of molten metal exists, the contaminants are removed by the use of fluxing agents. If welding is done in a vacuum, either by solid-state or fusion processes, the contaminants are removed much more easily, and coalescence is established with considerable ease. In outer-space conditions, mating parts may become welded together under light loads when such was not intended.

The various welding processes not only differ considerably in the manner in which temperature and pressure are combined and achieved, they also vary as to the attention that must be given to the cleanliness of the metal surface prior to welding and to possible oxidation or contamination of the metal during welding. If high temperatures are used, most metals are affected more adversely by the surrounding environment and, if actual melting occurs, serious modification of the metal may result. Also, the metallurgical structure and quality of the metal may be affected—usually adversely—as a consequence of the heating and cooling that accompany most welding processes, and these effects should be taken into account.

In summary, in order to obtain satisfactory welds it is desirable to have (1) a satisfactory heat and/or pressure source, (2) a means of protecting or cleaning the metal, and (3) avoidance of, or compensation for, harmful metallurgical effects.

A chart of welding processes. The various welding processes have been defined and classified by the American Welding Society as shown in Figure 29-1 and assigned letter symbols to facilitate their designation. These processes provide a variety of ways of meeting the three requirements stated previously. They provide a wide range of processes that make it possible to achieve effective and economical welds in nearly all metals and combinations of metals. The result has been that welding has replaced other types of permanent fastening to such a degree that a large proportion of manufactured products contain one or more welds.

FORGE WELDING

Although little *forge welding* (FOW), except in one type of application, is done today, it is the most ancient of the welding processes and a review of it is of

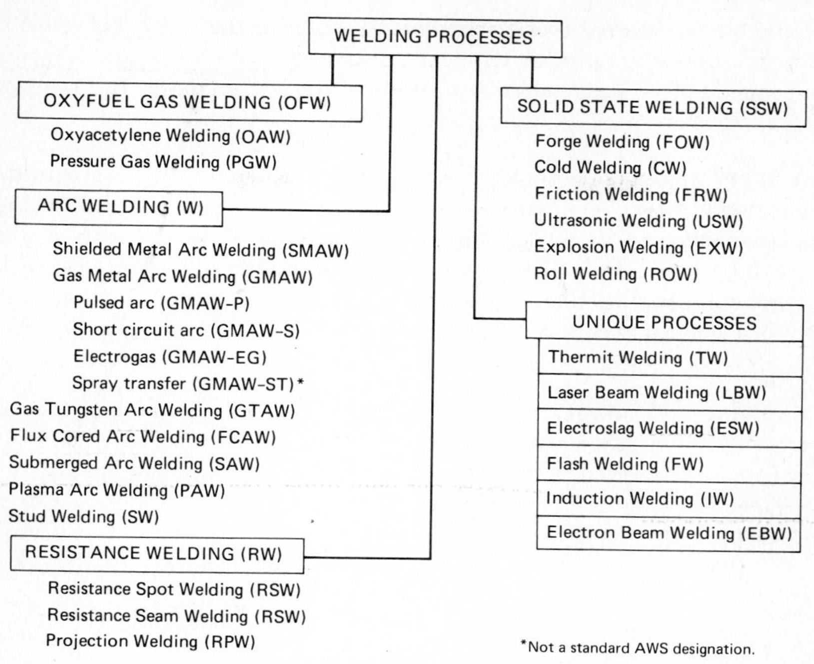

FIGURE 29-1. Chart of welding processes.

both historical and practical use in understanding how and why modern welding processes were developed. The armor makers of ancient times owed their positions of prominence in their societies as much to their ability to join two pieces of steel into a single strong piece as to their ability to harden and temper the metal. Perhaps the only undesirable social consequence that can be charged to modern welding is that it was partly responsible for the passing of the colorful village blacksmith, who, with his forge, hammer, and anvil, could fasten together pieces of steel by forge welding to form a variety of products.

The blacksmith used a charcoal forge as his source of heat. Pieces that were to be welded were heated to a forging temperature and the ends scarfed by hammering to permit them to be fitted together without undue thickness. The ends were again heated until they were nearly to the proper temperature. They then were withdrawn and dipped into some borax, which acted as a flux. Heating was then continued for a short interval until the blacksmith judged by their color that the workpieces were at the proper temperature for welding. He then withdrew them from the forge, struck them sharply against the anvil, or with his hammer, to knock off the scale and impurities, and then placed the ends on each other on the anvil and hammered them to force them into the necessary proximity to complete the weld. Thus the competent blacksmith did solid-state welding and was able to produce a welded joint that was as strong as the original metal. However, because of the crudeness of his heat source, the uncertainty of his temperature control, and the difficulty of maintaining metal cleanliness, a great amount of skill was required, and the results were variable.

Forge-seam welding. Although forge welding, as done by the blacksmith, is seldom done today, very large amounts of *forge-seam welding* are done in the manufacture of pipe, as was described in Chapter 14. A heated strip of steel is formed into a cylinder and the edges welded together in either a lap or butt joint by the pressure that is exerted as the formed metal is either pulled through a conical welding bell or passed between formed rolls (see Figure 14-22).

Cold welding. *Cold welding* (CW), a solid-state process, is a unique variation of forge welding, in that no heating is used and only a single blow or pressure application is employed. Coalescence results solely through the rapid application of pressure. It thus represents one extreme of the possible range of temperature–pressure possibilities.

To produce a cold weld the faying surfaces are cleaned, usually by wire brushing, placed in contact, and subjected to localized pressure sufficient to produce about 30 to 50 per cent cold working. This can be done in a punch press or a special tool. The result is as shown in Figure 29-2a. Undoubtedly, some heating does occur as a result of the severe cold working of the metal, but the high localized pressure is the primary factor in producing coalescence.

The use of cold welding is confined to fastening small parts, such as the electrical connections shown in Figure 29-2b.

FIGURE 29-2. (*Left*) Section through a cold weld. (*Right*) Small parts joined by cold welding. (*Courtesy Koldweld Corporation.*)

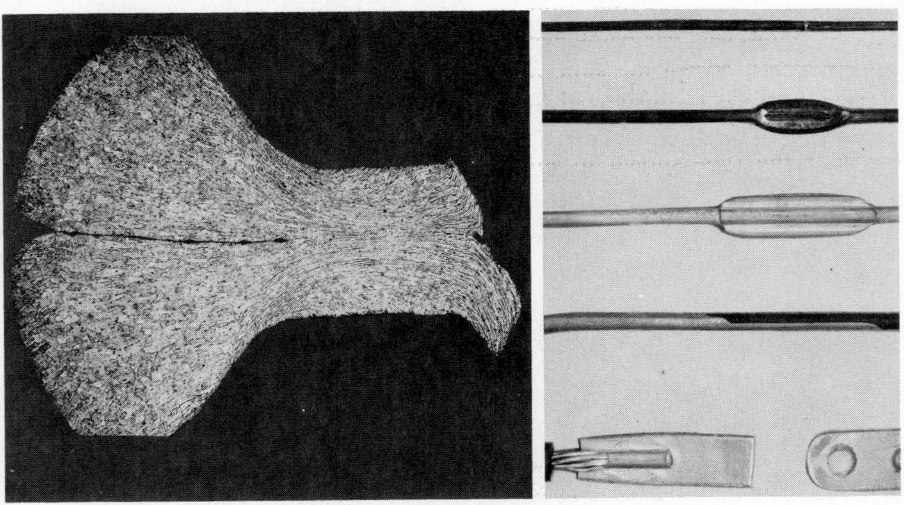

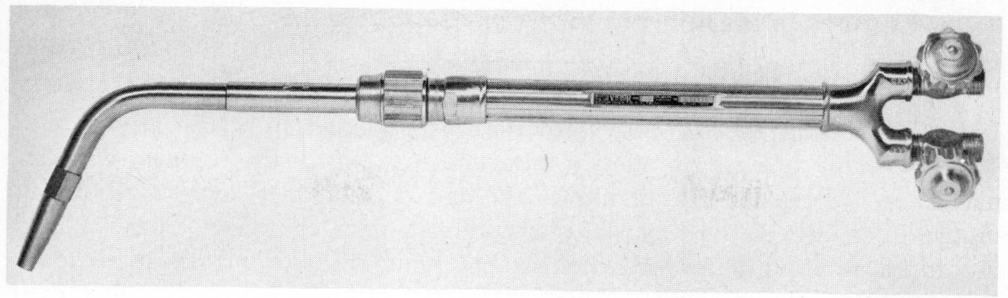

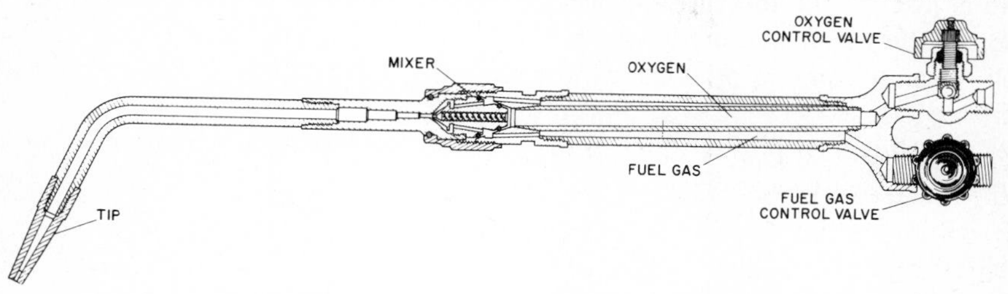

FIGURE 29-3. Oxyacetylene welding torch. (*Courtesy Victor Equipment Company.*)

OXYFUEL GAS WELDING[1]

Oxyfuel gas welding (OFW) covers a group of welding processes that utilize as the heat source a flame resulting from the burning of a fuel gas and oxygen, mixed in proper proportions. The oxygen usually is supplied in relatively pure form but may, in rare cases, come from air.

It was the development of a practical torch to burn acetylene and oxygen, shortly after 1900, that brought welding out of the blacksmith's shop, demonstrated its potential, and started its development as a manufacturing process. However, gas-flame welding has largely been replaced by other welding processes, except for some repair work and a few special applications, but acetylene still is the principal fuel gas employed in the process.

The combustion of oxygen and acetylene (C_2H_2), by means of a welding torch of the type shown in Figure 29-3, produces a temperature of about 3482°C (6300°F) in a two-stage reaction. In the first stage the oxygen and acetylene from the pressure tanks react in the manner

$$C_2H_2 + O_2 \rightarrow 2CO + H_2$$

This reaction occurs near the end of the torch tip. The second stage of the reaction—combustion of the CO and H_2—occurs just beyond the first combustion zone in two reactions:

[1] Why the AWS committee came up with this term, when they were clearly trying to imply "gas–oxygen flame welding," is a mystery.

$$2CO + O_2 \rightarrow 2CO_2$$

$$H_2 + \tfrac{1}{2}O_2 \rightarrow H_2O$$

The oxygen for these secondary reactions is obtained from the atmosphere. This two-stage process produces a flame having two distinct zones, with the maximum temperature occurring at the end of the inner cone, as shown in Figure 29-4, where the first stage of combustion is complete. Most welding should be done with the torch held so that this point of maximum temperature is just off the metal being welded. The outer zone of the flame serves to preheat the metal and at the same time provides some shielding from oxidation, because some oxygen from the surrounding air is consumed for secondary combustion.

As shown in Figure 29-4, three types of flames can be obtained by varying the oxygen-to-acetylene ratio. If the ratio is about 1:1 to 1.15:1, a *neutral* flame is obtained. A higher ratio produces an oxidizing flame, whereas a lower ratio produces a flame that is *carburizing.* The same three types of flames can be obtained with other fuel gases.

Most welding is done with a neutral flame, because such a flame has a minimum chemical effect on most heated metals. Oxidizing flames are used only in welding certain brasses and bronzes. Carburizing flames are used in welding Monel, very low carbon steels, a few alloy steels, and in applying hard-surfacing materials such as Stellite and Colmanoy.

Acetylene gas is produced by the reaction between calcium carbide and water in the reaction

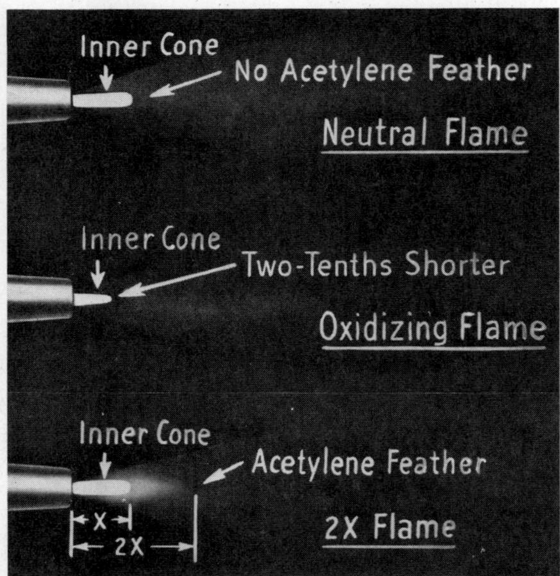

FIGURE 29-4. Neutral, oxidizing, and carburizing flame characteristics. (*Courtesy Linde Division, Union Carbide Corporation.*)

$$CaC_2 + 2H_2O \rightarrow Ca(OH)_2 + C_2H_2$$

For welding purposes, acetylene most often is obtained in portable storage tanks holding up to 8.5 cubic meters (300 ft^3) at 1.72-MPa (250-psi) pressure. Because acetylene is not safe when stored as a gas at over 0.1 MPa (15 psi), it must be dissolved in acetone. Consequently, the storage cylinders are filled with a porous filler, such as balsa-wood chips and infusorial earth, with acetone filling the voids in the filler material and, in turn, dissolving the acetylene.

When large quantities of acetylene are needed, acetylene generators can be employed. These operate on the principle of dropping pieces of calcium carbide into water to generate the acetylene.

Stabilized methylacetylene propadiene, known best under the trade name MAPP gas, has replaced ordinary acetylene to a considerable extent, particularly where portability is important. It is more dense, thus providing more energy for a given volume, and it can be stored safely in ordinary pressure tanks. Oxygen for gas-flame welding almost always is obtained from pressure tanks.

The pressures used in gas-flame welding ordinarily vary from 6.9 to 103.4 kPa (1 to 15 psi), pressure regulators being employed to reduce and maintain the desired pressure. Because mixtures of acetylene and oxygen or air are highly explosive, precautions must be taken to avoid mixing the gases improperly or by accident. All acetylene fittings have left-hand threads, whereas those for oxygen are equipped with right-hand threads, thus preventing improper connections from being made.

Uses, advantages, and limitations. Almost all oxyfuel gas welding is *fusion* welding; the metals being joined are melted at the point where welding occurs, and no pressure is involved. Because a slight gap usually exists between the pieces being joined, filler material usually must be added in the form of a wire or rod that is melted in the flame or in the pool of weld metal.

Good-quality welds can be obtained by the OFW process if proper technique and care are used. Control of the temperature of the work is easily accomplished. However, exposure of the heated and molten metal to various gases in the flame and the atmosphere, without effective shielding, makes it difficult to prevent contamination. In addition, because the heat source is not concentrated, considerable areas of the metal are heated and distortion is likely to occur. As a result, since the high development of shielded-arc and inert gas metal-arc welding, these processes have largely replaced flame welding.

Pressure gas welding. *Pressure gas welding* (PGW) is a process used to make butt joints between such objects as pipe and railroad rails. The ends are heated with a gas flame to a temperature below the melting point and then forced together under considerable pressure. The process thus is a type of solid-phase welding. Figure 29-5 shows this process being used to join sections of pipe.

FIGURE 29-5. Pressure welding of pipe. (*Courtesy Linde Division of Union Carbide Corporation.*)

ARC WELDING

Almost from the time electricity became a commercial reality, it was recognized that an electric arc between two electrodes was a concentrated heat source approaching 3871°C (7000°F) in temperature. As early as 1881, various persons attempted to use an arc between a carbon electrode and metal workpieces as the heat source for fusion welding, using the basic circuit shown in Figure 29-6. As in gas-flame welding, it was necessary to add filler metal from a metallic wire. Later, a bare metal wire was used as the electrode; it melted in the arc and thus automatically supplied the needed filler metal. However, the results were very uncertain. Because of the instability of the arc, a great amount of skill was required to maintain it, and contamination and oxidation of the weld metal resulted from its exposure to the atmosphere at such high temperatures. Furthermore, there was little or no understanding of the metallurgical effects and requirements. Consequently, although the great potential of the process was recognized, particularly as a result of experiences

FIGURE 29-6. Basic arc-welding circuit.

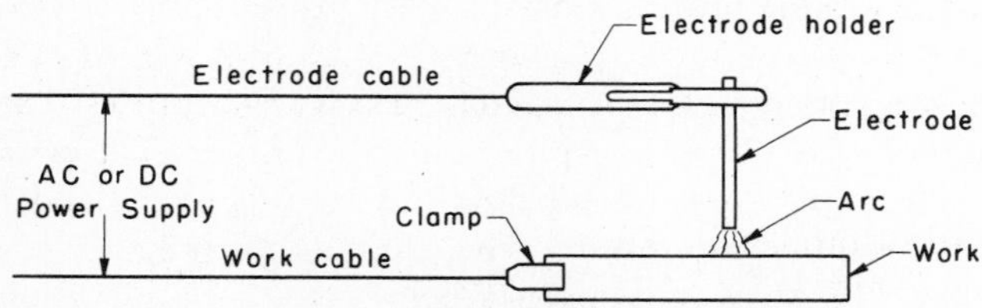

during World War I, very little use was made of the process until after that time. About 1920, shielded metal electrodes were developed, which provided a stable arc and shielding from the atmosphere, and some fluxing action for the molten pool of metal, thus solving the major problems related to arc welding. The use of the process expanded very rapidly and has continued to do so ever since, until we now have a considerable variety of arc-welding processes available.

All arc-welding processes employ essentially the same basic circuit depicted in Figure 29-6, except that alternating current is used at least as much as direct current. When direct current is used, if the work is made positive (the anode of the circuit) and the electrode is made negative, *straight polarity* is said to be used. When the work is negative and the electrode is positive, the polarity is *reversed.* When bare electrodes are used, greater heat is liberated at the anode. Certain shielded electrodes, however, change the heat conditions and are used with reverse polarity.

All arc welding is done with metal electrodes. In one type the electrode is consumed and thus supplies needed filler metal to fill in the voids in the joint and to speed the welding process. In this case the electrode has a melting point below the temperature of the arc. Small droplets are melted from the end of the electrode and pass to the parent metal. The size of these droplets varies greatly and the mechanism of the transfer varies with different types of electrodes and processes. As the electrode melts, the arc length and the resistance of the arc path vary, as a result of the change in the arc length and the presence, or absence, of metal particles in the path. This requires that the electrode be moved toward or away from the work to maintain the arc and satisfactory welding conditions.

No ordinary manual arc welding is done today with bare electrodes; shielded (covered) electrodes always are used. However, a large amount of automatic and semiautomatic arc welding is done in which the electrode is a continuous, bare metal wire, but always in conjunction with a separate shielding and arc-stabilizing medium and automatic feed-controlling devices that maintain the proper arc length.

In the other type of metal arc welding, the electrode is made of tungsten, which is not consumed by the arc except by relatively slow vaporization. In some applications a separate filler wire must be used to supply the needed metal.

Shielded metal arc welding. *Shielded metal arc welding* (SMAW) uses electrodes that consist of metal wire, usually from 1.59 to 9.53 mm (1/16 to 3/8 inch) in diameter, upon which is extruded a coating containing chemical components that add a number of desirable characteristics, including all or a number of the following:

1. Provide a protective atmosphere.
2. Stabilize the arc.

3. Act as a flux to remove impurities from the molten metal.
4. Provide a protective slag to accumulate impurities, prevent oxidation, and slow down the cooling of the weld metal.
5. Reduce weld-metal spatter and increase the efficiency of deposition.
6. Add alloying elements.
7. Affect arc penetration.
8. Influence the shape of the weld bead.
9. Add additional filler metal.

Coated electrodes are classified on the basis of the tensile strength of the deposited weld metal, the welding position in which they may be used, the type of current and polarity (if direct current), and the type of covering. A four- or five-digit system of designation is used, as indicated in Figure 29-7. As an example, type E7016 is a low-alloy steel electrode that will provide a deposit having a minimum tensile strength of 70,000 psi in the non-stress-relieved condition; it can be used in all positions, with either alternating current or reverse-polarity direct current, and it has a low-hydrogen-type coating.

In general, the cellulosic coatings contain about 50 per cent SiO_2; 10 per cent TiO_2; small amounts of FeO, MgO, and Na_2O; and about 30 per cent volatile matter. The titania coatings have about 30 per cent SiO_2, 50 per cent TiO_2; small amounts of FeO, MgO, Na_2O, and Al_2O_3; and about 5 per cent volatile material. The low-hydrogen coatings typically contain about 28 per cent TiO_2 plus ZrO_2 and 25 per cent CaO plus MgO. They eliminate dissolved hydrogen in the deposited weld metal and thus prevent microcracking. To be effective they must be baked just prior to use to assure the removal of all moisture from the coating.

All electrodes are marked with colors in accordance with a standard established by the National Electrical Manufacturers Association so they can be readily identified as to type.

FIGURE 29-7. Designation system for arc-welding electrodes.

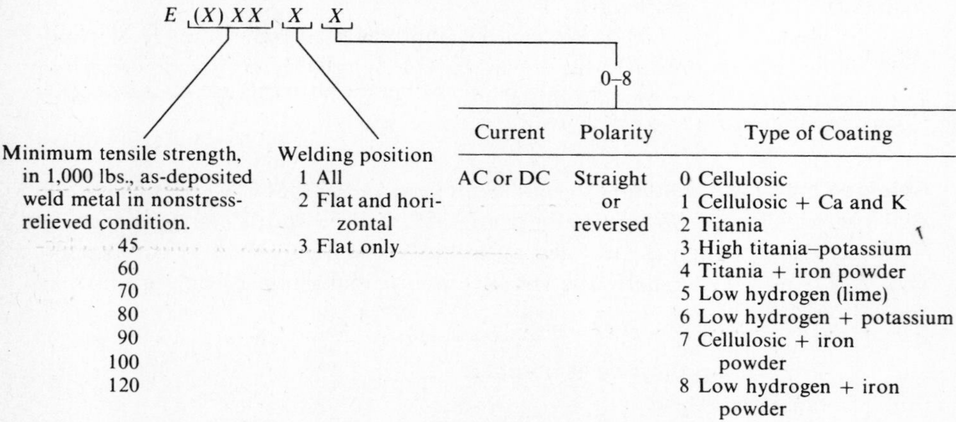

FIGURE 29-8. Artist's representation of the metal transfer in a shielded metal arc, and an actual photograph of arc welding. (*Courtesy General Electric Company.*)

As the coating on the electrode melts and vaporizes, it forms a protective atmosphere that stabilizes the arc and protects the molten and hot metal from contamination. Fluxing constituents unite with any impurities in the molten metal and float them to the surface to be entrapped in the slag coating that forms over the weld. This slag coating protects the cooling metal from oxidation and slows down the cooling rate to prevent hardening. The slag is easily chipped from the weld when it has cooled. Figure 29-8 depicts the way in which metal is deposited from a shielded electrode.

Electrodes having iron powder in the coating are used extensively, particularly in production-type welding. These greatly increase the amount of metal that can be deposited with a given size of electrode wire and current and thus reduce welding costs.

The insulating coating on one type of electrode melts sufficiently slowly that it protrudes slightly beyond the melting filler wire and thus, if dragged along the work, will maintain the proper arc length. These are called *contact* or *drag electrodes*.

Although a very large amount of welding still is done with ordinary shielded electrodes, in recent years there has been a great increase in the use of other methods of shielding, largely because they permit the use of continuous electrodes and automatic electrode-feeding devices.

Gas tungsten arc welding *Gas tungsten arc welding* (GTAW)[2] was one of the first major developments away from the use of ordinary shielded electrodes. Originally developed for welding magnesium, it employs a tungsten electrode, held in a special holder through which an inert gas is supplied with sufficient flow to form an inert shield around the arc and the molten pool of metal, thus shielding them from the atmosphere. This arrangement is shown

[2] This process formerly was known as TIG welding—for Tungsten, Inert-Gas.

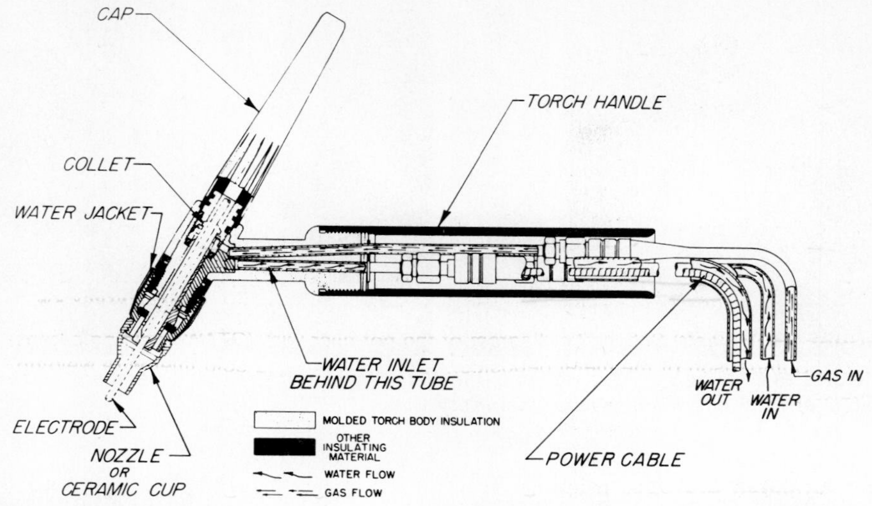

FIGURE 29-9. Welding torch used in nonconsumable metal-electrode, inert-gas (GTAW) welding. (*Courtesy Linde Division, Union Carbide Corporation.*)

in Figure 29-9. Argon or helium, or a mixture of them, is used as the inert shielding medium. Because the tungsten electrode is virtually not consumed at arc temperatures in these inert gases, the arc length remains constant so that the arc is stable and easy to maintain. The tungsten electrodes often are treated with thorium or zirconium to provide better current-carrying and electron-emission characteristics. A high-frequency, high-voltage current usually is superimposed on the regular ac or dc welding current to make it easier to start and maintain the arc.

If filler metal is required, it must be supplied by a separate wire, as in gas-flame welding. However, in many applications where a close fit exists between the parts being welded, no filler metal may be needed. The modification depicted in Figure 29-10 has come into some use. A fine, continuous filler wire is heated by the passage of an ac current so that it melts as it feeds into the weld puddle, just behind the arc, as the result of the I^2R effect. As shown, the deposition rate is several times what can be achieved with a cold wire, and it can be increased further by oscillating the filler wire from side to side when making a wide weld. The hot-wire process cannot be used in welding copper or aluminum, however, because of their low resistivities.

Gas tungsten arc welding produces very clean welds, and no special cleaning or slag removal is required because no flux is employed. With skillful operators, welds often can be made that are scarcely visible. However, the surfaces to be welded must be clean and free of oil, grease, paint, or rust, because the inert gas does not provide any cleaning or fluxing action.

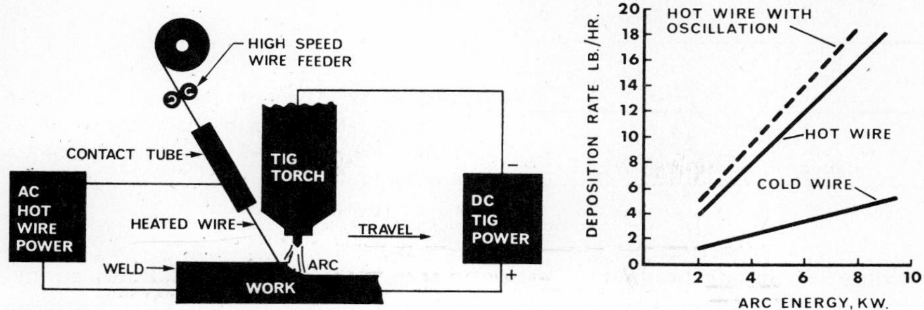

FIGURE 29-10. (*Left*) Schematic diagram of the hot-filler-wire (GTAW) welding process. (*Right*) Comparison of the metal deposition rates with hot and cold filler wire welding. (*Courtesy Welding Journal.*)

Gas tungsten arc spot welding. A variation of gas tungsten arc welding is employed for making spot welds between two pieces of metal without the necessity of having access to both sides of the joint. The basic procedure is shown in Figure 29-11. A modified and vented inert-gas, tungsten arc gun and nozzle are used, with the nozzle pressed against one of the two pieces of the joint. The workpieces must be sufficiently rigid to sustain the pressure that must be applied to one side to hold them in reasonably good contact. The arc between the tungsten electrode and the upper workpiece provides the necessary heat, and an inert gas, usually argon or helium, flows through the nozzle and provides a shielding atmosphere. Automatic controls move the electrode to make momentary contact with the workpiece to start the arc, and then withdraw and hold it at a correct distance to maintain the arc. The duration of the arc is timed automatically so that the two workpieces are heated sufficiently to form a spot weld under the pressure of the "gun" nozzle. The depth and size of the weld nugget are controlled by the amperage, time, and type of shielding gas, as shown in Figure 29-12.

Because access to only one side of the work is required, this type of spot

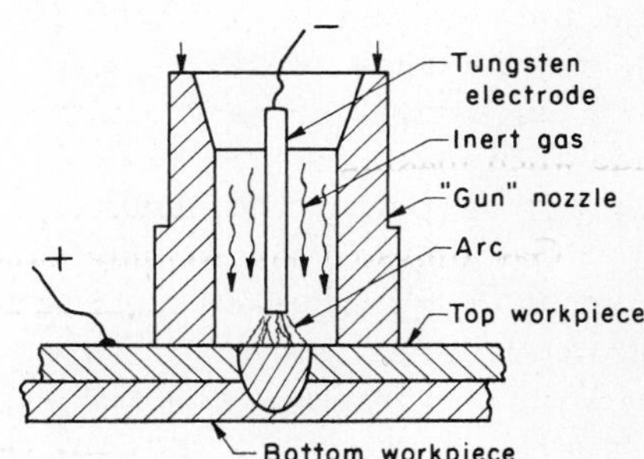

FIGURE 29-11. Schematic diagram of the method of making spot welds by the inert-gas-shielded tungsten arc process.

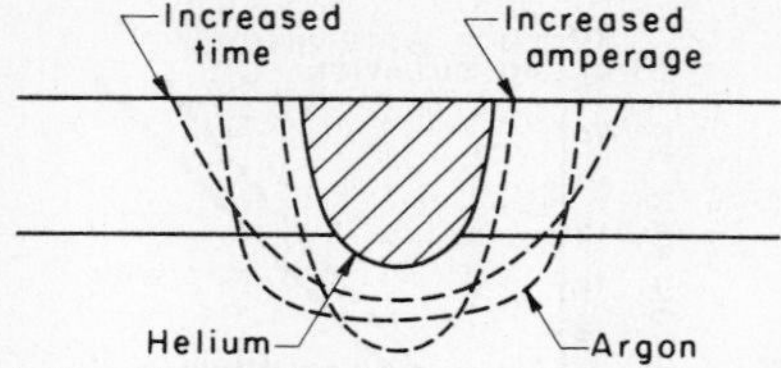

FIGURE 29-12. Effect of changes of current, time, and shielding gas on the shape of the weld nugget in inert-gas-shielded tungsten arc spot welding.

welding has an advantage over resistance spot welding in certain applications, as in fastening relatively thin sheet metal to a heavier framework, illustrated in Figure 29-13.

Gas metal arc welding *Gas metal arc welding* (GMAW) was a logical outgrowth of gas tungsten arc welding, differing in that the arc is maintained between an automatically fed, consumable wire electrode and the workpiece. It thereby automatically provides the additional filler. Formerly designated as MIG (metal, inert-gas) welding, the basic circuit and equipment are depicted in Figure 29-14. Although argon and helium, or mixtures of them, can be used for welding virtually any metal, they are used primarily for welding nonferrous metals. For welding steel, some O_2 or CO_2 usually is added to improve the arc stability and to reduce weld spatter. The cheaper CO_2 alone can be used for welding steel, provided that a deoxidizing electrode wire is employed.

The shielding gases have considerable effect on the nature of the metal

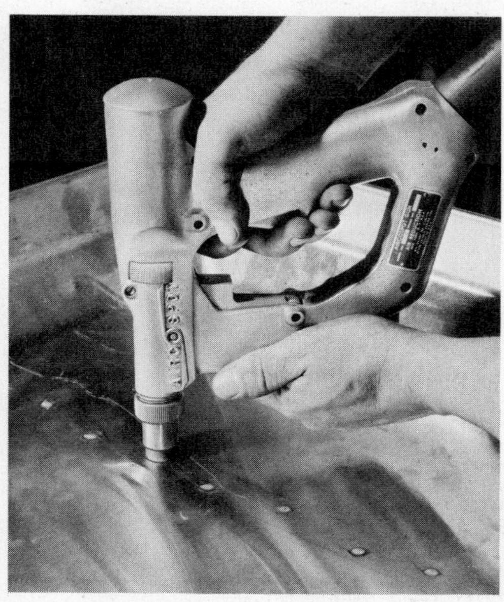

FIGURE 29-13. Making a spot weld by the inert-gas-shielded tungsten arc process. (*Courtesy Air Reduction Company, Inc.*)

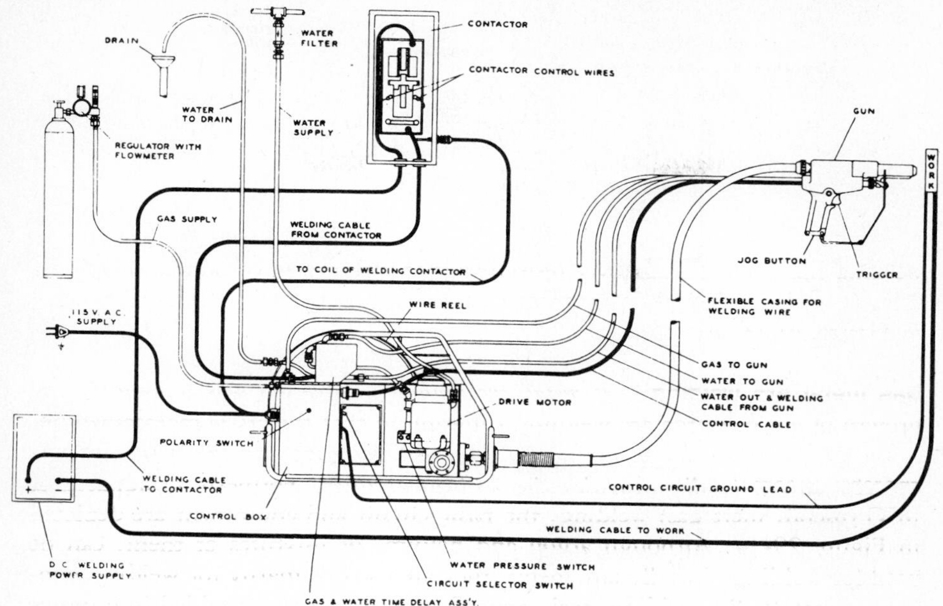

FIGURE 29-14. Circuit diagram and equipment for gas metal arc (GMAW) welding. (*Courtesy Air Reduction Company, Inc.*)

(drop) transfer from the electrode to the work and also affect the tendency for undercutting. Several types of electronic controls, which alter the wave form of the current, make it possible to vary the mechanism of metal transfer—by drops, spray, or short-circuiting drops. Some of these variations in the basic process are: *pulsed arc welding* (GNAW-P), *short circuiting arc welding* (GNAW-S), and *spray transfer welding* (GMAW-ST). *Buried arc welding* (GMAW-B) is another variation in which carbon dioxide–rich gas is used and the arc is buried in its own crater. Figure 29-15 illustrates one system for producing a series of peak current surges that preheat the electrode just before droplet formation.

Gas metal arc welding is fast and economical because there is no frequent changing of electrodes, as with stick-type electrodes. In addition, there is no slag formed over the weld, the process often can be automated and, if done manually, the welding head is relatively light and compact, as shown in Figure 29-16.

Flux cored arc welding. *Flux cored arc welding* (FCAW) basically utilizes a continuous electrode wire which contains a granular flux within its hollow core. However, quite frequently additional shielding is provided by a small flow of CO_2 around the arc zone, so it often is considered to be a variation of gas metal arc welding. It is used primarily for welding steel, and the basic ar-

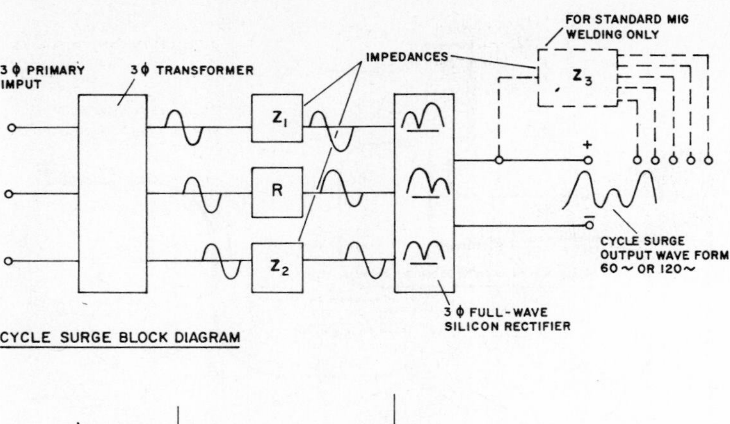

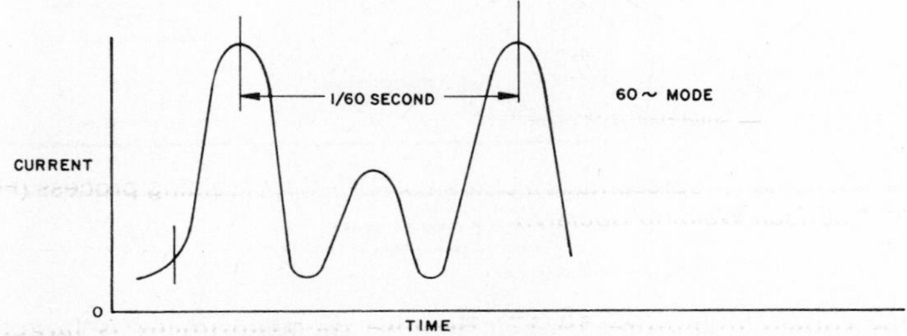

FIGURE 29-15. Method of producing peak-current surges to control droplet transfer. (*Courtesy Welding Journal.*)

FIGURE 29-16. Welding with a manually held GMAW welding gun. (*Courtesy Air Products and Chemicals, Inc.*)

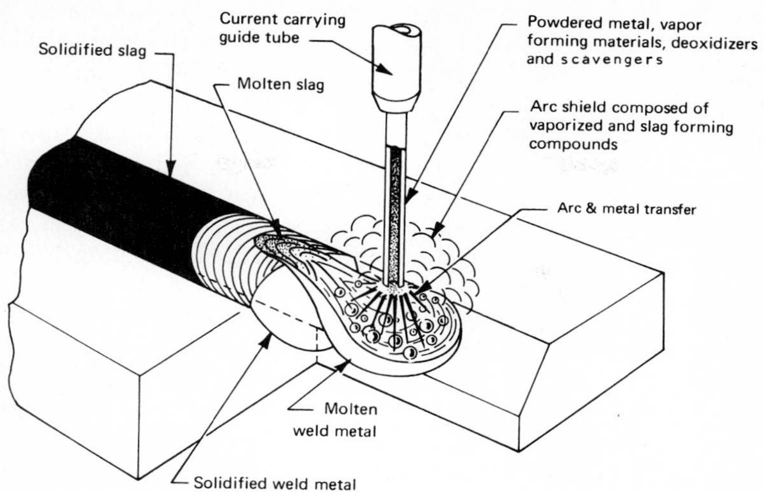

FIGURE 29-17. Schematic representation of the flux-cored arc welding process (FCAW). (*Courtesy The American Welding Society.*)

rangement is shown in Figure 29-17. Because the equipment is larger, it is not as readily portable and as easily manipulated as regular GMAW equipment. However, excellent welds can be made with the process.

Submerged arc welding. In *submerged arc welding* (SAW), as shown in Figure 29-18, the arc is maintained beneath a granular flux. Either ac or dc current can be used as the power source. The flux is deposited just ahead of the electrode, which is in the form of coiled wire—copper-coated to provide good

FIGURE 29-18. Basic features of the submerged arc welding process (SAW). (*Courtesy Linde Division, Union Carbide Corporation.*)

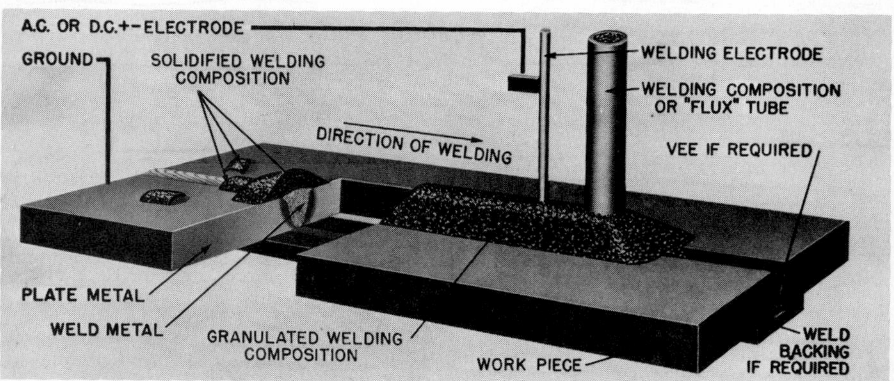

electrical contact. Because the arc is completely submerged in the flux, only a few small flames are visible. The granular flux provides excellent shielding of the molten metal and, because the pool of molten metal is relatively large, good fluxing action occurs, so as to remove impurities. Consequently, very high quality welds are obtainable. A portion of the flux is melted and solidifies into a glasslike covering over the weld. This, with the flux that is not melted, provides a thermal coating that slows down the cooling of the weld area and thus helps to produce a soft, ductile weld. The solidified flux cracks loose from the weld on cooling and is easily removed. Surplus unmelted flux is recovered with a vacuum cleaner.

Submerged arc welding is most suitable for making flat butt or fillet welds. The equipment used varies from the portable type shown in Figure 29-19, wherein the welding head is moved along the work automatically, to production setups, such as shown in Figure 29-20, or manual equipment, such as shown in Figure 29-21. In the latter, the electrode wire is fed through a funnel device that also contains the flux and the electrode guide, and the arc length is controlled automatically through the feeding of the wire. Such equipment is very useful for making short or nonlinear welds where the time required for setting up the usual equipment would not be justified.

High welding speeds and welds in thick plates can be achieved with the ordinary submerged arc process. Welding speeds of 762 mm (30 inches) per minute in 25.4-mm (1-inch)-thick steel plate, or 304.8 mm (12 inches) per minute in 38.1-mm (1½-inch) plate, are common. Because the metal is deposited in fewer passes than in the case of manual arc welding, there is less probability of entrapped slag or of voids, and higher quality welds are more consistently obtained.

Welding heads with a single electrode are most commonly used, but machines utilizing three electrodes, operating from three-phase alternating

FIGURE 29-19. Making a submerged arc weld using portable equipment, guided by the edge-preparation groove in the plates being welded. (*Courtesy The Lincoln Electric Company.*)

FIGURE 29-20. Production setup for welding large pipe by the submerged arc process. (*Courtesy Linde Division, Union Carbide Corporation.*)

current, often are employed in production work. The process is widely used for large-volume welding, as in the building of ships or the manufacture of large-diameter steel pipe or tanks.

In a modification of the basic process, iron powder is deposited in the prepared gap between the plates to be joined, ahead of the flux and in con-

FIGURE 29-21. Manual submerged arc welding. (*Courtesy The Lincoln Electric Company.*)

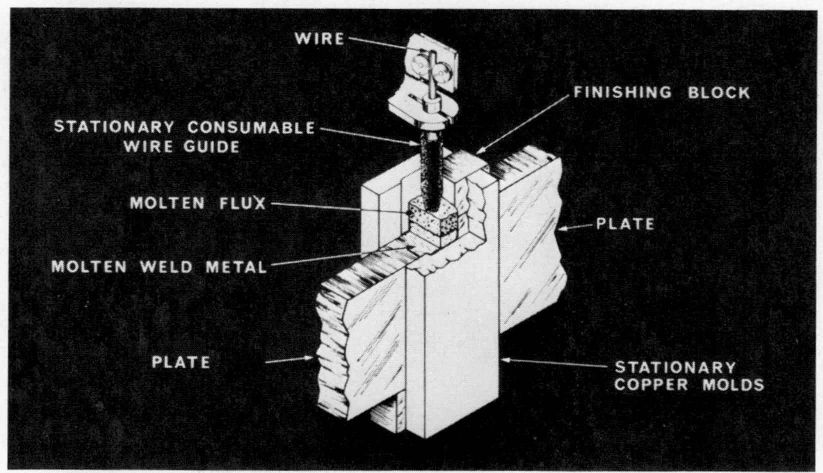

FIGURE 29-22. Setup for making vertical welds by the submerged arc process. (*Courtesy The Welding Journal.*)

junction with a backing strip. As with iron powder-coated electrodes in stick welding, this permits substantially higher deposition rates to be obtained.

By using the arrangement shown in Figure 29-22, vertical welds can be made by the submerged arc process. Stationary, copper side molds are employed, and a stationary, consumable wire guide is used. The consumable wire guide, coated with flux, melts as it enters the molten flux pool, thereby replacing the flux that solidifies at the copper–weld interface. Good-quality welds up to 102 mm (4 inches) thick can be made by this procedure, but for plate thicker than about 51 mm (2 inches), the electroslag process, which will be discussed later, is usually more economical.

Stud welding. *Stud welding* (SW) is an arc-welding process by which special types of metal studs can be attached to workpieces. A special gun is used, such as is shown in Figure 29-23, into which the stud is inserted. A dc arc is established between the end of the stud and the workpiece until sufficient temperature is produced, thereby welding the stud to the work. Automatic equipment controls the establishing of the arc, its duration, and the application of pressure to the stud.

A wide variety of studs is available, as shown in Figure 29-24. The recessed end of the stud is filled with flux. A ceramic ferrule, which is placed over the end of the stud before it is positioned in the gun, is an important factor in the process, acting to concentrate the arc heat and to protect the metal from the atmosphere. It also confines the molten or plastic metal to the weld area and shapes it round the base of the stud. After the weld is complete, the ferrule is broken from the stud.

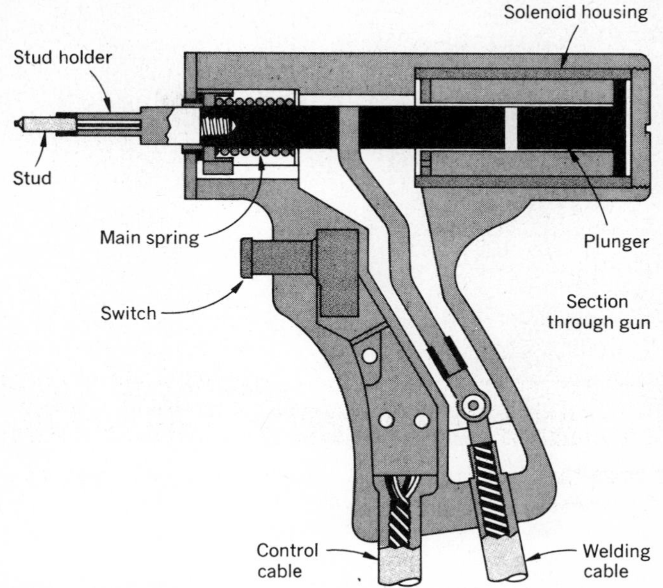

FIGURE 29-23. Schematic diagram of a stud-welding gun. (*Courtesy American Machinist.*)

FIGURE 29-24. (*Left*) Some of the types of available studs for stud welding. (*Center*) A stud and a ceramic ferrule. (*Right*) A stud after welding, and a section through a welded stud. (*Courtesy Nelson Stud Welding Co.*)

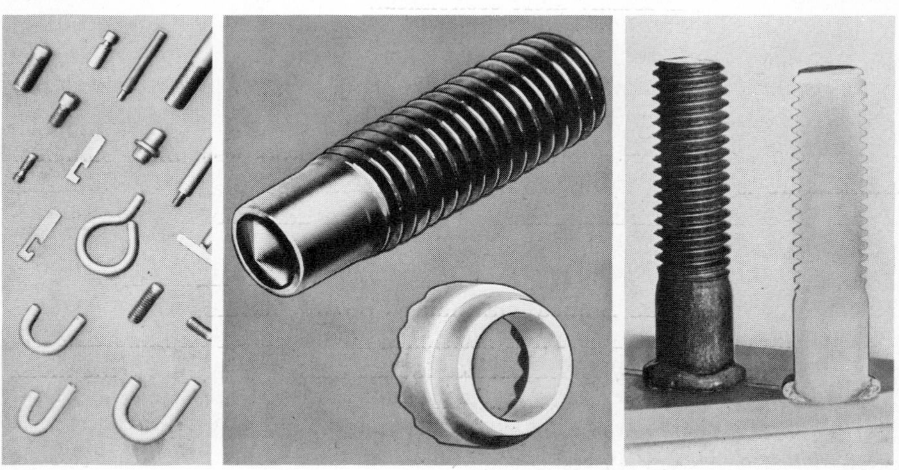

FIGURE 29-25. Production setup for making 720 stud welds per hour, fastening 22-mm (⅞-inch)-long studs to powder metallurgy parts. (*Courtesy Nelson Stud Welding Co.*)

The process requires almost no skill on the part of the operator; once the stud and ferrule are placed in the gun chuck and the gun is positioned on the work, all the operator has to do is pull the trigger. The remainder of the cycle is automatic, consuming less than 1 second. It thus is well suited to, and widely used in, manufacturing, eliminating the necessity for drilling and tapping many holes. Special production-type stud-welding machines, such as shown in Figure 29-25, are available.

Advantages and disadvantages of arc welding. Because of its great flexibility and the variety of processes that are available, arc welding is an extremely useful, versatile, and widely used process. However, except for gas-shielded tungsten arc spot welding, stud welding, and, to some degree, submerged arc welding, the various arc welding processes have one disadvantage—the quality of the weld depends ultimately on the skill and integrity of the worker who does the welding. Consequently, proper training, selection, and supervision of personnel are of great importance, even more so than adequate inspection.

Power sources for arc welding. Arc welding requires a large amount of current that does not change in magnitude as the voltage varies over a considerable range. The load voltage usually is from 30 to 40 volts, although the actual voltage across the arc ordinarily varies from about 12 to 30 volts, depending primarily on the arc length. Both dc and ac sources are available, having "drooping voltage" characteristics, as shown in Figure 29-26, and capacities from 150 to 1000 amperes. These characteristics assure that the cur-

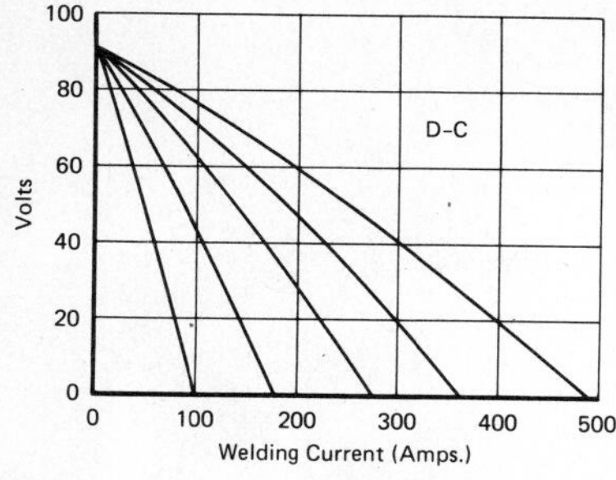

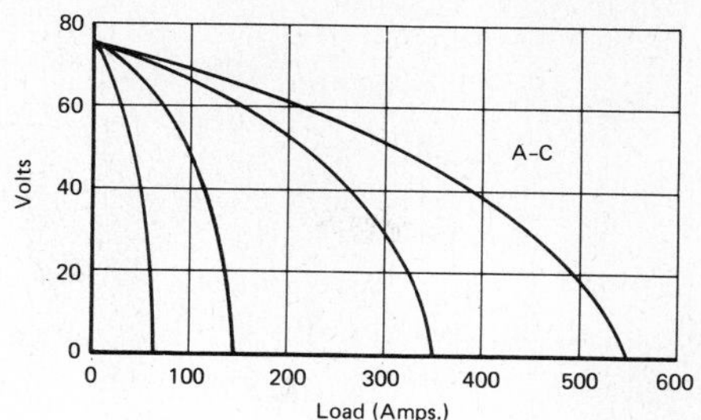

FIGURE 29-26. Drooping-voltage characteristics of typical arc-welding power sources: direct-current (*left*) and alternating-current (*right*).

rent does not vary greatly as the voltage fluctuates over the usual operating range.

In earlier years most direct current for welding was provided by motor-generator sets. However, today, solid-state transformer-rectifier machines, such as shown in Figure 29-27, are the most generally used power source. Using a three-phrase power supply, these usually can provide both ac and dc output. For field work where electric power is not available, gasoline-engine-driven dc generators are used.

FIGURE 29-27. Rectifier-type dc and ac welding machine, 300 amperes capacity, with cover removed to show interior. (*Courtesy The Lincoln Electric Company.*)

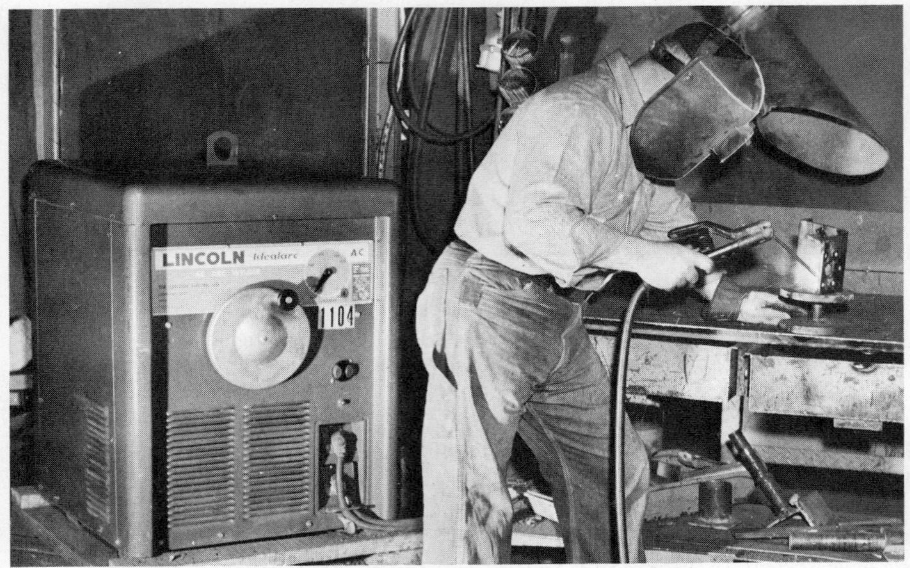

FIGURE 29-28. Ac welding machine being used on a typical welding job. (*Courtesy The Lincoln Electric Company.*)

If only ac welding is to be done, relatively simple transformer-type machines are available, such as shown in Figure 29-28. Usually, these are single-phase devices having low power factors, but when several machines are to be operated, as in production-type situations, they may be connected to the several phases of a three-phase supply and thus help to balance the load. Usually, they have internal capacitors to improve the power factor.

FUSION WELD TYPES AND JOINTS

There are four basic types of fusion welds, as illustrated in Figure 29-29. *Bead welds* require no edge preparation. However, because the weld is made on a

FIGURE 29-29. Four basic types of fusion welds.

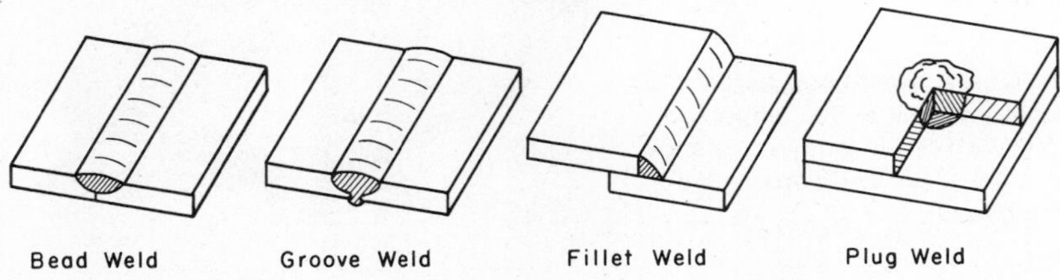

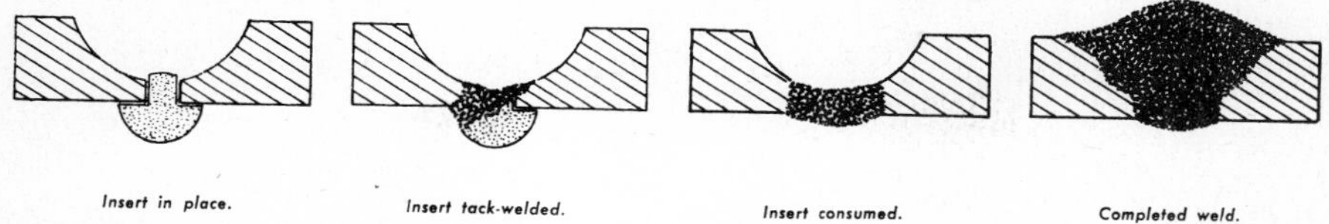

FIGURE 29-30. Use of a consumable backup insert in making arc welds. (*Courtesy Arcos Corporation.*)

flat surface and the penetration thus is limited, they are suitable only for joining thin sheets of metal, for building up surfaces, or for applying hard facing metals.

Groove welds are used where full-thickness strength is sought on thicker materials. These require some type of edge preparation to make a groove between the abutting edges. V, double V, U, and J configurations are most common, usually produced by oxyacetylene flame cutting. The type of groove configuration depends primarily on the thickness of the work, the welding process to be employed, and the position of the work, the primary consideration being to obtain a sound weld throughout the full thickness with a minimum deposit of weld metal. The weld may be made in either a single pass or by multiple-pass procedures, depending on the thickness of the material and the welding process used. As shown in Figure 29-30, special consumable insert rings, or strips, often are used to assist in obtaining proper spacing between the mating edges and to aid in assuring proper quality in the root pass. These are especially useful in pipeline welding, particularly under field conditions and where the welding must be done from only one side of the work.

Fillet welds are used for tee, lap, and corner joints. The size of fillet welds is measured by the leg of the largest 45° right triangle that can be inscribed within the contour of the weld cross section. This is shown in Figure 29-31, which also indicates the proper shape for fillet welds to avoid excess metal and to reduce stress concentration. Fillet welds require no special edge preparation. They may be continuous or made intermittently, spaces being left between short lengths of weld.

FIGURE 29-31. Preferred shape of fillet welds and the method of measuring fillet-weld size.

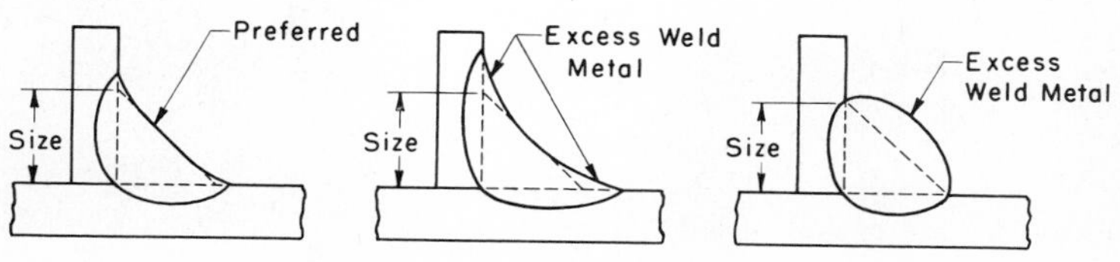

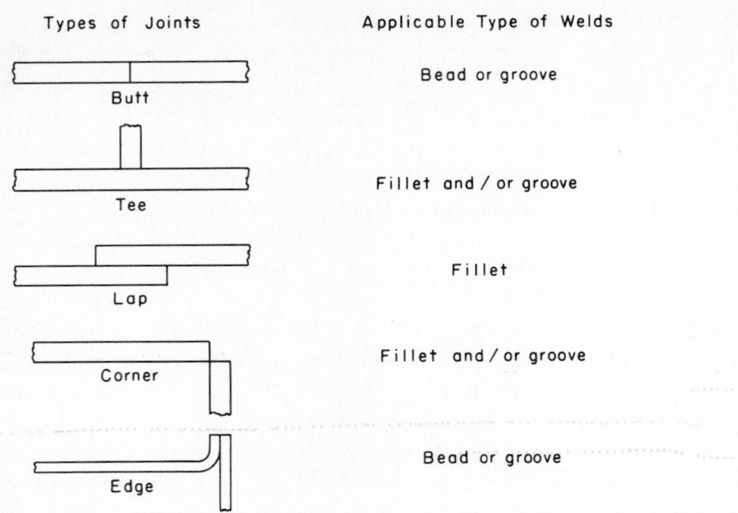

FIGURE 29-32. Basic types of fusion-weld joints, and types of welds used in making them.

Plug welds are used to attach one part on top of another, replacing rivets or bolts. A hole is made in the top plate, and welding is started at the bottom of this hole. They offer substantial saving in weight as compared with riveting or bolting.

Figure 29-32 shows the five basic types of joints that can be made through the use of bead, groove, and fillet welds. In selecting the type of weld

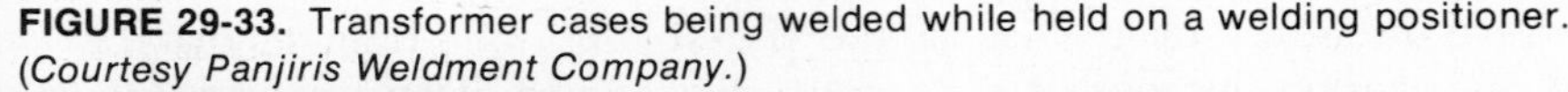

FIGURE 29-33. Transformer cases being welded while held on a welding positioner. (*Courtesy Panjiris Weldment Company.*)

joint to be used, the primary consideration should be the type of loading that will be applied. Too frequently this basic fact is neglected, and a large proportion of what are erroneously called "welding failures" are the result of such oversight. Cost and accessibility for welding are important, but secondary, factors in joint selection. Cost is affected by the required edge preparation, the amount of weld metal that must be deposited, the type of process and equipment that must be used, and the speed and ease with which the welding can be accomplished. Accessibility obviously will have considerable influence on several of these factors.

In production welding, extensive use is made of welding *positioners* to which the work is clamped. These make it possible for the work to be turned to permit the welding to be done in the most favorable position. Figure 29-33 shows a large weldment being fabricated on a positioner of moderate size. Special positioners are employed which are capable of holding large sections of ships weighing many tons.

Review questions

1. What is a weld?
2. What conditions must exist in order for a weld to be achieved?
3. What factors normally impede the formation of a weld? How may they be overcome?
4. Why is cold welding limited in its scope of usage?
5. Why are acetylene storage tanks filled with acetone?
6. What is the advantage of stabilized methylacetylene propadiene over acetylene as a heat source?
7. What is the effect of too much oxygen being supplied to an oxyacetylene torch used in welding?
8. What is the principal use for pressure gas-flame welding?
9. Give two reasons why arc welding has largely replaced oxyfuel gas welding.
10. How does the temperature of the heat source in arc welding compare with that in oxyfuel gas welding?
11. What functions does the electrode fulfill in consumable-electrode arc welding?
12. Why is less skill required for GTAW than for SMAW?
13. What functions are filled by the coating on an arc welding electrode?
14. What is indicated by the electrode designation E 7018?
15. Why is iron powder often added to electrode coatings?
16. What is the purpose of pulsing the current in arc welding?
17. Why is tungsten used as the electrode in gas tungsten arc welding?
18. Why are power sources for arc welding designed to having drooping characteristics?

19. Why is gas tungsten arc welding particularly advantageous for welding relatively delicate work where good apperance must be maintained?
20. Why did gas metal arc welding replace most gas tungsten arc welding?
21. Why is a flux-cored electrode usually used in conjunction with CO_2 when welding steel using GMAW?
22. Why are superior metallurgical properties usually obtained with submerged arc welding, compared with ordinary shielded metal arc welding?
23. The inventor of submerged arc welding has stated that during some of the early developmental work he sometimes was not sure whether he was developing a welding process or a casting process. Can you explain why?
24. Would you expect a weld in 50.8-mm (2-inch)-thick steel, made by submerged arc welding to be more ductile than if made by ordinary coated-electrode welding? Why?
25. What is the major difficulty in using submerged arc welding for making vertical welds?
26. How is the need for special shielding flux or gases eliminated in stud welding?
27. What is a major advantage of gas-shielded tungsten arc spot welding?
28. What are the four basic types of fusion welds?
29. What are five basic types of joints made by fusion welding?
30. What should be the primary factor determining the type of joint that should be used in a weldment?
31. What are two advantages of substituting plug welds for riveting?
32. A weldment, made from AISI 1025 steel, is being welded with E6012 electrodes. Some difficulty is being experienced with cracking in the weld beads and in the heat-affected zones. What possible corrective measures can you suggest?
33. A base for a special machine tool will weigh 635 kg (1,400 lb) if made as a gray iron casting. Pattern cost would be $450 and the foundry has quoted a price of $1.32 per kg ($0.60 per lb) for making the casting. If the part is made as a weldment, it will require 363 kg (800 lb) of steel costing $0.31 per kg ($0.14 per lb). Cutting, edge preparation, and setup time will require 30 hours at a rate of $10.00 per hour for labor and overhead. Welding time will be 55 hours at an hourly rate of $9.50. Ninety-one kg (200 lb) of electrode will be required, costing $0.37 per kg ($0.17 per lb). (a) Which method of fabrication will be more economical if only one part is required? (b) What number of parts would be required for welding and casting to break even?

CHAPTER 30

Resistance welding

RESISTANCE-WELDING THEORY

In resistance welding, both heat and pressure are utilized in producing coalescence. The heat is the consequence of the resistances of the workpieces and the interface between them to the flow of electrical current. A certain amount of pressure is applied initially to hold the workpieces in contact, thereby controlling the electrical resistance at the interface, and is increased when the proper temperature is attained, to facilitate the coalescence. Because of the pressure utilized, coalescence occurs at a lower temperature than with oxyfuel gas or arc welding. Consequently, in all modern resistance welding, no intentional melting of the metal occurs. Thus they essentially are solid-state processes, although not officially so classified by the American Welding Society.

In some of the resistance welding processes, additional pressure can be applied immediately after coalescence is achieved to provide a certain amount of forging action, with some accompanying grain refinement. Also, some additional heating can be induced, as a part of the process, to provide tempering and/or stress relief. Usually, the required temperature can be obtained and the weld completed in a few seconds or less. Consequently, resistance welding is a very rapid and economical process, extremely well suited to manufacturing, and it is widely used, particularly in mass production.

Heating. Heat for resistance welding is obtained by passing a large electrical current through the workpieces for a short interval, utilizing the basic relationship

$$H = I^2RT$$

where H is the heat, I the current, R the electrical resistance of the circuit, and T the time or duration of the current flow. In most cases, alternating current is used. As indicated in Figure 30-1, the work is a part of the electrical circuit. It is important to note that the total resistance of the assembly between the electrodes consists of three parts: (1) the resistance of the workpieces, (2) the contact resistance between the electrodes and the work, and (3) the resistance between the faying surfaces of the workpieces. Because it is desired to have the maximum temperature occur at the point where the weld is to be made—at the faying surfaces—it is essential to keep resistances (1) and (2) as low as possible with respect to resistance (3). Obviously, with materials having low electrical resistance, such as aluminum and copper, this condition is difficult to achieve, and they require much larger currents and more attention to the interface conditions than when welding steel.

The resistance of the workpieces is determined by the type and thickness of the metal. It usually is much less than the other two resistances because of the larger area involved and the relatively high electrical conductivity of most metals. The resistance between the work and the electrodes is minimized by using electrode materials that are excellent electrical conductors, by controlling the shape and size of the electrodes, and by using the proper pressure be-

FIGURE 30-1. Fundamental resistance-welding circuit.

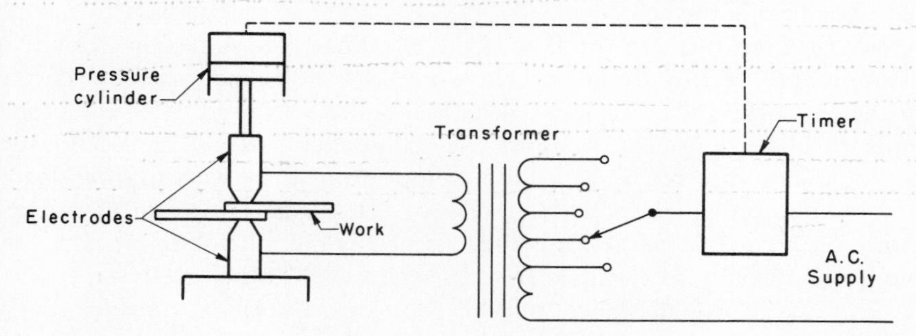

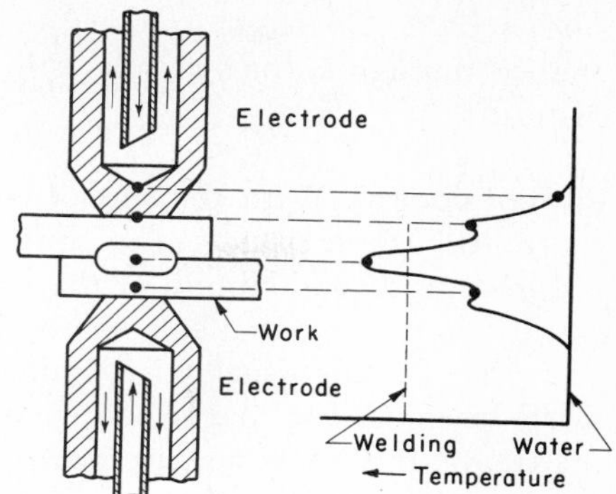

FIGURE 30-2. Desired temperature distribution across the electrodes and the workpiece in resistance welding.

tween the work and the electrodes. Because any change in the pressure between the work and the electrodes also tends to change the pressure between the faying surfaces, only limited control of the electrode-to-work resistance can be obtained in this manner.

The resistance between the faying surfaces is a function of (1) the quality of the surfaces; (2) the presence of nonconductive scale, dirt, or other contaminants; (3) the pressure; and (4) the contact area. These factors must be controlled to obtain uniform results.

As indicated in Figure 30-2, the objective is to bring the faying surfaces simultaneously to the proper elevated temperature while keeping the remaining material and the electrodes at much lower temperatures. The electrodes usually are water-cooled to keep their temperature low and to aid in maintaining them in proper condition. When metals of different thickness or of different conductivities are to be welded, they can be brought to the proper welding temperature simultaneously by using an electrode having higher conductivity against the thicker, or high-resistance, matrial than the one that is in contact with the thinner or lower-resistance material.

Pressure. Because pressure in resistance welding affects the contact resistance, permits welds to be made at lower temperatures, and provides a forging action, the control of both its magnitude and timing is very important. If too little pressure is used, the contact resistance is high and surface burning and pitting of the electrodes may result. On the other hand, if too high pressure is employed, molten or softened metal may be squirted or squeezed from between the faying surfaces, or the work may be indented by the electrodes. Ideally, a moderate pressure should be applied prior to and during the passage of the welding current, to establish proper contact resistance. It should then be increased considerably just as the proper welding heat is attained, to complete the coalescence and forge the weld to produce a fine grain structure.

On small, foot-operated machines only a single spring-controlled pressure is used. On larger, production-type welders, the pressures usually are applied through air or hydraulic cylinders that are controlled and timed automatically.

Current control. With the surface conditions held constant and the pressure controlled, the temperature in resistance welding is regulated by controlling the magnitude and timing of the welding current. Very precise and sophisticated controls are available for this purpose.

The current usually is obtained from a "step-down" transformer. On small machines the magnitude is controlled through taps on the primary of the transformer or by an autotransformer that varies the primary voltage supplied to the main transformer. On larger machines several methods are used. In *phase-shift control* the magnitude and wave shape of the primary current are altered. With *slope control* the current is permitted to rise gradually to full magnitude in from about 3 to 25 cycles.

In large, production-type welders, not only is the magnitude of the current controlled, but the extent of the current flow and the application of the current and pressure are carefully programmed. This is achieved by means of adjustable, electronic, synchronous timers that start and stop the current in proper synchronism with the voltage wave to avoid undesirable, high, transient currents that could damage electrodes and contactors and produce nonuniform welds. Figure 30-3 shows a relatively simple current and pressure cycle for spot welding that includes forging and postheating.

Power supply. The magnitude of the current required for resistance welding is so great that special types of circuits are employed in most machines to reduce the load on the power lines. Ordinary single-phase circuits are used only in smaller machines. Larger machines employ three-phase circuits. In one type, three-phase power is rectified and the energy stored in a special transformer. When the dc flow through the transformer is interrupted, the collapse of the

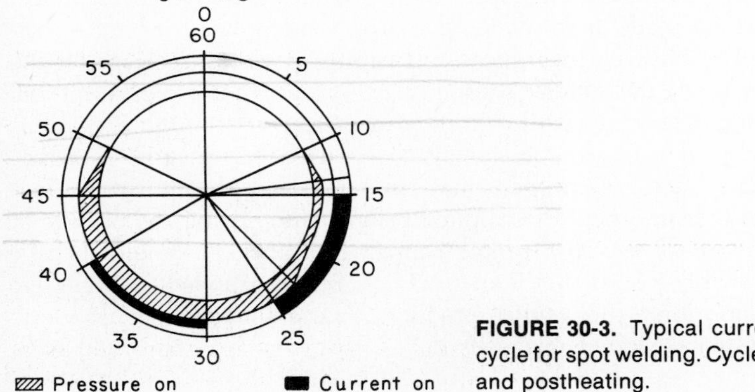

FIGURE 30-3. Typical current and pressure cycle for spot welding. Cycle includes forging and postheating.

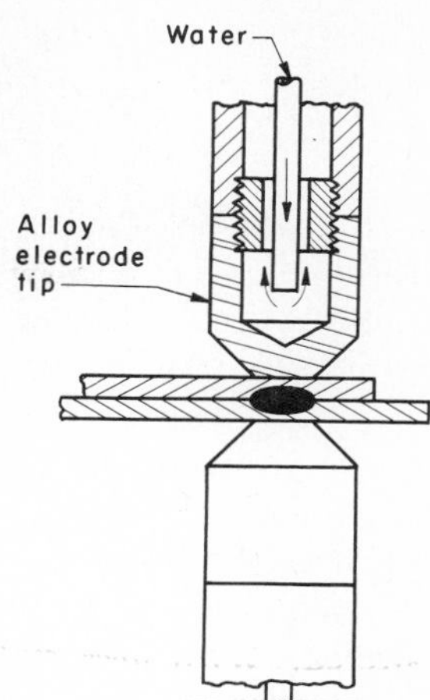

FIGURE 30-4. Arrangement of the electrodes and the work in spot welding.

field provides the required heavy current flow in the secondary circuit. Many modern resistance welders utilize dc welding current, obtained through solid-state rectification of three-phase power. Such machines reduce the current demand per phase, give a balanced load, and produce excellent welds.

RESISTANCE WELDING PROCESSES

Resistance spot welding. *Resistance spot welding* (RSW) is the simplest and most widely used of this type of welding. As shown in Figure 30-4, the overlapping work is positioned between water-cooled electrodes, which have reduced areas at the tips to produce welds that usually are from 1.6 to 12.7 mm ($^1/_{16}$ to ½ inch) in diameter. After the electrodes are closed on the work, a controlled cycle of pressure application and current flow occurs, producing a weld at the interface. The electodes then open, and the work is removed.

A satisfactory spot weld, such as is shown in Figure 30-5, consists of a

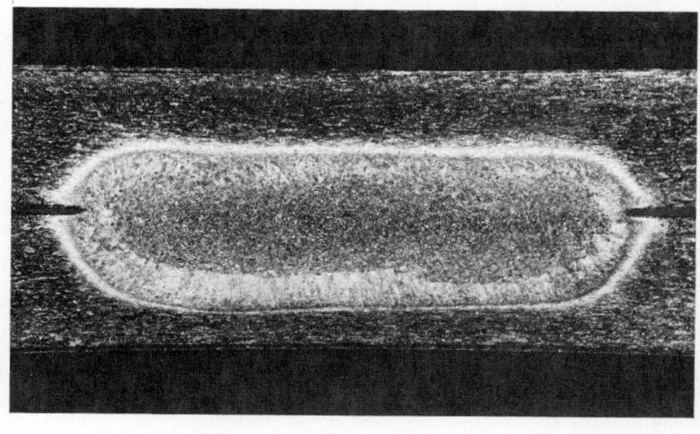

FIGURE 30-5. Section showing weld nugget between two sheets of 1.3-mm (0.051-inch) aluminum alloy. The radius of the upper electrode was greater than that of the lower electrode. (*Courtesy Lockheed Aircraft Corporation.*)

FIGURE 30-6. Tear test of a satisfactory spot weld, showing how failure occurs outside the weld.

nugget of coalesced metal formed with no melting of the material between the faying surfaces. There should be no, or only a very slight, indentation of the metal under the electrodes. The strength of the weld should be such that in a tensile or tear test the weld will remain intact and failure will occur in the heat-affected zone surrounding the nugget, as illustrated in Figure 30-6. If proper current density and timing, electrode pressure and shape, and surface conditions are maintained, sound spot welds can be obtained with excellent consistency.

Spot-welding machines. Three general types of spot-welding machines are available. For light-production-type work, primarily on steel, where complex current-pressure cycles are not required, the simple *rocker-arm* type, illustrated in Figure 30-7, often is used. On these machines the lower electrode arm is stationary, and the upper electrode, mounted on a pivoted arm, is brought

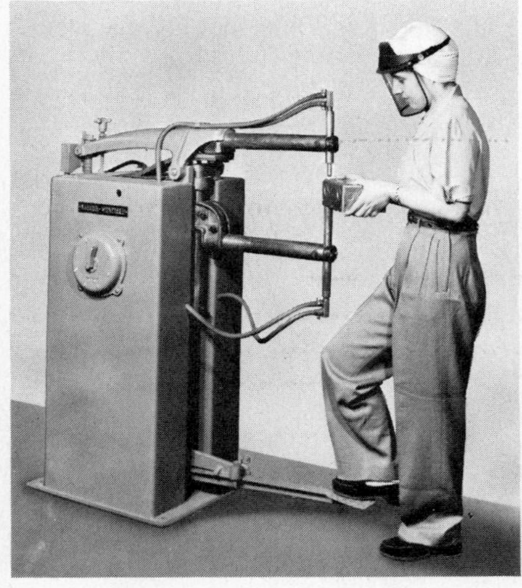

FIGURE 30-7. Part being welded on a foot-operated, rocker-arm, spot-welding machine. (*Courtesy The Taylor-Winfield Corporation.*)

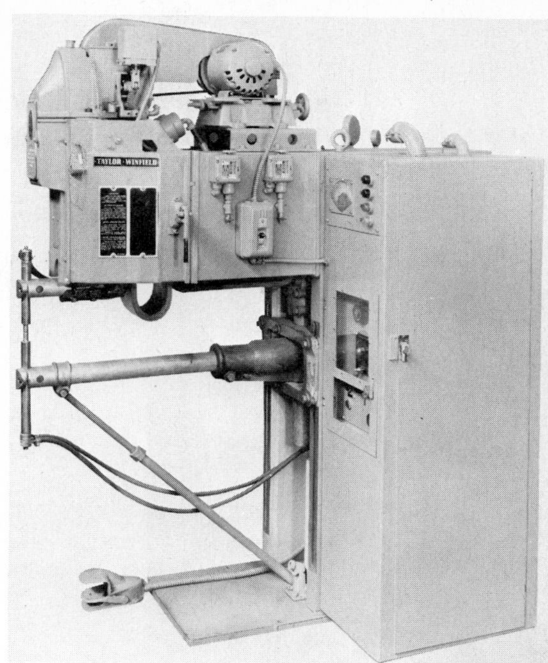

FIGURE 30-8. A 50-kVa, three-phase, press-type, dc spot welder. Rectifiers are used to provide the dc output. (*Courtesy The Taylor-Winfield Corporation.*)

down into contact with the work by means of a spring-loaded foot pedal. On larger machines and on machines for larger-volume work, this motion is obtained through an air cylinder or an electric motor. Machines of this type are available with throat depths up to about 1220 mm (48 inches) and transformer capacities up to 50 kVa.

Most large spot welders, and those used for greater production rates, are of the *press type,* such as is shown in Figure 30-8. In these the movable electrode has a straight-line motion, provided by an air or hydraulic cylinder. Such machines are adaptable to any type of pressure-controlled cycle that may be desired. Capacities up to 500 kVa and 1524-mm (60-inch) throat depth are common. Special-purpose spot welders of this type, employing multiple welding heads, are widely used in mass-production industries. Some, such as is shown in Figure 30-9, can make up to 200 spot welds in 60 seconds.

The application of spot welding is greatly extended through the use of *portable spot welding guns,* such as shown in Figure 30-10. Each gun is connected to the power supply and control units by flexible air hoses and electrical cables and by water hoses for cooling, where required. Such equipment permits bringing the welding unit and the operator to the work, thus greatly extending the use of spot welding in applications where work is too large to be moved to a welding machine. Such an application is shown in Figure 30-11. These units frequently make up to 200 spot welds per minute.

FIGURE 30-9. Large progressive-type spot welder which has 50 transformers and makes 200 spot welds on an automobile underbody in less than 6 seconds. (*Courtesy Progressive Welder Sales Company.*)

FIGURE 30-10. Typical portable spot-welding guns. (*Courtesy Progressive Welder Sales Company.*)

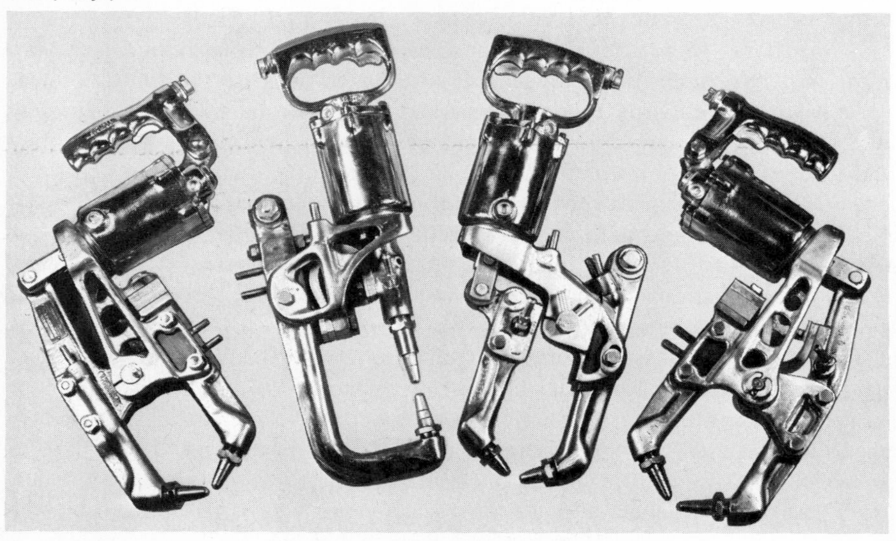

FIGURE 30-11. Portable spot-welding gun being used on an automobile body on an assembly line. (*Courtesy General Motors Corporation.*)

In some installations portable welding guns have been installed on automatic positioners[1] which position the gun in the desired three-dimensional location and permit one or more spot welds to be made automatically, without the need for an operator.

Spot-weldable metals. One of the great advantages of spot welding is the fact that virtually all the commercial metals can be spot-welded, and most of them can be spot-welded to each other. In only a few cases do the welds tend to be brittle. Table 30-1 shows combinations of metals that can be spot-welded satisfactorily.

Although the majority of spot welding is done on wrought sheet, other forms of metal also can be spot-welded. Sheets can be spot-welded to rolled shapes and steel castings, and some types of die castings can be welded without difficulty. Except for aluminum, most metals require no special preparation, except to be sure that the surface is free of corrosion and is not badly pitted. For best results, aluminum and magnesium should be cleaned immediately prior to welding by mechanical or simple chemical means. Metals that have high electrical conductivity require clean surfaces to assure that the electrode-to-metal resistance is low enough for adequate temperature to be devel-

[1] See Figure 38-15.

TABLE 30-1. Metal combinations that can be spot-welded

Metal	Aluminum	Brass	Copper	Galvanized Iron	Iron (wrought)	Monel	Nickel	Nickel Silver	Nichrome	Steel	Tin Plate	Zinc
Aluminum	X										X	X
Brass		X	X	X	X	X	X	X	X	X	X	X
Copper		X	X	X	X	X	X	X	X	X	X	X
Galvanized iron		X	X	X	X	X	X	X	X	X	X	
Iron (wrought)		X	X	X	X	X	X	X	X	X	X	
Monel		X	X	X	X	X	X	X	X	X	X	
Nickel		X	X	X	X	X	X	X	X	X	X	
Nickel silver		X	X	X	X	X	X	X	X	X	X	
Nichrome		X	X	X	X	X	X	X	X	X	X	
Steel		X	X	X	X	X	X	X	X	X	X	
Tin plate	X	X	X	X	X	X	X	X	X	X	X	
Zinc	X	X	X									X

oped in the metal itself. Silver and copper are difficult to weld because of their high heat conductivity. However, many copper alloys are readily resistance-welded. Water cooling adjacent to the spot-weld area can be used to aid in obtaining adequate welding temperature in the desired spot area of such materials.

The practical limit of thicknesses that can be spot-welded by ordinary processes is about 3.18 mm (⅛ inch), where each piece is of the same thickness. However, a thin piece can readily be welded to another piece that is much thicker than 3.18 mm. Spot welding of two 12.7-mm (½-inch) steel plates has been done satisfactorily as a replacement for riveting.

Resistance seam welding. *Resistance seam welds* (RSEW) are made by two distinctly different processes. In most cases, where the weld is between two sheets of metal, the seam actually is a series of overlapping spot welds, as shown in Figure 30-12. The basic equipment is the same as for spot welds, except that two rotating disks are used as electrodes, arranged as shown in Figure 30-13. As the metal passes between the electrodes, timed pulses of current pass through it to form the overlapping, elliptical welds. The timing of the welds and the movement of the work must be adjusted so that the workpieces do not get too hot. External cooling of the work, by air or water, often is used. A typical seam-welding machine is shown in Figure 30-14.

This type of seam welding is used primarily for the production of liquid- or pressure-tight vessels, such as gasoline tanks, automobile mufflers, or heat

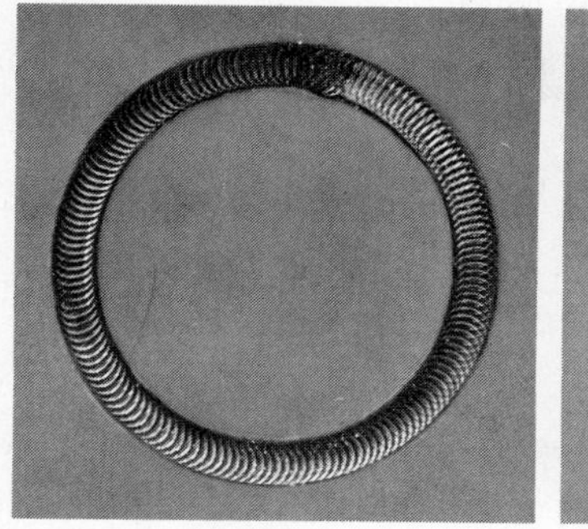

FIGURE 30-12. Seam welds made with overlapping spots of varied spacing. (*Courtesy The Taylor-Winfield Corporation.*)

exchangers. Special shapes of electrodes, such as the notched disks shown in Figure 30-15, can be used to produce intermittent seam welds and to adapt the processes to a wide variety of products.

The second type of resistance seam welding is being used increasingly to make butt welds between metal plates. In this process the electrical resistance of the abutting metal(s) is utilized, but a high-frequency current is employed—up to 450 kHz—that confines the heating adjacent to the surfaces to be joined. As they reach the welding temperature, the heated surfaces are progressively pressed together by passing through pressure rolls. The most extensive use of this process is in making pipe, using the arrangement illustrated in Figure 30-16. However, it is being used increasingly for making structural shapes from flat stock, as shown in Figure 30-17. Material from 0.13 mm (0.005 inch) to more than 12.7 mm (½ inch) in thickness can be welded at speeds up to 82 meters (250 feet) per minute. The combination of high-frequency current and high welding speed produces a very narrow heat-affected zone.

FIGURE 30-13. Schematic representation of seam welding.

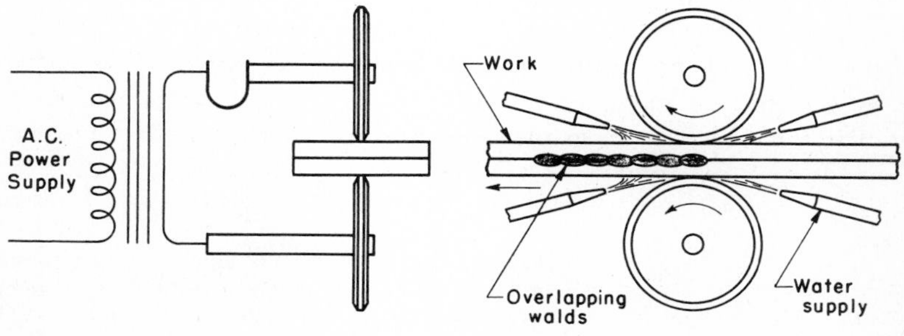

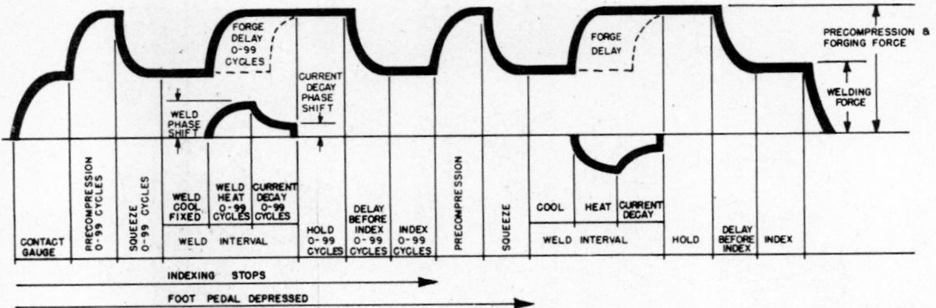

VARIABLE ELECTRODE FORCE — INTERMITTENT DRIVE — SINGLE IMPULSE WITH CURRENT DECAY FOR WELDING LIGHT ALLOYS

FIGURE 30-14. (*Left*) A 125-kVa seam welder. (*Top*) Typical cycle of electrode force versus current used in spot and seam welding. (*Courtesy Sciaky Bros., Inc.*)

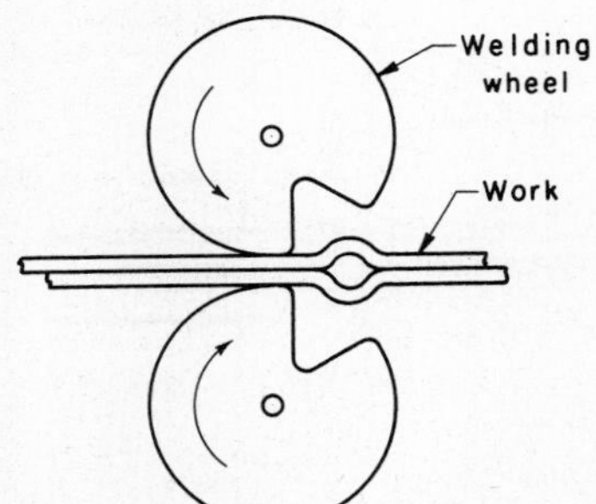

FIGURE 30-15. Method of using notched electrodes in seam welding to accommodate irregular sections in the work.

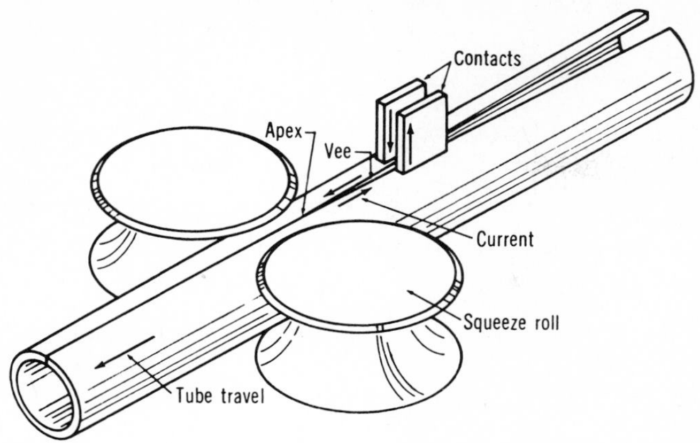

FIGURE 30-16. Method of making a seam weld in pipe by means of a high-frequency current as the heat source.

Projection welding. Two disadvantages of spot welding are that electrode maintenance is a considerable problem and, usually, only one spot weld is made at a time. If more strength, or attachment, is required than can be provided by one spot weld, several such welds must be made. *Projection welding* (RPW) provides a means of overcoming both of these disadvantages and thus is particularly well adapted to mass production.

The principle of projection welding is illustated in Figure 30-18. A *dimple* is embossed on one of the workpieces at the location where a weld is

FIGURE 30-17. Fabricating an I-beam from three plates by means of two simultaneous high-frequency resistance welds. (*Courtesy AMF Thermatool, Inc.*)

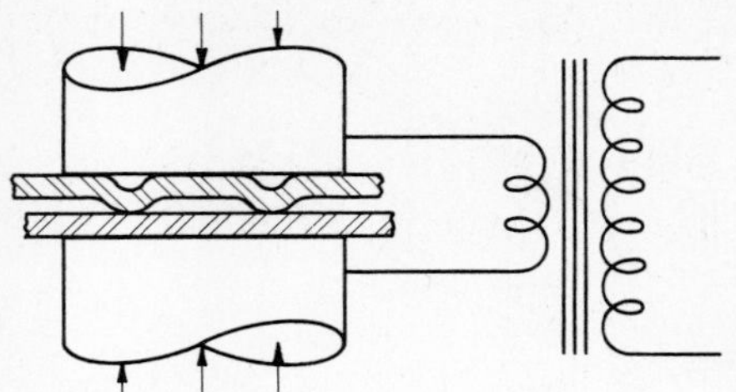

FIGURE 30-18. Principle of projection welding.

desired. The workpieces then are placed between plain, large-area electrodes in the projection-welding machine, and pressure and current are applied, as in spot welding. Because the contact area between the work and the electrodes is much greater than the area of the end of the dimple, nearly all the resistance of the circuit is at the dimple, and virtually all the heating occurs at the point where the weld is desired. As the metal in the dimple becomes plastic, due to the heating, the pressure causes the dimple to flatten as the weld is formed, and the workpieces are forced tightly together.

Several projection welds can be made at one time by having several dimples between the electrodes, the number being limited only by the capacity of the machine to provide the required current and pressure. Another important advantage is that the dimples, or projections, can be made almost any desired shape—such as round, oval, or circular rings—to produce welds of desired shapes as required for strength or location. Many spot-welding machines are convertible to projection welding by changing the electrodes.

Because the projections are formed on the work in punch presses, they often can be formed concurrently with other blanking or forming operations at virtually no cost. They should be shaped so that the weld forms outward from the center of each projection. Bolts and nuts are often attached to other metal parts by projection welding. Such bolts and nuts are available with the projections formed on them, ready for welding.

OTHER RESISTANCE WELDING PROCESSES

Electromagnetic solid-state welding. *Electromagnetic solid-state welding* is a special type of resistance welding in which, as in induction heating, high-frequency currents are induced in the pieces to be joined, by means of a magnetic field. This produces rapid heating at the interfaces. Large pulses of current then are passed through the workpieces, in parallel directions, to cause them to be pulled together as a consequence of the mutually attractive electromagnetic fields that are established. The combined heating and pressure produce solid-state coalescence. Although the process is not used extensively, it is very rapid and is useful for joining quite a range of similar and dissimilar metals, particularly combinations that are difficult to join by most welding processes.

Advantages and disadvantages of resistance welding. Resistance welding processes have a number of distinct advantages that account for their wide use, particularly in mass production:

1. They are very rapid.
2. The equipment is semiautomatic.
3. They are economical of material; no filler metal is required.
4. Skilled operators are not required.
5. Dissimilar metals can easily be joined.
6. A high degree of reliability and reproducibility can be achieved.

Resistance welding, of course, has some disadvantages, the principal ones being:

1. The equipment has a high initial cost.
2. There are limitations to the types of joints that can be made.
3. Skilled maintenance personnel are required to service the control equipment.
4. For some materials the surfaces must receive special preparation prior to welding.

However, it is evident that these disadvantages are not serious when resistance welding is to be done in considerable volume.

Review questions

1. Why is the interface resistance of such great importance in resistance welding?
2. How is shielding from the atmosphere provided in spot welding?
3. What problems are introduced in spot welding if the materials being joined have considerable difference in resistance?
4. What will occur if the temperature gets too high in spot welding?
5. Why must the shape and contour of the electrodes for spot welding be controlled?
6. Why should there be no melting of the material at the interface in making resistance welds?
7. Why is it difficult to spot-weld copper?
8. In what two ways may seam welds be produced by resistance welding?
9. What difficulties are encountered in spot welding a thin piece of material to a much thicker piece of the same material?
10. Why is the use of projection welding usually restricted to mass-production manufacturing?

11. What are three advantages of projection welding as compared with spot welding?

12. What would be the effect of using low-frequency current in the seam welding process depicted in Figure 30-16?

Case study 22.
THE DISPUTED BOLT FLANGE

The Weld-Rite Company received an order from the XO Corporation to fabricate two of the units shown in Figure CS-22. The order stated: "Units will be assembled together by XO Corp. by means of 3/4″-16 UNF-2 Hex. Hd. bolts and nuts."

After the XO Corporation received the units, it returned them to the Weld-Rite Company, stating that they were defective and refused to pay for them. The Weld-Rite Company rechecked the units and found they were completely within the specified tolerances and filed a lawsuit to collect payment. (1) Check to see if you can find anything wrong with the design of the units. (2) Should the Weld-Rite Company be paid for its work?

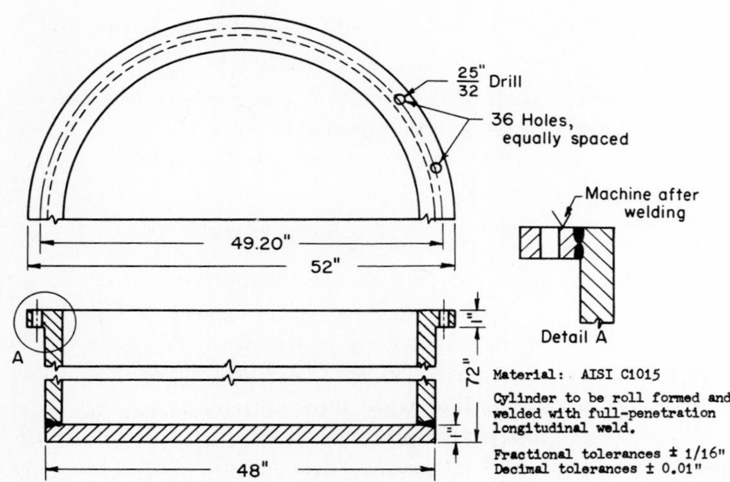

FIGURE CS-22. Design of bolting flange on processing vessel.

Miscellaneous welding and related processes

As indicated in Figure 29-1, there are a number of very useful welding processes that utilize heat sources other than an oxyfuel gas flame, electrical resistance, or an electric arc. Although some are quite old, others are among the newest of the welding processes. Most of them are suitable for only limited, but very important, applications.

THERMIT WELDING

Thermit welding (TW) is an old process and has been replaced by electroslag welding for some applications. However, it still is very effective and is used extensively for joining thick sections of steel, particularly where the contour varies, as in joining railroad rails or steel castings. It also is very useful in repairing large broken or cracked steel castings.

Heating and coalescence are produced by superheated molten metal and slag, obtained from the reaction between a metal oxide and aluminum. In addition to furnishing heat, the molten metal also supplies any required filler metal.

The thermit process utilizes a mechanical mixture of about one part of finely divided aluminum and three parts of iron oxide. When the mixture is ignited, it reacts according to the chemical equation

$$8Al + 3Fe_3O_4 = 9Fe + 4Al_2O_3 + \text{heat}$$

producing a temperature of over 2760°C (5000°F) in about 30 seconds. The ignition temperature of about 1150°C (2100°F) is supplied by a magnesium fuse.

The essential steps in welding by the thermit process are shown in Figure 31-1. The sections to be welded must be prepared to provide clearance between them. Wax is used to fill in the gap and is built up to form the desired shape of the weld, riser, and runner system—similar to the procedure of investment casting. A box is placed around the work and rammed with a material similar to molding sand to form a mold.

When the mold is completed, a heating torch is used to dry it and to melt the wax. Heating is continued until the faces of the work are at a red heat. One or more crucibles, filled with the thermit mixture, are set atop the mold, holes in their bottoms connecting with the runner system in the mold. The mixture in the crucibles is ignited, and the molten metal flows out of the crucible, filling the mold and at the same time supplying sufficient heat to raise the surfaces of the workpieces to a high-enough temperature to produce coalescence. After the deposited metal has cooled, the mold is removed. The weld then can be ground to final shape if desired.

ELECTROSLAG WELDING

Originated in Russia, *electroslag welding* (ESW) has been developed in the United States and several European countries to be a very effective process for welding thick sections of steel. The essential components are illustrated in Figure 31-2. The heat is derived from the electrical resistance of the molten slag into which the electrode wires dip. There is no arc involved. The process thus is entirely different from submerged arc welding. Also, the electrical resistance of the material being welded plays no part in producing heat. The temperature of the slag bath is up to 1760°C (3200°F) and thus is much higher than the melting points of the base metal and the electrodes. Consequently, the slag melts the edges of the pieces that are being joined and simultaneously melts the continuously fed electrodes, thereby supplying needed filler metal. Typically, there is about a 63.5 mm (2½ inch) depth of molten slag and 12.7 to 19 mm (½ to ¾ inch) of molten metal confined in the gap

FIGURE 31-1. Steps in repairing a large casting by thermit welding: (*left, top*) the broken casting; (*left, center*) crack gouged out ready for wax; (*left, bottom*) wax pattern in place; (*right, top*) thermit powder being melted and flowing into completed mold; (*right, bottom*) welding completed, gates and risers removed, and ready for finishing. (*Courtesy United Chromium Division, Metal & Thermit Corporation.*)

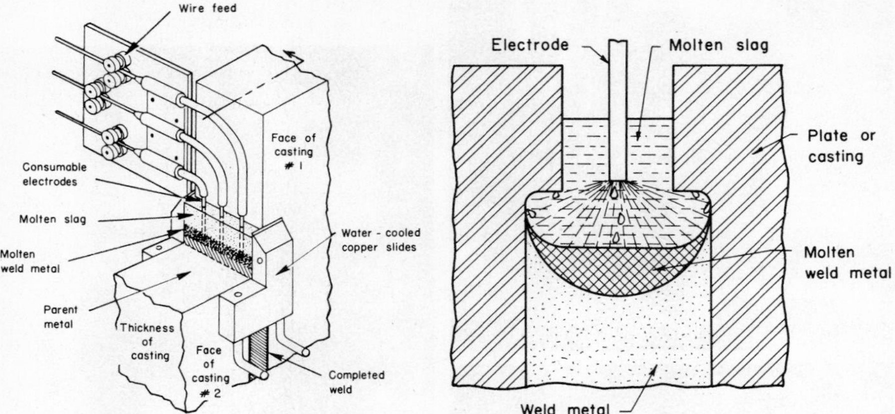

FIGURE 31-2. (*Left*) Arrangement of equipment and work for making a vertical weld by the electroslag process. (*Right*) Section through workpieces and weld during the making of an electroslag weld.

between the plates being welded by means of sliding, water-cooled plates. Obviously, the best conditions for maintaining a deep slag bath exist in vertical joints, so the process is used most frequently for this type of work. However, it also is used successfully for making circumferential joints in large pipe, using special curved slag-holder plates and rotating the pipe to maintain the area where welding is occurring in a vertical position.

Because very large amounts of weld metal and heat can be supplied, electroslag welding is the best of all the welding processes for making welds in thick plates. Thicknesses up to 457 mm (18 inches) in plain and alloy steels are welded without difficulty. Three electrodes commonly are used, often reciprocated in the direction of plate thickness.

Electroslag welding has virtually replaced thermit welding for welding plates or castings of uniform thickness, or if they taper uniformly. The choice between electroslag and submerged arc welding primarily is one of economics. The latter tends to be more economical for joints below about 63.5 mm (2½ inches), whereas the former is more economical for thicker joints.

FLASH WELDING

In *flash welding* (FW) the two surfaces to be welded are rapidly brought into light contact while connected to an electrical power source, as indicated in Figure 31-3. As they make contact, a large current passes through the joint and, because the interfaces are not absolutely smooth, some momentary arcing occurs. Consequently, the resistance to current flow and the incidental arcing

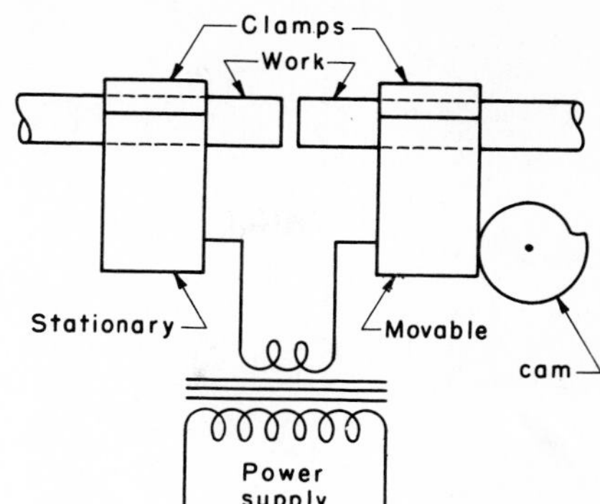

FIGURE 31-3. Principle of the flash-welding process.

produce rapid heating of the metals to a welding temperature, after which they are forced together with considerable pressure, resulting in simultaneous coalescence over the entire area of abutment. This action usually is accompanied by some upsetting of the softened metal. The current flow usually is sustained until the entire process is completed.

The process involves the following six steps:

1. Clamp the work in the machine.
2. Apply the welding voltage.
3. Bring the parts into light contact to establish current flow and cause *flashing* to occur.
4. Upset the parts by applying a high pressure when flashing has caused them to reach plastic temperature.
5. Cut off the welding current.
6. Unclamp the work.

The upsetting action should be sufficient to cause the plastic metal in the abutting surfaces to be displaced outward radially, so that all impure metal is squeezed out into the fin, or *flash,* that is formed. This assures sound metal in the weld. Figure 31-4 depicts the relationship among the current, work (or platen) travel, and time in making a flash weld. Figure 31-5 shows parts before and after being flash-welded. In addition to the six basic steps, preheat and postheat steps can be added in order to improve the metallurgical properties.

Flash welding can be employed for butt welding solid or tubular metals. It is widely used in manufacturing such products as tubular metal furniture, metal windows, and pipe. Except for very small sections, the equipment required is rather large and expensive, but excellent welds can be made at high production rates, making the process well suited to mass production. Although the resulting flash usually must be removed, in most cases no surface preparation is required.

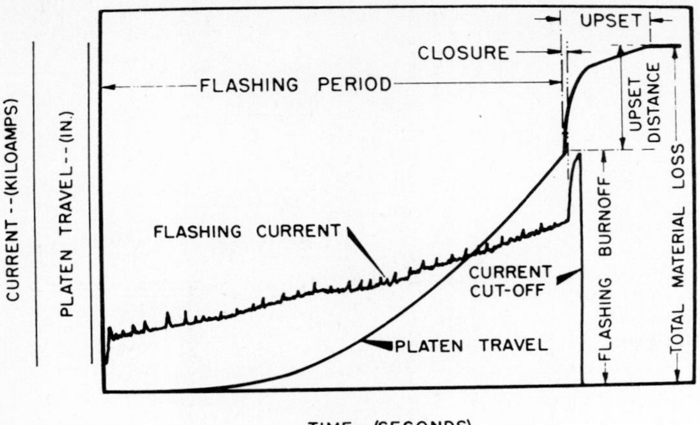

FIGURE 31-4. Relationship of current, platen travel, upset, and time in flash welding. (*Courtesy The Welding Journal.*)

ELECTRON BEAM WELDING

Electron beam welding (EBW) is a fusion welding process in which heating results from the impingement of a beam of high-velocity electrons on the metal to be welded. Originally developed for obtaining ultra-high-purity welds in reactive and refractory metals in the atomic energy and rocketry fields, its unique qualities have led to its substantial use in numerous applications.

The electron optical system employed is shown in Figure 31-6. The electrons must be generated and focused in a very high vacuum, and the welding usually is performed at a pressure of from 0.13 to 133 mPa. However, the process can be adopted to weld in pressures of from 0.13 to 13 Pa or even at atmospheric pressure, but the penetration of the beam and the depth-to-width ratio are reduced as the pressure increases.

A high-voltage current heats the tungsten filament to about 2204°C

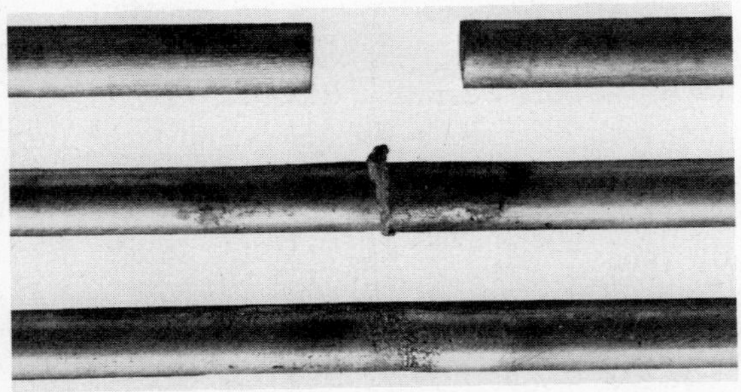

FIGURE 31-5. Flash-welded parts: (*top*) before welding; (*center*) welded, showing flash; (*bottom*) flash ground smooth.

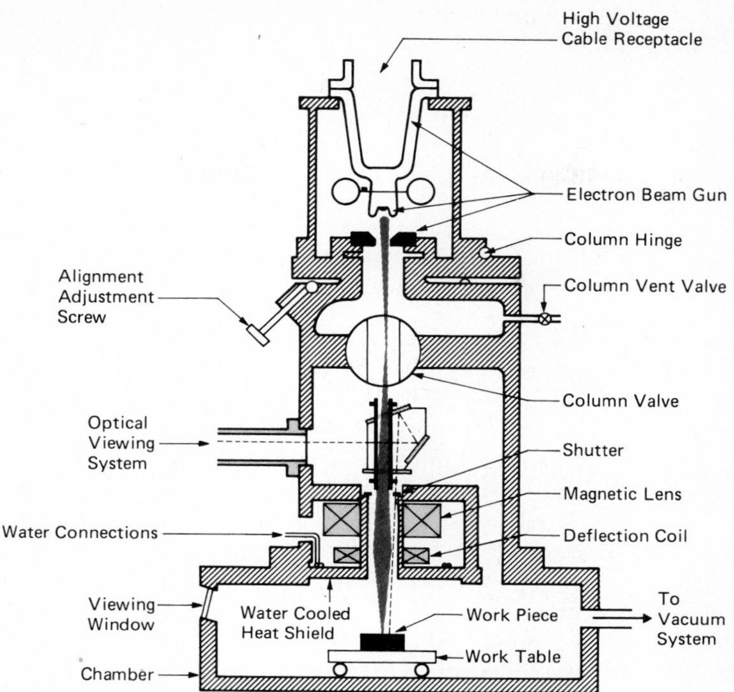

FIGURE 31-6. Schematic diagram of the electron beam welding process. (*Courtesy American Machinist.*)

(4000°F), causing it to emit high-velocity electrons. By means of a control grid, accelerating anode, and focusing coils, the electrons are converted into a concentrated beam and focused onto the workpiece in a spot from 0.79 to 3.18 mm (1/32 to 1/8 inch) in diameter. Usually the work is enclosed and moved under the electron gun in the vacuum chamber. Under these conditions, the vacuum assures degassification and decontamination of the molten weld metal, and very high quality welds can be obtained. However, the size of the vacuum chamber required naturally imposes serious limitations on the size of the workpiece that can be accommodated. As a consequence, electron-beam welding machines have been developed that permit the workpiece to remain outside the vacuum chamber. In these machines the electron beam emerges through a small orifice in the vacuum chamber to strike the adjacent workpiece. High-capacity vacuum pumps take care of the leakage through the orifice. Although these machines have some of the advantages of the total-vacuum types, because they do not operate at as low pressure they do not have as great penetrating power and, as shown in Figure 31-7, they produce considerably wider welds.

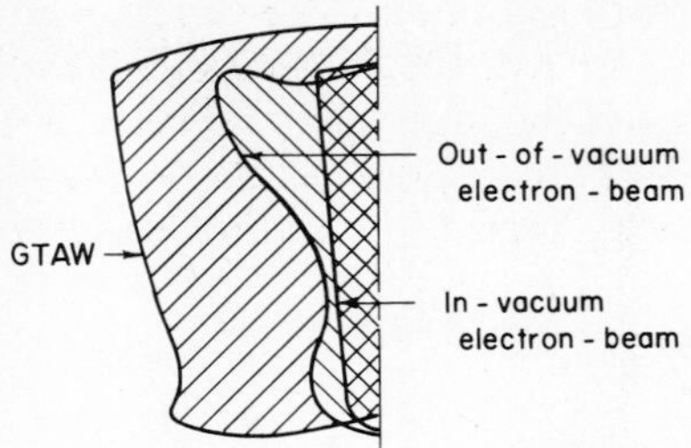

FIGURE 31-7. Comparison of weld widths of GTAW, out-of-vacuum electron beam, and in-vacuum electron beam welds.

In general, two ranges of voltages are employed in electron-beam welding. High-voltage equipment employs 50 to 100 kilovolts. This type provides a smaller spot size and greater penetration than does the lower-voltage type, which uses from 10 to 30 kilovolts. However, the high-voltage units, with their high electron velocities, emit considerable quantities of harmful X rays and thus require expensive shielding and indirect viewing systems for observing the work. Such X rays as are produced by the low-voltage types are sufficiently soft to be absorbed by the walls of the vacuum chamber, and they have adequate spot concentration and penetration for most applications. They also are less critical in adjustment. With them the work can be observed directly through viewing ports.

Materials that are difficult to weld by other processes, such as zirconium, beryllium, and tungsten, can be welded successfully by electron-beam welding, but the weld configuration should be simple and preferably flat. As shown in Figure 31-8, very narrow welds can be obtained as well as remark-

FIGURE 31-8. (*Left to right*) Electron beam welds in 7079 aluminum, thick stainless steel, and a multiple-tier weld in stainless steel tubing. (*Courtesy Hamilton Standard Division of United Aircraft Corp.*)

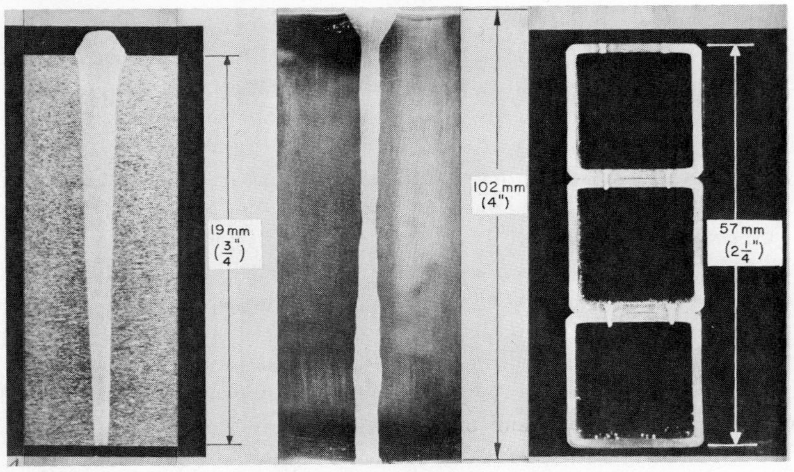

able penetrations, even through nonadjacent sections of metal. The equipment is quite expensive, and highly skilled labor is required. Obviously, the process is one to be employed primarily where extremely high quality welds are required or where other welding processes will not produce the required results. However, its unique capabilities have resulted in its routine use in a surprising number of applications.

ULTRASONIC WELDING

Ultrasonic welding (USW) is a solid-state process wherein coalescence is produced by localized application of very high frequency vibratory energy to the workpieces as they are held together under pressure. The basic components of the process are shown in Figure 31-9. Although there is some increase in temperature at the faying surfaces, it always is far below the melting points of the materials. It appears that the rapid reversals of stress along the bond interface play an important role in facilitating coalescence by breaking up and dispersing the mating-surface films.

The ultrasonic transducer used is essentially the same as is employed in ultrasonic machining and shown in Figure 25-49. It is coupled to a force-sensitive system that contains a welding tip on one end. The pieces to be welded are placed between this tip and a reflecting anvil, thereby concentrating the vibratory energy within the work. Either stationary tips, for spot welds, or rotating disks, for seam welds, can be used.

Ultrasonic welding is restricted to joining thin materials—sheet, foil, and wire; the maximum is about 2.54 mm (0.1 inch) for aluminum and 1.02 mm (0.04 inch) for harder metals. However, as indicated in Table 31-1, it is particularly valuable in that numerous dissimilar metals can readily be joined by the process. Because the temperatures involved are low, and no current

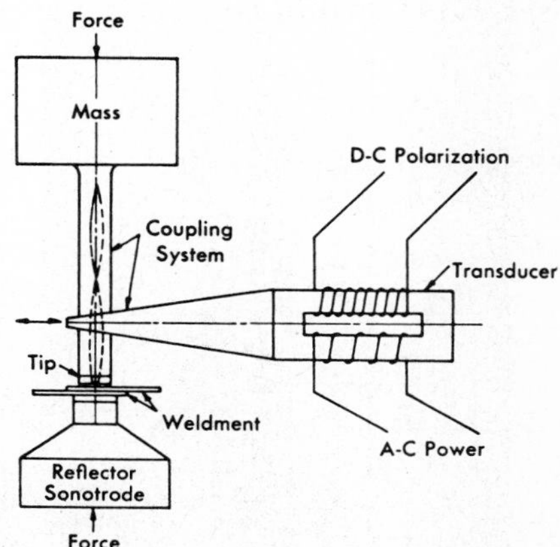

FIGURE 31-9. Schematic diagram of the equipment used in ultrasonic welding.

TABLE 31-1. Metal combinations weldable by ultrasonic welding

Metal	Aluminum	Copper	Germanium	Gold	Molybdenum	Nickel	Platinum	Silicon	Steel	Zirconium
Aluminum	X	X	X	X	X	X	X	X	X	X
Copper		X		X		X	X		X	X
Germanium			X	X		X	X	X		
Gold				X		X	X	X		
Molybdenum					X	X			X	X
Nickel						X	X		X	X
Platinum							X		X	
Silicon										
Steel									X	X
Zirconium										X

flow is involved and no arcing, intermetallic compounds seldom are formed, and there is no contamination of surrounding areas. The equipment is simple and reliable, and only moderate skill is required of the operator. However, it usually is not economically competitive when other processes can be used.

FRICTION WELDING

The heat for *friction welding* (FRW)—sometimes called *inertia welding*—is the result of mechanical friction between two abutting pieces of metal that are held together while one rotates and the other is held stationary. Two basic procedures are used. In one, as illustrated in Figure 31-10, the moving part is held and rotated in a motor-driven collet while the stationary part is pressed

FIGURE 31-10. Arrangement of equipment for inertia (friction) welding. (*Courtesy Materials Engineering.*)

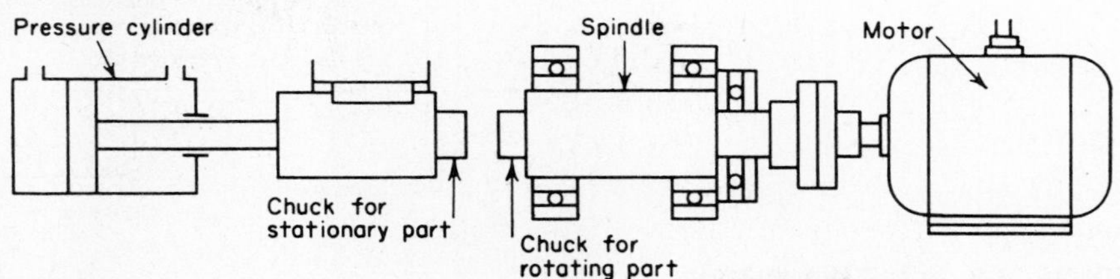

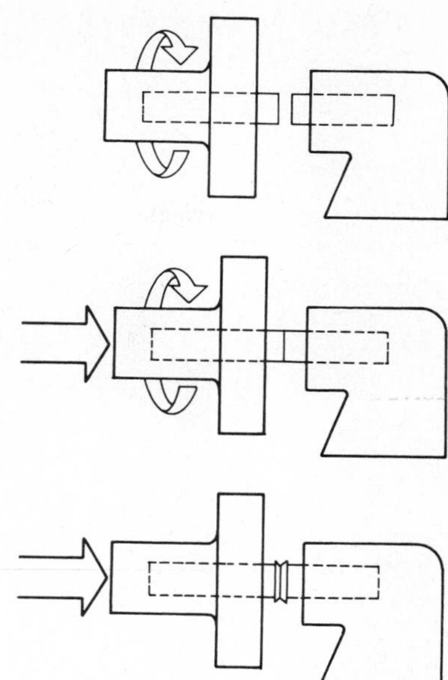

FIGURE 31-11. Schematic representation of the three steps in inertia welding.

against it with sufficient pressure so that the friction quickly generates enough heat to raise the abutting surfaces to welding temperature. As soon as the welding temperature is reached, rotation is stopped, and the pressure is maintained or increased until the weld is completed.

In the second (inertia) process, as indicated in Figure 31-11, the one workpiece is gripped in a rotating flywheel, with the kinetic energy of the flywheel being converted into heat by pressing the two workpieces together when the flywheel has attained the desired velocity. Figure 31-12 shows an example of the equipment used and the relationship between surface velocity, torque, and upset.

In both procedures the total cycle time for a weld usually is less than 25 seconds, whereas the actual time for heating and welding is about 2 seconds. No material is melted. Because of the very short period of heating—and thus the lack of time for the heat to flow away from the joint—the weld and heat-affected zones are very narrow. Surface impurities are displaced radially into a small flash that can be removed after welding if desired. Because virtually all the energy used is converted into heat, the process is very efficient. However, its use is restricted to joining round bars and tubes of the same size, or to joining bars and tubes to flat surfaces. The ends of the workpieces must be cut true and fairly smooth. It is used primarily for joining pieces of the same metal, but some dissimilar metals, such as aluminum and magnesium, also can be joined successfully.

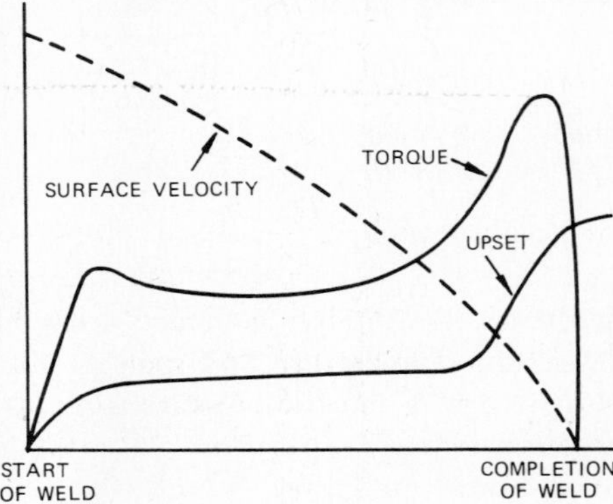

FIGURE 31-12. (*Top*) Inertia-type friction-welding machine and welded part. (*Bottom*) Relationship between surface velocity, torque, upset, and time in friction welding. (*Courtesy Production Technology, Inc.*)

LASER BEAM WELDING

The heat source in *laser beam welding* (LBW) is a focused laser beam, usually providing power intensities in excess of 10 kilowatts per square centimeter, but with low heat input—0.1 to 10 joules. The high-intensity beam produces a very thin column of vaporized metal, extending into the base metal. The column of vaporized metal is surrounded by a liquid pool, which moves along as welding progresses, resulting in welds having depth-to-width ratios greater than 4:1. Laser beam welds are most effective for autogenous welds, but filler metal can be added.

Deep-penetration welds produced by lasers are similar to electron-beam welds, but LBW has several advantages:

1. A vacuum environment is not required.
2. No X rays are generated.
3. The laser beam is easily shaped and directed with reflective optics.
4. Because only a light beam is involved, there does not need to be any

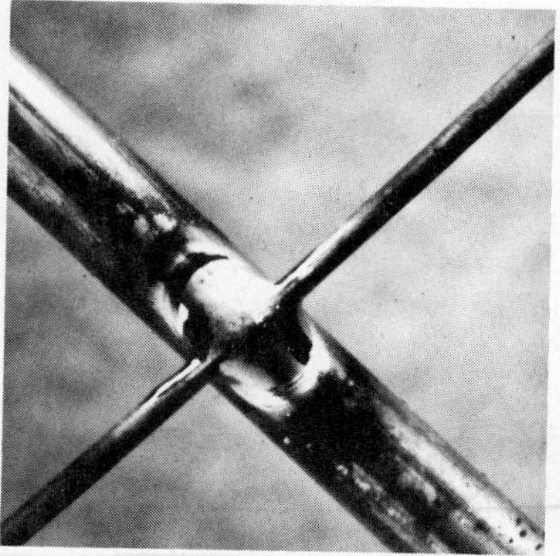
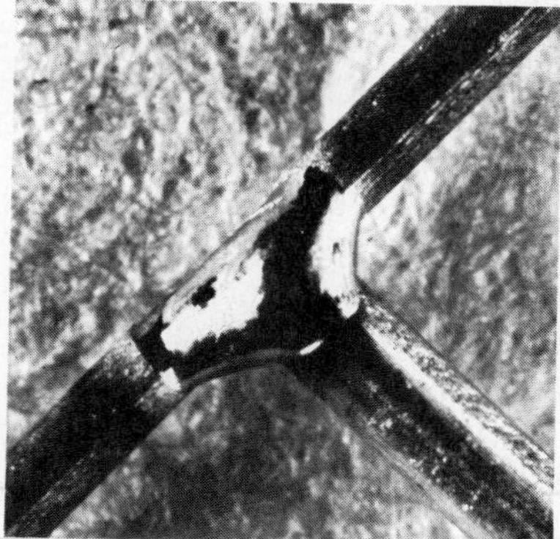

FIGURE 31-13. Typical welds made by laser welding. (*Courtesy Linde Division, Union Carbide Corporation.*)

physical contact between the workpieces and the welding equipment, and the beam will pass through transparent materials, permitting welds to be made inside transparent containers.

Because laser beams are concentrated, laser welds are small—usually less than 0.025 mm (0.001 inch). They thus are very useful in the electronics industry in connecting leads on small electronic components and in integrated circuitry. Lap, butt, tee, and cross-wire configurations are used. It also is possible to weld lead wires having polyurethane insulation without removing the insulation. The laser evaporates the insulation and completes the weld. Figure 31-13 shows examples of laser welds.

The equipment required for laser beam welding is high in cost but usually is designed for use by semiskilled workers. Thus its use is restricted to specialized applications.

EXPLOSION WELDING

Explosion welding (EXW) is used primarily for bonding sheets of corrosion-resistant metals to heavier plates of base metals (*cladding*), particularly where large areas are involved. An explosive material, usually in the form of a sheet, is placed on top of the two layers of metal and detonated progressively. A compressive stress wave, on the order of thousands of megapascals, progresses across the surface of the plates, so that a small open angle is formed between the two colliding surfaces. Surface films are liquefied or scarfed off the surfaces and jetted out of the interface, leaving clean surfaces which coalesce under the high pressure. The result is a cold weld having a wavy configuration at the interface.

DIFFUSION WELDING

Diffusion welding (DFW) occurs in the solid state when properly prepared surfaces are maintained in contact under sufficient conditions of pressure, temperature, and time. In one procedure, stacks of wrought sheets that have been cut to shape are stacked in a die cavity, and pressure is applied through a ram while the die is maintained at an elevated temperature—titanium, for example, at about 871°C (1600°F) at 13.8 MPa (2000 psi). By using the proper combination of temperature and pressure, grain growth is avoided and the wrought properties of the sheets are retained, permitting the resulting product to be substituted for a forging. Thus the need for extremely large forging presses is avoided.

Another procedure is to encapsulate the stacked sheets. Sometimes the sheets may be preplated with a filler metal to lower the required temperature or to avoid having to fill the capsule with an expensive atmosphere. In all cases, the joints must be tightly contacting surfaces.

SURFACING

Surfacing is the deposition of a layer of metal of one composition upon the surface or edge of a base metal of a different composition. The usual objectives are to obtain improved resistance to wear, abrasion, or chemical reactions. Gas-flame, arc, or plasma-arc methods are used. The process often is called *hard facing,* because the deposited surfaces usually are harder than the base metal. However, this is not always true. In some cases a softer metal, such as bronze, is applied to a harder base metal.

Surfacing metals. The metals most commonly used for surfacing fall into seven categories: (1) air-hardenable carbon and low-alloy steels, (2) high-alloy steels and irons, (3) cobalt-base alloys, (4) carbides, (5) copper-base alloys, (6) nickel-base alloys, such as Monel, Nichrome, and Hastelloy, and (7) stainless steels.

Surfacing methods and application. Nearly all the surfacing metals can be deposited by either gas-flame or arc welding. Arc welding is used more frequently, particularly for depositing the high-melting-point alloys. Submerged arc welding can be employed when large areas are to be surfaced, as illustrated in Figure 31-14.

The use of the plasma arc torch[1] greatly extends the capabilities of surfacing, because of its very high temperatures. To obtain true fusion of the surfacing material, a transferred arc is used. The surfacing metal is in the form of

[1] See Figure 32-9.

FIGURE 31-14. Applying a hard-surfacing layer to a cylindrical drum by the submerged arc process. (*Courtesy Linde Division, Union Carbide Corporation.*)

powder. A nontransferred arc can be used, but this provides only a mechanical bond. This is a form of *metallizing,* which will be discussed next.

METALLIZING

In metallizing, surfacing materials are melted and atomized in a special torch, or gun, and sprayed onto a base material. The process sometimes is called metal spraying. The gun, illustrated in Figure 31-15, usually employs an oxy-acetylene flame, but other gases sometimes are used. As a wire of the surfacing metal is fed automatically through the center of the flame, it melts and simultaneously is atomized by a stream of compressed air and blown onto the base

FIGURE 31-15. Schematic diagram of a metal-spraying gun. (*Courtesy Metallizing Engineering Company, Inc.*)

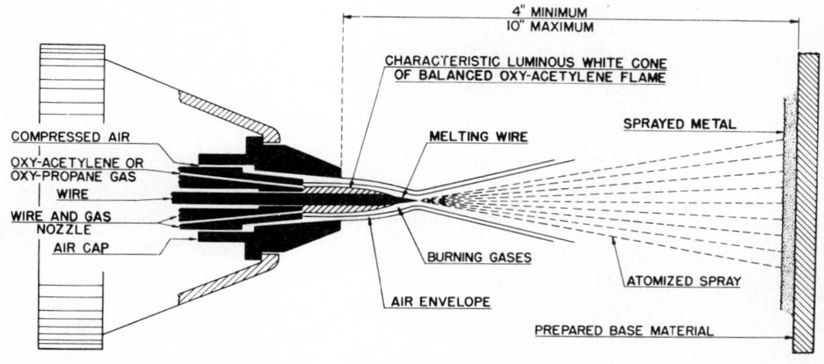

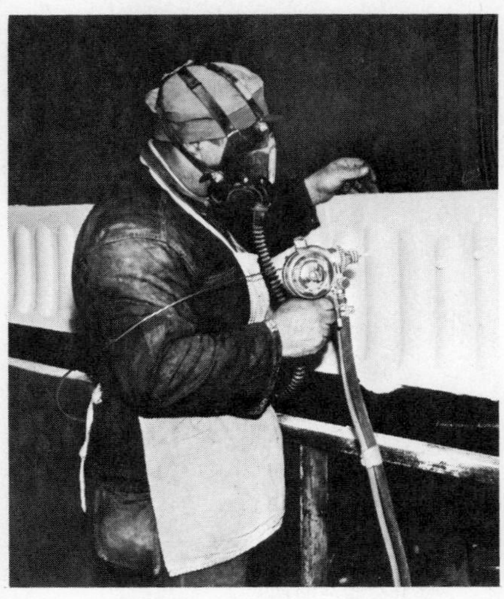

FIGURE 31-16. Spraying metal to construct a nest block for an airplane jig. (*Courtesy Metallizing Engineering Company, Inc.*)

material. Almost any kind of metal that can be made into a wire can be deposited by this procedure.

Another type of gun utilizes materials in the form of powders, which are blown through the flame. This has the advantage that not only metals but also cermets, oxides, and carbides can be sprayed.

The spray gun can be manipulated manually, as shown in Figure 31-16, or it can be held and fed mechanically, as shown in Figure 31-17. A distance of from 152 to 254 mm (6 to 10 inches) usually is maintained between the nozzle of the gun and the work.

Surface preparation. Because the bond obtained between the deposited and base materials by metallizing is purely mechanical, it is essential that the base

FIGURE 31-17. Spraying martinsitic stainless steel to improve the wear resistance of an integral piston rod, rotating the workpiece in a lathe. Gun shown uses metal wire. Other types use metal powder. (*Courtesy Wall Colmonoy Corporation.*)

FIGURE 31-18. Method of preparing surfaces for metal spraying by machining grooves and rolling the edges.

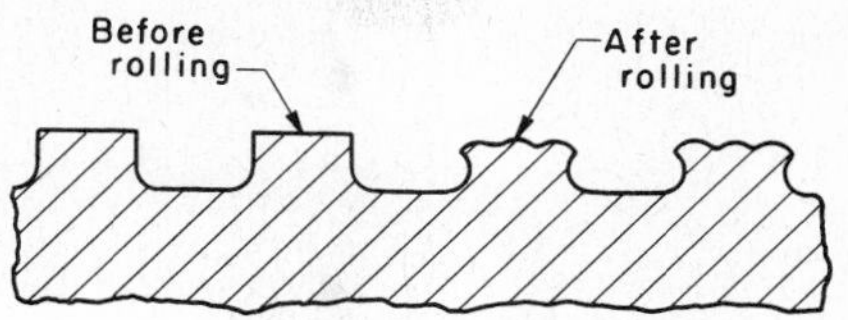

be prepared properly so that good mechanical interlocking can be obtained. With any method of surface preparation, the base surface must be clean and free of oil.

The most common method of surface preparation is grit blasting. For this purpose the grit should be sharp enough to produce a really rough surface. On cylindrical surfaces that can be rotated in a lathe, an effective method is to turn very rough threads and then roll the crests over slightly with a knurling tool. A modification that can be used on flat surfaces is to cut a series of parallel grooves, using a rounded gooving tool, and then roll over the lands between the grooves with a knurling tool, as indicated in Figure 31-18. Figure 31-19 shows a comparison of the surfaces obtained by grit blasting and rough machining. If the sprayed surface is to be machined, the base should be prepared by either rough machining or grooving so as to provide maximum interlocking.

Characteristics of sprayed metals. The properties of sprayed metals are what would be expected from the fact that they are broken into fine, molten particles, mixed with air, and then cooled rapidly upon hitting the base material. As a result, such coatings contain particles of oxidized metal and are harder, more porous, and more brittle than in the wrought state. In general, they

FIGURE 31-19. Surfaces obtained by two methods of preparation for metal spraying. (*Courtesy Metallizing Engineering Company, Inc.*)

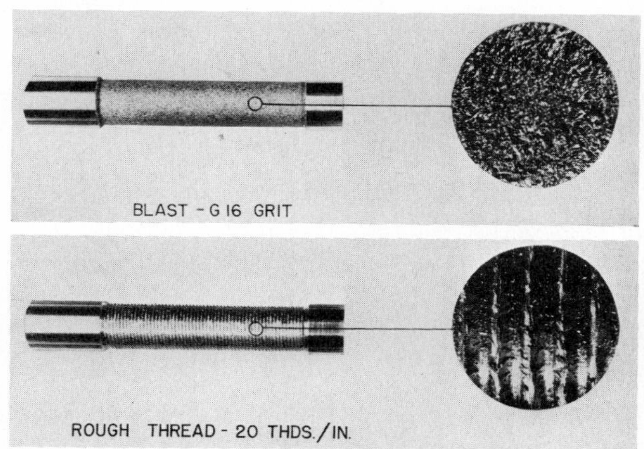

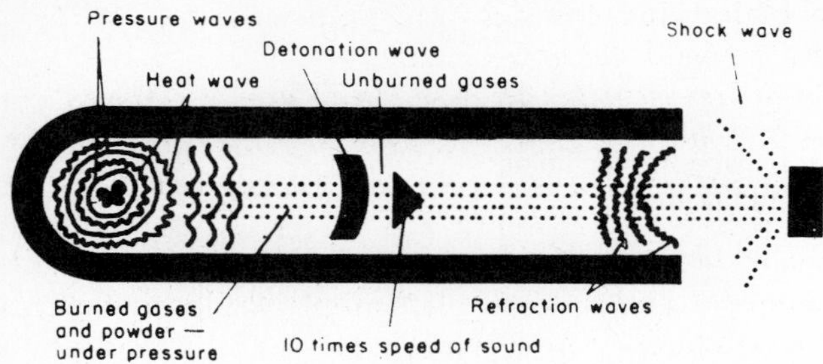

FIGURE 31-20. Schematic diagram of the flame-plating process by means of a detonation gun. (*Courtesy Linde Division, Union Carbide Corporation.*)

have many of the characteristics of cast metals. However, these properties make sprayed metals particularly well suited to resist abrasion and wear and to serve as bearing surfaces. The characteristic porosity retains lubricants, which adds to their quality as lubricated wearing surfaces.

Although sprayed coatings have only about 85 to 90 per cent the density and about one-third to one-half the strength of wrought metals, the electrical conductivity is nearly as good.

Flame plating. Another method of applying coating materials to base metal is through the use of the detonation gun, illustrated in Figure 31-20. A mixture of oxygen and acetylene and particles of the coating material, in powder form, is introduced into the gun and ignited. A detonation occurs and the resulting detonation wave hurls the powder particles out of the barrel and causes them to embed in the surface of the base material. Because some microscopic welding action occurs, producing a tenacious bond, the results are different from those obtained by the metallizing processes, in which only a mechanical bond is involved. This process commonly is called *flame plating.* The coatings have low porosity and a smoothness of about 3.2 μm (125 microinches), as coated, that can be improved by ordinary polishing methods. Tungsten carbide and aluminum oxide are the most common plating materials.

Applications of metal spraying. Metal spraying has a number of important applications.

1. *Protective coatings.* Zinc and aluminum are sprayed on steel and iron to provide corrosion resistance.

2. *Building up worn surfaces.* Worn parts can be salvaged by adding metal to desired areas by spraying.

3. *Hard surfacing.* Although metal spraying is not to be compared with hard surfacing by depositing weld metals, it is useful where thin coatings are adequate.

4. *Applying coatings of expensive metals.* Metal spraying provides a simple

method for applying thin coatings of noble metals to small areas where conventional plating would not be economic.

5. *Electrical conductivity.* Because metal can be sprayed on almost any surface, metal or nonmetal, it provides a simple means of supplying a conductive surface on a poor or nonconductor. Copper or silver frequently is sprayed on glass or plastics for this purpose.

6. *Reflecting surfaces.* Aluminum, sprayed on the back of glass by a special fusion process, makes an excellent reflecting surface that is used in traffic markers and for similar applications.

7. *Decorative effects.* One of the earliest and still important uses of metal spraying was to obtain decorative effects. Because sprayed metal can be treated in a variety of ways, such as buffed, wire-brushed, or left in the as-sprayed condition, it is used in both products and architectural works as a decorative device.

FLAME STRAIGHTENING

Basically, *flame straightening* is the creation of controlled, localized upsetting in order to straighten warped or buckled plates. The theory of the process is illustrated in Figure 31-21. If a straight piece of metal is heated in a localized area, as indicated by the curved path in the upper diagram, the material adjacent to *b* will be upset due to being restrained from lengthening by the remaining cool metal. When the upset portion cools, it will contract and the piece will be shortened on edge *b* and become bent, as shown in the lower diagram.

If the same procedure is used on a bent or warped piece of material, using the heated path shown in the lower diagram of Figure 31-21, the piece may be straightened because of the upsetting and subsequent contraction of the material near side *a*′. This procedure has been used to straighten various structures that have had members bent through accident.

A similar process can be used to flatten metal plates that have become dished due to buckling. In this case spots about 51 mm (2 inches) in diameter

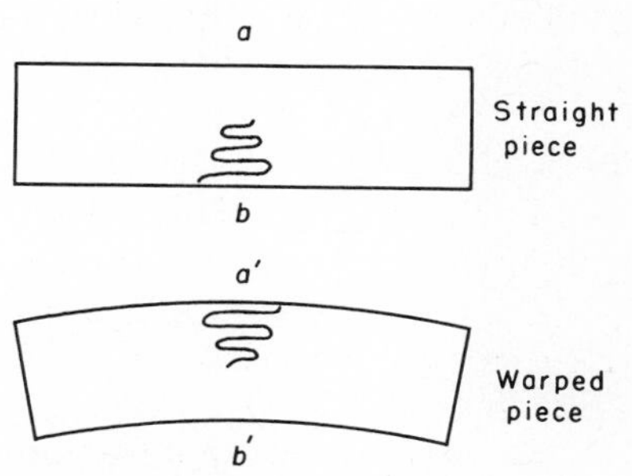

FIGURE 31-21. Theory of flame straightening.

are heated quickly to upsetting temperature so that the surrounding metal remains relatively cool. Cool water then is sprayed on the plate and the contraction of the upset spot straightens the buckle over a considerable area. To remove large buckles it is necessary to repeat the process on several spots.

Flame straightening cannot be used on thin material; the metal adjacent to the heated area must have sufficient rigidity to resist transferring the buckle from one area to another.

WELDING OF PLASTICS

Plastics of the thermoplastic type can be welded successfully using either a hot-gas torch of the type shown in Figure 31-22 or an electrically heated tool. In the hot-gas torch, a gas—usually air—is heated by an electrical coil as it passes through the torch. Electrical tools are similar to an electric soldering iron and are moved in contact with the material until the desired temperature is achieved. The heating is localized, at from 246 to 357°C (475 to 675°F), until the plastic softens. Some pressure then is applied to produce coalescence.

V-groove butt or fillet welds usually are employed. Because the plastic cannot be made to flow as in fusion welding of metals, some filler material usually has to be added. This is done by using a plastic filler rod that is heated simultaneously with the workpieces and then pressed or stretched into the joint, thus supplying some of the pressure needed to complete coalescence. This procedure is shown in Figure 31-22.

Heated-tool welding usually is employed for making lap-seam welds in

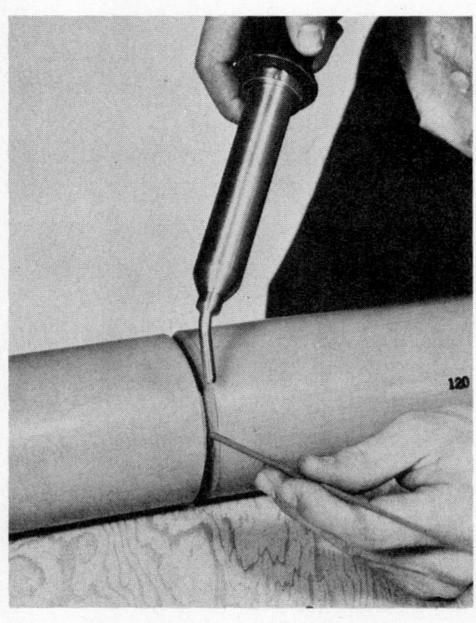

FIGURE 31-22. Using a hot-gas torch to make a weld in plastic pipe.

flexible plastic sheets. Pressure is applied by a roller, or other pressure device, after the material has been heated.

Butt welds sometimes are made in plastic pipe and rods by friction welding.

Review questions

1. What is the heat source for thermit welding?
2. For what type of work is thermit welding best suited?
3. One leg of a steel casting having a rectangular section 50 x 203 mm is cracked completely through. Why would it be necessary to vee-cut a considerable amount of metal adjacent to the crack before it could be repaired by thermit welding?
4. How does electroslag welding differ basically from a submerged arc welding?
5. Why can electroslag welding not be used where the plates to be joined vary irregularly in thickness?
6. What primarily determines the choice between using electroslag or submerged arc welding?
7. How is contamination and surface oxidation taken care of in flash welding?
8. What are two difficulties in the use of electron beam welding?
9. For what types of applications is electron beam welding best suited?
10. What is the principal role of the vibrations in ultrasonic welding?
11. On what types of material is ultrasonic welding usually employed?
12. Explain why friction welding is the most efficient welding process.
13. Would friction welding be suitable for joining two hexagonal rods of the same size? Why?
14. What are three advantages of laser beam welding?
15. Why can laser welds be made without contact between the workpieces and the equipment?
16. For what type of work is explosion welding most commonly used?
17. Why must the mating surfaces be carefully prepared in diffusion welding?
18. For what type of product can diffusion-welded products often be substituted?
19. How does surfacing differ from metallizing?
20. What general characteristics do sprayed metals usually have?
21. Explain why surfaces that are to be metallized should receive careful preparation.
22. What methods commonly are used to prepare surfaces for metallizing?
23. How does flame plating differ basically from metallizing?
24. What is the basic theory involved in flame straightening?
25. Why can flame straightening not be applied successfully to thin materials?
26. Why can thermosetting plastics not be welded?
27. Why is a regular gas-flame torch not suitable for welding plastics?

Case study 23.
THE CRACKING WING PLANKS

As shown in Figure CS-23(a), the Logjam Aircraft Company machined wing planks from aluminum-alloy extrusions and then joined the plank units as indicated at (b), using rows of spot welds and a sealant. The outer skin of the plank was the upper wing surface. As indicated, stress corrosion cracks were experienced, resulting in costly repairs. Develop a modified design for these wing planks that will eliminate this type of failure and should be no more costly to manufacture.

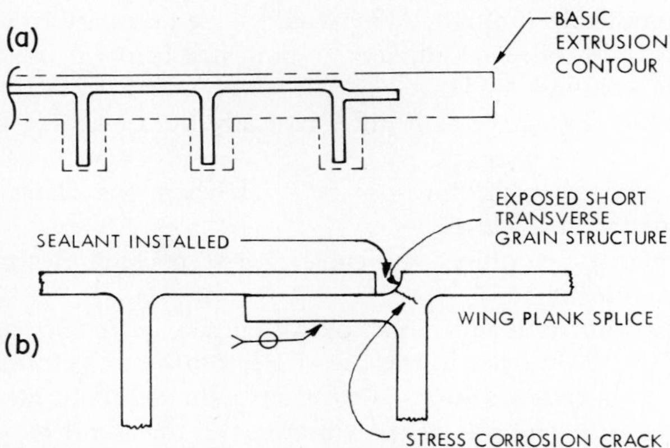

FIGURE CS-23. (a) Schematic showing method for making aircraft wing plank. (b) Method of splicing wing plank sections, and location of in-service cracks.

CHAPTER 32

Torch and arc cutting

For many years, metal sheets and plates have been cut by means of oxyfuel torches and electric arc equipment. Developed originally for use in salvage and repair work, then later for preparing plates for welding, these processes are now widely used for cutting metal sheets and plates into desired shapes for assembly and other processing operations. In recent years the development of laser and electron beam equipment has made possible the cutting of both metals and nonmetals, with cutting speeds up to 25.4 meters (1000 inches) per minute. Accuracies up to 0.254 mm (0.010 inch) are readily attainable, and speeds of 1.27 meters (50 inches) per minute are common. Figure 32-1 shows the commonly used torch and arc cutting processes with their AWS designations.

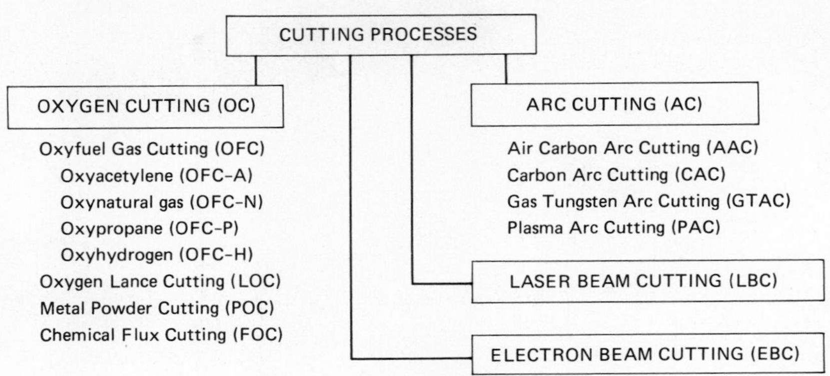

FIGURE 32-1. Common cutting processes and their AWS designations.

OXYGEN TORCH CUTTING

Oxyfuel gas cutting. By far the majority of metal cutting is done by *oxyfuel gas cutting* (OFC). In a few cases, primarily where the metal is not steel, the metal is merely melted by means of the flame of the oxyfuel gas torch and blown away to form a gap, or cut, in the metal. However, where steel is being cut, the process is one of rapid oxidation (burning) of iron at high temperatures according to the chemical equation

$$3\ Fe + 2O_2 = Fe_3O_4 + \text{heat}$$

Because this reaction does not occur until steel is at approximately 871°C (1600°F), a gas-flame torch is used to raise the metal to the temperature at which burning will start. After combustion starts, the torch continues to supply the oxygen required for the reaction. Theoretically, no further heat is required, but in most cases some additional heat must be supplied to compensate for heat losses into the atmosphere and the surrounding metal and thus to assure that the reaction progresses, particularly in the desired direction. Under some ideal conditions, where the area of burning is well confined so as to conserve the heat of combustion, as in cutting through a steel ingot, no supplementary heating is required, and a supply of oxygen through a small pipe will keep the cut progressing. This is known as *oxygen lance cutting* (LOC), and a temperature of about 1204°C (2200°F) has to be achieved in order for this procedure to be effective.

Fuel gases for oxyfuel gas cutting. Acetylene is by far the most common fuel used in oxyfuel gas cutting; thus the process often is called oxyacetylene cutting (OFC-A). The type of torch commonly used is shown in Figure 32-2. The tip contains a circular array of small holes through which the oxygen–acetylene mixture is supplied for the heating flame. A larger hole in

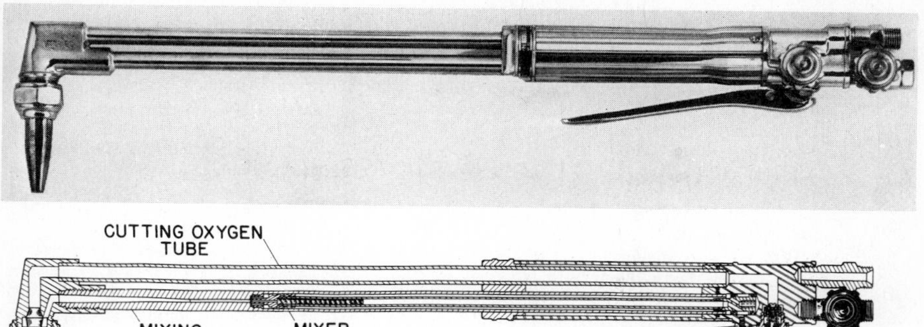

FIGURE 32-2. Oxyacetylene cutting torch. (*Courtesy Victor Equipment Company.*)

the center supplies a stream of oxygen, controlled by a lever valve. The rapid flow of the cutting oxygen not only produces rapid oxidation but also blows the oxides from the cut. If the torch is adjusted and manipulated properly, a smooth cut results, as shown in the top example of Figure 32-3.

The cutting torch can be manipulated manually. However, in most manufacturing it is moved over the desired path by mechanical means, since this procedure yields better accuracy and much smoother cut surfaces. Figure 32-4 shows the type of portable, electrically driven carriage that commonly is used. For straight cuts the device moves along a section of portable track. As shown, it also can be adjusted to cut circles of various radii. Where duplicate or more complex shapes are required, a pantograph-type machine, shown in Figure 32-5, is widely used. These have a tracer mechanism that follows a template or a line drawing. Some are now equipped with numerical tape or computer controls. All of these types produce desired shapes with remarkable accuracy. Accuracies of ±0.38 mm (0.015 inch) are possible, but accuracies of 0.76 to 1.0 mm (0.030 to 0.040 inch) are more common.

Fuel gases other than acetylene also are used for oxyfuel gas cutting, the most common being natural gas (OFC-N) and propane (OFC-P). Their use is a matter of economics due to special availability. For certain special work, hydrogen may be used (OFC-H).

In preparing plate edges for welding, two or three simultaneous cuts often are made, as shown in Figure 32-6.

Stack cutting. In order to cut a stack of thin sheets of steel successfully, two precautions must be observed. First, the sheets should be flat and smooth and be free of scale. Second, they must be clamped together tightly so there are no gaps between them that could interrupt uniform oxidation and permit slag

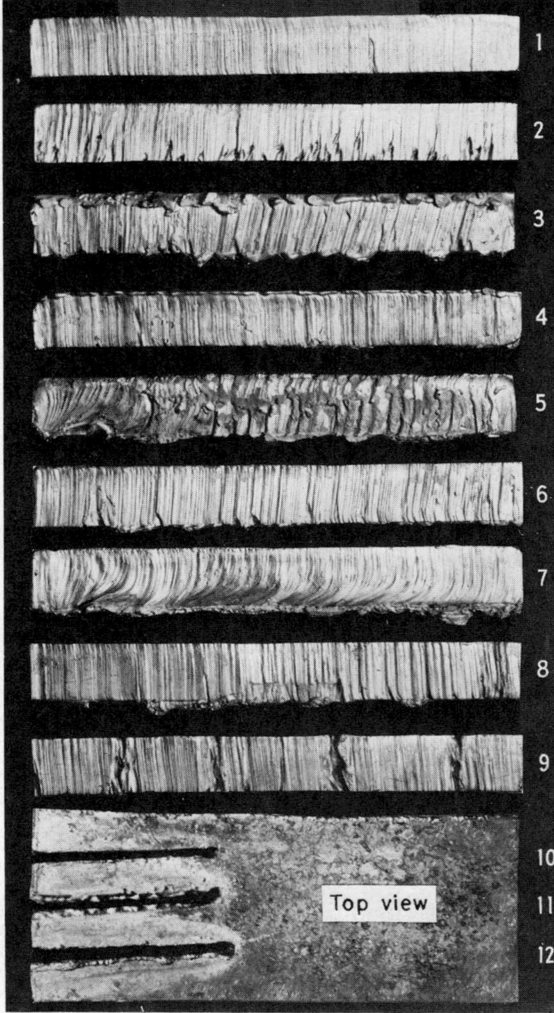

(1) Correct Procedure
Compare this cut in 1-in. plate with those below. The edge is square, the drag lines are vertical and not too pronounced.

(2) Preheat Flames Too Small
They are only about 1/8 in. long. Result: cutting speed was too slow, causing bad gouging effect at bottom.

(3) Preheat Flames Too Long
They are about 1/2 in. long. Result: surface has melted over, cut edge is irregular, and there is too much adhering slag.

(4) Oxygen Pressure Too Low
Result: top edge has melted over because of too slow cutting speed.

(5) Oxygen Pressure Too High
Nozzle size also too small. Result: entire control of the cut has been lost.

(6) Cutting Speed Too Slow
Result: irregularities of drag lines are emphasized.

(7) Cutting Speed Too High
Result: a pronounced rake to the drag lines and irregularities on the cut edge.

(8) Blowpipe Travel Unsteady
Result: the cut edge is wavy and irregular.

(9) Lost Cut Not Properly Restarted
Result: bad gouges where cut was restarted.

(10) Good Kerf
Compare this good kerf (viewed from the top of the plate) with those below.

(11) Too Much Preheat
Nozzle also is too close to plate. Result: bad melting of the top edges.

(12) Too Little Preheat
Flames also are too far from the plate. Result: heat spread has opened up kerf at top. Kerf is tapered and too wide.

FIGURE 32-3. Appearance of edges of metal cut properly and improperly by the oxyacetylene process. (*Courtesy Linde Division, Union Carbide Corporation.*)

and molten metal to be entrapped. *Stack cutting* is a useful technique where a modest quantity of duplicate parts is required but is insufficient in number to justify the construction of a blanking die. Obviously, the accuracy obtainable is not as good as can be obtained by blanking.

Metal powder cutting. Some ferrous metals, notably cast iron, stainless steel, and some high alloys, are not cut readily by the ordinary oxidation process.

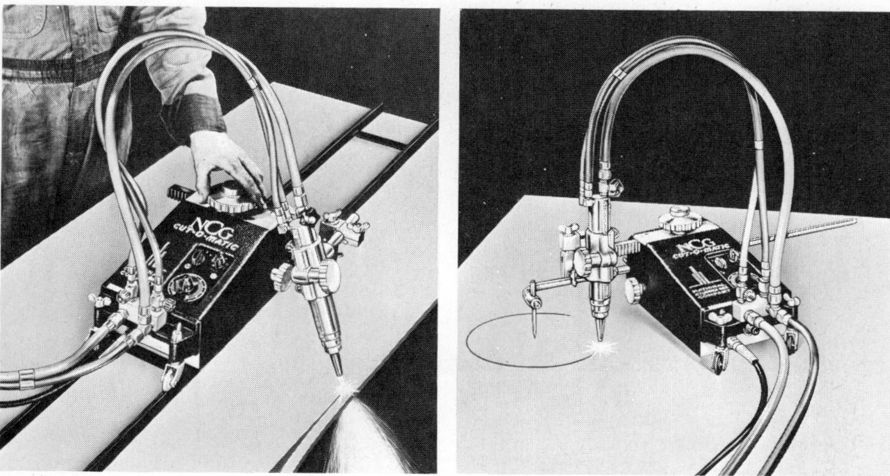

FIGURE 32-4. Oxyacetylene cutting with the torch carried by a small, portable, electrically driven carriage. (*Courtesy National Cylinder Gas Company.*)

FIGURE 32-5. Three parts being cut simultaneously from heavy steel plate, using a tracer-pantograph machine that is guided by the template at the bottom. (*Courtesy Linde Division, Union Carbide Corporation.*)

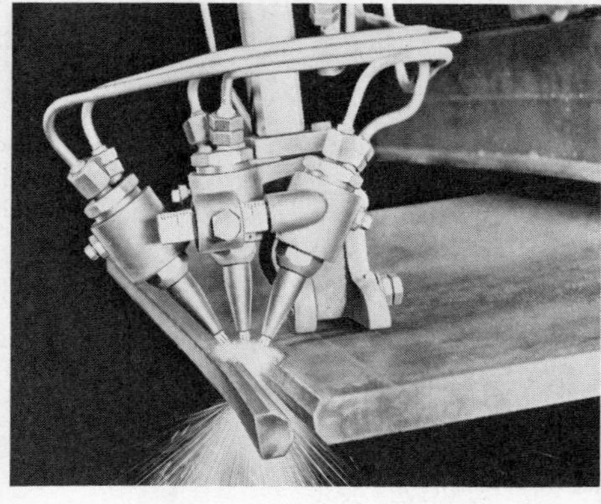

FIGURE 32-6. Plate edge being prepared by three simultaneous cuts with oxyacetylene cutting torches. (*Courtesy Linde Division, Union Carbide Corporation.*)

High-melting-point oxides are formed that are not oxidized or melted at the usual temperatures produced in oxyacetylene cutting. Also, they prevent the cutting oxygen from coming into contact with the iron in the material being cut. These difficulties are eliminated by *metal powder cutting* (POC), in which a stream of iron powder is blown into the cutting flame through a special opening in the torch tip. The rapid oxidation of this fine iron powder raises the temperature in the cutting area sufficiently to increase the oxidation of the oxides so that cutting proceeds satisfactorily.

Powder cutting also is effective in cutting certain high-melting-point, nonferrous metals, such as copper, which do not oxidize readily. To cut most nonferrous meterials it is necessary to melt them and then blow away the molten material and thereby form a kerf. These metals frequently have high heat conductivities that make it difficult to maintain a melting temperature. The higher temperature obtainable with the use of iron powder enables sufficiently rapid melting of the metal to be achieved so that a satisfactory cut can be made.

Chemical flux cutting. In the *chemical flux cutting* (FOC) used to cut hard-to-cut ferrous metals, a fine stream of a special flux is injected into the cutting oxygen as it leaves the tip of the torch. This flux increases the fluidity of the refractory oxides so that they are blown from the kerf, leaving the iron exposed to be oxidized more readily.

Arc-plasma torches, which will be discussed later, have replaced iron-powder and flux-injection cutting to a considerable extent.

Underwater torch cutting. Steel can be cut underwater by use of a special torch that supplies a flow of compressed air to provide the secondary oxygen for the oxyacetylene flame and to keep the water away from the zone where the burning of the metal occurs. Such a torch, shown in Figure 32-7, contains an auxiliary skirt, surrounding the main tip, through which the compressed air

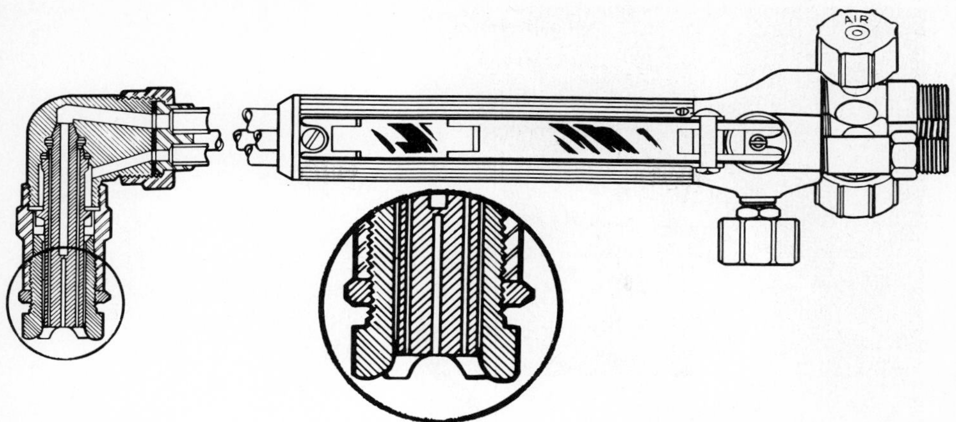

FIGURE 32-7. Underwater cutting torch. Note extra set of gas openings in the nozzle to provide an air bubble and the extra control valve. (*Courtesy The Bastian-Blessing Company.*)

flows. The torch is either ignited in the usual manner before descent or by an electric spark device after being submerged.

Acetylene gas usually is utilized for depths up to about 7.6 m (25 feet). For greater depths hydrogen is used, because the pressure involved is too great for safe operation with acetylene.

Flame machining. Although some use is made of flame cutting in removing metal as a rough machining process, it has proven practical in only very limited cases, primarily where the work can be rotated in a lathe. The blowpipe tip is mounted in order to make a small angle with the surface of the work, and as the work rotates, the torch is moved to cut away the desired amount of metal. The attainable accuracy is low, the resulting surface is quite rough, and difficulties resulting from thermal expansion often are encountered.

ARC CUTTING

Virtually all metals can be cut by electric arc procedures, wherein the metal is melted by the intense heat of the arc and then permitted, or forced, to flow from the kerf. With one process the resulting edges are considerably rougher than are obtained by oxyacetylene cutting, and the rate of cutting is considerably slower. However, one arc cutting process can produce edges that in some cases are smoother than can be obtained by flame cutting, and with much higher cutting speeds. Consequently, arc cutting processes are very effective and useful for certain applications.

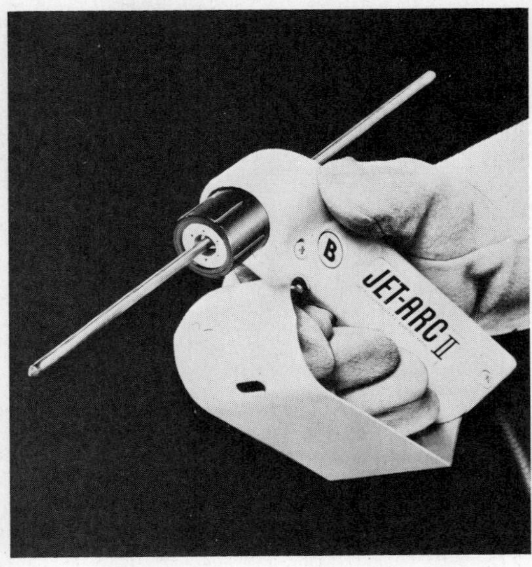

FIGURE 32-8. Gun used in the arc air process. Note air holes in holder surrounding electrode. (*Courtesy Jackson Products.*)

Air carbon arc cutting. *Air carbon arc cutting* (AAC) is the most widely used of the arc cutting processes. An arc is maintained between a carbon electrode and the workpiece, and high-velocity jets of air are directed at the molten metal from holes in the electrode holder, such as shown in Figure 32-8.[1] While there is some oxidation of molten metal, the primary effect of the air stream is to blow the molten metal and any oxides from the cut. The process is used primarily for cutting cast iron and for gouging grooves in steel plates preparatory to welding. Speeds up to 610 mm (24 inches) per minute are readily attained. Although the process can be used for cutting stainless steel and nonferrous metals, plasma arc cutting is more efficient. The disadvantages of arc air cutting are that it is quite noisy, and the hot metal particles tend to be blown out over a substantial area.

Plasma arc cutting. The torches used in *plasma arc cutting* (PAC) produce the highest temperature available from any practicable source. They thus are very useful for cutting metals, particularly nonferrous and stainless types that cannot be cut by the usual rapid oxidation induced by ordinary flame torches. Two types of torches are used, shown in Figure 32-9. Both are arranged so that the arc column is constricted within a small-diameter nozzle through which inert gas is directed. Because the arc fills a substantial part of the nozzle opening, most of the gas must flow through the arc and, as a consequence, is heated to a very high temperature, forming a plasma.

[1] In some types of equipment a single jet of air is used.

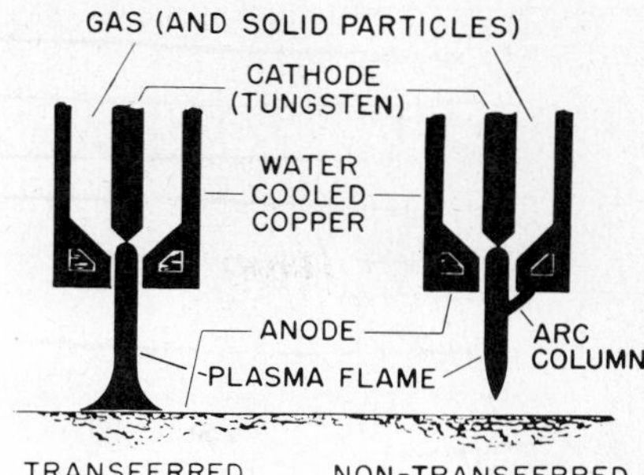

FIGURE 32-9. Principle of arc plasma torches. (*Courtesy Linde Division, Union Carbide Corporation.*)

With the nontransferred-type torch, wherein the arc column is completed within the nozzle, a temperature of about 16 649°C (30,000°F) is obtainable. With the transferred-type torch, wherein the arc column is between the electrode and the workpiece, the temperatures obtainable are estimated to be up to 33 316°C (60,000°F). Obviously, such high temperatures provide a means of very rapid cutting of any material by melting and blowing it from the cut. Cutting speeds up to 7620 mm (300 inches) per minute have been obtained in 6.35 mm (¼ inch) aluminum. The combination of extremely high temperature and jetlike action of the plasma produces narrow kerfs and remarkably smooth surfaces—nearly as smooth as can be obtained by sawing. Transferred-type torches usually are used for cutting metals, whereas the nontransferred type must be used for nonmetals.

Argon, helium, nitrogen, and mixtures of argon and hydrogen are used. Mixtures of 65 to 80 per cent argon and 20 to 35 per cent hydrogen are very common.

A limited amount of welding is done with plasma torches (PAW), but the temperature is too high for most work. Good results have been achieved with a torch that combines a small nontransferred arc within the torch, which heats the orifice gas, ionizing it and thereby forming a conductive path for the main transferred arc. This permits instant ignition of a low-current arc, which can be lower in magnitude, and more stable, and more readily controlled than in an ordinary plasma torch. Separate dc power supplies are used for the pilot and main arcs. An inert shielding gas usually is supplied through an outer cup surrounding the torch.

Gas tungsten arc cutting. *Gas tungsten arc cutting* (GTAC) is used primarily for making holes—up to about 9.5 mm (⅜ inch)—in sheet metal. A special tungsten electrode arc torch is used and a combination of argon and CO_2. Argon gas at low pressure is used while the metal is being melted by the arc; then CO_2, at a pressure of up to 0.56 MPa (80 psi), is used to blow away the molten metal and form the hole. Figure 32-10 shows holes being made by this process.

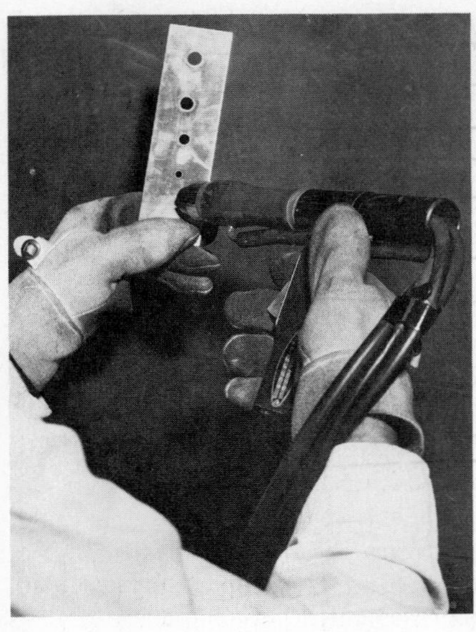

FIGURE 32-10. Making holes in sheet metal by the inert-gas arc process. (*Courtesy Hobart Brothers Company.*)

LASER BEAM CUTTING

Laser beam cutting (LBC) uses the intense heat from a laser beam to melt and/or evaporate the material being cut. Any known material can be cut by this process. For some nonmetallic materials the mechanism is purely evaporation, but for many metals a gas may be supplied, either inert to blow away the molten metal and provide a smooth, clean kerf or oxygen to speed the process through oxidation. The temperature achieved may be in excess of 11 093C (20,000°F), and cutting speeds of the order of 25.4 meters (1000 inches) per minute are not uncommon in nonmetals and 508 mm (20 inches) per minute in tough steels. As shown in Figure 32-11, very accurate cuts can be made. The kerf and heat-affected zone are narrower than with any other thermal cutting process, and it is quite easily adapted to computer or tape control.

Metallurgical and heat effects from flame and arc cutting. Flame and arc cutting obviously involve high temperatures, often localized. Consequently, it is always possible that the use of these processes could have harmful metallurgical and other heat effects, and such possibilities should not be overlooked. Fortunately, in most cases little or no difficulty is experienced, but in others, definite steps should be taken to avoid or overcome the harmful effects.

In low-carbon steel, below 0.25 per cent carbon, oxyacetylene cutting usually causes no serious metallurgical effects. Although there often is some minor hardening in a thin zone near the cut, and a small amount of grain growth, these effects usually will be wiped out if any subsequent welding of the cut edges is done. However, in steels of higher carbon content, these effects can be quite serious, and preheating and/or postheating may be required.

FIGURE 32-11. Cutting sheet metal with a plasma torch. (*Courtesy GTE Sylvania.*)

From a heat-effect viewpoint, arc air cutting is about the same as arc welding. If welding follows the use of this process, its heat effects will replace those left by the cutting process, and no special precautions are required. However, if no subsequent welding is to be done, consideration should be given to whether the resulting heat effects will be damaging in view of the load stresses that are anticipated.

Plasma arc cutting is so rapid, and the heat is so localized, that the heat-affected zone usually is less than 2.38 mm ($^3/_{32}$ inch), and the original hardness of the metal exists beyond 1.59 mm ($^1/_{16}$ inch) from the cut.

All these processes leave some residual stresses, with the cut surface in tension. Except in the case of thin sheet, gas or arc cutting usually will not produce warping in themselves. However, if subsequent machining removes only a portion of the total cut surface, or not all of the depth in residual tension, the resulting unbalancing of stresses may produce warping. Thus, if subsequent machining is to be done, it may be necessary to remove all cut surfaces to a substantial depth in order to achieve dimensional stability. Machining cuts should be sufficiently deep to get below the hardened surface and avoid dulling tools.

All flame- or arc-cut edges are rough, to varying degrees, and thus contain geometrical notches that can act as stress raisers and thus reduce the endurance strength. Consequently, if such edges are to be subjected to high or repeated tensile stressing, the cut surface and the heat-affected zone should be machined away or, as a minimum, stress-relief heat treatment be provided.

Review questions

1. Basically, what occurs when steel is cut by a torch?
2. Why is it usually necessary to supply some external heat from a torch during oxyacetylene cutting of steel?

3. What conditions determine whether steel can be cut with an oxygen lance?
4. What accuracy can typically be obtained with machine-guided cutting with an oxyacetylene torch?
5. Why can stainless steel not be cut with an ordinary cutting torch?
6. What precautions must be observed when cutting a stack of thin metal sheets with an oxyacetylene torch?
7. Why does the addition of iron powder to the cutting flame make it possible to cut stainless steel?
8. What is the principle upon which an underwater cutting torch operates?
9. Why should a torch-cut plate of steel not be used in an application where it will be subjected to repeated tensile stresses?
10. What is the basic action in cutting with an electric arc?
11. Why is the plasma arc torch used very little for welding?
12. What would you expect to result from cutting a 13-mm (½-inch) plate of AISI 4140 steel with an oxyacetylene cutting torch?
13. What difficulties might be experienced in subsequent machining of a 6.4-mm (¼-inch)-thick steel plate that had been torch-cut from a larger plate along all four edges?
14. What processes account for the removal of material in cutting with a laser beam?
15. Why is the heat-affected zone so narrow when cutting metal with a laser beam?

CHAPTER 33

Brazing, soldering, and adhesive bonding

BRAZING

Brazing differs from other welding processes in that a nonferrous filler metal is used, which has a melting temperature above 450°C (840°F) but below the solidus temperature of the base metal, or metals, and coalescence is between the filler and base metals, or betweeen the base metals and alloys that may be formed. Consequently, the base metals are not melted, and capillary attraction plays in important role in distributing the filler metal in the joint.

Five important characteristics and requirements result from the definition of brazing:

1. The 450°C (840°F) temperature is an arbitrary one, set to distinguish brazing from soldering.
2. Because the base metals are not melted, the brazing metal always is different in composition from the base metals.

3. For capillary attraction to exist, the clearance between the parts being joined must be quite small.
4. In order for capillary attraction to be effective, the base-metal surfaces must be clean so that the brazing metal can easily wet them. This usually requires the use of a flux or inert atmospheres in a vacuum furnace.
5. Heating a brazed joint above the melting point of the brazing metal, even followed by subsequent cooling, may destroy the integrity of the joint.

Brazing has several distinct advantages:

1. Virtually all metals can be joined by some type of brazing metals; coalescence always is between the base metal and a dissimilar brazing metal.
2. Less heating is required than for welding; the process can be done more quickly and more economically.
3. Lower temperatures are used; fewer difficulties due to distortion are encountered, and thinner and more complex assemblies can be joined successfully.
4. Many brazing operations can be mechanized.

A consideration of these advantages makes it apparent that brazing is particularly well suited for use in mass production and for delicate assemblies.

The major disadvantage of brazing is the fact that reheating can cause inadvertent melting of the braze metal, causing it to run out of the joint, thus weakening or destroying the joint. Too often this occurs when people apply heat to brazed parts in attempting to repair or straighten such devices as bicycles or motorcycles. Such a consequence, of course, is not a defect of brazing, but it can lead to most unfortunate results. Consequently, if brazing is specified for use in products that later might be subjected to such abuse, adequate warning should be given to those who will use the product.

The nature and strength of brazed joints. A brazed joint derives its strength from a combination of the braze metal and the base-metal alloy that is formed and the penetration of the low-viscosity brazing metal into the grain boundaries of the base metal. The strength of a properly brazed joint is between that of the base metal and that of the braze metal. However, the strength also is a function of the clearance between the parts being brazed. There must be sufficient clearance so that the braze metal will wet the joint and flow into it, but beyond this amount of clearance, the strength decreases rapidly to that of the braze metal. The proper amount of clearance varies considerably, depending primarily on the type of braze metal. Copper requires virtually no clearance when heated in a hydrogen atmosphere. In fact, a slight press fit of about 0.1 per cent per unit of diameter (0.001 inch per inch) is recommended. For silver-alloy brazing metals the proper absolute clearance is about 0.04 to 0.05

mm (0.0015 to 0.002 inch). When 60–40 brass is used for brazing iron and copper, a clearance of about 0.51 to 0.76 mm (0.02 to 0.03 inch) is desirable.

Brazing metals. The most commonly used brazing metals are copper, and copper, silver, and aluminum alloys. Table 33-1 lists some of the most frequently used brazing metals and their usages.

Copper is used only for brazing steel and other high-melting-point alloys, such as high-speed steel and tungsten carbide. Its use is confined almost exclusively to furnace heating in a protective hydrogen atmosphere in which the copper is extremely fluid and requires no flux. Copper brazing is used extensively for assemblies composed of low-carbon steel stampings, screw-machine parts, and tubing, such as are common in mass-produced products.

Copper alloys, in the form of "spelter" brass, were the earliest brazing ma-

TABLE 33-1. Commonly used brazing metals and their uses

Braze Metal	Composition	Brazing Process	Base Metals
Brazing brass	60% Cu, 40% Zn	Torch Furnace Dip Flow	Steel, copper, high copper alloys, nickel, nickel alloys, stainless steel
Manganese bronze	58.5% Cu, 1% Sn, 1% Fe, 0.25% Mn, 39.5% Zn	Torch	Steel, copper, high copper alloys, nickel, nickel alloys, stainless steel
Nickel silver	18% Ni, 55–65% Cu, 27–17% Zn	Torch Induction	Steel, nickel, nickel alloys
Copper silicon	1.5% Si, 0.25% Mn, 98.25% Cu, 1.5% Si, 1.00% Zn, 97.5% Cu	Torch	Steel
Silver alloys (no phosphorus)	5–80% Ag, 15–52% Cu, balance Zn + Sn + Cd	Torch Furnace Induction Resistance Dip	Steel, copper, copper alloys, nickel, nickel alloys, stainless steel
Silver alloys (with phosphorus)	15% Ag, 5% P, 80% Cu	Torch Furnace Induction Resistance Dip	Copper, copper alloys
Copper phosphorus	93% Cu, 7% P	Torch Furnace Induction Resistance	Copper, copper alloys

terials. Today, the most common use of copper alloys in brazing is in the repair of steel and iron castings. Tobin and manganese bronzes frequently are used for this purpose. Phos-copper is used extensively in brazing copper but should not be used on ferrous alloys.

Silver alloys are widely used in fabricating copper and nickel alloys. Although these brazing alloys are expensive, such a small amount is required that the cost per joint is low. One typical silver–phosphorus–copper alloy, known as Sil-Fos, is self-fluxing on clean copper and is used extensively for brazing this material.

The silver alloys also are used for brazing stainless steel. However, because the brazing temperatures are in the range of carbide precipitation, only stabilized stainless steels should be brazed with these alloys if continued corrosion resistance is desired.

Aluminum–silicon alloys, containing about 6 to 12 per cent silicon, are used for brazing aluminum and aluminum alloys. By using a braze metal that is not greatly unlike the base metal, the possibility of galvanic corrosion is reduced. However, because these brazing alloys have melting points of about 610°C (1130°F), when the melting temperature of commonly brazed aluminum alloys, such as 3003, is around 669°C (1290°F), control of the temperature used in brazing is quite critical. Thus, in brazing aluminum, good temperature control must be exercised, and proper fluxing action, surface cleaning, and/or use of a controlled-atmosphere or vacuum environment must be utilized to assure adequate flow of the braze metal and yet avoid damage to the base metal.

A commonly used procedure in connection with brazing aluminum is to use sheets that have one or both surfaces coated with the brazing alloy to a thickness of about 10 per cent of the total sheet thickness. These "brazing sheets" have sufficient coating to form adequate fillets, and joints are made merely by coating the joint area with suitable flux followed by heating.

Fluxes. *Fluxes* play a very important part in brazing by (1) dissolving oxides that may be on the surface prior to heating, (2) preventing the formation of oxides during heating, and (3) lowering the surface tension of the molten brazing metal and thus promoting its flow into the joint.

One of the primary factors affecting the quality and uniformity of brazed joints is cleanliness. Although fluxes will dissolve modest amounts of oxides, *they are not cleaners.* Before a flux is applied, dirt, particularly oil, should be removed from the surfaces that are to be brazed. The less the flux has to do prior to heating, the more effective it will be during heating.

Borax has been a commonly used brazing flux. *Fused* borax should be used because the water in ordinary borax causes bubbling when the flux is heated. Alcohol can be mixed with fused borax to form a paste.

Many modern fluxes are available that have lower melting temperatures than borax and are somewhat more effective in removing oxidation. Fluxes should be selected with reference to the base metal. Paste fluxes usually are utilized for furnace, induction, and dip brazing and either paste or powdered fluxes are used for torch brazing. In furnace, induction, and dip brazing the flux ordinarily is brushed onto the surfaces. In torch brazing it frequently is applied by dipping the heated end of the filler wire into the flux.

Fluxes for aluminum usually are mixtures of metallic hallide salts, the base typically being sodium and potassium chlorides and comprising from 15 to 85 per cent of the flux. Activators, such as fluorides or lithium compounds, are added. These fluxes do *not* dissolve the surface oxide film on aluminum.

Most brazing fluxes are corrosive, and the residue should be removed from the work after the brazing is completed. This is particularly important in the case of aluminum, and much, and successful, effort has been devoted to developing fluxless procedures for brazing aluminum, as will be discussed later.

Applying brazing metal. Brazing metal is applied to joints in three ways. The oldest, and a common method used in torch brazing, uses the brazing metal in the form of a rod or wire. When the joint has been heated to a sufficient temperature so that the base metal will melt the brazing wire or rod, the wire or rod is then melted by the torch and capillary attraction causes it to flow into the joint. Although the base metal should be hot enough to melt the braze metal, and assure its remaining molten and flowing into the joint, the actual melting should be done with the torch.

Obviously, this method of braze metal application requires considerable labor, and care is necessary to assure that it has flowed to the inner portions of the joint because it always is applied from the outside. To avoid these difficulties, the braze metal often is applied to the joint before heating, in the form of wire or shims. In cases where it can be done, rings or shims of braze metal often are fitted into internal grooves in the joint before the parts are assembled, as shown in Figure 33-1. When this procedure is employed, the parts usually must be held together by press fits, riveting, staking, tack welding, or a jig, to assure their proper alignment. In such preloaded joints, care must be exercised to assure that the filler metal is not pulled away from the intended surface by the capillary action of another surface with which it may be in contact. Capillary action always will pull the molten braze metal into the smallest clearance, whether or not such was intended.

Another precaution that must be observed is that the flow of filler metal not be cut off by the absence of required clearances or by no provision for the escape of trapped air. Also, fillets or grooves within the joint may act as reservoirs and trap the metal.

Special brazing jigs or fixtures often are used to hold the parts that are to be brazed in proper relationship during heating, especially in the case of complex assemblies. When these are used, it usually is necessary to provide springs that will compensate for expansion, particularly when two or more dissimilar metals are being joined. Figure 33-2 shows an excellent example of this procedure.

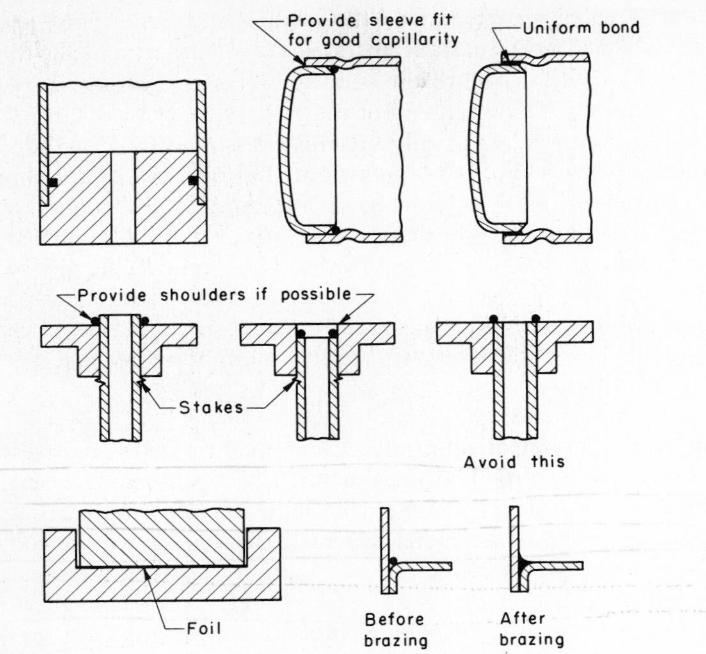

FIGURE 33-1. Methods of applying braze metal in sheet or wire form to assure proper flow into the joint.

FIGURE 33-2. Brazing fixture, with provision for expansion caused by brazing heat. (*Courtesy Aluminum Company of America.*)

FIGURE 33-3. Production setup for torch brazing school desk-chair frames. (*Courtesy The American Brass Company.*)

Heating methods used in brazing. A common source of heat for brazing is a gas-flame torch. In this *torch-brazing procedure,* oxyacetylene, oxyhydrogen, or other gas-flame sources can be used. Most repair brazing is done in this manner because of its flexibility and simplicity, but the process also is widely used in production brazing, as illustrated in Figure 33-3. Its major drawbacks are the difficulty in obtaining uniform heating, proper control of the temperature, and the requirement of costly skilled labor. In production-type torch brazing, specially shaped torches often are used to speed the heating and to aid in reducing the amount of skill required.

Large amounts of brazing are done in controlled-atmosphere furnaces. In such *furnace brazing,* the brazing metal must be preloaded into the work. If the work is not of such a nature that its preassembly will hold the parts in proper alignment and with adequate pressure, brazing jigs or fixtures must be used. Assemblies that are to be brazed usually can be designed so that such jigs or fixtures will not be needed if adequate consideration is given to this matter; often a light press fit will suffice. Figure 33-4 shows a number of typical furnace-brazed assemblies.

Because excellent control of brazing temperatures can be obtained and no skilled labor is required, furnace brazing is particularly well suited for mass production. Either box- or continuous-type furnaces can be used, the latter being more suitable for mass-production work. Furnace brazing is very economical, particularly when no flux is required, as in brazing steel parts with copper.

FIGURE 33-4. Furnace-brazed assemblies. (*Courtesy Pacific Metals Company.*)

In *salt-bath brazing* the parts are heated by dipping in a bath of molten salt that is maintained at a temperature slightly above the melting point of the brazing metal. This method has three major advantages: (1) the work heats very rapidly because it is completely in contact with the heating medium, (2) the salt bath acts as a protective medium to prevent oxidation, and (3) thin pieces can easily be attached to thicker pieces without danger of overheating because the temperature of the salt bath is below the melting point of the parent metal; this latter feature makes this process well suited for brazing aluminum.

It is essential that parts that are to be dip-brazed be held in jigs or fixtures or that they be prefastened in some manner and that the brazing metal be preloaded into the work. Also, to assure that the bath remains at the desired temperature, the volume must be rather large, depending on the weight and quantity of the assemblies that are to be brazed.

In *dip brazing,* the assemblies are immersed in a bath of molten brazing metal. The bath thus provides both the required heat and the braze metal for the joint. Because the braze metal usually will coat the entire work, it is wasteful of braze metal and is used primarily only for small parts, such as wire.

Induction brazing utilizes high-frequency induction currents for heating. The process has the following advantages, which account for its extensive use:

1. Heating is very rapid—usually only a few seconds being required for the complete cycle.
2. The operation can easily be made semiautomatic, so that semiskilled labor may be used.
3. Heating can be confined to the localized area of the joint, because of the shape of the heating coils and the short heating time. This reduces scale, discoloration, and distortion.
4. Uniform results are readily obtained.
5. By making new, and relatively simple, heating coils, a wide variety of work can be done with a single power unit.

The high-frequency power-supply units are available in both small and large capacities at very modest costs. The only other special equipment required for adapting induction brazing to a given job is a simple heating coil to fit around the joint to provide heating at the desired area. These coils are formed of copper tubing through which cooling water is carried. Although the filler material can be added to the joint manually after it is heated, the usual practice is to use preloaded joints to speed the operation and obtain more uniform joints. Figure 33-5 shows a typical induction-brazing operation.

FIGURE 33-5. Typical induction brazing operation. (*Courtesy Lepel High Frequency Laboratories, Inc.*)

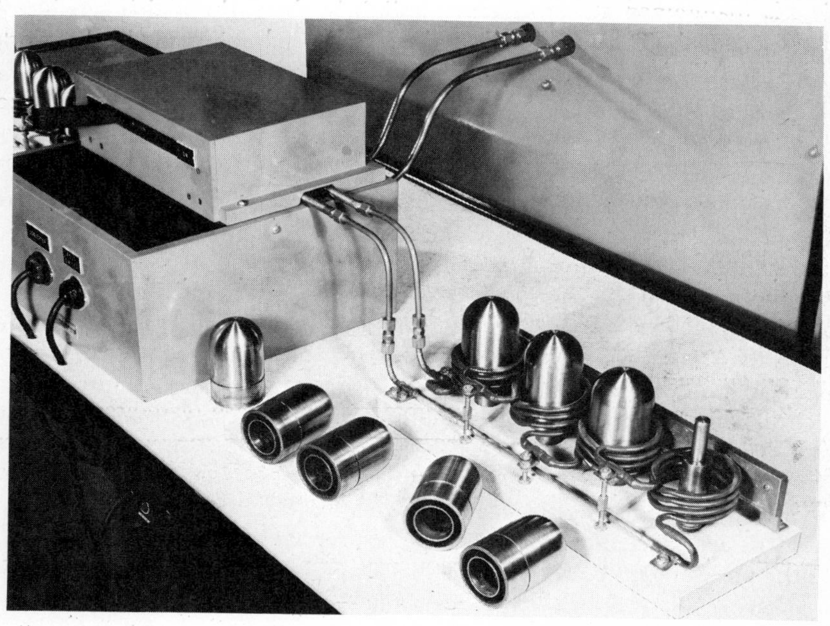

Induction brazing is so rapid that it can often be used to braze parts with high surface finishes, such as silver plating, without affecting the finish.

Some *resistance brazing* is done, in which the parts to be joined are held under pressure between two electrodes, similar to spot welding, and an electrical current is passed through them. However, unlike resistance welding, most of the resistance is provided by the electrodes, which are made of carbon or graphite. Most of the heating of the metal thus is by conduction from the hot electrodes. This process is used primarily in the electrical equipment manufacturing industry for brazing conductors, cable connections, and so on. Regular resistance welders and their timing equipment often are adapted for resistance brazing.

Flux removal. Although not all brazing fluxes are corrosive, most of them are. Consequently, flux residues usually must be completely removed. Most of the commonly used fluxes are soluble in hot water, so their removal is not difficult. In most cases immersion in a tank of hot water for a few minutes will give satisfactory results, provided that the water is kept really hot. Usually, it is better to remove the flux residue while it is still hot. Blasting with sand or grit is also an effective method of flux removal, but this procedure cannot be utilized if the surface finish must be maintained. Such drastic treatment seldom is necessary.

Fluxless brazing. Obviously, both the application of brazing flux and the removal of flux residues involve significant costs, particularly where complex joints and assemblies are involved. Consequently, a large amount of work has been devoted to the development of procedures whereby flux is not required, particularly for brazing aluminum. This work has been spurred by the obvious advantages of aluminum as a lightweight and excellent heat conductor for use in heat-transfer applications such as radiators in automobiles, where weight reduction is of increasing importance.

The brazing of aluminum is complicated by the high refractory oxide surface film, its low melting point, and its high galvanic potential. However, the successful fluxless brazing of aluminum has been achieved by employing rather complicated vacuum-furnace techniques, utilizing vacuums up to 0.0013 Pa (1×10^{-5} torr). Often a "getter" metal is employed to aid in absorbing the small amount of oxygen, nitrogen, and other occluded gases that remain in the "vacuum," or that may be evolved from the aluminum being brazed. The aluminum must be carefully degreased prior to brazing, and the design of the joint is quite critical. Sharp V-edge joints appear to give the best results.

Some success has been achieved in the fluxless induction brazing of aluminum in air. The induction heating coils are designed to act as clamps to hold the pieces being brazed together under pressure, and an aluminum braze metal having about 7 per cent silicon and 2.5 per cent magnesium is used. The resulting magnesium vapor apparently reduces some of the oxide on the surface of the aluminum and thus permits the braze metal to flow and cover the aluminum surface.

Design of brazed joints. Three types of brazed joints are used: *butt, scarf,* and *lap* or *shear*. These, together with some examples of good and poor design details, are shown in Figure 33-6. Because the basic strength of a brazed joint is somewhat less than that of the parent metals, desired strength must be obtained by utilizing sufficient joint area; this means some type of lap joint where maximum strength is required. If joints are made very carefully, a lap of 1 or 1¼ times the thickness of the metal can develop strength as great as that of the parent metal. However, for joints that are to be made in routine production, it is best to use a lap equal to three times the material thickness. Full electrical conductivity usually can be obtained with a lap about 1½ times the material thickness.

If maximum joint strength is desired, it is important to have some pressure applied to the parts during heating and until the braze metal has cooled sufficiently to attain most of its strength. In many cases, the needed pressure can be obtained automatically through proper joint selection and design.

In designing joints that are to be brazed, one must make sure that no gases can be trapped within the joint. Trapped gas may prevent the filler metal from flowing throughout the joint, owing to pressure developed during heating.

FIGURE 33-6. Examples of good and bad joint design for brazing.

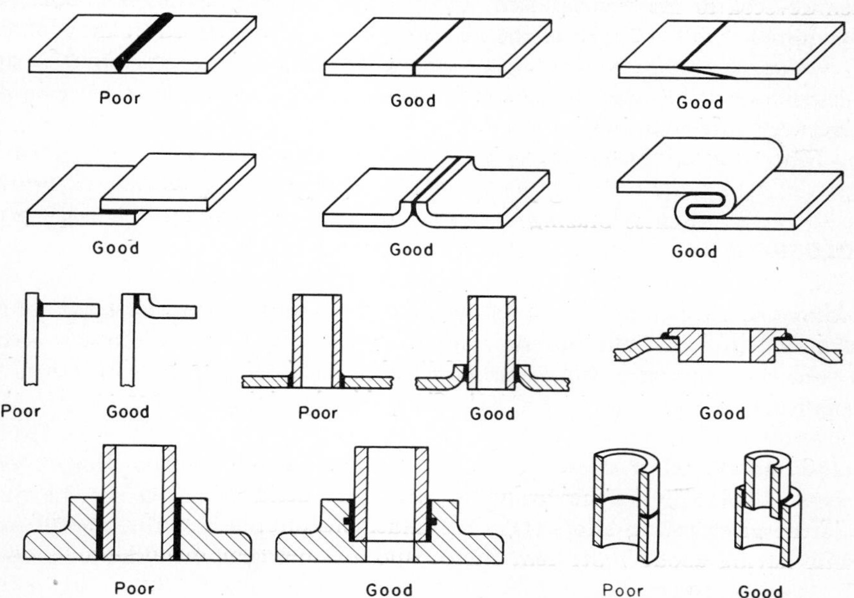

FIGURE 33-7. Large casting being repaired by braze welding. Four hours were required for preparation of the joint and 3 hours for welding. Twenty pounds of braze metal were used. (*Courtesy The Anaconda Brass Company.*)

Braze welding. *Braze welding* differs from brazing in that capillary attraction is not used to distribute the filler metal; the molten filler metal is deposited by gravity. Because relatively low temperatures are required, and thus there is less danger of causing warping than might result if welding were used, it is very effective for the repair of steel parts and ferrous castings and is used almost exclusively for this type of work. Obviously, the strength obtained is determined by the braze metal used, and considerable buildup of the braze metal is required if full strength of the repaired part is required.

Virtually all braze welding is done with an oxyacetylene torch. The surfaces should be "tinned" with a thin coating of the brazing metal before the remainder of the filler metal is added. Figure 33-7 shows an example of a casting being repaired by braze welding.

SOLDERING

Soldering is the joining of metals by means of a fusible metal or alloy, called *solder,* which has a melting temperature below 450°C (840°F). There is no coalescence, the joining being effected by adhesion between the solder and the parent metal.

Solder metals. Most solders are alloys of lead and tin with the addition of a very small amount of antimony—usually less than 0.5 per cent. The three most commonly used alloys contain 60, 50, and 40 per cent tin, and all melt below 241°C (465°F). Because tin is expensive, those having higher proportions of tin are used only where high fluidity is required. For *wiped* joints and for filling dents and seams, as in automobile body work where no strength is required, solder containing only 20 per cent tin is used.

For higher-temperature service, lead–silver solders are used. These contain 1.5 to 3.5 per cent silver and melt at about 310°C (590°F).

Soldering fluxes. As in brazing, soldering requires that the metal must be clean. Fluxes are used for this purpose, but it is essential that all dirt, oil, and grease be removed before the flux is applied. Soldering fluxes are not intended to, and will not, remove any appreciable amount of contamination.

Soldering fluxes are classified as *corrosive* or *noncorrosive.* A common noncorrosive flux is rosin in alcohol. This is suitable for copper and brass and for tin-, cadmium-, or silver-plated surfaces, if the surfaces are clean. Aniline phosphate is a more active noncorrosive flux, but it has limited use because it gives off toxic gases. In addition to being suitable for copper and brass, it can be used on aluminum, zinc, steel, and nickel.

The two most commonly used corrosive-type fluxes are muriatic acid and a mixture of zinc and ammonium chlorides. Acid fluxes are very active but are highly corrosive. Chloride fluxes are effective on aluminum, copper, brass, bronze, steel, and nickel if no oil is on the surface.

Heating for soldering. Although any method of heating that is suitable for brazing can be used for soldering, furnace and salt-bath heating are seldom used. Dip soldering is used extensively for soldering wire ends, particularly in electronics work, for automobile radiators, and for tinning. Induction heating is used extensively where large numbers of identical parts are to be soldered. However, a large amount of soldering still is done with electric soldering irons or guns. The principal requisites of these are that they have sufficient heat capacity and that the surface that is held against the work be flat and well tinned so as to assure good heat transfer.

As in brazing, the joints can be preloaded with solder, or the filler metal can be supplied from a wire. The method of heating usually determines which procedure is used.

Design and strength of soldered joints. Soldered joints seldom will develop shear strengths in excess of 1.72 MPa (250 psi). Consequently, if appreciable strength is required, soldered joints should not be used. If substantial strength is required, some type of mechanical joint, such as a rolled-seam lock joint, should be made prior to soldering. Butt joints should never be used, and designs where peeling action is possible should be avoided.

In making soldered joints, the parts must be held firmly so that no movement can occur until the solder has cooled well below the solidification temperature. Otherwise, the joint will be full of cracks and have very little strength.

Flux removal. The flux usually must be removed from soldered joints, either to prevent corrosion or for the sake of appearance. Flux removal usually is easily accomplished if the type of solvent used in the flux is known. Water-soluble fluxes can be removed with hot water and a brush. Alcohol will remove most rosin fluxes. When the flux contains a grease, as in most paste fluxes, a grease solvent can be used, followed by a hot-water rinse.

ADHESIVE BONDING

Tremendous advances have been made in recent years in the development, use, and reliability of adhesive bonding. Its use, even in such critical applications as automobiles and aircraft, is increasing rapidly. The adhesives used are thermoplastic and thermosetting resins, several artificial elastomers, and some ceramics. Both metals and nonmetals can be joined satisfactorily.

Adhesive materials and properties. Because of their distinctly different service characteristics and the differing requirements for their application, adhesives commonly are classified as being thermoplastic or thermosetting. The most commonly used adhesives thus are as follows:

Thermoplastic	Thermosetting
Nonvulcanized neoprene rubbers	Buna and neoprene rubbers
Polyimides	Epoxies
Vinyls	Isocyanates
	Phenolic-rubber
	Phenolic-vinyl

Because thermoplastic materials soften and lose strength when heated, they cannot be used for elevated-temperature applications. However, they are very easy to use in the following forms:

1. Air-drying dispersions, emulsions, or solutions that achieve their strength through evaporation of the solvent.
2. Fusible solids that liquefy when heated so as to fill the voids in the joint and then gain strength when they cool.
3. Pressure-sensitive tapes, which in reality are high-viscosity plastics on cloth or paper backings.

The thermosetting adhesives are activated by a chemical hardener and/or heat and pressure. Consequently, they can be used for elevated temperature service and where creep must be minimized. Their use has been greatly extended through the use of accelerating agents that are mixed with the basic resins just prior to use, thus reducing the need for heating the joint. In a few cases, thermoplastic and thermosetting adhesives are combined to obtain some of the properties of each type. Table 33-2 lists some of the adhesives that are

TABLE 33-2. Some common adhesives with their preparation, service temperatures, and strengths

Chemical Type	Cure Temperature °C (°F)	Service Temperature °C (°F)	Lap Shear Strength MPa at °C (psi at °F)	Peel Strength at Room Temperature N/cm (lb/in.)
Butyral-phenolic	135–177 (275–350)	−51–79 (−60–175)	17.2(2500) at RT[a] 6.9 at 79 (1000 at 175)	17.5 (10)
Epoxy; room-temperature cure	16–32 (60–90)	−51–82 (−60–180)	17.2(2500) at RT 10.3 at 82 (1500 at 180)	7.0 (4)
Epoxy; elevated-temperature cure	93–177 (200–350)	−51–177 (−60–350)	17.2(2500) at RT 10.3 at 177 (1500 at 350)	8.8 (5)
Epoxy-nylon	121–177 (250–350)	−251–82 (−420–180)	41.4(6000) at RT 13.8 at 82 (2000 at 180)	122.6 (70)
Epoxy-phenolic	121–177 (250–350)	−251–260 (−420–500)	17.2(2500) at RT 6.9 at 79 (1000 at 175)	17.5 (10)
Neoprene-phenolic	135–177 (275–350)	−51–82 (−60–180)	13.8(2000) at RT 6.9 at 82 (1000 at 180)	26.3 (15)
Nitrile-phenolic	135–177 (275–350)	−51–121 (−60–250)	27.6(4000) at RT 13.8 at 121 (2000 at 250)	105.1 (60)
Polyimide	288–343 (550–650)	−251–538 (−420–1000)	17.2(2500) at RT 6.9 at 538 (1000 at 1000)	5.3 (3)
Urethane	24–121 (75–250)	−251–79 (−420–175)	17.2(2500) at RT 6.9 at 79 (1000 at 175)	87.6 (50)

[a] RT, room temperature.

commonly used for structural purposes, with their service and curing temperatures and expected strengths.

Joint design and preparation. Adhesive-bonded joints often are classified as either continuous surface or core-to-face. In *continuous-surface bonds,* both of the adhered faying surface areas are relatively large and are of the same size and shape. *Core-to-face bonds* have one adhered area that is very small compared with the other, as in bonding lightweight honeycomb core structures to the face sheets. In designing a bonded joint, consideration must be given to the types of stress to which it will be subjected. As shown in Figure 33-8, these are tension, shear, cleavage, and peel. Because, as noted in Table 32-2, most of the adhesives are much weaker in peel and cleavage, it is important that as

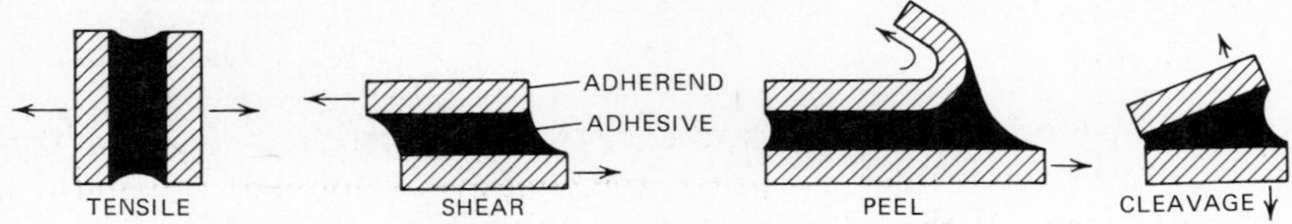

FIGURE 33-8. Types of stresses in adhesive-bonded joints.

much of the stress as possible be shear or tension. Adhesives tend to fall into two groups—those that are hard and brittle and have high tensile and shear strengths but low peel strengths, and those that are more ductile and have high peel strengths as well as good shear and tensile strengths. The first group usually requires only contact pressure during assembly and cure, and often can be cured at temperatures ranging from room to 149°C (300°F). The second group requires temperatures from 121 to 177°C (250 to 350°F) and pressures up to 1.38 MPa (200 psi) for 60 to 90 minutes for proper cure. Some of these give peel strengths of 123 N per centimeter (70 pounds per inch) of width.

As noted in Table 33-2, the lap shear strengths of some of the common adhesives range from 13.8 to 41.4 MPa (2000 to 6000 psi) at room temperature and their tensile strengths from 4.14 to 8.28 MPa (600 to 1200 psi). It thus is evident why joints should be designed to utilize the superior lap shear strengths whenever possible. Figure 33-9 shows some of the commonly used types of points.

To obtain satisfactory and consistent quality in adhesive-bonded joints, it is essential that the surfaces be prepared properly and that a standard procedure be established and adequate and frequent checks be made to assure that it is being followed. A four-step procedure customarily is used:

1. Cleaning: all contaminants and grease *must* be removed.
2. Etching: the surface must be made chemically receptive to the adhesive primer and provide maximum wetting characteristics.

FIGURE 33-9. Types of commonly used adhesive joints.

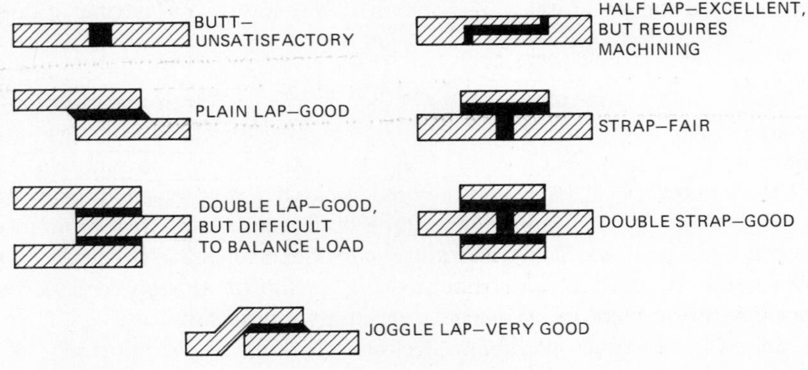

3. Rinsing.
4. Drying.

A low-viscosity primer may be applied, in one or more coats, by spraying or brushing. After the primer has dried, the adhesive is applied, usually in liquid or paste form. If the adhesive contains a solvent, most of it must be removed before the joint is closed.

When elevated temperatures are required for curing thermosetting adhesives, heat lamps, ovens, heated-platen presses, or autoclaves are employed, depending on the conditions.

Advantages and disadvantages of adhesive bonding. Adhesive bonding has a number of obvious advantages. Almost any material or combination of materials can be joined. For most adhesives, the curing temperatures are quite low, seldom exceeding 177°C (350°F), and a substantial number that cure at room temperature, or only slightly above, will provide adequate strength for many applications. Very thin and quite delicate materials, such as foils, can be joined to each other or to heavier sections. Because continuous bonding can be obtained throughout a joint, good load distribution and fatigue resistance can be obtained. Similarly, because of the large amount of contact area that usually can be obtained, the total joint strength often compares favorably with that resulting from other methods of attachment. Smooth contours are obtainable, and no holes have to be made, as with riveting or bolting. The adhesive can provide thermal and electrical insulation, act as a vibration dampener, and provide protection against galvanic action where dissimilar metals are joined. In most cases highly skilled labor is not required.

The major disadvantages of adhesive bonding are: (1) most of the adhesives are not stable above 177°C (350°F), although a few are quite good up to 260°C (500°F); (2) it is difficult to determine the quality of an adhesive-bonded joint by nondestructive means, although some methods give fairly good results for some types of joints; and (3) surface preparation, adhesive preparation, and curing procedures are quite critical if good and consistent results are to be obtained. However, the extensive and successful use of adhesive bonding is proof that these factors can be controlled if adequate quality-control procedures are adopted and followed.

While the unit strengths that can be obtained by adhesive bonding are relatively low, this is not a major disadvantage or limitation in most cases, since sufficient areas usually can be utilized if joints are designed properly.

Review questions

1. Why is brazing not considered to be a welding process?
2. Why is the fit between parts that are to be joined by brazing more critical than if welding were to be used?
3. What is the function of a brazing flux?

4. What are the advantages of brazing, as compared with welding?
5. What is a major disadvanage in the use of brazing?
6. In brazing a "closed" joint, what special precautions must be taken?
7. What metals and alloys most commonly are used as braze metals?
8. When dissimilar metals are furnace-brazed, what difficulties may be encountered?
9. For what conditions is dip brazing usually not a satisfactory process?
10. Why is salt-bath brazing particularly suitable for brazing aluminum parts?
11. What is the principal source of heat in resistance brazing?
12. Why does the circulation of cooling water through the coils used in induction brazing not slow the heating the workpieces?
13. What types of joints commonly are used for brazing?
14. Why must the residue from brazing flux usually be removed?
15. How does braze welding differ basically from ordinary brazing?
16. A vertical rod must be attached beneath a flat, horizontal surface. Would braze welding be a suitable process? Why?
17. What lap-to-thickness ratio usually is required in a brazed lap joint for the full strength of the parent metal to be equaled?
18. Why is soldering not classified as a welding process?
19. What are the components of soft solder?
20. What shear strength typically can be achieved in ordinary soldered joints?
21. What provisions should be made if a soldered joint must withstand substantial tension?
22. What agents are normally used as adhesives for metal?
23. How can adhesive-bonded joints be designed to have the full strength of the metals being joined?
24. What accounts for the increasing use of adhesive bonding?

Case study 24. CARGO-CONTAINER RACKS

On container ships, the cargo containers are held in racks on deck, stacked as many as six high. On one ship, the racks were designed and constructed of

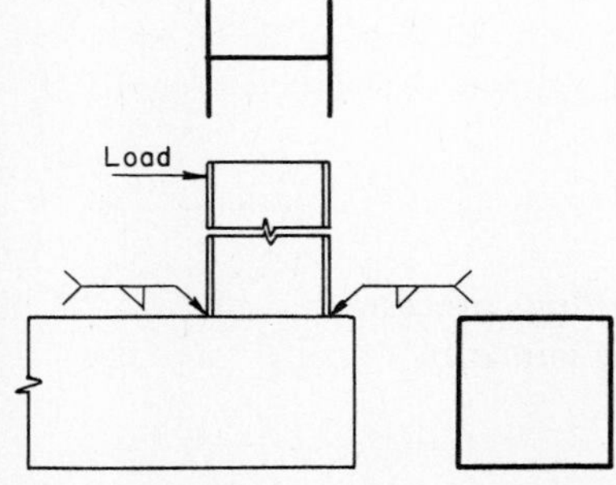

FIGURE CS-24. Method of attaching vertical legs to horizontal member of a cargo container rack.

four vertical 203-mm (8-inch) steel H sections, attached to transverse 254-mm (10-inch) box sections by means of 19-mm (¾-inch) fillet welds along the outside edges of the flanges, as shown schematically in Figure CS-24. On the ship's second voyage, during a Pacific storm, two of the racks failed at the attaching point of the vertical members, and six containers of cargo were lost overboard.

Was there a basic deficiency in the design? If so, what was it, and how could it have been corrected?

CHAPTER 34

Heat and design considerations in welding

HEAT EFFECTS

Heating and cooling are essential and integral components of all welding processes, except cold welding. In most cases the metal is subjected to temperatures above those at which phase transformations or grain growth occur. Because the amount of metal that is heated usually is small relative to the total mass of metal in the workpiece, at least some portions of the heated metal may be cooled quite rapidly. Also, the localized heating may result in residual stresses and/or warping of the workpiece. These may or may not adversely affect the service performance of the weldment. It should be obvious that these facts must be taken into consideration in designing weldments and in using welding if the potential benefits of the process are to be obtained and harmful side effects avoided. However, there are too many examples, in the form of failures in welded components, which show that designers and others associated with welding fail to give these factors proper consideration.

Flame and arc cutting also involve localized heating and subsequent cooling and thus also may give rise to some of the same heat effects. However, they are usually less severe and less extensive.

Welding metallurgy. Welding is probably the outstanding example of a manufacturing process wherein heating and cooling occur as essential steps and thus produce metallurgical effects that usually are not desired. Because in fusion welding some metal is melted and often cooled quite rapidly, the effects are most pronounced in connectjon with this type of welding. However, they also exist to a lesser degree in other types of welding wherein the heating–cooling cycle is less severe. If they are considered properly, these effects can be negated completely, and excellent service performance can be obtained through the use of welding. If they are not considered properly, the results can be disastrous—a failure of the designer or fabricator, not a failure of welding.

Because such a wide range of metals is welded and a variety of processes is used, welding metallurgy is an extensive subject. However, a few basic considerations will enable anyone who has a reasonable working knowledge of metallurgy to understand the changes in microstructure that occur. In fusion welding a pool of molten metal, either from the parent plate or from both the parent plate and an electrode or filler rod, is created. This pool is contained in a metal mold, formed by the parent plate. Ordinarily the molten pool is very small in comparison with the colder metal mold. Thus the process, metallurgically, is one of *casting* a small amount of molten metal into a metal mold. The resulting metallurgical and strength characteristics can be anticipated and explained on this basis.

Figure 34-1 shows a typical microstructure that results from a fusion weld. In the centre of the weld is a zone that is made up primarily of weld metal that has solidified from the molten state. Actually, it is a mixture of parent metal and electrode or filler metal, the ratio depending on the welding process used, the type of joint, and the edge preparation. This zone is cast metal of the particular composition of the mixture that has cooled; its microstructure reflects the cooling rate in the weld. Because it is cast metal, this zone cannot be expected to have the same properties as the *wrought* parent metal; it can achieve equal properties only through the addition of filler metal which, in the *as-cast* condition, has equal or superior properties. Thus one should use filler rods or electrodes which, in the *as-deposited* condition, have properties that equal, or are superior to, those of the parent metal. This requirement is the basis for the AWS specifications for electrodes and filler rods.

The grain structure in the weld metal zone may be fine or coarse and equiaxed or dendritic, depending on the type and volume of weld metal and the cooling rate. Most electrode and filler rod compositions tend to produce

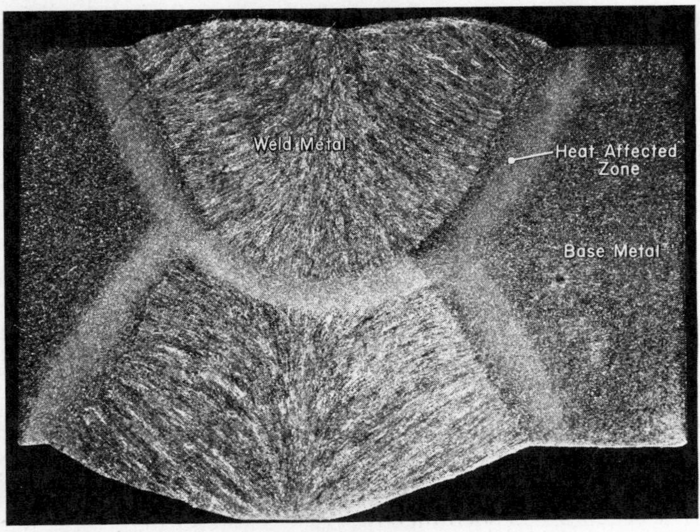

FIGURE 34-1. Grain structure and zones in a fusion weld.

fine, equiaxed grains, but the volume of weld metal and the cooling rate can easily defeat these objectives.

Adjacent to the weld metal is the ever-present, and undesirable, *heat-affected zone.* In this region, the parent metal was not melted, but it was heated to a high-enough temperature, and for a sufficient period of time, so that grain growth occurred. A variety of microstructures usually can be found in the heat-affected zone. In plain carbon steel these may range from very narrow regions of hard martensite to coarse perlite. It has lost the desirable characteristics that previously were imparted to the parent plate by hot working, and it does not have the superior properties of the weld metal. Consequently, the heat-affected zone is usually the weakest area in a weld, *in the as-welded condition.* Except where there are obvious defects in the weld deposit, most welding failures originate in the heat-affected zone.

Outside the heat-affected zone is the base metal that was not heated sufficiently to change its microstructure.

It is apparent that the microstructure of a weld is varied and complex. Much of the variation can be reduced or eliminated by suitable heat treatment following welding. However, it is also apparent that heat treatment cannot restore all the benefits of hot working. Another procedure that can reduce the variation in microstructure, particularly the sharpness of the variation, is to preheat the base metal adjacent to the weld just prior to welding. For plain carbon steels, a temperature of 93 to 204°C (200 to 400°F) usually is adequate. This procedure reduces the cooling rate of the weld deposit and the immediately adjacent metal in the heat-affected zone. Preheating also tends to reduce the sharpness of the boundaries between the weld and heat-affected zones, producing a more gradual change in microstructures and thereby eliminating a metallurgical stress raiser.

If the carbon content of plain carbon steels is greater than about 0.3 per cent, the cooling rates encountered in normal welding are sufficient to cause hardening, unless precautionary measures are taken. This also is true for many alloy steels, because of their high hardenability. Thus special pre- and post-welding heating cycles must be used when such steels are welded. It is for this reason that the various weldable, low-alloy steels have gained so much acceptance, since they usually can be welded without the necessity for preheating or postheating.

Where little or no melting occurs and there is considerable pressure applied to the heated metal, as in forge or resistance welding, the weld may retain some of the characteristics of wrought material.

The discussion thus far has considered the metallurgical effects related to the welding of steel. The effects of heating and cooling accompanying the welding of other metals will, of course, depend on the transformations and changes that can occur in such metals as they are subjected to the heating and cooling cycles that accompany the welding process.

Thermal stresses. As mentioned previously, another effect of the heating and cooling that accompany welding is the setting up of thermal stresses. These are of two types and are most pronounced in fusion welding, where maximum heating occurs.

Residual welding stresses are the result of restraint to thermal expansion and contraction *offered by the pieces being welded.* They may exist whether or not the pieces are attached to other portions of a structure or are restrained externally in any manner. The way in which residual stresses occur may be explained by means of Figure 34-2. As the material in *T*—the weld and the immediately adjacent metal—is heated, it expands. Its freedom to expand in the direction parallel with the weld is restrained by the much larger sections of relatively cold metal in the zones marked *C*. As a result, because the metal in zone *T* cannot expand freely in the lengthwise direction, it is *upset,* becoming slightly thicker in order to accommodate the increased volume. This upsetting involves permanent plastic deformation.

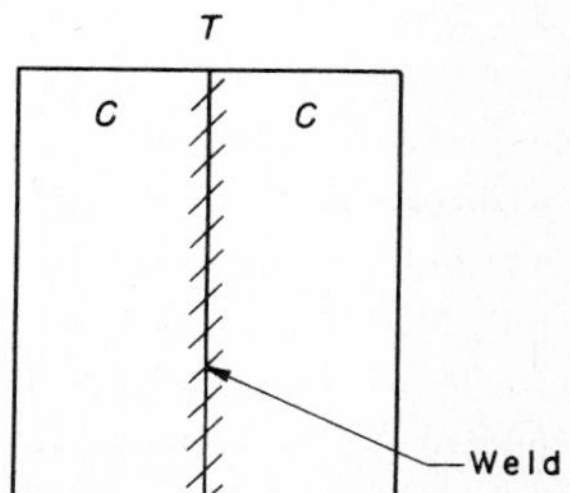

FIGURE 34-2. Conditions that cause residual welding stresses.

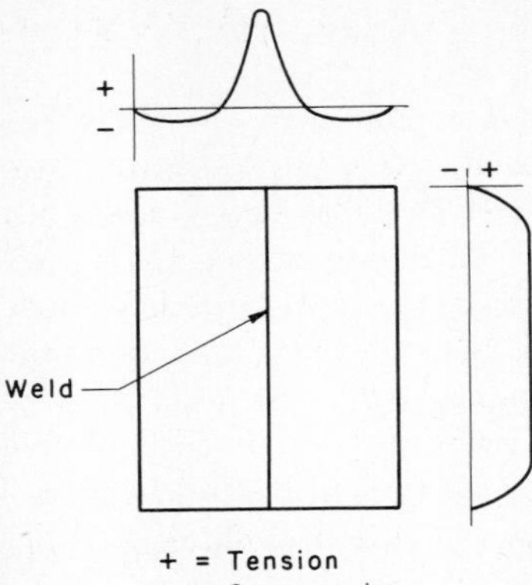

FIGURE 34-3. Typical pattern of residual stresses in a butt weld.

After the weld metal is deposited and solidifies, it cools and starts to contract. This contraction also is prevented by the metal in the zones marked *C* and, as a consequence, the metal in zone *T* is stretched, in attempting to have as great length as it had while hot and after being upset. Upon cooling, the metal in this zone is in a state of tension, balanced by low compression in the *C* zones, as shown in Figure 34-3.

The longitudinal residual stresses, parallel with the weld, are a maximum in the weld, being slightly above the yield strength of the parent plate, depending on the thickness of the metal and the amount and type of weld metal. For example, in a butt-fusion weld in 25.4-mm (1-inch) mild-steel plate, having a yield strength of 221 MPa (32,000 psi), the longitudinal residual stresses typically will have a maximum value of about 331 MPa (48,000 psi). In a similar 12.7-mm (½-inch) plate, these stresses will be about 276 MPa (40,000 psi). The reason these stresses can be greater than the yield strength of the plate is the fact that the weld metal, is the as-deposited condition, has a yield strength of about 359 MPa (52,000 psi), being designed to be stronger than the plate, and the resulting weld, being a mixture of weld metal and plate metal, has properties between those of the pure weld metal and the wrought plate metal. The longitudinal residual stresses decrease rapidly away from the weld and become compression stresses of low magnitude—usually not exceeding 34.5 MPa (5000 psi).

The second type of thermal stresses are called *reaction stresses.* These are due to the restraint to thermal expansion and contraction offered by other parts of the structure to which the pieces being welded are attached. Thus they exist only when two pieces that are being joined by welding are already attached to other portions of a structure.

The magnitude of reaction stresses is primarily an inverse function of the length of material that exists between the weld joint and a point of maximum rigidity. It can never exceed the yield strength of the parent material and, except where the restraint conditions are very "tight," these stresses are far below the yield-strength value, seldom exceeding 10.3 MPa (15,000 psi) in ordinary steel.

Effects of thermal stresses. The effects of thermal stresses resulting from welding have been widely misunderstood. The most obvious effect is that they can cause distortion, or warping, in weldments. Warping and buckling are the result of unbalanced stress conditions wherein the plates wanted to move but could not because of restraint. Warping and buckling tend to reduce the unbalanced stresses, but they usually are undesirable.

No fixed rules can be given for avoiding warping because the possible conditions that cause it are so varied. Several procedures are used. In general, a welding sequence should be used that permits the plates to have as much freedom as possible during the making of each weld. A general rule is to weld toward the point of greatest freedom. Sometimes the plates can be prepositioned so that the resulting distortion leaves them in the desired shape. Another common procedure is to completely restrain the plates during welding, thereby forcing some plastic flow in the plates and/or the cooling weld metal; this procedure can be utilized most readily on small weldments. Another procedure is to balance the resulting thermal stresses by depositing the weld metal in predetermined patterns and areas, often in short lengths. Warping frequently can be minimized by the use of *peening,* wherein each pass of multipass welds, except the first and last passes, is hammered with a peening tool to cause a small amount of localized plastic flow and thus permit the needed movement to occur.

As for residual stresses, there is no substantial evidence that they have a harmful effect on the strength performance of weldments *except in the presence of notches or in very rigid structures* where no plastic flow can occur. These are two conditions that should not exist in weldments *if they are properly designed and if proper workmanship is employed.* Unfortunately, welding makes it easy to inadvertently join heavy sections of steel and rigid configurations that will not permit the small amount of elastic or plastic movement required to reduce peak or concentrated stresses. Too often, geometric notches, such as sharp, interior corners, are incorporated in welded structures. Proper design can eliminate both rigidity and geometric notches. Other harmful notches, such as gas pockets, rough beads, porosity, and arc "strikes," can be triggers for weld failures, but these, too, can be avoided by proper welding procedures, good workmanship, and adequate supervision and inspection.

Residual stresses can cause warpage when weldments are machined, so as to unbalance the stress system. Consequently, weldments that are to undergo appreciable machining ordinarily are given a stress-relief heat treatment prior to machining.

Reaction stresses constitute a system of stresses that are imposed on the live-load system. They, too, can cause distortion. However, their most frequent effect is their tendency to cause cracking during or immediately following welding as the weld cools. This is particularly likely when welds are made under "tight" conditions where there is great restraint to the normal shrinkage that occurs in the direction transverse to the length of the weld. Often, where a multipass weld is being made, such a crack may occur in one of the early beads, when there is insufficient metal to withstand the shrinkage stresses. Such a condition may be serious if the crack is not noticed and chipped out and repaired or is not completely rewelded as subsequent beads are deposited. Preheating the adjacent metal for a few inches on each side of a weld that must be made under tight conditions will often eliminate this type of cracking.

Certain welding codes require some types of weldments to be stress-relieved by heat treatment prior to use. In many cases it appears that the improved performance that results from such treatment is due as much to the improvement in microstructure that results as it is to the reduction of residual stresses.

DESIGN CONSIDERATIONS

Welding is a unique process. It is not a process that can be substituted directly for another method of fastening without proper consideration being given to its peculiar characteristics and requirements. Unfortunately, welding is so easy and convenient to use that these two important facts often are overlooked—and the cause of many so-called welding failures can be traced to such negligence. Proper design must be employed.

As has been discussed previously, it is very easy to make structures too rigid by welding. Considerable thought may be required to design structures and joints that will permit sufficient flexibility that desirable readjustment of the stress can occur, but the multitude of successful welded structures attest to the fact that it can be done.

A very important factor that must be kept in mind in the use of welding is that it produces monolithic, or one-piece, structures. When two pieces are welded together, they become one piece. This can cause complications if not properly taken into account. For example, a crack in one piece of a multipiece structure may not be serious, because it seldom will progress beyond the single piece in which it occurs. However, when a large structure, such as a welded ship, a pipeline, storage tank, or a pressure vessel, consists of only one piece, a crack that starts in a single plate or weld may progress for a great distance and cause complete failure.[1] Obviously, such a situation is not the fault of welding but, rather, a failure to take a simple fact into account in designing the structure.

Another factor that relates to design is that a given material in small pieces may not behave as it does in a large piece. Welding makes it very easy

[1] Propagation velocities up to 1524 m (5000 feet) per second and distances of 304.8 m (1000 feet) have been recorded.

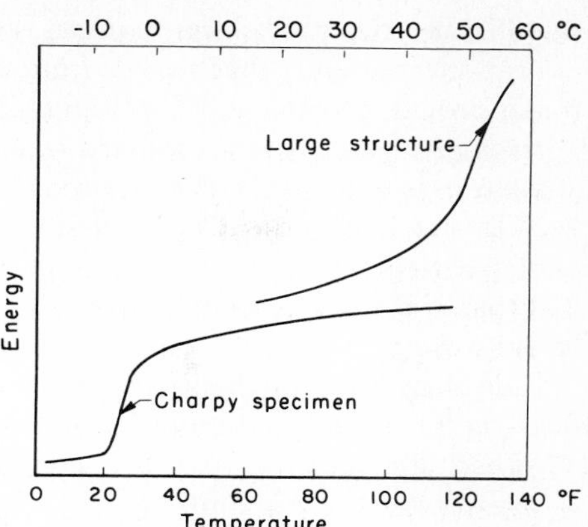

FIGURE 34-4. Effect of size on the energy-absorbing ability (*transition temperature*) of a steel specimen.

to obtain a structure that is a large piece. The importance of this fact is illustrated vividly in Figure 34-4, which shows the relationship between energy absorption and temperature for the same steel when tested in a Charpy impact specimen and in a large, welded structure. In a Charpy bar the material exhibited ductile behavior and good energy absorption at temperatures down to −4°C (25°F), but when welded into a large structure it exhibited ductile behavior down to only 43°C (110°F). Thus the notch-ductility characteristics of steel to be used in large welded structures may be of great importance, and ordinary specifications for steel frequently do not control this quality. More than one welded structure has failed because the designer did not take this fact into account.

Again, welding is an excellent fabricating process but, if satisfactory results are to be obtained, it cannot be used thoughtlessly, any more than can the rivets not be headed in a riveted connection or the nuts be left off bolts in a bolted connection. Before welding is adopted and used in manufacturing a product or structure, one should be sure that it has been designed properly for welding.

Review questions

1. What are the basic metallurgical characteristics of the deposited weld metal in a fusion weld?
2. Does the metal in an arc welding electrode have the same chemical composition as the parent material on which it is used? Why?
3. Why is the heat-affected zone usually the weakest part of a fusion weld?
4. Would you expect the weld "nugget" in a spot weld to have as high unit strength as the parent metal of which it is a part? Why?

5. What are two procedures that can be used to improve the properties in the heat-affected zone that results from a fusion weld?
6. An ardent proponent of welding claimed that the weld was always the strongest part of any ordinary steel structure in which it was used. To what extent was he right or wrong?
7. Why are longitudinal residual stresses not a direct function of the length of the weld?
8. What is the fundamental difference between residual and reaction stresses?
9. Why are the longitudinal residual stresses in thinner plate less in magnitude than those in thicker plate?
10. Is the presence of warping an indication of high residual stresses? Why?
11. What are the two primary conditions under which residual welding stresses may be harmful?
12. What are two reasons why a crack in a welded structure may be more serious than one in a riveted structure of similar design?
13. How is the energy-absorbing ability of many metals affected by temperature?
14. What two types of notches are likely to be found in welded structures unless proper precautions are taken?
15. Give three reasons why greater care should be taken in designing a structure that is to be welded than if it were to be riveted.

PART FIVE

Processes and techniques related to manufacturing

CHAPTER 35

Layout

After the configuration of a part is set forth on a drawing, or is clearly established in a designer's mind, in order to convert it into reality the dimensions that determine the size and location of its various geometrical surfaces must be established on the workpiece. The process of establishing these dimensions is called *layout.* It probably is most evident in the cases of machining and pattern making, but it also is involved in pressworking, welding, and other processes.

In most cases, layout involves transferring to a workpiece dimensions that are shown on a drawing. However, in some cases no drawings may be involved; the dimensions are only in someone's mind. Increasingly in modern manufacturing, the dimensions may be transferred from a person's mind to a punch-tape or a computer and then from the tape or computer memory to the machine tool. These procedures will be discussed in Chapter 39.

Designers and draftsmen should be aware of their relationship to layout.

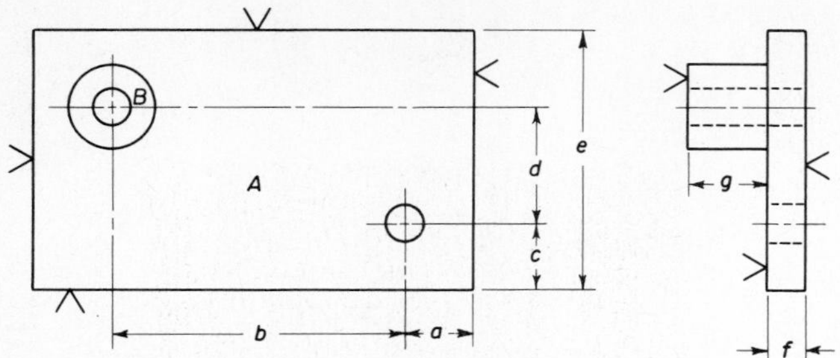

FIGURE 35-1. A part, showing dimensions that must be established on the workpiece in order to locate two holes.

A drawing that is designed and dimensioned so that it can easily be converted into reality can have a marked effect on reducing costs and, indirectly, quality. Unfortunately, sufficient examples exist to keep alive the belief among many shop people that the designers' motto is: "We design it; let the shop find a way to make it."

When only one or a few workpieces are involved, layout usually is accomplished manually, either directly on the workpiece or through the use of the controls available on the machine tools that are used. Digital controls make this latter procedure easier. When larger quantities are to be made, layout often is accomplished through the use of jigs and dies. These will be discussed in Chapter 36, but layout is an extremely important factor in their construction. If numerically controlled machines are used in manufacturing, even more attention may have to be given to the layout dimensions so that the required operations and sequences are compatible with the machine capabilities.

In order to understaand the problems associated with layout, consider the simple part depicted in Figure 35-1. All the required finished surfaces can be produced without difficulty, and dimensions *e, f,* and *g* can be obtained by direct surface-to-surface measurements, using ordinary measuring instruments. However, before the two holes can be machined, dimensions *a, b, c,* and *d* have to be established accurately on the workpiece. These are called *location dimensions* because they determine the location of geometrical shapes—in this case holes—with respect to another geometrical shape, the rectangular solid. Two methods for accomplishing this layout—manual and machine—will now be discussed.

MANUAL LAYOUT METHODS

The required dimensions *a, b, c,* and *d* can be established with a fair degree of accuracy by simple manual methods. In one method, surfaces *A* and *B* would be coated with a substance that could easily be scratched with a fine scribing

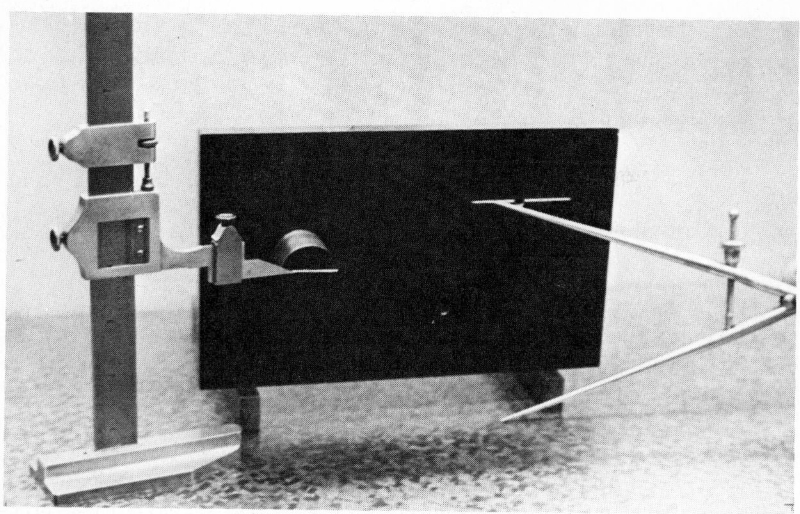

FIGURE 35-2. Scribing layout lines on a workpiece by means of a vernier height gage and a pair of dividers.

point. If only fair accuracy were required, chalk rubbed on the surface might be satisfactory. However, because the chalk is likely to be rubbed from the surface in handling the workpiece, a more permanent coating usually is used. Several commercial colored lacquers are available that can be brushed or sprayed on the surface to form a very thin coating that will remain until it is removed with a lacquer thinner. These lacquers dry quickly, and a scriber will readily scratch through the coating, leaving a fine, easily seen mark.

The next step usually would be to set the workpiece on a surface plate or on a pair of parallel bars, as shown in Figure 35-2. For accurate work, a vernier height gage with a scribing attachment would be set to the desired dimension c, or $c+d$, and slid along the surface plate with the scribing point in contact with the coated work surface, as shown in Figure 35-2, thus scribing a fine line at the desired distance from the edge that was resting on the surface plate or parallel bars. By thus using a vernier height gage, or a height-gage attachment with gage blocks, lines can be located to an accuracy of about $\pm$0.02 mm (0.001 inch) if care is exercised. If less accuracy is required, dividers can be used to establish the desired line, as also shown in Figure 35-2. If even less accuracy is required, a combination rule and square head may be used (see Figure 16-7), with the square head held against one finished side of the workpiece.

Dimensions a and $a+b$ could be established in a similar manner by standing the block on end.

Unfortunately, although the centers of the two holes can be established within 0.02 mm (0.001 inch) by the procedure described, this does not assure that the centers of the actual holes will be located this accurately. Consideration of the operations that are required to produce the holes makes it clear that such probably would not be true. To produce the holes to accurate size would require the following steps:

1. Center-punch the intersections of the layout lines.
2. Center-drill at the center-punched locations.
3. Drill the holes to approximate size.
4. Ream the drilled holes to final size.

Several factors in these steps tend to reduce the accuracy of the location of the centers. First, it is difficult to locate and hold the point of the center punch exactly at the intersection of the two fine, scribed layout lines. Second, when the center punch is struck, local surface irregularities and the nonhomogeneity of the metal in the workpiece may tend to cause the center punch to drift slightly from the desired location. It is true that a skilled worker can, after the first light striking of the center punch, by additional blows correct this drifting to some extent, but one cannot expect to have the final center-punch mark located more accurately than about 0.05 to 0.08 mm (0.002 to 0.003 inch).

Third, the center-drilling operation will usually decrease the accuracy of location to some extent. Again it is difficult to locate the mark from the center punch exactly under the center of the center drill. Also, the center drill may not be perfectly accurate in all respects, and it may wobble slightly.

The drilling operation may add to the inaccuracy. The drill may not start drilling exactly in the center of the center-drilled hole, and it may be deflected slightly as it drills into the workpiece. Thus, although the layout lines might be accurate to 0.02 mm (0.001 inch), the actual center of the hole may easily be off by as much as 0.13 mm (0.005 inch). Further, there is little assurance that the axis of the drilled hole is exactly normal to the surface on which the work was resting when the drilling was done, as a drill is slightly flexible and it can be deflected from the desired axis.

Similar difficulties may be encountered in producing other geometric surfaces at desired locations. Thus it is apparent that the problem of translating specified dimensions from a drawing into actual locations of component parts of a finished product is not only an important one but also a complicated one, and it is a matter that is of direct concern to the designer.

Toolmakers buttons. *Toolmaker's buttons* are a means for considerably improving the manual layout for holes. The essential features of these devices are shown in Figure 35-3. They are small, hollow, accurately ground cylinders, about 12.7 mm (½ inch) high and available in several diameters; the 12.7 mm (½ inch) size is most common.

To use toolmaker's buttons, the approximate location of the center of the desired hole is established by ordinary layout methods. The location does not have to be established exactly, but it should be within about 0.25 mm (0.010

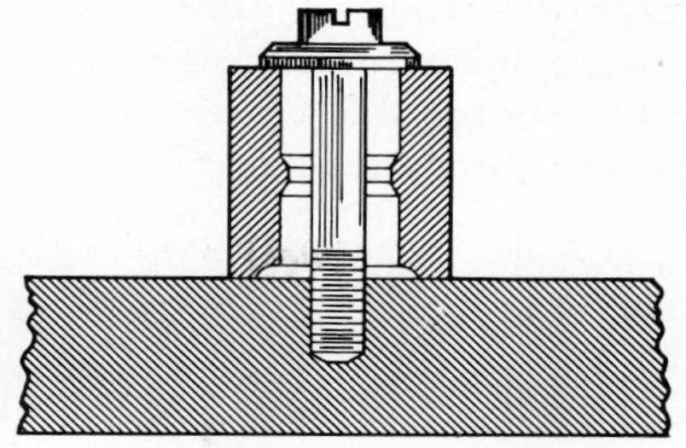

Sectional View of Button Applied

FIGURE 35-3. Toolmaker's button mounted on a workpiece, and sectional view showing the method of attachment.

inch). This location is center-punched, drilled, and tapped for a No. 5-40 screw. A toolmaker's button then is attached to the work by means of a No. 5-40 machine screw and a small washer, the screw being tightened sufficiently to hold the button firm, yet loose enough to permit it to be moved slightly under the washer by light tapping. Then, by using a dial gage or other suitable indicating device, the button is adjusted until its *axis* is exactly coincident with the axis of the desired hole. The attaching screw then is tightened to hold the button firmly in this position.

With the button properly located, the work is mounted in a lathe or other suitable machine tool, and located, by means of a dial or wiggler gage, so that the axis of the toolmaker's button is coincident with the axis of rotation of the work (or drill or boring tool if a rotating tool is to be used). The toolmaker's button then is removed. The hole is center-drilled and drilled somewhat undersize. Next, it is bored to approximate size. Because the axis of the bored hole is coincident and properly aligned with the axis of the toolmaker's button, it is exactly at the desired location, and all inaccuracies that might have existed in the drilled hole are eliminated. Such a hole often is finished to exact size by reaming. Although holes may be located to within ±0.013 mm (0.0005 inch) by means of toolmaker's buttons, it is obvious that the procedure is very laborious. It also is evident why machine layout methods, which require less skilled labor, are more economical and are used whenever possible.

MACHINE LAYOUT METHODS

By making use of the controlled movements of their worktables, layout can be accomplished on several machine tools in far less time and with greater accuracy than by hand methods. Referring to the part shown in Figure 35-1, if one of the longer edge surfaces were machined on a vertical milling machine, dimension *e* could readily be established merely by moving the machine table a distance equal to dimension e plus the diameter of the cutter. With the usual table controls the dimension could quickly be established to an accuracy of 0.02 mm (0.001 inch). On most machines equipped with digital readout systems, even greater accuracy could be obtained. Similarly, after completion of the lower edge (as pictured), dimension c could be established by moving the table a distance equal to c plus the radius of the milling cutter.

It is apparent that layout by means of the controlled motions available on a machine tool has a number of advantages over manual layout. It is both quicker and more accurate. Direct measurement is not required. As a consequence, vertical and horizontal milling machines, horizontal drilling, boring, and milling machines, and vertical drilling machines with controlled-movement tables are widely used in this manner. Since such machines tools usually can do several operations, such as milling, drilling, boring, and reaming, they are very flexible and useful because the necessary layout and all required machining operations often can be accomplished with a single setup.

Digital readout equipment. Digital readout devices, such as shown in Figure 35-4 and referred to previously, have revolutionized the controlling of table movements on machine tools, and have made layout on machines equipped with these devices much more rapid and accurate. These devices, usually electromechanical or optical-electrical, automatically indicate the position of a

FIGURE 35-4. Jig borer equipped with digital readout indicators, being used for boring a hole in a fixture. (*Courtesy Bendix Automation & Measurement Division.*)

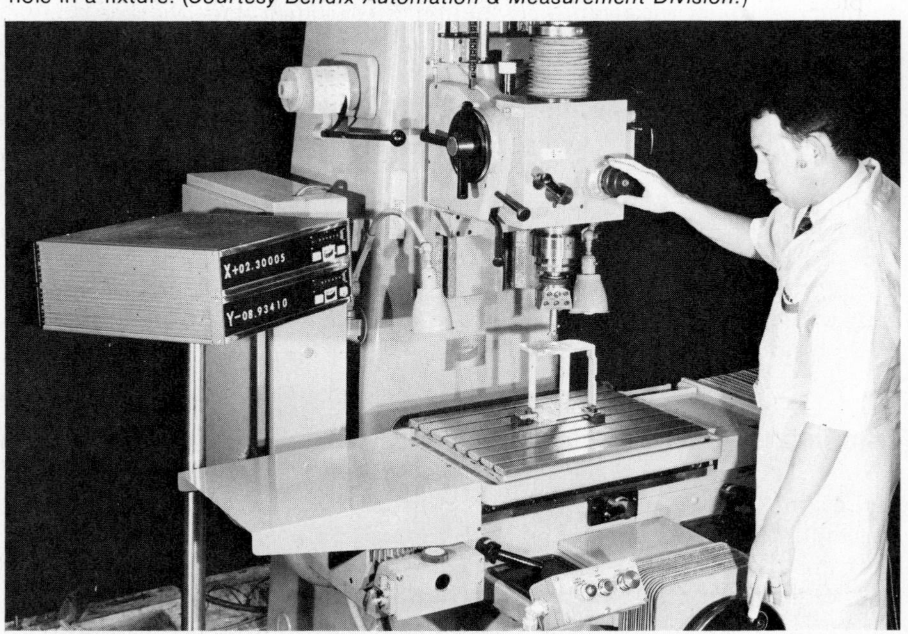

worktable in either two or three axis directions. Some operate through the table feed screws; others are connected between the table and the machine bed. Most systems have a "floating zero," meaning that the zero can be set for any position of the table merely by pushing a button or setting a switch. Accuracies and indications of 0.02 and 0.002 mm (0.001 and 0.0001 inch) are most common, although others can be obtained. Some are available that will read in either millimeters or inches by merely throwing a switch.

Such equipment can be used in several ways. One is to control all table movements from a single zero location. Referring to Figure 35-1, for example, the distances c and $c+d$, the locations of the two holes on one axis, would be obtained by moving the table the distance c and $c+d$, respectively, with the lower edge of the workpiece being at zero. Another method is to use an incremental procedure. With this procedure, after the dimension c was established, the zero would be reset, and the table could then be moved the required distance d for location of the center ot the second hole. Obviously, the procedure used is related to the manner in which the part drawing is dimensioned. Here, again, the designer and draftsman can facilitate the use of such equipment.

The use of digital readout equipment not only provides better accuracy and reduces errors in doing layout on a machine tool, but it also speeds the total machining process. The machinist can move the table, and thus the work, any desired distance more quickly without the necessity of counting the turns of a crank or reading graduated dials. He can know at all times the exact location of the machine spindle axis relative to some zero point on the workpiece. Most new machine tools are equipped with digital readout equipment, and many older machines are being retrofitted with such equipment.

Layout on jig borers. Where extremely accurate layout is required, as in making jigs, fixtures, and dies, special machine tools, called *jig borers,* are often used. Essentially, these are very accurately made vertical drilling machines that incorporate four special features:

1. Provision for accurately locating the worktable with respect to the cutting tool in at least two directions in a plane.
2. Oversized spindle, bearings, and quill so that light boring and grinding can be done, as well as drilling.
3. High accuracy in the movement of the critical parts, such as the rotation and feed of the spindle.
4. Provision to avoid the harmful effects of temperature changes.

Because the first three of these features are incorporated in a good many modern numerically controlled vertical milling machines and machining centers, permitting accuracies of ± 0.0025 mm (0.0001 inch) to be obtained readily, such machines are now used for much of the work that formerly was done only on jig borers. However, if greater accuracy is required, or if a considerable quantity of such high precision work must be done, jig borers generally are used, their use being restricted to high-precision jobs, usually not on large, heavy workpieces. In this manner there is less danger of the accuracy of the machine being lost through abuse.

In jig borers, accurate positioning of the worktable, in two mutually perpendicular directions, is achieved in several ways. One is the use of gage-block-quality end-measuring rods and a dial gage. Another is the use of very accurate lead screws. Most now utilize digital readout equipment that measures directly between the worktable and the machine bed. This method is employed in the jig borer shown in Figure 35-4 and provides rapid location to 0.0013 mm (0.00005 inch). Such a system does not require the table traversing screws to be made with extreme accuracy, and any wear or temperature effects in the screws do not affect the accuracy of the machine.

The jig borer shown in Figure 35-5 is equipped with both digital readout and tape control. It is arranged so that a tape may be prepared in the usual manner, or one can be prepared automatically by machining one part, using the digital readout system. The tape so prepared then can be used to control the machine automatically in machining subsequent, duplicate parts. This feature is particularly valuable where one may want to exactly duplicate a very accurate part at some later time.

Jig borers are built with the finest possible precision. Particular attention is given to the design of the spindle and spindle bearings to assure accu-

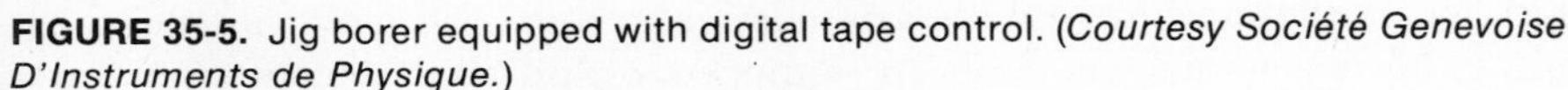

FIGURE 35-5. Jig borer equipped with digital tape control. (*Courtesy Société Genevoise D'Instruments de Physique.*)

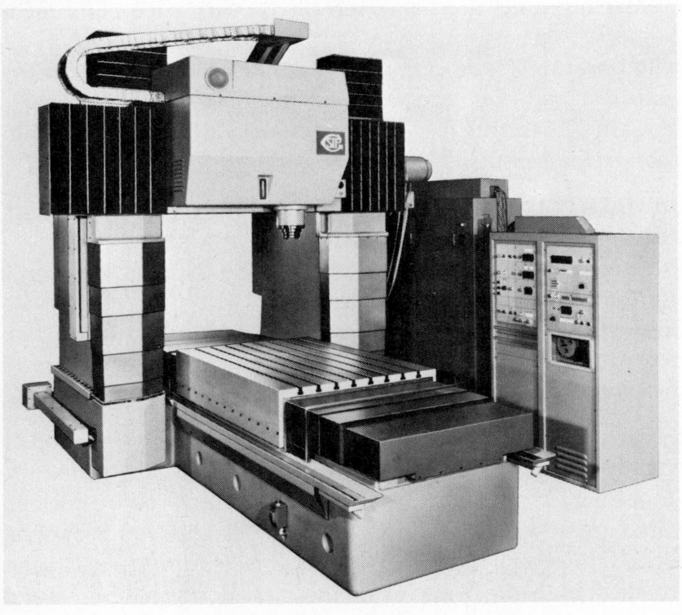

rate rotation and feed. The castings used are carefully annealed so that they will not warp with age. In several makes certain critical castings are made of Invar or other materials that have low-temperature coefficients, so that changes in temperature will not reduce the ultimate accuracy. Handwheels and other controls are designed to minimize the effect of the body heat of the operator.

When high accuracy is essential, it is desirable to locate jig borers in temperature-controlled rooms. In addition, some care may be necessary to provide coolant at a uniform temperature so that the work can be maintained at a fairly uniform temperature.

Although light milling can be done on some jig borers, such use is not recommended. They primarily are for center drilling, drilling, and boring. Some are equipped so that light grinding can be done, and similar machines are made specifically for jig grinding.

Layout on jig borers can be speeded materially if drawings of parts that are to be machined are dimensioned properly. In general, base-line, or reference-line, dimensioning should be used, as illustrated in Figure 35-6. The method shown, of enclosing the dimensions that the jig borer operator must use in boxes, is a good one; the other dimensions shown are for use in checking the work. This type of dimensioning is particularly helpful when digital readout equipment is used. However, digital readout equipment that has provision for a floating zero permits point-to-point dimensioning to be employed with less difficulty than would be experienced otherwise.

FIGURE 35-6. Metric drawing dimensioned for use with a jig borer.

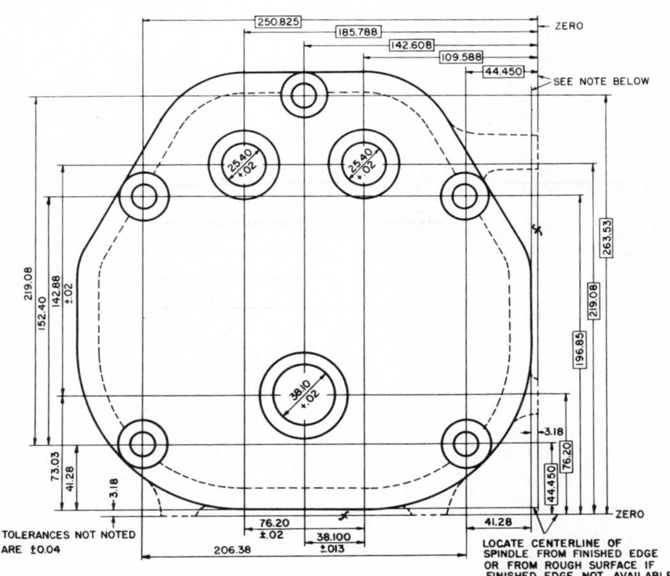

Because layout accuracy within 0.0025 mm (0.0001 inch) is readily obtainable on jig borers, in former years they often were used to produce small quantities of duplicate parts where extreme accuracy was required. At present, this seldom is done, because numerically and tape-controlled milling machines and machining centers routinely do such work at less cost.

Layout by photographic methods. Extensive use is made of photographic methods of layout, particularly in making large templates and in transferring location dimensions to large workpieces, such as are encountered in the aircraft industry. An accurate drawing is prepared, or a very accurate layout is made on a special type of plastic sheet. It then is photographed onto a sensitized glass plate, using a special template camera. After the glass plate has been developed, it can be used for projecting the layout image, full size, onto a sheet of metal that has been sensitized with a photographic emulsion. The exposed metal sheet then is developed in a large tank, thereby having the layout on it. In some cases the exposed and developed metal sheet is machined, according to the layout, and used as a template for layout of additional workpieces.

Review questions

1. What is the purpose of layout?
2. Explain why layout is directly related to the machine tool that will be used for production of a part.
3. Why should layout lines not be scratched directly into a workpiece?
4. Where may inaccuracy occur in laying out and drilling a hole at a desired location?
5. Why is it not necessary for the hole that is used for attaching a toolmaker's button to be located exactly?
6. What are three machine tools on which layout can readily be done?
7. What is the advantage of a "floating zero" in connection with a digital readout system?
8. Explain how a draftsman can facilitate the use of digital readout and control equipment on a machine tool.
9. To what machine tool is a jig borer most closely related?
10. What methods commonly are used to provide accurate table control on jig borers?
11. Why do most jig borers not use the table lead screws for dimensional control?
12. What special precautions would you suggest relative to the use of cutting fluids on a jig borer?
13. How is layout done photographically?
14. For what types of applications is photographic layout commonly used?

Case study 25.
THE BRONZE BOLT MYSTERY

You are employed by the Mountainous Irrigation Company, which had 1000 special bolts of phosphor bronze (10 per cent Sn) made to the design shown in Figure CS-25 for use in its various pumping plants. Within a few weeks after a number of these bolts were installed, the heads broke off several of them, the fracture being along the dashed line shown in Figure CS-25. The broken bolts were replaced, with special precautions being taken to assure that they were not overloaded. Again, after only a few days, several of these replacement bolts broke in the same manner. You have been assigned the task of determining the cause of the failures. Upon examining several unused bolts, you discover that many of them have fine cracks at the intersection of the head and body. Upon checking with the manufacturer, you learn that the bolts were made by ordinary heading procedures. (1) Outline the procedure you will use to determine the cause of the cracks. (2) What do you suspect is the cause? (3) How could the difficulty have been avoided?

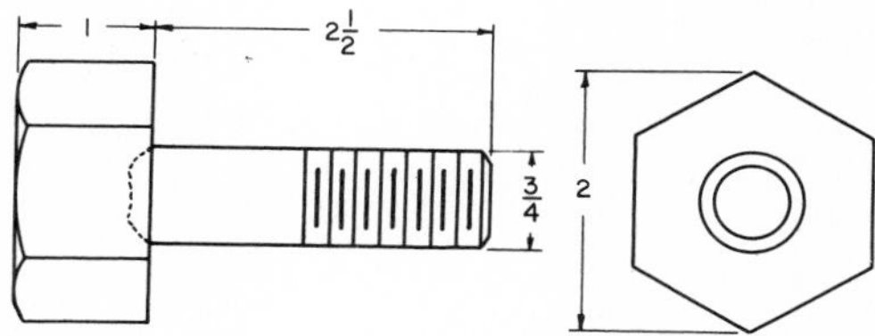

FIGURE CS-25. Special bronze bolt, showing location of fracture (dashed line).

CHAPTER 36

Jigs and fixtures

Four objectives are of prime importance in manufacturing: (1) maximizing the productivity of costly machine tools by minimizing the machine idle time required for setting up workpieces in the machine, (2) achieving and maintaining the required dimensions and quality, (3) minimizing the amount of labor required, and (4) reducing the degree of skill that is required. *Jigs* and *fixtures* are extremely important aids used to help achieve these objectives.

In previous chapters attention has been directed repeatedly to the manner in which workpieces are mounted and held in the various machines. In Chapter 35 the very important subject of layout was discussed, with special emphasis on the time and skill required to achieve accurate results. In this chapter jigs and fixtures will be considered as important production tools or adjuncts, with primary attention being directed toward their essential characteristics, the manner in which they accomplish layout and/or work holding, their relationship to the machine tools and manufacturing processes, and the economics of their use. Only minimal attention will be given to the details of their design. Just as the engineer who is concerned with manufacturing must understand the functioning, capabilities, and proper utilization of machine tools but almost never designs them, similarly he or she should have a thorough understanding of the basic principles of jigs and fixtures so as to utilize them effectively. However, in most cases, he can leave the design of their details to the tool designer.

Definitions. To understand and utilize jigs and fixtures effectively, it is important to define them clearly. Unfortunately, and particularly in earlier years, the terms "jigs" and "fixtures" sometimes are interchanged. This is not too surprising in that a jig quite often also performs the function of a fixture; the reverse is never true.

In defining jigs and fixtures, and in avoiding confusing the two, it is helpful to consider the subject of dimensioning as used in drafting practice. Dimensions are of two types—*size* and *location.* Size dimensions denote the size of geometrical shapes—cylinders, cubes, parallelopipeds, etc.—of which objects are composed. Location dimensions, on the other hand, determine the location of these geometrical shapes *with respect to each other.* With location dimensions in mind, one can precisely define a jig as follows: *a jig is a special device which, through built-in features, determines location dimensions that are produced by machining or fastening operations.* The key requirement of a jig is that it determine a location dimension. Thus jigs automatically accomplish layout.

In establishing location dimensions, jigs may do a number of other things. They frequently guide tools, as in drill jigs, and thus determine the location of a component geometrical shape. However, they do not always guide tools, as in the case of welding jigs where they hold component parts in a desired relationship with respect to each other while an unguided tool accomplishes the fastening. Thus the guiding of a tool is not a necessary requirement of a jig.

Similarly, jigs usually hold the work that is to be machined or fastened. However, this is not always true, because in certain cases the work actually supports the jig. Thus, although a jig *may* incidentally perform other functions, the basic requirement is that, through qualities that are built into it, certain critical dimensions of the workpiece are determined.

A fixture is a special device that holds work during machining or assembly operations. The key characteristic is that it is a *special* work-holding device, designed and built for a particular part or shape. A general-purpose device, such as a vise or clamp, is not a fixture. Thus a fixture has as its specific objective the facilitating of *setup,* or making holding easier.

Because many jigs hold the work while determining critical dimensions, it is apparent that in such cases they meet all the requirements of a fixture. But it is equally evident that fixtures never determine dimensions of parts which they hold—a basic requirement of a jig.

From their definition and function, it follows that jigs are associated with operations, whereas fixtures most commonly are related to specific machine tools. Thus the most common jigs are drilling jigs, reaming jigs, welding jigs, and assembly jigs. In these uses they often are not fastened to the machine table, being free to be moved so as to permit the proper registering of the work and the tool. Fixtures, on the other hand, most frequently are attached to a machine tool or table. Consequently, they are associated, in name, with the particular tool with which they are used. Because their function of holding is a broad one, there are many types of fixtures, such as milling fixtures, broaching fixtures, lathe fixtures, grinding fixtures, assembly fixtures, and plating fixtures.

Some basic factors in jig and fixture design. In order to assure their optimum functioning, the following six factors should be considered in designing jigs and fixtures:

1. Clamping of the work.
2. Support of the work against any forces that are imposed by the process.
3. Location of the work to provide required dimensional control.
4. Guiding of the tool, when required.
5. Provision for chips, if they are made.
6. Rapid and easy operation.

Items (3) and (4) apply only to jigs.

Clamping of the work is closely related to support of the work. However, there are certain principles that relate specifically to clamping. First, clamping *stresses* should be kept low. Any clamping, of course, induces some stresses that tend to cause some distortion of the workpiece, usually elastic. If this distortion is measurable, it will cause some inaccuracy in final dimensions, as illustrated in an exaggerated manner in Figure 36-1. The obvious solution is to spread the clamping forces over a sufficient area to reduce the stresses to a magnitude that will not produce appreciable distortion.

Second, the clamping forces should direct the work against the points of work support. This principle is illustrated in Figure 36-2. It is not safe to assume that if there is no *designed* clamping-force component either upward or downward, the work will be held properly and will not tend to rise from the

FIGURE 36-1. Exaggerated illustration of the manner in which too-large clamping forces can affect the final dimensions of a workpiece.

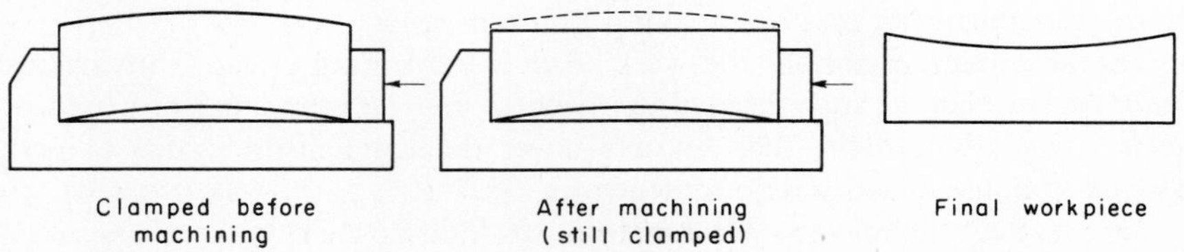

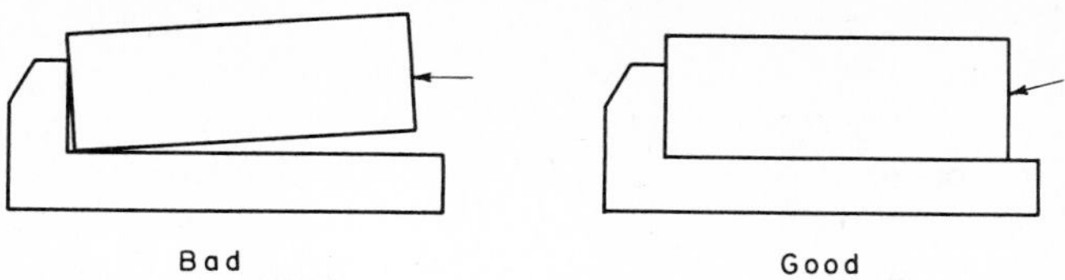

FIGURE 36-2. Effect of improperly directed clamping forces.

supports. Clamped surfaces often have some irregularities that may produce force components in an undesired direction. Consequently, clamping forces should be applied in directions that will assure that the work will remain in the desired position.

An important corollary to the second principle is that, whenever possible, jigs and fixtures should be designed so that the forces induced by the cutting tool aid in holding the workpiece in position against the supports. These forces are predictable, and proper utilization of them can materially aid in reducing the magnitude of the required clamping stresses.

The third principle is that as many operations as practicable should be performed with each clamping of the workpiece. This principle has both physical and economic aspects. Because some stresses result from each clamping, with the possibility of accompanying distortion, greater accuracy is probable if as many operations as practicable are performed with each clamping. Thus a jig or fixture should be designed to serve for several operations when possible. From the economic viewpoint it is obvious that if the number of jigs or fixtures is reduced, a smaller investment of capital usually will be required and fewer man-hours will be necessary for handling the workpiece in putting it in and removing it from the jigs or fixtures.

The locating of a workpiece in a jig or fixture requires adherence to the "three–two–one" principle—at least three locating points, or stops, are required to locate an object in the first plane, at least two points in a second plane to locate it in that plane, and at least one point to locate it in a third plane, assuming that the planes are not parallel and preferably are perpendicular.

In addition to locating the work properly, the stops or work-supporting areas must be arranged so as to provide adequate support against the forces imposed on the workpiece by the cutting tool. Such consideration of the cutting forces should be a routine step in the design of all jigs and fixtures that are used in connection with machining operations. Figure 36-3 illustrates this concept. As shown in Figure 36-3a, whenever possible the cutting force should always act against a fixed portion of the jig or fixture and not against a movable section. This aids materially in permitting lower clamping stresses to be utilized. Figure 36-3b illustrates the principle of keeping the points of clamping as nearly as possible in line with the action forces of the cutting tool so as to reduce their tendency to pull the work from the clamping jaws. Com-

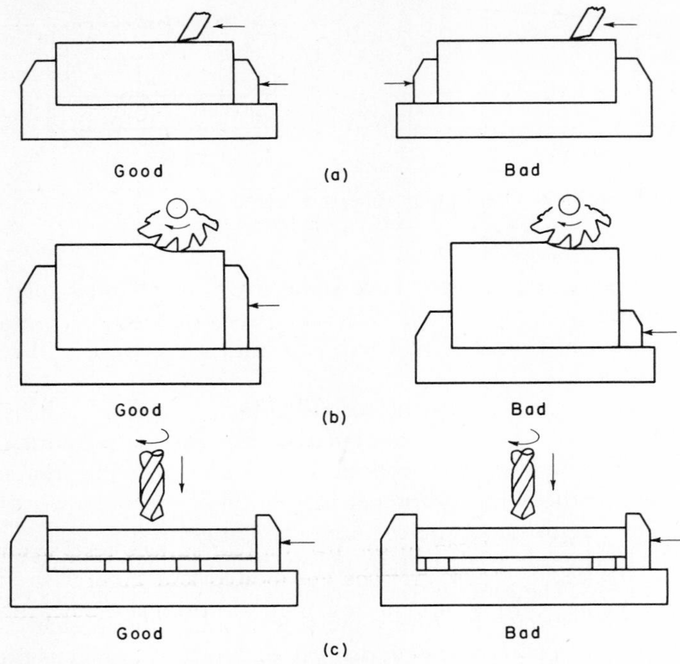

FIGURE 36-3. Proper work support to resist the forces imposed by cutting tools.

pliance with this principle results both in lower clamping stresses and less massive clamping devices. If the principle illustrated in Figure 36-3c is not followed, the action forces of the cutting tool may distort the work, with resulting inaccuracy, or broken tools. Compliance with these three principles also aids materially in reducing the possibility of workpiece vibration and tool chatter.

The principles of work location and tool guidance are illustrated in Figures 36-4 and 36-5. The two mounting holes in the base of the bearing block are located and drilled with the use of the drill jig shown in Figure 36-5. The dimensions *A, B,* and *C,* shown in Figure 36-4, are determined by the jig. There also is one other location dimension that must be controlled. This is the implied requirement that the axes of the mounting holes be at right angles to the bottom surface of the block.

The way in which these dimensions are determined in the finished block can be seen by reference to Figures 36-5 and 36-6. The bearing block rests in the jig on four buttons marked *x* in Figure 36-5. These buttons, made of hardened steel, are set into the bottom plate of the jig and are accurately ground so that their surfaces are in a single plane. The left-hand end of the block is held against another button marked *y* in Figure 36-5. This locating button is built into the jig so that its surface is at right angles to the plane of the *x* but-

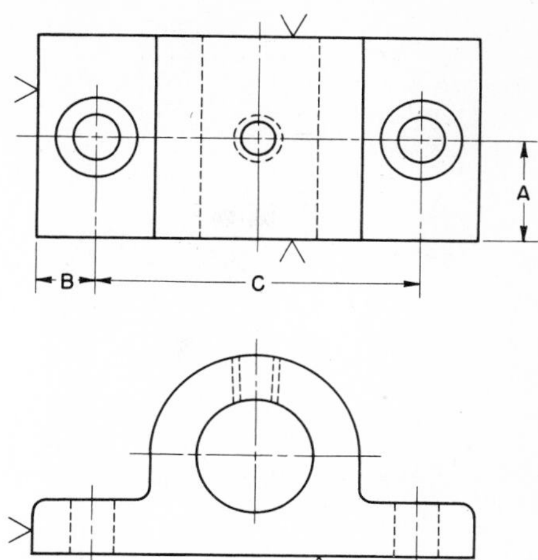

FIGURE 36-4. Bearing block.

tons. When the block is placed in the jig, its rear surface rests against three more buttons marked *z*. These buttons are located and ground so that their surfaces lie in a plane that is at right angles to the planes of both the *x* and *y* buttons.

The use of four buttons in the bottom of this jig (*x* buttons) appears not to adhere to the three–two–one principle stated previously. However, although only two *x* buttons would have been required for complete location, the use of only two buttons would not have provided adequate support adjacent to the holes that were to be drilled. This would have violated one of the fundamental design principles. Thus the three–two–one principle is a *minimum* concept, and often must be exceeded.

To assure that the mounting holes are drilled in their proper locations, the drill must be located and then guided during the drilling process. This is

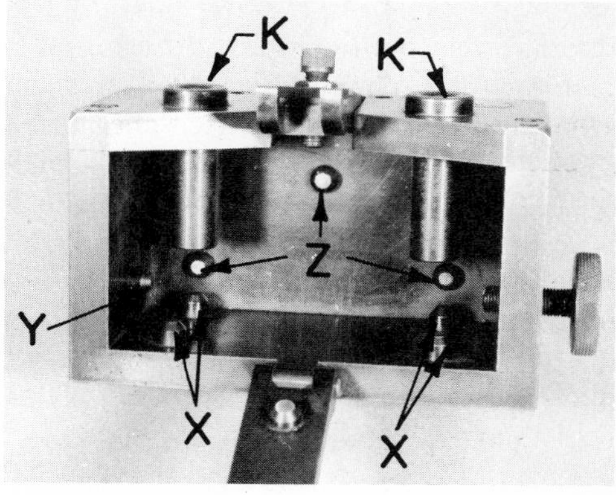

FIGURE 36-5. Box-type drill jig for drilling the mounting holes in the bearing block shown in Figure 36-4.

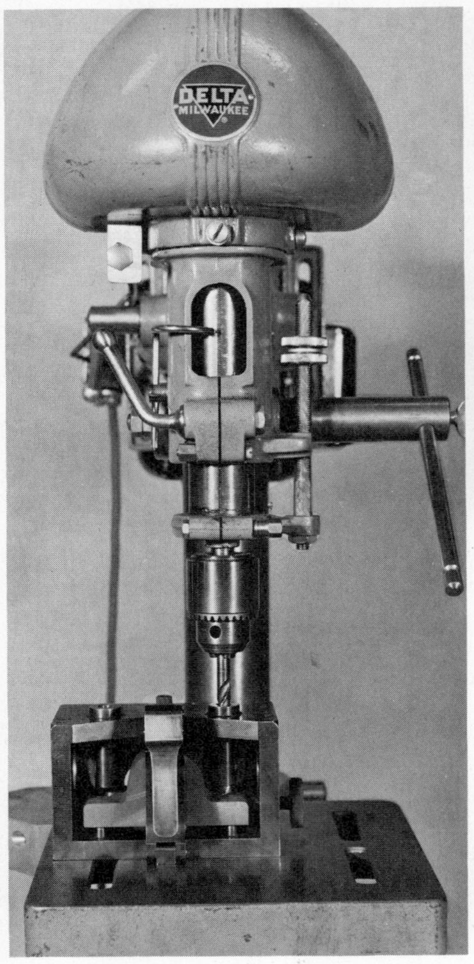

FIGURE 36-6. (*Left*) Drilling the mounting holes in a bearing block, using the drill jig shown in Figure 36-5. (*Right*) Close-up view, showing the block in the jig, resting on the locating buttons, and the drill being guided by the drill bushing.

accomplished by the two drill bushings, marked *K* in Figure 36-5. Such drill bushings are accurately made of hardened steel with their inner and outer cylindrical surfaces concentric. The inner diameter is made just enough larger than the drill that is to be used—usually 0.013 to 0.025 mm (0.0005 to 0.001 inch)—so that the drill can turn freely but not drift appreciably. The bushings are mounted in the upper plate of the jig by means of a press fit and positioned so that their axes are exactly perpendicular to the plane of the *x* buttons, at a distance *A* from the *z* buttons and at distances *B* and *C*, respectively, from the plane of the *y* button. Note that the bushings are sufficiently long that the drill is guided close to the surface where it will start drilling. Consequently, when the bearing block is properly placed and held in the jig,

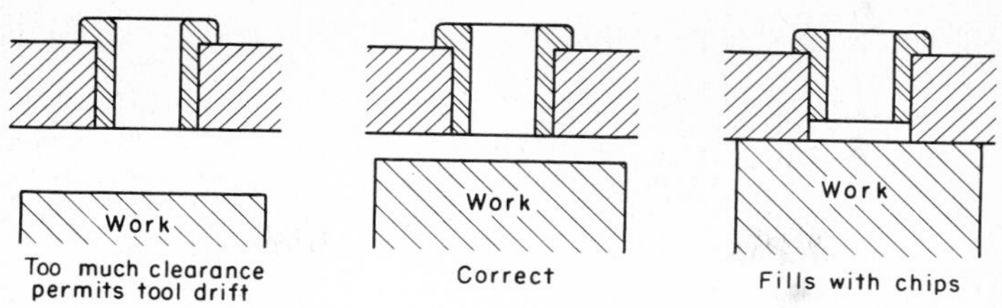

FIGURE 36-7. Importance of proper chip clearance.

the drill will be located and guided by the bushings so that the critical dimensions on the workpiece will be correct.

When jigs or fixtures are used in connection with chip-making operations, adequate provision must be made for the chips that are produced and for their easy removal. This is essential for several reasons. First, if chips become packed around the tool, heat will not be carried away, and tool life can be decreased or the tool actually broken. Figure 36-7 illustrates how insufficient clearance between the end of a drill bushing and the workpiece can prevent the chips from escaping, whereas too much clearance may not provide adequate drill guidance and can result in broken drills.

A second reason why chips must be considered relates to preventing them from interfering with the proper seating of the work in the jig or fixture. This is illustrated in Figure 36-8. Even though chips and dirt always have to be cleared from the locating and supporting surfaces by a worker or by automatic means, such as an air blast, the design details should be such that chips and other debris will not readily adhere to, or be caught in or on, the locating surfaces, corners, or overhanging elements and thereby prevent the work from seating properly. Such a condition results in distortion, high clamping stresses, and incorrect finished dimensions.

Another, and very important, factor in the design and selecting of jigs and fixtures is the ease and rapidity with which they can be utilized by the operating personnel. Jigs and fixtures are used only when considerable quantities of production are involved, and their primary purpose is to increase the

FIGURE 36-8. Bad and good methods of providing chip clearance to assure proper seating of the work.

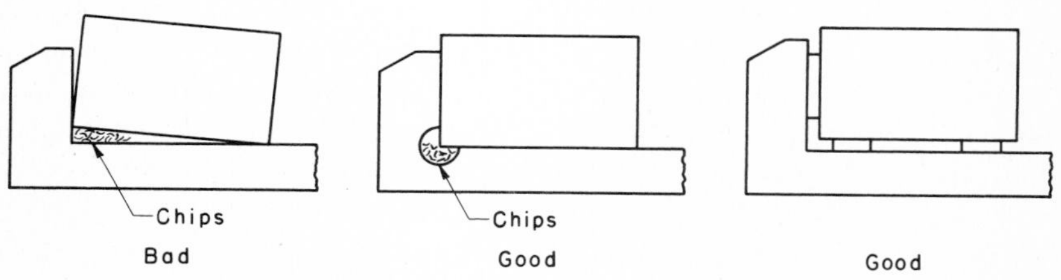

TABLE 36-1. Effects of reducing setup time by a constant amount for two production cycles

Case	Setup Time (minutes)	Machining Time (minutes)	Total Cycle Time (minutes)	Cost per Unit	Cost Saving per Unit (%)	Productivity Increase (%)
A (labor $6/hr; machine $3.50/hr)	3 2	10 10	13 12	$2.06 $1.90	=7.8	8.3
B (labor $4.00/hr; machine $6/hr)	3 2	5 5	8 7	$1.33 $1.17	= 12.0	14.1

productivity of workers and machines. While work is being put into or being taken out of jigs and fixtures, they, and usually also the machines with which they are used, are nonproductive. Although a small amount of time saved is of some importance when a machine of low productivity is involved, the same amount of saved time is of much greater importance when a machine of higher productivity is being utilized.

This important relationship is shown in Table 36-1. In this table the effect of reducing setup time by 1 minute is shown for two conditions—one where the machining time is 10 minutes and the other where it is only 5 minutes. In the first case, *A*, a skilled operator is used, who receives $6.00 per hour, and the cost for the machine is $3.50 per hour, which includes all overhead and interest costs. In the second case, *B*, a less-skilled operator, paid $4.00 per hour, can be employed because the machine is more nearly automatic, but this results in the cost for the machine being $6.00 per hour. The results of decreasing the setup time by a constant amount are very evident. Although the cost saving in each case happens to be the same ($0.16 per unit), the percentage saving and the percentage increase in productivity are much greater in the case of the more productive machine. This is typical of what can happen when machine-hour costs and productivity are high and emphasizes the importance of tooling adjuncts that reduce setup time when costly machine tools are utilized. Such tools are increasingly common.

There are several ways in which jigs and fixtures can be made easier and more rapid to use. An important one relates to the method in which the work is clamped. Some methods can be operated much more readily than others. For example, in the drill jig shown in Figure 36-5, a knurled clamping screw is used to hold the block against the buttons at the end of the jig. To clamp or unclamp the block in this direction requires several motions. On the other hand, a cam latch is used to close the jig and hold the block against both the bottom and rear locating buttons. This type of latch can be operated with a single motion.

Ease of operation of jigs and fixtures, as well as the controls on machine tools, not only directly increases the productivity of such equipment, it also does so indirectly through better satisfied workers and better treatment of the tools.

Transfer of skill. The construction of a jig, with the required locating surfaces, axes of the drill bushings, locating buttons, and so on, requires careful layout and machining. This requires the services of a skilled toolmaker and the use of precision tools, such as a jig borer. In the making of the jig, the highly skilled toolmaker transfers some of his skill to the jig, so that through the use of the jig a less skillful person can produce a workpiece with as high a degree of accuracy as if it had been made by the more highly skilled worker. This relationship is expressed by the equation

$$\text{total skill required for the job} = \text{skill built into the jig} + \text{skill required from the worker}$$

This relationship is an extremely important one in manufacturing. The greater the skill that is built into a jig or fixture, the less is the amount required of the worker who uses the tooling in producing a product. Similarly, the more of the critical setup that is built into a fixture or jig, the less is the time required to set up each piece in a machine tool. Consequently, well-designed jigs and fixtures not only reduce the requirement for highly skilled workers in production operations, they also increase the productivity of both workers and machines and thereby the availability of goods for society. This is especially important when very costly machines are utilized, because nonproductive time is reduced.

JIGS

Types of jigs. Because jigs are designed to facilitate certain processes, they are made in several basic forms and carry names that are descriptive of their general configurations or predominant features. Several of these are illustrated in Figure 36-9.

A *plate jig* is one of the simplest types, consisting only of a plate that contains the drill bushings, and a simple means of clamping the work in the jig, or the jig to the work. In the latter case, where the jig is clamped to the work, it sometimes is called a *clamp-on jig.* Such jigs frequently are used on large parts, where it is necessary to drill one or more holes that must be spaced accurately with respect to each other, or to a corner of the part, but that need not have an exact relationship with other portions of the work.

Channel jigs also are simple and derive their name from the cross-sectional shape of the main member. They can be used only with parts having fairly simple shapes.

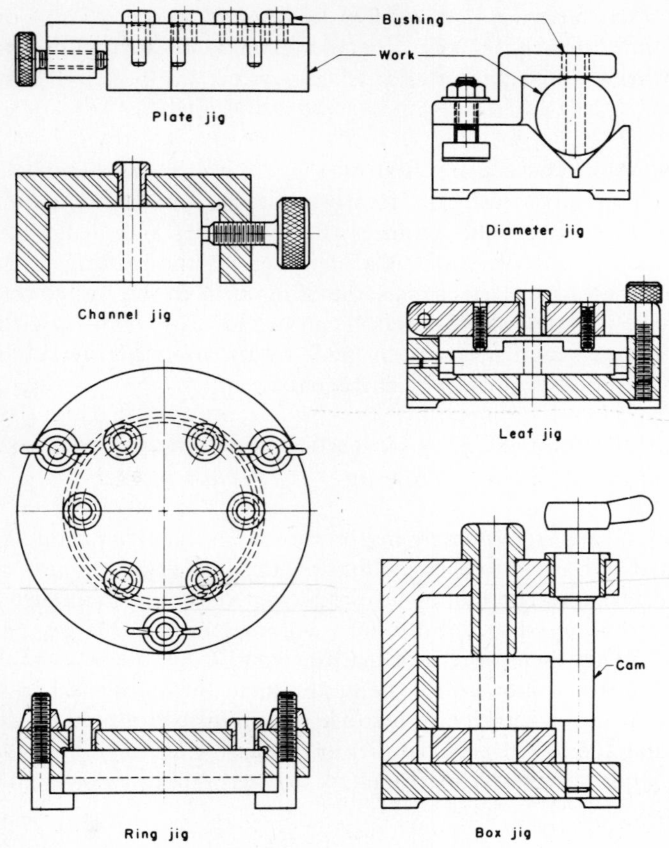

FIGURE 36-9. Common types of jigs.

Ring jigs are used only for drilling round parts, such as pipe flanges. The clamping must be sufficient to prevent the part from rotating in the jig.

Diameter jigs provide a means of locating a drilled hole exactly on a diameter of a cylindrical or spherical piece.

Leaf jigs derive their name from the hinged leaf or cover that can be swung open to permit the workpiece to be inserted and then closed to clamp the work in position. Drill bushings may be located in the leaf as well as in the body of the jig so as to permit locating and drilling holes on more than one side of the workpiece. Such jigs, which require turning to permit drilling from more than one side, are called *rollover jigs*.

Box jigs are very common, deriving their name from their boxlike construction. They have five fixed sides and a hinged cover or leaf, or, as shown in Figure 36-9, a cam that locks the workpiece in place. Usually, the drill bushings are located in the fixed sides to assure retention of their accuracy.

The fixed sides of the box usually are fastened by means of dowel pins and screws so that they can be taken apart and reassembled without loss of accuracy. Because of their more complex construction, box jigs are costly, but their inherent accuracy and strength make them justified where there is sufficient volume of production. They have two obvious disadvantages: (1) it usually is more difficult to put work into them than into more simple types, and (2) there is a greater tendency for chips to accumulate within them. Figure 36-5 shows a box-type jig.

Because jigs must be constructed very accurately and be made sufficiently rugged so as to maintain their accuracy despite the abuse to which they inevitably are subjected in use, they are expensive. Consequently, several methods have been devised to aid in lowering the cost of making jigs. One method involves the use of simple, standardized plate and clamping mechanisms, called *universal jigs,* such as shown in Figure 36-10. These can easily be equipped with suitable locating buttons and drill bushings to construct a jig for a particular job. Such universal jigs are available in a variety of configura-

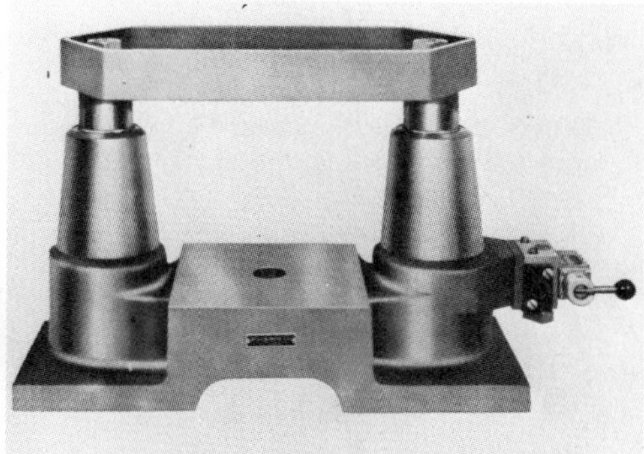

FIGURE 36-10. (*Top*) Two types of universal jigs, manual (*left*) and power-actuated (*right*). (*Right*) Completed jig, made from unit shown at upper right. (*Courtesy Cleveland Universal Jig Division, The Industrial Machine Company.*)

FIGURE 36-11. Typical standard components available for use in constructing jigs and fixtures. (*Courtesy Brown & Sharpe Manufacturing Co.*)

tions and sizes and, because they can be produced in quantities, their cost is relatively low. However, the variety of work that can be accommodated by such jigs obviously is limited.

A second method of reducing the cost of constructing jigs and fixtures is the use of standardized components, such as drill bushings, locating buttons, and clamping devices. As shown in Figure 36-11, a wide variety of such components is available.

Because *assembly jigs* usually must provide for the introduction of several component parts and the use of some type of fastening equipment, such as welding or riveting, they commonly are of the open-frame type. Such jigs are widely used in the automobile and aircraft industries. A very large jig of this type is shown in Figure 36-12.

Since fixtures are for the sole purpose of facilitating setup and clamping, they tend to be simpler than jigs and more open in design, to permit easy placement and removal of the workpiece. However, those details in jigs that involve support and clamping apply in the same manner to fixtures. In most instances, fixtures are attached to a machine table or to a workbench, so some attention must be given to this matter. If there is some possibility that the fixture may need to be used on more than one machine, the method of attachment should be designed so as to permit this to be done.

FIGURE 36-12. Large assembly jig used in an aircraft factory. This jig is constructed mainly of reinforced concrete. (*Courtesy Boeing Company.*)

Because fixtures are less complex than jigs, they are much less expensive to construct and thus can be economically justified for smaller production quantities. Although most fixtures are made specifically for given jobs, in many cases satisfactory results can be obtained by simple modification of standard devices. An example is the addition of special jaws to a standard vise, as shown in Figure 36-13. Frequently, minor modifications can be made to standard clamping components, such as those shown in Figure 36-11. By exercising a little ingenuity, very effective fixtures often can be constructed at very modest cost and thus make their use economical for quite small production volumes, with accompanying substantial gains in productivity.

ECONOMIC JUSTIFICATION OF JIGS AND FIXTURES

As has been discussed previously, jigs and fixtures are expensive, even when designed and constructed by using standard components. Obviously, their

FIGURE 36-13. Use of special adaptors for vise jaws.

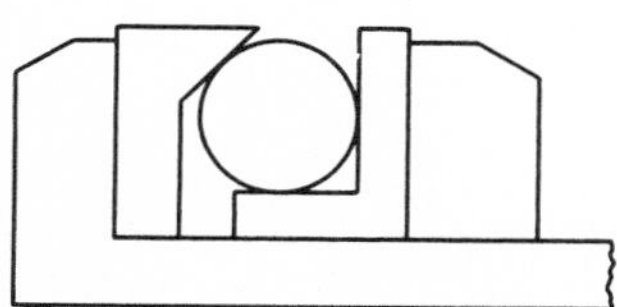

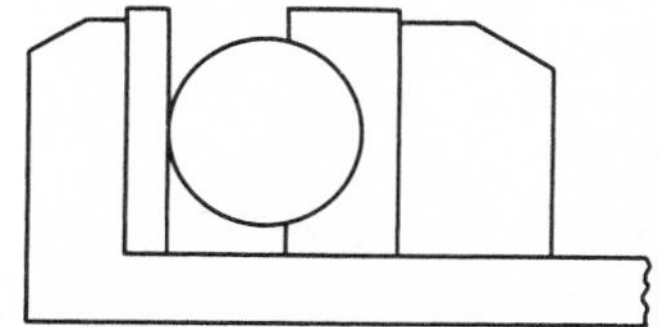

cost is a part of the total cost of production, and one must determine whether they can be justified economically by the savings in labor and machine cost that will result from their use.

In order to determine the economic justification of any special tooling, the following factors must be considered:

1. The cost of the special tooling.
2. Interest or profit charges on the tooling cost.
3. The savings in production labor cost resulting from the use of the tooling.
4. The savings in production machine cost due to increased productivity.
5. The number of units that will be produced using the tooling.

The economic relationship between these factors can be expressed in the following manner:

$$\underbrace{\underbrace{[\underbrace{(R)(t)}_{\substack{\text{labor cost}\\\text{per piece}\\\text{without}\\\text{tooling}}} + \underbrace{(R_m)(t)}_{\substack{\text{machine and}\\\text{overhead cost}\\\text{per piece}\\\text{without}\\\text{tooling}}}]}_{\text{total cost per piece without tooling}} - \underbrace{[\underbrace{(R_t)(t_t)}_{\substack{\text{labor cost}\\\text{per piece}\\\text{with}\\\text{tooling}}} + \underbrace{(R_m)(t_t)}_{\substack{\text{machine and}\\\text{overhead cost}\\\text{per piece}\\\text{with tooling}}}]}_{\substack{\text{total cost per piece}\\\text{using tooling (exclusive}\\\text{of tooling costs)}}}}_{\text{savings per piece (exclusive of tooling costs)}} \geqq \underbrace{\frac{\overbrace{C_t}^{\substack{\text{cost of}\\\text{tooling}}} + \overbrace{(C_t/2)(n)(i)}^{\substack{\text{interest on}\\\text{tooling cost}}}}{N}}_{\substack{\text{tooling cost}\\\text{per piece}}} \tag{36-1}$$

where R = labor rate per hour, without tooling
R_t = labor rate per hour, using tooling
t = hours per piece, without tooling
t_t = hours per piece, using tooling
R_m = machine cost per hour, including all overhead
C_t = cost of the special tooling
n = number of years tooling will be used
i = interest or profit rate invested capital is worth
N = number of pieces that will be produced with the tooling

This equation can be expressed in a simpler form:

$$(R + R_m)t - (R_t + R_m)t_t \geqq \frac{C_t}{N}\left(1 + \frac{n \times i}{2}\right) \tag{36-2}$$

This equation assumes straight-line depreciation and computes interest on the average amount of capital invested throughout the life of the tooling.[1] Where the time over which the tooling is to be used is less than 1 year, companies often do not include an interest cost. If this factor is neglected, the right-hand term of equation (36-2) reduces to C_t/N.

Equations (36-1) and (36-2) assume that the material cost will be the same regardless of whether special tooling is used. This is not always true. Although these equations are not completely accurate for all cases, they are satisfactory for determining tooling justification in most cases, because the life of such tooling seldom exceeds 5 years, and more frequently does not exceed 2 years. Furthermore, the savings usually should exceed the tooling costs by a substantial margin before special tooling is adopted.

The following example illustrates the use of equation (36-2) to determine tooling justification. In drilling a series of holes on a radial drill, the use of a drill jig will reduce the time from ½ hour per piece to 12 minutes. If a jig is not used, an A-grade machinist must be used, whose hourly rate is $6.00. If the jig is used, the job can be done by a B-grade machinist, whose rate is $4.50 per hour. The hourly rate for the radial drill is $3.75.

The cost of making the jig would include $150 for design, $75 for material, and 50 hours of toolmaker's labor, which is charged at the rate of $10 per hour to include all machine and overhead costs in the toolmaking department.

Investment capital is worth 8 per cent to the company. It is estimated that the jig would last 2 years and that it would be used for the production of 400 parts over this period. Is the jig justified? How many parts would have to be produced with the jig for it to "break even"?

To use equation (36-2), the cost of the jig, C_t, must be determined.

$$C_t = \$150 + \$75 + \$10 \times 50 = \$725$$

Substituting the values given in equation (36-2).

$$(\$6.00 + \$3.75)0.5 - (\$4.50 + \$3.75)0.2 \geqq \frac{\$725}{400}\left(1 + \frac{2 \times 0.08}{2}\right)$$

$$\$3.23 > \$1.96$$

Thus, use of the jig is justified. By omitting the value 400 in the solution above and solving for N, it is found to be 242+ pieces, so that at least 243 pieces would have to be produced with the jig for it to break even or pay out.

It should be noted that equations (36-1) and (36-2) assume that the production time of the machines that would be made available by the use of the special tooling can be used for other operations. If this is not the case,

[1] For the use of the slightly more complex concept of "average interest" or the use of more accurate sinking fund depreciation, see *Engineering Economy* by E. P. DeGarmo, J. R. Canada, and W. G. Sullivan, 6th ed. (New York: Macmillan Publishing Co., Inc. 1979).

there being no other use for the machines, the cost analysis should be altered to take this important fact into account. Otherwise, the tooling justification may be substantially in error.

Numerically controlled machine tools, discussed in Chapter 39, sometimes can eliminate the need for jigs, but again the choice should be based on an economic analysis.

Review questions

1. What distinguishes a jig from a fixture?
2. Jigs have been called "automatic layout devices." Explain.
3. An early treatise defined a jig as "a device that holds the work and guides a tool." Why was this definition incorrect?
4. Why would an ordinary vise not be considered to be a fixture?
5. What six basic factors should be considered in designing jigs and fixtures?
6. What difficulties can result from not keeping clamping stresses low in designing jigs and fixtures?
7. Explain the three–two–one concept related to jig design.
8. Which of the six basic design principles relating to jigs and fixtures would most often require the three–two–one concept to be exceeded?
9. Why is it desirable to have drill bushings extend as close to the workpiece as practicable?
10. What are two reasons for not having drill bushings come close to the workpiece?
11. Why does the use of down milling often make it easier to design a milling fixture than if up milling were used?
12. Explain what is meant by the expression, "a jig is a skill-transfer device."
13. A large assembly jig for an airplane-wing component gave difficulty when it rested on four points of support but was satisfactory when only three supporting points were used. Why?
14. Explain why the use of a given fixture may not be economical when used with one machine tool but may be economical when used in conjunction with another machine tool.
15. What are two reasons why rollover jigs are advantageous?
16. Using the following values, determine the number of pieces that would have to be made to justify the use of a jig costing \$3000:

$R = \$5.75$	$t_t = 1\frac{1}{4}$	$R_m = \$4.50$
$R_t = \$4.50$	$t = 2\frac{1}{4}$	$n = 3$
	$i = 10\%$	

Case study 26.
THE HOLES IN COLUMNS

To install some new equipment in one bay of an assembly plant, it is necessary to drill six holes, 9.53 mm (3/8 inch) in diameter, in a circular pattern 127 mm (5 inches) in diameter, on each of eight vertical steel columns at a height of 6.1 meters (20 feet) above the floor. It is essential that the holes be accurately spaced to within 5 minutes of arc and that their axes be parallel and normal to the face of each column. The face of each column is flush with a wall surface, so mechanical clamping or attachment cannot be used, and the building code will not permit welding. How would you go about this job? List the equipment that you would use.

CHAPTER 37

Decorative and protective surface treatments

A large proportion of all manufactured products must be given some type of decorative or protective surface treatments before they can be sold or used. Handling and the various manufacturing processes leave scratches, burrs, fins, fine pores, or other blemishes which detract from their appearance or present possible hazards to users. The most commonly used materials, such as most irons and steels, do not inherently possess the colors that customers, rightly or wrongly, want in products—especially in large-volume, consumer goods. Materials often are not adequately resistant to the environments in which they will be used. As materials become more scarce and more costly, there is a pressing need to substitute basically inferior materials by modifying their surfaces to withstand service conditions. As a consequence, after achieving their desired shape, most manufactured products require one or more additional operations to clean, protect, or color them.

These important decorative and protective surface treatments add to the cost of manufactured products. Further, as with other manufacturing processes, there is often a definite relationship between design and these finishing processes. In recent years, a large amount of attention has been devoted to developing finishing processes and equipment that enable them to be accomplished successfully in mass quantities at low cost. Consequently, designers should be knowledgeable about them. Through proper coordination of design and the shape-producing processes, occasionally the need for finishing operations can be eliminated, and invariably finishing costs can be reduced or better results achieved. In a few cases, processes that were developed for finishing have been adapted for producing desired shapes.

CLEANING AND SMOOTHING PROCESSES

The first step in finishing usually is cleaning and/or smoothing. Mechanical, chemical, and electrochemical methods are used.

Abrasive cleaning. One of the most common steps preliminary to the application of decorative or protective surface treatments is the removal of sand or scale from metal parts. Sand often adheres to certain types of castings, and scale often results when metal has been processed at elevated temperatures. In some cases, sand may be removable by simple vibratory shaking, but often some type of *abrasive cleaning* is employed to remove such foreign materials. Some type of abrasive, usually sand or steel grit or shot, is impelled against the surface to be cleaned. Fine glass shot may be used for some materials.

In *shot blasting,* a high-velocity air blast is used as the impelling agent. Air pressures from 0.4 to 0.69 MPa (60 to 100 psi) are used for ferrous metals and from 0.07 to 0.4 MPa (10 to 60 psi) for nonferrous metals. A nozzle with about a 9.5-mm (⅜-inch) diameter is often used. For large or only a few parts, the blast may be directed by hand, as shown in Figure 37-1. Quantities of small parts usually are moved past sets of stationary nozzles inside a hood. When done manually, protective clothing and breathing equipment must be provided and precautions taken to prevent the resulting dust from being spread. Ordinarily, a separate, well-vented room or booth is used with suitable collection equipment so as to avoid air pollution.

Equipment that impels the abrasive particles by mechanical means is being used increasingly. This type of equipment employs the centrifugal principle illustrated in Figure 15-19. It is more economical in the use of energy, and less difficulty is experienced in avoiding pollution.

If sand is used, it should be clean, sharp-edged, silica sand. Steel grit will clean much more rapidly than sand and causes much less dust. However, sand has a lower cost and somewhat greater flexibility.

Obviously, abrasive cleaning is effective only if the abrasive can reach all the areas of the surface that must be cleaned. Thus it may be difficult to use if the surface is very complex. Edges are rounded considerably during abrasive

FIGURE 37-1. Shot blasting in a special cleaning room. (*Courtesy Norton Company.*)

cleaning and, therefore, it cannot be used if sharp edges and corners must be maintained. If the nozzles are directed manually, the labor cost tends to become high.

Tumbling. Tumbling is a widely used method for mechanical cleaning. The parts are placed in a special barrel or drum until it is nearly full. Occasionally, the loaded barrel is rotated without the addition of any abrasive agents. However, in most cases metal slugs or jacks, or some abrasive, such as sand, granite chips, slag, or aluminum oxide pellets, are added. The rotation of the barrel causes the parts to be carried upward and then to tumble and roll over each other as they slide downward, as depicted in Figure 37-2. This produces a cutting action that usually will remove fins, flashes, scale, and sand. The

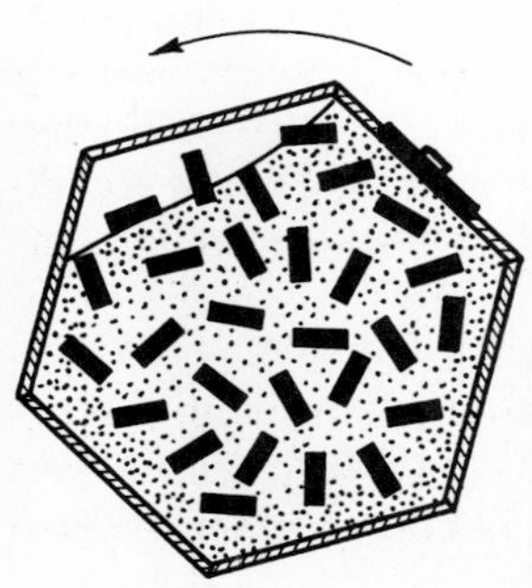

FIGURE 37-2. Principle of tumbling.

process can be used only on parts that are sufficiently rugged to withstand the tumbling action. However, by a suitable selection of abrasives, fillers, barrel size and speeds, and careful packing of the barrel, an amazing range of parts can be tumbled successfully. Delicate parts should not shift loosely during tumbling. In some cases such parts must be attached to racks within the barrel so that they will not strike each other.

Tumbling is an inexpensive cleaning method. Various shapes of slug materials are used. Several shapes often are mixed in a given load so that some will reach into all sections and corners to be cleaned. Tumbling usually is done dry, but it can also be done wet. Obviously, tumbled parts will have rounded edges and corners. The equipment can be arranged so that loading and unloading are accomplished readily and so that the slug material is separated from the workpieces by falling through suitable grid tables.

Wire brushing. A high-speed, rotary wire brush sometimes is used to clean surfaces; it also does a minor amount of smoothing. Wire brushing can be done by hand application of the workpiece to the brush, but more commonly automatic machines are used in which the parts are moved past a series of rotating brushes, similar to the procedure shown in Figure 37-9.

Wire brushing normally removes very little metal, except fine, sharp high spots. It produces a surface composed of fairly uniform, fine scratches that, for some purposes, is satisfactory as a final finish, or can easily be removed by barrel finishing or buffing.

Belt sanding. A simple and common method for obtaining smooth surfaces is by *belt sanding.* The workpieces are held against a moving, abrasive belt until the desired degree of finish is obtained. A series of belts of varying degrees of fineness can be used. If flat surfces are desired, the belt passes over a flat table at the point where the work is held against it.

Belt sanding, as used for finishing, cannot be considered a sizing operation, because only sufficient sanding is done to remove the sharp, high spots and thus produce a smoother surface. The resulting surface is composed of very fine scratches, their fineness depending on the grit of the belt.

Although greatly improved sanding belts now are available and fairly smooth surfaces can be obtained, belt sanding is a hand operation and therefore is quite slow and costly in labor. Consequently, it should not be used except where more economical methods cannot be utilized. It has the further disadvantage of not being effective where recesses or interior corners are involved. As a result, it is used primarily for delicate parts or where small quantities are involved.

Barrel finishing. Since specially shaped, artificial abrasive pellets became available, about 1946, great progress has taken place in *barrel finishing.* The advent of vibratory finishing equipment during recent years has given further impetus to this method of finishing.

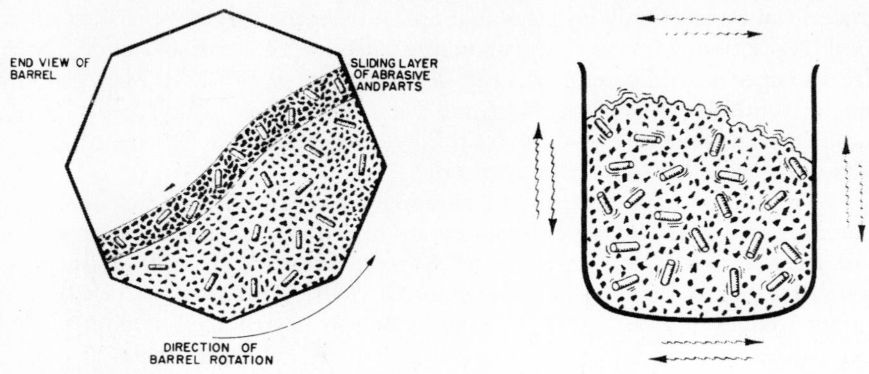

FIGURE 37-3. Comparison of actions in rotary and vibratory finishing. (*Courtesy Norton Company.*)

In this smoothing process, the workpieces and abrasive pellets are placed in a container and caused to move relative to each other, either by the container rotating, similar to barrel tumbling, or by the container vibrating through short strokes at from 900 to 3600 cycles per minute. The general actions involved are indicated in Figure 37-3.

Although the results obtained are essentially the same, vibratory finishing tends to be somewhat faster because the entire mass of workpieces and abrasive pellets are in constant, relative motion. Vibratory finishing tends to be superior to barrel finishing for smoothing and deburring interior surfaces.

Although some cleaning may occur incidentally in barrel finishing, the primary objective is to "cut down" the surface through the movement of the abrasive pellets across the surface of the workpiece. To assure this action, the work surface, the ratio of the work to abrasive, the speed of rotation or vibration, and the shape of the abrasive pellets relative to the workpiece configuration are very important. As shown in Figures 37-4 and 37-5, a considerable variety of pellet sizes and shapes is available to permit selection of shapes that will slide across all the required surfaces and not become lodged in restricted areas. Water and other compounds, such as dilute acids or soaps, frequently are added. Several natural abrasives, including slag, cinders, sharp sand, and granite chips, still are used, but in most cases the artificial abrasives have replaced these because of their greater uniformity of shape and cutting action.

Parts that are to be barrel or vibratory finished must be of such shape that they will not lock together, thus preventing them from moving freely. Some type of filler often is added to act as a carrying agent for the abrasive and to provide the necessary bulk to keep the workpieces separated from each other. Scrap punchings or mineral matter can be used in wet finishing, and in dry finishing, hardwood sawdust or leather scraps sometimes are employed. Heavy or intricate parts can be finished in a barrel by being racked in special

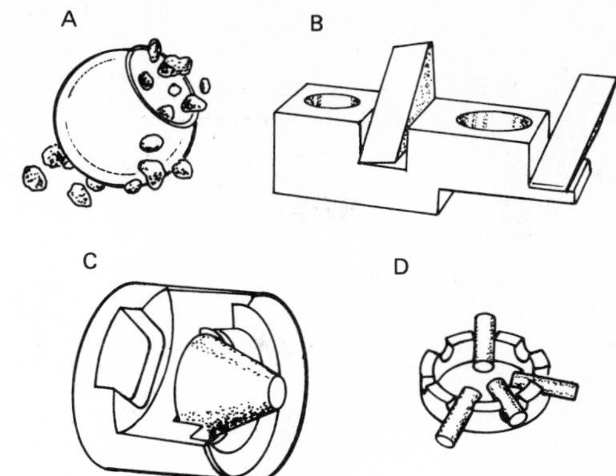

FIGURE 37-4. Some shapes of abrasives used for finishing various shapes of workpieces.

FIGURE 37-5. Various types and shapes of artificial abrasive pellets used for rotary and vibratory finishing. (*Courtesy Norton Company.*)

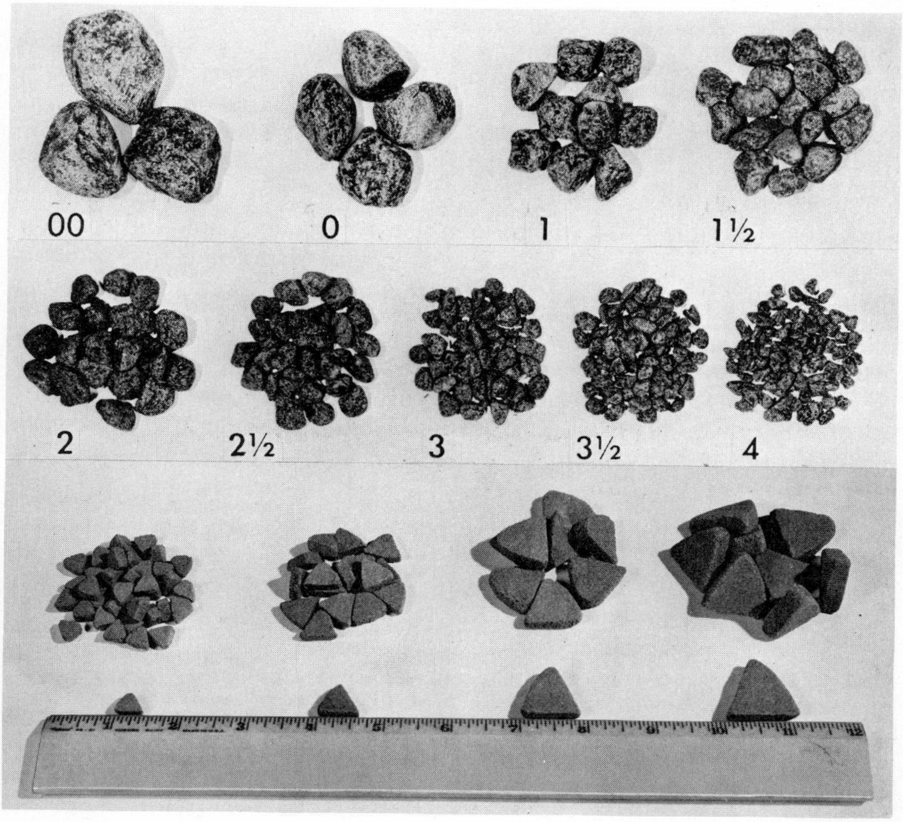

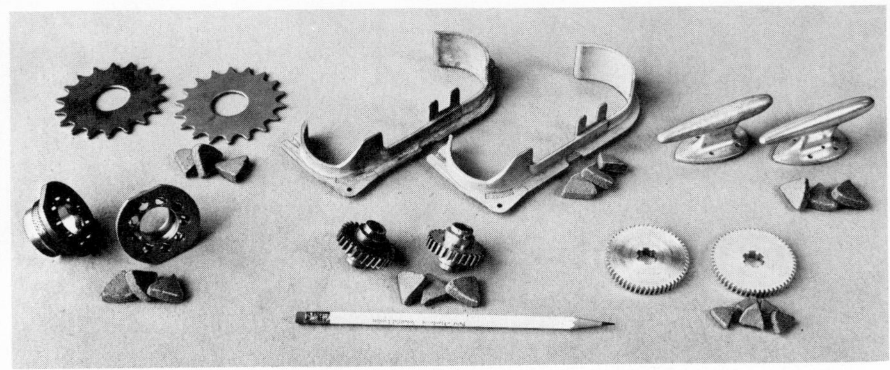

FIGURE 37-6. Examples of typical parts before and after rotary finishing. (*Courtesy Norton Company.*)

fixtures to prevent them from becoming nicked. When sharp edges must be maintained, or when certain areas are not to be smoothed, masking can be employed. Obviously, it is preferable for such restrictions to be avoided through design.

By using the proper abrasive, a wide range of finishes can be obtained, the best being virtually free of visible scratches. However, unusually deep scatches should be removed by wire brushing prior to barrel finishing if a uniform surface finish is desired. Figure 37-6 shows several examples of parts before and after barrel finishing. Finishing time varies from 10 minutes for soft, nonferrous parts to 2 or more hours for steel parts. Although they are batch processes, barrel and vibratory finishing are quite simple and economical. Sometimes the parts may be put through more than one barrel, using increasingly fine abrasives. Figure 37-7 shows a typical installation of several

FIGURE 37-7. Installation of several types of tumbling and rolling barrels. (*Courtesy Queen Products Division, King-Seeley Thermos Co.*)

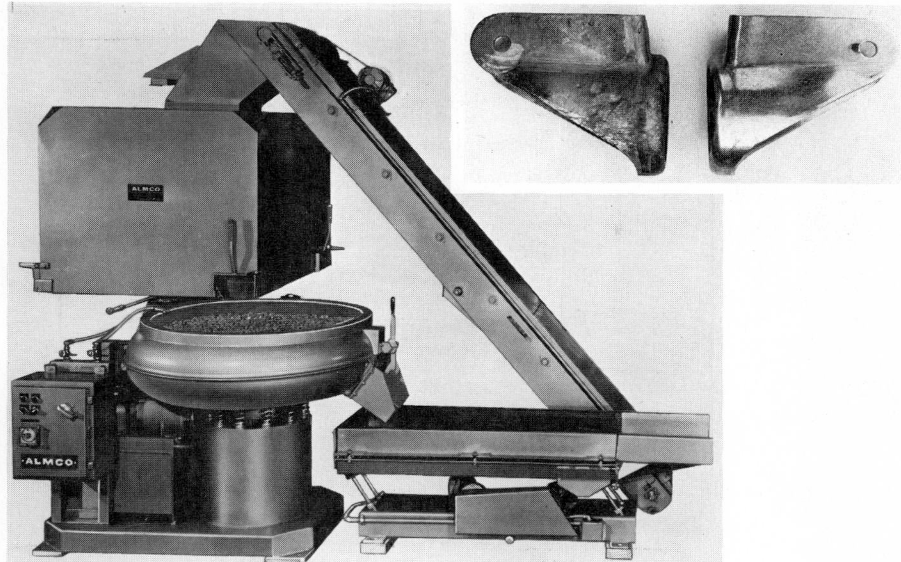

FIGURE 37-8. Vibratory-type finishing machine and close-up view of part before and after finishing. (*Courtesy Queen Products Division, King-Seeley Thermos Co.*)

tumbling and rolling barrels, and Figure 37-8 shows an example of a vibratory finishing machine with an automatic system for handling the work and the abrasive.

Buffing. *Buffing* is a polishing operation in which the workpiece is brought in contact with a revolving cloth buffing wheel that has been charged with a very fine abrasive, such as polishing rouge. Obviously, buffing is closely related to lapping in that the cloth buffing wheel is a carrying vehicle for the abrasive. The abrasive removes minute amounts of metal from the workpiece, thus eliminating fine scratch marks and producing a very smooth surface. When softer metals are buffed, there is some indication that a small amount of metal flow may occur that helps to reduce high spots and produce a high polish.

Buffing wheels are made of disks of linen, cotton, broadcloth, or canvas that are made more or less firm by the amount of stitching used to fasten the layers of cloth together. Buffing wheels for very soft polishing, or which can be used to polish into interior corners, may have no stitching, the cloth layers being kept in proper position by the centrifugal force resulting from the rotation of the wheel. Various types of polishing rouges are available, most of them being primarily ferric oxide in some type of binder.

Buffing should be used only to remove very fine scratches, or to remove oxide or similar coatings that may be on the work surface. If it is done by manually holding the work against the rotating buffing wheel, it is quite ex-

FIGURE 37-9. Parts being buffed on an automatic machine. (*Courtesy Murray-Way Corporation.*)

pensive because of the labor cost. However, semiautomatic buffing machines are available, such as is shown in Figure 37-9, in which the workpieces are held in fixtures on a rotating, circular worktable and moved past a series of individually driven buffing wheels. The buffing wheels can be adjusted to desired positions so as to buff different portions of specific workpieces. If the workpieces are not too complex, very good results can be obtained quite economically with such equipment. Obviously, part design plays an important role where automatic buffing is to be employed.

Barrel burnishing. In many cases, results comparable with those obtained by buffing can be obtained by barrel burnishing. In this pocedure, balls, shot, or round-ended pins are added to the work in a rotating barrel. No cutting action is involved. Instead, the slug material produces peening and rubbing actions, reducing the minute irregularities and producing an even surface. Burnishing will not remove visible scratches or pits, so in most cases the parts should first be rolled or vibrated with a fine abrasive. However, the resulting surface is smooth, uniform, and free of porosity.

Barrel burnishing is normally done wet, using water to which some lubricating or cleaning agent has been added, such as soap or cream of tartar. Because the rubbing action between the work and the shot material is very important, the barrel should not be loaded more than half full with work and shot, and the ratio of shot to work should be about 2 volumes to 1. The ratio should be such that the workpieces do not rub against each other. The speed of rotation of the barrel should be adjusted so that the workpieces will not be thrown out of the mass as they reach the top position and roll down the inclined surfaces.

It usually is necessary to use several sizes and shapes of shot material to assure that it can come in contact with inside corners and other recesses that must be rubbed. Balls from 3.2 to 6.4 mm (⅛ to ¼ inch) in diameter, pins, jacks, and ball cones commonly are used.

Parts that cannot be permitted to bump against each other can be burnished successfully by fastening them in racks inside the burnishing barrel. The shot material is then added, and burnishing is carried out in the usual manner.

ELECTROPOLISHING

Electropolishing, the reverse of electroplating, sometimes is used for polishing metal parts. The workpiece is made the anode in an electrolyte with a cathode added to complete the electrical circuit. In the resulting deplating, material is removed most rapidly from raised, rough spots, producing a very smooth, polished surface. It usually is not economical to remove more than about 0.025 mm (0.001 inch) of material, so the process is used primarily to produce mirrorlike surfaces, and the initial surface must be quite smooth. A final finish of less than 0.05 μm (2 microinches) can be obtained if the initial roughness does not exceed 0.18 to 0.020 μm (7 to 8 microinches) rms.

Electropolishing was originated for polishing metallurgical specimens and later was adapted for polishing stainless steel sheets and parts. It is particularly useful for polishing irregular shapes that would be difficult to buff. For best results the metal should be fine grain and free of surface defects exceeding the coarseness of a 180-grit scratch.

CHEMICAL CLEANING

At some stage in the finishing of virtually all metal products it is necessary to employ *chemical cleaning* to remove oil, dirt, scale, or other foreign material that may adhere to the surface so that subsequent painting or plating can be done successfully. One or more of three cleaning processes is used.

Alkaline cleaning. *Alkaline cleaning* employs such agents as sodium metasilicate or caustic soda with some type of soap to aid in emulsification. Wetting agents often are added to assist in obtaining thorough cleaning. The cleaning action is by emulsification of the oils and greases. Thus the solution must penetrate any dirt that covers them. It also is necessary to thoroughly rinse the cleansing solution from the work surface so as not to leave any residue.

The cleansing bath must be controlled to maintain a constant and proper pH value. Too high as well as too low pH levels may produce poor results.

Cleaning with emulsifiable solvents. *Cleaning with emulsifiable solvents* is done by combining an organic solvent with a hydrocarbon-soluble emulsifying agent such as sulfonated castor oil with water, or blending a soap and an organic solvent such as kerosene with a small amount of water. Cresylic acid or some other blending agent is used.

The work is dipped in the solvent solution and then rinsed once or twice. If the work is to be electroplated, it should have a subsequent treatment in an alkaline cleaner to remove any organic matter that remains on the metal. As a preparation for painting, solvent cleaning and rinsing usually are adequate. Solvent cleaning is used extensively for metals such as aluminum, lead, and zinc, which are chemically active and might be attacked by alkaline cleaners.

Vapor degreasing. *Vapor degreasing* is widely used to remove oil from ferrous parts and from such metals as aluminum and zinc alloys, which would be attacked by alkaline cleaners. A nonflammable solvent, such as trichlorethylene, is heated to its boiling point, and the parts to be cleaned are hung in its vapors. The vapor condenses on the work and washes off the grease and oil. Excess vapor is condensed by cooling coils in the top of the vapor chamber. Although grease and oil from the work are washed off into the liquid solvent, causing the bath to become dirty, because they are only slightly volatile at the boiling temperatures of the solvent, the vapor remains relatively clean at all times and so continues to clean effectively.

It must be remembered that vapor degreasing is effective only if the vapor condenses on the work. Thus the work must remain relatively cool. This offers no difficulty except in the case of thin sheets that contain considerable amounts of oil and do not have sufficient heat capacity to remain cool and thus condense enough vapor to bring about satisfactory cleaning.

Vapor degreasing has the advantages of being rapid and of having almost no visible effect on the surface. Its major disadvantages are that vapor alone does not remove solid dirt, and it frequently must be followed by alkaline cleaning to remove remaining organic matter. If the surface has substantial solid dirt in addition to oil, this may be removed by passing the work through a boiling liquid, thereby removing most of the dirt and some of the oil, then through cold liquid to cool the work, and finally through hot vapor to remove the remaining oil.

Pickling. *Pickling* involves dipping metal parts in dilute acid solutions to remove the oxides and dirt that are left on the surface by various processing operations. The most commonly used pickling solution is a 10 per cent sulfuric acid bath at temperatures of from 66 to 85°C (150 to 185°F). Muriatic acid also is used, either cold or hot. When used cold, pickling baths have approximately equal parts of acid and water. At temperatures ranging from 38 to 60°C (100 to 150°F), more dilute solutions are used.

It is very important that parts be throughly cleaned prior to pickling; the pickling solution will not act as a cleaner, and any dirt or oil on the surface will result in an uneven removal of the oxides. Alkaline cleaning is usually employed for this purpose. Pickling *inhibitors,* which decrease the attack of the acid on the metal but do not interfere with the action of the acid on the oxides, frequently are added to the pickling bath.

After the parts are removed from a pickling bath, they should be rinsed thoroughly, to remove all traces of acid, and then dipped in a bath that is slightly alkaline to prevent subsequent rusting. Where it will not interfere with further processing, a dip in cold milk of lime often is used for this purpose. Parts should not be overpickled, as this may result in roughening the surface.

Ultrasonic cleaning. *Ultrasonic cleaning* is used extensively where very high quality cleaning is required for relatively small parts. In this process the parts are suspended, or placed in wire baskets, in a cleaning bath, such as Freon, that contains an ultrasonic transducer operating at a frequency that causes cavitation in the liquid. Excellent results can be obtained in from 60 to 200 seconds in most cases. It usually is best to remove gross dirt, grease, and oil before doing ultrasonic cleaning.

FINISHES

Paints are by far the most widely used finishes on manufactured products, and a great variety is available to meet a wide range of requirements. Today, most paints and enamels are synthetic organic compounds that dry by polymerization or by a combination of polymerization and adsorption of oxygen. Water is frequently the carrying vehicle for the pigments. Moderate amounts of heat can be used to accelerate the drying, but many synthetic paints and enamels will dry in less than 1 hour without the use of heat. The older, oil-base paints and enamels require too much drying time for use in mass production and thus seldom are used.

Table 37-1 lists the more commonly used organic finishes and their important characteristics. *Nitrocellulose lacquers,* although very fast drying and capable of producing very beautiful finishes, are not sufficiently durable for most commercial applications. The *alkyds* are general-purpose paints but do not have sufficient durability for hard service conditions. The *acrylic enamels* are widely used for automobile finishes. *Silicones* and *fluoropolymers* are specialty finishes; their high cost is justified only where their special properties are important.

Asphaltic paints, which are solutions of asphalt in some type of solvent, such as benzine or toluol, are still used extensively, especially in the electrical industry, where resistance to corrosion is required but appearance is not of prime importance.

TABLE 37-1. Commonly used organic finishes and their qualities

Material	Durability (scale of 1–10)	Relative Cost (scale of 1–10)	Characteristics
Nitrocellulose lacquers	1	2	Fast drying; low durability
Epoxy esters	1	2	Good chemical resistance
Alkyd-amine	2	1	Versatile; low adhesion
Acrylic lacquers	4	1.7	Good color retention; low adhesion
Acrylic enamels	4	1.3	Good color retention; tough; high baking temperature
Vinyl solutions	4	2	Flexible; good chemical resistance; low solids
Silicones	4–7	5	Good gloss retention; low flexibility
Fluoropolymers	10	10	Excellent durability; difficult to apply

Paint application. In manufacturing, almost all painting is done by one of four methods: *dipping, hand spraying, automatic spraying,* or *electrocoating.* In most cases at least two coats of paint are required. The first, or prime, coat serves primarily to (1) assure adhesion, (2) provide a leveling effect by filling in minor porosity and other surface blemishes, and (3) improve corrosion resistance and thus prevent later coatings from being dislodged in service. These properties are less obtainable in the more highly pigmented paints that are used for final coats because of their better color and appearance. In using multiple coats, one must be sure that the carrying vehicles in the final coats do not unduly soften the previous coats.

Paint application by *dipping* is used extensively. The parts are either dipped manually into the paint or are passed down into the paint while on a conveyor. Obviously, all of the workpiece is coated, and thus it is a very simple and generally economical technique where all surfaces require painting. Consequently, it is used for prime coats and for small parts where the loss of paint due to overspray would be excessive if ordinary spray painting were used. On the other hand, the unnecessary amount of paint used can make the process uneconomical if only some surfaces actually require painting or in cases where very thin, uniform coatings of some of the modern primers are adequate, particularly on large objects such as automobile bodies. Other difficulties with dipping are the tendency of the paint to run, thus producing a wavy surface, and the final drop of paint that usually is left at the lowest drip point. It also is essential that the paint in the dip tanks be kept stirred at all times and be of uniform viscosity.

Spray painting is probably the most widely used painting process, owing to its versatility and economy in the use of paint. The paint is atomized by three methods: air, mechanical pressure, or electrostatically. Either manual or automatic application is used. When hand spraying is used, either air or mechanical atomization is employed, and the spray of paint is directed against the work by means of a hand-manipulated gun. The worker must exercise considerable skill in obtaining proper coverage without allowing the paint to "run" or "drape" downward. Consequently, only a very thin film can be deposited at one time—usually not over 0.025 mm (0.001 inch)—if conventional methods are used. As a result, several coats usually must be applied with some intervening time for drying. Somewhat thicker coatings can be applied in one operation by using a *hot spray* method. In this procedure, the paint is sprayed while hot.

Obviously, spray painting by hand, as illustrated in Figure 37-10, is costly from the viewpoint of labor expense and therefore is replaced by automatic methods whenever practicable. The simplest type of automatic equipment consists of a chain conveyor on which the parts are moved past a series of spray heads. However, if regular spray heads are used, results are often not satisfactory. A large part of the paint may be wasted, because a considerable portion will go into the spaces between the parts on the conveyor and thus not hit the surfaces needing painting. In addition, it is difficult to get uniform coverage.

Much better results are obtained from either manual or automatic spray painting by using the electrostatic principle. In the air process, the spray gun atomizes the paint, giving the particles an electrostatic charge and considerable velocity. The atomized particles are attracted to, and deposited on, the

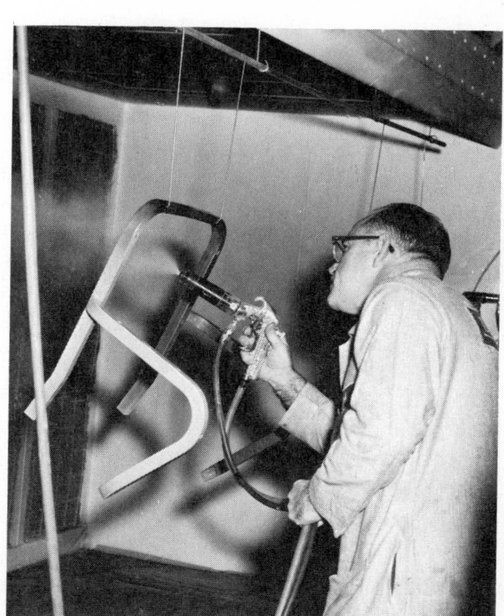

FIGURE 37-10. Spray painting a chair, using an electrostatic-type spray gun. (*Courtesy The DeVilbiss Company.*)

FIGURE 37-11. Electrostatically finishing aluminum extrusions, using two reciprocating disks. (*Courtesy Ransburg Corporation.*)

work, which is grounded electrically. Under proper control, painting efficiencies of from 75 to 95 per cent can be obtained.

In a second, airless electrostatic method, the paint is fed onto the interior of a rapidly rotating cone or disk that is one electrode of a high-potential electrostatic circuit. The rotation of the cone or disk causes the paint to flow outward to the edge by centrifugal force. As the thin film of paint reaches the edge and then is spun off, the particles are charged electrostatically and atomized without the need for any air pressure. With the workpiece being the other electrode of the circuit, the paint is transferred, as in the previously described method. The primary advantages of this method, which is illustrated in Figure 37-11, are that because no air pressure is used for atomization, there is less spray loss, much less extensive provision must be made to take care of fumes, and there is a higher efficiency in the use of paint—running as high as 99 per cent.

With automatic spray-painting systems it often is necessary to do some touch-up work manually where the coverage is not completely uniform. There also is a tendency for the paint to be drawn to the nearest edge or surface, making it difficult to get paint into deep recesses. Of course, the workpiece must be electrically conductive.

Electrocoating is the most recent basic development in paint application. It permits the economy of ordinary dip painting to be achieved, but overcomes its disadvantages while permitting thinner and more uniform coatings and superior coverage in interior recesses. The principle of the process is shown in Figure 37-12. The paint particles, in a water solvent, are given an

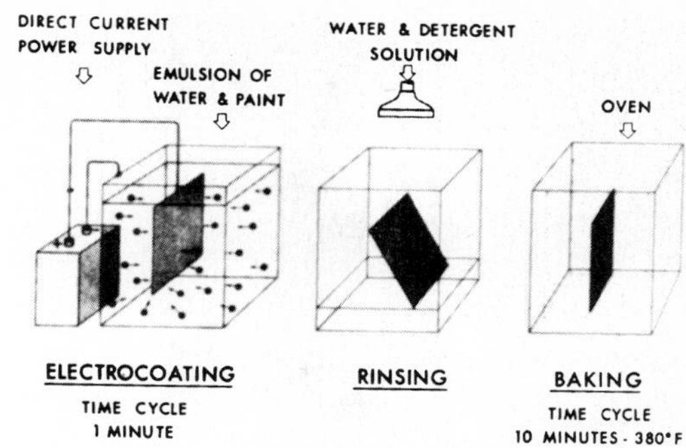

FIGURE 37-12. Principle of the electrocoating method of painting. (*Courtesy Materials Engineering.*)

electrostatic charge by applying a dc voltage between the tank (cathode) and the workpiece (anode). As the workpiece enters and passes through the tank, the paint particles are attracted to, and deposited on, it in a uniform, thin coating from 0.02 to 0.038 mm (0.0008 to 0.0015 inch) thick. When the coating reaches the desired thickness, determined by controlling the conditions, no more paint is deposited. The water in the deposited film is drawn away by electroosmosis, leaving a coating that is composed of more than 90 per cent resins and pigments. The workpiece is removed from the dip tank, rinsed by a water spray, and baked for about 25 minutes at about 190°C (375°F).

Electrocoating is especially suitable for applying the prime coat to complex metal structures, such as automobile bodies, where good corrosion resistance is necessary. The flow of paint to hard-to-reach areas can be improved by placing electrodes in the workpiece at strategic locations. In addition, because the solvent is water, there is no fire danger as exists when large-area tanks are employed with regular dipping primers. As shown in Figure 37-13, electrocoating is readily adapted to conveyor-line production.

A new development is the application of paint in powder form by an electrostatic spray process. Several coats, such as primer and finish, can be applied and then followed by a single baking, instead of requiring baking after each coat as in conventional spray processes.

Drying. Most paints and enamels used in manufacturing require from 2 to 24 hours to dry at normal room temperatures. This obviously is not practical. They can be dried satisfactorily in from 20 minutes to 1 hour at temperatures of from 135 to 232°C (275 to 450°F). Consequently, some drying at elevated temperatures usually is done, using either a baking oven or, more often, a tunnel or panel of infrared heat lamps. The latter involve relatively low investment, not much floor space, and are very flexible.

Although drying at elevated temperatures can be accomplished without difficulty on metal parts, this is not the case with wooden products. The temperatures are high enough to expand the gases, moisture, and sap that are in

FIGURE 37-13. Automobile bodies entering an electrocoating paint tank. (*Courtesy Ford Motor Company.*)

the wood, even though it is quite dry. These are forced to the surface after the paint has started to harden and form small bubbles that roughen the surface or, if they break, leave small holes in the painted surface.

Hot-dip coatings. Large quantities of metal parts are given corrosion-resistant coatings by being dipped into certain molten metals. Those most commonly used are zinc, tin, and an alloy of lead and tin.

Hot-dip galvanizing is the most widely used method of providing steel with a protective coating. After the parts, or sheets, have been cleaned, they are fluxed by dipping them into a solution of zinc chloride and hydrochloric acid. They then are dipped into a molten zinc bath. The resulting zinc coating is complex, consisting of a layer of $FeZn_2$ at the metal surface, an intermediate layer of $FeZn_7$, and an outer layer of pure zinc. Hot-dip galvanizing gives a good degree of corrosion resistance.

The coating thickness should be controlled; coatings that are too thick crack and peel. A wide variety of "spangle" patterns can be obtained by proper processing. When galvanizing is done properly, considerable subsequent bending and forming can be done without damaging the coating. However, rimmed steel should not be galvanized.

Tin "plating" also can be done by hot dipping. After the steel has been cleaned, it is dipped into the molten tin, the surface of which is covered with a layer of zinc chloride. In this manner the work passes through the zinc chloride before entering the molten tin. As the work leaves the tin bath it passes through rollers that are immersed in palm oil, thus removing the excess tin. However, most tin plate now is produced by an electrolytic process that gives more uniform coating with less tin being required.

Terne coating is similar to hot-dip tin coating, but an alloy of 15 to 20 per cent tin and the remainder lead is used in place of pure tin. This process is thus cheaper than tin coating and provides satisfactory corrosion resistance for some purposes.

Phosphate coatings. Two phosphate coating processes are used extensively to provide corrosion resistance, usually to steel. In these processes the surface of the metal is converted into an insoluble crystalline phosphate by treatment with a dilute acid phosphate solution.

Parkerizing produces a fairly corrosion-resistant coating from 0.004 to 0.008 mm (0.00015 to 0.0003 inch). The treatment requires about 45 minutes and provides quite good corrosion resistance for parts that are to be kept painted. *Bonderizing* is similar to Parkerizing, but its primary purpose is not to give corrosion resistance but to form a surface to which paint will adhere tightly. The coating is thinner than that obtained by Parkerizing, but it reduces the activity of the metal surface so that corrosion at the paint–metal interface is retarded. As a result, if the paint coat is scratched, there is less likelihood of rust starting and progressing and thereby causing the paint adjacent to the scratch to loosen.

Blackening. Many steel parts are treated to produce a black, lustrous surface that will be resistant to rusting when handled. Such coatings usually are obtained by converting the surface into black iron oxide. One method is to heat the parts in a closed box of spent carburizing compound at 649°C (1200°F) for about 1½ hours and then quench them in oil. Another method consists of immersing the parts in special blackening salts at 149°C (300°F) for about 15 minutes. A third method is to heat the parts in a rotary retort furnace to about 399°C (750°F). A small quantity of linseed or fish oil is then added. After a few minutes the parts are removed from the furnace, spread out, and allowed to cool. When they have cooled they are dipped into an oil that retards rust.

A *gun-metal finish* is obtained by heating the parts in a retort with a small amount of charred bone to 399°C (750°F). When the parts are oxidized, they are allowed to cool to about 343°C (650°F). A mixture of bone and some carbonic oil is then added and the heating continued for several hours. The work is then removed from the furnace and dipped in sperm oil.

ELECTROLYTIC FINISHES

Large quantities of both metal and plastic parts are electroplated to provide corrosion or wear resistance, improved appearance (such as color or luster), or an increase in dimensions. Plating is applied to virtually all base metals—copper, brass, nickel–brass, aluminum, steel, and zinc-base die castings—and also to plastics. In order to plate plastics, they first must be coated with some electrically conductive material.

The most common plating metals are tin, cadmium, chromium, copper, gold, platinum, silver, and zinc. Except for the making of tin plate for the container industry, chromium is by far the most common metal plated as the surface layer. However, in most cases a thin layer of both copper and nickel is deposited beneath the chromium. Gold, silver, and platinum are very important plating metals in the jewelry and electronics industries.

All electroplating processes essentially are the same, although the methods may vary somewhat in details. The basic process is indicated in Figure 37-14. The parts to be plated are made the cathode and suspended in a solution that contains dissolved salts of the metal to be deposited. The anode is a suspended slab of the metal to be deposited. Other materials may be added to the electrolyte to increase its conductivity. When a dc voltage is applied, metallic ions migrate to the cathode and, upon losing their charges, are deposited as metal upon it.

Successful plating depends greatly on (1) the preparation of the surface, (2) the ability of the bath to produce coatings in recessed areas, and (3) the crystalline character of the deposited metal. The actual deposition of the plating metal is governed by the bath composition and concentration, the bath temperature, and the current density. These are interdependent and must be carefully controlled to obtain satisfactory and consistent results.

Surfaces that are to be electroplated must be prepared properly to obtain satisfactory results. All defects, pin holes, and scratches must be removed if a smooth, lustrous finish is desired. The surface must be chemically clean. Proper combinations of degreasing, cleaning, and pickling are used to assure a clean surface to which the plating material will adhere.

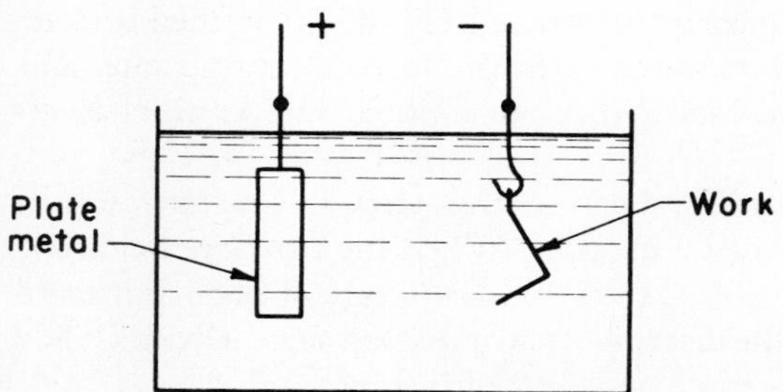

FIGURE 37-14. Basic electroplating circuit.

Plating solutions are chosen on the basis of their *throwing power,* referring to their ability to deposit sound metal in recesses. Cyanide solutions have better throwing power than acid solutions and, therefore, commonly are used, although they are more dangerous to handle.

The plating metal tends to be attracted to, and build up on, corners and protrusions (see Figure 40-7). This makes it difficult to obtain uniform plating thickness on parts of irregular shape containing recesses and interior corners. Improved results can often be obtained by using several properly spaced anodes, or anodes having a shape similar to the workpiece. Obviously, there are limitations in the use of such procedures.

Nickel plating provides good corrosion resistance but does not retain its luster and is expensive. Consequently, it has largely been replaced as an outer coating by chromium where appearance is important, and by cadmium for many applications where appearance is not of much importance and only moderate corrosion resistance is required.

As mentioned previously, chromium seldom is used alone. In modern practice, the first layer usually is bright acid copper, which produces a leveling effect and makes it possible to reduce the thickness of the nickel layer that follows. The nickel layer need not exceed 0.008 to 0.015 mm (0.0003 to 0.0006 inch). Chromium then is plated as the final layer to provide both protection and appearance.

A new procedure is to apply a very thin layer of stressed nickel, about 0.0005 mm (0.00002 inch), on top of the ordinary nickel layer, followed by chrome plating. The combined stresses produce fine microcracking in the chromium layer and result in improved corrosion resistance.

In mass production most electroplating is done as a continuous process, using the type of equipment shown in Figure 37-15. The parts to be plated are suspended on a conveyor and are lowered into successive plating, washing, and fixing tanks wherein the various operations are performed. Such methods make it possible to obtain economical plating where the volume of work is high. Ordinarily, only one type of workpiece can be plated at a time because the solutions, timing, and conditions of current density must be changed when different sizes and shapes are to be processed.

Hard chromium plating is used to build up worn parts to larger dimensions, to coat the face of cutting tools to reduce friction and wear, and to resist wear and corrosion. Hard chromium coatings always are applied directly to the base material and usually are much thicker than ordinary chrome plating, commonly from 0.07 to 0.25 mm (0.003 to 0.010 inch) thick. However, greater thicknesses—up to 0.76 mm (0.030 inch)—are used for such items as diesel cylinder liners. The hardness of hard chrome plating typically is from 66 to $70R_C$. Hard chrome plating does not have a leveling effect, and thus defects or roughness in the base surface will be amplified and made more apparent. Very smooth surfaces can be obtained by suitable grinding and polishing techniques.

FIGURE 37-15. Modern, automatic, continuous-plating installation. Slabs of plating metal can be seen on both sides of the tank in the foreground. Racks of parts are lowered into the tank at the far left. (*Courtesy The Udylite Corporation.*)

Cadmium–titanium plating is used successfully to provide corrosion resistance for high-strength steels that are highly subject to hydrogen embrittlement when they are plated with either zinc or cadmium alone.

As will be discussed in Chapter 40, the ease and cost of obtaining satisfactory electroplating on parts can be greatly affected through design.

Electroforming. *Electroforming* is an important modification of electroplating that is used to produce metal parts by electroplating metal onto an accurately made mandrel that has the inverse contour, dimensions, and surface finish desired on the finished product. When the desired thickness of deposited metal has been obtained, the workpiece is separated from the mandrel; the method of separation depends on the mandrel material and its shape.

The mandrels are made from a variety of materials, including plastics, glass, Pyrex, and various metals (most commonly aluminum or stainless steel). The mandrel must be made electrically conductive, by a coating, if it is not made from metal.

Electroformed parts most commonly are made of nickel, iron, copper, or silver, and thicknesses up to 16 mm (⅝ inch) have been deposited successfully. Metals deposited by electroforming have their own distinct properties. Dimensional tolerances are very good, often up to 0.0025 mm (0.0001 inch), and surface finishes of 0.05 μm (2 microinches) can be obtained quite readily if the mandrel is adequately smooth.

A wide variety of parts and shapes can be made by electroforming, the principal limitation being that it must be possible to remove the part from the mandrel. In some applications a relatively thin electroformed shell is backed up with other materials, often cast into the shell, to provide the required strength. However, in most cases the wall thickness is made sufficient to provide the necessary strength.

Electroforming is particularly useful for high-cost metals and, because of the low tooling cost, for low production quantities. Some care must be taken to minimize internal stresses, which commonly may be from 13.8 to 34.5 MPa (2000 to 5000 psi), and which can be much higher if plating temperatures and current densities are not carefully controlled.

Anodizing. *Anodizing* is a process that is widely used to provide corrosion-resistant and decorative finishes to aluminum. It is somewhat the reverse of electroplating in that (1) the work is made the anode in an electrolytic circuit, and (2) instead of a layer of material being added to the surface, the reaction progresses inward, increasing the thickness of the highly protective, but thin, aluminum oxide layer that normally exists on aluminum.

One of the common forms of anodizing uses a 3 per cent solution of chromic acid as the electrolyte at a temperature of about 38°C (100°F). The voltage is raised from about 0 to 40 at the rate of about 8 volts per minute, and then is maintained at full voltage for about 30 to 60 minutes with a current density of about 11 to 32 amperes per square meter (1 to 3 amp/ft^2). This treatment produces a converted layer that is about 0.0013 to 0.0025 mm (0.00005 to 0.0001 inch) thick. It is used primarily on aircraft materials. Because the coating is an integral part, the metal can be subjected to quite severe forming and drawing operations without destroying the coating or reducing its protective qualities. Parts that are anodized in this manner usually have a grayish-green color, resulting from the presence of reduced chromium in the coating. Other colors can be obtained by the use of suitable dye materials. The anodized surface also provides a good paint base.

More complex anodizing treatments are often used, some of which are *Alumilite* finishes. The most common of these uses a solution containing 15 to 25 per cent sulfuric acid. This produces a transparent coating on pure aluminum and an opaque coating on alloys. These coatings are submicroscopically porous. A wide variety of colors can be obtained by the use of suitable dyes that penetrate these pores. Some of the colors will not resist sunlight, but it is possible to use a special type of dye and a sealing process that will produce good sun resistance.

Inasmuch as anodyzing does not add to the dimensions, there is no necessity for providing any dimensional allowance, as must be done when electroplating is used.

ELECTROLESS PLATING

Electroless nickel plating. Because of the almost impossibility of obtaining uniform plating thickness on even moderately complex shapes with electroplating, and because of the large amounts of energy consumed, extensive work has been done in developing *electroless plating,* with major success in plating with nickel. Basically, electroless plating is accomplished by autocatalytic reduction of the metallic ion of the plating metal in an aqueous solution in which the workpiece acts as the catalyst. In the case of electroless nickel plating, sodium hypophosphite acts as the reducing agent, reducing nickel salts to nickel metal and, incidentally, supplying a small amount of phosphorus so that the resulting plated material is a solid solution of phosphorus (about 8 per cent) in nickel.

In addition to having as good corrosion resistance as electroplated nickel, electroless nickel has an as-deposited hardness of about 49 to $55R_c$, which can be increased to as high as $70R_c$ by suitable heat treatment. Also, because it is purely a chemical process, the coatings obtained are uniform in thickness, not being affected by part complexity.

Nickel-carbide plating. A very useful electroless plating process has been developed wherein minute particles of silicon carbide are codeposited in a nickel-alloy matrix. As shown in Figure 37-16, the particles of silicon carbide are 1 to 3 μm (0.000,04 to 0.0001 inch) in size and constitute about 25 per cent of the volume of the deposit. Not only is the coating as corrosion resistant as nickel, because of the very high hardness of the carbide particles (about 4500 on the Vickers scale, where tungsten carbide is 1300 and hardened steel about 900, or $62R_c$), the wear and abrasion resistance of such coatings is outstanding. As with electroless nickel plating, the thickness of the coating is not affected by part shape. Figure 37-17 shows an example.

Coating thicknesses typically are up to 0.20 mm (0.008 inch). The process has a wide range of applications, being of outstanding value for coating molding dies for plastics that contain substantial amounts of abrasive filler materials, such as glass fibers.

A modification of this process utilizes minute artificial polycrystalline

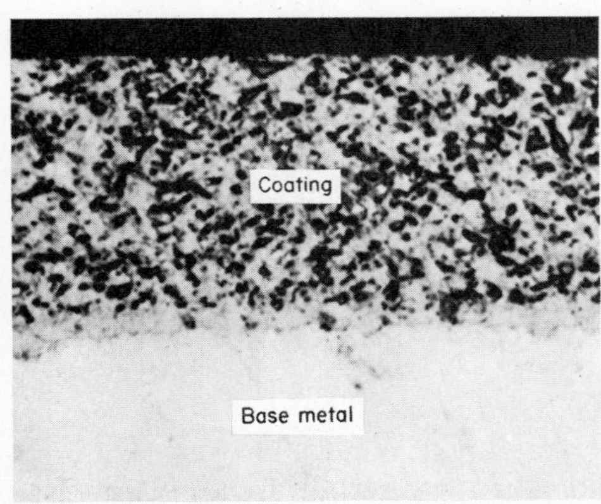

FIGURE 37-16. Photomicrograph of nickel-carbide coating obtained by electroless plating. (*Courtesy Electro-Coatings, Inc.*)

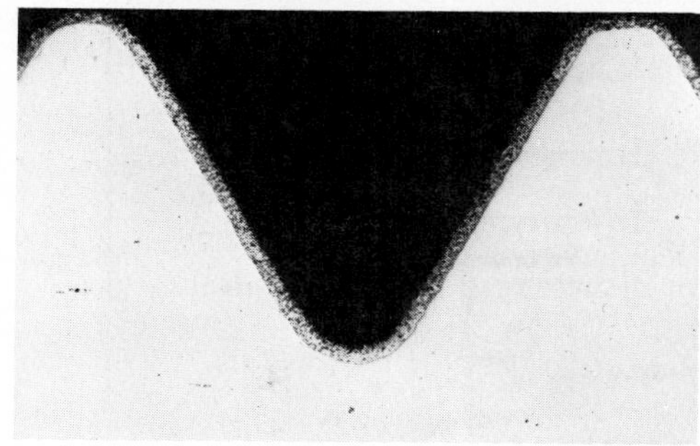

FIGURE 37-17. Photomicrograph showing uniform deposit obtained on irregularly shaped part by electroless plating. (*Courtesy Electro-Coatings, Inc.*)

diamond particles in place of silicon carbide. Such coatings usually do not exceed 0.025 mm (0.001 inch) in thickness, and they have outstanding wear resistance.

Impact plating, obtained by tumbling the parts in a tumbling barrel that contains a water slurry of very fine powder particles of the plating metal, glass spheres, and a "promoter" chemical, can be used to obtain a thin coating that is satisfactory for some purposes. The small glass balls peen the fine powder particles onto the workpieces, producing some cold welding. The deposited coatings are lamellar in structure and quite uniform in thickness. Any metal that can be obtained in a very fine powder form can be used as a plating material. One advantage of the process is that there is no danger of hydrogen embrittlement; therefore, it can be used on hardened steel.

VAPORIZED METAL COATINGS

Vacuum coating. *Vacuum coating* is widely used to deposit thin films of metal and metal compounds on various substrate materials. The process involves the evaporation of the metal or the compound in a high vacuum and the subsequent condensation of the vapor on the cool workpiece. A pressure (vacuum) of from 0.013 to 1.33 Pa usually is required. Such coatings are used as electrical conductors and resistors in the electronics industry, as decorative coatings, and, because of their outstanding properties, as reflective surfaces. Virtually any metal can be deposited by the vacuum-coating process—aluminum, chromium, gold, nickel, silver, germanium, and platinum being very common. The coatings are usually less than 0.5 μm (0.00002 inch) in thickness, and often are as little as 0.025 μm (1 microinch).

Vaporization is a surface phenomenon and does not constitute boiling. As the metal leaves the heated surface in atomic form, it travels in line-of-sight direction to the surface of the substrate. Thus, if an entire surface of a shape is to be coated, it must be rotated to expose all the surfaces. Fortunately, in most cases only a single surface must be coated.

Because of the very thin deposits required, the process is very economical

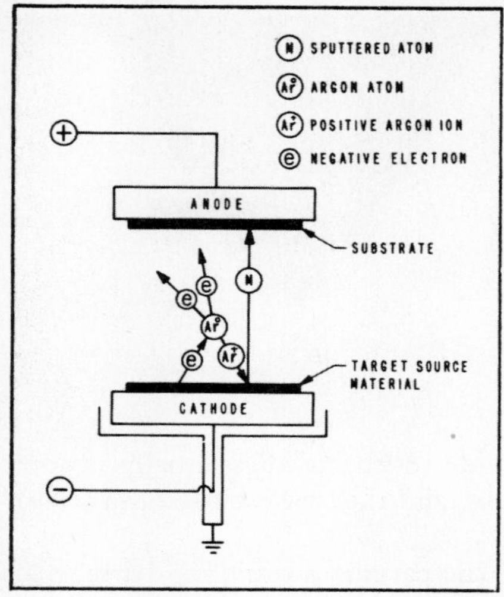

FIGURE 37-18. Schematic diagram of the radio-frequency sputtering process.

of metal, and expensive materials, such as gold or silver, often can be used economically over inexpensive part materials, such as plastics or steel.

Sputtering. For certain applications, *radio-frequency sputtering* is used as a substitute for electroplating and vapor-deposited coatings. Its most extensive use is for depositing thin films of metals in making solid-state devices and circuits. The basic process is indicated in Figure 37-18. The substrate, upon which metal is to be deposited, and the source metal are arranged as shown in a gas-filled chamber (often argon) that is evacuated to about 10 to 50 μm pressure. The substrate is made positive, relative to the source material, by a radio-frequency power source. When the applied potential reaches the ionization energy of the gas, electrons, generated at the cathode, collide with the gas atoms, ionizing them and creating a plasma. These positively charged ions, having high kinetic energy, are accelerated toward the cathode target, overcome the binding energy of the target material, and dislodge atoms that then travel across the electrode gap and are deposited on the substrate. Because of the energy of these atoms, usually between 15 and 50 electron volts, their adherence to the substrate is considerably better than if they were deposited by ordinary vacuum evaporation.

Review questions

1. Name three manufacturing processes that inherently result in the necessity for surface-cleaning and smoothing operations.
2. Why should a product designer be concerned about the finishing operations that may be required on the product?

3. What are three basic cleaning methods?
4. What materials commonly are used for abrasive cleaning?
5. What are the primary limiting factors in the use of abrasive cleaning?
6. Why should the barrel not be filled too full when tumbling is used for cleaning?
7. How may delicate parts be cleaned by tumbling?
8. What is the effect of wire brushing on a surface cleaned by that process?
9. On what type of surfaces is belt sanding effective?
10. What is the requisite condition for barrel finishing to be effective?
11. What is the basic difference in the action obtained in rolling and vibratory finishing barrels?
12. Basically, what type of process is buffing?
13. How is the stiffness of buffing wheels controlled?
14. Why is part design so important if buffing or belt sanding must be used to finish the part?
15. What type of surface is produced by barrel burnishing?
16. Explain how electropolishing produces a smooth surface.
17. What are three commonly used chemical cleaning methods?
18. Why is more than one of the three basic chemical cleaning processes often used on a given product?
19. Why may vapor degreasing not work well on parts made of thin aluminum?
20. What is the absolute condition that must be met for vapor degreasing to clean satisfactorily?
21. What is the purpose of pickling?
22. Under what conditions is ultrasonic cleaning usually employed?
23. Why have synthetic resin paints largely replaced oil-base paints?
24. Why are prime coats used in painting?
25. What four methods commonly are used for applying paints in manufacturing?
26. What is the reason for using the electrostatic principle in spray painting?
27. Why is rotating-disk atomization replacing air atomization in spray painting?
28. Why is dip painting limited in its use?
29. What are the principal advantages of electrocoating?
30. Why can infrared drying seldom be used for drying paints on wood products?
31. Why are paints that are excellent for finish coats usually not satisfactory for primers?
32. Why should the thickness of galvanizing be carefully controlled?
33. On what type of steel is galvanizing not satisfactory?
34. What are the differences in the purposes of Parkerizing and Bonderizing?
35. Why is it difficult to simultaneously put parts of widely differing shape through an automatic electroplating system?

36. What are two reasons why electroless plating is preferred over electroplating?
37. Why is part design especially important for parts that must be electroplated?
38. Why is vacuum coating usually preferred over electroplating when costly metals are the plating material?
39. What is the major advantage of sputtering over vapor vacuum evaporative coating?

Case study 27. THE DIRTY BRASS INSERTS

The alpha-brass insert in a rubber automobile valve stem (into which the valve core screws and onto which the air cap screws) will provide excellent adhesion to the rubber when the latter is vulcanized if there is no appreciable oxidation or other contaminants on the brass. In its operations within the United States, the Goodstone Tire Co. found that these inserts were sufficiently clean as they came from the supplier. However, because of the time lag, inserts that were shipped to its plant in southeast Asia required some type of cleaning, and a rather complex acid type of cleaning and quick drying were recommended. The company wishes to utilize an easier, dry type of cleaning, if possible. What economical and simple procedure would you recommend?

CHAPTER 38

Mass-production tools and techniques

Although the basic machining processes are utilized in manufacturing products in both large and small quantities, the general-purpose machine tools that have been discussed in previous chapters are not suitable for use in mass-production manufacturing. Because of the amount of time required for making setups and machine and tool adjustments, their productivity is too low, and they require too highly skilled operators. Consequently, ever since the birth of mass production, various approaches and techniques have been used to develop machine tools that would be highly effective in large-scale manufacture. Their effectiveness was closely related to the degree to which the design of the products was standardized and the time over which no changes in the design were permitted. If a part or product is highly standardized and will be manufactured in large quantities, it is quite easy to develop machines that will produce them with a minimum of skilled labor.[1] However, such special-

[1] A completely tooled screw machine is an example.

ized machines are very costly to design and build, and typically they are inflexible, not being capable of making any other product. Consequently, for them to be economical, they must be operated for considerable periods of time, making only the one product for which they were designed. Thus such machines, although highly efficient, can be utilized only to make products in very large volume, and desired changes of design in the products must be avoided or delayed because it would be too costly to scrap the machines.

Mass-produced products are manufactured to meet the demands of free-economy, mass-consumption markets, where there is a virtually constant demand for improvements and style changes. To meet these demands, it is essential that the machines that produce the products have considerable flexibility and be easily adapted for use in producing more than one product. Otherwise, they will not be economical, and the entire concept of low-cost manufacture is lost and, as a result, low-cost goods will not be available to the public.

A classic example of this situation existed in the United States a number of years ago in a highly mechanized plant that could turn out thousands of automobile frames for a given car each day, using only four operators. However, the cost of converting the equipment whenever a model change was made, and getting the "bugs" out of the modified or new equipment, was so great, and caused such delays, that the plant only operated for 1 year in actual production. Another plant, utilizing much less specialized equipment and many more workers, was built beside the old plant, and the old one was dismantled.

Such experiences led to the development of machine tools that provide the specialization required in mass production and yet retain flexibility. The most widely used concept is the construction of production machine tools from basic units that accomplish a *function,* rather than produce a specific product. A number of these units are combined and interconnected to produce completed parts or products. This has been made possible through two developments. The first was the development of self-contained, power-head production units. The second was the development of automatic transfer mechanisms that move workpieces from one production unit to the next. A third development is assuming substantial importance. This is the principle of automatic feedback control, which makes automation possible.

Powerhead production units. Many machining processes involve a rotating tool, in most cases fed either longitudinally or transversely with respect to the workpiece. Powerhead production units, such as shown in Figure 38-1, provide these basic, required motions. As shown, such a unit consists essentially of (1) a powered spindle, mounted in suitable bearings, that can be used to support and drive a variety of rotating cutting tools at selected speeds through a gear box, (2) a means for power feed, and (3) a frame or bed on which the other two components are mounted. Thus it is similar to the components in the upper column and spindle assembly of a heavy-duty, upright drilling machine, and it can easily be adapted to do drilling, milling, boring, ream-

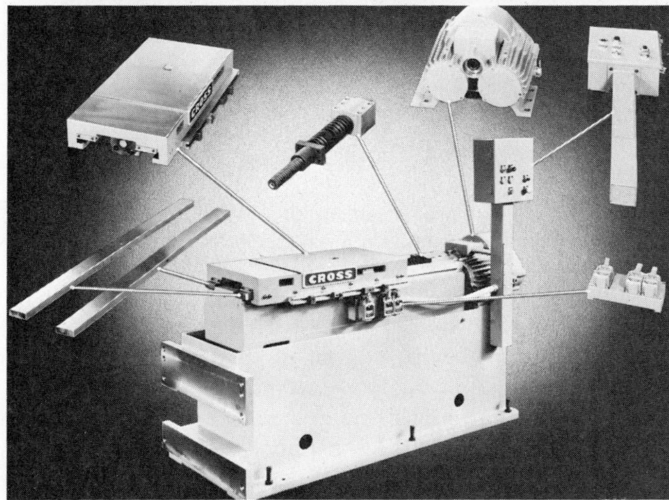

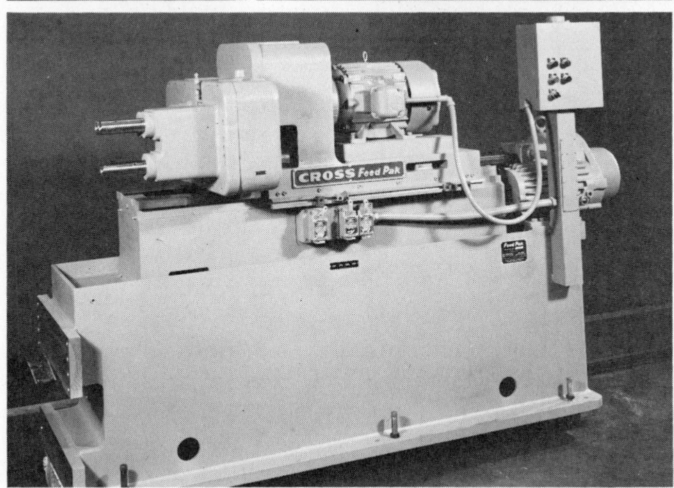

FIGURE 38-1. Powerhead production unit. (*Top*) Basic unit containing powered spindle. (*Center*) Exploded view, showing various components to provide base, spindle rotation, and feed. (*Bottom*) Complete unit. (*Courtesy The Cross Company.*)

FIGURE 38-2. Models of eight standard building-block units from which a wide variety of production machine tools can be built. (*Courtesy Heald Machine Company.*)

ing, tapping, grinding, and honing with high effectiveness. Some units, in modified form, are equipped with chucks so that simple turning operations also can be done.

Powerhead production units come in a substantial range of standardized sizes which, along with base and column units, serve as *building-block* components from which a wide variety of special production machines can be assembled. As an example, the eight standard building-block components shown in Figure 38-2 can be combined to form all the very different machining centers shown in Figure 38-3.

The application of this basic principle may be seen in Figures 38-4 through 38-7. In each case, although very different in size and arrangement, and for totally different products, most of each machine consists of standard components. When a design change occurs in a product, or the product is no longer to be made, the standard production units can be adapted to the new requirements. They can be regrouped, with new jigs, fixtures, and bases, if required. A unit previously used for drilling may now be used for tapping or milling on the same or entirely different product. Thus flexibility and adaptability are obtained along with high productivity. Also, an entirely new machine can be obtained in much less time because of the availability of standard components.

Transfer mechanisms. Means must be provided for moving workpieces from station to station on production machines, and for moving them from one machine to another. Machines that are provided with such automatic mechanisms are called *transfer machines.* Transferring usually is accomplished by one

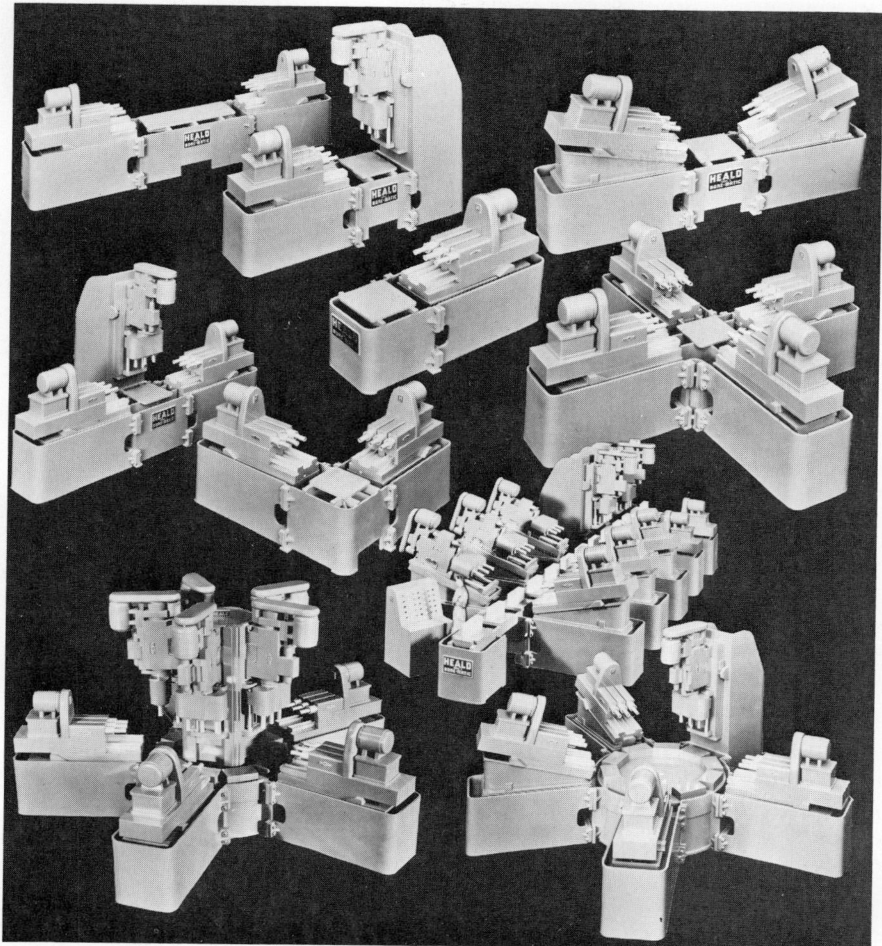

FIGURE 38-3. Examples of various types of machining units that can be built from the components shown in Figure 38-2. (*Courtesy The Heald Machine Company.*)

or more of four methods. Frequently, the work is pulled along supporting rails by means of an endless chain that moves intermittently as required. This method can be seen in Figure 38-4. In another method the work is pushed along continuous rails by air or hydraulic pistons. A third method, restricted to lighter workpieces, is to move them by an overhead chain conveyor, which may lift and deposit the work at the machining stations, as shown in Figure 38-6.

A fourth method often is employed when a relatively small number of operations—usually three to 10—are to be performed. The machining heads are arranged radially around a rotary indexing table, which contains fixtures in

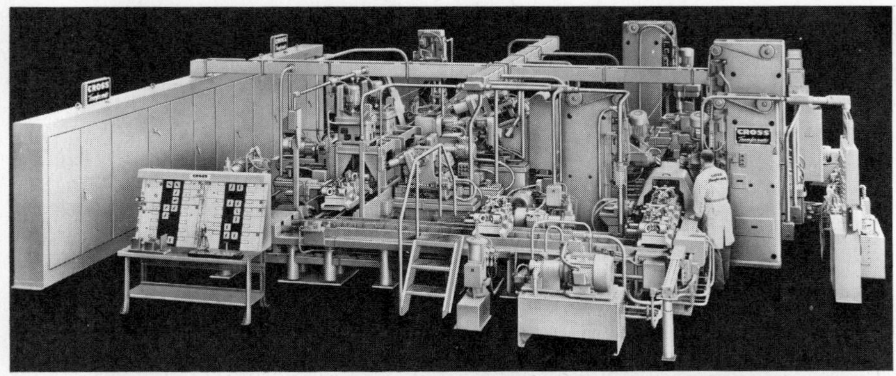

FIGURE 38-4. (*Top*) Transfer-type production machine for machining typewriter frame (*bottom left*). (*Bottom, right*) Pallet-type fixtures containing typewriter frames. (*Courtesy The Cross Company.*)

which the workpieces are mounted. Such units range in size from the rather small one shown in Figure 38-7 to quite large ones. The table movement may be continuous or intermittent. Face milling operations sometimes are performed by moving the workpieces past one or more vertical-axis milling heads. Such circular configurations have the advantages of being compact and of permitting the workpieces to be loaded and unloaded at a single station without having to interrupt the machining.

Means must be provided for positioning the workpieces correctly as they are transferred to the various stations. One method is to attach the work to carrier pallets or fixtures that contain locating holes or points that mate with retracting pins or fingers at each work station. The fixtures thus are located and then clamped in the proper positions. This method is used in the machine shown in Figure 38-4. Obviously, carrier fixtures are costly. When possible, they are eliminated and the workpiece transported directly and located by means of self-contained holes or surfaces. This procedure, used in the machine

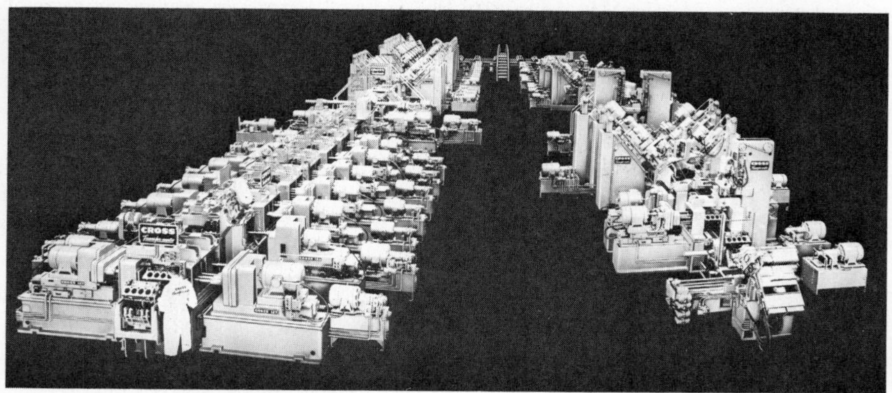

FIGURE 38-5. Complex transfer-type production unit arranged in a U-pattern. This unit machines V-8 cylinder blocks at the rate of 100 per hour. There are five independent sections that perform 265 drilling, 6 milling, 21 boring, 56 reaming, 101 counterboring, 106 tapping, and 133 inspection operations. These are performed at 104 stations, including 1 loading, 53 machining, 36 visual inspection, and 1 unloading. Provision is made for banking parts between each section. (*Courtesy The Cross Company.*)

FIGURE 38-6. Transfer-type production unit for grinding crankpins. Crankshafts are transferred between stations by an overhead conveyor. One crankshaft is completed and inspected each minute without requiring an operator. (*Courtesy Norton Company.*)

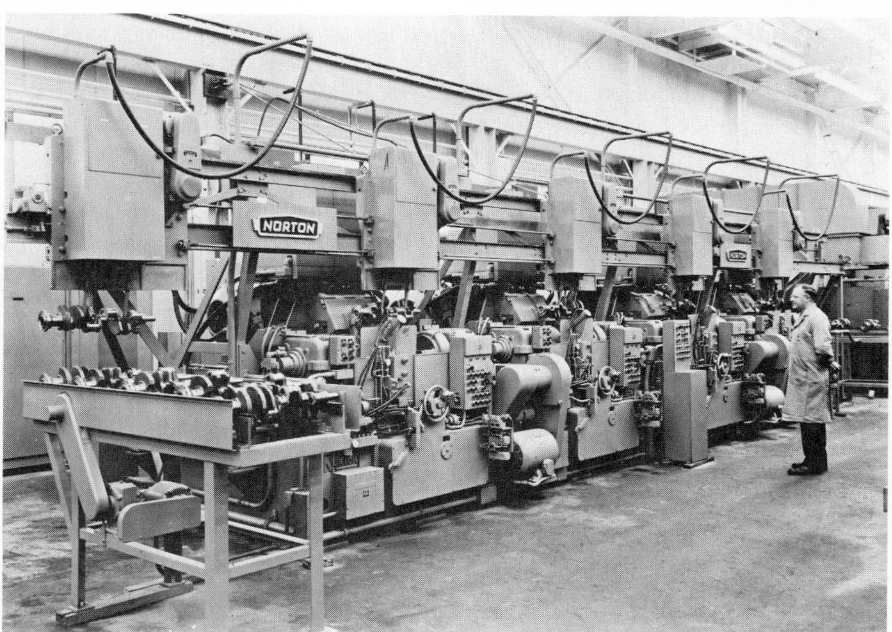

shown in Figure 38-5, also eliminates the labor required for fastening the workpieces to the carrier pallets.

Three problems are of considerable importance when large transfer machines are used. One is the matter of the geometric arrangement of the various production units. Whether or not transfer fixtures or pallets must be used is an important factor. As was shown in Figure 38-4, these fixtures and pallets usually are quite heavy. Consequently, when they are used, a closed, rectangular arrangement, such as that shown in Figure 38-4, is employed so that the fixtures are automatically returned to the loading point. If no fixtures or pallets are required, U or straight-line configurations can be employed, as shown in Figures 38-5 and 38-6, respectively. Thus, if the need for pallets or fixtures can be avoided, not only is their cost saved, but greater freedom of machine layout is provided. Whether pallets or fixtures must be used is dependent primarily on the size, rigidity, and design of the workpieces. The workpiece designer often can facilitate their elimination and should not fail to consider this matter. If no transfer pallet or fixture is to be used, locating bosses or points should be designed into the workpiece.

The matter of tool wear and replacement is of great importance when a large number of operations are incorporated in a single production unit. In such costly machines, having high production capabilities, it is essential that they be kept operating as much of the time as possible. At the same time, tools must be replaced before they become worn and produce defective parts. Transfer machines often have more than 100 cutting tools. If the entire, complex machine had to be shut down each time a single tool became dull and had to be replaced, the resulting productivity would be very low. This is avoided by designing the tooling so that certain groups have similar lives and then utilizing control panels that record tool wear in each group and shut down the machine before the tooling has deteriorated. All the tools in the affected group are then changed so that repeated shutdowns are not necessary.

Several methods have been developed for accurately presetting tools and for changing them rapidly. Figure 38-8 shows two versions of a quick-change tool holder. These are available for a wide range of tools, and they permit a large number of tools to be changed in a few minutes, thereby reducing machine downtime. Increasingly, tools are preset in standard, quick-change holders with excellent accuracy, often to within 0.005 mm (0.0002 inch). A simple and a more elaborate type of presetting equipment are shown in Figure 38-9.

A third problem encountered in the use of multiple-station machines is avoiding shutting down the entire machine in case only one or two stations should become inoperative. This usually is avoided by arranging the individual units in groups, or sections, and providing for a small amount of storage (*banking*) of workpieces between the sections. This permits production to continue on all remaining sections for a short time while one is shut down for tool changing or repair.

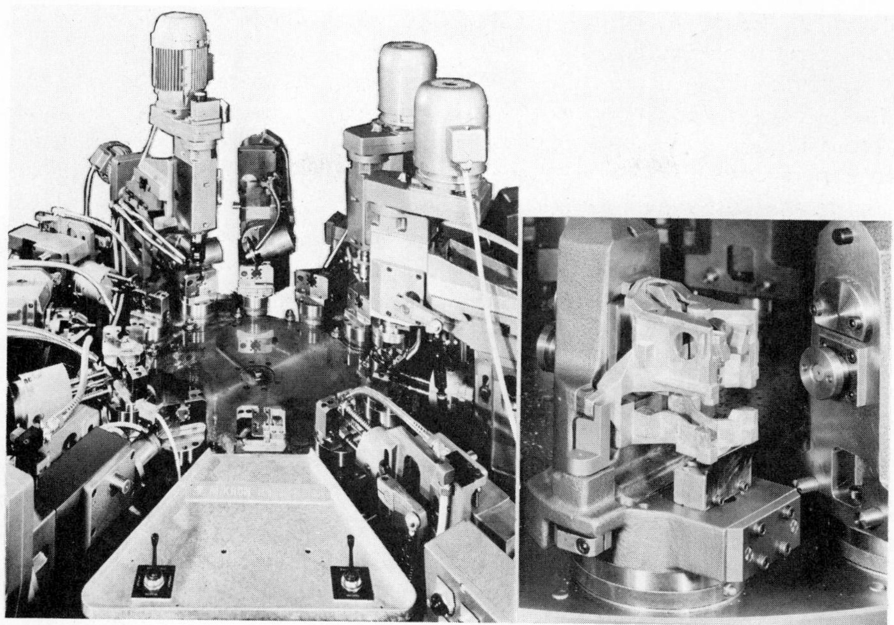

FIGURE 38-7. (*Left*) Automatic production unit employing a rotary indexing table for holding and transporting the workpieces. An operator loads and unloads the workpieces at the front station. (*Right*) Close-up view of workpiece in fixture at one station. (*Courtesy Mikron Haesler and Hamilton Technology, Inc.*)

Automation. Much has been written and discussed about automation—much of it misleading or erroneous owing to confusion between mechanization and automation. An *automated operation* is one that not only is mechanized but has the built-in capability of sensing when corrective action is required and making such corrections. For example, an automatic screw machine can carry out

FIGURE 38-8. Type of quick-change tool holder used in automatic machining units. (*Courtesy Bendix, Industrial Tool Division.*)

FIGURE 38-9. (*Top, left*) Preset tool in spindle holder. (*Top, right*) Simple device for presetting tools in holders. (*Courtesy The Weldon Tool Company.*) (*Bottom, left*) A more elaborate machine for presetting tools. (*Bottom, right*) Close-up view of the presetting equipment. (*Courtesy Ex-Cell-O Corporation.*)

operations at a high rate without a human operator. However, if a tool wears or breaks, the machine will not make corrective adjustments or shut down to prevent defective products from being produced. If the machine were automated, it would measure the part after each operation was performed, or as it was being performed, and make any adjustments required to assure that only parts meeting the design specifications were produced. Thus sensing and feedback control are essential requisites for automation.

Actually, rather few manufacturing processes are, as yet, really automated. Many are highly mechanized, and some elements of automation have been incorporated into them, as in the case of the transfer machine shown in Figure 38-10. This machine can machine the two types of flywheel housings shown; they are similar but require different machining operations. When the housings are fed into the machine, in mixed order, a sensing device contacts a distinguishing boss that is on one type but absent from the other. The sensing and feedback mechanism then sets the proper tooling for the housing, so as to omit the operations that are not required for the type involved.

It is common practice to equip transfer-type machines with automated gaging or probing heads which, after each operation, determine whether the operation was performed correctly and to detect whether any tool breakage has occurred that might cause damage in subsequent operations—checking a hole after drilling to make certain it is clear prior to tapping, for example.

Automation and transfer principles also are used very successfully for assembly operations. In addition to saving labor, automatic testing and inspection can be incorporated into such machines at as many points as desired. Not only does this assure better quality, but defective, partially completed assemblies are discovered and removed for rework or scrapping without addi-

FIGURE 38-10. Transfer-type production unit that can machine two different types of flywheel housings shown in the inset. (*Courtesy The Cross Company.*)

FIGURE 38-11. Automatic assembly machine for assembling the electronic unit and components shown in the inset. (*Courtesy Bendix Automation and Measurement Division.*)

FIGURE 38-12. Close-up view, showing a small resistor about to be inserted automatically into an assembly. (*Courtesy General Mills, Inc.*)

FIGURE 38-13. Machine for automatic assembly and inspection of two types of automobile differentials, and completed units shown in inset. (*Courtesy The Cross Company.*)

tional parts being added or additional assembly steps being completed on them.

The range of products now being completely or partially assembled automatically is very great. Figure 38-11 shows an automated assembly machine for assembling and testing electronic components. Figure 38-12 is a close-up view of one of the assembly heads on a similar machine. These machines are further examples of how standardized heads can be combined and converted for assembling different combinations of components. Figure 38-13 shows a

FIGURE 38-14. Large pallet and the components for two automobile differentials, assembled on the machine shown in Figure 38-13. (*Courtesy The Cross Company.*)

very different type of automated assembly machine for the assembly and testing of automobile differentials. In this case the components are carried through the machine on large pallets, one of which is shown in Figure 38-14. The tightening of many of the bolts is done to specified torque limits, and numerous inspections and tests are done automatically, including a noise test.

FIGURE 38-15. (*Bottom*) Automatic positioners feeding and removing parts in a large forming press. (*Top, left*) Motions provided by this type of positioner. (*Top, right*) Close-up view of the positioner holding the workpiece. (*Courtesy Unimation Inc.*)

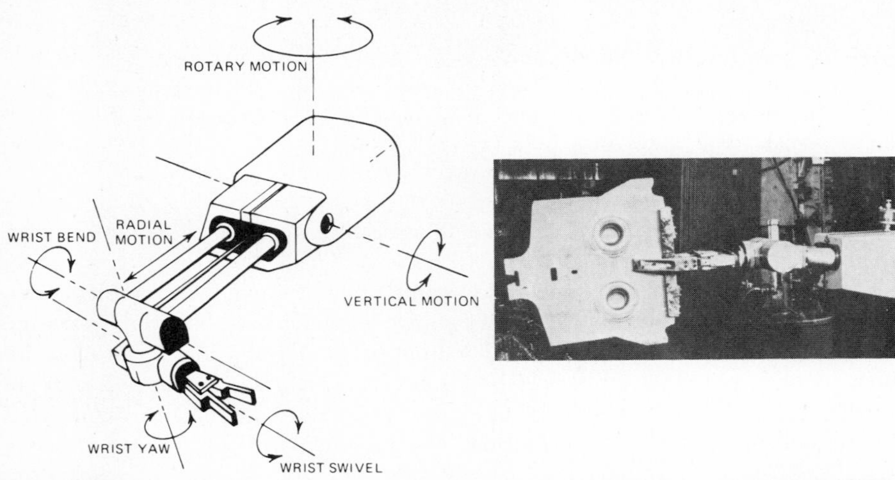

In many cases, some manual operations are combined with some automatic operations. For example, one transfer machine for assembling steering knuckle, front wheel hub, and disk-brake assemblies has 16 automatic and 5 manual work stations. As with machining operations, automatic assembly often can be greatly facilitated by proper consideration on the part of designers.

Automated positioners. *Automatic positioners,* shown in Figure 38-15, are playing very important roles in automating forming, machining, and assembly operations. As shown, some of these units have as many as six motions, which permit them to be used to perform not only heavy and difficult tasks, as in feeding forming and forging presses, but also to be programmed to perform difficult, complex tasks, such as welding automobile bodies. In addition, they often can be used to perform operations that tend to be hazardous or must be done under unpleasant conditions. Consequently, it is to be expected that automatic positioners are going to be used with increasing frequency.

Review questions

1. Why are general-purpose machine tools not suitable for use in mass production?
2. What are two distinctly different approaches in designing machine tools for use in mass production?
3. Why is it desirable for mass-production machines to be flexible?
4. What functional capabilities do most power-head production units have?
5. Explain what is meant by "building-block" construction as applied to machine tools.
6. What are five machining operations that could be performed by a single, typical power-head production unit?
7. What is a transfer-type machine?
8. What methods commonly are employed to transfer workpieces on production machines?
9. Why is it desirable to eliminate work pallets in connection with transfer machines?
10. Explain the relationship of work pallets to the geometric arrangement of the production units in a large transfer-type machine.
11. Why must the designer of a part that is to be made on a transfer-type machine take into account whether transfer pallets will be used?
12. What means are used for locating workpieces at individual work stations in transfer machines?
13. How is frequent shutdown of transfer-type machines for changing worn tools avoided?
14. What is the difference between automation and mechanization?

15. In mass production, why is it important to apply inspection before or during assembly rather than after assembly is completed?
16. Explain why product design is more important in mass production than in small-scale manufacturing.
17. What difficulties would you expect to encounter in attempting to use automated equipment in one of the developing countries?
18. Why is the use of automatic positioners likely to increase?

Case study 28.
MACHINING SMALL COLLARS

The Jo-Ko Company wants to consider manufacturing 2000 of the collars shown in Figure CS-17 (page 705) and has assigned you to determine whether they should be, or could be, made on its B & S screw machine, having a secondary-operation milling attachment (Figures 20-55 and 20-58), or on its recently acquired turning center, such as shown in Figure 17-22. (1) Could the collars be made on both machines? (2) List a feasible sequence of operations for making the collars on each machine (if they can be made on both machines). (3) Which machine do you believe would be more economical, and why?

Numerical, tape, and computer control

The most significant development in manufacturing during the past 20 years has been the advent and wide-scale adoption of numerically, tape-, and computer-controlled machine tools. As was pointed out in Chapter 38, there was a great need for machines that could bridge the gap between highly flexible, general-purpose machine tools and highly specialized, but inflexible, mass-production machines. Numerical control (N/C) of machine tools not only bridged this gap, but also created entirely new concepts in manufacturing, making routine certain operations that previously were very difficult to accomplish. And whereas in the earlier years highly trained programmers were required to do the programming required for their use, the development of low-cost, solid-state microprocessing chips has resulted in machines that can be programmed in a very short time, by personnel having only a few hours of training, using only simple machine shop language. As a consequence, there are few manufacturing facilities today, from the largest down to small job shops, that do not have, and routinely use, one or more numerically controlled machine tools.

Numerically controlled machines have the unique advantage of being economical for producing a single or a substantial number of units. Also, through the use of tape or computer control, there can be absolute assurance that a unit produced today can be duplicated exactly by the same machine at a later date.

A not-inconsiderable side result of the use of numerically controlled machines has been the increased realization of the importance and use of preproduction planning. Such planning is an absolute requirement in the use of numerically controlled machines, and a realization of its advantages has resulted in much better planning being done even when general-purpose tools are to be used.

Another side result has been the substantial decrease in the non-chip-producing time of machine tools, caused by the necessity for the operator to set speeds and feeds and locate the tool relative to the work. Quite simple forms of N/C and digital readout equipment have provided both greater productivity and increased accuracy.

A further, but not entirely unmixed, advantage of N/C has been the ability to routinely obtain greater accuracy than previously was obtainable. Most of the early N/C machine tools were developed for special types of work where accuracies of as much as 0.0013 mm (0.00005 inch) might be required, and most N/C machines were built to provide accuracies of at least 0.00254 mm (0.0001 inch), whether it was needed or not. While most N/C machine tools today will provide greater accuracy than is required for most jobs and the tendency to specify greater accuracy on parts than is required has diminished, the matter should be monitored.

Functions involved in machining. In machining a single part on an ordinary single-purpose machine tool, the operator usually will perform the following nine functions:

1. Plan the operation sequence.
2. Select the tools.
3. Set and change the tools.
4. Select the feeds and speeds.
5. Set the feeds and speeds.
6. Position the work relative to the tools.
7. Start and stop the machine operations.
8. Control the relative motion (path) between the work and the tool during cutting.
9. Cause the work (or tool) to move from the position at the end of one cut to that for the beginning of the next cut (usually at a rapid rate).

If the machine operator performs all these functions effectively, he or she must have a high degree of skill, yet during a substantial portion of the operation cycle she or he is idle—watching chips being made. At the same time, during a considerable portion of the total cycle the machine is not removing chips—its functional purpose.

As will be discussed in Chapter 40, items 1, 2, and 4 of the preceding list can be done before the operation begins, by someone other than the machine operator. This would lessen the skill and judgment requirements of the operator. If the number of identical parts to be produced is sufficient, and the motions required are relatively simple, items 3, 5, 6, 7, 8, and 9 can be done automatically or semiautomatically, as in a screw machine or a turret lathe. However, such a procedure is not economically feasible for a single or a few parts.

Item 6, positioning the work relative to the tool, can be greatly facilitated through the use of jigs, but this procedure is costly and cannot be justified when only one or a few parts are to be made.

Item 8, control of the relative motion between the work and the tool, cannot be done by an operator if the motion is at all complex. For example, turning a uniform taper on a lathe by simultaneously controlling the longitudinal and cross feeds is virtually impossible, and the accurate turning of a more complex surface is completely impossible. Although irregular surfaces can be produced by such machines as copying lathes and duplicating milling machines, these require rather expensive templates or models that make the cost very high if only one or a few parts are to be made. Furthermore, such machines are quite specialized and costly, and highly skilled labor usually is required to operate them.

It thus is apparent that functions 3, 5, 6, 7, and 9 of the list above cannot be done automatically or economically on the basic, general-purpose machine tools when a single or a few workpieces are involved. However, tape and computer control make it possible to do all these functions automatically and economically.

Basic principles of numerical control. As the name implies, *numerical control* is a method of controlling the motion of machine components by means of numbers. It can be illustrated in simple form by its application to item 6 of the previous list—positioning. Assume that the three 25.4-mm (1-inch) holes in the part shown in Figure 39-1 are to be drilled and bored on the Jigmil shown in Figure 39-2. The centers of these holes must be located relative to each other and with respect to the left-hand edge of the workpiece (*X* direction) and the bottom edge (*Y* direction). After the workpiece is set up on the table of the machine, the work and the tool must be brought into proper relationship for each of these holes. If this is to be done automatically, means must be available for precisely specifying, measuring, and controlling the relative motions of the machine table and the spindle carrier so that the location dimensions specified on the drwing will be reproduced on the workpiece. This requires signals that will command the driving motor for the machine component, such as a table, to move the component to the desired location (*open-loop*

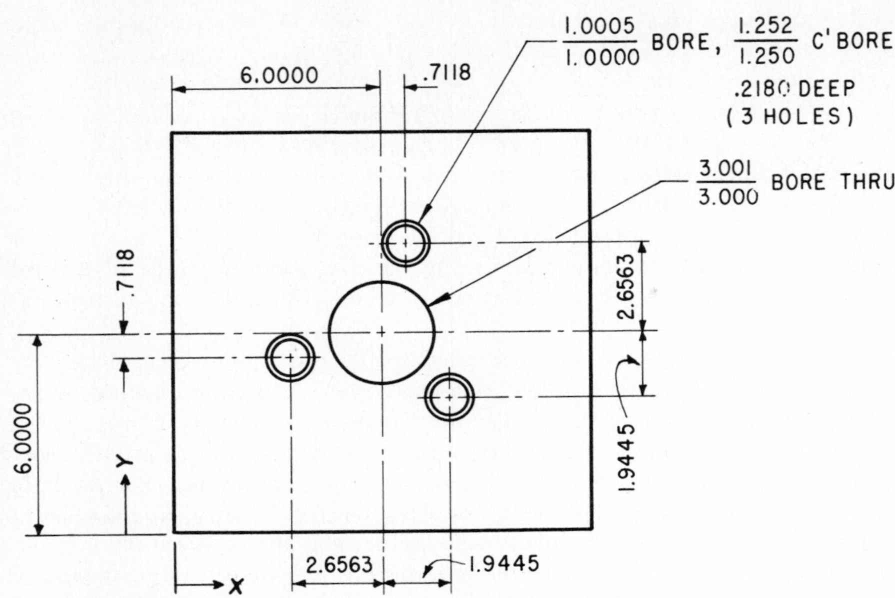

FIGURE 39-1. Part to be machined on a numerically controlled machine tool.

FIGURE 39-2. Tape-controlled Jigmil of the type that could be used for machining the holes in the part shown in Figure 39-1. Inset shows the setup of the head and the work for boring. (*Courtesy DeVlieg Machine Company.*)

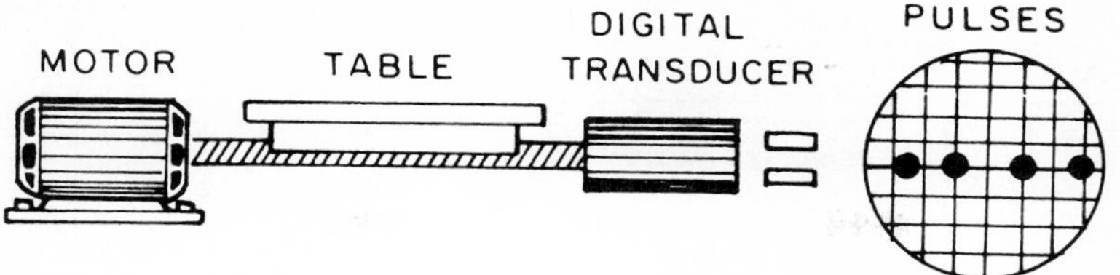

FIGURE 39-3. Schematic representation of a digital transducer and resulting output signal. (*Courtesy Norden Division of United Aircraft Corp.*)

system), or command and feedback signals that will command the driving motor to move the table and also tell the motor when the component has reached exactly the desired position (*closed-loop system*). Most N/C controls use the closed-loop system, with the feedback signals being supplied by transducers actuated either by the feed screw or by the actual movement of the component. The transducers may provide either *digital* or *analog* information (signals).

Digital information usually is in the form of electric pulses, illustrated schematically in Figure 39-3. Two basic types of digital transducers are used. One supplies *incremental* information and tells how much motion of the input shaft or table has occurred. The information supplied is similar to telling a newsboy that he is to deliver papers to the first, fourth, and eighth houses from a given corner on one side of a block. To follow the instructions, the newsboy would have to have a means of counting the houses (pulses) as he passes them and deliver papers when he has counted 1, 4, and 8. The second type of digital information is *absolute* in character, with each pulse corresponding to a specific location of the machine component. Using the newsboy analogy, this would correspond to telling him to deliver papers to the houses having house number 2400, 2406, and 2414. In this case it would only be necessary for the newsboy (machine component) to be able to read the house numbers (addresses) and stop and deliver a paper when he has arrived at a proper address. This "address" system is a common one in numerical-control systems, because it provides absolute location information relative to a zero point.

The analog type of information is illustrated in Figure 39-4. In this case the signal is usually in the form of an electric voltage that varies as the input shaft is rotated or the machine component is moved, the variable output being a function of movement. The movement is evaluated by measuring, or matching, the voltage, or by measuring the ratios between the applied and feedback voltages; this eliminates the effect of supply-voltage variations. Again using the newsboy analogy, this method is like telling the boy to deliver papers to those houses on one side of the block that are 7.6 m (25 feet), 68.6 m (225 feet), and 129.5 m (425 feet) from a given corner. For the boy to respond

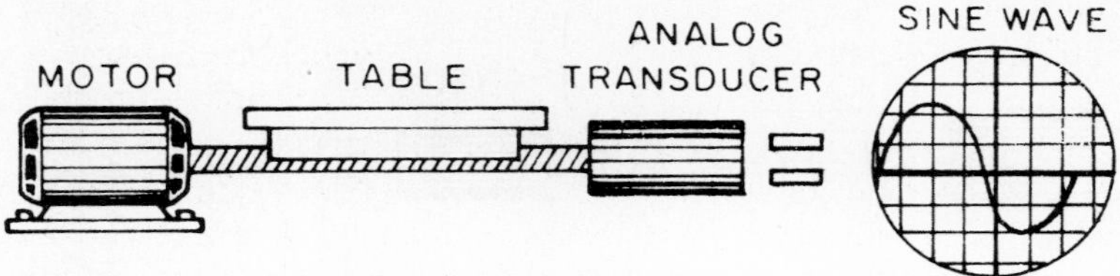

FIGURE 39-4. Schematic representation of an analog transducer and resulting output signal. (*Courtesy Norden Division of United Aircraft Corp.*)

properly, he would need a measuring device—a tape measure—with which to measure his movement from the corner. He then would deliver papers when his traveled distance was 7.6, 68.6, and 129.5 meters. Several types of N/C systems use analog information.

Figure 39-5 depicts the elements involved in a basic manual address-type N/C system for controlling one axis of motion of a machine tool. The operator sets the desired location for the workpiece for a given operation on the command module. When he or she actuates a control button, the system moves the table to the specified location, and its actual location is indicated on the actual-position readout display.[1] As many of these systems as desired can be combined to provide control in several axes—two- and three-axis controls are most common, but some machines have as many as seven. In many, conversion to either English or metric measurement is available by merely throwing a switch.

The components required for such a numerical-control system now are

FIGURE 39-5. Basic components and system for a manual address-type numerical control for one axis of a machine tool. (*Courtesy Norden Division of United Aircraft Corp.*)

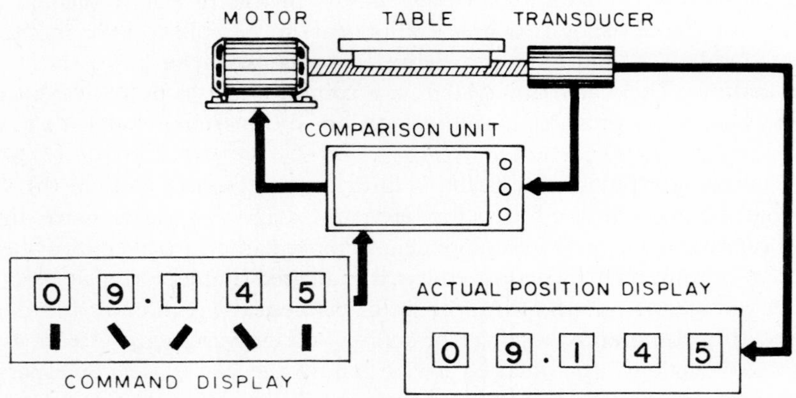

[1] In some systems the actual-position display can be set to read "zero" when the desired position has been reached.

well-standardized items of hardware. In most cases the driving motor is electric, but hydraulic systems also are used. They usually are capable of driving the machine elements, such as tables, at high rates of speed—up to 5080 mm (200 inches) per minute being common. Thus exact positioning can be achieved much more rapidly than by manual means. Several types of transducers are used, most of them being completely electrical. In one type a photoelectric sensor counts the optical fringes generated by superimposed ruled gratings that are slightly inclined to each other, one being attached to the machine table and the other fixed. In many systems the input of the transducer is connected directly to the lead screw, with special precautions being taken, such as the use of extra-large screws and ball nuts, to avoid backlash and to assure accuracy. Other systems drive the transducer from an accurate rack that is attached to the machine table. When pulse systems are used, the pulses are counted by common off–on electronic counters. Various degrees of accuracy are obtainable; guaranteed positioning accuracies of 0.025 mm or 0.0025 mm (0.001 inch or 0.0001 inch) are most common, but greater accuracies can be obtained at higher cost. Most N/C systems are built into the machines, but they can be retrofitted to some machine tools.

A basic N/C system of the type described usually performs only function 6 in the previous list—positioning the work relative to the tool. Although in such a system the remaining functions must be performed by the operator, there are many advantages relative to a machine without numerical control. Some of these are:

1. Greater accuracy and uniformity.
2. More rapid table movement.
3. Less chance of mistakes.
4. Eliminaton of the need for jigs, thus avoiding this cost and also speeding completion time for the first units.
5. Usually better preproduction planning, because economical use of such machines makes it imperative that better planning be done.

Tape control. Tape control of N/C machines provides a means for automatically accomplishing functions 3, 5, 7, 8, and 9, as well as function 6, of the previous list, thereby eliminating the necessity for the operator to perform these functions. They also can be performed more rapidly and, in the case of function 8, more accurately, permitting operations to be done readily that often could not be done effectively by manual control. In addition, functions 1, 2, and 4 can, by preproduction planning, be punched into the tape. Thus all nine of the required functions can be tape-controlled, permitting almost complete automation on a general-purpose tool, and for any production quantity.

Actually, punched tape is not a recent idea—the old player-piano roll was a form of tape control, and punched cards had been used for many years for controlling complicated weaving and business machines. Thus tape control

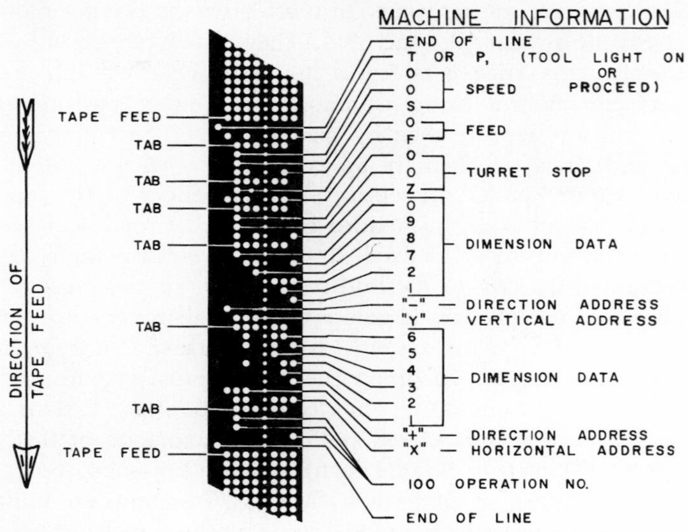

FIGURE 39-6. Block of control tape for controlling one operation on a turret-type drilling machine. (*Courtesy DeVlieg Machine Company.*)

of machine tools is an extension of an existing basic concept in which holes, representing information that has been punched into the tape, are "read" by sensing devices and used to actuate relays or other devices that control various electrical or mechanical mechanisms.

Most tape-controlled machine tools use a 1-inch-wide paper or Mylar tape containing eight information channels. Figure 39-6 shows an example of a "block" of such tape, punched with the information required for one operation on a turret-type drilling machine. By comparing Figures 39-7 and 39-5, it will be noted that the tape unit, in effect, replaces the manual command-display unit, although in most cases the manual command unit is retained so as to permit the operator to override the tape when necessary. One tape can actuate several axis-control systems, as well as the auxiliary control functions of a machine.

Although the majority of N/C and tape-controlled machine tools do not provide for machining contoured surfaces, many do. Most computer-assisted and direct-computer-controlled machines provide this feature. Such machines as the one shown in Figure 39-8 have the obvious advantages of the substitution of an inexpensive tape or computer program for expensive templates or three-domensional patterns. The required curves and contours are generated approximately by a series of very short, straight lines or segments of some type of regular curves, such as hyperbolas. Consequently, the program fed to the machine is arranged to approximate the required curve, within set limits. Figure 39-9 illustrates how a desired curve can be approximated by means of

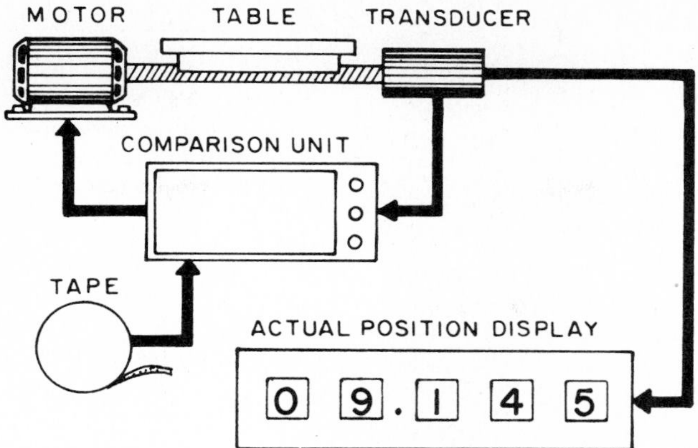

FIGURE 39-7. Basic components and system for tape control for one axis of a machine tool. (*Courtesy Norden Division of United Aircraft Corp.*)

FIGURE 39-8. Tape-controlled die-milling machine having control of motions in three axes. Similar machines are available with control of up to five axes. Inset shows a die being milled. (*Courtesy Giddings & Lewis Machine Tool Company.*)

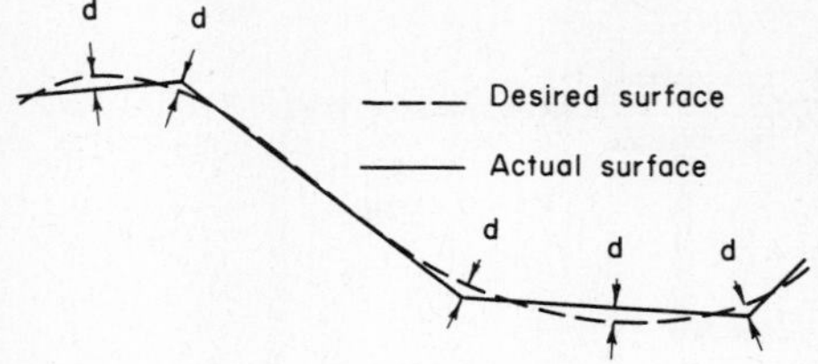

FIGURE 39-9. Method of approximating a curve with numerical control, with maximum deviation of the actual surface from the theoretical kept within distance *d*.

short, straight lines within a permissible deviation, *d*. It is apparent that the length of the straight-line or standard-curve segments must be varied in accordance with the deviation permitted and the shape of the curve. Most machine tools with contouring capability will produce a surface that is within 0.025 mm (0.001 inch) of the one desired, and many will provide considerably better performance. Most contouring machines have either two- or three-axis capability, but a good many have up to five-axis capability, as illustrated in Figure 39-10.

Obviously, contour machining requires that complex information be punched into the control tape; by ordinary procedures the number of straight-line or curved segments may be quite large, and manual programming of the tape can be quite laborious. This can be eliminated by the use of a computer—usually a mini computer built into the machine—that will translate simple commands into the complex information required by the machine.

Tape preparation. Obviously, the preparation of the tape for use in N/C is very important. In most cases the preparation is not difficult, since quite simple standard languages and programs have been developed.

Manually prepared tapes usually are punched on a typewriter-like machine, such as shown in Figure 39-11, or on more simple devices designed for

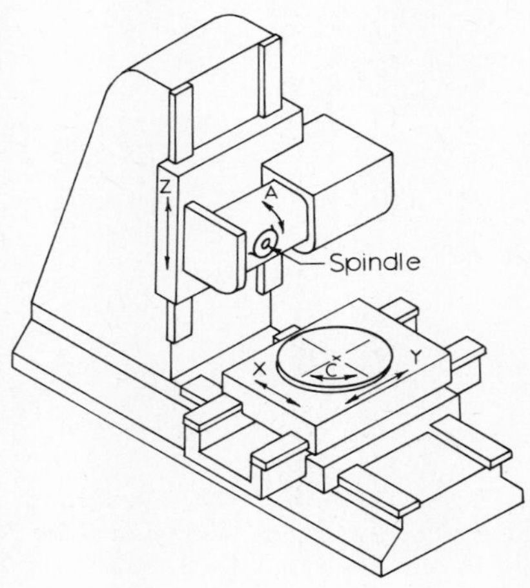

FIGURE 39-10. Five controlled axes in one type of numerically controlled machine tool.

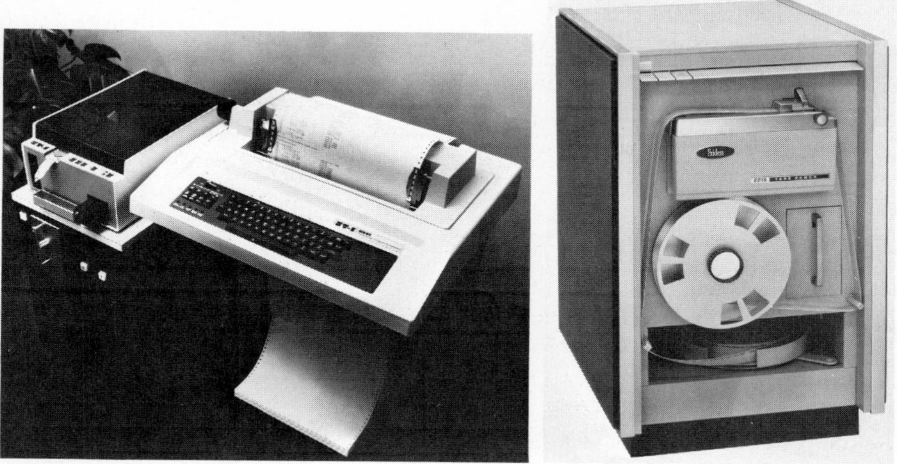

FIGURE 39-11. (*Left*) N/C tape preparation center. (*Right*) Tape-reading unit. (*Courtesy Manufacturing Data Systems Incorporated and Friden Division, Singer Business Machines, respectively.*)

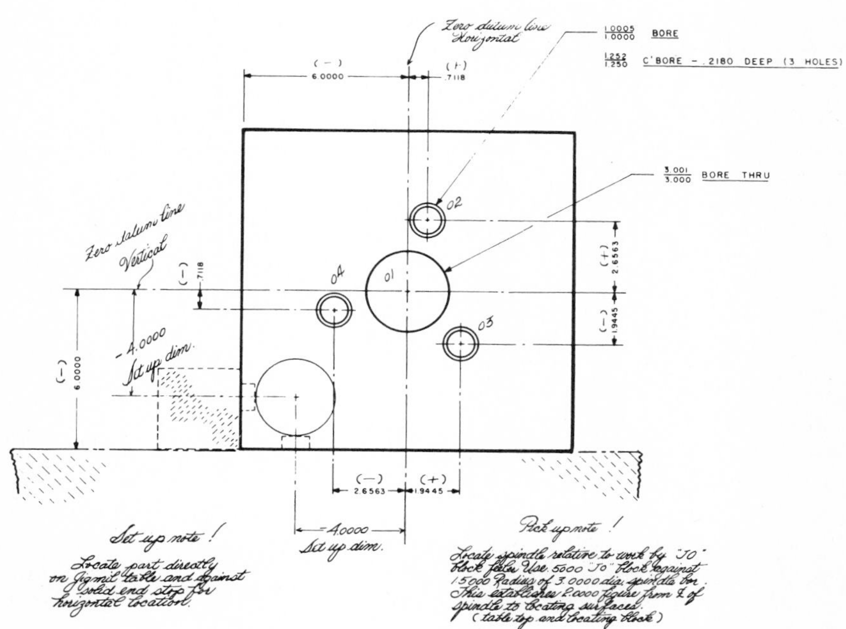

FIGURE 39-12. Drawing of a part, modified to provide information needed for preparing a tape program sheet and for setting up work on machine. (*Courtesy DeVlieg Machine Company.*)

PROGRAM SHEET FOR DEVLIEG *SPIRAMATIC* JIGMIL WITH Tapac III

Part No. ABC-1001A Part Name Plate Sheet 1 of 1

Oper. No.	X Dimension	Y Dimension	Z Turret	Spindle Feed No.	Spindle Speed No.	Tool Change Auto Step
000	X±000000	Y±000000	Z00	F0	S00	T/P

```
000      x-040000      y-040000                                    t
For set up purposes,first piece only,use .5000 Jo block against
spindle bar for horizontal(end stop) and vertical (table top).
Zero horizontal and vertical slides to this position and measurement.

001      x+000000      y+000000      z01      f6      s05          t
Core drill hole 2-7/8 dia. Hole #01 . 70 RPM .012 Feed

002                                  z02      f3      s13          t
Rough bore with tool #217 ,Hole #01;570RPM; .004 Feed

003                                  z02      f2      s13          t
Finish bore with tool # 217F; Hole #01; 570RPM; .003 FEED

004      x+007118      y+026563      z03      f6      s09          t
Drill 15/16 dia. thru 200 RPM .002Feed Hole #02

005      x+019445      y-019445                                    p
Same as operation #004; Hole #03

006      x-026563      y-007118                                    p
Same as operation #004; Hole #04

007      x+007118      y+026563      z04      f2      s15          t
Rough and finish bore and c'bore using tool #ct 10025. Hole #02
[illegible] RPM; .003 Feed

[illegible]  x+019445      y-019445                                p
Same as operation #007; Hole #03

[illegible]  x-026563      y-007118                                p
Same as operation #007; Hole #04
```

End of Line Carriage Ret.

NOTE: ...FOR MACHINE COMMAND, USE "PUNCH ON" CODE AT MARGIN
...FOR MANUSCRIPT MATTER, USE "PUNCH OFF" CODE AT MARGIN

FIGURE 39-13. Program sheet used in preparing the control tape for the part shown in Figure 39-1. (*Courtesy DeVlieg Machine Company.*)

specific machines. The basic steps can be illustrated by reference to the part shown in Figure 39-12.[2] The first step is to modify the drawing to establish the zero reference axes, the *X* and *Y* directions, the dimensioning with respect to the reference axes, and the setup instructions to establish the workpiece properly on the machine table with respect to the tool. These modifications are shown in Figure 39-12. Obviously, this step can be avoided if the original drawing is made in the required form, knowing that tape-controlled machines are to be utilized.

[2] The steps required by some machines often are considerably less complicated. The method described here is used to illustrate the basic principles.

TAPAC III
SPEED & FEED CONVERSION CHART

—BAR FEED—
PROGRAM NUMBERS AND ACTUAL FEED (INCH PER REV.)

PROGRAM NO.		F 1	F 2	F 3	F 4	F 5	F 6	F 7	F 8
ACTUAL FEED	3H & 4H	.002	.003	.004	.006	.008	.012	.017	.024
	5 H	.0034	.0042	.0059	.0085	.0132	.0183	.0256	.0365

—SPINDLE SPEED—
PROGRAM NUMBERS AND ACTUAL RPM

PROGRAM NO.			S01	S02	S03	S04	S05	S06	S07	S08	S09	S10	S11	S12	S13	S14	S15	S16				
SPEEDS	3 H	ST'D.	25	32	41	52	70	90	115	150	200	255	325	420	570	730	940	1200				
		HI-SP'D.	33	43	55	70	90	120	155	200	260	340	435	560	750	970	1250	1600				
	4 H	ST'D.	21	27	35	45	58	76	98	127	165	215	275	355	470	605	775	1000				
		HI-SP'D	26	34	44	56	70	95	120	150	210	275	355	440	575	750	950	1220				
	5 H	ST'D.	10	13	17	21	27	34	44	58	75	97	124	166	213	276	355	468	601	778	1000	N*
PROGRAM NO.			S01	S02	S03	S04	S05	S06	S07	S08	S09	S10	S11	S12	S13	S14	S15	S16	S17	S18	S19	S20

* N=NEUTRAL

FIGURE 39-14. Speed and feed conversion code chart used in preparing the control tape for part shown in Figure 39-1. (*Courtesy DeVlieg Machine Company.*)

The second step is to make a *program sheet,* such as shown in Figure 39-13. This sheet gives the coordinate dimensions for each operation, specifies the spindle traverse that determines the depth of the cut, the spindle speed and feed, and whether the same tool can continue the next operation or whether a tool change is required. The last four items are specified by code symbols obtained from the chart shown in Figure 39-14.

After the program sheet has been prepared, it is used by the operator to prepare the tape. Typing the information shown in Figure 39-13 produces the print copy shown in Figure 39-15 and simultaneously produces the punched tape, a section of which is shown in Figure 39-16.[3]

Before a tape is used, it usually is checked. This can be done by running it through the tape reader, shown in Figure 39-11, which is either connected to the tape writer so as to type a duplicate sheet showing the machine-information data (Figure 39-15) or to a special N/C plotting machine that will trace out all the tool–work paths as thy would occur on the machine tool.

[3] The entire tape for machining the part shown in Figure 39-12 measured 28 inches. The block shown in Figure 39-14 measured 3½ inches.

000	x−040000	y−040000				t
001	x+000000	y+000000	z01	f6	s05	t
002			z02	f3	s13	t
003			z02	f2	s13	t
004	x+007118	y+026563	z03	f6	s09	t
005	x+019445	y−019445				p
006	x−026563	y−007118				p
007	x+007118	y+026563	z04	f2	s15	t
008	x+019445	y−019445				p
009	x−026563	y−007118				p

FIGURE 39-15. Machine information data typed by tape-preparation unit in preparing the control tape for the part shown in Figure 39-1. (*Courtesy DeVlieg Machine Company.*)

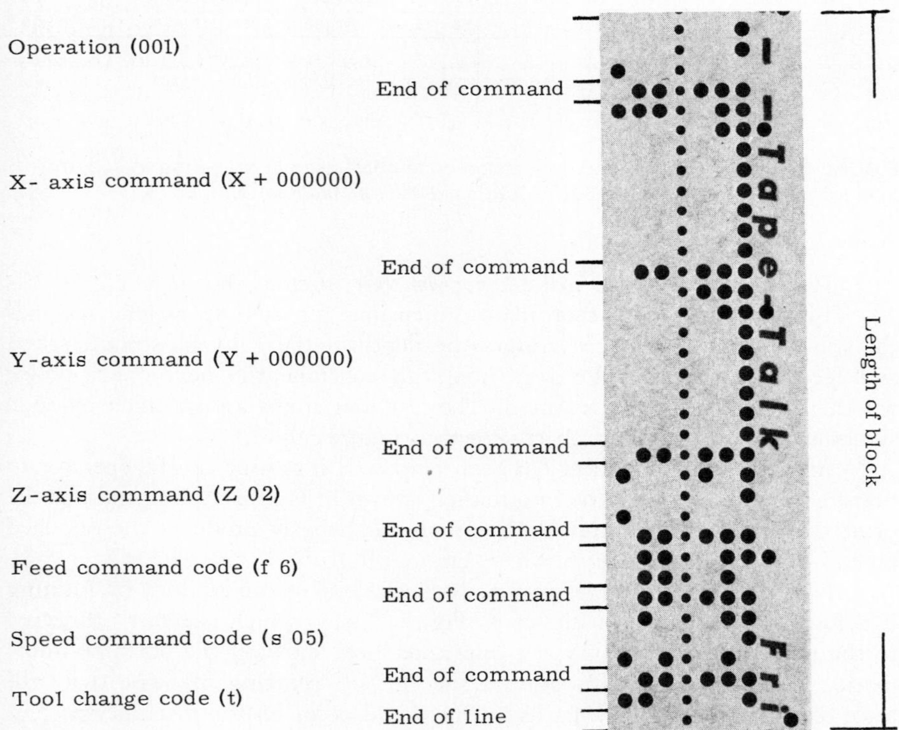

FIGURE 39-16. Block of control tape, corresponding to the machine information required for the second operation shown in Figure 39-15. (*Courtesy DeVlieg Machine Company.*)

If a part is to be duplicated a number of times, the tape may be made into a loop so that it will repeat the entire cycle of operations as often as is required. If only a single part is required, the tape usually is left in the form of a strip. Where a machine is capable of producing several different parts without retooling, the tapes corresponding to more than one part can be spliced together so that the machine will automatically change the processing as required. When it is desired to make a design change in a part, this often can be accomplished by splicing a new section in the existing tape.

Manual preparation of N/C tapes by the procedure just described can be quite time-consuming, especially for complex operations and for contouring. Consequently, such a procedure is rarely used today, being replaced by the use of special programming languages designed specifically for machine tool control, and by computer-assisted techniques.

Computer languages for N/C control. Unfortunately, the control mechanisms on N/C machines do not understand ordinary shop language that people use to describe what machining, or machine–work movements, must take place. In addition, the effects of various sizes and types of cutting tools, requiring different offsets as illustrated in Figure 39-17, and the capabilities of the machine tool regarding available power, speeds, feeds, table travel, and so on, must be taken into account. This barrier has been diminished by the use of computers and special, simplified programming languages which the computer can understand and convert into the commands required by N/C machine controls. One such language is APT II (Automatically Programmed Tools).

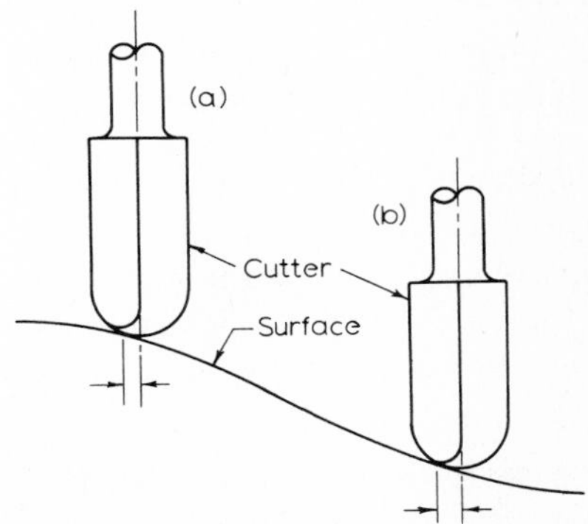

FIGURE 39-17. Effect of surface contour and cutter geometry on the relationship of cutter center line to point of contact.

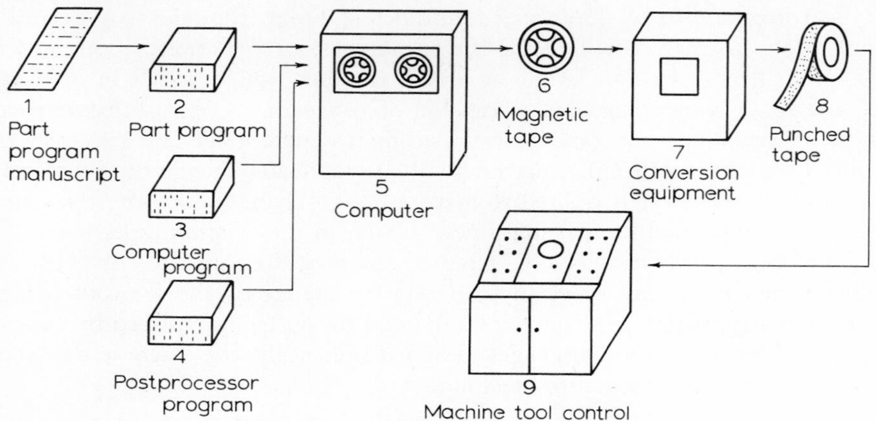

FIGURE 39-18. Schematic representation of the nine elements in a computer-assisted control program.

Figure 39-18 shows how nine elements can be combined to assist the programmer. The person doing the programming analyzes the part, dividing it into its geometric sections—straight lines, points, circles, ellipses, and so on—and determines their relative positions. He or she then writes a *part-program manuscript,* using the simplified programming language. Such languages usually have less than 200 words, with a limited number of characters and precise meanings. The part-program manuscript thus describes what operations the machine tool must perform in order to produce the part. This permits various standardized configurations to be described very briefly. For example, as shown in Figure 39-19, the locations of the centers of a circle of drilled holes can be described very quickly. Similarly, a *matrix*[4] of locations can be given merely by specifying the horizontal (*X*) distance between points in one line and the distance (*Y*) between parallel lines of points.

The part-program manuscript next is converted into the second element—the *part program.* This is done by a keypunch operator copying the part-program manuscript onto a deck of punched cards. Each card contains a specific machine-tool instruction.

The third element is the *computer program.* It is a previously prepared deck of punched cards, or tape, that contains instructions that the computer needs to understand and execute the part-program instructions and also those it will receive from the fourth element, the *post-processor program.* It is the job of the computer to calculate the coordinate points that the cutter must follow to execute the part-program instructions. At the same time the computer

[4] A matrix is a series of horizontal prallel lines of points with the distance between the points and lines specified.

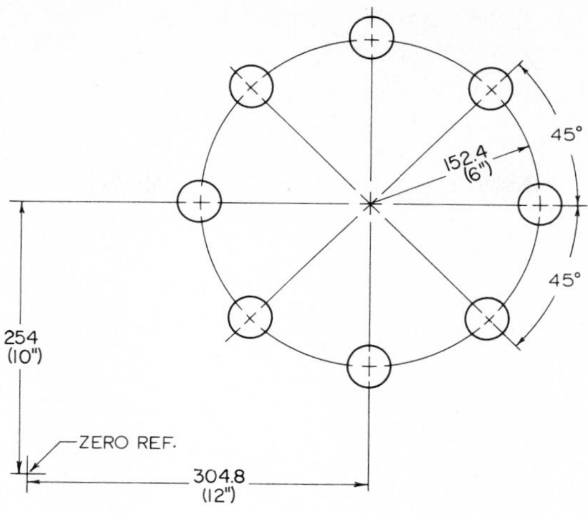

English Program:

```
INDEX/ GO TO / 18, 10, 1, 40
GO DELTA/ MINUS 1, 12
GO DELTA/ 1, 12
COPY/ 1 XY ROT, 45, 7
```

Metric Program:

```
INDEX/ GO TO / 457.2, 254, 25.4, 40
GO DELTA/ MINUS 25.4, 12
GO DELTA/ 25.4, 12
COPY/ 1 XY ROT, 45, 7
```

FIGURE 39-19. Circle of eight holes and APT program for machining these holes. Numerals 457.2, 254, and 25.4 (18, 10, and 1) are *X, Y,* and *Z* coordinates for table and tool. Numeral 40 is table movement rate. Numeral 12 is feed rate for drill. Numeral 45 is 45° of rotation. Numeral 7 is the instruction for seven duplicate holes to be drilled.

must take into account the individualities of the machine tool, relative to available speeds, feeds, accelerations and decelerations, and so forth, that are provided in the post-processor program, in either punched-card or magnetic-tape form.

A general-purpose computer is the fifth element of the system. The sixth element of the system is the output of the computer. It may be either a magnetic tape or a 1-inch-wide eight-channel perforated tape that is ready for use in the machine control.

The seventh element is necessary if the computer output is not in the form of 1-inch-wide, eight-channel tape. The *conversion unit* prepares the eighth element, the *punched tape,* from the computer output tape, if necessary. The ninth element is the *machine-tool control.*

By the use of such special languages and systems, control tapes can be prepared quite quickly and economically, even for complex parts and using the general-purpose procedure that has just been described. Fortunately, more simple and economical methods are available for most N/C machines, and these will now be discussed.

Computer-assisted tape preparation and control. As just discussed, the availability of computers has simplified the programming of N/C machines. Also, on any given machine tool, the number of operations that can be performed and the number of required commands are limited. Consequently, the capacity of the memory and processor elements that are required to convert a limited number of simple instructions into machine commands is relatively small. This permits a machine to be programmed in quite simple shop terminology, which will be processed by the computer into the commands required by the machine. A variety of such programs and controls now is provided by machine tool manufacturers. The CUTS program (Computer Utilized Turning System), provided by the The Warner & Swasey Co. for certain of its turning machines, is an example of this type of procedure and will be described in some detail.

FIGURE 39-20. Tapered sleeve. (*Courtesy The Warner & Swasey Co.*)

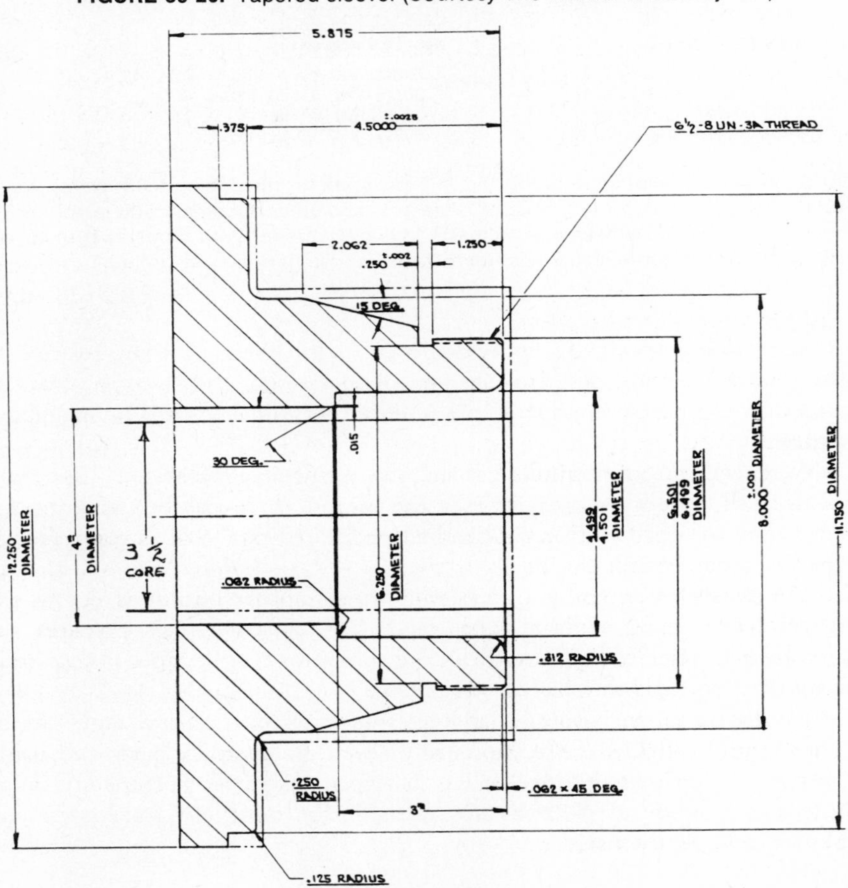

The operation of the CUTS program will be described in its application to the part shown in Figure 39-20. The programmer first fills out the two forms shown in Figure 39-21. Basically, the *header sheet* defines the material, so that the computer can select the speeds and feeds, and locates the part within an imaginary, three-dimensional box relative to control axes on the lathe, taking into account any interference offered by the chuck jaws.

On the *surface-description sheet* the programmer lists the operations to be performed, using a restricted list of ordinary machine-shop terms, such as turn, bore, face, and so forth, following a continuous, closed-path machining sequence and starting with the right-hand finished face. He or she also specifies the amount of material that must be removed and the surface finish desired. These enable the computer to decide on the number of cuts necessary and the feed. Very little specialized knowledge about programming is required, and a person with reasonably good knowledge of the machine and the processing can learn rather quickly to analyze the workpiece and fill out these two basic sheets.

The next step is to feed the data contained in the two programming sheets to the computer. This can be done directly through a teletype, or similar computer terminal, or by punched cards. This provides the typed copy shown in Figure 39-22. Making use of the input data and the CUTS program for the particular machine tool, which previously has been stored in the computer, the computer generates the *input and tooling sheet* shown in Figure 39-23 and, by connection to a tape-punch unit, the control tape. If desired, a *tape command* listing can also be obtained from the computer. A portion of this type of listing, showing the eight channels of tape commands for the illustrative example, is shown in Figure 39-24.[5] Such computer-assisted procedures greatly simplify the programming of single machines.

Direct numerical control. If suitable information regarding a machine tool is stored in the memory of a computer when input specifying the description of a workpiece and the operations that are to be performed on it is fed into the computer, it will automatically provide the output necessary to punch a control tape. The tape, in turn, can actuate the machine controller. If, on the other hand, the computer is connected directly to a suitable machine-control unit, the machine tool can be controlled by the user communicating directly with the computer either by a keyboard console or by punched cards. Because of the capabilities of modern computers, it is possible for several N/C machines to be controlled through a single computer. Such a system is called *Direct Numerical Control* (DNC). DNC also can provide current information to management regarding production output.

Although DNC systems have great potential and are used by a number of companies, their use has not been as widespread as once was thought would

[5] There are 275 tape commands for the total operation involving the part shown in Figure 39-20.

CUTS

WARNER & SWASEY COMPUTER UTILIZED TURNING SYSTEM

A — HEADER SHEET

Pg. 1 of 2

PART	PART NAME: TAPERED SLEEVE	PART NUMBER: 210725695	OPER.: 30
COMMENT	FOR W.T.S.C BY D.H.G.		
CLASS	MAT CLASS: 02 · DRILL SFPM · ROUGH SFPM · FINISH SFPM · BRINELL · RGH. FEED FACTOR · FIN. FEED FACTOR		
COMMENT	CAST IRØN CLASS 30		
MACHINE	MACHINE IDENTIFICATION — SIZE: SC28 · RPM · MTR · L		
GRIPDIAM	DIAMETER ON WHICH PART IS GRIPPED: 12.25		
SPIN-LOC	DIMENSION FROM SPINDLE NOSE FLANGE TO LOCATING POINT: 10.		
LOC-INT	DIMENSION FROM LOCATING POINT TO INTERFERENCE WITH HOLDING DEVICE: .875		
PRTLENTH	DIMENSION FROM FINISHED RIGHT END TO THE LEFT END OF THE PART: 5.875		
FRNTFACE	LOC. PT. TO FIN. RT. END: 5.875 · STOCK TO BE REMOVED: .375 · TOLERANCE: .005 · RMS FINISH: 125 · RGH. FEED FACTOR · FIN. FEED FACTOR · RGH. SFPM FACTOR · FIN. SFPM FACTOR · CUT		

Column positions printed on the form: 1, 8, 10, 15, 20, 25, 30, 35, 40, 45, 50, 55, 60, 65, 70, 73 (IDENT. SEQ.), 80.

Col. 1–8	Col. 10	Col. 43
PPIDEN	*	BØTH
EXTERNAL		

CUTS

WARNER & SWASEY COMPUTER UTILIZED TURNING SYSTEM

B — SURFACE DESCRIPTION

Pg 2 of 2

SURFACE DESCRIPTION		FIRST DIMENSION	SECOND DIMENSION	STOCK	RMS FIN.	CUT	ANGLE OR RADIUS	THIRD DIMENSION	FOURTH DIMENSION	FIFTH DIMENSION		IDENTIFICATION SEQUENCE
CHAMFER		.0620					45.					
DIAM		6.5010	6.4490	8.5000	125							
THREAD		1.3750	.1250									
NECK		.125	.25									
FACE		1.5000		.5000	125							
TAPER	1		2.0620				15.					
DIAM		8.0000	.0020	8.5000	063							
RADIUS							.250					
FACE		4.5000	.0050	.3750	125							
RADIUS							.1250					
DIAM		11.750		12.250	250							
FACE		4.875		.7500	250							
CUTSHARP												
DIAM		12.250		12.250								
INTERNAL												
RADIUS							.312					
DIAM		4.501	4.499	3.5	063							
RADIUS							.063					
FACE		3.0000		3.0000	125							
CHAMFER		.0150					30.					
DIAM		4.0000		3.5000	250							
FACE		5.8750										
CUT-PAST												
ENDPART												

FIGURE 39-21. Header and surface description sheets used in the CUTS system of programming for the part shown in Figure 39-20. (*Courtesy The Warner & Swasey Co.*)

```
PART,TAPERED SLEEVE,210725695,30
COMMENT,FOR W.T.S.C. BY D.H.G.
CLASS,02
COMMENT,CAST IRON CLASS 30
MACHINE,SC28
GRIPDIAM,12.25
SPIN-LOC,10
LOC-INT,.875
PRTLENTH,5.875
FRNTFACE,5.875,.375,.005,125
EXTERNAL
CHAMFER,.062,,,,,45.
DIAM,6.501,6.449,8.5,125
THREAD,1.375,.125
NECK,.125,.25
FACE,1.5,,1.5,125
TAPER,1,,2.062,,,,15.
DIAM,8.,.002,8.5,063
RADIUS,,,,,,.25
FACE,4.5,.005,.375,125
RADIUS,,,,,,.125
DIAM,11.75,,12.25,250
FACE,4.875,,.75,250
CUTSHARP
DIAM,12.25,,12.25
INTERNAL
RADIUS,,,,,,.312
DIAM,4.501,4.499,3.5,063
RADIUS,,,,,,.062
FACE,3.,,3.,125
CHAMFER,.015,,,,,30.
DIAM,4.,,3.5,250
FACE,5.875
CUT-PAST
ENDPART
```

FIGURE 39-22. Typed input copy to computer in CUTS system for the part shown in Figure 39-20. (*Courtesy The Warner & Swasey Co.*)

occur. This has been due largely to the advent of inexpensive minicomputers used in connection with single machines, which will be discussed next.

Computer numerical control. Just as they have made possible the extremely effective and versatile hand-held computers, so the development and availability of solid-state microprocessor and memory chips have greatly improved and simplified the programming and control of single N/C machines. The basic components in such circuitry are depicted in Figure 39-25.

Because the operations performed by a single machine tool are limited, and thus the machine processing information that must be available in the memory bank also is limited, only a relatively simple computer-memory unit is required to enable the machine to be programmed by quite simple input information from the user. Thus many modern N/C machines have small built-in computers, thereby providing what is called *Computer Numerical Control* (CNC). The advantages of DNC are largely obtained without the necessity of having a large computer; "software" (simple programs) is substituted for more elaborate and costly "hardware."

Applications of tape control. Tape control is used on a wide variety of machines. These range from single-spindle drilling machines, which often have only two-axis control, such as is shown in Figure 39-26, and can be obtained for about $10,000, to machining centers, such as is shown in Figure 39-27. The latter can do drilling, boring, milling, tapping, and so forth, with three-axis control. It can automatically select and change up to 60 or more preset tools. Such a machine costs over $250,000. Between these extremes are nu-

FIGURE 39-23. Input and tooling sheet provided by the computer for part shown in Figure 39-20. (*Courtesy The Warner & Swasey Co.*)

```
++INPUT AND TOOLING
WARNER & SWASEY CUTS    VERSION 2   MODIFICATION 2
PART     TAPERED SLEEVE              21Ø725695     3Ø
COMMENT  FOR W.T.S.C. BY D.H.G.
CLASS    Ø2
COMMENT  CAST IRON CLASS 3Ø
MACHINE  SC28
GRIPDIAM 12.25
SPIN-LOC 1Ø.
LOC-INT  .875
PRTLENTH 5.875
FRNTFACE 5.875   .375    .ØØ5 125
EXTERNAL
CHAMFER *    .Ø62Ø -Ø.       -Ø.        -Ø-Ø 45.ØØØØ
DIAM    *   6.5Ø1Ø  6.449Ø  8.5ØØØ 125-Ø
THREAD      1.375Ø   .125Ø -Ø           -Ø-Ø
NECK    *    .125Ø   .25ØØ -Ø           -Ø-Ø
FACE    *   1.5ØØØ -Ø.       1.ØØØ  125-Ø
TAPER   1 -Ø.        2.Ø62Ø -Ø.         -Ø-Ø 15.ØØØØ
DIAM    *   8.ØØØØ   .ØØ2Ø  8.5ØØØ  63-Ø
RADIUS  * -Ø.       -Ø.      -Ø.        -Ø-Ø    .25ØØ
FACE    *   4.5ØØØ   .ØØ5Ø   .375Ø 125-Ø
RADIUS  * -Ø.       -Ø.      -Ø.        -Ø-Ø    .125Ø
DIAM    *  11.75ØØ -Ø.       12.25ØØ 25Ø-Ø
FACE    *   4.875Ø -Ø.        .75ØØ 25Ø-Ø
CUTSHARP
DIAM    *  12.25ØØ -Ø.       12.25ØØ   -Ø-Ø
INTERNAL
RADIUS  * -Ø.       -Ø.      -Ø.        -Ø-Ø    .312Ø
DIAM    *   4.5Ø1Ø  4.499Ø  3.5ØØØ  63-Ø
RADIUS  * -Ø.       -Ø.      -Ø.        -Ø-Ø    .Ø62Ø
FACE    *   3.ØØØØ -Ø.       3.ØØØØ 125-Ø
CHAMFER *    .Ø15Ø -Ø.       -Ø.        -Ø-Ø 3Ø.ØØØØ
DIAM    *   4.ØØØØ -Ø.       3.5ØØØ 25Ø-Ø
FACE    *   5.875Ø -Ø.       -Ø.        -Ø-Ø
CUT-PAST
ENDPART
WARNER & SWASEY CUTS    VERSION 2 MODIFICATION 2        Ø9/Ø1/72    Ø8.41.46
 STATION  TOOL  OP NAME     HOLDER          INSERT   RADIUS   ZOFSET XOFSET
     1     1Ø    TURN    65Ø3 12L-HEAD TPG544     .Ø625   15.3125  -3.9833
     2     8Ø    ETRD    65Ø6 L-BLK     NT-4R     .Ø1ØØ   15.872Ø  -3.645Ø
     3     2Ø    FACE    65Ø3 12L-HEAD TPG544     .Ø625   15.2667  -4.1875
     4     43    RBOR    2835 3.5ØDBAR TPG433     .Ø469   15.Ø781   -.7938
           53    FBOR    2835 3.5ØDBAR CPG422     .Ø313   15.Ø9Ø6   2.59Ø6
     5     3Ø    FINI    65Ø3 12L-HEAD CPG632     .Ø313   15.34Ø6  -4.1594
     6     6Ø    EGRV    65Ø6 L-BLK1/4 NG-48R     .Ø1ØØ   15.99ØØ  -3.635Ø
START STATION NO. 1 , 16.Ø INCHES FROM THE STOPS, RADIAL SCALE Ø.Ø

 CAUTION - CHECK UNUSED STATIONS FOR INDEXING CLEARENCE.

                 TOTAL TIME    18.2 MINUTES
                 LAST BLOCK NUMBER'IS 275
                 TAPE LENGTH =   53 FEET    1Ø INCHES
```

++TAPELIST

WARNER & SWASEY CUTS VERSION 2 MODIFICATION 1 09/01/72

WARNER AND SWASEY SC-28

+++++++++ TAPE COMMAND LISTING +++++++++ PAG

N	G	X	Z	I	K	F	S	T	M
1	1		-3.0000		.99989	250.00			7
2							30		3
3	4	3.5000						11	
4									8
5	1	-.0992		.99999		25.00			6
6			-6.5500		.99989	250.00			7
7							30		3
8	4	3.0000							
9	1		-4.7000		.99999	4.89			6
10		-.1000	.1000	.70711	.70711	25.00			
11		-.1000	4.6000	.02173	.99976	250.00			7
12		.7000		.99989		125.00			
13							27		3

FIGURE 39-24. First 13 of the tape command listings provided by the computer through the CUTS system for the part shown in Figure 39-20. (*Courtesy The Warner & Swasey Co.*)

merous machine tools that do less varied work than the highly sophisticated machining centers but which combine high output, minimum setup time in changing from one job to another and remarkable flexibility because of the number of tool motions that are provided and controlled, usually with continuous-path contouring capabilities. The machines shown in Figures 39-28 and 39-29 are examples of this trend of providing great versatility along with high productivity. The versatility of such machining centers is being further increased by combining both rotary-work and rotary-tool operations—turning and milling—in a single machine, as in the one shown in Figure 17-22. There also are numerous tape-controlled machines that provide four- and five-axis contouring capability and highly accurate jig borers, such as is shown in Figure 35-5. Table (or tool) movements commonly occur at speeds up to 6350 mm (250 inches) per minute, and tools are changed in 6 seconds or less. It

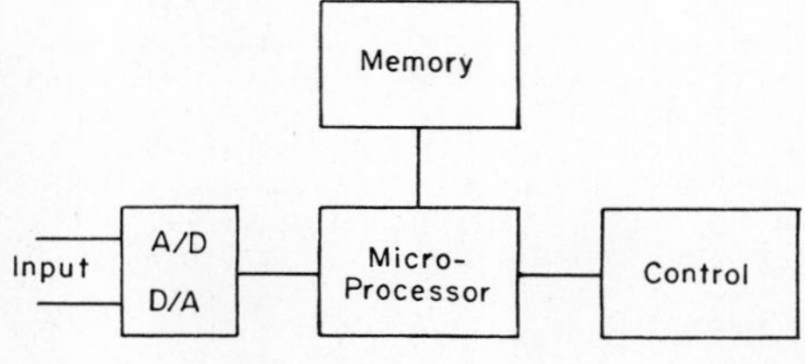

FIGURE 39-25. Basic components of the circuitry in a modern solid-state control system.

FIGURE 39-26. Tape-controlled, single-spindle drilling machine. (*Courtesy Colt Industries, Pratt & Whitney Machine Toll Division.*)

FIGURE 39-27. Large tape-controlled machining center. This machine has a 72-position index table and a pallet shuttle. It can select and automatically change tools in 7 seconds from a storage magazine that holds up to 60 preset tools. (*Courtesy Kearney & Trecker Corporation.*)

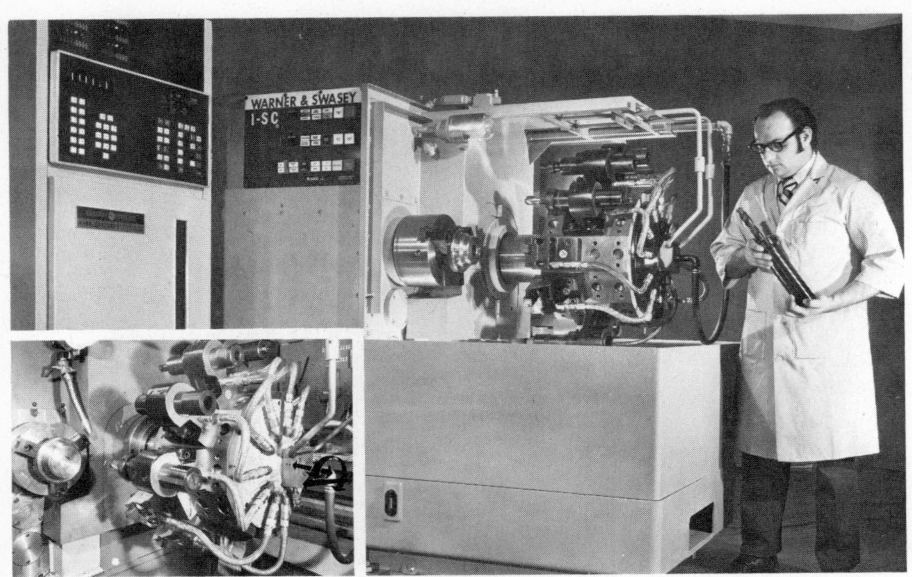

FIGURE 39-28. Tape-controlled turret lathe, showing a close-up view of the turret, tooling, and spindle and the motions provided. (*Courtesy The Warner & Swasey Co.*)

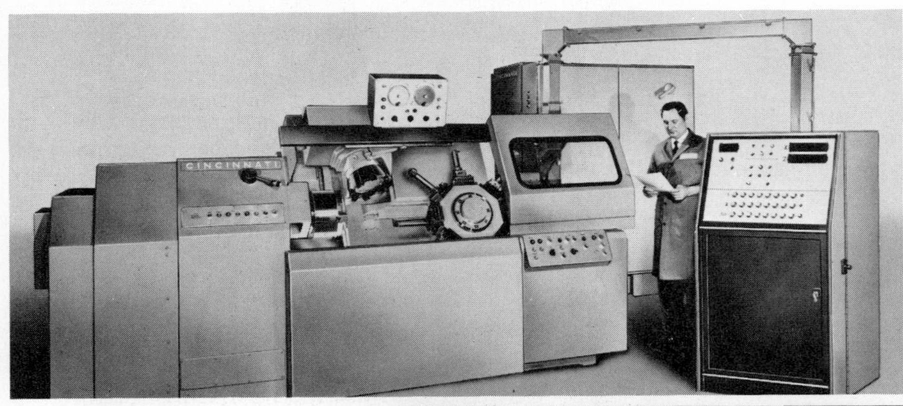

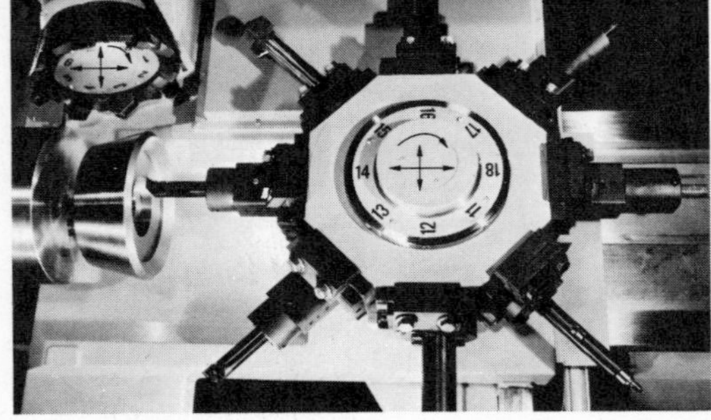

FIGURE 39-29. Tape-controlled turning machine having two indexing turrets, each capable of contouring. (*Courtesy Cincinnati Milacron Inc.*)

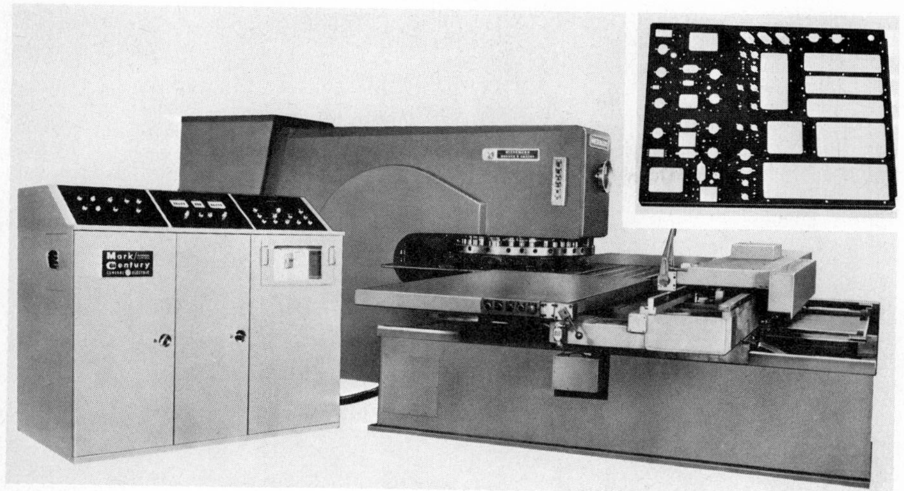

FIGURE 39-30. Tape-controlled turret-type punch press. Inset shows a typical sample of work done. (*Courtesy The Warner & Swasey Co., Wiedemann Division.*)

also is common to provide two worktables, permitting work to be set up on one while machining is done on work mounted on the other (Figure 39-27), with the tables being interchanged automatically in 12 seconds or less. Consequently, the productivity of such machines is very high, the chip-producing time often exceeding 75 per cent of the total.

Tape control also has been applied to a wide variety of production tools other than those that produce chips. One very important application is equipment for punching sheet metal. Figure 39-30 shows an example of such a machine and one type of work done. A tape-controlled spot welding machine is shown in Figure 39-31.

Economic considerations in tape and numerical control. N/C and tape-controlled machines are costly, but their use usually can be justified economically in from 1 to 3 years if their use factor is reasonably high, primarily through the substantial savings in setup and machining time and in tooling costs. Studies have shown that setup time is reduced by 20 to 70 per cent, handling time by 20 to 50 per cent, and scrap by as much as 45 per cent. Overall production gains of more than 100 per cent are common. Three examples of relative times and costs are shown in Table 39-1.

Advantages and disadvantages. The advantages of tape-controlled machine tools can be summarized as follows:

1. Greater accuracy. The precision built into the machine is utilized to the fullest extent by the control system.

FIGURE 39-31. Tape-controlled spot-welding machine. (*Courtesy General Electric Company.*)

2. Product quality improvement. A high order of repeatability is provided.
3. High production rates. Optimum feeds and speeds for each operation.
4. Lower tooling costs. Expensive jigs and templates are not needed.
5. Less lead time. Tape can be prepared in less time than conventional jigs and templates, and less setup time is required.
6. Fewer setups per workpiece. More operations can be done at each setup of the workpiece.
7. Better machine utilization. There is less machine idle time, owing to more efficient table or tool movement between successive operations and fewer setups. Cycle time is reduced.

TABLE 39-1. **Comparison of setup and cycle times and tooling costs for three typical parts machined on conventional and tape-controlled machine tools**

	Setup Time (hr)		Cycle Time (hr)		Tooling Costs	
Part	Conventional	Tape-controlled	Conventional	Tape-controlled	Conventional	Tape-controlled
Toolholder	5.25	None	1.86	1.30	$3140	$2350
Motor base	4.48	None	0.65	0.40	2350	1890
Bracket	4.63	None	0.64	0.21	835	260

8. Reduced inventory. Less inventory needs to be carried because parts can be run economically in smaller quantities.
9. Reduction in space required. Greater productivity and reduced tooling lessen floor and storage space, and smaller economic lot sizes reduce storage space required for inventories.
10. Less scrap. Operator errors are substantially reduced.
11. Less skill required of the operator. Program planning in preparing tapes reduces the necessity for operator decisions.

The only major disadvantage of tape-controlled machine tools is their cost. This means that they must have sufficient use to justify the investment and enable the previously indicated savings to be obtained. The control equipment now is virtually all made of solid-state modules, and the reliability is excellent. Availabilities exceeding 95 per cent are common. The programming has been greatly simplified. Consequently, N/C machines have provided a much needed solution for small- and medium-quantity production, and it is easy to see why they have been so widely adopted.

Review questions

1. Of the nine functions that are listed as being required for machining a part, which can be affected by the part designer?
2. Which of the nine functions can be considered to be truly productive, considered from the viewpoint of the functional purpose of a machine tool?
3. Which of the nine functions usually are transferred from the operator when tape control is used?
4. What is a disadvantage of having the feedback transducer driven by the table feed screw on a machine tool?
5. Explain what is meant by an address system of location specification.
6. What are the advantages provided by a manually operated digital-control system on a machine tool?
7. Assuming that a computer is not used, what steps would be required to prepare a control tape from a drawing?
8. Assume that the two mounting holes in the part shown in Figure 35-1 are to be machined on the 3H Jigmill that was used for the part shown in Figure 39-12 and that the following dimensions apply:

$a = 1.750$ in.	$e = 6.500$ in.
$b = 4.388$ in.	$f = 0.850$ in.
$c = 1.750$ in.	$g = 1.650$ in.
$d = 2.375$ in.	hole diameters $= 0.500$ in.

Prepare the program sheet, using suitable feeds and speeds for gray cast iron and using the data shown in Figure 39-14 for coding.

9. Explain why minicomputers with memories often are built into N/C control systems.
10. What is the function of a post-processor program in numerical control?
11. Explain how the use of tape control may reduce the economic lot size.
12. Explain how the use of tape control may reduce the lead time for producing a quantity of parts.

Case study 29. SHAFT AND CIRCULAR CAM

The Cab-Con Corporation makes the part shown in Figure CS-29 in large quantities. (Only the essential dimensions are shown.) Currently, it is machining the part from AISI 1120 bar stock that is 76.2 mm (3 inches) in diameter and then carburizing the outer periphery of the disk cam, which must have a hardness of at least $55R_c$, although the depth of the hardness need not exceed 0.15 mm (0.006 inch). Obviously, the machining cost and the wastage of material are excessive. The chief engineer has assigned you to devise a more economical procedure for producing this part. How would you propose to produce the part? (Assume that the volume is sufficient to justify any required new equipment, or that some operations can be subcontracted, if that would be more economical.)

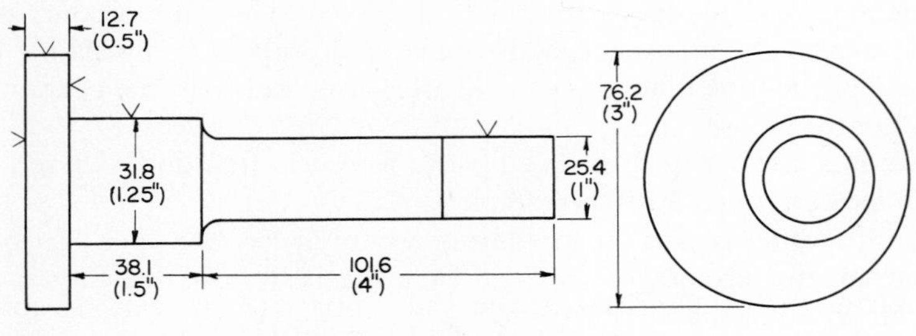

FIGURE CS-29. Shaft with eccentric circular cam.

CHAPTER 40

Planning manufacturing operations

Manufacturing is the utilization and management of materials, people, equipment, and money to produce products. The previous chapters have been devoted primarily to a discussion of materials, processes, and equipment. This chapter deals with an important aspect of the management of manufacturing—*planning*. Special emphasis will be given to the relationship between design, planning, and production.

For successful and economical manufacturing to be achieved, planning must start at the design stage and continue through the selection of materials, processes, equipment, and the scheduling of production. As materials and energy become more scarce and costly, and recycling and antipollution requirements become more stringent, so will good planning become more imperative. Investment in plant and equipment, now averaging over $30,000 per worker in most manufacturing industries, will increase as more complex, automated, and numerically controlled equipment and antipollution devices are installed. Consequently, if the cost of producing products is to remain low, so as to assure their consumption in the large volume that makes mass-production possible, good planning in manufacturing is an absolute requirement. Thus design and manufacturing engineers must understand how their activities fit into the planning process.

Not too many years ago, many designers felt that manufacturing engineers were their natural enemies and expressed their views as: "We design it; you make it." In one large plant, the manufacturing groups retaliated by adopting a policy of omitting from the product any design detail that caused manufacturing difficulties and waiting to see if anybody noticed it or if any difficulties ensued. Not too surprisingly, in many cases nothing further was heard about the matter. Clearly, this is not the best way to handle such important matters, and now most design and manufacturing engineers recognize the importance of using a "systems" approach, extending from the design stages through manufacturing, packaging, distribution, and use of the product. If the design engineer is not concerned with all the problems related to converting his or her ideas from a design on paper into "hardware" in the hands of the user, an otherwise brilliant design may be worthless because it cannot be manufactured and marketed economically. Adequate planning of manufacturing operations can do much to avoid such situations.

The relationship of design to production. As was mentioned in Chapter 10, good design usually occurs in three phases. In the *conceptual* or *idea* phase, the designer conceives of an idea for a device that will accomplish some function. This stage establishes the functional requirements that must be met by the device.

In the second, *functional-design* stage, a device is designed that will achieve the functional requirements established in the conceptual stage. Often more than one functional design will be made, suggesting alternative ways in which the functions can be met. However, little attention is given to the manufacturing operations required to make the device. The designer should, of course, keep in mind whether the device can be made, but she or he is more concerned with materials than with processes at this stage.

The third phase of design is called *production design.* Although attention should also be given to the appearance of the product at this stage, particularly if sales appeal is important, the major emphasis is on providing a design that can be manufactured economically. The design engineer must, of course, know that certain manufacturing processes and operations exist that can produce the desired product. However, merely knowing that feasible processes exist is not sufficient. He also needs to know their limitations and possibilities as to cost, accuracy, tolerance requirements, and so forth. An excellent example of such a situation occurred in the designing of a new automatic transmission for an automobile. Four multiple-disk clutches were required—two large and two small. By knowing not only that a blanking process could be used for producing both the friction and metal disks, but also understanding thoroughly the blanking process and the tolerance requirements, the engineers were able to design each clutch with the same number of friction and metal disks and with the small and large clutches of such diameters that each piece blanked from the center of a large clutch element was exactly the correct diameter for a small clutch element. As a result, very substantial savings in material and blanking costs were affected. This likely would not have occurred if the designers have been concerned only with function, and if the production department had then been required to manufacture whatever the designers had called for. Thus, if maximum economy is to be achieved, the designer should be aware of the intimate relationship between design details and production operations.

It is extremely important that the relationship between production and design be given careful consideration throughout the production-design phase, and it should not be forgotten even at the functional-design stage. Changes can be made for pennies in the design room that would cost hundreds and thousands of dollars to effect later in the factory. This type of consideration should be an integral and routine part of planning for manufacturing. One outstandingly successful company follows the practice of having the design engineers make a working prototype of every new product before any production drawings are made. If the model performs in accordance with the conceptual requirements, a second model is made, using, insofar as possible, the same manufacturing methods that will be used later for production. Any changes that will permit easier and more economical production are incorporated into this second model. If the second model meets the functional requirements of the engineering design group, it is then sent to the drafting room, and production drawings are made from it. The company feels that this practice has virtually eliminated costly-to-produce details from its products.

The designer plays a key role in determining what processes and equipment must be used to manufacture the product he or she designs, although often indirectly. If a die casting is specified, obviously a die-casting machine must be used. If welding is specified, welding equipment must be available. This type of direct relationship is quite clear. However, other equipment and processes may be specified just as certainly in not-so-obvious ways.

One of the most common ways in which equipment and processes may be specified indirectly is through dimensional tolerances placed on a drawing. If a tolerance of ±0.005 mm (0.0002 inch) is shown, a grinding operation may be specified just as definitely as if the word *grind* were placed on the drawing. Designers often fail to realize this fact, and unnecessarily close tolerances and expensive and unnecessary operations result. A classic example of this type occurred in an airplane factory where a redrawing operation was being performed on a sheet-metal part to bring a main indentation in the piece within a general requirement on the drawing that stated "Tolerances on all dimensions ±0.001 inch." Investigation revealed that the only thing that had to fit into this particular indentation was the toe of the shoe of a pilot. A similar situation may be brought about through the careless or unnecessary distribution of finish marks on drawings, or the use of the phrase "Finish all over" when there is no need for having all surfaces machined.

On the other hand, indefinite dimensions and tolerances can lead to important requirements being ignored by the factory. In a recent large court suit, one important drawing contained the statement "Approximately not to scale," and another had an important dimension reading "128$^5/_{16}$ inches approx." It was not surprising that the shop ignored certain other dimensions that caused it some inconvenience but which were critical. Consequently, designers should realize that the dimensions and tolerances they place on a drawing may have implications and results far beyond what they anticipated.

Design details are directly related to the processing that will be used,

FIGURE 40-1. Good and bad details in casting design.

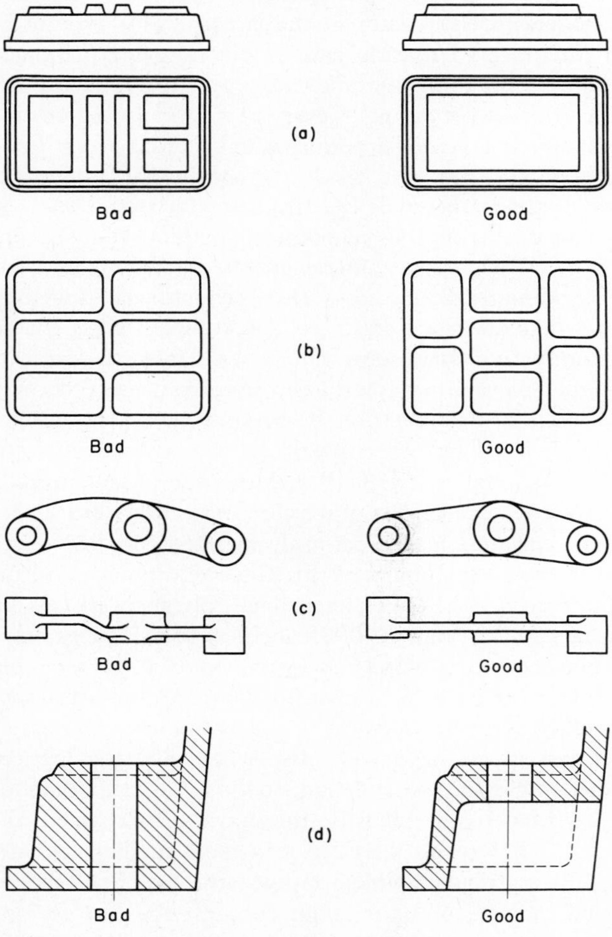

(a) Bad Good

(b) Bad Good

(c) Bad Good

(d) Bad Good

making the processing easy, difficult, or impossible, and affecting the cost and/or quality. Some of these will now be considered.

Design details related to casting. Some design details that relate to casting are shown in Figure 40-1. These emphasize the basic fact that castings shrink during solidification and cooling, and that difficulties may be experienced unless uniform sections are employed and adequate provisions is made to avoid excessive contraction stresses. Also, by remembering this principle, one often may reduce the weight and cost of castings, as illustrated at Figure 40-1d. Figure 40-1c emphasizes that mold parting lines should be in a single plane, if practicable. This also applies to forgings. All of these, and similar good design details, can be followed in most situations.

Most of the design details relating to sand castings also apply to permanent-mold castings, such as die castings. However, if permanent cores are utilized, as most often is the case, other unique problems are encountered. Figure 40-2 illustrates how relatively simple design changes can have a major effect on the cost of the die and the resulting casting.

Design details related to lettering. Lettering or decorative detail is often included on the surface of products. As shown in Figure 40-3, this should be handled differently for various processes. When letters or decorative devices are included on castings or forgings, the design should call for them to be above the surface of the part. On sand castings they can be produced easily in this manner by applying stock letters to the surface of the pattern, which will in turn result in their being above the surface of the casting. To have the lettering below the surface of the casting requires that the pattern be engraved. In the case of die castings and forgings, the lettering can be engraved in the dies, requiring only a minimum of machining and producing the desired,

FIGURE 40-2. Good and bad design details in die castings.

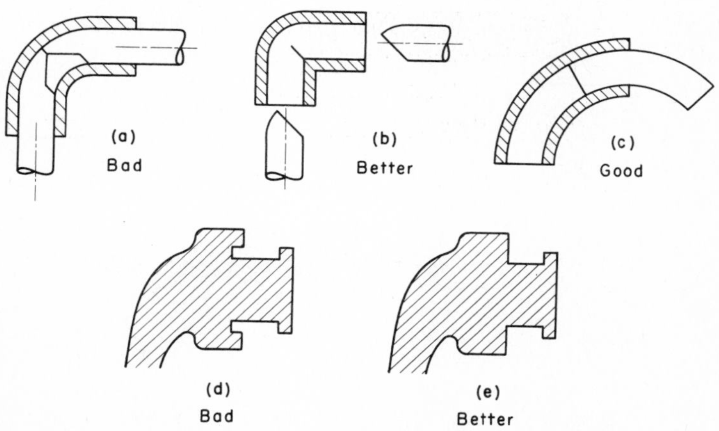

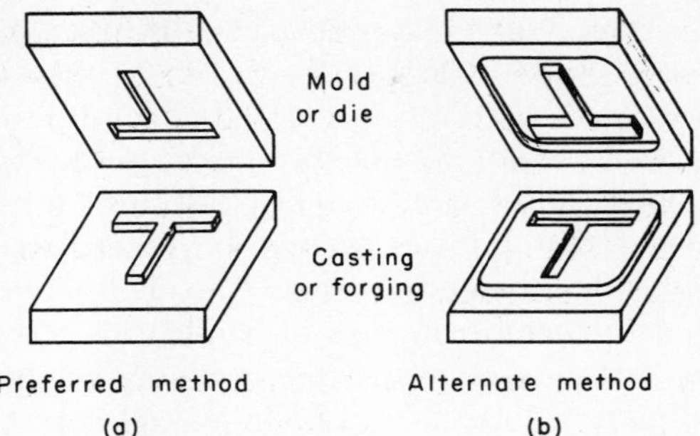

FIGURE 40-3. Preferred methods of providing lettering for different processes.

raised letters on the finished parts. On the other hand, if the design calls for depressed lettering on die castings and forgings, all the surface in the dies, except the letters, must be machined away. This is an expensive process and can be done only by use of special machines. If depressed lettering must be used on die castings and forgings, the method illustrated in Figure 40-3b sometimes can be employed, the lettering being depressed in a small, raised panel. This requires only the small amount of material in the panel, surrounding the lettering, to be removed from the dies.

When lettering must be included on a machined surface, it should be engraved into the surface. If raised lettering is essential, it should be made on a separate, small plate and attached to the machined surface. In many cases this can be done very satisfactorily by means of modern adhesives.

Design details related to forgings. When designing forgings, two primary factors should be kept in mind: (1) metal flow, and (2) minimizing the number of operations and dies. These two often are closely related. As illustrated in Figure 40-4a, a shape that would be very satisfactory for casting may

FIGURE 40-4. Examples of good and bad design details for forgings.

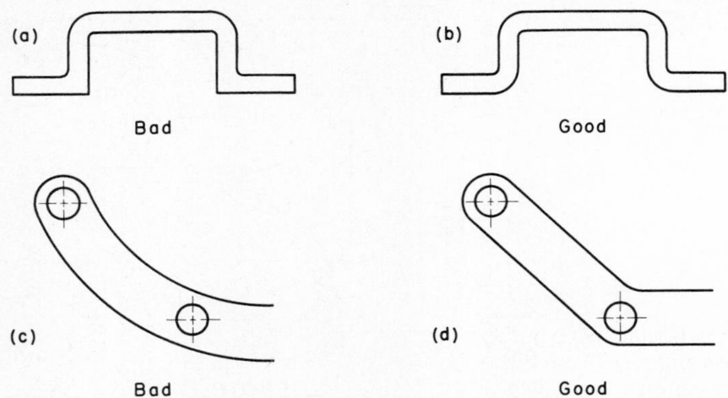

be quite poor for forging, requiring more than one operation. This difficulty can be eliminated by the simple change shown in Figure 40-4b. Figure 40-4c and d illustrates how a shape that is easy to draw on the drafting board may be excessively costly to produce because of added difficulties in machining the forging die. If upset forging is involved, the rules illustrated in Figure 14-18 should be observed. Quite often the combining of two processes, such as forging and inertia welding, can provide an economical solution.

Design details related to machining. The ways in which design details can affect machining are almost unlimited. Figure 40-5 illustrates just a few. It is far better for the designer to visualize how the workpiece will be machined and make minor modifications that will permit easy and economical machining than to force the manufacturing department to find some way to machine a needlessly bothersome detail, usually at excessive cost. Too often the manufacturing departments will take for granted that the part has to be made as

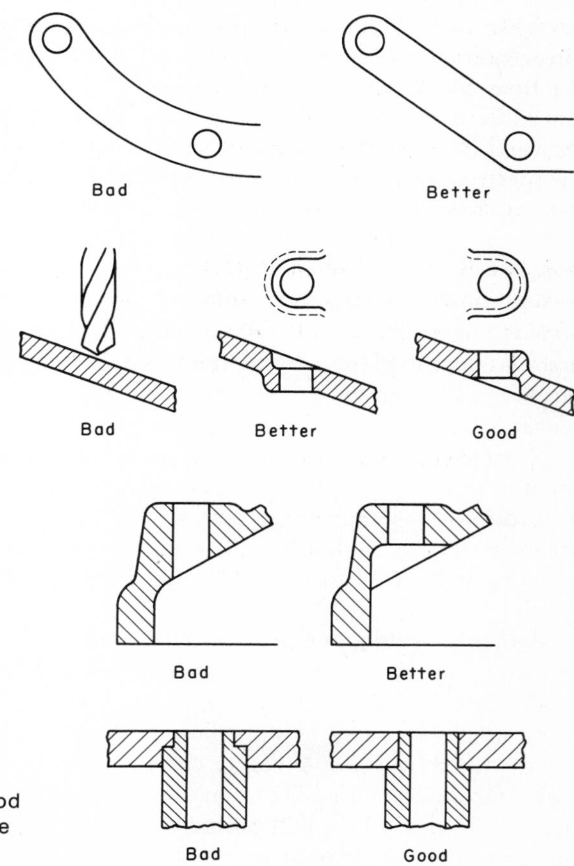

FIGURE 40-5. Examples of good and bad design details from the viewpoint of machining.

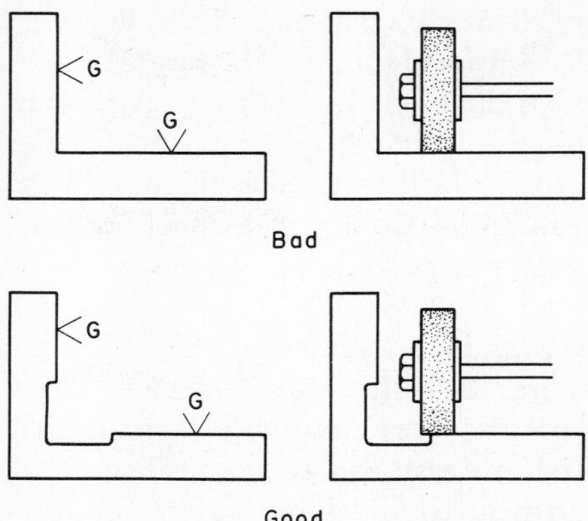

FIGURE 40-6. Method of providing proper clearance for grinding-wheel mounting and for overtravel.

designed, rather than take the trouble to contact the designer to ascertain whether some modification can be made that will facilitate machining.

Figure 40-6 illustrates how required grinding may be made difficult, or virtually impossible, through poor design. With no provision for entry of the grinding wheel or its supporting arbor (top views), the required surfaces cannot be ground by ordinary procedures. A simple change in the design corrects the difficulty. Another common deficiency in the design of parts that require the grinding of external cylindrical surfaces is the failure to provide any means to grip or hold the workpiece during grinding. This happens most frequently when only one or a few pieces are involved.

Design details related to finishing. As was discussed in Chapter 37, the cost of required finishing operations—cleaning, smoothing, plating, and painting—is greatly affected by design. This is particularly important when large quantities of a product are to be manufactured. As has been pointed out in previous chapters, some manufacturing processes result in less need for finishing than others. Thus the choice of processing methods, and design to permit the use of such processes, may be very important where fine finish is required. The various finishing and decorative treatments often require special considerations and some modification of design details. Figure 40-7, for example, illustrates good and bad design details where plating is to be used.

Design, processing, and product liability. More and more, poor design and its resulting consequence in the form of manufacturing defects are factors in product-liability claims. Unfortunately, with the experience of 20–20 hindsight, it often is easy to show how a manufacturing defect could easily have been avoided by a simple design change. Thus designers are called upon to exercise foresight in relating their designs to the required processing so that defects and difficulties will not occur. Figure 40-8 shows an example where a designer failed to take into account a very simple and elementary fact regard-

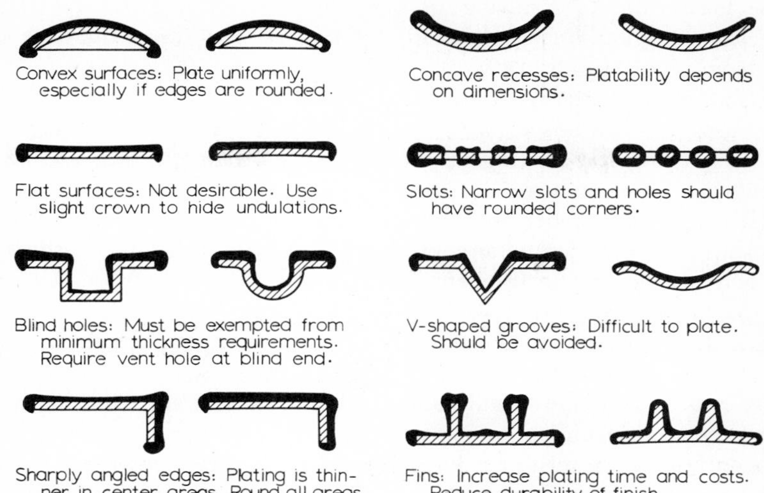

Convex surfaces: Plate uniformly, especially if edges are rounded.
Concave recesses: Platability depends on dimensions.
Flat surfaces: Not desirable. Use slight crown to hide undulations.
Slots: Narrow slots and holes should have rounded corners.
Blind holes: Must be exempted from minimum thickness requirements. Require vent hole at blind end.
V-shaped grooves: Difficult to plate. Should be avoided.
Sharply angled edges: Plating is thinner in center areas. Round all areas.
Fins: Increase plating time and costs. Reduce durability of finish.

FIGURE 40-7. Good and bad design details for electroplating.

ing casting. The company initially manufactured and marketed a relatively small woodworking tool that was supported at the end of a cast-aluminum beam having the size shown in Figure 40-8a. The product was very successful, and larger and larger models were made, each having a larger beam with an exactly proportional cross section. Finally, the beam was the size shown in Figure 40-8b, but almost the entire center was a large shrinkage cavity. One

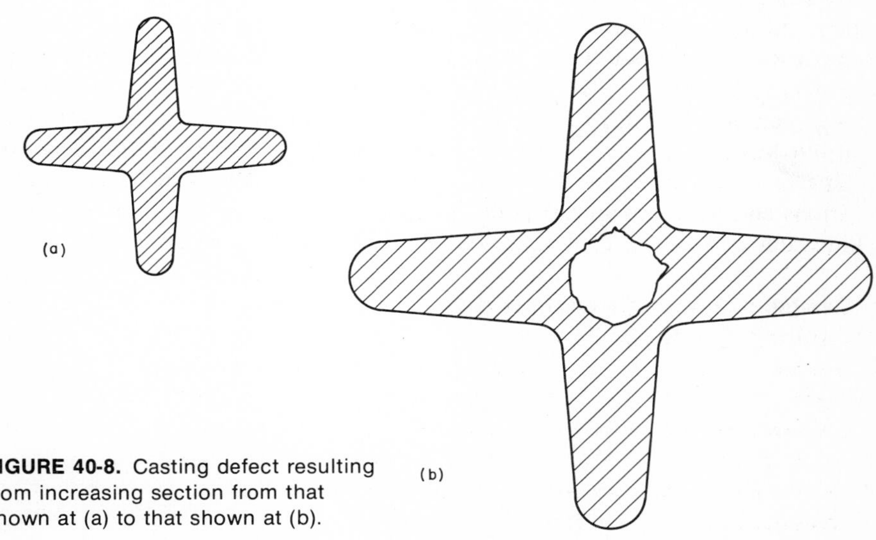

FIGURE 40-8. Casting defect resulting from increasing section from that shown at (a) to that shown at (b).

of these beams broke, resulting in a serious accident and a large damage award. The design engineer admitted that he was unaware of the problem of hot spots and shrinkage in castings!

Although, admittedly, it often is difficult for a designer to foresee all the possible uses and misuses to which a product will be subjected, there is no excuse for not properly relating the design to the processing that will be used to manufacture the product.

Although the number of examples that could be given to show the close relationship between design and production is almost limitless, those presented here should demonstrate the fact that design is an important phase of production planning and that all designers should have an intimate knowledge of production processes and use this knowledge in carrying out their work.

Quantity–process–design relationships. Most processes are not equally suitable and economical for producing a range of quantities for a given product. Consequently, the quantity to be produced should be considered, and the design should be adjusted to the process that actually is to be used before it is "finalized." As an example, consider the part shown in Figure 40-9. Assume that, *functionally,* brass, bronze, a heat-treated aluminum alloy, or ductile iron would be suitable materials. What material and process would be most economical if 1, 100, or 1000 parts were to be made?

If only one were to be made, it is most likely that contour sawing, followed by drilling the 19 mm (3/4 inch) hole, would be most economical. The irregular surface would be difficult to produce by other machining processes, and any casting process would require the making of a pattern, which would be about as costly to produce as to saw the desired part. It would be unlikely that a suitable piece of ductile iron or heat-treated aluminum alloy would be readily available. Because brass would be considerably cheaper than bronze, brass, contour sawing, and drilling most likely would be the best combination. For only one part the excess cost of brass over ductile iron would not be

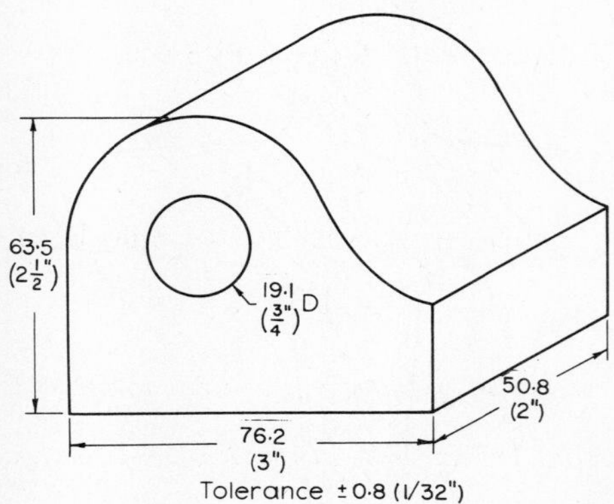

FIGURE 40-9. Part to be analyzed for production.

great, and this combination would require no special consideration on the part of the designer.

For a quantity of 100 parts, it is apparent that the excess cost for brass would be appreciable, and machining costs would have to be minimized. Consequently, casting undoubtedly would be the most economical process, and ductile iron would be cheaper than any of the other permissible materials. Although the design requirements for casting this simple shape would be minimal, the designer would want to consider them, particularly as to whether the hole should be cored.

For 1000 parts, entirely different solutions become feasible. In this case the use of an aluminum extrusion, with the individual 50.8 mm (2 inch) units being sawed off, might be the most economical solution.[1] All other machining would be eliminated. About 55 meters (180 feet) of extrusion would be required, including sawing allowance. The cost of the extrusion die would not be very high; thus the per-piece cost would not be great and likely would be more than offset by the savings in machining costs. So for this quantity, this method of production should be investigated. If it were to be used, the designer should make sure that any tolerances specified were well within commercial extrusion tolerances.

This simple example clearly illustrates how quantity can affect both material and process selection, and the selection of the process may require special considerations and design revisions on the part of the designer. Obviously, if the dimensional tolerances were changed, entirely different solutions might result. When more complex products are involved, these relationships become more complicated, but they also usually are more important and require detailed consideration by the designer.

Part analysis for basic requirements. After a satisfactory production design has been completed, the next step in planning is to determine the basic job requirements that must be satisfied. These usually are determined by analysis of the drawings and the job orders. They involve consideration and determination of the following:

1. Size and shape of the geometric components of the workpiece.
2. Tolerances.
3. Material from which the part is to be made.
4. Number of pieces to be produced.

Such an analysis for the "threaded shaft" shown in Figure 40-10 would be as follows[2]:

[1] This assumes the time required to obtain the extrusions—probably requiring the making of a special die—was not a factor.

[2] Because Figures 40-10 and 40-11 are dimensioned in English units, no metric equivalences are given in the discussion.

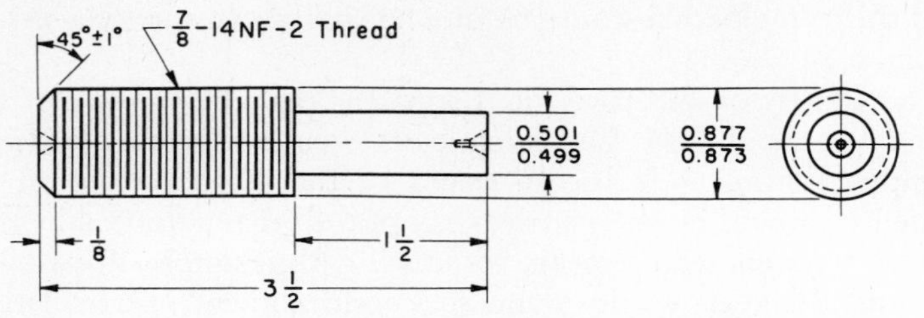

FIGURE 40-10. Threaded shaft.

1. a. Two concentric and adjacent cylinders, having diameters of 0.877/0.873 and 0.501/0.499' respectively, and lengths of 2 inches and 1½ inches.
   b. Three parallel, plain surfaces forming the ends of the cylinders.
   c. A 45° × ⅛-inch bevel on the outer end of the ⅞-inch cylinder.
   d. A ⅞-inch NF-2 thread cut on the entire length of the ⅞-inch cylinder.
2. The closest tolerance is 0.002 inch, and the single angular tolerance is ±1°.
3. The material is AISI 1020 cold-rolled steel.
4. The job order calls for 15 parts.

A number of conclusions regarding the processing can be drawn from this analysis. First, because concentric, external, cylindrical surfaces are involved, it is apparent that turning operations are required and that the piece should be made on some type of lathe. Second, because only 15 pieces are to be made, the use of a turret lathe probably would not be justified, so an ordinary engine lathe will be used. Third, because the maximum, required diameter is approximately ⅞-inch, 1-inch-diameter cold-rolled stock will be satisfactory; it will provide about $^1/_{16}$ inch for rough and finish turning of the large diameter. From this information, decisions can be made regarding the equipment and personnel that will be used and the time that will be required to accomplish the task.

Routing sheets. After the production requirements have been determined, the next step is to set up a *routing sheet.* This lists the operations that must be performed in order to produce the part, in their sequential order, and the machines or work stations and the tooling that will be required for each operation. For example, Figure 40-11 shows the drawing of a small, round "punch," and Figure 40-12 is a routing sheet for making this part. Once the routing of a part has been determined, the planning of each operation in the processing can then be done.

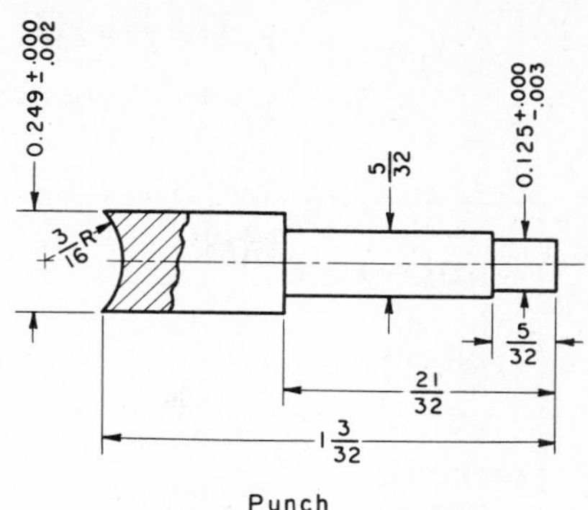

Matl. - 0.250 dia. AISI 1040
H.T. to 50 R.C. on 0.249 dia.

FIGURE 40-11. Drawing of a punch.

Operation sheets. Although routing sheets are very useful for the general planning of manufacturing, they usually do not give sufficiently detailed information to act as instructions for the individual machine operator in carrying out an operation or for scheduling machines and personnel. Such information can be provided on *operation sheets,* such as shown in Figure 40-13, which lists, in sequence, the operations required for machining the threaded shaft shown in Figure 40-10. Commonly, a single operation sheet lists the operations that are done in sequence on a single machine. However, they may cover all the operations for a given part even though more than one machine is required.

Operation sheets vary greatly as to details. The simpler types often list only the required operations and the machines to be used; speeds and feeds may be left to the discretion of the operator, particularly where skilled workers and small quantities are involved. However, it is common practice for complete details to be given regarding tools, speeds, and often the time allowed for completing each operation. Such data are necessary if the work is to be done on N/C machines, and experience has shown that these preplanning steps are advantageous where ordinary machine tools are used.

The determination of standard times for manufacturing operations is not within the scope of this text, but it is a well-established technique of industrial engineering. When an operation is machine-controlled, as in making a lathe cut of a certain length with power feed, the required time can be determined by simple mathematics. For example, if a cut is 254 mm (10 inches) in length and the turning speed is 200 rpm, with a feed of 0.127 mm (0.005 inch) per revolution, the time required will be 10 minutes. Several procedures are available for determining the time required for people-controlled machining elements, such as moving the carriage of a lathe by hand back from the end of one cut to the starting point of a following cut. Actual time studies, accumulated data from past operations, or some type of motion-time data,

DARVIC INDUSTRIES

ROUTING SHEET

NAME OF PART Punch PART NO. 2

QUANTITY 1,000 MATERIAL SAE 1040

OPERATION NUMBER	DESCRIPTION OF OPERATION	EQUIPMENT OR MACHINES	TOOLING
1	Turn $\frac{5}{32}$, 0.125, and 0.249 diameters	J & L turret lathe	#642 box tool
2	Cut off to $1\frac{3}{32}$ length	"	#6 cutoff in cross turret
3	Mill $\frac{3}{16}$ radius	#1 Milwaukee	Special jaws in vise $\frac{3}{16}$ form cutter x 4" D
4	Heat treat. 1,700° F for 30 minutes, oil quench	Atmosphere furnace	
5	Degrease	Vapor degreaser	
6	Check hardness	Rockwell tester	

FIGURE 40-12. Routing sheet for making the punch shown in Figure 40-11.

such as MTM,[3] can be employed for estimating such times. Each can provide accurate results that can be used for establishing standard times for use in planning or as a basis for worker compensation. Various handbooks and books on machine-shop estimating contain tables of average times for a wide variety

[3] Methods–time–measurement.

DARVIC INDUSTRIES

OPERATION SHEET

PART NAME: Threaded Shaft

OPER. NO.	NAME OF OPERATION	MACH. TOOL	CUTTING TOOL	CUTTING SPEED		FEED ipr	DEPTH OF CUT Inches	REMARKS
				ft/min	rpm			
10	Face End	Engine Lathe		90	318	Hand		Use 3-jaw Universal chuck
20	Center Drill End	"	Combination center drill		750	Hand		
30	Cut off to $3\frac{9}{16}$ length	"	Parting tool	90	318	Hand		To prevent chattering, keep overhang of work and tool at a minimum and feed steadily. Use lubricant.
40	Face to length	"	RH facing tool (small radius point)	90	318	Hand	(R) $\frac{1}{8}$ max. (F) .005	Before replacing part in 3-jaw chuck, scribe a line marking the $3\frac{1}{2}$ inch length.
50	Center Drill End	"	Combination center drill		750	Hand		
60	Place between centers, turn $\frac{.501}{.499}$ diameter, and face shoulder	"	RH turning tool (small radius point)	90 100	(R)318 (F)492	(R) .0089 (F) .0029	(R) .081(3) (F) .007	
70	Remove and replace end for end and turn $\frac{.877}{.873}$ diameter	"	RH tools (R) (small radius point) (F) Round nose tool	90 120	(R)318 (F)492	(R) .0089 (F) .0029	(R) .057 (F) .005	
80	Produce 45°-chamfer	"	RH round nose tool	120	492	Hand	(R) $\frac{1}{8}$ max. (F) .005	
90	Cut $\frac{7}{8}$-14 NF-2 thread	"	Threading tool	60	208		(R) .004 (F) .001	(1) Swivel compound rest to 30 degrees. (2) Set tool with thread gage. (3) When tool touches outside diameter of work set cross slide to zero. (4) Depth of cut for roughing = .004. (5) Engage thread dial indicator on any line. (6) Depth of cut for finishing = .001 Use compound rest.
	Remove burrs and sharp edges	"	Hand file					

Date ______________ ______________________

FIGURE 40-13. Operation sheet for the threaded shaft shown in Figure 40-10.

of elemental operations for use in estimating and setting standards. However, such data should be used with great caution, even for planning purposes, and they should never be used as a basis of wage payment. The conditions under which they were obtained may have been very different from those for which a standard is being set.

N/C, automation, planning, and organization. As was discussed in Chapter 39, some type of planning always precedes the start of manufacturing. One of the incidental side effects of numerically and computer-controlled machines has been a better realization of the benefits of preproduction planning. When such equipment is used, preproduction planning has to be done in great detail, and the designer usually has to give careful consideration to the requirements that the use of such machines imposes. This forces the acquiring of familiarity with the capabilities and limitations of the equipment that is to be utilized. It thus is likely that more attention will be given to reducing minor variations between designs and to the use of standard details that can be programmed and produced more readily on automatic equipment. Such procedures aid greatly in increasing the utilization factor of very costly machines.

At the same time, the experience gained from the increased use of numerically controlled equipment has made it clear that when ordinary machine tools are utilized there also is much benefit to be obtained by a certain amount of preproduction planning, particularly at the design stage. As has been said, someone has to do the planning. The real question is: Is the planning being done at the proper time, by the proper person, and is it planning in the real sense, or is it planning how to overcome previously imposed difficulties?

A key factor in assuring that the proper amount of planning is done, by the right persons, is the organization within a plant. Usually, a number of people are involved, and their efforts must be coordinated. Furthermore, even after adequate planning is done, the operations must be scheduled and *controlled* to assure that they actually are carried out as planned.

Because design, material selection, and process selection are directly related, engineers engaged in design and manufacturing play key roles in optimizing this relationship that is so essential in producing quality products at reasonable cost. Their efforts and cooperation are vital in improving productivity and the standard of living.

Review questions

1. Explain how preproduction planning and manufacturing processes are interrelated.
2. How does production design differ from functional design?
3. Give three examples of how a designer of a part may indirectly dictate what machines must be used to produce the part.
4. Give an example of a design detail that would be satisfactory if the part

were to be made by casting but would be bad if it were to be made by machining.

5. Give an example of a poor design detail that may lead to a defective product.
6. What are two ways in which the design engineer can contribute to product failure?
7. What process is available to correct the condition depicted in the bottom examples shown in Figure 40-7?
8. What are two advantages to be derived by making the change in casting design depicted in the bottom example of Figure 40-1?
9. Assume that the part shown in Figure 40-9 is to be made from ductile cast iron. What production processes would you recommend for: (a) 1 part; (b) 5 parts; (c) 1000 parts?
10. What is the purpose of a routing sheet?
11. Why will the increased use of automation require designers to be more familiar with the properties of materials and with manufacturing processes?
12. If the hole in the part shown in Figure 40-9 were square instead of round, would this change your answer to question 9? How?

Case study 30. THE COMPONENT WITH THE TRIANGULAR HOLE

The Johnstone Company estimates that its annual requirements for the socket component shown in Figure CS-30 will be at least 50,000 units, and that this volume will continue for at least 5 years. Consequently, it wants to consider all practicable methods for making the component and has assigned you the task of determining these methods and of recommending which should be explored in detail to determine the most effective and economical process. The specifications call for a lightweight metal, such as an aluminum alloy, and that it must have a tensile strength of at least 138 Mpa and an elongation of at least 3 per cent.

Determine at least five practicable methods for producing the part and suggest which two appear most likely to be most economical and thus should be investigated fully. (Give the reasons for your selections.)

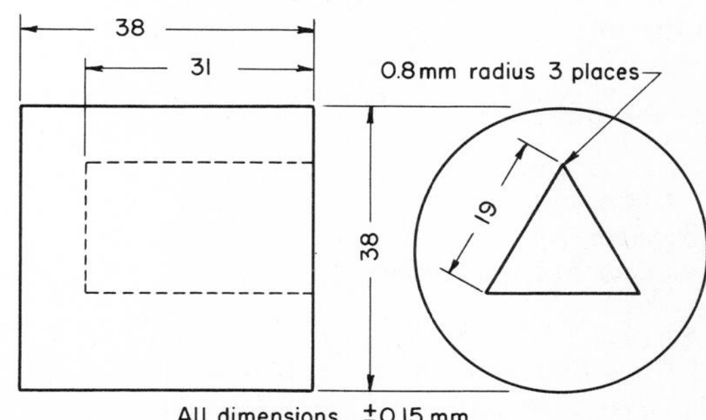

FIGURE CS-30. Component containing dead-end triangular hole.

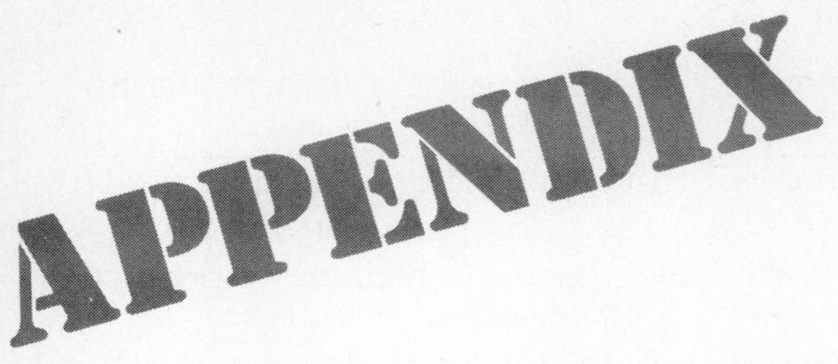

Selected references for additional study

The following is a list of selected and reasonably available books that will be useful to those who wish additional information regarding the subject matter of this textbook. The table on pages 1061–1062 indicates the works considered to be especially helpful for the subjects indicated.

Title	Author(s)	Publisher
1. *A Concise Guide to Plastics*	Simonds and Church	Van Nostrand Reinhold Company
2. *A Treatise on Milling Machines* (3rd ed.)		Cincinnati Milacron Inc.
3. *American Malleable Iron*		Malleable Founders Society
4. *Cast Metals Handbook* (4th ed.)		American Foundrymen's Society
5. *Chip Formation, Friction, and Finish*	Ernst and Merchant	Cincinnati Milacron Inc.

Title	Author(s)	Publisher
6. *Closed-Die Forgings*		Bethlehem Steel Corporation
7. *Cold Finished Steel Bar Handbook*		American Iron and Steel Institute
8. *Deformation and Fracture Mechanics of Engineering Materials*	Hertzberg	John Wiley & Sons, Inc.
9. *Die Casting for Engineers*		The New Jersey Zinc Co.
10. *Dimensional Control*		Bendix Automation and Measurement Division
11. *Elements of Materials Science and Engineering* (3rd ed.)	Van Vlack	Addison-Wesley Publishing Company, Inc.
12. *Forging of Powder Metallurgy Preforms*	Hauser	American Powder Institute
13. *Foundry Engineering*	Taylor, Fleming, and Wulff	John Wiley & Sons, Inc.
14. *Foundry Engineering*	Beeley	Halstead Press Division, John Wiley & Sons, Inc.
15. *Fundamental Principles of Polymeric Materials for Practicing Engineers*	Rosen	Barnes and Noble
16. *Fundamentals of Metal Casting*	Flinn	Addison-Wesley Publishing Company, Inc.
17. *Gear Handbook*	Dudley	McGraw-Hill Book Company
18. *Grinding Practice* (3rd ed.)	Colvin and Stanley	Van Nostrand Reinhold Company
19. *Handbook of Industrial Metrology*		Prentice-Hall, Inc.
20. *High-Velocity Forming of Metals*		Society of Manufacturing Engineers
21. *ISO System of Limits and Fits, General Tolerances and Deviations*		American National Standards Institute
22. *Jigs and Fixtures*	Colvin and Haas	McGraw-Hill Book Company
23. *Machining Data Handbook* (2nd ed.)		Machinability Data Center
24. Manufacture of Plastics	Smith	Van Nostrand Reinhold Company
25. *Manufacturing Processes* (7th ed.)	Amstead, Ostwald, and Begemen	John Wiley & Sons, Inc.
26. *Materials* ("Scientific American" book)		W. H. Freeman and Company
27. *Materials Science*	Ruoff	Prentice-Hall, Inc.
28. *Mechanical Metallurgy*	Dieter	McGraw-Hill Book Company
29. *Mechanical Processing of Metals*	Kalpakjian	Van Nostrand Reinhold Company
30. *Mechanical Working of Metals*	Crane	Macmillan Publishing Co., Inc.

Title	Author(s)	Publisher
31. *Mechanics of Plastic Deformation in Metal Processing*	Thomsen, Yang, and Kobayashi	Macmillan Publishing Co., Inc.
32. *Metals Handbook* (7 vols.) Vol. 1, *Properties and Selection of Metals* Vol. 2. *Heat Treating, Cleaning, and Finishing* Vol. 3. *Machining* Vol. 4. *Forming* Vol. 5. *Forging and Casting* Vol. 6. *Welding and Brazing*		American Society for Metals
33. *Metalworking with Aluminum*		The Aluminum Association, Inc.
34. *Nature and Properties of Engineering Materials*	Jastrzebski	John Wiley & Sons, Inc.
35. *Non-Traditional Machining Processes*	Springborn	American Society of Tool & Manufacturing Engineers
36. *Numerical Control for Machine Tools*	Borron	McGraw-Hill Book Company
37. *Physics of Metal Cutting*	Ernst	American Society for Metals
38. *Powder Metallurgy*	Sands and Shakespeare	Newnes (London)
39. *Powder Metallurgy*	Wulff	American Society for Metals
40. *Powder Metallurgy for Engineers*	Dixon and Clayton	Machinery Publishing Co.
41. *Principles of Engineering Materials*	Barrett, Nix, and Tetelman	Prentice-Hall, Inc.
42. *Precision Investment Castings*	Cady	Van Nostrand Reinhold Company
43. *Processing and Materials of Manufacture*	Lindberg	Allyn and Bacon, Inc.
44. *Statistical Quality Control* (4th ed.)	Grant and Leavenworth	McGraw-Hill Book Company
45. *Steel Castings Handbook* (3rd ed.)		Steel Founders Society of America
46. *The Closed Die Forging Process*	Kyle	Macmillan Publishing Co., Inc.
47. *The Making, Shaping, and Treating of Steel* (9th ed.)	McGannin	United States Steel Corp.
48. *Theory and Flow and Fracture of Solids*	Nadai	McGraw-Hill Book Company
49. *The Oxy-Acetylene Handbook*		Linde Division of Union Carbide Corp.
50. *The Science of Precision Measurement*		The DoALL Company
51. *Tool Engineers Handbook* (3rd ed.)	Society of Manufacturing Engineers	McGraw-Hill Book Company

Title	Author(s)	Publisher
52. *Weldability of Steels*	Stout	Welding Research Council
53. *Welding and Welding Technology*	Little	McGraw-Hill Book Company
54. *Welding for Engineers*	Udin, Funk, and Wulff	John Wiley & Sons, Inc.
55. *Welding Handbook* (6th ed.) (in 5 sections; 6 vols.)		American Welding Society

Chapter Number	Subject	References as Listed on pp. 1058–1061
1	Introduction	25, 43
2	Properties of Materials	11, 28, 32(1), 41
3	The Nature of Metals and Alloys	8, 11, 23, 26, 27, 41
4	Production and Properties of Common Engineering Metals	43, 47
5	Equilibrium Diagrams	11, 28
6	Heat Treatment	32(2), 47
7	Alloy Irons and Steels	27, 32(1), 34
8	Nonferrous Alloys	11, 32(1), 41
9	Nonmetallic Materials	1, 11, 15, 24, 26
10	Material Selection	32(1)
11	Casting	3, 4, 9, 13, 14, 16, 25, 32(5), 42, 45
12	Powder Metallurgy	12, 38, 39, 40
13	The Theoretical Basis for Metal Forming	28, 29, 31, 48
14	Hot-Working Processes	6, 12, 25, 29, 32(5), 43, 46, 47
15	Cold-Working Processes	7, 20, 25, 29, 30, 32(4), 33, 43
16	Measurement, Gaging, and Quality Control	10, 19, 21, 25, 44, 51
17	Metal Cutting	5, 23, 32(3), 37, 51
18	Shaping and Planing	25, 43
19	Drilling and Reaming	23, 25, 43
20	Turning and Related Operations	25, 51
21	Boring	25, 43
22	Milling	2, 25, 43, 51
23	Sawing and Filing	25, 43
24	Broaching	25, 43, 51
25	Abrasive Machining Processes	18, 23, 25
26	Chipless Machining Processes	35, 43
27	Thread Cutting and Forming	25, 43
28	Gear Manufacturing	17, 43, 51
29	Forge, Oxyfuel Gas, and Arc Welding	32(6), 49, 53, 54, 55

Chapter Number	Subject	References as Listed on pp. 1058–1061
30	Resistance Welding	32(6), 53, 54, 55
31	Miscellaneous Welding and Related Processes	32(6), 53, 54, 55
32	Torch and Arc Cutting	32(6), 49, 55
33	Brazing, Soldering, and Adhesive Bonding	32(6), 43, 54, 55
34	Heat and Design Considerations in Welding	52, 54, 55
35	Layout	51
36	Jigs and Fixtures	22, 51
37	Decorative and Protective Surface Treatments	32(2), 43
38	Mass-Production Tools and Techniques	25, 43, 51
39	Numerical, Tape, and Computer Control	25, 36, 43, 51
40	Planning Manufacturing Operations	51

INDEX